Carbon Nanomaterials Based on Graphene Nanosheets

Carbon Nanomaterials Based on Graphene Nanosheets

Editor
Ling Bing Kong
School of Materials Science and Engineering
Nanyang Technological University
Singapore

CRC Press is an imprint of the
Taylor & Francis Group, an **informa** business
A SCIENCE PUBLISHERS BOOK

CRC Press
Taylor & Francis Group
6000 Broken Sound Parkway NW, Suite 300
Boca Raton, FL 33487-2742

First issued in paperback 2020

ISBN-13: 978-1-4987-2504-0 (hbk)
ISBN-13: 978-0-367-78255-9 (pbk)

Library of Congress Cataloging-in-Publication Data

Names: Kong, Ling Bing, editor.
Title: Carbon nanomaterials based on graphene nanosheets / editor, Ling Bing Kong, School of Materials Science and Engineering, Nanyang Technological University, Singapore.
Description: Boca Raton, FL : CRC Press, [2016] | Includes bibliographical references and index.
Identifiers: LCCN 2016044382| ISBN 9781498725040 (hardback : acid-free paper) | ISBN 9781498725057 (e-book)
Subjects: LCSH: Carbon nanofibers. | Nanostructured materials. | Graphene.
Classification: LCC TA455.C3 C357 2016 | DDC 620.1/15--dc23
LC record available at https://lccn.loc.gov/2016044382

Visit the Taylor & Francis Web site at
http://www.taylorandfrancis.com

and the CRC Press Web site at
http://www.crcpress.com

Preface

Since the discovery of graphene, it has become one of the most widely and extensively studied materials. This book was aimed to summarize the progress in synthesis, processing, characterization and applications, of a special group of nanocarbon materials derived from graphene or graphene related derivatives, by using various strategies, in different forms. More specifically, three forms of macrosized materials will be presented, i.e., one-dimension or 1D (fibers, wires, yarns, streads, etc.), two-dimension or 2D (films, membranes, papers, sheets, etc.) and three-dimension or 3D (bulk, hydrogels, aerogels, foams, sponges, etc.). Seven chapters are included, with the first chapter (this chapter) serves to introduce the concept, definition, nomenclature of graphene, graphene oxide and their derivatives. The main topics are covered in Chapters 2–7. Although they have coherent connections, each chapter of them is designed to independent, so that they can be used independently.

In Chapter 2, various synthetic routes, including chemical and physical routes, top-down (e.g., exfoliation) and bottom-up (chemical vapor deposition or CVD), will be discussed in a relatively detailed way. Although the most potential route with large-scale production is wet-chemical exfoliation, due to the comparatively simple process, cheap chemicals and non-expensive facility requirement, other methods could be used as alternatives, especially in case that special requirements are required.

Chapter 3 and Chapter 4 are used to discuss the progress in synthesis, preparation and characterization of various hybrids based on graphene, graphene oxide and their derivatives, with various inorganic components. The inorganic components include nanosized oxides (Chaper 3) and noble metallic nanoparticles, normal metals, non-graphene nanocarbons (carbon nanotubes or carbon nanofibers), semiconductor nanoparticles and so on (Chapter 4).

Macrosized one-dimensional (1D), two-dimensional (2D) and three-dimensional (3D) structures, based on graphene or graphene related materials will be systematically presented and discussed in Chapters 5–7. In Chapter 5, 1D macrosized materials, also called fibers, yarns, wires and so on, will be elaborated, focusing on the inter-relationship among processing, microstructure, properties and performances. Chapter 6 covers 2D materials, including thin films (deposited substrates), free-standing papers, membranes, sheets and so on. Chapter 7 describes all the 3D structured materials based on graphene or graphene related items, which include hydrogels, aerogels, foams, sponges and so on.

The target readers of this book include undergraduate students, postgraduate students, researchers, designers, professors, and program/project managers, from areas/fields of materials science and engineering, applied physics, chemical engineering, biomaterials, materials manufacturing and design, institutes, research founding agencies, etc. It can also be used as a text book for postgraduate students and training materials for engineers and workers.

Contents

CHAPTER 1

Introduction

Ling Bing Kong

1.1. Graphene and Graphene Oxide

Graphene is one of the most popular two-dimensional (2D) materials, which has attracted considerable attention all around the world. Graphene was initially used to refer to an isolated single atom thick carbon sheet [1], while it is used now to represent all 2D sheet-like or flake-like carbon materials [2, 3]. Besides monolayer (single atomic layer) graphene, ultrathin multilayered carbon materials derived from graphite through exfoliation are also called graphene or graphene-based materials. Moreover, graphene oxide and various chemically modified graphenes have also emerged in the open literature [3]. Graphene and graphene oxide, as atomically-thin precursors, can be stacked, folded, crumpled or pillared into various three-dimensional (3D) architectures [4, 5]. There are also one-dimensional (1D) and 2D (different from the 2D mentioned before) carbon materials based on graphene or graphene oxide. Therefore, it is necessary to have a widely accepted definition for graphene.

1.2. Principles for Definition

There are several principles that have been used to define graphene and graphene related nanocarbons, which have been described in detail in Ref. [2]. A brief discussion on each principle will be presented in this section. The first principle is to use specific names to describe the different types of graphene related materials, instead of simply using the word-graphene. However, the fact is that different graphene-based materials are simply named as graphene, without differentiation. The second issue is distinguishing between graphene and graphene layer. Graphene layer was initially used to refer to the carbon atomic sheets with sp^2 hybridization [6, 7], but is now being used in the literature for isolated graphene [1].

The next principle is to take into consideration the lateral dimensions, which could range from tens of nanometers to micrometers and even to macroscopic dimensions. The size of lateral dimensions has an important effect when the graphene or graphene related materials are used in composites to achieve desired electrical properties, such as electrical conductivity, i.e., the percolation thresholds. Other properties that are closely related to lateral size include band gaps, cell interactions, and so on. Thickness that is actually the number of atomic carbon layers is also a very important variable, which therefore should be stated as accurately as possible, in order to ensure that the related results or data will be more meaningful.

School of Materials Science and Engineering, Nanyang Technological University, Singapore.
E-mail: elbkong@ntu.edu.sg

1.3. Definitions of Graphene and Graphene Related Items

Accordingly, there have been various terms for graphene and graphene related materials, which also follow Ref. [2].

Graphene is defined as a single-layer atomic carbon sheet, with a perfect hexagonal configuration of sp^2 carbon atoms, which is either isolated free-standing or deposited on a substrate. The lateral dimensions can vary in range from several nanometers to micrometers for free-standing graphene and even to the macroscale for deposits on substrates.

Graphene layer means a single-layer atomic carbon sheet with hexagonal configuration of sp^2 carbon atoms, which is only present within a given carbon material structure, having either 3D ordering (graphitic) or other (e.g., turbostratic or rotationally faulted) characteristics. Therefore, it is a building block (unit) of a certain structure.

Turbostratic carbon is used to describe three-dimensional configurations, with sp^2-bonded carbon atoms, but without clearly defined atomic carbon layers. The name is a combination of the words "turbo" (rotated) and "strata" (layer), therefore it is also known as rotationally faulted.

Bilayer graphene and trilayer graphene stand for 2D or sheet-like layered materials that contain two (2) and three (3) stacked graphene layers, respectively, which could be further specified as "AB-stacked bilayer graphene" or "rotationally faulted trilayer graphene".

Multilayer graphene (MLG) is similar to the above graphenes, but the number of atomic carbon sheets is between 2 and 10, which are well-defined and tightly stacked with an extended lateral dimension.

Few-layer graphene (FLG) is actually a group of multilayer graphene, with 2–5 layers of graphene.

Graphite nanoplates, graphite nanosheets and graphite nanoflakes include 2D graphite materials with stackings of ABA or ABCA, with a thickness and/or lateral dimension of less than 100 nm.

Exfoliated graphite refers to multilayer structured items, which are made by partially exfoliating graphite into thin multilayer sheets, while still retaining the 3D crystal stacking of graphite, by using thermal, chemical or mechanical methods.

Graphene nanosheet refers to a single atomic carbon sheet, with carbon atoms arranged hexagonally, with sp^2 hybridization, which is either free-standing or deposited on substrates. According to the definition of nanomaterials, it should have a lateral dimension of < 100 nm. However, graphene nanosheet or nanosheets have been widely used in the literature to represent all 2D graphene materials, and are used similarly in this book.

Graphene microsheet refers to a single-layer atomic carbon sheet with sp^2-bonded carbon atoms arranged hexagonally, which is either free-standing or deposited on substrates, with a lateral dimension ranging from 100 nm to 100 μm.

Graphene nanoribbon is a single-layer atomic carbon strip, with sp^2-bonded carbon atoms having a hexagonal configuration, which is either free-standing or deposited on substrates. The width should be less than 100 nm.

Graphene quantum dots (GQD) is an alternative name for graphene nanosheets or few-layer graphene nanosheets, with lateral dimensions of < 10 nm (or average of 5 nm).

Graphene oxide (GO) stands for all the graphenes, which are chemically derived through oxidation and exfoliation, in which the basal planes contain a large amount of oxygen, in the form of oxygen containing functional groups.

Graphite oxide is a bulk solid made by oxidizing graphite through functionalization of the basal planes, with the interlayer spacing expanded to a certain degree. It is also known as expanded graphite.

Reduced graphene oxide (rGO) is the product of graphene oxide that is reduced by using various reducing methods, such as chemical, thermal, microwave, photo-chemical, photo-thermal and microbial/bacterial; the graphene oxide is reduced by eliminating oxygen or oxygen containing functional groups. rGO has characteristics similar to that of graphene.

Graphenization is used to describe the processes that are used to prepare graphene layers, either in 2D or 3D forms. Two related terms are (i) carbonization, meaning the conversion of organic materials into a carbonaceous solid, and (ii) graphitization, referring to the crystallization of carbonaceous materials into 3D ordered structures.

Graphene materials, graphene-based materials, graphene nanomaterials, graphene-family nanomaterials are also used to imply more generally the collection of the above discussed 2D materials that contain the word graphene.

Graphenic carbon materials represent the broadest class of carbonaceous solids that are formed by elemental carbon with bounding of sp^2-hybridization, in either 2D or 3D forms.

There are also numerous ways that have been used in the open literature to represent newly emerged materials based on or related to graphene, in one way or another. Examples are briefly discussed below.

Graphene oxide nanosheet is graphene oxide (monolayer) with a lateral dimension of < 100 nm, according to the definition of nanotechnology. However, in the literature, graphene oxide nanosheet or graphene oxide nanosheets are also used to stand for those with a lateral size of > 100 nm.

Few-layer graphene nanoribbons belong to the category of graphene nanoribbons, with 2–5 atomic carbon layers.

Multilayer graphene oxide film refers to a multilayer structure with restacking of graphene oxide monolayer sheets that are deposited on substrates. A similar term is graphene film.

There are also some terms and usages that appear in the literature but do not follow the above discussed principles in the nomenclature, which are presented as follows.

Graphite layer is a term that is not recommended because graphite is usually used to refer to the 3D crystals. Therefore, graphene layer is a more appropriate term to replace graphite layer.

Graphene nanosheet is similar to graphene oxide nanosheet or reduced graphene nanosheet. Although it is not recommended to use this term in Ref. [2], its usage can be quite frequently found in the literature, while it will be used in this book.

Graphene nanoplates, graphene nanoplatelets: these terms are used for some industry products with microscale lateral dimension, but are not recommended in our scientific nomenclature for reasons given above. “Graphene” does not need the prefix “nano” to indicate thinness, and instead “nano” used in this way should indicate the lateral dimension.

Graphitic is also not recommended in Ref. [2], which is not adopted in this book.

Graphene materials can also be incorporated or hybridized in various ways with various other components, thus leading to additional terms, such as folded, wrinkled, activated, decorated or functionalized [8]. One group of such materials, i.e., hybrids with inorganic components, will be discussed in detail in Chapter 2 and Chapter 3 in this book.

References

[1] Novoselov KS, Geim AK, Morozov SV, Jiang D, Zhang Y, Dubonos SV et al. Electric field effect in atomically thin carbon films. Science. 2004; 306: 666–9.

[2] Bianco A, Cheng HM, Enoki T, Gogotsi Y, Hurt RH, Koratkar N et al. All in the graphene family - A recommended nomenclature for two-dimensional carbon materials. Carbon. 2013; 65: 1–6.

[3] Geim AK, Novoselov KS. The rise of graphene. Nature Materials. 2007; 6: 183–91.

[4] Compton OC, Nguyen ST. Graphene oxide, highly reduced graphene oxide, and graphene: versatile building blocks for carbon-based materials. Small. 2010; 6: 711–23.
[5] Luo JY, Jang HD, Sun T, Xiao L, He Z, Katsoulidis AP et al. Compression and aggregation-resistant particles of crumpled soft sheets. ACS Nano. 2011; 5: 8943–9.
[6] Boehm HP, Setton R, Stumpp E. Nomenclature and terminology of graphite-intercalation compounds. Carbon. 1986; 24: 241–5.
[7] Boehm HP, Setton R, Stumpp E. Nomenclature and terminology of graphite-intercalation compounds. Pure and Applied Chemistry. 1994; 66: 1893–901.
[8] Zhu YW, Murali S, Stoller MD, Ganesh KJ, Cai WW, Ferreira PJ et al. Carbon-based supercapacitors produced by activation of graphene. Science. 2011; 332: 1537–41.

CHAPTER 2

Synthesis of Graphene Nanosheets

Ling Bing Kong,[1,a,*] *Freddy Boey,*[1,b] *Yizhong Huang,*[1,c] *Zhichuan Jason Xu,*[1,d] *Kun Zhou,*[2] *Sean Li,*[3] *Wenxiu Que,*[4] *Hui Huang*[5] and *Tianshu Zhang*[6]

2.1. Introduction

One of the critical challenges in using graphene nanosheets for practical applications is the lack of an efficient way to reliably produce them in both large quantities and of high qualities. Also, it is still not possible to control the synthesis of graphene nanosheets with precisely defined size, shape and edge chemistry, which is crucial for most applications, especially when one or more of these parameters are important. For example, it is necessary to open up the energy band gap of graphene, so that it can be used as an active material in field-effect transistors (FETs) [1]. For electrode applications, electrical conductivity is more important than other parameters. In this case, the graphene nanosheets should have high purity and perfect structure. In this respect, the aim of this chapter is to summarize the progress in the synthesis of graphene nanosheets and their related materials. However, due to the huge number of references, it is impossible to provide a complete list; therefore representative examples will be presented for each synthetic method.

[1] School of Materials Science and Engineering, Nanyang Technological University, Singapore.
[a] E-mail: elbkong@ntu.edu.sg
[b] E-mail: mycboey@ntu.edu.sg
[c] E-mail: yzhuang@ntu.edu.sg
[d] E-mail: xuzc@ntu.edu.sg
[2] School of Mechanical & Aerospace Engineering, Nanyang Technological University, Singapore; E-mail: kzhou@ntu.edu.sg
[3] School of Materials Science and Engineering, The University of New South Wales, Australia; E-mail: sean.li@unsw.edu.au
[4] Electronic Materials Research Laboratory, School of Electronic and Information Engineering, Xi'an Jiaotong University, P. R. China; E-mail: wxque@xjtu.edu.cn
[5] Singapore Institute of Manufacturing Technologies (SIMTech), Singapore; E-mail: hhuang@SIMTech.a-star.edu.sg
[6] Anhui Target Advanced Ceramics Technology Co. Ltd., Hefei, Anhui, P. R. China; E-mail: 13335516617@163.com
* Corresponding author

2.2. Synthetic Methods

Two main strategies have been employed to synthesize graphene: top-down (exfoliation of graphite) and bottom-up (chemical synthesis) [2]. The top-down methods typically include mechanical exfoliation of HOPG [3, 4], chemical oxidation/exfoliation of graphite followed by reduction of graphene oxide (GO) [5–9] and solution-based exfoliation of graphite intercalation compounds (GICs) [10]. The bottom-up approaches for graphene synthesis comprise epitaxial growth on metallic substrates by means of CVD [11–13], thermal decomposition of SiC [14], and organic synthesis [15, 16] based on precursor molecules [17–20].

2.2.1. Liquid-Phase Exfoliation of Graphite

2.2.1.1. Oxidation Exfoliation

Graphite oxide (GO), an oxygen-rich carbonaceous layered material, is produced by the controlled oxidation of graphite [5, 6]. Each layer of GO is essentially an oxidized graphene sheet commonly referred to as graphene oxide [21, 22]. It has been widely accepted that GO consists of intact graphitic regions interspersed with sp^3 hybridized carbons containing hydroxyl and epoxide functional groups on the top and bottom surfaces of each sheet and sp^2 hybridized carbons containing carboxyl and carbonyl groups mostly at the sheet edges [23–26]. Therefore, GO is hydrophilic and thus can be readily dispersed in water to form stable colloidal suspensions [27, 28].

Figure 2.1 summarizes the process in general, which has been modified in many alternative ways in the open literature [29, 30]. The process starts from the preparation of bulk quantities of chemically

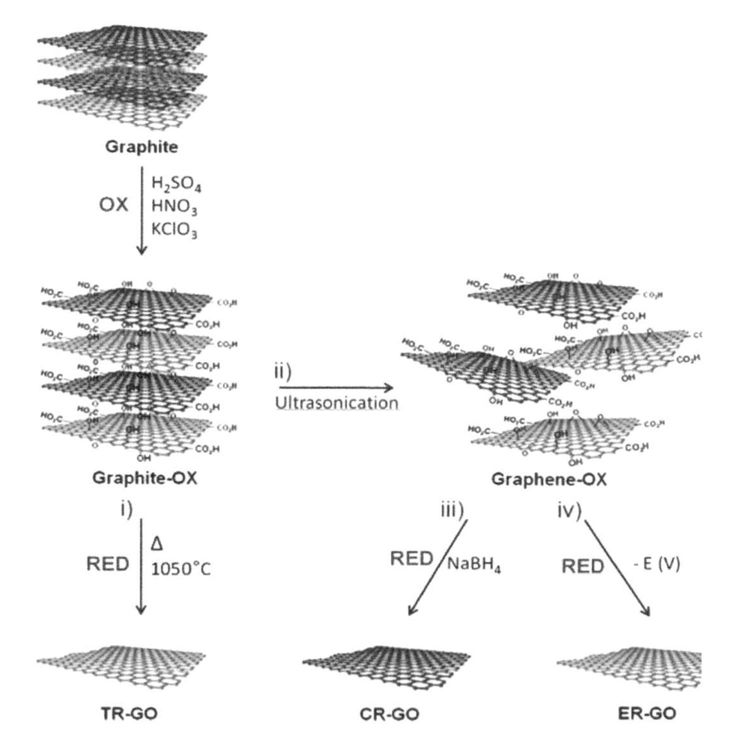

Fig. 2.1. Schematic of the procedure followed for the production of chemically modified graphenes using graphite as the starting material. An oxidative treatment is initially performed to generate graphite oxide. This is followed by: (i) thermal reduction/exfoliation of graphite oxide to produce TR-GO, (ii) exfoliation by ultrasonication generating graphene oxide, (iii) chemical reduction of graphene oxide by using $NaBH_4$ to produce CR-GO, (iv) electrochemical reduction of graphene oxide producing ER-GO. "OX" and "RED" stand for oxidation and reduction, respectively. Reproduced with permission from [30], Copyright © 2011, Wiley-VCH Verlag GmbH & Co.

modified graphite (CMG), which is derived from graphite microparticles that are oxidized by a mixture of HNO_3, H_2SO_4 and $KClO_3$ [31]. The oxidizing agents can be different. The resulting graphite oxide can be treated in several ways: (i) rapid heating of the sample in N_2 atmosphere up to 1050°C resulting in the formation of TR-GO [32], (ii) ultrasonication of the graphite oxide dispersion in dimethylformamide (DMF) or ultrapure water leading to a dispersion of graphene oxide [33], (iii) chemical reduction of the water-dispersed graphene oxide with sodium borohydride ($NaBH_4$, 50 mM) for 1 h resulting in CR-GO [8, 34, 35] and (iv) modification of the surface to form electrochemically modified graphene [36, 37].

Brodie's method was developed more than 150 years ago, and is still widely used today [5]. In a typical process of this method, 10 g of natural graphite powder and 85 g of $NaClO_3$ were mixed in a flask in an ice-bath. 60 ml of fuming HNO_3 was added drop wisely [28]. Another portion of acid (40 ml) was added after 16 h and the slurry was heated to 60°C and kept for 8 h. A slow heating rate of 1.5°C/min was used to avoid strong deflagration. The reaction was terminated by transferring the mixture into 1 l of distilled water. The suspension was washed with 5 × 200 ml of 3 M HCl solution and with a copious amount of distilled water until the supernatant had a specific conductivity of 10 $\mu S \cdot cm^{-1}$. The residual graphite oxide was decanted and dried. The first specimen of the graphite oxide series (GO-1) was oxidized forward (to GO-2), with the same procedure, except that a triple quantity (30 g) of GO-1 was used instead of graphite. Then, the whole oxidation procedure was repeated twice more in the same fashion to obtain the samples with the highest degree of oxidation. The four samples will be denoted as GO-1, GO-2, GO-3 and GO-4 where 1–4 correspond to the number of subsequent oxidation steps.

Dilute aqueous suspension of GO-1, 0.1 $g \cdot l^{-1}$, pH = 4.5, was examined visually and by means of electron microscopy [28]. The partial disaggregation of GO-1 still formed a fine colloid, but coarse particles could be observed, which settled down in a short time. After 1 day aging, a clear supernatant and compact sediment were formed, as shown in the left image of Fig. 2.2. The platelet aggregates could not be cleaved. However, rapid and spontaneous exfoliation was observed, when NaOH was added, leading to the formation of a stable suspension, with more intensive light scattering, as shown in the right image of Fig. 2.2. The exfoliation effect of NaOH is schematically demonstrated in the figure as well. Figure 2.3 (a) shows TEM image of the GO at pH = 4.5, in which both coarse slabs and finer lamellae with micron-sized or smaller lateral dimensions could be observed. With the addition of NaOH (pH = 10), only thin carbon foils in random spatial arrangement and crumpled conformation could be found, which was indicative of the delaminated GO structure, as shown in Fig. 2.3 (b).

A typical process of Hummers' method is described as follows [6]. 100 g of flake graphite powder and 50 g of sodium nitrate were mixed in 2.3 liters of sulfuric acid. The mixing was carried out in a container had been cooled to 0°C in an ice-bath for the purpose of safety. While maintaining vigorous agitation, 300 g of potassium permanganate was added to the suspension. The rate of addition was controlled carefully to prevent the temperature of the suspension from exceeding 20°C. The ice-bath was then removed and the temperature of the suspension was increased to 35 ± 3°C, where it was maintained for 30 min. As the reaction progressed, the mixture gradually thickened with a diminishing in effervescence. At the end of 20 min, the mixture became pasty with the evolution of only a small amount of gas. The paste was

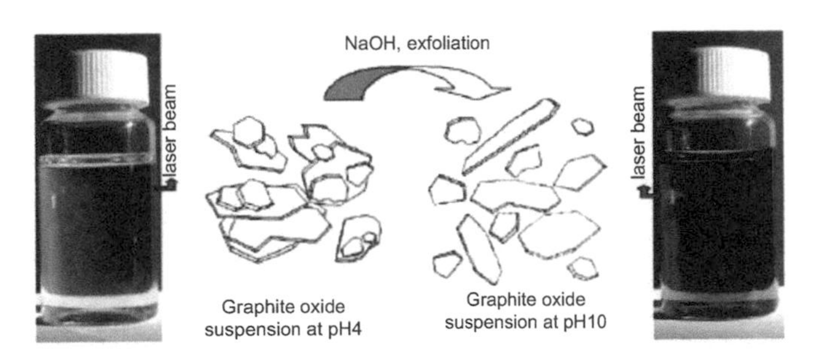

Fig. 2.2. Stability and laser light scattering of the GO suspensions at pH = 4 and pH = 10. Reproduced with permission from [28], Copyright © 2006, Elsevier.

Color image of this figure appears in the color plate section at the end of the book.

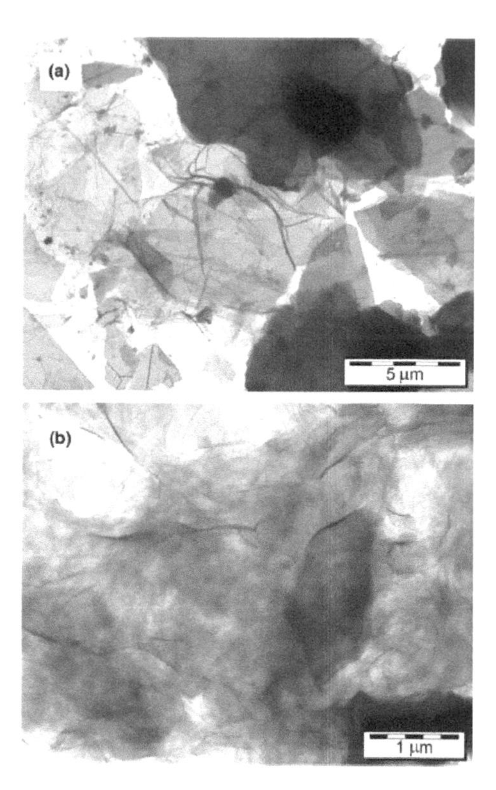

Fig. 2.3. TEM images of GO-1 suspension at pH = 4.5 (a) and at pH = 10 (b). Reproduced with permission from [28], Copyright © 2006, Elsevier.

brownish-gray in color. At the end of 30 min, 4.6 liters of water were slowly stirred into the paste, causing violent effervescence and an increase in temperature to 98°C. The diluted suspension, brown in color, was maintained at this temperature for 15 min. The suspension was then further diluted to about 14 liters with warm water and treated with 3% hydrogen peroxide to reduce the residual permanganate and manganese dioxide to colorless soluble manganese sulfate. Upon treatment with the peroxide, the suspension turned bright yellow. The suspension was then filtered, resulting in a yellow-brown filter cake. The filtering was conducted while the suspension was still warm in order to avoid precipitation of the slightly soluble salt of mellitic acid formed as a side reaction. After washing the yellowish-brown filter cake three times with a total of 14 liters of warm water, the graphitic oxide residue was dispersed in 32 liters of water to 0.5% solids. The remaining salt impurities were removed by treating with resinous anion and cation exchangers. The dry form of graphitic oxide was obtained by using centrifugation, followed by dehydration at 40°C over phosphorus pentoxide. This method has been widely used in the open literature.

A modified Hummers' method has been developed to synthesize graphite oxide nanosheets in high yield of 122 wt% based on the raw graphite and 68% based on the recovery of carbon [21]. The GO nanosheets had an average thickness of several nanometers and an average size of about 20 μm. Furthermore, excellent flexibility of the nanosheets was observed. When embedded in polymer matrix polymer, they were present in two types of secondary conformations, i.e., (i) lamination-layer-aggregate and (ii) random-shape-aggregate. Figure 2.4 shows the various categories of the GO nanosheets: oxidized form, partially oxidized form and reduced form.

Figure 2.5 shows SEM image of the product reduced with nascent hydrogen from aluminium powder and hydrochloric acid. They were random shaped aggregates. There were both singular and plural aggregates of the GO nanosheets, which were irregularly bent and deformed like crumpled paper [38]. This shows that, when the affinity between the nanosheets and the dispersion medium was very low, the nanosheets tended to aggregate like a linear flexible polymer. It was also found that the XRD peak corresponding to

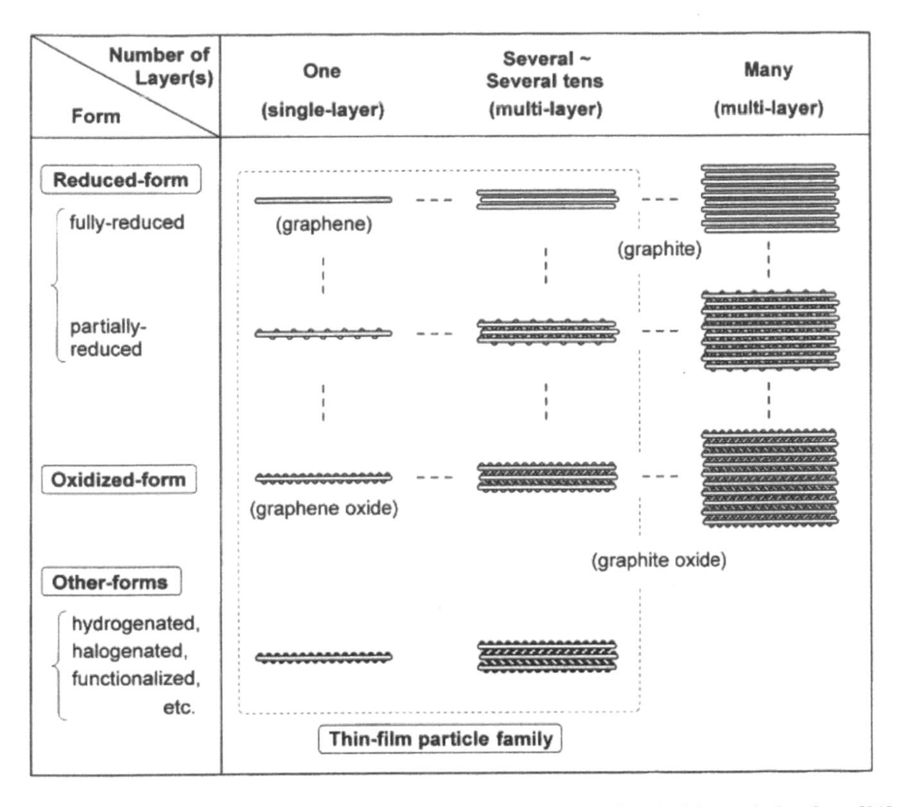

Fig. 2.4. Classification of the graphene and graphene oxide nanosheets. Reproduced with permission from [21], Copyright © 2004, Elsevier.

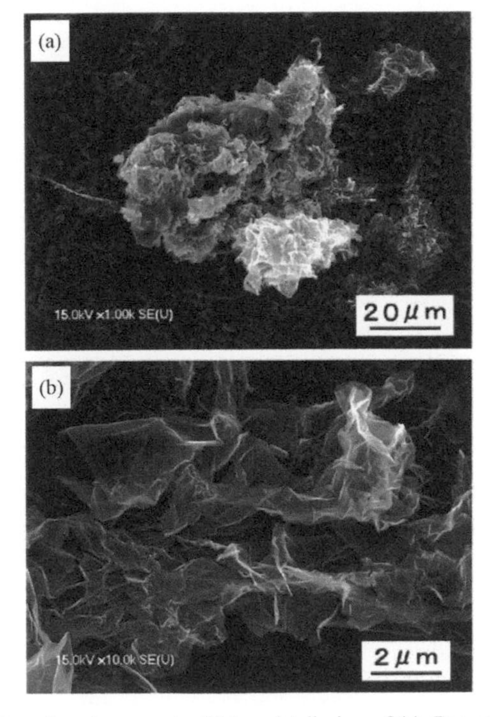

Fig. 2.5. SEM images of a random shaped aggregate: (b) is a detail view of (a). Reproduced with permission from [21], Copyright © 2004, Elsevier.

the spacing of 0.83 nm disappeared, but it was present in the original nanosheets in the oxidized form. This implied that the degree of both the interparticular orientation and interlayer orientation inside each nanosheet was reduced due to the deformation and aggregation.

Figure 2.6 (a) shows SEM image of a part of the random shaped aggregate, while Fig. 2.6 (b) demonstrates TEM image of a center of bend in the random shaped aggregate. Although GO nanosheets possessed a dense carbonaceous skeleton, they were very flexible, due to their extremely small thinness and very high aspect ratios. In this case, each of the fundamental layers of the particle is also flexible. The single-layered nanosheets were similar to a two-dimensional flexible macromolecule.

Dispersion behaviors of GO nanosheets in various solvents have been well-studied in the open literature [33, 39, 40]. For example, as-prepared graphite oxide can be dispersed in N,N-dimethylformamide, N-methyl-2-pyrrolidone, tetrahydrofuran and ethylene glycol. In all these solvents, full exfoliation of the graphene oxide into individual single-layer graphene oxide sheets was achieved with the aid of sonication [33]. The graphene oxide dispersions exhibited long-term stability, consisting of sheets with sizes ranging from hundreds of nm to a few μm, similar to the dispersion of graphene oxide in water. These results provide guidance while selecting solvents to disperse graphene-based materials.

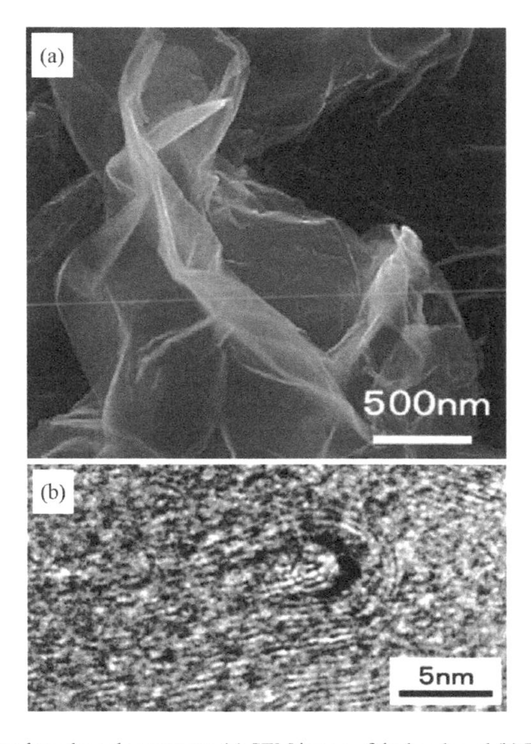

Fig. 2.6. Fine structure of the random shaped aggregate: (a) SEM image of the bends and (b) TEM image of the center of one bend. Reproduced with permission from [21], Copyright © 2004, Elsevier.

The as-prepared graphite oxide material was dispersed in water and 13 organic solvents with a nominal concentration of 0.5 $mg \cdot ml^{-1}$ with the aid of ultrasonication. The dispersions were then allowed to settle for several weeks. Figure 2.7 shows photographs of all the dispersions immediately after the sonication (top) and 3 weeks after the sonication (bottom). For the just dispersed samples, graphene oxide could be well-dispersed in most of the solvents, while it was not stable in dichloromethane and *n*-hexane, and was dispersed in methanol and *o*-xylene to a lesser extent. However, some dispersions were stable only for a short while, from hours to days, such as those in acetone, ethanol, 1-propanol, DMSO and pyridine. Four solvents, including ethylene glycol, DMF, NMP and THF, offered long-term stability, which is comparable to that of dispersion in water. In the five most stable dispersions, only a small amount of precipitate was observed during the first few days after the sonication and they were kept stable without further precipitation.

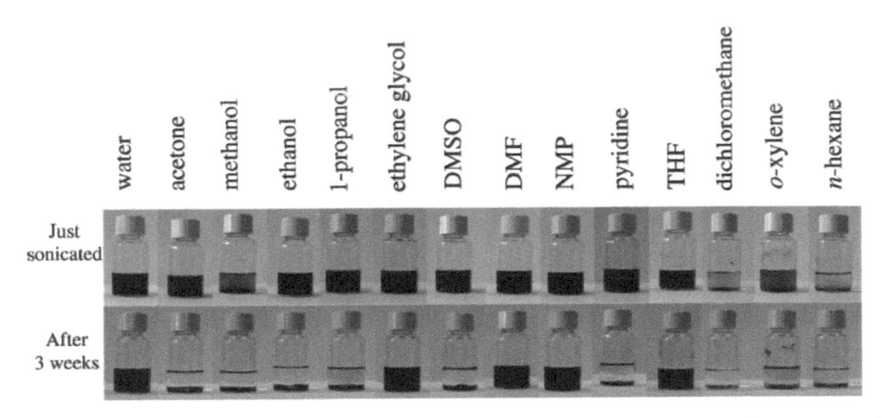

Fig. 2.7. Photographs of the as-prepared graphene oxide dispersed in water and 13 organic solvents prepared with bath ultrasonication for 1 h. Top: dispersions immediately after the sonication. Bottom: dispersions 3 weeks after the sonication. Reproduced with permission from [33], Copyright © 2008, American Chemical Society.

Color image of this figure appears in the color plate section at the end of the book.

Those precipitates were attributed to graphite oxide particles that were not sufficiently exfoliated. In other words, if these graphite oxide particles are properly exfoliated, they can be dispersed.

To disperse graphene oxide, the organic solvents should be polar molecules, because the graphene oxide sheets contain a large quantity of polar oxygen-containing groups, such as hydroxyl, carbonyl, carboxyl, etc., which would trigger a strong graphene oxide sheet–solvent interaction. Water and the four good organic solvents have high electrical dipole moment values, while the relatively poor solvents possess lower dipole moment values.

The GO exfoliated by using the above methods has strong interlayer hydrogen bonds between the oxygen functional groups of adjacent graphene oxide layers. Therefore, GO is usually hydrophilic, so the direct exfoliation of GO in non-aqueous solvents is not effective. This is because organic solvents are unable to penetrate the interlayer spaces of GO and disrupt these hydrogen bonds, leading to poor exfoliation. One idea is to reduce the density of the hydrogen bond groups, such as hydroxyls, through chemical functionalization, so that the graphene oxide layers become less hydrophilic and the strength of interlayer hydrogen bonding is decreased. In this case, it is possible to exfoliate GO in organic solvents. It has been shown that graphite oxide can be functionalized to form chemically modified graphite oxide derivatives [41]. Examples include the silylation of butyl amine-intercalated graphite oxide [42, 43].

Functionalized graphite oxides were prepared by treating them with organic isocyanates [44]. The isocyanate-treated GOs (iGOs) could be exfoliated into functionalized graphene oxide nanoplatelets, so that stable dispersions in polar aprotic solvents could be readily achieved. According to FT-IR spectroscopy and elemental analysis, the isocyanate treatment led to the functionalization of the carboxyl and hydroxyl groups in the GO through the formation of amides and carbamate esters, respectively, as shown in Fig. 2.8. The degree of the functionalization could be controlled through either the reactivity of the isocyanate or the reaction time.

Figure 2.9 shows the dispersions of the parent GO in DMF (left), phenyl isocyanate-treated GO in water (middle) and phenyl isocyanate-treated GO in DMF (right), all at a concentration of 1 $mg \cdot ml^{-1}$. The vials with parent GO in DMF and phenyl isocyanate-treated GO in water contained visible precipitates, indicating their poor dispersing stability. In contrast, the dark brown dispersion of the phenyl isocyanate-treated GO in DMF had no visible precipitate, and was stable for weeks.

It is expected that the Brodie, Staudenmaier and Hummers' methods and their modified versions will be continuously used to produce GO nanosheets [45–49]. However, due to the use of strong acids and oxidants during the processing, the structural defects caused by the chemical reactions disrupted the electronic structure of graphene, so that GO usually has a sheet resistance that is several orders of magnitude higher than that of pristine graphene or graphite. Therefore, chemical or thermal reduction

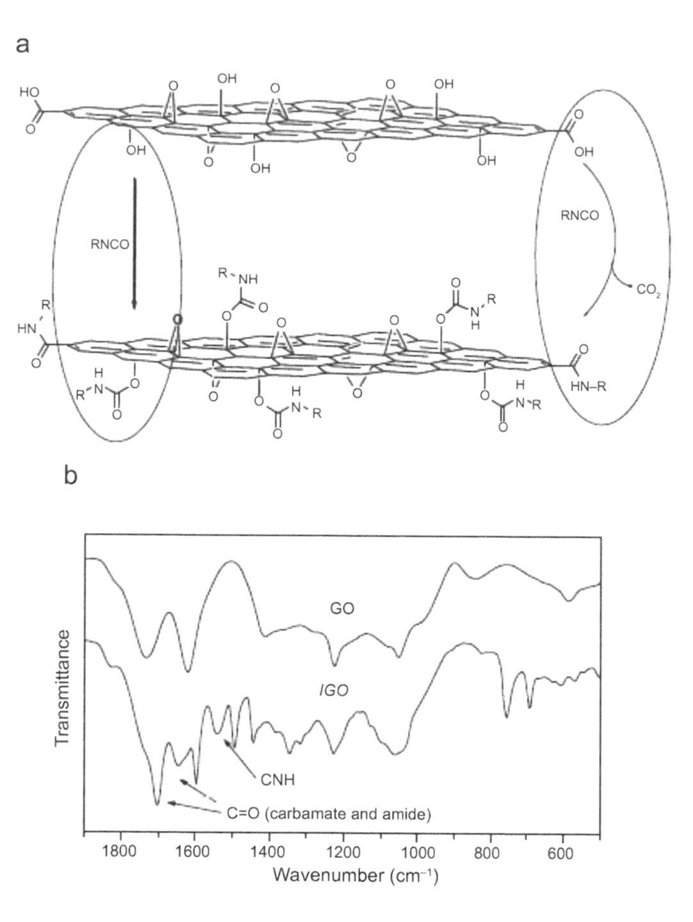

Fig. 2.8. (a) Proposed reactions during the isocyanate treatment of GO, where the organic isocyanates reacted with the hydroxyl (left oval) and carboxyl groups (right oval) of the GO sheets to form carbamate and amide functionalities, respectively. (b) FT-IR spectra of GO and phenyl isocyanate-treated GO. Reproduced with permission from [44], Copyright © 2006, Elsevier.

of GO is necessary in order to recover electrical conductivity for some applications, such as electrodes. However, the complete reduction of GO to defect-free high quality graphene is still a challenge [50–54]. Reduction of GO has been mainly carried out by using chemical methods, by using various reducing agents. Among them, hydrazine is the most widely used one, including hydrazine monohydrate [55–57], dimethylhydrazine [58, 59] and anhydrous hydrazine [8]. Efficient chemical reduction of GO has mostly been achieved in solutions, because both sides of the sheets can be in contact with the reducing agents. Hydrazine is able to effectively remove the in-plane functional groups, like epoxy and hydroxyls, but is unable to eliminate the edge moieties, including carboxyl and carbonyl [7].

Due to their hydrophobic characteristics, graphite or graphene sheets cannot be stably dispersed in water without using any dispersing agents. Therefore, when GO is reduced to graphene, precipitation occurs. A strategy, known as electrostatic stabilization, has been proposed to disperse chemically reduced graphene nanosheets from graphite derived GO in water [7]. This discovery makes it possible to develop large-scale production of aqueous graphene dispersions without the need for polymeric or surfactant stabilizers.

In this process, the as-synthesized graphite oxide was suspended in Ultrapure Milli-Q® water to form a brown dispersion, with the residual salts and acids being completely removed through dialysis. The as-purified graphite oxide suspensions were then dispersed in water at a dispersion concentration of 0.05 wt%. In the chemical reduction of graphite oxide to graphene, the homogeneous dispersion (5.0 ml) was mixed with 5.0 ml of water, 5.0 ml of hydrazine solution (35 wt% in water) and 35.0 ml of ammonia solution (28 wt% in water). The weight ratio of hydrazine to GO was about 7:10. After being vigorously

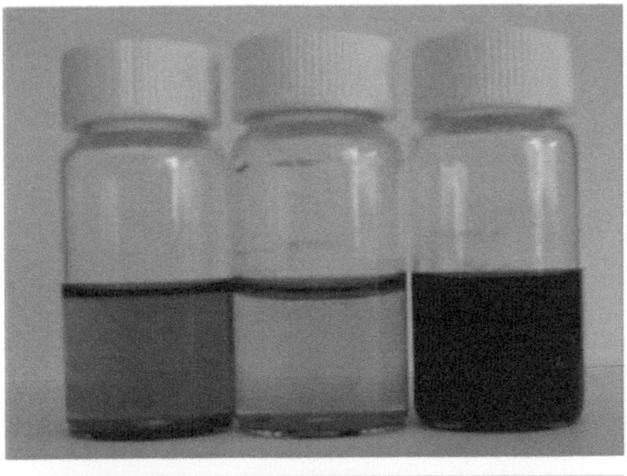

Fig. 2.9. Vials containing the dispersions (1 mg·ml^{-1}) of GO in DMF (left), phenyl isocyanate-treated GO in water (middle) and phenyl isocyanate-treated GO in DMF (right). The top image shows the dispersions 24 h after preparation. The bottom image shows the inverted dispersions with the precipitate clearly shown on the bottom of the left and middle vials. Reproduced with permission from [44], Copyright © 2006, Elsevier.

shaken or stirred, the vial was put in a water bath at 95°C for 1 h. The concentration of hydrazine in the reduction mixture can be varied from 0.0175 wt% to 1.75 wt%. In order to obtain stable dispersions with concentrations of > 0.0175 wt%, the excessive hydrazine should be removed by dialysis against a 0.5% ammonia solution after the reduction is finished. A $R_{N2H4/GO}$ ratio of 7:10 was the optimal ratio to obtain stable dispersions of highly conducting graphene nanosheets.

Figure 2.10 shows a schematic diagram demonstrating the solution-based route to obtain hydrophilic graphite oxide, which is exfoliated as individual graphene oxide (GO) sheets by ultrasonication in water. However, chemically reduced graphene nanosheets obtained through this method precipitate as irreversible agglomerates due to their hydrophobic nature. The resulting graphene agglomerates are insoluble in water and organic solvents, which cannot be used for further processing.

Surface charge or zeta potential studies have indicated that the as-prepared GO nanosheets are negatively charged when they are dispersed in water, as shown in Fig. 2.11 (a). This is attributed to the ionization of the carboxylic acid and phenolic hydroxyl groups on surfaces of the GO nanosheets [23, 60]. Therefore, the formation of stable GO colloids in water is attributed to the electrostatic repulsion, rather than just the hydrophilicity of GO. Because it is difficult to reduce the carboxylic acid groups with hydrazine, these groups still remain after the reduction, which has been confirmed by FT-IR analysis, as demonstrated in Fig. 2.11 (b). This result implied that the surfaces of the graphene nanosheets in aqueous solution are still charged after the chemical reduction. As a result, the electrostatic repulsion mechanism could be used to form well-dispersed graphene colloids.

It has been shown that, in most colloids, the colloidal stability of an electrostatically stabilized dispersion is strongly dependent on pH, the concentration of electrolytes and the electrolytes of dispersed particles [61]. Based on these facts, chemically reduced graphene nanosheet stable colloids through electrostatic stabilization have been achieved, as shown schematically in Fig. 2.10. Graphene oxide

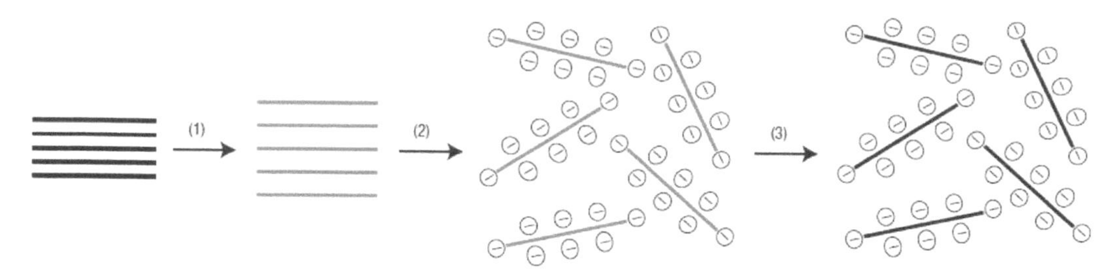

Fig. 2.10. Scheme showing the chemical route to the synthesis of aqueous graphene dispersions. (1) Oxidation of graphite (black blocks) to graphite oxide (lighter colored blocks) with greater interlayer distance. (2) Exfoliation of graphite oxide in water by sonication to obtain GO colloids that are stabilized by electrostatic repulsion. (3) Controlled conversion of GO colloids to conducting graphene colloids through deoxygenation by hydrazine reduction. Reproduced with permission from [7], Copyright © 2008, Nature Publishing Group.

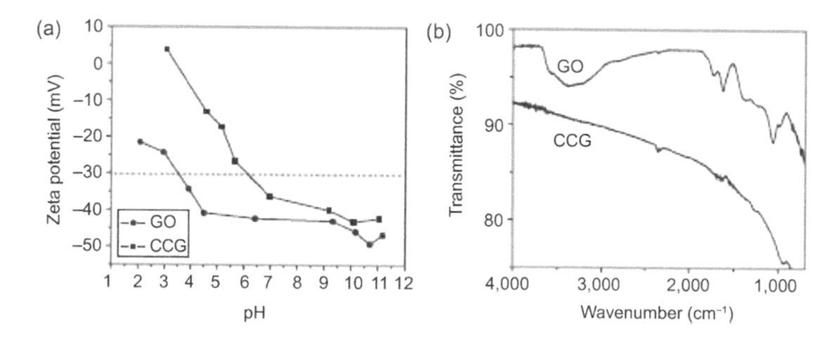

Fig. 2.11. Surface properties of the GO and chemically converted graphene (CCG). (a) Zeta potential of GO and CCG as a function of pH, in aqueous dispersions at a concentration of 0.05 mg·ml^{-1}. (b) FT-IR spectra of the GO and CCG. The absorption band at 1,700 cm^{-1} is attributed to carboxyl groups. The absorption of the CCG sheets at this range is observable, but not as pronounced as that observed for the GO, due to the overlapping of the strong absorption of graphene sheets in this region. Reproduced with permission from [7], Copyright © 2008, Nature Publishing Group.

nanosheets in the dispersions can be directly converted to graphene nanosheets that can be dispersed in water, without precipitation. In this case, the metal salts and acids that generally remained in the starting graphite oxide should be completely removed, because they can neutralize the charges on the graphene nanosheets and thus destabilize the dispersions. Ammonia was used to control the pH of the suspensions to be about 10. Volatile ammonia can be easily removed after the processing of the graphene nanosheets.

As the GO dispersions had concentrations of < 0.5 mg·ml^{-1}, reduction with hydrazine under optimized conditions caused no increase in the particle size of the resulting graphene nanosheets, as shown in Fig. 2.12 (a). Figure 2.12 (b) shows an atomic force microscopy (AFM) image of the graphene nanosheet on silicon substrates. The flat graphene nanosheets had a thickness of 1 nm. In the graphene nanosheet suspensions, ammonia and hydrazine dissociated into ionic items with behaviors like electrolytes. If the amount of hydrazine exceeded the optimal level, i.e., hydrazine:GO = 7:10 by weight, the stability of the dispersion decreased with the increasing concentration of hydrazine. Therefore, excessive hydrazine must be immediately removed from the dispersions after the reduction. Also, the concentrations of the GO dispersions should be < 0.5 mg/ml.

Besides hydrazine, other reducing agents, such as hydrides, hydroquinone [62, 63], and *p*-phynylene diamine [64], vitamin C and amino acid [65, 66], wild carrot root [67], potassium iodide [68], iron (Fe) [69], zinc (Zn) [70], and so on, have also been used for GO reduction. Hydrides include sodium borohydride [34, 62, 71, 72] and sodium hydride [73]. It has been reported that these residual edge groups can be removed by using concentrated H_2SO_4 after the initial reduction step [74]. As an alternative to chemical methods, hydrogen plasma treatment has also been shown to result in efficient reduction [56]. Some examples are discussed in the following part.

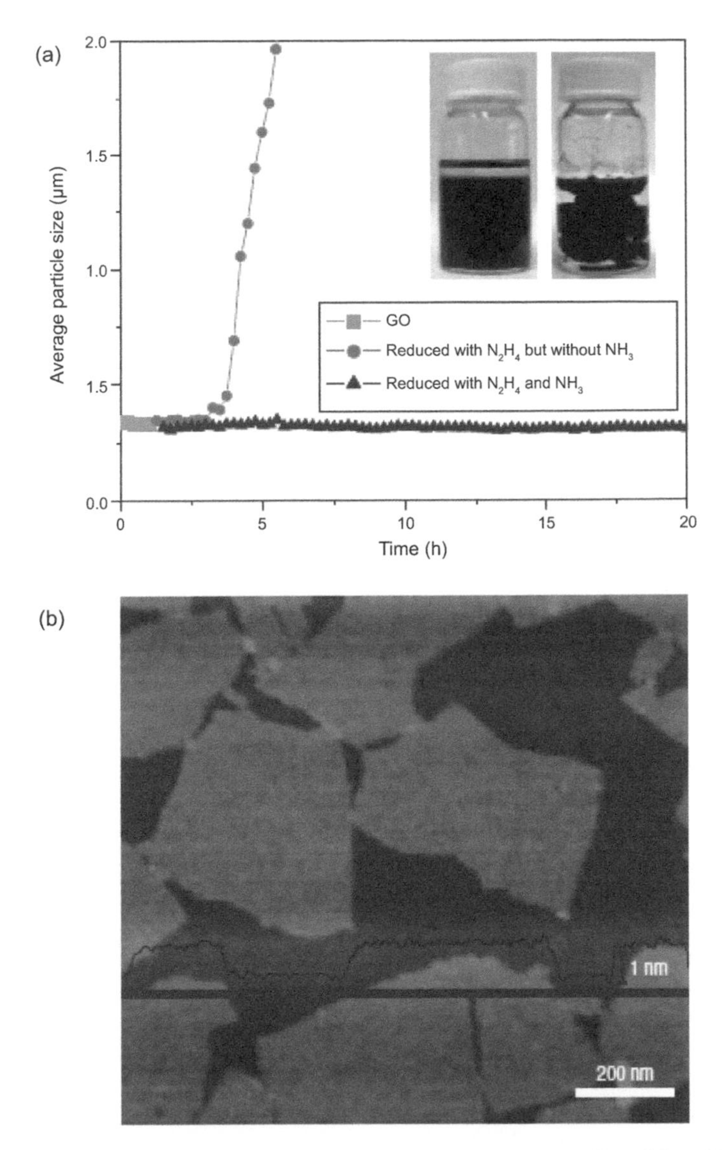

Fig. 2.12. Colloidal and morphological characterization of the CCG dispersions. (a) Effect of the addition of ammonia on the dispersion state of the CCG sheets, characterized by measuring average particle sizes over a long period of time. The photographs shown in the inset were taken two days after the reduction reaction was complete with (left) and without (right) the addition of ammonia. The concentration of the starting GO solution is 0.5 $mg \cdot ml^{-1}$. (b) Tapping mode AFM image of the CCG sheets with a height profile (blue curve with scale bar of 1 nm) taken along the red line. The sample was prepared by drop-casting a dilute CCG dispersion onto a silicon wafer. Reproduced with permission from [7], Copyright © 2008, Nature Publishing Group.

The reduction of GO to reduced GO (rGO) by using sodium hydride (NaH) is described [73]. When GO dispersion in methanol was treated with fresh sodium hydride powder at a dosage of 100 mg/ml, instantaneous reduction of GO to rGO occurred in less than 1 min. Accordingly, the yellow colored GO suspension became dark black. After being kept stationary for 10 min, the reduced suspension centrifuged at a high speed of 11000 g for 60 min at 20°C, this was followed by washing with DI water to remove the sodium methoxide ($NaOCH_3$) deposits on the rGO nanosheet surface. Stable rGO dispersion was formed by redispersing the rGO powder with adsorbed $NaOCH_3$ in pure methanol.

For the as-prepared GO, the addition of the electronegative oxygen and the removal of unsaturated π electrons caused the GO sheets to be distorted, due to the introduction of nonplanar sp3 bonds, as shown in Fig. 2.13 (top). When NaH powder was dropped into the GO suspension, the immediate reduction was accompanied by the generation of hydrogen gas bubbles. At the same time, methanol was deprotonated into methoxy ions by sodium hydride [75], which in turn stabilized the rGO nanosheets in methanol, as shown schematically in Fig. 2.13. With respect to the initial mass of the graphite flakes, the process offered a high yield of about 68%, with most rGO nanosheets having 1–4 layers. Nearly 100% GO could be reduced to rGO in this process. Due to the electrostatic repulsion between individual rGO nanosheets in methanol, which was induced by the adsorbed methoxide ion (Na^+CH3O^-), the rGO suspension was highly stable with no visible settling.

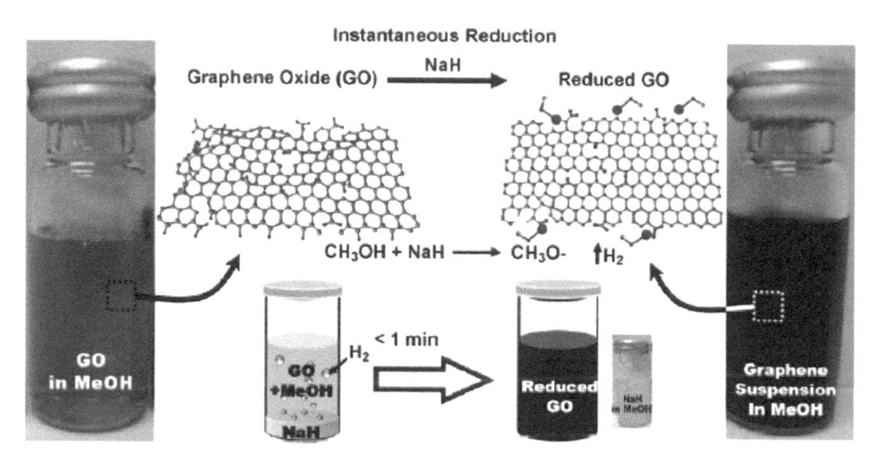

Fig. 2.13. Schematic diagram of the hydride reduction process. Photographs of glass vials containing the dispersion of GO in methanol (left) and the stable dispersion of rGO in methanol (right). Center right, bottom: picture of a vial with NaH in methanol. The dark black color of the stabilized rGO in contrast to the yellow color of the GO indicates the partial restoration of the interlayer π network of the rGO nanosheets. The cartoon and the three-dimensional (3D) chemical structures, with the gray, red and blue balls representing carbon, oxygen and sodium atoms, respectively, in the ball-and-stick model, show the reduction of GO to rGO with hydride, together with the release of hydrogen gas and the stabilization of the rGO suspension by sodium methoxide ions. Reproduced with permission from [73], Copyright © 2010, John Wiley & Sons.

Microstructural analysis results have indicated that sodium methoxide with dendritic structures was deposited on the surface of the resulting rGO nanosheets. The rGO suspension was destabilized when deionized (DI) water was added, due to the removal of the stabilizing methoxide ions from surface of the rGO nanosheets. It has been confirmed that sodium was completely removed, so that the rGO nanosheets had a smooth surface, as demonstrated by SEM image in the inset of Fig. 2.14 (b). When a 300-nm thick silicon dioxide substrate was contacted with an rGO solution for 10 min, Fig. 2.14 (c) shows that the rGO had a large areal coverage on SiO_2/Si substrates. AFM observation indicated that the rGO had a thickness of about 0.6 nm, which was thinner than the original GO (about 1.2 nm), as illustrated in Fig. 2.14 (a, b). This observation could be attributed to the presence of (i) partially unreduced nonplanar oxy-functional groups, (ii) remnant sp^3 C–C bonds or (iii) gas/solvent molecules trapped between the substrate and the rGO nanosheets [76]. In addition, the rGO deposits exhibited a higher degree of folding, due to the π–π interaction energy within the nanosheets. A possible folding mechanism is shown in Fig. 2.14 (c, left). Also due to the negatively charged methoxide adsorbed on the rGO nanosheet surfaces, the rGO deposits had no agglomeration during deposition.

Figure 2.15 shows Raman spectra of GO and rGO. GO had an I_D/I_G ratio of 1.88 ± 0.25, while the I_D/I_G ratio of the rGO was 1.08 ± 0.15, which indicated that the hydride reduction decreased the relative content of the sp3 carbon atoms and other related defects. In addition, the reduction of GO to rGO also led to a decrease in $I_{D'}/I_G$ ratio, which is usually used to quantify the content of weak defects, induced intravalley scattering (D' peak) and the graphenic region (G peak).

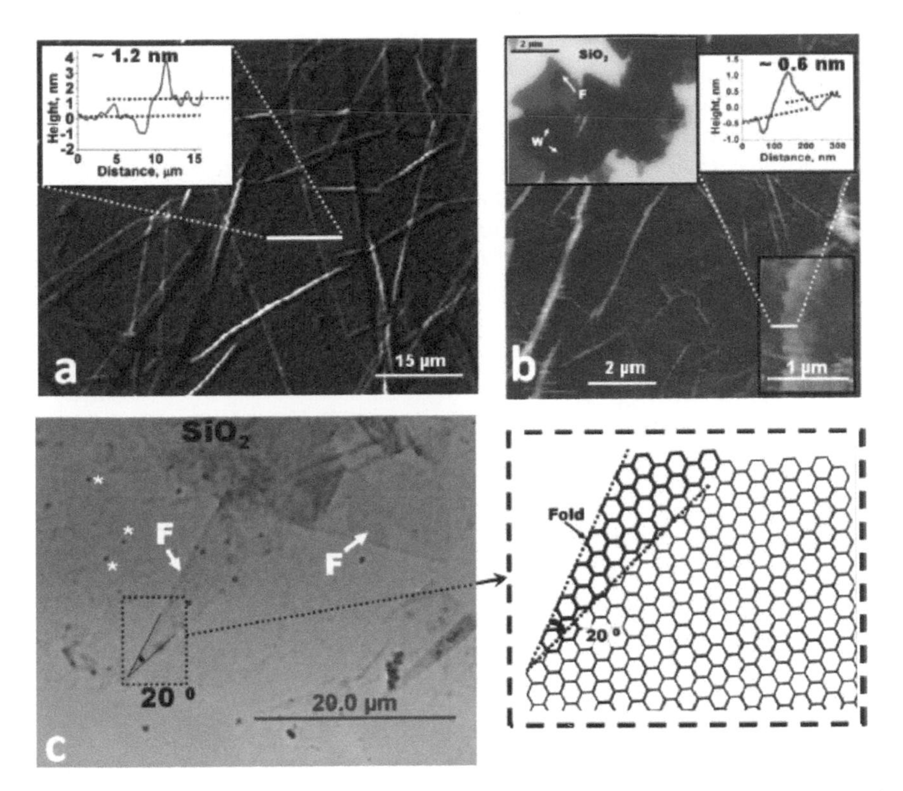

Fig. 2.14. (a, b) Atomic force microscopy (AFM) images of the GO nanosheets (a) and rGO nanosheets (b) spin-coated on 300-nm-thick silica wafers. The height profiles shown as the insets indicate that the thickness of the GO monolayer is about 1.2 nm, while that of the rGO monolayer is 0.6 nm. The top left inset in (b) shows an SEM image of the washed rGO nanosheets deposited on a 300-nm-thick silica substrate. The nanosheets had wrinkles (W) and folds (F). (c) Optical image of the rGO nanosheets deposited on a 300-nm-thick silica substrate, which shows large-area coverage of sporadic folding (F). The stars indicate the possible residual sodium methoxide deposits on the rGO nanosheets and on the substrate. The schematic diagram of the carbon structure (right) depicts a possible mechanism of folding of the rGO nanosheets. Reproduced with permission from [73], Copyright © 2010, John Wiley & Sons.

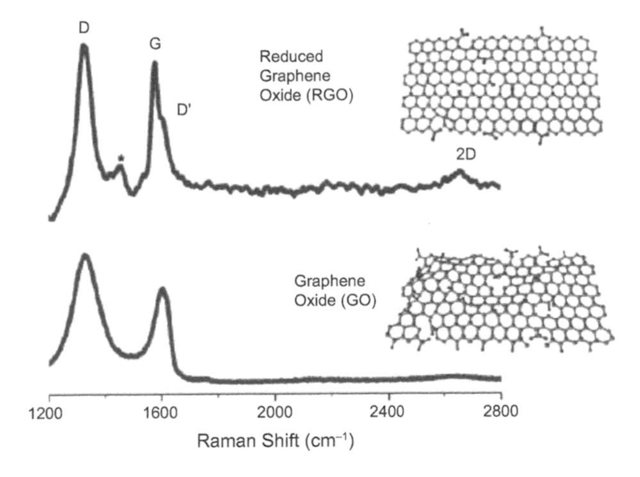

Fig. 2.15. Raman spectra of the GO and rGO nanosheets. The I_D/I_G and $I_{D'}/I_G$ ratios for rGO were decreased after the reduction process, which suggested the presence of the long-range crystallographic order of the sp^2 carbon atoms in the rGO nanosheets. The peak marked (*) in the rGO spectrum was attributed to residual surface-adsorbed sodium methoxide molecules. 3D chemical structures of the GO and rGO nanosheets are shown on the right. Reproduced with permission from [73], Copyright © 2010, John Wiley & Sons.

Fe is a cheap and widely available element. Fe powder has been used to reduce exfoliated graphite oxide at room temperature, with the principle being shown schematically in Fig. 2.16 [69]. Due to the presence of residual Fe, the resulting graphene nanosheets exhibited a high adsorption capacity of 111.62 mg·g^{-1} for methylene blue at room temperature. More importantly, the items could be recycled from the suspensions by using magnetic separation after the absorption. In a representative experiment, 1 g of Fe powder with an average particle size of 10 μm and 20 ml of HCl (35 wt%) were directly added into 100 ml of GO suspension at room temperature. The mixture was stirred for 30 min and then kept stationary for a period of time. After reduction, 15 ml of HCl (35 wt%) was added into the suspension in order to fully remove the excessive Fe powder. The resulting graphene nanosheet powder was collected with filtration.

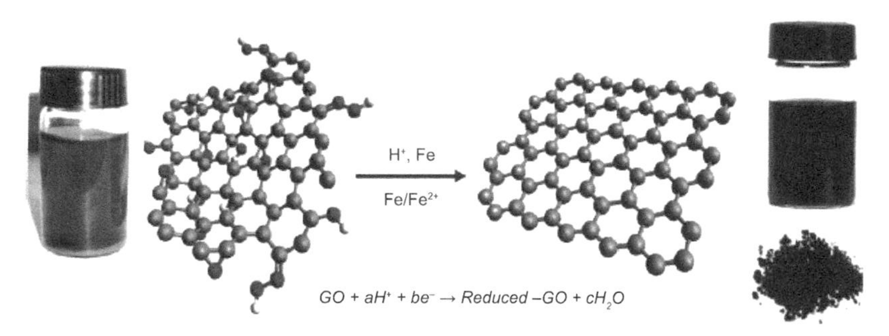

Fig. 2.16. Illustration of preparation of the graphene nanosheets with Fe powder as the reducing agent. Reproduced with permission from [69], Copyright © 2011, American Chemical Society.

Figure 2.17 (A) shows that the brown colored GO suspension was rapidly darkened in 60 min upon the addition of Fe and H^+. Due to the hydrophilic characteristics of the oxygenated graphene layers, GO is easily exfoliated in aqueous media [27, 28]. As a result, GO sheets with a thickness of 1 nm can be readily dispersed in water to form a stable colloidal suspension, as shown in Fig. 2.17 (A). With the presence of H^+, Fe reacted with H^+ to form Fe^{2+} ions at the surface of the Fe particles. In this case, the GO nanosheets with negative charges (–37 mV) could be attracted onto the surface of the positively charged Fe particles to form spherical structures, as shown in Fig. 2.17 (D, E). Therefore, the Fe particles were covered by the GO nanosheets, in close contact, so that reduction of the GO to rGO became possible, through the fast electron transport from Fe/Fe^{2+} to GO [37].

After reduction for 360 min, rGO powder was obtained, which comprised of randomly aggregated crumpled thin sheets, as illustrated in Fig. 2.17 (F, H). Figure 2.17 (I) shows cross-sectional TEM image of the rGO nanosheets. The thickness of the rGO nanosheets was between 1–5 nm, consisting of approximately 2–10 stacked monatomic graphene layers. It was found that, without HCl, the reduction of GO was very slow. The introduction of HCl dissolved the passive film of the Fe particles and promoted the reduction potential of Fe/Fe^{2+}.

The magnetization hysteresis loop of the rGO powder is shown in Fig. 2.18, which is an S-like curve with a saturation magnetization value of 1.66 emu·g^{-1}. Moreover, the rGO exhibited typical superparamagnetic behavior, without coercivity and remanence magnetization. The rGO powder exhibited a strong adsorption to MB in aqueous solution. More importantly, the magnetic rGO could be readily separated from the solution by using a magnetic field. In summary, this is a cost-effective, environmentally friendly and large-scale method to produce graphene nanosheets.

Another interesting example is the reduction of GO in strong alkaline solutions [77]. It was incidentally found that a stable graphene suspension could be quickly prepared by simply heating an exfoliated GO suspension under strongly alkaline conditions at moderate temperatures of 50–90°C, as shown in Fig. 2.19 (a). The finding originated from an experiment using free-radical addition to exfoliate GO through the introduction of functional groups, which was demonstrated in CNTs [78]. NaOH was added to the GO suspension in order to increase the solubility of the alkyl free-radical initiator, with carboxyl-termination. However, the suspension experienced a rapid color change from yellow-brown to dark black,

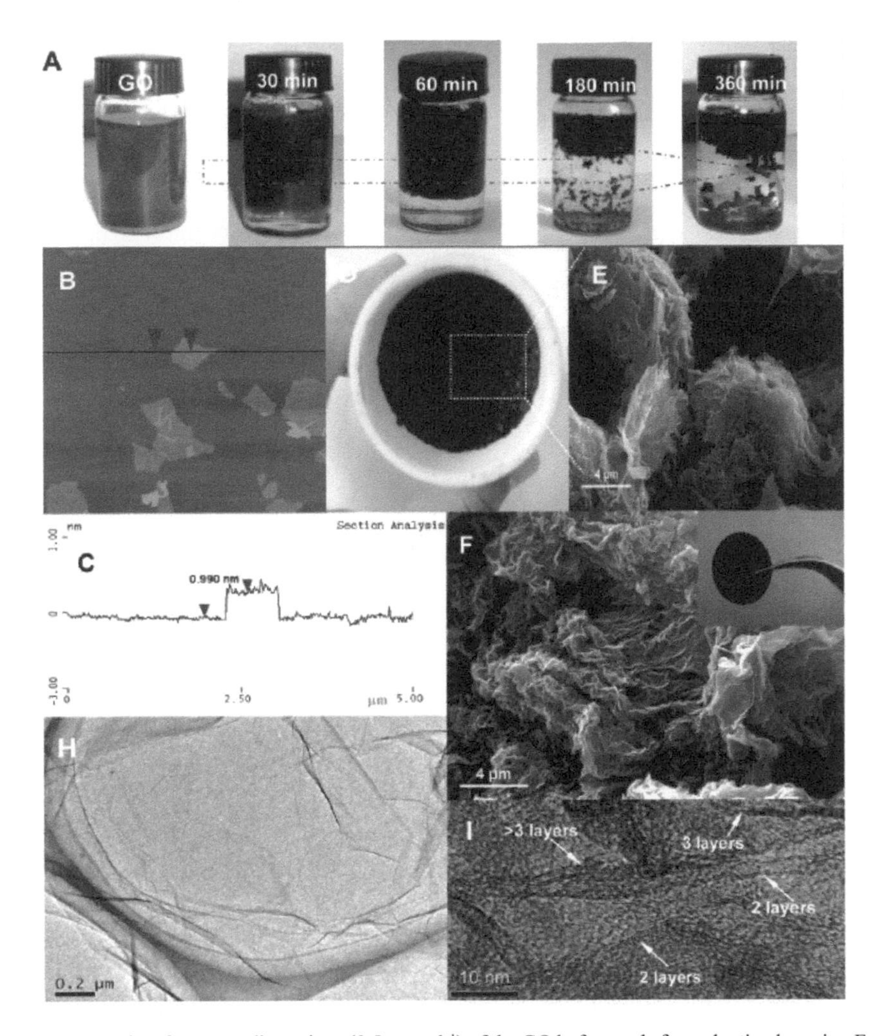

Fig. 2.17. (A) Photographs of aqueous dispersions (0.5 mg·ml^{-1}) of the GO before and after reduction by using Fe powder for different reaction times. (B) AFM image of the GO dispersed on mica and (C) the corresponding line profile. (D) Photograph and (E) SEM image of the rGO after reduction for 30 min without the presence of HCl. (F) SEM and (H, I) TEM images of the rGO after reduction for 360 min. The inset of panel F shows a piece of pressed graphene paper. Reproduced with permission from [69], Copyright © 2011, American Chemical Society.

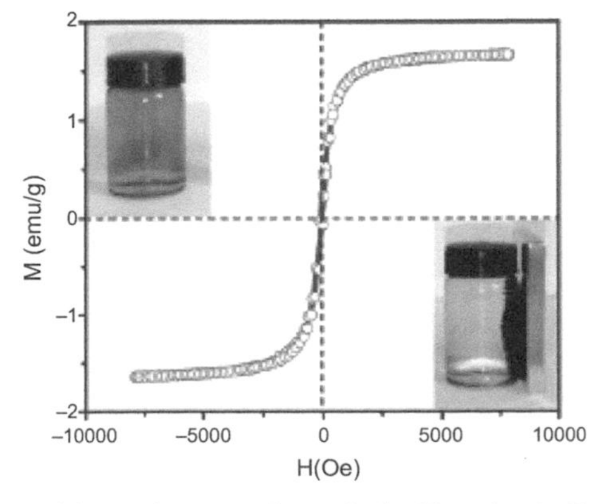

Fig. 2.18. Magnetic hysteresis loop of the graphene nanosheets obtained by using the Fe reduction for 360 min without acid treatment at room temperature. Reproduced with permission from [69], Copyright © 2011, American Chemical Society.

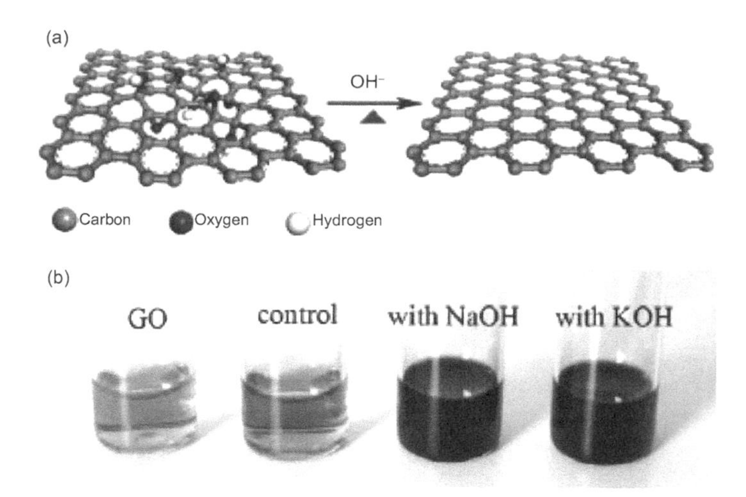

Fig. 2.19. (a) Illustration of the deoxygenation (reduction) of exfoliated GO under alkaline conditions and (b) photographs of the exfoliated-GO suspensions (0.5 mg·ml^{-1}) before and after the reaction. The control sample in (b) was prepared by heating the pristine exfoliated-GO suspension without the presence of NaOH or KOH at 90ºC for 5 h with the aid of sonication. No obvious color change was observed in the control suspension, even when it was heated for a prolonged time period at relatively higher temperatures. Reproduced with permission from [77], Copyright © 2008, John Wiley & Sons.

which implied that the exfoliated GO was deoxygenated (reduced), leading to stable aqueous graphene suspensions, as demonstrated in Fig. 2.19 (b). At a suitable temperature, e.g., 80ºC, the yellow-brown exfoliated GO suspension became black in a few minutes.

The graphene suspensions obtained in this way exhibited very high stability, which made it possible for further processing, such as spin-coating, casting or spraying to take place. Because the negatively charged oxide functional groups were not completely removed, the stability of the graphene suspension was attributed to a strengthened electrostatic stabilization caused by the strong alkaline conditions. This was because the repulsion between negatively charged graphene nanosheets was increased at higher pH values. The deoxygenation (reduction) of the exfoliated GO under the strong alkaline conditions could be understood to be the reverse process of the oxidation reaction of graphite in strong acids [6]. This hypothesis was supported by the dependency of the deoxygenation reaction on pH value, i.e., the higher the pH value of the exfoliated-GO suspension, the faster the reaction would be. More importantly, if the pH value of the suspension was sufficiently high, the reaction could take place at room temperature or even lower. On the other hand, it is not surprising that all the reducing agents in the chemical reductions of exfoliated GO are intrinsically strong alkalis, such as hydrazine and dimethylhydrazine, as discussed previously [7, 56, 58, 79].

Other routes, including electrochemical reduction [37, 80, 81], photocatalytic reduction [82] and flash conversion [83], have been explored to reduce exfoliated GO to rGO nanosheets. However, these methods are more suitable for the reduction of thin films than for bulk or solutions.

A simple, efficient, low-cost and environmentally friendly electrochemical method has been reported to fabricate electrochemically reduced graphene oxide (ER-GO) films with an O/C ratio of less than 6.25% [37]. When coupled with a spray coating technique, the reduction method can be used to prepare large-area patterned ER-GO films with thicknesses ranging from monolayer to several microns on various conductive and insulating substrates. GO slurry was prepared by using the modified Hummers' method [84], which was used to fabricate GO films on substrates by using spray coating, as demonstrated schematically in Fig. 2.20. Each spraying cycle consisted of spraying for 1 s and drying for 20 s. By controlling the concentration of the dispersion and the number of the spray coating cycles, the thickness of GO films could be readily controlled. By applying templates, patterned GO films could be obtained, as shown in Fig. 2.20.

The GO film could be directly reduced to rGO by using the potentiostatic method. Digital camera photographs showed that as the electrochemical reaction proceeded, a yellow–brown/black and near circular

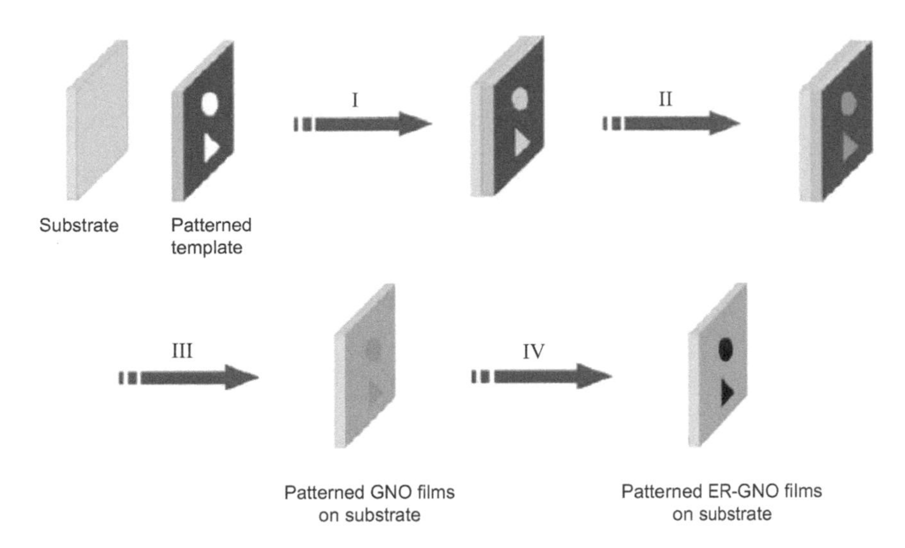

Fig. 2.20. Schematic diagram of the four-step procedure to fabricate patterned ER-GO films with desired thickness on a given substrate. (I) A template (mask) with circular and triangular patterns (holes) was put onto the target substrate. By using templates with the desired patterns, GO films with complex shapes on different substrates could be fabricated. (II) A solution of GO was spray coated onto the substrate through the template. The thickness of the deposited GO film could be controlled by varying the concentration of GNO in solution and the number of spray-coatings. (III) After the template was removed, GO films with circular and triangular patterns were obtained. (IV) By electro-reducing the GO films, patterned ER-GO films on the substrate were obtained. Reproduced with permission from [37], Copyright © 2009, John Wiley & Sons.

area grew around the tip of the electrode, as demonstrated in Fig. 2.21 (a–c) [37]. The color change is an indicator that the GO was reduced and the π network was restored, i.e., rGO was obtained. SEM images, which are shown in Fig. 2.21 (d–g), indicate that the morphology of the rGO films was similar to that of the GO films, suggesting that the electrochemical reduction is a non-destructive method, which is important for various applications. After electrochemical reduction the thickness of the film was decreased from 7 to 5 μm, which was attributed to the elimination of the oxide groups from the GO [7, 8, 57, 59]. At the same time, the rGO film had an electrical conductivity of 8.50×10^3 S·m^{-1}, which was nearly eight orders of magnitude higher than that of the original GO film 2.80×10^{-5} S·m^{-1}. This implied that the conjugated sp^2 carbon network was present in the rGO film, which was responsible for the circular area that was brighter than the unreduced area in Fig. 2.21 (e) and (f).

The process of flash reduction is discussed as a demonstration in this part. The flash experiments could be carried out either in air or in N_2 atmosphere inside a glovebox [83]. It was found that lower flash energy, measured as f/stops on the flash units, was needed to achieve reduction in N_2 than in air. This is because in N_2 the water content in the GO was lower, leading to more effective heating. Free-standing GO films could be flash reduced with a single close-up of < 1 cm flash by using a Xenon lamp equipped on a common digital camera. Interdigitated electrode arrays on GO/polystyrene thin film could be flash patterned. The typical flash energies applied to the samples were 0.1–2 J·cm^{-2}. The uniformity and contrast of the patterns could be improved by using multiple flashes of lower energy. High powder flashes were not suggested, in order to avoid the formation of cracks, due to large volume shrinkage and rapid degassing. GO/polystyrene blends were found to offer better defined patterns and more durable rGO films, due to the heat sink effect of the nonabsorbing polystyrene spheres, which were melted to fuse with the rGO sheets.

As demonstrated in Fig. 2.22 (a) and (b), a dramatic change in color was observed in the GO thin film before and after flash reduction [83]. Upon a close camera flash within 1 cm, the transparent brown film instantaneously turned black and opaque, which was accompanied by a loud popping sound. The strong photoacoustic response was attributed to the rapid expansion of air near the surface of the film, due to a heat pulse generated by photothermal energy conversion, during which thermal deoxygenating reactions were effectively triggered to reduce the GO to rGO. The resulting black film was more hydrophobic, as shown in the inset of Fig. 2.22 (a) and (b). The water contact angle of the GO film was increased from about

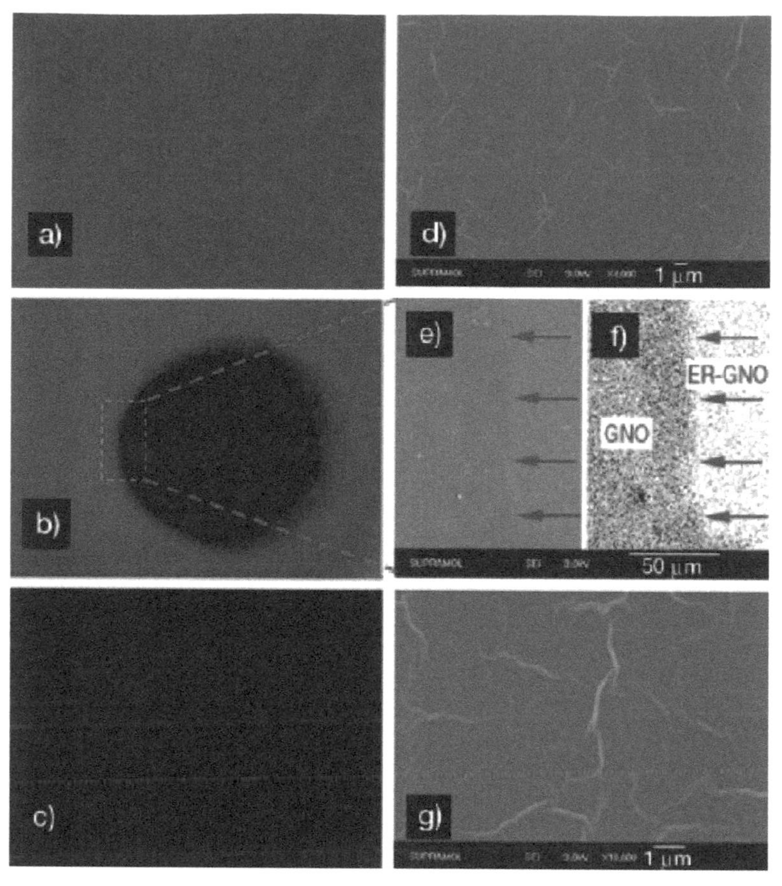

Fig. 2.21. Optical photographs (a–c) and SEM images (d–g) of the 7-mm-thick GO films on quartz of 5 × 4 cm^2, before (0 s, a, d), during (1000 s, b, e, f) and after (5000 s, c and g) the electrolysis. Image (f) was obtained from image (e) with a contrast enhancement of 100. The blue arrows indicate the boundary between the circular ER-GO area and the unreduced GO in the image (b). Reproduced with permission from [37], Copyright © 2009, John Wiley & Sons.

45° to about 78° after the flash reduction, approaching the value measured on a piece of highly ordered pyrolytic graphite (HOPG) of 81°. This implied that the oxygen-containing groups had been removed.

Figure 2.22 (c) shows TGA curves of the GO films before and after flash reduction in N_2 atmosphere [83]. GO showed a first mass loss of about 15% near 100°C due to water removal and a second mass loss of about 25% near 220°C, which was attributed to the loss of oxygen containing groups [55]. The thermal behavior of the flash reduced GO was similar to that produced by using high temperature annealing, with < 2% mass loss in the same temperature range. FTIR spectra of the GO before and after the flash reduction are shown in Fig. 2.22 (d). The spectrum of the GO film showed transmission bands of carboxylic acid at wavelengths of 1630–1730 cm^{-1}, phenyl hydroxyl at about 1100 cm^{-1} and epoxide groups at 930 cm^{-1} [55, 85]. All these groups disappeared in the flash reduced rGO. The flash reduced rGO film also showed much lower transmission in the IR spectrum, demonstrating the electrically conducting characteristic of the carbon [86]. The C/O atomic ratios of the GO and rGO were 1.15 and 4.23, respectively.

The GO films generally exhibited an expansion by more than ten times in volume after the flash reduction, due to the rapid degassing, as shown in Fig. 2.22 (e). This observation is different from that of the rGO film obtained by using the electrochemical reduction method, as discussed above [37]. The expansion of the flashed film implied that the rGO nanosheets were completely exfoliated, which was confirmed by the XRD results shown in Fig. 2.22 (f). The GO film possessed only one sharp peak centered at $2\theta = 9.12°$ in the XRD pattern, corresponding to a spacing of 9.7 Å between the original GO nanosheets, which is

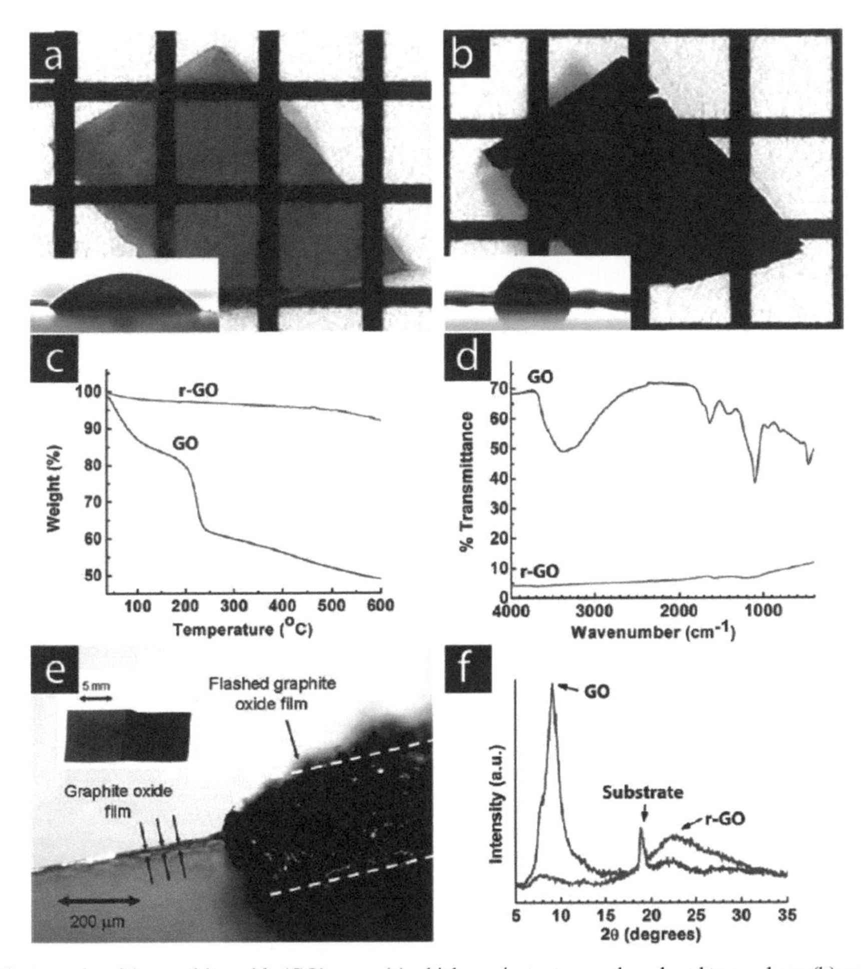

Fig. 2.22. Photographs of the graphite oxide (GO) paper (a) which was instantaneously reduced to graphene (b) upon exposure to a photographic camera flash, with the grids in the background having a dimension of 1 mm × 1 mm. The flash reduction of GO was evident by the significant change in color (a, b), water contact angle (insets), TGA (c) and FTIR (d). In (c, d, f) the blue lines correspond to GO and the red lines correspond to the flash reduced GO (rGO). (e) Cross-sectional view of a GO paper showing large thickness expansion after the flash reduction. Only the right half of the GO sample was flashed. The left part of the picture shows the cross-sectional view of the light brown colored GO film. The thickness was increased by nearly 2 orders of magnitude, leading to a very fluffy and potentially high surface area film. (f) The lack of a graphitic peak in the XRD pattern of the rGO suggested the disordered packing of the rGO nanosheets, consistent with the large volume expansion observed in (e). Reproduced with permission from [83], Copyright © 2009, American Chemical Society.

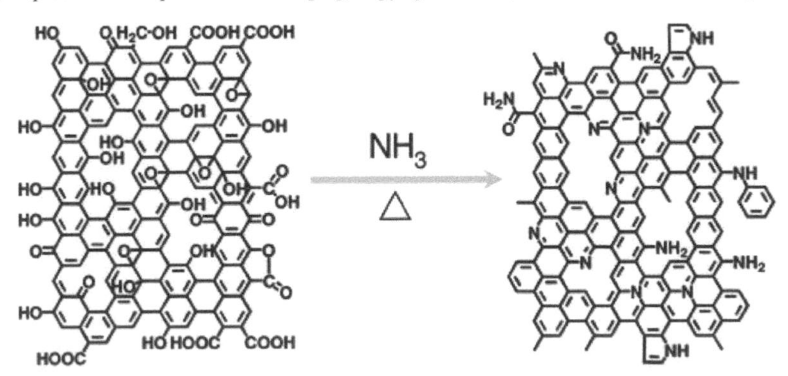

Fig. 2.23. Schematic of the transition from GO to N-doped rGO due to the annealing in NH_3 atmosphere. Reproduced with permission from [93], Copyright © 2009, American Chemical Society.

close to the typical thickness value of a single GO layer, i.e., about 1 nm [57, 87]. After flash reduction, the GO peak disappeared, whereas a broad peak centered at about 22.5° was observed, which is slightly lower than the peak for bulk graphite. The absence of the graphite peak indicated that the flash reduced rGO nanosheets were disorderly packed [88]. The flash reduced rGO films had a density of about 0.14 $g \cdot cm^{-3}$, which is about 6% of the density of bulk graphite. They exhibited an electrical conductivity of about 1000 $S \cdot m^{-1}$. The open structure of the flash reduced rGO thin films could be useful for applications requiring high surface area, such as electrodes of supercapacitors [89, 90].

One of the most striking advantages of the flash reduction is its capability of photopatterning, which is not possible when using the conventional chemical reduction methods. By using photomasks, conductive rGO patterns can be created on the insulating GO films. Furthermore, the exposed areas can also be removed directly by using flashing with enhanced photothermal effect of rGO. In other words, both patterning and etching can be realized in one experimental setup, by simply varying the dose of the flash irradiation, which has been demonstrated [83].

As stated earlier, thermal reduction has also been widely used to reduce GO. Thermal reduction of GO is usually carried out at annealing temperatures of > 200°C in inert or reducing environments. Generally, the reduction efficiency is increased with increasing temperature [85, 91, 92]. It should be mentioned that in the presence of oxygen, GO is oxidized gradually at lower temperatures (< 200°C) and very quickly at high temperatures [55]. Therefore, a tight control of annealing environment is very important for thermal reduction of GO.

One study indicated that annealing GO in NH_3 atmosphere at temperatures of > 300°C led to reduced graphene nanosheets doped with N. The N doping was achieved through the formation of C–N bonds, as shown schematically in Fig. 2.23 [93]. According to X-ray photoelectron spectroscopy (XPS) results, the highest doping level of N was about 5%, which was obtained at 500°C. The original GO had an oxygen level of about 28%, which was decreased to about 2% after annealing at 1100°C. The N binding configuration of the doped GO was pyridinic N. The oxygen groups in the as-prepared GO reacted with NH_3 to form C–N bonds. If the GO was prereduced by thermal annealing in H_2, the reactivity with NH_3 was largely decreased and thus the N-doping level became very limited, because fewer oxygen groups were available to facilitate the reaction of GO and NH_3. The rGO annealed in NH_3 exhibited higher electrical conductivity than that annealed in H_2, implying that the reduction in NH_3 was more effective than in H_2. The N-doped rGO demonstrated n-type conduction behavior with the Dirac point (DP) at negative gate voltages in the three terminal devices. It is believed that this method could be used to synthesize N-doped rGO nanosheets for various practical applications.

The most widely used reduction methods, involving the use of hydrazine and high-temperature treatment, are not ideal for environmental and technological reasons. Green methods are more desirable and will be the research subject of this field in the future. Although several methods have claimed to be green and environmentally friendly, more efforts should be directed towards developing new ways for GO reduction, especially in large-scale production. In addition, it has also been found that multistep reduction through the combination of different processes could be more effective in removing specific functional groups that cannot be eliminated by using the one-step reduction routes [74].

2.2.1.2. Solvent Exfoliation

Solvent exfoliation is a method used to directly produce graphene nanosheets from graphite, without the presence of the intermediate phase, graphene oxide. Graphite has been exfoliated by using a range of organic solvents, such as N-methylpyrrolidone (NMP) [94] and N,N-dimethylformamide (DMF) [95]. The exfoliation of graphite is attributed to the strong interaction between the graphitic basal planes and the solvents, i.e., the system energy after exfoliation and subsequent dispersion could be minimized. In this case, solvents used for the exfoliation of graphite should have a surface tension of about 40 $mJ \cdot m^{-2}$. Transmission electron microscopy (TEM) and electron diffraction (ED) characterizations indicated that monolayer graphene sheets with high crystal quality can be produced through such exfoliations.

Large-scale graphene dispersions with concentrations of up to 0.01 $mg \cdot ml^{-1}$, produced by dispersion and exfoliation of graphite in organic solvents, including N-methyl-pyrrolidone, have been reported [94].

The presence of individual graphene sheets was confirmed by using Raman spectroscopy, transmission electron microscopy (TEM) and electron diffraction (ED). A monolayer yield of about 1 wt% was achieved, which could potentially be increased to 7–12 wt% through further processing optimization. More importantly, the graphene sheets prepared in this way had no defects or oxides, as evidenced by the results of X-ray photoelectron, infrared and Raman spectroscopies. The monolayer graphene suspensions can be used to prepare semi-transparent conductive films and conductive polymer composites.

Sieved graphite powder was put into NMP (spectrophotometric grade, > 99.0%), which was treated in bath sonication. After sonication, gray liquid was obtained, which consisted of a homogeneous phase and large numbers of macroscopic aggregates. These aggregates could be removed through mild centrifugation, resulting in homogeneous dark dispersions [96, 97]. The dispersions derived from different graphite concentrations are shown in Fig. 2.24 (a). Highly stable dispersions remained for up to five months after

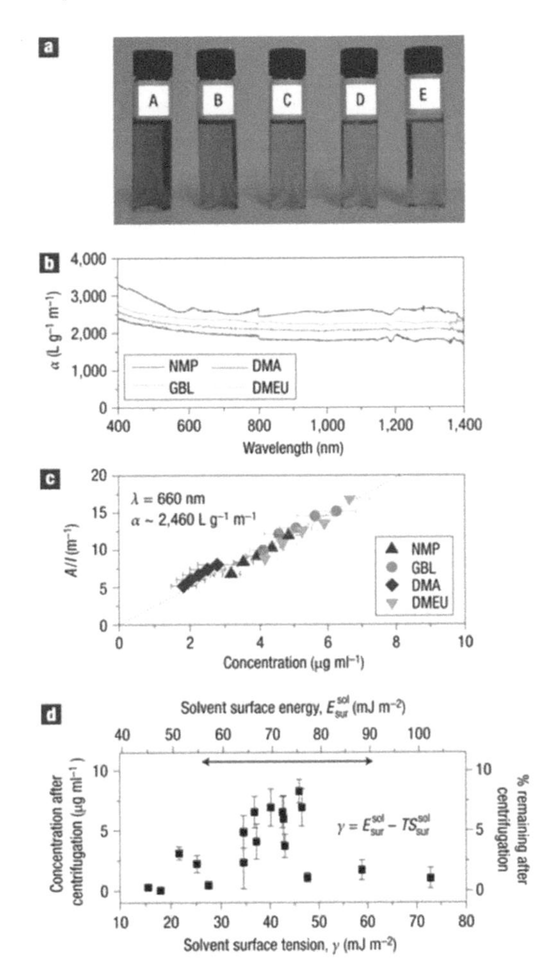

Fig. 2.24. (a) Dispersions of graphite flakes in NMP, at a range of concentrations ranging from 6 μg·ml^{-1} (A) to 4 μg·ml^{-1} (E) after centrifugation. (b) Absorption spectra for graphite flakes dispersed in NMP, GBL, DMA and DMEU at concentrations of 2–8 μg·ml^{-1}. (c) Optical absorbance (λ_{ex} = 660 nm) divided by cell length (*A*/*l*) as a function of concentration for graphene in the four solvents NMP, GBL, DMA and DMEU showing Lambert–Beer behavior with an average absorption coefficient of $<\alpha_{660}>$ = 2,460 L·g^{-1}·m^{-1}. The *x*-axis error bars come from the uncertainty in measuring the mass of graphene/graphite in solution. (d) Graphite concentration measured after centrifugation for a range of solvents plotted versus solvent surface tension. The data were converted from absorbance (660 nm) using $A/l = <\alpha_{660}>C$ with $<\alpha_{660}>$ = 2,460 L·g^{-1}·m^{-1}. The original concentration, before centrifugation, was 0.1 mg/ml. The *y*-axis error bars represent the standard deviation calculated from five measurements. Shown on the right axis is the percentage of material remaining after centrifugation. On the top axis, the surface tension has been transformed into surface energy using a universal value for surface entropy of $S^{sol}_{sur} \approx 0.1$ mJ·K^{-1}·m^{-2}. The horizontal arrow shows the approximate range of the reported literature values for the surface energy of graphite. Reproduced with permission from [94], Copyright © 2008, Macmillan Publishers Limited.

the preparation. Other solvents that have been used to disperse nanotubes were also studied, including N,N-Dimethylacetamide (DMA), γ-butyrolactone (GBL) and 1,3-dimethyl-2-imidazolidinone (DMEU) [98]. The dispersions were characterized by using UV–vis–IR absorption spectroscopy, the absorption coefficient plotted versus wavelength has been shown in Fig. 2.24 (b). The absorbances of the suspensions at 660 nm divided by the cell length, as a function of concentration, are shown in Fig. 2.24 (c), all of which showed the Lambert–Beer behavior. By measuring the optical absorbance after mild centrifugation and using the absorption coefficient at 660 nm to transform absorbance into concentration, the amount of graphite flakes dispersed as a function of solvent surface energy could be quantified, as shown in Fig. 2.24 (d).

It was found that the enthalpy of mixing for graphite dispersed in the selected solvents is close to zero and that the solvent–graphite interaction is van der Waals force instead of covalent bonding. Figure 2.25 (a) shows SEM of the starting powder graphite, which consists of flakes with lateral size of < 500 μm and thickness of < 100 μm. After ultrasonication, the flakes were broken into smaller pieces, with lateral size of tens of μm and thicknesses of μm, as shown in Fig. 2.25 (b). Representative bright-field TEM images of the items that could be observed are shown in Fig. 2.25 (c–g). The items could be classified into three types: (i) monolayer graphene (Fig. 2.25 (c–e)), (ii) folded graphene layers (Fig. 2.25 (f)) and (iii) bilayer and multilayer graphene (Fig. 2.25 (g)). They possessed lateral sizes of a few μm. Scrolled and slightly folded edges were occasionally observed. The results indicated that the graphite powder was exfoliated into monolayer and few-layer graphene nanosheets. Figure 2.25 (h) shows distribution of the thickness of the graphene nanosheets, demonstrating that the number fraction of monolayer graphene, i.e., the number of monolayers over the total number of flakes observed, in NMP dispersions was about 28%. In addition, the sediment could be recycled to produce dispersions of monolayer graphene, implying that the eventual graphene yield could be 7–12 wt% relative to the starting graphite mass.

A graphene monolayer and a graphene bilayer are shown in Fig. 2.26 (a) and (b), respectively. Figure 2.26 (c) shows the normal-incidence electron diffraction pattern of the flake in Fig. 2.26 (a). This pattern is the typical sixfold symmetry of graphite/graphene [99]. Normal-incidence selected area diffraction patterns of the flake in Fig. 2.26 (b) are shown in Fig. 2.26 (d, e), with one pattern (Fig. 2.26 (d)) for the monolayer graphene and one (Fig. 2.26 (e)) for the multilayer graphene. The presence of the monolayer and multilayer graphene was further confirmed by the diffraction intensities shown in Fig. 2.26 (f–h) and the intensity ratio of (1100) and (2110) shown in Fig. 2.26 (i).

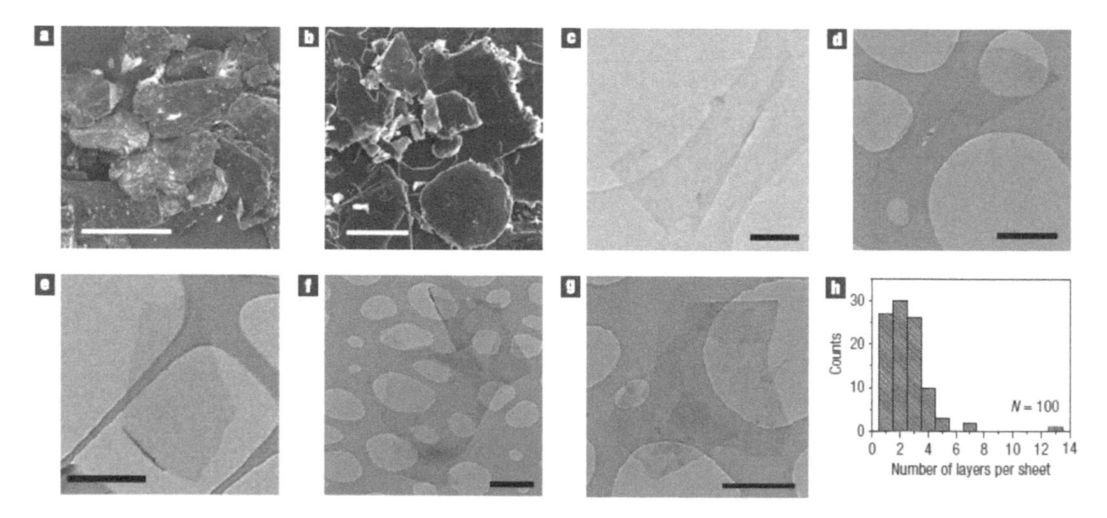

Fig. 2.25. (a) SEM image of the sieved pristine graphite (scale bar: 500 μm). (b) SEM image of the sediment after centrifugation (scale bar: 25 μm). (c–e) Bright-field TEM images of the monolayer graphene flakes deposited from GBL (c), DMEU (d) and NMP (e), respectively (scale bars: 500 nm). (f, g) Bright-field TEM images of a folded graphene sheet and multilayer graphene, both deposited from NMP (scale bars: 500 nm). (h) Histogram of the number of visual observations of flakes as a function of the number of monolayers per flake for the NMP dispersions. Reproduced with permission from [94], Copyright © 2008, Macmillan Publishers Limited.

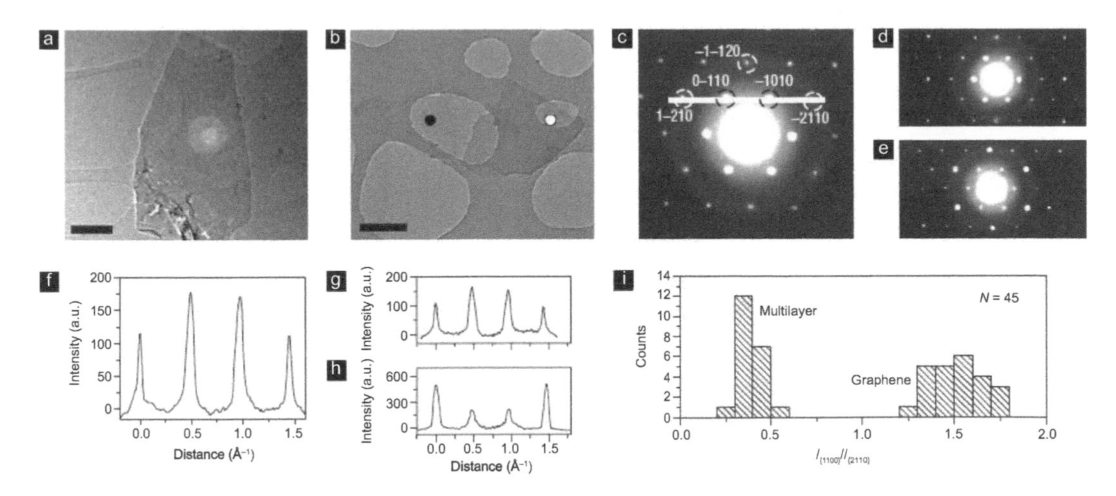

Fig. 2.26. (a, b) High-resolution TEM images of the solution-cast monolayer (a) and bilayer (b) graphene (scale bar: 500 nm). (c) Electron diffraction pattern of the sheet in (a), with the peaks labeled by Miller–Bravais indices. (d, e) Electron diffraction patterns taken from the positions of the black (d) and white spots (e), respectively, of the sheet shown in (b), using the same labels as in (c). The graphene is clearly one layer thick in (d) and two layers thick in (e). (f–h) Diffracted intensity taken along the (1–210) to (–2110) axis for the patterns shown in (c–e), respectively. (i) Histogram of the ratios of the intensity of the (1100) and (2110) diffraction peaks for all the diffraction patterns collected. A ratio > 1 is the signature of graphene. Reproduced with permission from [94], Copyright © 2008, Macmillan Publishers Limited.

Graphene can also be produced in the liquid phase by sonicating tetrabutylammonium hydroxide (TBA) and oleum-intercalated graphite in DMF to obtain high quality graphene sheets, which can be fabricated into transparent conducting films by using Langmuir-Blodgett deposition [100]. High-quality graphene sheets (GS) were prepared by exfoliating commercial expandable graphite by briefly heating for 60 s to 1000°C in forming gas. The exfoliated graphite was reintercalated by grinding with oleum (fuming sulphuric acid with 20% free SO_3). After that, the oleum-intercalated graphite was inserted with tetrabutylammonium hydroxide (TBA, 40% solution in water) in N,N-dimethylformamide (DMF), as shown in Fig. 2.27 (a). The TBA-inserted oleum-intercalated graphite was sonicated in a DMF solution of 1,2-distearoyl-sn-glycero-3-phosphoethanolamine-N-[methoxy(polyethyleneglycol)-5000] (DSPE-mPEG) for 60 min to form a homogeneous suspension, as illustrated in Fig. 2.27 (b). Large pieces of items were removed with centrifugation from the supernatant, as demonstrated in Fig. 2.27 (c). High concentration GS suspensions in DMF were transferrable to other solvents, such as water and organic solvents.

AFM observation indicated that the deposits on substrates from the supernatant contained about 90% of single-layer GS with various shapes and sizes, as shown in Fig. 2.27 (d). The average size of the single-layer GS was about 250 nm and the average topographic height was about 1 nm. TEM and electron diffraction (ED) pattern of representative single-layer GS are shown in Fig. 2.27 (e) and (f), respectively. The single GS was well-crystallized, because its ED pattern was similar to that of peeled-off graphene [99].

IR spectra indicated that room-temperature oleum treatment was much less oxidative than the Hummers' method. The as-prepared GS had much fewer functional groups, as shown in Fig. 2.28 (a), while the as-prepared Hummers' GO possessed more functional groups, as seen in Fig. 2.28 (d). Figure 2.28 (b) shows X-ray photoelectron spectroscopy (XPS) of the as-prepared GS, showing weak signals at higher binding energy. That binding energy corresponded to a small amount of C–O, which could be removed by annealing at 800°C in H_2. XPS spectrum of the annealed GS was same as that of a pristine highly oriented pyrolytic graphite (HOPG) crystal, confirming the high crystallinity and quality of the GS.

Structures of the intermediate and final products of the GS and Hummers' GO are schematically shown in Fig. 2.28 (c) and (f), respectively. Because the oxidization experienced by the intermediate and the as-prepared GS was relatively mild, the covalently bound functional groups, such as the carboxylic and hydroxyl groups, were most observed at the edges. This has been supported by the fact that the

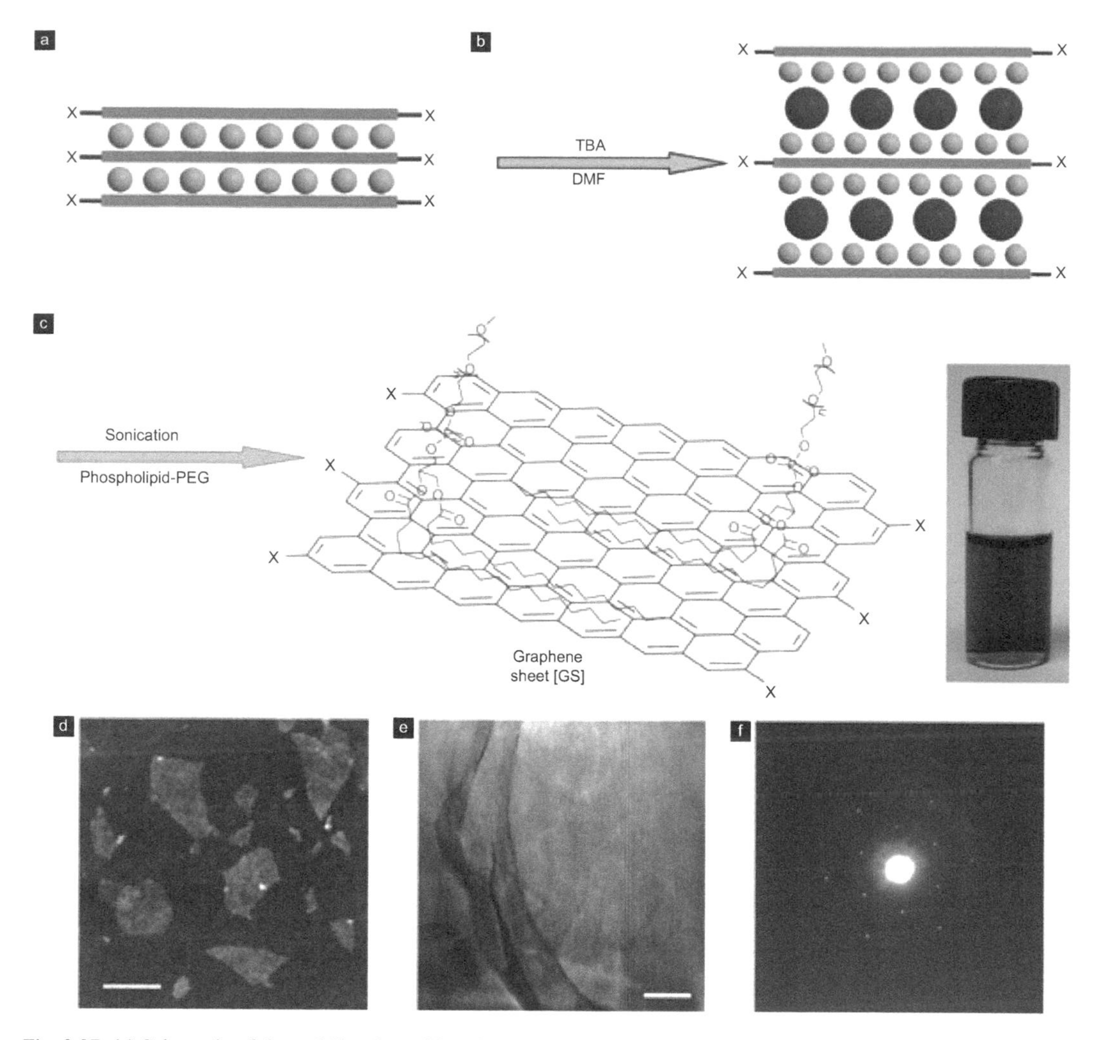

Fig. 2.27. (a) Schematic of the exfoliated graphite reintercalated with sulphuric acid molecules (teal spheres) between the layers. (b) Schematic of TBA (blue spheres) insertion into the intercalated graphite. (c) Schematic of GS coated with DSPE-mPEG molecules and a photograph of a DSPE-mPEG/DMF solution of GS. (d) AFM image of the typical GS with a size of hundred nm and topographic height of ~1 nm (scale bar: 300 nm). (e) Low-magnification TEM images of the typical GS with a size of hundred nm (scale bar: 100 nm). (f) Electron diffraction (ED) pattern of the as-obtained GS as in panel (e), showing its excellent crystallinity. Reproduced with permission from [100], Copyright © 2008, Macmillan Publishers Limited.

electrical conductivity of the as-prepared GS was similar to that of the vacuum-annealed GS at 800°C shown in Fig. 2.29 (b, c). In contrast, the Hummers' GO was significantly oxidized, with the presence of disrupted conjugation, missing carbon atoms and abundant functional groups, including epoxide, hydroxyl, carbonyl and carboxyl at both the edges and in the plane, as demonstrated in Fig. 2.28 (f). In this case, thermal annealing removed only some of the functional groups, while all defects of the Hummers' GO sheets could not be entirely eliminated. Therefore, the weak oleum treatment condition is important to obtain high-quality GS without the presence of excessive chemical functional groups, so that the property degradation could be effectively prevented.

The weakened van der Waals interaction in intercalated graphite compounds allows for easier dispersion in organic solvents. A solution-phase technique has been developed to produce large-area bilayer and trilayer graphene from graphite, with controlled stacking behaviors [101]. The ionic compounds, iodine chloride (ICl) or iodine bromide (IBr), intercalated graphite at every second or third layer, leading to second- or third-stage controlled graphite intercolation compounds, respectively. The resulting solution dispersions

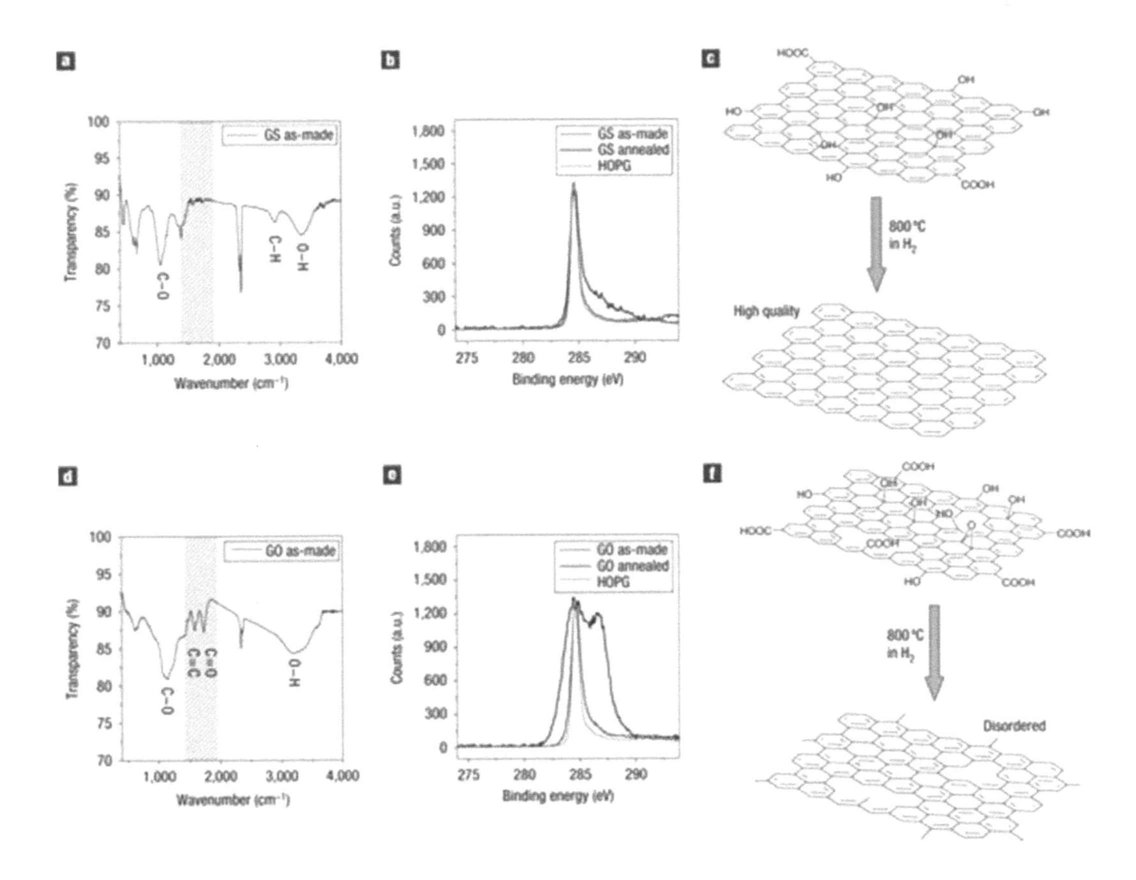

Fig. 2.28. (a) IR spectrum (400–4,000 cm^{-1}) of the as-prepared GS. The shaded region of ~1,400–1,900 cm^{-1} is attributed to the carboxylic groups. (b) XPS spectra of the as-prepared, annealed GS and a HOPG crystal. The spectra of the annealed GS and HOPG are very similar. (c) Schematic of the atomic structure of as-prepared (top) and the annealed GS. (d) IR spectrum (400–4,000 cm^{-1}) of as-prepared GO. The shaded region of ~1,400–1,900 cm^{-1} is attributed to the carboxylic groups. (e) XPS spectra of as-prepared, annealed GO and a HOPG crystal. (f) Schematic of the atomic structure of as-prepared (top) and annealed GO (bottom). Reproduced with permission from [100], Copyright © 2008, Macmillan Publishers Limited.

contained large amounts of bilayer or trilayer graphene nanosheets accordingly. Because the process was conducted with mild sonication, it could provide graphene nanosheets with areas of as large as 50 μm^2. As a result, the graphene nanosheets exhibited high electrical properties, with unannealed samples having resistance of as low as ~1 kΩ and hole mobilities of as high as 400 cm^2/V·s.

HOPG was used as a high-quality graphite source, halogen ICl (to form Stage-2 ionic graphite intercalation compounds or GIC) and IBr (to form Stage-3 ionic GIC) were used as the intercalants. The staging phenomenon was related to the effect of long-range lattice strain. Once the exposed surface of the graphite planes interact with the intercalated species, the intercalant molecules are introduced into the host material from the exposed end surfaces, so that layered structures are formed. In this case, Stage-2 and Stage-3 GICs have every second or third layer of the graphite lattice that are intercalated, respectively, as shown in Fig. 2.30 (a). The stage of the GIC stands for the crystallographic arrangement of the intercalant and the graphite layers, with a superlattice structure along the *c*-axis. The top three profiles in Fig. 2.30 (b) illustrate the XRD patterns of HOPG, IBr Stage-3 GIC and ICl Stage-2 GIC. The results confirmed that halogen molecules intercalated into every second or third layer of graphite.

The derived Stage-2 and Stage-3 GICs were used as precursors of expanded graphite, which were annealed at 800°C for 5 min in an Ar. The IBr or ICl intercalants between the graphene layers were volatilized rapidly and thus removed completely. The substantial anisotropic volume changes for the Stage-3 and Stage-2 expanded samples are shown in Fig. 2.30 (c) and (d), respectively. During the thermal

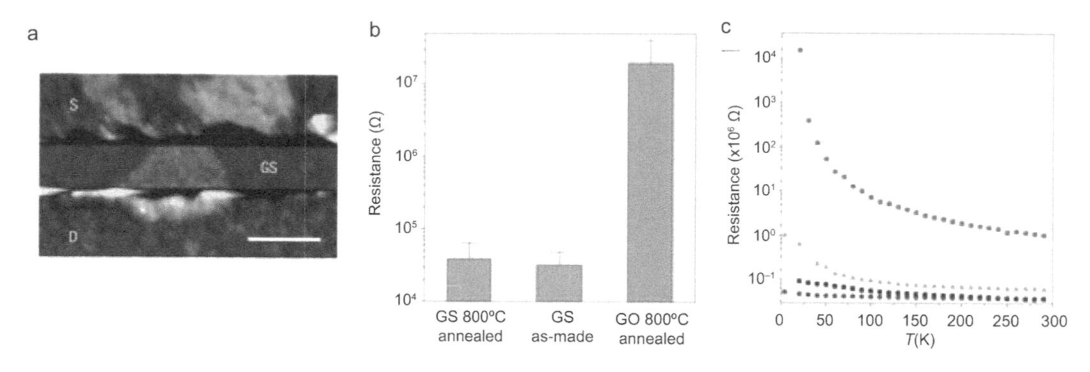

Fig. 2.29. (a) AFM image of a typical device with a single GS (single layer thickness of about 1 nm) bridging the channel (channel length $L \approx 100$ nm GS with titanium/gold contacts and silicon backgate) between the source (S) and drain (D) electrodes (scale bar: 200 nm). (b) Mean resistance histograms for 10 devices each of the as-prepared GS, annealed GS and annealed GO. The resistances of as-prepared GS and the annealed GS were similar, indicating the high quality of the as-prepared GS. (c) Resistances of the as-prepared GS (green curve), GS annealed at 800°C with a titanium/gold contact (black), GS annealed at 800°C with a palladium contact (blue) and GO annealed at 800°C (red). Reproduced with permission from [100], Copyright © 2008, Macmillan Publishers Limited.

Color image of this figure appears in the color plate section at the end of the book.

expansion, the bilayer and trilayer crystal domains were isolated and cross-linked, with stacking faults formed during the intercalation process. The symmetry-breaking of graphite lattice for the Stage-3 and Stage-2 expanded materials is demonstrated in the XRD patterns. In natural graphite, the graphene layers are strongly bound with van der Waals forces, while in the expanded items, bilayer and trilayer crystallites were cross-linked by the crystalline defects, which were easily broken up in a liquid-phase dispersion by using mild ultrasonication. The dispersion process for expanded graphite is shown in Fig. 2.30 (e).

Full Raman spectra and 2D peaks for representative graphene flakes are shown in Fig. 2.31 (a) and (b), respectively. The 2D peaks corresponding to specific numbers of stacked layers in the exfoliated graphene nanosheets exhibited the same layering dependence as that observed in micromechanically cleaved graphene [102, 103]. This implies that the solution-processed graphene flakes are AB stacked. The monolayer graphene exhibited a single Lorentzian at 2,660 cm^{-1}, while the component at 2,690 cm^{-1} became dominant with the increasing number of layers. However, the 2D signature in Raman spectroscopy cannot be used to unambiguously differentiate 3-layer and 4-layer graphene nanosheets. According to the D-peak at ~1,340 cm^{-1} and the fact that the monolayers had a similar size as that of the laser spot, it was concluded that the exfoliated graphene did not contain a large quantity of defects or functionalization.

Figure 2.31 (c) shows high-resolution TEM (HRTEM) images of edges of the bilayer and trilayer flakes. Figure 2.20 (d) shows a TEM image of representative few-layer graphene with an area of ~50 μm^2. Electron diffraction patterns indicated typical Bragg reflection intensity ratios for the monolayer and few-layer graphene. AFM images of the isolated bilayer and 3–4-layer graphene flakes are shown in Fig. 2.31 (e) and (f), respectively.

On the other hand, theoretically understanding the solution phase dispersion of pristine unfunctionalized graphene is important for the production of conducting inks and top-down approaches for applications in electronics [104]. A theoretical framework has been established, in which molecular dynamics simulations and the kinetic theory of colloid aggregation were employed to elucidate the mechanisms of stabilization of liquid phase exfoliated graphene nanosheets in N-methylpyrrolidone (NMP), N,N'-dimethylformamide (DMF), dimethyl sulfoxide (DMSO), γ-butyrolactone (GBL) and water. According to the calculations of the potential of mean force between two solvated graphene nanosheets using molecular dynamics (MD) simulations, it was found that the dominant barrier hindering the aggregation of graphene was the last layer of confined solvent molecules between the graphene nanosheets, which had resulted from the strong affinity of the solvent molecules for graphene. The source of the energy barrier responsible for repelling the nanosheets were the steric repulsions between solvent molecules and graphene, before desorption of

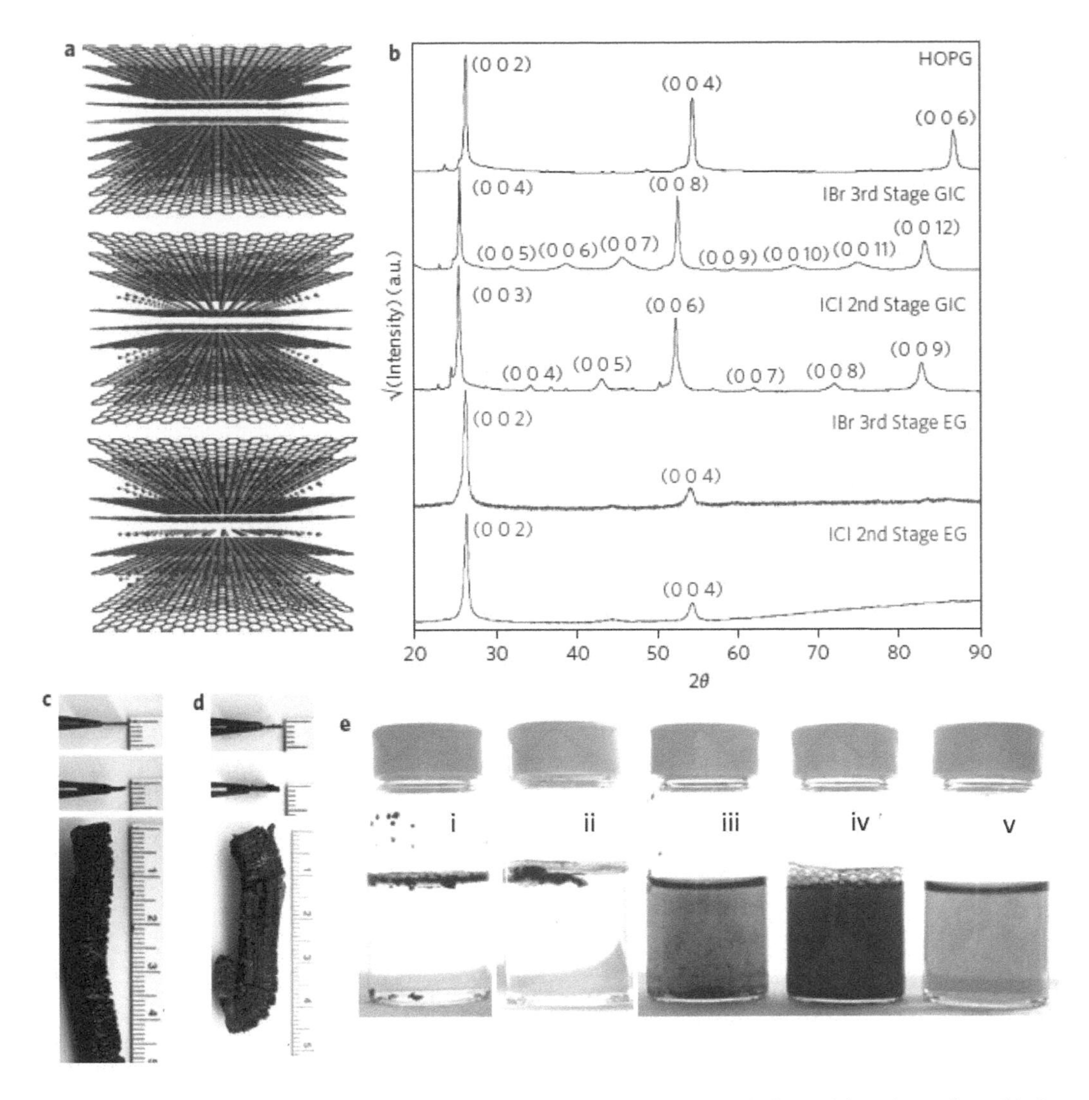

Fig. 2.30. (a) Three-dimensional computer-generated molecular models (carbon–gray, iodine–red, bromine–yellow, chlorine–cyan) of HOPG (top), IBr Stage-3 GIC (middle) and ICl Stage-2 GIC (bottom). (b) XRD patterns of HOPG, IBr Stage-3 GIC, ICl Stage-2 GIC, IBr Stage-3 expanded graphite (EG) and ICl Stage-2 EG. For IBr Stage-3 GIC, an additional three peaks appear between two of the three main peaks, corresponding to the (002), (004) and (006) planes of HOPG, which is a clear signature of Stage-3 GIC, with the intercalant layer inserted between every three graphite layers. Similar XRD patterns were observed for the ICl Stage-2 GIC. Both Stage-3 and Stage-2 expanded graphites showed much weaker (004) peaks than the HOPG, while the (006) peak became unobservable. (c) Photographs of HOPG (top), Stage-3 GIC (middle) and Stage-3 EG (bottom). (d) Photographs of HOPG (top), Stage-2 GIC (middle), and Stage-2 EG (bottom). (e) Photograph of steps involved in forming the suspensions of EGs in 2 wt% sodium cholate aqueous solution. As-prepared Stage-3 (i) and Stage-2 (ii) expanded materials floating on the solution, followed by 30 min homogenization (iii) and 10 min sonication (iv). (v) Clear and gray graphene solutions after centrifugation at 2,000 rpm. Reproduced with permission from [101], Copyright © 2011, Macmillan Publishers Limited.

the confined single layer of solvent occurred. A kinetic theory of colloid aggregation has been formulated to model the aggregation of graphene nanosheets in the liquid phase, in order to predict the stability using the potential of mean force. The theory could be used to describe the experimentally observed degradation of the single-layer graphene fraction in NMP. With these results, the potential solvents according to their ability to disperse pristine unfunctionalized graphene are in an order of NMP ≈ DMSO > DMF > GBL > H_2O, which is consistent with the fact that the first three solvents have been most widely used in the open literature.

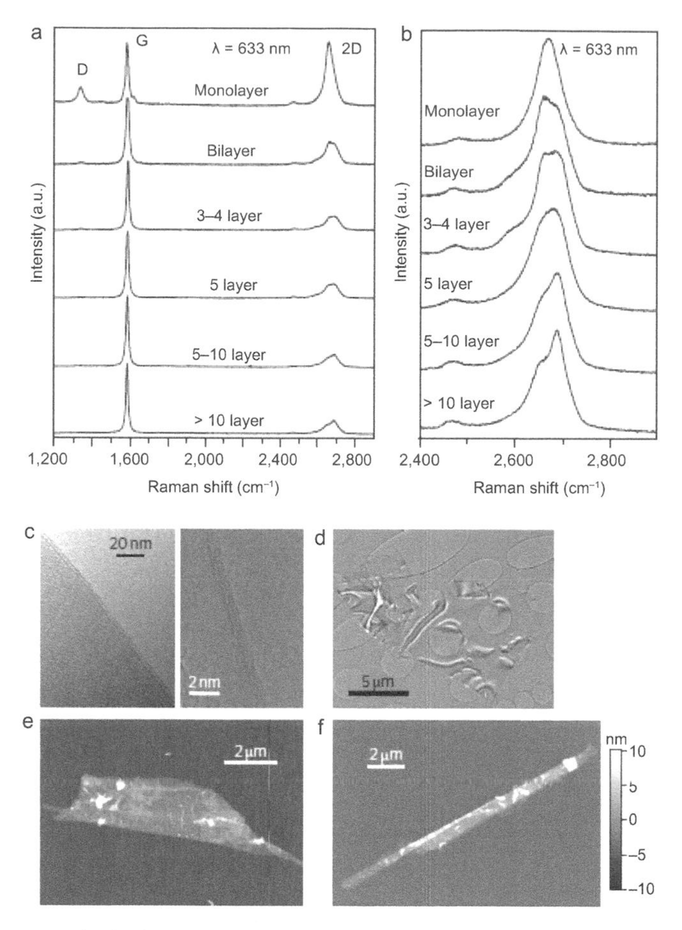

Fig. 2.31. (a, b) Raman spectra (excitation wavelength $\lambda = 633$ nm) for graphene flakes with different numbers of stacked layers. Note that for multilayer graphene flakes, the D-peaks at 1340 cm^{-1} (a) are absent, but for the relatively smaller monolayer graphene flake, the weak D-peak arises due to edge effects. 2D Raman spectra (b) corresponding to specific numbers of stacked layers of our graphene flakes show the same features observed using the Scotch-tape method on the same substrate. (c) HRTEM images of the edges of bilayer (left) and trilayer (right) graphene flakes. (d) TEM image of a representative multilayer graphene flake with an area of ~50 μm^2. (e, f) Representative AFM images of bilayer (e) and 3–4 layer graphene flakes (f). Reproduced with permission from [101], Copyright © 2011, Macmillan Publishers Limited.

2.2.1.3. Surfactant Exfoliation

Graphite has also been exfoliated in aqueous surfactant solutions, which was reported to be more convenient for further solution processing. Sodium dodecyl benzene sulfonate (SDBS) and sodium cholate (SC) are promising candidates for the exfoliation of graphite, when they are used at concentrations that are below their critical micelle concentrations (CMCs) [105–108]. According to zeta potential measurements, the dispersed graphene sheets are electrostatically charged, so that they are more sensitive to the change in pH value. In particular, sodium cholate allows exfoliation of graphite at high yields, with a concentration of up to 0.04 $mg \cdot ml^{-1}$. Similar to those derived from organic solvent exfoliation, the graphene layers exfoliated in aqueous surfactant dispersions are also defect-free. The high processability of this approach allows the deposition of graphene layers on insulating substrates and the fabrication of transparent electrodes by vacuum filtration.

By using the surfactant, sodium cholate, graphene dispersions stabilized in water have been prepared at concentrations of up to 0.3 $mg \cdot ml^{-1}$ [108]. The suspensions were obtained with low power sonication for long times of up to 400 h, followed by centrifugation. The dispersed concentration increased with increasing sonication time, while the quality of dispersions could be controlled through the centrifugation rates at 500–2000 rpm. TEM analysis results indicated that the graphene nanosheets consisted of 1–10 stacked monolayers, with one layer graphene sheets to be up to 20%. Most of the sheets contained 4 stacked monolayers, with the length and width of 1 μm and 400 nm, respectively. As the centrifugation rate was increased from 500 to 5000 rpm, the average flake length was decreased to 0.5–1 μm. According to Raman spectroscopies, the graphene nanosheets derived at centrifugation rates of < 2000 rpm were nearly defect-free. The dispersions can be used to fabricate high-quality free-standing graphene films.

Natural flake graphite was used as starting material. Sodium cholate (NaC) was dissolved in Millipore water at concentrations of 0.01–20 $mg \cdot ml^{-1}$. Graphene dispersions were prepared by adding the graphite powder at an initial concentration of $C_{G,i}$ = 5 $mg \cdot ml^{-1}$ to 400 ml NaC solution in 500 ml capped round-bottomed flasks. Various surfactant concentrations (C_{NaC}) were used. Ultrasonication was carried out in a low power sonic bath. The nominal power output for the bath was 80 W, while real application was about 16 W. Due to the long time sonication, it is necessary to continuously refill bath water in order to maintain sonication efficiency and prevent overheating.

Figure 2.32 (A) shows the measured absorbance per unit cell length, A/l, as a function of configuration rate for a given set of experiments [108]. It is understandable that the absorption decreased with increasing configuration rate, due to the decreasing content of graphene. The inset in Fig. 2.32 (A) indicates that there was an optimized concentration of NaC, which is C_{NaC} = 0.1 $mg \cdot ml^{-1}$. Figure 2.32 (B) shows that the samples became clearer gradually as the centrifugation was increased, in consistence with the observation in Fig. 2.32 (A). The effect of sonication time on absorption and the calculated concentration of graphene are demonstrated in Fig. 2.32 (C).

Figure 2.33 shows TEM images of four exfoliated graphene samples. Both well-exfoliated graphene multilayers (Fig. 2.33 (B)) and monolayers (Fig. 2.33 (C, D)) were observed [108]. Thick objects that were opaque to the TEM beam were also observed across the entire TEM grid for the samples prepared at the lowest centrifugation rate of 500 rpm, due to the presence of some graphite nanoparticles with lateral sizes of 4–15 μm. At centrifugation rates of ≥ 1500 rpm, no such thick objects could be detected. It is interesting to note that. Figure 2.33 (D) indicates that some monolayers had quite large aspect ratios. Free-standing graphene sheets could be fabricated with the suspensions. One example is shown in Fig. 2.34 (A), which could be mildly bent and flexed without breaking. Such films exhibited densities of 1000–1440 kg/m^3. SEM analysis showed the films consisted of graphene nanosheets that were well-aligned in the plane of the film, as shown in Fig. 2.34 (B) [108].

Sodium dodecylbenzene sulfonate (SDBS) has been used to disperse and exfoliate graphite to produce high concentration and high quality graphene suspensions in water-surfactant solutions [109]. It was found that more than 40% of the exfoliated graphene nanosheets had < 5 layers, with 3% of them consisting of monolayers. At the same time, the monolayers were most likely free of defects. The graphene dispersions were highly stable against re-aggregation by Coulomb repulsion, due to the adsorbed surfactant. The stabilization could be explained by using the DLVO and Hamaker theory. The dispersions could be used to fabricate free-standing graphene papers through vacuum filtration. Raman and IR spectroscopic analysis indicated the graphene papers consisted of nanosheets that were free of defects, while a trace of oxide was detected by using X-ray photoelectron spectroscopy (XPS).

Commercial graphite powder and SDBS were used as starting materials in the experiment. The graphite powder was sieved through a 0.5 mm mesh to remove the large particles. SDBS solutions with concentrations of 5–10 $mg \cdot ml^{-1}$ were prepared in Millipore water by stirring. A typical sample was prepared by dispersing graphite in the desired SDBS concentration by using sonication in a low power sonic bath for 30 min. The resulting dispersion was left for about 24 h to allow the aggregates to precipitate, which was then centrifuged for 90 min at 500 rpm. After centrifugation (CF), the top 15 ml of the dispersion was decanted by pipet and retained. Various graphene concentrations were prepared for optical characterization.

Absorption coefficient, R, is related to the absorbance, A, through the Lambert-Beer law, i.e., $A = \alpha Cl$, where C is the concentration and l is the path length, which is usually used to characterize dispersions. An absorption spectrum is shown in the inset of Fig. 2.35 [109]. The strong absorption band was attributed to the SDBS. The volume of graphene contained in a given sample was precisely measured by filtering under

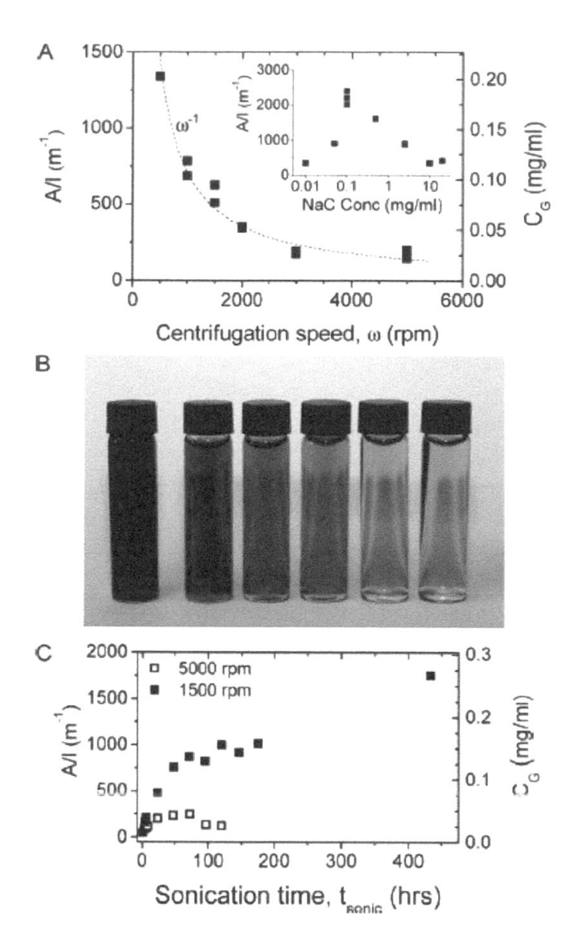

Fig. 2.32. (A) A/l as a function of centrifugation speed ($C_{G,i}$ = 5 mg·ml^{-1}, CNaC 0.1 mg·ml^{-1}, t_{sonic} = 24 h, centrifugation for 90 min). Inset: Absorbance/cell length, A/l, as a function of surfactant concentration (C_G = 5 mg·ml^{-1}, t_{sonic} = 24 h, centrifugation at 1000 rpm for 30 min). Subsequently, a surfactant concentration of 0.1 mg·ml^{-1} was used for all dispersions. (B) Photographs of the surfactant-stabilized graphene dispersions,with $C_{G,i}$ = 5 mg/ml, C_{NaC} = 0.1 mg·ml^{-1} and t_{sonic} = 24 h. Left to right: uncentrifuged, centrifuged for 90 min at 1000 rpm, 1500 rpm, 2000 rpm, 3000 rpm, and 5000 rpm, respectively. Note that the centrifuged dispersions have been diluted by a factor of 10 to highlight the color change. (C) A/l as a function of sonication time ($C_{G,i}$ = 5 mg·ml^{-1}, C_{NaC} = 0.1 mg·ml^{-1}, centrifugation for 90 min) for centrifugation speeds of 5000 and 1500 rpm. Note that in both A and C the right axis shows the graphitic concentration calculated using an absorption coefficient of 6600 L·(g·m)$^{-1}$. Reproduced with permission from [108], Copyright © 2010, American Chemical Society.

high vacuum onto an alumina membrane with known mass. Thermogravimetric (TGA) analysis indicated that the dried films were 64 ± 5% graphitic, while the remainder was the residual surfactant. With this value, the concentrations of the dispersions could be estimated. Figure 2.35 shows the absorbance per unit length (A/l) versus concentration of graphite after centrifugation (C_G) of a dispersion, which was serially diluted with 0.5 mg·ml^{-1} SDBS solution. A straight line fitting led to the absorption coefficient at 660 nm to be R = 1390 ml·(mg·m)$^{-1}$, which was in agreement with the content of graphite/graphene.

Figure 2.36 shows representative TEM images of the different types of graphene nanosheets obtained by using the surfactant exfoliation method [109]. A small quantity of monolayer graphene flakes was observed, as shown in Fig. 2.36 (A). A larger proportion of flakes were few-layer graphene sheets, including both bilayers and trilayers, as shown in Fig. 2.36 (B) and (C). Additionally, some disordered flakes with multiple layers were also observed, which were similar to the one shown in Fig. 2.36 (D). The disordering characteristics implied that such flakes were formed due to the re-aggregation of smaller flakes. There

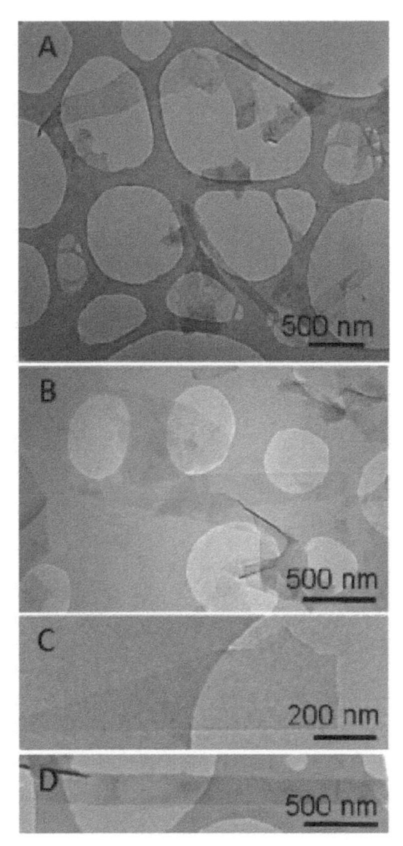

Fig. 2.33. (A) Wide field and (B) close-up TEM images of a highly exfoliated graphene monolayer and multilayer flakes, (C) a monolayer graphene flake, and (D) a graphene ribbon. These samples were deposited from dispersions prepared with the following parameters: (A) $C_{G,i} = 5$ mg·ml^{-1}, CNaC = 0.1 mg·ml^{-1}, $t_{sonic} = 72$ h, centrifugation at 1500 rpm for 90 min, (B) $C_{G,i} = 5$ mg·ml^{-1}, $C_{NaC} = 2.5$ mg·ml^{-1}, $t_{sonic} = 48$ h, centrifugation at 1000 rpm for 30 min, (C and D) $C_{G,i} = 5$ mg·ml^{-1}, $C_{NaC} = 0.1$ mg·ml^{-1}, $t_{sonic} = 144$ h, centrifugation at 1500 rpm for 90 min. Reproduced with permission from [108], Copyright © 2010, American Chemical Society.

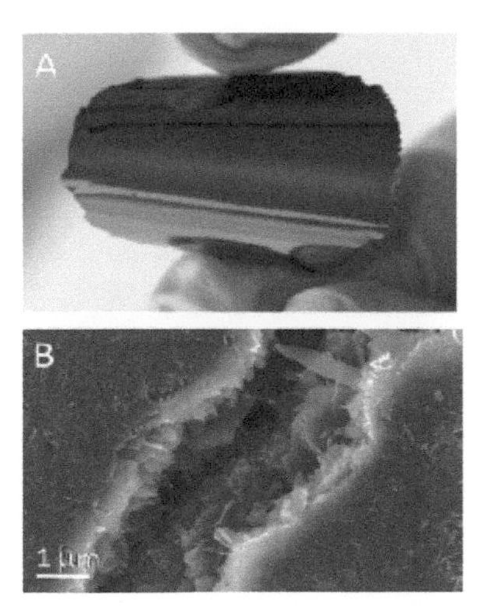

Fig. 2.34. (A) Free-standing graphene film derived from the NaC/graphene dispersion ($C_{G,i} = 5$ mg·ml^{-1}, $C_{NaC} = 0.1$ mg·ml^{-1}, $t_{sonic} = 171$ h, centrifugation at 1000 rpm for 30 min). (B) SEM image of a crack on the surface of a free-standing graphene film ($C_{G,i} = 5$ mg·ml^{-1}, $C_{NaC} = 0.1$ mg·ml^{-1}, $t_{sonic} = 24$ h, centrifugation at 1500 rpm for 90 min). Reproduced with permission from [108], Copyright © 2010, American Chemical Society.

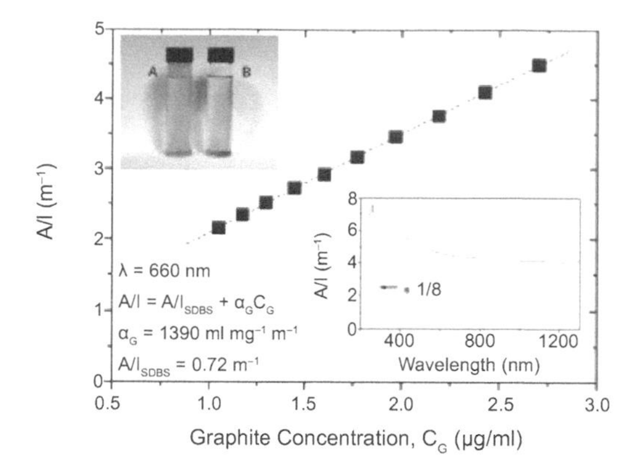

Fig. 2.35. Absorbance per unit length at $\lambda = 660$ nm as a function of graphite concentration (after centrifugation) for an SDBS concentration, $C_{SDBS} = 0.5$ mg·ml^{-1}. Graphite concentration before centrifugation was $C_{G,i} = 0.1$ mg·ml^{-1}. NB, the curve does not go through the origin due to the presence of a residual SDBS absorbance. (Intercept of $A/l = 0.72$ m^{-1} compares with the residual absorbance of $A/l \approx 0.5$ m^{-1} for SDBS at $C_{SDBS} = 0.5$ mg·ml^{-1}.) Bottom inset: Absorption spectrum for a sample with $C_{SDBS} = 0.5$ mg·ml^{-1} and $C_G = 0.0027$ mg·ml^{-1}. The portion below 400 nm is dominated by surfactant absorption and has been scaled by a factor of 1/8 for clarity. The portion above 400 nm is dominated by graphene/graphite with some residual SDBS absorption. Top inset: Surfactant-stabilized graphite dispersions (A) before and (B) immediately after centrifugation. The dispersions are almost transparent due to the low concentration of graphite. Reproduced with permission from [109], Copyright © 2009, American Chemical Society.

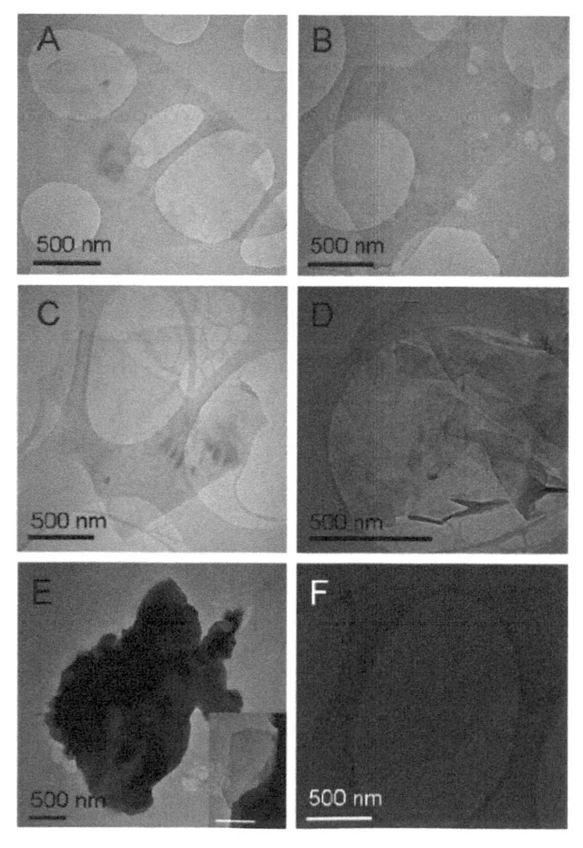

Fig. 2.36. Representative TEM images of the flakes prepared by using the surfactant processing method: (A) monolayer, (B) bilayer, (C) trilayer, (D) disordered multilayer and (E) very large flake. Inset: close-up of an edge of a very large flake showing a small multilayer graphene flake protruding. (F) A monolayer from the sample prepared by sediment recycling. Reproduced with permission from [109], Copyright © 2009, American Chemical Society.

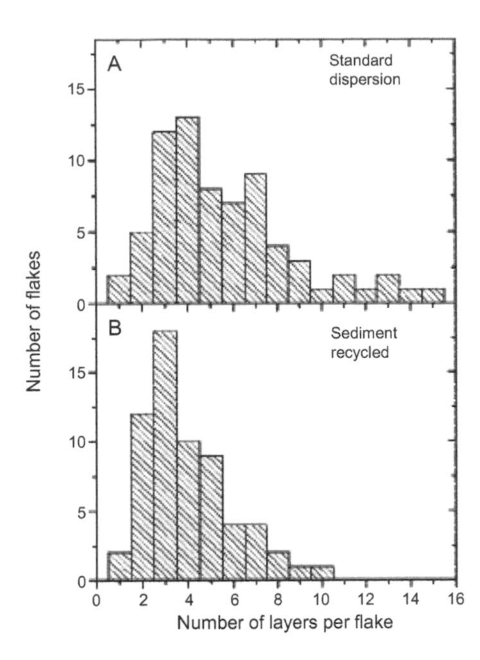

Fig. 2.37. Histogram of the number of layers per flake for dispersions from original sieved graphite and from recycled sediment. This histogram does not include the two very large flakes of the type shown in Fig. 2.25 (E). Reproduced with permission from [109], Copyright © 2009, American Chemical Society.

were also a limited number of flakes with large sizes, as shown in Fig. 2.36 (E). The inset of Fig. 2.36 (E) indicates that they were graphites of thin multilayers protruding from their edges. Figure 2.37 (A) shows a histogram of the large flakes for the standard dispersions.

By recycling the sediment remaining after centrifugation, the overall yield of graphene exfoliation was effectively increased. About 3% monolayer graphene could be obtained from the recycled sediment, as shown in Fig. 2.36 (F). The flake thickness distribution shifted towards thinner ones, with more bilayers and trilayers being present, i.e., 67% of flakes observed had layers < 5, as illustrated in Fig. 2.37 (B). Figure 2.38 (A) shows a HRTEM image of a graphene monolayer similar to that shown in Fig. 2.36 (A). The presence of residual surfactant led to the nonuniform distribution of the graphene nanosheets. A fast Fourier transform (FFT) of the image is shown in the inset. The (1100) spots can clearly be seen, while the (2110) ones are too faint. This difference in intensity confirmed the presence of monolayer graphene [94, 99]. A HRTEM image of a multilayer graphene nanosheet is shown in Fig. 2.38 (B), where both the (1100) and (2110) spots were visible, with the (2110) spots being more intense.

Figure 2.38 (C) shows an aberration corrected HRTEM image of a monolayer graphene nanosheet, as confirmed by the FFT in the inset. The nonuniformity in the phase contrast transfer was due to the presence of residual surfactant. A filtered image of a part of the region enclosed by the white square in Fig. 2.38 (C) is shown in Fig. 2.38 (D), clearly revealing the hexagonal nature of graphene. Figure 2.38 (E) shows the intensity analysis along the left dashed line, demonstrating a hexagon width of 2.4 Å, close to the expected value of 2.5 Å of graphene. Analysis of the intensity profile along the right dashed line is shown in Fig. 2.38 (F), giving rise to a C-C bond length of 1.44 Å, close to the expected value of 1.42 Å. All the imaged areas were free of structural defects, implying that the exfoliation technique was nondestructive to the graphene nanosheets.

2.2.1.4. Spontaneous Exfoliation

Spontaneous exfoliation of graphite in chlorosulfonic acid has been reported, with achieved suspensions having concentrations of up to 2 $mg \cdot ml^{-1}$, which was nearly one order of magnitude higher than those

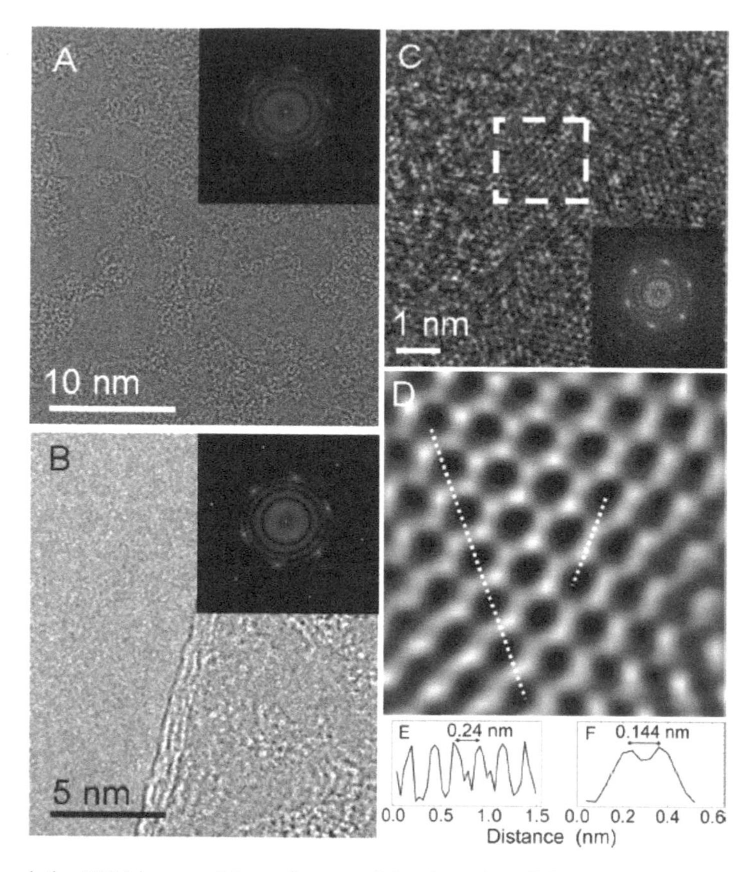

Fig. 2.38. High-resolution TEM images of the surfactant exfoliated graphene flakes: (A) HRTEM image of a section of a graphene monolayer with the inset showing the fast Fourier transform (equivalent to an electron diffraction pattern) of the image, (B) HRTEM image of a section of a trilayer with the inset showing the fast Fourier transform of the image and (C) HRTEM image of part of a graphene monolayer with the inset showing the fast Fourier transform of the region enclosed by the white square (scale bar: 1 nm). (D) A filtered image of a part of the region in the white square. (E) Intensity analysis along the left white dashed line shows a hexagon width of 2.4 Å. (F) Intensity analysis along the right white dashed line shows a C-C bond length of 1.44 Å. Reproduced with permission from [109], Copyright © 2009, American Chemical Society.

prepared by other methods [110]. More importantly, this method works without the use of covalent functionalization, surfactant stabilization or sonication, which could either compromise the properties of the derived graphene or reduce the flake sizes [111]. Liquid-crystalline phases at concentrations of as high as 20–30 $mg{\cdot}ml^{-1}$ have been achieved by using the spontaneous exfoliation method. Transparent conducting films could be fabricated by using these dispersions, with electrical conductivity of 1000 $\Omega{\cdot}\square^{-1}$ and optical transparency of ~80%. High-concentration suspensions, both isotropic and liquid crystalline, could be used to make flexible electronic devices and multifunctional fibers.

Superacids have been used to effectively dissolve single-walled carbon nanotubes (SWCNTs), showing great promise for bulk processing of SWCNT-based materials [112]. Three sources of graphite were studied: (i) graphoil (commercial graphitic material used in seals), (ii) microcrystalline graphite and (iii) highly ordered pyrolytic graphite (HOPG). The concentration of the dispersed phase was determined by centrifuging the dispersion, extracting the top (isotropic) phase and then measuring absorbance by using UV–vis–NIR (near-infrared) spectra. Figure 2.39 (a) indicates that graphene dispersions could be readily achieved by using the acid from different graphite sources without using sonication. Concentrations of the suspensions were ten to a hundred times higher than those obtained by organic solvent exfoliation or surfactant exfoliation, as discussed earlier.

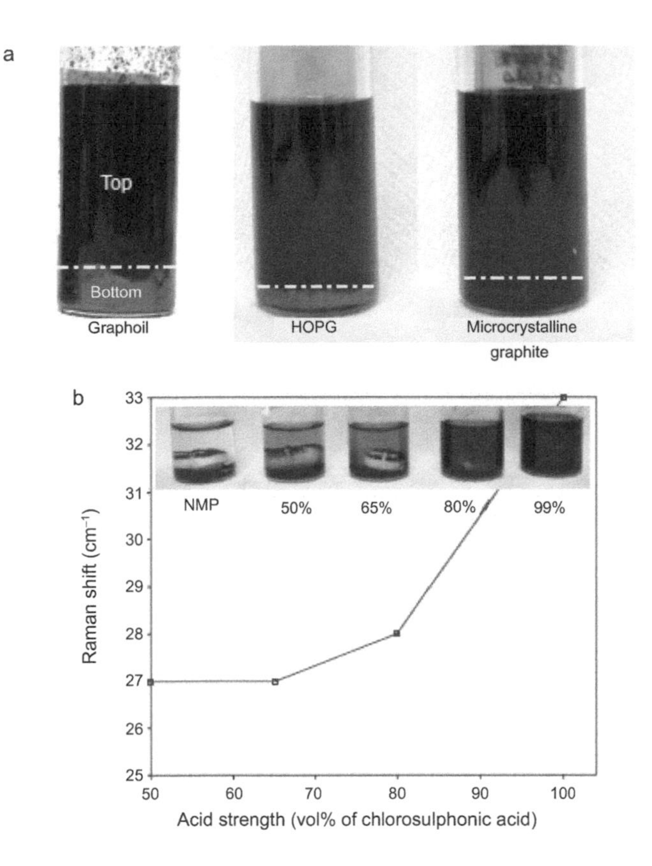

Fig. 2.39. Solubility and solvent quality of the graphite dispersions. (a) Chlorosulphonic acid dispersions of graphite at an initial concentration of 25 mg·ml^{-1}, derived from different sources as indicated below the vials. A dark upper portion (top) was obtained for all the sources after 12 h of centrifugation at 5,000 rpm, with a gray-colored lower portion (bottom). The yellow line on the vials indicated the interface between the top and bottom phases. The soluble portion was removed and isolated for solubility examination. (b) Acid-induced shifts in the liquid-phase Raman G-peak for the graphite dispersed in the same mixtures of chlorosulphonic acid in sulphuric acid. The G-peak shift, denoted as dG, is a quantitative measure of the degree of protonation. The image in the insert shows a qualitative comparison between graphite dissolution in different solvents, showing graphite in the vials with a Teflon-coated stir bar to promote dissolution. Starting from the left, graphite is dissolved in NMP, 50, 65 and 80 vol% chlorosulphonic acid (HSO_3Cl) in sulphuric acid (H_2SO_4) and pure chlorosulphonic acid. The dispersions had a concentration of 10 mg·ml^{-1}. The acid dispersions were centrifuged at 5,000 rpm for 12 h, while the NMP dispersion was centrifuged for 3 h. The amount of centrifugation time was different for the two cases because the settling time was linearly dependent on the density differential between the particles and solvent. Thus, the centrifugation time was scaled by this density difference. Reproduced with permission from [110], Copyright © 2010, Macmillan Publishers Limited.

It was found that the quality of the dispersion was affected by the strength of the acid, which could be well-controlled by mixing chlorosulphonic acid and concentrated (98%) sulphuric acid [113]. The insert in Fig. 2.39 (b) shows that, the solubility of graphite was decreased significantly, as the acidity was reduced to 80% chlorosulphonic acid. The dispersion of SWCNT in superacids was due to the protonation of the SWCNT sidewalls, thus leading to electrostatic repulsion and debundling. The degree of protonation is measured as the positive charge per carbon atom in graphitic materials, which is characterized by the acid-induced shift (dG) in the location of the G peak in the Raman spectra [113–115]. Figure 2.39 (b) shows dG of the acid mixtures. The Raman shift dG decreased with decreasing acidity, corresponding to a decrease in solubility. In addition, the Raman spectra of the graphite powder before and after acid dissolution were almost the same, which implied that no defect was introduced in the starting materials after the acid treatment.

Low- and high-magnification TEM images of a single-layered graphene flake are shown in Fig. 2.40 (a) and (b), while Fig. 2.40 (c) shows HRTEM image of a few-layered graphite flake. The SWCNT grid network is clearly visible through the flake. Electron diffraction is illustrated in Fig. 2.40 (d), displaying typical Bragg reflections, with the intensity ratio of single-layer graphene, as shown in Fig. 2.40 (e) [116]. Figure 2.40 (f) shows an AFM image, with the height profile of one flake having a step height of 0.5 nm, which is consistent with that of single-layer graphene. It was found that 70% of the dispersed graphene nanoflakes were single-layer. The graphene nanosheets made from microcrystalline graphite and graphoil had an average flake size of 300 and 900 nm, respectively.

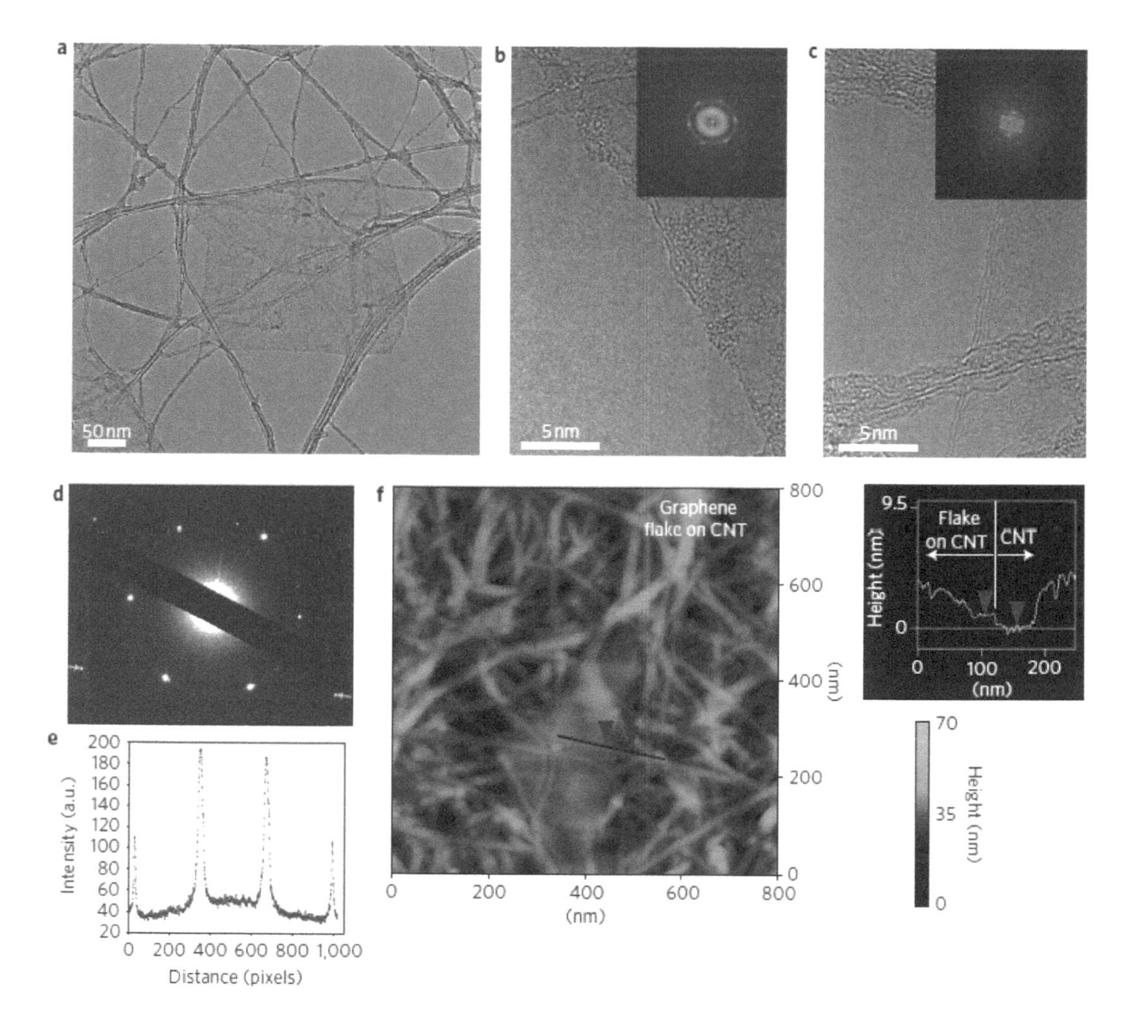

Fig. 2.40. Evidence for single-layer graphene dissolution. (a) Low-magnification TEM showing a small flake of graphene. The CNT grid network beneath the graphene flake is clearly visible. (b) HRTEM of a single-layer edge, with fast Fourier transform (FFT) shown as in insert. (c) HRTEM of the edge of a few-layered graphene. Adsorbates of unknown origin are clearly visible at high magnification, similar to flakes obtained by micromechanical cleavage. (d) Electron diffraction showing the typical intensity profile along the line delimited by the two yellow arrows (e). (f) AFM image and its height profile showing a flake with a height of 0.5 nm. Reproduced with permission from [110], Copyright © 2010, Macmillan Publishers Limited.

Suspensions of rigid anisotropic molecules, such as nanoclays, usually experience an isotropic/liquid-crystalline transition as their concentration is increased [117–120]. In order to prove this concept, the top phase was extracted and quenched after centrifugation. The powder was then redispersed at a concentration of 20 mg·ml^{-1} and centrifuged to induce phase separation. The top phase was structureless with a concentration of 1.8 mg·ml^{-1}. The bottom phase showed birefringence under a cross-polarizer, which was a typical feature of liquid crystals, as shown in Fig. 2.42 (a–c). Therefore, it was confirmed that the dispersions of soluble graphene underwent an isotropic/liquid-crystalline phase separation. The liquid crystalline Schlieren texture in Fig. 2.41 (b, c) was very similar to the typical discotic nematic samples, which is a promising characteristic for fabricating fibers and films. Similarly, oxidized and reduced GNRs in chlorosulphonic acid at a high concentration of 3.4 mg/ml were also entirely soluble and formed liquid crystals at high concentration, as shown in Fig. 2.41 (d–f). Therefore, chlorosulphonic acids are promising candidates for exfoliation and dispersion of single-layer graphene at very high concentrations.

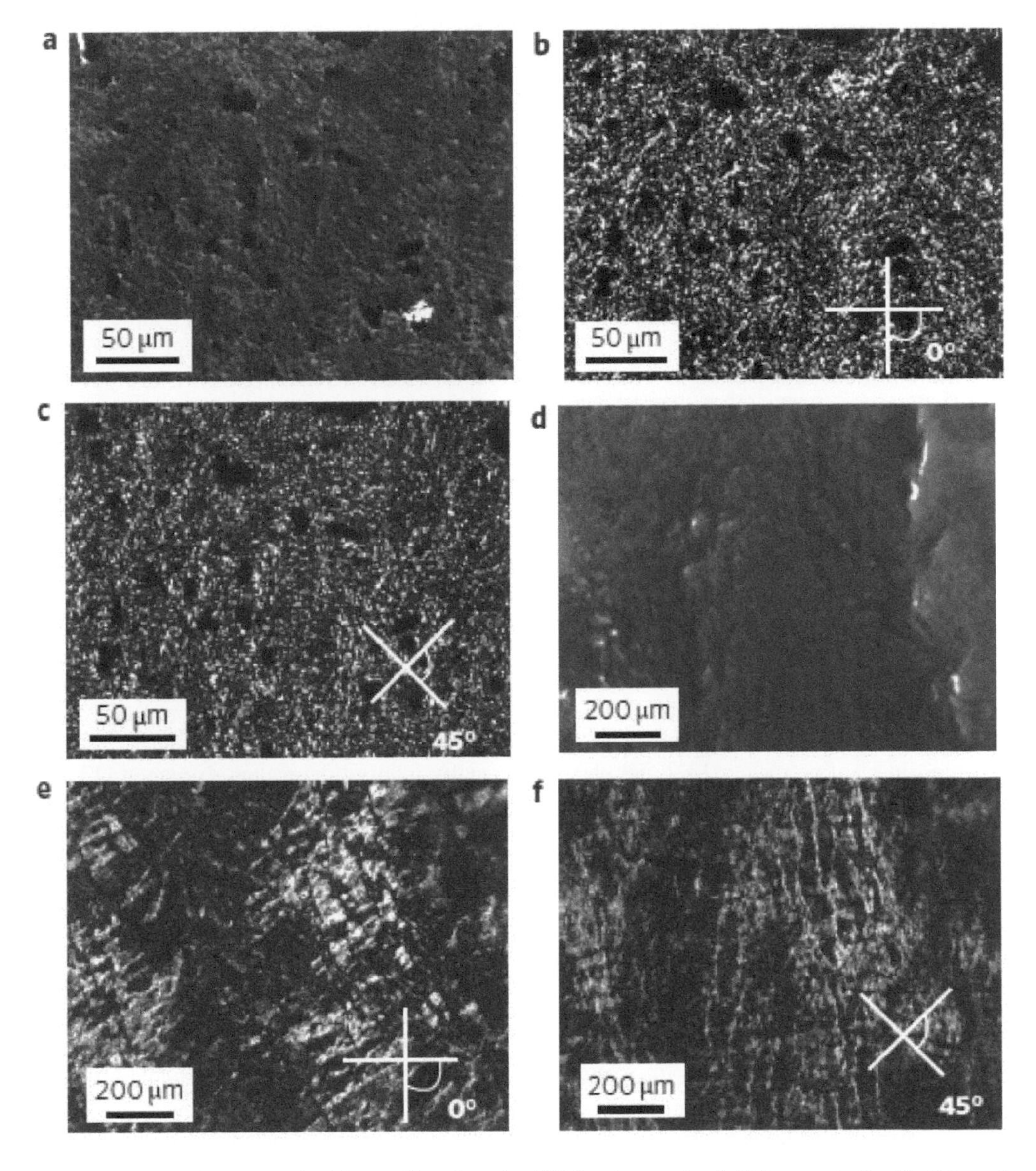

Fig. 2.41. Evidence for the graphene liquid-crystalline phase. (a–c) Light micrographs of a high concentration (2 wt%) graphene dispersion in chlorosulphonic acid (scale bar: 50 μm): transmitted light (a), transmitted polarized light with analyser and polarizer crossed at 90° (b), crossed analyser and polarizer rotated by 45° with respect to image of (b) (c). (d–f) Light micrographs of high concentration (2 wt%) oxidized nanoribbon dispersions in chlorosulphonic acid (scale bar: 200 μm): transmitted light (d), transmitted polarized light with analyser and polarizer crossed at 90°, crossed analyser (e) and polarizer rotated by 45° with respect to image of (e) (f). Reproduced with permission from [110], Copyright © 2010, Macmillan Publishers Limited.

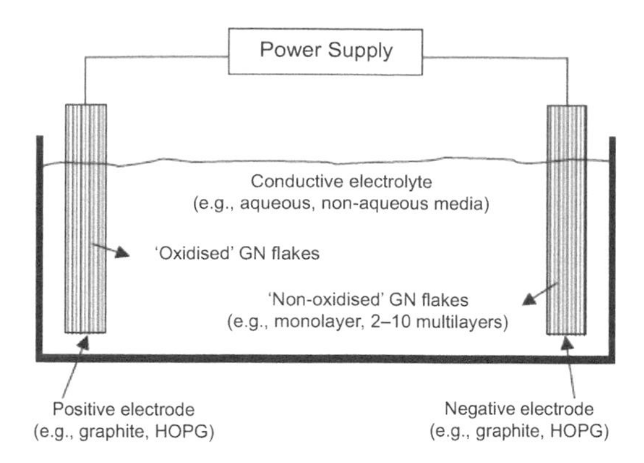

Fig. 2.42. Schematic diagram of an electrochemical cell used for electrochemical intercalation and exfoliation of HOPG to produce graphene nanosheets. Reproduced with permission from [121], Copyright © 2013, Elsevier.

2.2.1.5. Electrochemical Exfoliation

The electrochemical approach to exfoliate graphite has several advantages, including simple processing (single-step), being environmentally friendly (when using ionic liquid electrolytes or aqueous surfactants) and operation at ambient conditions. Strongly volatile solvents or reducing agents are not used in electrochemical exfoliation [121, 122]. The feasibility of electrochemical method to produce graphene nanosheet powder from milligram and gram quantities has been widely demonstrated [123–127]. With the well-established principles of electrochemical cell design and engineering, this method can be easily scaled up [128]. The process can be finished in a controlled time duration, fast processing of large quantities of graphene nanosheet is also possible [129]. The graphene nanosheets are produced in a liquid solution of electrolyte, in which an electrical current is applied to drive structural expansion, either oxidation or reduction, intercalation and exfoliation, of graphite in the forms of rod, plate or wire [123]. Yield, productivity and properties of the graphene nanosheets can be maximized by controlling the electrolysis parameters and electrolytes [126].

Electrochemical methods originated from studies on the intercalation of pure sulphuric acid in graphite particulates [130], electrochemical intercalation of F ions into graphite in aqueous [131] and anhydrous HF media [132], conversion of a $NiCl_2$-based GIC into a more useful $Ni(OH)_2$-based GIC by electrochemical polarization in alkaline media [133] and electrochemically induced intercalant exchange [134, 135]. Highly oriented pyrolytic graphite (HOPG) and vitreous carbon electrode materials have also been used for electrochemical exfoliation [136–140]. Electrochemical exfoliation methods involve the application of cathodic or anodic potentials, and currents or voltage, in either aqueous (acidic or other media) or non-aqueous solutions of electrolytes. The scaling-up of the electrochemical exfoliation is determined by the yield of the graphene nanosheets produced [124]. The ideal procedure should be able to produce monolayer graphene nanosheets, with high conductivity and large size.

An experimental set-up of electrochemical intercalation and exfoliation of graphite into graphene nanosheets consists mainly of the following parts: (i) a graphite working electrode (WE), (ii) a standard reference electrode or a quasi-reference electrode that could vary depending on the electrolyte used, (iii) a counter electrode that is used in a three-electrode arrangement, (iv) an electrolyte solution that could be aqueous (e.g., acidic or surfactant) or non-aqueous (e.g., organic or ionic liquid) and (v) a DC power supply or a potentiostat. Figure 2.42 shows a schematic diagram of an electrochemical exfoliation set-up for a two-electrode undivided electrochemical cell [121].

Generally, a constant potential is applied to the WE or the potential could be cycled between pre-determined values as in cyclic voltammetry. Sometimes, a constant current could be applied. In this case, the WE corrodes and a black precipitate or sludge will gradually be formed at bottom of the reactor. During the electrochemical exfoliation process, the host graphite WE is oxidized or reduced, so that cations or anions from the electrolyte cause intercalation of the graphite, as shown in Fig. 2.43 [121]. The intercalation induces structural expansion of the graphite matrix [136]. Once the intercalation is complete and a GIC is produced, a different potential or current, but with reverse polarity to the level applied during the intercalation is used to exfoliate graphite flakes into the electrolyte solution. The selection of ions or surfactants is done to ensure that graphene nanosheets remain dispersed rather than aggregating back into graphite [126].

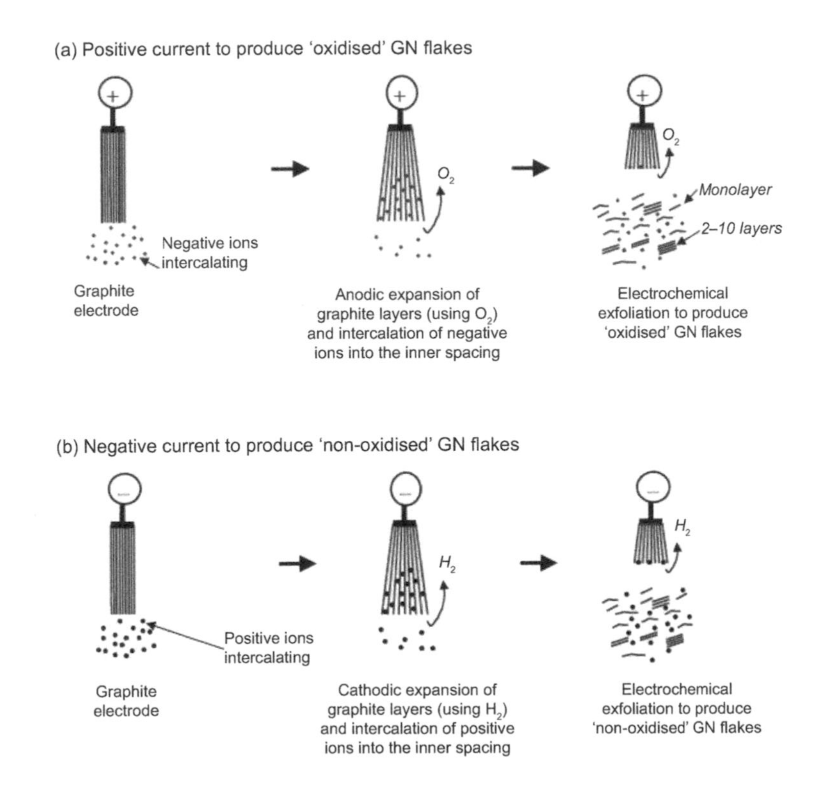

Fig. 2.43. Electrochemical approaches: (a) oxidation, intercalation and exfoliation (negative ions are shown in red color) and (b) reduction, intercalation and exfoliation to produce single and multilayer graphene nanosheets (positive ions are shown in deep blue color). Reproduced with permission from [121], Copyright © 2013, Elsevier.

For anodic oxidation, positive potentials are applied to the graphite WE (or anode) in order to oxidize it, so as to trigger the intercalation of anions, leading to structural expansion and ultimate exfoliation of graphene nanosheets [123]. High purity graphite rods with a diameter of 6 mm (99.999% purity) were used as electrodes. Poly(sodium-4-styrenesulfonate) (PSS, M_w = 70,000) was dissolved in de-ionized (DI) water to form the electrolyte solution with a concentration of 0.001 M. Figure 2.44 shows schematic diagram of the synthesis set-up. In the experiment, two graphite rods were placed in the electrolysis cell filled with the electrolyte. A constant potential of 5 V (DC voltage) was applied to the two electrodes. After 20 min of electrolysis, a black product was gradually observed at the positive electrode (anode). The exfoliation continued for 4 h. Then the product dispersion was collected from the electrolysis cell. The dispersion was centrifuged at low speed (1000 rpm) to remove large agglomerates. The top of the dispersion was decanted. This graphene–PSS suspension was very stable, without precipitation after 6 months storage. Dry graphene nanosheet powders were obtained from the dispersion by washing and drying. The yield of the graphene nanosheet powder was about 15 wt%.

Figure 2.45 shows Raman spectrum of the dried graphene nanosheet powder, together with that of the graphite rod as the inset [123]. A broad D band (1350 cm^{-1}) and a sharp G band (1580 cm^{-1}) were observed in the Raman spectrum of the graphene nanosheet powder. The G line corresponds to the in-plane

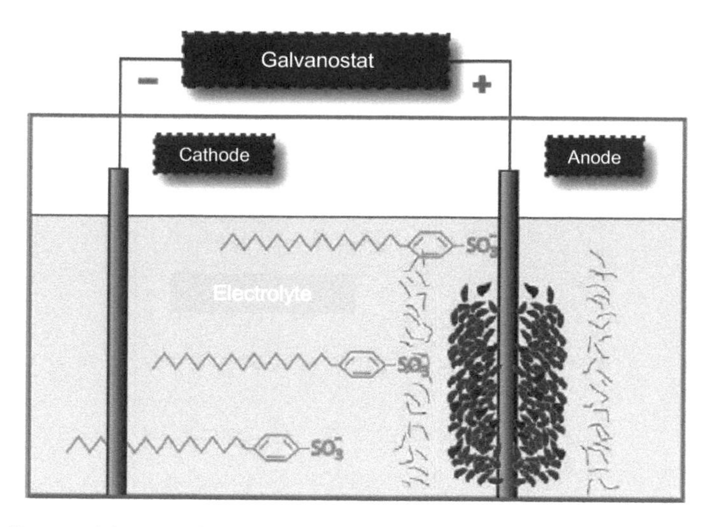

Fig. 2.44. Schematic diagram of the set-up for synthesis of graphene by using electrochemical exfoliation. Reproduced with permission from [123], Copyright © 2009, Elsevier.

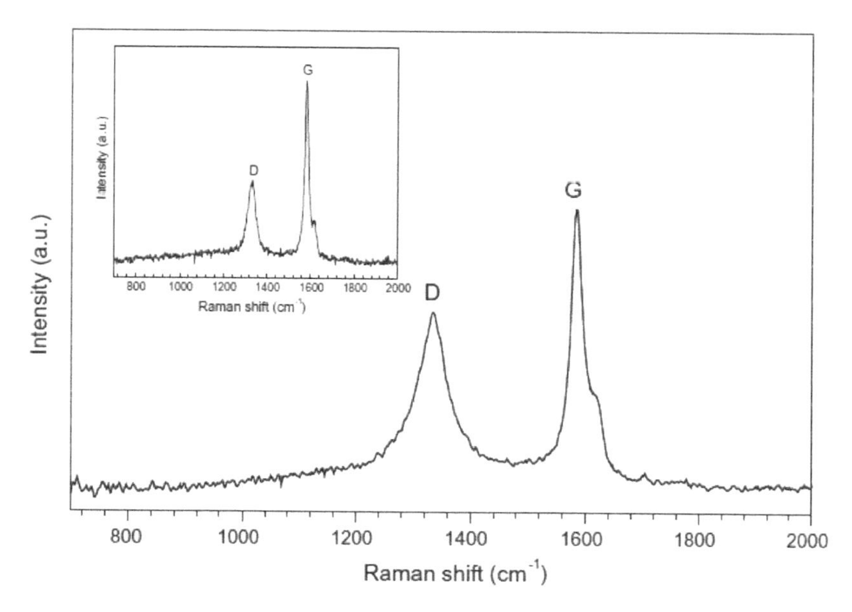

Fig. 2.45. Raman spectrum of the graphene nanosheet powders, together with the Raman spectrum of the pristine graphite rod shown in the inset. Reproduced with permission from [123], Copyright © 2009, Elsevier.

bond-stretching motion of the pairs of C sp^2 atoms, i.e., the E_{2g} phonons, whereas the D line represents the breathing modes of rings or κ-point phonons of A_{1g} symmetry [141]. It has been observed that the graphene nanosheets produced by using the chemical exfoliation approaches usually show a strong D band in the Raman spectrum, characterized by a high intensity ratio of $I_D/I_G > 1$, which is attributed to the defects and partially disordered crystal structure in graphene nanosheets [109]. Therefore, the electrochemically exfoliated graphene nanosheets had a low content of defects, because the intensity of the G band is much higher than that of the D band.

Figure 2.46 (a) shows SEM image of the graphene nanosheet powder [123], which exhibited typical crumpling and scrolling characteristics of graphene nanosheets [99]. Figure 2.46 (b) demonstrates low magnification TEM image of the graphene nanosheets, which were stacked into multilayered structures,

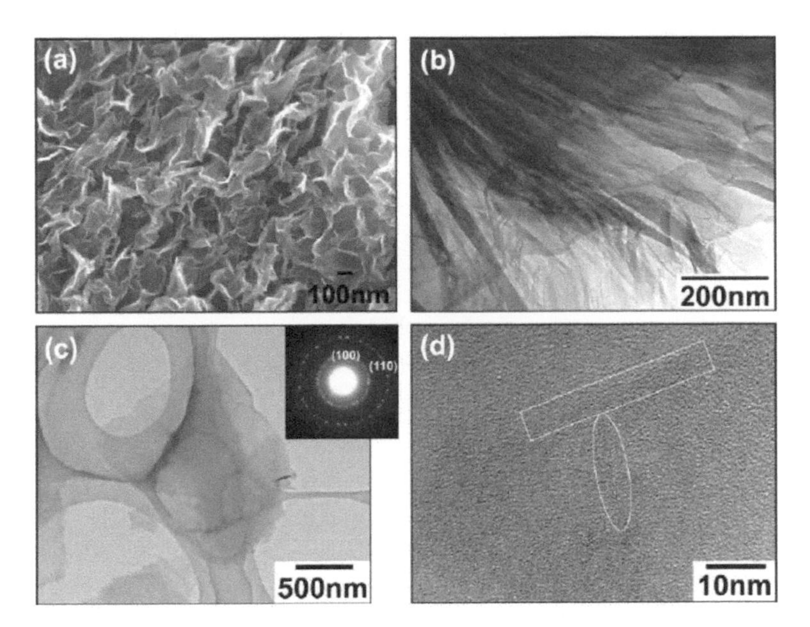

Fig. 2.46. (a) SEM image of the graphene nanosheet powders. (b) TEM image of a collection of the graphene nanosheets. (c) TEM image of a large graphene flake. The inset is the corresponding SAED pattern. (d) HRTEM image of the graphene sheets, showing the featureless basal planes and a cross-sectional view of the edges of folded graphene nanosheets. Reproduced with permission from [123], Copyright © 2009, Elsevier.

exhibiting a rippled and entangled profile. Figure 2.46 (c) illustrates TEM image of a few flat graphene nanosheets with larger sizes, where 2–3 layers of graphene nanosheets were overlapping. Selected area electron diffraction (SAED) of the graphene nanosheets is shown as the inset in Fig. 2.46 (c). The diffraction rings could be indexed to the hexagonal graphitic structure, indicating the crystalline nature of the graphene nanosheets. A high resolution TEM (HRTEM) image of the graphene nanosheets is shown in Fig. 2.46 (d). Most regions were featureless basal plane, while cross-sectional views due to folding of the graphene nanosheets were occasionally observed.

Anodic intercalation of graphite GO and the subsequent reduction of GO to rGO nanosheet is efficient in terms of time duration, but the sp^3 defects cannot be completely eliminated to restore the sp^2 configuration [127]. To address this problem, it is thus proposed to use cathodic reduction of the host graphite electrode, which can be conducted in either aqueous or non-aqueous electrolytes [124, 126].

One example is a solution route that was inspired by the lithium rechargeable battery, which offered high-yield (> 70%) exfoliation of graphite into conductive few-layer graphene nanosheets, with an average thickness < 5 layers [124]. In this case, the negative graphite electrode was electrochemically charged and expanded in an electrolyte of Li salts and organic solvents at a relatively high current density, which was further exfoliated into few-layer graphene nanosheets by using sonication. Figure 2.47 shows a schematic diagram to illustrate the exfoliation mechanism. The highly dispersible graphene nanosheets could be used to make conductive inks, leading to conductive films coated on commercial papers, with a resistivity of $15\ \Omega\cdot\square^{-1}$, at a graphene loading of $< 1\ mg\cdot cm^{-2}$.

X-ray photoelectron spectroscopy (XPS) spectra of the exfoliated graphene nanosheets are shown in Fig. 2.47 (a, b). Li and PC were observed on the graphene nanosheets. At high charging voltages, Li-PC complexes accelerated the expansion of the electrode. Figure 2.47 (c) shows capacitive currents of the expanded graphite electrode, which were increased with the increasing charging voltage. Thermogravimetric analysis (TGA) of the expanded graphite electrode indicated that a large amount of PC was present at the graphite electrode, as shown in Fig. 2.47 (d). Obviously, the amount of PC intercalated into the graphite electrode was increased with the increasing charging voltage. It thus implied that the organic intercalants were from the electrochemical process and not due to organic contaminants. Representative SEM images of the thermally annealed electrode are shown in Fig. 2.47 (e, f, inset), illustrating the presence of graphene

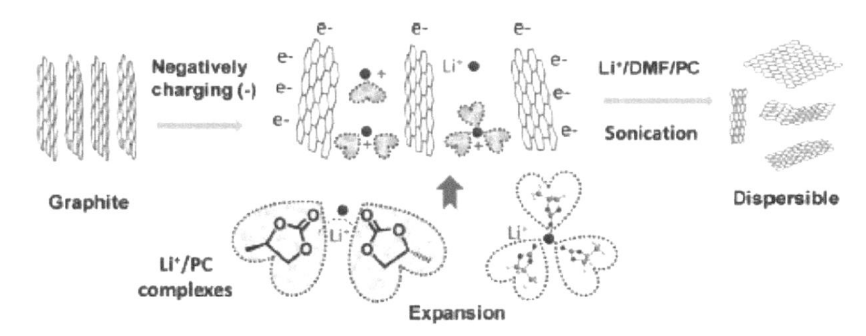

Fig. 2.47. Exfoliation of graphite into few-layer graphene nanosheets through the intercalation of Li^+ complexes. Graphite was electrochemically charged in Li^+/PC at a high voltage. The expanded graphite was then exfoliated by power sonication in LiCl in a DMF/PC mixed solvent. Reproduced with permission from [124], Copyright © 2011, American Chemical Society.

nanosheets. The graphene nanosheets were derived from the expansion of the graphite layers caused the releasing of the gas due to the decomposition of PC. The black appearance and high conductivity of the expanded graphite particles confirmed that the electrochemical exfoliation was a nonoxidative process.

The hydrophobic few-layer graphene nanosheets could be floated on the surface of water and dispersed in dichlorobenzene and diphenyl ether at 1 mg/ml, as shown in Fig. 2.48 (a, right). Figure 2.49 (b) shows SEM images of the few-layer graphene nanosheets, which exhibited an average lateral dimension of 1–2 µm. The size histogram is shown in Fig. 2.49 (c). The diffraction spot intensity in TEM images suggested that the thickness of the graphene nanosheets was 1–3 layers, as shown in Fig. 2.50 (a, b). According to the statistical sampling of the few-layer graphene nanosheets by AFM, it was found that 50% of the graphene nanosheets consisted of 2–3 layers of graphene, as shown in Fig. 2.49 (c) and Fig. 2.50 (c).

Figure 2.50 (d, right) shows Lorentzian peak fittings of the 2D peak using deconvoluted peaks characteristic of the few-layer graphene nanosheets with two and three layers. As shown in Fig. 2.50 (d, left), the I_D/I_G intensity ratio was < 0.1, indicating the presence of a very low level of defects. All the data allowed the conclusion that $> 70\%$ of the few-layer graphene nanosheets had a thicknesses of < 5 layers. The few-layer graphene films transferred onto plastic substrates showed good electrical conductivity without the requirement of thermal annealing.

Another example is the electrochemical intercalation of sodium dodecyl sulfate (SDS) into graphite, which was followed by electrochemical exfoliation of the SDS-intercalated graphite electrode [126]. These electrochemical processes resulted in a stable colloidal suspension of graphene/SDS. It was found that the structural ordering, size and number of layers of the final graphene sheets were determined by the potential value for the intercalation of SDS into the graphite. According to the Raman spectroscopy and transmission electron microscopy results, a relatively high intercalation potential led to graphene nanosheets with higher structural ordering and less number of layers.

Figure 2.51 shows a schematic diagram of the electrochemical approach for graphite intercalation and exfoliation [126]. In this experiment, electrochemical intercalation of SDS into graphite was conducted in 0.1 M SDS solution, with graphite used as the anode. The SDS-intercalated graphite electrode was electrochemically exfoliated to form stable graphene/SDS suspension, when the electrode was used as the cathode. The graphene/SDS suspensions were centrifuged at 1000 rpm for 10 min to remove large agglomerates. The top suspension was separated by decantation, which was demonstrated to be long term stable without precipitation. Graphene nanosheet powder could be derived from the graphene/SDS suspension through long time high speed centrifugation at 20,000 rpm, followed by washing and drying.

Figure 2.52 shows TEM images of the graphene nanosheets that were processed at intercalation potentials of 1.4, 1.6, 1.8 and 2.0 V [126]. The intercalation potential showed obvious influence on transmission and shape of the graphene nanosheets. In most cases, the graphene nanosheets exhibited multilayered structures. Disordered and relatively thick graphene flakes were observed in the sample processed at the intercalation potential of 1.4 V, while the intercalation potential of 1.6 V led to relatively transparent, more ordered, much larger and thin graphene flakes. At higher intercalation potentials of 1.8

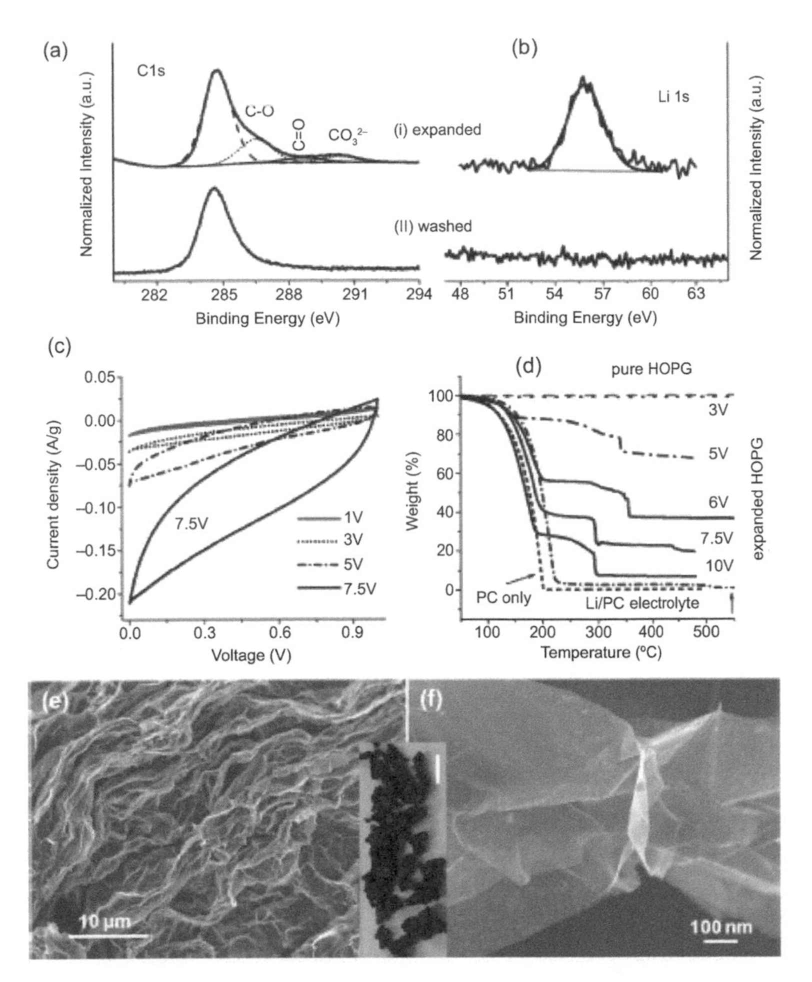

Fig. 2.48. (a, b) XPS narrow scans of C 1s and Li 1s for (i) the expanded HOPG electrode and (ii) after washing of (i) with HCl ($1.0\ mol{\cdot}l^{-1}$) and water. (c) Cyclic voltammograms recorded between 0 and 1.0 V, at a scan rate of $0.05\ V{\cdot}s^{-1}$ after the graphite electrode as the negative electrode had been precharged at potentials of 1, 3, 5 and 7.5 V for 60 min. The electrochemical cell consisted of the graphite electrode, a Li metal electrode and $30\ mg{\cdot}ml^{-1}$ $LiClO_4$ in PC solution. (d) TGA curves of expanded graphite samples in N_2 atmosphere at different charging voltages. TGA data for graphite, PC and $30\ mg{\cdot}ml^{-1}$ $LiClO_4$ in PC (Li/PC) are included for comparison. (e, f) SEM images of the expanded HOPG electrode (charged at 7.5 V) after performing TGA (annealed at the rate of $10°C{\cdot}min^{-1}$ to 500°C in N_2 gas flow). The inset shows a photograph of the expanded HOPG after annealing, with a scale bar of 5 mm. Reproduced with permission from [124], Copyright © 2011, American Chemical Society.

and 2.0 V, highly ordered hexagonal graphite lattices of multilayered graphene flakes could be observed, as shown in Fig. 2.52 (c, d). The buckled or wrinkled parts seen in Fig. 2.52 (c, d) was related to the thermodynamic stability of 2-D structure of graphene [99].

The graphene/SDS suspensions could be used to fabricate graphene films by using electrophoresis deposition, as shown in Fig. 2.53 [126]. The electrophoresis experiment was performed at 3 V for 3 h, after which both cathode and anode Pt electrodes were rinsed with ultra-pure water, followed by drying with stream of N_2. Low and high magnification SEM images of the Pt surface are shown in Fig. 2.53 (a) and (b), respectively. Similarly, Fig. 2.53 (c) and (d) illustrate low and high magnification SEM images of the Pt anode electrode after the electrophoresis process. The surface structure of Pt anode exhibited almost no change before and after the electrolysis, i.e., no graphene film was deposited on the surface. In contrast, Fig. 2.53 (e) clearly indicates that the surface of Pt cathode was covered with nanosized and microsized particles after the electrophoresis process. High magnification image revealed that the particles were agglomerates of graphene nanosheets, with sizes of as large as 1.3 μm, as shown in Fig. 2.53 (f). Such

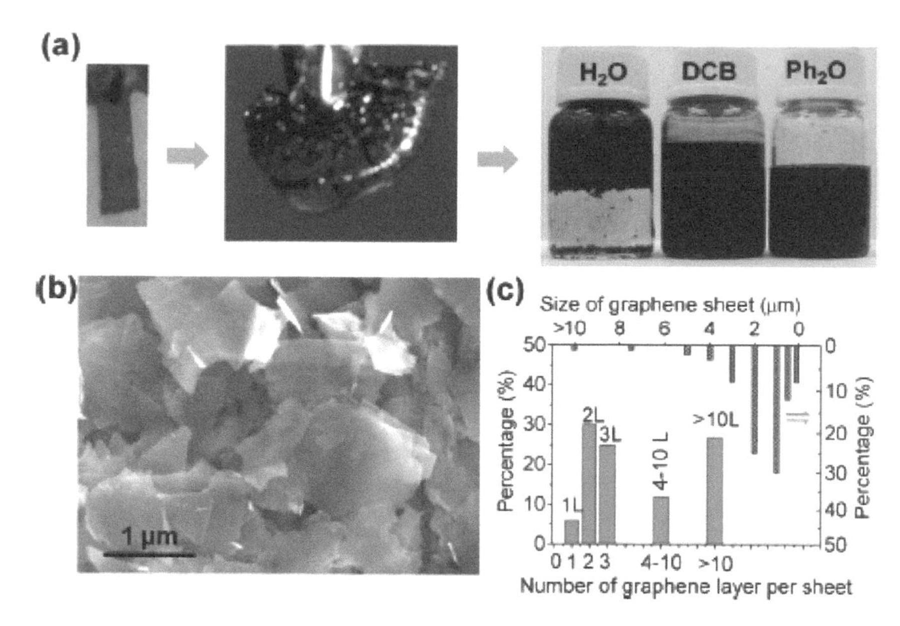

Fig. 2.49. (a) Photographs of (left) the HOPG used as a cathode, (center) HOPG after electrochemical intercalation, where it expanded by > 50 times in volume and (right) after sonication of the expanded HOPG to produce few-layer graphene nanosheet powder, which was dispersed at 1 mg·ml^{-1} in water, dichlorobenzene (DCB) and diphenyl ether (Ph_2O). (b) Low magnification SEM image. (c) Thickness and size distribution histograms of the few-layer graphene nanosheets, estimated from AFM analysis. Reproduced with permission from [124], Copyright © 2011, American Chemical Society.

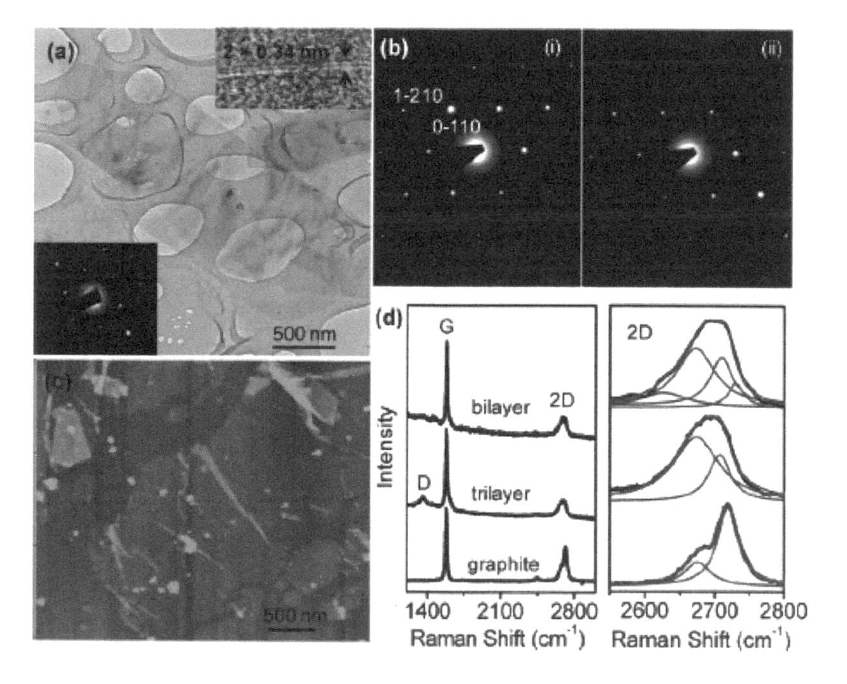

Fig. 2.50. (a) TEM images and electron diffraction pattern of the few-layer graphene nanosheets. (b) Electron diffraction patterns of (i) single and (ii) bilayer nanosheets. (c) AFM image of the few-layer graphene nanosheets, spin-coated onto a Si substrate. The thickness was about 1.5 nm, corresponding to a bilayer. (d) (left) Raman spectra (532 nm laser) of the few-layer graphene nanosheets on Si substrates, compared with the spectrum of graphite. (right) Lorentzian peak fitting of the 2D bands of the bilayer and trilayer. Reproduced with permission from [124], Copyright © 2011, American Chemical Society.

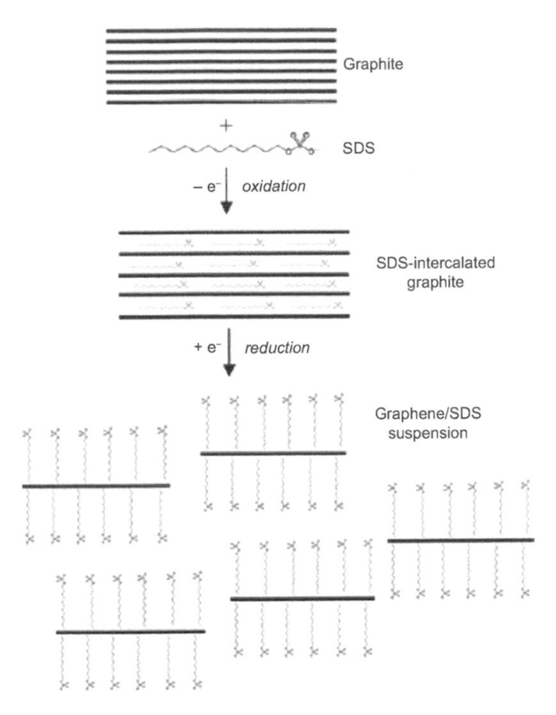

Fig. 2.51. Schematic diagram of the electrochemical approach to produce graphene/SDS suspension. Reproduced with permission from [126], Copyright © 2012, Elsevier.

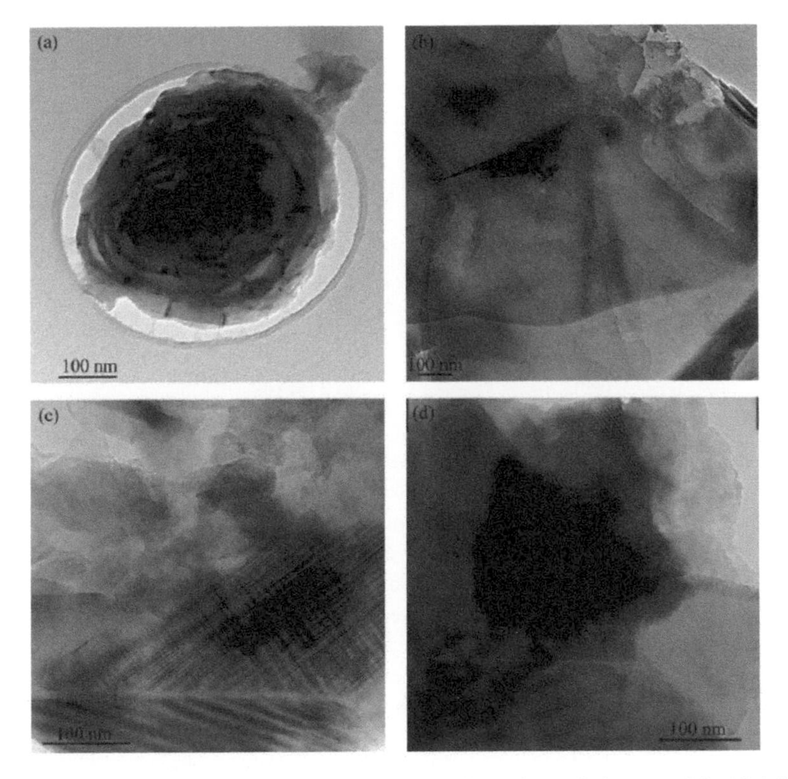

Fig. 2.52. TEM images of the graphene nanosheets synthesized at different intercalation potentials: (a) 1.4, (b) 1.6, (c) 1.8 and (d) 2.0 V. Reproduced with permission from [126], Copyright © 2012, Elsevier.

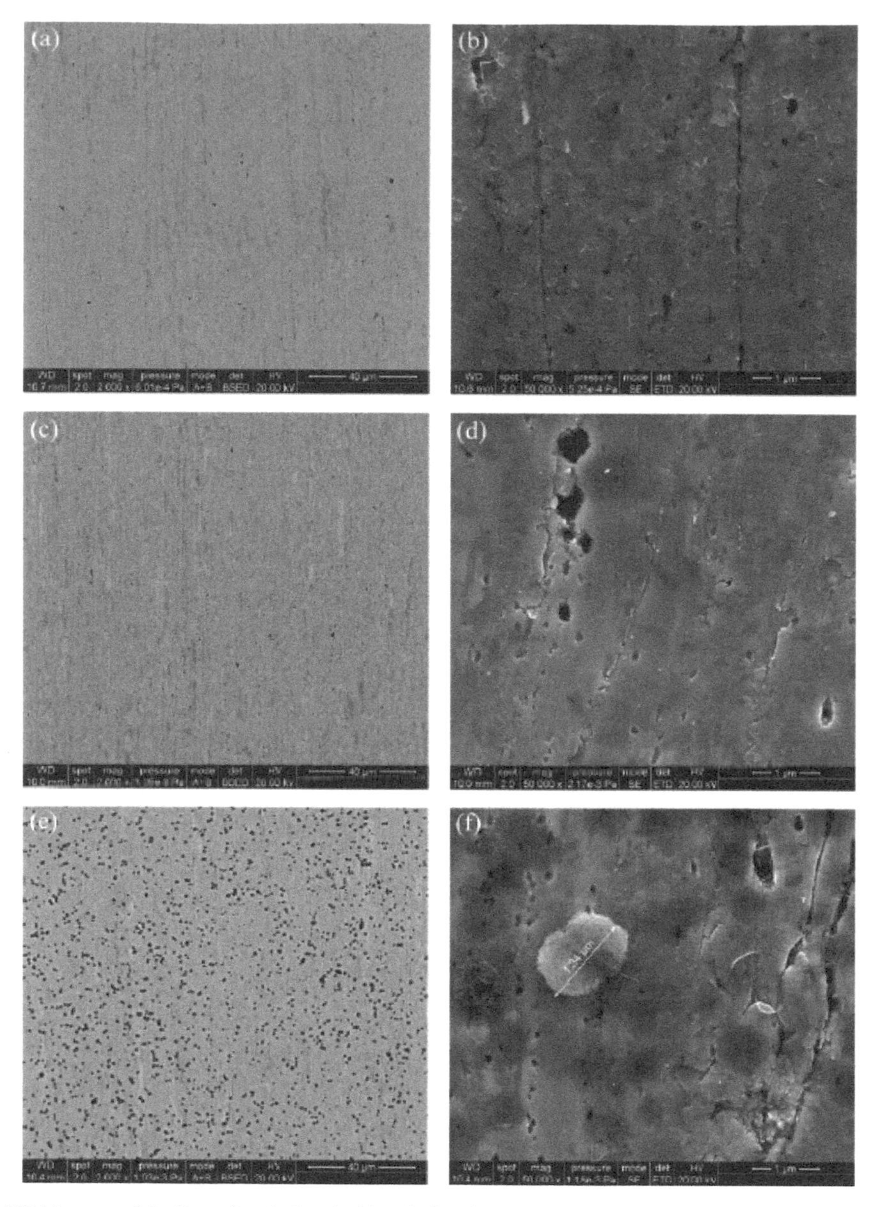

Fig. 2.53. SEM images of the Pt surface before (a, b) and after the electrophoresis process at 3 V for 3 h in the graphene/SDS suspension (c, d at anode and e, f at cathode). Reproduced with permission from [126], Copyright © 2012, Elsevier.

graphene nanosheets were formed due to electrochemical reduction of the GO flakes, which are similar to the GO films electrochemically deposited on glassy carbon electrodes [142].

A simple and fast electrochemical exfoliation of HOPG crystals in sulfuric acid to produce single and multilayer graphene nanosheets has been reported [143]. The graphene nanosheets developed in this way mainly had an AB-stacked bilayer structure, with lateral sizes ranging from several μm to tens of μm. Electrical properties of the electrochemically exfoliated nanosheets were higher than those of the commonly reduced graphene oxide nanosheets. Stable suspensions could be derived by dissolving the graphene nanosheets in dimethyl formamide (DMF). When water was added into the suspensions, self-aggregation occurred at the air-DMF interfaces, due to the strong surface hydrophobicity. This occurrence led to continuous films, which exhibited ultratransparency of ~96% transmittance and relatively low sheet resistance of < 1 k$\Omega\cdot\square^{-1}$ after a simple treatment with HNO_3. According to Raman and STM characterization results, the graphene nanosheets exfoliated by using this electrochemical method preserved the intrinsic structure of graphene.

In this method, natural graphite flakes (NGF) with average sizes of 5–20 mm and highly oriented pyrolytic graphite (HOPG) with a dimension of $1.5 \times 1.5 \times 0.3$ mm^3 were used as the electrode and source of graphene for the electrochemical exfoliation. The graphite flake was adhered to a tungsten wire by using a silver pad, which was then inserted into the ionic solution as an anode in such a way that only the graphite was immersed into the solution. A grounded Pt wire was placed parallel to the graphite flake at a distance of 5 cm. The ionic solution was formed by dissolving 4.8 g sulfuric acid (H_2SO_4, 98%) in 100 ml DI water. DC bias voltages were applied on graphite electrode from −10 to +10 V to trigger the electrochemical exfoliation process. The exfoliated graphene nanosheets were collected with a 100 nm porous filter and washed with DI water by using vacuum filtration. The dried graphene nanosheets were dispersed in DMF solution by using gentle sonication in a water-bath for 5 min to form graphene suspension. High speed centrifugation at 2500 rpm was used to remove the large graphite particles.

The experimental setup of the electrochemical exfoliation is shown in Fig. 2.54 (a) [143]. Static bias of +1 V was first applied to the graphite for 5–10 min, which was then increased to +10 V for 1 min. The low bias voltage was to wet the sample and cause gentle intercalation of SO_4^{2-} ions into the grain boundary of the graphite [144]. As the high bias of +10V was applied, the graphite was dissociated into small pieces, which were spreading onto surface of the solution, as shown in Fig. 2.54 (b). Figure 2.54 (c) shows the photograph of the dispersed graphene sheets in DMF solution. Although the electrochemical exfoliation of graphite using H_2SO_4 solution was very efficient, the graphene nanosheets contained a high level of defects, because H_2SO_4 is a strong oxidation agent. This problem could be addressed by adding KOH solution into the H_2SO_4 solution. For example, 2.4 g 98% H_2SO_4 in 100 ml DI water was mixed with 11 ml 30% KOH solution to have a pH value of about 1.2. It was also found that the exfoliation conditions could be optimized by adjusting the applied voltages.

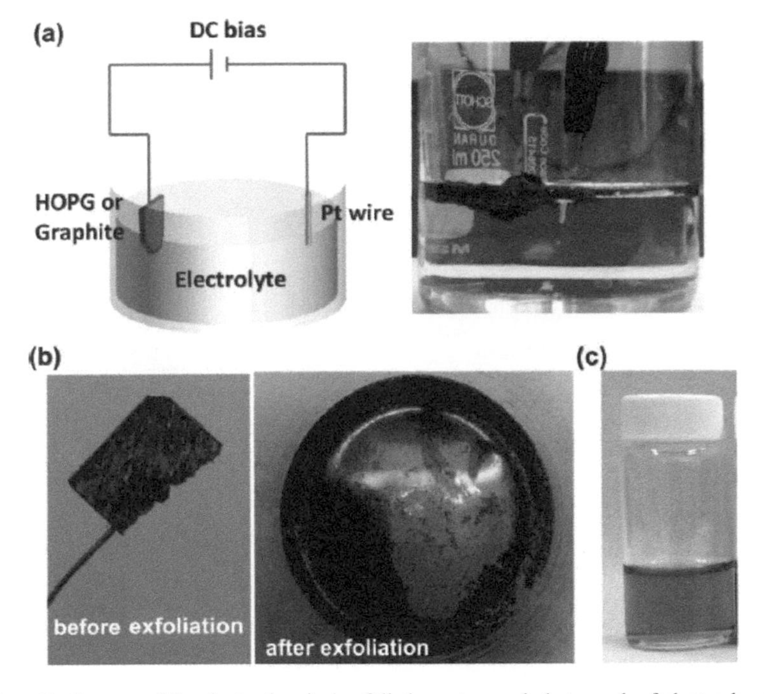

Fig. 2.54. (a) Schematic diagram of the electrochemical exfoliation set-up and photograph of electrochemically exfoliated graphite. (b) Photographs of the graphite before and after the electrochemical exfoliation. (c) Photograph of the dispersed graphene nanosheets in DMF. Reproduced with permission from [143], Copyright © 2011, American Chemical Society.

Figure 2.55 (a) shows AFM image of the electrochemically exfoliated graphene nanosheet film with a thickness of about 1.5 nm, which was drop-cast on a SiO_2 substrate [143]. The graphene nanosheets had a thickness of < 3 nm and more than 65% of them were < 2 nm, as shown in Fig. 2.55 (b). Their lateral size was in the range of 1–40 μm, which was larger than that of those exfoliated in liquid phases with extensive sonication. The total yield of the graphene nanosheets was 5–8 wt%. Figure 2.55 (c) shows representative TEM image of the exfoliated graphene nanosheets, most of which were bi-layered. The interlayer distance was about 0.45 nm, which was larger than 0.335 nm in ordinary graphite. Figure 2.55 (d) shows STM image in a selected area of the graphene nanosheets. The bright lattice pattern marked with circles and spots indicated that the bilayer graphene nanosheet exhibited an A-B stacking, i.e., Bernal stacking [145]. Figure 2.55 (e) Attenuated Total Reflection Fourier transform infrared (ATR-FTIR) spectrum of the graphene nanosheet films, which confirmed the presence of free or nonbonded SO_4^{2-}, C-O-C and C-OH [146, 147].

Another electrochemical exfoliation of graphite was able to produce graphene nanosheets, which exhibited a high yield (> 80%) of one- to three-layer graphene nanosheets, a high C/O ratio of 12.3 and low sheet resistance of 4.8 kΩ·□$^{-1}$ for a single graphene nanosheet [148]. Experimental setup for the electrochemical exfoliation of graphite is shown in Fig. 2.56 (a). Natural graphite flakes, platinum wires and 0.1 M H_2SO_4 solution were used as working electrodes, counter electrodes and electrolytes, respectively. When a positive voltage of +10 V was applied to the graphite electrode, the graphite flakes began to expand, dissociate and spread into the solution, as shown in Fig. 2.56 (b) and (c). The exfoliation

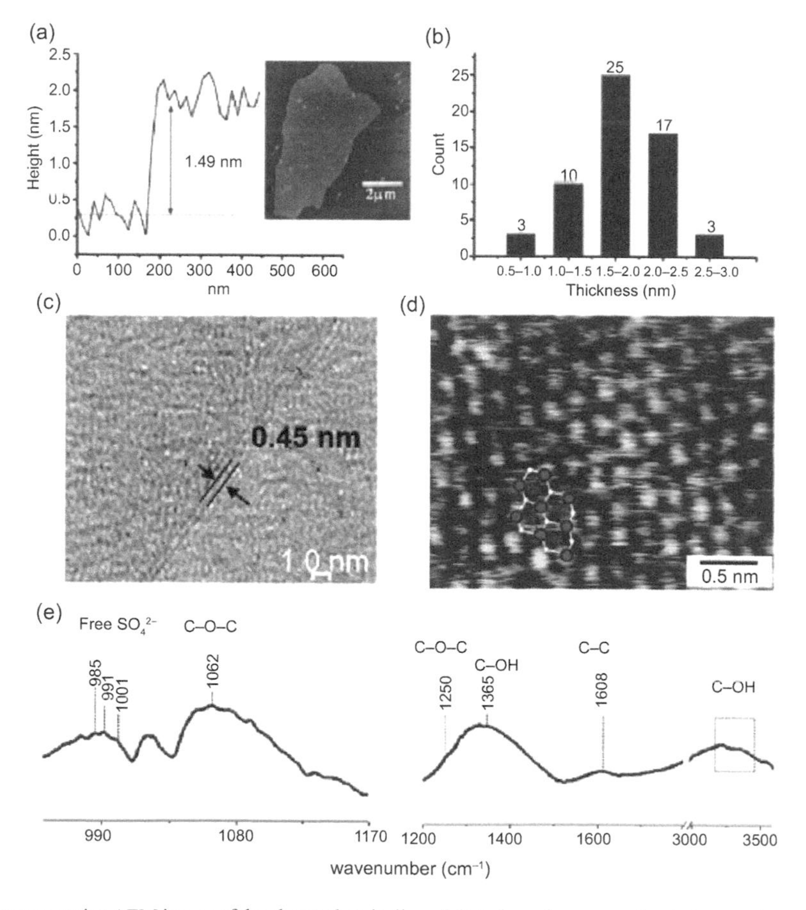

Fig. 2.55. (a) Representative AFM image of the electrochemically exfoliated graphene nanosheets drop-cast on a SiO_2 substrate. (b) Statistical thicknesses of the graphene nanosheet films. (c) Representative TEM image of the exfoliated bilayered graphene nanosheets. (d) STM image of a bilayered graphene, with the hexagons indicating the atom configuration of the two layer graphene nanosheets. (e) ATR-FTIR spectrum of the graphene nanosheet film. Reproduced with permission from [143], Copyright © 2011, American Chemical Society.

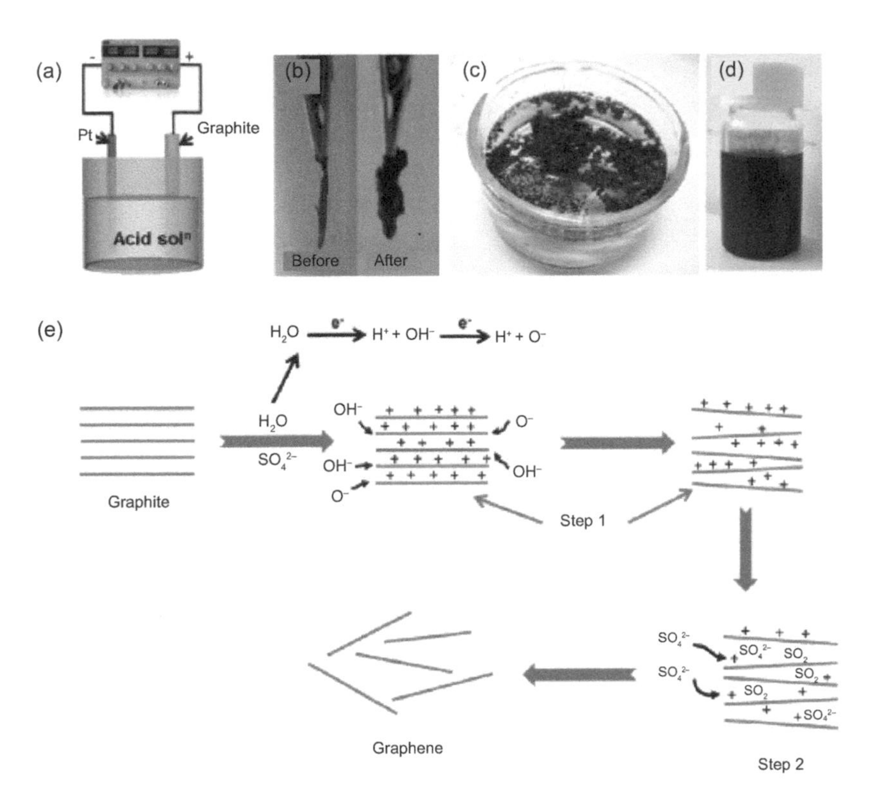

Fig. 2.56. (a) Schematic illustration of the electrochemical exfoliation of graphite, (b) photographs of graphite flakes before and after the exfoliation, (c) exfoliated graphene nanosheets floating on the surface of water, (d) dispersed graphene nanosheets in DMF at a concentration of about 1 mg/ml and (e) schematic illustration of the proposed mechanism of the electrochemical exfoliation process. Reproduced with permission from [148], Copyright © 2013, American Chemical Society.

process was completed in 2 min. The exfoliated graphene nanosheets could be collected by using the method described above and dispersed in DMF to form stable suspensions at concentrations of up to 1.0 mg/ml, as shown in Fig. 2.56 (d).

Figure 2.56 (e) shows mechanism of electrochemical exfoliation graphite in 0.1 M H_2SO_4 solution, involving (i) the oxidation of water to produce hydroxyl and oxygen radicals, 33 (ii) oxidation or hydroxylation at edge sites and grain boundaries of the graphite electrode by the radicals, (iii) opening up of the defective sites at the edges or grain boundaries due to oxidation and (iv) intercalation of SO_4^{2-} to cause expansion of the graphite [144, 149]. Further experiment results suggested that the presence of water and sufficiently high concentration of H_2SO_4 are the most important factors that determine the exfoliation efficiency of the electrochemical exfoliation method.

Figure 2.57 (a) shows AFM image of the graphene nanosheets cast on silicon substrates [148]. The average thickness of the graphene nanosheets was about 0.86 nm, which is thicker than that of pristine graphene, implying that the graphene nanosheets had oxygen-containing groups on their surfaces. The distribution of thickness of the graphene nanosheets according to the AFM height profiles is shown in Fig. 2.57 (b). More than 80% of them consisted of 1–3 layers, while bilayer graphene nanosheets were ~35%. The sizes of the nanosheets were in the range of 5–10 μm. High-resolution TEM (HRTEM) images confirmed that the graphene nanosheets had 1–4 layers, as shown in Fig. 2.57 (c–e). Figure 2.57 (d) shows an interlayer distance of ~0.34 nm in a bilayer graphene nanosheet. Selected area electron diffraction (SAED) pattern of the graphene sheets exhibited typical 6-fold symmetry, as shown in Fig. 2.57 (f), confirming the hexagonal crystalline structure of bilayer graphene nanosheet [150].

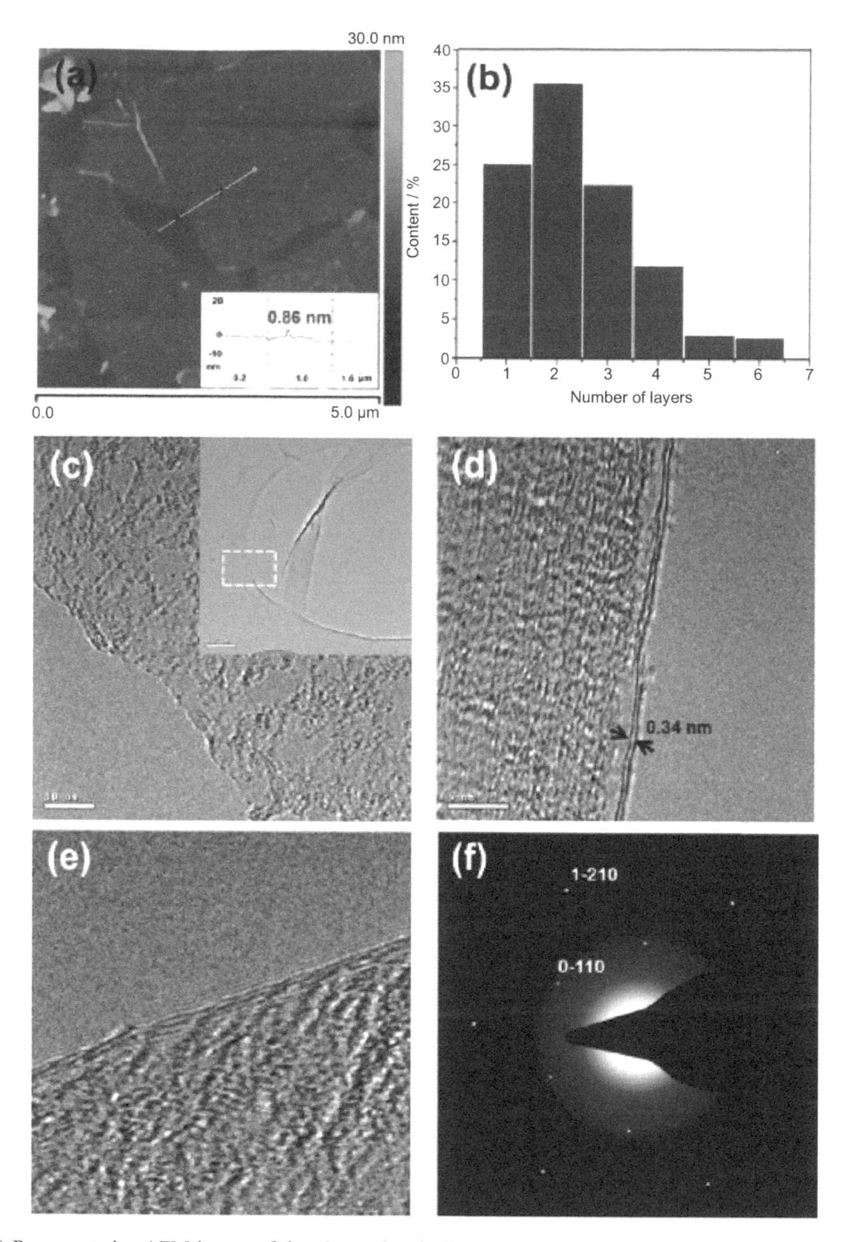

Fig. 2.57. (a) Representative AFM image of the electrochemically exfoliated graphene nanosheets on SiO_2. (b) Statistical thickness analysis of the graphene nanosheets according to AFM height. (c, d, e) HRTEM images of the single-, bi- and four-layer graphene nanosheets. The inset in (c) is the low magnification image of the graphene nanosheets. (f) SAED pattern of a bilayer graphene nanosheet. Reproduced with permission from [148], Copyright © 2013, American Chemical Society.

Electrochemical exfoliation of graphite assisted by ionic liquids (ILs) has been shown to be a high-throughput, green and scalable graphene production technique [151]. By using an IL/acetonitrile electrolyte, a significantly low level of ionic liquids was necessary to maintain acceptable exfoliation efficiencies. Figure 2.58 shows schematic diagram of the experimental setup used for electro-exfoliation of graphite in the presence of ionic liquid electrolytes. High purity iso-molded graphite rod (6.35 mm diameter with 4 cm effective length exposed to the electrolyte) was used as the anode, while a platinum wire was used as the counter-electrode (1.6 mm diameter and 4 cm effective length). Both IL/water solutions and IL/acetonitrile solutions were employed as electrolytes. A constant voltage of 7 V was applied for the electrochemical exfoliation experiments, which were conducted at room temperature. 0.1 M ionic liquid solutions in acetonitrile were used as the electrolyte, with an IL/solvent volume ratio of 1:50. The exfoliated graphene nanosheets were dispersed in N-methyl-2-pyrrolidone (NMP).

Four ILs were assessed, with results being shown in Fig. 2.59 [151]. It is necessary to mention that the solution coloration was not correlated to the graphite exfoliation yield. For example, about 86% of the graphite electrode was exfoliated when BMPyrr BTA was used as IL, while the final solution was virtually colorless. In contrast, the exfoliation rate was the lowest, i.e., 29%, for EMIM BF_4, but the solution was darkest in color.

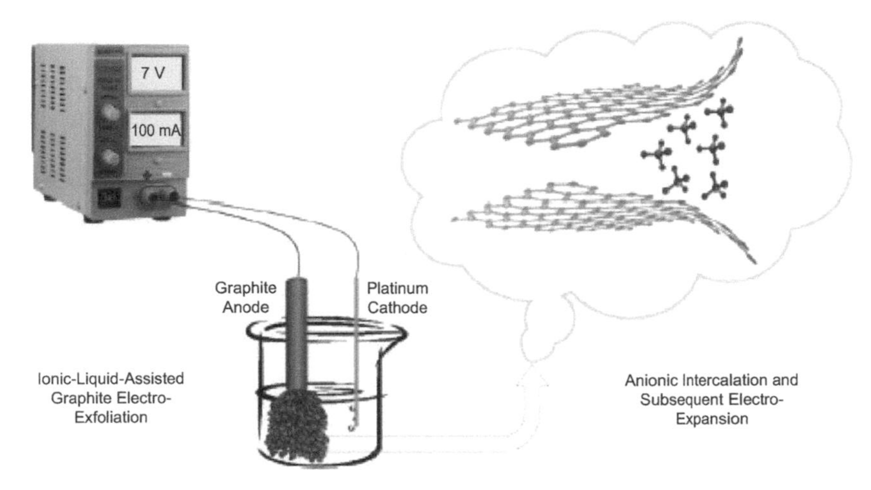

Fig. 2.58. Experimental setup used for graphite anodic exfoliation in the presence of ionic liquids. Reproduced with permission from [151], Copyright © 2014, Elsevier.

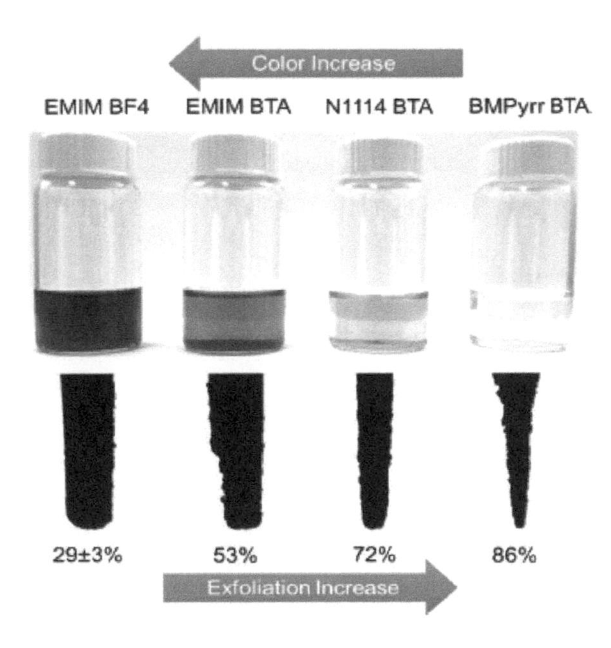

Fig. 2.59. Color changes and the exfoliation yield after 4 h of ionic-liquid-assisted electrochemical graphite exfoliation at 7 V in 0.1 M IL/acetonitrile with an IL/solvent volume ratio of 1:50 at 293 K. Reproduced with permission from [151], Copyright © 2014, Elsevier.

After electrochemical exfoliation for 4 h, the precipitates were collected and thoroughly washed using deionized water and ethanol. Figure 2.60 shows TEM and FESEM images of the representative samples of the exfoliated graphene nanosheets and the by-products. A variety of structures, ranging from ultrathin graphene nanosheets (Fig. 2.60 (a–d)) to semi-transparent carbonaceous particles (Fig. 2.60 (e)) and rolled sheets (Fig. 2. 60 (f)), were observed. The majority of the produced materials were the ultrathin irregularly shaped sheets. The average length of the nanosheets was in the order of 500 nm. According to the folded edges shown in Fig. 2.60 (b) and (c), most of the graphene nanosheets obtained were monolayers or bilayers.

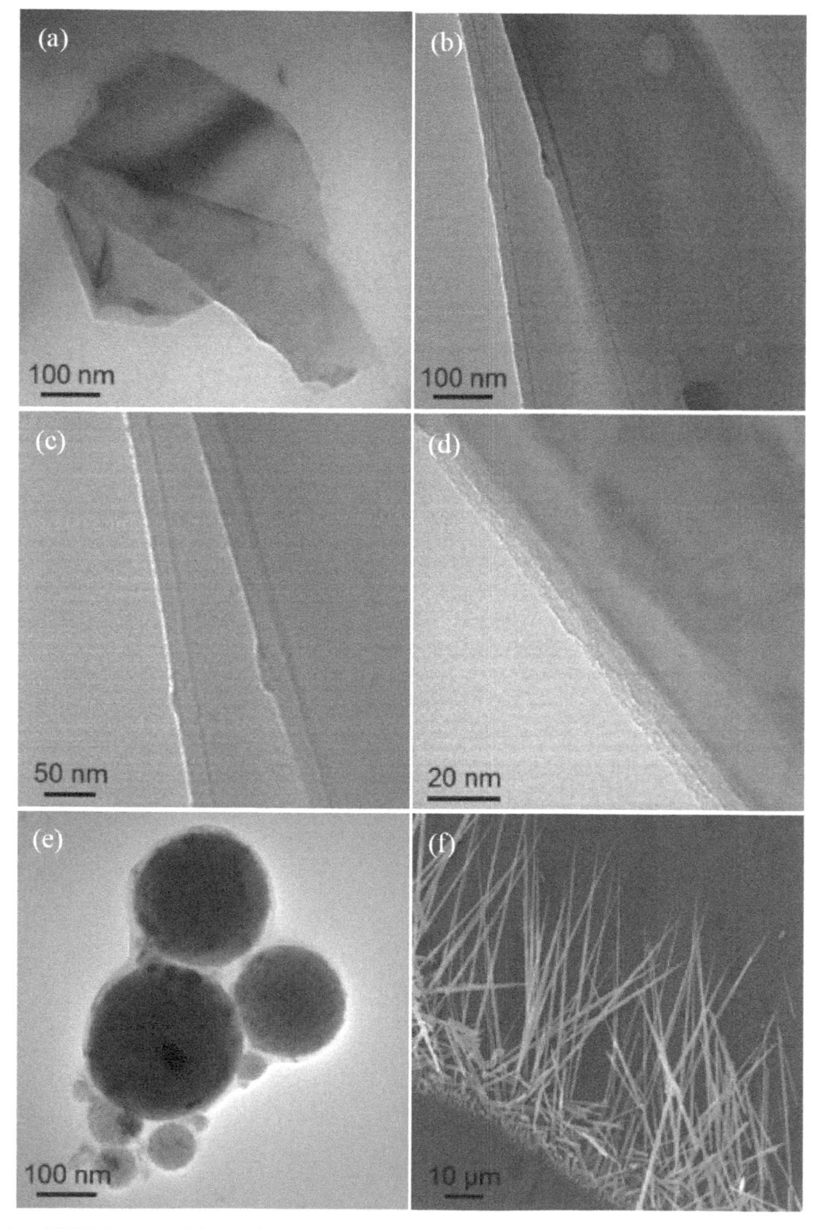

Fig. 2.60. TEM and SEM images of the products generated by using the electrochemical exfoliation of iso-molded graphite in EMIM BF_4/acetonitrile electrolyte at 7 V and 293 K: (a–d) crumpled/folded sheets, (e) semi-transparent carbonaceous particles and (f) rolled nanosheets. Reproduced with permission from [151], Copyright © 2014, Elsevier.

2.2.1.6. Other Solution Exfoliation Methods

Several other solution exfoliation methods that have not been covered previously will be discussed in this part, such as the direct exfoliation of graphite in ionic liquids (ILs), supercritical fluidic (SCF) exfoliation, hydrothermal reduction and solvothermal synthesis.

Direct exfoliation of graphite by using ionic liquid, 1-butyl-3-methyl-imidazolium bis(trifluoromethanesulfonyl)imide ([Bmim][Tf_2N]), with a molecular structure shown in Fig. 2.61, into micrometer-sized few-layer graphene nanosheets, with concentrations of up to 0.95 mg/ml, has been reported [152]. Natural graphite flakes were dispersed in [Bmim][Tf_2N], which was then subjected to tip ultrasonication for a total of 60 min with cycles of 5–10 min. The resulting dispersion was centrifuged at 10000 rpm for 20 min to obtain graphene sheets in IL. Yield of the exfoliation was measured quantitatively by using vacuum filtration. Figure 2.61 shows photographs of the samples before and after the ultrasonication of the graphite flakes in [Bmin][Tf_2N]. Dark black dispersion was formed after the ultrasonication, which was still retained after centrifugation. The dispersion was very stable with small levels of sedimentation and aggregation. Tyndall scattering effect was observed for a diluted dispersion as shown in the figure.

Most of the exfoliated graphene nanosheets had a thickness of < 5 layers according to the STEM and bright field TEM images, as shown in Fig. 2.62 [152]. Figure 2.62 (A) and (B) show a monolayer with folded edges and a bilayer graphene nanosheet, respectively. The exfoliated graphene nanosheets had a lateral size at the micrometer scale, together with a few smaller graphene fragments. The presence of the fragments was attributed to the irregular crystalline flakes of the natural graphite or the breaking of large flakes caused by the ultrasonication. The atomic ordering of the carbon atoms was confirmed by the bright field (BF) TEM and selected area electron diffraction (SAED) from a region of few-layer graphene (FLG), with orientation along the [0001] zone axis, as shown in Fig. 2.62 (C) and (D), respectively. The majority of the graphene nanosheets exhibited a very low level of defects, while dislocations and stacking faults were also observed occasionally, according to the diffraction contrast BF TEM imaging conditions. Small graphite particles and multilayered structures were formed due to the partial restacking of the FLG domains, which were present in the final suspensions. As compared with the electrochemical exfoliation in ILs, this method does not consume electricity, which could be one of its advantages. However, the quality of the graphene nanosheet suspension needs further improvement.

Graphene nanosheets have been prepared by using supercritical fluidic (SCF) exfoliation of graphite directly in organic solvents and ethanol solutions containing the sodium salt of 1-pyrenesulfonic acid [153, 154]. A rapid one-pot supercritical fluid (SCF) exfoliation process was developed to produce high-quality large-scale processable graphene nanosheets in SCFs, including ethanol, N-methyl-pyrrolidone (NMP), and DMF [153]. Direct high-yield exfoliation of graphite into graphene nanosheets was possible in SCFs conditions, due to their high diffusivity and solvating power. The graphene nanosheets prepared in this way contained 90–95% sheets of < 8 layers, 6–10% monolayer and 5–10% ones of ≥ 10 layers.

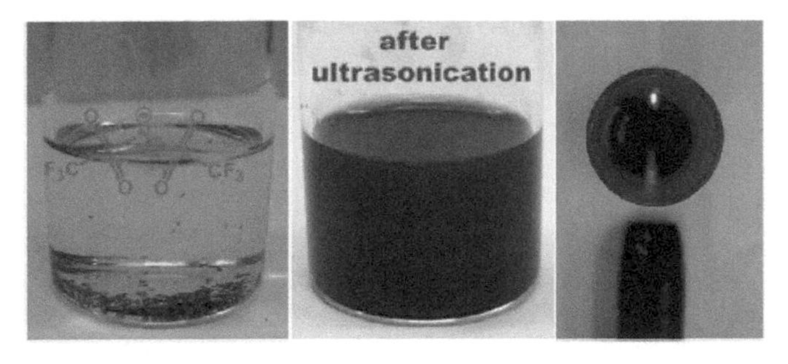

Fig. 2.61. Photographs of the exfoliation of graphite in [Bmim][Tf_2N] before (left) and after (middle) ultrasonication and the Tyndall effect of a diluted graphene suspension with a laser pointer (right). Reproduced with permission from [152], Copyright © 2010, The Royal Society of Chemistry.

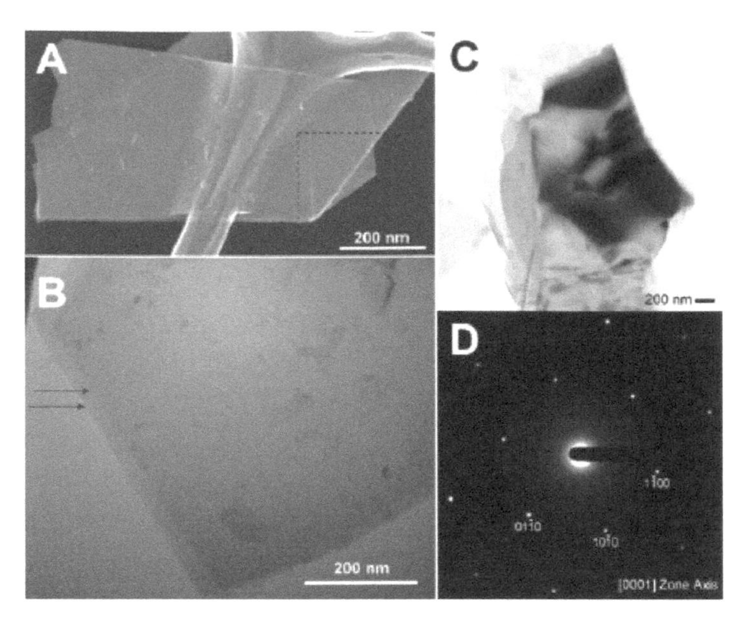

Fig. 2.62. STEM images of the dispersed graphene under the SE (A) and TE (B) mode, respectively. The copper grid (A) was found to be sandwiched by a few-layer graphene (bottom) and a monolayer graphene (top) with folded edges, as highlighted in a red rectangle. The arrows in (B) indicate the edges of a bilayer graphene. Bright field TEM image (C) of the multilayer graphene with respective SAED pattern (D) collected from the dark area in (C). Reproduced with permission from [152], Copyright © 2010, The Royal Society of Chemistry.

The SCF exfoliations were conducted in a stainless-steel reactor with a volume of 10 ml. In the experiment, graphite powder (10–20 mg) was put into the stainless-steel reactor vessel, together with a solvent (5 ml), ethanol, DMF or NMP, and was subjected to low-power sonication for 10 min. After that, the sealed reactor vessel was heated to 300–400°C for 15–60 min in a specially designed tube furnace, while the reactor reached the optimum temperature in 3 min. The reaction pressure was maintained at 38–40 MPa, by adjusting the loading of the reactor vessel and furnace temperature. The reaction was terminated by submerging the hot reactor in an ice-water bath. The exfoliated graphene nanosheets were collected accordingly by washing, centrifuging and drying. Photographs of the black solutions obtained by 5 min sonication of the graphene nanosheets in different fresh solvents, including DMF, ethanol and NMP, are shown in Fig. 2.63 (a–c), together with their dried powders shown in Fig. 2.64 (d–f). The concentration of dispersed graphene nanosheet solutions was 2–4 mg/ml.

Figure 2.63 (g) shows a schematic diagram of the SCF exfoliation of graphite into graphene nanosheets [153]. SCFs have unique features, including low interfacial tension, excellent wetting of surfaces, and high diffusion coefficients [155, 156]. Therefore, the SCF solvents could rapidly penetrate the interlayers of graphite with high diffusivity and solvation power that was much higher than the interlayer energy of graphite, thus leading to a one-pot rapid and high-yield exfoliation.

TEM analysis indicated that graphene nanosheets exhibited different types and a quite wide range of sizes, as shown in Fig. 2.64 [153]. The average size was in the range of hundreds nm to 2 μm. Most of the nanosheets were agglomerated, consisting of different types of layer structures, from monolayer to multilayer, as illustrated in Fig. 2.64 (a–c). Some of them were only monolayer graphene, as demonstrated in Fig. 2.64 (d–f). The TEM images were consistent with the Raman analysis, i.e., a large proportion of sheets were < 8 layers. The electron diffraction (ED) pattern of the graphene suggested that the nanosheets were well-crystallized, with one to a few-layered graphene structures, as shown in Fig. 2.64 (g–i). The unoxidized graphene nanosheets possessed high conduction and good electron-carrier capacity.

A simple, clean and controlled hydrothermal dehydration route to convert graphene oxide (GO) into stable graphene suspension has been reported [157]. 25 ml GO aqueous suspension with a concentration of 0.5 mg·ml^{-1} was added to a Teflon lined autoclave, which was then heated at 180°C for 6 h. The autoclave

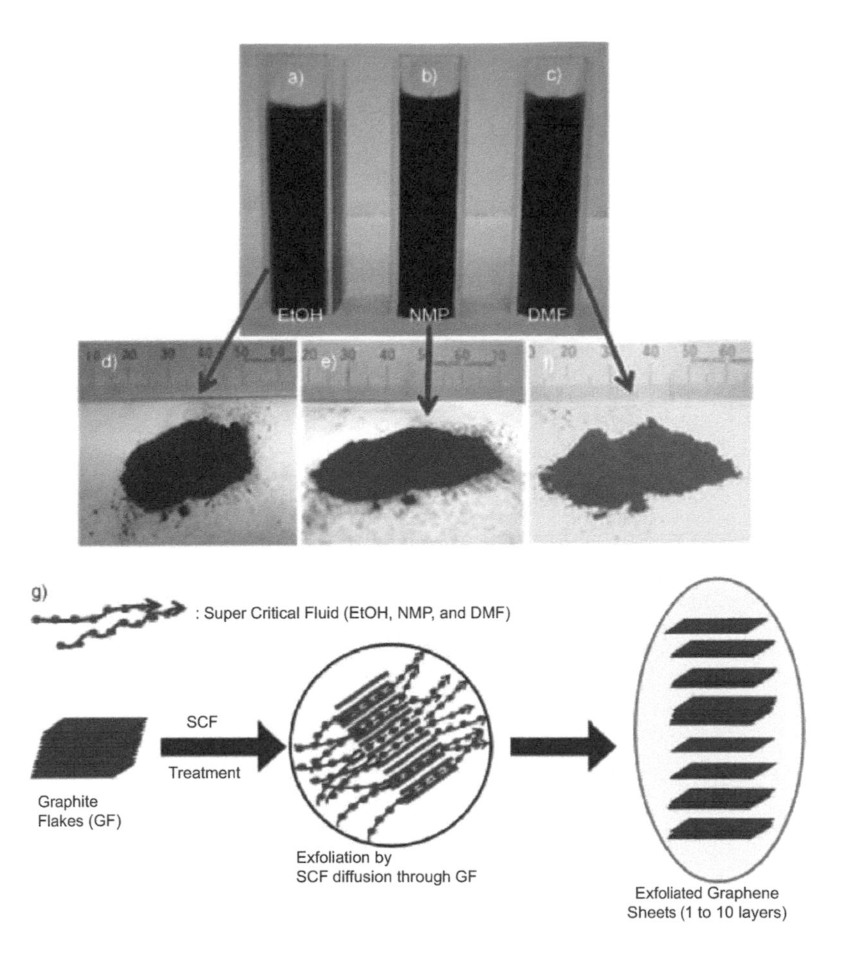

Fig. 2.63. Photographs of the exfoliated graphene nanosheets dispersed in ethanol (a), NMP (b) and DMF (c), together with about 1 g of the respective dried graphene nanosheet powders (d–f). (g) Schematic diagram showing the supercritical fluidic (SCF) exfoliation of graphite crystals to graphene. Reproduced with permission from [153], Copyright © 2010, John Wiley & Sons.

was then cooled to room temperature. After the hydrothermal treatment, the GO was reduced to graphene, which precipitated at bottom of the autoclave as a black powder, due to the low solubility graphene. The graphene could be dispersed by ultrasonification in water. The pH value of the solution was adjusted with hydrochloric acid and ammonia solution. Hydrazine-reduced GO was obtained by adding 10 ml of 98% hydrazine solution into 10 ml of 0.5 mg·ml^{-1} GO solution. The reduction process was carried out at 50°C for 12 h. The graphene nanosheet powder was collected with filtration, washed and dried at 90°C.

It has been acknowledged that superheated H_2O promotes acid-catalyzed reaction of organic compounds, due to the sufficiently high H^+ concentration, as compared with the normal liquid phase. The hydrothermal reduction was first carried out in neutral water. UV-vis absorption spectra of the product showed that the absorption peak at 227 nm of the GO corresponding to $\pi \rightarrow \pi^*$ transitions of aromatic C=C bonds was red-shifted to 254 nm after hydrothermal reduction treatment at 180°C for 6 h, as shown in Fig. 2.65. The absorption in the whole spectral region of > 238 nm was increased correspondingly, which indicated the presence of the π-conjugation network within the graphene nanosheets, i.e., the GO had been well-reduced during the hydrothermal treatment. The inset shows photographs of the samples before and after the hydrothermal treatment, demonstrating a color change of the GO solution from yellow brown to black.

The effect of pH value on the hydrothermal reaction was examined. It was found that the hydrothermal dehydration of GO solution at pH value of 11 led to stable graphene solution, which could be used directly

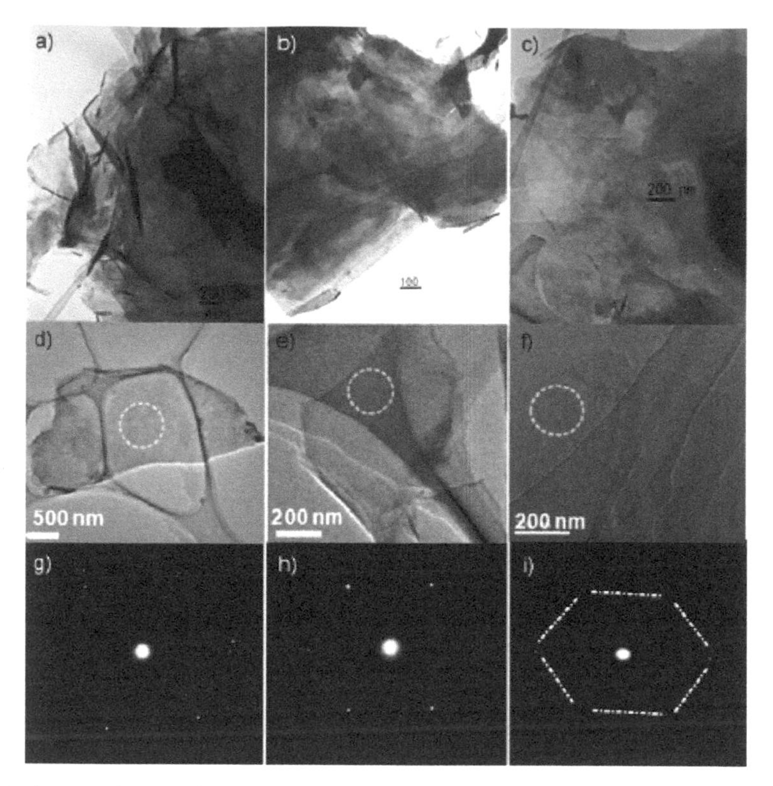

Fig. 2.64. HR-TEM images of the graphene nanosheets exfoliated in (a) DMF, (b) EtOH and (c) NMP. (d–f) Monolayer of the graphene nanosheets and (g–i) selected area electron diffraction (SAED) patterns of the graphene nanosheets showing crystalline graphene structures. Reproduced with permission from [153], Copyright © 2010, John Wiley & Sons.

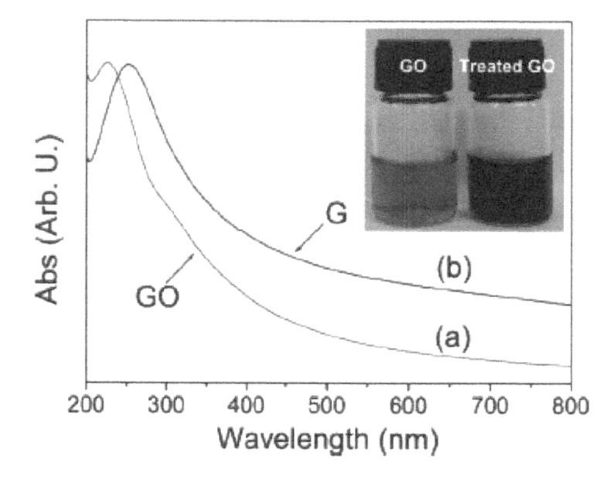

Fig. 2.65. UV-vis absorption spectra of the GO (a) before and (b) after hydrothermal treatment at 180°C for 6 h. Inset shows the color change of 0.5 mg/ml GO solution before and after hydrothermal treatment. Reproduced with permission from [157], Copyright © 2009, American Chemical Society.

to deposit single-layer graphene film through spin coating, as shown in Fig. 2.66 (a) [157]. In contrast, hydrothermal treatment of GO solution at pH value of 3 produced aggregation of graphene, which could not be dispersed even in concentrated ammonia solution. The reduction process was considered to be similar to the H^+-catalyzed dehydration of alcohol, where water was provided with H^+ for the protonation of OH. As illustrated in Fig. 2.67, both intramolecular and intermolecular dehydrations took place on the edges or basal planes of the graphene nanosheet. The edges of the GO, and some parts of its basal plane, were terminated with hydrogen, hydroxyl, ether and carboxylic groups. For intramolecular dehydration, the

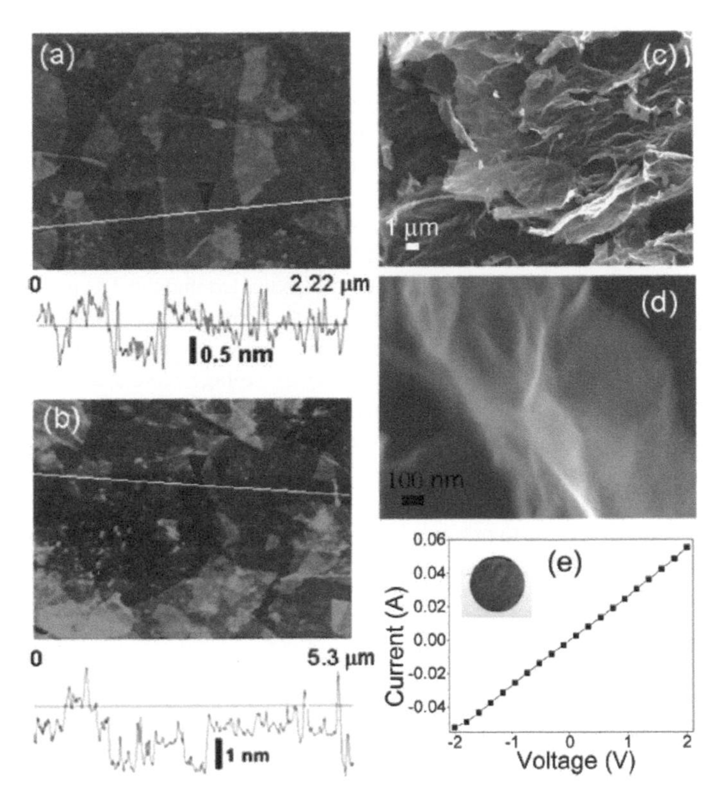

Fig. 2.66. (a, b) AFM images of the G and GO after and before hydrothermal treatment at 180°C for 6 h, respectively, and the corresponding section analyses. (c, d) SEM images at different magnifications of the G and (e) room-temperature I-V curve of the G disk, exhibiting Ohmic characteristics. The G disk was prepared by pressing the powder into the pellet with a hydraulic press. Reproduced with permission from [157], Copyright © 2009, American Chemical Society.

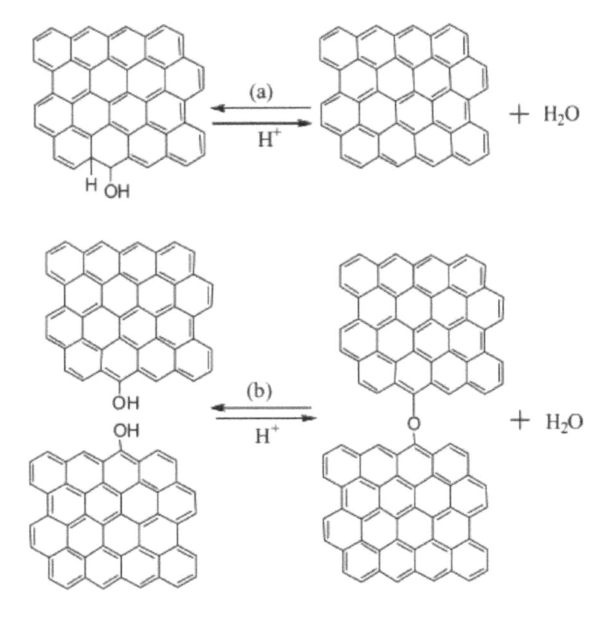

Fig. 2.67. (a) Intramolecular dehydration of G catalyzed by H^+ and (b) intermolecular dehydration of G observed at high pH values, leading to aggregated products. Symbolic H and OH functional groups were drawn terminating the edges of G in (a). Reproduced with permission from [157], Copyright © 2009, American Chemical Society.

elimination of OH and H occurred at the edge sites of graphene, followed by the recovery of π-bonding. However, the acid-catalyzed dehydration was reversible, i.e., acid-catalyzed hydration. Therefore, not all oxygen groups could be removed from the graphene nanosheets after the reaction, so that the residual negatively charged oxygen functional groups allowed the graphene nanosheets to be dispersed in water.

Figure 2.66 (a) shows AFM image of the graphene nanosheet after the hydrothermal treatment. The graphene nanosheets had a thickness of about 0.8 nm, which was smaller than that of the GO of about 1.5 nm, as shown in Fig. 2.66 (b), due to the removal of the functional groups on surfaces of the GO. The thickness was also slightly less than that of chemically converted graphene nanosheets, which is usually > 1.0 nm. SEM images of the hydrothermally reduced GO indicated that the 2-D molecular structure was well-retained as the reaction was conducted at neutral or alkaline pH levels, as shown in Fig. 2.66 (c). The graphene nanosheets were sufficiently thin to be flexible and were restacked when freeze-drying, as demonstrated in Fig. 2.66 (d). Figure 2.66 (e) shows the current-voltage curve of a graphene disk, indicating that the hydrothermally reduced graphene was highly conductive.

A direct chemical synthesis method was demonstrated to directly produce graphene nanosheets in gram-scale quantity as a bottom-up approach [158]. Ethanol and sodium reacted to form an intermediate solid that was then pyrolized, leading to a fused array of graphene nanosheets that could be dispersed by mild sonication. This method is able to produce bulk graphene nanosheets from nongraphitic precursors. The solvothermal reactions were conducted in a Teflon-lined reactor, with a volume of 23 ml. In typical synthesis, 2 g sodium (Na) and 5 ml ethanol, i.e., with molar ratio of 1:1, were sealed in the reactor vessel,

Fig. 2.68. Example of the bulk quantity of graphene product. The image consists of about 2 g of graphene powder. Reproduced with permission from [158], Copyright © 2008, Macmillan Publishers Limited.

which was heated at 220°C for 72 h. The solvothermal reaction yielded a solid product, which was the precursor of graphene nanosheets. This precursor was rapidly pyrolized and then washed with deionized water (100 ml). The suspended solid was finally filtered and dried in vacuum at 100°C for 24 h. 0.1 g per ml of ethanol, or 0.5 g per round of solvothermal reaction, could be obtained, with examples as shown in Fig. 2.68. Representative TEM images of the as-synthesized graphene nanosheet samples after post-sonication are shown in Fig. 2.69, while SEM images of the sample before sonication are shown in Fig. 2.70.

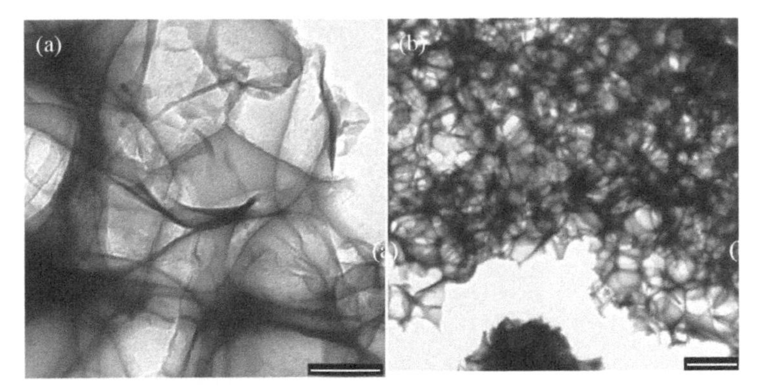

Fig. 2.69. TEM images of the agglomerated graphene nanosheets at different magnifications, with scale bars of 200 nm in (a) and 1 μm in nm (b). The extent of sheet formation and the tendency for sheets to coalesce into overlapped regions were observed. An inherent sheet-like structure displaying an intricate long-range array of folds was evident. As the images were taken in transmission mode, the relative opacity of individual sheets was a result of interfacial regions with overlap between individual nanosheets. The nanosheets extended in lateral dimensions over micrometer length scales, ranging from 1×10^{-7} to greater than 1×10^{-5} m. Reproduced with permission from [158], Copyright © 2008, Macmillan Publishers Limited.

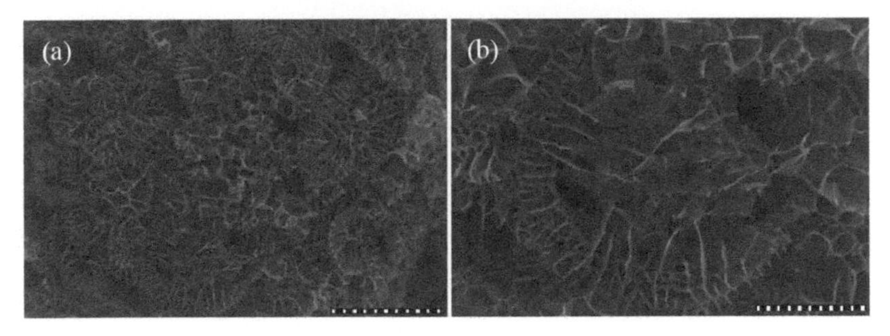

Fig. 2.70. SEM images of the as-synthesized graphene structures. (a, b) The bulk graphene product obtained from the pyrolysis of the solvothermal product was highly porous, consisting of individual nanosheets, with scale bars of 15 μm in (a) and 6 μm in (b). The entire image consisted of individual graphene sheets held in a porous structure that typically extended over more than 1×10^{-4} m, with the presence of numerous cavities or holes. The graphene therefore initially obtained as fused nanosheets, weakly held into a foam-like structure that was then easily separated into individual nanosheets by sonication in ethanol for several minutes. Reproduced with permission from [158], Copyright © 2008, Macmillan Publishers Limited.

According to previous discussion, the solution exfoliation methods are unable to control the number of layers of the exfoliated graphene nanosheets, i.e., the resulting dispersions usually exhibit polydispersity in properties. However, it has been shown that bilayer graphene has a tunable bandgap in the infrared range [159, 160] and exhibits unique quantum mechanical behavior [161], as compared with the zero bandgap single-layer graphene, while trilayer graphene is a semimetal material, with a band overlap that can be controlled by the application of an external electric field [162, 163]. Therefore, it is desirable to synthesize graphene suspensions with monodisperse thickness, so as to fabricate graphene films with unique properties for given applications, by using appropriate film deposition methods [100, 164].

In this regard, density gradient ultracentrifugation (DGU) has been used to separate graphene nanosheets with different thickness, due to their different buoyant densities [165]. This technique has been scaled up to produce commercially available graphene in aqueous solutions. Graphene dispersions were prepared by using horn ultrasonication of naturally occurring graphite flakes in an aqueous solution containing 2% w/v SC. The highly hydrophobic graphene nanosheets exfoliated in this way were stabilized by the amphiphilic SC molecules, with the hydrophobic faces associated with the graphene leaving the opposing hydrophilic faces to interact with the surrounding aqueous environment, as shown in Fig. 2.71 (A). Horn ultrasonication facilitated the formation of thin graphene nanosheets and thicker graphite flakes, leading to unstable gray-black graphene/graphite slurries. To remove the fast sedimenting

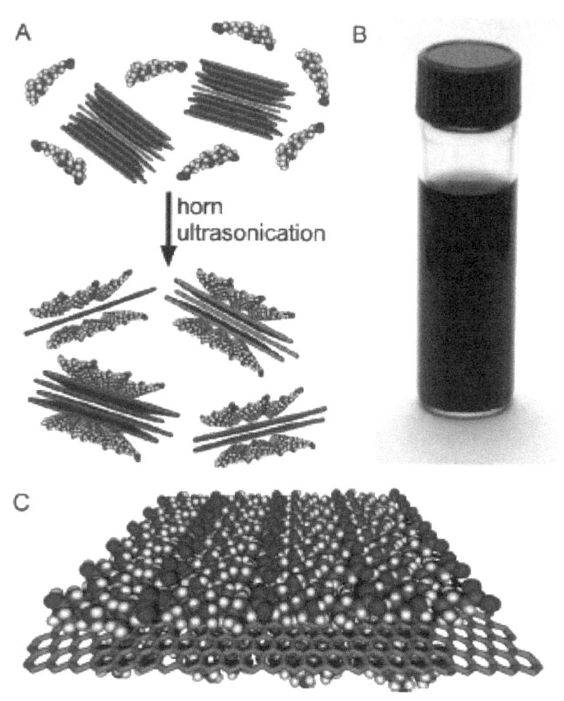

Fig. 2.71. (A) Schematic diagram of the graphene exfoliation process. Graphite flakes were combined with sodium cholate (SC) in aqueous solution. Horn ultrasonication triggered the exfoliation of the graphite into few-layer graphene nanosheets that were encapsulated by the SC micelles. (B) Photograph of a 90 μg·ml^{-1} graphene dispersion in SC with long-term stability. (C) Schematic diagram illustrating an ordered SC monolayer on graphene. Reproduced with permission from [165], Copyright © 2009, American Chemical Society.

thick graphite items, sedimentation centrifugation was conducted at 15 krpm for 60 min, resulting in a black supernatant that contained few-layer graphene nanosheets. The sedimented graphene dispersions had concentrations of up to 90 μg·ml^{-1}, with an example shown in Fig. 2.71 (B).

SC has been successfully used to establish differences in buoyant density for carbon nanotubes, according to their wall number, diameter and chiral handedness [166–168]. Therefore, SC can be employed to encapsulate carbon nanotubes uniformly and reproducibly. Due to the planar structure of SC, it is also able to encapsulate graphene, by well-ordered assemblies of SC in aqueous solution, as illustrated schematically in Fig. 2.71 (C). Similarly, the graphene-SC complexes exhibited different buoyant densities for the thicknesses of the graphene nanosheets encapsulated in the complex, which could be separated by using the DGU-based sorting approach.

Figure 2.72 (A) shows a photograph of the centrifuge tube following the separation, with clearly visible multiple gray bands at different locations inside the gradient [165]. Representative AFM images and line profiles of the items from fractions f4 and f16 are shown in Fig. 2.72 (B–D). The flakes from fraction f4 showed an average thickness of 1.1 nm. Single-layer graphene nanosheet on SiO_2 usually has a thickness of about 1 nm, due to the adsorption of water, so that the slightly thicker graphene nanosheet observed in the work could be attributed to the possible presence of residual sodium cholate molecules on the graphene surface. In contrast, the graphene nanosheets from f16 had an average thickness of 1.5 nm.

Thickness histograms of representative graphene fractions and the concentrated and sedimented graphene dispersions are shown in Fig. 2.73 (A, B). According to the thickness distributions of the sedimented, concentrated and f4 graphene solutions in Fig. 2.73 (A), it was found that the distributions became gradually sharpening with increasing buoyant density refinement. 37% of the sedimented graphene solution had thicknesses of > 2 nm, while these nanosheets constituted only 2.6% of the dispersion, following the concentration step gradient processing. Following DGU, 80% of the graphene nanosheets from the

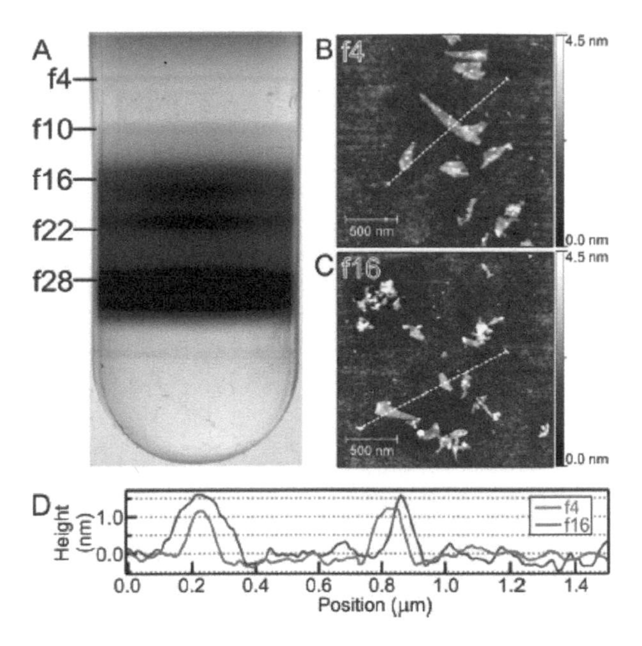

Fig. 2.72. (A) Photograph of a centrifuge tube following the first iteration of density gradient ultracentrifugation (DGU). The concentrated graphene was diluted by a factor of 40 to ensure that all graphene bands could be clearly resolved in the photograph. The lines marked the positions of the sorted graphene fractions within the centrifuge tube. (B, C) Representative AFM images of the graphene deposited using the fractions f4 (B) and f16 (C) onto SiO_2. (D) Height profile of the regions marked in panels B (blue curve) and C (red curve), demonstrating the different thicknesses of the graphene nanosheets obtained from different DGU fractions. Reproduced with permission from [165], Copyright © 2009, American Chemical Society.

fraction f4 exhibited thicknesses of $\leq$ 1.2 nm, very close to that of single-layer graphene. Comparatively, only 24% of the concentrated graphene nanosheets were single-layered. The thickness distributions of the graphene sorted with DGU showed a monotonic increase in the average sheet thickness with increasing buoyant density, as shown in Fig. 2.73 (B).

A geometrical model of the buoyant density of the graphene-SC complex has been proposed, as shown in Fig. 2.73 (C), in which the thickness of the graphene nanosheet was defined as N, representing the average number of graphene layers inside the sheet separated by the graphene interlayer distance t_{gr} = 0.34 nm. On both sides of the graphene nanosheet, there was an anhydrous layer, with a thickness of t_A, containing the SC encapsulation layer. The SC molecules in this region coated the graphene surface with surface packing density σ. The SC layer was surrounded by an electrostatically bound hydration shell, with a thickness of t_H. The hydration layer had the lowest density of any of the components in the complex and thus served to decrease the buoyant density of the graphene-SC assembly. The model was in a good agreement with the experimental data, as shown in Fig. 2.73 (D) [165].

Development of graphene suspensions with high concentrations has been an important topic, due to the requirements for various applications. Functionalization has been acknowledged to be the most effective approach to improve the dispersion capability of graphene nanosheets in different solvents [169–172]. An example of chemical graphene functionalization was demonstrated by the reaction of aryl diazonium compounds with *in situ* activated, exfoliated and reduced graphene [173].

Natural graphite flake with diameters of 2–3 mm was used as the starting carbon source after being degassed *in vacuo*, while being heated to 300°C. All synthetic steps involving NaK alloy were carried out with extreme caution, by using blast shields at all times, under strict exclusion of air or moisture, with argon as the protection gas. Argon was streamed through a purification/drying column equipped with blue gel molecular sieves of 0.3 nm, potassium hydroxide and phosphorus pentoxide before use. 1,2-Dimethoxyethane was degassed, refluxed over sodium wire in an argon atmosphere and freshly distilled from NaK alloy directly into a flame-dried, argon-filled round-bottomed flask. An excess (0.45 ml,

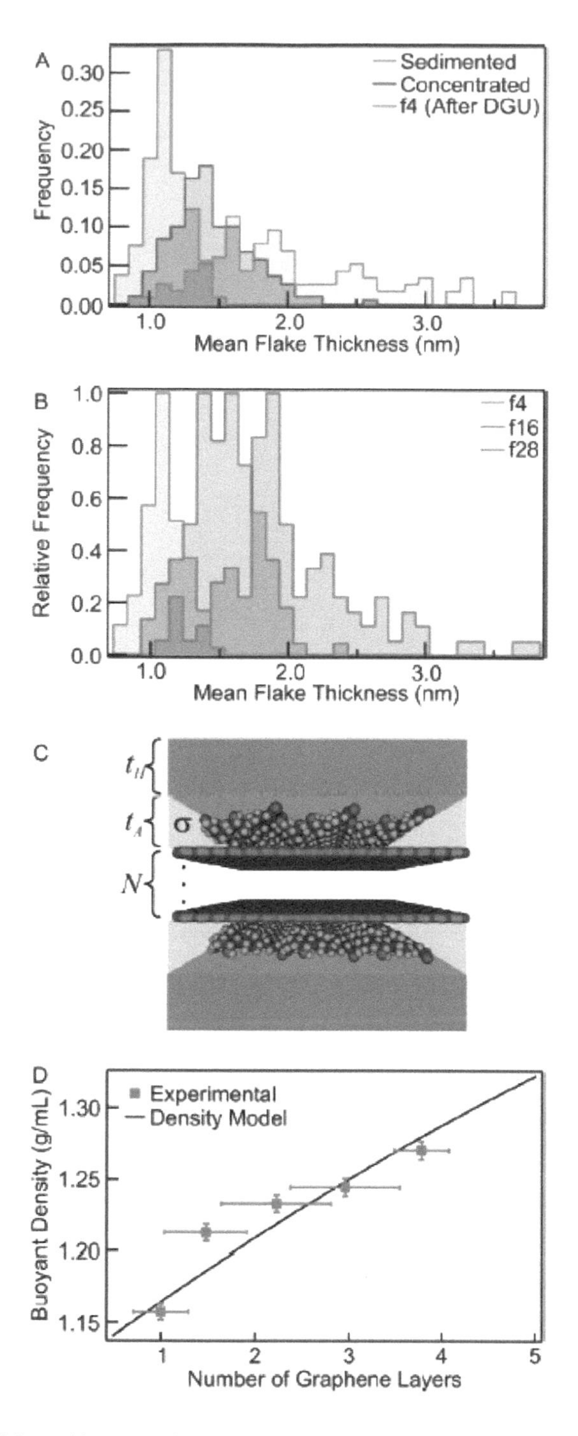

Fig. 2.73. (A) Mean flake thickness histograms for the sedimented (gray), concentrated (purple) and DGU fraction f4 (blue) graphene solutions, estimated by using AFM. (B) Mean flake thickness histograms plotted by relative frequency (mode thickness scaled to unity) for the DGU fractions f4 (blue), f16 (red) and f28 (green). (C) Buoyant density model for the SC-encapsulated graphene, in which a flake of thickness N was coated by a surfactant with packing density σ and effective thickness t_A and a hydration layer with a thickness of t_H. (D) Fit of the geometrical density model to the experimental data, showing fairly good agreement. Reproduced with permission from [165], Copyright © 2009, American Chemical Society.

Color image of this figure appears in the color plate section at the end of the book.

9 mmol K and 2 mmol Na) of freshly prepared liquid sodium potassium alloy was added and dispersed by a short ultrasonication step. Graphite flakes (23 mg, 1.9 mmol) were immersed in the deep-blue solution and stirred for three days until the blue color disappeared. The reaction solution was quenched by the addition of the solid diazonium compound (4.6 mmol or 12 mmol, respectively) to obtain medium or highly functionalized material, while for low diazonium concentrations the solidified alkali metal was partially removed before the addition.

The reaction sequence for the wet chemical bulk functionalization of graphene is shown in Fig. 2.74 [173]. The initial step was the reduction of graphite with solvated electrons. Liquid alloy of sodium (Na) and potassium (K) was used as the electron source and 1,2-dimethoxyethane (DME) as an inert solvent. Dissolution of the alkali metals in DME was indicated by the solution becoming deep blue. As compared to liquid ammonia, DME was more advantageous, due to several reasons, such as the possibility to work at room temperature and its inertness against diazonium salts. After the addition of dried natural graphite flakes, the deep blue solution became graphene suspension. During this process, the alloy was solidified, because the more electropositive K was consumed, leading to an increase in the melting point of the alloy. This implied that further reduction was continued until saturation was reached, after the initially formed solvated electrons were consumed by graphite. The negative charges on the graphite were balanced by the solvated and intercalated K^+ ions, forming graphite intercalation compounds (GICs) with alkali metals [174–176].

As the diazonium salts, 4-*tert*-butylphenyldiazonium tetrafluoroborate (BPD) and 4-sulfonylphenyldiazonium chloride (SPD), were added to the DME dispersion, vigorous evolution of N_2

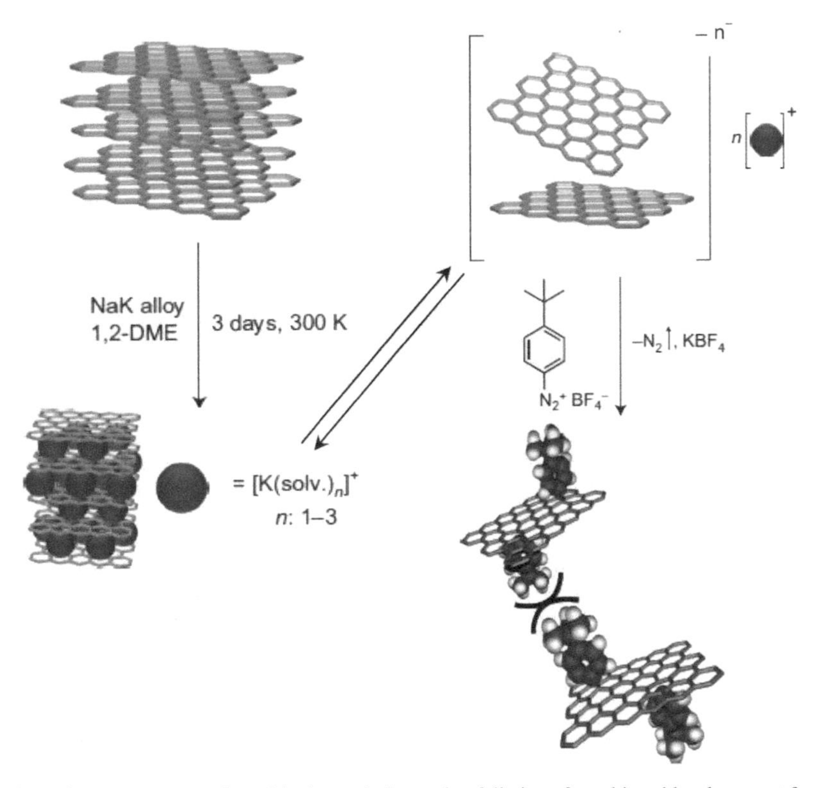

Fig. 2.74. Reaction scheme. Representation of the intercalation and exfoliation of graphite with subsequent functionalization of intermediately generated reduced graphene yielding 4-tert-butylphenyl functionalized graphene (double bonds in the basal planes have been omitted for clarity). 1,2-DME, 1,2-dimethoxyethane. Reproduced with permission from [173], Copyright © 2011, Macmillan Publishers Limited.

and heating of the reaction mixture were observed, indicating the reoxidation and functionalization of the charged graphene by the diazonium cations $Ar–N_2^+$. Due to the electron uptake of $Ar–N_2^+$ in the immediate vicinity of the charged graphene, N_2 and highly reactive aryl radicals were formed, which was similar to the single-electron transfer (SET) reaction mechanisms observed for the reactions of carbon nanotubes and comparable carbon-based materials [177]. The covalent addition to the conjugated π-electron system of the graphene would consume the *in situ* formed aryl radicals. The functional group would effectively prevent the formation of scroll through the intramolecular π–π stacking or graphite through the intermolecular π–π stacking [178].

Figure 2.75 shows Raman and AFM experimental results of the samples at different stages of the reduction, exfoliation and functionalization sequence with BPD [173]. A significant increase in volume was observed for the defunctionalized material. Raman spectra of the foam-like defunctionalized sample confirmed the complete reversibility of the reaction and full restoration of the pristine planar sp^2 carbon network, which was found to be more effective than thermal decarboxylation [179, 180]. The graphite was intercalated into few-layer graphene nanosheets, as shown in Fig. 2.75 (a, b). The reduced graphene was of single-layer characteristic in the Raman spectrum, as demonstrated in Fig. 2.75 (d). The absence of a D-band implied the defect-free nature of the graphene before the addition of the diazonium salt. Figure 2.75 (j) shows that the highly functionalized material exhibited significantly broadened lines, with I_D/I_G ratios of > 1.2. The highly functionalized graphene was also single-layer, with a thickness of 2 nm, as illustrated in Fig. 2.75 (k, l).

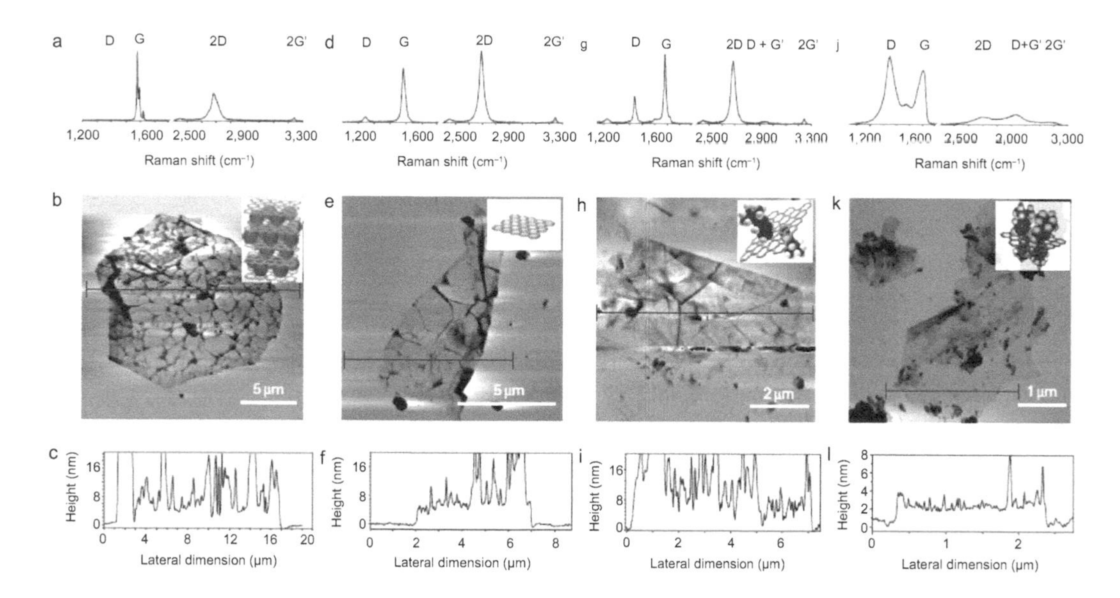

Fig. 2.75. Raman and AFM experiments conducted at different stages of the reduction, exfoliation and functionalization sequence with BPD. (a) Raman spectrum of a defect-free nano-GIC, exhibiting typical G-band splitting due to the guest incorporation between the graphene sheets. The 2D-band shape strongly suggests residual Bernal stacked graphene sheets, as shown by peak deconvolution into four single Lorentzians. (b) AFM image from the nano-GIC. (c) AFM profile along the red line indicated in (b), revealing heights of 2–6 nm. (d) Raman spectrum of chemically exfoliated graphene without G-band splitting and typical single Lorentzian lineshape for the 2D-band (FWHM = 38.5 cm^{-1}). (e) AFM image from chemically exfoliated single-layer graphene. (f) Height profile extracted along the red line in (e), which is in good agreement with the heights measured for micromechanically cleaved graphene on the same instrument. (g) Raman spectrum of a medium functionalized monolayer flake, exhibiting an intensive D-band at 1350 cm^{-1} (I_D/I_G = 0.57) and a symmetrical 2D-band at 2696 cm^{-1} (FWHM = 27.6 cm^{-1}). (h) AFM image of the functionalized single-layer graphene. (i) Height profile extracted along the red line in (h), yielding 3 nm averaged at flat regions. (j) Raman spectrum of a highly functionalized single-layer graphene exhibiting a spectral shape comparable to graphene oxide. The disappearance of the 2D resonance as well as the overall broadening of the line widths is indicative of very high functionalization densities. (k) AFM image from the highly functionalized flake. (l) Height profile along the red line in (k). Reproduced with permission from [173], Copyright © 2011, Macmillan Publishers Limited.

Figure 2.76 shows HRTEM results, which provided further evidence of the covalent functionalization of graphene nanosheets [173]. Functionalized graphene nanosheets were readily observed at low magnifications, as shown in Fig. 2.76 (a, e). At low functionalization densities, typical hexagonal pattern of the crystal lattice was well-retained, as evidenced by the HRTEM images and the corresponding fast Fourier transforms (FFTs) in Fig. 2.76 (b, f). When the graphene nanosheets were highly covalently functionalized, their surfaces were covered by the randomly distributed domains of immobilized phenyl groups, as demonstrated in Fig. 2.76 (c, g). Such a reversible functionalization is expected to be an effective strategy to tailor the properties of graphene nanosheets for targeted applications.

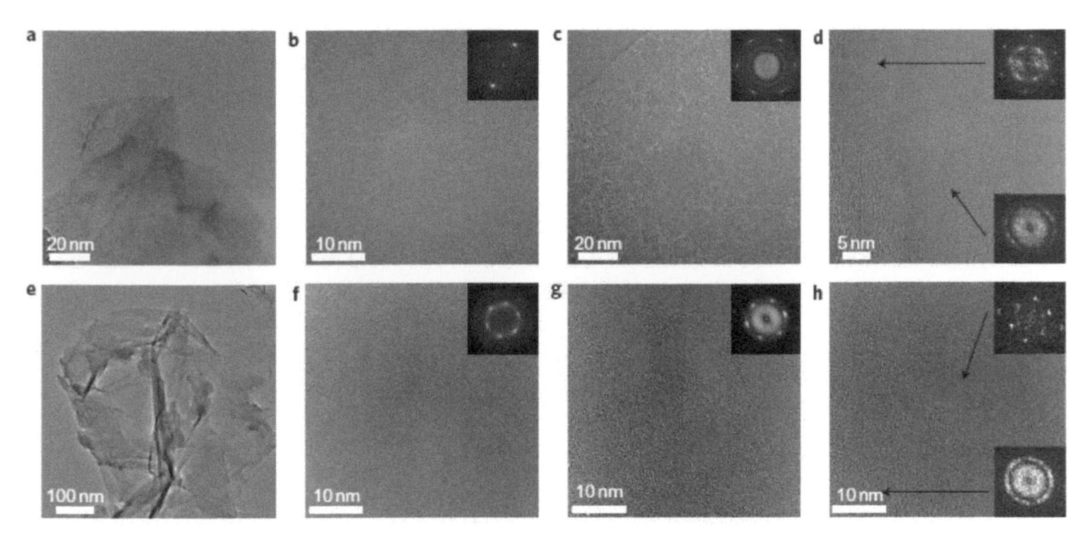

Fig. 2.76. HRTEM micrographs of functionalized graphene derivatives. (a) HRTEM image of a functionalized graphene sheet. (b) Low BPD functionalization density with preserved crystal lattice. (c) High functionalization density leading to the formation of amorphous regions among crystalline domains, as observed in the HRTEM image and in the FFT pattern. (d) Coexisting areas of preserved graphene lattice and highly functionalized domains. (e) Low magnification of a functionalized graphene nanosheet bearing heteroatom (sulfur) markers. (f) Crystalline graphene lattice for low functionalization densities. (g) Highly functionalized sulfur-marked graphene flake exhibiting a strong amorphous contribution to the FFT patterns. Functional group bearing domains are visible. (h) Spatially confined areas of high and low functionalization density and their respective FFT patterns. Reproduced with permission from [173], Copyright © 2011, Macmillan Publishers Limited.

2.2.1.7. Mechanochemical Exfoliation

Mechanochemical exfoliation offers a simple, cheap and rapid method to fabricate graphene nanosheets directly from graphite, in which mechanical energy is used to exfoliate graphite or reduce graphene oxide (GO) [181–189]. Selected examples will be presented in this part to demonstrate the advantages and feasibility of this method.

Few-layer graphene (FLG) nanosheets can be prepared from graphite powder with melamine in a nitrogen atmosphere by using ball milling [183]. The exfoliated graphene nanosheets can be dispersed in water or DMF to form suspensions at concentrations of up to 0.2 mg/ml. During the ball-milling, graphite powder interacted with melamine (2,4,6-triamine-1,3,5-triazine) and exfoliation of graphite was facilitated, as schematically shown in Fig. 2.77. Because graphite is insoluble, but graphene is suspendable in DMF, the efficiency of the exfoliation could be estimated. If melamine was absent or the graphite powder was sonicated in DMF with the presence of melamine, no dispersions were obtained. XPS results indicated that no difference was observed before and after mechanochemical treatment, which meant that the intrinsic structure of graphite remained intact when subject to the ball milling, i.e., the mechanochemical milling is a non-oxidation process. Figure 2.78 shows a representative TEM image of the graphene nanosheets, with the wrinkles confirming the large size and high flexibility of the graphene.

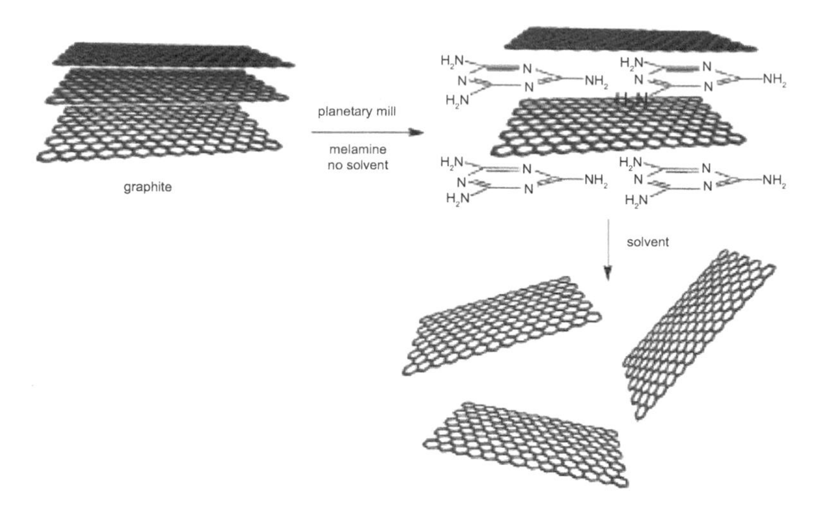

Fig. 2.77. Schematic diagram of the exfoliation of graphite by using the ball-milling method. Reproduced with permission from [183], Copyright © 2011, The Royal Society of Chemistry.

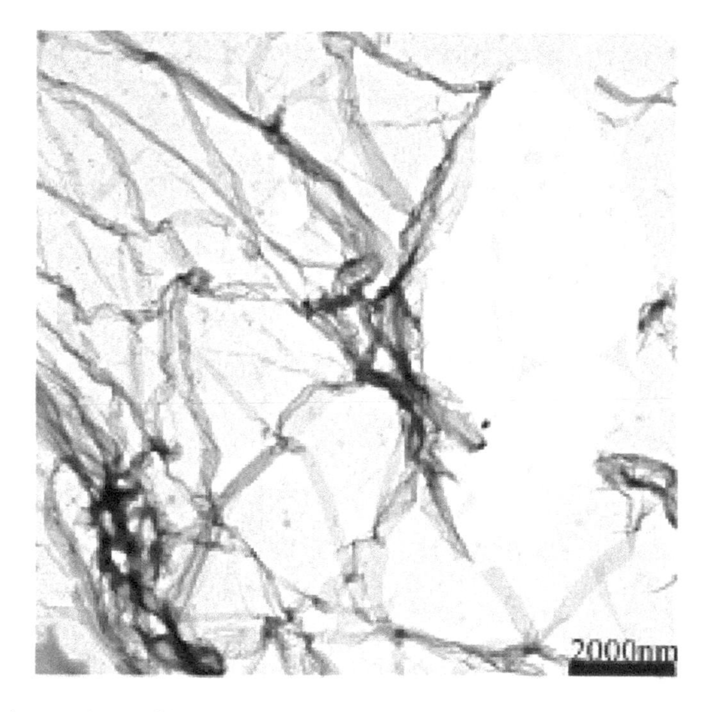

Fig. 2.78. Representative TEM image of the exfoliated graphene through the ball milling method. Reproduced with permission from [183], Copyright © 2011, The Royal Society of Chemistry.

A stirred media milling, as compared with a shaking plate for fast screening of favorable process parameters, has been employed as a wet media delamination approach to develop surfactant-free few-layer-graphene (FLG) suspensions from unmodified isostatic graphite in organic solvents [184]. The wet delamination in selected solvents is similar to ultrasound-assisted exfoliation methods, as discussed earlier. With N-methylpyrrolidone (NMP) as the dispersing medium, the influence of delamination process parameters on FLG concentration was studied, which led to an optimal set of processing parameters. Among the various delamination methods, the delamination by enhanced shear resulted in the largest nanosheet structures. The highest value for the FLG yielding rate, as a ration of mass/time, was found in a stirred media mill.

Unmodified isostatic graphite, with a purity of > 99.5% and a median particle size of $x_{50,3}$ = 15–20 μm, was used as feed material. N-methylpyrrolidone (NMP), cyclohexanone (CHO) and i-propanol (IPA), all with a purity of > 99.5%, were selected as the organic solvents. In all the delamination experiments, 1 wt% graphite powder was dispersed in each solvent. A shaking plate and a lab-scale stirred media mill were used as the delamination tools. Yttria stabilized ZrO_2 (YSZ) beads, with diameters of 30–200 μm, were used as delamination media. For the delamination experiment using the shaking plate, 0.2 g graphite, 20 ml solvent and 30 g ZrO_2 beads were placed in a 30 ml screw-cap bottle, which was mounted on a shaking plate. For the delamination experiment using the stirred media mill, the chamber of the mill, with a volume of 600 ml, was loaded with 170 ml solvent, 1.7 g graphite and 1.5 kg ZrO_2 beads. All the delamination experiments were conducted at room temperature. The milled samples were centrifuged at a high speed to obtain graphene suspensions.

The delamination of graphite in the stirred media mill was realized by shear forces, as well as direct friction between the beads and the graphite particles. The moving ZrO_2 beads displaced the solvent and therefore a viscous energy transfer occurred if a graphite particle was in between two beads, as shown in Fig. 2.79 [184]. The Reynolds number in the milling chamber was assessed to be in the range of 1–4 × 10^6 for the stirrer rotation speed range. Therefore, the flow was turbulent, which led to Reynolds-shear stress, due to the intensive velocity fluctuations [190, 191]. The van der Waals forces between the graphene layers in the graphite stack were overcome by the shear stress, so that the FLG nanosheets were pealed from the graphite particles. At the same time, the solvent NMP acted as a stabilizing agent to prevent the FLG nanosheets from agglomerating or re-stacking.

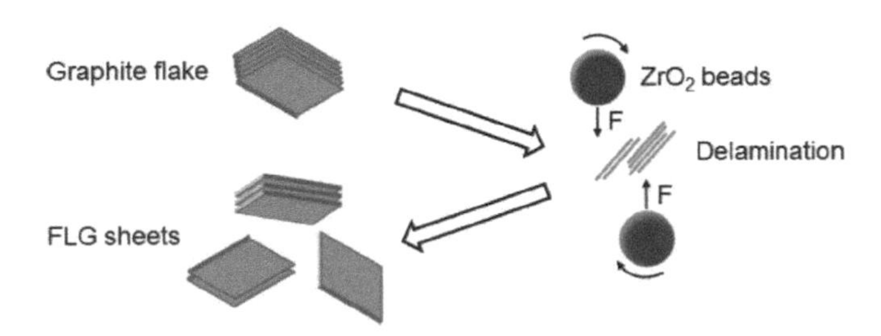

Fig. 2.79. Schematic diagram showing the principle of graphite exfoliation by using the stirred media milling in organic solvents. Reproduced with permission from [184], Copyright © 2015, Elsevier.

With 100 μm ZrO_2 beads as delamination media, after shaking for 1 h, the contents of graphene nanosheets were 0.044 $g \cdot l^{-1}$ and 0.055 $g \cdot l^{-1}$, in CHO and NMP, respectively, while no carbon was detected in the supernatant of IPA, as shown in Fig. 2.80 (A). For stirred media milling at a stirrer rotation speed of 1000 rpm, the graphene concentrations obtained after 1 h of processing were higher by a factor of about 10, as compared with those obtained with the shaking plate. Even in IPA, carbon could be detected, as illustrated in Fig. 2.80 (B). Therefore, the stirred media milling was much more effective than the shaking plate.

Raman spectroscopy has been demonstrated to be one of the most powerful techniques to distinguish between single-layer graphene (SLG), few-layer graphene (FLG) and graphitic nanoparticles [192–196]. Generally, the FWHM of the 2D-Raman peak (2D-FWHM) at 2690 cm^{-1} is narrower than 39 cm^{-1} for LSG, between 39 and 65 cm^{-1} for FLG and broader than 65 cm^{-1} for graphite-like nanoparticles. Statistical Raman investigations by mapping have been performed to evaluate the quality of the samples shown in Fig. 2.80 (A, B). In Fig. 2.81, a map of 2D-FWHM is shown to be exemplary for the sample processed in the stirred media mill in NMP as the dispersing agent. From the color code in Fig. 2.81 it can be seen that in most regions of the mapped area, Raman spectra with 2D-FWHM < 65 cml are obtained proving a high content of FLG.

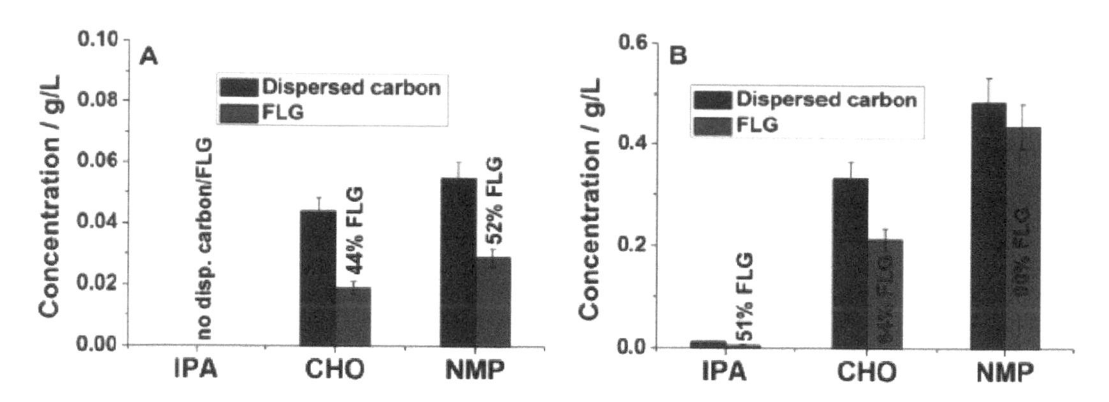

Fig. 2.80. Dispersed carbon and FLG concentrations obtained after 1 h of processing in different solvents by using 100 lm ZrO_2 beads: (A) shaking plate and (B) stirred media mill with stirrer rotation speed of 1000 rpm. The results were obtained for the supernatants of the centrifuged samples. Reproduced with permission from [184], Copyright © 2015, Elsevier.

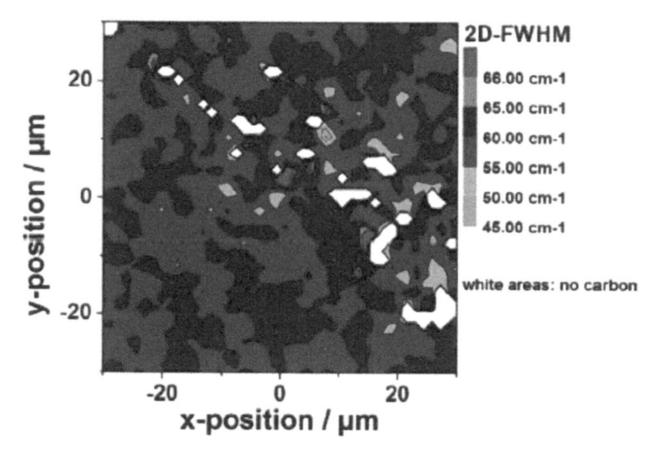

Fig. 2.81. Map for 2D-FWHM measured for the sample processed for 1 h by using the stirred media milling in NMP as the dispersing medium. 100 μm ZrO_2 beads were used as delamination media and the stirrer tip speed was 3.4 $m{\cdot}s^{-1}$. Reproduced with permission from [184], Copyright © 2015, Elsevier.

Color image of this figure appears in the color plate section at the end of the book.

A versatile and eco-friendly approach, simple solid-state mechanochemical ball-milling, was developed to reduce graphene oxide (GO) into high-quality graphene nanosheets in the presence of hydrogen (H_2) [185]. The ball-milling process was able to effectively recover the graphitic structure in the resultant graphene nanosheets, with enhanced electrical conductivities of up to 3400 S/m, after a heat treatment at 900°C for 2 h in N_2.

The ball-milling of GO to produce graphene oxide (BMRGO) was conducted by using a planetary micro ball-mill machine at a speed of 900 rpm. In the experiment, GO powder and N_2 gas were charged into a stainless steel capsule, together with stainless steel balls of 5 mm in diameter. The container was fixed in the planetary ball-mill machine to be milled for time durations of 30–240 min. The samples were BMRGO30, BMRGO60, BMRGO120, BMRGO180 and BMRGO240 for milling times of 30, 60, 120, 180 and 240 min, respectively. The milled powders were black, as shown in Fig. 2.82, which were collected and purified by Soxhlet extraction with 1 M HCl solution to remove the metallic impurities [185]. The purified powders were washed with water and methanol and then dried in vacuum at 80°C to obtain BMRGO powders.

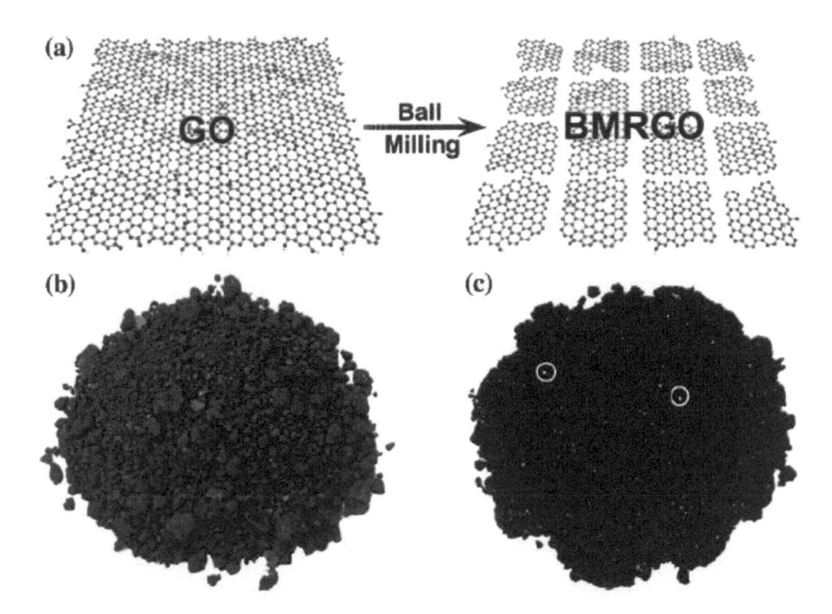

Fig. 2.82. (a) A schematic diagram of the solvent-free mechanochemical reduction of graphene oxide (GO) in the presence of hydrogen, together with photographs of (b) GO and (c) BMRGO. The color change from GO (brown) to BMRGO30 (dark black) is a visual indication of the mechanochemical reduction and the shiny light reflection (white circles) of BMRGO30 implied its high crystallinity. Reproduced with permission from [185], Copyright © 2014, Elsevier.

Figure 2.83 (a) shows that the conversion from GO to BMRGOs was proportional to the ball milling time up to 120 min, which was leveled off thereafter [185]. Correspondingly, the carbon/oxygen (C/O) ratio was increased from 1.51 of GO to 6.17 of BMRGO120 and reached 6.80 of BMRGO240. At the same time, the BET surface area was also increased from 3.65 $m^2 \cdot g^{-1}$ for GO to 10.30 $m^2 \cdot g^{-1}$ for BMRGO120, as shown in Fig. 2.83 (a). The relatively low surface areas of the BMRGOs implied that the graphitic structure had been effectively recovered by the ball-milling. Figure 2.83 (b) shows TGA curves of all samples conducted in air. Thermal stability of the BMRGOs was enhanced due to the elimination of the oxygenated groups during the ball-milling in hydrogen.

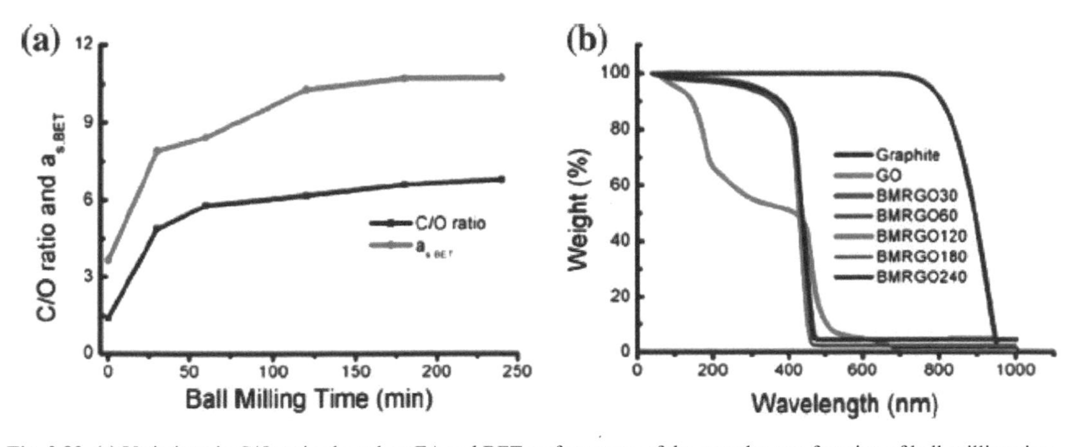

Fig. 2.83. (a) Variations in C/O ratios based on EA and BET surface areas of the samples as a function of ball-milling time. (b) TGA thermograms of the samples with a heating rate of 10°C·min^{-1} in air. Reproduced with permission from [185], Copyright © 2014, Elsevier.

Color image of this figure appears in the color plate section at the end of the book.

Representative TEM images and SAED pattern of the BMRGO30 powder are shown in Fig. 2.84 (a–c) [185]. The sample exhibited a typical wrinkled sheet-like morphology of graphene nanosheets (Fig. 2.84 (a)). The high degree structural restoration by mechanochemical milling was confirmed by the high magnification TEM image, as shown in Fig. 2.84 (b). The high crystallinity of the ball-milled powder was further evidenced by the SAED pattern with a hexagonal symmetry, as demonstrated in Fig. 2.84 (c).

Electrical conductivities of the samples were measured by using pellets made from the powders [185]. The pellets had a diameter of 2.5 cm and thickness of about 200 μm, as shown in the inset of Fig. 2.84 (d). Electrical conductivity of the GO sample was 0.2 S/m, which was significantly increased by three orders of magnitude, after it was reduced to BMRGOs. The BMRGO samples exhibited conductivities in the range of 13–120 S/m, which was attributed to the restoration of π-conjugated networks. The decrease in the electrical conductivity of the samples with prolonged ball milling time was due to the decrease in the size of the graphene nanosheets. It was found that the electrical conductivity of the BMRGO30 sample could be further increased by thermally annealing at 900°C for 2 h in N_2. Figure 2.84 (e) shows that the BMRGO pellet remained intact after the thermal annealing. However, the GO pellet became rough and was cracked after the thermal treatment, as shown in Fig. 2.84 (f), and caused gas evolution during annealing.

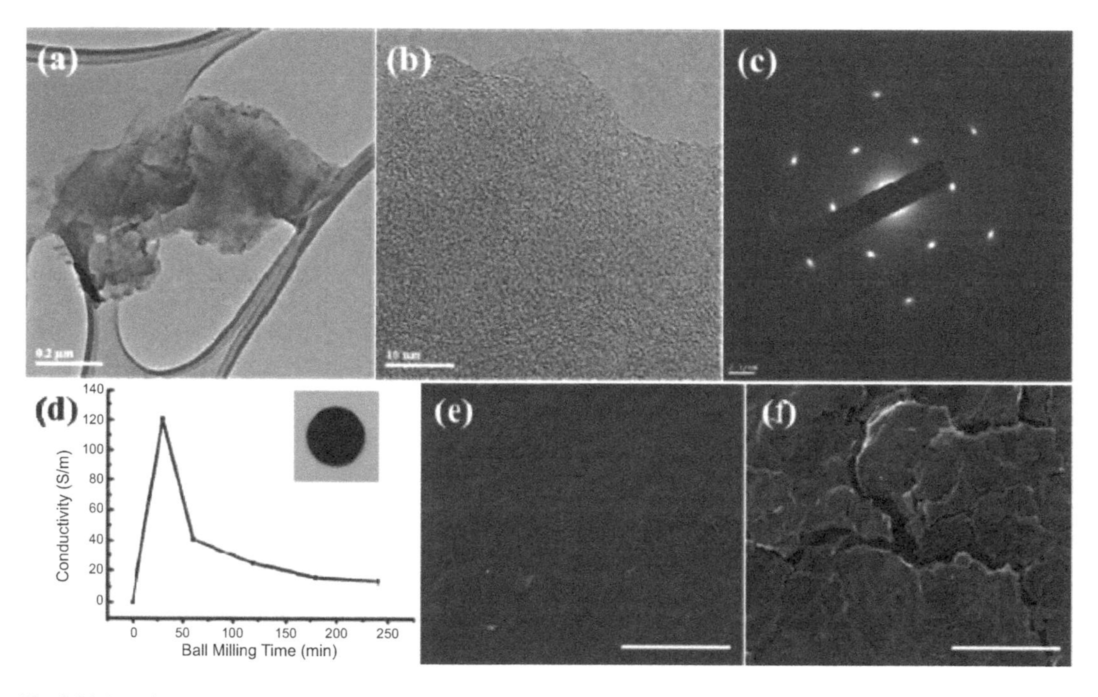

Fig. 2.84. TEM images and selected area electron diffraction (SAED) pattern of the BMRGO30 sample: (a) low magnification, (b) high-magnification at the edge and (c) SAED pattern. (d) Conductivity plots of the GO and BMRGOs as a function of ball-milling time, with the inset showing a photograph of the BMRGO30 pellet with a diameter of 2.5 cm used for the measurement. SEM images obtained from the surface of sample pellets after heat treatment at 900°C for 2 h in nitrogen: (e) BMRGO30 and (f) GO, with scale bars of 100 μm. Reproduced with permission from [185], Copyright © 2014, Elsevier.

2.2.1.8. Strategies to Improve the Quality and Quantity of Graphene

Almost all the methods discussed above use strong ultrasonication to facilitate the exfoliation, which is inevitably detrimental to the produced graphene nanosheets, because most applications require that the graphene nanosheets should be as large as possible. Meanwhile, the low production yield is another drawback of the solution exfoliation methods. Fortunately, various strategies have been proposed and demonstrated to produce large-sized and highly productive graphene nanosheets, for practical applications [197, 198].

A mild sonication exfoliation approach, through bromine-intercalation of graphite, has been reported to produce large-sized graphene nanosheets [199]. Few-layer graphene (FLG) nanosheets with sizes of up to 30 μm were obtained, which was close to the grain size of the starting graphite powder.

Molecular bromine (Br_2) intercalates into graphite and the effects of the molecule on graphite matrix have been extensively studied [200]. At high concentration, C_8Br_2 is formed, with the molecules being oriented parallelly to the graphene planes. In this case, the bromine atoms are centered on top of the hexagons, as shown in Fig. 2.85 (a), with a calculated equilibrium interlayer distance of 0.64 nm [201–203]. As the concentration is increased to an intermediate level, $C_{18}Br_2$ is obtained, with the Br_2 molecules being oriented perpendicularly to the graphene planes, so that the atoms situated stand on top of a carbon atom, as illustrated in Fig. 2.85 (b). In this case, the theoretical interlayer distance is 0.77 nm, only slightly smaller than the experimental value of 0.88 nm. At the lowest concentration, $C_{32}Br_2$ is stable, the optimal structure is similar to the case of high concentration, as demonstrated in Fig. 2.85 (c), where the interlayer distance is 0.62 nm. The interlayer binding energy is related to the charge transfer from Br_2 to the graphene layers, with values of 0.08e, 0.04e and 0.01–0.02e for high, medium and low bromine concentrations, respectively. Therefore, a low concentration of bromine is favorable for graphite exfoliation. Also, at low concentrations, the cohesive energy of Br_2 molecule to the graphene was much smaller than that of the –OH groups in graphite oxide [204], so that the removal of Br_2 from graphene is not a problem.

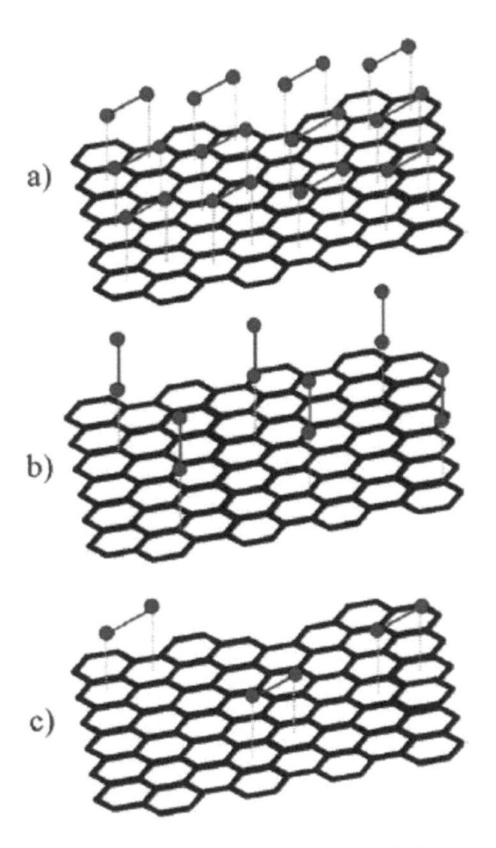

Fig. 2.85. Schematic images of the Br_2 molecule orientation on graphite layers at different concentrations: (a) C_8Br_2, (b) $C_{16}Br_2$ and (c) $C_{32}Br_2$. Reproduced with permission from [199], Copyright © 2009, IOP Publishing.

For the experiment, bromine–graphite in deionized water was sonicated for 10 min at 45 kHz and 100 W. SEM images of the graphite and bromine–graphite samples are shown in Fig. 2.86 (a) and (b), respectively. Intercalation of bromine obviously promoted the exfoliation of graphite. A large size distribution was observed. It was found that sonication times of > 45 min at 100 W led to damaging of

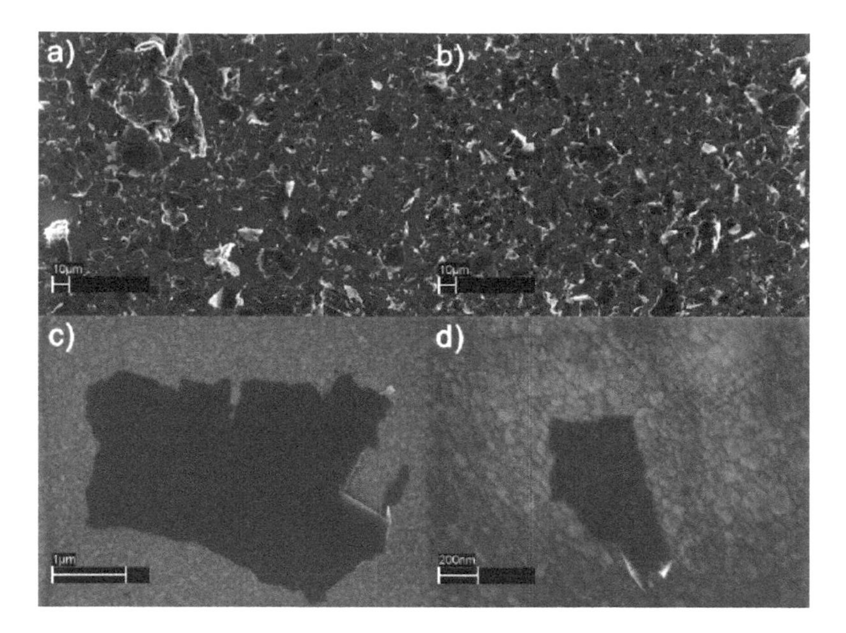

Fig. 2.86. Overview SEM images of the samples deposited on silicon substrates from (a) suspensions prepared from graphite and (b) suspensions prepared from bromine–graphite. SEM images of (c) a large flake and (d) a small flake deposited on platinized silicon, in which the structure of the underlying substrate is clearly visible through the graphene nanosheets. Reproduced with permission from [199], Copyright © 2009, IOP Publishing.

the graphene nanosheets in the suspensions. TEM images by collecting the nanosheets on grids with a carbon support film are shown in Fig. 2.87 (a–c). Quite a number of nanosheets were folded. Figure 2.87 (d) shows a representative thickness map of the exfoliated samples.

A modified solution-phase method has been demonstrated for high yield production of graphene oxide and graphene nanosheets with ultra-large sizes of up to 200 μm [205]. The experiment was started with natural graphite of 50 mesh size, corresponding to 300 μm in average size. Graphene oxide was obtained by exfoliating the graphite in water, by using a gentle shaking treatment, instead of the conventional strong ultrasonication. The exfoliated GO nanosheets were reduced to graphene by using hydrazine hydrate in the presence of a nonionic polymeric surfactant.

After exfoliation, the suspension exhibited anisotropy textures when it was shaken, which implied that the obtained nanosheets had large sizes with high aspect ratios. Figure 2.88 (a) shows a representative SEM image of the GO nanosheets [205]. They had lateral sizes dominantly in the order of 100 μm, while some of them were as large as 200 μm. Only a small number of fragments with sizes of several micrometers were observed, which were mainly derived from the micrometer-sized graphite flakes. In other words, this method showed no damage to the graphite, so that the size of the graphene nanosheets could be further increased by using large graphite particles and the small-sized nanosheets could be avoided by controlling the particle size of the starting graphite flakes.

Figure 2.88 (b) shows an AFM image of a GO nanosheet, with an average thickness of about 1 nm, suggesting the unilamellar nature of the nanosheets [205]. Representative SEM image of the as-reduced graphene nanosheets is shown in Fig. 2.88 (c), which had the similar large size feature to the parent GO. AFM image indicated that the nanosheets had an average thickness of about 0.5 nm, close to the theoretical thickness of 0.335 nm of single-layer graphene, as shown in Fig. 2.88 (d).

Both the GO and graphene materials were made into solid foams from their colloidal suspensions, by using a freeze-drying method, as demonstrated in Fig. 2.89 (a) and (b). Their SEM images are shown in Fig. 2.89 (c) and (d), respectively. The foams exhibited an ultra-hollow framework structure, consisting of the large nanosheets. The graphene foam could be further purified by annealing in an inert atmosphere.

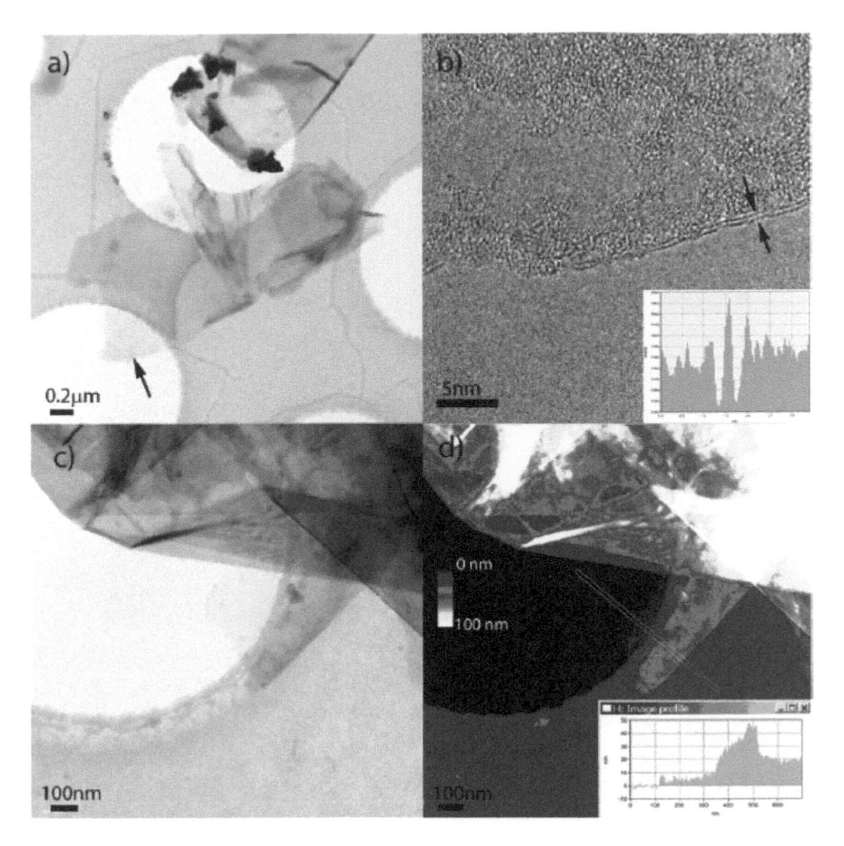

Fig. 2.87. (a) TEM image of two graphene nanosheets. (b) High resolution TEM image of the folded edge indicated by the arrow in image (a). The flake is only 2 or 3 layers thick, as indicated by the intensity profile in the inset. (c) An overview TEM image of a folded nanosheet and (d) a thickness map constructed from the measured intensities in the overview image. The inset shows a thickness profile of the marked region in (d). Reproduced with permission from [199], Copyright © 2009, IOP Publishing.

Color image of this figure appears in the color plate section at the end of the book.

The annealed graphene foam was highly conductive, but still behaving like a sponge with high elasticity and flexibility. It was found that the nanosheets were highly cross-linked after the thermal annealing. The suspensions could also be filtrated to make GO and graphene papers with high electrical conductivities [205].

A kitchen blender has been used to produce high-quality few-layer graphene (FLG) nanosheets, with an average thickness of 1.5 nm (20% $\leq$ 1 nm) and no oxidation, which can be considered as a potential large-scale production method [191]. As a kitchen blender is equipped with a rotating impeller, it can induce multiple fluid dynamic events, including shear, turbulence and collisions, which can be used to exfoliate graphite into graphene nanosheets, due to the lateral self-lubricating characteristics of the layer structured graphite, with the mechanism schematically shown in Fig. 2.90.

During the experiment, graphite power was dispersed in N,N-dimethylformamide (DMF) at an initial concentration of 3 mg/ml to form 400 ml graphite dispersions. The exfoliation can be carried out similarly in other any other solvents that have suitable surface tension [94] or Hansen solubility parameters [206], as well as water/surfactant solutions, which have been used in the conventional chemical exfoliation processes, as discussed before. The kitchen blender used in this experiment was equipped with a five-blade impeller. The exfoliation was conducted at 5000 rpm for up to 8 h, during which six samples were collected at 0.5, 1, 2, 3, 5, 8 h, respectively. The collected samples were centrifuged for 45 min at 500 rpm to obtain FLG dispersions.

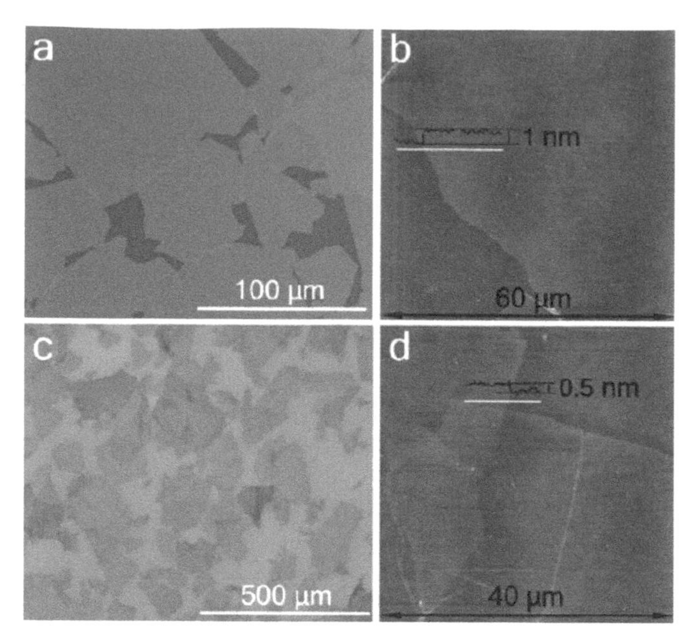

Fig. 2.88. (a) SEM and (b) AFM images of GO sheets that were deposited on a Si substrate by Langmuir–Blodgett assembly at a constant surface pressure of 10 mN·m^{-1}. (c) SEM and (d) AFM images of graphene sheets deposited on a polyethyleneimine (PEI) pre-coated Si substrate. Reproduced with permission from [205], Copyright © 2010, The Royal Society of Chemistry.

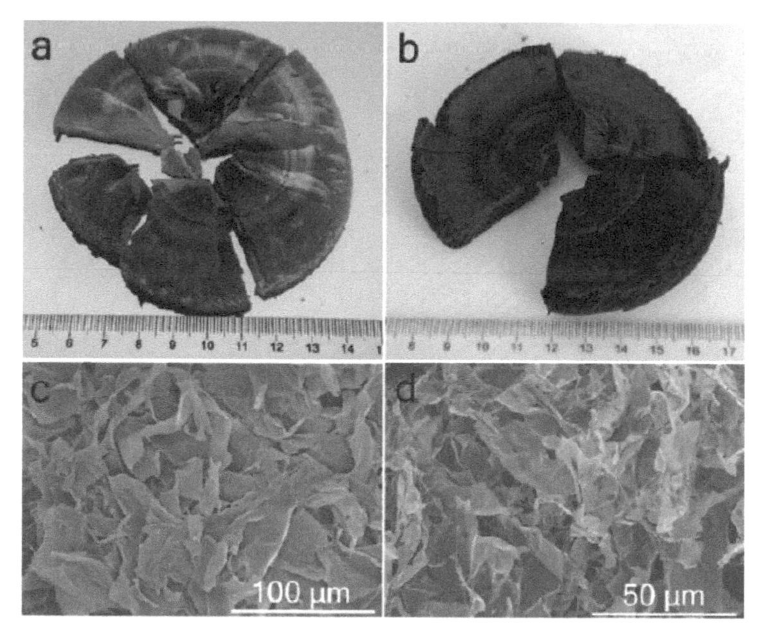

Fig. 2.89. Photographs and SEM images of the freeze-dried solid foams: (a, c) GO sheets and (b, d) graphene sheets. Reproduced with permission from [205], Copyright © 2010, The Royal Society of Chemistry.

Figure 2.91 (a) shows Raman mapping of a representative sample of the exfoliated graphene nanosheets. As stated earlier, the defect content of a graphene nanosheet is characterized by the intensity ratio of the D band to the G band, I_D/I_G [94]. There are two types of defects, i.e., (i) basal-plane defects and (ii) edge defects. Basal plane defects usually result in the broadening of G bands, which are observed in most graphene nanosheets exfoliated by using the chemical exfoliation routes. The presence of edge defects is

Fig. 2.90. Schematic of a kitchen blender that was used to produce FLG nanosheets. Reproduced with permission from [191], Copyright © 2014, Elsevier.

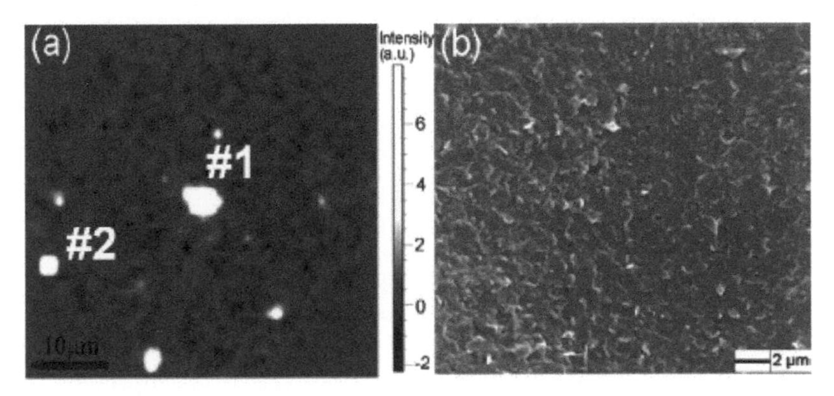

Fig. 2.91. (a) A Raman mapping image of representative samples of the exfoliated graphene nanosheets. The Raman map plots the intensity integral of the spectra between 2600 and 2800 cm^{-1}. The excitation wavelength was 532 nm. (b) SEM image of the vacuum filtered film of the graphene nanosheets. Reproduced with permission from [191], Copyright © 2014, Elsevier.

generally related to the fragmentation effect. Because the narrow G band was not broadened, basal-plane defects were not dominant in the blender exfoliated graphene nanosheets. Therefore, the weak and narrow D band in the filtered film was attributed to the edge defects. In other words, no basal-plane defects were produced during the kitchen blender facilitated exfoliation of graphite. Moreover, I_D/I_G for the filtered film was less than 0.12, much lower than those of graphene oxide and the chemically reduced graphene nanosheets. The high-quality of FLG was also confirmed by the high conductivity. It was found that a 20 μm thick film obtained by using vacuum filtering the FLG dispersions showed a sheet resistance 2.3 Ω/□, corresponding to a high DC conductivity of 2.2×10^4 $S \cdot m^{-1}$.

Two types of forces were responsible for the exfoliation of graphite in this case, i.e., (i) normal force and (ii) lateral force. The lateral force promoted the relative motion of the graphene layers, due to the self-lubricating behavior of graphite. The exfoliation mechanism was understood from the fluid dynamics point of view in the kitchen blender. The Reynolds number of the flow in the tank is $R_e = \rho ND^2/\mu \approx 2 \times 10^6$, corresponding to a full turbulent flow [190]. As mentioned above, four fluid dynamics events could be responsible for the exfoliation and fragmentation of graphite: (I) velocity gradient to induce viscous shear stress, (II) intensive velocity fluctuations in turbulence to induce Reynolds shear stress (III), viscous forces as the inertial forces related to the very large Reynolds number in the turbulence to enhance collisions of graphite particles and (IV) possible pressure difference induced by the turbulent pressure fluctuations to provide normal force to exfoliate graphite, as shown schematically in Fig. 2.92 (a–c). These fluid dynamics events could also lead to exfoliation as fragmentation. The lateral exfoliation was evidenced by the TEM images of the partially exfoliated flakes, as shown in Fig. 2.92 (d) and (f). The stacked flakes exhibited a slipped configuration with lateral relative displacement of either translation or rotation, as illustrated in Fig. 2.92 (e) and (g), respectively.

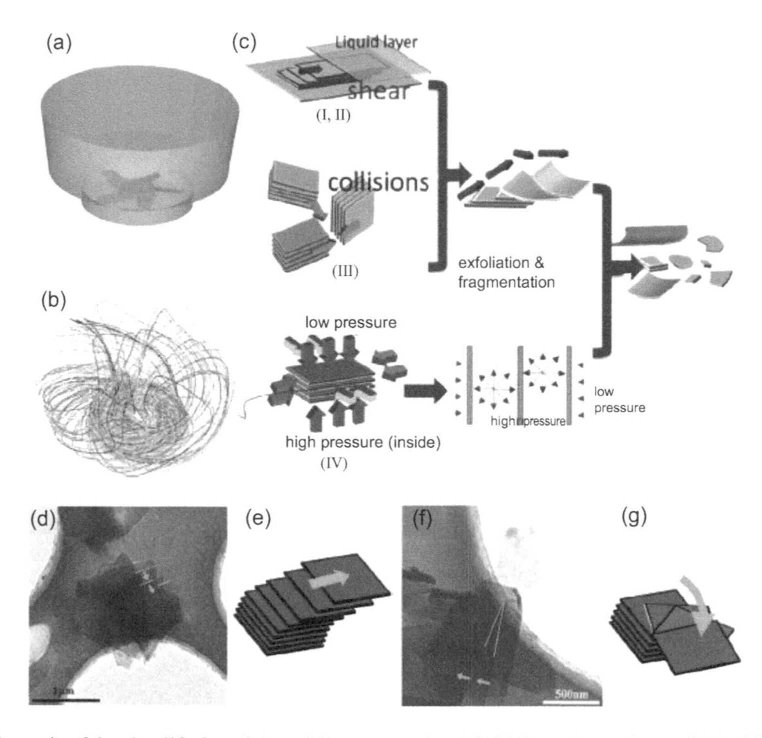

Fig. 2.92. (a) Schematic of the simplified model used for computational fluid dynamics analyses. (b) Pathlines released form the impeller surface colored by particle ID. (c) Illustration for the exfoliation mechanism. Deliberately captured partially exfoliated FLG flakes with translational (d, e) and rotational (f, g) lateral exfoliation. Reproduced with permission from [191], Copyright © 2014, Elsevier.

An attempt has been made to promote the exfoliation of graphite from laboratory small-scale synthesis to industrially large-scale production, by using blender-like shear milling [207]. The method led to graphene nanosheets that are unoxidized and have no basal plane defects. Simple model indicated that as the local shear rate exceeded 10^4 per second, exfoliation of graphite was started. It was confirmed that, by fully characterizing the scaling behavior of the graphene production rate, large-scale exfoliation could be readily achieved. A high-shear laboratory mixer was used in the experiment, with the main component of the mixing head to be a 4-blade rotor that was fixed within a stator. Commercial graphite powders were used as the starting materials. The mixer was run at the desired speeds for predetermined mixing time durations. The main controllable mixing parameters included diameter of the rotors, D = 12–110 mm, the initial graphite concentration (C_i = 1–100 mg·ml^{-1}), mixing time (t = 3–540 min), liquid volume (V = 0.2–300 l) and rotor speed (N = 500–7000 rpm). Solvent N-methyl-2-pyrrolidone (NMP) and aqueous solutions of surfactant sodium cholate (NaC) and polymer polyvinyl alcohol (PVA) were used to form graphene suspensions.

Figure 2.93 (a) shows the mixer that was used to generate high shear with a closely spaced rotor/stator combination and a range of rotor diameters, as shown in Fig. 2.93 (b, c). Large volume suspensions in NMP and NaC solutions are demonstrated in Fig. 2.93 (d), which were obtained with the processing parameters of D = 32 mm, C_i = 50 mg/ml, t = 20 min, V = 4.5 l and N = 4500 rpm. After centrifugation, the suspensions contained large quantities of high-quality graphene nanosheets, including some monolayers, as shown in Fig. 2.93 (e–h). Large-scale trial facilities are shown in Fig. 2.94 (f–h).

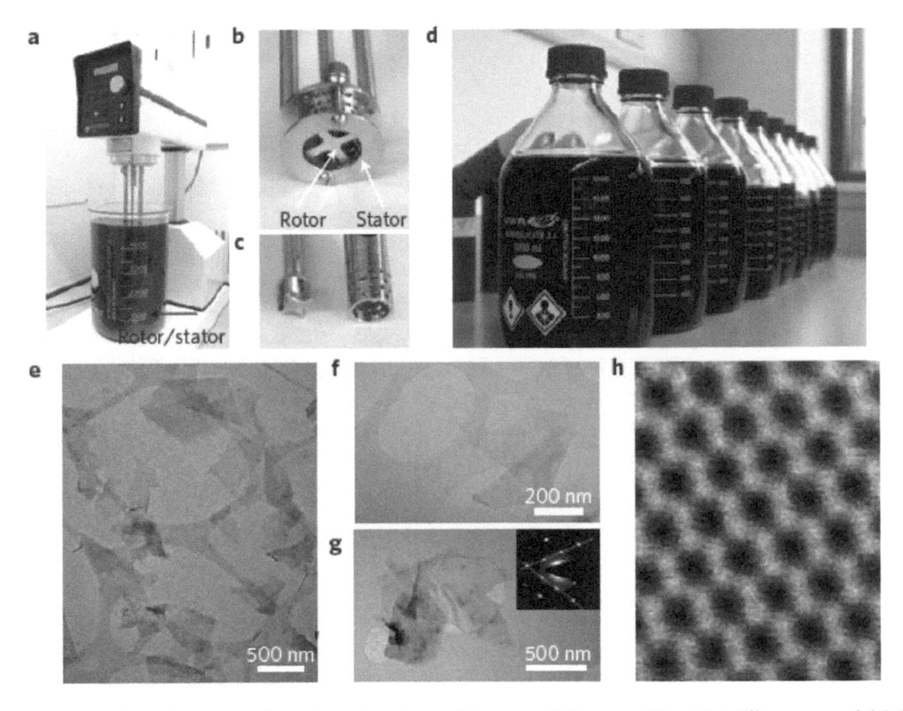

Fig. 2.93. Production of graphene nanosheets by using shear mixing to exfoliate graphite. (a) A Silverson model L5M high-shear mixer with mixing head in a 5 l beaker of graphene dispersion. (b, c) Close-up view of a D = 32 mm mixing head (b) and a D = 16 mm mixing head with rotor (left) separated from stator (c). (d) Graphene–NMP dispersions produced through shear exfoliation. (e) Wide-field TEM image of the graphene nanosheets after centrifugation. (f–h) TEM images of individual nanosheets (f), a multilayer (g, bottom left) and monolayer (g, right) as evidenced by its electron diffraction pattern (g, inset) and a monolayer (h, imaged by high-resolution scanning TEM). Reproduced with permission from [207], Copyright © 2014, Macmillan Publishers Limited.

Fig. 2.94. Scaling of the graphene production with a shear mixer. (a, b) 11-cm-diameter rotor (f) and stator (g) used during large-scale trials. (c) Shear exfoliation of graphite in water–surfactant solution at the scale of V = 100 l. Reproduced with permission from [207], Copyright © 2014, Macmillan Publishers Limited.

2.2.2. Mechanical Exfoliation

Mechanical exfoliation of graphite is an alternative method to create graphene nanosheets [208]. Because the inter-layer van der Waals interaction energy of graphite is about 2 eV·nm^{-2}, it is expected that graphite can be exfoliated by using a force with the magnitude of 300 nN·µm^{-2} [209]. Due to the extremely low force, an adhesive tape can be used to peel off graphene nanosheets from graphite crystals [210, 211]. Obviously, mechanical exfoliation offers graphene with high electrical and structural quality, due to the nondestructive character of the method. It is also a straightforward approach to produce graphene nanosheets directly from graphite. However, it is difficult to provide very thin sheets of a few nanometers or monolayers using this method. Also, large-scale production and reproducibility are another two critical problems of mechanical exfoliation of graphite. In addition, glue residues could remain in the samples, which would have negative effects on their properties, such as carrier mobility [212, 213], although the problem can be addressed by using post-thermal treatments [76, 214].

The simplest and most effective method to exfoliate graphite and other layer structured materials by using common adhesive tape has been widely demonstrated [3, 79]. In this method, the adhesive tape was allowed to repeat the process of alternative sticking and peeling, so that graphene sheets with thicknesses from 1 µm down to even monolayer thickness could be obtained. The graphene nanosheets or graphite pieces could be simply transferred onto suitable substrates by gently pressing the tapes. The apparent contrast of graphene monolayer on SiO_2/Si substrates, with the thicknesses of the oxide layer of either 300 or 90 nm, was maximized, due to the Fabry–Perot multilayer cavity, in which the optical path added by graphene to the interference of the SiO_2/Si system was maximized for specific oxide thicknesses [215–217]. Hence, in practice, thicker graphite flakes deposited on a 300 nm SiO_2 will appear yellow to bluish as the thickness decreases, as shown in Fig. 2.95 (a). However, when thicknesses came to be < 10 nm, darker to lighter shades of purple color would appear, indicating the presence of graphene with a few layers or even one single monolayer, as demonstrated in Fig. 2.95 (b). The coloring appearance of the graphene nanosheets is attributed to the interference phenomenon between the graphene and the substrates, which can thus be changed by adapting the thickness of the dielectric layer.

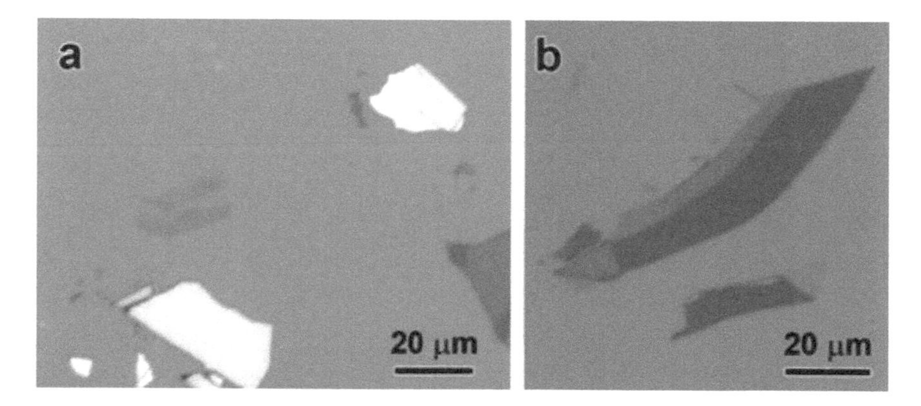

Fig. 2.95. Micromechanically exfoliated graphene nanosheets dispersed on substrates. Optical images of (a) thin graphite and (b) few-layer graphene (FLG) and single-layer graphene (lighter purple contrast) on a 300 nm SiO_2 covered Si substrate. Yellow-like color indicates thicker samples (> 100 nm), while bluish and lighter contrast indicates thinner samples. Reproduced with permission from [208], Copyright © 2010, Elsevier.

It has been reported that a graphene nanosheet with a thickness of only 1 atom could be obtained by mechanical exfoliation [218]. By using aberration-correction in combination with a monochromator, 1-Å resolution was achieved at an acceleration voltage of only 80 kV. Therefore, every individual carbon atom in the field of view was detected and resolved. A highly crystalline lattice along with occasional point defects was observed. The formation and annealing of Stone-Wales defects were observed *in situ*. Figure 2.96 (a) shows an optical micrograph of a large graphene sample on the support grid, with the 1-µm grid perforation holes being clearly visible, whereas Fig. 2.96 (b) shows a low magnification TEM image of

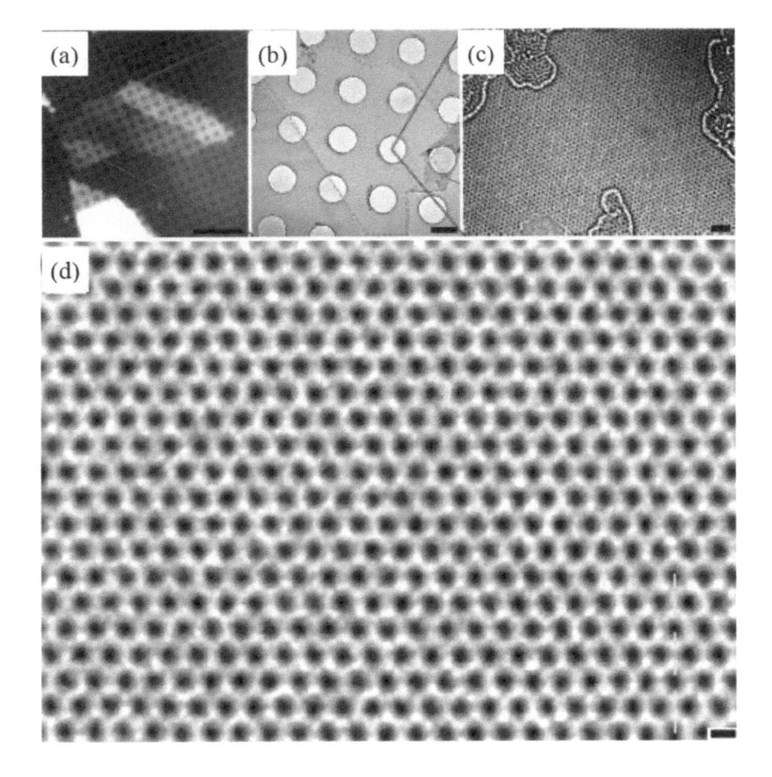

Fig. 2.96. (a) Optical micrograph and (b) low-magnification TEM image of the graphene sheets on perforated carbon film. A single-layer region is outlined by the red dashed line. (c) Unfiltered CCD exposure (1 s) of a single-layer graphene nanosheet. The structures near the edge of the image are adsorbates and a hole (formed after prolonged irradiation) is observed near the lower-edge left. Scale bars are 10 μm (a), 1 μm (b) and 1 nm (c). (d) Direct image of a single-layer graphene nanosheet, with carbon atoms being white. Reproduced with permission from [218], Copyright © 2008, American Chemical Society.

a single-layer graphene nanosheet, covering several holes. Figure 2.96 (c) shows a high-resolution TEM image that was obtained by zooming in on one of the suspended nanosheet regions. The image was a single unfiltered CCD exposure, in which the intensity profile was a direct representation of the carbon atomic structure in graphene, with the carbon atoms being white. The large area shown in Fig. 2.96 (c) was structurally perfect. A direct image of a single-layer graphene nanosheet is shown in Fig. 2.96 (d).

2.2.3. Graphene Ribbons

Graphene nanoribbons form a special group of nanocarbon materials [219–221]. Graphene nanoribbons have borders that can exhibit edge states and different electronic, chemical and magnetic properties, depending on their size and type of the border. The most studied chiral edge configurations include (i) 0° (armchair) and (ii) 30° (zigzag), thus leading to armchair (A-GNRs) and zigzag nanoribbons (Z-GNRs), respectively [222]. In addition, there have also been reports on edge reconstruction with pentagonal and heptagonal carbon rings [223]. Methods to produce graphene nanoribbons mainly include (i) unzipping of carbon nanotubes, (ii) chemical exfoliation of graphite and (iii) chemical vapor deposition (CVD). The first two methods will be discussed in this part, the third one will be included in next part.

2.2.3.1. Unzipping of CNTs

If carbon nanotubes can be considered as rolled up graphene nanosheets, unrolling (unzipping) CNTs would result in graphene. Various unzipping methods have been developed to obtain graphene from CNTs, mainly including (i) intercalation–exfoliation of MWCNTs with chemical treatments in liquid NH_3 with Li, followed by exfoliation with HCl and thermal treatment [224], (ii) chemical unzipping through acid reactions to break the carbon-carbon bonds with oxidizing agents, such as H_2SO_4 and $KMnO_4$ [225], (iii) catalytic cutting with metallic nanoparticles [226], (iv) electrical bombardment [227] and (v) physicochemical removal of the unembedded part of the CNTs in a polymer matrix, followed by Ar plasma treatment [228].

The unzipping mechanism of the first method is shown in Fig. 2.97 [224]. The strong electrostatic attraction between the negatively charged MWCNTs and NH_3-solvated Li^+ facilitated the intercalation of the MWCNTs. The unzipping started from the opened tips with structural defects, due to the insertion of Li-NH_3. At the same time, defects on the walls, induced by the oxidation of acids, triggered the intercalation and unzipping, leading to the detachment of graphene nanosheets. The initial unzipping resulted in edges that were terminated with hydrogen and/or amino groups. HCl reacted with Li^+ and neutralized NH_3, resulting in exfoliation, so that the CNTs were unzipped.

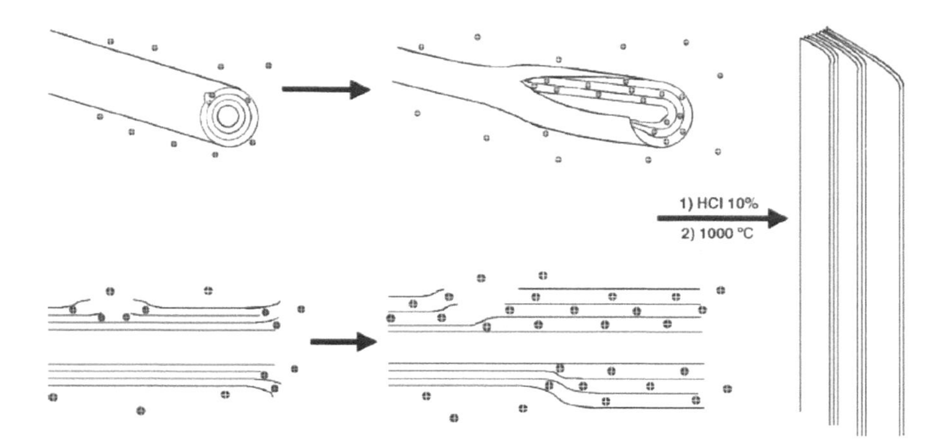

Fig. 2.97. Possible mechanism for the intercalation and unzipping of MWCNTs, with three-dimensional (above) and cross-sectional (below) views of the initial stages. The dots represent NH_3-solvated Li^+, which were attracted to the negatively charged MWCNTs. The intercalation expanded interlayer spacing almost twice and the stress ruptured the walls. After the treatment with HCl, the tubes were exfoliated and disordered. After thermal treatment, further exfoliation occurred together with recrystallization of some graphitic planes, leading to nanoribbons. Reproduced with permission from [224], Copyright © 2009, American Chemical Society.

Four stages were observed in the unzipping process, as shown in Fig. 2.98 [224]. According to their appearances, the nanotubes were classified into (i) intact CNTs, (ii) damaged CNTs with only the outer layers being exfoliated, (iii) partially exfoliated ones with one or both ends being opened or walls being unopened but damaged and (iv) completely exfoliated MWCNTs. Further evidence of exfoliated MWCNTs is shown in Fig. 2.99. Figure 2.99 (A) shows an unzipped MWCNT and a curved fully exfoliated tube formed by stacked graphene layers. More unzipped samples are shown in Fig. 2.99 (B) and (C). The fast Fourier transform hexagonal pattern shown in the inset of Fig. 2.99 (B) confirmed the presence of graphene layers stacked with an ABAB... configuration. Some rippled graphene nanosheets caused by the chemical treatment were also observed, as demonstrated in Fig. 2.99 (C). Figure 2.99 (D) indicates that porous microstructures were present in some exfoliated MWCNTs, which could be attributed to the detachment of graphene segments.

Therefore, this method cannot exfoliate all the MWCNTs entirely, with the maximum content of nanoribbons to be 60%, which could be attributed to several reasons. Firstly, because the formation of the defects discussed above was an arbitrary process, it is difficult to maintain the defects to be formed in such a way that they would be well-aligned along the length of the MWCNTs, especially when the tubes are too long. Secondly, the Li^+-NH_3 complexes and the Li^+ without being combined with NH_3 could be removed during the thermal treatment before the occurrence of the exfoliation. Also, some of the functional

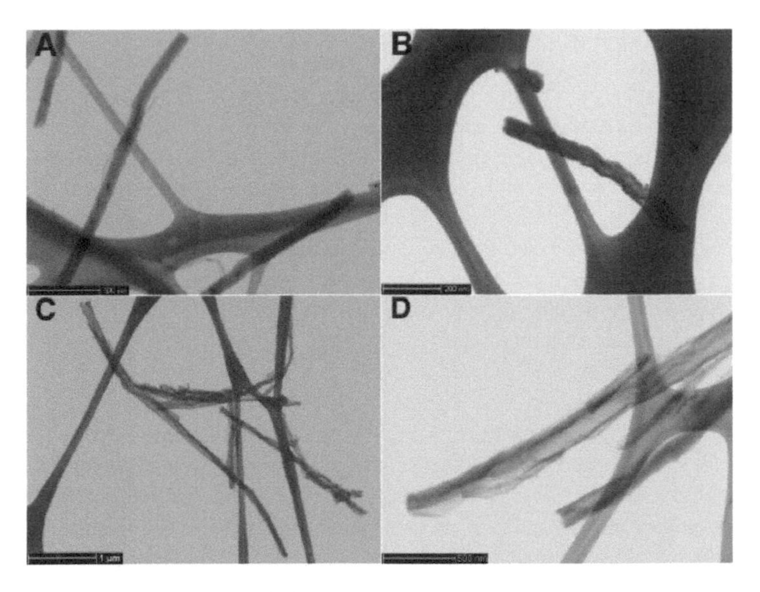

Fig. 2.98. Representative STEM images of the MWCNTs at the four states after the exfoliation treatments: (A) intact, (B) damaged, (C) a long tube partially exfoliated, along with shorter damaged tubes and (D) exfoliated MWCNTs. Reproduced with permission from [224], Copyright © 2009, American Chemical Society.

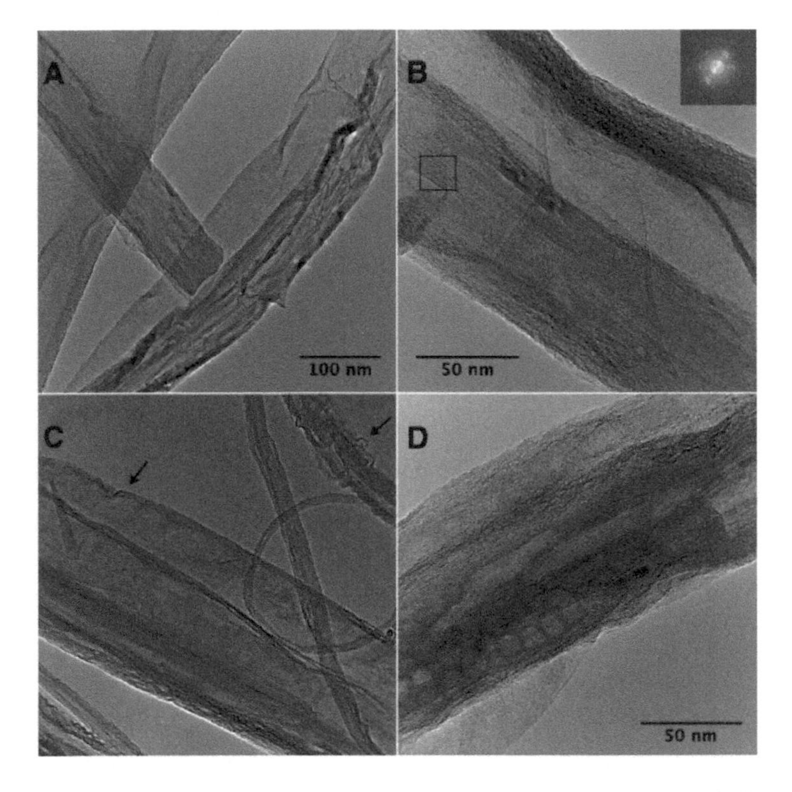

Fig. 2.99. Representative TEM images of the exfoliated MWCNTs. (A) Some nanotubes were unzipped with a layered structure, while others appeared as ribbons. (B) An example of an unwrapped nanotube, with the inset showing the fast Fourier transform of the marked area, demonstrating the hexagonal pattern of ABAB... graphite. (C) Rippled graphene sheets (arrows) on the exfoliated MWCNT and damaged MWCNT. (D) Graphene nanosheets with holes in some regions, which exhibited irregular shapes. Reproduced with permission from [224], Copyright © 2009, American Chemical Society.

groups responsible for the exfoliation would be on the surface of the tubes, which had no contribution to the exfoliation. In addition, precise controlling of the dimension of the nanoribbons is also a critical problem. As a result, further improvement is necessary for this method to achieve full unzipping of the CNTs and controlling of the nanoribbons, for practical applications.

The second method is a simple solution-based oxidative process, which can be used to produce nanoribbons with a nearly 100% yield, by lengthwise cutting and unzipping the side walls of multi-walled carbon nanotubes (MWCNTs) [225]. The derived nanoribbon exhibited high water solubility, which could be reduced chemically to restore the electrical conductivity of graphite. The graphene nanoribbons could be used in the fields of electronics and composite materials.

The direct oxidation experiment was started with commercially available MWCNTs, which were suspended in concentrated sulphuric acid (H_2SO_4) for time durations of 1–12 h and then treated with 500 wt% potassium permanganate ($KMnO_4$). The strong H_2SO_4 conditions were required to exfoliate the nanotube. The reaction mixture was stirred at room temperature for 1 h and then heated to 55–70°C for 1 h. When all the $KMnO_4$ was consumed, the mixture was quenched by pouring ice with a small amount of hydrogen peroxide (H_2O_2). The solution was filtered over a polytetrafluoroethylene (PTFE) membrane and the solids were washed with acidic water, ethanol and ether. The exfoliation of the MWCNTs to graphene nanoribbons was a stepwise oxidation. The reaction procedure was roughly the same, except that 100 wt% $KMnO_4$ was added five times to 500 wt%. When the $KMnO_4$ was consumed at every step, a portion of the reaction solution was collected for analysis. Nanoribbon suspension in water with a concentration of 200 mg/l was reduced by using 1 vol% concentrated ammonium hydroxide (NH_4OH) and 1 vol% hydrazine monohydrate ($N_2H_4 \cdot H_2O$), with or without 1 wt% SDS surfactant. The solution was covered with a thin layer of silicon oil and then heated at 95°C for 1 h.

The unzipping of the nanotubes occurred along a line, thus leading to straight-edged nanoribbons. It could be a linear longitudinal cutting or a spiralling opening, depending on the initial site of the attack and the chiral angle of the nanotubes. A representative mechanism is shown in Fig. 2.100 (a), where the opening was on the mid-section of the nanotube [225]. However, the actual location of the initial attack was not clarified, which could occur at an arbitrary site.

The mechanism of opening was proposed, based on the oxidation of alkenes by permanganate in acid. The first step was the formation of manganate ester, as shown in Fig. 2.100 (b, 2), which was the rate controlling step [225]. Further oxidation resulted in dione in the dehydrating medium, as demonstrated in Fig. 2.100 (b, 3). Juxtaposition of the buttressing ketones distorted the b,c-alkenes (red in 3), so that they were prone to further be attached by permanganate. As the process was continued, the buttressing-induced strain on the b,c-alkenes was reduced, owing to the creation of more space for the carbonyl projection. However, on the other hand, the b,c-alkenes became more reactive, because of the bond-angle strain induced by the enlarged hole, as illustrated in Fig. 2.100 (b, 4). Therefore, once an opening was initiated, the opening would be continued, until the nanotube was unzipped. Moreover, the ketones could be converted into the carboxylic acids through their O-protonated forms, which then would line the edges of the nanoribbons. Eventually, the bond-angle strain was relieved as the nanotube was completely unzipped to the graphene nanoribbon, i.e., Fig. 2.100 (b, 5). The same unzipping process was also applicable to single-walled carbon nanotubes (SWCNTs), to produce narrow nanoribbons, but with a problem of entanglement.

TEM analysis confirmed that nanoribbons were produced from MWCNTs, with a starting diameter of 40–80 nm and 15–20 inner nanotube layers, as shown in Fig. 2.100 (c). After reaction, the width of the carbon nanostructures was increased to about 100 nm, which had linear edges with little pristine MWCNTs side-wall structure remaining. Figure 2.100 (d) shows AFM image of the graphene nanosheets, revealing the presence of single atomic layers. Figure 2.100 (e) shows SEM image of the nanoribbons on a silicon substrate. The ribbons could be as long as 4 μm if they were not cut by the tip sonication. The nanoribbons could be dispersed as single or thin layers, with uniform widths and predominantly straight edges over their entire length.

The degree of consecutive tube opening in the MWCNTs could be controlled by adjusting the amount of oxidizing agent. TEM results indicated that 80–100% of the MWCNTs were completely unzipped on the side walls into nanoribbons, when 500 wt% $KMnO_4$ was used. The successive opening reaction was demonstrated in five iterations, in which there was a stepwise increase in the amount of $KMnO_4$

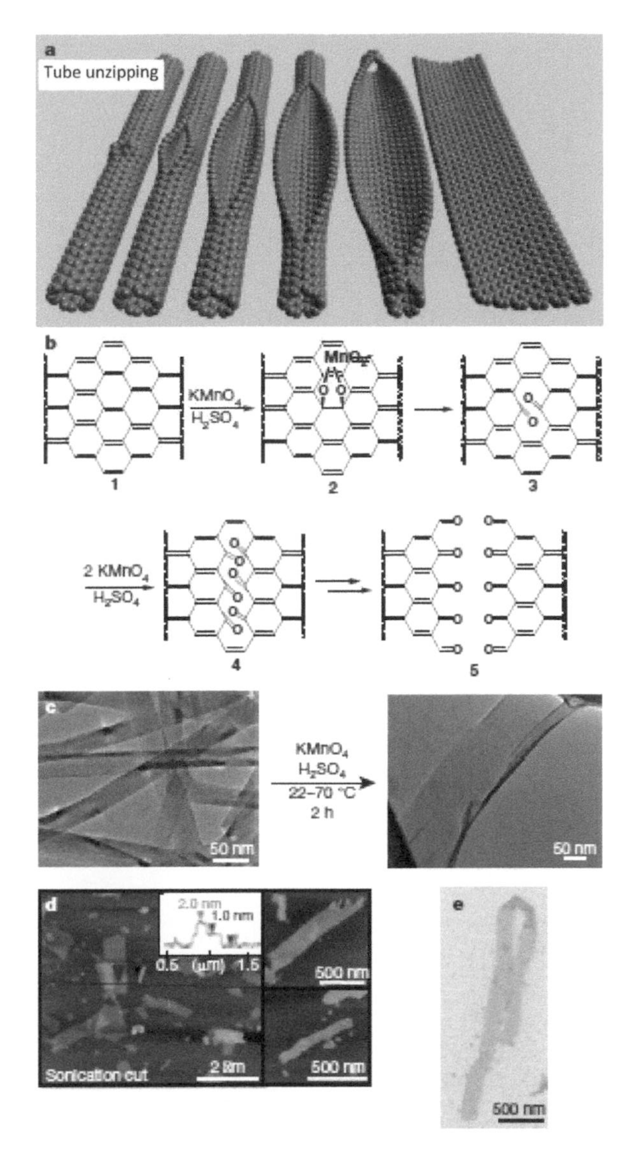

Fig. 2.100. Formation mechanism and images of the graphene nanoribbons derived from MWCNTs. (a) Representation of the gradual unzipping of one wall of a carbon nanotube to form a nanoribbon, without showing the oxygenated sites. (b) Chemical unzipping mechanism of the nanotube. The manganate ester in (b, 2) could also be protonated. (c) TEM images depicting the transformation of MWCNTs (left) into oxidized nanoribbons (right). The right-hand side of the ribbon is partly folded onto itself. The dark structures are part of the carbon imaging grid. (d) AFM images of the partly stacked multiple short fragments of nanoribbons that were horizontally cut by tip-ultrasonic treatment of the original oxidation product to facilitate spin-casting onto the mica surface. The height data (inset) indicates that the ribbons are generally single-layered. The two small images on the right show some other characteristic nanoribbons. (e) SEM image of a folded, 4-μm-long single-layer nanoribbon on a silicon surface. Reproduced with permission from [225], Copyright © 2009, Macmillan Publishers Limited.

at 100 wt%. For the first iteration (sample I), 100 wt% $KMnO_4$ was used, while 200 wt% $KMnO_4$ was used for the second iteration (sample II), until the final iteration (sample V), when 500 wt% $KMnO_4$ was used. This resulted in consecutive unencapsulation of the different layers by unzipping the successive MWCNTs. TEM images of the representative samples are shown in Fig. 2.101 (a–e). It was found that the walls of the MWCNTs were opened to a higher degree as the content of the oxidation agent was increased. Figure 2.101 (f) shows statistical plot of average diameter of the remaining MWCNTs, which decreased with the increasing degree of oxidation.

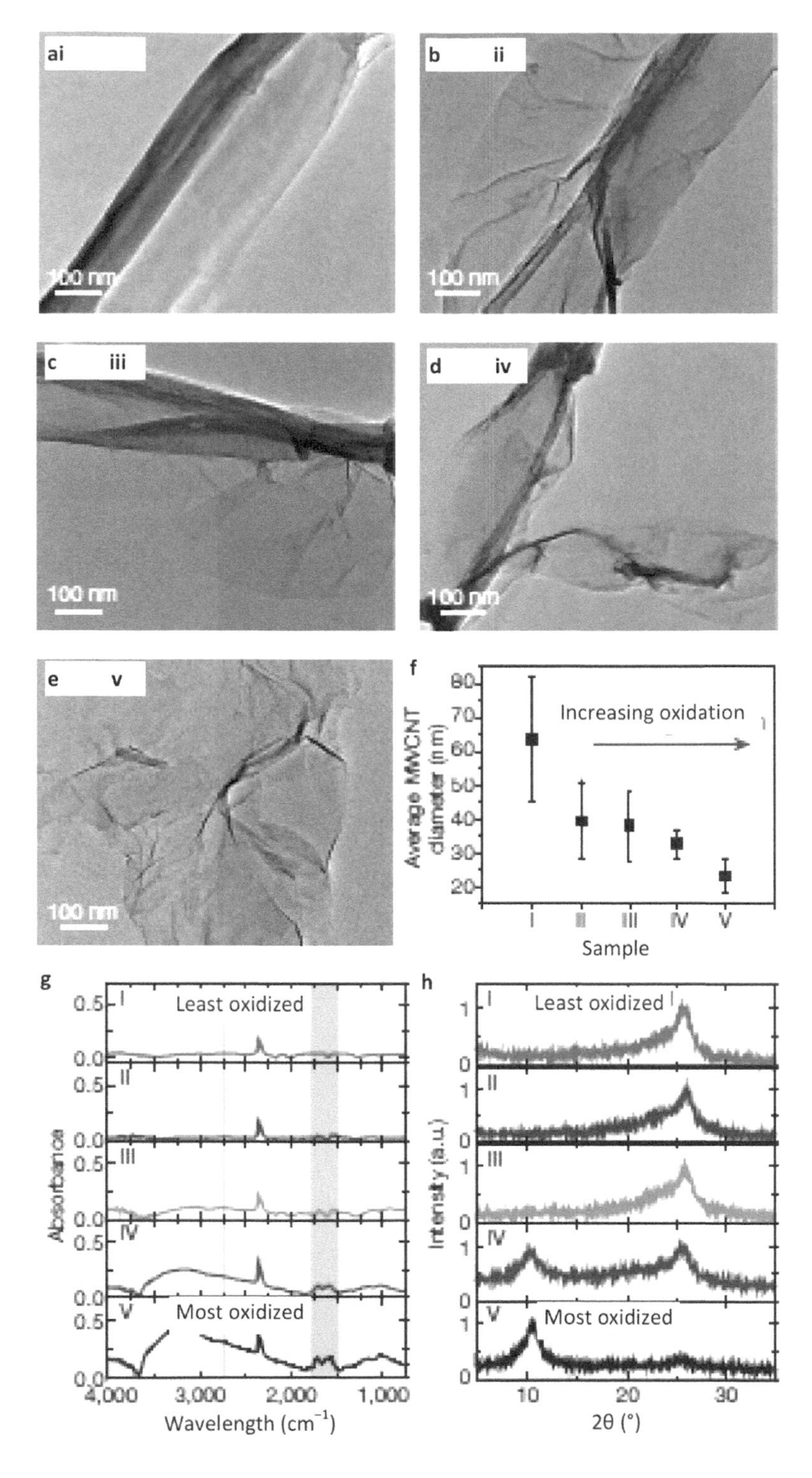

Fig. 2.101. Stepwise opening of MWCNTs to form nanoribbons. (a–e) TEM images of the stepwise opening of MWCNTs, representing the incremental exposure of the system to $KMnO_4$, with the least oxidized sample (sample I) shown in (a) and the most oxidized sample (sample V) in (e). (f) Scatter plot showing change in average diameter of the MWCNTs with increasing exposure to $KMnO_4$. (g) ATR–IR spectroscopy of stepwise opening/oxidation of the MWCNTs. (h) XRD patterns of the stepwise opening of the nanotube. Reproduced with permission from [225], Copyright © 2009, Macmillan Publishers Limited.

The degrees of oxidation of the samples for the five iterative $KMnO_4$ treatment steps, with partly and/or completely unzipped MWCNTs, were examined by using attenuated-total reflection infrared (ATR–IR) spectroscopy and thermogravimetric analysis (TGA). Figure 2.101 (g) shows ATR–IR spectroscopies of the five samples. The C=O stretch (purple region in Fig. 2.101 (g)) increased from 1,690 cm^{-1} in sample III (green line) to 1,710 cm^{-1} in sample V (black line), consistent with the declining conjugation. The COO–H/O–H stretch at 3,600–2,800 cm^{-1} (yellow region of Fig. 2.101 (g)) was observed in sample III, and continued to increase through the series. It implied that the number of carboxyl and hydroxyl functionalities was increased and trapped water could be present. TGA results indicated that the total weight losses were 20% and 49% for samples I and V, respectively. This was because the number of volatile side-wall functionalities present was increased with the increasing level of $KMnO_4$ or increasing degree of oxidation.

Figure 2.101 (h) shows XRD patterns of the five samples. The graphite (002) spacing increased with increasing degree of oxidation. Samples I–III exhibited 2θ values of about 25.8°, corresponding to a d-spacing of 3.4 Å. Sample IV had two peaks, one at 10.8° and one at 25.4°, corresponding to d-spacings of 8.2 Å and 3.5 Å, respectively. Sample V displayed a predominant peak at 10.6°, corresponding to a d-spacing of 8.3 Å, with minimal signal from the MWCNTs at $2\theta = 25.8°$.

Transition metal nanoparticles of Ni or Co have been employed to longitudinally unzip MWCNTs to produce graphene nanoribbons [226]. The process involved catalytic hydrogenation of carbon, in which the sp^2 hybridized carbon atoms were cut by the metal particles along the nanotubes, resulting in the liberation of the hydrocarbon species. Partially unzipped carbon nanotubes were also observed, which have been predicted to have magnetoresistance (MR) effects [229]. The nanoribbons produced were usually 15–40 nm in width and 100–500 nm in length. In the experiments, MWCNTs and CN_xMWCNTs (nitrogen-doped) were synthesized with an aerosol pyrolysis method [230–232].

MWCNTs or CN_xMWCNTs were dispersed ultrasonically in a solution of 3 wt% $CoCl_2$ in methanol (CH_3OH), with a high power sonication tip for 1 min. About 30 μl of the dispersion was dropped on a clean Si substrate, which was then dried at room temperature. The sample was heated to 500°C for 1 h in a quartz tube at the center of a tubular furnace, so that nucleation of the metal nanoparticles was realized. The catalytic hydrogenation of carbon was conducted at 850°C for 30 min, in the flow of Ar/H_2 with a volume ratio of 90:10 by volume as carrier gas. At the same time, metal nanoparticles were also deposited on the carbon nanotubes by using magnetron sputtering. First, pristine CNTs were dispersed in methanol at a concentration of 100 μg/ml with an ultrasonic bath for 1 h. 50 ml of the dispersion was evaporated at room temperature. MWCNT or CN_xMWCNT films were formed by placing a Si wafer vertically against the wall of the vial through the capillary action during the evaporation of the solvent. A 2 nm film of Ni was then deposited on the substrate by using pulsed DC magnetron sputtering for 5 s with an Ar plasma, at a chamber pressure of 10^{-7} Torr and a frequency of 50 kHz, with an input power of 60 W. The sputtering deposition was carried out at 32 mTorr at room temperature. The experimental conditions for the unzipping with Ni nanoparticles were the same as those of Co nanoparticles, i.e., 850°C for 30 min.

Figure 2.102 shows SEM images of the unzipping process. Metal nanoparticles were nucleated on the surface of the CN_xMWCNTs, as indicated by the encircled areas in Fig. 2.102 (a, b and d). The nanoparticles travelled on the surface of the nanotubes, forming random opening of the outer graphitic layers, through catalytic hydrogenation of the graphene nanosheets [233, 234] (Fig. 2.102 (a)). Larger metal nanoparticles with a size of about 40 nm travelled along the axis of some carbon nanotubes, leading to deeper opening of the MWCNTs and CN_xMWCNTs, as shown in Fig. 2.102 (b–d). As a result, graphene nanoribbons and nanosheets were created, as illustrated in Fig. 2.102 (e). Further SEM and STEM studies indicated that there were partially unzipped CNTs, graphene nanoribbons and graphene nanosheets, as shown in Fig. 2.103. For example, when a MWCNT was nearly completely unzipped, but short segments of carbon nanotubes still remained at both ends, thus forming a graphene nanoribbon with CNT leads, which was about 20 nm wide and 500 nm long, as demonstrated in Fig. 2.103 (a). Figure 2.103 (b) shows that, the nanoribbons derived from the CN_xMWCNTs could have different sizes, with widths of 15–40 nm and lengths of 100–500 nm. Longer nanoribbons were occasionally observed. SEM and bright field (BF) STEM images obtained from the same twisted nanoribbon from CN_xMWCNTs cut with Ni nanoparticles are shown in Fig. 2.103 (c) and (d). There were also large and irregular shaped nanosheets, as shown in Fig. 2.102 (e) and (f).

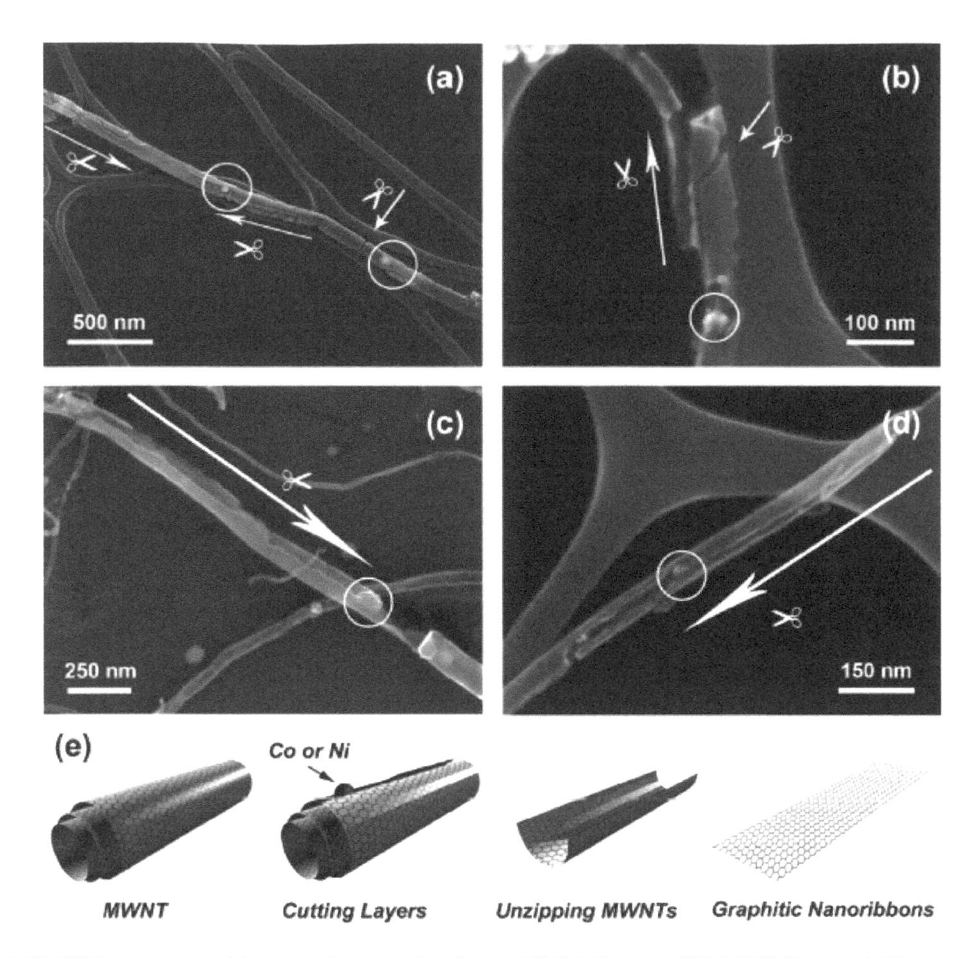

Fig. 2.102. SEM micrographs of the prepared samples. ((a), (b), and (d)) SEM images of CN_xMWCNTs cut with Ni nanoparticles. (c) SEM image of a MWCNT on a Si substrate, with Co nanoparticles. Arrows guide the eye along the nanocutting and circles mark the presence of metal nanoparticles. (e) Schematic diagram representing different stages of cutting MWCNTs along their axes with metal nanoparticles of Ni or Co. The scheme shows the unzipping process of an armchair MWCNT so as to form a zigzag graphitic nanoribbon. Reproduced with permission from [226], Copyright © 2010, American Chemical Society.

According to low-magnification TEM images, it was found that the unzipping effect of the catalyst nanoparticles occurred on the surface of the nanotubes. Figure 2.104 (a) shows a short CN_xMWCNT that was unzipped by the effect of carbon hydrogenation, together with a drawing model of the unzipped CN_xMWCNT. Figure 2.104 (c) shows that low contrast graphene layers were present, as indicated by the arrows. Representative high-resolution TEM image is shown in Fig. 2.104 (d), clearly demonstrating that the nanotubes used in the experiments exhibited high crystallinity. This method is somehow similar to most of the chemical unzipping approaches.

Electrical current combined with nanomanipulation has been used to unzip a portion of the MWCNTs, so as to produce graphene nanoribbons with desired widths and lengths [227]. During the unzipping process, the structure and electrical transport characteristics of the derived nanoribbons could be characterized simultaneously. As expected, high quality graphene nanoribbons had high current-carrying capacity. Figure 2.105 shows schematic diagrams showing the unzipping mechanism of the MWCNTs into nanoribbons. By using a movable electrode, a selected individual MWCNT could be contacted, which was then unzipped on the outer walls induced by the applied electrical current through the contact and nanotube. By controlling the voltage bias, only part of the outer wall of the MWCNT was attached, so that a graphene nanoribbon could be produced which was clinging to the remaining MWCNT inner core, as shown in Fig.

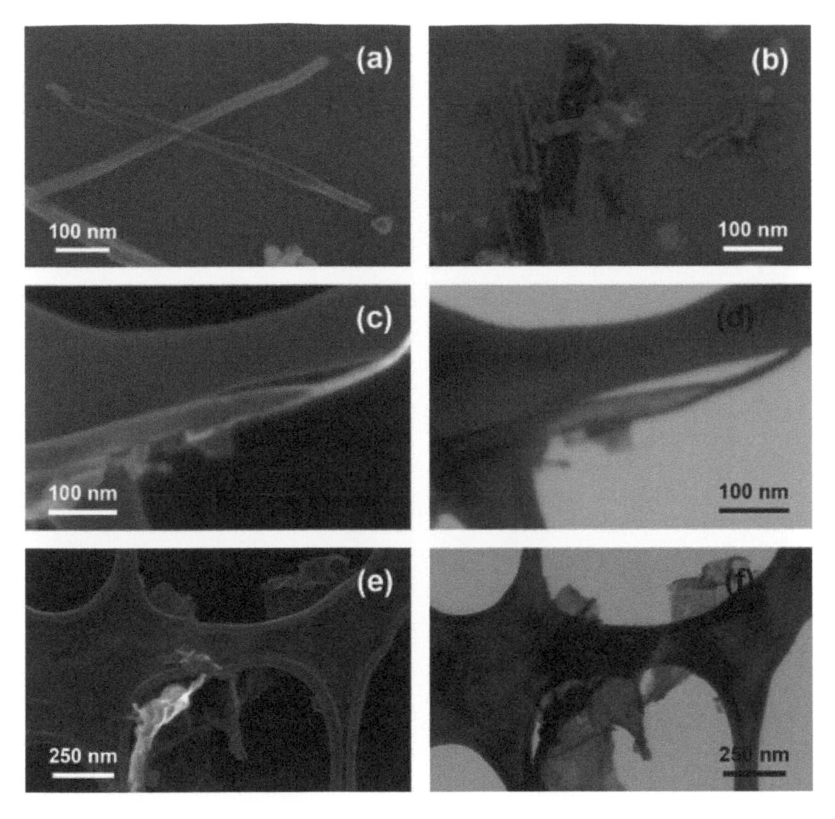

Fig. 2.103. SEM and STEM micrographs of the produced graphitic nanoribbons and nanosheets. (a) SEM image of a MWCNT almost completely unzipped. (b) SEM image of different sized nanoribbons, as a result of the cutting experiments on CN_xMWCNTs with Ni sputtered nanoparticles. SEM (c) and bright field (BF) STEM (d) images of a twisted nanoribbon, from CN_xMWCNT cutting experiments. SEM (e) and BF-STEM (f) images of a representative nanosheet obtained from an ultrasonic dispersed CN_xMWCNT cut sample. Reproduced with permission from [226], Copyright © 2010, American Chemical Society.

2.105 (c). The nanoribbon was then isolated from the MWCNT by sliding out the MWCNT inner core, as shown in Fig. 2.105 (d). Careful manipulation made it possible to have a graphene nanoribbon with each end being electrically and mechanically attached to a conducting electrode, where the remaining portion of the MWCNT was used as one electrode.

A representative SEM image of the graphene nanoribbons is shown in Fig. 2.105 (e). The GNR is fully suspended in vacuum, with each end electrically and mechanically attached to a conducting electrode. The original MWCNT with a diameter of 30 nm was located on the right side of the nanoribbon. The length of the graphene nanoribbon was about 300 nm, while the width was 45 nm. The uniform feature along the ribbon axis implied that about half of the MWCNAT outermost shells were removed during the electrical unzipping process. The unzipping of the MWCNTs was so fast that no intermediate steps could be observed in the TEM experiments. Obviously, this method cannot be used for large-scale production, but is very convenient when it is used to obtain graphene nanoribbon for *in situ* observation and evaluation.

The last example in this part is the production of graphene nanoribbons through the unzipping of MWCNTs by using plasma etching nanotubes that were partly embedded in a polymer film [228]. The nanoribbons obtained in this way exhibited smooth edges and a narrow width distribution in the range of 10–20 nm. The high quality of the graphene nanoribbons was evidenced by Raman spectroscopy and electrical transport properties. Graphene nanoribbons with controlled widths, edge structures, placement and alignment could be fabricated by unzipping CNTs with well-defined structures in arrays for device integration applications.

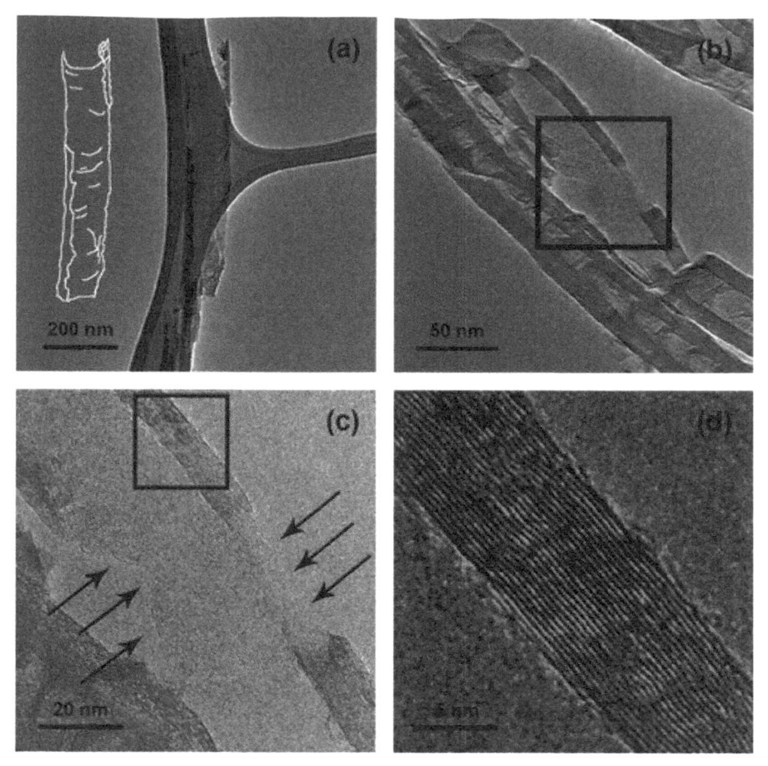

Fig. 2.104. TEM images of the unzipped $CN_xMWCNTs$. (a) and (b) Low magnification images of the $CN_xMWCNTs$ after the cutting process. (c) Higher magnification TEM image that corresponds to the box in (b), where few-layer graphene is bonding two parts of a CN_xMWCNT, as the arrows point out. (d) High-resolution TEM image corresponding to the box in (c), which exhibits the (002) G planes present in carbon nanotubes. Reproduced with permission from [226], Copyright © 2010, American Chemical Society.

In the typical experiment, 1 mg of MWCNTs was dispersed in 10 ml 1% Tween 20 aqueous solution with sonication for 5 min, followed by centrifugation at 16,400 g for 10 min to remove aggregates. The MWCNT suspension was deposited onto a Si substrate that was pre-treated with 3-aminopropyltriethoxysilane (APTES, 12 ml in 20 ml H_2O), rinsed with water and then blow-dried. The sample was then calcined at 350°C for 10 min to remove the Tween 20. A PMMA solution was spin-coated on the MWCNTs on the substrate at 3,000 rpm for 1 min, which was then heated at 170°C for 2 h on a hot plate. The PMMA–MWCNT film was peeled off in 1 M KOH solution at 80°C [235]. After that, the film was rinsed with water and placed closely onto a Si substrate, through heating at 80°C for 10 min. The PMMA–MWCNT film was etched by using 10 W Ar plasma at 40 mTorr. The etched film was lifted with water and then adhered to an APTES-treated 500-nm SiO_2/Si substrate with a deposited Pt/W marker array. After the PMMA was removed with acetone vapor, the obtained sample was calcined at 300°C for 10 min to remove the residual PMMA. The process is shown in Fig. 2.106 [228].

AFM image and the G-band image of a 0.9-nm-thick graphene nanoribbon are shown in Fig. 2.107 (a) and (b), respectively. The 2D peak could be well-fitted by a sharp and symmetric Lorentzian peak, as shown in the inset of Fig. 2.107 (c), indicating that the graphene nanoribbon was single-layered. AFE image, G-band image and 2D peak of a 1.3-nm-thick nanoribbon are shown in Fig. 2.107 (d–f). The 2D peak could be fitted by four Lorentzians (the inset of Fig. 2.107 (f)), suggesting the bilayer characteristic of an AB-stacked graphene. There were also bilayer nanoribbons that exhibited 2D peaks with different line shapes from that of the AB-stacked graphene, due to the different stacking structures between the layers in the starting MWCNTs. Typical trilayer nanoribbons displayed an ABA-stacked graphene structure. The 2D band was broadened and up shifted with increasing thickness. The average I_D/I_G values were 0.38, 0.30 and 0.28 for single-, bi- and trilayer nanoribbons, all with 10–20-nm widths, respectively.

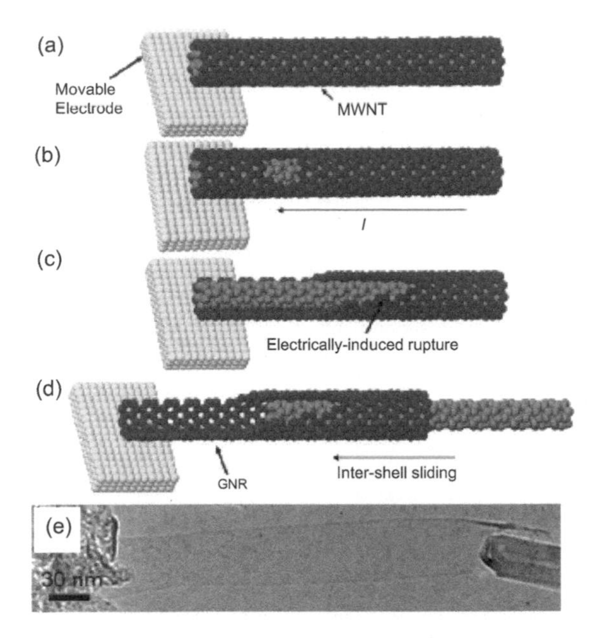

Fig. 2.105. Schematic diagrams of the unzipping process from MWCNTs to graphene nanoribbon, by using the electrical rupture method, where a double-wall carbon nanotube (DWCNT) was used for simplicity: (a) MWCNT before the partial wall rupture, (b) electrical current induced rupture of the outer wall of the nanotube, (c) partial outer-wall rupture of the nanotube leading to a precursor graphene nanoribbon under the MWCNT inner core and (d) inter-shell sliding between the nanoribbon and the inner core resulting in a suspended electrically contacted nanoribbon. (e) TEM image of a nanoribbon derived from a MWCNT through the electrical rupture and unzipping method. The original MWCNT is shown on the right side, which served as a mechanical support and electrical contact for the nanoribbon. Reproduced with permission from [227], Copyright © 2010, American Chemical Society.

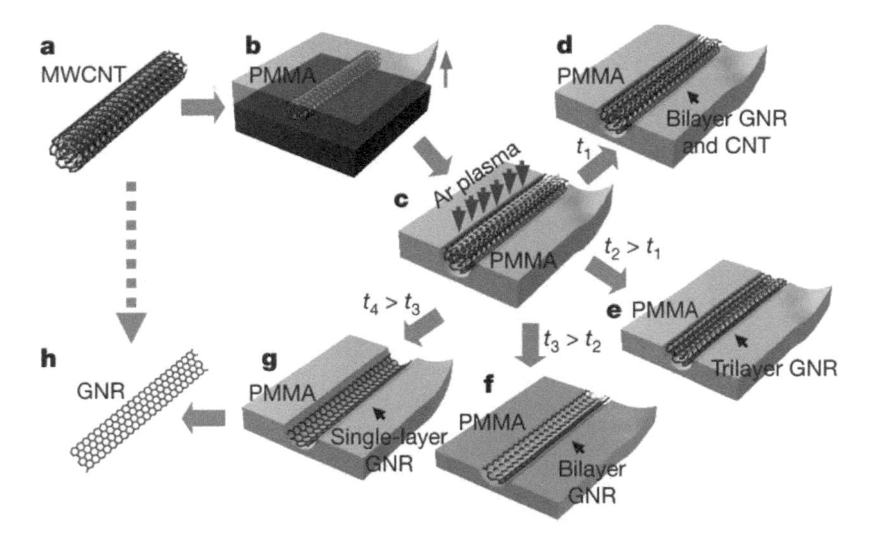

Fig. 2.106. Fabrication of graphene nanoribbons from MWCNTs. (a) A pristine MWCNT used as the starting raw material. (b) The MWCNT deposited on a Si substrate and then coated with a PMMA film. (c) The PMMA–MWCNT film peeled from the Si substrate, turned over and then exposed to an Ar plasma. (d–g) Several possible products generated after etching for different times, including nanoribbons with CNT cores produced after etching for a short time t_1 (d), tri-, bi- and single-layer nanoribbons produced after etching for times t_2, t_3 and t_4, respectively ($t_4 > t_3 > t_2 > t_1$, e–g). (h) Removal of the PMMA to release the graphene nanoribbons. Reproduced with permission from [228], Copyright © 2009, Macmillan Publishers Limited.

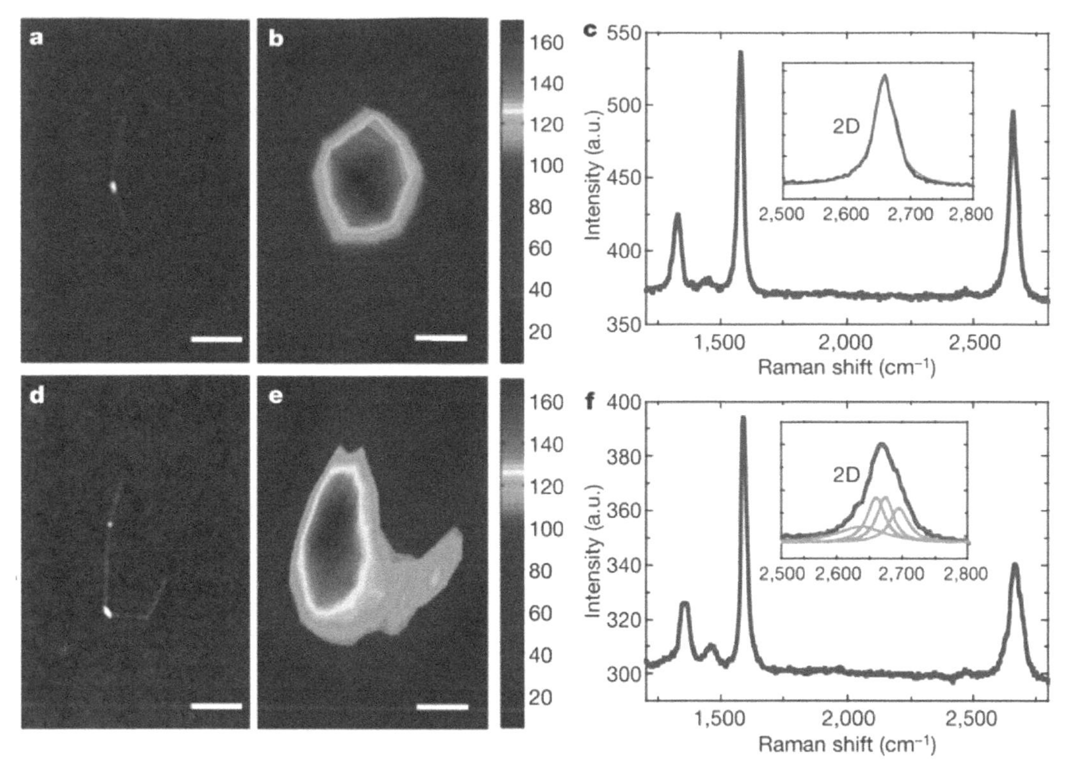

Fig. 2.107. AFM images, Raman imaging and spectra of the graphene nanoribbons. (a–c) An AFM image (a), G-band Raman image (b) and Raman spectrum (c) of a single-layer nanoribbon. Inset of (c) is a 2D-band spectrum (blue) of the nanoribbon and a single-Lorentzian fit (red). (d–f) An AFM image (d), a G-band Raman image (e) and the Raman spectrum (f) of a bilayer nanoribbon. Inset of (f) is 2D band spectrum (blue) of the nanoribbon and the best-fit curve (red), which is a combination of four Lorentzians (green). The peaks at about 1,440 cm^{-1} in (c) and (f) are attributed to a trace of PMMA left on the SiO_2. Scale bars are 200 nm. Reproduced with permission from [228], Copyright © 2009, Macmillan Publishers Limited.

Color image of this figure appears in the color plate section at the end of the book.

In this approach, embedding the MWCNTs in the PMMA film as an etching mask has been the essential step to protect parts of the shells of the MWCNTs from being etched by the plasma. Due to its relatively low viscosity and promising wetting capability, PMMA has been acknowledged to be a suitable conformal coating on the MWCNTs. The intensity of the plasma source also played a critical role in ensuring the longitudinal unzipping of the CNTs. Because chemical etching of MWCNTs by O_2 plasma was too rapid along the nanotube circumference, it cannot be used to produce graphene nanoribbons by using this method. In contrast, the bombardment effect of Ar plasma resulted in anisotropic physical etching to remove atoms only at the unprotected sites along the longitudinal direction of the CNTs [236]. Both power and time duration of the etching were important. Prolonged etching duration could increase the yield of single- and few-layer nanoribbons, but caused breaking of the nanoribbons. In addition, a narrow distribution in diameter and a fewer number of layers of the MWCNTs would allow to produce uniform nanoribbons. The productivity of this method lies between the chemical routes and the electric rupture approach, while its advantage is the ability to more precisely control the dimension of the graphene nanoribbons.

2.2.3.2. Other Methods for Graphene Ribbons

Besides the methods discussed above, several other strategies have been emerged, some of which will be described in this part. For example, a two-step method has been reported to produce few-layer nanoribbons by unzipping mildly gas-phase oxidized MWCNTs, by using mechanical sonication in an organic solvent

[237]. The nanoribbons exhibited very high quality, with smooth edges, low ratios of disorder to graphitic Raman bands and high electrical conductance and mobility.

The unzipping steps are shown in Fig. 2.108 (a) [237]. Raw soot materials containing pristine MWCNTs synthesized by using arc discharge were first calcined in air at 500°C, which is a mild condition that has been used to remove impurities. It can also etch or oxidize MWCNTs at the defect sites and ends, while the pristine sidewalls of the nanotubes are not affected [238]. The calcined nanotubes were then dispersed in a 1,2-dichloroethane (DCE) organic solution of poly(*m*-phenylenevinylene-co-2,5-dioctoxy-*p*-phenylenevinylene) (PmPV) by applying sonication. During this process, the calcined nanotubes were effectively unzipped into graphene nanoribbons. The remaining nanotubes and graphitic carbon nanoparticles were removed through ultracentrifugation. The yield of the nanoribbons was about 60%, corresponding to a 2% of the starting raw nanotube soot material, which could be further improved by optimizing the processing parameters.

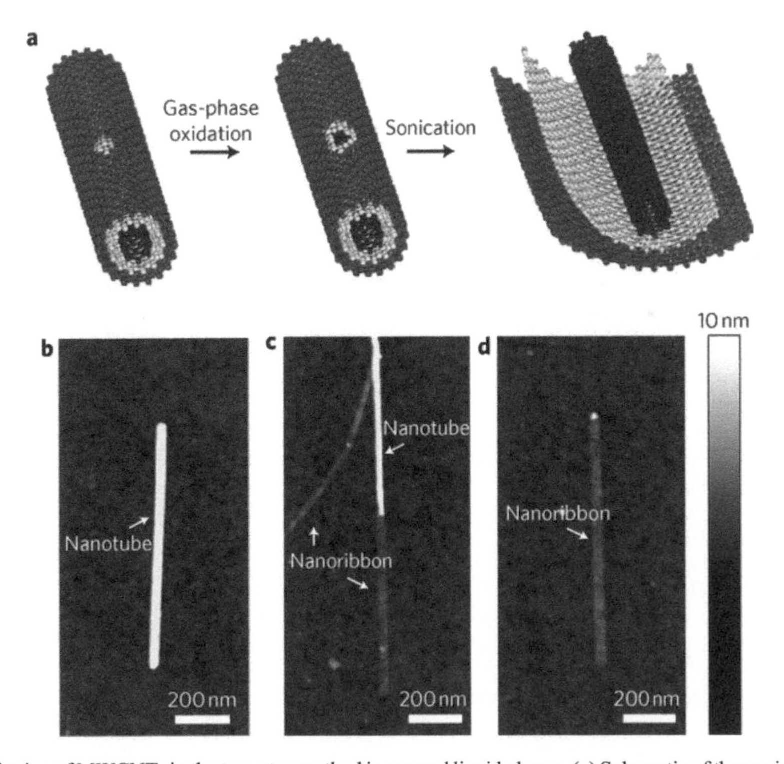

Fig. 2.108. Unzipping of MWCNTs in the two-step method in gas and liquid phases. (a) Schematic of the unzipping processes. In the mild gas-phase oxidation step, oxygen reacted with the pre-existing defects on the nanotubes to form etch pits on the sidewalls. In the solution-phase sonication step, sonochemistry and hot gas bubbles enlarged the pits and unzipped the tubes. (b–d) AFM images of pristine, partially and fully unzipped nanotubes, respectively. The heights of the nanoribbons in (c) and (d) are 1.4 and 1.6 nm, respectively, much lower than that of the pristine nanotube shown in (b) with a height of 9 nm. Reproduced with permission from [237], Copyright © 2009, Macmillan Publishers Limited.

AFM images of representative graphene nanoribbons are shown in Fig. 2.108 (b–d). Both the MWCNTs and the unzipped products were deposited on SiO_2/silicon substrates. The graphene nanoribbons had heights of 1–2.5 nm, which were much lower than those of the MWCNTs. It was found that most of the nanoribbons were single-, bi- or trilayered, with widths in the range of 10–30 nm.

A precise unzipping of CNTs into regular graphene nanoribbons has been realized by using flattened nanotubes (F-CNTs) as the starting material, which were cut along the folded edges, by using acid, because of the high energy sites that were preferentially attacked [239]. From flattened double-wall CNTs (F-DWCNTs) using mixed acids of H_2SO_4 and $KMnO_4$, all the graphene nanoribbons were of bilayer structure, with a narrow uniform width of 8–12 nm and straight edges.

The flattened DWCNTs were synthesized by using a CVD gas-flow reaction with mixed ethanol and acetone as the carbon source and ferrocene as catalyst in a hydrogen flow at 1150°C. In this process, large diameter DWCNTs were synthesized through the transient growth over the iron particles in the gas flow under suitable conditions [240]. The DWCNTs could be flattened, bundled or interconnected into a continuous assembly in the gas flow, which were collected as yarns, films or cotton-like products. A collection of yarn-like F-DWCNTs was used for unzipping.

In a typical experiment, the F-DWCNTs (1.20 mg) were dispersed in 4 ml concentrated H_2SO_4 (98%) in a flask, which was kept at 60°C in a water bath and stirred at 200 rpm for 30 min for pre-oxidation. After that, 4.8 mg $KMnO_4$ powder was added and stirred at 60°C for 90 min for oxidation. After the reaction was finished, 25 ml distillated water was added and stirred for 20 min for hydrolysis. 3.5 ml H_2O_2 was then added, before the reaction mixture was centrifuged at 10000 rpm for 5 min. The oxidized product was purified with a 3.0 wt% aqueous solution of HCl, distilled water and ethanol through centrifugation. The oxidized product was heated at 60°C for 4 h to obtain black powder. Different amounts of $KMnO_4$ were used to test its effect on the oxidation reaction, with weight ratios of $M_{KMnO4}/M_{F\text{-}DWCNTs}$ = 2, 4, 5, 10, 15, 20 and 25, while the effect temperature (20, 40, 60 and 80 C) was also examined, with $M_{KMnO4}/M_{F\text{-}DWCNTs}$ = 5.

TEM image in Fig. 2.109 (a) shows that the F-DWCNTs were regular tapes with straight folded edges at the two sides [239]. The F-DWCNTs had uniform widths of 8–12 nm. They were stacked in bundles, consisting of 10–15 FDWCNTs. The collapsed portions of the F-DWCNTs at the two sides were heavily bent into small arcs with diameters of 2 nm, equivalent to small diameter SWCNTs, as shown in Fig. 2.109 (b–d). Such small arcs at the folded edge were in a highly bent state, which were vulnerable to physical and chemical attacks. The unzipping process was similar to the expanding of graphite to graphene oxide nanosheets and unzipping MWCNTs to GNRs, as discussed before. It was found that $M_{KMnO4}/M_{F\text{-}DWCNTs}$ = 4 at a temperature of 60°C was the optimized condition.

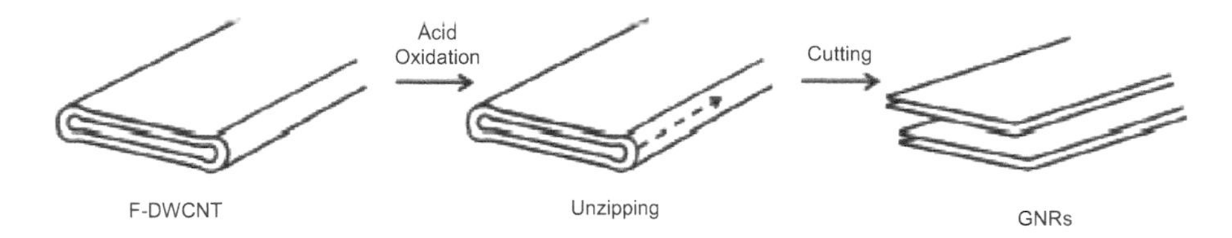

Fig. 2.109. Schematic of directional unzipping of flattened carbon nanotubes. Reproduced with permission from [239], Copyright © 2012, The Royal Society of Chemistry.

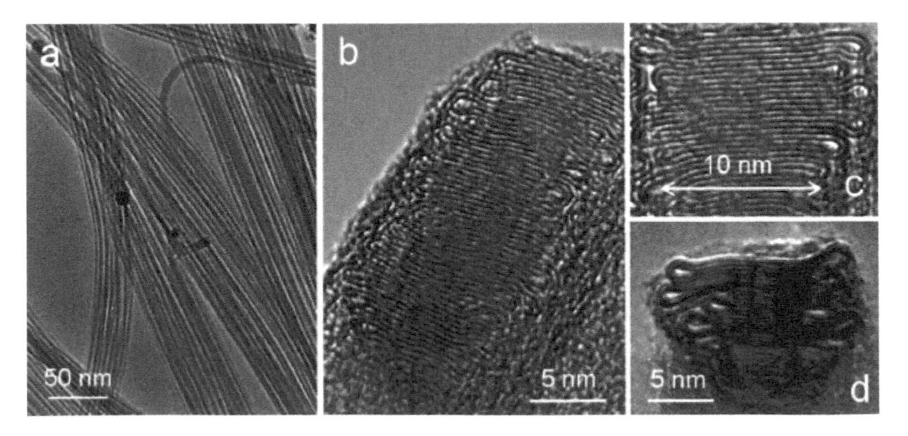

Fig. 2.110. TEM images of the F-DWCNTs used for unzipping: (a) typical morphology, (b and c) typical ends of F-DWCNT bundles showing the stacked F-DWCNTs and (d) a bundle of stacked F-DWCNTs with displacement. Reproduced with permission from [239], Copyright © 2012, The Royal Society of Chemistry.

Representative TEM image of the unzipped product is shown in Fig. 2.111 (a), confirming the formation of thin nanoribbons [239]. The graphene nanoribbons appeared as regular tapes with a double-layer structure and straight edges, as clearly illustrated in Fig. 2.111 (b). The typical width of thc nanoribbons was about 10 nm, which was half of the circumference of the F-DWCNTs, while the length was > 200 nm. Figure 2.111 (c) shows two nanoribbons, which were overlapped and slightly displaced perpendicularly to the length direction. They were most likely from one completely unzipped F-DWCNT. Figure 2.111 (d) shows a nanoribbon at an intermediate stage of unzipping, which clearly indicated that the unzipping was along the folded edge of the F-DWCNT. The unzipped nanoribbons were partially exfoliated from the bundle, with the cutting front stopped at the bundle surface. All the results suggested that the unzipping of the F-DWCNT proceeded unidirectionally from one side to the other. It is believed that this method can be applied to other CNTs.

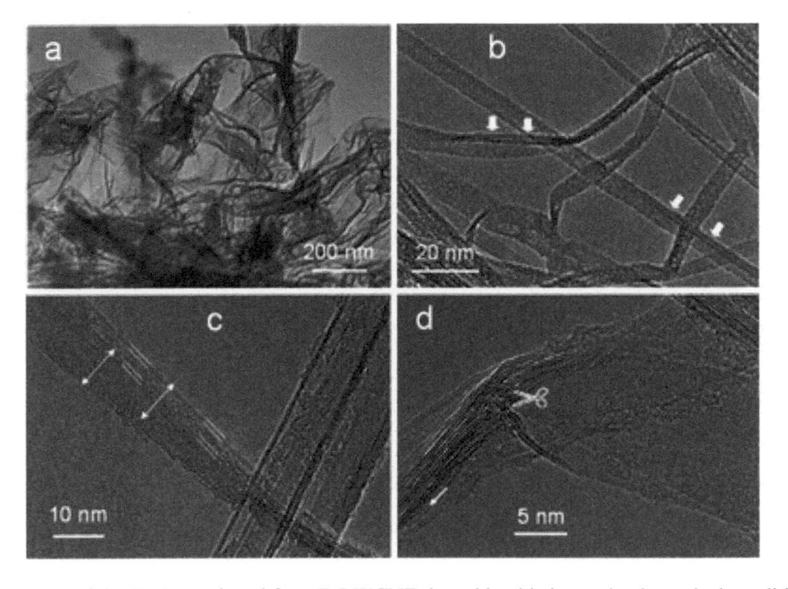

Fig. 2.111. TEM images of the GNRs unzipped from F-DWCNTs by acid oxidation under the typical conditions of $M_{KMnO4}/M_{F\text{-}DWCNTs} = 4$ at 60°C: (a) low magnification image showing typical morphology, (b) graphene nanoribbons formed by unzipping the F-DWCNTs, (c) a completely unzipped nanoribbon and (d) a nanoribbon at an intermediate stage of unzipping, with the scissors to indicate the cutting from the edge of the F-DWCNT. Reproduced with permission from [239], Copyright © 2012, The Royal Society of Chemistry.

A microwave method to derive graphene and a few-graphene nanosheets from turbostratic carbon fibers has been reported [241]. The microwave radiation as an additional energy source was in conjunction with a mild oxidizing agent, hydrogen peroxide. This method addressed the problems associated with the excessive oxidation and considerably shortened the overall processing time. Commercial polyacrylonitrile (PAN)-based carbon fibers and carbon nanofibers were used for the exfoliation experiments. The microwave treatment was a two-step process, i.e., (i) initial step in the presence of hydrogen peroxide and (ii) subsequent dry expansion of the exfoliated CF leading to expanded carbon fibers. The carbon fibers were finely chopped into segments with lengths of 5–10 μm, as shown in Fig. 2.112 (a), which were immersed in hydrogen peroxide and subjected to microwave radiation for 30 min. During this process, the outer sheath of the carbon fibers was oxidized, so that it could be removed and partially expanded into graphenic layers, as illustrated in Fig. 2.112 (b). Under microwave irradiation, the carbon fibers were rapidly oxidized and thus became more and more porous [242]. Figure 2.112 (c) shows XRD pattern of the expanded carbon fibers, with a sharp peak at 12° and the graphitic peak at 26°, due to the expansion of the graphitic layers. The sharp peak implied the presence of stacked graphene layers together with locally folded graphene

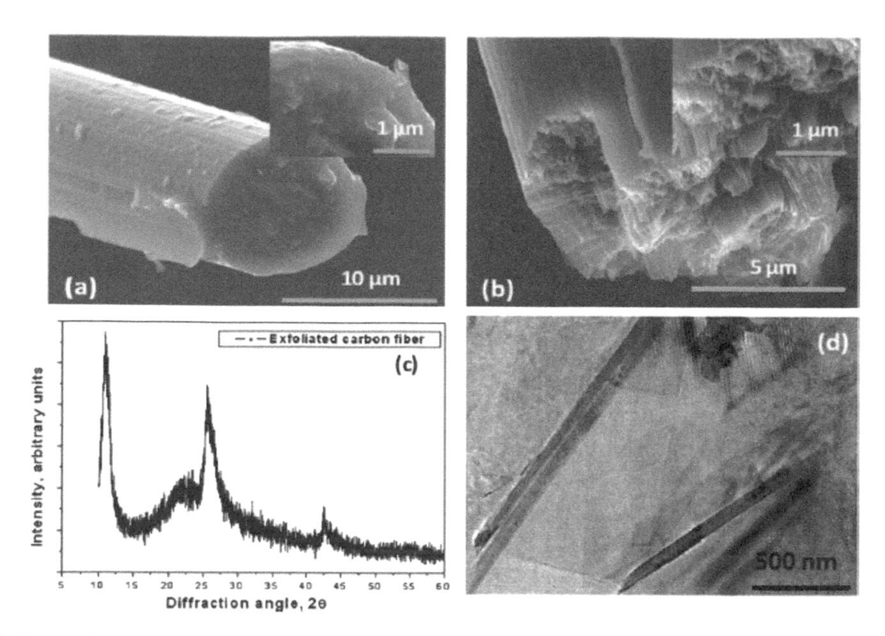

Fig. 2.112. SEM micrograph of carbon fiber (a), peroxide-treated carbon fiber (b), XRD of exfoliated carbon fiber (c), and TEM of exfoliated carbon fiber (d). Insert in (a) and (b) shows high-resolution micrographs. Reproduced with permission from [241], Copyright © 2012, The Royal Society of Chemistry.

stacks, as confirmed by the TEM image in Fig. 2.112 (d). After this exfoliation process, the dried product remaining in the container could be reused for microwave radiation and exfoliation. It means that the carbon fibers can be used repeatedly.

Because the graphene layers of the carbon fibers were expanded, weakly adhering graphene layers were formed accordingly. Microwave radiation caused rapid boiling of the peroxide trapped within the graphene stacks, which also partially removed the residual peroxide, thus resulting in expansion of the graphenic structures. Figure 2.113 (a) shows SEM image of the microwave expanded carbon fibers, in which curled-up structures, together with some free graphene nanosheets, were observed, especially at the tip of the sample. Usually, sharp edges and facets are less stable, as compared with the structures having smooth curvature, while curling or formation of spheres can be attributed to the requirement of minimizing the surface area [243]. The surface stress-induced strains resulted in the formation of ellipsoidal structures, with weakly adhering graphene platelets, as shown in Fig. 2.113 (b). The graphene nanosheets were then detached under ultrasonication. Representative SEM and TEM images of the supernatant are shown in Fig. 2.113 (c) and (d), respectively, demonstrating typical features of graphene nanosheets. The as-derived graphene nanosheets could be further reduced with 5 wt% hydrazine solution to remove any unreacted oxide groups. One of the advantages of this finding is that carbon fibers are much cheaper than CNTs. Therefore, it could be a cost-effective method towards large-scale production of graphene nanosheets.

Ab initio calculations have revealed that, under a uniaxial external tensile strain, O atoms adsorbed on graphene could form parallel epoxy chains, so that the graphene nanosheet would be cut into nanoribbons by oxygen attack, whereas quantum dots would be obtained if the graphene layer was not strained, as illustrated in Fig. 2.114 [244].

It has been recognized that, due to the high symmetry of the honeycomb lattice of graphene, the formation of epoxy chains on graphene nanosheet is random, so that oxygen cutting usually leads to the formation of graphene quantum dots (QDs), i.e., following the route of Fig. 2.114 (a→b→c→d) [32, 245–248]. By applying a uniaxial tensile strain, that symmetry could be broken, resulting in an orientation-selective cutting of graphene, i.e., following the route of Fig. 2.114 (a→e→f→g), so that graphene nanoribbons could be obtained, instead of graphene QDs [244]. The ab initio calculations indicated that the epoxy chains tended to be aligned along a zigzag direction that was nearly perpendicular to the direction

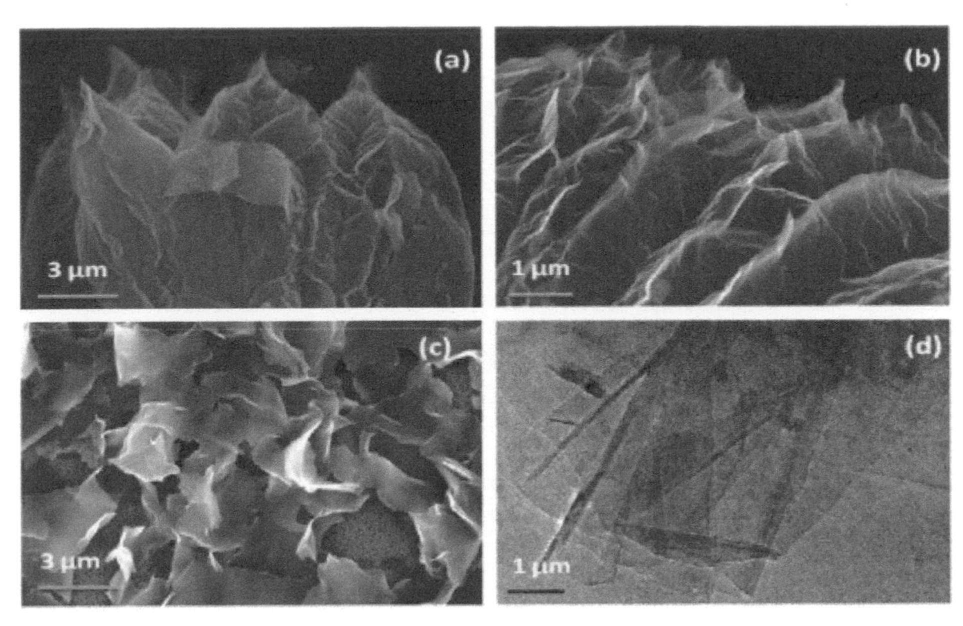

Fig. 2.113. SEM micrographs of microwave expanded carbon fibers (a) and (b); SEM (c) and TEM (d) micrographs of graphene sheets after centrifugation. Reproduced with permission from [241], Copyright © 2012, The Royal Society of Chemistry.

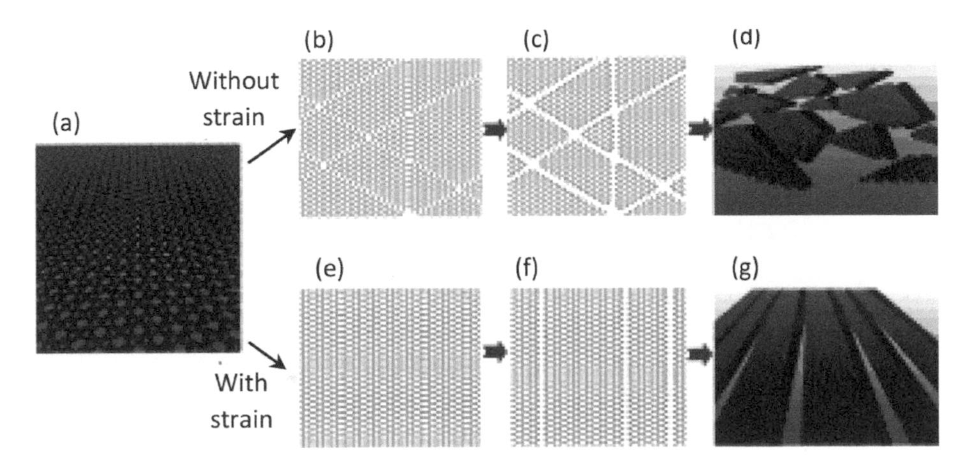

Fig. 2.114. (a) Perfect graphene nanosheet that can be cut in two ways by oxygen attack: (b) → (c) → (d) random cutting of graphene into quantum dots and (ii) (e) → (f) → (g) orientation-selective cutting of strained graphene into graphene nanoribbons. Reproduced with permission from [244], Copyright © 2012, John Wiley & Sons.

of the strain. Furthermore, the external strain could decrease both the reaction barrier and the enthalpy of reaction, as the graphene was cut along that direction. Therefore, graphene nanoribbons with large and controllable aspect ratios could be produced readily by using this method.

Because the strain in a freestanding graphene was completely released on cutting, a potential experiment to apply external tensile strain to cut graphene into nanoribbons was proposed, as shown in Fig. 2.115 [244]. It consisted of four steps: (i) transferring graphene onto a flexible polymer film with suitable binding between graphene and polymer film, (ii) bending or stretching the polymer film to produce strain on the graphene, (iii) adsorption of O atoms on the graphene to form parallel epoxy chains and (iv) further oxidation to unzip the graphene layer into nanoribbons with desired dimension. The first two steps have been well-developed [249–253], while nanosized graphene has been produced by cutting preoxidized graphene nanosheet with a scanning probe microscopic [245]. In addition, both stretching of graphene on

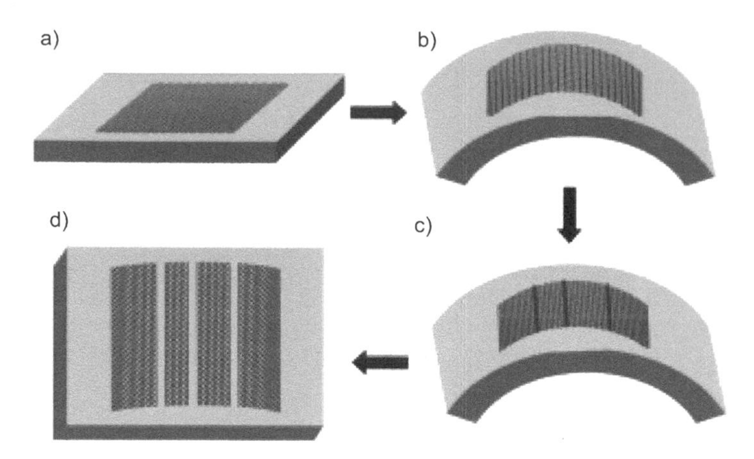

Fig. 2.115. (a) Graphene monolayer transferred to a flexible polymer substrate. (b) The polymer substrate is bent to create a strain on the attached graphene sheet. (c) Annealing of adsorbed O atoms to form parallel epoxy chains on the graphene. (d) Unzipping of the graphene layer into nanoribbons through further oxidation. Reproduced with permission from [244], Copyright © 2012, John Wiley & Sons.

polymer film [249] and oxidation cutting of graphene [245] have been demonstrate in single- and few-layer graphene nanosheets. Therefore, it is expected that experimental results will be presented to confirm this theoretical study in the near future.

2.2.4. Graphene through Chemical Vapor Deposition (CVD)

The CVD method has been used extensively for the growth of carbon nanotubes in high yields in the presence of catalytic amounts of nanoparticles [254–256]. The chemical nature of the metal catalysts and carbon sources plays a key role in the growth of carbon nanotubes. This technique has also proven to be efficient for producing large-area graphene layers on transition-metal surfaces [257], including platinum [258], ruthenium [12, 259–261], nickel [262], and copper [263, 264]. CVD-grown graphene, on copper, can be produced in large areas with very few structural defects [265, 266]. Significant progress on CVD graphene synthesis has been achieved in the last two years. Monolayer graphene was formed on 30-inche copper [266]. However, CVD growth of graphene nanosheets is still limited by the requirement of substrates, which needs to be further developed so as to synthesize graphene on a large-scale from various carbon sources.

One example is the CVD synthesis of few-layer graphene ribbons by using ZnS substrates as templates [267]. The few-layer graphene nanoribbons (FLGNRs), with a thickness of about 3.4 nm, widths of 0.5–5 μm and lengths of several microns, were obtained through the decomposition of CH_4 at 750ºC on the ZnS templates that were later dissolved with acid treatment. In a typical experiment, ZnS ribbons were synthesized on a Si substrate by using a physical vapor deposition (PVD) method, with Au as the catalyst. The Si substrate with the ZnS ribbons could be used to grow graphene nanoribbons at 750°C and 100 sccm CH_4.

Figure 2.116 (a) shows a schematic diagram, demonstrating the formation mechanism of the graphene nanosheets, with the ZnS ribbons as the template. SEM and TEM images of the samples are shown in Fig. 2.116 (b) and (c), respectively. The ZnS ribbons were single crystals, with a uniform morphology, together with an average width of 0.5–5 μm and a length of tens to hundred micrometers. Over the ZnS ribbons, uniform carbon layer was deposited. Figure 2.116 (d) shows that almost all the ZnS ribbons were entirely coated with a uniform nanosheet, which was confirmed by the TEM image, as shown in Fig. 2.116 (c). The graphene nanoribbons followed the shape of the ZnS ribbons, as confirmed by the SEM and TEM images shown in Fig. 2.116 (f) and (g). Therefore, the morphologies of the graphene ribbons can be controlled by controlling the morphologies of the ZnS ribbons. In addition, Fig. 2.116 (h) shows a photograph of the samples, implying that this method has great potential for large-scale production, by simply using larger substrates.

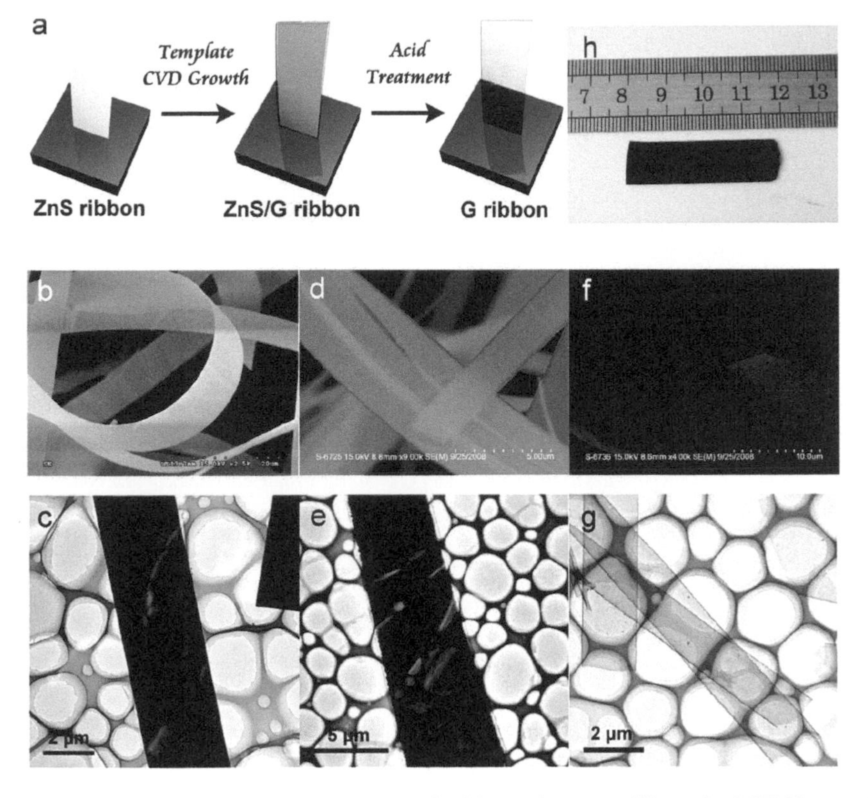

Fig. 2.116. (a) Schematic diagram of the template CVD growth of the graphene nanoribbons. (b–g) SEM images (b, d, f) and TEM images (c, e, g) of the ZnS ribbons (b, c), the ZnS/G ribbons (d, e) and the graphene nanoribbons (f, g). (h) Photograph of the graphene nanoribbon sample on a Si substrate. Reproduced with permission from [267], Copyright © 2009, American Chemical Society.

2.3. Summary

Various strategies have been developed to synthesize graphene nanosheets and their related materials. It is believed that more and more strategies will appear in the open literature from the research point of view. Therefore, a huge database has been formed by the references in the open literature and the content of the database will be increased even at an unexpected rate. Among the synthetic methods, wet-chemical solution route seems to be the only way used for large-scale production. In addition, there are unlimited ways to reduce the products derived from the wet-chemical solution synthetic routes, i.e., graphene oxides (GO), but no reference has been available on which method or methods should be used for practical large-scale production. In other words, standards and regulations should be established for this issue. Also, as stated in the beginning of this chapter, Synthetic route, which can be used to produce graphene with controlled thickness (number of atomic carbon layers), lateral dimensions, high purity, desired edge chemistries, and so on, should be either identified from the currently available methods or the newly developed ones in the near future.

References

[1] Lin YM, Dimitrakopoulos C, Jenkins KA, Farmer DB, Chiu HY, Grill A et al . 100-GHz transistors from wafer-scale epitaxial graphene. Science. 2010; 327: 662.

[2] Englert JM, Hirsch A, Feng XL, Muellen K. Chemical methods for the generation of graphenes and graphene nanoribbons. Angewandte Chemie-International Edition. 2011; 50: A17–A24.

[3] Novoselov KS, Geim AK, Morozov SV, Jiang D, Zhang Y, Dubonos SV et al . Electric field effect in atomically thin carbon films. Science. 2004; 306: 666–9.
[4] Pang SP, Englert JM, Tsao HN, Hernandez Y, Hirsch A, Feng XL et al. Extrinsic corrugation-assisted mechanical exfoliation of monolayer graphene. Advanced Materials. 2010; 22: 5374–7.
[5] Brodie BC. Sur le poids atomique du graphite. Annales de Chimie et de Physique. 1860; 59.
[6] Hummers WS, Offeman RE. Preparation of graphitic oxide. Journal of the American Chemical Society. 1958; 80: 1339.
[7] Li D, Mueller MB, Gilje S, Kaner RB, Wallace GG. Processable aqueous dispersions of graphene nanosheets. Nature Nanotechnology. 2008; 3: 101–5.
[8] Tung VC, Allen MJ, Yang Y, Kaner RB. High-throughput solution processing of large-scale graphene. Nature Nanotechnology. 2009; 4: 25–9.
[9] He P, Sun J, Tian SY, Yang SW, Ding SJ, Ding GQ et al. Processable aqueous dispersions of graphene stabilized by graphene quantum dots. Chemistry of Materials. 2015; 27: 218–26.
[10] Li XL, Wang XR, Zhang L, Lee SW, Dai HJ. Chemically derived, ultrasmooth graphene nanoribbon semiconductors. Science. 2008; 319: 1229–32.
[11] Sutter E, Acharya DP, Sadowski JT, Sutter P. Scanning tunneling microscopy on epitaxial bilayer graphene on ruthenium (0001). Applied Physics Letters. 2009; 94: 133101.
[12] Sutter PW, Flege JI, Sutter EA. Epitaxial graphene on ruthenium. Nature Materials. 2008; 7: 406–11.
[13] Berger C, Song ZM, Li XB, Wu XS, Brown N, Naud C et al. Electronic confinement and coherence in patterned epitaxial graphene. Science. 2006; 312: 1191–6.
[14] Berger C, Song ZM, Li TB, Li XB, Ogbazghi AY, Feng R et al. Ultrathin epitaxial graphite: 2D electron gas properties and a route toward graphene-based nanoelectronics. Journal of Physical Chemistry B. 2004; 108: 19912–6.
[15] Doessel L, Gherghel L, Feng XL, Muellen K. Graphene nanoribbons by chemists: nanometer-sized, soluble, and defect-free. Angewandte Chemie-International Edition. 2011; 50: 2540–3.
[16] Fogel Y, Zhi LJ, Rouhanipour A, Andrienko D, Raeder HJ, Muellen K. Graphitic nanoribbons with dibenzo[e,l] pyrene repeat units: Synthesis and self-assembly. Macromolecules. 2009; 42: 6878–84.
[17] Palma CA, Samori P. Blueprinting macromolecular electronics. Nature Chemistry. 2011; 3: 431–6.
[18] Treier M, Pignedoli CA, Laino T, Rieger R, Muellen K, Passerone D et al. Surface-assisted cyclodehydrogenation provides a synthetic route towards easily processable and chemically tailored nanographenes. Nature Chemistry. 2011; 3: 61 7.
[19] Fuhrer MS. Graphene: Ribbons piece-by-piece. Nature Materials. 2010; 9: 611–2.
[20] Cai JM, Ruffieux P, Jaafar R, Bieri M, Braun T, Blankenburg S et al. Atomically precise bottom-up fabrication of graphene nanoribbons. Nature. 2010; 466: 470–3.
[21] Hirata M, Gotou T, Horiuchi S, Fujiwara M, Ohba M. Thin-film particles of graphite oxide 1: High-yield synthesis and flexibility of the particles. Carbon. 2004; 42: 2929–37.
[22] Hirata M, Gotou T, Ohba M. Thin-film particles of graphite oxide. 2: Preliminary studies for internal micro fabrication of single particle and carbonaceous electronic circuits. Carbon. 2005; 43: 503–10.
[23] Lerf A, He HY, Forster M, Klinowski J. Structure of graphite oxide revisited. Journal of Physical Chemistry B. 1998; 102: 4477–82.
[24] Hontorialucas C, Lopezpeinado AJ, Lopezgonzalez JDD, Rojascervantes ML, Martinaranda RM. Study of oxygen-containing groups in a series of graphite oxide-physical and chemical characterization. Carbon. 1995; 33: 1585–92.
[25] He HY, Riedl T, Lerf A, Klinowski J. Solid-state NMR studies of the structure of graphite oxide. Journal of Physical Chemistry. 1996; 100: 19954–8.
[26] He HY, Klinowski J, Forster M, Lerf A. A new structural model for graphite oxide. Chemical Physics Letters. 1998; 287: 53–6.
[27] Titelman GI, Gelman V, Bron S, Khalfin RL, Cohen Y, Bianco-Peled H. Characteristics and microstructure of aqueous colloidal dispersions of graphite oxide. Carbon. 2005; 43: 641–9.
[28] Szabo T, Tombacz E, Illes E, Dekany I. Enhanced acidity and pH-dependent surface charge characterization of successively oxidized graphite oxides. Carbon. 2006; 44: 537–45.
[29] Goh MSH, Bonanni A, Ambrosi A, Sofer Z, Pumera M. Chemically-modified graphenes for oxidation of DNA bases: analytical parameters. Analyst. 2011; 136: 4738–44.
[30] Ambrosi A, Bonanni A, Sofer Z, Cross JS, Pumera M. Electrochemistry at chemically modified graphenes. Chemistry-A European Journal. 2011; 17: 10763–70.
[31] Staudenmaier L. Verfahren zur darstellung der graphitsaure. Berichte der Deutschen Chemischen Gesellschaft 1898; 31: 1481.
[32] McAllister MJ, Li JL, Adamson DH, Schniepp HC, Abdala AA, Liu J et al. Single sheet functionalized graphene by oxidation and thermal expansion of graphite. Chemistry of Materials. 2007; 19: 4396–404.
[33] Paredes JI, Villar-Rodil S, Martinez-Alonso A, Tascon JMD. Graphene oxide dispersions in organic solvents. Langmuir. 2008; 24: 10560–4.
[34] Shin HJ, Kim KK, Benayad A, Yoon SM, Park HK, Jung IS et al. Efficient reduction of graphite oxide by sodium borohydride and its effect on electrical conductance. Advanced Functional Materials. 2009; 19: 1987–92.

[35] Zhou TN, Chen FR, Liu K, Deng H, Zhang Q, Feng JW et al. A simple and efficient method to prepare graphene by reduction of graphite oxide with sodium hydrosulfite. Nanotechnology. 2011; 22: 045704.
[36] Guo HL, Wang XF, Qian QY, Wang FB, Xia XH. A green approach to the synthesis of graphene nanosheets. ACS Nano. 2009; 3: 2653–9.
[37] Zhou M, Wang YL, Zhai YM, Zhai JF, Ren W, Wang FA et al. Controlled synthesis of large-area and patterned electrochemically reduced graphene oxide films. Chemistry-A European Journal. 2009; 15: 6116–20.
[38] Wen X, Garland CW, Hwa T, Kardar M, Kokufuta E, Li Y et al. Crumpled and collapsed conformations in graphite oxide membranes. Nature. 1992; 355: 426–8.
[39] Dubin S, Gilje S, Wang K, Tung VC, Cha K, Hall AS et al. A one-step, solvothermal reduction method for producing reduced graphene oxide dispersions in organic solvents. ACS Nano. 2010; 4: 3845–52.
[40] Kim DH, Yun YS, Jin HJ. Difference of dispersion behavior between graphene oxide and oxidized carbon nanotubes in polar organic solvents. Current Applied Physics. 2012; 12: 637–42.
[41] Lerf A, He HY, Riedl T, Forster M, Klinowski J. C-13 and H-1 MAS NMR studies of graphite oxide and its chemically modified derivatives. Solid State Ionics. 1997; 101: 857–62.
[42] Matsuo Y, Tabata T, Fukunaga T, Fukutsuka T, Sugie Y. Preparation and characterization of silylated graphite oxide. Carbon. 2005; 43: 2875–82.
[43] Matsuo Y, Fukunaga T, Fukutsuka T, Sugie Y. Silylation of graphite oxide. Carbon. 2004; 42: 2117–9.
[44] Stankovich S, Piner RD, Nguyen ST, Ruoff RS. Synthesis and exfoliation of isocyanate-treated graphene oxide nanoplatelets. Carbon. 2006; 44: 3342–7.
[45] Allen MJ, Tung VC, Kaner RB. Honeycomb carbon: A review of graphene. Chemical Reviews. 2010; 110: 132–45.
[46] Zhu YW, Murali S, Cai WW, Li XS, Suk JW, Potts JR et al. Graphene and graphene oxide: Synthesis, properties, and applications. Advanced Materials. 2010; 22: 3906–24.
[47] Dreyer DR, Park S, Bielawski CW, Ruoff RS. The chemistry of graphene oxide. Chemical Society Reviews. 2010; 39: 228–40.
[48] Dreyer DR, Todd AD, Bielawski CW. Harnessing the chemistry of graphene oxide. Chemical Society Reviews. 2014; 43: 5288–301.
[49] Park SJ, Ruoff RS. Chemical methods for the production of graphenes. Nature Nanotechnology. 2009; 4: 217–24.
[50] Liang YY, Frisch J, Zhi LJ, Norouzi-Arasi H, Feng XL, Rabe JP et al. Transparent, highly conductive graphene electrodes from acetylene-assisted thermolysis of graphite oxide sheets and nanographene molecules. Nanotechnology. 2009; 20: 434007.
[51] Su Q, Pang SP, Alijani V, Li C, Feng XL, Muellen K. Composites of graphene with large aromatic molecules. Advanced Materials. 2009; 21: 3191–5.
[52] Cai MZ, Thorpe D, Adamson DH, Schniepp HC. Methods of graphite exfoliation. Journal of Materials Chemistry. 2012; 22: 24992–5002.
[53] Yang H, Hernandez Y, Schlierf A, Felten A, Eckmann A, Johal S et al. A simple method for graphene production based on exfoliation of graphite in water using 1-pyrenesulfonic acid sodium salt. Carbon. 2013; 53: 357–65.
[54] You SJ, Luzan SM, Szabo T, Talyzin AV. Effect of synthesis method on solvation and exfoliation of graphite oxide. Carbon. 2013; 52: 171–80.
[55] Stankovich S, Dikin DA, Piner RD, Kohlhaas KA, Kleinhammes A, Jia YY et al. Synthesis of graphene-based nanosheets via chemical reduction of exfoliated graphite oxide. Carbon. 2007; 45: 1558–65.
[56] Gomez-Navarro C, Weitz RT, Bittner AM, Scolari M, Mews A, Burghard M et al. Electronic transport properties of individual chemically reduced graphene oxide sheets. Nano Letters. 2007; 7: 3499–503.
[57] Gilje S, Han S, Wang MS, Wang KL, Kaner RB. A chemical route to graphene for device applications. Nano Letters. 2007; 7: 3394–8.
[58] Stankovich S, Dikin DA, Dommett GHB, Kohlhaas KM, Zimney EJ, Stach EA et al. Graphene-based composite materials. Nature. 2006; 442: 282–6.
[59] Eda G, Fanchini G, Chhowalla M. Large-area ultrathin films of reduced graphene oxide as a transparent and flexible electronic material. Nature Nanotechnology. 2008; 3: 270–4.
[60] Szabo T, Berkesi O, Forgo P, Josepovits K, Sanakis Y, Petridis D et al. Evolution of surface functional groups in a series of progressively oxidized graphite oxides. Chemistry of Materials. 2006; 18: 2740–9.
[61] Li D, Kaner RB. Processable stabilizer-free polyaniline nanofiber aqueous colloids. Chemical Communications. 2005; 3286–8.
[62] Bourlinos AB, Gournis D, Petridis D, Szabo T, Szeri A, Dekany I. Graphite oxide: Chemical reduction to graphite and surface modification with primary aliphatic amines and amino acids. Langmuir. 2003; 19: 6050–5.
[63] Wang GX, Yang J, Park JS, Gou XL, Wang B, Liu H et al. Facile synthesis and characterization of graphene nanosheets. Journal of Physical Chemistry C. 2008; 112: 8192–5.
[64] Chen Y, Zhang X, Yu P, Ma YW. Stable dispersions of graphene and highly conducting graphene films: a new approach to creating colloids of graphene monolayers. Chemical Communications. 2009: 4527–9.
[65] Gao J, Liu F, Liu YL, Ma N, Wang ZQ, Zhang X. Environment-friendly method to produce graphene that employs vitamin C and amino acid. Chemistry of Materials. 2010; 22: 2213–8.
[66] Tran DNH, Kabiri S, Losic D. A green approach for the reduction of graphene oxide nanosheets using non-aromatic amino acids. Carbon. 2014; 76: 193–202.

[67] Kuila T, Bose S, Khanra P, Mishra AK, Kim NH, Lee JH. A green approach for the reduction of graphene oxide by wild carrot root. Carbon. 2012; 50: 914–21.
[68] Das AK, Srivastav M, Layek RK, Uddin ME, Jung DS, Kim NH et al. Iodide-mediated room temperature reduction of graphene oxide: a rapid chemical route for the synthesis of a bifunctional electrocatalyst. Journal of Materials Chemistry A. 2014; 2: 1332–40.
[69] Fan ZJ, Kai W, Yan J, Wei T, Zhi LJ, Feng J et al. Facile synthesis of graphene nanosheets via Fe reduction of exfoliated graphite oxide. Acs Nano. 2011; 5: 191–8.
[70] Dey RS, Hajra S, Sahu RK, Raj CR, Panigrahi MK. A rapid room temperature chemical route for the synthesis of graphene: metal-mediated reduction of graphene oxide. Chemical Communications. 2012; 48: 1787–9.
[71] Shen JF, Hu YZ, Shi M, Lu X, Qin C, Li C et al. Fast and facile preparation of graphene oxide and reduced graphene oxide nanoplatelets. Chemistry of Materials. 2009; 21: 3514–20.
[72] Si YC, Samulski ET. Synthesis of water soluble graphene. Nano Letters. 2008; 8: 1679–82.
[73] Mohanty N, Nagaraja A, Armesto J, Berry V. High-throughput, ultrafast synthesis of solution-dispersed graphene via a facile hydride chemistry. Small. 2010; 6: 226–31.
[74] Gao W, Alemany LB, Ci LJ, Ajayan PM. New insights into the structure and reduction of graphite oxide. Nature Chemistry. 2009; 1: 403–8.
[75] Brown HC, Krishnamurthy S. 40 years of hydride reductions. Tetrahedron. 1979; 35: 567–607.
[76] Ishigami M, Chen JH, Cullen WG, Fuhrer MS, Williams ED. Atomic structure of graphene on SiO_2. Nano Letters. 2007; 7: 1643–8.
[77] Fan XB, Peng WC, Li Y, Li XY, Wang SL, Zhang GL et al. Deoxygenation of exfoliated graphite oxide under alkaline conditions: A green route to graphene preparation. Advanced Materials. 2008; 20: 4490–3.
[78] Peng HQ, Alemany LB, Margrave JL, Khabashesku VN. Sidewall carboxylic acid functionalization of single-walled carbon nanotubes. Journal of the American Chemical Society. 2003; 125: 15174–82.
[79] Novoselov KS, Jiang D, Schedin F, Booth TJ, Khotkevich VV, Morozov SV et al. Two-dimensional atomic crystals. Proceedings of the National Academy of Sciences of the United States of America. 2005; 102: 10451–3.
[80] Wang ZJ, Zhou XZ, Zhang J, Boey F, Zhang H. Direct electrochemical reduction of single-layer graphene oxide and subsequent functionalization with glucose oxidase. Journal of Physical Chemistry C. 2009; 113: 14071–5.
[81] Ramesha GK, Sampath S. Electrochemical reduction of oriented graphene oxide films: An *in situ* Raman spectroelectrochemical study. Journal of Physical Chemistry C. 2009; 113: 7985–9.
[82] Williams G, Seger B, Kamat PV. TiO_2-graphene nanocomposites. UV-assisted photocatalytic reduction of graphene oxide. Acs Nano. 2008; 2: 1487–91.
[83] Cote LJ, Cruz-Silva R, Huang JX. Flash reduction and patterning of graphite oxide and its polymer composite. Journal of the American Chemical Society. 2009; 131: 11027–32.
[84] Kovtyukhova NI, Ollivier PJ, Martin BR, Mallouk TE, Chizhik SA, Buzaneva EV et al. Layer-by-layer assembly of ultrathin composite films from micron-sized graphite oxide sheets and polycations. Chemistry of Materials. 1999; 11: 771–8.
[85] Becerril HA, Mao J, Liu ZF, Stoltenberg RM, Bao ZN, Chen YS. Evaluation of solution-processed reduced graphene oxide films as transparent conductors. Acs Nano. 2008; 2: 463–70.
[86] Leitner T, Kattner J, Hoffmann H. Infrared reflection spectroscopy of thin films on highly oriented pyrolytic graphite. Applied Spectroscopy. 2003; 57: 1502–9.
[87] Cote LJ, Kim F, Huang JX. Langmuir-blodgett assembly of graphite oxide single layers. Journal of the American Chemical Society. 2009; 131: 1043–9.
[88] Fujimoto H. Theoretical X-ray scattering intensity of carbons with turbostratic stacking and AB stacking structures. Carbon. 2003; 41: 1585–92.
[89] Luo JY, Jang HD, Huang JX. Effect of sheet morphology on the scalability of graphene-based ultracapacitors. Acs Nano. 2013; 7: 1464–71.
[90] Stoller MD, Park SJ, Zhu YW, An JH, Ruoff RS. Graphene-based ultracapacitors. Nano Letters. 2008; 8: 3498–502.
[91] Matuyama E. Pyrolysis of graphitic acid. Journal of Physical Chemistry. 1954; 58: 215–9.
[92] Wang X, Zhi LJ, Muellen K. Transparent, conductive graphene electrodes for dye-sensitized solar cells. Nano Letters. 2008; 8: 323–7.
[93] Li XL, Wang HL, Robinson JT, Sanchez H, Diankov G, Dai HJ. Simultaneous nitrogen doping and reduction of graphene oxide. Journal of the American Chemical Society. 2009; 131: 15939–44.
[94] Hernandez Y, Nicolosi V, Lotya M, Blighe FM, Sun ZY, De S et al. High-yield production of graphene by liquid-phase exfoliation of graphite. Nature Nanotechnology. 2008; 3: 563–8.
[95] Blake P, Brimicombe PD, Nair RR, Booth TJ, Jiang D, Schedin F et al. Graphene-based liquid crystal device. Nano Letters. 2008; 8: 1704–8.
[96] Giordani S, Bergin SD, Nicolosi V, Lebedkin S, Kappes MM, Blau WJ et al. Debundling of single-walled nanotubes bydilution: Observation of large populations of individual nanotubes in amide solvent dispersions. Journal of Physical Chemistry B. 2006; 110: 15708–18.
[97] Hasan T, Scardaci V, Tan PH, Rozhin AG, Milne WI, Ferrari AC. Stabilization and “debundling” of single-wall carbon nanotube dispersions in N-methyl-2-pyrrolidone (NMP) by polyvinylpyrrolidone (PVP). Journal of Physical Chemistry C. 2007; 111: 12594–602.

[98] Bergin SD, Nicolosi V, Streich PV, Giordani S, Sun ZY, Windle AH et al. Towards solutions of single-walled carbon nanotubes in common solvents. Advanced Materials. 2008; 20: 1876–81.
[99] Meyer JC, Geim AK, Katsnelson MI, Novoselov KS, Booth TJ, Roth S. The structure of suspended graphene sheets. Nature. 2007; 446: 60–3.
[100] Li XL, Zhang GY, Bai XD, Sun XM, Wang XR, Wang E et al. Highly conducting graphene sheets and Langmuir-Blodgett films. Nature Nanotechnology. 2008; 3: 538–42.
[101] Shih CJ, Vijayaraghavan A, Krishnan R, Sharma R, Han JH, Ham MH et al. Bi- and trilayer graphene solutions. Nature Nanotechnology. 2011; 6: 439–45.
[102] De Marco P, Nardone M, Del Vitto A, Alessandri M, Santucci S, Ottaviano L. Rapid identification of graphene flakes: alumina does it better. Nanotechnology. 2010; 21: 255703.
[103] Ferrari AC, Meyer JC, Scardaci V, Casiraghi C, Lazzeri M, Mauri F et al. Raman spectrum of graphene and graphene layers. Physical Review Letters. 2006; 97: 187401.
[104] Shih CJ, Lin SC, Strano MS, Blankschtein D. Understanding the stabilization of liquid-phase-exfoliated graphene in polar solvents: Molecular dynamics simulations and kinetic theory of colloid aggregation. Journal of the American Chemical Society. 2010; 132: 14638–48.
[105] Moore VC, Strano MS, Haroz EH, Hauge RH, Smalley RE, Schmidt J et al. Individually suspended single-walled carbon nanotubes in various surfactants. Nano Letters. 2003; 3: 1379–82.
[106] O'Connell MJ, Boul P, Ericson LM, Huffman C, Wang YH, Haroz E et al. Reversible water-solubilization of single-walled carbon nanotubes by polymer wrapping. Chemical Physics Letters. 2001; 342: 265–71.
[107] Bergin SD, Nicolosi V, Cathcart H, Lotya M, Rickard D, Sun ZY et al. Large populations of individual nanotubes in surfactant-based dispersions without the need for ultracentrifugation. Journal of Physical Chemistry C. 2008; 112: 972–7.
[108] Lotya M, King PJ, Khan U, De S, Coleman JN. High-concentration, surfactant-stabilized graphene dispersions. ACS Nano. 2010; 4: 3155–62.
[109] Lotya M, Hernandez Y, King PJ, Smith RJ, Nicolosi V, Karlsson LS et al. Liquid phase production of graphene by exfoliation of graphite in surfactant/water solutions. Journal of the American Chemical Society. 2009; 131: 3611–20.
[110] Behabtu N, Lomeda JR, Green MJ, Higginbotham AL, Sinitskii A, Kosynkin DV et al. Spontaneous high-concentration dispersions and liquid crystals of graphene. Nature Nanotechnology. 2010; 5: 406–11.
[111] Schwamb T, Burg BR, Schirmer NC, Poulikakos D. An electrical method for the measurement of the thermal and electrical conductivity of reduced graphene oxide nanostructures. Nanotechnology. 2009; 20: 405704.
[112] Davis VA, Parra-Vasquez ANG, Green MJ, Rai PK, Behabtu N, Prieto V et al. True solutions of single-walled carbon nanotubes for assembly into macroscopic materials. Nature Nanotechnology. 2009; 4: 830–4.
[113] Rai PK, Pinnick RA, Parra-Vasquez ANG, Davis VA, Schmidt HK, Hauge RH et al. Isotropic-nematic phase transition of single-walled carbon nanotubes in strong acids. Journal of the American Chemical Society. 2006; 128: 591–5.
[114] Ramesh S, Ericson LM, Davis VA, Saini RK, Kittrell C, Pasquali M et al. Dissolution of pristine single walled carbon nanotubes in superacids by direct protonation. Journal of Physical Chemistry B. 2004; 108: 8794–8.
[115] Sumanasekera GU, Allen JL, Fang SL, Loper AL, Rao AM, Eklund PC. Electrochemical oxidation of single wall carbon nanotube bundles in sulfuric acid. Journal of Physical Chemistry B. 1999; 103: 4292–7.
[116] Meyer JC, Geim AK, Katsnelson MI, Novoselov KS, Obergfell D, Roth S et al. On the roughness of single- and bi-layer graphene membranes. Solid State Communications. 2007; 143: 101–9.
[117] Onsager L. The effects of shape on the interaction of colloidal particles. Annals of the New York Academy of Sciences. 1949; 51: 627–59.
[118] Bates MA, Frenkel D. Nematic-isotropic transition in polydisperse systems of infinitely thin hard platelets. Journal of Chemical Physics. 1999; 110: 6553–9.
[119] van der Kooij FM, Kassapidou K, Lekkerkerker HNW. Liquid crystal phase transitions in suspensions of polydisperse plate-like particles. Nature. 2000; 406: 868–71.
[120] van der Beek D, Lekkerkerker HNW. Liquid crystal phases of charged colloidal platelets. Langmuir. 2004; 20: 8582–6.
[121] Low CTJ, Walsh FC, Chakrabarti MH, Hashim MA, Hussain MA. Electrochemical approaches to the production of graphene flakes and their potential applications. Carbon. 2013; 54: 1–21.
[122] Hilder M, Winther-Jensen B, Li D, Forsyth M, MacFarlane DR. Direct electro-deposition of graphene from aqueous suspensions. Physical Chemistry Chemical Physics. 2011; 13: 9187–93.
[123] Wang GX, Wang B, Park JS, Wang Y, Sun B, Yao J. Highly efficient and large-scale synthesis of graphene by electrolytic exfoliation. Carbon. 2009; 47: 3242–6.
[124] Wang JZ, Manga KK, Bao QL, Loh KP. High-yield synthesis of few-layer graphene flakes through electrochemical expansion of graphite in propylene carbonate electrolyte. Journal of the American Chemical Society. 2011; 133: 8888–91.
[125] Paredes JI, Villar-Rodil S, Fernandez-Merino MJ, Guardia L, Martinez-Alonso A, Tascon JMD. Environmentally friendly approaches toward the mass production of processable graphene from graphite oxide. Journal of Materials Chemistry. 2011; 21: 298–306.
[126] Alanyalioglu M, Segura JJ, Oro-Sole J, Casan-Pastor N. The synthesis of graphene sheets with controlled thickness and order using surfactant-assisted electrochemical processes. Carbon. 2012; 50: 142–52.

[127] Morales GM, Schifani P, Ellis G, Ballesteros C, Martinez G, Barbero C et al. High-quality few layer graphene produced by electrochemical intercalation and microwave-assisted expansion of graphite. Carbon. 2011; 49: 2809–16.
[128] Sires I, Low CTJ, Ponce-de-Leon C, Walsh FC. The deposition of nanostructured β-PbO2 coatings from aqueous methanesulfonic acid for the electrochemical oxidation of organic pollutants. Electrochemistry Communications. 2010; 12: 70–4.
[129] Wei D, Grande L, Chundi V, White R, Bower C, Andrew P et al. Graphene from electrochemical exfoliation and its direct applications in enhanced energy storage devices. Chemical Communications. 2012; 48: 1239–41.
[130] Jnioui A, Metrot A, Storck A. Electrochemical production of graphite salts using a 3-dimensional electrode of graphite particles. Electrochimica Acta. 1982; 27: 1247–52.
[131] Berlouis LEA, Schiffrin DJ. The electrochemical formation of graphite-bisulfate intercalation compounds. Journal of Applied Electrochemistry. 1983; 13: 147–55.
[132] Noel M, Santhanam R, Flora MF. Effect of polypyrrole film on the stability and electrochemical activity of fluoride based graphite-intercalation compounds in HF media. Journal of Applied Electrochemistry. 1994; 24: 455–9.
[133] Inagaki M, Iwashita N, Wang ZD, Maeda Y. Electrochemical synthesis of graphite-intercalation compounds with nickel and hydroxides. Synthetic Metals. 1988; 26: 41–7.
[134] Stumpp E, Schubert P, Ehrhardt C. Preparation of metal halide graphite-intercalation compounds by intercalate exchange. Synthetic Metals. 1990; 34: 73–8.
[135] Noel M, Santhanam R. Electrochemistry of graphite intercalation compounds. Journal of Power Sources. 1998; 72: 53–65.
[136] Hathcock KW, Brumfield JC, Goss CA, Irene EA, Murray RW. Incipient electrochemical oxidation of highly oriented pyrolytic-graphite-correlation between surface blistering and electrolyte anion intercalation. Analytical Chemistry. 1995; 67: 2201–6.
[137] Skowronski JM. Electrochemical intercalation of $HClO_4$ into graphite and CrO_3-graphite intercalation compound. Synthetic Metals. 1995; 73: 21–5.
[138] Alliata D, Haring P, Haas O, Kotz R, Siegenthaler H. Anion intercalation into highly oriented pyrolytic graphite studied by electrochemical atomic force microscopy. Electrochemistry Communications. 1999; 1: 5–9.
[139] Toyoda M, Sedlacik J, Inagaki M. Intercalation of formic acid into carbon fibers and their exfoliation. Synthetic Metals. 2002; 130: 39–43.
[140] Toyoda M, Yoshinaga A, Amao Y, Takagi H, Soneda Y, Inagaki M. Preparation of intercalation compounds of carbon fibers through electrolysis using phosphoric acid electrolyte and their exfoliation. Journal of Physics and Chemistry of Solids. 2006; 67: 1178–81.
[141] Ferrari AC, Robertson J. Interpretation of Raman spectra of disordered and amorphous carbon. Physical Review B. 2000; 61: 14095–107.
[142] Chen LY, Tang YH, Wang KL, Liu CB, Luo SL. Direct electrodeposition of reduced graphene oxide on glassy carbon electrode and its electrochemical application. Electrochemistry Communications. 2011; 13: 133–7.
[143] Su CY, Lu AY, Xu YP, Chen FR, Khlobystov AN, Li LJ. High-quality thin graphene films from fast electrochemical exfoliation. ACS Nano. 2011; 5: 2332–9.
[144] Kang FY, Leng Y, Zhang TY. Influences of H_2O_2 on synthesis of H_2SO_4-GICs. Journal of Physics and Chemistry of Solids. 1996; 57: 889–92.
[145] Varchon F, Mallet P, Magaud L, Veuillen JY. Rotational disorder in few-layer graphene films on 6H-SiC(000-1): A scanning tunneling microscopy study. Physical Review B. 2008; 77: 165415.
[146] Krauss B, Lohmann T, Chae DH, Haluska M, von Klitzing K, Smet JH. Laser-induced disassembly of a graphene single crystal into a nanocrystalline network. Physical Review B. 2009; 79.
[147] Guo X, Xiao HS, Wang F, Zhang YH. Micro-Raman and FTIR spectroscopic observation on the phase transitions of $MnSO_4$ droplets and ionic interactions between Mn^{2+} and SO_4^{2-}. Journal of Physical Chemistry A. 2010; 114: 6480–6.
[148] Parvez K, Li RJ, Puniredd SR, Hernandez Y, Hinkel F, Wang SH et al. Electrochemically exfoliated graphene as solution-processable, highly conductive electrodes for organic electronics. ACS Nano. 2013; 7: 3598–606.
[149] Lu J, Yang JX, Wang JZ, Lim AL, Wang SL, Loh KP. One-pot synthesis of fluorescent carbon nanoribbons, nanoparticles, and graphene by the exfoliation of graphite in ionic liquids. Acs Nano. 2009; 3: 2367–75.
[150] Sun ZZ, Yan Z, Yao J, Beitler E, Zhu YW, Tour JM. Growth of graphene from solid carbon sources. Nature. 2010; 468: 549–52.
[151] Najafabadi AT, Gyenge E. High-yield graphene production by electrochemical exfoliation of graphite: Novel ionic liquid (IL)-acetonitrile electrolyte with low IL content. Carbon. 2014; 71: 58–69.
[152] Wang XQ, Fulvio PF, Baker GA, Veith GM, Unocic RR, Mahurin SM et al. Direct exfoliation of natural graphite into micrometre size few layers graphene sheets using ionic liquids. Chemical Communications. 2010; 46: 4487–9.
[153] Rangappa D, Sone K, Wang MS, Gautam UK, Golberg D, Itoh H et al. Rapid and direct conversion of graphite crystals into high-yielding, good-quality graphene by supercritical fluid exfoliation. Chemistry-A European Journal. 2010; 16: 6488–94.
[154] Jang JH, Rangappa D, Kwon YU, Honma I. Direct preparation of 1-PSA modified graphene nanosheets by supercritical fluidic exfoliation and its electrochemical properties. Journal of Materials Chemistry. 2011; 21: 3462–6.

[155] Johnston KP, Shah PS. Materials science-Making nanoscale materials with supercritical fluids. Science. 2004; 303: 482–3.
[156] Serhatkulu GK, Dilek C, Gulari E. Supercritical CO_2 intercalation of layered silicates. Journal of Supercritical Fluids. 2006; 39: 264–70.
[157] Zhou Y, Bao QL, Tang LAL, Zhong YL, Loh KP. Hydrothermal dehydration for the "green" reduction of exfoliated graphene oxide to graphene and demonstration of yunable optical limiting properties. Chemistry of Materials. 2009; 21: 2950–6.
[158] Choucair M, Thordarson P, Stride JA. Gram-scale production of graphene based on solvothermal synthesis and sonication. Nature Nanotechnology. 2009; 4: 30–3.
[159] Ohta T, Bostwick A, Seyller T, Horn K, Rotenberg E. Controlling the electronic structure of bilayer graphene. Science. 2006; 313: 951–4.
[160] Zhang YB, Tang TT, Girit C, Hao Z, Martin MC, Zettl A et al. Direct observation of a widely tunable bandgap in bilayer graphene. Nature. 2009; 459: 820–3.
[161] Novoselov KS, McCann E, Morozov SV, Fal'ko VI, Katsnelson MI, Zeitler U et al. Unconventional quantum Hall effect and Berry's phase of 2_π in bilayer graphene. Nature Physics. 2006; 2: 177–80.
[162] Koshino M, McCann E. Gate-induced interlayer asymmetry in ABA-stacked trilayer graphene. Physical Review B. 2009; 79.
[163] Craciun MF, Russo S, Yamamoto M, Oostinga JB, Morpurgo AF, Tarucha S. Trilayer graphene is a semimetal with a gate-tunable band overlap. Nature Nanotechnology. 2009; 4: 383–8.
[164] Vijayaraghavan A, Sciascia C, Dehm S, Lombardo A, Bonetti A, Ferrari AC et al. Dielectrophoretic assembly of high-density arrays of individual graphene devices for rapid screening. Acs Nano. 2009; 3: 1729–34.
[165] Green AA, Hersam MC. Solution phase production of graphene with controlled thickness via density differentiation. Nano Letters. 2009; 9: 4031–6.
[166] Green AA, Hersam MC. Processing and properties of highly enriched double-wall carbon nanotubes. Nature Nanotechnology. 2009; 4: 64–70.
[167] Arnold MS, Green AA, Hulvat JF, Stupp SI, Hersam MC. Sorting carbon nanotubes by electronic structure using density differentiation. Nature Nanotechnology. 2006; 1: 60–5.
[168] Green AA, Duch MC, Hersam MC. Isolation of single-walled carbon nanotube enantiomers by density differentiation. Nano Research. 2009; 2: 69–77.
[169] Englert JM, Roehrl J, Schmidt CD, Graupner R, Hundhausen M, Hauke F et al. Soluble graphene: generation of aqueous graphene solutions aided by a perylenebisimide-based bolaamphiphile. Advanced Materials. 2009; 21: 4265–9.
[170] Craciun MF, Khrapach I, Barnes MD, Russo S. Properties and applications of chemically functionalized graphene. Journal of Physics-Condensed Matter. 2013; 25: 423201.
[171] Wei WL, Qu XG. Extraordinary physical properties of functionalized graphene. Small. 2012; 8: 2138–51.
[172] Sreeprasad TS, Berry V. How do the electrical properties of graphene change with its functionalization? Small. 2013; 9: 341–50.
[173] Englert JM, Dotzer C, Yang G, Schmid M, Papp C, Gottfried JM et al. Covalent bulk functionalization of graphene. Nature Chemistry. 2011; 3: 279–86.
[174] Rudorff W, Schulze E. Uber alkaligraphitverbindungen. Zeitschrift Fur Anorganische Und Allgemeine Chemie. 1954; 277: 156–71.
[175] Ginderow D. Preparation of intersticial compounds in graphite from liquid reagents. Annales De Chimie France. 1971; 6: 5–16.
[176] Ginderow D, Setton R. Intercalation compounds in graphite via reactions in solution. Comptes Rendus Hebdomadaires Des Seances De L Academie Des Sciences Serie C. 1970; 270: 135–7.
[177] Allongue P, Delamar M, Desbat B, Fagebaume O, Hitmi R, Pinson J et al. Covalent modification of carbon surfaces by aryl radicals generated from the electrochemical reduction of diazonium salts. Journal of the American Chemical Society. 1997; 119: 201–7.
[178] Viculis LM, Mack JJ, Kaner RB. A chemical route to carbon nanoscrolls. Science. 2003; 299: 1361.
[179] Schniepp HC, Li JL, McAllister MJ, Sai H, Herrera-Alonso M, Adamson DH et al. Functionalized single graphene sheets derived from splitting graphite oxide. Journal of Physical Chemistry B. 2006; 110: 8535–9.
[180] Erickson K, Erni R, Lee ZH, Alem N, Gannett W, Zettl A. Determination of the local chemical structure of graphene oxide and reduced graphene oxide. Advanced Materials. 2010; 22: 4467–72.
[181] Jeon IY, Shin YR, Sohn GJ, Choi HJ, Bae SY, Mahmood J et al. Edge-carboxylated graphene nanosheets via ball milling. Proceedings of the National Academy of Sciences of the United States of America. 2012; 109: 5588–93.
[182] Jeon IY, Choi HJ, Jung SM, Seo JM, Kim M, Dai LM et al. Large-scale production of edge-selectively functionalized graphene nanoplatelets via ball milling and their use as metal-free electrocatalysts for oxygen reduction reaction. Journal of the American Chemical Society. 2013; 135: 1386–93.
[183] Leon V, Quintana M, Antonia Herrero M, Fierro JLG, de la Hoz A, Prato M et al. Few-layer graphenes from ball-milling of graphite with melamine. Chemical Communications. 2011; 47: 10936–8.
[184] Damm C, Nacken TJ, Peukert W. Quantitative evaluation of delamination of graphite by wet media milling. Carbon. 2015; 81: 284–94.

[185] Chang DW, Choi HJ, Jeon IY, Seo JM, Dai LM, Baek JB. Solvent-free mechanochemical reduction of graphene oxide. Carbon. 2014; 77: 501–7.

[186] Yan L, Lin MM, Zeng C, Chen Z, Zhang S, Zhao XM et al. Electroactive and biocompatible hydroxyl-functionalized graphene by ball milling. Journal of Materials Chemistry. 2012; 22: 8367–71.

[187] Zhao WF, Fang M, Wu FR, Wu H, Wang LW, Chen GH. Preparation of graphene by exfoliation of graphite using wet ball milling. Journal of Materials Chemistry. 2010; 20: 5817–9.

[188] Chen JF, Duan M, Chen GH. Continuous mechanical exfoliation of graphene sheets via three-roll mill. Journal of Materials Chemistry. 2012; 22: 19625–8.

[189] Posudievsky OY, Khazieieva OA, Koshechko VG, Pokhodenko VD. Mechanochemical delamination of graphite in the presence of various inorganic salts and formation of graphene by its subsequent liquid exfoliation. Theoretical and Experimental Chemistry. 2014; 50: 103–9.

[190] Bakker A, Gates LE. Properly choose mechanical agitators for viscous liquids. Chemical Engineering Progress. 1995; 91: 25–34.

[191] Yi M, Shen ZG. Kitchen blender for producing high-quality few-layer graphene. Carbon. 2014; 78: 622–6.

[192] Lee DS, Riedl C, Krauss B, von Klitzing K, Starke U, Smet JH. Raman spectra of epitaxial graphene on SiC and of epitaxial graphene transferred to SiO_2. Nano Letters. 2008; 8: 4320–5.

[193] Casiraghi C, Hartschuh A, Qian H, Piscanec S, Georgi C, Fasoli A et al. Raman spectroscopy of graphene edges. Nano Letters. 2009; 9: 1433–41.

[194] Malard LM, Pimenta MA, Dresselhaus G, Dresselhaus MS. Raman spectroscopy in graphene. Physics Reports-Review Section of Physics Letters. 2009; 473: 51–87.

[195] Engiert JM, Vecera P, Knirsch KC, Schaefer RA, Hauke F, Hirsch A. Scanning-Raman-microscopy for the statistical analysis of covalently functionalized graphene. Acs Nano. 2013; 7: 5472–82.

[196] Eigler S, Hof F, Enzelberger-Heim M, Grimm S, Mueller P, Hirsch A. Statistical Raman microscopy and atomic force microscopy on heterogeneous graphene obtained after reduction of graphene oxide. Journal of Physical Chemistry C. 2014; 118: 7698–704.

[197] Segal M. Selling graphene by the ton. Nature Nanotechnology. 2009; 4: 611–3.

[198] Tour JM. Scaling up exfoliation. Nature Materials. 2014; 13: 545–6.

[199] Widenkvist E, Boukhvalov DW, Rubino S, Akhtar S, Lu J, Quinlan RA et al. Mild sonochemical exfoliation of bromine-intercalated graphite: a new route towards graphene. Journal of Physics D-Applied Physics. 2009; 42: 112003.

[200] Dresselhaus MS, Dresselhaus G. Intercalation compounds of graphite. Advances in Physics. 2002; 51: 1–186.

[201] Sasa T, Takahash.Y, Mukaibo T. Crystal structure of graphite bromine lamellar compounds. Carbon. 1971; 9: 407–16.

[202] Erbil A, Kortan AR, Birgeneau RJ, Dresselhaus MS. Intercalate structure, melting, and the commensurate-incommensurate transition in bromine-interlalated graphite. Physical Review B. 1983; 28: 6329–46.

[203] Benedict LX, Chopra NG, Cohen ML, Zettl A, Louie SG, Crespi VH. Microscopic determination of the interlayer binding energy in graphite. Chemical Physics Letters. 1998; 286: 490–6.

[204] Boukhvalov DW, Katsnelson MI. Modeling of graphite oxide. Journal of the American Chemical Society. 2008; 130: 10697–701.

[205] Zhou XF, Liu ZP. A scalable, solution-phase processing route to graphene oxide and graphene ultralarge sheets. Chemical Communications. 2010; 46: 2611–3.

[206] Yi M, Shen ZG, Zhang XJ, Ma SL. Achieving concentrated graphene dispersions in water/acetone mixtures by the strategy of tailoring Hansen solubility parameters. Journal of Physics D-Applied Physics. 2013; 46: 025301.

[207] Paton KR, Varrla E, Backes C, Smith RJ, Khan U, O'Neill A et al. Scalable production of large quantities of defect-free few-layer graphene by shear exfoliation in liquids. Nature Materials. 2014; 13: 624–30.

[208] Soldano C, Mahmood A, Dujardin E. Production, properties and potential of graphene. Carbon. 2010; 48: 2127–50.

[209] Zhang YB, Small JP, Pontius WV, Kim P. Fabrication and electric-field-dependent transport measurements of mesoscopic graphite devices. Applied Physics Letters. 2005; 86: 073104.

[210] Lu XK, Huang H, Nemchuk N, Ruoff RS. Patterning of highly oriented pyrolytic graphite by oxygen plasma etching. Applied Physics Letters. 1999; 75: 193–5.

[211] Dayen JF, Mahmood A, Golubev DS, Roch-Jeune I, Salles P, Dujardin E. Side-gated transport in focused-ion-beam-fabricated multilayered graphene nanoribbons. Small. 2008; 4: 716–20.

[212] Bolotin KI, Sikes KJ, Jiang Z, Klima M, Fudenberg G, Hone J et al. Ultrahigh electron mobility in suspended graphene. Solid State Communications. 2008; 146: 351–5.

[213] Chen JH, Jang C, Xiao SD, Ishigami M, Fuhrer MS. Intrinsic and extrinsic performance limits of graphene devices on SiO_2. Nature Nanotechnology. 2008; 3: 206–9.

[214] Moser J, Barreiro A, Bachtold A. Current-induced cleaning of graphene. Applied Physics Letters. 2007; 91: 163513.

[215] Blake P, Hill EW, Castro Neto AH, Novoselov KS, Jiang D, Yang R et al. Making graphene visible. Applied Physics Letters. 2007; 91: 063124.

[216] Jung I, Pelton M, Piner R, Dikin DA, Stankovich S, Watcharotone S et al. Simple approach for high-contrast optical imaging and characterization of graphene-based sheets. Nano Letters. 2007; 7: 3569–75.

[217] Gao LB, Ren WC, Li F, Cheng HM. Total color difference for rapid and accurate identification of graphene. ACS Nano. 2008; 2: 1625–33.

[218] Meyer JC, Kisielowski C, Erni R, Rossell MD, Crommie MF, Zettl A. Direct imaging of lattice atoms and topological defects in graphene membranes. Nano Letters. 2008; 8: 3582–6.

[219] Terrones M, Botello-Mendez AR, Campos-Delgado J, Lopez-Urias F, Vega-Cantu YI, Rodriguez-Macias FJ et al. Graphene and graphite nanoribbons: Morphology, properties, synthesis, defects and applications. Nano Today. 2010; 5: 351–72.

[220] Terrones M. Nanotubes unzipped. Nature. 2009; 458: 845–6.

[221] Dutta S, Pati SK. Novel properties of graphene nanoribbons: A review. Journal of Materials Chemistry. 2010; 20: 8207–23.

[222] Jia XT, Hofmann M, Meunier V, Sumpter BG, Campos-Delgado J, Romo-Herrera JM et al. Controlled formation of sharp zigzag and armchair edges in graphitic nanoribbons. Science. 2009; 323: 1701–5.

[223] Girit CO, Meyer JC, Erni R, Rossell MD, Kisielowski C, Yang L et al. Graphene at the edge: stability and dynamics. Science. 2009; 323: 1705–8.

[224] Cano-Marquez AG, Rodriguez-Macias FJ, Campos-Delgado J, Espinosa-Gonzalez CG, Tristan-Lopez F, Ramirez-Gonzalez D et al. Ex-MWNTs: Graphene sheets and ribbons produced by lithium intercalation and exfoliation of Carbon nanotubes. Nano Letters. 2009; 9: 1527–33.

[225] Kosynkin DV, Higginbotham AL, Sinitskii A, Lomeda JR, Dimiev A, Price BK et al. Longitudinal unzipping of carbon nanotubes to form graphene nanoribbons. Nature. 2009; 458: 872–6.

[226] Laura Elias A, Botello-Mendez AR, Meneses-Rodriguez D, Jehova Gonzalez V, Ramirez-Gonzalez D, Ci LJ et al. Longitudinal cutting of pure and doped carbon nanotubes to form graphitic nanoribbons using metal clusters as nanoscalpels. Nano Letters. 2010; 10: 366–72.

[227] Kim KP, Sussman A, Zettl A. Graphene nanoribbons obtained by electrically unwrapping carbon nanotubes. ACS Nano. 2010; 4: 1362–6.

[228] Jiao LY, Zhang L, Wang XR, Diankov G, Dai HJ. Narrow graphene nanoribbons from carbon nanotubes. Nature. 2009; 458: 877–80.

[229] Santos H, Chico L, Brey L. Carbon nanoelectronics: Unzipping tubes into graphene ribbons. Physical Review Letters. 2009; 103: 086801.

[230] Terrones M, Kamalakaran R, Seeger T, Ruhle M. Novel nanoscale gas containers: encapsulation of N_2 in CNx nanotubes. Chemical Communications. 2000: 2335–6.

[231] Botello-Mendez AR, Campos-Delgado J, Morelos-Gomez A, Romo-Herrera JM, Rodriguez AG, Navarro H et al. Controlling the dimensions, reactivity and crystallinity of multiwalled carbon nanotubes using low ethanol concentrations. Chemical Physics Letters. 2008; 453: 55–61.

[232] Pinault M, Mayne-L'Hermite M, Reynaud C, Beyssac O, Rouzaud JN, Clinard C. Carbon nanotubes produced by aerosol pyrolysis: growth mechanisms and post-annealing effects. Diamond and Related Materials. 2004; 13: 1266–9.

[233] Baker RTK, Sherwood RD, Derouane EG. FURTHER-STUDIES OF THE NICKEL-GRAPHITE-HYDROGEN REACTION Further studies of the nickel-graphite-hydrogen reaction. Journal of Catalysis. 1982; 75: 382–95.

[234] Ci LJ, Xu ZP, Wang LL, Gao W, Ding F, Kelly KF et al. Controlled nanocutting of graphene. Nano Research. 2008; 1: 116–22.

[235] Jiao LY, Fan B, Xian XJ, Wu ZY, Zhang J, Liu ZF. Creation of nanostructures with poly(methyl methacrylate)-mediated nanotransfer printing. Journal of the American Chemical Society. 2008; 130: 12612-+.

[236] Winters HF, Coburn JW, Chuang TJ. Surface process in plasma-assisted etching environments. Journal of Vacuum Science & Technology B. 1983; 1: 469–80.

[237] Jiao LY, Wang XR, Diankov G, Wang HL, Dai HJ. Facile synthesis of high-quality graphene nanoribbons. Nature Nanotechnology. 2010; 5: 321–5.

[238] Colbert DT, Zhang J, McClure SM, Nikolaev P, Chen Z, Hafner JH et al. Growth and sintering of fullerene nanotubes. Science. 1994; 266: 1218–22.

[239] Kang YR, Li YL, Deng MY. Precise unzipping of flattened carbon nanotubes to regular graphene nanoribbons by acid cutting along the folded edges. Journal of Materials Chemistry. 2012; 22: 16283–7.

[240] Zhong XH, Li YL, Liu YK, Qiao XH, Feng Y, Liang J et al. Continuous multilayered carbon nanotube yarns. Advanced Materials. 2010; 22: 692–6.

[241] Sridhar V, Jeon JH, Oh IK. Microwave extraction of graphene from carbon fibers. Carbon. 2011; 49: 222–6.

[242] Nabais JMV, Carrott PJM, Carrott M, Menendez JA. Preparation and modification of activated carbon fibres by microwave heating. Carbon. 2004; 42: 1315–20.

[243] Banhart F, Fuller T, Redlich P, Ajayan PM. The formation, annealing and self-compression of carbon onions under electron irradiation. Chemical Physics Letters. 1997; 269: 349–55.

[244] Ma L, Wang JL, Ding F. Strain-induced orientation-selective cutting of graphene into graphene nanoribbons on oxidation. Angewandte Chemie-International Edition. 2012; 51: 1161–4.

[245] Fujii S, Enoki T. Cutting of oxidized graphene into nanosized pieces. Journal of the American Chemical Society. 2010; 132: 10034–41.

[246] Gao XF, Wang LL, Ohtsuka Y, Jiang DE, Zhao YL, Nagase S et al. Oxidation unzipping of stable nanographenes into joint spin-rich fragments. Journal of the American Chemical Society. 2009; 131: 9663–9.

[247] Liu J, Rinzler AG, Dai HJ, Hafner JH, Bradley RK, Boul PJ et al. Fullerene pipes. Science. 1998; 280: 1253–6.

[248] Li JL, Kudin KN, McAllister MJ, Prud'homme RK, Aksay IA, Car R. Oxygen-driven unzipping of graphitic materials. Physical Review Letters. 2006; 96: 176101.
[249] Ni ZH, Yu T, Lu YH, Wang YY, Feng YP, Shen ZX. Uniaxial strain on graphene: Raman spectroscopy study and band-gap opening. ACS Nano. 2008; 2: 2301–5.
[250] Huang MY, Yan HG, Chen CY, Song DH, Heinz TF, Hone J. Phonon softening and crystallographic orientation of strained graphene studied by Raman spectroscopy. Proceedings of the National Academy of Sciences of the United States of America. 2009; 106: 7304–8.
[251] Gong L, Kinloch IA, Young RJ, Riaz I, Jalil R, Novoselov KS. Interfacial stress transfer in a graphene monolayer nanocomposite. Advanced Materials. 2010; 22: 2694-+.
[252] Mohiuddin TMG, Lombardo A, Nair RR, Bonetti A, Savini G, Jalil R et al. Uniaxial strain in graphene by Raman spectroscopy: G peak splitting, Gruneisen parameters, and sample orientation. Physical Review B. 2009; 79: 205433.
[253] Huang MY, Yan HG, Heinz TF, Hone J. Probing strain-induced electronic structure change in graphene by Raman spectroscopy. Nano Letters. 2010; 10: 4074–9.
[254] Ren ZF, Huang ZP, Xu JW, Wang JH, Bush P, Siegal MP et al. Synthesis of large arrays of well-aligned carbon nanotubes on glass. Science. 1998; 282: 1105–7.
[255] Yudasaka M, Kikuchi R, Matsui T, Ohki Y, Yoshimura S, Ota E. Specific conditions for Ni catalyzed carbon nanotube growth by chemical vapor deposition. Applied Physics Letters. 1995; 67: 2477–9.
[256] Che GL, Lakshmi BB, Fisher ER, Martin CR. Carbon nanotubule membranes for electrochemical energy storage and production. Nature. 1998; 393: 346–9.
[257] Kim KS, Zhao Y, Jang H, Lee SY, Kim JM, Kim KS et al. Large-scale pattern growth of graphene films for stretchable transparent electrodes. Nature. 2009; 457: 706–10.
[258] Vitchev R, Malesevic A, Petrov RH, Kemps R, Mertens M, Vanhulsel A et al. Initial stages of few-layer graphene growth by microwave plasma-enhanced chemical vapour deposition. Nanotechnology. 2010; 21: 095602.
[259] Jang WJ, Kim HW, Jeon JH, Yoon JK, Kahng SJ. Recovery and local-variation of Dirac cones in oxygen-intercalated graphene on Ru(0001) studied using scanning tunneling microscopy and spectroscopy. Physical Chemistry Chemical Physics. 2013; 15: 16019–23.
[260] Liao Q, Zhang HJ, Wu K, Li HY, Bao SN, He P. Oxidation of graphene on Ru(0001) studied by scanning tunneling microscopy. Applied Surface Science. 2010; 257: 82–6.
[261] Marchini S, Guenther S, Wintterlin J. Scanning tunneling microscopy of graphene on Ru(0001). Physical Review B. 2007; 76: 075429.
[262] Reina A, Jia XT, Ho J, Nezich D, Son HB, Bulovic V et al. Large area, few-layer graphene films on arbitrary substrates by chemical vapor deposition. Nano Letters. 2009; 9: 30–5.
[263] Li JY, Ren ZY, Zhou YX, Wu XJ, Xu XL, Qi M et al. Scalable synthesis of pyrrolic N-doped graphene by atmospheric pressure chemical vapor deposition and its terahertz response. Carbon. 2013; 62: 330–6.
[264] Wei DC, Liu YQ, Wang YH, Zhang HL, Huang LP, Yu G. Synthesis of N-doped graphene by chemical vapor deposition and its electrical properties. Nano Letters. 2009; 9: 1752–8.
[265] Li XS, Cai WW, An JH, Kim SY, Nah JH, Yang DX et al. Large-area synthesis of high-quality and uniform graphene films on copper foils. Science. 2009; 324: 1312–4.
[266] Bae SK, Kim HK, Lee YB, Xu XF, Park JS, Zheng Y et al. Roll-to-roll production of 30-inch graphene films for transparent electrodes. Nature Nanotechnology. 2010; 5: 574–8.
[267] Wei DC, Liu YQ, Zhang HL, Huang LP, Wu B, Chen JY et al. Scalable synthesis of few-layer graphene ribbons with controlled morphologies by a template method and their applications in nanoelectromechanical switches. Journal of the American Chemical Society. 2009; 131: 11147–54.

CHAPTER 3

Graphene-Inorganic Hybrids (I)

Ling Bing Kong,[1,a,*] *Freddy Boey,*[1,b] *Yizhong Huang,*[1,c] *Zhichuan Jason Xu,*[1,d] *Kun Zhou,*[2] *Sean Li,*[3] *Wenxiu Que,*[4] *Hui Huang*[5] and *Tianshu Zhang*[6]

3.1. Introduction

As stated in Chapter 1, hybrids of graphene or graphene related materials have formed a new group of materials, which have properties that are either better than their individual components or not found in the components. To date, a wide range of components have been used to produce hybrids with graphene or graphene related materials. In this book, only inorganic components will be discussed. Due to the huge number of references on these new hybrid materials, two chapters will be used to present them. This chapter will cover hybrids based on oxides, while those derived from other inorganic components, including metals, nanocarbons (non-graphene, e.g., carbon nanotubes or CNTs) and semiconductors, will be discussed in the next chapter.

Almost all metallic oxides have been explored to create hybrids with graphene or graphene related materials. There have been review papers regarding graphene–oxide hybrids. Two of the most extensive reviews are necessarily mentioned, i.e., one is about graphene-TiO_2 hybrids for photocatalytic applications [1], while the other one is about graphene-oxides for energy storage applications [2]. However, these two reviews have been given their respective emphasis and no more general and extensive reviews have been found in the open literature. In this chapter, efforts have been made to include all metallic oxides that have been employed to form hybrids with graphene.

There are six main types of combinations of graphene nanosheets with the inorganic particles in general and metallic oxides in particular: (i) nano-sized particle anchored on graphene, (ii) graphene-wrapped

[1] School of Materials Science and Engineering, Nanyang Technological University, Singapore.
[a] E-mail: elbkong@ntu.edu.sg
[b] E-mail: mycboey@ntu.edu.sg
[c] E-mail: yzhuang@ntu.edu.sg
[d] E-mail: xuzc@ntu.edu.sg
[2] School of Mechanical & Aerospace Engineering, Nanyang Technological University, Singapore; E-mail: kzhou@ntu.edu.sg
[3] School of Materials Science and Engineering, The University of New South Wales, Australia; E-mail: sean.li@unsw.edu.au
[4] Electronic Materials Research Laboratory, School of Electronic and Information Engineering, Xi'an Jiaotong University, P. R. China; E-mail: wxque@xjtu.edu.cn
[5] Singapore Institute of Manufacturing Technologies (SIMTech), Singapore; E-mail: hhuang@SIMTech.a-star.edu.sg
[6] Anhui Target Advanced Ceramics Technology Co. Ltd., Hefei, Anhui, P. R. China; E-mail: 13335516617@163.com
* Corresponding author

inorganic particles, (iii) graphene-encapsulated particles, (iv) two-dimensional sandwich-like configuration, (v) graphene/particle layered structure consisting of aligned layers of inorganic particles and (vi) three-dimensional construction of graphene conductive networks containing inorganic particles. Figure 3.1 shows schematic drawings of possible combinations between graphene nanosheet and inorganic particles, which were previously used to describe hybrids of graphene and oxide nanoparticles [2]. Graphene–oxide hybrids can be classified into one or more of these combinations, which may not be specifically mentioned in the context of the discussion.

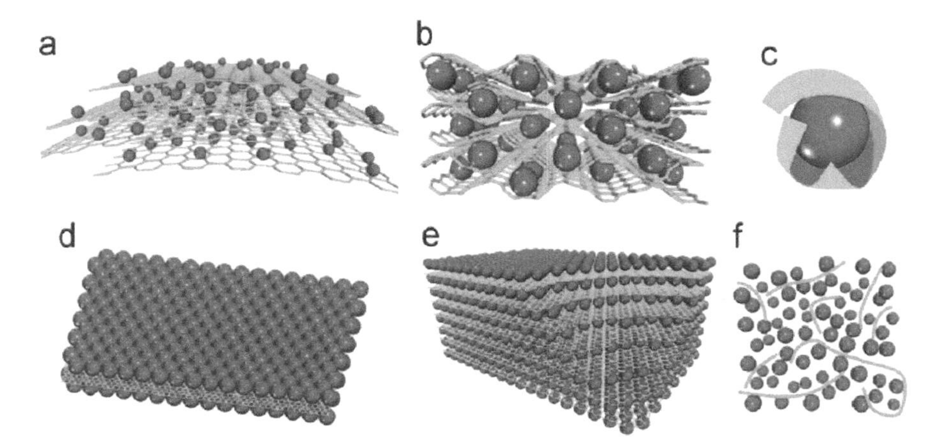

Fig. 3.1. Schematic diagrams of the structural models of graphene/inorganic particles (metallic oxides) hybrids: (a) anchored model in which nanosized inorganic particles are anchored on the surface of the graphene nanosheets, (b) wrapped model where inorganic particles are wrapped by the graphene nanosheets, (c) encapsulated model with inorganic particles being encapsulated by the graphene, (d) sandwich-like model in which graphene nanosheets are used as a template to form particle/graphene/particle sandwich-like structures, (e) layered model composed of alternating layers of inorganic nanoparticles and graphene and (f) mixed model of graphene and inorganic particles that are mechanically mixed, with graphene forming a conductive network. Red items are inorganic particles and blue items are graphene nanosheets. Reproduced with permission from [2], Copyright © 2012, Elsevier.

Color image of this figure appears in the color plate section at the end of the book.

3.2. Graphene-Oxide Hybrids and Preparation

3.2.1. Graphene—Transitional Metal Oxides (TMOs)

3.2.1.1. Graphene-TiO_2

TiO_2 has been extensively studied for various applications, including as a photocatalyst, in lithium ion batteries, due to its high stability, low cost, relatively low toxicity and excellent photocatalytic and electrochemical performances [3]. However, pure TiO_2 exhibited several disadvantages for practical applications. For example, it showed limited photoactivity in the visible range of the Earth's solar spectrum, while its electrical conductivity is too low when it is used as an electrode of lithium ion batteries. Therefore, it is necessary to modify TiO_2 in order to improve its performance, by using doping and form composites with other components [4, 5].

Among various materials, nanocarbon materials have been demonstrated to be promising candidates in forming composites with TiO_2, because of their various advantages, including high chemical stability and tunability, relatively high electrical conductivity and tunable microstructure. Nanocarbon materials, such as carbon nanotubes (CNTs), fullerenes, graphene nanosheets, have been widely used for such purpose, due to their unique structural and electrical properties [6–8]. Among these carbon materials, graphene has recently emerged as a rising star in forming composite materials, because of its excellent electrical properties, large specific surface area, flexible structure, high transparency and stable thermal properties.

Various methods and strategies have been proposed to synthesize graphene or graphene oxide–TiO_2 (G–TiO_2 or GO–TiO_2) hybrids, such as (i) simple mixing or mixing with sonication [9, 10], (ii) sol–gel process [11, 12], (iii) liquid-phase deposition [13], (iv) hydrothermal and solvothermal synthesis [12, 14, 15]. It is well-known that graphene nanosheets have a strong tendency to form aggregates in solutions or suspensions, due to the strong van der Waals interactions and hydrogen bonds in polar solvents. As a result, methods such as electrostatic stabilization and chemical functionalization have been developed to prepare stable graphene suspensions [16].

The mixing and sonication method is the simplest route to prepare GO–TiO_2 and G–TiO_2 hybrids. However, it is expected that the interaction between the two components is relatively weak, because chemical bonding is not formed during the mixing process. Guo et al. used ultrasonication method to incorporate TiO_2 nanoparticles on the graphene layers [9]. The average size of the TiO_2 nanoparticles was controlled to be 4–5 nm on the graphene nanosheets without using any surfactant. In this method, the pyrolysis and condensation of the dissolved $TiCl_4$ into TiO_2 were effectively and efficiently triggered due to the application of the ultrasonic waves. The photocatalytic activity of the G–TiO_2 composites containing 25 wt.% TiO_2 was higher than that of commercial pure TiO_2, which was readily attributed to the extremely small size of the TiO_2 nanoparticles and the graphene–TiO_2 composite microstructure in which the crystalline TiO_2 nanoparticles were homogeneously dispersed on the graphene nanosheets. Because the graphene nanosheets in the composites had a very close contact with the TiO_2 nanoparticles, the photo-electron conversion of TiO_2 was greatly enhanced, by reducing the recombination of photo-generated electron–hole pairs.

GO–TiO_2 composite was synthesized first by sonochemical reaction of $TiCl_4$ in GO suspensions [9]. TiO_2 precursors were prepared with a molar ratio of ethanol:H_2O:$TiCl_4$ = 35:11:1, in which 0.25 g of GO was added and thoroughly mixed. The suspensions were then treated with ultrasonication at room temperature for 3 h. The composites were recovered with centrifugation and thoroughly washed with ethanol. G–TiO_2 hybrid was formed by using a hydrazine chemical reduction method to convert the GO–TiO_2 hybrid [17].

A possible mechanism was proposed to describe the formation of the uniformly distributed TiO_2 nanoparticles on graphene nanosheets by using the sonochemical method, as shown in Fig. 3.2. Firstly, Ti ions were physically adsorbed on the graphene nanosheets, because the GO contained oxidized graphene nanosheets, with their basal planes heavily decorated with epoxide and hydroxyl groups, as well as carboxyl groups. The carboxylic moieties then interacted with the Ti cations through physisorption, electrostatic

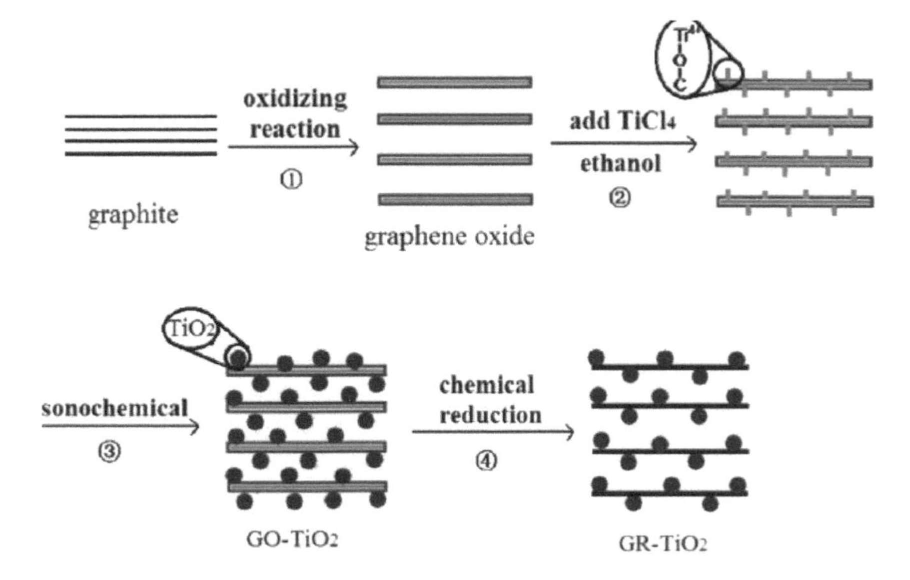

Fig. 3.2. Possible formation mechanism of the TiO_2 nanoparticles on the graphene nanosheets by using the sonochemical method. Reproduced with permission from [9], Copyright © 2011, Elsevier.

binding, and/or charge-transferring. When the Ti ions were attached to the carboxyl groups, reaction to TiO_2 would occur on the surface of the exfoliated nanosheets. Since the number of ionizable carboxyl groups at the edge might not be large enough to trap the cations and stabilize the metal oxide nanoparticles once formed, nonuniform size distribution and large-sized particles would be observed. One way to achieve homogeneous dispersion is to use surface functionalization [18, 19]. In this case, surface modification or additional agents were not used due to the application of ultrasonication.

Ultrasonic agitation promoted the homogenization of the solutes and increased the diffusion rate of $TiCl_4$ to the GO surface. The aggregation of $TiCl_4$ on surfaces of the GO nanosheets could be prevented, when the small bubbles were collapsed. Therefore, a thin layer coating, with a thickness of a few nanometers, was formed on the GO surfaces, so that uniform dispersion of fine TiO_2 could be readily achieved, without using surface functionalization. Figure 3.3 shows XRD patterns of pure TiO_2, graphene (GR), graphene (GR)–TiO_2 and graphene (GR)–TiO_2–T (calcined at 450°C in N_2) [9]. The quite broad peak of the GR–TiO_2 at $2\theta = 25.5$ implied that the TiO_2 nanoparticles in the sample were of amorphous nature. After calcination at 450°C, the peak was sharpened, due to increase in crystallinity of the TiO_2 particles.

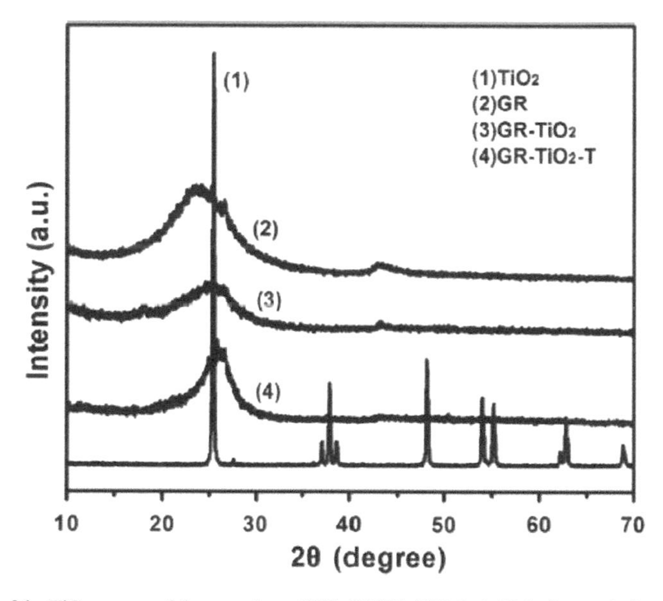

Fig. 3.3. XRD patterns of the TiO_2 nanoparticles, graphene (GR), G(GR)–TiO_2 hybrid, before and after calcinations in nitrogen. Reproduced with permission from [9], Copyright © 2011, Elsevier.

Pure graphene exhibited a layered structure with a very smooth surface, while wrinkles were observed in the hybrid sample both at the edge of the graphene sheets and on the interlayer nanosheets. The TiO_2 nanoparticles were uniformly dispersed on both the graphene surface and the interlayers. No significant agglomeration was formed in the hybrid.

Figure 3.4 shows TEM images of the G–TiO_2 hybrids before (GR–TiO_2) and after (GR–TiO_2–T) the calcination. As shown in Fig. 3.4 (a), the as-synthesized GR–TiO_2 exhibited wrinkles on the 2D graphene nanosheets. The edges of the sheets were folded. No TiO_2 nanoparticles were identified on the graphene nanosheets, as illustrated in Fig. 3.4 (b) and (c), Ti and O were detected with EDX. Selected area electron diffraction (SAED) pattern of the hybrid consisted of diffraction spots of graphite and diffraction rings of TiO_2. After calcination, crystallized TiO_2 particles uniformly scattered on the graphene nanosheets could be readily observed, as shown in Fig. 3.4 (d). Figure 3.4 (f) indicates that the particle size was in the range of 4–5 nm. The crystal lattice of TiO_2 nanoparticles was demonstrated in Fig. 3.4 (e).

Williams et al. used a similar method to prepare GO–TiO_2 hybrid, which was further reduced to rGO-TiO_2 hybrid [20]. TiO_2 colloidal suspension in ethanol at a concentration of 10 mM was prepared

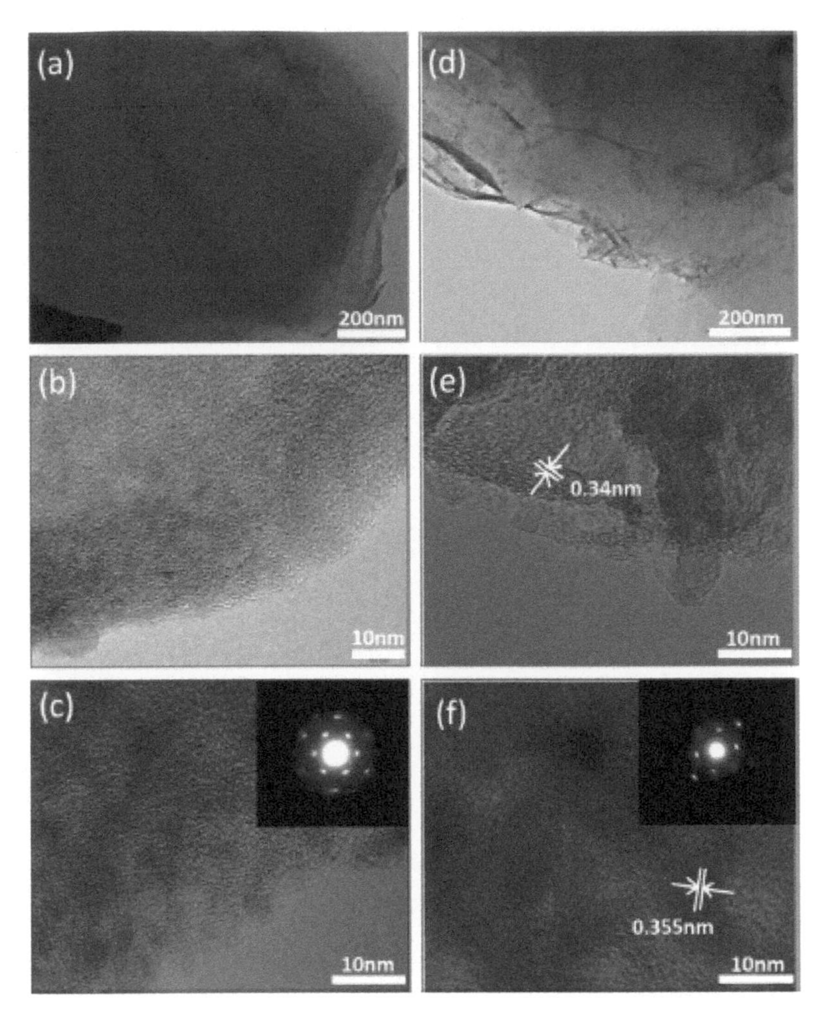

Fig. 3.4. TEM images of G–TiO_2 before (a–c) and after calcination (GR–TiO_2–T) (d–f) at different resolutions. Reproduced with permission from [9], Copyright © 2011, Elsevier.

separately through hydrolysis of titanium isopropoxide, in which titanium isopropoxide was added dropwise into a vigorously stirred solution of ethanol. The colloidal suspension was continuously stirred to avoid agglomeration of the nanoparticles. The nanoparticles had a size range of 2–7 nm. GO–TiO_2 hybrid was formed by adding GO powder into the TiO_2 suspension.

Zhang and Pan used a mixing and sonication method to obtain rGO–TiO_2 hybrid [21]. Commercial TiO_2 (P25) nanoparticles (80% anatase + 20% rutile) were used directly. 10 mg of GO and 90 mg of P25 were added into 100 ml of distilled water. After 1 h of sonication, the mixture was centrifuged and then dried at 30°C. The obtained product was finally calcined at 300°C for 2 h in Ar. G–P25 hybrid was then prepared by reducing the GO. XRD patterns of pure P25 and the G–P25 hybrid are shown in Fig. 3.5 (a). The two samples had a similar XRD pattern, without the signal from graphene due to the small amount of graphene contained in the hybrid. Figure 3.5 (b) shows TEM image of the G–P25 hybrid sample, in which the graphene nanosheets and the TiO_2 nanoparticles were obviously demonstrated. Figure 3.5 (c) indicates that the TiO_2 nanoparticles were encapsulated within the graphene nanosheets with a thickness of a few layers. FT-IR spectra, as shown in Fig. 3.5 (d), implied that all the oxygen-containing functional groups were completely eliminated during the thermal reaction.

The sol–gel process has widely employed to synthesize GO–TiO_2 hybrids with chemical interaction between the two components. As stated earlier, GO is dissolvable in water, so that it can be easily blended

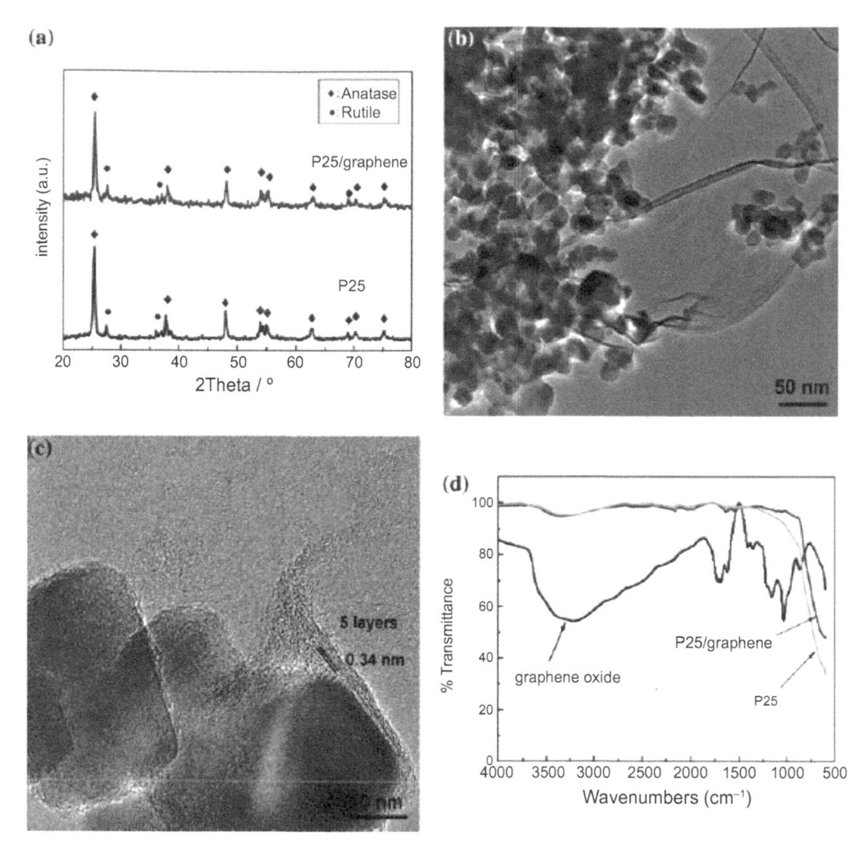

Fig. 3.5. (a) XRD patterns of pure P25 and the G–P25(TiO_2) hybrid. (b, c) TEM images of the G–P25 hybrid sample. (d) Fourier transform infrared (FTIR) spectra of the GO, P25 and G–P25 hybrid. Reproduced with permission from [21], Copyright © 2011, Springer.

with titanium alkoxide precursors. The hydroxyl groups on the surfaces of GO nanosheets can facilitate oxo- or hydroxo-bridges with metal centers. TiO_2 phase is formed in sol–gel processes, during the hydrolysis of the alkoxides, through the oxo- or hydroxo-bridges. The hydroxo-connected GO nanosheets are homogeneously dispersed in the sols, which leads to biphasic gels and thus GO–TiO_2 hybrids [22]. Various titanium compounds, not only alkoxides but also chlorides and fluoride, including $Ti(OPri)_4$ [23], $Ti(OBu)_4$ [24], titanium peroxo complexes [25, 26], $TiCl_4$ [27], $TiCl_3$ [28], TiF_4 [29], $(NH4)_2TiF_6$ [30], $Ti(SO_4)_2$ [31], etc., have been used to prepare GO–TiO_2 hybrids.

Peng et al. developed a sol-gel process to synthesize GO–TiO_2 with $Ti(OPri)_4$ as the precursor [23]. They obtained lamellar GO nanosheets dispersed with one-dimensional titania nanotubes (TNT) or nanorods (TNR). The GO–TNT and G–TNR composites could find potential application in photocatalysis, sensors and photovoltaic solar cells. The formation process of the GO–TNT and G–TNR hybrids is schematically illustrated in Fig. 3.6. In the first step, Ti containing organic species were intercalated into the inter-layers of GO to form a Ti species-pillared GO (GOTi). The GO was synthesized from natural graphite by using the Hummers' and Offeman's method, with a chemical formula of C8O4.4H1.4 and an interlayer distance of 0.76 nm. Intercalation with titanium tetra-isopropoxide (TTIP, 0.1 M) led to the expansion of GO, resulting in an inter-layer distance of 1.3–1.4 nm. The obtained precipitate was subjected to hydrothermal treatment in 10 M NaOH solution at 423 K for 24 h, in order to obtain GO–TNT hybrid. The hydrothermal product was finally calcined at 823–1023 K in vacuum to get G–TNR hybrid.

Figure 3.7 (a) shows SEM image of the GOTi, which had a stacking structure of platelets without particle aggregation. Figure 3.7 (b) shows TEM image of a representative transparent platelet, with a

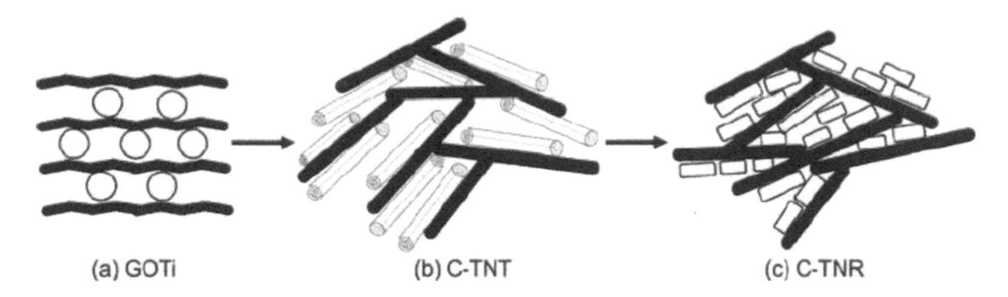

Fig. 3.6. Schematic diagram to describe the formation of the C(GO)-TNT (titania nanotube) and C(GO)-TNR (titania nanorod) hybrids. Reproduced with permission from [23], Copyright © 2008, Royal Society of Chemistry.

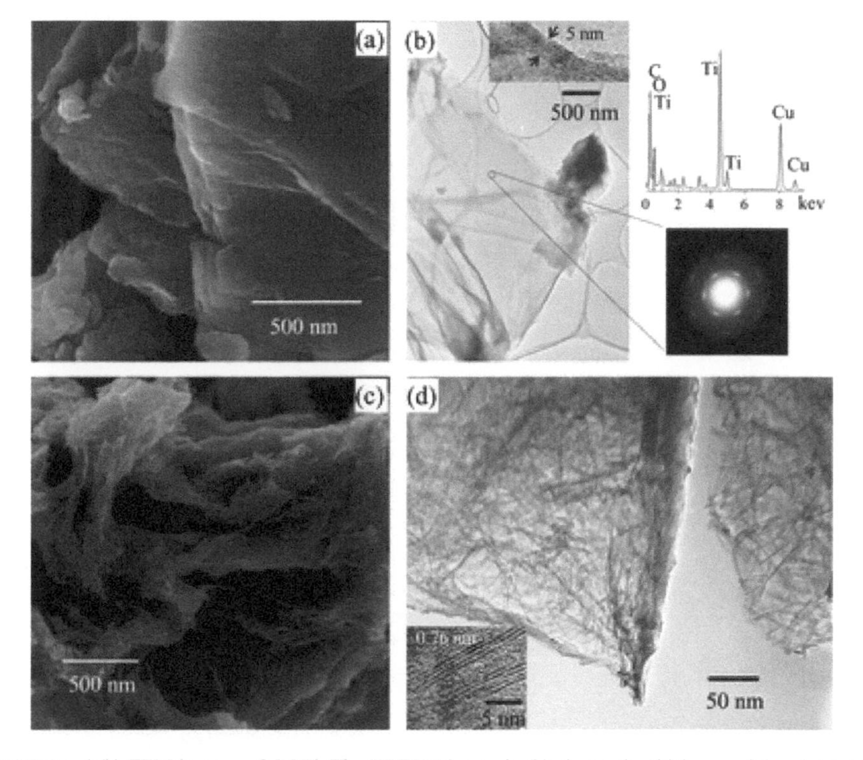

Fig. 3.7. (a) SEM and (b) TEM images of GOTi. The HRTEM image in (b) shows the thickness of the platelet to be 5 nm, with the 'right-up' and 'right-bottom' insets being EDX result and SAD pattern, respectively. (c) SEM and (d) TEM images of C(GO)-TNT (titania nanotube), with the inset being HRTEM image of a single nanotube. Reproduced with permission from [23], Copyright © 2008, Royal Society of Chemistry.

thickness of 5 nm. HRTEM image indicated that, at the curved edges, a large number of carbon fringes were observed, as shown in Fig. 3.7 (b). Selected area electron diffraction (SAED) result was a hexagonal single crystal pattern with bright diffraction spots. The inter-plane spacings of the spots (0.246 nm and 0.143 nm) could be indexed with the in-plane (hk0) lattices of GO. EDX analysis revealed the composition of the platelet, with 4.24 mol% Ti and 4.96 mol% O and 85.9 mol% C, as shown in the 'right-up' inset of Fig. 3.7 (b), which confirmed the formation of the Ti species–pillared GO structure.

Figure 3.7 (c) shows SEM image of the hydrothermal derived G-TNT hybrid. An assembly of delaminated graphene nanosheets was observed, with their surfaces decorated with tubular materials, without agglomeration. The tubular materials were distributed randomly on the surface of graphene nanosheets with the tubular axes being parallel to the graphene nanosheets. However, the packed structure of the graphene nanosheets was looser than that of GOTi, because of the growth of tubular materials in between the graphene nanosheets. Figure 3.7 (d) indicated that two-dimensional (2D) deposits and hollow tubes with an external diameter of 6–9 nm were formed along either flat or curved surfaces of the graphene nanosheets. HRTEM image (inset of Fig. 3.7 (d)) confirmed the presence of the hollow tubular materials, which had asymmetric layer numbers of walls and an interlayer distance of 0.76 nm, corresponding to $d_{(200)}$ of hydrogen trititanate [32, 33]. It can be understood that, during the hydrothermal treatment in concentrated alkali environment, the intercalated Ti species in the GOTi were converted into titanate nanotubes, where the graphene nanosheets acted as templates that directed the growth of the one-dimensional nanotubes in a 2D confinement.

A sol-gel-like anionic sulfate surfactant is used to assist the stabilization of graphene in aqueous solutions and facilitate the self-assembly of *in situ* grown nanocrystalline TiO_2, rutile and anatase, together with graphene nanosheets [28]. These nanostructured TiO_2-graphene hybrid materials were evaluated as anode materials of Li-ion batteries. The hybrid materials showed significantly improved Li-ion insertion/extraction capabilities. The specific capacity of the hybrid was doubled at high charge rates, as compared with the pure TiO_2, which was attributed to the increased conductivity of the hybrid electrode due to the formation of a percolated graphene network.

Figure 3.8 shows a schematic illustration of this approach, in which graphene was first dispersed with the aid of an anionic sulfate surfactant, i.e., sodium dodecyl sulfate (SDS). Self-assembly of the surfactants together with the metal oxide precursor then occurred, leading to the *in situ* crystallization of the precursors. Hybrids with both rutile and anatase TiO_2 have been prepared by using this method. The preparation of rutile TiO_2–G hybrid with 0.5 wt% graphene is described as an example. 2.4 mg of graphene was dispersed in 3 ml of SDS aqueous solution (0.5 mol·l^{-1}). The suspension was diluted to 15 ml and sonicated for 10–15 min. 25 ml of $TiCl_3$ (0.12 mol/L) aqueous solution was mixed with the as-prepared SDS–G suspension. Then, 2.5 ml of H_2O_2 (1 wt%) was added dropwise, followed by the addition of deionized water under vigorous stirring until a total volume of 80 ml was reached. The process was repeated to prepare hybrid materials with 0.17, 5 and 10 wt% graphene. For anatase hybrid, 5 ml of 0.6 M Na_2SO_4 was added as an additional item, while the rest of the steps were almost the same.

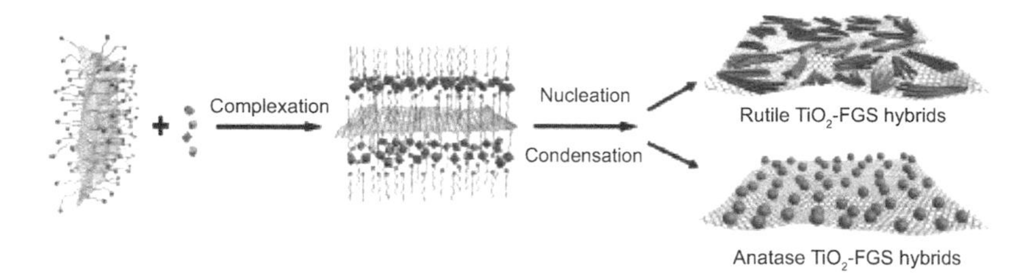

Fig. 3.8. Anionic sulfate surfactant mediated stabilization of graphene and growth of self-assembled G–TiO_2 hybrid nanostructures. Reproduced with permission from [28], Copyright © 2009, American Chemical Society.

Figure 3.9 shows XRD patterns of representative TiO_2–G hybrids with nanocrystalline rutile and anatase phases. Their crystal sizes were estimated to be 6 and 5 nm, respectively.

Figure 3.10 (a) shows TEM image of the graphene, with intrinsic microscopic roughening and out-of-plane deformations known as wrinkles. More than 80% of the graphene nanosheets were single layer. TEM and SEM images of the as-grown rutile TiO_2–G hybrid nanostructures are shown in Fig. 3.10 (b–e). As demonstrated in Fig. 3.10 (b) and (c), the graphene nanosheets were densely covered with TiO_2 nanoparticles. The nanostructured TiO_2 consisted of rod-like rutile nanocrystals that were formed in parallel interspaced with the SDS surfactants. The SEM image of Fig. 3.10 (d) revealed that the rod-like nanostructured rutile particles were randomly oriented on the graphene nanosheets. The nanostructured rutile particles were mostly on the graphene nanosheets with the rod length being parallel to the graphene surface, which was confirmed by the cross-sectional TEM image, as illustrated in Fig. 3.10 (e). Representative TEM images of the anatase TiO_2–G hybrid nanostructures are shown in Fig. 3.10 (f) and (g). Spherical aggregated anatase TiO_2 nanoparticles were homogeneously distributed on the graphene nanosheets.

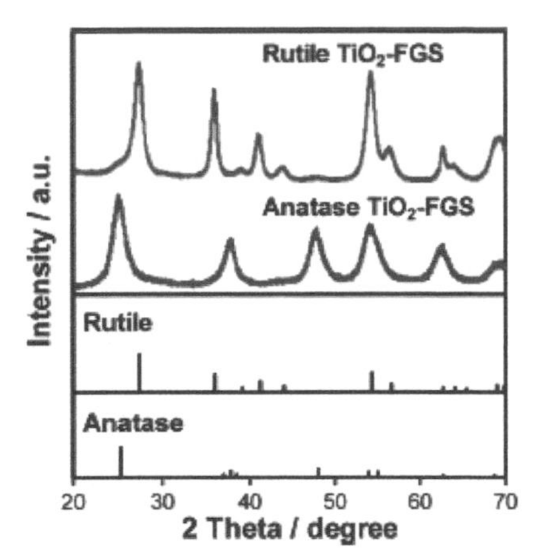

Fig. 3.9. XRD patterns of the G–TiO_2 (anatase) and rutile G–TiO_2 (rutile) hybrid materials. Standard diffraction peaks of anatase TiO_2 (JCPDS No. 21-1272) and rutile TiO_2 (JCPDS No. 21-1276) are shown as vertical bars. Reproduced with permission from [28], Copyright © 2009, American Chemical Society.

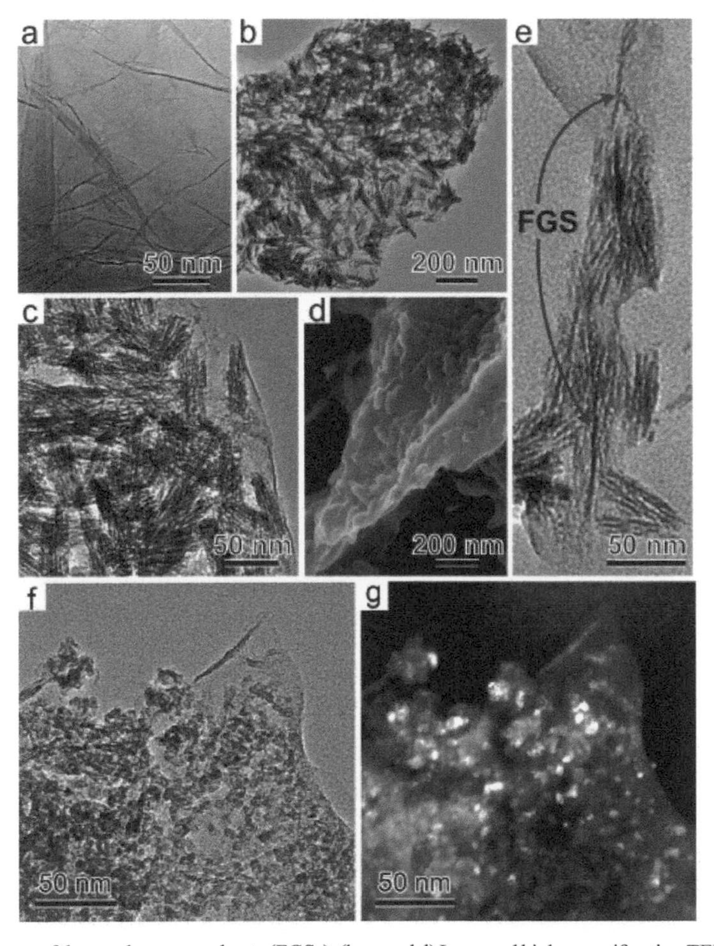

Fig. 3.10. (a) TEM image of the graphene nanosheets (FGSs). (b, c, and d) Low- and high-magnification TEM and SEM images of the self-assembled rutile TiO_2–G hybrids. (e) Cross-sectional TEM image of the rutile TiO_2–G hybrid with nanostructured rutile TiO_2 on the graphene nanosheets (FGS). (f) Plane-view TEM image of anatase TiO_2–G hybrid nanostructures. (g) Dark-field TEM image of the anatase TiO_2–G hybrid nanostructure. Reproduced with permission from [28], Copyright © 2009, American Chemical Society.

In this case, the SDS surfactant played an important role in determining the interfacial interactions between graphene and TiO_2 nanoparticles. The surfactant molecules were adsorbed onto the graphene nanosheets through the hydrophobic tails, so that they could be homogeneously dispersed in the suspensions and thus effectively interacted with the TiO_2 precursor through the hydrophilic head groups. Therefore, the three components were thoroughly mixed. Upon formation, the TiO_2 nanoparticles were attached to the graphene surfaces, because the surfactant sulfate head groups would be strongly bonded with TiO_2 [34]. Although the graphene nanosheets might also provide surface functional sites, e.g., carboxylate, epoxy, and hydroxyl groups, to bond with TiO_2 nanoparticles, the number of them was not sufficient to cater to all the oxide species. As a consequence, without the presence of surfactant, phase separation would take place.

Titanium oxide nanoparticle-graphene oxide (TiO_2–GO) and titanium oxide nanoparticle-reduced graphene oxide (TiO_2–rGO) hybrids were prepared by using TiF_4 as the precursor that was hydrolyzed at 60°C for 24 h [29]. In this approach, highly faceted anatase nanocrystals with petal-like morphologies were formed on surface and embedded in between the graphene nanosheets. At sufficiently high concentrations of GO, long-range ordered TiO_2–GO sheets were obtained due to self-assembly, if the reaction media was not stirred. TiO_2–GO hybrids could be well-dispersed at low concentrations of < 0.75 mg·ml^{-1} in water and ethanol. At too high titania concentrations, graphene papers could not be prepared with filtration through anodic membranes. The TiO_2–GO hybrid was reduced to TiO_2–rGO either chemically or thermally.

To prepare the hybrids, GO was dissolved in deionized water under sonication to yield GO solution. Titanium fluoride (TiF_4) was then added into the solution with vigorous stirring at various concentrations. The resulting suspensions were ultrasonicated and then heated at 60°C for 24 h. To obtain TiO_2–rGO hybrid, TiO_2–GO was suspended in deionized water with sonication and then reduced with hydrazine hydrate. The precipitates were collected with filtration and washed with water and methanol.

Figure 3.11 shows SEM images of the TiO_2–GO hybrids derived from solutions with different concentrations of TiF_4 [29]. At low concentrations of TiF_4, almost all TiO_2 nanoparticles possessed a highly faceted flower-like morphology, with cluster diameters of 200–400 nm, as shown in Fig. 3.11 (a). At high concentrations of TiF_4, elongated particles appeared, with sizes in the range of 150–250 nm in length and 100–150 nm in thickness. The number of the elongated particles increased gradually with the increasing concentration of TiF_4.

TiO_2 nanoparticles were embedded in between the GO layers. It was found that the presence of GO was a necessary requirement for the hydrolysis reaction to form TiO_2, which suggested the GO served as a template for the growth of TiO_2. The highly oxygenated GO, with hydroxyls and epoxides on the basal planes and carboxylic acids along the edges and at defect sites, acted as ligands for the hard Ti^{4+} Lewis acid, as discussed before [23]. An EDS spectrum is shown in Fig. 3.11 (f), which indicates that Ti and F were present in the hybrid. Therefore, Ti-F could have been formed, due to the bounding of F to the nanocrystalline TiO_2 surface.

As the concentration of GO was increased from 0.75 to 1.5 mg·ml^{-1}, long-range ordered assembly of TiO_2-GO hybrid was observed, if the reaction solutions were not stirred. This could be attributed to the self-assembly of the dispersed nanosheets of GO, GO-Ti^{4+}, or GO-TiO_2. Previous experiments indicated, that at concentrations of ≤ 3 mg·ml^{-1}, GO tended to self-assemble into flake- or sheet-like structures [35–38]. Larger strings of TiO_2 clusters, with diameters of 250–500 nm, were grown at the edges of the stacked graphene nanosheets. The stacks had spacing distances of 200–300 nm. For the reactions that were stirred, the staked sheet aggregates were not observed, even at significantly high GO concentrations. However, stirring promoted the conversion of TiF_4 to TiO_2.

Figure 3.12 shows representative TEM images of the TiO_2-GO hybrids [29]. The TiO_2 nanoparticles were found to have both flower-like morphologies of 250–320 nm in length and seed-shaped structures, together with structures between the two extremes, as shown in Fig. 3.12 (a–c). Dark-field image revealed that each single seed was a single crystal, as demonstrated in Fig. 3.12 (d). HRTEM images indicated that each nanoflower/particle consisted of very small crystallites, with crystal sizes of 2–4 nm, which were crystallographically aligned with same orientation. The TiO_2 nanoparticles possessed a porous microstructure, although no surfactant was used. Similar mesoporous crystallographically aligned anatase TiO_2 structures from TiF_4 hydrolysis were observed when using CNTs as template [39]. TGA/DTA experiments suggested that the GO was stabilized by the embedded TiO_2 [40]. It has been reported that

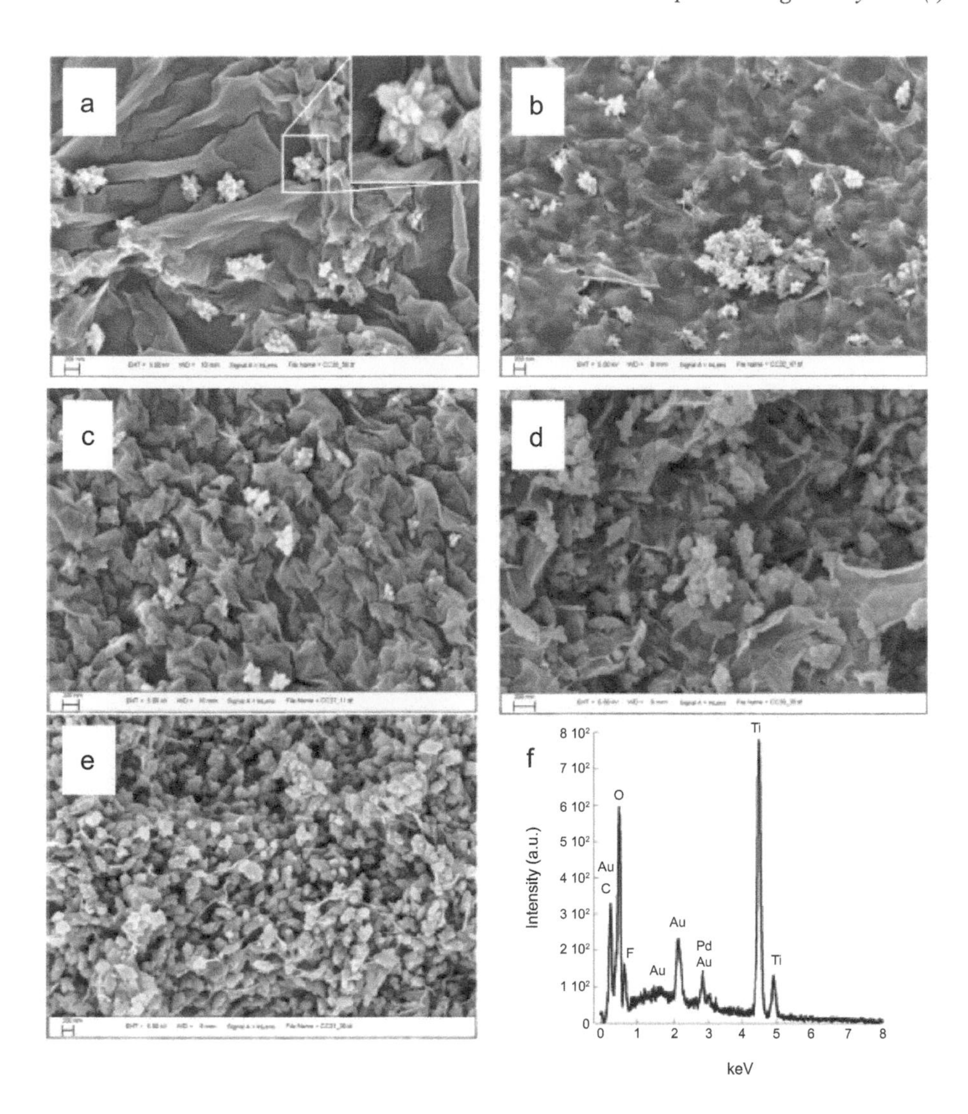

Fig. 3.11. SEM images of the TiO_2–GO hybrids from the solutions with different concentrations (M) of TiF_4: (a) 0.01, (b) 0.02, (c) 0.04, (d) 0.08, and (e) 0.16. (f) EDX pattern of the 0.08 M TiF_4 and TiF_4/GO weight ratio = 13.2 hybrid. All were from reactions with aqueous GO dispersion at 0.75 mg·ml^{-1}. Scale bars of (a–e) are 200 nm. Reproduced with permission from [29], Copyright © 2009, American Chemical Society.

GO can contain up to 25% of water [37]. Accordingly, the sample with a 2.1 weight ratio of TiO_2/GO was formed from an initial TiF_4/GO weight ratio of 13.2.

When heated at high temperatures, GO is reduced to graphene, or thermally reduced graphene oxide (TRGO) [40]. SEM analysis of the hybrid after TGA/DTA study showed that the TiO_2 nanoparticles on the surfaces of the hybrid had been sintered into elongated networks, with a diameter of about 1 μm, while the TiO_2 nanoparticles with seed or flower morphology inside the hybrid were transformed into a cube-like aggregate structure, with cube sizes of 50–200 nm, as demonstrated in Fig. 3.13 (a) and (b), respectively. This observation implies that the sintering behavior of the TiO_2 nanoparticles in the hybrid was affected by the graphene, which could be further explored from the ceramic sintering point of view. EDS results revealed that no F was detected in the TiO_2–TRGO hybrid, due to its loss during the heating process. Figure 3.13 (d) shows XRD pattern of the TiO_2–TRGO hybrid, indicating that the anatase TiO_2 was transferred to rutile.

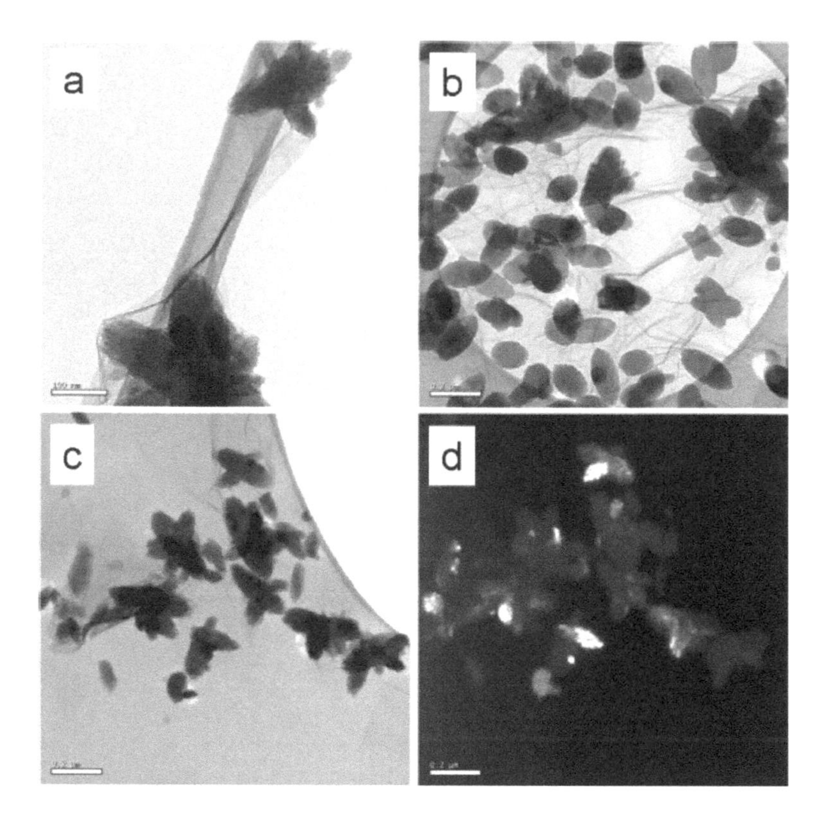

Fig. 3.12. TEM images of the TiO_2–GO hybrid from solution of 0.08 M TiF_4 and TiF_4/GO weight ratio = 13.2 and with aqueous GO dispersion at 0.75 $mg \cdot ml^{-1}$. (a-c) Bright-field images showing morphologies ranging from TiO_2 nanoflowers to TiO_2 seeds and graphene oxide. (d) Dark-field image showing that each petal is a single crystalline. Scale bar of (a) is 100 nm and scale bar of (b–d) is 0.2 μm. Reproduced with permission from [29], Copyright © 2009, American Chemical Society.

It was found that the reduction of GO in TiO_2–GO hybrid is faster than that of pure GO. The morphology of the TiO_2–rGO hybrid was similar to that of the original TiO_2–GO hybrid. SEM images revealed that the TiO_2 exhibited a certain degree of aggregation and that the surface of the graphene became smoother, as shown in Fig. 3.14 (a) and (b). Similarly, no F was detected in the TiO_2–RGO hybrid with EDS (Fig. 3.14 (c)). The presence of Si was attributed to potential contamination. Figure 3.14 (d) shows XRD pattern of the TiO_2–CRGO (chemically reduced graphene oxide) hybrid, confirming that the anatase TiO_2 experienced on phase transformation during the reduction.

One of the problems which have been encountered when using the sol-gel method with alkoxide precursors is the difference in dissolving behavior between the alkoxides and GO. Titanium alkoxides are usually dissolved in ethanol. When they are mixed with GO containing aqueous solutions, fast hydrolysis will take place in an uncontrolled way, resulting in the rapid precipitation of TiO_2. As a result, homogeneous hybrids cannot be obtained. However, this problem has been partly addressed by using precursors with a slow rate of hydrolysis in the presence of water, such as $Ti(OBu)_4$ [41]. In this respect, developing titanium alkoxides with a low hydrolysis rate could be a meaningful subject, in order to prepare homogeneous GO–TiO_2 hybrids by using the sol-gel process.

In summary, the sol–gel method has been employed to prepare various hybrids of TiO_2 and graphene or graphene oxide. Various modifications have also been introduced to further increase the efficiency of the sol-gel method. For example, Chen et al. [42] used prepared rGO–TiO_2 hybrids with different kinds of semiconductors, which could be controlled by adjusting the GO content in the starting solutions. Manga et al. [22] used an ionic salt, titanium (IV) bis(ammonium lactate) dihydroxide, as the precursor

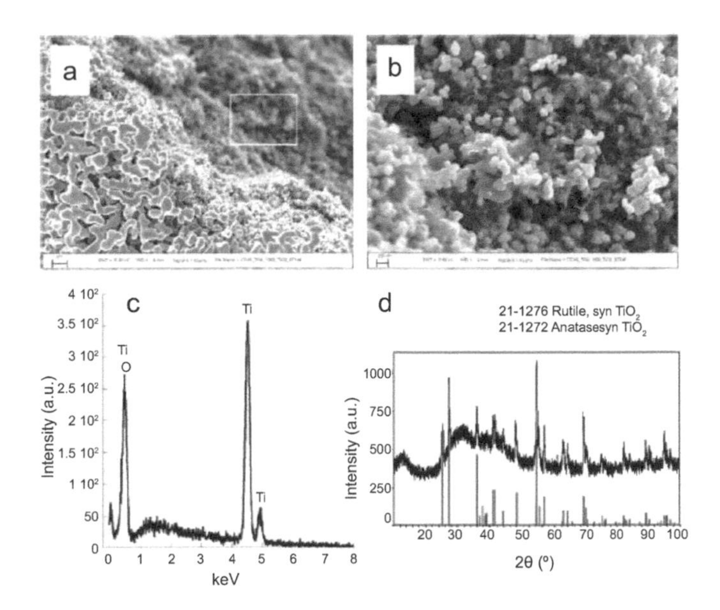

Fig. 3.13. SEM images of the thermally reduced graphene oxide-titania (TiO_2–TRGO) hybrid heated at 1000°C: (a) image showing surface TiO_2 sintered into about 1 μm wide globular networks (scale bar of 1 μm), (b) magnification of approximate area outlined in (a) showing morphology of the TiO_2 on inner layers of the hybrid (scale bar is 200 nm). Initial untreated sample was TiO_2–GO from solution of 0.08 M TiF_4 and TiF_4/GO weight ratio = 13.2. (c) EDX pattern of the TiO_2–TRGO. (d) XRD pattern of the TiO_2–TRGO hybrid, showing partial phase transformation to rutile phase. Reproduced with permission from [29], Copyright © 2009, American Chemical Society.

Color image of this figure appears in the color plate section at the end of the book.

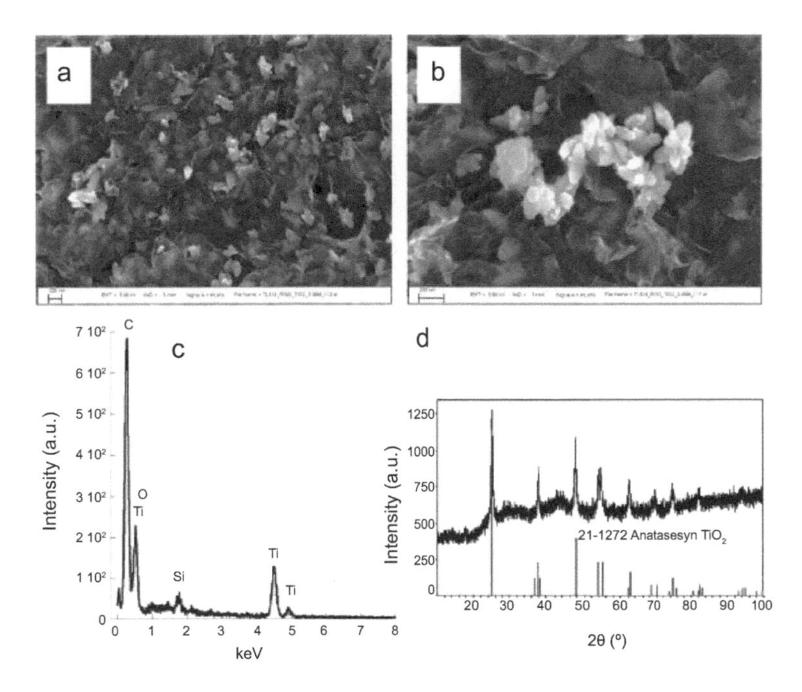

Fig. 3.14. (a, b) SEM images of the chemically reduced graphene oxide-titania (TiO_2–CRGO) hybrid – initial TiO_2 morphologies were retained (scale bars = 200 nm). Initial untreated sample was TiO_2–GO from solution of 0.08 M TiF_4 and TiF_4/GO weight ratio = 13.2. (c) EDX pattern of the TiO_2–CRGO hybrid, in which the presence of silicon was due to possible contamination. (d) XRD pattern of the TiO_2–CRGO, showing the retention of the anatase phase. Reproduced with permission from [29], Copyright © 2009, American Chemical Society.

of TiO_2, which had good water solubility and thus could be well mixed with GO solutions. It is expected that solution-based sol-gel method will be continuously popular in synthesizing TiO_2–G hybrid materials.

In hydrothermal and solvothermal synthesis, reactions are conducted at high temperature and high pressure. When GO–TiO_2 hybrids are synthesized by using these methods, GO is usually reduced to rGO. Complete reduction of GO into graphene can be achieved, if chemical reagents, such as hydrazine and $NaBH_4$, are included. Also, phase transformation of TiO_2 is generally observed. One of the advantages of these methods is their capability to offer various unique morphologies of the products. Safety is a critical requirement when using these high temperature and high pressure synthesis technologies. Representative examples will be described in the following section.

Zhang et al. [43] used a hydrothermal method to prepare a rGO–P25 hybrid, with crystalline structure and surface area similar to that of pure P25. At the same time, the original GO was reduced. GO was dissolved in the mixture of distilled H_2O and ethanol by using ultrasonic treatment, in which P25 was added and stirred to form a homogeneous suspension. The suspension was then placed in a Teflon-sealed autoclave and heated at 120°C for 3 h to prepare the hybrid. Figure 3.15 shows a representative TEM image

Fig. 3.15. Representative TEM image of the P25-G, with P25 loading on the surface of graphene and concentrating along the wrinkles. Reproduced with permission from [43], Copyright © 2010, American Chemical Society.

of the rGO–P25 hybrid. The hybrid exhibited a two-dimensional nanosheet structure, with micrometer-long wrinkles. Due to the presence of carboxylic acid groups on the GO nasnosheets, the P25 nanoparticles were dispersed on the graphene support and were prone to accumulate along the wrinkles and edges.

Liang et al. [44] proposed a two-step method to prepare a TiO_2–GO hybrid, in which TiO_2 was first dispersed on GO nanosheets through slow hydrolysis of the precursor $Ti(BuO)_4$ and then crystallized into anatase nanocrystals through hydrothermal treatment. The slow hydrolysis was realized in an EtOH–H_2O (15:1) solution. H_2SO_4 was used to direct the selective growth of TiO_2 on GO nanosheets. In the hydrothermal step, a H_2O–dimethylformamide (DMF) solution was used, where DMF was used to increase the dispersion of the graphene nanosheets and decrease their aggregation. The functional groups on the GO nanosheets provided reactive and anchoring sites for nucleation and growth of the TiO_2 nanoparticles, as shown in Fig. 3.16. In the first hydrolysis step, nearly amorphous TiO_2 fine particles were coated on GO nanosheets. The selective growth prevented the presence of TiO_2 in the solution, as shown in Fig. 3.16 (b)). Without H_2SO_4, the hydrolysis reaction was too fast, so that all TiO_2 nanoparticles were formed in the solution, but not on the GO nanosheets. Figure 3.16 (c) shows that the hydrothermal treatment promoted the crystallization of the amorphous TiO_2 nanoparticles into anatase nanocrystals. The crystalline characteristics of the TiO_2 are also confirmed by the TEM and XRD results, as shown in Fig. 3.17 (b–d).

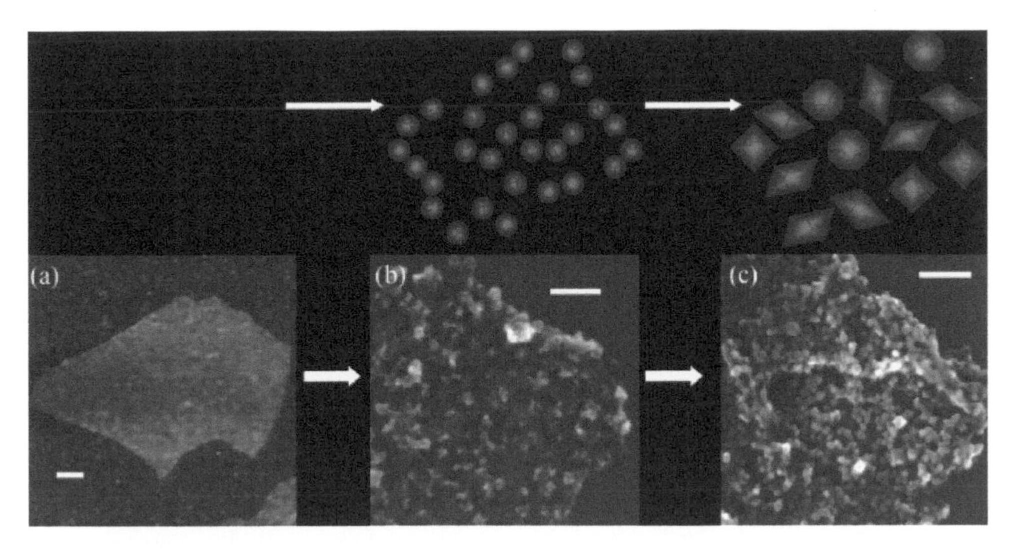

Fig. 3.16. Synthesis of TiO_2 nanocrystals on GO sheets. Top panel: Reaction scheme: (a) AFM image of a starting GO nanosheet; (b) SEM image of particles grown on a GO nanosheet after the first hydrolysis reaction step; (c) SEM image of TiO_2 nanocrystals on GO after hydrothermal treatment in the second step. Scale bars = 100 nm. Reproduced with permission from [44], Copyright © 2010, Springer.

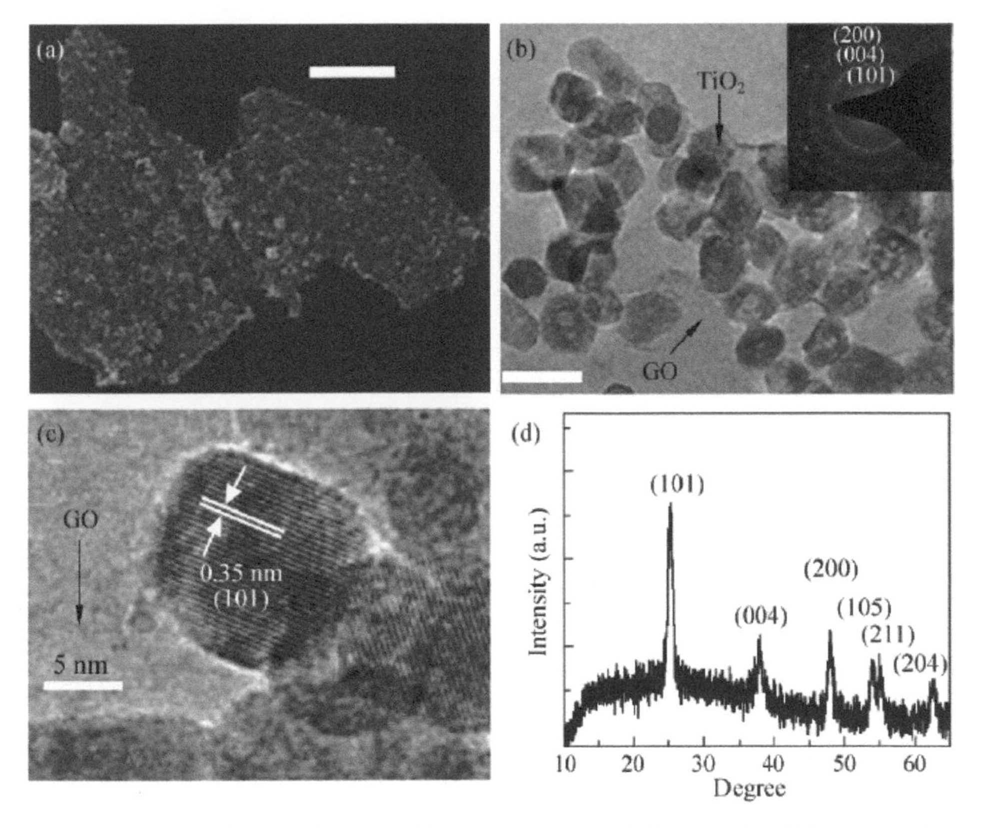

Fig. 3.17. (a) SEM image, (b) low magnification and (c) high magnification TEM images of the TiO_2 nanocrystals grown on GO nanosheets. The scale bar is 400 nm for the SEM image in (a) and 20 nm for the TEM image in (b). (d) An XRD pattern of the graphene–TiO_2 hybrid. Reproduced with permission from [44], Copyright © 2010, Springer.

The density of the TiO_2 nanocrystals coated on the graphene nanosheets could be controlled by the loading ratio of $Ti(BuO)_4$/GO. Figure 3.18 indicates that, with increasing content of $Ti(BuO)_4$, the density of the TiO_2 nanoparticles was increased. The average size of the TiO_2 nanoparticles could be controlled through the EtOH/water ratio in the first step. It was found that the average size was increased from about 15 nm in EtOH/water of 15/1 to about 30 nm in EtOH/water of 3/1. This observation was attributed to the fact that the rate of the hydrolysis reaction increased with the increasing content of water. However, higher content of water resulted in a decrease in the surface area of the TiO_2–GO hybrids. In such TiO_2–GO hybrids, the TiO_2 nanocrystals were directly grown on the graphene nanosheets, so that there were strong interactions between the two components. This strong coupling is expected to be an advantage for various applications.

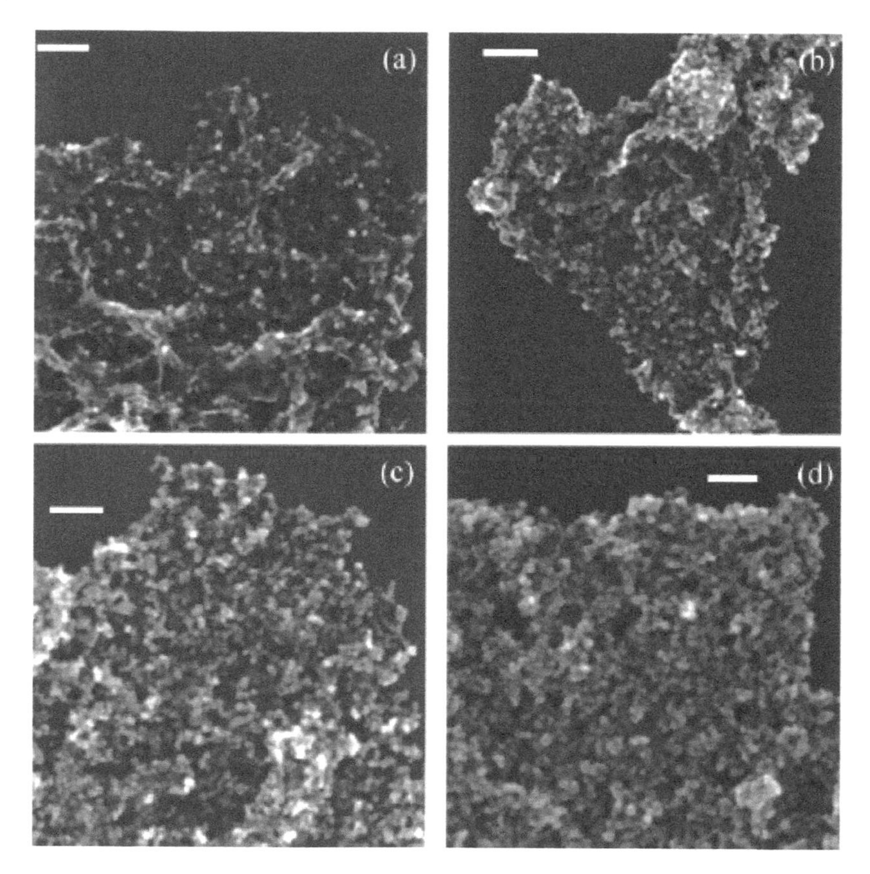

Fig. 3.18. SEM images of the graphene–TiO_2 hybrids formed with various GO/TiO_2 mass ratios: (a) 1:1, (b) 1:3, (c) 1:9 and (d) 1:18. The scale bars are 100 nm. The TiO_2 coating density was dependent on the loading ratio of $Ti(BuO)_4$/GO. Reproduced with permission from [44], Copyright © 2010, Springer.

TiO_2 nanospindles have also used to prepare graphene-based hybrids, by using a hydrothermal process with TiO_2 nanotubes as self-sacrificing precursors [45]. GO in the as-synthesized hybrids was reduced by using thermal treatment in ammonia atmosphere. Furthermore, the TiO_2 surface was effectively nitrated into TiN_xO_y or even TiN. The hybrids have been evaluated for lithium ion battery applications. An overview of the synthesis mechanism is illustrated in Fig. 3.19.

To synthesize titanate nanotube precursor, TiO_2 (Degussa P25, 20% rutile + 80% anatase, with particle sizes of 20–30 nm) was dispersed in 10 M NaOH aqueous solution. The resulting suspension was hydrothermally treated in a Teflon-lined stainless steel autoclave at 150°C for 20 h. The resulting titanate nanotube precipitate was washed with 0.1 M HCl solution until pH = 1–2. TiO_2 nanospindles with different sizes were also derived from the titanate nanotube precursor. The precursor was re-dispersed in

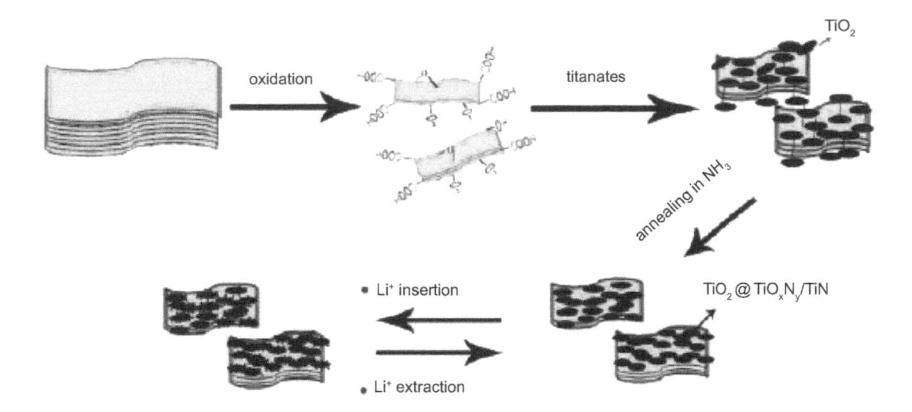

Fig. 3.19. Schematic diagram demonstrating the synthesis and the subsequent lithiation-delithiation processes of the TiO_2@TiN_xO_y/TiN–graphene nanohybrids. Reproduced with permission from [45], Copyright © 2010, American Chemical Society.

mixtures with different compositions of water and ethanol or ethylene glycol, after which dimethylamine was added and mixed. The resulting mixture was hydrothermally treated at 180°C for 12 h. To synthesize TiO_2 nanospindles–GO hybrids, TiO_2 nanospindle powder was dispersed in GO solution, which was then hydrothermally reacted in a similar way. The TiO_2 nanospindles–GO hybrid powders were thermally treated in NH_3 to produce TiO_2@TiO_xN_y/TiN nanospindles–graphene hybrids.

Figure 3.20 shows SEM (a) and TEM (b) images of the precursor. It exhibited a tubular morphology, with an average diameter of 10 nm and lengths ranging from hundreds of nm to a few μm. XRD patterns indicated that the titanate precursor was transformed into anatase TiO_2 after the hydrothermal treatment.

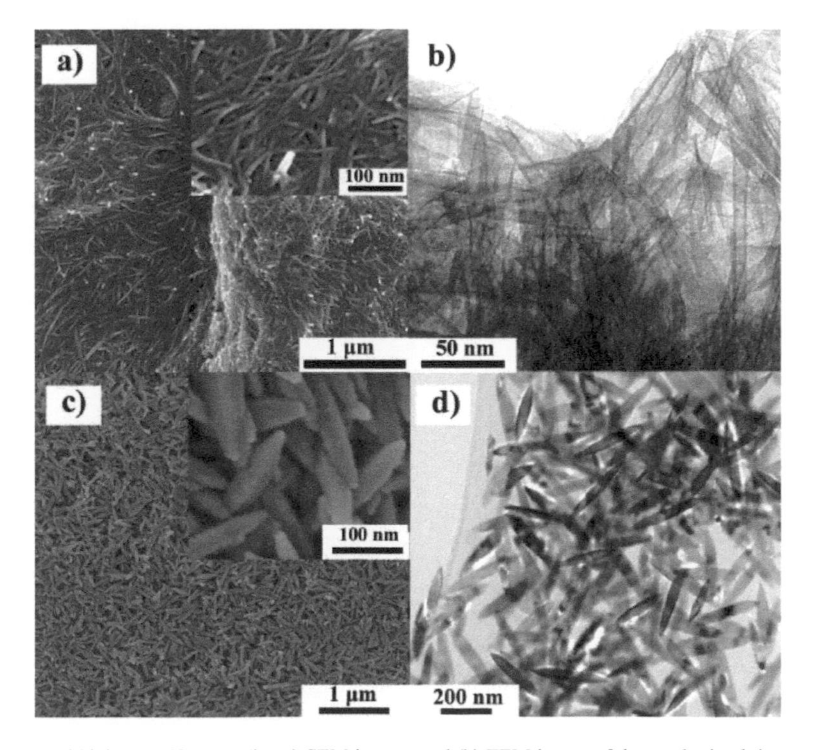

Fig. 3.20. (a) Low and high magnification (inset) SEM images and (b) TEM image of the as-obtained titanate precursor. (c) Low and high magnification (inset) SEM images and (d) TEM image of the as-synthesized TiO_2 nanospindles. Reproduced with permission from [45], Copyright © 2010, American Chemical Society.

The as-obtained anatase TiO_2 possessed quite high crystallinity and high purity. As shown in Fig. 3.20 (c), the anatase TiO_2 had a spindle-like morphology with a uniform size distribution. According to Fig. 3.20 (d), the TiO_2 spindles were 30–50 nm in diameter and 200–300 nm in length. HRTEM results revealed that the TiO_2 spindles were single crystal and grew preferably along the direction of the [001] axis. The tip was confined by one set of (101) outer surfaces, which tended to adsorb dimethylamines [46, 47].

Figure 3.21 shows TEM images of the precursor hydrothermally treated with the presence of alkyl amines in a mixture of ethanol and ethylene glycol with a volume ratio of 1:1 for different time durations. It was found that, some tiny nanospindles were formed at the initial stage. The nanospindles were attached perpendicularly on the precursor nanotubes. With increasing reaction time duration, the number of the tiny nanospindles was increased. Gradually, the tubular-structured precursor disappeared, whereas densely distributed nanospindles were formed. More importantly, the size of the nanospindles was dependent on the viscosity of the reaction solvent.

Fig. 3.21. TEM images of the products obtained after different periods of hydrothermal reaction time durations: (a) 0, (b) 0.5, (c) 1.5 and (d) 3 h. Reproduced with permission from [45], Copyright © 2010, American Chemical Society.

In the formation of the TiO_2 nanospindles, the roles played by water and alkyl amines were important. It has been accepted that the layer in the layered titanate consists of corrugated ribbons of edge-shared TiO_6 octahedra, with a width of about three TiO_6 octahedra, which are laterally linked through corner sharing and separated by H ions from the neighboring layers [48–50]. The crystal structure of the titanate $H_xTi_3O_7$ is shown in Fig. 3.22. Because of this layered structure, the titanate was exfoliated into monomer or oligomer units, which were assembled into the anatase nanospindles. During the exfoliation process, anatase phase was formed, due to the nucleation and growth from the monomer or oligomer building blocks, as described in Fig. 3.22.

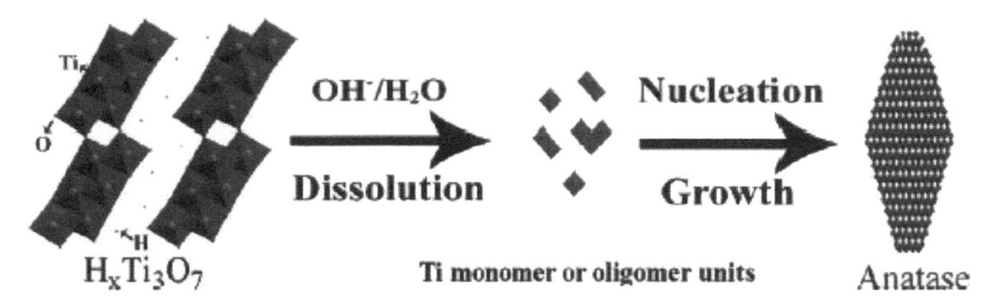

Fig. 3.22. Possible formation mechanism of the anatase nanospindles. Reproduced with permission from [45], Copyright © 2010, American Chemical Society.

It was proposed that the dissolution of the titanate nanotubes started from their outer surface, while the nucleation of the dissolved units occurred at the same time. In this case, the undissolved sites served as seeds for the growth of the spindle-shaped TiO_2 nanocrystals. Accordingly, the nanospindles were grown in the direction perpendicular to the precursor nanotubes. Because of its high viscosity, ethylene glycol could reduce the growth rate of the spindle crystals to a certain degree. This hypothesis has been experimentally confirmed, i.e., if pure ethylene glycol was used as the reaction solvent, only very small TiO_2 nanoparticles could be obtained [45].

Lee et al. used the hydrothermal method to prepare graphene-wrapped anatase TiO_2 nanoparticles (NPs) with enhanced photocatalytic performance under visible light irradiation [51]. Amorphous TiO_2 spherical NPs were synthesized by using a sol-gel process with titanium (IV) isopropoxide as the precursor [52]. The TiO_2 NPs were dispersed in ethanol with sonication, in which 3-aminopropyl-trimethoxysilane (APTMS) was added. After that, the solution was heated and refluxed. The APTMS-treated TiO_2 NPs were thoroughly washed with ethanol to remove the APTMS moiety. GO suspension with a concentration of 0.2 mg·ml^{-1} was added into the amine-functionalized TiO_2 NP dispersion under vigorous stirring at pH 6. Various weight ratios of GO to TiO_2 NPs were examined. The mixtures were then centrifuged and washed with deionized water. The amorphous TiO_2 was converted to the anatase phase through hydrothermal treatment in a solution of ethanol and deionized water at 180°C for 16 h. The hydrothermally reacted samples were calcined at 400°C in Ar to remove all the organic components and obtain highly crystalline graphene-TiO_2 NPs. Graphene-TiO_2 NPs were obtained by using a two-step hydrothermal process. Crystallized anatase TiO_2 NPs were first wrapped by GO nanosheets and then the GO was reduced to graphene by the same hydrothermal treatment and calcination. The process and representative results are shown in Fig. 3.23.

There are other examples of TiO_2–GO hybrids synthesized by using the hydrothermal method. For instance, a two-phase approach was used to develop hybrids with a uniform distribution of TiO_2 nanorods on GO nanosheets [53, 54]. TiO_2 nanorods were dispersed in toluene and stabilized with oleic acid to form phase one. GO nanosheets were dispersed in deionized water to make phase two. The self-assembly of the two components took place at the water/toluene interface. Glucose has been used to replace hydrazine in the hydrothermal method, as an ecofriendly chemical, to reduce GO to rGO. For instance, an rGO–TiO_2 hybrid was prepared by using glucose as the reducing agent in a one-step hydrothermal synthesis. However, glucose is not able to completely reduce GO under hydrothermal conditions, while the reduction mechanism has not been convincingly clarified [55]. Nevertheless, the hydrothermal method has been acknowledged to be an effective route in synthesizing GO–TiO_2 hybrids, especially in terms of GO reduction, as compared with the UV-assisted photocatalytic approach and hydrazine method [56].

Solvothermal method has been used to synthesize GO–TiO_2 hybrids to avoid the uncontrollable hydrolysis of titanium alkoxide precursors [57–59]. For example, Zhou et al. reported a solvothermal route to produce a rGO–TiO_2 hybrid, from a GO dispersion and $Ti(OBu)_4$ solution in isopropanol [57]. Figure 3.24 shows SEM and TEM images of the G–TiO_2 hybrid. As illustrated in Fig. 3.24 (a), the graphene sheets were decorated with a dense layer of TiO_2 nanoparticles. The TiO_2 nanoparticles could be clearly identified due to the absence of aggregation. The fringe of graphene is shown in Fig. 3.24 (c), confirming the presence of graphene nanosheets. The TiO_2 nanoparticles had an equiaxed geometry, with a narrow size distribution of about 15 nm. HRTEM image (Fig. 3.24 (d)) revealed that the crystalline TiO_2 was anatase.

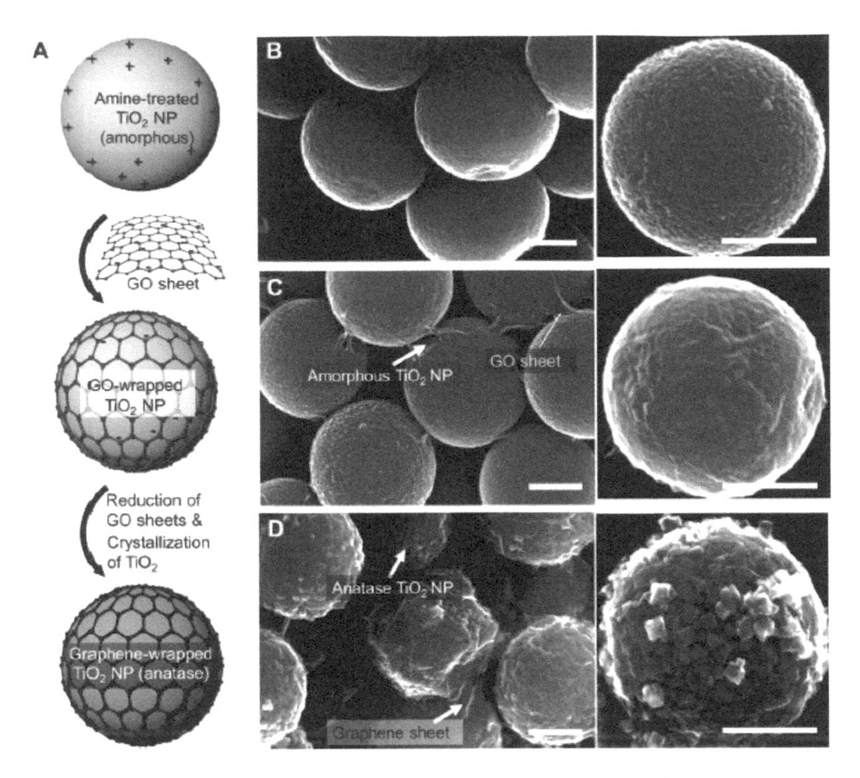

Fig. 3.23. Schematic illustration of the synthesis steps for graphene wrapped anatase TiO_2 NPs and the corresponding SEM images. (A) Synthesis steps of graphene-wrapped TiO_2 NPs; the surface of amorphous TiO_2 NPs was modified by APTMS and then wrapped by graphene oxide (GO) nanosheets through electrostatic interaction. Graphene-wrapped anatase TiO_2 NP was synthesized through one-step hydrothermal GO reduction and TiO_2 crystallization. (B) SEM images of bare, amorphous TiO_2 NPs prepared by sol-gel method. (C) SEM images of GO-wrapped amorphous TiO_2 NPs. The weight ratio of GO to TiO_2 was 0.02:1. (D) SEM images of graphene-wrapped anatase TiO_2 NPs. Scale bar: 200 nm. Reproduced with permission from [51], Copyright © 2012, John Wiley & Sons.

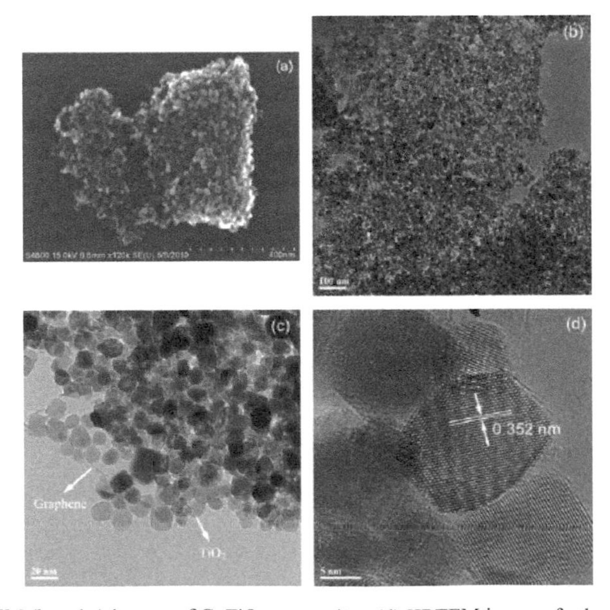

Fig. 3.24. SEM (a) and TEM (b and c) images of G–TiO_2 composites. (d) HRTEM image of selected area. Reproduced with permission from [57], Copyright © 2012, The Royal Society of Chemistry.

3.2.1.2. Graphene-VO$_x$

Due to their variable chemical valence, vanadium oxides include mainly V_2O_5 and VO_2. Vanadium oxides have been intensively studied because of their wide range applications. Bulk vanadium oxides have exhibited a variety of fascinating electronic, magnetic and structural properties [60]. The nanostructured V_2O_5 and VO_2 have found potential applications in various areas, such as gas sensors [61–63], electromechanical actuators [64], room-temperature ferromagnetic materials [65], catalysts [66] and electrode material for lithium–ion batteries [67–69]. Graphene–VO$_x$ hybrids include G–V_2O_5 [70–73] and G–VO_2 [74–79]. G–V_2O_5 hybrids have been mainly used as cathode materials for lithium ion batteries, while G–VO_2 hybrids have unique thermochromic properties. Selected samples will be discussed in this section to show the preparation and characterization of this group of hybrid materials.

Rui et al. reported a solvothermal approach to prepare reduced graphene oxide (rGO) supported highly porous polycrystalline V_2O_5 spheres, V_2O_5–rGO hybrid [70]. During the solvothermal process, reduced vanadium oxide (rVO) nanoparticles with sizes in the range of 10–50 nm were deposited on rGO nanosheets through heterogeneous nucleation. The rVO nanoparticles were subsequently oxidized to V_2O_5 after annealing in air at 350°C, and were assembled into polycrystalline porous spheres with sizes of 200–800 nm. The weight ratio between the rGO and V_2O_5 could be well-controlled by controlling the weight ratio of the precursors. Morphology of the V_2O_5–rGO hybrids was dependent on the weight ratio. The V_2O_5–rGO hybrids exhibited promising performances as cathode materials of lithium ion batteries. Figure 3.25 shows synthetic procedure of the V_2O_5–rGO hybrids, starting with vanadium isopropoxide (VO(OiPr)$_3$) as the precursor.

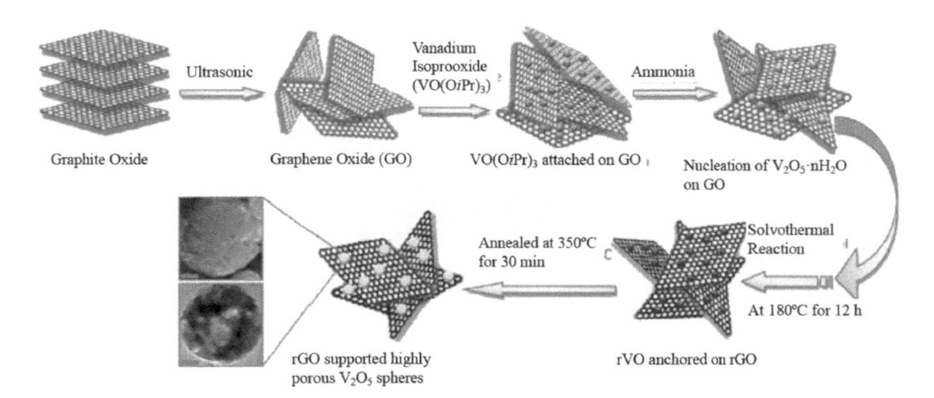

Fig. 3.25. Schematic diagram of the production of highly porous V_2O_5 spheres on the surface of rGO nanosheets. Scale bar in SEM and TEM images is 100 nm. Reproduced with permission from [70], Copyright © 2011, The Royal Society of Chemistry.

Figure 3.26 shows SEM and TEM images of the V_2O_5–rGO (46 wt%) hybrid [70]. The V_2O_5 spheres, with diameters in the range of 200–800 nm, were deposited onto the rGO nanosheets, as shown in Fig. 3.26 (a). High magnification SEM image of Fig. 3.26 (b) revealed that the porous V_2O_5 spheres contained pores with sizes of about 50 nm. As shown in Fig. 3.26 (c), some V_2O_5 spheres were broken, probably due to their hollow structure. Porous structure of the V_2O_5 spheres was confirmed by the TEM images of Fig. 3.26 (d) and (e). The lattice fringes of Fig. 3.26 (f) with a spacing of 0.34 nm could be indexed to the (110) planes of orthogonal V_2O_5. The highly porous V_2O_5 spheres were formed from rVO nanoparticles during the annealing process. The V_2O_5 spheres were attached on the surface of the rGO nanosheets so strongly that they could withstand long time sonication. Such a strong adhesion would ensure charge transport, due to the highly conductive rGO nanosheets. It was found that the diameter of the V_2O_5 spheres in the V_2O_5–rGO hybrids increased as the content of rGO was decreased.

Another example was reported by Liu and Yang [72], who developed a hybrid material consisting of graphene nanosheets and V_2O_5 nanowires. The ultra-long V_2O_5 single crystal nanowires were deposited on transparent graphene substrate, which possessed high electrochemical properties, as cathode of lithium ion

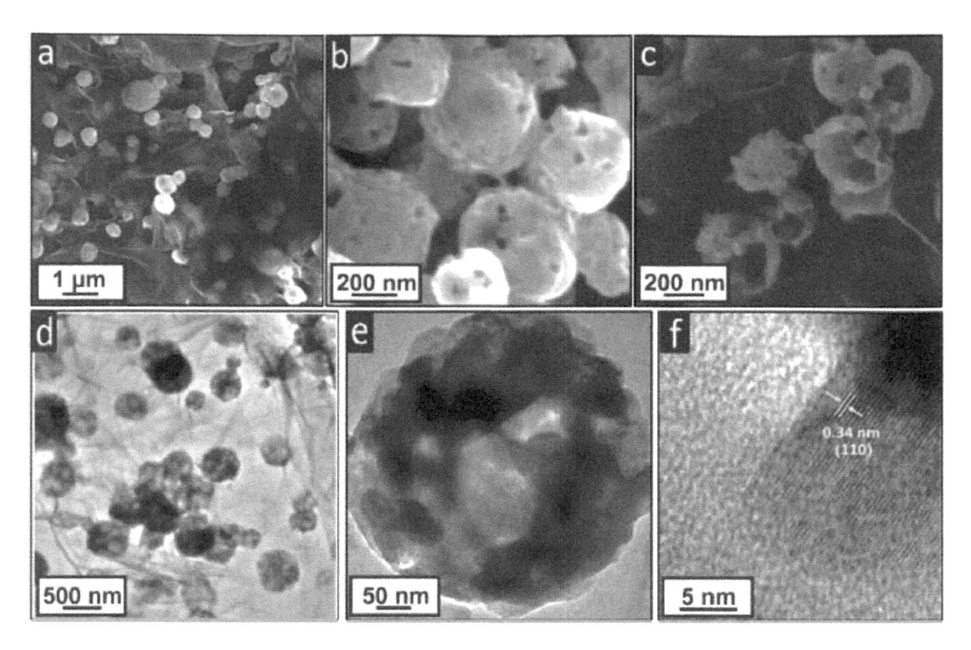

Fig. 3.26. Representative SEM (a–c) and TEM images (d–e) of the as-synthesized V_2O_5–rGO (46 wt%) hybrid, with the V_2O_5 spheres being highly porous. The SEM image in (c) shows some broken V_2O_5 spheres, which implies that they could have hollow structures. (f) HRTEM image of the V_2O_5 in the V_2O_5–rGO (46 wt%) hybrid. Reproduced with permission from [70], Copyright © 2011, The Royal Society of Chemistry.

batteries, with high initial discharge capacities and excellent rate capacities. The concurrent effects of the unique nanostructure of the V_2O_5 wires with short diffusion pathway for lithium ions and the high electrical conductivity of the graphene nanosheets in the hybrid were responsible for its outstanding electrochemical performances. More importantly, the fabrication process was environmentally friendly, because strong reduction and oxidation reagents were not used and no toxic gases were released. Detailed procedures and potential electron transfer behavior of the hybrid materials are illustrated in Fig. 3.27.

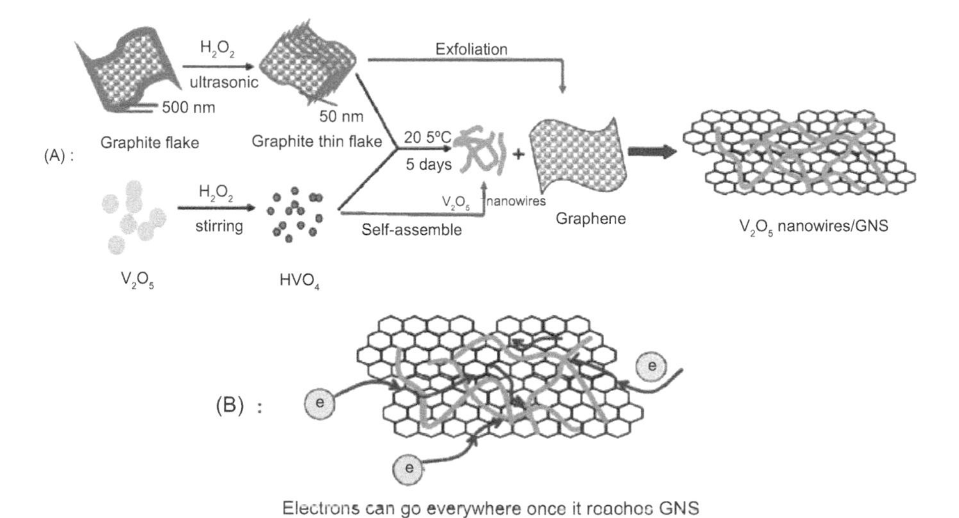

Fig. 3.27. (A) Schematic illustration of the synthesis procedure of the V_2O_5 nanowire–graphene nanosheet hybrid. (B) Ideal electron-transfer pathway of the V_2O_5 nanowire–graphene hybrid. Reproduced with permission from [72], Copyright © 2011, The Royal Society of Chemistry.

Low magnification TEM images of the hybrid are shown in Fig. 3.28 (a) and (b). The broken V_2O_5 nanowires were caused by the ultrasonic treatment during the TEM sample preparation, while graphene nanosheet layers were clearly visible. Figure 3.28 (c) shows a cross-sectional TEM image of a thick graphene layer, demonstrating an interlayer spacing of 0.34 nm. In selected areas, the graphene platelet thickness was 3–7 nm, corresponding to 10–20 graphene layers, as shown in Fig. 3.28 (d). HRTEM images and SAED patterns are shown in Fig. 3.28 (e) and (f), indicating the single crystal states of the V_2O_5 nanowires. The inter-planar d-spacing of 0.41 nm corresponded to the (101) lattice planes of orthorhombic V_2O_5.

Hybrids with sandwich-like structures are shown in Fig. 3.28 (g) and (h). Some nanowires were wrapped by GNS, whereas other wires were attached on the graphene surface. The graphene layers that covered the individual V_2O_5 nanowire in Fig. 3.28 (h) were 2–5 nm thick, consisting of 6–15 graphene layers. Such sandwich structures make it easier for the electrons to transfer through the hybrid.

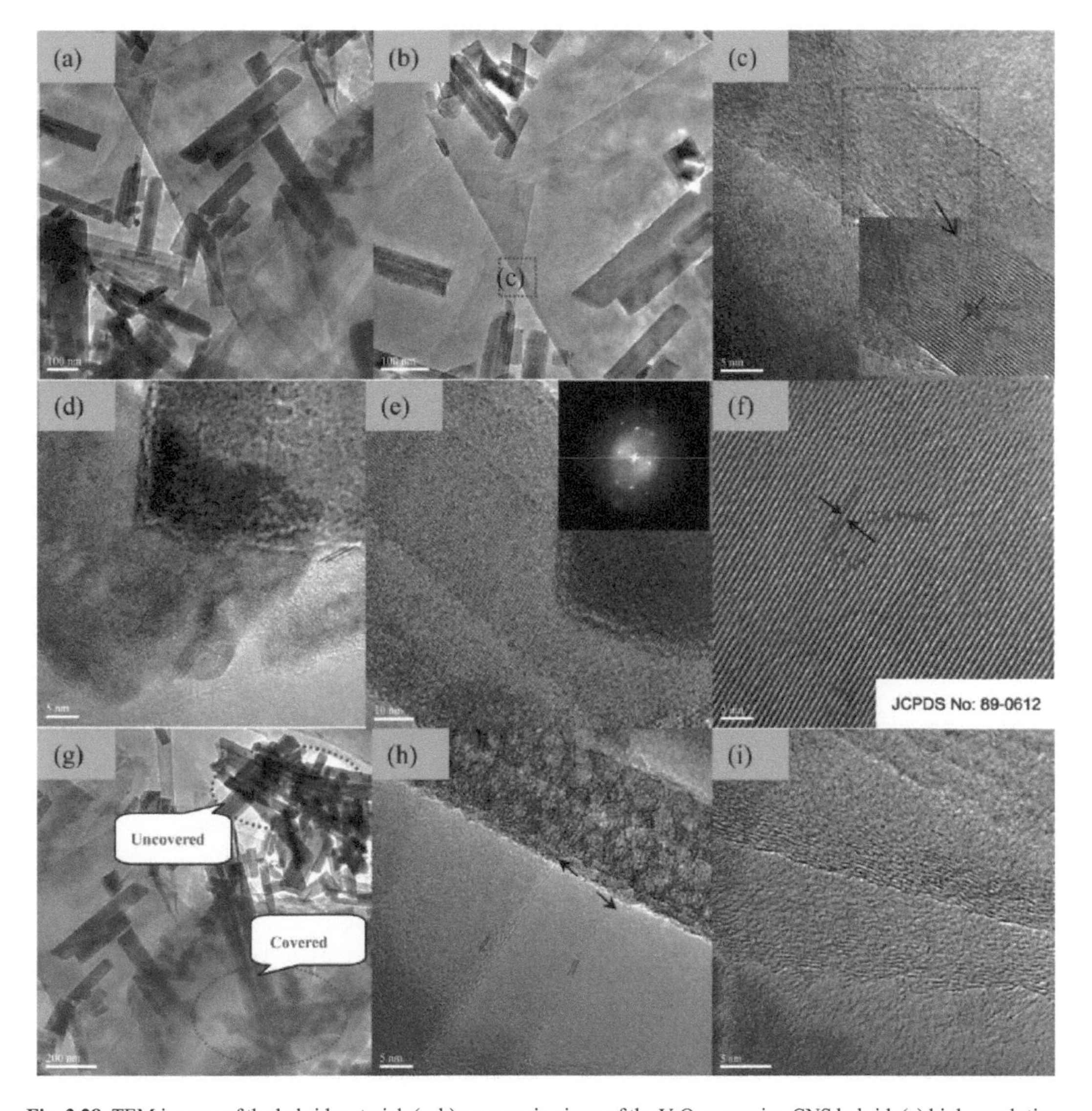

Fig. 3.28. TEM images of the hybrid material: (a, b) panoramic views of the V_2O_5 nanowire–GNS hybrid, (c) high-resolution TEM image showing the direct and visible graphene nanosheet stacking along the (002) with a layer spacing of 0.34 nm, (d) a closed view of an individual V_2O_5 nanowire hybrid with a very thin graphene layer, (e, f) lattice fringes of the V_2O_5 nanowires, (g, h) selected sandwich-like structures showing that the V_2O_5 nanowires were wrapped by graphene, (i) cross-sectional HRTEM image of selected graphene layers. Reproduced with permission from [72], Copyright © 2011, The Royal Society of Chemistry.

Figure 3.28 (i) shows a cross-sectional TEM of a stacking sheet structure of the graphene, in which individual graphene monolayers were clearly observed. The content of graphene in the hybrid was about 3.89% [72].

A simple bottom-up approach was reported to prepare hybrids of single crystalline VO_2 ribbons with graphene nanosheets [79]. The VO_2-graphene ribbon hybrids could be used with electrode properties for high-power lithium ion batteries. The VO_2-graphene ribbon hybrids were synthesized by using hydrothermal synthesis and simultaneous chemical reduction, as schematically illustrated in Fig. 3.29 (a). V_2O_5 was initially dissolved in water and dispersed onto the surface of GO nanosheets, during which V_2O_5 was gradually reduced to VO_2 ribbons, as shown in Fig. 3.29 (b) and (c). The resulting ribbons and residual graphene oxide nanosheets were assembled into 3D interpenetrating architectures. Compositions of the VO_2-graphene hybrids were controlled by adjusting the ratio of V_2O_5 and GO.

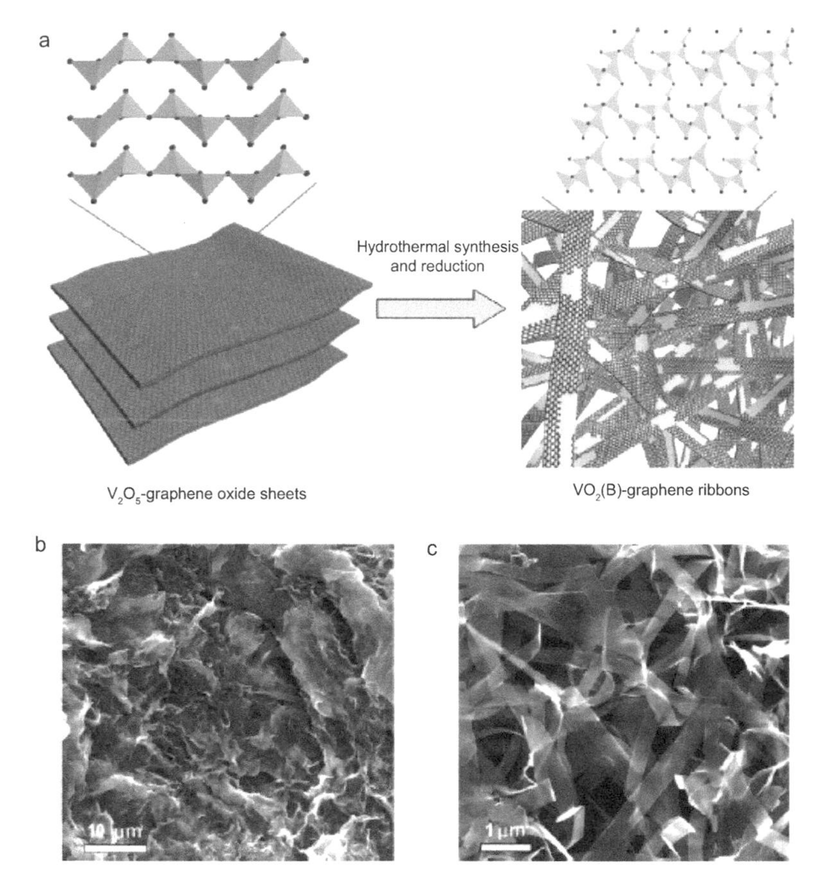

Fig. 3.29. Synthesis of VO_2–graphene hybrid ribbons. (a) Formation of the VO_2–graphene ribbons through hydrothermal synthesis and simultaneous reduction of layered V_2O_5–graphite oxide at 180°C. Structural model of layered, orthorhombic V_2O_5 phase projected along [001] facet is shown on the top left of (a), while the ideal model of the resulting monoclinic VO_2(B) phase projected along [010] facet is shown on the top right of (a), where the edge-sharing VO_6 octahedra is visible. Representative SEM images of (b) V_2O_5–graphite oxide hybrid and (c) VO_2–graphene sample hydrothermally treated for 1.5 and 12 h. Reproduced with permission from [79], Copyright © 2013, American Chemical Society.

The ribbons were 200–600 nm in width and tens of μm in length, as shown in Fig. 3.29 (c) and Fig. 3.30 (a). AFM images indicated that the ribbons exhibited a uniform thickness of ~10 nm. HRTEM image and SAED pattern are shown in Fig. 3.30 (b-d), demonstrating that the ribbons were single crystals. The lattice fringes with a spacing of 0.21 nm corresponded to the spacing of (003) the planes of VO_2(B). The nanosheets were tightly covered with graphene layers, which were not continuous. It was found that

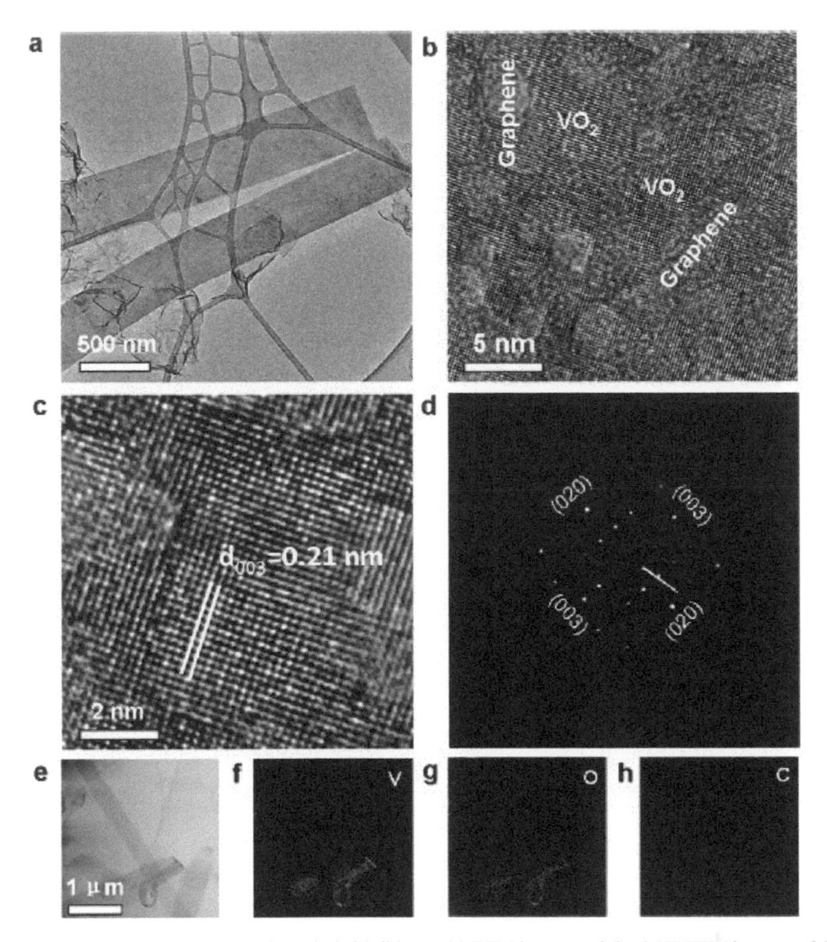

Fig. 3.30. Characterization of the VO_2–graphene hybrid ribbons. (a) TEM image and (b, c) HRTEM images of the VO_2 ribbons, demonstrating the graphene nanosheets on the surface of the crystalline ribbons. The exposed lattice fringes shown in (c) have a spacing of 0.21 nm, corresponding to the (003) plane of VO_2(B). (d) Diffraction patterns with well-defined arrays of dots, demonstrating the single crystalline feature of the VO_2(B) ribbons. (e) STEM image and corresponding elemental mappings of (f) vanadium, (g) oxygen and (h) carbon, indicating homogeneous dispersion of all elements in the ribbons. Reproduced with permission from [79], Copyright © 2013, American Chemical Society.

all V, O and C elements were homogeneously distributed in the ribbons, as shown in Fig. 3.30 (e–h). Nitrogen adsorption–desorption isotherms of the VO_2–graphene hybrid exhibited a typical II hysteresis loop at a relative pressure between 0.6 and 0.9, with characteristic pores of different sizes.

3.2.1.3. Graphene-Cr_2O_3

Cr_2O_3 is one of the potential anode materials due to its high theoretical capacity (1058 mA h g^{-1}) and relatively low electromotive force value (1.085 V) [80–82]. Reports on graphene–Cr_2O_3 hybrids are relatively less, as compared with those based on other oxides. Currently, the graphene–Cr_2O_3 hybrids reported in the open literature are mainly used as anode materials for lithium ion batteries [83–85]. A representative example reported by Yue et al. [83] is discussed as follows.

CrOOH was first grown on the surface of graphene nanosheets by using a hydrothermal method, and was then thermally decomposed to Cr_2O_3 [83]. More significantly, carbon layers were formed from the glucose on the CrOOH particles that prevented their aggregation. Specifically, $Cr(NO_3)_3{\cdot}9H_2O$ was dissolved in a GO solution with a concentration of 0.5 g l^{-1}. After stirring for 0.5 h, urea was added into the solution. Hydrothermal treatment of the solution was conducted at 120°C for 5 h to form G–CrOOH. The black powder was collected and then thermally treated in N_2 at 650°C for 2 h, leading to G–Cr_2O_3 hybrid. To produce carbon-coated G–Cr_2O_3 hybrids, either G–CrOOH or G–Cr_2O_3 was dispersed in distilled water by ultrasonication, in which glucose was added. The solutions were hydrothermally treated at 160°C for 10 h. After that, the precipitates were collected and then heated in N_2 at 650°C for 2 h to obtain G–Cr_2O_3–C1 and G–Cr_2O_3–C2, respectively. The formation procedure of the hybrids is shown in Fig. 3.31. The formation of CrOOH and the reduction of GO into G were confirmed by XRD patterns. The CrOOH was of rhombohedral structure of $HCrO_2$. After thermal annealing, the CrOOH was converted into rhombohedral Cr_2O_3.

Figure 3.32 shows representative TEM and SEM images of the G–CrOOH, G–Cr_2O_3, G–Cr_2O_3–C1 and G–Cr_2O_3–C2 samples [83]. As shown in Fig. 3.32 (A), a wrinkled graphene nanosheet was observed in the G–CrOOH hybrid, but no particles were found on the surface of the sample. EDX confirmed the presence of the three elements, C, O and Cr. The calcination procedure in N_2 led to the formation of crystalline Cr_2O_3 nanoparticles (Fig. 3.32 (B)). However, the problem of agglomeration was addressed as carbon coating was used, as shown in TEM image of the sample G–Cr_2O_3–C1 (Fig. 3.32 (C)). Lattice spacings of 0.216 and 0.248 nm, corresponding to the (113) and (110) planes of Cr_2O_3, were

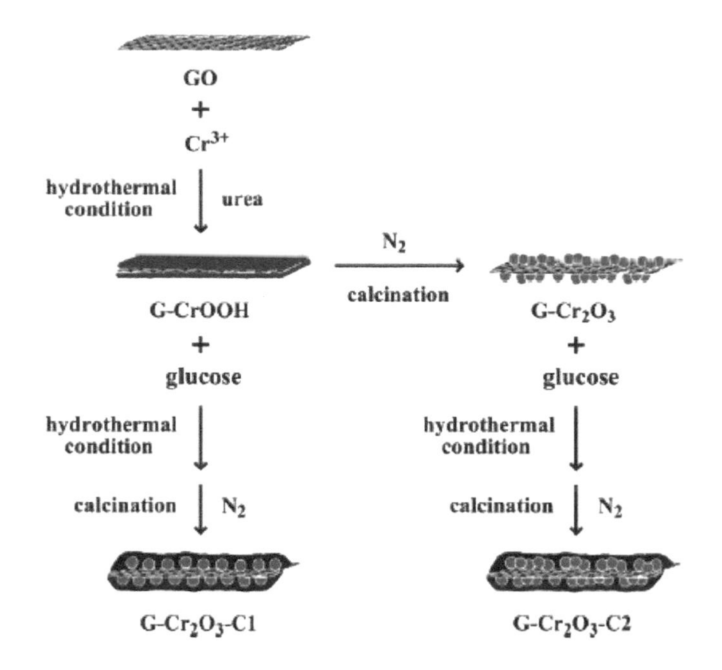

Fig. 3.31. Illustration of the synthesis routes for G–Cr_2O_3, G–Cr_2O_3–C1 and G–Cr_2O_3–C2. Reproduced with permission from [83], Copyright © 2013, Elsevier.

clearly observed, as shown in Fig. 3.32 (E) and (F). The Cr_2O_3 nanoparticles with diameters of 5–10 nm were randomly deposited on the graphene nanosheets, while aggregation was occasionally observed (Fig. 3.32 (E)). In contrast, the particle size of the carbon coated sample was decreased to 2–5 nm and the particles were more homogeneously dispersed (Fig. 3.32 (F)). According to the TEM image of G–Cr_2O_3–C2 (Fig. 3.32 (D)) and G–Cr_2O_3 (Fig. 3.32 (B)), the carbon layers formed after the formation of Cr_2O_3 had a very slight effect on the size and aggregation of the Cr_2O_3 particles.

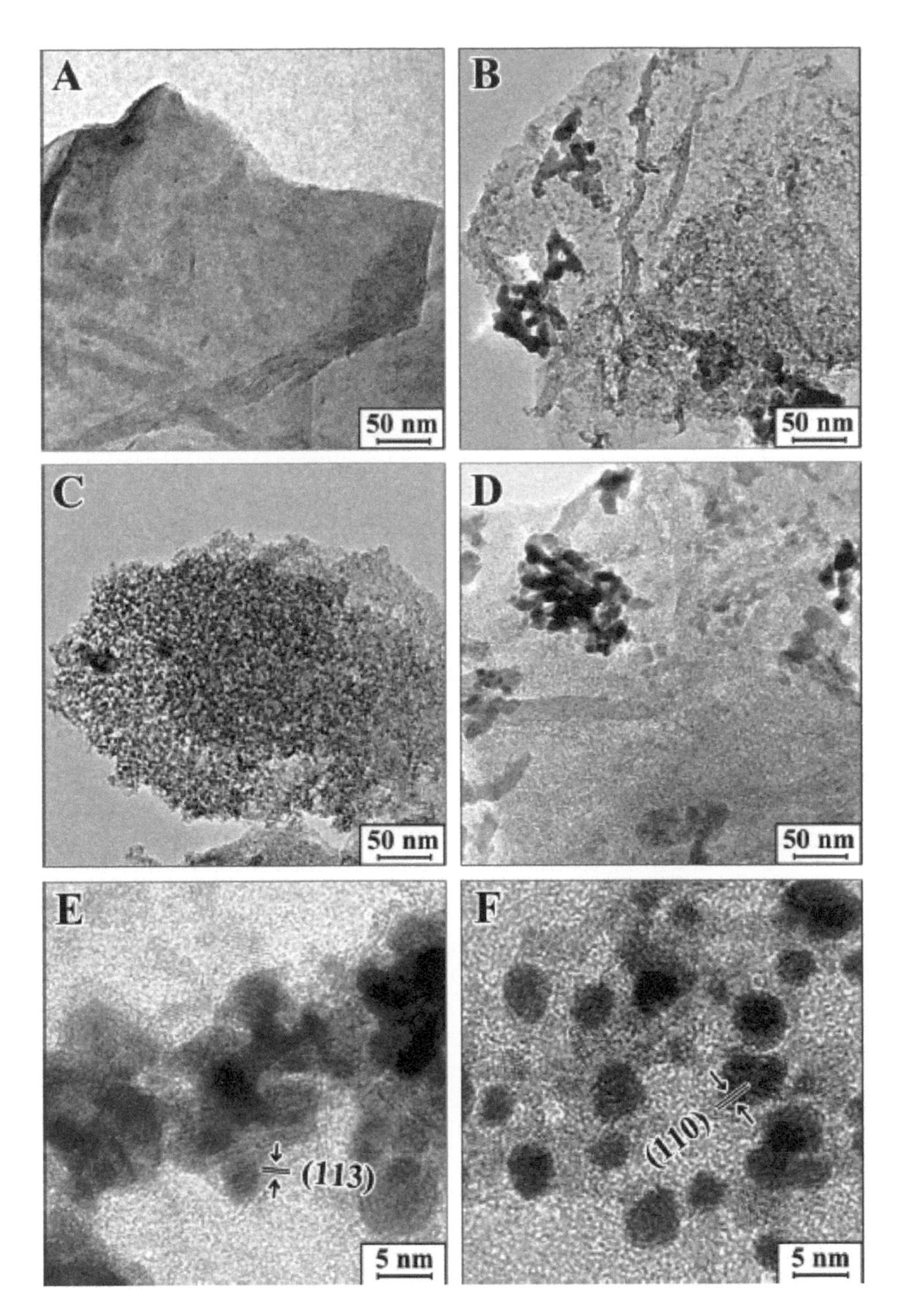

Fig. 3.32. TEM images of (A) G–CrOOH, (B) G–Cr_2O_3, (C) G–Cr_2O_3–C1, (D) G–Cr_2O_3–C2 and HRTEM images of (E) G–Cr_2O_3, (F) G–Cr_2O_3–C1. Reproduced with permission from [83], Copyright © 2013, Elsevier.

3.2.1.4. Graphene-Manganese Oxides

Due to their variable valences, a number of stable manganese oxides, including MnO, Mn_3O_4, Mn_2O_3, MnO_2, with different types of crystal structures, can be formed [86, 87]. The diversity of crystal structures, together with the presence of defects, morphologies, porosity and textures, has made manganese oxides potential candidates for various applications, especially in electrochemical energy storage applications [88, 89]. One of the problems in using manganese oxides as electrodes in energy storage devices is their poor conductivity. Incorporation with conductive graphene nanosheets to form nanocomposites or nanohybrids has been recognized as an effective strategy to address this problem.

MnO_2–G and MnO_2–GO hybrids have been prepared by using various methods, such as the chemical synthetic method [90–111], hydrothermal treatment [112–122], electrodeposition [123–127], microwave irradiation [128–130], for predominant applications as supercapacitors [90–92, 94, 95, 97–99, 101–104, 106, 107, 109, 110, 113, 114, 118, 120–123, 125–129], as well as for application in lithium ion batteries [96, 105, 108, 112, 115, 117, 119], biosensors [93, 111, 124], water splitting [100, 130], as catalysts [116] and so on. Different methods have also been used to synthesize Mn_3O_4–G, MnO–G and MnO_x–G hybrids, which are used for similar types of applications [131–144]. Representative samples of the synthesis and characterization of this group of hybrids are discussed as follows.

Chen et al. developed a simple chemical synthetic method to obtain MnO_2–G nanohybrids [90]. The GO was prepared from powdered flake graphite (400 mesh) by using a modified Hummers' method. Nanohybrids with different MnO_2/GO ratios were synthesized, with the synthesized GO and $MnCl_2 \cdot 4H_2O$, which were dispersed in isopropyl alcohol with the aid of ultrasonication. The suspensions were heated to about 83°C in a water cooled condenser under vigorous stirring, into which the $KMnO_4$ aqueous solution was added rapidly. After refluxing for 0.5 h, the mixtures were cooled to room temperature. To obtain nanocomposite samples, the reacted suspensions were centrifuged, washed and dried in air at 60°C. The samples were labeled according to the CMG_{ratio}, i.e., CMG_3, CMG_{15} and CMG_{20}, respectively. Pure MnO^2 without the presence of GO was also prepared for comparison. In addition, the effect of isopropyl alcohol on the reduction of GO was studied, with the sample being referred to as GOI, i.e., GO treated with isopropyl alcohol. The reaction between GO and $KMnO_4$ was also examined [90].

Figure 3.33 shows representative TEM images of the samples [90]. The GO sheets were made of multilayer structures, with diameters of a few micrometers, as shown in Fig. 3.33 (a). Pure nanosized MnO_2 exhibited a needle-like morphology, with typical diameters of 20–50 nm and lengths of 200–500 nm, as demonstrated in Fig. 3.33 (b). TEM images of the nanohybrid powders were from CMG_3. As shown in Fig. 3.33 (c), the hybrids consisted of GO sheets and MnO_2 nanoneedles, with the latter being randomly dispersed on the former. The presence of MnO_2 was confirmed by XRD results. There was no obvious difference in morphology between the MnO_2 particles in the hybrids and the pure one, which means that the growth behavior of the MnO_2 particles was not affected by GO. Some MnO_2 nanoneedles were brighter than others, because they were enveloped by a thin layer of graphene. TEM images at the tip of a MnO_2 nanoneedle are shown in Fig. 3.33 (e) and (f). The nanoneedles consisted of nanorods that were aggregated along the lateral faces, as shown in Fig. 3.33 (e). The nanorods at the center of the nanoneedles were longer than the rest, which could imply that the formation of the MnO_2 nanoneedles was related to the oriented attachment mechanism. The lattice fringes had an interplanar distance of about 0.7 nm shown in Fig. 3.33 (f) which was very close to that of the (110) plane of the tetragonal structured α-MnO_2 [145, 146].

The formation process of the GO–MnO_2 nanohybrids was studied, by taking samples from the mixtures before the reactions were finished, with CMG_3 as example [90]. The samples were analyzed by using TEM, UV–vis and Raman spectra. TEM images of the samples with different reaction time durations are shown in Fig. 3.34. As shown in Fig. 3.34 (a), the sample contains GO nanosheets with disordered precursors on their surfaces, after reaction for 10 s. TEM images of the CMG_3 samples with the reaction times from 1 to 10 min are shown in Fig. 3.34 (b–e), illustrating that the dispersants on the GO nanosheets became more and more ordered and gradually nanoneedles were observed, so that the growth of the MnO_2 nanoneedle followed a dissolution-crystallization mechanism. The morphology of the sample shown in Fig. 3.34 (f) was almost the same as that in Fig. 3.34 (e), which means that the growth of MnO_2 was nearly completed after reaction for 10 min.

The development of the GO–MnO_2 hybrids has been explained by using a schematic diagram as shown in Fig. 3.35 [90]. The basal planes of GO nanosheets are usually rich with epoxy and hydroxyl groups, while their edges are attached with carbonyl and carboxyl groups, which usually serve as active sites for nanostructures to attach [147]. MnO_2 is composed of MnO_6 octahedron units as building blocks, leading to MnO_2 crystals with different crystallographic structures and morphologies, when they are linked in

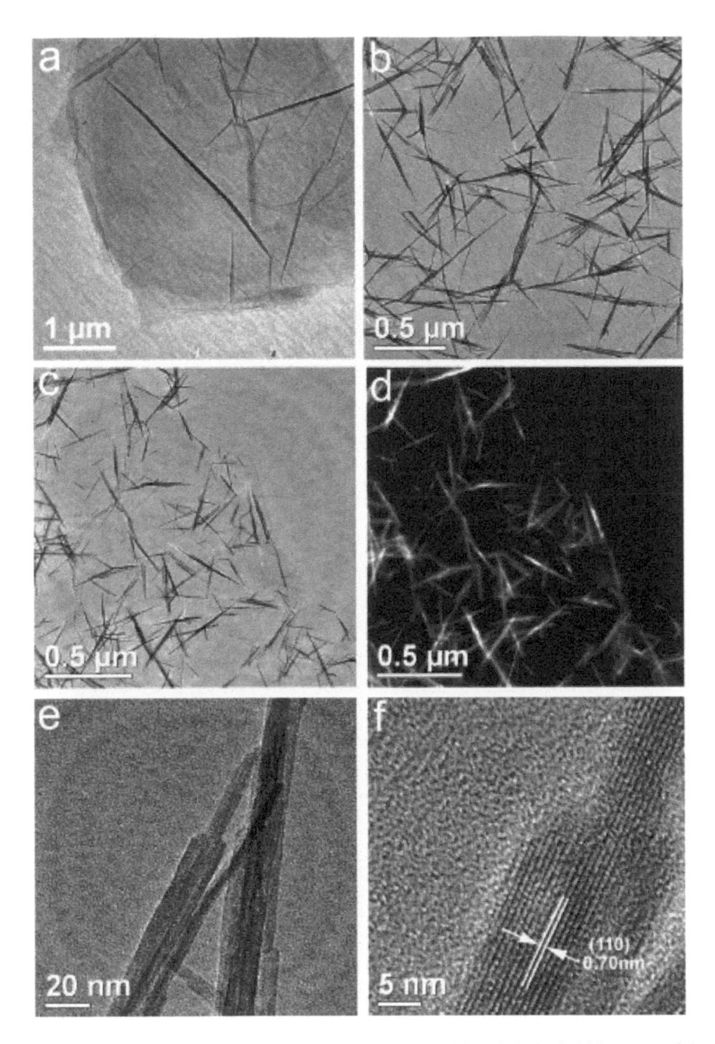

Fig. 3.33. (a, b) TEM images of the GO and pure MnO_2, (c, d) bright-field and dark-field images of the CMG_3, and (e, f) the HRTEM images of a MnO_2 nanoneedle. Reproduced with permission from [90], Copyright © 2010, American Chemical Society.

different ways [148]. During the early stage of the reactions, Mn^{2+} ions from $MnCl_2 \cdot 4H_2O$ solution would be bonded with the O atoms of the negatively charged oxygen-containing functional groups on the GO nanosheets through electrostatic forces. When $KMnO_4$ solution was added, the redox reaction occurring between Mn^{2+} and Mn^{7+} at a relatively higher temperature (83°C) would quickly produce a large number of nuclei. Through intermolecular hydrogen bonds or covalent coordination bonds, the Mn atoms of the MnO_6 octahedron could be attached to the O atoms of the functional groups on the GO nanosheets.

The formation of nanoneedle-like MnO_2 had a close relation to the ratio of DI-water/isopropyl alcohol and the way that the $KMnO_4$ solution was added [149]. In MnO_2 crystals, the (001) faces are most energetic [146]. Due to its higher polarity, less steric hindrance and stronger electrostatic effect, the H_2O molecule coordinates with the MnO_6 octahedrons more easily than isopropyl alcohol [150]. As a result, the interaction of the H_2O molecules with the O atoms of the MnO_6 units would be in the (001) direction preferably. Due to the small quantity of water, i.e., small volume ratio of H_2O: isopropyl alcohol = 5:50 ml, O atoms only in the (001) direction would bind with the H_2O molecules, while those in other directions would be combined with the isopropyl alcohol molecules. In addition, the way to add the $KMnO_4$ solution affected the crystal growth process through kinetics. A rapid introduction of $KMnO_4$ led to quick formation of nuclei and fast growth of crystals. Crystal growths through dissolution-crystallization and oriented

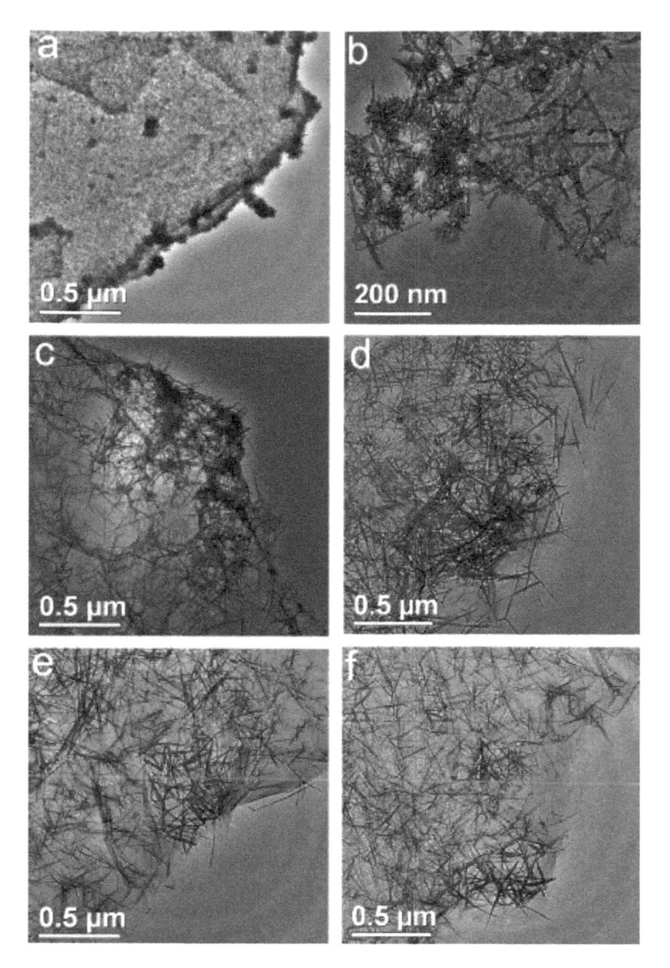

Fig. 3.34. TEM images of the CMG_3 samples collected after reaction for different time durations: (a) 10 s, (b) 1 min, (c) 3 min, (d) 5 min, (e) 10 min and (f) 30 min. Reproduced with permission from [90], Copyright © 2010, American Chemical Society.

attachment occurred at the same time. The nanorods at the center grew faster, thus leading to the presence of needle-like structures. It was also found that exfoliation of the GO nanosheets and growth of the MnO_2 crystals took place simultaneously.

Controlled experimental results indicated that MnO_2 nanoneedles could not be well-obtained, when using GO and $KMnO_4$ as the precursors, without $MnCl_2{\cdot}4H_2O$, while needle-like morphology was observed when GO was not used [90]. Therefore, it was suggested that the reaction between $MnCl_2{\cdot}4H_2O$ and $KMnO_4$ in the water-isopropyl alcohol mixture was responsible for the formation of the nanoneedle-like MnO_2 [149]. The GO–MnO_2 nanohybrids exhibited promising electrochemical properties when they were used as electrode materials of supercapacitors. The hybrid materials are also expected to have potential applications as catalysts, absorbents, and electrodes for other energy storage devices.

Another chemical synthetic route is called the co-assemble method [92]. By using this method, monodisperse MnO_2 nanospheres wrapped by graphene nanosheets, with a honeycomb-like "opened" structure, were obtained. The graphene-wrapped MnO_2 nanohybrids (shorten as GW-MnO_2) were formed with the positively charged honeycomb MnO_2 nanospheres and negatively charged graphene nanosheets, through electrostatic interactions. The GW-MnO_2 nanohybrids possessed promising electrochemical performance as electrode materials of supercapacitors.

To synthesize honeycomb MnO_2 nanospheres, $KMnO_4$ was dissolved in distilled water, under vigorous stirring. Then, oleic acid was added into the mixture, to form a steady emulsion. After the emulsion was

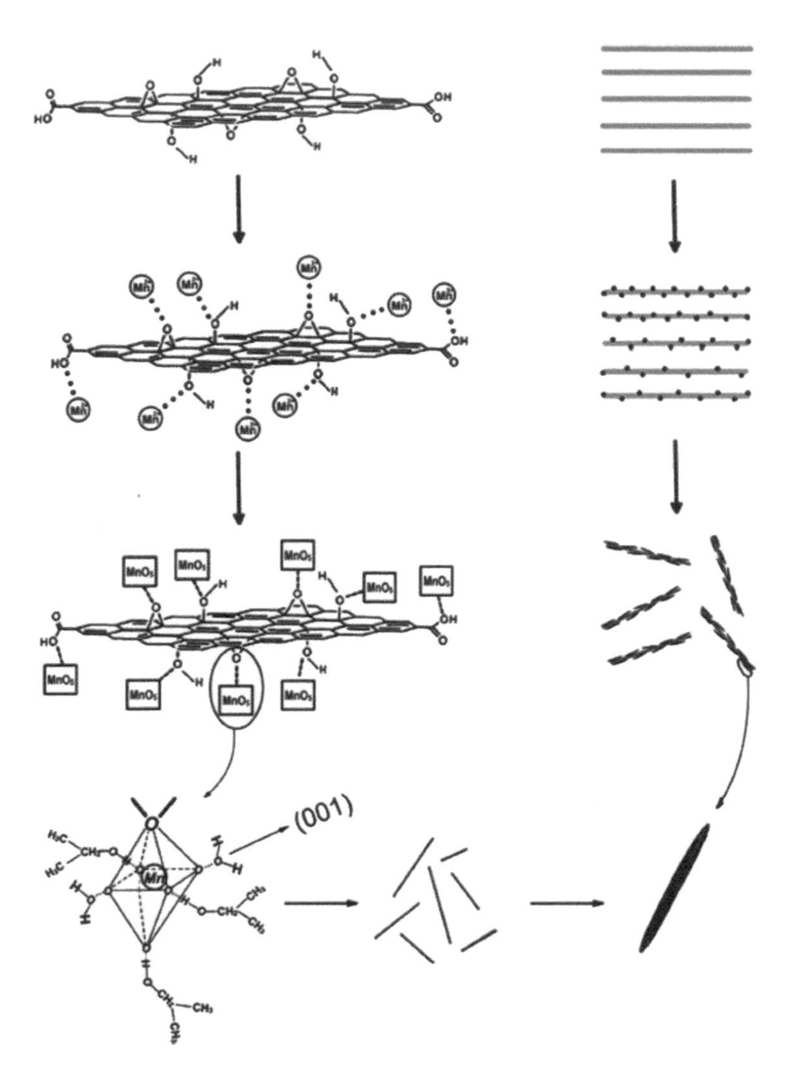

Fig. 3.35. Formation mechanism of the GO-MnO_2 nanohybrids. Reproduced with permission from [90], Copyright © 2010, American Chemical Society.

aged at room temperature for 24 h, a brown-black product was collected. The product was then washed several times with distilled water and alcohol to remove all of the residual reactants. Finally, the sample was dried in air at 60°C for 12 h. Electrostatic assembly towards graphene-wrapped MnO_2 nanohybrids is demonstrated schematically in Fig. 3.36 [92]. The powder of honeycomb MnO_2 nanospheres was first reacted with aminopropyltrimethoxysilane (APTS), so that the surface of the nanospheres was modified with NH_2 groups [151]. The mixtures were formed by adding the honeycomb MnO_2 nanospheres, APTS and toluene successively into a round-bottomed flask, which were then stirred and refluxed at 120°C under N_2 for 6 h. After that, the modified MnO_2 nanospheres were washed and dried in air at 60°C. Suspensions of the modified honeycomb MnO_2 nanospheres dispersion (3 mg·ml^{-1}) and graphene nanosheets (0.3 mg·ml^{-1}) were mixed and reacted under vigorous stirring at room temperature for 1 h. GW-MnO_2 nanohybrids were obtained after centrifugation, washing with water and drying. A typical sample contained fractions of MnO_2 and graphene of 90.9% and 9.1%, respectively.

Figure 3.37 shows TEM images of the honeycomb MnO_2 nanospheres, graphene nanosheets and the GW-MnO_2 nanohybrids [92]. The MnO_2 nanospheres with an average diameter of 100 nm had a honeycomb-like structure, as shown in Fig. 3.37 (a). The individual MnO_2 nanosphere consisted of nanoplatelets that

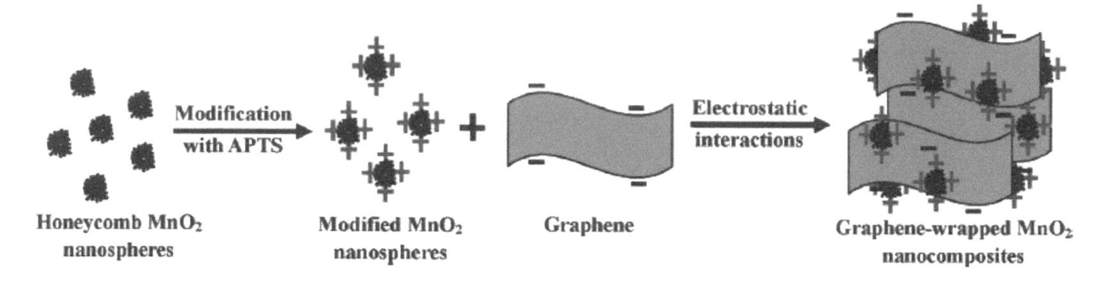

Fig. 3.36. Schematic illustration to show the formation process of the GW-MnO_2 nanohybrids. Reproduced with permission from [92], Copyright © 2012, American Chemical Society.

were self-aligned perpendicularly to the nanosphere surface and emanated from the center, similar to the structure of a honeycomb, as demonstrated in Fig. 3.37 (b). The graphene nanosheets appeared as transparent thin layers, with an average thickness of about 0.86 nm. The electronic conjugation within the graphene nanosheets was restored after reduction in hydrazine. Such reduced GO would acted as a flexible and conductive agent of the nanohybrids [152]. The GW-MnO_2 nanohybrids exhibited crinkled and rough textured structures, due to the presence of flexible and ultrathin graphene sheets. The honeycomb MnO_2 nanospheres were actually wrapped by the flexible and ultrathin graphene nanosheets. This kind of structure would both ensure high electrical conductivity of the nanohybrids and effectively prevent the MnO_2 nanoparticles from electrochemical dissolution. Figure 3.37 (c) indicated that the honeycomb MnO_2 nanospheres were firmly attached to the graphene nanosheets, which withstood the strong ultrasonic forces.

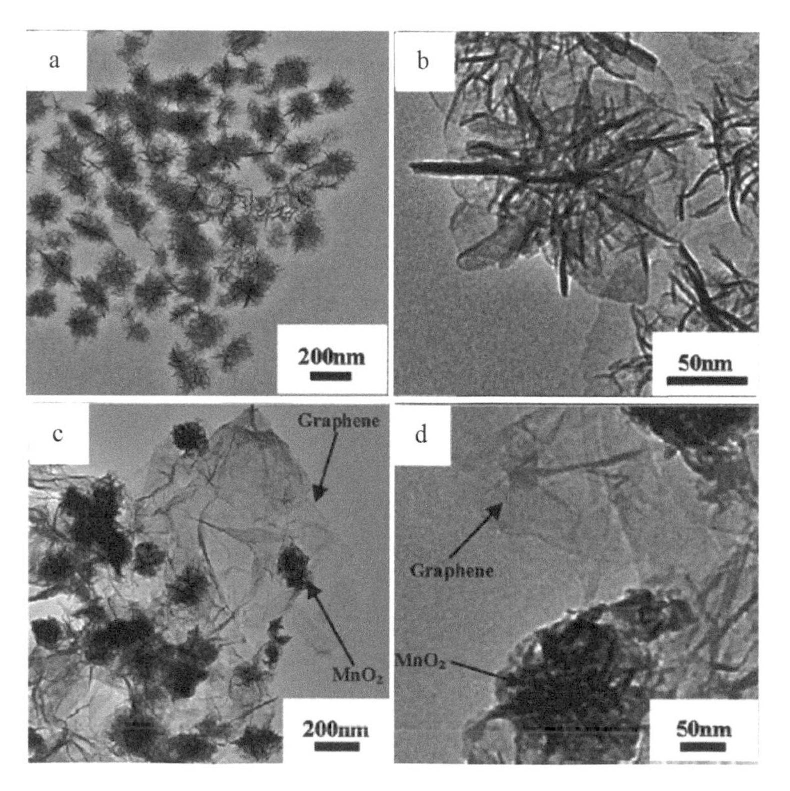

Fig. 3.37. Representative TEM images of the samples: (a, b) honeycomb MnO_2 nanospheres and (c, d) GW-MnO_2 nanohybrids. Reproduced with permission from [92], Copyright © 2012, American Chemical Society.

With the modification of APTS moieties, the honeycomb MnO_2 nanospheres were positively charged on their surface [153]. These positively charged honeycomb MnO_2 nanospheres were readily assembled with the negatively charged graphene nanosheets due to electrostatic interactions. During the modification of the MnO_2 nanospheres, pH value was very important, with an optimized value of 2, which was necessary to graft the APTS moieties onto the surface of the nanoparticles, whereas the graphene dispersion should be adjusted to pH = 9. The composition of the nanohybrids was confirmed by using TGA analysis.

Three examples of using the hydrothermal method to synthesize G–MnO_2 hybrids are discussed in this section. The first example was reported by Yu et al. [117], who prepared α-MnO_2 nanorods separately by using a hydrothermal method. The reaction solution contained $KMnO_4$ and $MnSO_4 \cdot H_2O$ with a molecular ratio of 2:1. The solution was hydrothermally treated at 180°C for 6 h with a Teflon-lined stainless steel autoclave. The products were washed with DI water and dried at 80°C. Figure 3.38 shows a schematic diagram demonstrating the process of synthesis of the rGO–α-MnO_2 hybrids. The as-synthesized α-MnO_2 nanorods were added in DI water and sonicated to form a dispersion, which was then mixed with GO dispersion with a weight ratio of 1:1 under strong stirring. After that, the GO nanosheets were reduced by using concentrated $NH_3 \cdot H_2O$ and hydrazine solution, which was added stepwise. The reduction was conducted at 90°C for 2 h. The reacted suspension was cooled to room temperature, filtered, washed and dried to obtain nanohybrids. The final contained 37.5 wt% rGO.

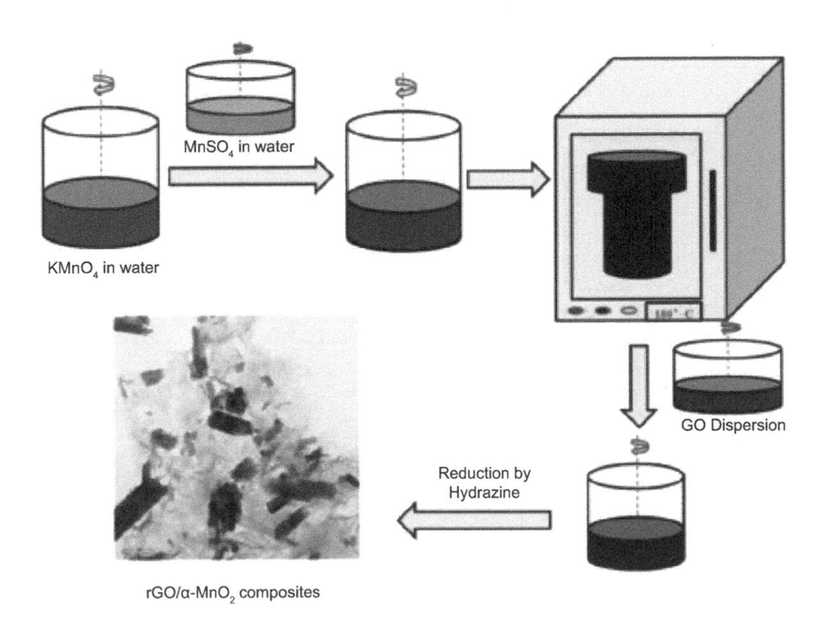

Fig. 3.38. Schematic diagram to show the synthesis process of the rGO–α-MnO_2 hybrids. Reproduced with permission from [117]. Copyright © 2013, The Royal Society of Chemistry.

Figure 3.39 shows representative SEM and TEM images of the as-synthesized α-MnO_2 nanorods and the final rGO–α-MnO_2 hybrids [117]. It was found that the morphology and crystal structure of manganese oxides could be tremendously affected by the ratio of $KMnO_4$ and $MnSO_4$ [154]. In this case, the α-MnO_2 nanorods exhibited a diameter of about 50 nm and a length of about 500 nm, as shown in Fig. 3.39 (a, b). Due to less agglomeration, the α-MnO_2 nanorods could be readily dispersed in GO solutions, as illustrated in Fig. 3.39 (c). The rGO nanosheets acted as a support and electrically conducting medium of the nanohybrids. Slightly different from the one discussed above, the α-MnO_2 nanorods were merely attached onto surface of the rGO nanosheets rather than wrapped by them. This combination would ensure full exposure of both the rGO nanosheets and the α-MnO_2 nanorods to Li ions and oxygen during the charge/discharge process, which is excellent for applications such as the electrode materials for lithium ion batteries. HRTEM image (Fig. 3.39 (d)) indicated that the α-MnO_2 nanorods were well-crystallized. The

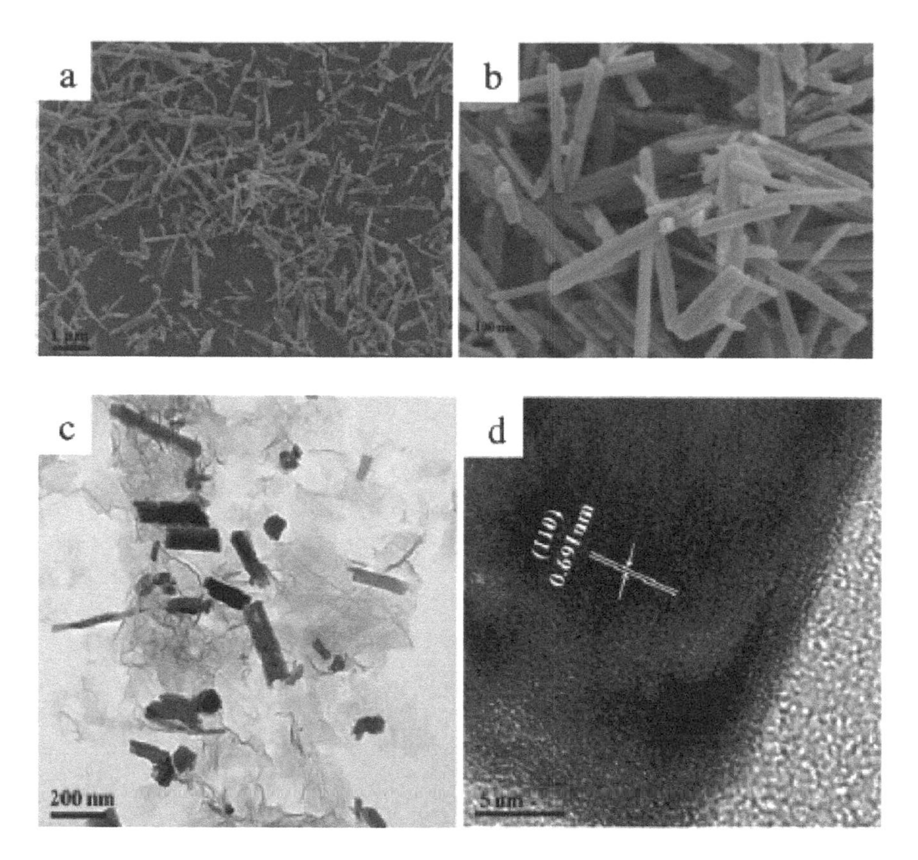

Fig. 3.39. (a and b) SEM images of the α-MnO_2 nanorods, (c) TEM image of the rGO–α-MnO_2 hybrids, and (d) crystallographic dimension of the α-MnO_2 nanorods. Reproduced with permission from [117]. Copyright © 2013, The Royal Society of Chemistry.

rGO–α-MnO_2 hybrid has been used as the cathode for Li-air batteries, which offered a specific capacity of 558.4 $mAh \cdot g^{-1}$. This performance was much better than that of pure α-MnO_2 electrode, due to the high electrical conductivity and homogeneous microstructure.

A three-dimensional (3D) hybrid of MnO_2 and graphene foam has been developed by using the hydrothermal method [113]. The morphology of the MnO_2 nanostructures can be readily controlled by the acidity of the solution. Such 3D hybrids can be used as free-standing electrodes of supercapacitors. Due to its 3D structure, the hybrid possessed a relatively high specific capacitance of 560 $F \cdot g^{-1}$ at current density of 0.2 $A \cdot g^{-1}$, together with high cycling stability.

Firstly, 3D graphene foams were prepared by using chemical vapor deposition (CVD) on nickel foam that was used as the substrate and support, with ethanol as the carbon precursor [155–157]. Free-standing 3D graphene foams were obtained by etching the Ni substrate away with 3 M HCl at 80°C. MnO_2 nanostructures were deposited *in situ* on the graphene foams by using the hydrothermal method. To prepare the hydrothermal solutions, $KMnO_4$ was dissolved into DI-water or diluted HCl solution under vigorous stirring for 10 min at room temperature. The $KMnO_4$ solution was then transferred into a Teflon-lined stainless steel autoclave. The 3D graphene foams fixed on glass slides were immersed into the solution, followed by thermal treatment at 150°C for 6 h. MnO_2–graphene 3D foam hybrids were obtained after thorough washing and appropriate drying process.

Because of conformal growth during the CVD, the 3D graphene foams exhibited a structure identical to that of the Ni foams, with a honeycomb structure having pore sizes of 100–300 μm, as shown in Fig. 3.40 (a) and (b) [113]. After removing the Ni substrate, the free-standing foam architecture was still retained, indicating the strong mechanical strength of the graphene nanosheets. The CVD derived graphene was defect-free, which was evidenced by the Raman spectra [158]. According to the integral ratio of 2D and G (Fig. 3.40 (d)), the graphene foams were constructed with single-layer and few-layer nanosheets. Such graphene foams had a large surface area of 670 $m^2 \cdot g^{-1}$. MnO_2 nanoparticles were deposited on the surface of the 3D graphene architecture, with a uniform distribution, with thicknesses of 5–8 μm, as demonstrated in Fig. 3.40 (c).

In this study, morphologies of the MnO_2 nanoparticles were controlled by adjusting the experimental conditions, especially the pH value. Figure 3.41 shows SEM and TEM images of the samples that were obtained under different experimental conditions [113], i.e., (i) without the presence of HCl, homogeneous reticular structure was produced (sample A), (ii) with the addition of 0.03 M HCl, the sample exhibited a microstructure consisting of crumpled flower-like structures and spherical particles (sample B), (iii) with the presence of 0.1 M HCl, the microstructure was characterized by homogeneous and regular flower-like structures (sample C) and (iv) with the addition of 0.3 M HCl, the MnO_2 nanostructure became uniform hollow nanotubes of 150–170 nm in diameter (sample D). This is a unique achievement in synthesizing MnO_2 nanostructures on defect-free graphene nanosheets, by simply controlling the acidity of the hydrothermal solution, without using templates or surfactants. It means that the nucleation and the growth of MnO_2 nanostructures in this case were sensitive to pH value. In strong acidic solutions, hollow nanotubes were formed because the MnO_2 crystal anisotropically grew along the [001] direction, while the interior along the [001] axis. At the same time, the metastable growths in other directions were completely avoided, due to higher surface energy [159, 160].

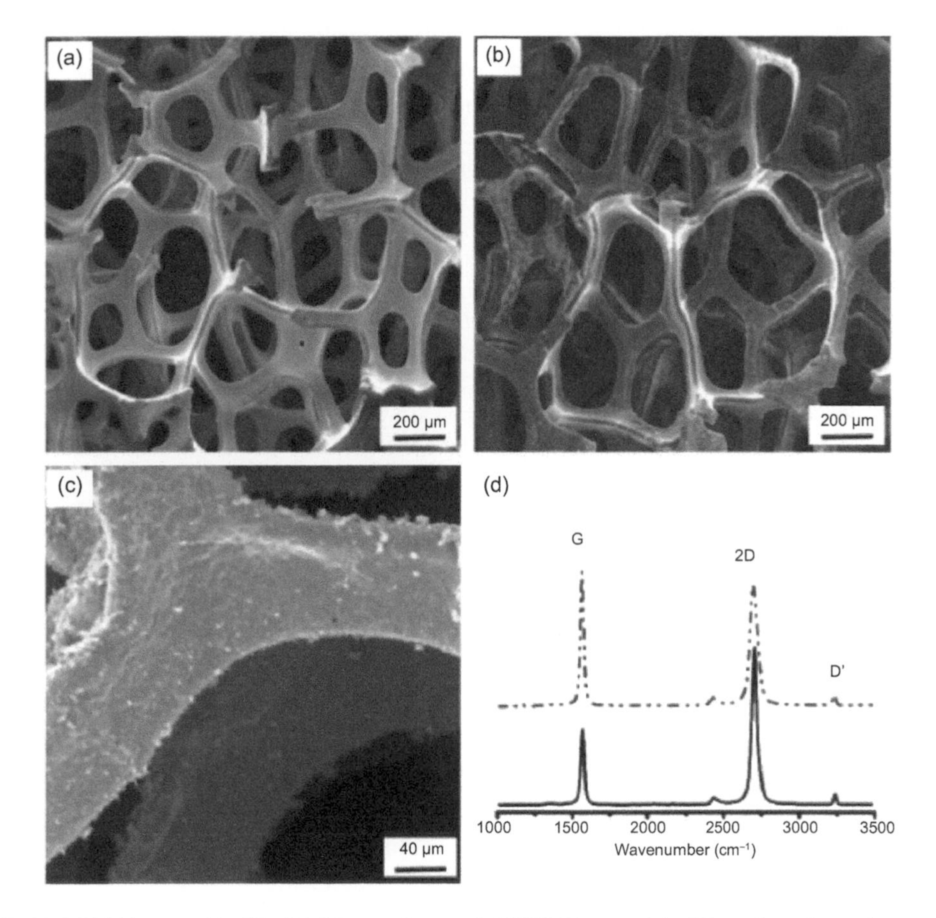

Fig. 3.40. (a) FESEM image of the 3D nickel foam substrate. (b) FESEM image of the 3D graphene foam after removal of the nickel substrate. (c) Low magnification FESEM image of the MnO_2–graphene 3D foam hybrid. (d) Raman spectra of the 3D graphene foam obtained at different locations. Reproduced with permission from [113]. Copyright © 2012, Elsevier.

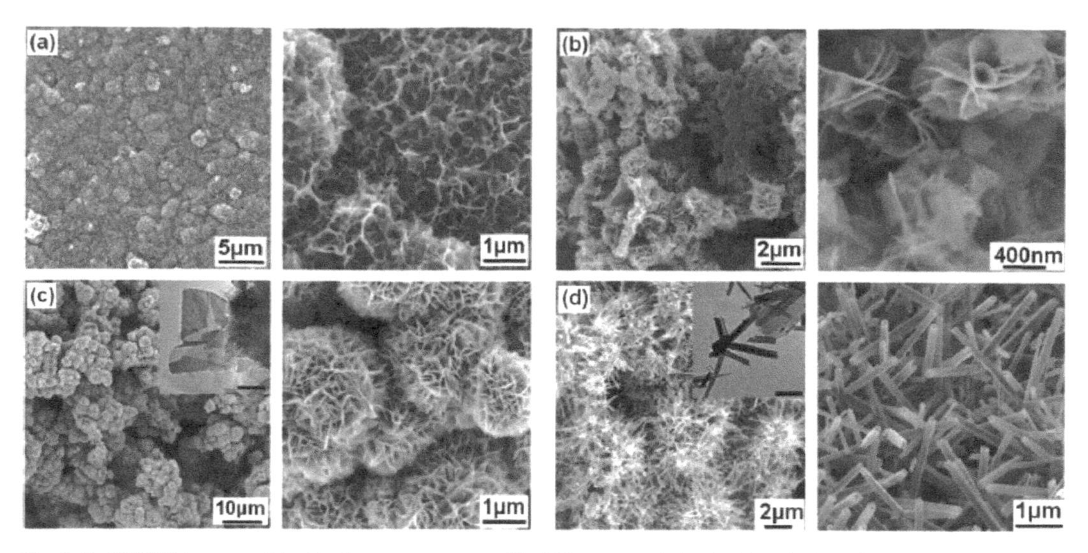

Fig. 3.41. FESEM images of the MnO_2 nanostructures with different morphologies deposited on the 3D graphene foams at various magnifications: (a) Sample A, (b) Sample B, (c) Sample C and (d) Sample D. In (a–d), left and right panels are the images at different magnifications. The insets in (c) and (d) are corresponding TEM images of the MnO_2 nanostructures, with scale bars of 0.2 μm. Reproduced with permission from [113]. Copyright © 2012, Elsevier.

A new type of MnO_2–nitrogen-doped graphene hybrid aerogel composites (MNGAs) have been fabricated by using hydrothermal synthetic technique [119]. It was facilitated by a redox reaction between $KMnO_4$ and the carbon in nitrogen-doped graphene hydrogels (NGHs). e The as-prepared MNGAs demonstrated a quite high performance, when they were used as anodes of lithium ion batteries, with a discharge capacity of 909 $mAh \cdot g^{-1}$ after 200 cycles at a current density of 400 mA g^{-1}, which was better than those of similar composite electrodes. They also exhibited excellent cycling stability and high rate capability, due to the synergistic effects of the homogeneously dispersed MnO_2 nanoparticles and the highly conductive three-dimensional networks based on the porous nitrogen-doped graphene nanosheets.

To fabricate the MNGAs, NGHs were immersed in a solution of 0.1 M $KMnO_4$ + 0.1 M Na_2SO_4. A shaker was used to speed up the diffusion of $KMnO_4$ and Na_2SO_4 into NGHs at room temperature, during which spontaneous reaction between $KMnO_4$ and carbon in NGHs took place. After the reaction, the hybrid hydrogels were thoroughly dialyzed in deionized water. Finally, MNGAs were obtained by freeze-drying the hybrid hydrogels in vacuum. The loading of MnO_2 was controlled by varying the immersion time (i.e., 30, 60 and 120 min, designated as MNGAs-x with x = 30, 60 and 120 min). NGAs and pure MnO_2 were prepared similarly for comparison.

NGHs were obtained by hydrothermally treating the mixtures of aqueous GO dispersion and ammonia at 180°C for 12 h, as shown in Fig. 3.42 [119], where the ammonia acted as the nitrogen source to dope the graphene nanosheets during the hydrothermal process [161]. The formation of NGHs was driven by hydrogen bonding, p–p stacking and hydrophobic interactions between nitrogen-doped graphene nanosheets. MnO_2 was formed due to the spontaneous redox deposition, $4KMnO_4 + 3C + H_2O \rightarrow 4MnO_2 + K_2CO_3 + 2KHCO_3$ [159]. In the final MNGAs, the loading of MnO_2 was 16.1, 27.3 and 42.5 wt% in the MNGAs-30, MNGAs-60 and MNGAs-120, respectively.

The NGAs exhibited a 3D network structure with interconnected pores, which could be retained after the deposition of MnO_2 nanoparticle for 30 min (MNGA-30). However, as the deposition time was increased to 60 and 120 min, the interconnected porous structure was compromised to a certain degree. It was found that this kind of porous network structures effectively prevent the MnO_2 nanoparticles from agglomeration or aggregation, ensuring the high dispersion behavior of the active materials. Figure 3.43 shows representative TEM images of the samples [119]. The number of MnO_2 nanoparticles was increased with increasing immersion and reaction time, while no aggregates were observed. Figure 3.43 (e) indicated

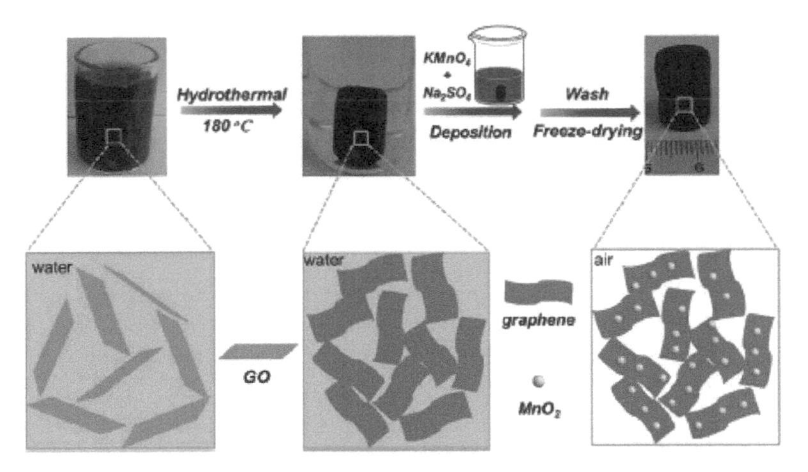

Fig. 3.42. Schematic diagram showing the procedures to fabricate the MnO_2–nitrogendoped graphene hybrid aerogel composites (MNGAs). Reproduced with permission from [119]. Copyright © 2015, The Royal Society of Chemistry.

that the MnO_2 nanoparticles were of an average particle size of 4–8 nm. HRTEM image in Fig. 3.43 (f) confirmed the formation of MnO_2 [162].

The electrodeposition process has been shown to be an effective technique to synthesize manganese oxide-based hybrids with graphene nanosheets. One example has been demonstrated to prepare MnO_2–graphene hybrids used as electrode materials of supercapacitors [127]. Firstly, MnO_2 was deposited on a gold plate by using an electrodeposition method. Graphene synthesized by using thermal CVD were then transferred onto the MnO_2 layers. The two alternative processes were repeated to produce multilayered structures, in which thicknesses or the number of graphene layers was controlled intentionally. The procedure for the preparation of the MnO_2/graphene composite electrode is represented in Fig. 3.44.

CVD derived graphene nanosheets were deposited on Cu foil, which were transferred onto the MnO_2 films with the aid of poly(methyl methacrylate) (PMMA) [163]. Before deposition, one side of the Cu foils

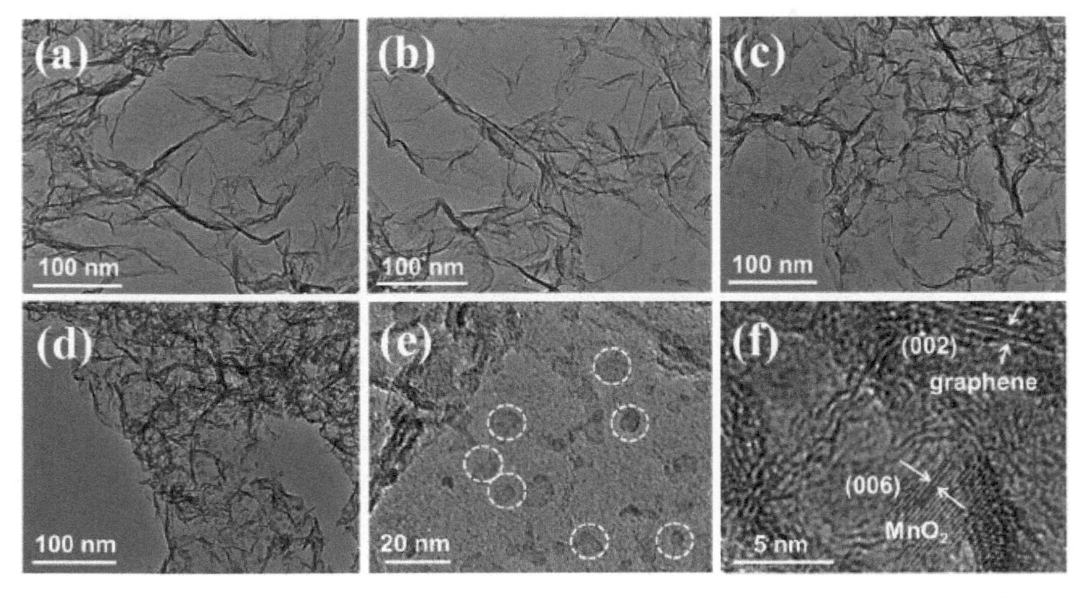

Fig. 3.43. TEM images of NGAs (a), MNGAs-30 (b), MNGAs-60 (c) and MNGAs-120 (d) and TEM images of MNGAs-120 at high magnifications (e and f). Reproduced with permission from [119]. Copyright © 2015, The Royal Society of Chemistry.

Color image of this figure appears in the color plate section at the end of the book.

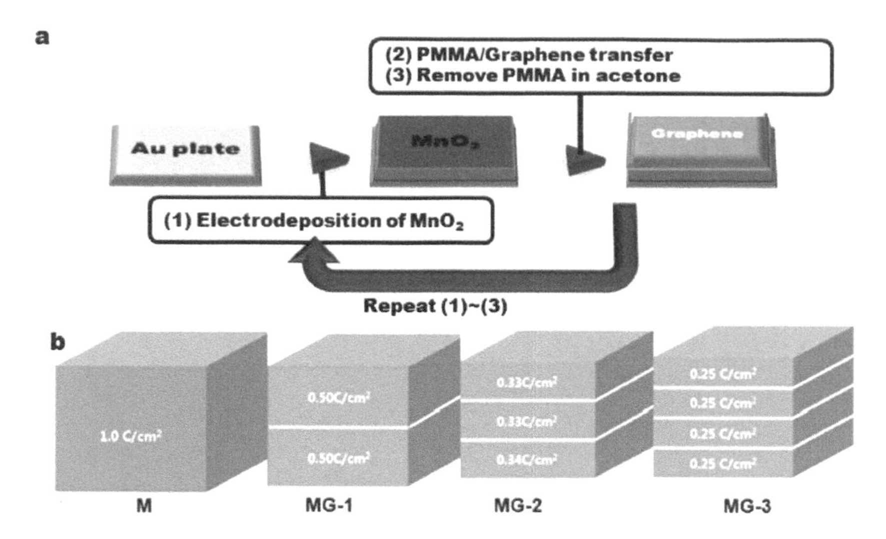

Fig. 3.44. (a) Schematic diagram to show fabrication procedure of the MnO_2–graphene hybrid electrode. (b) Types of the MnO_2–graphene hybrid as a function of intercalation. Reproduced with permission from [127]. Copyright © 2011, The Royal Society of Chemistry.

was coated with a layer of PMMA, so that the graphene layers were deposited on the PMMA on one side and on Cu on the other side. After deposition, the graphene on the Cu/graphene side was removed through mechanical polishing. Then, the Cu substrate was etched away by an aqueous solution of iron nitrate ($Fe(NO_3)_3$, 0.05 g ml^{-1}). The PMMA/graphene stacks were washed with deionized water and placed on target substrates and then dried. After the transfer was finished, a small amount of liquid PMMA solution was dropped onto the PMMA/graphene to dissolve the pre-coated PMMA, forming a new PMMA film, which was then slowly cured at room temperature for about 30 min before it was dissolved by acetone. Structures of the hybrids were controlled by adjusting the thicknesses of the graphene and MnO_2 layers, as shown in Fig. 3.44 (b).

Figure 3.45 shows SEM and TEM images of the MnO_2–graphene–MnO_2 sandwich-structured composite. Cross-sectional SEM image (Fig. 3.45 (a)) of the fractured surface of the composite film revealed that the graphene layers were stacked in between the MnO_2 layers with uniform thickness. Exposed graphene layers could be observed in the composite film, as illustrated in Fig. 3.45 (b). The MnO_2 and graphene layers were contacted intimately, without the presence of cracks and voids, suggesting the strong adhesion between the two components. Figure 3.45 (c) further confirmed the visibility of the graphene interlayers. Such a sandwich-structured MnO_2–graphene hybrid exhibited promising electrochemical performances. The sample MG-3 (Fig. 3.44 (b)) exhibited a specific capacitance of 399.82 $F \cdot g^{-1}$ at 10 mV s^{-1}, with a high stability after 5000 redox cycles. In this type of hybrids, graphene layers greatly improved the electrical conductivity of the MnO_2-based electrode and enhanced the electrolyte-electrode interactions.

Another example was reported by Xiao et al. [124]. In this work, a hybrid consisting of MnO_2 nanowire networks embedded in freestanding reduced graphene oxide paper (rGOP), on which high-density Pt nanoparticles were deposited, was developed to be used as a new type of flexible electrochemical sensor. In the ternary system, rGOP offered strong mechanical strength and high electrical conductivity, the MnO_2 nanowire networks provided a large surface area, while the well-dispersed and small-sized Pt nanoparticles were of high catalytic activity. It exhibited high sensitivity and selectivity to H_2O_2, without the presence of enzyme.

Freestanding GOP was fabricated using a mold-casting method with aqueous dispersion of GO nanosheets in a casting mold made of polytetrafluoroethylene (PTFE). Size of the GOP could be easily controlled by using casting molds with the desired sizes, while thickness of the paper was simply controlled by controlling the volumes of the GO dispersion. rGOPs were obtained by chemical reduction of GOP with HI solution [164]. MnO_2 nanowires were anodically deposited on the rGOPs by using an *in situ*

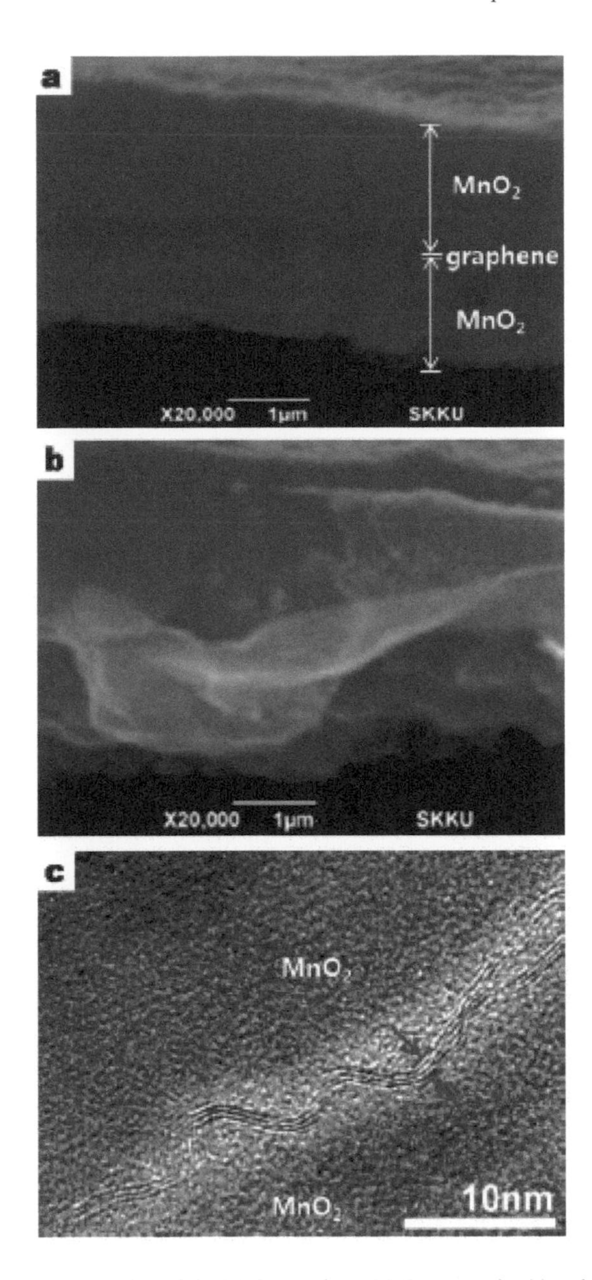

Fig. 3.45. SEM images of the cross-section of the MnO_2–graphene–MnO_2 composite (a) and exposed graphene from the composite (colorized for clarity) (b), and a TEM image of the MnO_2–graphene–MnO_2 (c). Reproduced with permission from [127]. Copyright © 2011, The Royal Society of Chemistry.

electrodeposition method. The deposition was carried out in 0.1 M $NaSO_4$ containing 50 mM $MnSO_4$, leading to MnO_2–rGOP hybrid. An ultrasonic-electrodeposition method was used to deposit Pt Nanoparticles on MnO_2–rGOP. This potentiostatic deposition was conducted in a 0.1 M KCl solution containing 1.0 mM K_2PtCl_6 at a deposition potential of –0.2 V for 150 s. During the deposition process, a continuous ultrasonic irradiation was applied to the electrolyte to minimize the aggregation and reduce the size of the Pt nanoparticles. The obtained paper electrode was denoted as Pt–MnO_2–rGOP. Figure 3.46 shows a schematic diagram of detailed procedures to fabricate the nanohybrid electrode [124].

Figure 3.47 shows representative SEM images of the samples at different fabrication stages. The GOP had a uniform thickness, with GO nanosheets being aligned in parallel to form a layered structure,

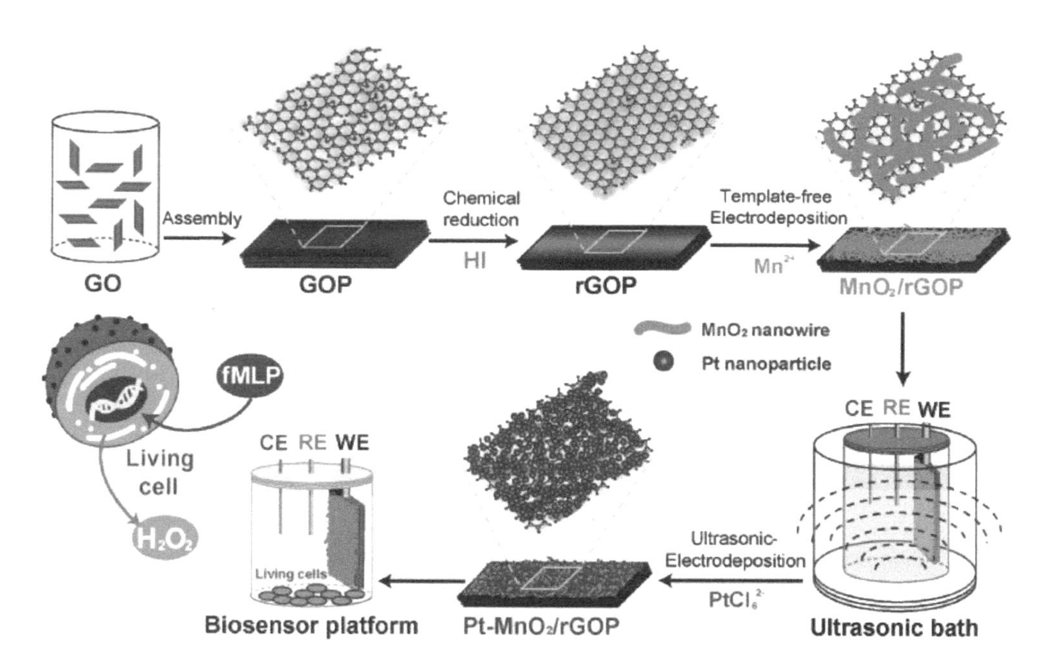

Fig. 3.46. Schematic diagram showing fabrication process of the nanohybrid paper electrode for the biosensor. Reproduced with permission from [124]. Copyright © 2012, WILEY-VCH Verlag GmbH & Co. KGaA, Weinheim.

as shown in Fig. 3.47 (A–C). After the chemical reduction by HI, the layered structure of the nanosheets and the wrinkle feature of the paper surface were retained in the rGOP, as demonstrated in Fig. 3.47 (D–F). Due to the elimination of the oxygen containing groups on the GO nanosheets through chemical reduction, the nanosheets were packed in a more dense way in the rGOP. Figure 3.47 (G) shows that the MnO_2 nanowires possessed an average diameter of 10–15 nm and an average length of 100–200 nm. The Pt nanoparticles covered the MnO_2 nanowires and also filled in the void space of the MnO_2 nanowire network, owing to the high affinity of Pt to MnO_2, as illustrated in Fig. 3.47 (H). However, on bare rGOP substrates, the density of Pt nanoparticles deposited with the same process was much lower, as shown in Fig. 3.47 (I). It was also found that ultrasonication played a significant role in producing a uniform distribution of the Pt nanoparticles.

Fan et al. reported a method to prepare graphene–MnO_2 hybrid by depositing MnO_2 nanoparticles on graphene nanosheets with microwave irradiation [128]. The graphene–MnO_2 hybrid was used as the positive electrode, while activated carbon nanofibers (ACN) were used as the negative electrode, to construct an asymmetric supercapacitor. High energy density was achieved in a neutral aqueous Na_2SO_4 electrolyte. At optimized conditions, the asymmetric supercapacitor could be cycled reversibly in a voltage range of 0–1.8 V, with maximum energy density of 51.1 $Wh\cdot kg^{-1}$ and an excellent cycling durability. The graphene–MnO_2 hybrids were directly formed through redox reaction between graphene and $KMnO_4$ under microwave irradiation. Graphene nanosheet powder was first dispersed with the aid of ultrasonication, into which $KMnO_4$ powder was added. The resultant suspension was heated by using a household microwave oven (2.45 GHz, 700 W) for 5 min and then cooled to room temperature naturally. After that, graphene–MnO_2 hybrid powder was obtained by filtering, washing and drying the black deposits. Birnessite-type MnO_2 was produced through the reaction between carbon and $KMnO_4$, as discussed above [159].

The deposition of MnO_2 nanoparticles prevented graphene nanosheets from agglomerating, as shown in Fig. 3.48 (a) and (b) [128]. MnO_2 nanoparticles were uniformly distributed on the surface of the graphene nanosheets. In comparison, pure MnO_2 had a size of 500 nm, with a flower-like structure, as shown in Fig. 3.48 (c). Phase composition was confirmed by the XRD results of Fig. 3.48 (d). The MnO_2 nanoparticles were adhered so strongly on the graphene nanosheet surface that they could withstand long

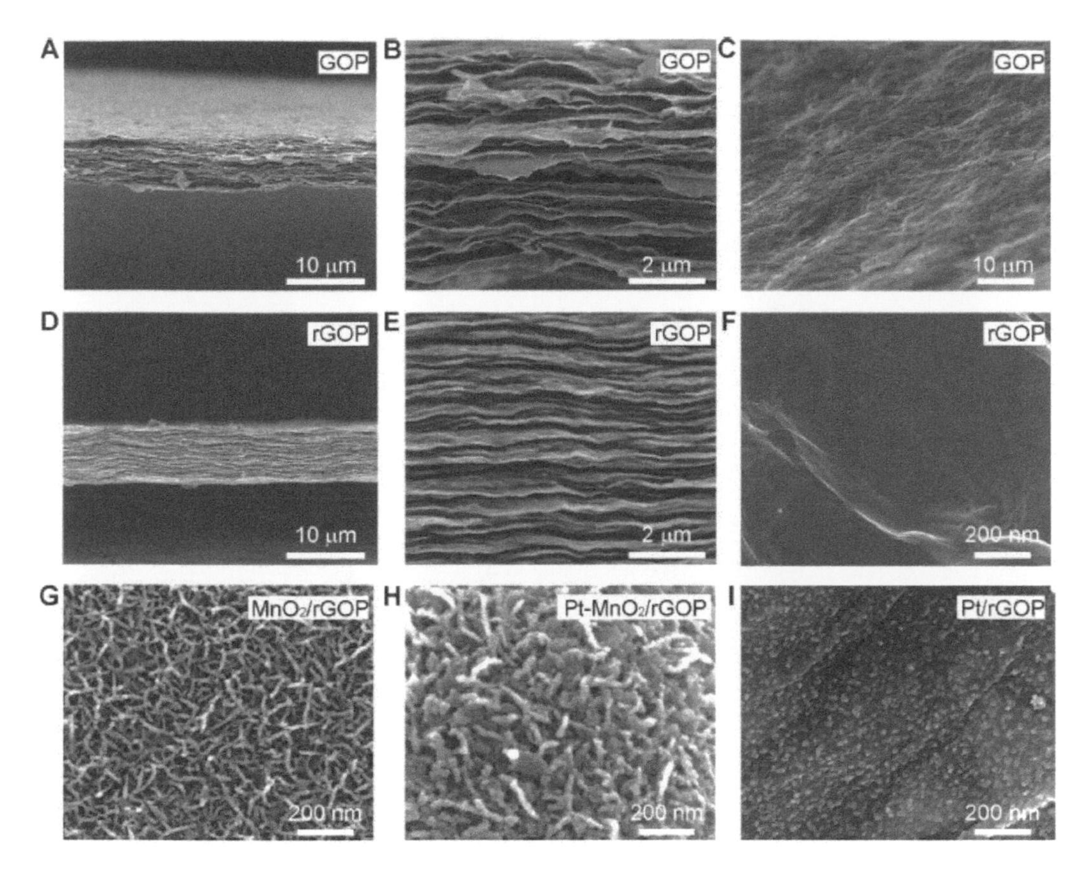

Fig. 3.47. Representative SEM images of the samples at different fabrication stages. Reproduced with permission from [124]. Copyright © 2012, WILEY-VCH Verlag GmbH & Co. KGaA, Weinheim.

time sonication. This strong adhesion ensured the high recycle performances of the hybrids as electrode materials of supercapacitors. The effectively separated graphene nanosheets facilitated a short diffusion path for electrolyte ions during the charge/discharge process, so that the electrochemical function of MnO_2 could be entirely utilized. Furthermore, the strong adhesion of the MnO_2 nanoparticles on the graphene nanosheets also enabled a fast electron transport, thus leading to enhanced electrochemical performance.

Duan et al. developed a method to synthesize Mn_3O_4–G hybrids, the Mn_3O_4 nanoparticles having three different morphologies, i.e., spheres, cubes and ellipsoids, were selectively deposited on nitrogen-doped graphene nanosheets, by using a two-step liquid-phase procedure [131]. These hybrid materials exhibited a promising oxygen reduction reaction (ORR) activity and high durability. The mesoporous microstructure, nitrogen doping and strong bonding between the oxide species and the nitrogen-doped graphene were readily responsible for the ORR catalytic activity. Among these three types of Mn_3O_4 nanoparticles, the one with ellipsoidal shape demonstrated the highest ORR activity, with a more positive onset-potential of –0.13 V and a higher kinetic limiting current density (J_K) of 11.69 mA cm^{-2} at –0.60 V. It was suggested that ORR activity of the hybrid materials was closely related to the exposed crystalline facets of the catalytic components, associated with the shape of the Mn_3O_4 nanocrystals.

The synthesis process is illustrated in Fig. 3.49 [131]. GO nanosheets were mixed with $Mn(AC)_2$ and ammonia in different solvents, which were first treated in a water bath at 80°C for 10 h. During this treatment process, irregular Mn_3O_4 nuclei were created on the GO nanosheets through heterogeneous nucleation. The GO-Mn nuclei solution was further solvothermally treated, in order to reduce the GO and dope it with nitrogen, due to the high pressure and high temperature [165]. At the same time, the Mn_3O_4

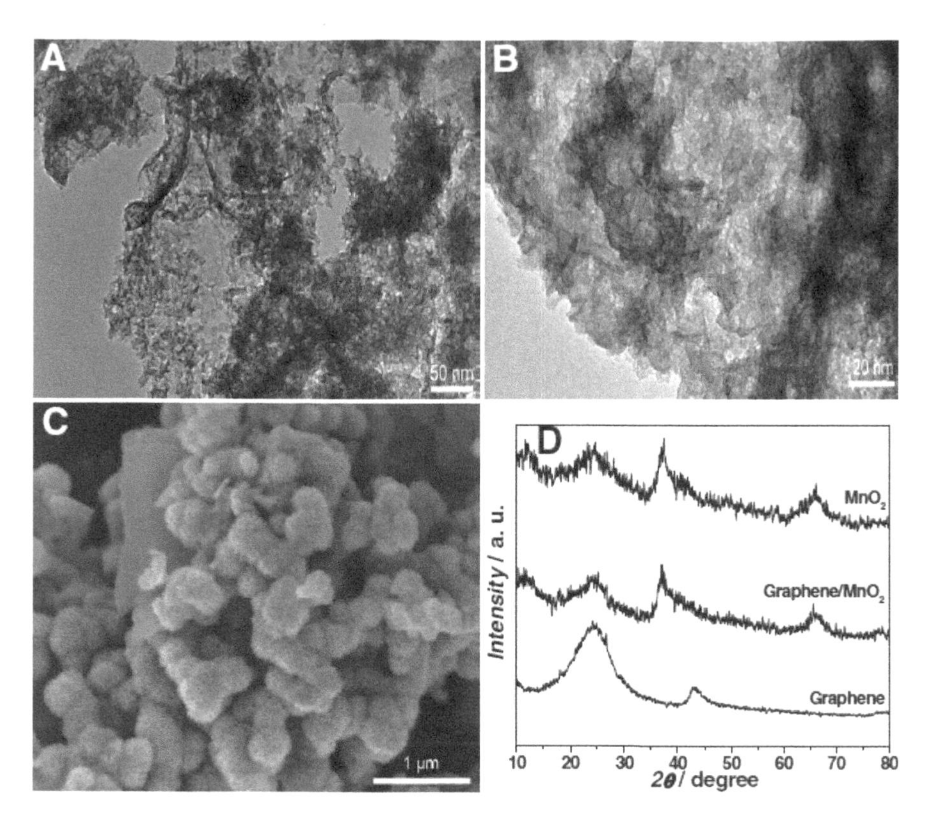

Fig. 3.48. (a) Low and (b) high magnification TEM images of the graphene–MnO_2 hybrid, showing that nanoscopic MnO_2 particles were closely deposited on graphene nanosheets. (c) SEM image of pure MnO_2 particles with a flower-like morphology. (d) XRD patterns of the graphene, MnO_2 and graphene–MnO_2 hybrid powder. Reproduced with permission from [128]. Copyright © 2011, WILEY-VCH Verlag GmbH & Co. KGaA, Weinheim.

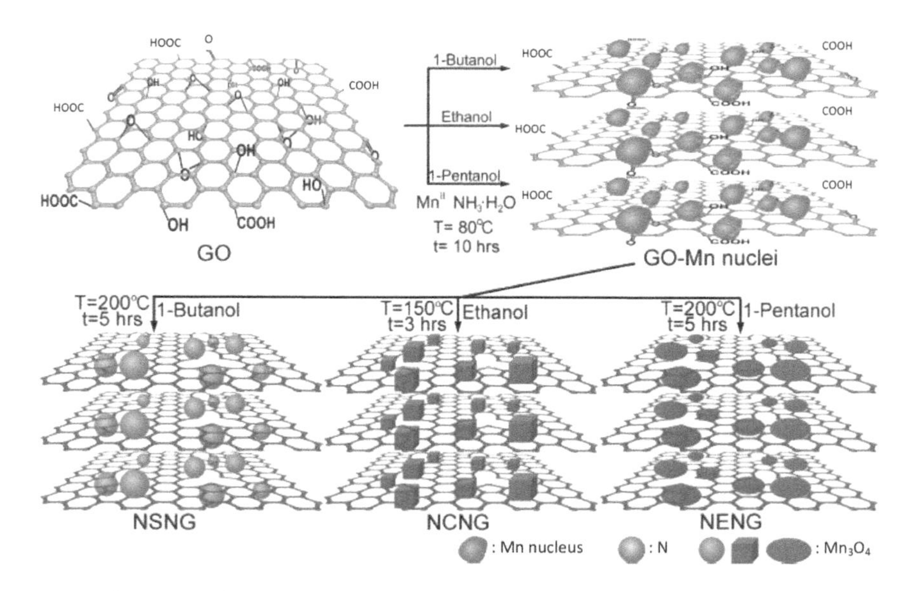

Fig. 3.49. Schematic diagram showing preparation procedure of the Mn_3O_4 nanoparticles with three morphologies on nitrogen-doped graphene nanosheets. Reproduced with permission from [131]. Copyright © 2013, WILEY-VCH Verlag GmbH & Co. KGaA, Weinheim.

nuclei grew into nanoparticles with sphere, cube and ellipsoid-like morphologies, when different reaction conditions were applied.

Figure 3.50 shows TEM images of the three groups of nanoparticles [131]. The Mn_3O_4 quasi-nanospheres on nitrogen-doped graphene, or NSNG in short, were synthesized in 1-butanol at 200°C for 5 h (Fig. 3.50 (a)). TEM image revealed that the nanospheres were of sizes ranging from several nm to 20 nm. HRTEM image (Fig. 3.50 (b)) indicated that each individual Mn_3O_4 nanosphere was of single-crystal characteristic, with an inter-planar spacing of 0.272 nm, corresponding to the (103) plane of hausmannite Mn_3O_4 [87]. The hausmannite Mn_3O_4 was also confirmed by electron diffraction (ED) pattern, as illustrated in the inset of Fig. 3.50 (b). Mn_3O_4 nanoparticles with a quasi-nanocube morphology and particle sizes of 5–20 nm were obtained after solvothermal treatment in ethanol at 150°C for 3 h, as shown in

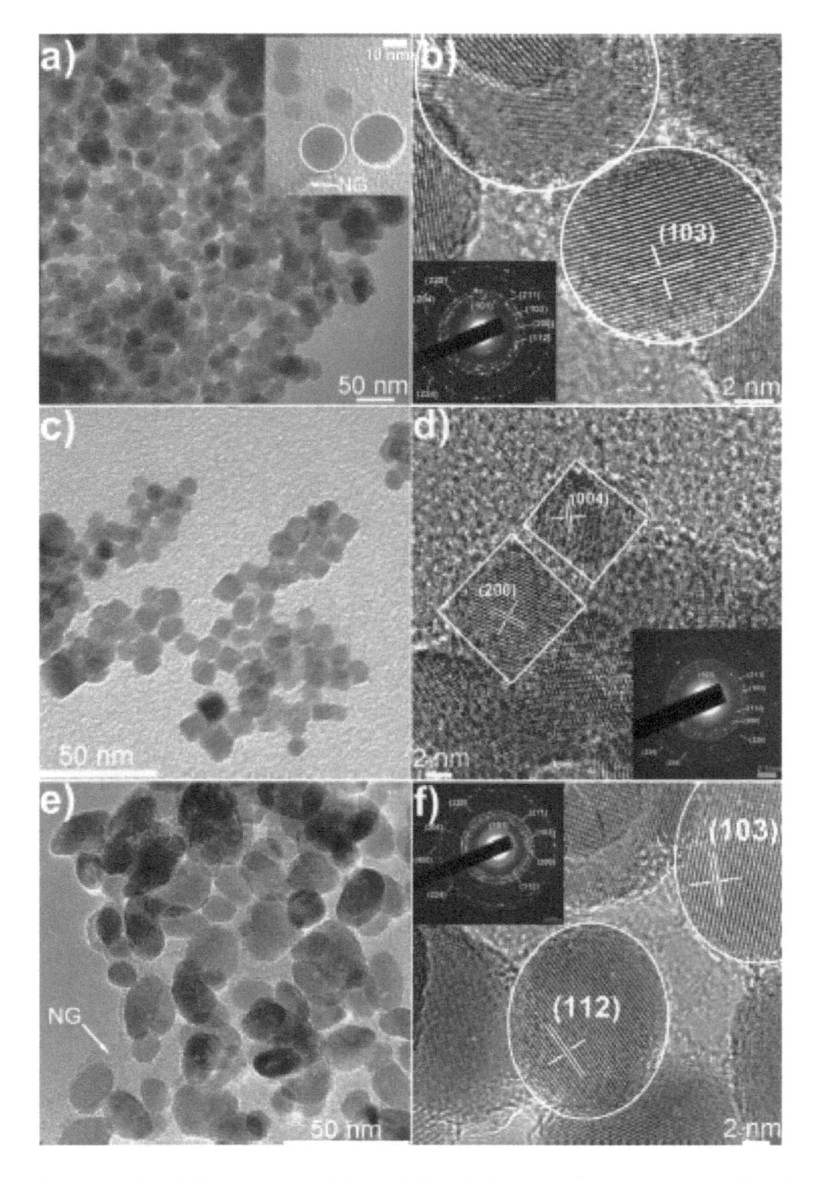

Fig. 3.50. (a) TEM images and (b) HRTEM image of the NSNG, with the inset to be the ED pattern. (c) TEM image and (d) HRTEM image of the NCNG, with the inset to be the ED pattern. (e) TEM image and (f) HRTEM image of the NENG, with the inset to be the ED pattern. Reproduced with permission from [131]. Copyright © 2013, WILEY-VCH Verlag GmbH & Co. KGaA, Weinheim.

Fig. 3.50 (c) (NCNG). In HRTEM image (Fig. 3.50 (d)) of the NCNG, the two inter-planar spacings of 0.284 nm and 0.234 nm corresponded to (200) and (004) planes of hausmannite Mn_3O_4, in agreement with the ED pattern, i.e., the inset of Fig. 3.50 (d). If the solvothermal treatment was conducted in 1-pentanol at 200°C for 5 h Mn_3O_4 quasi-nanoellipsoids were obtained (NENG), as demonstrated in Fig. 3.50 (e). The nanoellipsoids were 5 nm in width and 20 nm in length. The two inter-planar distances of 0.307 nm and 0.272 nm observed in the HRTEM image (Fig. 3.50 (f)) of NENG could be assigned to (112) and (103) planes of hausmannite, which was supported by the ED pattern shown as inset in the figure.

The fact that the Mn_3O_4-based hybrid nanomaterials possessed extraordinary high ORR activity was ascribed to a synergistic effect of the mesoporous microstructure, nitrogen dopant, the presence of asymmetry charge and spin density, as well as the strong coupling between the metal ions and the nitrogen-doped graphene [133]. The mesoporous microstructure facilitated fluent mass transport during the catalytic reaction, while unbalanced charge distribution was created due to the nitrogen dopant which promoted the adsorption of O_2 molecules. The strong bonding between the nitrogen-doped graphene and the metal ion was beneficial to electron transfer, leading to high durability. The ORR performance was related to the shape of the Mn_3O_4 nanoparticles. It is widely accepted that, due to the high oxidation state and strong tendency to form a redox couple, (001) facets with Mn_2O_4 termination (Mn^{IV}) of Mn_3O_4 nanocrystals are of strong catalytic activity [166]. Therefore, Mn_3O_4 nanoparticles with an ellipsoid-like morphology, with most exposed (001) facets, possessed the highest ORR performance.

3.2.1.5. Graphene-Iron Oxides

Iron oxides that have been involved in hybrids with graphene or graphene oxide mainly include Fe_3O_4 and Fe_2O_3. These hybrids have found various applications in different areas. Fe_3O_4–G hybrids have been synthesized by using various methods, including physical encapsulation or mixing [167–170], chemical precipitation or reaction [171–214], sol-gel process [215], hydrothermal or solvothermal treatment [216–241], for applications in anode materials of Li ion batteries or electrodes of supercapacitors [173, 174, 184, 185, 188, 189, 192, 193, 201, 205, 210, 213, 215, 216, 220, 221, 224, 229, 231-236, 239], microwave absorbers or electromagnetic shielding [175, 177, 178, 190, 199, 202, 203, 208, 212, 228, 230, 237], water treatment and purification [169, 171, 172, 176, 180–183, 187, 195, 196, 198, 204, 211, 223, 226, 227, 238], medicinal therapy or biosensors [167, 191, 200, 214, 241], as catalysts [197, 218, 222, 225], electrochemical actuators [168] and high dielectric constant materials [219].

Physical mixing is a simple way to prepare Fe_3O_4–G hybrids. One example was demonstrated by Liang et al. [170], who created a water-soluble graphene–Fe_3O_4 hybrid, which could be used to fabricate free-standing graphene–Fe_3O_4 hybrid papers. Such hybrid papers exhibited high electrical conductivity, promising mechanical strength, extraordinary flexibility and superparamagnetism, which could be readily controlled through the contents of the Fe_3O_4 nanoparticles. A special magnetically controlled switch behavior was realized by using these hybrid papers.

Water soluble Fe_3O_4 nanoparticles were synthesized by using a chemical precipitation method, from $FeCl_2 \cdot 4H_2O$ (in 1.0 M HCl) and $FeCl_3 \cdot 6H_2O$, with $NH_3 \cdot H_2O$ as the precipitant in N_2 atmosphere. Black precipitates were obtained after filtration, washing and drying, which were purged with N_2 and treated with 2 M $HClO_4$ at room temperature. After centrifugation, the precipitates were re-dispersed in distilled water, after which thorough dialysis was conducted to make the dispersions reach neutrality. The Fe_3O_4 nanoparticle aqueous solutions had a concentration of 7 $mg \cdot ml^{-1}$. It was interesting to find that water-soluble Fe_3O_4 nanoparticles could be used to prevent the agglomeration of graphene nanosheets from GO through chemical reduction. Graphene–Fe_3O_4 hybrid papers were prepared by using a two-step approach, as shown in Fig. 3.51, i.e., (i) physical mixing and (ii) chemical reduction [170]. The minimum amount of the Fe_3O_4 nanoparticle that could compensate the hydrophobic interaction between graphene nanosheets was as low as 2 wt% of the GO, in order to produce homogeneous solutions. In that study, two graphene–Fe_3O_4 samples were investigated, i.e., graphene–Fe_3O_4-1 with 3.4 wt% Fe_3O_4 and graphene–Fe_3O_4-2 with 6.5% Fe_3O_4.

The graphene–Fe_3O_4 hybrid water suspensions were made into free-standing papers, as shown in Fig. 3.52 (a, b). After thermal annealing, the hybrid papers were still smooth. The hybrid papers exhibited a well-stacked structure, with Fe_3O_4 nanoparticles distributed homogeneously, as demonstrated in Fig. 3.52

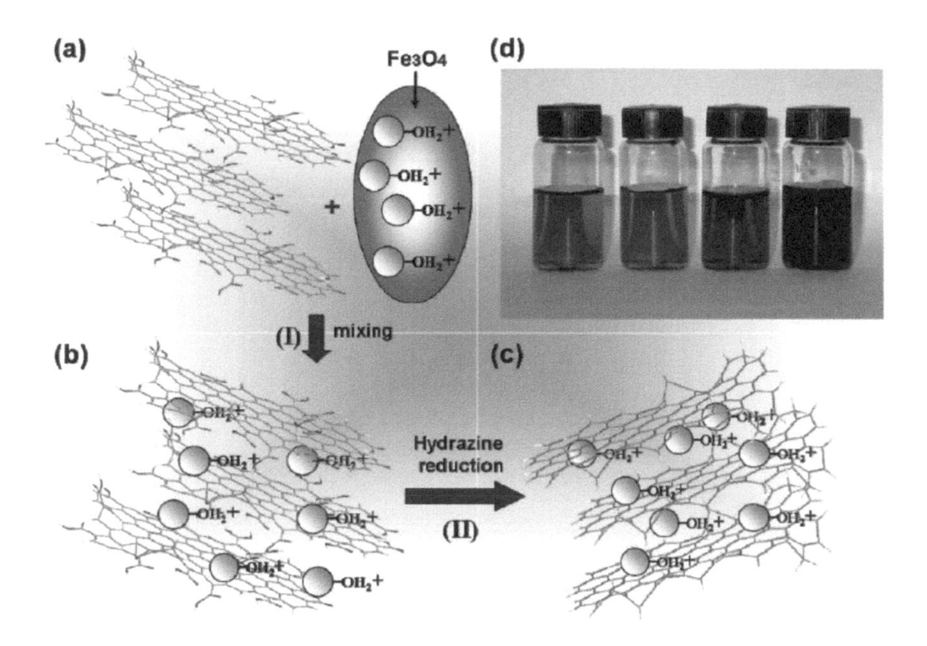

Fig. 3.51. Two-step method to synthesize the graphene–Fe_3O_4 hybrid papers: (I) mixing of GO aqueous solution with water-soluble Fe_3O_4 nanoparticle solution (a, b) and (II) chemical reduction of the mixed suspension with hydrazine (c, d). The photographs (d) of water dispersions (from left to right) include: (1) water-soluble Fe_3O_4 nanoparticles, (2) mixture of water-soluble Fe_3O_4 nanoparticles and GO nanosheets, (3) reduced graphene without Fe_3O_4 nanoparticles and (4) reduced graphene with water-soluble Fe_3O_4 nanoparticles, with a concentration of 0.1 mg·ml^{-1}. Reproduced with permission from [170]. Copyright © 2010, American Chemical Society.

(b, c, f). The graphene–Fe_3O_4-2-annealed papers possessed a similar structure (Fig. 3.52 (e)) to that of general graphene papers (Fig. 3.52 (d)). These results indicated that the layered structure of the graphene was not affected by the presence of the Fe_3O_4 nanoparticles (Fig. 3.52 (f)). This result corresponds to the observed XRD patterns (Fig. 3.52 (c)). It is expected that such layered structures should have promising mechanical strength and electrical conductivity.

A one-step method was developed to prepare G–Fe_3O_4 hybrid hydrogels [172]. The hybrid had a 3D interconnected network, which was created through the synergistic effects of self-assembly of GO nanosheets and *in situ* deposition of metal oxide nanoparticles, including α-FeOOH nanorods and magnetic Fe_3O_4 nanoparticles on the graphene nanosheets. The GO nanosheets were reduced by the ferrous ions. Compositions of the graphene hydrogels could be readily controlled by adjusting the pH value of the GO suspensions. After the oil-saturated α-FeOOH–G superhydrophobic aerogels were heated, 3D hematite α-Fe_2O_3 monoliths with porous microstructures were derived. The G–metal oxide hydrogels and aerogels demonstrated strong capability to remove heavy metal ions and oils from water.

For self-assembly of G–FeOOH and G–Fe_3O_4 hydrogels, GO suspensions with a concentration of 2 mg·ml^{-1} were first prepared with Hummers' method, in which $FeSO_4$ was added. The pH value of the GO suspensions was adjusted from 3 to 11 with ammonia aqueous solution. Reaction was realized in an oil bath at 90°C to form 3D black monoliths, which were washed with distilled water and freeze-dried into aerogels. Time-dependent photographs of the samples are shown in Fig. 3.53 (a). After reaction for 0.5 h, black reduced GO nanosheets were formed and dispersed in water, slightly floating from the bottom of the container. Gradually, the assembled graphene monolith was floated and shrunk, due to the decrease in the diameter of the columnar hydrogel. Finally, black hydrogel with a columnar shape was obtained. This method can be easily scaled up, as demonstrated in Fig. 3.53 (b). The as-obtained graphene hydrogels had a very high content of water, with a highly interconnected 3D network microstructure. The network was formed by micrometered pores, as shown in Fig. 3.53 (c). Figure 3.53 (d) indicates that a large number

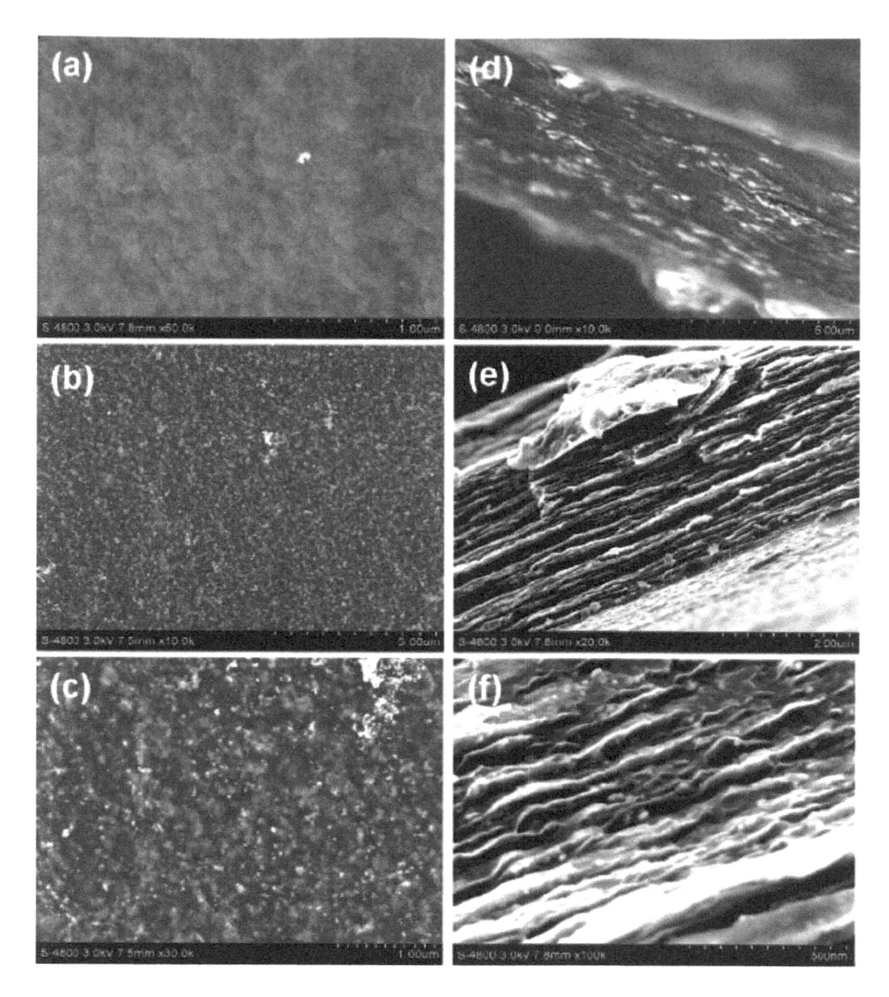

Fig. 3.52. Top-view SEM images of the annealed graphene paper (a) and graphene–Fe_3O_4-2 paper (b). (c) Enlarged SEM image of (b). Cross-sectional SEM images of the annealed graphene paper (d) and graphene–Fe_3O_4-2 paper (e). (f) High-resolution SEM image of the annealed graphene–Fe_3O_4-2 paper of (e). The white spots in the images (b, c, e and f) are Fe_3O_4 nanoparticles. Reproduced with permission from [170]. Copyright © 2010, American Chemical Society.

of nanoparticles were homogenously enwrapped in the graphene nanosheets. Figure 3.53 (e) shows a representative TEM image of the hydrogels, in which rod-like nanoparticles with a size of about 60 nm were attached on thin graphene nanosheets. Figure 3.53 (f) shows representative XRD pattern of the hybrids, indicating that the particles were orthorhombic phase α-FeOOH.

SEM and TEM images of the hybrid samples are shown in Fig. 3.54 (a–c). It was found that the properties of the G–α-FeOOH hydrogels were closely related to the quantity of Fe(II). The higher the concentration of $FeSO_4$ in the GO suspension, the higher the quality of the 3D network would be. For example, in the samples with Fe(II) contents of less than 0.0625 mmol, the pores possessed a size of tens of micrometers, thus leading to large 3D gel cylinders, which means that the network of the hydrogel was formed with a weak interaction, as shown in Fig. 3.54 (d–f). Moreover, the aerogels after freeze-drying with low contents of $FeSO_4$ even collapsed, as illustrated in Fig. 3.54 (g, h). The collapse of the samples with low contents of Fe(II) was attributed to the presence of GO.

It was discovered that the pH value of the initial GO suspensions exhibited a significant effect on the compositions of the as-prepared graphene hydrogels. In the samples derived from the GO suspensions with pH values of 7–10, α-FeOOH was the dominant crystalline phase, but also with the presence of diffraction peaks of $Fe(OH)_3$. At pH = 11, 3D graphene hydrogels with well-developed magnetic properties could be

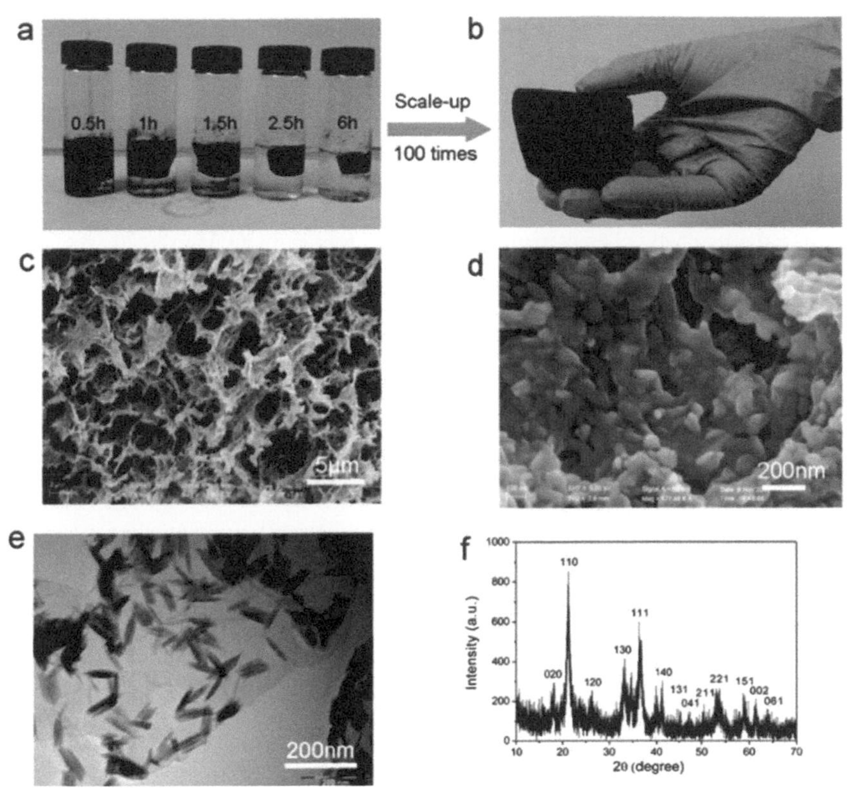

Fig. 3.53. (a) Photographs of the hydrogels with 10 ml of GO suspensions (2 mg·ml^{-1}) at pH = 3 in the presence of 0.5 mmol of $FeSO_4$ after reaction for different time durations. (b) Representative photograph of a scaled-up hydrogel sample by using 1000 ml of GO and 50 mmol of $FeSO_4$. (c, d) Low- and high-magnified SEM images of the aerogels. (e) TEM image of the freeze-dried G–α-FeOOH hydrogels. (f) XRD pattern of the G–α-FeOOH aerogels. Reproduced with permission from [172]. Copyright © 2012, American Chemical Society.

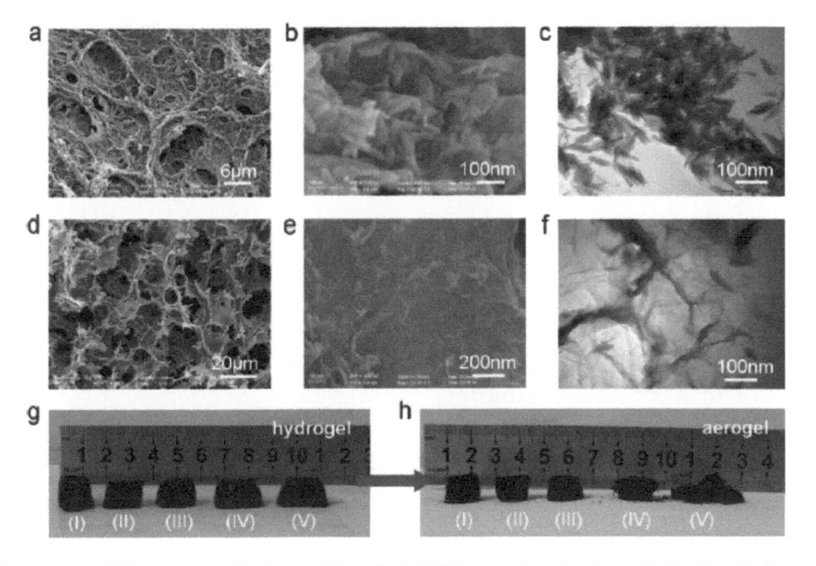

Fig. 3.54. SEM images at different magnifications of the G–FeOOH aerogels dried from the hydrogels derived from 10 ml GO (2 mg·ml^{-1}) suspension at pH = 3, with different amounts of $FeSO_4$: (a–c) 1 mmol and (d–f) 0.0625 mmol. Photographs of the G–FeOOH hydrogels (g) and corresponding aerogels (h) dried from the hydrogels obtained with different amounts of $FeSO_4$. The concentrations of $FeSO_4$ used to prepare the samples (I)–(V) in (g) and (h) were 1, 0.5, 0.25, 0.125, and 0.0625 mmol, respectively. Reproduced with permission from [172]. Copyright © 2012, American Chemical Society.

obtained. The magnetic properties were attributed to the formation of Fe_3O_4 particles, which were uniformly decorated on the thin graphene nanosheets with an average size of 30 nm.

Figure 3.55 shows formation mechanism of the G–iron oxide hydrogels [172]. Fe^{2+} ions were attached to the GO nanosheets due to electrostatic interactions. The Fe^{2+} ions were oxidized into Fe^{3+} by the oxygen containing functional groups on the GO surface. At low pH values, α-FeOOH nanorods were deposited *in situ* on the surfaces of the reduced GO nanosheets through the hydrolysis of Fe^{3+} ions. If the pH value of the initial GO suspension was too high, Fe_3O_4 was formed, due to the coprecipitation of the Fe^{3+} and Fe^{2+} ions. Due to the magnetic characteristic of Fe_3O_4, the reduced GO nanosheets were self-assembled into 3D hydrogels with interconnected networks. At the same time, the magnetic nanoparticles separated the graphene nanosheets and stabilized the graphene hydrogels. This effect was also applicable in case of other metal ions, such as Mn(II) and Ce(III).

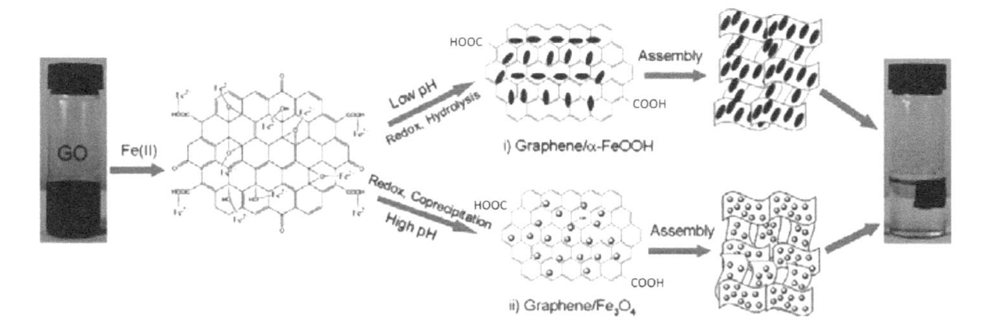

Fig. 3.55. Schematic diagram of formation mechanism of the G–iron oxide hydrogels. Reproduced with permission from [172]. Copyright © 2012, American Chemical Society.

The G–α-FeOOH aerogels were superhydrophobic with high porosity, so that they adsorbed oils and nonpolar organic solvents without suctioning water. Therefore, these materials could be used for marine oil-spill recovery applications. Figure 3.56 (a) shows an example of the aerogels to selectively absorb gasoline on water surface. More importantly, the aerogels could be readily regenerated, as shown in Fig. 3.56 (b). It was found that, after eight gasoline absorbing-drying recycles, adsorption capacity of the G–α-FeOOH aerogels adsorbent was retained to be above 90%. This high retention capacity was mainly attributed to the strongly interconnected network porous structure. They could absorb a wide range of nonpolar organic solvents and oils, as shown in Fig. 3.56 (c). For most adsorbates, the maximum uptake capacity was up to nearly 30 times the weight of the absorbent. The superhydrophobicity of the hybrid aerogels was closely related to the hydrophobic π–π stacking of the reduced GO, while the presence of the Fe_3O_4 nanoparticles on the graphene nanosheets increased the surface roughness of the materials. After the oil-saturated G–α-FeOOH aerogels were burned, hematite α-Fe_2O_3 was formed, as shown in Fig. 3.56 (d). Microstructures of the α-Fe_2O_3 containing blocks are demonstrated in Fig. 3.56 (e, f). Figure 3.56 (f) shows that the porous nanosheets of nanoparticles were well-retained after the graphene nanosheets were burned off. In other words, the superhydrophobic G–α-FeOOH aerogels can be used as precursor to obtain metal oxides with 3D hierarchical microstructures. The hybrid aerogels exhibited promising performances in removing toxic pollutant ions, including Cr(IV) and Pb(II) [172].

He and Gao reported a simple one-step method to synthesize G–Fe_3O_4 hybrid, consisting of graphene nanosheets attached with magnetite NPs [175]. The hybrid possessed multifunctional properties, due to the electrically conductive graphene and the superparamagnetic Fe_3O_4 NPs. In this case, G–Fe_3O_4 hybrid was formed through the *in situ* reduction of GO and simultaneous formation of Fe_3O_4 NPs. The surfaces of the graphene nanosheets were uniformly decorated with a layer of Fe_3O_4 NPs. The hybrid powder could be dispersed in polar solvents, leading to feasible solution processing. Because of its unique properties, this G–Fe_3O_4 hybrid could find various applications. To prepare the G–Fe_3O_4 hybrid, NaOH was mixed with DEG, which was heated at 120°C for 1 h in the presence of nitrogen and then cooled down to 70°C,

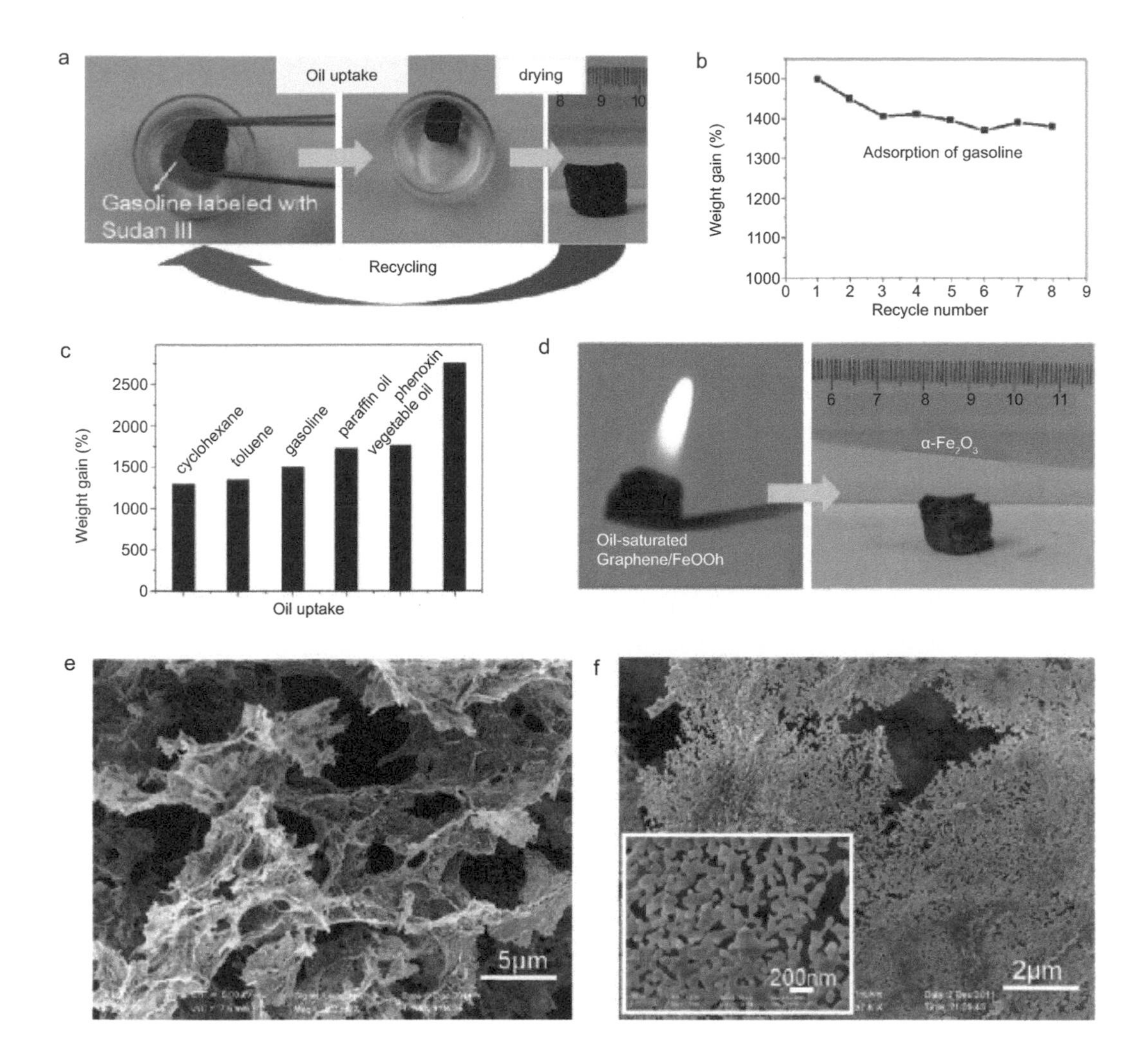

Fig. 3.56. (a) Photographs showing gasoline adsorbing process of the G–α-FeOOH aerogels. The G–α-FeOOH aerogels were obtained from the hydrogels derived from 10 ml of GO (2 mg·ml^{-1}) suspension at pH = 3 with 0.5 mmol of $FeSO_4$. (b) Regeneration capacity of the aerogels to adsorb gasoline, which was measured by removing the gasoline at 100°C. (c) Adsorption capacities of the aerogels for different organic solvents and oils. (d) Photographs of an oil-saturated G–α-FeOOH aerogel sample before and after burning. (e, f) SEM images of the 3D microstructures of the burned oil-saturated G–α-FeOOH aerogel containing α-Fe_2O_3. Reproduced with permission from [172]. Copyright © 2012, American Chemical Society.

forming NaOH/DEG solution with NaOH concentration of 10 mg·ml^{-1}. After that, GO power was dispersed in DEG, into which $FeCl_3$ was added to produce Fe_3O_4 precursor solution. The mixture was heated to 220°C for 30 min in nitrogen flow, into which 5 ml NaOH/DEG solution was injected rapidly. The newly formed mixture was kept at 220°C for 1 h. Variable conditions included mass feed ratio of $FeCl_3$ to GO (R_{feed}-$FeCl_3$) and mass feed ratio of NaOH to GO (R_{feed}-NaOH). The dispersible G–Fe_3O_4 hybrid could be further made into Graphene–Fe_3O_4-C and G–Fe_3O_4-polyurethane composites.

Representative TEM images of the G–Fe_3O_4 hybrid are shown in Fig. 3.57 (a–c) [175]. It was observed that surfaces of the graphene nanosheets were covered by a dense layer of Fe_3O_4 NPs, with a narrow size distribution and an average size of 6.3 nm, as shown in Fig. 3.57 (d). HRTEM image (the inset in Fig. 3.57 (c)) shows lattice spacing of (311) plane of cubic magnetite. AFM images indicated that the graphene nanosheets in the hybrid had a lateral dimension of several micrometers. Figure 3.58 shows SEM images of the graphite, graphene and G–Fe_3O_4 hybrid samples [175]. According to the SEM images of the G–

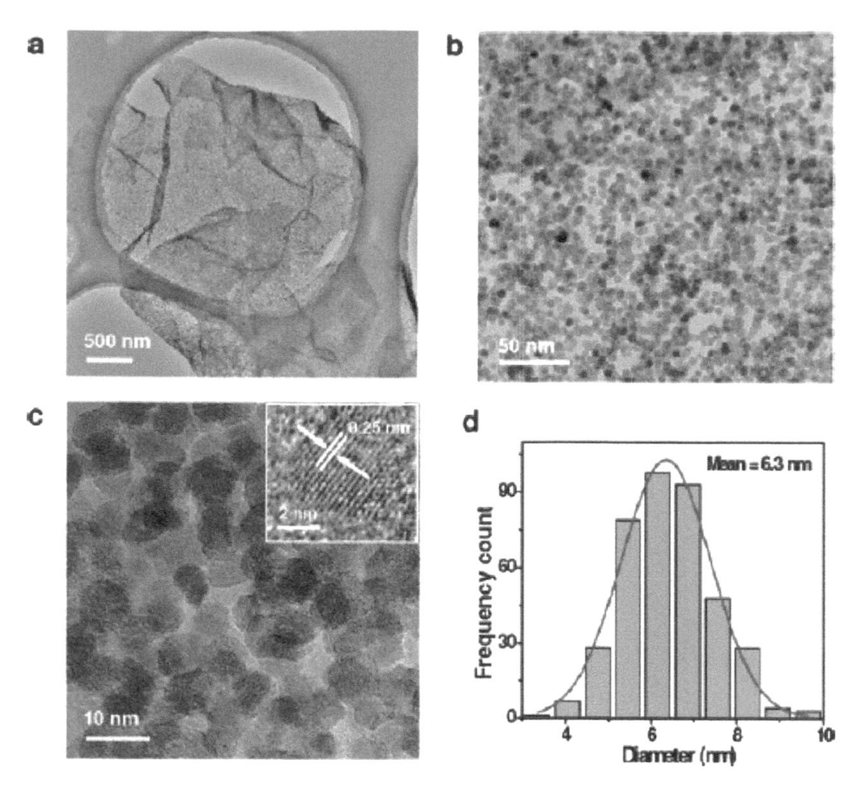

Fig. 3.57. TEM images at different magnifications (a–c) and size distribution of the Fe_3O_4 nanoparticles (d) of the G–Fe_3O_4, which was made under conditions of R_{feed}-$FeCl_3$ = 4, R_{feed}-NaOH = 5/3 and t = 1 h. Reproduced with permission from [175]. Copyright © 2010, American Chemical Society.

Fe_3O_4 samples prepared with various feed ratios, the Fe_3O_4 NPs (bright dots) were uniformly distributed on the slightly wrinkled surface of the graphene nanosheets, as illustrated in Fig. 3.58 (c–e). As shown in Fig. 3.58 (f), vacuum filtered G–Fe_3O_4 hybrid film possessed a layered structure of thin graphene nanosheets. The crystal structure of the Fe_3O_4 NPs was confirmed by XRD results.

Figure 3.59 shows a schematic diagram of the formation mechanism of the G–Fe_3O_4 hybrid, together with representative TEM images of the samples reacted for different time durations [175]. It was found that the formation of the Fe_3O_4 nanocrystals on graphene nanosheets could explained with the La Mer model. As NaOH solution was injected into the precursor mixture, a temporally discrete nucleation took place, which was then accelerated due to the high temperature of the reaction system. As a result, a burst nucleation was triggered within a very short while. Such a quick nucleation led to the formation of magnetite nanocrystals with a relatively small size and narrow size distribution. The nuclei grew quickly into larger crystals during the first 5 min after the initial nucleation.

The G–Fe_3O_4 hybrid film obtained by using vacuum-filtration was about 0.7 $S \cdot m^{-1}$, which indicated that the insulating GO was effectively reduced to conductive graphene. Depending on the content of the hybrid powder, the G–Fe_3O_4-polyurethane composite sheets were gradually darkened, as shown in Fig. 3.60 (a). Owing to the presence of the Fe_3O_4 NPs and the polymer matrix, the composite films were magnetic and mechanically flexible, as demonstrated in Fig. 3.60 (b). Figure 3.60 (c) shows electrical conductivity of the G–Fe_3O_4-polyurethane composites. It is reasonably expected that such multifunctional G–Fe_3O_4 polymer composites could be used for applications in electromagnetic interference (EMI) shielding, microwave absorbing and flexible electronic materials.

A one-step hydrothermal method has been developed to produce a G–Fe_3O_4 hybrid with superparamagnetic property [217]. Anhydrous $FeCl_3$ was used as the source of iron, while ethylene

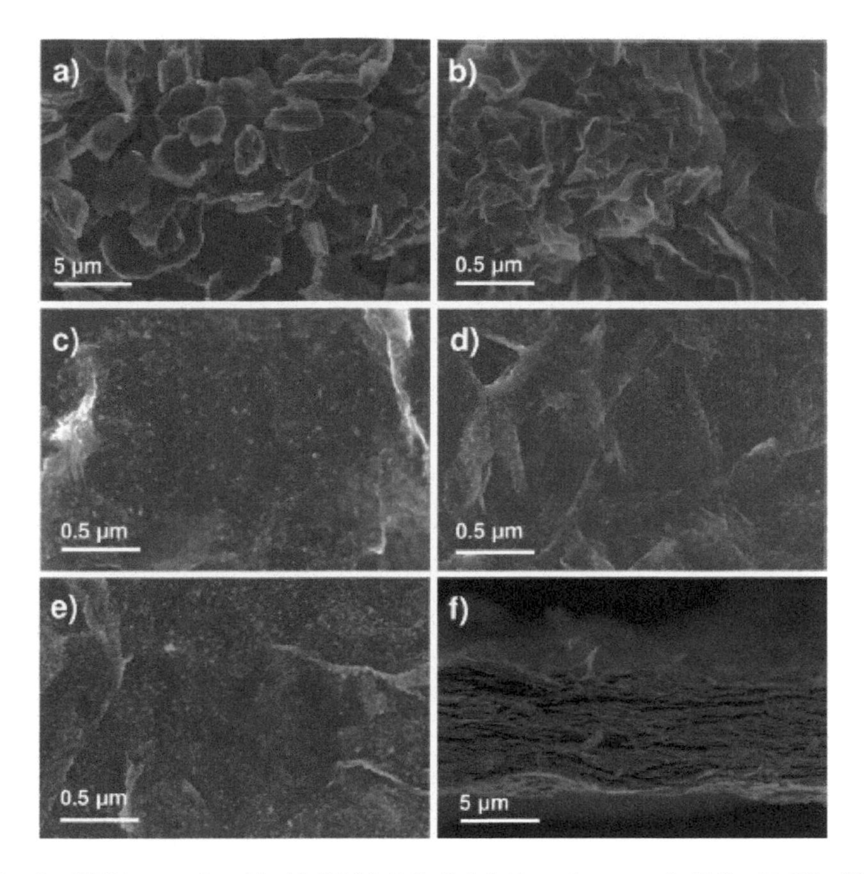

Fig. 3.58. Top-view SEM images of graphite (a), GO (b), G–Fe_3O_4 hybrid samples prepared with R_{feed}-NaOH = 5/3 for t = 1 h but with different R_{feed}-$FeCl_3$ of (c) 2, (d) 3 and (e) 4, together with cross-sectional SEM image of the vacuum filtrated G–Fe_3O_4 hybrid sheets (f). Reproduced with permission from [175]. Copyright © 2010, American Chemical Society.

glycol (EG) or mixture of EG and diethylene glycol (DEG) was used as the solvent and reducing agent. In these hybrids, sizes of the magnetite nanocrystals and nanocrystal clusters could be well-controlled by controlling the reaction time and ratio of EG and DEG. Figure 3.61 shows a schematic diagram of the hydrothermal synthesis process, in which the formation of the NPs and reduction of GO to graphene nanosheets occurred simultaneously.

To synthesize GO–Fe_3O_4 hybrids, GO powder was dispersed in DEG with the aid of ultrasonication to prepare solution A, while $FeCl_3$ and NaOAc were dissolved in DEG to form solution B. Then, solution A and solution B were thoroughly mixed, and the resultant solution was hydrothermally treated at 220°C for different time durations, in order to obtain GO–Fe_3O_4 hybrid powders. It was found that the crystal size of the Fe_3O_4 NPs increased gradually with the increasing reaction time. When using a mixture solvent of DEG and EG, the hydrothermal reaction temperature was 205°C. In this case, particle size of Fe_3O_4 was dependent on the volume ratio of DEG/EG (v/v in ml). With increasing DEG/EG ration, the average particle size was increased.

Figure 3.62 shows TEM images of the Fe_3O_4 nanocrystals and the G–Fe_3O_4 hybrid samples synthesized with different hydrothermal reaction times [217]. After reaction for 2, 4 and 6 h, the Fe_3O_4 nanocrystals were homogeneously distributed on the surface of the graphene nanosheets, with average sizes of 2.3, 6.3 and 8.5 nm, respectively, as shown in Fig. 3.62 (b–d). Selected area electron diffraction (SAED) patterns indicated that the Fe_3O_4 NPs were highly crystallized. The Fe_3O_4 NPs were strongly attached onto the graphene nanosheets, so that the G–Fe_3O_4 hybrids could withstand the strong sonication, although there was no chemical bonding between the Fe_3O_4 nanocrystals and the graphene nanosheets.

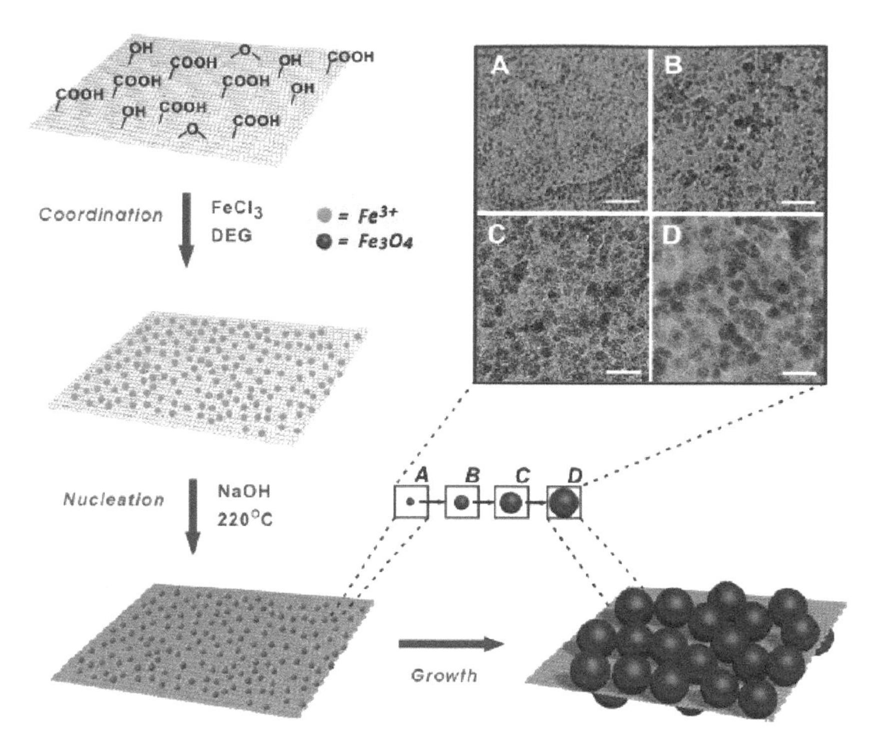

Fig. 3.59. Schematic illustration showing the formation mechanism of the G–Fe_3O_4 hybrid. The inset shows TEM images of G–Fe_3O_4 hybrid samples after reaction for different time durations: (A) 0, (B) 1, (C) 5 and (D) 60 min. Scale bars = 20 nm. Reproduced with permission from [175]. Copyright © 2010, American Chemical Society.

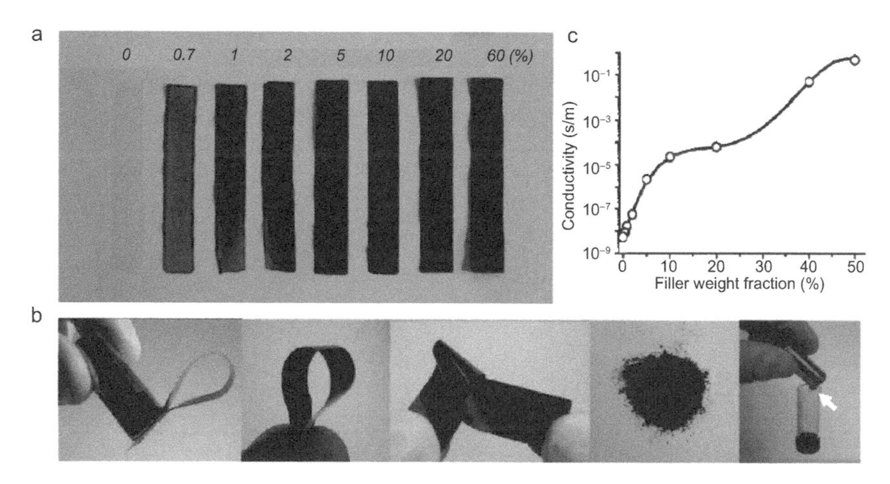

Fig. 3.60. (a) Photographs of the G–Fe_3O_4-polyurethane composite sheets with different contents of the hybrid powder (about 40 μm in thickness). (b) From left to right: photographs of a strip of the G Fe$_3$O$_4$-polyurethane composites (50 wt%) lifted with a magnet, held and twisted with hands, dried G–Fe_3O_4 powder and 6.0 mg G–Fe_3O_4 powder that can lift a 5.05 g glass bottle with a magnet, respectively. (c) Electrical conductivity of the G–Fe_3O_4-polyurethane composites as a function of the weight fraction of the hybrid powder. Reproduced with permission from [175]. Copyright © 2010, American Chemical Society.

Similar to hydrothermal synthesis, the solvothermal method has also been widely used to prepare G–Fe_3O_4 hybrids. One example was reported by Zhou et al. starting with GO and $FeCl_3$ as the precursors [238]. During the solvothermal reaction, GO was reduced to graphene, while Fe_3O_4 particles were grown on

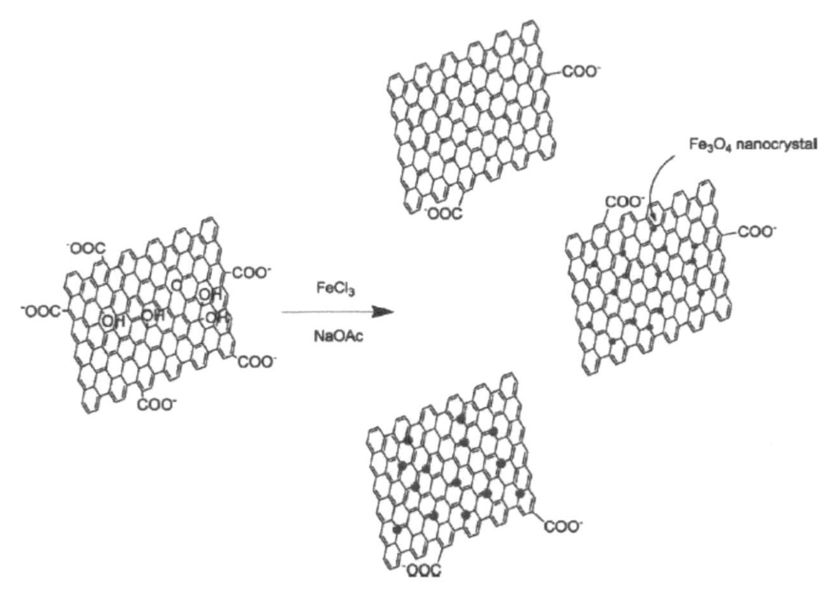

Fig. 3.61. Schematic diagram showing the formation of G–Fe_3O_4 hybrids for different reaction times. Reproduced with permission from [217]. Copyright © 2011, Elsevier.

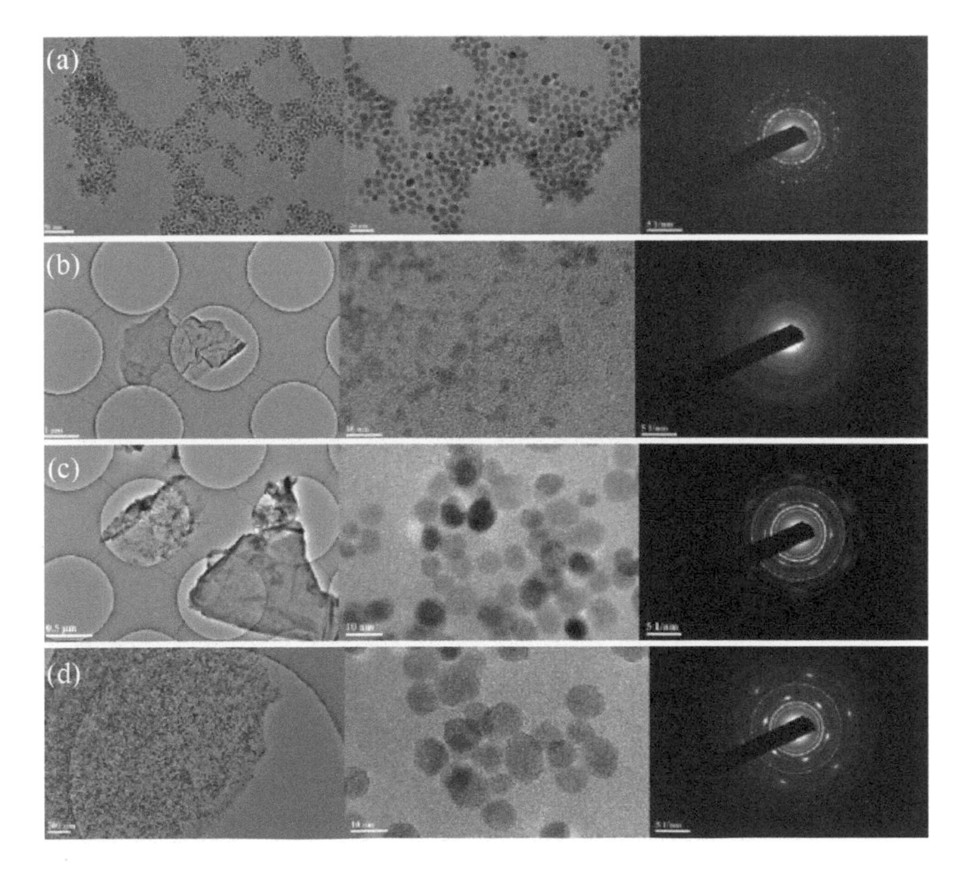

Fig. 3.62. TEM images of the Fe_3O_4 nanocrystals (a) and the G–Fe_3O_4 hybrids synthesized after reaction for different times (b–d). Average particle sizes of the Fe_3O_4 nanocrystals are 7.8 nm (a), 2.3 (b, 2 h), 6.3 (c, 4 h) and 8.5 nm (d, 6 h). Reproduced with permission from [217]. Copyright © 2011, Elsevier.

the surfaces of the graphene nanosheets at the same time. The Fe_3O_4 particles exhibited a very narrow size distribution. GO powder was dispersed in ethylene glycol to form GO suspension, into which $FeCl_3 \cdot 6H_2O$ was added and thoroughly mixed through strong stirring. Then, NaAc and polyethylene glycol were added into the mixture. The mixture was finally used for solvothermal reaction at 200°C for 16 h in order to obtain G–Fe_3O_4 hybrid powders.

Figure 3.63 (a) shows a representative AFM image of the GO used in the solvothermal reaction, with the GO nanosheets having a thickness of about 1.1 nm [238]. Typical TEM images of the G–Fe_3O_4 hybrid are shown in Fig. 3.63 (b-c). Fe_3O_4 particles with a microspherical morphology of uniform size were evenly distributed on the surface of the graphene nanosheets, as demonstrated in Fig. 3.63 (b). Due to the extremely thin monolayer, the transparent graphene nanosheets could only be observed from the fringe and wrinkles. The Fe_3O_4 particles exhibited an average diameter of about 200 nm. Figure 3.63 (c) indicated that the Fe_3O_4 microspherical particles were of a porous structure, implying that they consisted of a large number of Fe_3O_4 NPs. The small Fe_3O_4 NPs had an average size of 15 nm. Stoichiometric composition of the Fe_3O_4 NPs was confirmed by EDS results. Figure 3.63 (d) shows a high-resolution TEM (HRTEM) image of a selected area in the Fe_3O_4 microspherical particle, revealing a well-developed crystallinity. The lattice spacing values of 0.48 and 0.29 nm were attributed to the (111) and (202) planes of Fe_3O_4. SAED pattern (inset of Fig. 3.63 (d)) further confirmed the high crystalline characteristic of the Fe_3O_4 NPs. Because GO nanosheets usually contain a large quantity of hydroxyl and carboxyl groups, they are negatively charged in solutions or suspensions. Therefore, during the solvothermal reaction, positive Fe^{3+} ions would be easily attached onto the surface of the GO nanosheets through electrostatic interaction. The attached Fe^{3+} ions acted as nuclear sites to form Fe_3O_4 NPs, when the Fe^{3+} ions were partially reduced. The

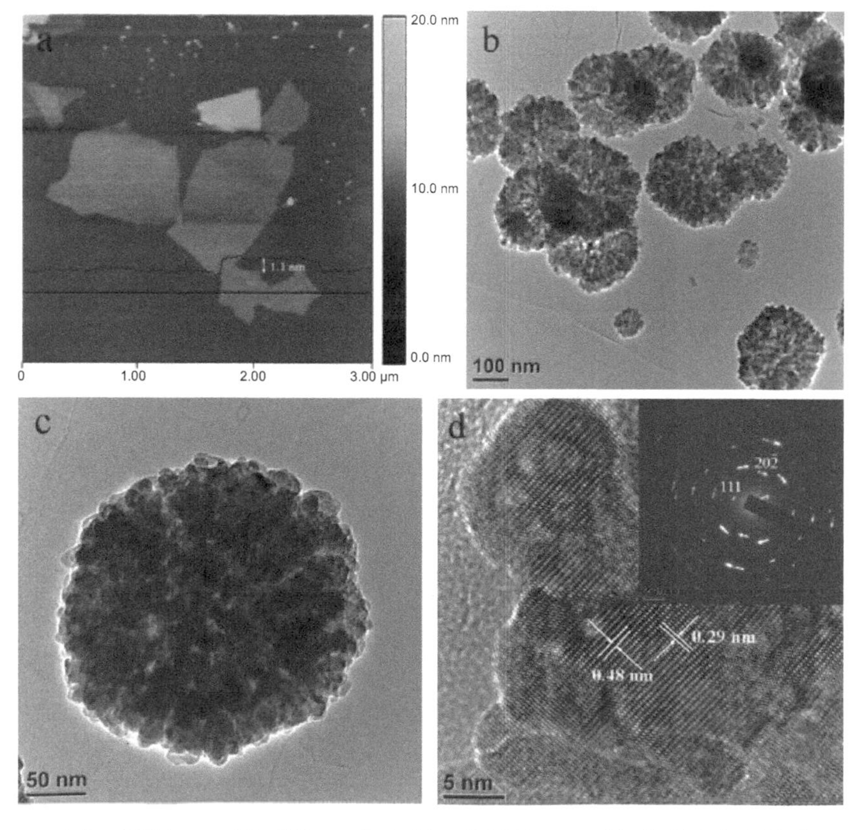

Fig. 3.63. Representative AFM image of the GO nanosheets (a) and TEM images of the G–Fe_3O_4 hybrids: low magnification image (b), (c) high magnification image of an individual Fe_3O_4 microspherical particle and (d) selected area HRTEM image of the Fe_3O_4 microspherical particle, with inset to be SAED pattern of Fe_3O_4. Reproduced with permission from [238]. Copyright © 2010, The Royal Society of Chemistry.

small Fe_3O_4 NPs were agglomerated into microspherical particles. Figure 3.64 shows a detailed diagram of synthesis procedure of the G–Fe_3O_4 hybrids.

It was found that the morphology of the G–Fe_3O_4 hybrids was dependent on the initial concentration of Fe^{3+}. Figure 3.65 shows SEM images of the G–Fe_3O_4 hybrid samples obtained with different amounts of $FeCl_3 \cdot 6H_2O$ used in the precursors [238]. Although grape-like Fe_3O_4 microspherical particles were present in all samples, the density of Fe_3O_4 particles was increased with the increasing concentration of Fe^{3+}. As the starting amounts of $FeCl_3 \cdot 6H_2O$ were 0.1 and 0.5 g, the difference in particle size between the two samples was not very significant, as seen in Fig. 3.65 (a, b), while there was sharp increase in particle size,

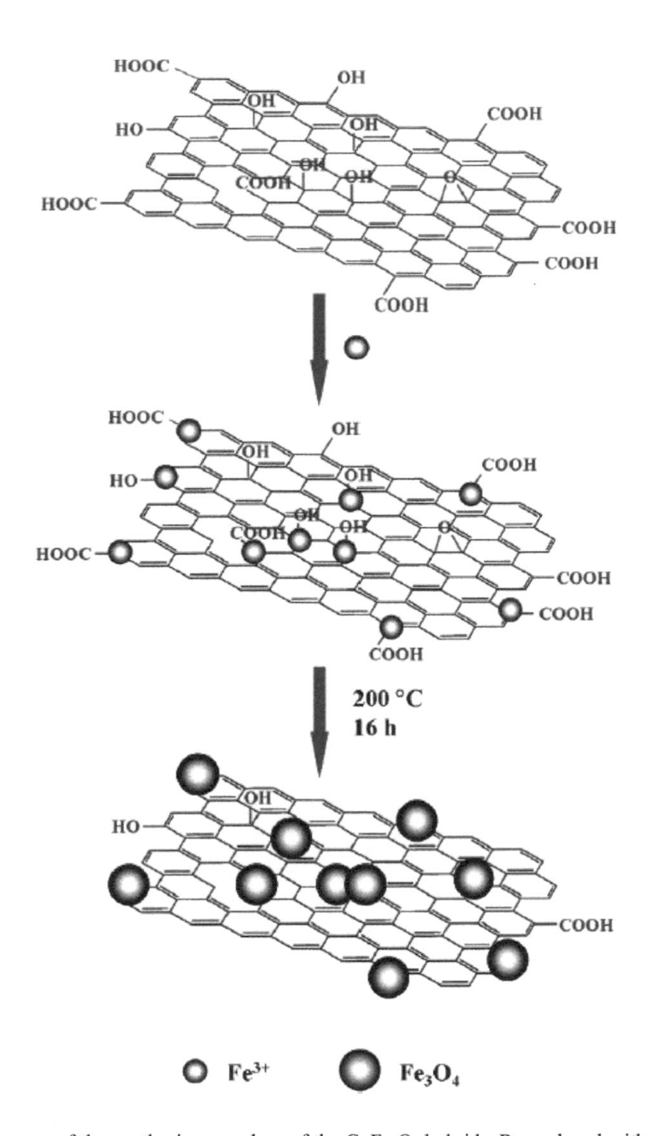

Fig. 3.64. Schematic diagram of the synthesis procedure of the G–Fe_3O_4 hybrids. Reproduced with permission from [238]. Copyright © 2010, The Royal Society of Chemistry.

as the initial amount of $FeCl_3 \cdot 6H_2O$ was increased to 1 g (Fig. 3.65 (d)). In other words, concentration of the Fe^{3+} can be used as a factor to control the density and size of the Fe_3O_4 NPs in the G–Fe_3O_4 hybrids.

Fe_2O_3–G hybrids have been prepared by using chemical routes [242–257], hydrothermal synthesis [258-271], sol-gel process [272], which are predominantly used in lithium ion batteries and supercapacitors

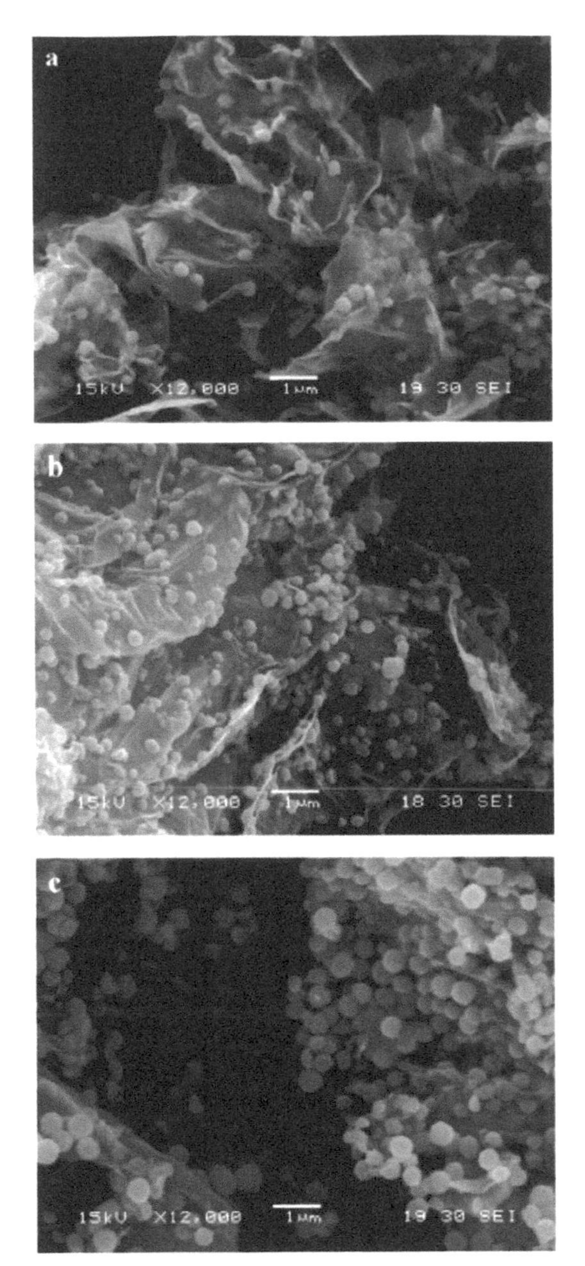

Fig. 3.65. SEM images of the G–Fe_3O_4 hybrids prepared with different amounts of $FeCl_3{\cdot}6H_2O$ in the precursors: (a) 0.1 g, (b) 0.5 g and (c) 1 g. Reproduced with permission from [238]. Copyright © 2010, The Royal Society of Chemistry.

[242–247, 251, 254, 257, 258, 260–270, 272], with minor applications as photocatalysts [259], in water treatment [248, 255], tribolotical enhancement [250], microwave absorbers [252], gas sensors [253] and biocompatible materials [256]. Selected samples with focus on the synthesis and characterization of Fe_2O_3–G hybrids are presented in this section.

A simple two-step homogeneous precipitation method was reported to synthesize G–Fe_2O_3 hybrids by using $FeCl_3$ as the source of iron [242]. The precipitation was carried out in GO suspension with urea as the precipitant. After precipitation reaction, the GO was reduced into graphene (rGO) with hydrazine under microwave irradiation. At the same time, Fe_2O_3 NPs were formed, which were distributed on the surface of the rGO nanosheets. Such hybrids exhibited encouraging performances as the anode materials of

Li-ion batteries, with discharge and charge capacities of 1693 and 1227 mAh·g^{-1}, which were normalized with respect to the mass of Fe_2O_3 to be 1355 and 982 mAh·g^{-1}.

Figure 3.66 shows a schematic diagram of the synthesis procedure of the rGO–Fe_2O_3 hybrids [242]. $FeCl_3$ was hydrolysed in the GO suspension containing urea at 90°C, with the molar ratio of $FeCl_3$ to urea to be 1:30. During the hydrolysis process, urea gently released hydroxyl ions, so as to form $Fe(OH)_3$ precipitates. The $Fe(OH)_3$ particles were attached onto the surface of the GO nanosheets through oxygen-containing functional groups, including carboxyl, hydroxyl and epoxyl. The reacted suspension was cooled to room temperature, into which a certain amount of hydrazine was added. After that, the suspension was irradiated with microwaves. By doing this, the GO nanosheets were reduced to graphene nanosheets, while the $Fe(OH)_3$ dehydrated to form Fe_2O_3 NPs. The presence of Fe_2O_3 was confirmed by XRD, XPS and Raman results. TGA analysis indicated that the hybrids contained 20 wt% rGO.

Figure 3.67 shows representative SEM and TEM images of the rGO–Fe_2O_3 hybrids [242]. Low magnification SEM image indicated that the hybrid powder possessed a curled morphology, leading to a paper-like structure, as shown in Fig. 3.67 (a). High magnification SEM image revealed that the Fe_2O_3 NPs were uniformly distributed on the surface of the rGO nanosheets, as illustrated in Fig. 3.67 (b). The Fe_2O_3 NPs had an average diameter of 60 nm, as evidenced by the TEM image in Fig. 3.67 (c). There were interfaces between the rGO and the Fe_2O_3 NPs, which was clearly demonstrated by the HRTEM image in Fig. 3.67 (d). The interlayer spacing of 0.25 nm corresponded to that of (110) planes of crystalline Fe_2O_3.

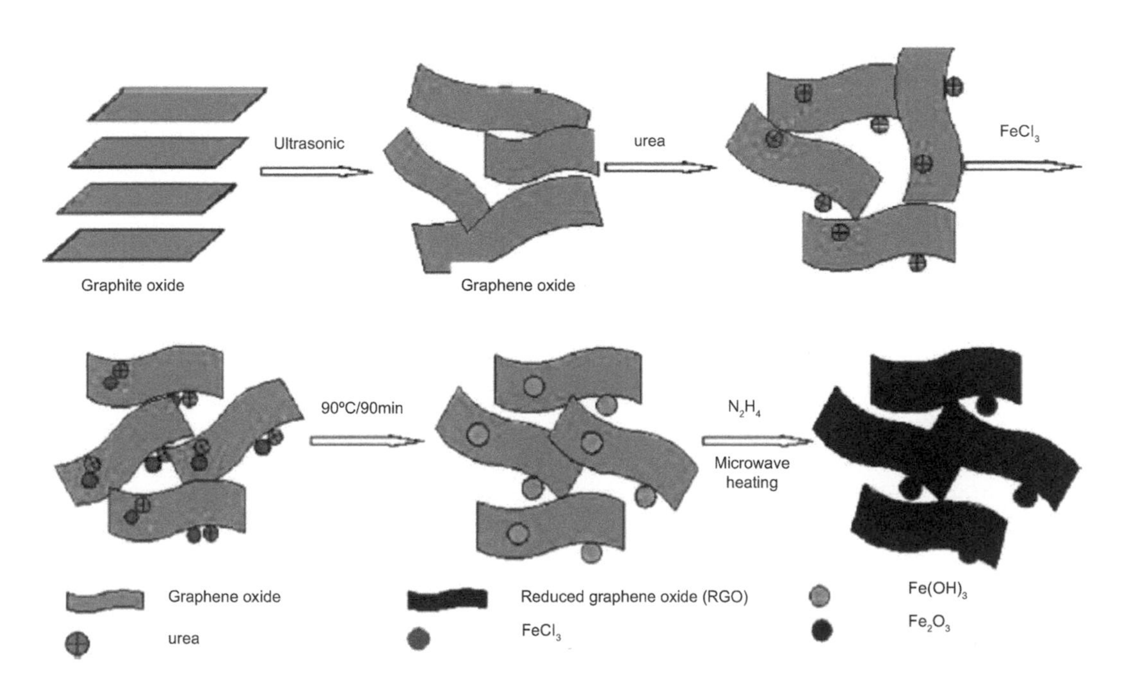

Fig. 3.66. Schematic diagram showing the formation mechanism of the rGO–Fe_2O_3 hybrids. Reproduced with permission from [242]. Copyright © 2011, American Chemical Society.

Another chemical route to synthesize G–Fe_2O_3 hybrids was reported by Zhu et al. [246]. The special aspect of the G–Fe_2O_3 hybrids was that the graphene nanosheets were synthesized by using the chemical vapor deposition (CVD) method, rather than the wet-chemical exfoliation method. Moreover, the graphene nanosheets were highly porous, with one or two layers and an extremely high specific surface area (SSA) of up to 2038 m^2·g^{-1}. The CVD derived graphene was named nanomesh graphene (NMG). The NMG–Fe_2O_3 hybrid was selected as an example in that study. The Fe_2O_3 NPs were prepared by using an adsorption-precipitation process, which was claimed by the authors to be applicable in case of other oxides as well. Figure 3.68 shows a schematic diagram demonstrating the formation mechanism

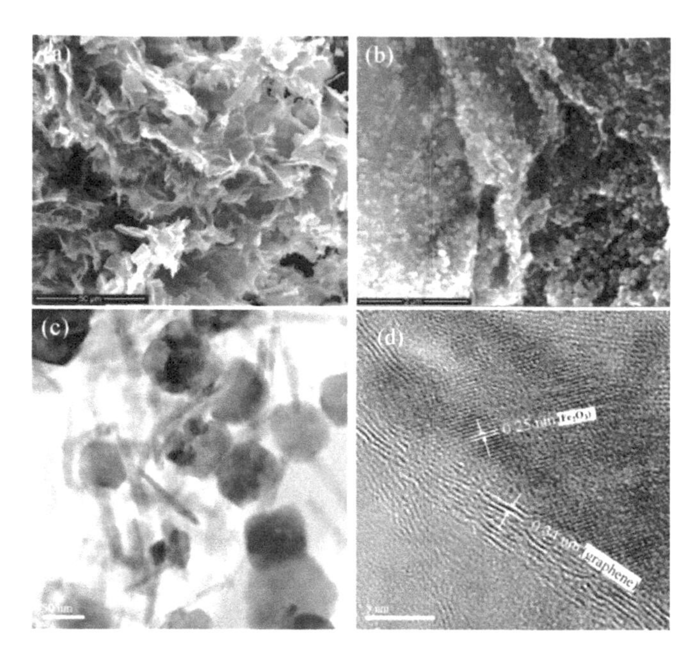

Fig. 3.67. Representative SEM and TEM images of the rGO–Fe_2O_3 hybrids. Reproduced with permission from [242]. Copyright © 2011, American Chemical Society.

of the NMG–Fe_2O_3 hybrid, starting with the NMG nanosheets and $FeCl_3$. Due to the high absorbability of the NMG nanosheets, $FeCl_3$ aqueous solution was readily adsorbed into the pores of the NMG. When $NH_3{\cdot}H_2O$ aqueous solution was added into the suspension, NMG–$Fe(OH)_3$ was formed, which led to the formation of the NMG–Fe_2O_3 hybrid as a result of calcination. Therefore, the Fe_2O_3 NPs were filled into the pores of the NMG nanosheets. Composition of the NMG–Fe_2O_3 hybrids could be readily controlled by controlling the amount of the $FeCl_3$.

Figure 3.69 shows TEM images of the NMG–Fe_2O_3 hybrid samples with different loadings of Fe_2O_3 [246]. The Fe_2O_3 NPs had sizes of 1.5−5 nm, and were uniformly distributed on the NMG nanosheets, as the content of Fe was 10% (Fig. 3.69 (a–c)). The Fe_2O_3 NPs were attached on walls of the meshes, as shown in Fig. 3.69 (c). A special core/void/shell structure was formed in the NMG–Fe_2O_3 hybrid. The presence of the void could buffer the volume change causing the intercalation-deintercalation processes of the lithium ions. The formation of the core/void/shell structure was closely related to the adsorption-precipitation process. The mesh pores would have confined the drops of the $FeCl_3$ solution. Because the $FeCl_3$ precursor drops were isolated individually, continuous growth of the precipitated nanoparticles was effectively prevented. However, as the amount of $FeCl_3$ was increased, the particle size of the Fe_2O_3 NPs was also increased. It was found that the particle sizes of the 20%Fe-NMG and 70%Fe-NMG hybrids were 20−30 nm and 20–200 nm, respectively, as shown in Fig. 3.69 (d, e). Therefore, the Fe precursor at high

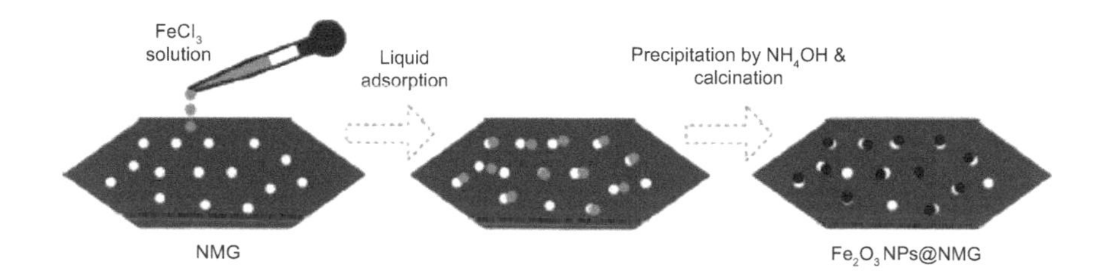

Fig. 3.68. Schematic diagram showing the synthesis procedure of the Fe_2O_3–NMG composites. Reproduced with permission from [246]. Copyright © 2014, American Chemical Society.

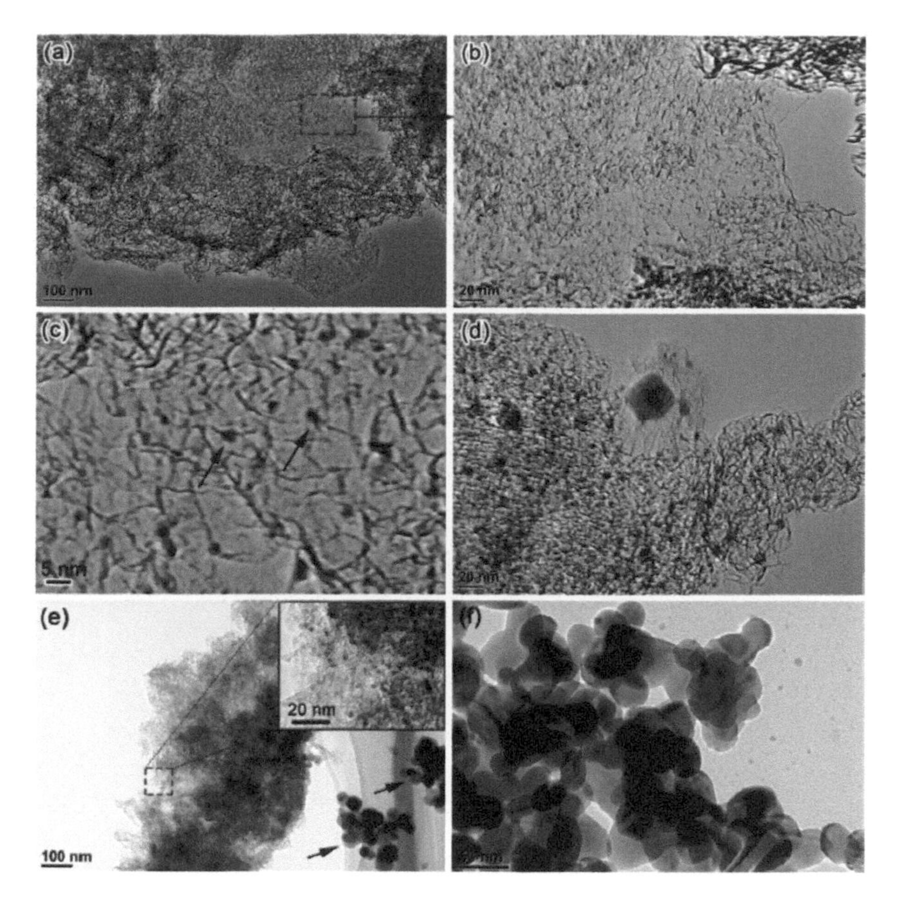

Fig. 3.69. (a–c) TEM images of 10%Fe-NMG, (d) 20%Fe-NMG, (e) 70%Fe-NMG, and (f) pure Fe_2O_3. The inset of (e) a high-magnification TEM image of the 70%Fe-NMG hybrid sample. Reproduced with permission from [246]. Copyright © 2014, American Chemical Society.

concentrations was precipitated in the liquid phase, instead of on the surface of the NMG nanosheets. In comparison, Fig. 3.69 (f) indicated that pure Fe_2O_3 NPs without NMG had sizes of 30–50 nm.

The uniform distribution of the Fe_2O_3 NPs in the NMG matrix was further confirmed by the EDS results of the 10%Fe–NMG sample. The representative HRTEM images and SAED pattern of the 10%Fe–NMG sample are shown in Fig. 3.70 (a, b), confirming the presence of Fe_2O_3 NPs and indicating that the disordered graphene nanosheet stacks had 1–3 layers [246]. The lattice fringe of 0.16 nm was attributed to the (211) planes of crystalline Fe_2O_3. Near the Fe_2O_3 NPs, (002) lattice plane of graphene with a spacing of about 0.38 nm could be identified easily, implying that intimate interfacial contact was formed between the two components. Particle size analysis result indicated that the Fe_2O_3 NPs of the 10%Fe–NMG hybrid sample possessed a narrow size distribution of 1.5–5 nm centered at about 3 nm. In comparison, the pores of pure NMG showed a slightly larger average size of about 4 nm, so that there was still space after the pores were filled with the Fe_2O_3 NPs. This space could well accommodate the expansion of the active materials during the intercalation and deintercalation of Li^+ ions.

Electrochemical performance and other characterization results demonstrated that the 10%Fe–NMG sample possessed thinner solid-electrolyte interface (SEI) films and thus lower electrolyte resistance, which led to higher lithium storage capacity and rate capability [246]. Typical HRTEM images of the 10%Fe–NMG sample after long cycles of discharge/charge cycles are shown in Fig. 3.70 (c, d). Figure 3.70 (a) revealed that the interfaces between the Fe_2O_3 NPs and the NMG nanosheets became less pronounced

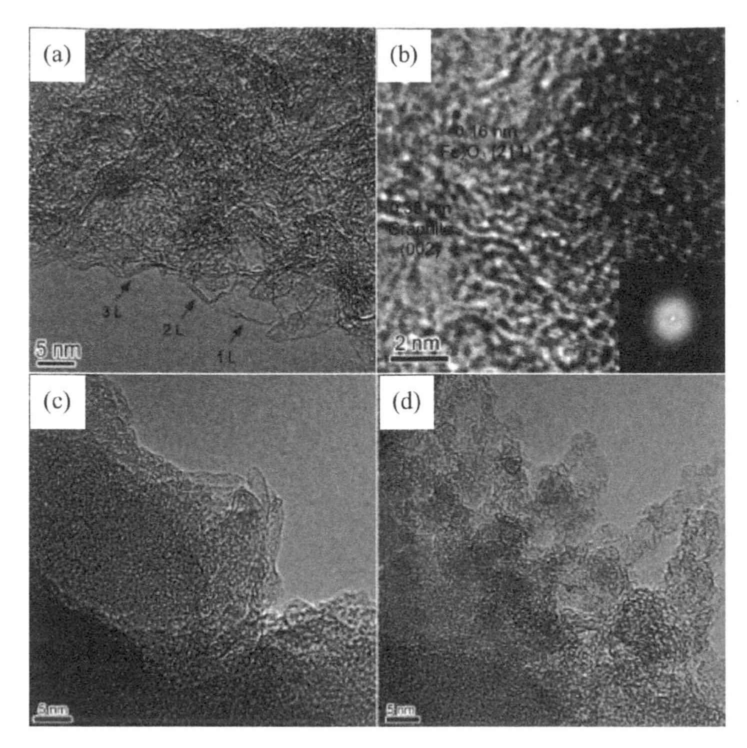

Fig. 3.70. (a, b) HRTEM images of the 10%Fe-NMG hybrid. The inset of (b) is SAED pattern of the sample. (c, d) HRTEM images of the 10%Fe-NMG sample after 50 discharge/charge cycles at a current density of 150 mA g^{-1}. Reproduced with permission from [246]. Copyright © 2014, American Chemical Society.

as compared with the pristine sample, which could be readily attributed to the presence of SEI films. It was also found that some columnar structures were occasionally formed on surface of the Fe_2O_3 NPs, as illustrated in Fig. 3.70 (b), which implied that the formation and decomposition processes of the SEI films in the presence of Fe_2O_3 NPs were different from those on NMG nanosheets. Therefore, the presence of the Fe_2O_3 NPs could trigger the decomposition of the SEI films, which is an advantage in terms of improving the electrochemical performance of electrode materials for practical applications.

A simple hydrothermal method has been reported to synthesize porous α-Fe_2O_3–G hybrids [260]. $FeSO_4 \cdot 7H_2O$ aqueous solution (40 mg ml^{-1}) and homogeneous GO solution (1 mg ml^{-1}) were mixed for hydrothermal synthesis at 150°C for 12 h. The hydrothermal reaction was combined with a slow annealing route, the pore size of the hybrid and crystallinity of the oxide could be controlled by controlling the heating rate. The porous hybrids exhibited promising performances as electrodes of supercapacitors.

Figure 3.71 shows XRD patterns of the hybrid samples before and after thermal annealing [260]. The as-synthesized sample contained rGO and goethite FeOOH. Thermal annealing led to the formation of hematite α-Fe_2O_3. The α-Fe_2O_3 phase in the slowly annealed hybrid sample had a lower crystallinity than that in the quickly heated sample, with XRD crystal sizes of 15.3 and 33.2 nm, respectively. Therefore, the crystallinity of the α-Fe_2O_3 NPs in this method was closely related to the heating rate, which is of significance specifically for application as electrode materials. It was found that the electrode materials with an appropriate crystallite size could be much higher than those of amorphous and highly crystallized ones [273].

Representative TEM images of the samples are shown in Fig. 3.72 [260]. In the as-synthesized sample, the FeOOH nanoparticles possessed a short rod-like morphology, and were uniformly distributed on the surface of the graphene nanosheets, as shown in Fig. 3.72 (a). Such morphology and distribution of the NPs were not altered by the thermal annealing, while the FeOOH was transformed to porous α-Fe_2O_3, as illustrated in Fig. 3.72 (b). The graphene nanosheets were found to have wrinkles and folds. Figure 3.72

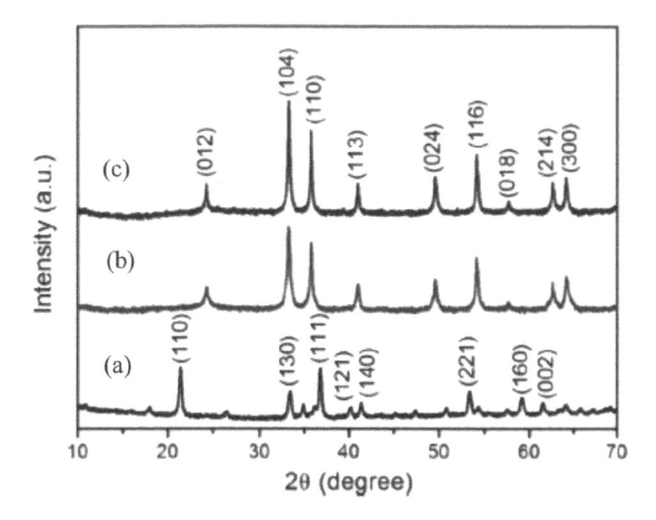

Fig. 3.71. XRD patterns of the rGO-FeOOH hybrid (a), slowly annealed hybrid sample (b) and quickly heated sample (b). Reproduced with permission from [260]. Copyright © 2015, American Chemical Society.

(c) confirmed that the α-Fe_2O_3 NPs were highly porous, with slit-like open pores having widths of 3–4 nm. In the HRTEM image of Fig. 3.72 (d), the fringe spacing of 0.27 nm was attributed to the (104) plane of α-Fe_2O_3. The porous α-Fe_2O_3 NPs and the graphene nanosheets in the hybrids prepared in this way were highly integrated. Such a strong integration is beneficial to the reduction of ion diffusion paths, while the porous and active materials could accommodate more electrolytes.

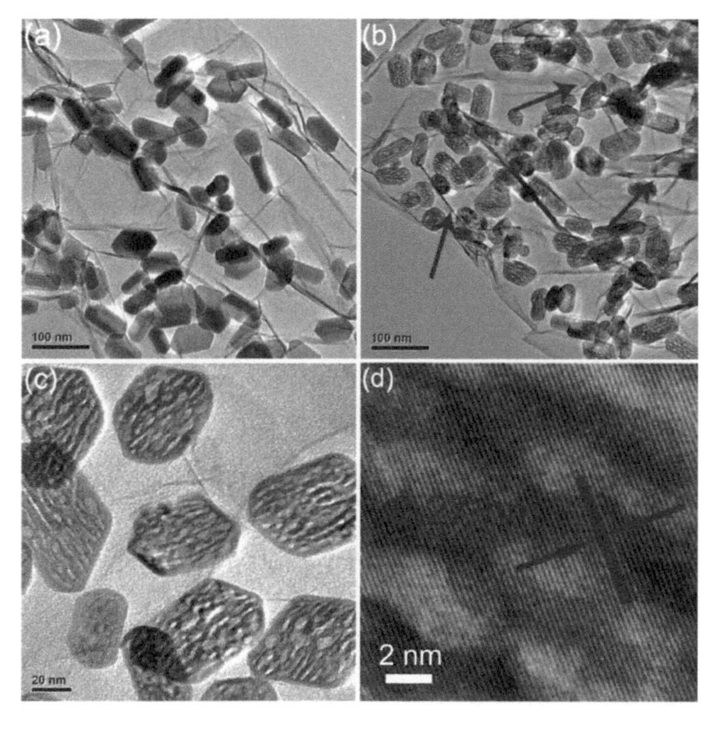

Fig. 3.72. (a) TEM image of the rGO-FeOOH hybrid, (b, c) TEM images of the slowly annealed hybrid and (d) HRTEM image of the porous α-Fe_2O_3 NPs. Reproduced with permission from [260]. Copyright © 2015, American Chemical Society.

3.2.1.6. Graphene-Cobalt Oxides

Co_3O_4–G hybrids can be synthesized by using the chemical thermal decomposition or precipitation methods [147, 274–290] and hydrothermal processes [291–301]. They are mainly used for applications in energy storage devices [274–278, 280, 281, 283, 284, 286, 288, 291–293, 295–300], water treatment [294], water splitting [279, 285, 289, 290], field emission [282], and microwave absorbers [301]. Another cobalt oxide that has been used to form hybrids with graphene or graphene oxide is CoO, thus producing CoO–G hybrids for various applications [302–306]. Typical examples of the two groups of hybrids are discussed as follows.

Yang et al. developed a method to prepare the G–Co_3O_4 hybrid, with the Co_3O_4 to be in the form of nanofibers fabricated by using electrospinning, as shown schematically in Fig. 3.73 [274]. To prepare the Co_3O_4 nanofibers with the electrospinning technique, polyvinylpyrrolidone (PVP) polymer was dissolved in ethanol to form a homogeneous solution, in which cobalt(II) acetate tetrahydrate ($Co(Ac)_2 \cdot 4H_2O$) was added to obtain a spinning solution. Composite nanofibers were then electrospun out. The as-spun PVP/$Co(Ac)_2$ composite nanofibers were dried at 60°C to remove the solvent residuals. The dried samples were annealed at 450°C for 5 h in air to form Co_3O_4 nanofibers, while the PVP was burnt out. In order to be integrated with the graphene nanosheets, the Co_3O_4 nanofibers were modified by 3-aminopropyltrimethoxysilane (APS). GO suspensions were diluted to 4 mg ml^{-1} with water. pH value was found to be a critical factor in order to obtain high quality G–Co_3O_4 hybrid paper through vacuum filtration. The GO solution and the modified Co_3O_4 nanofibers were mixed at pH = 8, while the solution was adjusted to have pH = 4 during the filtration. The content of Co_3O_4 in the G–Co_3O_4 hybrid paper was 78.3%.

A typical SEM image of the electrospun $Co(Ac)_2$/PVP composite nanofibers is shown in Fig. 3.74 (a) [274]. The composite nanofibers had an average diameter in the range of 700–800 nm, which was reduced to about 600 nm after thermal annealing, as illustrated in Fig. 3.74 (b). Figure 3.74 (c) revealed that the Co_3O_4 nanofibers consisted of Co_3O_4 NPs, with a mesoporous structure. The periodic fringe spacing of 0.46 nm observed in the HRTEM image of Fig. 3.74 (d) was attributed to the (111) lattice plane of cubic Co_3O_4.

Figure 3.75 shows SEM and TEM images of the G–Co_3O_4 hybrid paper [274]. The hybrid paper had a uniform thickness of about 20 μm, as shown in Fig. 3.75 (a). The graphene nanosheets became more crumpled, so that the hybrid was highly porous, due to the presence of the Co_3O_4 nanofibers. The hybrid sample exhibited a BET specific surface area of 43.2 $m^2 \cdot g^{-1}$. The presence of the pores was a significant advantage in maintaining fast diffusion of Li^+ ions, when the hybrid was used as the electrode of lithium

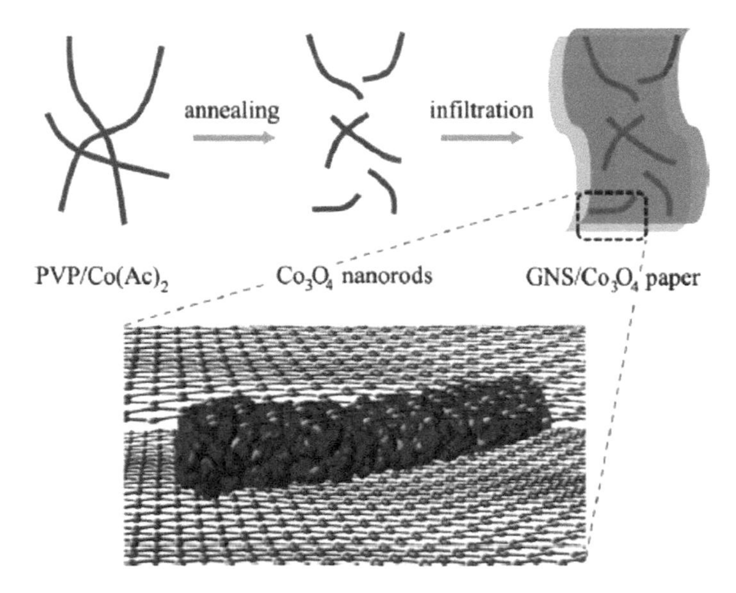

Fig. 3.73. Schematic diagram showing the fabrication process of the G–Co_3O_4 nanofiber hybrid paper. Reproduced with permission from [274]. Copyright © 2013, American Chemical Society.

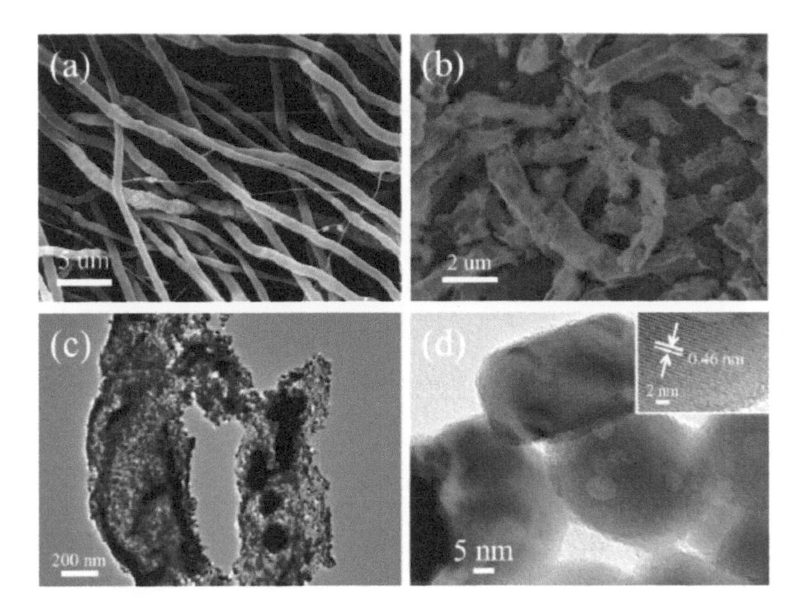

Fig. 3.74. (a) SEM image of the as-electrospun $Co(AC)_2$/PVP nanofibers. (b) SEM image of the Co_3O_4 nanofibers after annealing at 450°C. (c) TEM image of the Co_3O_4 nanofibers. (d) HRTEM image, with the inset showing the lattice fringe of the cubic Co_3O_4. Reproduced with permission from [274]. Copyright © 2013, American Chemical Society.

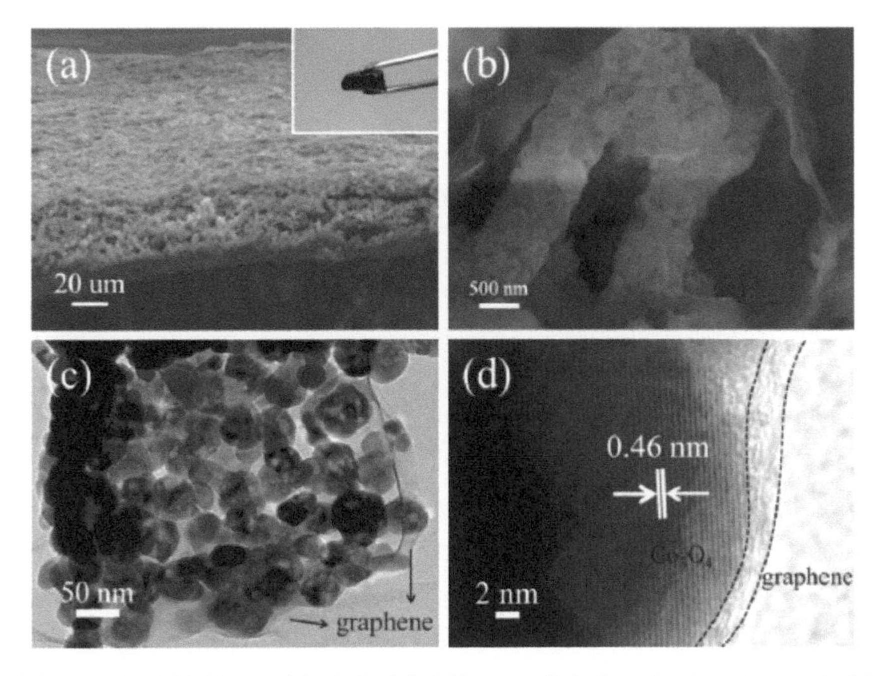

Fig. 3.75. (a) Cross-sectional SEM image of the G–Co_3O_4 hybrid paper, with the inset showing a photograph of the paper. (b) Enlarged view of the paper. (c) TEM image of the sample. (d) HRTEM image of hybrid paper. Reproduced with permission from [274]. Copyright © 2013, American Chemical Society.

ion batteries. As demonstrated in Fig. 3.75 (b), the high magnification SEM image indicated that the Co_3O_4 nanofibers were strongly attached on the graphene nanosheets, while the Co_3O_4 nanofibers were coated with graphene nanosheets. This highly integrated structure made the hybrid highly conductive.

Figure 3.75 (c) shows a TEM image of the G–Co_3O_4 hybrid, confirming the strong interaction between the graphene nanosheets and the Co_3O_4 nanofibers [274]. Figure 3.75 (d) shows a HRTEM image at the edge of the graphene-coated Co_3O_4 nanofibers. The lattice fringes were along the (111) zone axis of crystalline Co_3O_4, whereas the interplanar spacing of about 0.37 nm corresponded to the (002) lattice planes of graphite. Because the graphene nanosheets were not connected in a continuous way, Li^+ ions could penetrate the structure and react with the Co_3O_4 NPs. In addition, the graphite structure was likely formed in the hybrid, which effectively increased its electrical conductivity.

Chen and Wang reported a G–Co_3O_4 hybrid, with a sheet-on-sheet nanostructure, in which Co_3O_4 porous nanosheets were used to separate graphene nanosheets [275]. In this nanostructure, the 2D Co_3O_4 nanosheets were in close contact with the 2D graphene nanosheets. Figure 3.76 shows a schematic diagram demonstrating the formation process of the nanolayered G–Co_3O_4 hybrid. Firstly, GO-$Co(OH)_2$ was prepared from $Co(NO_3)_2{\cdot}6H_2O$ and GO suspension in hexamethylenetetramine (HMT, $(CH_2)_6N_4$) aqueous solution. The mixture was then sealed in a specialized glass tube by using microwave irradiation at 180°C for 5 min. G–Co_3O_4 hybrid was finally obtained by calcining the GO-$Co(OH)_2$ at 300°C for 2 h in N_2, during which GO was reduced while $Co(OH)_2$ was decomposed. The sheet-on-sheet structure was attributed to the interaction between the $Co(OH)_2$ and the GO nanosheets with the presence of carboxyl and hydroxyl groups. After reduction, the graphene nanosheets were well-separated and stabilized by the Co_3O_4 nanosheets.

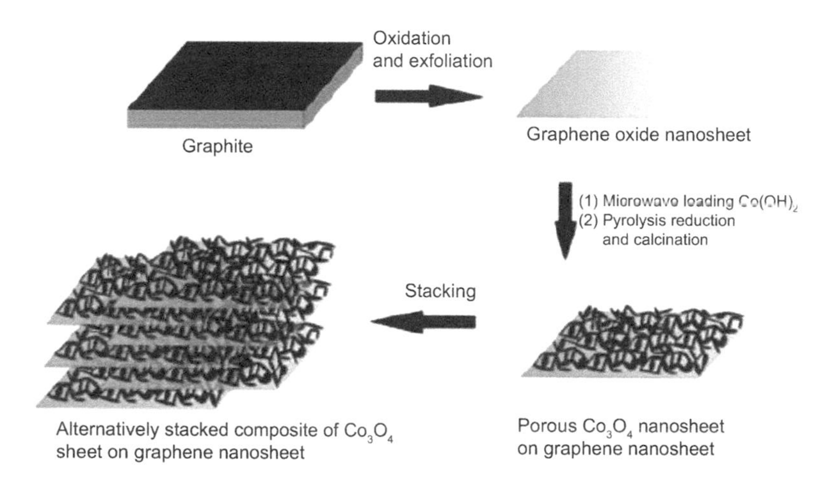

Fig. 3.76. Schematic diagram showing the growth process of the G–Co_3O_4 hybrid with a sheet-on-sheet nanostructure. Reproduced with permission from [275]. Copyright © 2010, The Royal Society of Chemistry.

Figure 3.77 shows TEM images of the graphene nanosheet, Co_3O_4 porous nanosheet and G–Co_3O_4 sheet-on-sheet hybrid [275]. Due to the few-layered structure, the graphene nanosheets were transparent with rippled and crumpled features, as shown in Fig. 3.77 (a). Figure 3.77 (b) shows a porous Co_3O_4 nanosheet, with a lateral dimension of several micrometers. The Co_3O_4 nanosheet had a thickness of about 100 nm and contained pores with sizes of 60–100 nm. TEM images of Fig. 3.77 (c, d) revealed that a few porous Co_3O_4 nanosheets were deposited on graphene nanosheets, thus leading to the unique sheet-on-sheet structure of the G–Co_3O_4 hybrid. As shown in the cross-sectional HRTEM image in Fig. 3.77 (e), the thickness of the graphene nanosheets was about 5 nm, consisting of about 14 layers of single carbon atomic sheet. The lattice fringe with spacing of 0.24 nm shown in Fig. 3.77 (f) was in a good agreement with the (311) plane of cubic Co_3O_4.

Figure 3.78 shows representative SEM images of the two components and their G–Co_3O_4 hybrid [275]. The graphene nanosheets were curled, with the thickness of a few nanometers (Fig. 3.78 (a)). The porous structure of the Co_3O_4 nanosheets was clearly demonstrated in Fig. 3.78 (b). SEM images of the G–Co_3O_4 hybrid from different angles are shown in Fig. 3.78 (c, d). The Co_3O_4 nanosheets and the

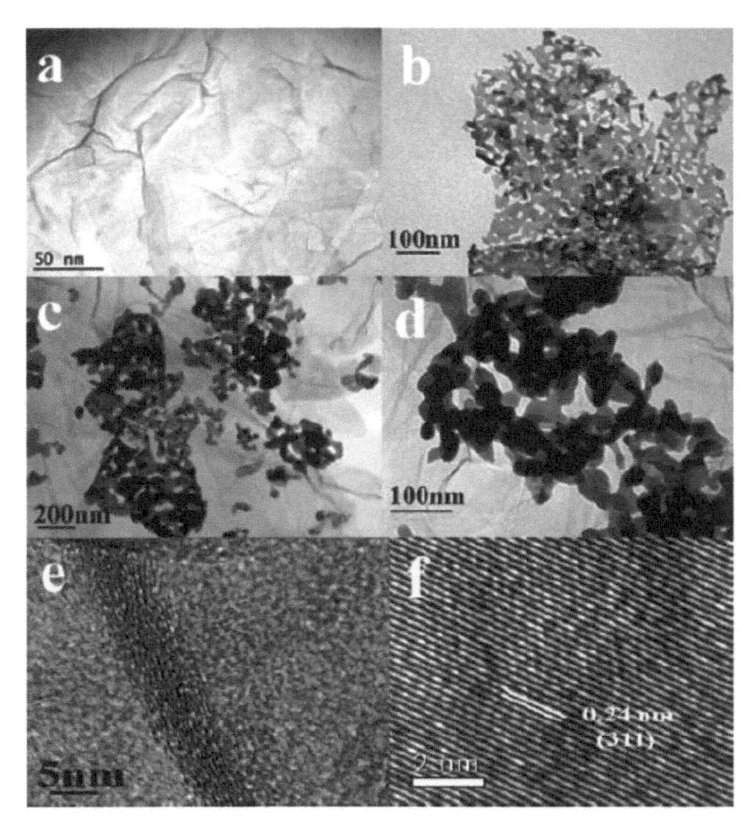

Fig. 3.77. TEM images of the graphene nanosheets (a), porous Co_3O_4 nanosheets (b), G–Co_3O_4 sheet-on-sheet hybrid (c–d), cross-sectional HRTEM image of a graphene nanosheet (e) and HRTEM image of a Co_3O_4 nanosheet (f). Reproduced with permission from [275]. Copyright © 2010, The Royal Society of Chemistry.

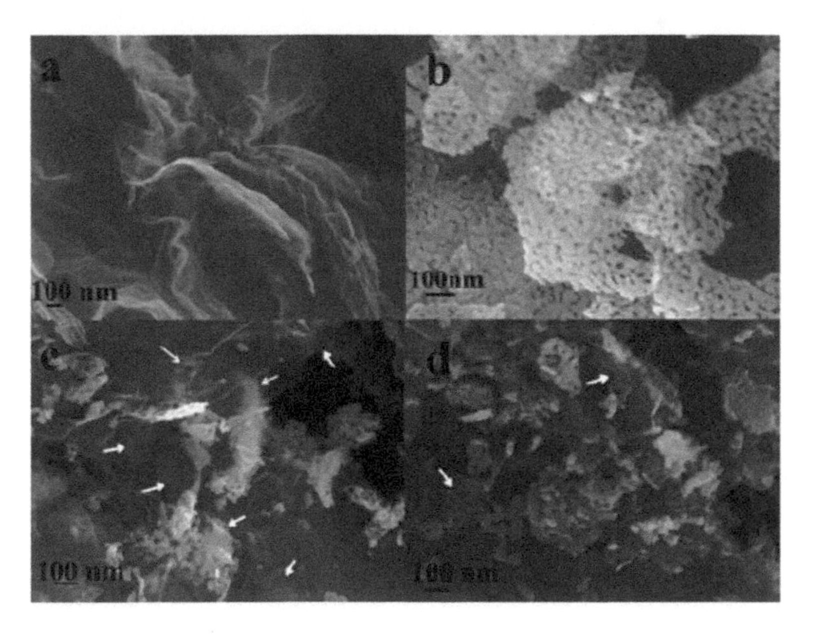

Fig. 3.78. SEM images of the graphene nanosheets (a), porous Co_3O_4 nanosheets (b) and G–Co_3O_4 sheet-on-sheet hybrid (c–d). Reproduced with permission from [275]. Copyright © 2010, The Royal Society of Chemistry.

graphene nanosheets were stacked alternatively to form a sheet-on-sheet layered structure. Therefore, the graphene nanosheets were effectively separated by the Co_3O_4 nanosheets, so that potential aggregation or restacking of the graphene nanosheets could be avoided, which facilitated the promising electrochemical performances of the hybrid for use as the anode electrodes of lithium ion batteries.

A hydrothermal method was reported to synthesize a Co_3O_4–rGO–CNTs hybrid, with Co_3O_4 microsphere arrays to be grown on free-standing rGO–CNTs paper [293]. GO was suspended in water as a brown dispersion, which was dialysed to completely remove the residual salts and acids. Purified CNTs were directly mixed with the GO suspension. The mixed suspension was used to fabricate GO–CNT paper through vacuum filtration with a 0.2 μm porous polytetrafluoroethylene (PTFE) membrane. $Co(NO_3)_2{\cdot}6H_2O$ and NH_4NO_3 were dissolved in the solution of H_2O and $NH_3{\cdot}H_2O$. The homogeneous solution, together with the GO–CNTs paper, was subjected to hydrothermal reaction at 90°C for 14 h, in order to obtain Co_3O_4–rGO–CNTs hybrid paper. Other precipitants, including urea ($CO(NH_2)_2$), L-lysine and KOH, were also used, for comparison. The alkaline hydrothermal process triggered the formation of Co_3O_4 and the reduction of the GO.

Representative top-view SEM images of the Co_3O_4–rGO–CNTs hybrid paper are shown in Fig. 3.79 (a, b) [293]. Co_3O_4 microspheres with a size of about 1 μm were formed after the hydrothermal reaction, and were uniformly distributed on the surface of the rGO–CNTs paper. Due to the presence of oxygen-containing functional groups, i.e., hydroxyl and carbonyl groups, Co_3O_4 crystals could be linked onto the surface of the GO–CNTs paper. The Co_3O_4–rGO–CNTs hybrid paper exhibited a layered structure, with an average thickness of about 50 μm, as demonstrated in Fig. 3.79 (c). The Co_3O_4 microspheres were closely packed to form an interconnected monolayer array. The Co_3O_4–rGO–CNTs hybrid paper was highly flexible, as illustrated by the photograph in Fig. 3.79 (d). The inset in Fig. 3.79 (c) revealed that the CNTs were sandwiched between the rGO nanosheets. The skeleton of the array grown on the layer-structured rGO–CNTs paper was formed by interconnected Co_3O_4 microsphere building blocks (Fig. 3.79 (e, f)). The content of Co_3O_4 in the as-obtained hybrid paper was about 48 wt%. Figure 3.79 (g) shows TEM image of the Co_3O_4 microspheres, further confirming that their diameter was about 1 μm. The interplanar spacing of 0.24 nm was attributed to the (311) lattice plane of Co_3O_4 crystals (inset of Fig. 3.79 (d)), in agreement with the atomic structure shown in Fig. 3.79 (h). The hybrid flexible paper could be used as an electrode of supercapacitors, with a specific capacitance of 378 $F{\cdot}g^{-1}$ at 2 $A{\cdot}g^{-1}$ and 297 $F{\cdot}g^{-1}$ at 8 $A{\cdot}g^{-1}$. Such promising electrochemical performances could be readily attributed to the unique hierarchical microstructure of the hybrid paper.

An electrostatic-induced spread growth method was reported to homogeneously and completely coat β-$Co(OH)_2$ nanosheets on graphene nanosheets, which were coated on Cu foil, thus leading to a binder-free CoO–G hybrid that could be used as an electrode [303]. Figure 3.80 shows a schematic diagram demonstrating the synthesis procedure of the binder-free CoO–G hybrid, which was derived from β-$Co(OH)_2$–G with the β-$Co(OH)_2$ nanosheets being uniformly distributed on the graphene nanosheets. α-$Co(OH)_2$ ultra-thin nanosheets were synthesized from ammonia water and $Co(NO_3)_2$. The as-derived green-blue precipitate was washed with deionized water and then hydrothermally treated in solvents of water/methanol (16 ml/16 ml) at 180°C for 30 min. The obtained α-$Co(OH)_2$ nanosheet powder was mixed with GO suspension, into which hydrazine monohydrate was added. The mixture was then refluxed at 95°C for 24 h to form the β-$Co(OH)_2$–G hybrid. Finally, the synthesized β-$Co(OH)_2$–G hybrid sample was dispersed in DMF with concentrations of 6–8 $mg{\cdot}ml^{-1}$. The suspension was dropped on Cu foil and then dried at 80°C for 30 min. After the DMF was vaporized, the sample was calcined at 350°C for 3 h in Ar/H_2 (5 vol% H_2) at pressures of 3–5 MPa to form the CoO–G hybrid.

SEM observation indicated that the ultrathin α-$Co(OH)_2$ nanosheets had a thickness of about 2.6 nm, with a hexagonal crystal structure. The as-synthesized α-$Co(OH)_2$ was positively charged, while the GO is negatively charge, thus facilitating assembly through electrostatic interactions, as shown schematically in Fig. 3.80. Due to the high surface area of the GO nanosheets, the α-$Co(OH)_2$ nanosheets could be adsorbed in a relatively large quantity. The α-$Co(OH)_2$ was transferred into β-$Co(OH)_2$, while the GO was reduced into graphene nanosheets. The final CoO–G hybrid contained about 13 wt% carbon.

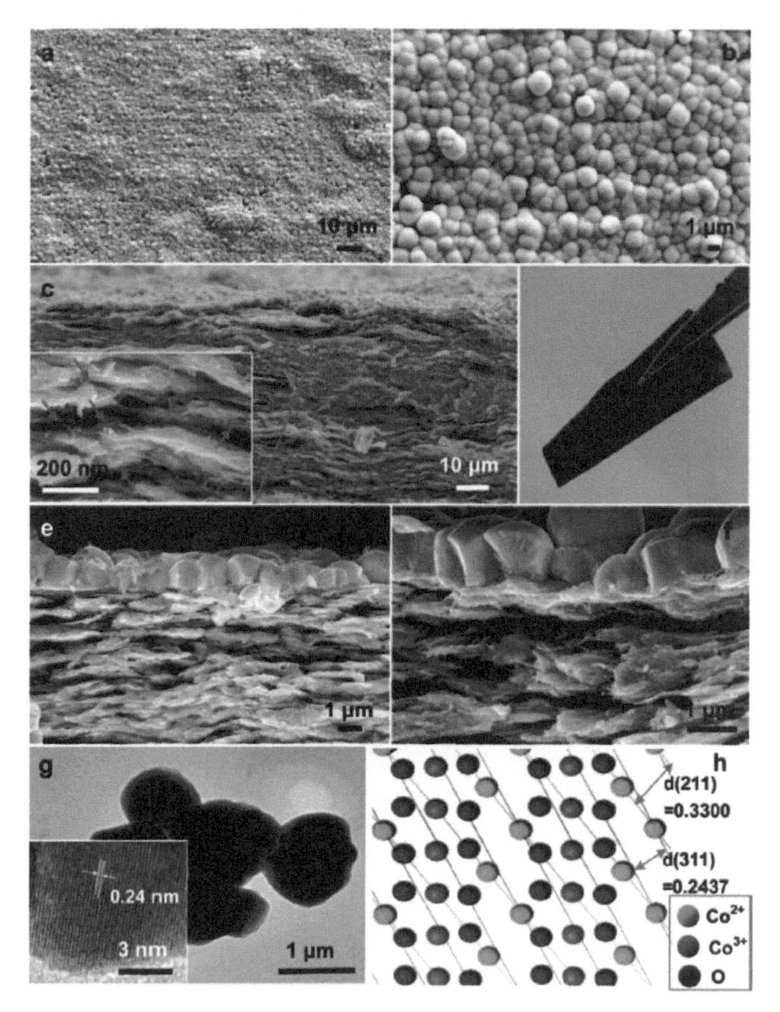

Fig. 3.79. Typical top-view (a, b) and side-view (c) SEM images of the flexible Co_3O_4–rGO–CNTs hybrid paper. The inset in (c) is the enlarged view of the region enclosed by the green rectangle. The photograph in (d) demonstrates the flexibility of the paper. (e, f) Enlarged cross-sectional views of the paper. (g) TEM and HRTEM images of the Co_3O_4 microspheres. (h) Atomic structure of the cubic phase Co_3O_4. Reproduced with permission from [293]. Copyright © 2012, Wiley-VCH Verlag GmbH & Co. KGaA, Weinheim.

The hybrid was strongly attached to the Cu foil after calcination, as shown in Fig. 3.81 (a, b) [303]. Figure 3.81 (c) shows XRD pattern of the CoO–G hybrid, corresponding to cubic CoO. SAED pattern of the CoO–G hybrid exhibited two sets of diffraction patterns, belonging to cubic CoO and graphene. The weak diffraction ring of graphene was due to its multilayer structure. A large number of mesopores were produced in the CoO–G hybrid after calcination, which were 2–6 nm in size. Some larger pores with sizes of 10–140 nm were also observed, corresponding to the voids between the different layers. This multiple porous structure made the hybrid suitable for application as an electrode. When used as the anode of lithium ion batteries, it offered a promising rate capability of 172 $mAh \cdot g^{-1}$ at a high current density of 20 $A \cdot g^{-1}$, with a high cycle stability up to 5000 cycles at 1 $A \cdot g^{-1}$. More importantly, the high capacity and cycling stability were retained over a wide temperature range from 0 to 55°C.

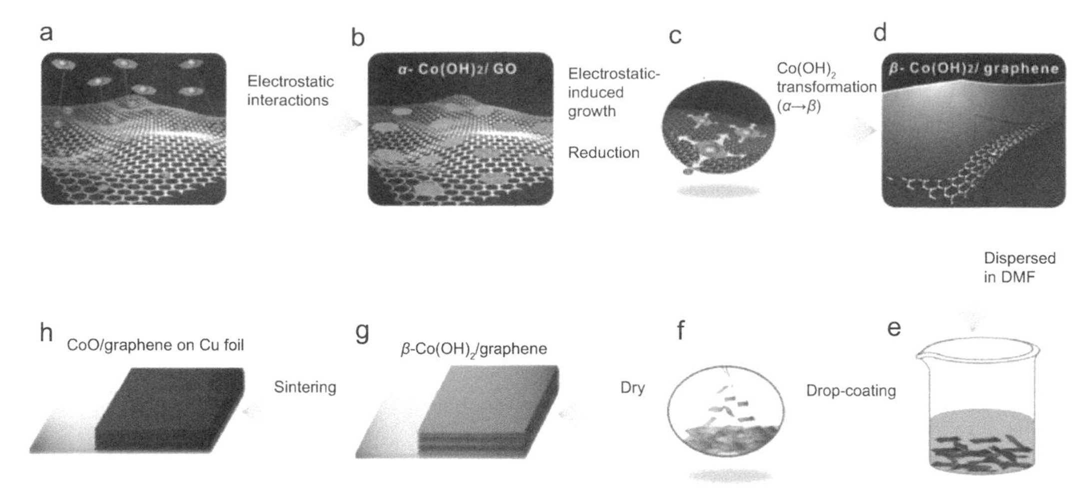

Fig. 3.80. (a) Schematic diagram illustrating the GO nanosheets with negative charge and α-$Co(OH)_2$ with positive charge. (b) Electrostatic interactions between the GO and α-$Co(OH)_2$. (c) Growth process of the electrostatic-induced spread growth method. (d) Homogeneous β-$Co(OH)_2$ on graphene. (e) β-$Co(OH)_2$–G hybrid dispersed in N,N-dimethyl formamide (DMF). (f) Coating of hybrid suspension. (g) β-$Co(OH)_2$–G hybrid on Cu foil. (h) CoO–G hybrid on Cu foil. Reproduced with permission from [303]. Copyright © 2013, Wiley-VCH Verlag GmbH & Co. KGaA, Weinheim.

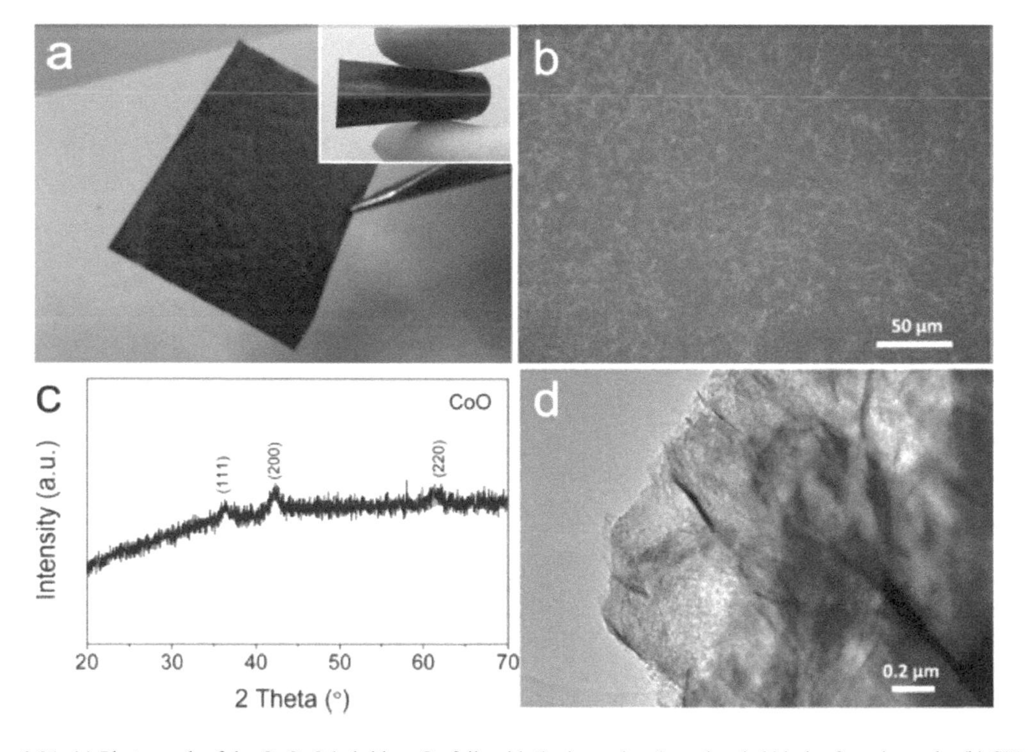

Fig. 3.81. (a) Photograph of the G–CoO hybrid on Cu foil, with the inset showing a bended binder-free electrode. (b) SEM image of the as-prepared electrode. (c) XRD pattern of the G–CoO hybrid. (d) TEM image of the G–CoO hybrid. Reproduced with permission from [303]. Copyright © 2013, Wiley-VCH Verlag GmbH & Co. KGaA, Weinheim.

3.2.1.7. Graphene-Nickel Oxide

Nickel oxide–graphene hybrids are mainly based on NiO, with major applications in energy storage devices, i.e., supercapacitors [307–313] and lithium ion batteries [312, 314–318], and have been synthesized by using various methods. One example is discussed in the following section.

A multi-step strategy was proposed to fabricate the NiO–G hybrid [309]. Figure 3.82 shows a schematic diagram of the detailed assembly process, including (1) adsorption of nitrate ions onto GO nanosheets in the mixture of GO and nitrate salt, (2) stacking of nitrate-adsorbed GO nanosheets into a membrane-like layered structure (GO/nickel nitrate) by slow vacuum drying and (3) transformation from GO/nickel nitrate into NiO–G by using a thermal treatment at a low temperature of about 500°C in vacuum. At the same time, the GO was reduced to graphene, due to the elimination of the chemisorbed oxygen, while the nickel nitrate was decomposed to NiO. The heating rate should be $< 5°C\ min^{-1}$, in order to avoid exfoliation of the layered structure and hence maintain the membrane-like feature. The aqueous dispersible GO allowed for uniform adsorption of the nickel ions into the GO layers, which ensured the formation of the sandwich structure. The sandwich structure in turn confined the NiO NPs, in order to prevent them from growing into large particles. The use of soluble nickel nitrate offered two advantages, i.e., (i) high solubility leading to high content and (ii) relatively low calcination temperature to form NiO NPs.

Figure 3.83 (a) shows a photograph of a NiO–G paper [309]. Such hybrid papers exhibited a relatively high electrical conductivity of 15 Ω/□ at a thickness of 20 μm. Figure 3.83 (b) shows XRD patterns of the NiO–G hybrid and the components. The graphite was characterized by a sharp peak at 26.5°, corresponding to the diffraction peak of (002) with a lattice spacing of 0.335 nm. After oxidation and expansion, the peak was shifted to 12.5°, due to the presence of the intercalated oxygen. A further downshift to about 10.5°, together with a broadening, was observed in the diffraction peak, after the GO nanosheets adsorbed nickel

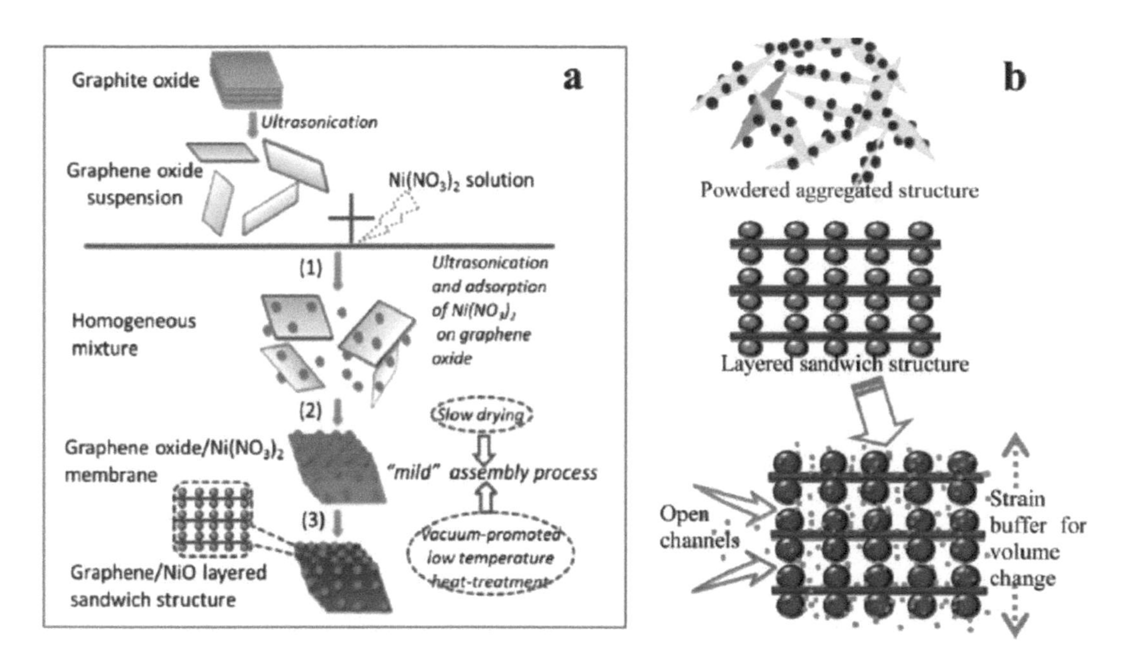

Fig. 3.82. (a) Schematic diagram of the self-assembly strategy to form NiO–G sandwich structures. (b) Schematic diagram showing the formation of powder aggregate structured (upper panel) and layer sandwich-structured (middle panel) oxide–G hybrids. The layered structure offered open channels for ion transport and served as a strain buffer for volume changes during the electrochemical reaction (bottom panel). Reproduced with permission from [309]. Copyright © 2011, The Royal Society of Chemistry.

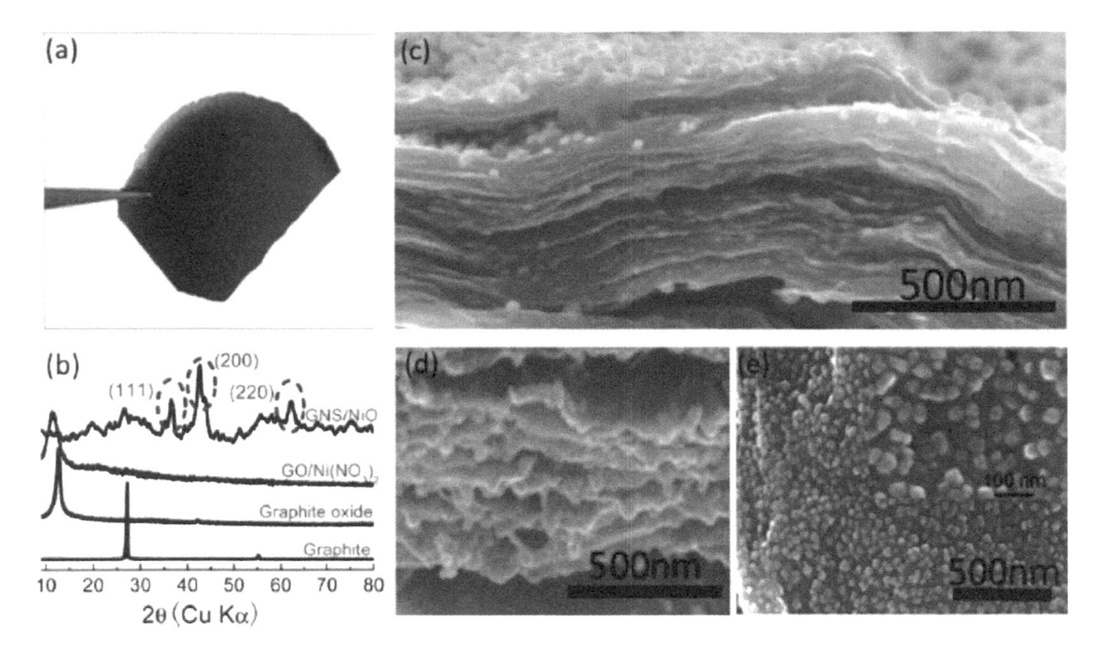

Fig. 3.83. Structure characterization of the sandwich-structured membrane: (a) photograph of a NiO–G sandwich paper, (b) XRD patterns of the GO–$Ni(NO_3)_2$ and NiO–G papers, together with graphite and graphite oxide, (c, d, e) cross-sectional and top-view SEM images of a NiO–G paper. The sample of (d) possessed a slightly exfoliated structure due to the rapid heating rate. Reproduced with permission from [309]. Copyright © 2011, The Royal Society of Chemistry.

nitrate to form GO/nitrate composite. Thermal annealing triggered the transformation from amorphous nickel nitrate to NiO NPs, accompanied by an increase in size from less than 1 nm of nickel nitrate to ~25 nm of NiO. The NiO NPs filled the spaces of the GO nanosheets, so that no XRD diffraction peaks corresponding to the layer-to-layer spacing were detected in the NiO–G hybrids.

Figure 3.83 (c) shows cross-sectional SEM image of the free-standing NiO–G hybrid paper [309]. It possessed a nice layered nanostructure, in which NiO NPs could be observed both on the surface of the stacked structure and in the space between the parallel layers of the graphene nanosheets. When a higher heating rate of 20°C·min^{-1} was used, a slightly exfoliated layered structure could be obtained, as shown in Fig. 3.83 (d). It further confirmed that the NiO NPs were distributed uniformly on the graphene layers and were filled in the layer spaces as well. As shown in Fig. 3.83 (e), the NiO NPs on the sample surface had sizes of 20–40 nm. The uniform distribution of the NiO NPs on graphene nanosheets was also evidenced by using TEM observation. The NiO–G hybrid demonstrated a strong interaction between the two components, which could withstand strong sonication. The hybrids exhibited promising electrochemical performances as electrodes of supercapacitors.

3.2.1.8. Graphene-Copper Oxides

There are two types of hybrids based on copper oxides and graphene, i.e., CuO–G [319–326] and Cu_2O–G [327–331]. Their main applications are also in energy storage, including as the anodes of lithium ion batteries and electrodes of supercapacitors. One example for each type of hybrid will be discussed in this section.

A simple chemical route was employed to deposit CuO NPs on GO nanosheets in a water–isopropanol reaction system without using any additives [326]. Such a CuO–GO hybrid was a promising catalyst for the decomposition of AP. It has been accepted that the morphology and texture of the as-synthesized CuO–GO hybrid samples were affected by polarity, steric hindrance and electrostatic effects. The different coordination capabilities of water and isopropanol offered a potential flexibility in controlling the properties of the products.

$Cu(Ac)_2 \cdot H_2O$ was used as the precursor to synthesize CuO–GO hybrids with different mass ratios of CuO:GO (0.5:1, 1:1, 2:1 and 3:1). GO was suspended in isopropanol to form a homogeneous brown suspension, into which $Cu(Ac)_2 \cdot H_2O$ powder was added. The mixture was then refluxed at 83°C for 30 min. After that, a certain amount of deionized (DI) water was rapidly added and the mixture was refluxed for another 30 min. CuO–GO hybrids were obtained as black precipitates. The samples with different mass ratios of CuO:GO were labeled as CuO:GO_{ratio}. XRD analysis results revealed that diffraction peaks were clearly observed in the samples with CuO: GO mass rations of > 1.

Figure 3.84 shows TEM images of the CuO:$GO_{0.5}$ hybrid [326]. Spindle-like CuO NPs, with diameters of 40–60 nm and lengths of 80–120 nm, were uniformly distributed on the GO nanosheets, as shown in Fig. 3.84 (a). Figure 3.84 (b) indicated that the spindle-like CuO particles consisted of a large number of small nanorods that were 7 nm thick and 40 nm long. In the HRTEM image of a CuO spindle, the interplanar spacing of about 0.252 nm was attributed to $(\bar{1}11)$ plane of monoclinic CuO (inset of Fig. 3.84 (b)). By comparing the bright- and dark-field TEM images in Fig. 3.84 (c, d), it was found that the GO nanosheet was of a single carbon atomic layer. Some CuO spindles were embedded by the graphene nanosheets.

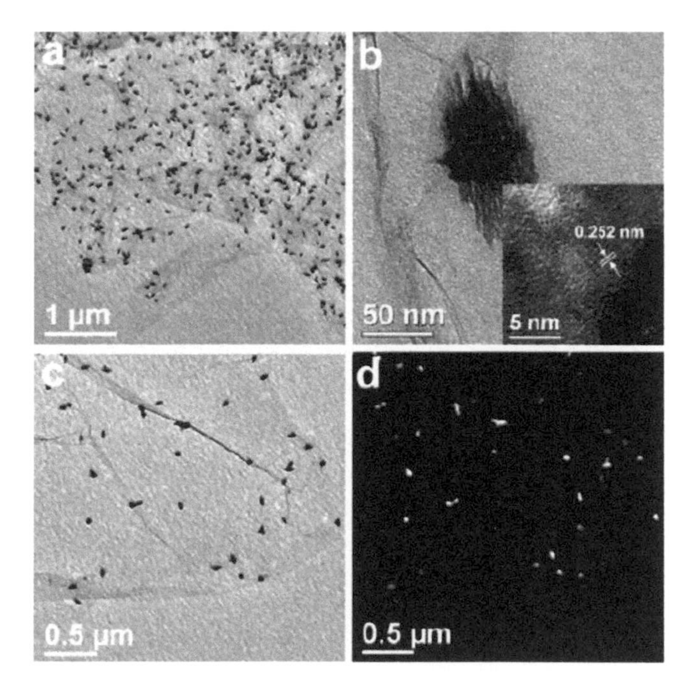

Fig. 3.84. TEM images of the CuO:$GO_{0.5}$ hybrid: (a, b, c) bright-field TEM images, with the inset of (b) to be a HRTEM image of a CuO particle, and (d) dark-field TEM image. Reproduced with permission from [326]. Copyright © 2010, The Royal Society of Chemistry.

Figure 3.85 shows TEM images of the CuO–GO hybrids with different CuO:GO mass ratios [326]. The CuO particles in the CuO:GO_1 hybrid exhibited a similar spindle-like morphology to the CuO:$GO_{0.5}$ (Fig. 3.85 (a)). As the mass ratio of CuO:GO was increased, spherical particles were formed. For example, in CuO:GO_2, the CuO spherical particles had sizes of 60–80 nm (Fig. 3.85 (b)). However, these spherical CuO particles also consisted of small nanorods (Fig. 3.85 (c)). Due to the high content of CuO, some CuO particles were detached from the GO nanosheets, as shown in Fig. 3.85 (d).

In completely converted GO nanosheets from graphite, the ideal C:O ratio is 2:1, with epoxy and hydroxyl on basal planes, as well as carbonyl and carboxyl groups at the edges [332, 333]. These oxygen-containing functional groups act as anchor sites to link various nanoparticles, so as to form hybrids. In this study, at a low concentration of $Cu(Ac)_2$, CuO NPs occupied only part of the anchor sites, so that they were highly dispersed on the GO nanosheets. As the mass ratio of CuO:GO was increased from 0.5:1 to

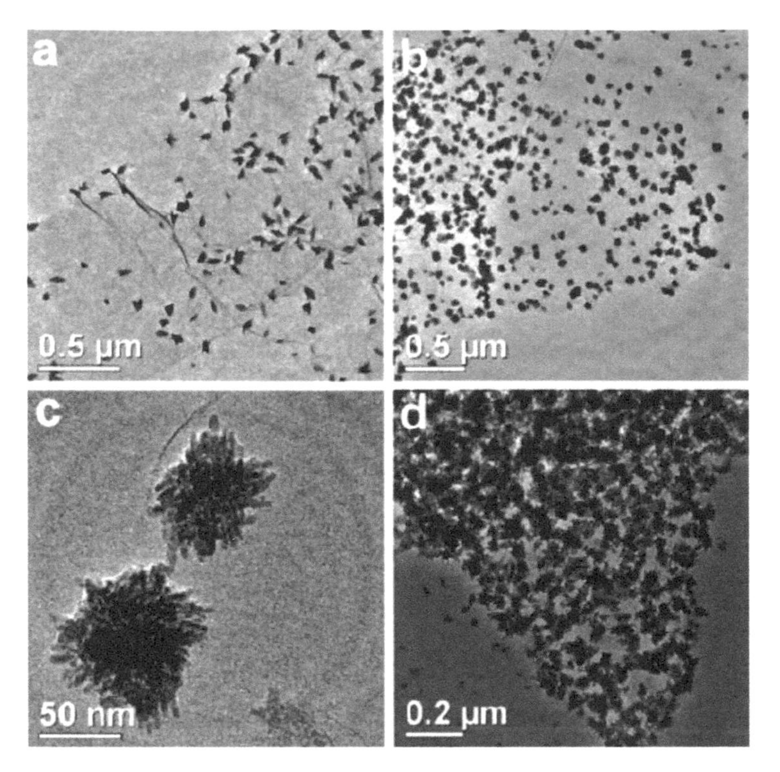

Fig. 3.85. TEM images of the CuO–GO hybrid samples with different CuO:GO mass ratios: (a) CuO:GO1, (b, c) $CuO:GO_2$ and (d) $CuO:GO_3$. Reproduced with permission from [326]. Copyright © 2010, The Royal Society of Chemistry.

2:1, the density of the CuO nanoparticles was increased accordingly. As the mass ratio of CuO:GO was increased to 3:1, all the anchor sites were taken by the CuO NPs. As a result, some CuO NPs were not attached on the GO nanosheets. This observation was further confirmed by the SEM results.

A schematic diagram to demonstrate the formation mechanism of the CuO NPs in the presence of water is illustrated in Fig. 3.84 [326]. According to the authors, Cu^{2+} in solution usually has a coordination number of six, forming an octahedron configuration. With a small amount of water, water molecules were only located at the top and bottom sites. Because the binding energies of H_2O molecules with the Cu^{2+} were lower than those of coordinating atoms located at the midplane sites, CuO NPs would grow preferentially along the axis with H_2O molecules. As a result, the growth rates of CuO crystals along that axis were higher than the growth rates of those in the plane directions, so that rod-like crystals were formed. Due to the presence of oxygen atoms, it is easy to form hydrogen bonds in isopropanol, which was responsible for the aggregation of the rod-like CuO nanocrystals. At the same time, small crystallites always tend to aggregate in order to decrease their surface Gibbs free energies. As the amount of water was increased, some H_2O molecules would occupy coordination sites in the equatorial direction of the octahedron, as shown in Fig. 3.84 (b), so that anisotropic grain growth was hindered. As a consequence, spherical CuO nanocrystals were produced. The spherical particles could be dispersed in a more uniform way as compared to their rod-like counterparts.

It was observed that the use of isopropanol together with a small amount of water played a dominant role in the production of the CuO–GO hybrids. If no water molecules were present, CuO particles could not be obtained, because water facilitated the hydrolization of $Cu(Ac)_2$. However, if there were no isopropanol molecules, it was hard to deposit CuO NPs on the GO nanosheets. According to the steric hindrance and electrostatic effects, H_2O molecules are stronger than isopropanol molecules when coordinating with Cu^{2+} and GO. Isopropanol was used to disperse the GO nanosheets, due to its molecules forming bonds with the oxygen-containing groups of GO through weak hydrogen bonds. When Cu^{2+} ions were added into

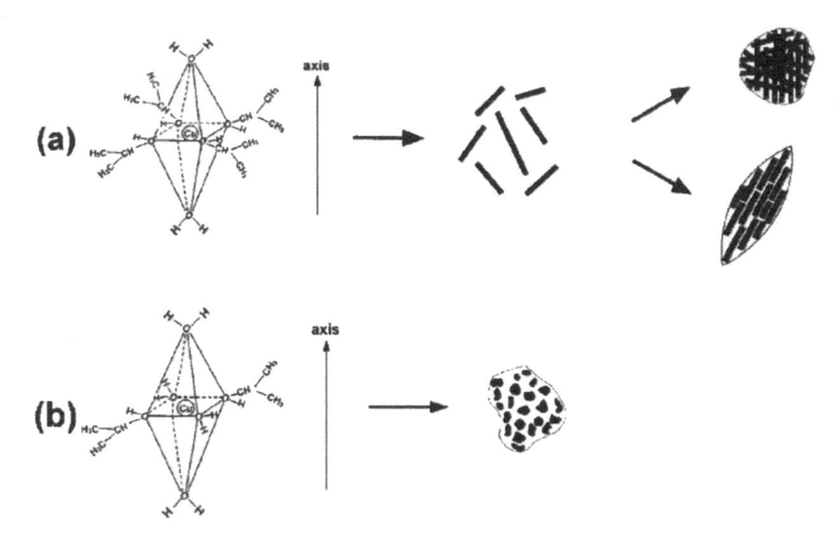

Fig. 3.86. Schematic diagram showing the formation mechanism of the CuO NPs with the presence of water: (a) a small amount of water and (b) a large amount of water. Reproduced with permission from [326]. Copyright © 2010, The Royal Society of Chemistry.

the GO suspension, they could replace some isopropanol molecules, forming bonds with GO through electrostatic forces. Also, isopropanol molecules would coordinate with the Cu^{2+} ions to form a distorted octahedron. These coordinated isopropanol molecules were gradually replaced by H_2O molecules, when water was added, so as to affect the growth behaviors of the CuO NPs. After the addition of water into the reaction system, Cu^{2+} were easily coordinated with H_2O; with the introduction of a small amount of water, the H_2O molecules might occupy the axial positions of the CuO growth units, leaving isopropanol in the equatorial positions. Moreover, the interaction of the H_2O molecules with Cu^{2+} in other directions became possible when more water was introduced, resulting in CuO nanocrystals with other shapes. If only water was used, the interactions between Cu^{2+} and GO were largely weakened, due to the strong bonding of H_2O molecules with both Cu^{2+} and GO, so that CuO NPs were hardly deposited onto the GO nanosheets. The obtained CuO–GO hybrid showed strong catalytic activity for the thermal decomposition of AP. They are also expected to have other applications, such as in gas sensors, magnetic phase transitions and nanoelectronic devices [326].

A one-step hydrothermal reaction method was used to develop Cu_2O–rGO hybrids, from $Cu(Ac)_2$ and GO nanosheets [329]. In these Cu_2O–rGO hybrids, Cu_2O was present in the form of nanowire mesocrystals. The mesocrystals were formed with nanowires as building blocks, characterized by the presence of octahedral crystal faces. 3D-oriented attachment of octahedrons led to the formation of larger mesocrystals, through a "nanoparticle-by-nanoparticle" growth. The presence of GO played a specific role in the transformation of Cu_2O mesocrystals to hierarchical structures. Other factors, including the type of organic additives, precursor concentration and pH value, also had effects on the nanowire architecture of mesocrystals. With increasing content of GO, 3D networks were formed from the Cu_2O mesocrystals that were deposited on the rGO nanosheets. The Cu_2O–rGO hybrids demonstrated high sensitivity to NO_2.

To synthesize the Cu_2O–rGO hybrids, $Cu(Ac)_2$, *o*-anisidine, acetic acid, 4,4′,4″,4‴-(porphine-5,10,15,20-tetrayl)tetrakis(benzoic acid) (porphyrin-COOH), perylene-3,4,9,10-tetracarboxylic dianhydride (PTCDA) and L-(+)-tartaric acid were used as the starting materials. $Cu(Ac)_2$ and GO were dissolved in deionized water, into which *o*-anisidine was added, forming the reaction solution. Hydrothermal reaction was conducted at 200°C for 15 h.

Figure 3.87 shows SEM images and representative XRD pattern of the Cu_2O nanowire mesocrystals and Cu_2O–rGO hybrids [329]. The mesocrystals possessed octahedron morphologies and distinct triangular external faces, with sizes of 10–100 μm. XRD (Fig. 3.87 (b)) revealed that the mesocrystals were cubic Cu_2O. The items were constructed by branched nanowires with diameters of 80–110 nm

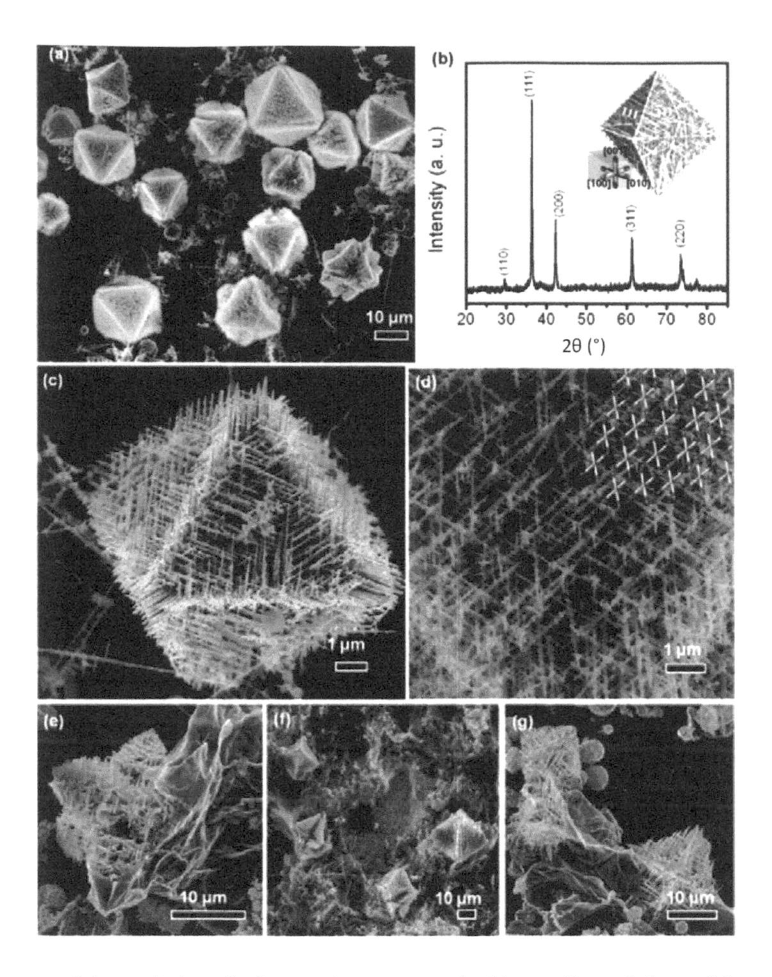

Fig. 3.87. SEM images of the octahedron Cu_2O nanowire mesocrystals: (a) overall morphology of the octahedron Cu_2O nanowire mesocrystals, (b) XRD pattern of the Cu_2O mesocrystals, with the inset showing a schematic illustration of the crystal structure of the octahedron Cu_2O nanowire mesocrystal, together with the crystal orientations of the hexapod branching, (c) an octahedron Cu_2O mesocrystal along the [111] view and (d) interior morphology with the overlaid hexapod grid. (e–g) SEM images of the Cu_2O–rGO hybrids with higher GO contents. Reproduced with permission from [329]. Copyright © 2012, American Chemical Society.

(Fig. 3.87 (c, d)). The octahedron mesocrystal exhibited a similar face centered cubic (FCC) crystal unit structure, as illustrated in the inset of Fig. 3.87 (b). The interior nanowire architecture was likely an octahedron constructed by fractal growth in the format of a hexapod (Fig. 3.87 (d)). The hexapod subunit with six arms was formed due to the diffusion-limited conditions, leading to branches growing along the ⟨100⟩, ⟨001⟩, and ⟨010⟩ directions of the cubic Cu_2O (inset of Fig. 3.87 (b)). Cu_2O mesocrystals were produced due to the repeated growth of the hexapod branches. At low content of GO, a continuous network of rGO was not formed. However, the mesocrystals were interconnected by the rGO network in the samples with high contents of GO, as shown in Fig. 3.87 (e–g).

Figure 3.88 shows TEM images of the Cu_2O nanowire mesocrystals and Cu_2O–rGO hybrids [329]. Figure 3.88 (a) indicated that the nanowire mesocrystal appeared as a dense network consisting of well-organized dendritic nanowires. Figure 3.88 (b) shows an entire plane of the dendritic nanowires, which was single-crystalline, as evidenced by the SAED pattern (inset of Fig. 3.88 (b)). The hexapods/dendrites were crystallographically aligned one another along certain main crystallographic axes of the Cu_2O. HRTEM images and SAED patterns of the main stem and nanowire branch are shown in Fig. 3.88 (c, e), confirming that their growth directions were [001] and [100], respectively. The anisotropic

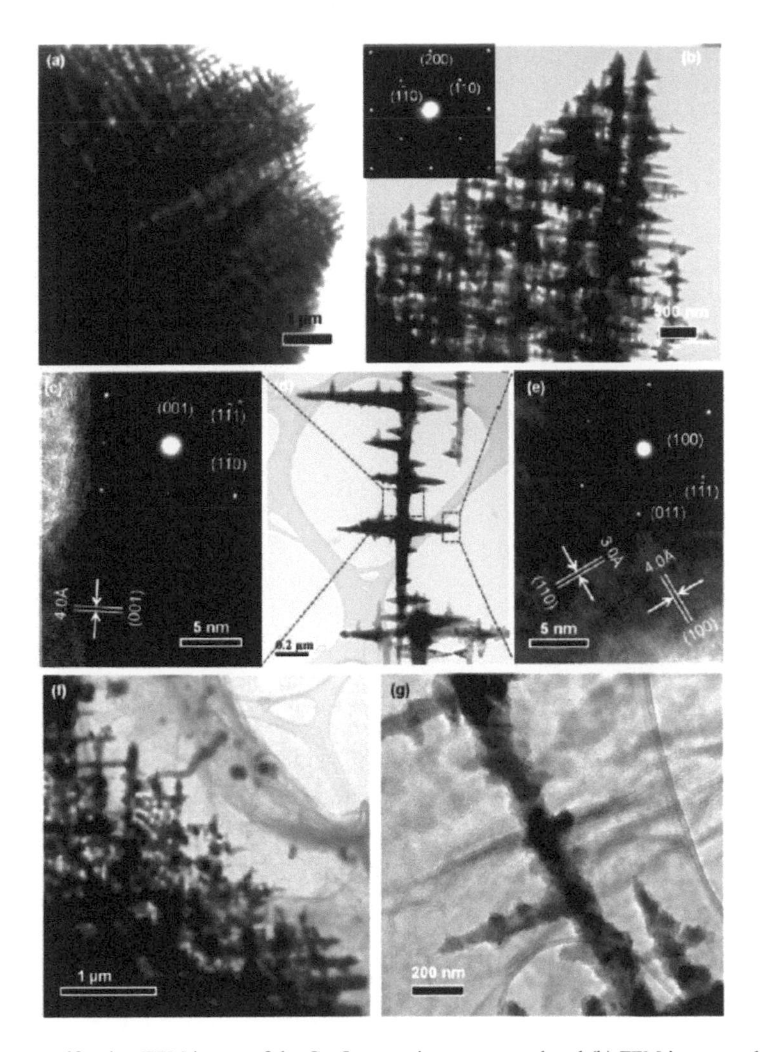

Fig. 3.88. (a) Low-magnification TEM image of the Cu_2O nanowire mesocrystal and (b) TEM image at the edge of the 3D mesocrystal, with the inset showing SAED of the entire plane. (d) TEM image of a fragment of the nanowire mesocrystal. (c, e) HRTEM images of the stem and branch of the nanowire indicated in panel (d), respectively, with the insets to be their corresponding SAED patterns. (f) TEM and (g) HRTEM image of the Cu_2O–rGO hybrids formed in the presence of higher GO contents. Reproduced with permission from [329]. Copyright © 2012, American Chemical Society.

growth of the branched Cu_2O nanowires was realized through the dissolution-recrystallization process or Ostwald ripening mechanism [334]. TEM and HRTEM images of the Cu_2O–rGO hybrids are in Fig. 3.88 (f, g). Wrinkled rGO nanosheets were observed, which covered the surface of the Cu_2O nanostructures.

Morphological evolution of the Cu_2O nanowire mesocrystal has been systematically studied by the authors [329]. As the initial nucleation was started, amorphous microspheres were agglomerated with rough surfaces, no free-standing Cu_2O nanowires could be observed. Gradually, faceted mesocrystals with relatively small sizes were formed, due to the transformation from the amorphous spheres to the crystalline mesocrystals. Finally, 3D mesocrystals with distinct octahedron configuration and relatively large sizes were obtained. Without the presence of GO, only nanowires were present. With the addition of GO, the nanowires coexisted with the amorphous microspheres, especially in the samples with the intermediate reaction time durations. Further studies indicated that the Cu_2O mesocrystals were derived from the intermediate amorphous microspheres. The mesocrystals initially resembled the morphology of the amorphous microsphere, inside which nanowires were crystallized. As a result, the amorphous microspheres were gradually transferred into spherical mesocrystal intermediates, with hierarchical structures consisting of nanowires, thus leading to the final faceted octahedral mesocrystals. EDX analysis results revealed that the C content of the mesocrystals was lower than that of the intermediate microspheres, due to the decrease in content of the polymeric phase caused by the reaction.

The addition of GO played a critical role in the formation of the Cu_2O nanowire mesocrystals. Without the presence of GO, only single-crystal particles were obtained. In the reaction system, water molecules, Cu^{2+} ions, amine groups of *o*-anisidine and the oxygen-containing groups on the GO were interconnected through chemical and hydrogen bonds. In addition, oxo- or hydroxobridges could be formed between the functional groups of the GO and the Cu^{+}/Cu^{2+} ions. Therefore, the reaction mixture became the colloidal solution. In this case, Cu_2O nuclei appeared, which grew and precipitated as the nanoparticles that served as the seeds. At a critical point, the seed particles aggregated to form amorphous microspheres, in order to minimize the energy of the system. The *o*-anisidine-coordinated GO acted as a polyelectrolyte to stabilize the amorphous Cu_2O nanoparticles. After that, transformation of the amorphous microspheres started, leading to the final Cu_2O mesocrystals and Cu_2O–rGO hybrids.

3.2.1.9. Graphene-ZnO

ZnO-G hybrids have been prepared by using the thermal decomposition or chemical methods [335–347], hydrothermal synthesis methods [348–353], with applications mainly as photocatalysts [335–338, 340, 341, 345, 346, 349–352], as well as in photodetectors [344, 353], white light emitting diodes [347], supercapacitors [342, 343] and lithium ion batteries [348].

A simple chemical method was reported to synthesize ZnO–G quasi core–shell structured hybrid quantum dots (QDs) [339]. The hybrid QDs were then used to derive high quality graphene nanosheets, by etching out the ZnO cores. GO suspension in dimethyl formamide (DMF) and $Zn(Ac)_2 \cdot 2H_2O$ solution in DMF were mixed and heated to 95°C for 5 h, leading to ZnO–G hybrid core–shell QDs. After drying, the ZnO cores were removed by using HCl acid solution, with graphene nanosheets being left behind. The whole process is schematically illustrated in Fig. 3.89. The formation of ZnO was confirmed by XRD analysis result.

Representative TEM images of the ZnO–G hybrid quasi core–shell structures are shown in Fig. 3.90. As shown in Fig. 3.90 (a), the outer graphene layers over the ZnO core particles are indicated with red arrows. Figure 3.90 (b) shows an image taken from the layer encircling the ZnO QDs. The graphene layer had a monolayer structure, with uniform contrast for visible hexagonal atomic lattice. The distance between the carbon atoms was measured to be about 0.14 nm (inset of Fig. 3.90 (b)). Curved graphene layers were also observed, as shown in the blue dot covered regions of Fig. 3.90 (c). Due to their corrugation and scrolling characteristics, 2D structured graphene nanosheets usually tend to bend in order to achieve a thermodynamically stable state. Therefore, graphene nanosheets with different layers, i.e., monolayer, bilayer, trilayer, quadlayer, and pentalayer, were observed at the edge with isolated fragments. Figure 3.90 (d) shows one of such examples.

An example of the hydrothermal method to synthesize ZnO–G hybrids was reported by Zhang et al. [350]. The ZnO–G hybrid was derived through an ethanol-thermal reaction with a Zn–EG (ethylene glycol)–Ac complex as the precursor. The authors claimed that the synthesis process had several advantages. Firstly, the Zn–EG–Ac possessed a large amount of functional groups, which facilitated strong interaction with the functional groups of GO, so that ZnO NPs could be uniformly distributed on the surfaces of the GO nanosheets after hydrolysis reaction, without the requirement of other precipitants. Secondly, because the Zn–EG–Ac was stable in ethanol, ZnO NPs would be grown onto the graphene nanosheets, instead of within the solution, thus making intimate contact in the final hybrid. Additionally, the GO was reduced by ethanol during the reaction, so that no other reducing agents were needed. The Zn–EG–Ac complex was derived from $Zn(Ac)_2$ and EG, which was mixed with GO in ethanol. The mixture was solvothermally treated at 150°C for 8 h to obtain the ZnO–G hybrid. Parameters, including the ratio of GO and Zn–EG–Ac, the type of Zn source and the type of the solvent, were studied. A schematic diagram to demonstrate the formation of the ZnO–G hybrid is shown in Fig. 3.91.

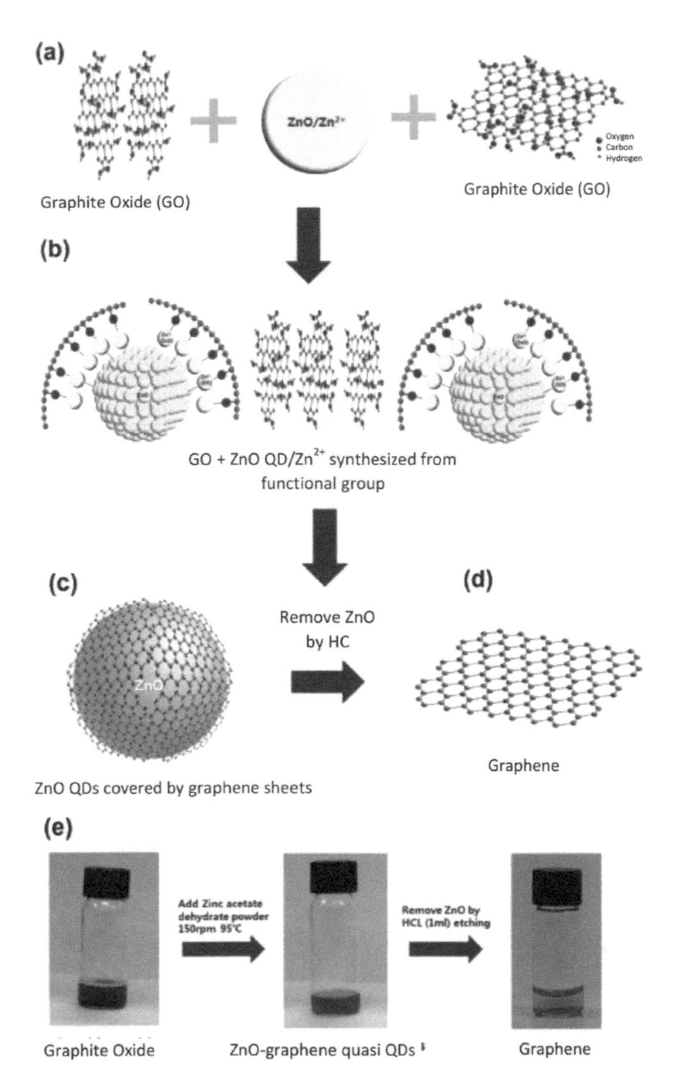

Fig. 3.89. Schematic diagram of the synthesis process: (a) formation of functional groups on the GO surface after treatment with an acid, (b) embryo ZnO QDs formed under the chemical reaction and formation of ZnO–G quasi core–shell QDs. The graphene layers wrapped the ZnO cores due to the chemical reactions between the functional groups (carboxyl, hydroxy, and epoxy) on the GO and Zn^{2+} ions, (c) schematic view of the ZnO–G quasi core–shell QDs and (d) schematic view of the final graphene nanosheet. Reproduced with permission from [339]. Copyright © 2013, Elsevier.

Representative SEM and TEM images of the ZnO–G hybrid are shown in Fig. 3.92 [350]. As demonstrated in Fig. 3.92 (a), the ZnO NPs had an average size of about 10 nm, and were uniformly distributed on the surfaces of the graphene nanosheets. The largest graphene nanosheets possessed a dimension of 2 μm. The uniform distribution of the ZnO NPs on the graphene nanosheets was further confirmed by TEM images, as shown in Fig. 3.92 (b, c). Because the graphene nanosheets were sufficiently thin, they were transparent under SEM and TEM observations. The lattice fringes of 0.248 nm were attributed to the (101) crystal planes of wurtzite ZnO (Fig. 3.92 (d)). The formation of the ZnO NPs in the hybrid was evidenced by XRD results. It was found that the presence of GO was critical to obtain ZnO NPs with small crystal size and narrow size distribution. If no GO was used during the synthesis, the ZnO NPs tended to agglomerate.

The type of Zn precursor also had a significant effect on the microstructure of the ZnO–G hybrids. If $Zn(Ac)_2{\cdot}2H_2O$ was used instead of the Zn–EG–Ac, the resultant product consisted of ZnO balls with

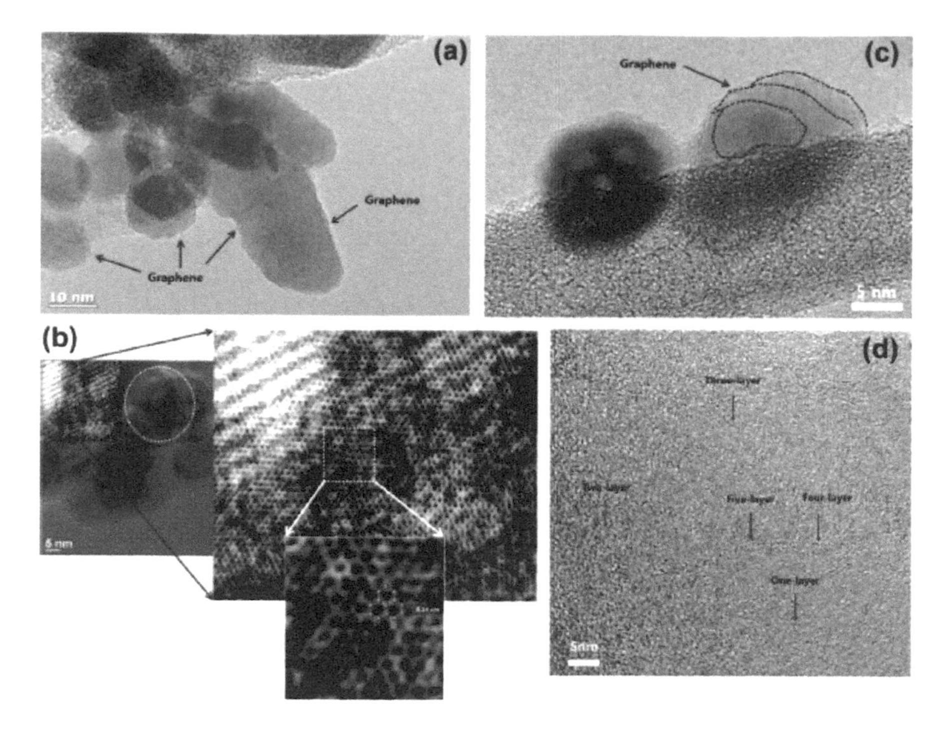

Fig. 3.90. (a) HRTEM image of the ZnO–G hybrid quasi QDs. (b) HRTEM image of the ZnO cores covered by graphene layer (left). White dot circle is an image of the ZnO–G quasi core–shell QDs. Red squared region is the magnified image of one ZnO QD covered by a monolayer graphene. An enlarged view of the monolayer graphene is indicated by the red lines. The area in the white dotted square box is magnified to disclose the atomic image. In the single layer, the gray-colored atoms formed the hexagon and the hexagon center is black. (c) HRTEM image showing the graphene layer areas over the ZnO core. (d) High magnification TEM image showing the edges of the graphene film consisting of one, two, three, four, and five layers. The cross-sectional view was obtained by the inherent curving characteristic at the edges. In-plane lattice fringes suggest a local stacking order of the graphene layers. Reproduced with permission from [339]. Copyright © 2013, Elsevier.

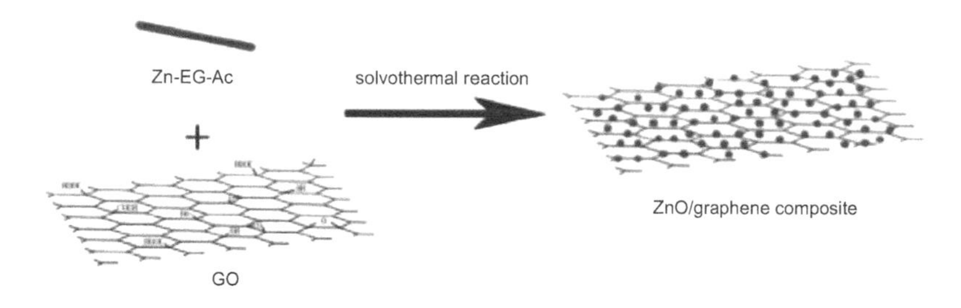

Fig. 3.91. Schematic diagram of the fabrication process of the ZnO–G hybrid. Reproduced with permission from [350]. Copyright © 2012, The Royal Society of Chemistry.

an average size of 1 μm. The ZnO balls were prone to agglomerate. Similar results were obtained for other inorganic Zn salts. Therefore, the rich EG and Ac groups in the precursor promoted the interactions with the –COOH and –OH groups of the GO, so that ZnO NPs were highly dispersed on the graphene nanosheets. Additionally, the use of ethanol as the solvent was also very important. For example, if water was used, homogeneous ZnO–G hybrids could not be produced, because the ZnO NPs not only tended to agglomerate but also were detached from the graphene nanosheets. This was because the reaction between the Zn–EG–Ac and water was so fast that the ZnO particles grew quickly.

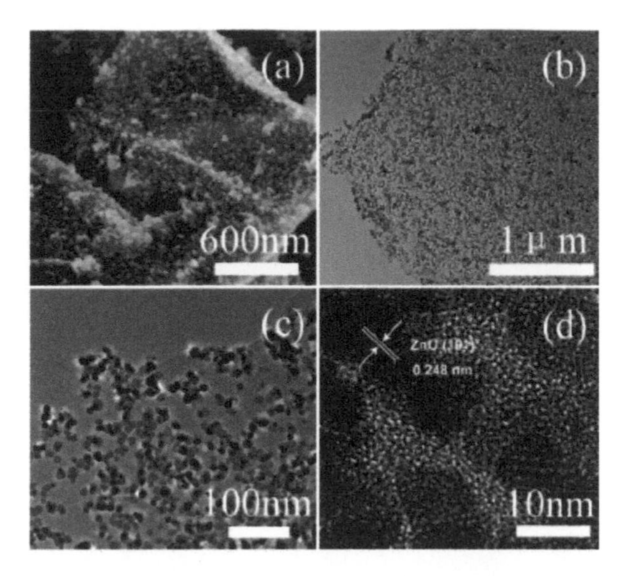

Fig. 3.92. (a) SEM image and (b, c) TEM images of the ZnO–G hybrid. (d) HRTEM image of the ZnO NPs derived from 0.04 g GO and 0.5 g Zn precursor at 150°C for 8 h. Reproduced with permission from [350]. Copyright © 2012, The Royal Society of Chemistry.

3.2.1.10. Graphene—Other Transitional Metal Oxides

Graphene-based hybrids with other transitional metal oxides include ZrO_2–G [354–356], MoO_2–G [357, 358], MoO_3–G [359, 360], WO_3–G [361–366] and RuO_2–G [367], which have been made with different methods for various potential applications. Three examples are discussed as follows.

Liu et al. used the atomic layer deposition (ALD) technique to prepare ZrO_2–G hybrids, with tetrakis(dimethylamido)zirconium(IV) [$Zr(NMe_2)_4$] and H_2O as the precursors [355]. The properties of the ZrO_2 component, such as morphology and crystallinity, could be well-controlled. Graphene powder was loaded into the ALD reactor that was preheated to a temperature first. $Zr(NMe_2)_4$ and deionized water were then alternatively introduced. $Zr(NMe_2)_4$ was at 75°C, while H_2O was kept at room temperature. Delivery lines were kept at 100°C to avoid condensation of the precursors. High purity N_2 gas was used as the carrier gas at a flow rate of 20 sccm, while the ALD reactor was maintained at a low pressure of 0.3–0.4 Torr. One ALD cycle consisted of six steps: (1) supply of $Zr(NMe_2)_4$ of 0.5 s, (2) extended exposure of $Zr(NMe_2)_4$ in the ALD reactor for 0.3 s, (3) purge of excessive $Zr(NMe_2)_4$ and byproducts for 30 s, (4) supply of H_2O for 1.0 s, (5) extended exposure of H_2O in the ALD reactor for 3.0 s and (6) purge of excess H_2O and byproducts for 30 s. ZrO_2–G hybrids were obtained by repeating the cycle at different reaction temperatures (150°C, 200°C and 250°C).

SEM images and XRD patterns of the ZrO_2–G hybrids obtained after ALD reaction for 100 cycles at 150°C, 200°C and 250°C are shown in Fig. 3.93 [355]. Smooth thin films had been deposited on the wrinkles of the graphene nanosheets, with wrinkle thicknesses to be 25 ± 0.5, 22 ± 0.7 and 19 ± 0.4 nm, for reaction temperatures of 150°C, 200°C and 250°C, respectively (Fig. 3.93 (a–c)). Based on the thickness of the pristine graphene wrinkle of about 4 nm, the growth rates were 1.05, 0.90 and 0.75 nm per cycle, at 150°C, 200°C and 250°C, respectively. The sample prepared at 250°C exhibited high crystallinity, as evidenced by the strong diffraction peaks in the XRD pattern (Fig. 3.93 (d)). With decreasing deposition temperature, the crystallinity of the ZrO_2 phase was gradually decreased. The deposition of the ZrO_2 on graphene nanosheets followed an "island growth" mode at the early stage. Once the "islands" coalesced into continuous layers, a "layer-by-layer growth" mode was started. The deposition rate was increased with increasing temperature.

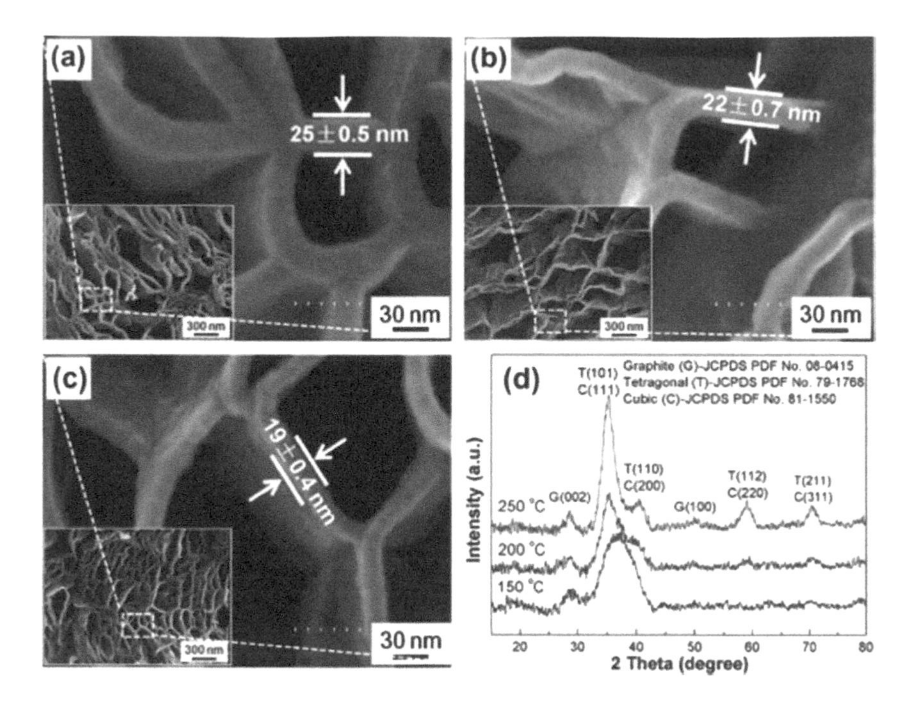

Fig. 3.93. SEM images of the ZrO_2–G hybrids after reaction for 100 cycles at different temperatures: (a) 150°C, (b) 200°C and (c) 250°C, together with their corresponding XRD patterns (d). Reproduced with permission from [355]. Copyright © 2013, Elsevier.

To realize ALD process on substrates, their surface should be terminated with hydroxyl groups in order to provide reactive sites for the nucleation and growth of deposited materials [368]. In this light, GO containing abundant hydroxyl groups is an ideal substrate for ALD of ZrO_2. The deposition sequence and crystallinity of the ZrO_2–G hybrids are schematically illustrated in Fig. 3.94. In the first half-cycle, $Zr(NMe_2)_4$ reacted with –OH groups on the graphene nanosheets. The chemically bonded $Zr(NMe_2)_4$ then reacted with H_2O in the second half-cycle. Eventually, ZrO_2 phase was formed after the reaction cycle was repeated. The hydroxyl groups on the graphene nanosheets were responsible for the formation of the ZrO_2 islands at the earlier stages. The effect of reaction temperature morphology and crystallinity of the ZrO_2 phase could be attributed to the temperature-dependent nature of the hydroxyl groups. For example, high temperature could lead to dehydroxylation [355].

The second example is that of the MoO_2–G hybrid, with hierarchical nanoarchitectures, which was synthesized by a solution-based method combined with a subsequent reduction process [357]. In this hybrid, the component MoO_2 possessed hierarchical nanostructures, while it was highly conductive due to the presence of the graphene nanosheets. The MoO_2–G hybrid exhibited promising electrochemical performances when it was used as the electrode of lithium ion batteries. Figure 3.95 shows a schematic diagram describing the fabrication process of the hierarchical MoO_2–G hybrid. As starting materials, phosphomolybdic acid (PMA) and GO were mixed, and then reacted with hydrazine hydrate solution to prepare the precursor of the hybrid. After annealing at 500°C for 5 h in H_2/Ar, hierarchical MoO_2–G hybrids were produced. Formation of the MoO_2–G hybrid was realized through three steps: (i) assembly of graphene and "heteropoly blue" (HPB) clusters, (ii) HPB rods wrapped with graphene nanosheets and (iii) *in situ* production of graphene-wrapped hierarchical MoO_2 rods in a reducing environment. The morphology and structure of the hybrids were significantly affected by the annealing temperature.

SEM analysis results indicated that the hierarchical MoO_2 rods had diameters of 1–3 μm and lengths of 5–10 μm, as well as rough surfaces that were embedded in the matrix of the graphene nanosheets [357]. There were also graphene nanosheets without the attachment of the MoO_2 nanorods, thus serving as a

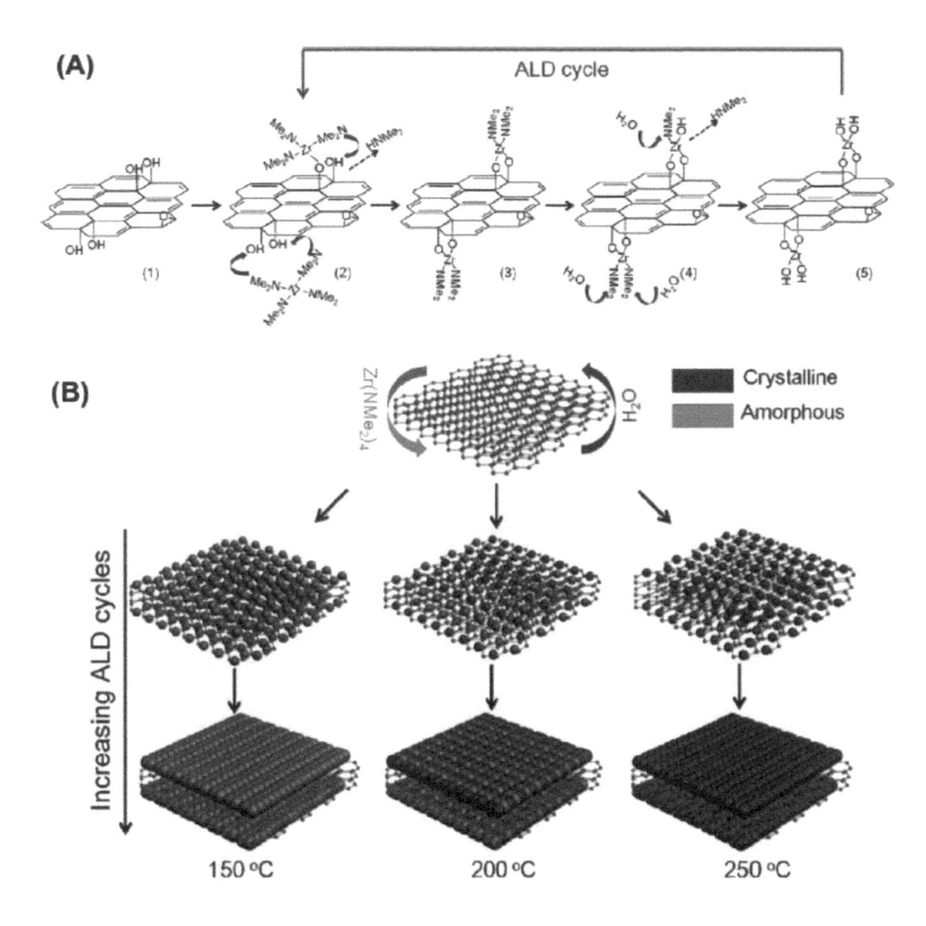

Fig. 3.94. (A) Schematic diagram of one ALD cycle with $Zr(NMe_2)_4$ and H_2O as the precursors. (B) Schematic diagram showing the formation of the ZrO_2–G hybrids with different crystallinity obtained at different temperatures. Reproduced with permission from [355]. Copyright © 2013, Elsevier.

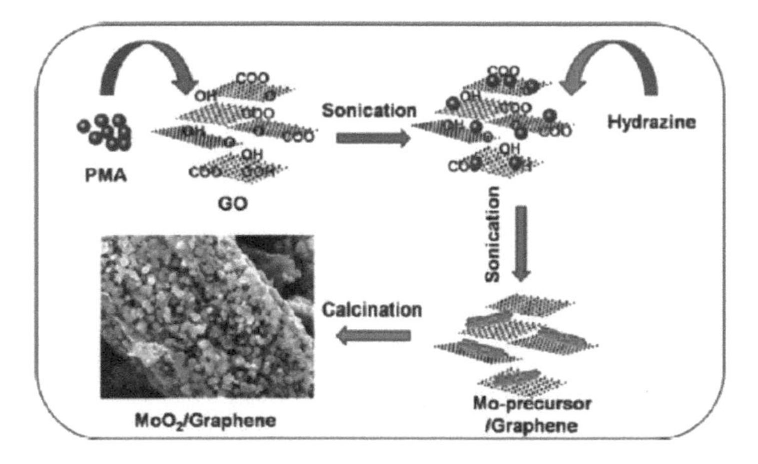

Fig. 3.95. Schematic diagram of the fabrication process of the MoO_2–G hybrid. Reproduced with permission from [357]. Copyright © 2011, American Chemical Society.

connecting agent for the hybrid. The hierarchical MoO_2 rods, as a secondary structure, were comprised of primary MoO_2 nanocrystallites, with average sizes of 30–80 nm. The formation of the fine-structured MoO_2 nanocrystallites was attributed to the confinement effect of the graphene nanosheets. Without the presence of the graphene nanosheets, the MoO_2 particles had a much larger size.

Representative TEM images and SAED pattern of the MoO_2–G hybrid [357]. As shown in Fig. 3.96 (a), the rod-like assemblies consisted of a large number of MoO_2 nanocrystals. The SAED rings confirmed that the hierarchical MoO_2 rods had polycrystalline characteristics, as demonstrated in Fig. 3.96 (b). High magnification TEM image (Fig. 3.96 (c)) revealed that the MoO_2 crystallites were tightly covered by the graphene nanosheets. Figure 3.96 (d) shows a HRTEM image at the edge of an individual MoO_2–G unit, in which the lattice fringes were along the [113] zone axis of the MoO_2. The spacings of ~1.7 and ~1.8 Å corresponded to the interplanar spacings of (220) and (121) planes of monoclinic MoO_2. The single-layer of graphene nanosheets covering the surface of the MoO_2 nanocrystals was about several nanometers in thickness. The interplanar spacing of ~0.37 nm was attributed to the (002) lattice plane of graphite.

An et al. used a one-step hydrothermal method to synthesize a WO_3–G hybrid [362]. The WO_3 component was present in the form of nanorods, while 3.5 wt% graphene was included in the hybrid. To prepare the hydrothermal reaction solutions, GO, $Na_2WO_4 \cdot 2H_2O$ and NaCl were dissolved in water, with pH values to be adjusted with HCl solution. The hydrothermal reaction was conducted at 180°C for 24 h. The WO_3–G hybrid demonstrated a significantly enhanced visible-light photocatalytic activity, as compared to pure WO_3 nanorods and commercial WO_3. It also possessed a promising sensitivity and high selectivity to NO_2. The enhancement in catalytic and gas sensing performances was attributed to several factors, including increased molecular adsorption, extended light absorption, efficient charge separation and improved conductivity.

Figure 3.97 shows TEM images of the WO_3–G hybrid, together with GO nanosheets [362]. Expectedly, the GO was of a crumpled layered structure, with several stacking layers, as demonstrated in Fig. 3.97 (a). Figure 3.97 (b) indicated that 2D graphene nanosheets were present in the hybrid sample. On surface of the graphene nanosheets, rod-like structured WO_3 nanocrystals were uniformly distributed. The uniform distribution of the nanorods on the graphene nanosheets and the intimate contact between the

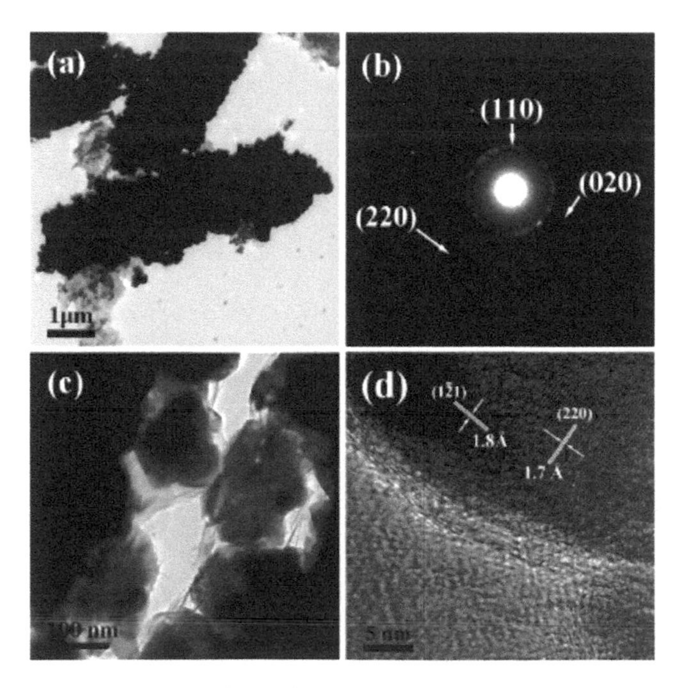

Fig. 3.96. (a) TEM image, (b) SAED pattern, (c) TEM image (high magnification) and (d) HRTEM image of the MoO_2–G hybrid. Reproduced with permission from [357]. Copyright © 2011, American Chemical Society.

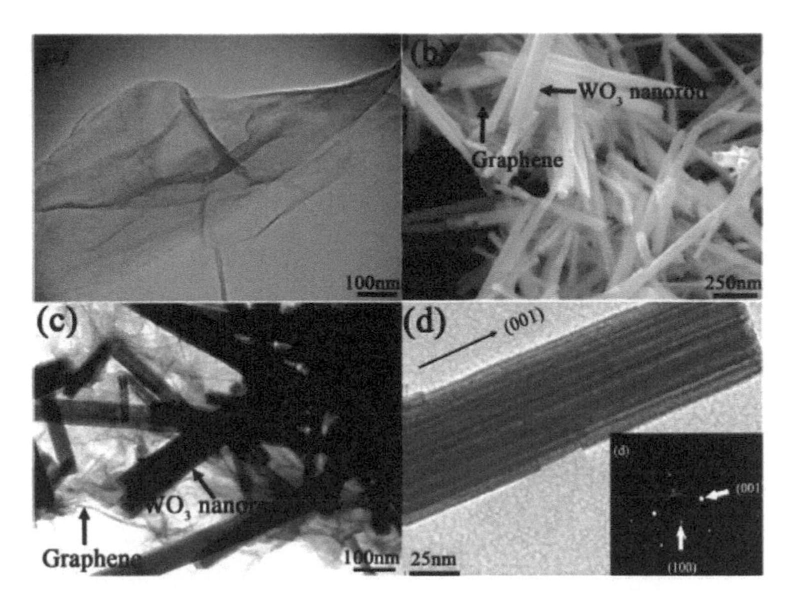

Fig. 3.97. (a) TEM image of the GO nanosheets, (b) SEM image of the WO_3–G hybrids, (c) TEM image of the WO_3–G hybrids and (d) TEM image of a single WO_3 nanorod, together with a SAED pattern. Reproduced with permission from [362]. Copyright © 2012, The Royal Society of Chemistry.

two components was clearly illustrated by the TEM image in Fig. 3.97 (c). The SAED pattern (inset in Fig. 3.97 (d)) revealed a set of well-defined spots, corresponding to the [010] zone axis of hexagonal WO_3, with a preferred growth direction along the (001) surface.

The presence of NaCl played a significant role in the arrangement of 1D WO_3 nanorods on the graphene nanosheets. Without the presence of NaCl, the WO_3 nanorods were seriously agglomerated. This was because NaCl was strongly adsorbed onto certain crystal planes of the WO_3 nucleus, weakening the interactions among the nanorods and thus preventing agglomeration. The level of pH value also had an effect on the morphology of the hybrids. Too low pH value led to WO_3 particles with a wider size distribution, while too high pH value resulted in large agglomerated particles. The morphology of the WO_3 particles was also influenced by the growth temperature.

A possible growth mechanism of the WO_3–G hybrid has been proposed [362]. At the initial stage, H_2WO_4 was formed near the GO nanosheets, due the addition of HCl solution. WO_3 crystal nuclei were then created from the H_2WO_4 during the hydrothermal reaction. Furthermore, due to the presence of NaCl, the nucleation and preferential growth were along the c-axes. The functional groups on the GO nanosheets served as anchor sites to facilitate the *in situ* growth of small WO_3 nanorods. Meanwhile, GO was reduced to graphene through solvothermal reduction.

3.2.2. Graphene—Main Group Oxides (MGOs)

Main group oxides, including Al_2O_3, SiO_2, In_2O_3, SnO_2 and Sb_2O_3, have been used to form hybrids with graphene or GO, among which SnO_2 has been studied more widely than other oxides. Therefore, only SnO_2–G hybrids are included in this section. SnO_2–G hybrids are mainly used as anode materials of lithium ion batteries, as well as materials for other applications such as gas sensors, photocatalysts, biosensors, water purification and electric field emission. Various methods have been employed to synthesize SnO_2–G hybrids, including conventional wet-chemical routes [369–401], hydrothermal or solvothermal treatment [402–421], thermal decomposition process [422], high-energy milling [423] and sputtering techniques [424]. Two examples of SnO_2–G hybrids prepared by using the wet chemical route and hydrothermal reaction technique are discussed as follows.

A two-step method was reported to prepare SnO_2–G hybrids by dispersing SnO_2 NPs on both sides of single-layer graphene nanosheets, with the procedure shown schematically in Fig. 3.98 [381]. At the first step, GO nanosheets were attached with Sn^{4+} through electrostatic force, from which SnO_2 NPs of about 3 nm in size were obtained. At the second step, the GO was reduced to graphene, while the SnO_2 NPs were still on the nanosheets. The SnO_2–G hybrid contained 60 wt% SnO_2 NPs. The high content of SnO_2 made the hybrid suitable for applications as the anode material of Li-ion batteries.

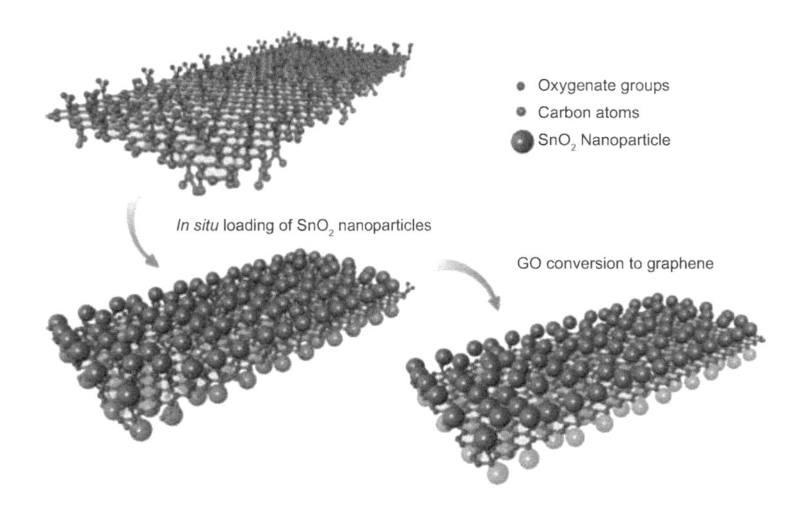

Fig. 3.98. Schematic diagram of the two-step method to disperse SnO_2 NPs on the surface of single-layer graphene nanosheets to form the SnO_2–G hybrid. Reproduced with permission from [381]. Copyright © 2010, The Royal Society of Chemistry.

Figure 3.99 shows representative TEM images of the SnO_2–GO and SnO_2–G hybrids [381]. It was clearly demonstrated that the SnO_2 NPs were uniformly dispersed on the surface of the GO nanosheets, as shown in Fig. 3.99 (a). Low magnification TEM image of the SnO_2–GO hybrid was transparent. SAED pattern (upper left inset of Fig. 3.99 (a)) indicated that, two sets of diffraction signals, i.e., (i) one set of isolated dots from single-layer GO and (ii) one set of rings from the SnO_2 NPs, were observed. As illustrated in the HRTEM image (Fig. 3.99 (b)), the grain size of the SnO_2 NPs was about 3 nm. The lattice spacings of 0.33 nm and 0.26 nm corresponded to (110) and (101) planes of rutile SnO_2. More importantly, the SnO_2 NPs were highly dispersed. Because of the high dispersion of the SnO_2 NPs on both sides of the GO nanosheets, aggregation of the graphene nanosheets during the reduction was effectively prevented. During the thermal treatment, oxygenate groups on the GO nanosheets were eliminated, so as to form the SnO_2–G hybrid. The high dispersion of the SnO_2 NPs after reduction was confirmed by the TEM image shown in Fig. 3.99 (a). At the same time, the crystal structure and particle size of the SnO_2 NPs were not affected, as illustrated in Fig. 3.99 (b). The presence of crystalline SnO_2 NPs and graphene was confirmed by the SAED pattern (inset of Fig. 3.99 (b)).

A solvothermal strategy was developed to synthesize a SnO_2–G hybrid, with the SnO_2 NPs having a mesoporous structure [403]. The mesoporous SnO_2–G hybrid exhibited a large surface area, high electrical conductivity and stable pore structure, making it suitable for various applications. To prepare the mesoporous hybrid, GO powder was first dispersed in a mixed solvent of ethanol and water, into which cetyltrimethylammonium bromide (CTAB) was added. After that, NaOH and $SnCl_4$ were dissolved to form the solvothermal reaction. The solvothermal synthesis was carried out at 160°C for 12 h. For comparison, mesoporous SnO_2 NPs and nonporous SnO_2–G hybrid were synthesized at the same time, but without the presence of GO and CTAB, respectively. The final products were denoted as M-SnO_2, SnO_2–G and M-SnO_2–G, for mesoporous SnO_2 NPs, nonporous SnO_2 NPs and graphene hybrid and mesoporous SnO_2 NPs and graphene hybrid, respectively.

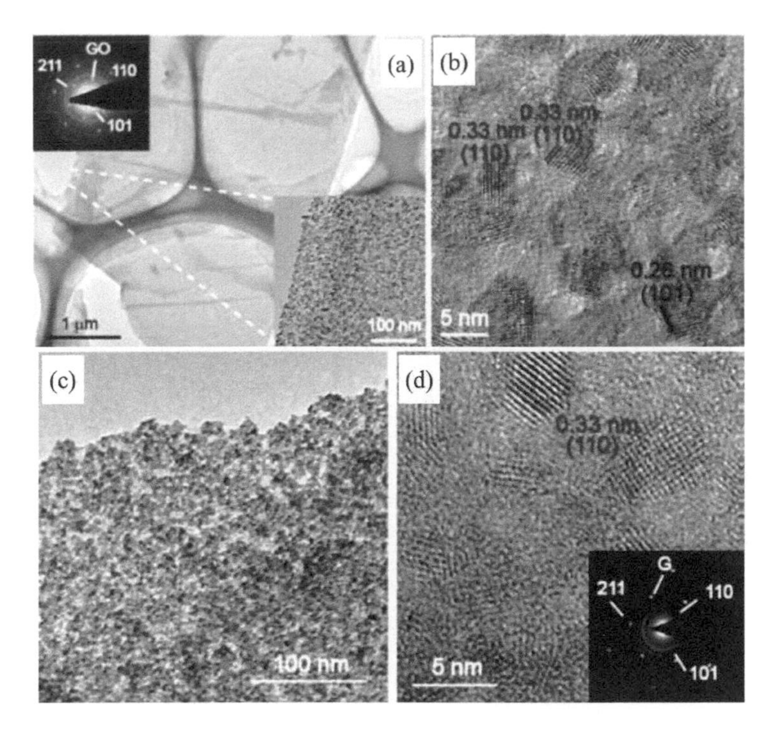

Fig. 3.99. TEM images of the hybrids: (a) TEM image of the SnO_2 NPs dispersed on the GO nanosheets, with upper left and bottom right insets showing SAED pattern and higher magnification TEM image of the SnO_2–GO hybrid, (b) HRTEM image of the SnO_2–GO hybrid, (c) TEM image of the SnO_2–G hybrid and (d) HRTEM image of (c), with the inset showing the SAED pattern of the SnO_2–G. Reproduced with permission from [381]. Copyright © 2010, The Royal Society of Chemistry.

The as-prepared M-SnO_2 possessed disordered wormhole-like pores. The mesoporous walls consisted of connected SnO_2 NPs with sizes of 3–4 nm. Figure 3.100 shows TEM images of the SnO_2–G and M-SnO_2–G hybrids [403]. As shown in Fig. 3.100 (a), SnO_2 NPs were densely loaded onto the graphene nanosheets, with a certain degree of agglomeration, due to the high content of ~79.5%. The mesoporous G–M-SnO_2 hybrid contained nearly the same level of SnO_2 NPs (~78.8%). Comparatively, the degree of agglomeration of the G–M-SnO_2 hybrid was less serious, as illustrated in Fig. 3.100 (c). The lattice spacing of ~0.332 nm (Fig. 3.100 (b)) was attributed to the (110) planes of SnO_2. Aggregation of the nonporous SnO_2 NPs was also observed in the HRTEM image, while the mesoporous SnO_2 NPs were highly dispersed (Fig. 3.100 (d)). The mesoporous SnO_2 NPs had pores of 3.6–3.8 nm in diameter, with a wall thickness of about 3 nm. There was no difference in the crystal structure between the two types of SnO_2 NPs.

The SnO_2 layers with disordered pore arrays of 4–5 nm were uniformly distributed on surface of the graphene nanosheets. The small pores in the M-SnO_2–G hybrid were formed when the SnO_2 NPs were grown on the CTAB template instead of random nanoparticle aggregation. Figure 3.101 shows a schematic diagram to understand the possible mechanism of the formation of the M-SnO_2–G [403]. Because the GO nanosheets were negatively charged in highly basic solutions, CTA^+ ions were strongly adsorbed through electrostatic interactions, thus leading to the formation of micelle assemblies. When the Sn^{4+} ions were mixed with the GO-CTA^+ items, they quickly reacted with OH^- ions to form negatively charged $Sn(OH)_x$ at interfaces of the micelles. Due to dehydration and polymerization of the $Sn(OH)_x$, the micelles on the GO surface were linked together. At the same time, the spherical micelles could be transferred into cylindrical ones, so that mesoporous SnO_2 NPs were formed on the GO surface during the solvothermal treatment, while GO was reduced to graphene in the meantime. The mesoporous structure of the M-SnO_2–G hybrid was present after the CTAB surfactants were removed after thorough washing with warm ethanol.

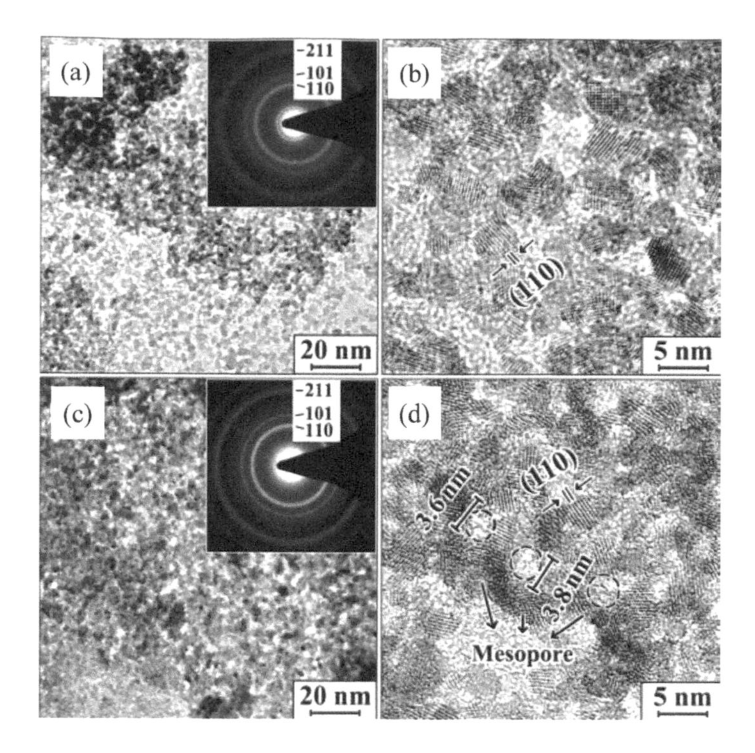

Fig. 3.100. TEM and HRTEM images of SnO_2–G (a, b) and M-SnO_2–G (c, d). The insets of (a, c) are SAED patterns of the SnO_2–G and M-SnO_2–G hybrids. Reproduced with permission from [403]. Copyright © 2013, Wiley-VCH Verlag GmbH & Co. KGaA, Weinheim.

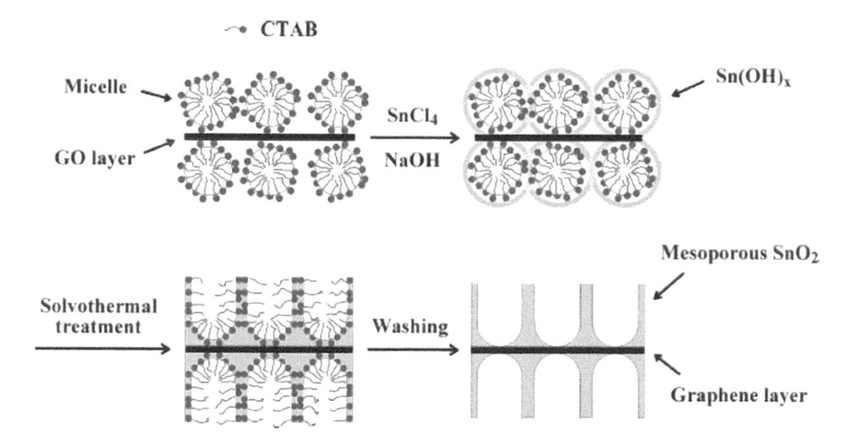

Fig. 3.101. Schematic diagram showing the formation mechanism of the M-SnO_2–G hybrid. Reproduced with permission from [403]. Copyright © 2013, Wiley-VCH Verlag GmbH & Co. KGaA, Weinheim.

3.2.3. Graphene—Complex Oxides and Multiple Oxides

Hybrids of graphene or graphene oxides with various complex oxides and multiple oxides have been reported. For complex oxides, three groups will be discussed. The first group is formed by ferrites [425-440], while the second group consists of cobaltites [441-448]. The last group introduced here contains mainly cathode materials of lithium ion batteries, such as $LiMn_{1/3}Ni_{1/3}Co_{1/3}O_2$ [449, 450],

$LiFePO_4$ [451–457], $Li_4Ti_5O_{12}$ [458, 459] and $Li_3V_2PO_4$ [460, 461]. In addition, more and more other binary oxide systems have been used to form hybrids with graphene nanosheets [462–472]. Selected examples are discussed in the following section.

A simple one-pot solvothermal reaction method was used to prepare rGO–MFe_2O_4 (M = Mn, Zn, Co, and Ni) hybrids with GO and metal chlorides as the starting materials [427]. These rGO–MFe_2O_4 hybrids could be used to remove dye pollutants in water, due to their high adsorption capacity, strong photocatalytic activity and convenient magnetic recovery. During the solvothermal reaction, GO was reduced, while MFe_2O_4 microspheres were grown on the rGO nanosheets at the same time, as shown schematically in Fig. 3.102. $FeCl_3 \cdot 6H_2O$, $MCl_2 \cdot nH_2O$ (M = Mn, Zn, Co and Ni) and NaAc were mixed in EG and PEG, during which positively charged Fe^{3+} and M^{2+} cations were attached to the negatively charged surface of the GO nanosheets through electrostatic interaction. The homogeneous dispersions were solvothermally treated at 200°C for 10 h. In this reaction, EG and PEG were used as the solvent and reducing agent, while NaAc served to adjust the alkalinity of the suspensions to facilitate the formation of MFe_2O_4 NPs from the Fe^{3+} and M^{2+} ions.

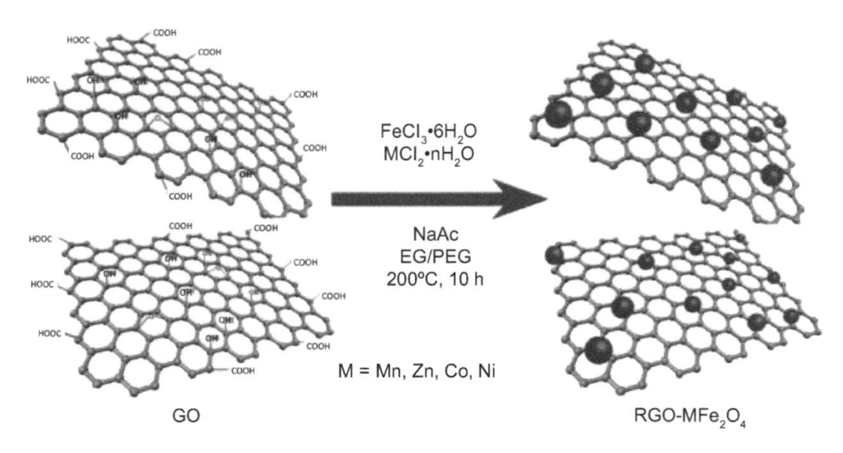

Fig. 3.102. Illustration for the synthesis of the rGO–MFe_2O_4 (M = Mn, Zn, Co, Ni) hybrids. Reproduced with permission from [427]. Copyright © 2013, Elsevier.

In all samples, MFe_2O_4 microspherical particles were uniformly dispersed on silk-like rGO nanosheets to form rGO–MFe_2O_4 hybrids. Figure 3.103 shows representative TEM images of the rGO–$MnFe_2O_4$ hybrids. It was observed that flake-like rGO nanosheets were strongly attached with the $MnFe_2O_4$ microspheres, as demonstrated in Fig. 3.103 (a). Figure 3.102 (b) indicated that the $MnFe_2O_4$ microspheres had an average diameter of 150 nm. An individual $MnFe_2O_4$ microsphere is shown in Fig. 3.103 (c). The particle consisted of a large number of small $MnFe_2O_4$ NPs, with an average size of about 8 nm, so that it was characterized by a porous structure. As illustrated in Fig. 3.103 (d), the $MnFe_2O_4$ NPs were highly crystallized, and had well-defined lattice spacings of 0.21, 0.24 and 0.30 nm, corresponding to the (400), (222) and (220) planes of the $MnFe_2O_4$ crystal.

The size and loading density of the MFe_2O_4 microspheres on rGO nanosheets could be readily controlled by controlling the concentrations of the starting metal ions. Figure 3.104 shows SEM images of the rGO–$MnFe_2O_4$ hybrids and bare $MnFe_2O_4$ particles, in which rGO–$MnFe_2O_4$-1/2 and rGO–$MnFe_2O_4$-2 represented the samples synthesized with half and twice the amount of metal ions as compared with the standard sample. The $MnFe_2O_4$ microspheres in rGO–$MnFe_2O_4$-1/2 and RGO–$MnFe_2O_4$ samples possessed a similar particle size of about 150 nm, while the average size of the $MnFe_2O_4$ microspheres in the RGO–$MnFe_2O_4$-2 hybrid was about 250 nm. This is because more metal ions would generate more $MnFe_2O_4$ NPs which led to more $MnFe_2O_4$ microspheres with larger size. Without the presence of GO nanosheets, the bare $MnFe_2O_4$ microspheres had a very narrow size distribution, with particle size slightly larger than that of their counterparts in the rGO–$MnFe_2O_4$ hybrid (Fig. 3.104 (d)).

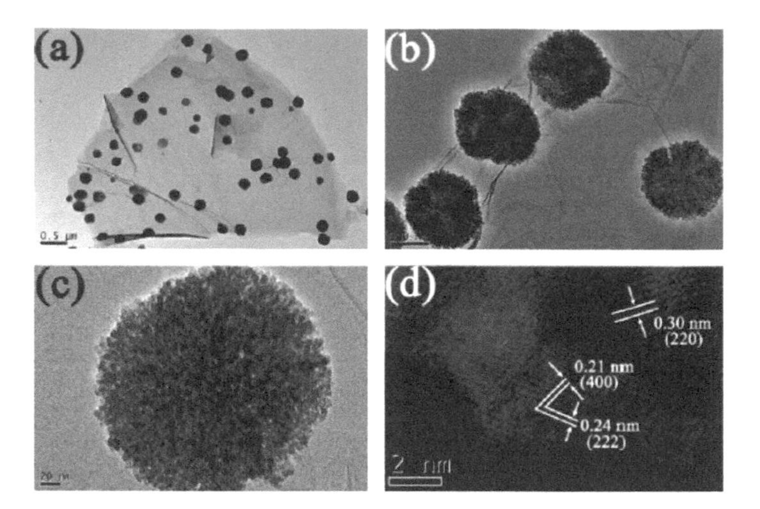

Fig. 3.103. TEM/HRTEM images of the rGO–$MnFe_2O_4$ hybrids. (a) Low- and (b) high-magnification TEM images, (c) TEM image showing the structure of individual $MnFe_2O_4$ microsphere on rGO nanosheets, (d) HRTEM image of the $MnFe_2O_4$ microsphere. Reproduced with permission from [427]. Copyright © 2013, Elsevier.

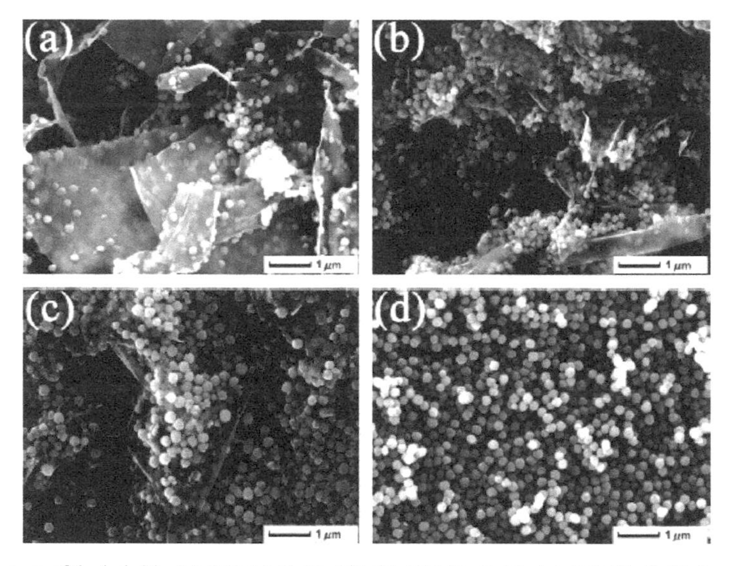

Fig. 3.104. SEM images of the hybrids: (a) rGO–$MnFe_2O_4$-1/2, (b) rGO–$MnFe_2O_4$, (c) rGO–$MnFe_2O_4$-2 and (d) bare $MnFe_2O_4$ microspheres. Reproduced with permission from [427]. Copyright © 2013, Elsevier.

The rGO–$MnFe_2O_4$ hybrids exhibited pretty high saturation magnetization, relatively low remanence and coercivity. They had a strong effect on the efficiency of removing dye pollutants in water. Due to their strong magnetic properties, the rGO–$MnFe_2O_4$ hybrids could be easily separated from the solutions by using a magnet after the decontamination reactions.

A 3D structured catalyst has been developed with N-doped graphene films containing in- and out-of-plane pores [441]. The in-plane porous graphene nanosheets were used as precursors to produce out-of-plane pores through the introduction of $NiCo_2O_4$ NPs between the graphene nanosheets. Such 3D hybrid films were made of porous N-doped graphene–$NiCo_2O_4$ (PNG–NiCo), which exhibited a strong catalytic activity for oxygen evolution reaction (OER). The high catalytic performance was attributed to the synergistic effects of the graphene nanosheets and the $NiCo_2O_4$ NPs.

Figure 3.105 shows a schematic diagram of the fabrication process. Graphene nanosheets were oxidized and etched with $KMnO_4$ and hydrochloric acid (HCl) to produce in-plane pores. Free-standing films were prepared through filtration of the porous graphene (PG). Nitrogen (N) atoms were introduced onto the graphene nanosheets as dopants by thermally treating them in ammonia to obtain N-doped porous graphene film (PNG). To prepare the PNG-NiCo hybrid, $Ni(NO_3)_2 \cdot 6H_2O$, $Co(NO_3)_2 \cdot 6H_2O$ and urea were dissolved in a mixed solvent of ethanol and H_2O, forming an intermediate complex, i.e., $[M(H_2O)_{6-x}(urea)_x]^{2+}$. PNG film was emerged in the solution and then heated at 90°C for 2 h. During the heating process, the metal complex $[M(H_2O)_{6-x}(urea)_x]^{2+}$ was decomposed to form nuclei of $NiCo_2O_4$. In this case, the functional groups of the graphene nanosheets, such as –COOH, served as anchor sites to host the nuclei through hydrogen bonding, van der Waals forces or covalent bonding. Due to the presence of pores, the precursors were sucked by the graphene film, so that $NiCo_2O_4$ NPs were grown and attached on the surfaces of the graphene nanosheets, thus forming a sandwich-like structure. The hybrid films were obtained by drying and subsequent annealing at 250°C for 2 h. In addition, N-doped graphene–$NiCo_2O_4$ hybrid film (NG-NiCo) was prepared similarly by replacing PGO with GO and porous graphene–$NiCo_2O_4$ hybrid film (PG-NiCo) was fabricated without using $NH_3 \cdot H_2O$ for comparison.

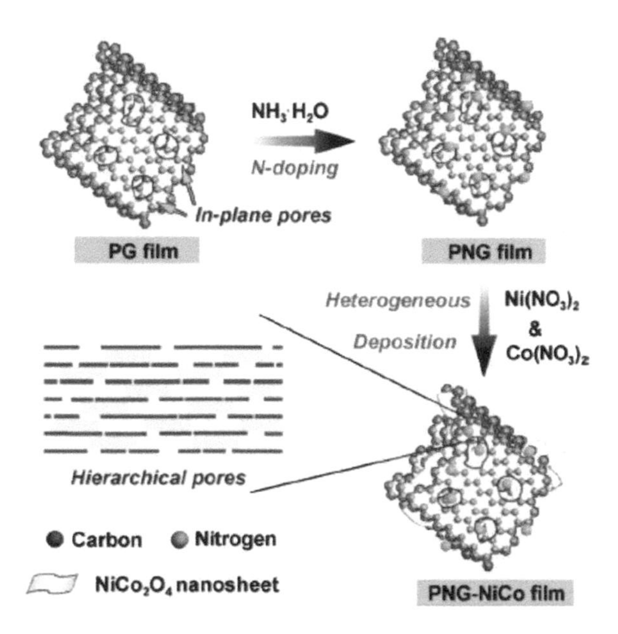

Fig. 3.105. Fabrication of the 3D hybrid catalyst, with PG, PNG and PNG–NiCo standing for porous graphene, porous N-doped graphene and porous N-doped graphene–$NiCo_2O_4$ hybrid, respectively. Reproduced with permission from [441]. Copyright © 2013, American Chemical Society.

The PNG-NiCo hybrid film was highly flexible, with a lateral size of several centimeters and a thickness of ~17 μm, as shown in Fig. 3.106 (a, b). The relatively oriented assembly of the graphene nanosheets in the PNG-NiCo hybrid film resulted in large out-of-plane macropores, with sizes ranging from tens of nanometers to several micrometers, as demonstrated in Fig. 3.106 (c). Due to the relatively ordered lamellar structure, the PNG-NiCo could promote effective transport during the catalytic reaction. Figure 3.106 (d) indicated that the in-plane pores of the PNG-NiCo were small mesopores with sizes of tens of nanometers that were randomly distributed on the graphene nanosheets. The presence of the out-of-plane macropores and in-plane mesopores in the PNG-NiCo hybrid film was confirmed by the results of nitrogen adsorption analyses. The pore size distribution showed the presence of mesopores and small macropores in the range of 10–100 nm.

Although the $NiCo_2O_4$ and PNG-NiCo exhibited a similar pore size distribution, there was a significant difference in their originality. The pores of the $NiCo_2O_4$ particles were formed due to the restacking of the

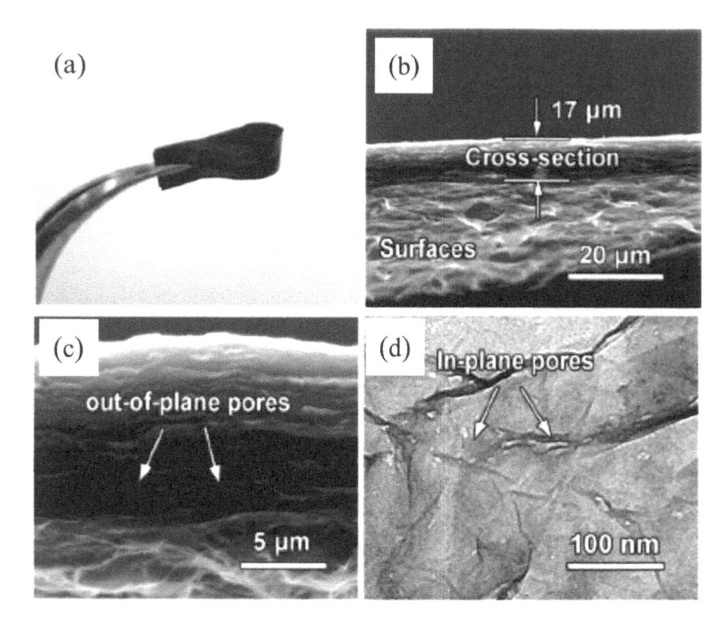

Fig. 3.106. (a) Photograph and (b, c) SEM images of the PNG–NiCo. (c) TEM image of PNG. Reproduced with permission from [441]. Copyright © 2013, American Chemical Society.

NPs, whereas those of the PNG-NiCo hybrid were comprised of both the in-plane pores of the graphene nanosheets and the restacking pores of the $NiCo_2O_4$ NPs. It was also found that the porosity was mainly contributed by the porous graphene nanosheets. The presence of the hierarchical porous structures of the PNG-NiCo hybrid ensured highly accessible surfaces for catalysts.

A simple hydrothermal method has been used to synthesize hybrids of graphene with $LiFePO_4$ (LFP) [451]. Polyethylene glycol (PEG) was used as a surfactant in the hydrothermal reaction. The LFP architectures were constructed with ordered LFP nanocubes. The micro/nanostructured LFP with different morphologies, i.e., cube-cluster-like, dumbbell-like, rod-like and rugby-like, were controlled by controlling the pH value of the precursor. The LFP/C–rGO hybrids exhibited higher electrochemical performance than bare LFP. It was found that, as the cathode material of lithium-ion batteries, the rugby-like LFP/C–rGO sample showed the highest electrochemical performance, with a discharge specific capacity of about 150 $mA{\cdot}h{\cdot}g^{-1}$ after 100 cycles and a high reversible specific capacity of 152 $mA{\cdot}h{\cdot}g^{-1}$ at 0.1 C.

To synthesize LFP, lithium hydroxide monohydrate ($LiOH{\cdot}H_2O$), iron sulfate heptahydrate ($FeSO_4{\cdot}7H_2O$) and phosphoric acid (H_3PO_4) were mixed in a molar ratio of 2:1:1. Firstly, PEG-400 (mean molecular weight of 380–420) was dissolved in DI water to form a homogeneous transparent solution. $FeSO_4{\cdot}7H_2O$ and H_3PO_4 were dissolved in DI water, which was then quickly mixed with the PEG solution. After that, $LiOH{\cdot}H_2O$ was slowly added into the mixed solution. The pH value of the precursor solution was controlled with $NH_3{\cdot}nH_2O$ solution. The solutions were subject to hydrothermal reaction at 190°C for 12 h, in order to form LFP powders.

The LFP powders were used to prepare LFP/C–rGO hybrids. Firstly, GO was dispersed in DI water by using ultrasonication and centrifugation to obtain clear suspension. Emulsifier OP-10 ($C_{34}H_{62}O_{11}$) was dispersed in DI water to form a solution, into which the LFP powders were added. After stirring for 6 h, the GO solution was added, with the GO/LFP weight ratio to be 7%. Hydrazine was then dropped into the mixture solution to reduce the GO, which was heated at 80°C under constant stirring to evaporate the solvent. The obtained powders were thermally treated at 650°C for 2 h in N_2/H_2 (95%/5%) to form the LFP/C–rGO hybrids.

Figure 3.107 shows SEM images of the LFP powders derived from the precursors at different pH values [451]. The LFP powder exhibited a microstructure consisting of cubes as the pH was at 8.2 (Fig. 3.107 (a)). The cube units with lengths ranging from 1 μm to 2 μm were assembled into cube clusters with a length of about 6 μm. A hierarchical microstructure was then formed by the self-assembled and

Fig. 3.107. SEM images of the LFP samples derived from the precursors with different pH values: (a) pH = 8.2, (b) pH = 8.7, (c) pH = 9.0 and (d) pH = 9.8. Reproduced with permission from [451]. Copyright © 2014, American Chemical Society.

cross-linked cubes. As the pH value was increased to 8.9, the LFP particles had a dumbbell-like shape, with a length of about 15 μm and widths of 6–8 μm, as shown in Fig. 3.107 (b). The middle section of dumbbell-like microstructure contained many nanosized particles while the two ends were formed with microsized particles. If the reaction was conducted at pH = 9.0, powders of rod-like microstructures with lengths of 6–7 μm and widths of 2–3 μm were obtained, as demonstrated in Fig. 3.107 (c). The rod-like nanostructured LFP was hierarchically constructed with a large number of nanocubes, which were assembled side by side. Typical length and width of the rugby-like nanostructures were in the ranges of 3–4 μm and 0.4–2 μm, respectively. The rugby-like nanostructure was derived from irregular nanocubes with sizes of 200–300 nm (Fig. 3.107 (d)). Therefore, pH value of precursor solution can be used to control the morphology and grain size of LFP powders.

Figure 3.108 shows an evolving mechanism explaining the formation of the self-assembled micro/nanostructured LFP powders at different pH values [451]. The formation process followed a dissolution-recrystallization mechanism. When the reactants, i.e., Li_3PO_4 precipitate, Fe^{2+} and PEG-400, were mixed during the hydrothermal reaction, LFP crystalline nuclei were formed from the Li_3PO_4 precipitate particles. Possibly, PEG would be adsorbed on surface of the LFP nuclei, so that cubic-shaped LFP nanocrystals were grown from the precursor solution. Due to driving force to minimize the surface area, the tiny primary nanocrystals were self-assembled into more stable cubes with larger sizes. As the reaction proceeded, the LFP cubes continued to aggregate into 3D hierarchical structures. At different pH values, through the dissolution and recrystallization process, the LFP cubes were assembled in a way of either side-by-side or cross-linking. As a result, cube-clusters-like, dumbbell-like, rod-like and rugby-like morphologies were derived at different pH values. The difference in morphology among the four samples was attributed to the difference in the rate of crystal growth in the precursors with different pH values. Generally, the rate of crystal growth in solution is closely related to the solubility of the solutes. At relatively low pH values, e.g., pH = 8.2 and 8.7, the solubility of the precursors was high, thus leading to a high crystal growth rate. As a consequence, crystals and assemblages with relatively large sizes were obtained. At high pH values, i.e., pH = 9 and 9.8, the presence of a large number of OH^- ions at the interface limited the growth rates of the faces. In this case, the nucleation process was dominant over the crystal growth, so that the crystals and assemblages had relatively smaller sizes.

In the LFP/C–rGO hybrids, the LFP particles were co-modified by rGO nanosheets and carbon coating layers, so that an effective conducting network was formed. The LFP particles and rGO were connected

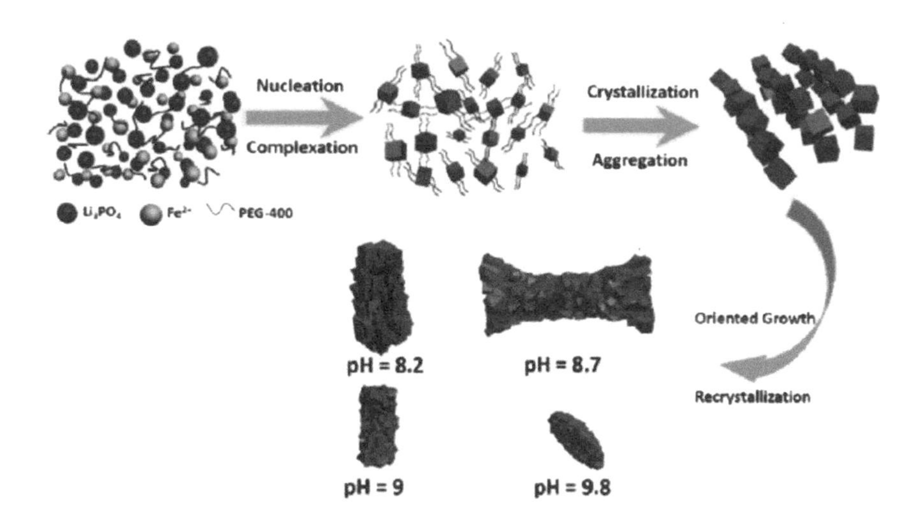

Fig. 3.108. Schematic diagram showing the formation mechanism of various self-assembled LFP samples during the hydrothermal reaction. Reproduced with permission from [451]. Copyright © 2014, American Chemical Society.

by carbon, while the LFP particles were covered by rGO nanosheets. The cube-cluster-like and dumbbell-like LFP particles were incompletely wrapped by the C/RGO, whereas the rod-like and rugby-like LFP particles with the smaller sizes were entirely wrapped by the C/RGO. For cube-cluster-like LFP, the surface of the powder was coated by a layer of carbon with thicknesses of 5–8 nm. In dumb-bell-like LFP/C–rGO hybrid, an amorphous carbon coating layer with an uneven thickness was formed on the surface of the LFP particles. In the case of rod-like LFP/C–rGO hybrid, the amorphous carbon layer had a thickness of 5 nm. The rod-like and rugby-like LFP/C–rGO hybrids contained thinner and more uniform carbon coating layers as compared with the other two hybrids. During the formation of the carbon coating layers, emulsifier OP-10 acted not only as a surfactant but also the source of carbon. Therefore, the LFP particles and rGO nanosheets were effectively connected by the carbon layers, so that a conducting network was formed which thus increased the conductivity of the hybrids.

3.3. Summary

As mentioned in the beginning, almost all metallic oxides have been used to produce hybrids with graphene or graphene oxide nanosheets, by using various approaches, which lead to materials with a wide range of applications. Although it is expected that different materials will have different characteristics, so that different methods should be used, there should be a general principle that can be used to govern their synthesis procedures. The two most commonly used methods are the wet-chemical precipitation and hydrothermal/solvothermal reactions. When using precipitation or coprecipitation approaches, the type of precipitating agents, concentration of the agents, way of addition/mixing, temperature, concentration of the parent solution, composition (cation/graphene ratio) of the solution, as well as other parameters, such as the washing conditions, calcination temperature/time and even heating rate, all will have influences on the properties, microstructure and thus performance of the final hybrids. For hydro/solvothermal synthetic routes, there are more factors that can be used to tailor the properties of the final products, such as the reaction temperature and time, type and concentration of the precursors, additives and so on. Therefore, there is still work to do in order to regulate the processing and characterization of these types of newly emerged hybrid materials.

References

[1] Morales-Torres S, Pastrana-Martinez LM, Figueiredo JL, Faria JL, Silva AMT. Design of graphene-based TiO_2 photocatalysts-A review. Environmental Science and Pollution Research. 2012; 19: 3676–87.

[2] Wu ZS, Zhou GM, Yin LC, Ren W, Li F, Cheng HM. Graphene/metal oxide composite electrode materials for energy storage. Nano Energy. 2012; 1: 107–31.

[3] Leary R, Westwood A. Carbonaceous nanomaterials for the enhancement of TiO_2 photocatalysis. Carbon. 2011; 49: 741–72.

[4] Wang P, Han L, Zhu CZ, Zhai YM, Dong SJ. Aqueous-phase synthesis of Ag-TiO_2-reduced graphene oxide and Pt-TiO_2-reduced graphene oxide hybrid nanostructures and their catalytic properties. Nano Research. 2011; 4: 1153–62.

[5] Liu G, Wang LZ, Yang HG, Cheng HM, Lu GQ. Titania-based photocatalysts-crystal growth, doping and heterostructuring. Journal of Materials Chemistry. 2010; 20: 831–43.

[6] Sampaio MJ, Silva CG, Marques RRN, Silva AMT, Faria JL. Carbon nanotube-TiO_2 thin films for photocatalytic applications. Catalysis Today. 2011; 161: 91–6.

[7] Yu JG, Ma TT, Liu G, Cheng B. Enhanced photocatalytic activity of bimodal mesoporous titania powders by C-60 modification. Dalton Transactions. 2011; 40: 6635–44.

[8] Wang F, Zhang K. Physicochemical and photocatalytic activities of self-assembling TiO_2 nanoparticles on nanocarbons surface. Current Applied Physics. 2012; 12: 346–52.

[9] Guo JJ, Zhu SM, Chen ZX, Li Y, Yu ZY, Liu QL et al. Sonochemical synthesis of TiO_2 nanoparticles on graphene for use as photocatalyst. Ultrasonics Sonochemistry. 2011; 18: 1082–90.

[10] Bell NJ, Ng YH, Du AJ, Coster H, Smith SC, Amal R. Understanding the enhancement in photoelectrochemical properties of photocatalytically prepared TiO_2-reduced graphene oxide composite. Journal of Physical Chemistry C. 2011; 115: 6004–9.

[11] Mishra AK, Ramaprabhu S. Functionalized graphene-based nanocomposites for supercapacitor application. Journal of Physical Chemistry C. 2011; 115: 14006–13.

[12] Wang F, Zhang K. Reduced graphene oxide-TiO_2 nanocomposite with high photocatalystic activity for the degradation of rhodamine B. Journal of Molecular Catalysis A-Chemical. 2011; 345: 101–7.

[13] Jiang GD, Lin ZF, Chen C, Zhu LH, Chang Q, Wang N et al. TiO_2 nanoparticles assembled on graphene oxide nanosheets with high photocatalytic activity for removal of pollutants. Carbon. 2011; 49: 2693–701.

[14] Fan Y, Lu HT, Liu JH, Yang CP, Jing QS, Zhang YX et al. Hydrothermal preparation and electrochemical sensing properties of TiO_2-graphene nanocomposite. Colloids and Surfaces B-Biointerfaces. 2011; 83: 78–82.

[15] Ding SJ, Chen JS, Luan DY, Boey FYC, Madhavi S, Lou XW. Graphene-supported anatase TiO_2 nanosheets for fast lithium storage. Chemical Communications. 2011; 47: 5780–2.

[16] Niyogi S, Bekyarova E, Itkis ME, McWilliams JL, Hamon MA, Haddon RC. Solution properties of graphite and graphene. Journal of the American Chemical Society. 2006; 128: 7720–1.

[17] Wang Y, Li YM, Tang LH, Lu J, Li JH. Application of graphene-modified electrode for selective detection of dopamine. Electrochemistry Communications. 2009; 11: 889–92.

[18] Si YC, Samulski ET. Exfoliated graphene separated by platinum nanoparticles. Chemistry of Materials. 2008; 20: 6792–7.

[19] Cui YM, Shan DJ, Zhu YR. Studies on photocatalytic oxidation of I^- over TiO_2 thin film. Chinese Journal of Inorganic Chemistry. 2001; 17: 401–6.

[20] Williams G, Seger B, Kamat PV. TiO_2-graphene nanocomposites. UV-assisted photocatalytic reduction of graphene oxide. ACS Nano. 2008; 2: 1487–91.

[21] Zhang YP, Pan CX. TiO_2/graphene composite from thermal reaction of graphene oxide and its photocatalytic activity in visible light. Journal of Materials Science. 2011; 46: 2622–6.

[22] Manga KK, Wang SL, Jaiswal M, Bao QL, Loh KP. High-gain graphene-titanium oxide photoconductor made from inkjet printable ionic solution. Advanced Materials. 2010; 22: 5265–70.

[23] Peng WQ, Wang ZM, Yoshizawa N, Hatori H, Hirotsu T. Lamellar carbon nanosheets function as templates for two-dimensional deposition of tubular titanate. Chemical Communications. 2008: 4348–50.

[24] Wojtoniszak M, Zielinska B, Chen XC, Kalenczuk RJ, Borowiak-Palen E. Synthesis and photocatalytic performance of TiO_2 nanospheres-graphene nanocomposite under visible and UV light irradiation. Journal of Materials Science. 2012; 47: 3185–90.

[25] Stengl V, Popelkova D, Vlacil P. TiO_2-graphene nanocomposite as high performace photocatalysts. Journal of Physical Chemistry C. 2011; 115: 25209–18.

[26] Zou F, Yu Y, Cao N, Wu LZ, Zhi JF. A novel approach for synthesis of TiO_2-graphene nanocomposites and their photoelectrical properties. Scripta Materialia. 2011; 64: 621–4.

[27] Akhavan O, Ghaderi E. Photocatalytic reduction of graphene oxide nanosheets on TiO_2 thin film for photoinactivation of bacteria in solar light irradiation. Journal of Physical Chemistry C. 2009; 113: 20214–20.

[28] Wang DH, Choi DW, Li JB, Yang ZG, Nie ZM, Kou R et al. Self-assembled TiO_2-graphene hybrid nanostructures for enhanced Li-ion insertion. ACS Nano. 2009; 3: 907–14.

[29] Lambert TN, Chavez CA, Hernandez-Sanchez B, Lu P, Bell NS, Ambrosini A et al. Synthesis and characterization of titania-graphene nanocomposites. Journal of Physical Chemistry C. 2009; 113: 19812–23.

[30] Zhang HJ, Xu PP, Du GD, Chen ZW, Oh K, Pan DY et al. A facile one-step synthesis of TiO_2/graphene composites for photodegradation of methyl orange. Nano Research. 2011; 4: 274–83.
[31] Zhang Q, He YQ, Chen XG, Hu DH, Li LJ, Yin T et al. Structure and photocatalytic properties of TiO_2-graphene oxide intercalated composite. Chinese Science Bulletin. 2011; 56: 331–9.
[32] Bavykin DV, Friedrich JM, Walsh FC. Protonated titanates and TiO_2 nanostructured materials: Synthesis, properties, and applications. Advanced Materials. 2006; 18: 2807–24.
[33] Chen Q, Du GH, Zhang S, Peng LM. The structure of trititanate nanotubes. Acta Crystallographica Section B-Structural Science. 2002; 58: 587–93.
[34] Wang DH, Liu J, Huo QS, Nie ZM, Lu WG, Williford RE et al. Surface-mediated growth of transparent, oriented, and well-defined nanocrystalline anatase titania films. Journal of the American Chemical Society. 2006; 128: 13670–1.
[35] Park SJ, Lee KS, Bozoklu G, Cai WW, Nguyen ST, Ruoff RS. Graphene oxide papers modified by divalent ions—Enhancing mechanical properties via chemical cross-linking. ACS Nano. 2008; 2: 572–8.
[36] Park SJ, Ruoff RS. Chemical methods for the production of graphenes. Nature Nanotechnology. 2009; 4: 217–24.
[37] Stankovich S, Dikin DA, Piner RD, Kohlhaas KA, Kleinhammes A, Jia YY et al. Synthesis of graphene-based nanosheets via chemical reduction of exfoliated graphite oxide. Carbon. 2007; 45: 1558–65.
[38] Stankovich S, Piner RD, Chen XQ, Wu NQ, Nguyen ST, Ruoff RS. Stable aqueous dispersions of graphitic nanoplatelets via the reduction of exfoliated graphite oxide in the presence of poly(sodium 4-styrenesulfonate). Journal of Materials Chemistry. 2006; 16: 155–8.
[39] Liu B, Zeng HC. Carbon nanotubes supported mesoporous mesocrystals of anatase TiO_2. Chemistry of Materials. 2008; 20: 2711–8.
[40] Subrahmanyam KS, Vivekchand SRC, Govindaraj A, Rao CNR. A study of graphenes prepared by different methods: characterization, properties and solubilization. Journal of Materials Chemistry. 2008; 18: 1517–23.
[41] Li HB, Zhang W, Zou LD, Pan LK, Sun ZZ. Synthesis of TiO_2-graphene composites via visible-light photocatalytic reduction of graphene oxide. Journal of Materials Research. 2011; 26: 970–3.
[42] Chen C, Cai WM, Long MC, Zhou BX, Wu YH, Wu DY et al. Synthesis of visible-light responsive graphene oxide/ TiO_2 composites with p/n heterojunction. Acs Nano. 2010; 4: 6425–32.
[43] Zhang H, Lv XJ, Li YM, Wang Y, Li JH. P25-graphene composite as a high performance photocatalyst. ACS Nano. 2010; 4: 380–6.
[44] Liang YY, Wang HL, Casalongue HS, Chen Z, Dai HJ. TiO_2 nanocrystals grown on graphene as advanced photocatalytic hybrid materials. Nano Research. 2010; 3: 701–5.
[45] Qiu YC, Yan KY, Yang SH, Jin LM, Deng H, Li WS. Synthesis of size-tunable anatase TiO_2 nanospindles and their assembly into anatase@titanium oxynitride/titanium nitride-graphene nanocomposites for rechargeable lithium ion batteries with high cycling performance. ACS Nano. 2010; 4: 6515–26.
[46] Liu CM, Yang SH. Synthesis of angstrom-scale anatase titania atomic wires. ACS Nano. 2009; 3: 1025–31.
[47] Liu CM, Sun H, Yang SH. From nanorods to atomically thin wires of anatase TiO_2: Nonhydrolytic synthesis and characterization. Chemistry-A European Journal. 2010; 16: 4381–93.
[48] Xu CY, Zhang Q, Zhang H, Zhen L, Tang J, Qin LC. Synthesis and characterization of single-crystalline alkali titanate nanowires. Journal of the American Chemical Society. 2005; 127: 11584–5.
[49] Feist TP, Davies PK. The soft chemical synthesis of TiO_2 (B) from layered titanates. Journal of Solid State Chemistry. 1992; 101: 275–95.
[50] Paek MJ, Ha HW, Kim TW, Moon SJ, Baeg JO, Choy JH et al. Formation efficiency of one-dimensional nanostructured titanium oxide affected by the structure and composition of titanate precursor: A mechanism study. Journal of Physical Chemistry C. 2008; 112: 15966–72.
[51] Lee JS, You KH, Park CB. Highly photoactive, low bandgap TiO_2 nanoparticles wrapped by graphene. Advanced Materials. 2012; 24: 1084–8.
[52] Chen DH, Huang FZ, Cheng YB, Caruso RA. Mesoporous anatase TiO_2 beads with high surface areas and controllable pore sizes: A superior candidate for high-performance dye-sensitized solar cells. Advanced Materials. 2009; 21: 2206–10.
[53] Liu JC, Bai HW, Wang YJ, Liu ZY, Zhang XW, Sun DD. Self-Assembling TiO_2 nanorods on large graphene oxide sheets at a two-phase interface and their anti-recombination in photocatalytic applications. Advanced Functional Materials. 2010; 20: 4175–81.
[54] Liu JC, Liu L, Bai HW, Wang YJ, Sun DD. Gram-scale production of graphene oxide-TiO_2 nanorod composites: Towards high-activity photocatalytic materials. Applied Catalysis B-Environmental. 2011; 106: 76–82.
[55] Shen JF, Yan B, Shi M, Ma HW, Li N, Ye MM. One step hydrothermal synthesis of TiO_2-reduced graphene oxide sheets. Journal of Materials Chemistry. 2011; 21: 3415–21.
[56] Fan WQ, Lai QH, Zhang QH, Wang Y. Nanocomposites of TiO_2 and reduced graphene oxide as efficient photocatalysts for hydrogen evolution. Journal of Physical Chemistry C. 2011; 115: 10694–701.
[57] Zhou KF, Zhu YH, Yang XL, Jiang X, Li CZ. Preparation of graphene-TiO_2 composites with enhanced photocatalytic activity. New Journal of Chemistry. 2011; 35: 353–9.
[58] Jiang BJ, Tian CG, Pan QJ, Jiang Z, Wang JQ, Yan WS et al. Enhanced photocatalytic activity and electron transfer mechanisms of graphene/TiO_2 with exposed (001) facets. Journal of Physical Chemistry C. 2011; 115: 23718–25.

[59] Wang P, Zhai YM, Wang DJ, Dong SJ. Synthesis of reduced graphene oxide-anatase TiO_2 nanocomposite and its improved photo-induced charge transfer properties. Nanoscale. 2011; 3: 1640–5.

[60] Surnev S, Ramsey MG, Netzer FP. Vanadium oxide surface studies. Progress in Surface Science. 2003; 73: 117–65.

[61] Liu JF, Wang X, Peng Q, Li YD. Vanadium pentoxide nanobelts: Highly selective and stable ethanol sensor materials. Advanced Materials. 2005; 17: 764–7.

[62] Yu HY, Kang BH, Pi UH, Park CW, Choi SY, Kim GT. V_2O_5 nanowire-based nanoelectronic devices for helium detection. Applied Physics Letters. 2005; 86: 253102.

[63] Serier H, Achard MF, Babot O, Steunou N, Maquet J, Livage J et al. Designing the width and texture of vanadium oxide macroscopic fibers: Towards tuning mechanical properties and alcohol-sensing performance. Advanced Functional Materials. 2006; 16: 1745–53.

[64] Gu G, Schmid M, Chiu PW, Minett A, Fraysse J, Kim GT et al. V_2O_5 nanofibre sheet actuators. Nature Materials. 2003; 2: 316–9.

[65] Krusin-Elbaum L, Newns DM, Zeng H, Derycke V, Sun JZ, Sandstrom R. Room-temperature ferromagnetic nanotubes controlled by electron or hole doping. Nature. 2004; 431: 672–6.

[66] Ponzi M, Duschatzky C, Carrascull A, Ponzi E. Obtaining benzaldehyde via promoted V_2O_5 catalysts. Applied Catalysis A-General. 1998; 169: 373–9.

[67] Liu P, Lee SH, Tracy CE, Yan YF, Turner JA. Preparation and lithium insertion properties of mesoporous vanadium oxide. Advanced Materials. 2002; 14: 27–30.

[68] Takahashi K, Limmer SJ, Wang Y, Cao GZ. Synthesis and electrochemical properties of single-crystal V_2O_5 nanorod arrays by template-based electrodeposition. Journal of Physical Chemistry B. 2004; 108: 9795–800.

[69] Cao AM, Hu JS, Liang HP, Wan LJ. Self-assembled vanadium pentoxide (V_2O_5) hollow microspheres from nanorods and their application in lithium-ion batteries. Angewandte Chemie-International Edition. 2005; 44: 4391–5.

[70] Rui XH, Zhu JX, Sim DH, Xu C, Zeng Y, Hng HH et al. Reduced graphene oxide supported highly porous V_2O_5 spheres as a high-power cathode material for lithium ion batteries. Nanoscale. 2011; 3: 4752–8.

[71] Cheng JL, Wang B, Xin HL, Yang GC, Cai HQ, Nie FD et al. Self-assembled V_2O_5 nanosheets/reduced graphene oxide hierarchical nanocomposite as a high-performance cathode material for lithium ion batteries. Journal of Materials Chemistry A. 2013; 1: 10814–20.

[72] Liu HM, Yang WS. Ultralong single crystalline V_2O_5 nanowire/graphene composite fabricated by a facile green approach and its lithium storage behavior. Energy & Environmental Science. 2011; 4: 4000–8.

[73] Wang ZL, Xu D, Huang Y, Wu ZY, Wang LM, Zhang XB. Facile, mild and fast thermal-decomposition reduction of graphene oxide in air and its application in high-performance lithium batteries. Chemical Communications. 2012; 48: 976–8.

[74] Kim HK, Kim YN, Kim KS, Jeong HY, Jang AR, Han SH et al. Flexible thermochromic window based on hybridized VO_2/graphene. ACS Nano. 2013; 7: 5769–76.

[75] Lee MS, Wee BH, Hong JD. High performance flexible supercapacitor electrodes composed of ultralarge graphene sheets and vanadium dioxide. Advanced Energy Materials. 2015; 5.

[76] Nethravathi C, Rajamathi CR, Rajamathi M, Gautam UK, Wang X, Golberg D et al. N-doped graphene-VO_2(B) nanosheet-nuilt 3D flower hybrid for lithium ion battery. ACS Applied Materials & Interfaces. 2013; 5: 2708–14.

[77] Nethravathi C, Viswanath B, Michael J, Rajamath M. Hydrothermal synthesis of a monoclinic VO_2 nanotube-graphene hybrid for use as cathode material in lithium ion batteries. Carbon. 2012; 50: 4839–46.

[78] Wang HW, Yi H, Chen X, Wang XF. One-step strategy to three-dimensional graphene/VO_2 nanobelt composite hydrogels for high performance supercapacitors. Journal of Materials Chemistry A. 2014; 2: 1165–73.

[79] Yang SB, Gong YJ, Liu Z, Zhan L, Hashim DP, Ma LL et al. Bottom-up approach toward single-crystalline VO_2-graphene ribbons as cathodes for ultrafast lithium storage. Nano Letters. 2013; 13: 1596–601.

[80] Li H, Balaya P, Maier J. Li-storage via heterogeneous reaction in selected binary metal fluorides and oxides. Journal of the Electrochemical Society. 2004; 151: A1878–A85.

[81] Grugeon S, Laruelle S, Dupont L, Chevallier F, Taberna PL, Simon P et al. Combining electrochemistry and metallurgy for new electrode designs in Li-ion batteries. Chemistry of Materials. 2005; 17: 5041–7.

[82] Hu J, Li H, Huang XJ, Chen LQ. Improve the electrochemical performances of Cr_2O_3 anode for lithium ion batteries. Solid State Ionics. 2006; 177: 2791–9.

[83] Yue WB, Tao SS, Fu JM, Gao ZQ, Ren YM. Carbon-coated graphene-Cr_2O_3 composites with enhanced electrochemical performances for Li-ion batteries. Carbon. 2013; 65: 97–104.

[84] Zhao GX, Wen T, Zhang J, Li JX, Dong HL, Wang XK et al. Two-dimensional Cr_2O_3 and interconnected graphene-Cr_2O_3 nanosheets: synthesis and their application in lithium storage. Journal of Materials Chemistry A. 2014; 2: 944–8.

[85] Wang F, Li WS, Hou MY, Li C, Wang YG, Xia YY. Sandwich-like Cr_2O_3-graphite intercalation composites as high-stability anode materials for lithium-ion batteries. Journal of Materials Chemistry A. 2015; 3: 1703–8.

[86] Jankovsky O, Sedmidubsky D, Simek P, Sofer Z, Ulbrich P, Bartunek V. Synthesis of MnO, Mn_2O_3 and Mn_3O_4 nanocrystal clusters by thermal decomposition of manganese glycerolate. Ceramics International. 2015; 41: 595–601.

[87] Chen ZW, Jiao Z, Pan DY, Li Z, Wu MH, Shek CH et al. Recent advances in manganese oxide nanocrystals: rabrication, characterization, and microstructure. Chemical Reviews. 2012; 112: 3833–55.

[88] Wei WF, Cui XW, Chen WX, Ivey DG. Manganese oxide-based materials as electrochemical supercapacitor electrodes. Chemical Society Reviews. 2011; 40: 1697–721.
[89] Xu CJ, Kang FY, Li BH, Du HD. Recent progress on manganese dioxide based supercapacitors. Journal of Materials Research. 2010; 25: 1421–32.
[90] Chen SQ, Zhu JW, Wu XD, Han QF, Wang X. Graphene oxide-MnO_2 nanocomposites for supercapacitors. ACS Nano. 2010; 4: 2822–30.
[91] Sumboja A, Foo CY, Wang X, Lee PS. Large areal mass, flexible and free-standing reduced graphene oxide/manganese dioxide paper for asymmetric supercapacitor device. Advanced Materials. 2013; 25: 2809–15.
[92] Zhu JY, He JH. Facile synthesis of graphene-wrapped honeycomb MnO_2 nanospheres and their application in supercapacitors. ACS Applied Materials & Interfaces. 2012; 4: 1770–6.
[93] Zhang YB, Su M, Ge L, Ge SG, Yu JH, Song XR. Synthesis and characterization of graphene nanosheets attached to spiky MnO_2 nanospheres and its application in ultrasensitive immunoassay. Carbon. 2013; 57: 22–33.
[94] Li ZP, Mi YJ, Liu XH, Liu S, Yang SR, Wang JQ. Flexible graphene/MnO_2 composite papers for supercapacitor electrodes. Journal of Materials Chemistry. 2011; 21: 14706–11.
[95] Sathish M, Mitani S, Tomai T, Honma I. MnO_2 assisted oxidative polymerization of aniline on graphene sheets: Superior nanocomposite electrodes for electrochemical supercapacitors. Journal of Materials Chemistry. 2011; 21: 16216–22.
[96] Li JX, Zhao Y, Wang N, Ding YH, Guan LH. Enhanced performance of a MnO_2-graphene sheet cathode for lithium ion batteries using sodium alginate as a binder. Journal of Materials Chemistry. 2012; 22: 13002–4.
[97] Chen H, Zhou SX, Chen ML, Wu LM. Reduced graphene Oxide-MnO_2 hollow sphere hybrid nanostructures as high-performance electrochemical capacitors. Journal of Materials Chemistry. 2012; 22: 25207–16.
[98] Lee MT, Fan CY, Wang YC, Li HY, Chang JK, Tseng CM. Improved supercapacitor performance of MnO_2-graphene composites constructed using a supercritical fluid and wrapped with an ionic liquid. Journal of Materials Chemistry A. 2013; 1: 3395–405.
[99] Zhang BW, Yu B, Zhou F, Liu WM. Polymer brush stabilized amorphous MnO_2 on graphene oxide sheets as novel electrode materials for high performance supercapacitors. Journal of Materials Chemistry A. 2013; 1: 8587–92.
[100] Yuan WY, Shen PK, Jiang SP. Controllable synthesis of graphene supported MnO_2 nanowires via self-assembly for enhanced water oxidation in both alkaline and neutral solutions. Journal of Materials Chemistry A. 2014; 2: 123–9.
[101] Shao YL, Wang HZ, Zhang QH, Li YG. High-performance flexible asymmetric supercapacitors based on 3D porous graphene/MnO_2 nanorod and graphene/Ag hybrid thin-film electrodes. Journal of Materials Chemistry C. 2013; 1: 1245–51.
[102] Kim MJ, Hwang YS, Kim JH. Super-capacitive performance depending on different crystal structures of MnO_2 in graphene/MnO_2 composites for supercapacitors. Journal of Materials Science. 2013; 48: 7652–63.
[103] Zhang JT, Jiang JW, Zhao XS. Synthesis and capacitive properties of manganese oxide nanosheets dispersed on functionalized graphene sheets. Journal of Physical Chemistry C. 2011; 115: 6448–54.
[104] Li Y, Zhao NQ, Shi CS, Liu EZ, He CN. Improve the supercapacity performance of MnO_2-decorated graphene by controlling the oxidization extent of graphene. Journal of Physical Chemistry C. 2012; 116: 25226–32.
[105] Yu AP, Park HW, Davies A, Higgins DC, Chen ZW, Xiao XC. Free-standing layer-by-layer hybrid thin film of γ-MnO_2 nanotube as anode for lithium ion batteries. Journal of Physical Chemistry Letters. 2011; 2: 1855–60.
[106] Deng LJ, Zhu G, Wang JF, Kang LP, Liu ZH, Yang ZP et al. Graphene-MnO_2 and graphene asymmetrical electrochemical capacitor with a high energy density in aqueous electrolyte. Journal of Power Sources. 2011; 196: 10782–7.
[107] Zhang GN, Ren LJ, Deng LJ, Wang JF, Kang LP, Liu ZH. Graphene-MnO_2 nanocomposite for high-performance asymmetrical electrochemical capacitor. Materials Research Bulletin. 2014; 49: 577–83.
[108] Xing LL, Cui CX, Ma CH, Xue XY. Facile synthesis of alpha-MnO_2/graphene nanocomposites and their high performance as lithium-ion battery anode. Materials Letters. 2011; 65: 2104–6.
[109] Peng LL, Peng X, Liu BR, Wu CZ, Xie Y, Yu GH. Ultrathin two-dimensional MnO_2/graphene hybrid nanostructures for high-performance, flexible planar supercapacitors. Nano Letters. 2013; 13: 2151–7.
[110] Huang HJ, Wang X. Graphene nanoplate-MnO_2 composites for supercapacitors: a controllable oxidation approach. Nanoscale. 2011; 3: 3185–91.
[111] Li LM, Du ZF, Liu S, Hao QY, Wang YG, Li QH et al. A novel nonenzymatic hydrogen peroxide sensor based on MnO_2/graphene oxide nanocomposite. Talanta. 2010; 82: 1637–41.
[112] Li LJ, Raji ARO, Tour JM. Graphene-wrapped MnO_2-graphene nanoribbons as anode materials for high-performance lithium ion batteries. Advanced Materials. 2013; 25: 6298–302.
[113] Dong XC, Wang XW, Wang L, Song H, Li XG, Wang LH et al. Synthesis of a MnO_2-graphene foam hybrid with controlled MnO_2 particle shape and its use as a supercapacitor electrode. Carbon. 2012; 50: 4865–70.
[114] Yang WL, Gao ZQ, Wang J, Wang B, Liu Q, Li ZS et al. Synthesis of reduced graphene nanosheet/urchin-like manganese dioxide composite and high performance as supercapacitor electrode. Electrochimica Acta. 2012; 69: 112–9.
[115] Cao Y, Wei ZK, He J, Zang J, Zhang Q, Zheng MS et al. α-MnO_2 nanorods grown *in situ* on graphene as catalysts for Li-O_2 batteries with excellent electrochemical performance. Energy & Environmental Science. 2012; 5: 9765–8.

[116] Nie RF, Shi JJ, Xia SX, Shen L, Chen P, Hou ZX et al. MnO_2/graphene oxide: a highly active catalyst for amide synthesis from alcohols and ammonia in aqueous media. Journal of Materials Chemistry. 2012; 22: 18115–8.

[117] Yu Y, Zhang BL, He YB, Huang ZD, Oh SW, Kim JK. Mechanisms of capacity degradation in reduced graphene oxide/α-MnO_2 nanorod composite cathodes of Li-air batteries. Journal of Materials Chemistry A. 2013; 1: 1163–70.

[118] Feng XM, Yan ZZ, Chen NN, Zhang YB, Ma YW, Liu XF et al. The synthesis of shape-controlled MnO_2/graphene composites via a facile one-step hydrothermal method and their application in supercapacitors. Journal of Materials Chemistry A. 2013; 1: 12818–25.

[119] Sui ZY, Wang CY, Shu KW, Yang QS, Ge Y, Wallace GG et al. Manganese dioxide-anchored three-dimensional nitrogen-doped graphene hybrid aerogels as excellent anode materials for lithium ion batteries. Journal of Materials Chemistry A. 2015; 3: 10403–12.

[120] Li ZP, Wang JP, Liu S, Liu XH, Yang SR. Synthesis of hydrothermally reduced graphene/MnO_2 composites and their electrochemical properties as supercapacitors. Journal of Power Sources. 2011; 196: 8160–5.

[121] Zhai T, Wang FX, Yu MH, Xie SL, Liang CL, Li C et al. 3D MnO_2–graphene composites with large areal capacitance for high-performance asymmetric supercapacitors. Nanoscale. 2013; 5: 6790–6.

[122] Kalubarme RS, Ahn CH, Park CJ. Electrochemical characteristics of graphene/manganese oxide composite catalyst for Li-oxygen rechargeable batteries. Scripta Materialia. 2013; 68: 619–22.

[123] He YM, Chen WJ, Li XD, Zhang ZX, Fu JC, Zhao CH et al. Freestanding three-dimensional graphene/MnO_2 composite networks as ultra light and flexible supercapacitor electrodes. ACS Nano. 2013; 7: 174–82.

[124] Xiao F, Li YQ, Zan XL, Liao K, Xu R, Duan HW. Growth of metal-metal oxide nanostructures on freestanding graphene paper for flexible biosensors. Advanced Functional Materials. 2012; 22: 2487–94.

[125] Gao HC, Xiao F, Ching CB, Duan HW. High-performance asymmetric supercapacitor based on graphene hydrogel and nanostructured MnO_2. ACS Applied Materials & Interfaces. 2012; 4: 2801–10.

[126] Cheng Q, Tang J, Ma J, Zhang H, Shinya N, Qin LC. Graphene and nanostructured MnO_2 composite electrodes for supercapacitors. Carbon. 2011; 49: 2917–25.

[127] Lee HS, Kang JM, Cho MS, Choi JB, Lee YK. MnO_2/graphene composite electrodes for supercapacitors: the effect of graphene intercalation on capacitance. Journal of Materials Chemistry. 2011; 21: 18215–9.

[128] Fan ZJ, Yan J, Wei T, Zhi LJ, Ning GQ, Li TY et al. Asymmetric supercapacitors based on graphene/MnO_2 and activated carbon nanofiber electrodes with high power and energy density. Advanced Functional Materials. 2011; 21: 2366–75.

[129] Yan J, Fan ZJ, Wei T, Qian WZ, Zhang ML, Wei F. Fast and reversible surface redox reaction of graphene-MnO_2 composites as supercapacitor electrodes. Carbon. 2010; 48: 3825–33.

[130] Zhang JT, Guo CX, Zhang LY, Li CM. Direct growth of flower-like manganese oxide on reduced graphene oxide towards efficient oxygen reduction reaction. Chemical Communications. 2013; 49: 6334–6.

[131] Duan JJ, Chen SQ, Dai S, Qiao SZ. Shape control of Mn_3O_4 nanoparticles on nitrogen-doped graphene for enhanced oxygen reduction activity. Advanced Functional Materials. 2014; 24: 2072–8.

[132] Li N, Geng ZF, Cao MH, Ren LJ, Zhao XY, Liu B et al. Well-dispersed ultrafine Mn_3O_4 nanoparticles on graphene as a promising catalyst for the thermal decomposition of ammonium perchlorate. Carbon. 2013; 54: 124–32.

[133] Duan JJ, Zheng Y, Chen SQ, Tang YH, Jaroniec M, Qiao SZ. Mesoporous hybrid material composed of Mn_3O_4 nanoparticles on nitrogen-doped graphene for highly efficient oxygen reduction reaction. Chemical Communications. 2013; 49: 7705–7.

[134] Lee JW, Hall AS, Kim JD, Mallouk TE. A facile and template-free hydrothermal synthesis of Mn_3O_4 nanorods on graphene sheets for supercapacitor electrodes with long cycle stability. Chemistry of Materials. 2012; 24: 1158–64.

[135] Lee JS, Lee TM, Song HK, Cho J, Kim BS. Ionic liquid modified graphene nanosheets anchoring manganese oxide nanoparticles as efficient electrocatalysts for Zn-air batteries. Energy & Environmental Science. 2011; 4: 4148–54.

[136] Wang L, Li YH, Han ZD, Chen L, Qian B, Jiang XF et al. Composite structure and properties of Mn_3O_4/graphene oxide and Mn_3O_4/graphene. Journal of Materials Chemistry A. 2013; 1: 8385–97.

[137] Yang SH, Song XF, Zhang P, Gao LB. Crumpled nitrogen-doped graphene-ultrafine Mn_3O_4 nanohybrids and their application in supercapacitors. Journal of Materials Chemistry A. 2013; 1: 14162–9.

[138] Kim HY, Kim SW, Hong JH, Park YY, Kang KS. Electrochemical and *ex-situ* analysis on manganese oxide/graphene hybrid anode for lithium rechargeable batteries. Journal of Materials Research. 2011; 26: 2665–71.

[139] Zhang X, Sun XH, Chen Y, Zhang DC, Ma YW. One-step solvothermal synthesis of graphene/Mn_3O_4 nanocomposites and their electrochemical properties for supercapacitors. Materials Letters. 2012; 68: 336–9.

[140] Fan YF, Zhang XD, Liu YS, Cai Q, Zhang JM. One-pot hydrothermal synthesis of Mn_3O_4/graphene nanocomposite for supercapacitors. Materials Letters. 2013; 95: 153–6.

[141] Qu JY, Gao F, Zhou Q, Wang ZY, Hu H, Li BB et al. Highly atom-economic synthesis of graphene/Mn_3O_4 hybrid composites for electrochemical supercapacitors. Nanoscale. 2013; 5: 2999–3005.

[142] Kwon OS, Kim TJ, Lee JS, Park SJ, Park HW, Kang MJ et al. Fabrication of graphene sheets intercalated with manganese oxide/carbon nanofibers: Toward high-capacity energy storage. Small. 2013; 9: 248–54.

[143] Seredych M, Bandosz TJ. Evaluation of GO/MnO_2 composites as supercapacitors in neutral electrolytes: role of graphite oxide oxidation level. Journal of Materials Chemistry. 2012; 22: 23525–33.

[144] Su YZ, Li S, Wu DQ, Zhang FB, Liang HW, Gao PF et al. Two-dimensional carbon-coated graphene/metal oxide hybrids for enhanced lithium storage. ACS Nano. 2012; 6: 8349–56.

[145] Villegas JC, Garces LJ, Gomez S, Durand JP, Suib SL. Particle size control of cryptomelane nanomaterials by use of H_2O_2 in acidic conditions. Chemistry of Materials. 2005; 17: 1910–8.
[146] Portehault D, Cassaignon S, Baudrin E, Jolivet JP. Morphology control of cryptomelane type MnO_2 nanowires by soft chemistry. Growth mechanisms in aqueous medium. Chemistry of Materials. 2007; 19: 5410–7.
[147] Xu C, Wang X, Zhu JW, Yang XJ, Lu LD. Deposition of Co_3O_4 nanoparticles onto exfoliated graphite oxide sheets. Journal of Materials Chemistry. 2008; 18: 5625–9.
[148] Devaraj S, Munichandraiah N. Effect of crystallographic structure of MnO_2 on its electrochemical capacitance properties. Journal of Physical Chemistry C. 2008; 112: 4406–17.
[149] Chen S, Zhu JW, Han QF, Zheng ZJ, Yang YG, Wang X. Shape-controlled synthesis of one-dimensional MnO_2 via a facile quick-precipitation procedure and its electrochemical properties. Crystal Growth & Design. 2009; 9: 4356–61.
[150] Kamlet MJ, Abboud JLM, Abraham MH, Taft RW. Linear solvation energy relationships, 23: A comprehensive collection of the solvatochromic parameters, π^*, α and β and some methods for simplifying the generalized solvatochromic equation. Journal of Organic Chemistry. 1983; 48: 2877–87.
[151] Yang SB, Feng XL, Ivanovici S, Muellen K. Fabrication of graphene-encapsulated oxide nanoparticles: Towards high-performance anode materials for lithium storage. Angewandte Chemie-International Edition. 2010; 49: 8408–11.
[152] Fan ZJ, Kai W, Yan J, Wei T, Zhi LJ, Feng J et al. Facile synthesis of graphene nanosheets via Fe reduction of exfoliated graphite oxide. Acs Nano. 2011; 5: 191–8.
[153] Liu JW, Zhang Q, Chen XW, Wang JH. Surface assembly of graphene oxide nanosheets on SiO_2 particles for the selective isolation of hemoglobin. Chemistry-A European Journal. 2011; 17: 4864–70.
[154] Guo CX, Wang M, Chen T, Lou XW, Li CM. A hierarchically nanostructured composite of MnO_2/conjugated polymer/graphene for high-performance lithium ion batteries. Advanced Energy Materials. 2011; 1: 736–41.
[155] Dong XC, Li B, Wei A, Cao XH, Chan-Park MB, Zhang H et al. One-step growth of graphene-carbon nanotube hybrid materials by chemical vapor deposition. Carbon. 2011; 49: 2944–9.
[156] Dong XC, Wang P, Fang WJ, Su CY, Chen YH, Li LJ et al. Growth of large-sized graphene thin-films by liquid precursor-based chemical vapor deposition under atmospheric pressure. Carbon. 2011; 49: 3672–8.
[157] Maiyalagan T, Dong XC, Chen P, Wang X. Electrodeposited Pt on three-dimensional interconnected graphene as a free-standing electrode for fuel cell application. Journal of Materials Chemistry. 2012; 22: 5286–90.
[158] Graf D, Molitor F, Ensslin K, Stampfer C, Jungen A, Hierold C et al. Spatially resolved raman spectroscopy of single- and few-layer graphene. Nano Letters. 2007; 7: 238–42.
[159] Li BX, Rong GX, Xie Y, Huang LF, Feng CQ. Low-temperature synthesis of α-MnO_2 hollow urchins and their application in rechargeable Li^+ batteries. Inorganic Chemistry. 2006; 45: 6404–10.
[160] Zhou M, Zhang X, Wei JM, Zhao SL, Wang L, Feng BX. Morphology-controlled synthesis and novel microwave absorption properties of hollow urchinlike α-MnO_2 nanostructures. Journal of Physical Chemistry C. 2011; 115: 1398–402.
[161] Hill JP, Alam S, Ariga K, Anson CE, Powell AK. Nanostructured microspheres of MnO_2 formed by room temperature solution processing. Chemical Communications. 2008: 383–5.
[162] Mao L, Zhang K, Chan HSO, Wu JS. Nanostructured MnO_2/graphene composites for supercapacitor electrodes: the effect of morphology, crystallinity and composition. Journal of Materials Chemistry. 2012; 22: 1845–51.
[163] Li XS, Zhu YW, Cai WW, Borysiak M, Han BY, Chen D et al. Transfer of large-area graphene films for high-performance transparent conductive electrodes. Nano Letters. 2009; 9: 4359–63.
[164] Xiao F, Song JB, Gao HC, Zan XL, Xu R, Duan HW. Coating graphene paper with 2D-assembly of electrocatalytic nanoparticles: A modular approach toward high-performance flexible electrodes. ACS Nano. 2012; 6: 100–10.
[165] Deng DH, Pan XL, Yu L, Cui Y, Jiang YP, Qi J et al. Toward N-doped graphene via solvothermal synthesis. Chemistry of Materials. 2011; 23: 1188–93.
[166] Biswal M, Dhas VV, Mate VR, Banerjee A, Pachfule P, Agrawal KL et al. Selectivity tailoring in liquid phase oxidation over MWNT-Mn_3O_4 nanocomposite catalysts. Journal of Physical Chemistry C. 2011; 115: 15440–8.
[167] Chen YT, Guo F, Qiu YC, Hu H, Kulaots I, Walsh E et al. Encapsulation of particle ensembles in graphene nanosacks as a new route to multifunctional materials. ACS Nano. 2013; 7: 3744–53.
[168] Liang JJ, Huang Y, Oh JY, Kozlov M, Sui D, Fang SL et al. Electromechanical actuators based on graphene and graphene/Fe_3O_4 hybrid paper. Advanced Functional Materials. 2011; 21: 3778–84.
[169] Sohn KN, Na YJ, Chang HK, Roh KM, Jang HD, Huang JX. Oil absorbing graphene capsules by capillary molding. Chemical Communications. 2012; 48: 5968–70.
[170] Liang JJ, Xu YF, Sui D, Zhang L, Huang, Ma YF et al. Flexible, magnetic, and electrically conductive graphene/Fe_3O_4 paper and its application For magnetic-controlled switches. Journal of Physical Chemistry C. 2010; 114: 17465–71.
[171] Chandra V, Park JS, Chun Y, Lee JW, Hwang IC, Kim KS. Water-dispersible magnetite-reduced graphene oxide composites for arsenic removal. ACS Nano. 2010; 4: 3979–86.
[172] Cong HP, Ren XC, Wang P, Yu SH. Macroscopic multifunctional graphene-based hydrogels and aerogels by a metal ion induced self-assembly process. ACS Nano. 2012; 6: 2693–703.
[173] Qu QT, Yang SB, Feng XL. 2D sandwich-like sheets of iron oxide grown on graphene as high energy anode material for supercapacitors. Advanced Materials. 2011; 23: 5574–80.

[174] Chen WF, Li SR, Chen CH, Yan LF. Self-assembly and embedding of nanoparticles by *in situ* reduced graphene for preparation of a 3D graphene/nanoparticle aerogel. Advanced Materials. 2011; 23: 5679–83.
[175] He HK, Gao C. Supraparamagnetic, conductive, and processable multifunctional graphene nanosheets coated with high-density Fe_3O_4 nanoparticles. ACS Applied Materials & Interfaces. 2010; 2: 3201–10.
[176] Li J, Zhang SW, Chen CL, Zhao GX, Yang XK, Li JX et al. Removal of Cu(II) and fulvic acid by graphene oxide nanosheets decorated with Fe_3O_4 nanoparticles. ACS Applied Materials & Interfaces. 2012; 4: 4991–5000.
[177] Shen B, Zhai WT, Tao MM, Ling JQ, Zheng WG. Lightweight, multifunctional polyetherimide/graphene@Fe_3O_4 composite foams for shielding of electromagnetic pollution. ACS Applied Materials & Interfaces. 2013; 5: 11383–91.
[178] Guan PF, Zhang XF, Guo J. Assembled Fe_3O_4 nanoparticles on graphene for enhanced electromagnetic wave losses. Applied Physics Letters. 2012; 101: 153108.
[179] Li Y, Chu J, Qi JY, Li X. An easy and novel approach for the decoration of graphene oxide by Fe_3O_4 nanoparticles. Applied Surface Science. 2011; 257: 6059–62.
[180] He FA, Fan JT, Ma D, Zhang LM, Leung CW, Chan HL. The attachment of Fe_3O_4 nanoparticles to graphene oxide by covalent bonding. Carbon. 2010; 48: 3139–44.
[181] Narayanan TN, Liu Z, Lakshmy PR, Gao W, Nagaoka Y, Kumar DS et al. Synthesis of reduced graphene oxide-Fe_3O_4 multifunctional freestanding membranes and their temperature dependent electronic transport properties. Carbon. 2012; 50: 1338–45.
[182] Ma YX, Li YF, Zhao GH, Yang LQ, Wang JZ, Shan X et al. Preparation and characterization of graphite nanosheets decorated with Fe_3O_4 nanoparticles used in the immobilization of glucoamylase. Carbon. 2012; 50: 2976–86.
[183] Prakash A, Chandra S, Bahadur D. Structural, magnetic, and textural properties of iron oxide-reduced graphene oxide hybrids and their use for the electrochemical detection of chromium. Carbon. 2012; 50: 4209–19.
[184] Behera SK. Enhanced rate performance and cyclic stability of Fe_3O_4-graphene nanocomposites for Li ion battery anodes. Chemical Communications. 2011; 47: 10371–3.
[185] Dong YC, Ma RG, Hu MJ, Cheng HH, Yang QD, Li YY et al. Thermal evaporation-induced anhydrous synthesis of Fe_3O_4-graphene composite with enhanced rate performance and cyclic stability for lithium ion batteries. Physical Chemistry Chemical Physics. 2013; 15: 7174–81.
[186] Xue YH, Chen H, Yu DS, Wang SY, Yardeni M, Dai QB et al. Oxidizing metal ions with graphene oxide: the *in situ* formation of magnetic nanoparticles on self-reduced graphene sheets for multifunctional applications. Chemical Communications. 2011; 47: 11689–91.
[187] Bhunia P, Kim GY, Baik CW, Lee HY. A strategically designed porous iron-iron oxide matrix on graphene for heavy metal adsorption. Chemical Communications. 2012; 48: 9888–90.
[188] Zhou GM, Wang DW, Li F, Zhang LL, Li N, Wu ZS et al. Graphene-wrapped Fe_3O_4 anode material with improved reversible capacity and cyclic stability for lithium ion batteries. Chemistry of Materials. 2010; 22: 5306–13.
[189] Zhang M, Jia MQ. High rate capability and long cycle stability Fe_3O_4-graphene nanocomposite as anode material for lithium ion batteries. Journal of Alloys and Compounds. 2013; 551: 53–60.
[190] Wang TS, Liu ZH, Lu MM, Wen B, Ouyang QY, Chen YJ et al. Graphene-Fe_3O_4 nanohybrids: Synthesis and excellent electromagnetic absorption properties. Journal of Applied Physics. 2013; 113.
[191] Yang XY, Zhang XY, Ma YF, Huang Y, Wang YS, Chen YS. Superparamagnetic graphene oxide-Fe_3O_4 nanoparticles hybrid for controlled targeted drug carriers. Journal of Materials Chemistry. 2009; 19: 2710–4.
[192] Zhang M, Lei DN, Yin XM, Chen LB, Li QH, Wang YG et al. Magnetite/graphene composites: microwave irradiation synthesis and enhanced cycling and rate performances for lithium ion batteries. Journal of Materials Chemistry. 2010; 20: 5538–43.
[193] Shi WH, Zhu JX, Sim DH, Tay YY, Lu ZY, Zhang XJ et al. Achieving high specific charge capacitances in Fe_3O_4/reduced graphene oxide nanocomposites. Journal of Materials Chemistry. 2011; 21: 3422–7.
[194] Ji JY, Zhang GH, Chen HY, Li Y, Zhang GL, Zhang FB et al. A general strategy to prepare graphene-metal/metal oxide nanohybrids. Journal of Materials Chemistry. 2011; 21: 14498–501.
[195] Xie GQ, Xi PX, Liu HY, Chen FJ, Huang L, Shi YJ et al. A facile chemical method to produce superparamagnetic graphene oxide-Fe_3O_4 hybrid composite and its application in the removal of dyes from aqueous solution. Journal of Materials Chemistry. 2012; 22: 1033–9.
[196] Geng ZG, Lin YH, Yu XX, Shen QH, Ma L, Li ZY et al. Highly efficient dye adsorption and removal: a functional hybrid of reduced graphene oxide-Fe_3O_4 nanoparticles as an easily regenerative adsorbent. Journal of Materials Chemistry. 2012; 22: 3527–35.
[197] Huo X, Liu J, Wang BD, Zhang HL, Yang ZY, She XG et al. A one-step method to produce graphene-Fe_3O_4 composites and their excellent catalytic activities for three-component coupling of aldehyde, alkyne and amine. Journal of Materials Chemistry A. 2013; 1: 651–6.
[198] Zhang WJ, Shi XH, Zhang YX, Gu W, Li BY, Xian YZ. Synthesis of water-soluble magnetic graphene nanocomposites for recyclable removal of heavy metal ions. Journal of Materials Chemistry A. 2013; 1: 1745–53.
[199] Wang RH, Xu CH, Sun J, Gao LB, Lin CC. Flexible free-standing hollow Fe_3O_4/graphene hybrid films for lithium-ion batteries. Journal of Materials Chemistry A. 2013; 1: 1794–800.
[200] Fan XJ, Jiao GZ, Gao LB, Jin PF, Li X. The preparation and drug delivery of a graphene-carbon nanotube-Fe_3O_4 nanoparticle hybrid. Journal of Materials Chemistry B. 2013; 1: 2658–64.

[201] Zhuo LH, Wu YQ, Wang LY, Ming J, Yu YC, Zhang XB et al. CO_2-expanded ethanol chemical synthesis of a Fe_3O_4@graphene composite and its good electrochemical properties as anode material for Li-ion batteries. Journal of Materials Chemistry A. 2013; 1: 3954–60.
[202] Chen DZ, Wang GS, He SJ, Liu J, Guo L, Cao MS. Controllable fabrication of mono-dispersed RGO-hematite nanocomposites and their enhanced wave absorption properties. Journal of Materials Chemistry A. 2013; 1: 5996–6003.
[203] Wang LN, Jia XL, Li YF, Yang F, Zhang LQ, Liu LP et al. Synthesis and microwave absorption property of flexible magnetic film based on graphene oxide/carbon nanotubes and Fe_3O_4 nanoparticles. Journal of Materials Chemistry A. 2014; 2: 14940–6.
[204] Yu HT, Li YC, Li XH, Fan LZ, Yang SH. Highly dispersible and charge-tunable magnetic Fe_3O_4 nanoparticles: facile fabrication and reversible binding to GO for efficient removal of dye pollutants. Journal of Materials Chemistry A. 2014; 2: 15763–7.
[205] Lu XY, Wang RH, Bai Y, Chen JJ, Sun J. Facile preparation of a three-dimensional Fe_3O_4/macroporous graphene composite for high-performance Li storage. Journal of Materials Chemistry A. 2015; 3: 12031–7.
[206] Shen JF, Hu YZ, Shi M, Li N, Ma HW, Ye MX. One step synthesis of graphene oxide-magnetic nanoparticle composite. Journal of Physical Chemistry C. 2010; 114: 1498–503.
[207] Bai LZ, Zhao DL, Xu Y, Zhang JM, Gao YL, Zhao LY et al. Inductive heating property of graphene oxide-Fe_3O_4 nanoparticles hybrid in an AC magnetic field for localized hyperthermia. Materials Letters. 2012; 68: 399–401.
[208] Ma EL, Li JJ, Zhao NQ, Liu EZ, He CN, Shi CS. Preparation of reduced graphene oxide/Fe_3O_4 nanocomposite and its microwave electromagnetic properties. Materials Letters. 2013; 91: 209–12.
[209] Hu AP, Chen XH, Tang YH, Yang L, Xiao HH, Fan BB. A facile method to synthesize Fe_3O_4/graphene composites in normal pressure with high rate capacity and cycling stability. Materials Letters. 2013; 91: 315–8.
[210] Luo JS, Liu JL, Zeng ZY, Ng CF, Ma LJ, Zhang H et al. Three-dimensional graphene foam supported Fe_3O_4 lithium battery anodes with long cycle life and high rate capability. Nano Letters. 2013; 13: 6136–43.
[211] Zhang YB, Chen B, Zhang LM, Huang J, Chen FH, Yang ZP et al. Controlled assembly of Fe_3O_4 magnetic nanoparticles on graphene oxide. Nanoscale. 2011; 3: 1446–50.
[212] Singh K, Ohlan A, Viet Hung P, Balasubramaniyan R, Varshney S, Jang JH et al. Nanostructured graphene/Fe_3O_4 incorporated polyaniline as a high performance shield against electromagnetic pollution. Nanoscale. 2013; 5: 2411–20.
[213] Dong YC, Ma RG, Hu MJ, Cheng H, Yang QD, Li YY et al. Thermal evaporation-induced anhydrous synthesis of Fe_3O_4 graphene composite with enhanced rate performance and cyclic stability for lithium ion batteries. Physical Chemistry Chemical Physics. 2013; 15: 7174–81.
[214] Cong HP, He JJ, Lu Y, Yu SH. Water-soluble magnetic-functionalized reduced graphene oxide sheets: *In situ* synthesis and magnetic resonance imaging applications. Small. 2010; 6: 169–73.
[215] Yu SH, Conte DE, Baek SH, Lee DC, Park SK, Lee KJ et al. Structure-properties relationship in iron oxide-reduced graphene oxide nanostructures for Li-ion batteries. Advanced Functional Materials. 2013; 23: 4293–305.
[216] Wei W, Yang SB, Zhou HX, Lieberwirth I, Feng XL, Muellen K. 3D graphene foams cross-linked with pre-encapsulated Fe_3O_4 nanospheres for enhanced lithium storage. Advanced Materials. 2013; 25: 2909–14.
[217] Ren LL, Huang S, Fan WQ, Liu TX. One-step preparation of hierarchical superparamagnetic iron oxide/graphene composites via hydrothermal method. Applied Surface Science. 2011; 258: 1132–8.
[218] Lin SX, Shen CM, Lu DB, Wang CM, Gao HJ. Synthesis of Pt nanoparticles anchored on graphene-encapsulated Fe_3O_4 magnetic nanospheres and their use as catalysts for methanol oxidation. Carbon. 2013; 53: 112–9.
[219] He FA, Lam KH, Ma D, Fan JT, Chan LH, Zhang LM. Fabrication of graphene nanosheet (GNS)-Fe_3O_4 hybrids and GNS-Fe_3O_4/syndiotactic polystyrene composites with high dielectric permittivity. Carbon. 2013; 58: 175–84.
[220] Dong YC, Yung KC, Ma RG, Yang XK, Chui YS, Lee JM et al. Graphene/acid assisted facile synthesis of structure-tuned Fe_3O_4 and graphene composites as anode materials for lithium ion batteries. Carbon. 2015; 86: 310–7.
[221] Li BJ, Cao HQ, Shao J, Qu MZ. Enhanced anode performances of the Fe_3O_4-carbon-rGO three dimensional composite in lithium ion batteries. Chemical Communications. 2011; 47: 10374–6.
[222] Wu ZS, Yang SB, Sun YP, Parvez K, Feng XL, Muellen K. 3D nitrogen-doped graphene aerogel-supported Fe_3O_4 nanoparticles as efficient eletrocatalysts for the oxygen reduction reaction. Journal of the American Chemical Society. 2012; 134: 9082–5.
[223] Ai LH, Zhang CY, Chen ZL. Removal of methylene blue from aqueous solution by a solvothermal-synthesized graphene/magnetite composite. Journal of Hazardous Materials. 2011; 192: 1515–24.
[224] Chen Y, Song BH, Tang XS, Lu L, Xue JM. One-step synthesis of hollow porous Fe_3O_4 beads-reduced graphene oxide composites with superior battery performance. Journal of Materials Chemistry. 2012; 22: 17656–62.
[225] Zeng T, Zhang XL, Ma YR, Niu HY, Cai YQ. A novel Fe_3O_4-graphene-Au multifunctional nanocomposite: green synthesis and catalytic application. Journal of Materials Chemistry. 2012; 22: 18658–63.
[226] Cheng G, Liu YL, Wang ZG, Zhang JL, Sun DH, Ni JZ. The GO/rGO-Fe_3O_4 composites with good water-dispersibility and fast magnetic response for effective immobilization and enrichment of biomolecules. Journal of Materials Chemistry. 2012; 22: 21998–2004.
[227] Fan W, Gao W, Zhang C, Tjiu WW, Pan JS, Liu TX. Hybridization of graphene sheets and carbon-coated Fe_3O_4 nanoparticles as a synergistic adsorbent of organic dyes. Journal of Materials Chemistry. 2012; 22: 25108–15.

[228] Sun X, He JP, Li GX, Tang J, Wang T, Guo YX et al. Laminated magnetic graphene with enhanced electromagnetic wave absorption properties. Journal of Materials Chemistry C. 2013; 1: 765–77.

[229] Liu DQ, Wang X, Wang XB, Tian W, Liu JW, Zhi CY et al. Ultrathin nanoporous Fe_3O_4-carbon nanosheets with enhanced supercapacitor performance. Journal of Materials Chemistry A. 2013; 1: 1952–5.

[230] Chang YH, Li J, Wang B, Luo H, He HY, Song QJ et al. Synthesis of 3D nitrogen-doped graphene/Fe_3O_4 by a metal ion induced self-assembly process for high-performance Li-ion batteries. Journal of Materials Chemistry A. 2013; 1: 14658–65.

[231] Jiang YB, Jiang ZJ, Yang LF, Cheng S, Liu ML. A high-performance anode for lithium ion batteries: Fe_3O_4 microspheres encapsulated in hollow graphene shells. Journal of Materials Chemistry A. 2015; 3: 11847–56.

[232] Meng XF, L. XY, Sun XF, Wang J, L. XL, Du XF et al. Graphene oxide sheets-induced growth of nanostructured Fe_3O_4 for a high-performance anode material of lithium ion batteries. Journal of Materials Chemistry A. 2015; 3: 12938–46.

[233] Su J, Cao MH, Ren L, Hu CW. Fe_3O_4-graphene nanocomposites with improved lithium storage and magnetism properties. Journal of Physical Chemistry C. 2011; 115: 14469–77.

[234] Li XY, Huang XL, Liu DP, Wang X, Song SYl, Zhou L et al. Synthesis of 3D hierarchical Fe_3O_4/graphene composites with high lithium storage capacity and for controlled drug delivery. Journal of Physical Chemistry C. 2011; 115: 21567–73.

[235] Hu AP, Chen XH, Tang QL, Liu Z, Zeng B. One-step synthesis of Fe_3O_4@C/reduced-graphite oxide nanocomposites for high-performance lithium ion batteries. Journal of Physics and Chemistry of Solids. 2014; 75: 588–93.

[236] Wang GC, Liu T, Xie XL, Ren ZY, Bai JB, Wang H. Structure and electrochemical performance of Fe_3O_4/graphene nanocomposite as anode material for lithium-ion batteries. Materials Chemistry and Physics. 2011; 128: 336–40.

[237] Xue WD, Zhao R, Du X, Xu FW, Xu M, Wei KX. Graphene-Fe_3O_4 micro-nano scaled hybrid spheres: Synthesis and synergistic electromagnetic effect. Materials Research Bulletin. 2014; 50: 285–91.

[238] Zhou KF, Zhu YH, Yang X, Li C. One-pot preparation of graphene/Fe_3O_4 composites by a solvothermal reaction. New Journal of Chemistry. 2010; 34: 2950–5.

[239] Chen Y, Song BH, Lu L, Xue JM. Ultra-small Fe_3O_4 nanoparticle decorated graphene nanosheets with superior cyclic performance and rate capability. Nanoscale. 2013; 5: 6797–803.

[240] Liu YW, Guan MX, Feng LL, Deng SL, Bao JF, Xie SY et al. Facile and straightforward synthesis of superparamagnetic reduced graphene oxide-Fe_3O_4 hybrid composite by a solvothermal reaction. Nanotechnology. 2013; 24.

[241] Wate PS, Banerjee SS, Jalota-Badhwar A, Mascarenhas RR, Zope KR, Khandare J et al. Cellular imaging using biocompatible dendrimer-functionalized graphene oxide-based fluorescent probe anchored with magnetic nanoparticles. Nanotechnology. 2012; 23.

[242] Zhu XJ, Zhu YW, Murali S, Stollers MD, Ruoff RS. Nanostructured reduced graphene oxide/Fe_2O_3 composite as a high-performance anode material for lithium ion batteries. ACS Nano. 2011; 5: 3333–8.

[243] Cao XH, Zheng B, Rui XH, Shi WH, Yan QY, Zhang H. Metal oxide-coated three-dimensional graphene prepared by the use of metal-organic frameworks as precursors. Angewandte Chemie-International Edition. 2014; 53: 1404–9.

[244] Lin J, Raji ARO, Nan KW, Peng ZW, Yan Z, Samuel ELG et al. Iron oxide nanoparticle and graphene nanoribbon composite as an anode material for gigh-performance Li-ion batteries. Advanced Functional Materials. 2014; 24: 2044–8.

[245] Lee JY, Lee KH, Kim YJ, Ha JS, Lee SS, Son JG. Sea-urchin-inspired 3D crumpled graphene balls using simultaneous etching and reduction process for high-density capacitive energy storage. Advanced Functional Materials. 2015; 25: 3606–14.

[246] Zhu X, Song XY, Ma XL, Ning GQ. Enhanced electrode performance of Fe_2O_3 nanoparticle-decorated nanomesh graphene as anodes for lithium-ion batteries. ACS Applied Materials & Interfaces. 2014; 6: 7189–97.

[247] Zhu JH, Chen MJ, Qu HL, Luo ZP, Wu SJ, Colorado HA et al. Magnetic field induced capacitance enhancement in graphene and magnetic graphene nanocomposites. Energy & Environmental Science. 2013; 6: 194–204.

[248] Vadahanambi S, Lee SH, Kim WJ, Oh IK. Arsenic removal from contaminated water using three-dimensional graphene-carbon nanotube-iron oxide nanostructures. Environmental Science & Technology. 2013; 47: 10510–7.

[249] Xu C, Chen ZX, Fu XZ. Graphene oxide-mediated formation of freestanding, thickness controllable metal oxide films. Journal of Materials Chemistry. 2011; 21: 12889–93.

[250] Song HJ, Jia XH, Li N, Yang XF, Tang HQ. Synthesis of α-Fe_2O_3 nanorod/graphene oxide composites and their tribological properties. Journal of Materials Chemistry. 2012; 22: 895–902.

[251] Zhang WY, Zeng Y, Xiao N, Hng HH, Yan QY. One-step electrochemical preparation of graphene-based heterostructures for Li storage. Journal of Materials Chemistry. 2012; 22: 8455–61.

[252] Zhang H, Xie AJ, Wang CP, Wang HS, Shen YH, Tian XY. Novel rGO/α-Fe_2O_3 composite hydrogel: synthesis, characterization and high performance of electromagnetic wave absorption. Journal of Materials Chemistry A. 2013; 1: 8547–52.

[253] Jiang ZX, Li J, Aslan H, Li Q, Li Y, Chen ML et al. A high efficiency H_2S gas sensor material: paper like Fe_2O_3/graphene nanosheets and structural alignment dependency of device efficiency. Journal of Materials Chemistry A. 2014; 2: 6714–7.

[254] Xiao LS, Schroeder M, Kluge S, Balducci A, Hagemann U, Schulzad C et al. Direct self-assembly of Fe_2O_3/reduced graphene oxide nanocomposite for high-performance lithium-ion batteries. Journal of Materials Chemistry A. 2015; 3: 11566–74.

[255] Nandi D, Gupta K, Ghosh AK, De A, Banerjee S, Ghosh UC. Manganese-incorporated iron(III) oxide-graphene magnetic nanocomposite: synthesis, characterization, and application for the arsenic(III)-sorption from aqueous solution. Journal of Nanoparticle Research. 2012; 14: 1272.

[256] Mendes RG, Bachmatiuk A, El-Gendy AA, Melkhanova S, Klingeler R, Buechner B et al. A facile route to coat iron oxide nanoparticles with few-layer graphene. Journal of Physical Chemistry C. 2012; 116: 23749–56.

[257] Yang SB, Sun Y, Chen L, Hernandez Y, Feng XL, Muellen K. Porous iron oxide ribbons grown on graphene for high-performance lithium storage. Scientific Reports. 2012; 2: 427.

[258] Xiao L, Wu DQ, Han S, Huang YS, Li S, He MZ et al. Self-assembled Fe_2O_3/graphene aerogel with high lithium storage performance. ACS Applied Materials & Interfaces. 2013; 5: 3764–9.

[259] Pradhan GK, Padhi DK, Parida KM. Fabrication of α-Fe_2O_3 nanorod/RGO composite: A novel hybrid photocatalyst for phenol degradation. ACS Applied Materials & Interfaces. 2013; 5: 9101–10.

[260] Yang SH, Song XF, Zhang P, Lian Gao L. Heating-rate-induced porous α-Fe_2O_3 with controllable pore size and crystallinity grown on graphene for supercapacitors. ACS Applied Materials & Interfaces. 2015; 7: 75–9.

[261] Li LJ, Zhou GM, Weng Z, Shan XY, Li F, Cheng HM. Monolithic Fe_2O_3/graphene hybrid for highly efficient lithium storage and arsenic removal. Carbon. 2014; 67: 500–7.

[262] Wang G, Liu T, Luo YJ, Zhao Y, Ren ZY, Bai JB et al. Preparation of Fe_2O_3/graphene composite and its electrochemical performance as an anode material for lithium ion batteries. Journal of Alloys and Compounds. 2011; 509: L216–L20.

[263] Wang Z, Ma CY, Wang HL, Liu ZH, Hao ZP. Facilely synthesized Fe_2O_3-graphene nanocomposite as novel electrode materials for supercapacitors with high performance. Journal of Alloys and Compounds. 2013; 552: 486–91.

[264] Du M, Xu CH, Sun J, Gao LB. Synthesis of α-Fe_2O_3 nanoparticles from $Fe(OH)_3$ sol and their composite with reduced graphene oxide for lithium ion batteries. Journal of Materials Chemistry A. 2013; 1: 7154–8.

[265] Zhang M, Qu BH, Lei DN, Chen YJ, Yu XZ, Chen LB et al. A green and fast strategy for the scalable synthesis of Fe_2O_3/graphene with significantly enhanced Li-ion storage properties. Journal of Materials Chemistry. 2012; 22: 3868–74.

[266] Xia XF, Hao QL, Lei W, Wang WJ, Sun DP, Wang X. Nanostructured ternary composites of graphene/Fe_2O_3/polyaniline for high-performance supercapacitors. Journal of Materials Chemistry. 2012; 22: 16844–50.

[267] Zou YQ, Kan J, Wang Y. Fe_2O_3-graphene rice-on-sheet nanocomposite for high and fast lithium ion storage. Journal of Physical Chemistry C. 2011; 115: 20747–53.

[268] Zhu JX, Zhu T, Zhou XZ, Zhang YY, Lou XW, Chen XD et al. Facile synthesis of metal oxide/reduced graphene oxide hybrids with high lithium storage capacity and stable cyclability. Nanoscale. 2011; 3: 1084–9.

[269] Zhang WY, Zeng Y, Xu C, Tan HT, Liu WL, Zhu JX et al. Fe_2O_3 nanocluster-decorated graphene as O_2 electrode for high energy Li-O_2 batteries. RSC Advances. 2012; 2: 8508–14.

[270] Kan J, Wang Y. Large and fast reversible Li-ion storages in Fe_2O_3-graphene sheet-on-sheet sandwich-like nanocomposites. Scientific Reports. 2013; 3.

[271] Xue XY, Ma CH, Cui CX, Xing LL. High lithium storage performance of α-Fe_2O_3/graphene nanocomposites as lithium-ion battery anodes. Solid State Sciences. 2011; 13: 1526–30.

[272] Kim IT, Magasinski A, Jacob K, Yushin G, Tannenbaum R. Synthesis and electrochemical performance of reduced graphene oxide/maghemite composite anode for lithium ion batteries. Carbon. 2013; 52: 56–64.

[273] Wang GP, Zhang L, Zhang JJ. A review of electrode materials for electrochemical supercapacitors. Chemical Society Reviews. 2012; 41: 797–828.

[274] Yang XL, Fan KC, Zhu YH, Shen JH, Jiang X, Zhao P et al. Electric papers of graphene-coated Co_3O_4 fibers for high-performance lithium-ion batteries. ACS Applied Materials & Interfaces. 2013; 5: 997–1002.

[275] Chen SQ, Wang Y. Microwave-assisted synthesis of a Co_3O_4-graphene sheet-on-sheet nanocomposite as a superior anode material for Li-ion batteries. Journal of Materials Chemistry. 2010; 20: 9735–9.

[276] Cheng MY, Ye YS, Cheng JH, Yeh YJ, Chen BH, Hwang BJ. Defect-free graphene metal oxide composites: formed by lithium mediated exfoliation of graphite. Journal of Materials Chemistry. 2012; 22: 14722–6.

[277] Yang XL, Fan KC, Zhu YH, Shen JH, Jiang X, Zhao P et al. Tailored graphene-encapsulated mesoporous Co_3O_4 composite microspheres for high-performance lithium ion batteries. Journal of Materials Chemistry. 2012; 22: 17278–83.

[278] Pan LY, Zhao HB, Shen WC, Dong XW, Xu JQ. Surfactant-assisted synthesis of a Co_3O_4/reduced graphene oxide composite as a superior anode material for Li-ion batteries. Journal of Materials Chemistry A. 2013; 1: 7159–66.

[279] Suryanto BHR, Lu XY, Zhao C. Layer-by-layer assembly of transparent amorphous Co_3O_4 nanoparticles/graphene composite electrodes for sustained oxygen evolution reaction. Journal of Materials Chemistry A. 2013; 1: 12726–31.

[280] Hsieh CT, Lin JS, Chen YF, Teng HS. Pulse microwave deposition of cobalt oxide nanoparticles on graphene nanosheets as anode materials for lithium ion batteries. Journal of Physical Chemistry C. 2012; 116: 15251–8.

[281] Wang HW, Hu ZA, Chang YQ, Chen YL, Zhang ZY, Yang YY et al. Preparation of reduced graphene oxide/cobalt oxide composites and their enhanced capacitive behaviors by homogeneous incorporation of reduced graphene oxide sheets in cobalt oxide matrix. Materials Chemistry and Physics. 2011; 130: 672–9.

[282] Baby TT, Sundara R. A facile synthesis and field emission property investigation of Co_3O_4 nanoparticles decorated graphene. Materials Chemistry and Physics. 2012; 135: 623–7.

[283] Ryu WH, Yoon TH, Song SH, Jeon SK, Park YJ, Kim ID. Bifunctional composite catalysts using Co_3O_4 nanofibers immobilized on nonoxidized graphene nanoflakes for high-capacity and long-cycle Li-O_2 batteries. Nano Letters. 2013; 13: 4190–7.

[284] Park KH, Lee DJ, Kim JM, Song JC, Lee YM, Kim HT et al. Defect-free, size-tunable graphene for high-performance lithium ion battery. Nano Letters. 2014; 14: 4306–13.

[285] Liang YY, Li YG, Wang HL, Zhou JG, Wang J, Regier T et al. Co_3O_4 nanocrystals on graphene as a synergistic catalyst for oxygen reduction reaction. Nature Materials. 2011; 10: 780–6.

[286] Choi BG, Chang SJ, Lee YB, Bae JS, Kim HJ, Huh YS. 3D heterostructured architectures of Co_3O_4 nanoparticles deposited on porous graphene surfaces for high performance of lithium ion batteries. Nanoscale. 2012; 4: 5924–30.

[287] Zhang HJ, Bai YJ, Feng YY, Li X, Wang Y. Encapsulating magnetic nanoparticles in sandwich-like coupled graphene sheets and beyond. Nanoscale. 2013; 5: 2243–8.

[288] Wang RH, Xu CH, Sun J, Liu YQ, Gao L, Lin CC. Free-standing and binder-free lithium-ion electrodes based on robust layered assembly of graphene and Co_3O_4 nanosheets. Nanoscale. 2013; 5: 6960–7.

[289] Xiao JW, Kuang Q, Yang SH, Xiao F, Wang SL, Guo L. Surface structure dependent electrocatalytic activity of Co_3O_4 anchored on graphene sheets toward oxygen reduction reaction. Scientific Reports. 2013; 3: 2300.

[290] Zhao YF, Chen SQ, Sun B, Su DW, Huang XD, Liu H et al. Graphene-Co_3O_4 nanocomposite as electrocatalyst with high performance for oxygen evolution reaction. Scientific Reports. 2015; 5.

[291] Dong XC, Xu H, Wang XW, Huang YX, Chan-Park MB, Zhang H et al. 3D graphene-cobalt oxide electrode for high-performance supercapacitor and enzymeless glucose detection. ACS Nano. 2012; 6: 3206–13.

[292] Liao QY, Li N, Jin SX, Yang GW, Wang CX. All-solid-state symmetric supercapacitor based on Co_3O_4 nanoparticles on vertically aligned graphene. ACS Nano. 2015; 9: 5310–7.

[293] Yuan CZ, Yang L, Hou LR, Li JY, Sun YX, Zhang XG et al. Flexible hybrid paper made of monolayer Co_3O_4 microsphere arrays on rGO/CNTs and their application in electrochemical capacitors. Advanced Functional Materials. 2012; 22: 2560–6.

[294] Zhang Z, Hao JH, Yang WS, Lu BP, Ke X, Zhang BL et al. Porous Co_3O_4 nanorods-reduced graphene oxide with intrinsic peroxidase-like activity and catalysis in the degradation of methylene blue. ACS Applied Materials & Interfaces. 2013; 5: 3809–15.

[295] Kim HY, Seo DH, Kim SW, Kim JS, Kang KS. Highly reversible Co_3O_4/graphene hybrid anode for lithium rechargeable batteries. Carbon. 2011; 49: 326–32.

[296] Wu ZS, Sun Y, Tan YZ, Yang SB, Feng XL, Muellen K. Three-dimensional graphene-based macro- and mesoporous frameworks for high-performance electrochemical capacitive energy storage. Journal of the American Chemical Society. 2012; 134: 19532–5.

[297] Sun CW, Li F, Ma C, Wang Y, Ren YL, Yang W et al. Graphene-Co_3O_4 nanocomposite as an efficient bifunctional catalyst for lithium-air batteries. Journal of Materials Chemistry A. 2014; 2: 7188–96.

[298] Yue WB, Jiang SH, Huang WJ, Gao ZQ, Li J, Ren Y et al. Sandwich-structural graphene-based metal oxides as anode materials for lithium-ion batteries. Journal of Materials Chemistry A. 2013; 1: 6928–33.

[299] Zhu JX, Sharma YK, Zeng ZY, Zhang XJ, Srinivasan M, Mhaisalkar S et al. Cobalt oxide nanowall arrays on reduced graphene oxide sheets with controlled phase, grain size, and porosity for Li-ion battery electrodes. Journal of Physical Chemistry C. 2011; 115: 8400–6.

[300] He GY, Li JH, Chen HQ, Shi J, Sun XQ, Chen S et al. Hydrothermal preparation of Co_3O_4@graphene nanocomposite for supercapacitor with enhanced capacitive performance. Materials Letters. 2012; 82: 61–3.

[301] Liu PB, Huang Y, Wang L, Zong M, Zhang W. Hydrothermal synthesis of reduced graphene oxide-Co_3O_4 composites and the excellent microwave electromagnetic properties. Materials Letters. 2013; 107: 166–9.

[302] Peng CX, Chen BD, Qin Y, Yang SH, Li CZ, Zuo YH et al. Facile ultrasonic synthesis of CoO quantum dot/graphene nanosheet composites with high lithium storage capacity. ACS Nano. 2012; 6: 1074–81.

[303] Huang XL, Wang RZ, Xu D, Wang ZL, Wang HG, Xu JJ et al. Homogeneous CoO on graphene for binder-free and ultralong-life lithium ion batteries. Advanced Functional Materials. 2013; 23: 4345–53.

[304] Mao S, Wen ZH, Huang TH, Hou Y, Chen JH. High-performance bi-functional electrocatalysts of 3D crumpled graphene-cobalt oxide nanohybrids for oxygen reduction and evolution reactions. Energy & Environmental Science. 2014; 7: 609–16.

[305] Qi YY, Zhang H, Du N, Yang DR. Highly loaded CoO/graphene nanocomposites as lithium-ion anodes with superior reversible capacity. Journal of Materials Chemistry A. 2013; 1: 2337–42.

[306] Sun YM, Hu XL, Luo W, Huang YH. Ultrathin CoO/graphene hybrid nanosheets: A highly stable anode material for lithium-ion batteries. Journal of Physical Chemistry C. 2012; 116: 20794–9.

[307] Wu CH, Deng SX, Wang H, Sun YX, Liu JB, Yan HG. Preparation of novel three-dimensional NiO/ultrathin derived graphene hybrid for supercapacitor applications. ACS Applied Materials & Interfaces. 2014; 6: 1106–12.

[308] Deng P, Zhang HY, Chen YM, Li ZH, Huang ZK, Xu XF et al. Facile fabrication of graphene/nickel oxide composite with superior supercapacitance performance by using alcohols-reduced graphene as substrate. Journal of Alloys and Compounds. 2015; 644: 165–71.

[309] Lv W, Sun F, Tang DM, Fang HT, Liu CB, Yang QH et al. A sandwich structure of graphene and nickel oxide with excellent supercapacitive performance. Journal of Materials Chemistry. 2011; 21: 9014–9.
[310] Zhao B, Song JS, Liu P, Xu WW, Fang T, Jiao Z et al. Monolayer graphene/NiO nanosheets with two-dimension structure for supercapacitors. Journal of Materials Chemistry. 2011; 21: 18792–8.
[311] Wu MS, Lin YP, Lin CH, Lee JT. Formation of nano-scaled crevices and spacers in NiO-attached graphene oxide nanosheets for supercapacitors. Journal of Materials Chemistry. 2012; 22: 2442–8.
[312] Huang Y, Huang XL, Lian JS, Xu D, Wang LM, Zhang XB. Self-assembly of ultrathin porous NiO nanosheets/graphene hierarchical structure for high-capacity and high-rate lithium storage. Journal of Materials Chemistry. 2012; 22: 2844–7.
[313] Luan F, Wang GM, Ling YC, Lu XH, Wang HY, Tong YX et al. High energy density asymmetric supercapacitors with a nickel oxide nanoflake cathode and a 3D reduced graphene oxide anode. Nanoscale. 2013; 5: 7984–90.
[314] Mai YJ, Shi SJ, Zhang D, Lu Y, Gu CD, Tu JP. NiO-graphene hybrid as an anode material for lithium ion batteries. Journal of Power Sources. 2012; 204: 155–61.
[315] Qiu DF, Xu ZJ, Zheng MB, Zhao B, Pan LJ, Pu L et al. Graphene anchored with mesoporous NiO nanoplates as anode material for lithium-ion batteries. Journal of Solid State Electrochemistry. 2012; 16: 1889–92.
[316] Qiu DF, Bu G, Zhao B, Lin ZX, Pu L, Pan LJ et al. *In situ* growth of mesoporous NiO nanoplates on a graphene matrix as cathode catalysts for rechargeable lithium-air batteries. Materials Letters. 2015; 141: 43–6.
[317] Huang XH, Zhang P, Wu JB, Lin Y, Guo RQ. Porous NiO/graphene hybrid film as anode for lithium ion batteries. Materials Letters. 2015; 153: 102–5.
[318] Tao LQ, Zai JT, Wang KX, Wan YH, Zhang HJ, Yu C et al. 3D-hierarchical NiO-graphene nanosheet composites as anodes for lithium ion batteries with improved reversible capacity and cycle stability. RSC Advances. 2012; 2: 3410–5.
[319] Liu Y, Wang W, Gu L, Wang YW, Ying YL, Mao YY et al. Flexible CuO nanosheets/reduced-graphene oxide composite paper: Binder-free anode for high-performance lithium-ion batteries. ACS Applied Materials & Interfaces. 2013; 5: 9850–5.
[320] Seo SD, Lee DH, Kim JC, Lee GH, Kim DW. Room-temperature synthesis of CuO/graphene nanocomposite electrodes for high lithium storage capacity. Ceramics International. 2013; 39: 1749–55.
[321] Wang B, Wu XL, Shu CY, Guo YG, Wang CR. Synthesis of CuO/graphene nanocomposite as a high-performance anode material for lithium-ion batteries. Journal of Materials Chemistry. 2010; 20: 10661–4.
[322] Lu LQ, Wang Y. Sheet like and fusiform CuO nanostructures grown on graphene by rapid microwave heating for high Li-ion storage capacities. Journal of Materials Chemistry. 2011; 21: 17916–21.
[323] Zhao B, Liu P, Zhuang H, Jiao Z, Fang T, Xu WW et al. Hierarchical self-assembly of microscale leaf-like CuO on graphene sheets for high-performance electrochemical capacitors. Journal of Materials Chemistry A. 2013; 1: 367–73.
[324] Baby TT, Sundara R. Synthesis and transport properties of metal oxide decorated graphene dispersed nanofluids. Journal of Physical Chemistry C. 2011; 115: 8527–33.
[325] Qiu DF, Zhao B, Lin ZX, Pu L, Pan LJ, Shi Y. *In situ* growth of CuO nanoparticles on graphene matrix as anode material for lithium-ion batteries. Materials Letters. 2013; 105: 242–5.
[326] Zhu JW, Zeng GY, Nie FD, Xu XM, Chen S, Han QF et al. Decorating graphene oxide with CuO nanoparticles in a water-isopropanol system. Nanoscale. 2010; 2: 988–94.
[327] Wu SX, Yin ZY, He QY, Lu GQ, Zhou XZ, Zhang H. Electrochemical deposition of Cl-doped n-type Cu_2O on reduced graphene oxide electrodes. Journal of Materials Chemistry. 2011; 21: 3467–70.
[328] Yan XY, Tong XL, Zhang YF, Han XD, Wang YY, Jin GQ et al. Cuprous oxide nanoparticles dispersed on reduced graphene oxide as an efficient electrocatalyst for oxygen reduction reaction. Chemical Communications. 2012; 48: 1892–4.
[329] Deng SZ, Tjoa V, Fan HM, Tan HR, Sayle DC, Olivo M et al. Reduced graphene oxide conjugated Cu_2O nanowire mesocrystals for high-performance NO_2 gas sensor. Journal of the American Chemical Society. 2012; 134: 4905–17.
[330] Hong CL, Jin X, Totleben J, Lohrman J, Harak E, Subramaniam B et al. Graphene oxide stabilized Cu_2O for shape selective nanocatalysis. Journal of Materials Chemistry A. 2014; 2: 7147–51.
[331] Xu C, Wang X, Yang LC, Wu YP. Fabrication of a graphene-cuprous oxide composite. Journal of Solid State Chemistry. 2009; 182: 2486–90.
[332] Kovtyukhova NI, Ollivier PJ, Martin BR, Mallouk TE, Chizhik SA, Buzaneva EV et al. Layer-by-layer assembly of ultrathin composite films from micron-sized graphite oxide sheets and polycations. Chemistry of Materials. 1999; 11: 771–8.
[333] Szabo T, Berkesi O, Forgo P, Josepovits K, Sanakis Y, Petridis D et al. Evolution of surface functional groups in a series of progressively oxidized graphite oxides. Chemistry of Materials. 2006; 18: 2740–9.
[334] Wang WZ, Xu CK, Wang GH, Liu YK, Zheng CL. Preparation of smooth single-crystal Mn_3O_4 nanowires. Advanced Materials. 2002; 14: 837–40.
[335] Yang YG, Ren LL, Zhang C, Huang S, Liu TX. Facile fabrication of functionalized graphene sheets (FGS)/ZnO nanocomposites with photocatalytic property. ACS Applied Materials & Interfaces. 2011; 3: 2779–85.
[336] Wang JF, Tsuzuki T, Tang B, Hou XL, Sun LW, Wang XG. Reduced graphene oxide/ZnO composite: Reusable adsorbent for pollutant management. ACS Applied Materials & Interfaces. 2012; 4: 3084–90.

[337] Yang YG, Liu TX. Fabrication and characterization of graphene oxide/zinc oxide nanorods hybrid. Applied Surface Science. 2011; 257: 8950–4.
[338] Kavitha T, Gopalan AI, Lee KP, Park SY. Glucose sensing, photocatalytic and antibacterial properties of graphene-ZnO nanoparticle hybrids. Carbon. 2012; 50: 2994–3000.
[339] Son DI, Kwon BW, Kim HH, Park DH, Angadi B, Choi WK. Chemical exfoliation of pure graphene sheets from synthesized ZnO-graphene quasi core-shell quantum dots. Carbon. 2013; 59: 289–95.
[340] Liu XJ, Pan LK, Lv T, Lu T, Zhu G, Sun ZZ et al. Microwave-assisted synthesis of ZnO-graphene composite for photocatalytic reduction of Cr(VI). Catalysis Science & Technology. 2011; 1: 1189–93.
[341] Roy P, Periasamy AP, Liang CT, Chang HT. Synthesis of graphene-ZnO-Au nanocomposites for efficient photocatalytic reduction of nitrobenzene. Environmental Science & Technology. 2013; 47: 6688–95.
[342] Lu T, Pan LK, Li HB, Zhu G, Lv T, Liu XJ et al. Microwave-assisted synthesis of graphene-ZnO nanocomposite for electrochemical supercapacitors. Journal of Alloys and Compounds. 2011; 509: 5488–92.
[343] Zhang YP, Li HB, Pan LK, Lu T, Sun Z. Capacitive behavior of graphene-ZnO composite film for supercapacitors. Journal of Electroanalytical Chemistry. 2009; 634: 68–71.
[344] Zhan ZY, Zheng LX, Pan YZ, Sun GZ, Li LJ. Self-powered, visible-light photodetector based on thermally reduced graphene oxide-ZnO (rGO-ZnO) hybrid nanostructure. Journal of Materials Chemistry. 2012; 22: 2589–95.
[345] Herring NP, Almahoudi SH, Olson CR, El-Shall MS. Enhanced photocatalytic activity of ZnO-graphene nanocomposites prepared by microwave synthesis. Journal of Nanoparticle Research. 2012; 14.
[346] Luo QP, Yu XY, Lei BX, Chen HY, Kuang DB, Su CY. Reduced graphene oxide-hierarchical ZnO hollow sphere composites with enhanced photocurrent and photocatalytic activity. Journal of Physical Chemistry C. 2012; 116: 8111–7.
[347] Son DI, Kwon BW, Park DH, Seo WS, Yi YJ, Angadi B et al. Emissive ZnO-graphene quantum dots for white-light-emitting diodes. Nature Nanotechnology. 2012; 7: 465–71.
[348] Shuvo MAI, Khan MAR, Karim H, Morton P, Wilson T, Lin YR. Investigation of modified graphene for energy storage applications. ACS Applied Materials & Interfaces. 2013; 5: 7881–5.
[349] Zhou XB, Shi TJ, Zhou HO. Hydrothermal preparation of ZnO-reduced graphene oxide hybrid with high performance in photocatalytic degradation. Applied Surface Science. 2012; 258: 6204–11.
[350] Zhang Q, Tian CG, Wu AP, Tan TX, Sun L, Wang L et al. A facile one-pot route for the controllable growth of small sized and well-dispersed ZnO particles on GO-derived graphene. Journal of Materials Chemistry. 2012; 22: 11778–84.
[351] Kumar R, Singh RK, Singh J, Tiwari RS, Srivastava ON. Synthesis, characterization and optical properties of graphene sheets-ZnO multipod nanocomposites. Journal of Alloys and Compounds. 2012; 526: 129–34.
[352] Saravanakumar B, Mohan R, Kim SJ. Facile synthesis of graphene/ZnO nanocomposites by low temperature hydrothermal method. Materials Research Bulletin. 2013; 48: 878–83.
[353] Yin SY, Men XJ, Sun H, She P, Zhang W, Wu CF et al. Enhanced photocurrent generation of bio-inspired graphene/ZnO composite films. Journal of Materials Chemistry A. 2015; 3: 12016–22.
[354] Pang H, Lu QY, Gao F. Graphene oxide induced growth of one-dimensional fusiform zirconia nanostructures for highly selective capture of phosphopeptides. Chemical Communications. 2011; 47: 11772–4.
[355] Liu J, Meng XB, Hu YH, Geng DS, Banis MN, Cai M et al. Controlled synthesis of zirconium oxide on graphene nanosheets by atomic layer deposition and its growth mechanism. Carbon. 2013; 52: 74–82.
[356] Du D, Liu J, Zhang XY, Cui XL, Lin YH. One-step electrochemical deposition of a graphene-ZrO_2 nanocomposite: Preparation, characterization and application for detection of organophosphorus agents. Journal of Materials Chemistry. 2011; 21: 8032–7.
[357] Sun YM, Hu XL, Luo W, Huang YH. Self-assembled hierarchical MoO_2/graphene nanoarchitectures and their application as a high-performance anode material for lithium-ion batteries. ACS Nano. 2011; 5: 7100–7.
[358] Seng KH, Du GD, Li L, Chen ZX, Liu HK, Guo ZP. Facile synthesis of graphene-molybdenum dioxide and its lithium storage properties. Journal of Materials Chemistry. 2012; 22: 16072–7.
[359] Xia XF, Hao QL, Lei W, Wang WJ, Wang HL, Wang X. Reduced-graphene oxide/molybdenum oxide/polyaniline ternary composite for high energy density supercapacitors: Synthesis and properties. Journal of Materials Chemistry. 2012; 22: 8314-20.
[360] Yang XF, Lu CY, Qin JL, Zhang RX, Tang H, Song HJ. A facile one-step hydrothermal method to produce graphene-MoO_3 nanorod bundle composites. Materials Letters. 2011; 65: 2341–4.
[361] Qin JW, Cao MH, Li N, Hu CW. Graphene-wrapped WO_3 nanoparticles with improved performances in electrical conductivity and gas sensing properties. Journal of Materials Chemistry. 2011; 21: 17167–74.
[362] An XQ, Yu JC, Wang Y, Hu YM, Yu XL, Zhang GJ. WO_3 nanorods/graphene nanocomposites for high-efficiency visible-light-driven photocatalysis and NO_2 gas sensing. Journal of Materials Chemistry. 2012; 22: 8525–31.
[363] Huang H, Yue ZK, Li G, Wang XM, Huang J, Du YK et al. Ultraviolet-assisted preparation of mesoporous WO_3/reduced graphene oxide composites: superior interfacial contacts and enhanced photocatalysis. Journal of Materials Chemistry A. 2013; 1: 15110–6.
[364] Gui YH, Zhao JB, Wang WM, Tian JF, Zhao M. Synthesis of hemispherical WO_3/graphene nanocomposite by a microwave-assisted hydrothermal method and the gas-sensing properties to triethylamine. 2015; 155: 4–7.

[365] Srivastava S, Jain K, Singh VN, Singh S, Vijayan N, Dilawar N et al. Faster response of NO_2 sensing in graphene-WO_3 nanocomposites. Nanotechnology. 2012; 23: 205501.
[366] Shao DL, Yu MP, Lian J, Sawyer S. An ultraviolet photodetector fabricated from WO_3 nanodiscs/reduced graphene oxide composite material. Nanotechnology. 2013; 24: 295701.
[367] Kim JY, Kim KH, Yoon SB, Kim HK, Park SH, Kim KB. *In situ* chemical synthesis of ruthenium oxide/reduced graphene oxide nanocomposites for electrochemical capacitor applications. Nanoscale. 2013; 5: 6804–11.
[368] George SM. Atomic layer deposition: An overview. Chemical Reviews. 2010; 110: 111–31.
[369] Lin J, Peng ZW, Xiang CS, Ruan GD, Yan Z, Natelson D et al. Graphene nanoribbon and nanostructured SnO_2 composite anodes for lithium ion batteries. ACS Nano. 2013; 7: 6001–6.
[370] Li XF, Meng XB, Liu J, Geng DS, Zhang YB, Banis MN et al. Tin oxide with controlled morphology and crystallinity by atomic layer deposition onto graphene nanosheets for enhanced lithium storage. Advanced Functional Materials. 2012; 22: 1647–54.
[371] Wang X, Cao XQ, Bourgeois L, Guan H, Chen SM, Zhong YT et al. N-doped graphene-SnO_2 sandwich paper for high-performance lithium-ion batteries. Advanced Functional Materials. 2012; 22: 2682–90.
[372] Li XL, Qi W, Mei DH, Sushko ML, Aksay I, Liu J. Functionalized graphene sheets as molecular templates for controlled nucleation and self-assembly of metal oxide-graphene nanocomposites. Advanced Materials. 2012; 24: 5136–41.
[373] Prabakar SJR, Hwang YH, Bae EG, Shim SD, Kim DW, Lah MS et al. SnO_2/graphene composites with self-assembled alternating oxide and amine layers for high Li-storage and excellent stability. Advanced Materials. 2013; 25: 3307–12.
[374] Liang JF, Wei W, Zhong D, Yang QL, Li LD, Guo L. One-step *in situ* synthesis of SnO_2/graphene nanocomposites and its application as an anode material for Li-ion batteries. ACS Applied Materials & Interfaces. 2012; 4: 454–9.
[375] Liang JF, Zhao Y, Guo L, Li LD. Flexible free-standing graphene/SnO_2 nanocomposites paper for Li-ion battery. ACS Applied Materials & Interfaces. 2012; 4: 5742-8.
[376] Jiang YZ, Yuan TZ, Sun WP, Yan M. Electrostatic spray deposition of porous SnO_2/graphene anode films and their enhanced lithium-storage properties. ACS Applied Materials & Interfaces. 2012; 4: 6216-20.
[377] Chen MX, Zhang CC, Li LZ, Liu Y, Li XC, Xu XY et al. Sn powder as reducing agents and SnO_2 precursors for the synthesis of SnO_2-reduced graphene oxide hybrid nanoparticles. ACS Applied Materials & Interfaces. 2013; 5: 13333–9.
[378] Wang XY, Zhou XF, Yao KX, Zhang JG, Liu ZP. A SnO_2/graphene composite as a high stability electrode for lithium ion batteries. Carbon. 2011; 49: 133–9.
[379] Lim HN, Nurzulaikha R, Harrison I, Lim SS, Tan WT, Yeo MC et al. Preparation and characterization of tin oxide, SnO_2 nanoparticles decorated graphene. Ceramics International. 2012; 38: 4209–16.
[380] Yao J, Shen XP, Wang B, Liu HK, Wang GX. In situ chemical synthesis of SnO_2-graphene nanocomposite as anode materials for lithium-ion batteries. Electrochemistry Communications. 2009; 11: 1849–52.
[381] Zhang LS, Jiang LY, Yan HJ, Wang WD, Wang W, Song WG et al. Mono dispersed SnO_2 nanoparticles on both sides of single layer graphene sheets as anode materials in Li-ion batteries. Journal of Materials Chemistry. 2010; 20: 5462–7.
[382] Zhang M, Lei DN, Du ZF, Yin XM, Chen LB, Li QH et al. Fast synthesis of SnO_2/graphene composites by reducing graphene oxide with stannous ions. Journal of Materials Chemistry. 2011; 21: 1673–6.
[383] Zhu CZ, Fang YX, Wen D, Dong SJ. One-pot synthesis of functional two-dimensional graphene/SnO_2 composite nanosheets as a building block for self-assembly and an enhancing nanomaterial for biosensing. Journal of Materials Chemistry. 2011; 21: 16911–7.
[384] Vinayan BP, Ramaprabhu S. Facile synthesis of SnO_2 nanoparticles dispersed nitrogen doped graphene anode material for ultrahigh capacity lithium ion battery applications. Journal of Materials Chemistry A. 2013; 1: 3865–71.
[385] Cui SM, Wen ZH, Mattson EC, Mao S, Chang JB, Weinert M et al. Indium-doped SnO_2 nanoparticle-graphene nanohybrids: simple one-pot synthesis and their selective detection of NO_2. Journal of Materials Chemistry A. 2013; 1: 4462–7.
[386] Le NH, Seema H, Kemp KC, Ahmed N, Tiwari JN, Park SJ et al. Solution-processable conductive micro-hydrogels of nanoparticle/graphene platelets produced by reversible self-assembly and aqueous exfoliation. Journal of Materials Chemistry A. 2013; 1: 12900–8.
[387] Wang YX, Lim YG, Park MS, Chou SL, Kim JH, Liu HK et al. Ultrafine SnO_2 nanoparticle loading onto reduced graphene oxide as anodes for sodium-ion batteries with superior rate and cycling performances. Journal of Materials Chemistry A. 2014; 2: 529–34.
[388] Tang JJ, Yang J, Zhou LM, Xie J, Chen GH, Zhou XY. Layer-by-layer self-assembly of a sandwich-like graphene wrapped SnO_x@ graphene composite as an anode material for lithium ion batteries. Journal of Materials Chemistry A. 2014; 2: 6292–5.
[389] Fu W, Du FH, Wang KX, Ye TN, Wei XD, Chen JS. *In situ* growth of ultrafine tin oxide nanocrystals embedded in graphitized carbon nanosheets for use in high-performance lithium-ion batteries. Journal of Materials Chemistry A. 2014; 2: 6960–5.
[390] Botas C, Carriazo D, Singha G, Rojo T. Sn– and SnO_2–graphene flexible foams suitable as binder-free anodes for lithium ion batteries. Journal of Materials Chemistry A. 2015; 3: 13402–10.

[391] Yang AK, Xue Y, Zhang Y, Zhang XF, Zhao H, Li XJ et al. A simple one-pot synthesis of graphene nanosheet/SnO_2 nanoparticle hybrid nanocomposites and their application for selective and sensitive electrochemical detection of dopamine. Journal of Materials Chemistry B. 2013; 1: 1804–11.

[392] Meng XB, Geng DS, Liu J, Banis MN, Zhang Y, Li RY et al. Non-aqueous approach to synthesize amorphous/crystalline metal oxide-graphene nanosheet hybrid composites. Journal of Physical Chemistry C. 2010; 114: 18330–7.

[393] Wei Y, Gao C, Meng FL, Li HH, Wang L, Liu JH et al. SnO_2/reduced graphene oxide nanocomposite for the simultaneous electrochemical detection of cadmium(II), lead(II), copper(II), and mercury(II): An interesting favorable mutual interference. Journal of Physical Chemistry C. 2012; 116: 1034–41.

[394] Zhu XJ, Zhu YW, Murali S, Stoller MD, Ruoff RS. Reduced graphene oxide/tin oxide composite as an enhanced anode material for lithium ion batteries prepared by homogenous coprecipitation. Journal of Power Sources. 2011; 196: 6473–7.

[395] Du ZF, Yin XM, Zhang M, Hao QY, Wang YG, Wang TH. *In situ* synthesis of SnO_2/graphene nanocomposite and their application as anode material for lithium ion battery. Materials Letters. 2010; 64: 2076–9.

[396] Wang L, Wang D, Dong ZH, Zhang FX, Jin J. Interface chemistry engineering for stable cycling of reduced GO/SnO_2 nanocomposites for lithium ion battery. Nano Letters. 2013; 13: 1711–6.

[397] Zhu CZ, Wang P, Wang L, Han L, Dong SJ. Facile synthesis of two-dimensional graphene/SnO_2/Pt ternary hybrid nanomaterials and their catalytic properties. Nanoscale. 2011; 3: 4376–82.

[398] Zhu J, Lei DN, Zhang GH, Li QH, Lu BG, Wang TH. Carbon and graphene double protection strategy to improve the SnO_x electrode performance anodes for lithium-ion batteries. Nanoscale. 2013; 5: 5499–505.

[399] Seema H, Kemp KC, Chandra V, Kim KS. Graphene-SnO_2 composites for highly efficient photocatalytic degradation of methylene blue under sunlight. Nanotechnology. 2012; 23.

[400] Sladkevich S, Gun J, Prikhodchenko PV, Gutkin V, Mikhaylov AA, Novotortsev VM et al. Peroxide induced tin oxide coating of graphene oxide at room temperature and its application for lithium ion batteries. Nanotechnology. 2012; 23.

[401] Huang YS, Wu DQ, Wang JZ, Han S, Lv L, Zhang F et al. Amphiphilic polymer promoted assembly of macroporous graphene/SnO_2 frameworks with tunable porosity for high-performance lithium storage. Small. 2014; 10: 2226–32.

[402] Zhou XS, Wan LJ, Guo YG. Binding SnO_2 nanocrystals in nitrogen-doped graphene sheets as anode materials for lithium-ion batteries. Advanced Materials. 2013; 25: 2152–7.

[403] Yang S, Yue WB, Zhu J, Ren Y, Yang XJ. Graphene-based mesoporous SnO_2 with enhanced electrochemical performance for lithium-ion batteries. Advanced Functional Materials. 2013; 23: 3570–6.

[404] Wang RH, Xu CH, Sun J, Gao L, Yao HL. Solvothermal-induced 3D macroscopic SnO_2/nitrogen-doped graphene aerogels for high capacity and long-life lthium storage. ACS Applied Materials & Interfaces. 2014; 6: 3427–36.

[405] Liu HD, Huang JM, Li XL, Liu J, Zhang YX, Du K. Flower-like SnO_2/graphene composite for high-capacity lithium storage. Applied Surface Science. 2012; 258: 4917–21.

[406] Zhang CF, Peng X, Guo ZP, Cai CB, Chen ZX, Wexler D et al. Carbon-coated SnO_2/graphene nanosheets as highly reversible anode materials for lithium ion batteries. Carbon. 2012; 50: 1897–903.

[407] Lim SP, Huang NM, Lim HN. Solvothermal synthesis of SnO_2/graphene nanocomposites for supercapacitor application. Ceramics International. 2013; 39: 6647–55.

[408] Liu J, Huang JM, Hao LL, Liu HD, Li XL. SnO_2 nano-spheres/graphene hybrid for high-performance lithium ion battery anodes. Ceramics International. 2013; 39: 8623–7.

[409] Wang DN, Yang JL, Li XF, Geng DS, Li RY, Cai M et al. Layer by layer assembly of sandwiched graphene/SnO_2 nanorod/carbon nanostructures with ultrahigh lithium ion storage properties. Energy & Environmental Science. 2013; 6: 2900–6.

[410] Xie J, Liu SY, Chen XF, Zheng YX, Song WT, Cao GS et al. Nanocrystal-SnO_2-loaded graphene with improved Li-storage properties prepared by a facile one-pot hydrothermal route. International Journal of Electrochemical Science. 2011; 6: 5539–49.

[411] Song HJ, Zhang LC, He CL, Qu Y, Tian YF, Lv Y. Graphene sheets decorated with SnO_2 nanoparticles: *in situ* synthesis and highly efficient materials for cataluminescence gas sensors. Journal of Materials Chemistry. 2011; 21: 5972–7.

[412] Zhang ZY, Zou RJ, Song GS, Yu L, Chen ZG, Hu JQ. Highly aligned SnO_2 nanorods on graphene sheets for gas sensors. Journal of Materials Chemistry. 2011; 21: 17360–5.

[413] Liang RL, Cao HQ, Qian D, Zhang JX, Qu MZ. Designed synthesis of SnO_2-polyaniline-reduced graphene oxide nanocomposites as an anode material for lithium-ion batteries. Journal of Materials Chemistry. 2011; 21: 17654–7.

[414] Xu CH, Sun J, Gao L. Direct growth of monodisperse SnO_2 nanorods on graphene as high capacity anode materials for lithium ion batteries. Journal of Materials Chemistry. 2012; 22: 975–9.

[415] Park SK, Yu SH, Pinna N, Woo SH, Jang BC, Chung YH et al. A facile hydrazine-assisted hydrothermal method for the deposition of monodisperse SnO_2 nanoparticles onto graphene for lithium ion batteries. Journal of Materials Chemistry. 2012; 22: 2520–5.

[416] Li BJ, Cao HQ, Zhang JX, Qu MZ, Lian F, Kong XH. SnO_2-carbon-RGO heterogeneous electrode materials with enhanced anode performances in lithium ion batteries. Journal of Materials Chemistry. 2012; 22: 2851–4.

[417] Wen ZH, Cui SM, Kim HJ, Mao S, Yu KH, Lu GH et al. Binding Sn-based nanoparticles on graphene as the anode of rechargeable lithium-ion batteries. Journal of Materials Chemistry. 2012; 22: 3300–6.

[418] Chen BB, Qian H, Xu JH, Qin LL, Wu QH, Zheng MS et al. Study on SnO_2/graphene composites with superior electrochemical performance for lithium-ion batteries. Journal of Materials Chemistry A. 2014; 2: 9345–52.

[419] Wang DN, Li XF, Wang JJ, Yang JL, Geng DS, Li RY et al. Defect-rich crystalline SnO_2 immobilized on graphene nanosheets with enhanced cycle performance for Li-ion batteries. Journal of Physical Chemistry C. 2012; 116: 22149–56.

[420] Jiang YZ, Xu Y, Yuan TZ, Yan M. Phase-tailored synthesis of tin oxide-graphene nanocomposites for anodes and their enhanced lithium-ion battery performance. Materials Letters. 2013; 91: 16–9.

[421] Liu LL, An MZ, Yang PX, Zhang JQ. Superior cycle performance and high reversible capacity of SnO_2/graphene composite as an anode material for lithium-ion batteries. Scientific Reports. 2015; 5.

[422] Mao S, Wen ZH, Kim HJ, Lu GH, Hurley P, Chen JH. A general approach to one-pot fabrication of crumpled graphene-based nanohybrids for energy applications. ACS Nano. 2012; 6: 7505–13.

[423] Chen ZX, Zhou M, Cao YL, Ai XP, Yang HX, Liu J. *In situ* generation of few-layer graphene coatings on SnO_2-SiC core-shell nanoparticles for high-performance lithium-ion storage. Advanced Energy Materials. 2012; 2: 95–102.

[424] Ding JJ, Yan XB, Li J, Shen BS, Yang J, Chen JT et al. Enhancement of field emission and photoluminescence properties of graphene-SnO_2 composite nanostructures. ACS Applied Materials & Interfaces. 2011; 3: 4299–305.

[425] Zeng GB, Shi N, Hess M, Chen X, Cheng W, Fan TX et al. A general method of fabricating flexible spinel-type oxide/reduced graphene oxide nanocomposite aerogels as advanced anodes for lithium-ion batteries. ACS Nano. 2015; 9: 4227–35.

[426] Zhang XJ, Wang GS, Cao WQ, Wei YZ, Liang JF, Guo L et al. Enhanced microwave absorption property of reduced graphene oxide (RGO)-$MnFe_2O_4$ nanocomposites and polyvinylidene fluoride. ACS Applied Materials & Interfaces. 2014; 6: 7471–8.

[427] Bai S, Shen XP, Zhong X, Liu Y, Zhu GX, Xu X et al. One-pot solvothermal preparation of magnetic reduced graphene oxide-ferrite hybrids for organic dye removal. Carbon. 2012; 50: 2337-46.

[428] Liu SQ, Xiao B, Feng LR, Zhou SS, Chen ZG, Liu CB et al. Graphene oxide enhances the Fenton-like photocatalytic activity of nickel ferrite for degradation of dyes under visible light irradiation. Carbon. 2013; 64: 197–206.

[429] Meidanchi A, Akhavan O. Superparamagnetic zinc ferrite spinel-graphene nanostructures for fast wastewater purification. Carbon. 2014; 69: 230-8.

[430] Zong M, Huang Y, Zhang N, Wu HW. Influence of (RGO)/(ferrite) ratios and graphene reduction degree on microwave absorption properties of graphene composites. Journal of Alloys and Compounds. 2015; 644: 491–501.

[431] Yao YJ, Cai YM, Lu F, Wei FY, Wang XY, Wang SB. Magnetic recoverable $MnFe_2O_4$ and $MnFe_2O_4$-graphene hybrid as heterogeneous catalysts of peroxymonosulfate activation for efficient degradation of aqueous organic pollutants. Journal of Hazardous Materials. 2014; 270: 61–70.

[432] Liu SY, Xie J, Fang CC, Cao GS, Zhu TJ, Zhao XB. Self-assembly of a $CoFe_2O_4$/graphene sandwich by a controllable and general route: towards a high-performance anode for Li-ion batteries. Journal of Materials Chemistry. 2012; 22: 19738–43.

[433] Hao JH, Zhang Z, Yang WS, Lu BP, Ke X, Zhang BL et al. *In situ* controllable growth of $CoFe_2O_4$ ferrite nanocubes on graphene for colorimetric detection of hydrogen peroxide. Journal of Materials Chemistry A. 2013; 1: 4352–7.

[434] Fu M, Jiao QZ, Zhao Y. Preparation of $NiFe_2O_4$ nanorod-graphene composites via an ionic liquid assisted one-step hydrothermal approach and their microwave absorbing properties. Journal of Materials Chemistry A. 2013; 1: 5577–86.

[435] Wang Z, Zhang X, Li Y, Liu ZT, Hao ZP. Synthesis of graphene-$NiFe_2O_4$ nanocomposites and their electrochemical capacitive behavior. Journal of Materials Chemistry A. 2013; 1: 6393–9.

[436] Shi YQ, Zhou KQ, Wang BB, Jiang SH, Qian XD, Gui Z et al. Ternary graphene-$CoFe_2O_4$/CdS nanohybrids: preparation and application as recyclable photocatalysts. Journal of Materials Chemistry A. 2014; 2: 535–44.

[437] Fu M, Jiao QZ, Zhao Y, Li HS. Vapor diffusion synthesis of $CoFe_2O_4$ hollow sphere/graphene composites as absorbing materials. Journal of Materials Chemistry A. 2014; 2: 735–44.

[438] Heidari EK, Zhang BL, Sohi MH, Ataie A, Kim JK. Sandwich-structured graphene-$NiFe_2O_4$-carbon nanocomposite anodes with exceptional electrochemical performance for Li ion batteries. Journal of Materials Chemistry A. 2014; 2: 8314–22.

[439] Chen XC, Hou CY, Zhang QH, Li YG, Wang HZ. One-step synthesis of Co-Ni ferrite/graphene nanocomposites with controllable magnetic and electrical properties. Materials Science and Engineering B-Advanced Functional Solid-State Materials. 2012; 177: 1067–72.

[440] Xia H, Qian YY, Fu YS, Wang X. Graphene anchored with $ZnFe_2O_4$ nanoparticles as a high-capacity anode material for lithium-ion batteries. Solid State Sciences. 2013; 17: 67–71.

[441] Chen S, Qiao SZ. Hierarchically porous nitrogen-doped graphene-$NiCo_2O_4$ hybrid paper as an advanced electrocatalytic water-splitting material. ACS Nano. 2013; 7: 10190–6.

[442] Mitchell E, Jimenez A, Gupta RK, Gupta BK, Ramasamy K, Shahabuddin M et al. Ultrathin porous hierarchically textured $NiCo_2O_4$-graphene oxide flexible nanosheets for high-performance supercapacitors. New Journal of Chemistry. 2015; 39: 2181–7.

[443] Zhang GQ, Xia BY, Wang X, Lou XW. Strongly coupled $NiCo_2O_4$-rGO hybrid nanosheets as a methanol-tolerant electrocatalyst for the oxygen reduction reaction. Advanced Materials. 2014; 26: 2408–12.

[444] Chen YJ, Zhuo M, Deng JW, Xu ZJ, Li QH, Wang TH. Reduced graphene oxide networks as an effective buffer matrix to improve the electrode performance of porous $NiCo_2O_4$ nanoplates for lithium-ion batteries. Journal of Materials Chemistry A. 2014; 2: 4449–56.

[445] Das AK, Layek RK, Kim NH, Jung DS, Lee JH. Reduced graphene oxide (RGO)-supported $NiCo_2O_4$ nanoparticles: an electrocatalyst for methanol oxidation. Nanoscale. 2014; 6: 10657–65.

[446] Ma LB, Shen XP, Ji ZY, Cai XQ, Zhu GX, Chen KM. Porous $NiCo_2O_4$ nanosheets/reduced graphene oxide composite: Facile synthesis and excellent capacitive performance for supercapacitors. Journal of Colloid and Interface Science. 2015; 440: 211–8.

[447] Wang HW, Hu ZA, Chang YQ, Chen YL, Wu HY, Zhang ZY et al. Design and synthesis of $NiCo_2O_4$-reduced graphene oxide composites for high performance supercapacitors. Journal of Materials Chemistry. 2011; 21: 10504–11.

[448] Wang JP, Wang SL, Huang ZC, Yu YM, Liu JL. Synthesis of long chain-like nickel cobalt oxide nanoneedles-reduced graphene oxide composite material for high-performance supercapacitors. Ceramics International. 2014; 40: 12751–8.

[449] Rao CV, Reddy ALM, Ishikawa Y, Ajayan PM. $LiNi_{1/3}Co_{1/3}Mn_{1/3}O_2$-graphene composite as a promising cathode for lithium-ion batteries. ACS Applied Materials & Interfaces. 2011; 3: 2966–72.

[450] Ding YH, Ren HM, Huang YY, Chang FH, He X, Fen JQ et al. Co-precipitation synthesis and electrochemical properties of graphene supported $LiMn_{1/3}Ni_{1/3}Co_{1/3}O_2$ cathode materials for lithium-ion batteries. Nanotechnology. 2013; 24: 375401.

[451] Lin M, Chen YM, Chen BL, Wu X, Kam KF, Lu W et al. Morphology-controlled synthesis of self-assembled $LiFePO_4$/C/RGO for high-performance Li-ion batteries. ACS Applied Materials & Interfaces. 2014; 6: 17556–63.

[452] Oh SW, Huang ZD, Zhang B, Yu Y, He YB, Kim JK. Low temperature synthesis of graphene-wrapped $LiFePO_4$ nanorod cathodes by the polyol method. Journal of Materials Chemistry. 2012; 22: 17215-21.

[453] Guo XK, Fan Q, Yu L, Liang JY, Ji WX, Peng LM et al. Sandwich-like $LiFePO_4$/graphene hybrid nanosheets: *in situ* catalytic graphitization and their high-rate performance for lithium ion batteries. Journal of Materials Chemistry A. 2013; 1: 11534–8.

[454] Ding YH, Ren HM, Huang YY, Chang FH, Zhang P. Three-dimensional graphene/$LiFePO_4$ nanostructures as cathode materials for flexible lithium-ion batteries. Materials Research Bulletin. 2013; 48: 3713–6.

[455] Hu LH, Wu FY, Lin CT, Khlobystov AN, Li LJ. Graphene-modified $LiFePO_4$ cathode for lithium ion battery beyond theoretical capacity. Nature Communications. 2013; 4: 1687.

[456] Wang L, Wang HB, Liu ZH, Xiao C, Dong SM, Han PX et al. A facile method of preparing mixed conducting $LiFePO_4$/graphene composites for lithium-ion batteries. Solid State Ionics. 2010; 181: 1685–9.

[457] Kim HY, Kim HS, Kim SW, Park KY, Kim JS, Jeon SK et al. Nano-graphite platelet loaded with $LiFePO_4$ nanoparticles used as the cathode in a high performance Li-ion battery. Carbon. 2012; 50: 1966–71.

[458] Tang YF, Huang FQ, Zhao W, Liu ZQ, Wan DY. Synthesis of graphene-supported $Li_4Ti_5O_{12}$ nanosheets for high rate battery application. Journal of Materials Chemistry. 2012; 22: 11257–60.

[459] Han SY, Kim IY, Jo KY, Hwang SJ. Solvothermal-assisted hybridization between reduced graphene oxide and lithium metal oxides: A facile route to graphene-based composite materials. Journal of Physical Chemistry C. 2012; 116: 7269–79.

[460] Liu HD, Yang G, Zhang XF, Gao P, Wang L, Fang JH et al. Kinetics of conventional carbon coated-$Li_3V_2(PO_4)_3$ and nanocomposite $Li_3V_2(PO_4)_3$/graphene as cathode materials for lithium ion batteries. Journal of Materials Chemistry. 2012; 22: 11039–47.

[461] Wu KL, Yang JP. Synthesis of carbon coated $Li_3V_2(PO_4)_3$/reduced graphene oxide composite for high-performance lithium ion batteries. Materials Research Bulletin. 2013; 48: 435–9.

[462] Pan LJ, Zhu XD, Xie XM, Liu YT. Smart hybridization of TiO_2 nanorods and Fe_3O_4 nanoparticles with pristine graphene nanosheets: Hierarchically nanoengineered ternary heterostructures for high-rate lithium storage. Advanced Functional Materials. 2015; 25: 3341–50.

[463] Ren YL, Wu HY, Lu MM, Chen YJ, Zhu CL, Gao P et al. Quaternary nanocomposites consisting of graphene, Fe_3O_4@Fe core@shell, and ZnO nanoparticles: Synthesis and excellent electromagnetic absorption properties. ACS Applied Materials & Interfaces. 2012; 4: 6436–42.

[464] Liu Q, Shi JB, Cheng MT, Li GL, Cao D, Jiang GB. Preparation of graphene-encapsulated magnetic microspheres for protein/peptide enrichment and MALDI-TOF MS analysis. Chemical Communications. 2012; 48: 1874–6.

[465] Lian PC, Liang SZ, Zhu XF, Yang WS, Wang HH. A novel Fe_3O_4-SnO_2-graphene ternary nanocomposite as an anode material for lithium-ion batteries. Electrochimica Acta. 2011; 58: 81–8.

[466] Liu S, Wang RH, Liu MM, Luo JQ, Jin XH, Sun J et al. Fe_2O_3@SnO_2 nanoparticle decorated graphene flexible films as high-performance anode materials for lithium-ion batteries. Journal of Materials Chemistry A. 2014; 2: 4598–604.

[467] Yang XL, Chen W, Huang JF, Zhou Y, Zhu YH, Li CZ. Rapid degradation of methylene blue in a novel heterogeneous Fe_3O_4@rGO@TiO_2-catalyzed photo-Fenton system. Scientific Reports. 2015; 5.

[468] Zhu JX, Lu ZY, Oo MO, Hng HH, Ma J, Zhang H et al. Synergetic approach to achieve enhanced lithium ion storage performance in ternary phased SnO_2-Fe_2O_3/rGO composite nanostructures. Journal of Materials Chemistry. 2011; 21: 12770–6.

[469] Zhu JX, Yin ZY, Li H, Tan HT, Chow CL, Zhang H et al. Bottom-up preparation of porous metal-oxide ultrathin sheets with adjustable composition/phases and their applications. Small. 2011; 7: 3458–64.

[470] Lu T, Zhang YP, Li HB, Pan LK, Li YL, Sun Z. Electrochemical behaviors of graphene-ZnO and graphene-SnO_2 composite films for supercapacitors. Electrochimica Acta. 2010; 55: 4170–3.
[471] Li S, Ling M, Qiu JX, Han JS, Zhang SQ. Anchoring ultra-fine TiO_2-SnO_2 solid solution particles onto graphene by one-pot ball-milling for long-life lithium-ion batteries. Journal of Materials Chemistry A. 2015; 3: 9700–6.
[472] Wang Y, Huang ZX, Shi YM, Wong JI, Ding M, Yang HY. Designed hybrid nanostructure with catalytic effect: beyond the theoretical capacity of SnO_2 anode material for lithium ion batteries. Scientific Reports. 2015; 5: 9164.

CHAPTER 4

Graphene-Inorganic Hybrids (II)

Ling Bing Kong,[1,a,*] *Freddy Boey,*[1,b] *Yizhong Huang,*[1,c] *Zhichuan Jason Xu,*[1,d] *Kun Zhou,*[2] *Sean Li,*[3] *Wenxiu Que,*[4] *Hui Huang*[5] and *Tianshu Zhang*[6]

4.1. Introduction

Graphene-based nanohybrids with metallic oxides have been discussed in the last chapter. In this chapter, hybrids with other inorganic components, including metallic nanoparticles, other nanocarbon (non-graphene) items, semiconductor nanoparticles and so on, will be elaborated upon. It has been acknowledged that the physical and chemical properties of noble metal nanostructures are highly dependent on their size, shape, composition and microstructure [1, 2]. Significant progress has been achieved in the synthesis and characterization of noble metal nanocrystals, which could be potentially used in various areas, including catalysis, sensors and medicine [3–7]. Due to their scarcity on the Earth and thus their high cost, the incorporation of noble metals with other materials has become one of the most effective ways to make full use of these special materials [8–10], and exploring graphene-based hybrids is one of them [11–13]. Moreover, graphene-based hybrids with other carbon nanostructures are also of special interest, because of their similarity in chemical composition and difference in physical morphology [14].

4.2. Graphene-Noble Metals

4.2.1. Graphene-Ag

It has been shown that the localized surface plasmon resonance (LSPR) of Ag nanocrystals can be readily tuned from the visible to the near infra-red (NIR) regime, by controlling their crystal size and

[1] School of Materials Science and Engineering, Nanyang Technological University, Singapore.
[a] E-mail: elbkong@ntu.edu.sg
[b] E-mail: mycboey@ntu.edu.sg
[c] E-mail: yzhuang@ntu.edu.sg
[d] E-mail: xuzc@ntu.edu.sg
[2] School of Mechanical & Aerospace Engineering, Nanyang Technological University, Singapore; E-mail: kzhou@ntu.edu.sg
[3] School of Materials Science and Engineering, The University of New South Wales, Australia; E-mail: sean.li@unsw.edu.au
[4] Electronic Materials Research Laboratory, School of Electronic and Information Engineering, Xi'an Jiaotong University, P. R. China; E-mail: wxque@xjtu.edu.cn
[5] Singapore Institute of Manufacturing Technologies (SIMTech), Singapore; E-mail: hhuang@SIMTech.a-star.edu.sg
[6] Anhui Target Advanced Ceramics Technology Co. Ltd., Hefei, Anhui, P. R. China; E-mail: 13335516617@163.com
* Corresponding author

morphology [2, 15]. As a result, Ag nanostructures have been widely used for SERS, optical sensing and cell imaging. Furthermore, Ag NPs are promising catalysts of various oxidation and oxidative coupling reactions [6]. To synthesize G–Ag hybrids, $AgNO_3$ is the most commonly used precursor, because it is cheap and widely available. More importantly, it can be easily reduced to Ag by using various reducing agents, such as amines, $NaBH_4$ and ascorbic acid [16–22]. Various synthetic approaches, such as photochemical reduction, substrate induced galvanic reaction and thermal evaporation, have been adopted to produce G–Ag hybrids [18–20, 23–33].

A simple method was developed to deposit Ag NPs on the surface of GO or rGO single-layer nanosheets, which were immobilized on a 3-aminopropyltriethoxysilane (APTES)-modified-Si/SiO_2 substrate. The Ag NPs were derived from $AgNO_3$ aqueous solution by heating at 75°C without using any reducing agent [16]. Figure 4.1 shows the experimental process to synthesize Ag NPs on GO and r-GO surfaces. GO nanosheets with micro-scale lateral size were first adsorbed on a APTES-modified SiO_x substrate (step 1). The GO nanosheets had a thickness of about 1.3 nm, implying that they were single layer. The density of GO on the APTES-modified SiO_x substrate could be controlled by adjusting the concentration of the GO aqueous suspension and the incubation time. The rGO substrate was obtained by reducing the GO substrate in hydrazine vapor (step 2). After the rGO substrate was heated in a 0.1 M $AgNO_3$ aqueous solution at 75°C for 30 min (step 3), Ag NPs were deposited on the surface of the rGO nanosheets, as shown in Fig. 4.2 (a). In this case, the rGO substrate served as not only a template to host the Ag NPs but also as a reducing agent. The Ag particles exhibited sizes ranging from tens of nanometers to 1 μm.

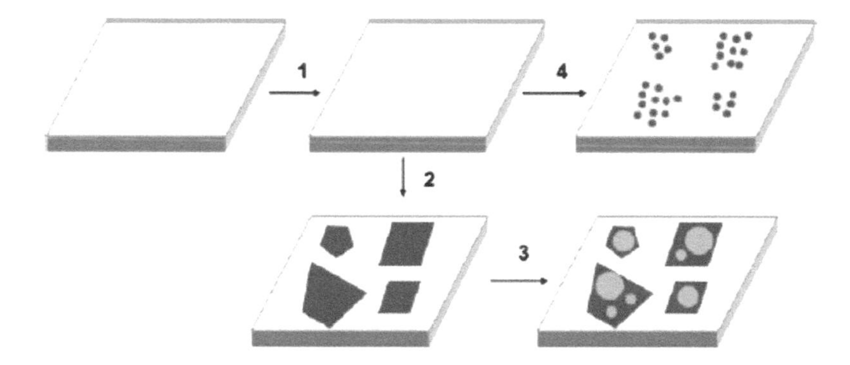

Fig. 4.1. (1) GO adsorbed on a APTES-modified SiO_x substrate. (2) GO reduced to rGO. (3) Growth of Ag particles by heating the rGO substrate in 0.1 M $AgNO_3$ at 75°C for 30 min. (4) Growth of Ag NPs by heating the GO substrate in 0.1 M $AgNO_3$ at 75°C for 30 min. Reproduced with permission from [16], Copyright © 2009, American Chemical Society.

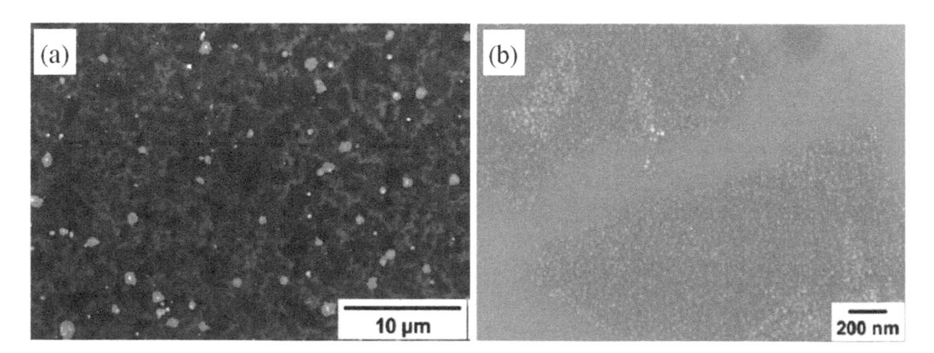

Fig. 4.2. SEM images of the Ag particles grown on the rGO surface (a) and GO surface (b). Reproduced with permission from [16], Copyright © 2009, American Chemical Society.

As the GO substrate was immersed in 0.1 M $AgNO_3$ aqueous solution at 75°C for 30 min (step 4), a large number of NPs were formed on the surface of the GO nanosheets, as illustrated in Fig. 4.2 (b). The NPs were about 6.0 nm. Comparatively, the Ag NPs on the GO surface had a much higher density and a smaller size than those on the rGO surface. This was because the GO contained much more functional groups to provide nucleation sites. At the same time, the GO also had many small aromatic conjugated domains to serve as electron-donors to reduce Ag^+ ions to form Ag NPs [34, 35]. Therefore, the carboxylic acid, hydroxyl and epoxide groups on the surface of GO could have served as nucleation sites for growth of the Ag particles. In contrast, once the GO was reduced to rGO, the number of functional groups was significantly decreased, resulting in the formation of fewer Ag particles. Also, the large-restored π-conjugated network would provide more electrons to reduce Ag^+ ions, thus leading to the formation of larger Ag particles. It was found that no NPs were observed on bare APTES-modified SiO_x, showing the importance of the template effect of the GO or rGO.

A solution-based method was reported to prepare GO–Ag hybrids, by embedding Ag NPs between the GO nanosheets [20]. The Ag NPs were stabilized without the requirement of functionalization on the particle surface. The hybrids exhibited optoelectronic properties that could be tuned over several orders of magnitude, by controlling the content of the Ag NPs, making them potential flexible and transparent semiconductors or semimetals. The GO–Ag hybrids were reduced into G–Ag ones through a chemical process. Figure 4.3 shows a schematic diagram of the synthesis process. To prepare the GO–Ag hybrid, Ag NPs were deposited on GO nanosheets through chemical reduction of Ag^+ by GO in KOH aqueous solution. The GO–Ag hybrid suspension could be used for deposition on any substrate, while the hybrid powder could be obtained by evaporating the suspensions at reduced pressure, as shown in Fig. 4.3 (a). Reduction of GO–Ag into G–Ag was conducted in hydrazine environment for 24 h.

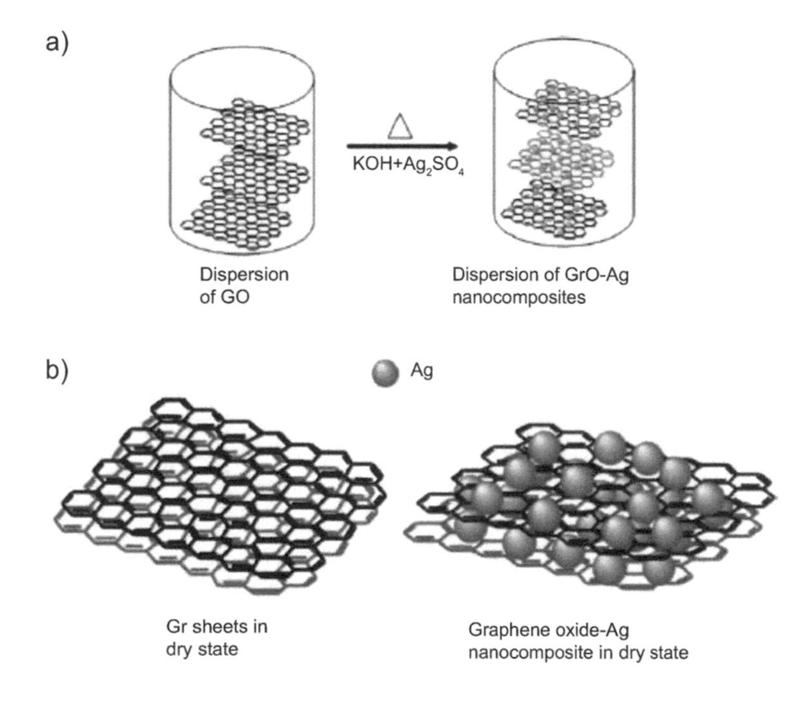

Fig. 4.3. Steps involved in the synthesis of the G–Ag hybrids. (a) Schematic diagram of the GO nanosheets and Ag nanoparticle-modified GO nanosheets in dispersion. (b) GO nanosheets without the spacers and those modified with the Ag NPs. The Ag NPs acted as nanoscale spacers to increase the interlayer spacing of the GO nanosheets, so that both faces of the GO nanosheets were available for NP deposition. Reproduced with permission from [20], Copyright © 2009, Wiley-VCH Verlag GmbH & Co. KGaA, Weinheim.

Figure 4.4 shows representative SEM and TEM images of the hybrids [20]. SEM images of the graphene films without and with the decoration of Ag NPs are shown in Fig. 4.4 (a) and (b), respectively. Evidently, the 2D structured GO nanosheets served as a template to disperse the Ag NPs. Individual nanosheets were observed, while the absence of charge during the SEM imaging implied that the network of the graphene nanosheets and the individual nanosheets were electrically conductive. As illustrated in Fig. 4.4 (b), the GO–Ag hybrid was free of agglomeration. Figure 4.4 (c) indicated that the GO nanosheets had a smooth finish but with wrinkles due to the thin sheet nature. The long wrinkles were observed on the larger GO nanosheets. The edges of the nanosheets were folded, thus appearing as ribbons. The measured lattice-fringe spacing of 0.34 nm in the ribbons corresponded to the (111) crystal plane of the nanosheets (Fig. 4.4 (d)).

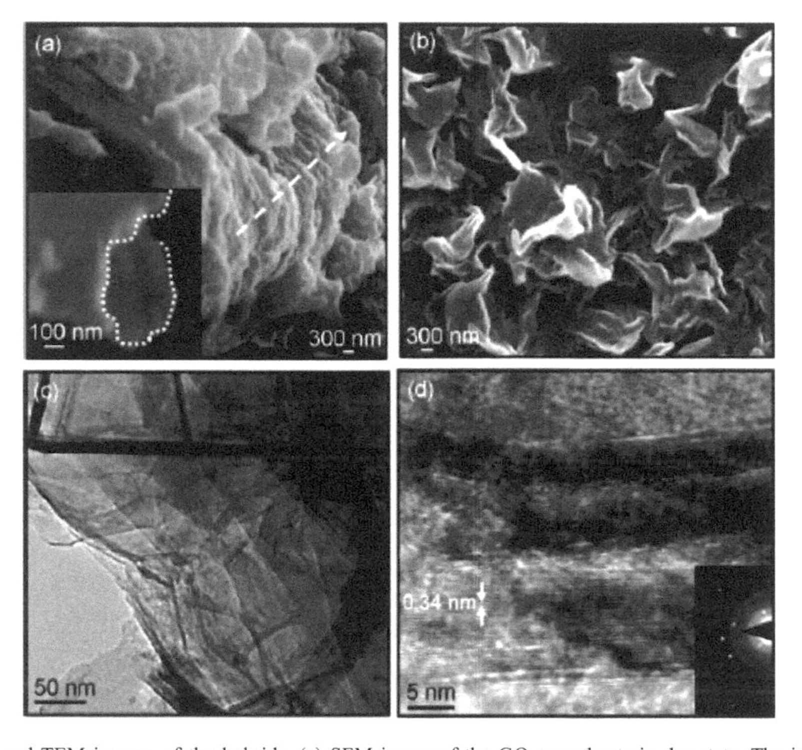

Fig. 4.4. SEM and TEM images of the hybrids. (a) SEM image of the GO nanosheets in dry state. The inset shows the magnified folded regions of the exfoliated nanosheets, demonstrating the large and isometric shapes of the 2D GO. A layered structure was observed at the edge of the agglomerates. (b) SEM image of the Ag–GO hybrid, in which an exfoliated bundle was formed with randomly dispersed, thin and crumpled nanosheets. (c) TEM images of a GO flake folded over itself, with sizes of several hundred nanometers. (d) HRTEM image showing a lattice-fringe spacing of 0.34 nm corresponding to the (111) crystal plane of the nanosheets, with the inset showing an electron diffraction pattern of as-made GO with excellent crystallization. Reproduced with permission from [20], Copyright © 2009, Wiley-VCH Verlag GmbH & Co. KGaA, Weinheim.

A wet-chemical method was presented to develop a new class of gauze-like hybrids, which consisted of GO nanosheets that enwrapped Ag/AgX (X = Br, Cl) NPs [32]. The GO/Ag/AgX hybrids showed high efficiency in harvesting visible light through plasmonic excitation. To synthesize the hybrids, GO aqueous solution and $AgNO_3$ solution were added to a surfactant solution of cetyltrimethylammonium bromide (CTAB) or cetyltrimethylammonium chloride (CTAC) in chloroform. The mixture was strongly stirred for a certain while, so that milky dispersions containing Ag/AgX/GO hybrids were obtained. After that, the white yellowish solids were collected and washed. Ag/AgX samples without the presence of GO were also prepared similarly. The synthesis process was carried out without being protected from the ambient light. In contrast, if the experiments were conducted in a strictly darkened room, with only a weak infrared lamp, the samples exhibited nearly no plasmon resonance absorptions in the visible region, which implied that the no Ag was produced. In other words, the formation of metal Ag NPs was triggered by the ambient light.

Figure 4.5 shows SEM images of the samples [32]. The Ag/AgX NPs had an average size of about 500 nm. With the presence of GO, the average particle size was decreased to about 200 nm. The NPs were entirely enwrapped with gauze-like GO nanosheets, thus forming Ag/AgX/GO hybrids. The smaller size of the Ag/AgX NPs in the Ag/AgX/GO hybrids, as compared with that of the corresponding Ag/AgX NPs without GO, was attributed to the effect of the GO nanosheets. This is because GO is an unconventional amphiphile, which could act as a surfactant in an oil-water system, so as to further improve the dispersibility of water phase in the chloroform solution of surfactants [36].

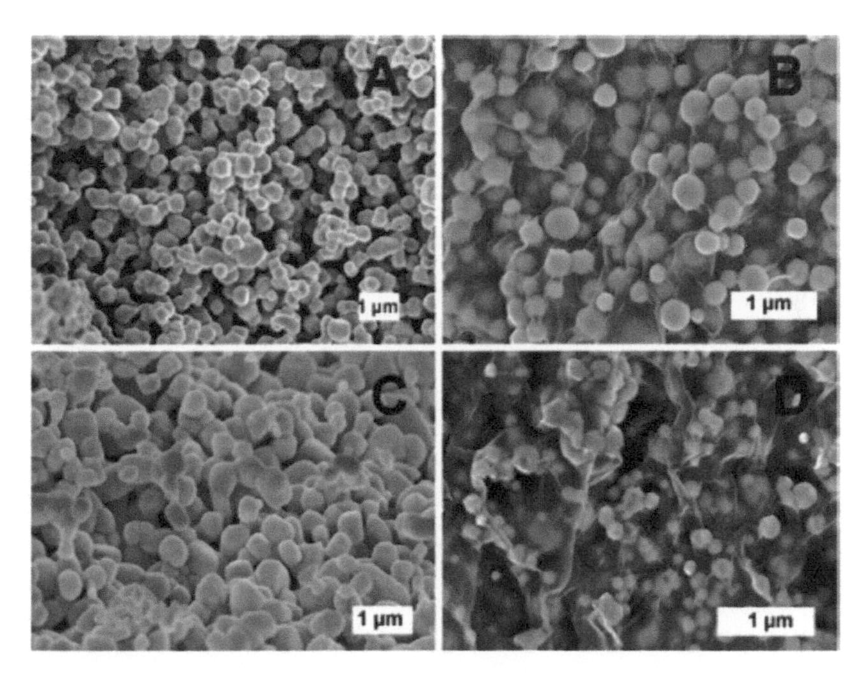

Fig. 4.5. Representative SEM images of the samples without and with GO (hybrids): (A) Ag/AgBr, (B) Ag/AgBr/GO, (C) Ag/AgCl and (D) Ag/AgCl/GO. Reproduced with permission from [32], Copyright © 2011, American Chemical Society.

4.2.2. Graphene-Au

Due to their unique properties, Au NPs have drawn significant attention in recent years with potential applications in biosensing, intracellular gene regulation, photonics and catalysis [1–7, 15]. Especially, these applications were highly dependent on size and morphology, which has triggered considerable interest in the synthesis of Au NPs with different methods. In this case, 2D structured graphene and its derivatives, with residual oxygen-containing functional groups on the surface, have been shown to be an excellent platform to deposit and grow Au NPs, i.e., G–Au hybrids [37, 38]. Up to now, G–Au hybrids have found a range of applications, such as SERS, catalysis, and electrochemical sensors [39–55]. The most straightforward method to prepare G–Au hybrids is through the direct chemical reduction of the Au precursor (e.g., $HAuCl_4$), in the presence of GO or rGO nanosheets, by using various reducing agents, including amines, $NaBH_4$ and ascorbic acid [40, 42, 44, 46, 48, 53, 55]. Various methods have been developed to synthesize hybrids, consisting of Au NPs on the GO or rGO nanosheets [39, 41, 43, 45, 47, 49, 52, 54, 56–59]. For instance, a G–Au hybrid has been produced by reducing $AuCl_4$ with $NaBH_4$ in rGO octadecylamine (ODA) solution [50, 51].

A simple electrochemical deposition method was used to prepare the G–Au hybrid [47]. Au NPs were deposited on the surface of graphene nanosheets, with morphologies that could be readily tailored by controlling the electrodeposition time and concentration of the precursor ($AuCl_4^-$). The Au-G hybrids exhibited promising electrocatalytic activities in oxygen reduction and glucose oxidation. The performance of the hybrid was much better than that of the mere Au NPs or graphene, demonstrating the synergistic effect of the two components.

Graphene nanosheet powder was dispersed in double distilled water to form a black suspension with a concentration of 1 $mg \cdot ml^{-1}$. To prepare G–Au/GC (glassy carbon) electrode, the suspension was cast on the surface of the pretreated GC electrode by using a microsyringe. After that, the solvent was evaporated at ambient temperature, so that a uniform graphene film was formed on the GC electrode (graphene/GC). Au NPs were electrodeposited on the graphene/GC electrodes in aqueous solutions of 0.5 $mol \cdot l^{-1}$ H_2SO_4 containing 1 $mmol \cdot l^{-1}$ $HAuCl_4$ as the precursor at a potential of −0.5 V (versus SCE) for 10 min. The solutions were deaerated by purging high-purity N_2 gas and were then kept in N_2 environment to isolate them from oxygen. The concentration of the precursor (from 0.1 $mmol \cdot l^{-1}$ to 10 $mmol \cdot l^{-1}$) and deposition time (from 1 min to 30 min) were used to control the morphology of the deposited Au NPs. After deposition, the samples were thoroughly rinsed with double distilled water.

Figure 4.6 shows SEM images of the G–Au hybrid electrodes and the Au NPs obtained after deposition for different times [47]. It was found that the density of the Au NPs increased gradually, as the deposition time was increased from 1 min to 30 min. After deposition for 30 min, the graphene/GC electrode surface was nearly entirely covered by the Au NPs, as shown in Fig. 4.6 (d). The size of the Au NPs was not significantly changed as the deposition time was increased from 1 min to 10 min (Fig. 4.6 (a–c)), while a sharp increase in particle size was observed after deposition for 30 min (Fig. 4.6 (d)). Furthermore, aggregation of the Au NPs occurred, which led to a decrease in catalytic efficiency. As a result, the sample with the deposition time of 10 min was the optimized one in terms of catalytic effect for glucose oxidation.

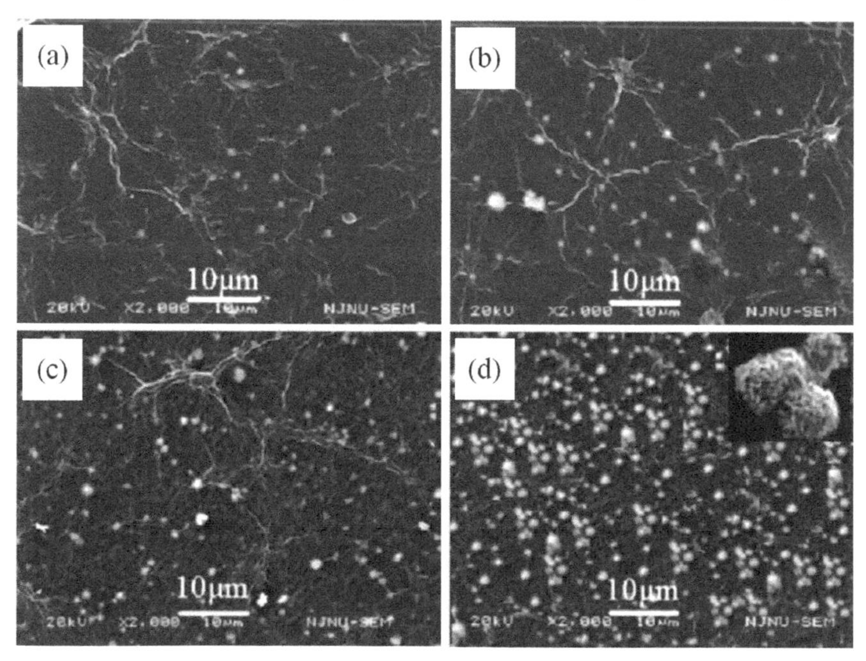

Fig. 4.6. SEM images of the Au–graphene/GC electrode fabricated through electrodeposition for 1 min (a), 5 min (b), 10 min (c) and 30 min (d) at a potential of −0.5 V in 0.5 $mol \cdot l^{-1}$ H_2SO_4 solution containing 1 $mmol \cdot l^{-1}$ $HAuCl_4$. The inset in panel (d) shows an image of the Au NPs at a high magnification. Reproduced with permission from [47], Copyright © 2010, Elsevier.

Figure 4.7 shows SEM images of the Au NPs obtained after deposition in solutions with different precursor concentrations [47]. Obviously, the density of the Au NPs increased with increasing precursor concentration, as demonstrated in Fig. 4.7 (a, c, e). Meanwhile, the precursor concentration also exhibited

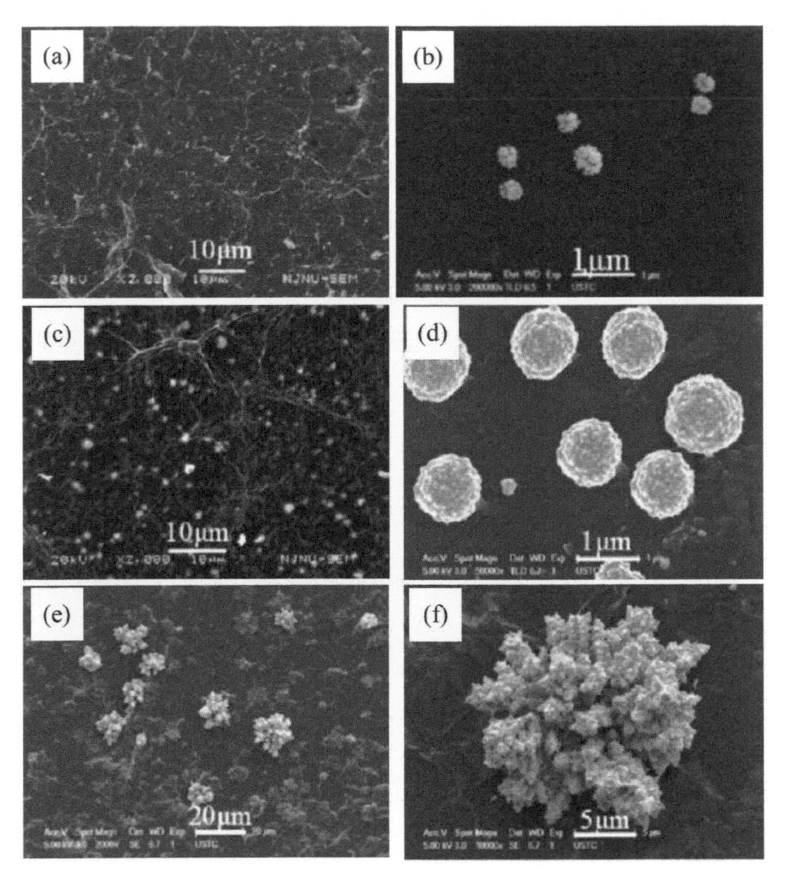

Fig. 4.7. SEM images of the Au NPs electrodeposited from 0.5 mol·l^{-1} H_2SO_4 solution containing 0.1 mmol·l^{-1} (a, b), 1 mmol·l^{-1} (c, d), and 10 mmol·l^{-1} $HAuCl_4$ (e, f), at a potential of −0.5 V for 10 min. Reproduced with permission from [47], Copyright © 2010, Elsevier.

a significant influence on the morphology and size of the Au NPs, as illustrated in Fig. 4.7 (b, d, f). After deposition in 0.1 mmol·l^{-1} $AuCl_4^-$ solution, the Au NPs appeared as 3D dendritic structures, with diameters of 250–300 nm. As the precursor concentration was increased to 1 mmol·l^{-1}, the Au NPs were spherical, with a diameter of 1.3 ± 0.2 μm, which consisted of small primary NPs with an average diameter of about 50 nm. At the highest precursor concentration of 10 mmol·l^{-1}, flower-like Au NPs with sizes of 15–20 μm were formed, which were constructed with small flakes with a size of about 200 nm. These results suggested that the density, size and morphologies of the Au NPs could also be well-controlled by adjusting the precursor concentration. The electrocatalytic activity of the hybrids for glucose oxidation was synergistically influenced by these factors.

Besides spherical Au NPs, various anisotropic Au nanostructures have also been prepared on graphene-based materials [40, 42, 44, 54]. As an example, the synthesis and characterization of snowflake-shaped Au nanostructures (SFGNs) on GO nanosheets are discussed here [54]. This study has demonstrated that graphene chemical derivatives (GCDs) could be used as excellent substrates for metal nuclei seeding and subsequent growth into nanodendritic structures. For GO nanosheets, the process was controlled by diffusion limited transport of the Au^{3+} ions, with anisotropic lattice incorporation of the gold atoms, during the seeding growth, which was governed by the mass-transfer rates, as shown in Fig. 4.8 (a). The highly anisotropic SFGNs, which were templated on the GO nanosheets, with thicknesses of 1–4 nm (1–5 single atomic carbon layers) and areas of 25–200 μm^2, had five primary branches with several sharp-edged secondary branches. The GO–SFGN hybrids had increased electrical conductivity Raman signals from GO. Moreover, as the GO–SFGN hybrids were reduced to G–SFGN hybrids, an apparent band gap of 164.24 meV and a

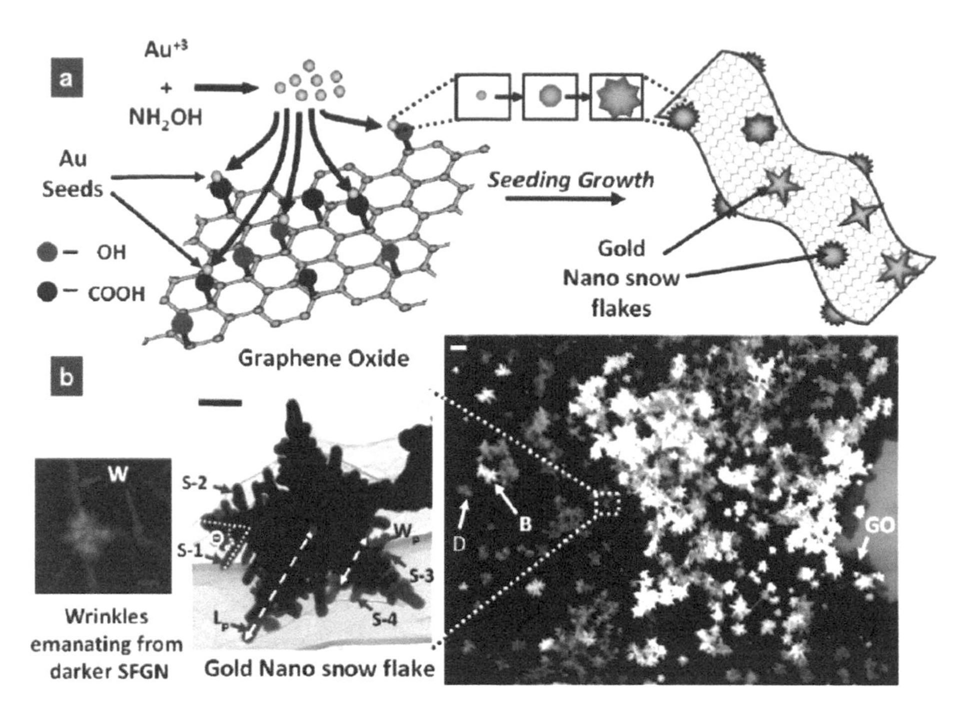

Fig. 4.8. Formation mechanism of the snowflake-shaped gold nanostructures (SFGNs) on GO nanosheets. (a) Interfacing the –COOH and –OH groups on the GO nanosheets with a freshly prepared solution of Au nuclei, formed during hydroxylamine assisted reduction of Au^{3+}, resulting in nuclei attachment and seed-mediated formation of SFGNs on the GO surface. (b) Right: FESEM of the SFGNs templated on GO lying on the silica surface. The SFGNs appearing darker (labeled as D) were on the rear surface, while the SFGNs appearing brighter (labeled as B) were on the front surface of the immobilized GO nanosheet (scale bar = 500 nm). Center: HRTEM image showing the detailed characteristics and the structural parameters of the SFGN exhibiting a dendritic morphology. Scale bar = 100 nm. Left: FESEM image showing wrinkles (labeled as W) on GO associated with a darker SFGN. Scale bar = 200 nm. Reproduced with permission from [54], Copyright © 2009, American Chemical Society.

Schottky barrier with a height of 38.98 meV were observed. Such a graphene–metal interface could find applications for the next generation graphene-based electronics, because the semiconductor properties of graphene could be modulated and integrated with other GCDs to make graphene logic devices and bioimaging/diagnostic systems. The surface-enhanced Raman spectroscopy (SERS) from SFGNs could be used for sensing biocomponents and bioprocesses. The optical properties of the SFGNs could be employed with the electrical sensitivity of graphene to construct solar cells and optical sensors. The IR absorption due to the anisotropic gold nanostructures could be used in thermal energy generators or nanocatalysts.

The oxy-functional groups on the GO nanosheets served to stabilize the gold nuclei, which were derived through *in situ* hydroxyl-amine assisted reduction of gold salt. The gold nuclei grew into the GO–SFGN hybrids. The experiment was started with the mixture of hydroxylamine (NH_2OH, 50% w/v) and 0.275 mM $HAuCl_4 \cdot 3H_2O$, together with GO suspension (80 mM carboxylic acid). After constant agitation at room temperature for 1 h, the GO–SFGN hybrid nanosheets were immobilized on an amine-silanized silica substrate. As shown in Fig. 4.8 (b, right), the GO served as an excellent template to deposit SFGNs on [54]. The relatively dark SFGNs in FESEM image (labeled as "D") were attributed to the fact that they were attached on the rear GO surface, thus giving lesser average surface electron density. The brighter SFGNs (labeled as "B") were deposited on the exposed GO surfaces. This is because the SFGNs were nucleated on both sides of the GO nanosheets that were exposed to the solution, as evidenced by the wrinkles of the GO nanosheets from the darker SFGNs (Fig. 4.8 (b left)).

The SFGNs could have either five or six primary branches (N_p), which contained secondary or side branches. Figure 4.8 (b, center) shows TEM image of an SFGN, with five primary branches (N_p = 5).

The primary branch had an average length (L_p) of 260 nm and an average width (W_p) of 120 nm. Each primary branch backbone structure consisted of parallel secondary or side branches with the same angles of emergence (Θ_e), i.e., the angle between a side branch and its primary branch backbone, ranging from 35° to 90° from branch to branch. This implied that the growth direction of secondary-branches was influenced by the crystal-lattice of their common primary branch, instead of random nucleation. The length of the secondary branches decreased from the end to the tip, which indicated that those near the center grew for a longer time. In other words, the primary branches were grown in a progressive way. Such secondary branching became more and more pronounced with increasing concentration of Au^{3+} ions in the solutions. Even ternary branches were observed in the SFGNs grown from high concentration solutions.

The SFGNs demonstrated a strong interaction with the surface of the GO nanosheets, which was confirmed by the Raman scattering signal analysis results. The Raman spectrum of GO usually has two peaks, i.e., D-band line (~1340 cm^{-1}) and the G-band line (~1590 cm^{-1}). In the GO–SFGN hybrids, their intensities were increased by more than two times. This surface enhancement in Raman signals was attributed to the presence of chemical interaction or bonding between the SFGNs and the GO nanosheets [60, 61]. As the density of the Au nanostructures was increased, the enhancement could be further enlarged. It was found that the SFGNs grew more favorably in the lateral direction than in the vertical direction to the GO surface, due to the template effect of the GO nanosheets in the solutions. Without the presence of GO, in the seeding solution of NH_2OH and Au salt, irregular aggregates with large sizes appeared as a precipitate.

A modified two-step model was proposed to explain the formation mechanism of the SFGNs [54]. Firstly, Au nuclei were formed on the GO nanosheets through the NH_2OH-assisted reduction of Au^{3+}, and were stabilized by the negatively charged $-COO^-$ and $-OH^-$ groups on the surface of the GO nanosheets. This step was confirmed by growing Au nuclei on GO, that were reduced with sodium borohydride in the presence of sodium citrate, as shown in Fig. 4.9 (a). In the second step, the Au nuclei started to grow, and their growth was sustained through hydroxyl-amine assisted Au^{3+} ion reduction, due to the catalytic effect of the Au surface [62].

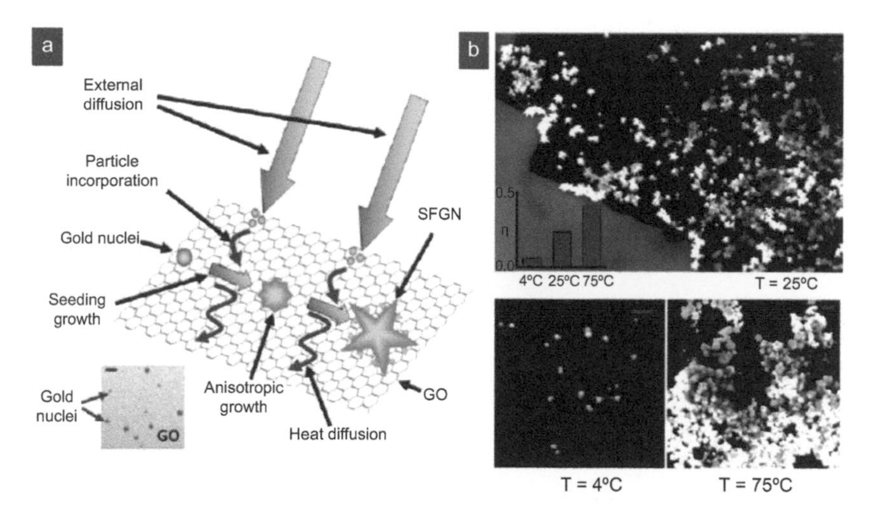

Fig. 4.9. Growth mechanism of the SFGNs on GO nanosheets and their structural dependence on the reaction temperature. (a) Schematic showing the elementary steps involved in the seeding growth of the SFGNs on GO template. Au ions diffused from the solution to the GO nanosheets where they were catalytically reduced and incorporated into the growing Au nuclei. Bottom inset shows the seed particles on GO that were prepared through sodium borohydride-assisted reduction of Au^{3+} in the presence of sodium citrate and GO. Scale bar = 10 nm. (b) Morphology and density of the GNs obtained at different reaction temperatures. GNs synthesized at room temperature (25°C) exhibited dendritic "snowflake" morphology with a high coverage on GO. At low temperature (4°C), the GNs had a spherical morphology with less coverage, while at higher temperature (75°C), the GNs exhibited a random cluster morphology with very dense coverage. Scale bars = 500 nm. The inset shows surface coverage index for the GO–Au hybrids synthesized versus reaction temperature. Reproduced with permission from [54], Copyright © 2009, American Chemical Society.

The formation of the gold nanostructures (GNs) with dendritic morphology was attributed to the diffusion limited kinetics during the seeding growth process. As shown in Fig. 4.9 (a), the seeding growth of the Au nuclei on GO nanosheets experienced two elementary processes, i.e., (i) external diffusion of the Au^{3+} ions from the solution to the nuclei and (ii) incorporation of the Au atoms into the crystal structure by hydroxylamine-induced catalysis. The heat related to the formation of the SFGNs at nanoscale was dissipated into the solution. Therefore, the net resistance to SFGNs' growth consisted of those from the two elemental steps, i.e., $k_F^{-1} = k_D^{-1} + k_G^{-1}$, where k_F is the rate-constant for the formation of the SFGN, k_D is the rate-constant for diffusion of the Au ions from the solution to the surface of the GO nanosheets and k_G is the rate-constant of incorporation of the Au atoms into the SFGN nanostructures. Because the incorporation step was catalytic, k_G was relatively higher; the external diffusion became the rate controlling step. This suggestion has been evidenced by the dependence of the SFGN nanostructures on the reaction temperature.

Figure 4.9 (b) shows FESEM images of the nanostructures synthesized at 4°C, 25°C and 75°C. SFGNs were only available at 25°C, implying their sensitivity to the synthesis temperature. The size and density of the nanoparticles increased with the increasing reaction temperature. The surface density of the GNs was evaluated by using a surface coverage index, i.e., the fraction of the GO surface covered by the Au nanostructures. The values were 0.06, 0.24 and 0.41, for the three temperatures, respectively, as shown in the inset of Fig. 4.9 (b).

At low temperature (4°C), as compared with the diffusive resistance, $k_D \propto T^{-1.5}$, the particle incorporation resistance, $k_G \propto \exp(E_A/(RT))$, was relatively larger, so that smaller particles with a lower density were formed. At 25°C (moderate temperature), the diffusive resistance was increased faster and became larger than the particle incorporation resistance, thus resulting in the formation of SFGNs. Because there was a difference in the surface chemical potential (μ), anisotropic particle growth took place on the surface of the SFGNs. Irregular surfaces have higher μ than blunt surfaces, thus growing more preferentially [63, 64]. The average rate of Au influx for the formation of an SFGN ($N_p = 5$) was about 0.135 g·cm^{-2}·h^{-1} at 25°C. At high temperature (75°C), both the mass-transfer rates and the chemical potentials were increased, which triggered the quick formation of disordered clusters but a high surface coverage, as shown in Fig. 4.10 (c). However,

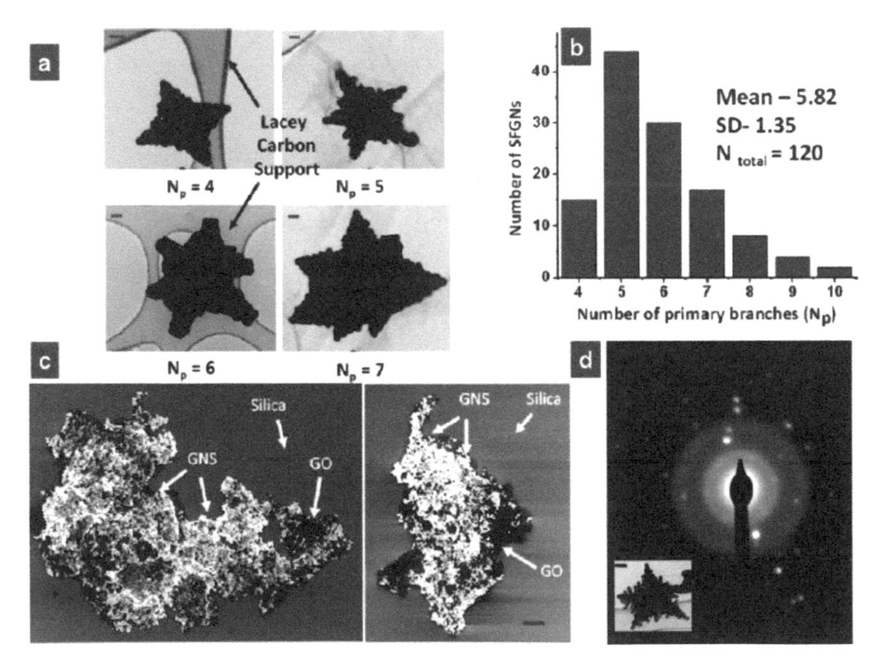

Fig. 4.10. Structures of the SFGNs. (a) TEM images of the SFGNs with 4, 5, 6, and 7 primary branches. Scale bar = 100 nm. (b) Distribution of primary branches in the SFGNs synthesized at 25°C. (c) FESEM images of the Au nanostructure obtained at 70°C with a high surface coverage density and high selectivity of Au on GO. Scale bar = 5 μm. (d) Selected area electron diffraction (SAED) pattern of an SFGN with $N_p = 5$ (inset), indicating the presence of defects and multiple crystal domains. Scale bar = 100 nm. Reproduced with permission from [54], Copyright © 2009, American Chemical Society.

the GO–Au nanohybrids obtained at 75°C possessed a higher enhancement factor in SERS signals (800%), as compared with that of the 25°C sample (250%). At higher concentrations of Au^{3+} ions in the solution, the growth of the secondary branches could be further promoted, with ternary branches being present.

Representative TEMs of the SFGNs obtained at 25°C, with N_p = 4, 5, 6 and 7, and the distribution of N_p, are shown in Fig. 4.10 (a) and (b) [54]. N_p had a relatively wide range of 4–12, with nearly half of the SFGNs (~49.2%) having N_p of 5 and 6, while only a small fraction of them had N_p values of ≥8 (~11.8%). Figure 4.10 (d) shows a selected area electron diffraction (SAED) pattern of the SFGNs (inset) with N_p = 5, exhibiting a mixed diffraction pattern, which indicated that the SFGNs contained defects and multiple crystal domains. If the SFGNs had incompletely formed secondary branches, the number of crystal defects was reduced. The secondary branches possessed similar angles, which suggested that they were grown on the primary branch due to similar crystal defects.

Other types of Au nanostructures, such as nanodots (NDs), nanorods (NRs), nanowires (NWs) and nanosheets, have been synthesized on GO or rGO nanosheets [40, 42, 44, 46, 59]. A one-pot photochemical method was developed to synthesize Au NDs on thiol-modified GO/rGO nanosheets through the reduction of $HAuCl_4$ [40]. In this case, 1-octadecanethiol (ODT) was preabsorbed on the GO/rGO nanosheets. The Au NDs *in situ* self-assembled into ordered chain-like structures, along the <100> direction of the rGO surface.

The Au NDs had average sizes of 1.2 ± 0.3 nm, with strong UV absorption at a peak of about 230 nm. The solution of Au NDs on rGO showed an absorption shoulder at ~260 nm, corresponding to rGO. The Au NDs were fluorescent with red emission (λ_{ex} = 360 nm and λ_{em} = 608 nm), which was quenched when the NDs were *in situ* synthesized on the rGO nanosheets. As shown in Fig. 4.11, the Au NDs on rGO were assembled into a regular pattern. HRTEM image of an individual Au ND is shown in the inset of Fig. 4.11 (a), in which the lattice fringe of 2.4 Å corresponded to the inter-plane spacing of face-centered cubic (fcc) Au (111). The particle chains formed with the Au NDs had lengths of 10–40 nm. The majority orientation of the chains was in the equi-anglar directions, rotating one another by about 60°. SAED pattern in Fig. 4.11 (b) revealed the reflection spots for graphene G{100} and G{110}, associated with the directions along which the Au ND chains were oriented. The distance between two adjacent and parallel Au ND chains was about 4.4 nm, as shown in Fig. 4.11 (b), which was comparable with the period of the linearly localized thiol groups of ODT [65].

Figure 4.12 shows a schematic diagram demonstrating the formation and self-assembly of the Au NDs [40]. It was found that the assembly of the Au NDs required three conditions. First of all, the ligand molecules had to be self-assembled into ordered patterns on the template. For example, if GO could be used as a template to form Au ND chains. Secondly, the size of the Au NDs had to be smaller than the period of the ordered thiol groups. As shown in Fig. 4.12, the periodical unit of the paired thiol groups of ODT (about 4.4 nm) was larger than the size of the synthesized Au NDs (about 1.2 nm). Lastly, both the formation and assembly of the Au NDs had to be *in situ* on the rGO surface. In summary, the process

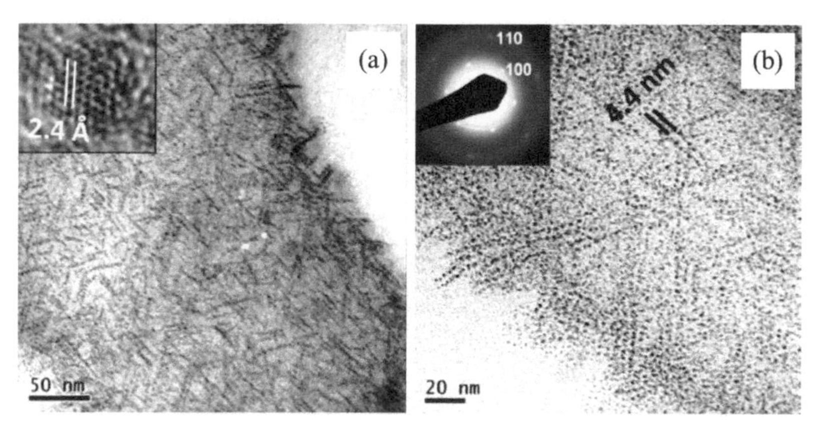

Fig. 4.11. (a) TEM and (b) magnified TEM images of the ODT-capped Au NDs synthesized *in situ* and assembled on the rGO nanosheet surface. Inset of (a) is a HRTEM image of an individual Au ND and inset of (b) is a SAED image of the area in (b). Reproduced with permission from [40], Copyright © 2010, Wiley-VCH Verlag GmbH & Co. KGaA, Weinheim.

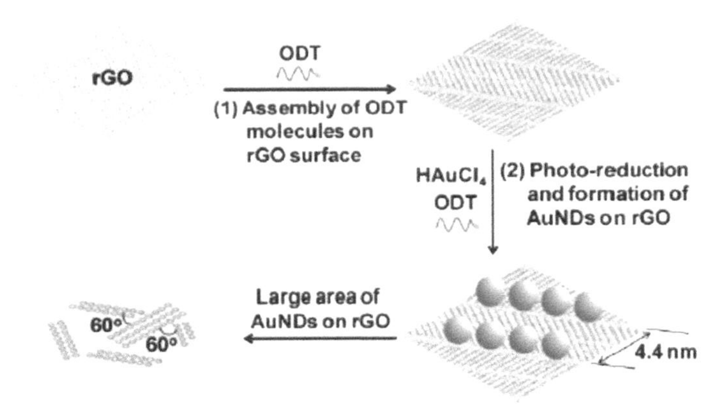

Fig. 4.12. Schematic diagram of *in situ* formation and assembly of the Au NDs on rGO nanosheets. Reproduced with permission from [40], Copyright © 2010, Wiley-VCH Verlag GmbH & Co. KGaA, Weinheim.

included (i) self-assembly of ODT molecules on the rGO nanosheets, (ii) reduction of Au^{3+} to Au on surface of the rGO and (iii) attachment of the nucleated Au NDs to the thiol groups. Experimental results indicated that if ODT-capped Au NDs were synthesized separately and later mixed with rGO nanosheets, then regular patterns could not be formed through assembly of the Au NDs.

Au NWs with alternating hcp and fcc domains were prepared by using GO as a template in the presence of 1-amino-9-octadecene [46]. Figure 4.13 shows TEM images and SAED pattern of the Au NWs, which were obtained by heating the mixture of $HAuCl_4$ and 1-amino-9-octadecene at 55°C for 36 h in the presence of GO nanosheets. The Au NWs had an average diameter of about 1.6 nm, as shown in Fig. 4.13 (a). HRTEM image revealed the hcp segment of an individual Au NW (Fig. 4.13 (b)). It was found that the Au NWs experienced a fast phase transformation and splitting under the irradiation of the electron beam, so that their long-range lattice structure could not be characterized. However, the hcp structure of the Au NWs was confirmed by XRD analysis. Meanwhile, the fcc structure was also observed in the XRD pattern, which was probably attributed to the presence of some spherical Au NPs as a byproduct, as demonstrated in Fig. 4.13 (a). All the Au NWs grew along the $<002>_h$ direction (or the $<111>_f$ direction for fcc phase), with a lattice spacing of 2.4 Å (Fig. 4.13 (b)). Discrete rings for Au$\{002\}_h$ plane (or Au$\{111\}_f$) and G{100} and G{110} of GO were observed in a SAED pattern of the Au NWs (Fig. 4.13 (c)).

As the reaction was conducted for only 10 h, with two-day ageing at room temperature, tadpole-shaped NWs were obtained with one end enlarged like a 'head' with sizes of 12–16 nm, as shown in Fig. 4.14 (a) [46]. Figure 4.14 (b) shows HRTEM image of a tadpole-shaped Au NW at the middle section, which was about 20 nm away from the 'head' with a diameter of about 7 nm. It consisted of hcp and fcc phases arranged alternatively. Fast Fourier transform (FFT) diffraction pattern of this area revealed streaks along the $[111]_f$ direction in the main fcc phase viewed along $[101]_f$ direction (inset of Fig. 4.14 (b)), due to the presence of stacking faults or hcp planes along the $[002]_h$ direction. The 'head' of the tadpole-shaped Au NW possessed a defect-free fcc crystal structure, as shown in Fig. 4.14 (c), supported by the FFT diffraction pattern oriented along the $[101]_f$ zone axis (inset).

In this case, the 1-amino-9-octadecene molecules acted as both the reducing agent for the reduction of Au^{3+} and the dispersant for the dispersion of GO in nonpolar hexane. Figure 4.15 shows a schematic diagram to explain the formation of the Au NWs [46]. After reaction for 10 h, Au seeds of 2 nm were formed on the GO nanosheets, which started to self-assemble into discontinuous chain-like structures, from the chain-like structured (1-amino-9-octadecene)AuCl complex through the reduction of Au^+ (Step (a)). The small Au seeds then continued to grow into short rod-like structures and finally Au NWs were formed (Step (b)). It was suggested that the oxidative etching and thinning of the Au NWs requires a sufficiently high temperature (i.e., 55°C). Therefore, short time reaction resulted in the formation of tadpole-shaped Au NWs.

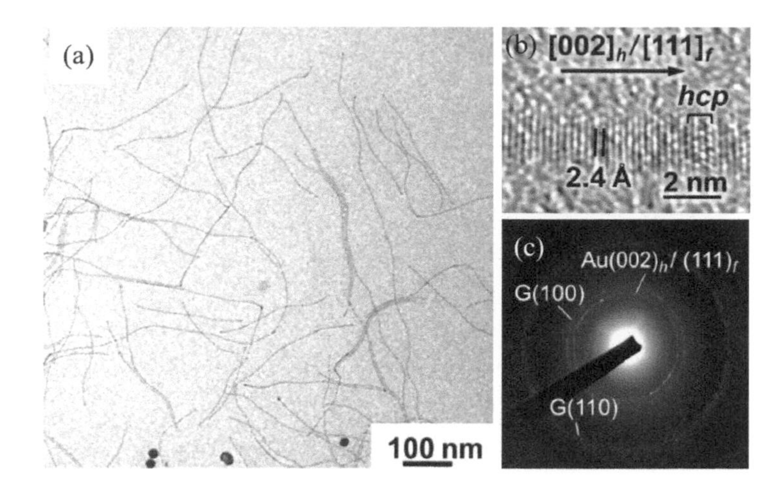

Fig. 4.13. (a) TEM image of the Au NWs grown on GO nanosheets, with the large black dots being Au NPs as byproducts. (b) HRTEM image of a section of an individual Au NW. (c) SAED pattern of a few Au NWs on GO nanosheets. Reproduced with permission from [46], Copyright © 2012, Wiley-VCH Verlag GmbH & Co. KGaA, Weinheim.

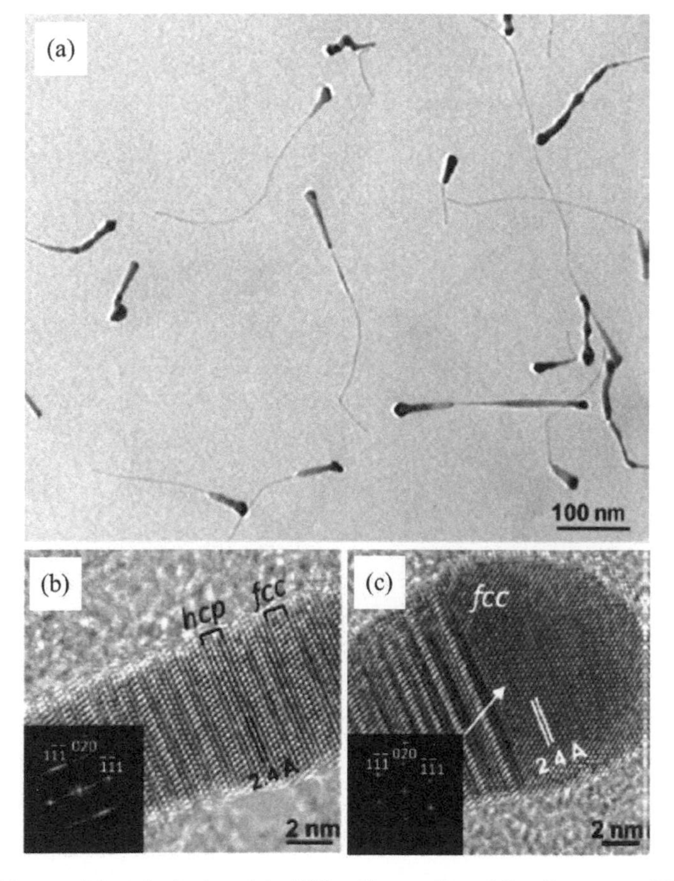

Fig. 4.14. (a) TEM image of the tadpole-shaped Au NWs with one enlarged 'head' grown on GO nanosheets. (b) HRTEM image of the middle section (about 20 nm away from the enlarged 'head') of an individual tadpole as shown in (a), with the inset to the corresponding FFT diffraction pattern. (c) HRTEM image of the enlarged 'head' of the tadpole-shaped Au NW, with the inset being the corresponding FFT diffraction pattern. Reproduced with permission from [46], Copyright © 2012, Wiley-VCH Verlag GmbH & Co. KGaA, Weinheim.

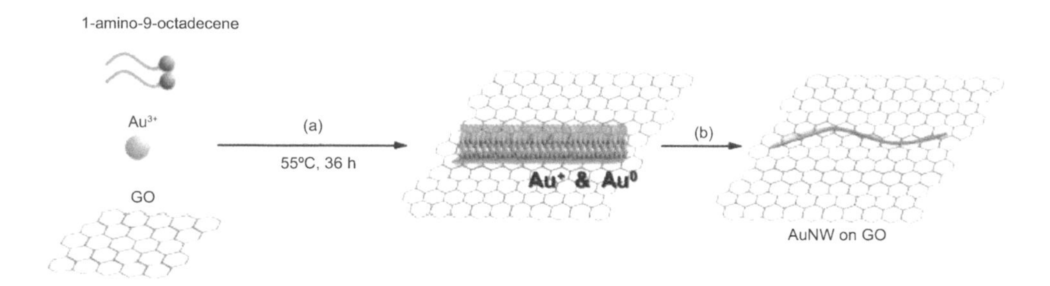

Fig. 4.15. Schematic diagram illustrating the formation process of the Au NWs through the reduction of the [(1-amino-9-octadecene)AuCl] complex on GO nanosheets. Reproduced with permission from [46], Copyright © 2012, Wiley-VCH Verlag GmbH & Co. KGaA, Weinheim.

Very interesting square-like Au nanosheets (SSs) were obtained on GO nanosheets by using a simple wet-chemical method [42]. The Au nanosheets had a hexagonal close-packed (hcp) structure that was stable under ambient conditions. The Au SSs were derived from a solution containing GO nanosheets, $HAuCl_4$ (4 mM) and 1-amino-9-octadecene (140 mM), $CH_3(CH_2)_7CH{=}C\ H(CH_2)_8NH_2$, in a mixture of hexane and ethanol with a volume ratio of 23:2, after heating at 55°C for 16 h. The Au SSs had lengths of 200–500 nm, as shown in Fig. 4.16 (a). Average thickness of the Au SSs was 2.4 ± 0.7 nm. Selected area electron diffraction (SAED) results indicated that the Au SSs were of hcp (2H type, with a symmetry of P63/mmc) crystal structure, as illustrated in Fig. 4.16 (c, d). Figure 4.16 (b) shows a high-resolution TEM (HRTEM) image of a small section of an individual Au SS, indicating the lattice pattern of a 2H-hcp Au along the $[110]_h$ zone axis. Crystal models, with 2D and 3D views, of the Au SSs have been proposed, as illustrated in the inset of Fig. 4.16 (a). The square basal plane in each nanostructure was normal to the

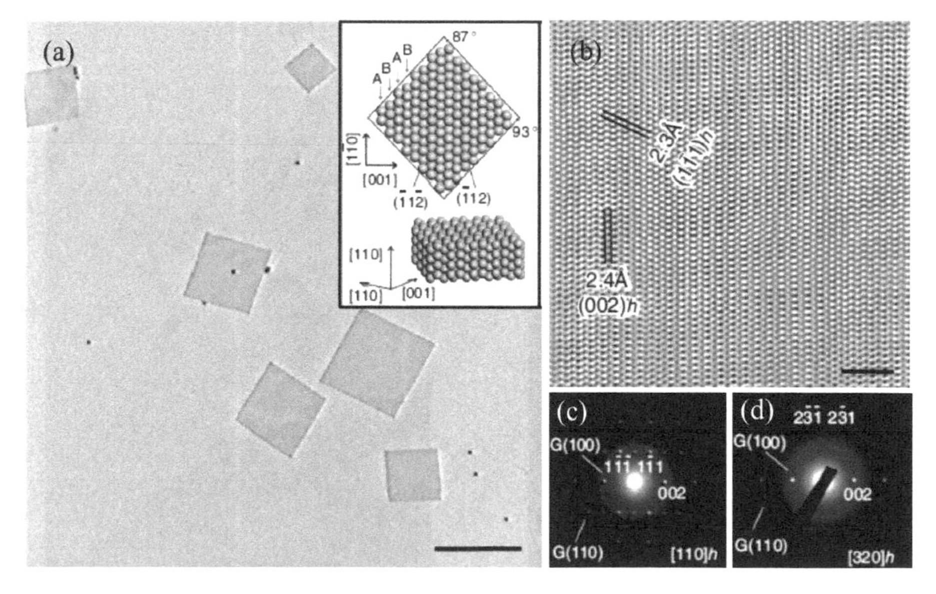

Fig. 4.16. TEM results of the Au SSs. (a) TEM image of the 2.4 nm thick Au SSs on GO nanosheets (scale bar = 500 nm). The black dots are Au nanoparticles (10–25 nm in diameter). The inset shows crystallographic models of the Au SSs with basal plane along the $[110]_h$ zone axis, showing ABAB stacking along the $[001]_h$ direction. (b) HRTEM image of a small region of an individual Au SS oriented normal to $[110]_h$, as indicated by the SAED in (c) (scale bar = 2 nm). (c) SAED pattern of an individual Au SS, showing diffraction rings of GO and spots of $[110]_h$ zone axis of the Au SS. (d) SAED of $[320]_h$ zone axis collected by tilting an Au SS from the $[110]_h$ zone axis along the $(110)_h$ reflection by about 19°. Reproduced with permission from [42], Copyright © 2010, Macmillan Publisher Ltd.

$[110]_h$ direction, with ABAB stacking along the $[001]_h$ direction and four sides enclosed by the $\{112\}_h$ planes. It was found that, if the heating time was increased, the thickness of the Au SSs also increased, while their square shape characteristic was not affected. However, the hcp phase became less stable as the thickness of the Au SSs was increased. Without the presence of the GO nanosheets, the Au SSs would have a wider size distribution and more spherical particles would be present. Also, the use of the mixed solvent of ethanol and hexane was an important requirement to produce Au SSs, due mainly to the effect of solvent polarity.

When the Au SSs were irradiated by an electron beam, a phase transformation from hcp to fcc occurred [42]. At the same time, after irradiation for about 20 s, the Au SSs gradually became porous, as shown in Fig. 4.17 (a, b). Figure 4.17 (c) revealed that the porous Au SS had a SAED pattern similar to that of twinned fcc phase. Initially, small domains of fcc packing were formed within the hcp matrix, whose long-range order was gradually destroyed, as a consequence of the rearrangement of the Au atoms. During the phase transformation, the planes of hexagonal packing were still observable as stacking faults and twin boundaries in the fcc phase, as shown in Fig. 4.17 (d). Finally, hcp was transformed into fcc structure, accompanied by stacking faults and twin defects. The phase transformation was attributed to the possible localized high temperature and removal of the amine molecules.

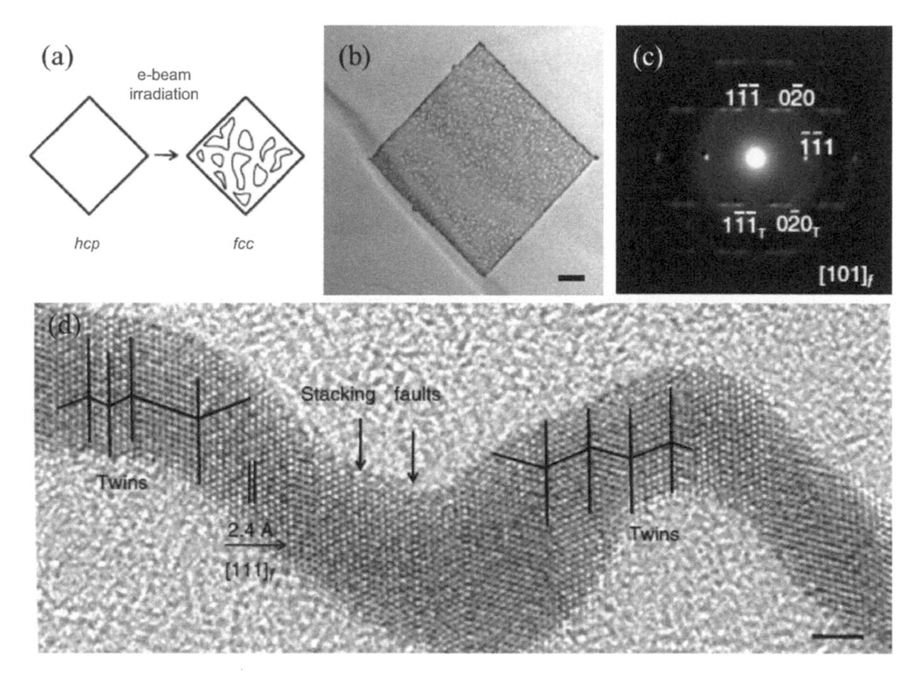

Fig. 4.17. E-beam-induced phase transformation of the Au SSs. (a) Schematic diagram of the e-beam-induced phase transformation of an Au SS. (b) TEM image of a porous Au SS after e-beam irradiation for about 20 s (scale bar = 50 nm). (c) SAED pattern of a porous Au SS after e-beam irradiation, showing the twinned structure viewed along the $[101]_f$ direction. (d) HRTEM image of a section of a porous Au SS, with marked twins and stacking faults (scale bar = 2 nm). Reproduced with permission from [42], Copyright © 2010, Macmillan Publisher Ltd.

Figure 4.18 shows a schematic diagram explaining the formation mechanism of the Au SSs [42]. It was suggested that the 1-amino-9-octadecene molecules played a significant role in the formation of the Au SSs. After mixing with 1-amino-9-octadecene, Au^{3+} was first reduced to Au^{+} to form 1-amino-9-octadecene–AuCl complex, which was adsorbed onto the GO nanosheets. The 1-amino-9-octadecene–AuCl complex was self-assembled into square-like structures through self-organization. Dendritic Au structures with a square-like morphology were gradually formed. These dendritic structures had a high density of defects, indicating the presence of the interfaces between adjacent misoriented Au nanocrystals, which were formed due to the inter-particle attachment. Through oxidative etching and smoothing, the thickness

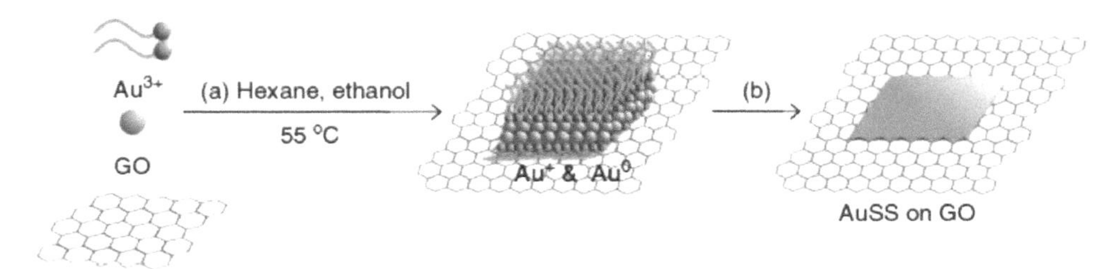

Fig. 4.18. Schematic illustration of the formation process of the Au SSs. Step (a): 1-amino-9-octadecene, Au^{3+} and GO nanosheets in mixed solvent of hexane and ethanol, heated at 55°C. Au^{3+} was partially reduced to Au^{+} and complexed with 1-amino-9-octadecene. Step (b): 1-amino-9-octadecene–AuCl complex was adsorbed on the GO nanosheets, leading to final Au SS. Reproduced with permission from [42], Copyright © 2010, Macmillan Publisher Ltd.

of the structures was reduced, thus leading to the formation of the square sheets from the center of the dendritic structures, after ageing for a certain while [66, 67]. The square-sheet center was thinner than the dendritic edges. Eventually, well-defined Au SSs were produced on the GO nanosheets.

A secondary growth-induced phase transformation from hcp to fcc was observed in the Au nanosheets, accompanied by an increase in their thickness [44]. The secondary growth led to structures with alternating hcp/fcc domains in the center section and defect-free fcc domains at one pair of the opposite thick edges, which were called by the authors, Au square-like plates (SPs). Firstly, ultrathin Au SSs were synthesized by using the similar method as discussed above with a minor modification [42], but with the differences including solutions of $HAuCl_4$ of 4 mM and 1-amino-9-octadecene of 170 mM, heating temperature of 58°C and reaction time of 14 h [44].

Figure 4.19 (a) shows TEM image of the Au SSs, which were redispersed in a fresh solution containing 3 mM $HAuCl_4$ and 140 mM 1-amino-9-octadecene [44]. The second reaction was carried out at 58°C for 10 h. The secondary growth step resulted in the formation of Au SPs, with side lengths of 200–400 nm,

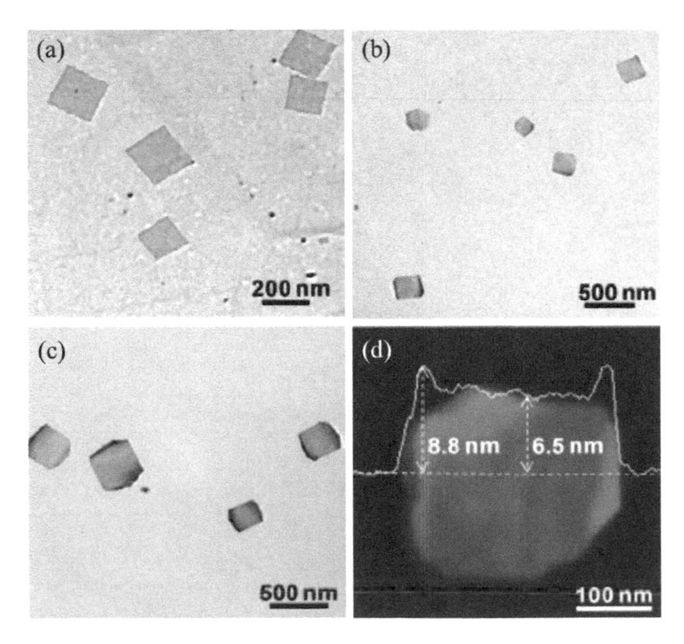

Fig. 4.19. (a) TEM image of the Au SSs synthesized on GO. (b, c) TEM images of Au SPs on GO synthesized from the secondary growth of Au SSs. (d) AFM image and section analysis of a typical Au SP. Reproduced with permission from [44], Copyright © 2011, Wiley-VCH Verlag GmbH & Co. KGaA, Weinheim.

as shown in Fig. 4.19 (b, c). The thickness of the Au SPs was not uniform over their basal plane, with the center area being thinner and one pair of opposite edges being thicker, as illustrated in Fig. 4.19 (b–d) and Fig. 4.20 (a). The thicknesses of the center region and edge were 6.5 and 8.8 nm, respectively, which included the surfactants (1-amino-9-octadecene) adsorbed on both sides of the Au SPs. It has been accepted that, as the dimension of a nanostructure is decreased to a critical value, surface energy becomes dominant in the total energy of the nanostructured system, so that the crystal structure can be different from that of the bulk [42, 68–70].

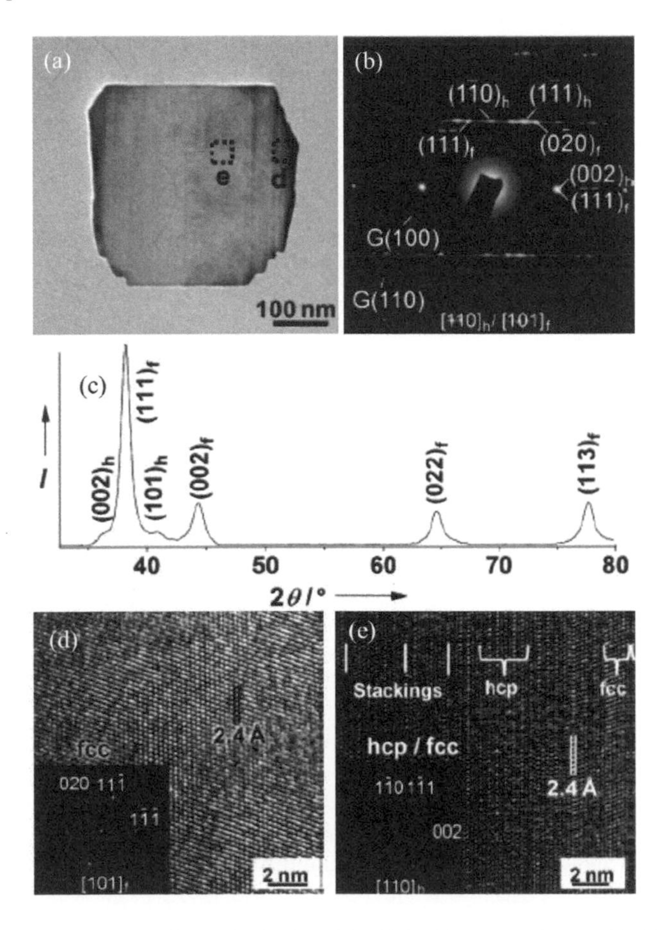

Fig. 4.20. (a) TEM image of a typical Au SP synthesized on GO. (b) SAED pattern of an Au SP. (c) XRD pattern of the Au SPs deposited on a glass substrate. (d, e) HRTEM images of the areas designated in (a). Insets in (d, e): fast Fourier transform (FFT) generated SAED patterns of the corresponding HRTEM images in (d, e). Reproduced with permission from [44], Copyright © 2011, Wiley-VCH Verlag GmbH & Co. KGaA, Weinheim.

Figure 4.20 (b) shows SAED pattern of the individual Au SP in Fig. 4.20 (a) [44]. The SAED patterns were a superposition of the fcc [110] and hcp [110] zone axes and streaks along the $[111]_f$ (or $[002]_h$) direction, due to the random stackings of the fcc and hcp phases. The coexistence of the two phases was also confirmed by the XRD result, as demonstrated in Fig. 4.20 (c). High-resolution TEM (HRTEM) images of the Au SPs are shown in Fig. 4.20 (d, e), corresponding to the two areas labeled in Fig. 4.20 (a). The edge area of the Au SP possessed a defect-free fcc crystal structure, as confirmed by the FFT generated SAED pattern (the inset of Fig. 4.20 (d)). In contrast, the center area exhibited faulted stacking planes of both the hcp and fcc phases, which was demonstrated by the FFT generated SAED pattern, as shown in the inset of Fig. 4.20 (e).

Figure 4.21 shows a schematic diagram illustrating the relative orientations of the Au SS and Au SP during transformations of shape and phase [44]. Despite the similar square-like morphologies, the sides of the Au SS and Au SP corresponded to different crystal facets. When the Au SS was used as the seed for the secondary growth, it was found that the secondary growth on the Au SS seeds was in the directions of $[1\bar{1}0]_h$, $[001]_h$ and $[110]_h$. The growth in the $[110]_h$ direction resulted in an increase in the thickness of the Au structure. As the thickness was increased, fcc domains were formed, due to the energy preference (Step (a) in Fig. 4.21), thus leading to the presence of faulted stackings. As a result, fcc segments were formed. The fcc stacking faults were randomly distributed over the original hcp structure. In this case, elastic strain was generated at the phase interfaces, which boosted the growth in the $[001]_h$ (or $[1\bar{1}\bar{1}]_f$) direction, as compared with that in the $[1\bar{1}0]_h$ (or $[1\bar{2}\bar{1}]_f$) direction. Consequently, the top and bottom edges were formed by the flat $[1\bar{1}0]_h/[1\bar{2}\bar{1}]_f$ planes, because the growth rate normal to these planes was relatively slow. Meanwhile, hexagon-like structures were observed at certain intermediate reaction stages.

The fact that one pair of the opposite edges of the Au SPs was thicker than the center region was attributed to the presence of the fcc packing domains. Once they appeared, they tended to continue to expand in both the lateral and vertical directions. Because the center area consisted of both hcp and fcc domains, the increase in thickness required additional energy to facilitate the phase transformation from hcp to fcc. As a consequence, the increase in thickness at the edges was faster than that in the center area. This is a dimension-related phase transition observed in nanostructured materials [71–73].

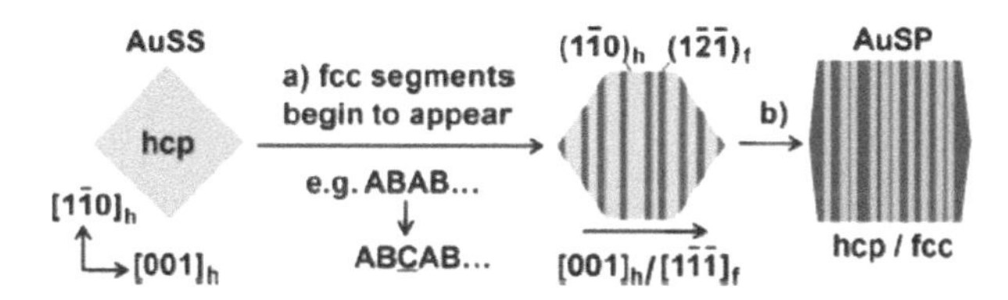

Fig. 4.21. Schematic diagram demonstrating the shape and phase evolution of the Au SS to Au SP (hcp = light gray, fcc = dark gray). Reproduced with permission from [44], Copyright © 2011, Wiley-VCH Verlag GmbH & Co. KGaA, Weinheim.

4.2.3. Graphene-Pd

Pd has a remarkable capacity for hydrogen (H_2) storage at room temperature and atmospheric pressure [2]. It is also a highly efficient catalyst for organic reactions and fuel cells [6]. Various G–Pd hybrids have been synthesized for different applications [53, 74–89]. A number of synthetic methods have been developed to synthesize graphene-templated Pd nanostructures (G–Pd hybrids), such as the laser reduction method [86], plasma-assisted reduction [87] and sacrificial Cu template-mediated method [82]. Details on the synthesis and characterization of selected examples are discussed in the following section.

The first example proposes a novel method for using supercritical carbon dioxide ($scCO_2$) to synthesize G–Pd hybrids, which exhibited promising sensing performances as non-enzymatic electrodes to ascorbic acid (AA), dopamine (DA) and uric acid (UA) [76]. Figure 4.22 shows a schematic diagram demonstrating the formation process of the hybrids. To deposit Pd NPs on graphene nanosheets, 40 mg graphene and 50 mg of palladium(II) hexafluoroacetylacetonate were loaded in a high-pressure stainless steel autoclave (500 ml). Methanol solution (50 ml) was used as the solvent, which was miscible with $scCO_2$. Dimethyl amineborane was used as the reducing agent. The reactor was pressurized with CO_2 up to 10 MPa at 50°C to maintain the supercritical state. The mixture was stirred vigorously for 2 h before depressurization. The products were collected and thoroughly washed with deionized water and methanol to obtain G–Pd@$scCO_2$. Hybrid without the presence of $scCO_2$, i.e., G Pd@air, was also prepared for comparison. In both cases, the content of Pd was 20 wt%.

Figure 4.23 shows TEM images of the two types of G–Pd hybrids [76]. Debundling of the graphene nanosheets can be observed by comparing the two images. Obviously, the entanglement of the graphene nanosheets in the G–Pd@air sample was more pronounced. It has been demonstrated that $scCO_2$ is an

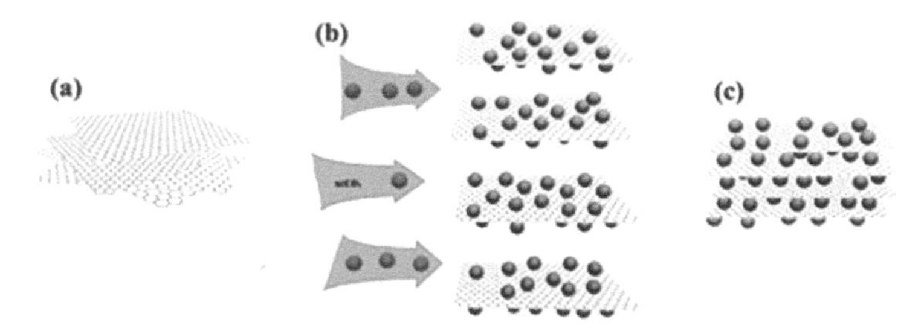

Fig. 4.22. Schematic diagram showing the formation process of the G–Pd@scCO_2 hybrid: (a) graphene nanosheets with agglomeration, (b) scCO_2 assisted debundling of the nanosheets to disperse the Pd NPs and (c) Pd NPs sandwiched between the highly conductive nanosheets which facilitated high electrochemical utilization and could prevent the graphene from restacking. Reproduced with permission from [76], Copyright © 2012, The Royal Society of Chemistry.

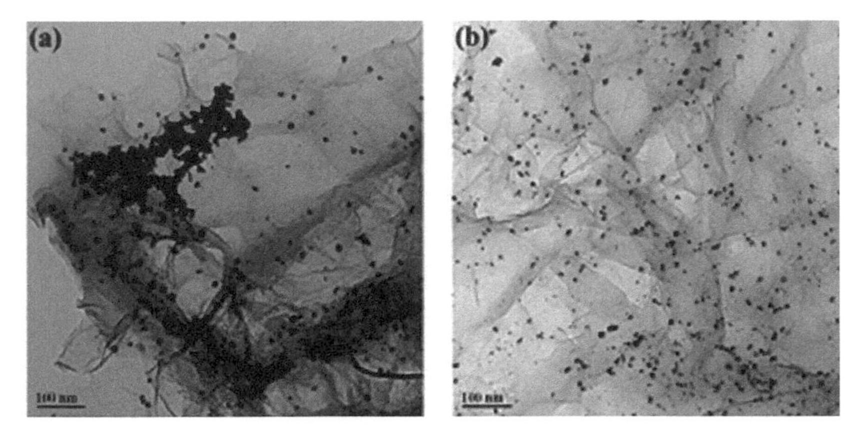

Fig. 4.23. TEM planar-view images of (a) G–Pd@air and (b) G–Pd@scCO_2 hybrids. Reproduced with permission from [76], Copyright © 2012, The Royal Society of Chemistry.

effective technique to produce graphene nanosheets by directly exfoliating graphite. TEM analysis indicated that the Pd NPs in the G–Pd@scCO_2 hybrid were sandwiched between the graphene nanosheets. Therefore, the NPs were encapsulated in the highly conductive graphene nanosheets. As a result, the G–Pd@scCO_2 hybrid exhibited much higher ESA than the G–Pd@air sample. In the unique structure of the G–Pd@ scCO_2 hybrid, both graphene and Pd were fully utilized, so that the electrochemical sensing activity of the electrode was significantly increased.

A soft chemical method was employed to prepare well-dispersed Pd NPs on low-defect graphene (LDG) nanosheets [77]. In this method, damaging of the graphene framework was effectively prevented, because neither oxidation of graphite in advance nor subsequent reduction of the LDG nanosheets was required, due to the lower oxidation degree. The Pd NPs with diameters of 1–5 nm were homogeneously distributed on the graphene nanosheets. Figure 4.24 shows a schematic diagram of the preparation process.

Few-layer graphene nanosheets were derived from the dispersion and exfoliation of graphite powder in N-methyl-pyrrolidone (NMP) by using moderate ultrasonic treatment. The exfoliation of graphite was realized because the energy required to exfoliate layer structured graphite was well-balanced by the solvent–graphene interaction. In this process, H^+ derived from the hydrolysis of palladium nitrate ($Pd(NO_3)_2$), together with the NO^{3-}, would oxidize the carbon nanosheets, so that a sufficient amount of oxygen-containing groups were produced on the surface of the exfoliated graphene nanosheets. The negatively charged groups served as anchor sites to facilitate the *in situ* formation of Pd NPs on the graphene nanosheets. Because the concentrations of the H^+ and NO^{3-} in this case were much lower than those used in the traditional oxidation process, excessive interruption of the basic graphene structure was effectively

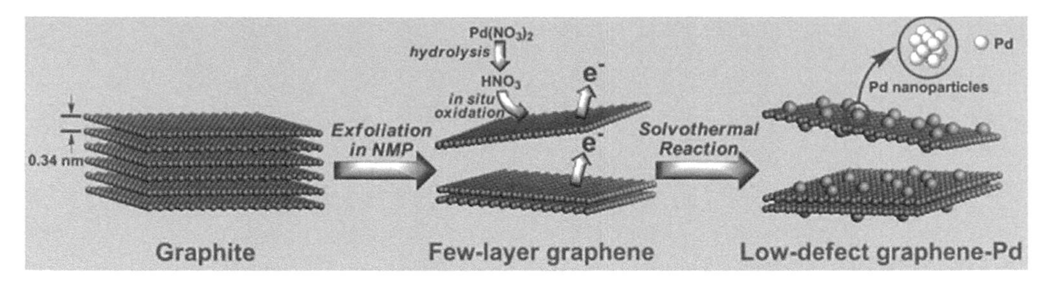

Fig. 4.24. Schematic diagram illustrating the synthesis process of the low-defect graphene–Pd hybrid catalyst with the soft chemistry method. Reproduced with permission from [77], Copyright © 2012, The Royal Society of Chemistry.

prevented. Therefore, the obtained graphene nanosheets contained a very low density of defects and thus the graphene-based catalyst possessed very high electrocatalytic activity for the oxidation of formic acid and methanol.

Figure 4.25 shows TEM images and particle size distributions of the G–Pd hybrids and nanocomposites based on other nanocarbons [77]. Figure 4.25 (a) revealed that the carbon nanosheets were nearly transparent, which implied that graphite had been effectively exfoliated into graphene nanosheets in NMP, although the ultrasonic treatment was moderate. At the same time, Pd NPs were uniformly attached on the LDG nanosheets. HRTEM image (the inset of Fig. 4.25 (a)) indicated the lattice fringes of both (111) plane of the face-centered cubic (fcc) Pd and (0002) plane of the graphene stacks. In comparison, rGO–Pd hybrid exhibited a wider particle size distribution than LDG–Pd, evidenced by the fact that a quite number of Pd particles were larger than 5 nm, as shown in Fig. 4.25 (b).

Two reasons have been considered to be responsible for the better dispersion of Pd NPs on the LDG nanosheets. Firstly, the amount of the oxygen-containing groups on the surface of the LDG nanosheets

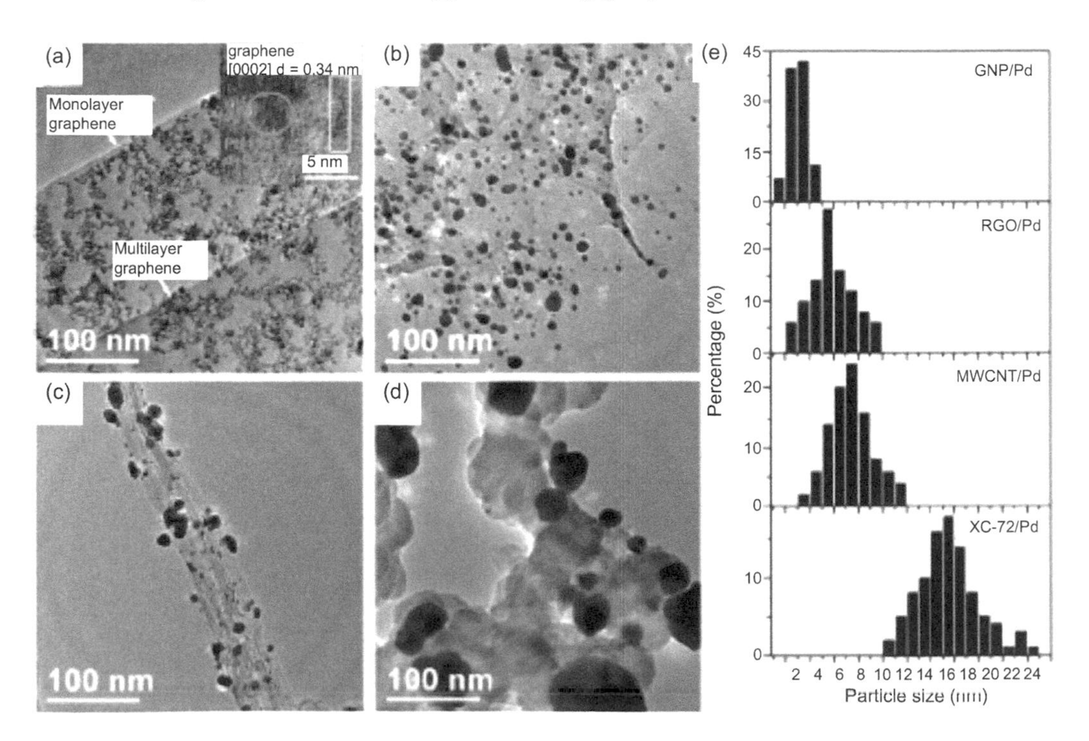

Fig. 4.25. TEM images of the hybrids: (a) LDG–Pd, (b) rGO–Pd, (c) MWNTs–Pd and (d) XC-72–Pd. The inset of (a) is a HRTEM image of the LDG–Pd. (e) Histograms of Pd particle size distribution of the hybrids with different carbon supporting materials. Reproduced with permission from [77], Copyright © 2012, The Royal Society of Chemistry.

was just sufficient due to the moderate synthetic conditions used in the process. In contrast, in the rGO–Pd hybrid, the graphene nanosheets had very rough surfaces and the amount of the groups was excessive with uneven distribution, thus resulting in the agglomeration of Pd nanoparticles. Secondly, the LDG contained more carbon-carbon unsaturated double bonds thus providing C_π sites, which were more efficient to anchor the Pd particles by forming Pd–C hybrids. In this case, Pd NPs were distributed on the carbon surfaces by forming stable π-complexes with the unsaturated carbon-carbon bonds. Comparatively, the content of such unsaturated carbon-carbon bonds in rGO was pretty low, due to the high functionalization level of the graphene nanosheets, thus leading to poor dispersion and wider size distribution of the Pd particles. Poor performances even in terms of particle size and size distribution were observed in the MWNTs–Pd and XC-72–Pd samples, as shown in Fig. 4.25 (c) and (d), respectively. This problem was caused by the relatively lower specific surface area of the two carbon forms. The effectiveness and efficiency of the different carbon materials for the dispersion of Pd NPs are shown in Fig. 4.25 (e), corresponding to average particle sizes of 3.4 nm, 5.8 nm, 7.6 nm and 15.9 nm, for the LDG–Pd, rGO–Pd, MWNTs–Pd and XC-72–Pd, respectively. This is the reason why the LDG–Pd hybrid had high electrochemical catalytic performance.

Another example is the synthesis of G–Pd hybrids by using a low temperature plasma-assisted approach without the presence of surfactants and reducing agents [87]. The plasma-assisted hybrid (PL-G–Pd) demonstrated promising catalytic performance for the catalytic HDS reaction of carbonyl sulfide, as compared with the conventional Pd on activated carbon, the G–Pd hybrids prepared by using impregnation and hydrogen reduction and the G–Pd derived from ethylene glycol through *in situ* reduction.

To prepare the G–Pd hybrids, Pd chloride ($PdCl_2$) was dissolved in ethanol. The solution was then added dropwise into GO suspension in ethanol, with the pH value being 10. The mixture was heated at 78°C for 6 h in a water bath under magnetic stirring. After the reaction, the product was collected by using centrifugation, and was dried at 60°C for 12 h in vacuum to form G–Pd^{2+}. The G–Pd^{2+} powder was treated with a dielectric barrier generated plasma discharge in hydrogen (H_2-DBD), so that both the GO and Pd^{2+} were reduced, resulting in PL-G–Pd hybrid. Meanwhile, hybrid samples of Pd NPs on graphene nanosheets were also made by using a liquid-phase ethylene glycol (EG) reduction route and a traditional two-step method (wet impregnation of graphene synthesized with H_2-DBD plasma and subsequent hydrogen reduction at 250°C), called EG-G–Pd and H_2-G–Pd. Conversional Pd/activated carbon (H_2-AC–G) catalyst was included for comparison.

Figure 4.26 shows FESEM and TEM images of the graphene nanosheets and G–Pd hybrid [87]. Morphology of the Pd NPs in the hybrids was significantly affected by the way of reduction. The graphene exfoliated by using DBD plasma exhibited a typical 2D structure with wrinkles, as shown in Fig. 4.26 (a, b). Potential re-staking of the graphene nanosheets was effectively prevented due to the presence of the wrinkles. SEM and TEM images of the 2% PL-G–Pd sample revealed that tiny Pd NPs were uniformly distributed on the wrinkled graphene nanosheets (Fig. 4.26 (c, d)). The Pd NPs had an average size of 2 nm, which was smaller than that of the Pd particles in the 2% EG-G–Pd sample (Fig. 4.26 (e)). The large Pd particle in the EG process was due to aggregation. Figure 4.26 (f) shows TEM image of the H_2-G–Pd hybrid made by using the conventional two-step method, as mentioned above. Obviously, the particles exhibited aggregation, although they were smaller than those in the EG-G–Pd sample. Therefore, it can be concluded that DBD plasma was a critical factor for uniform dispersion of the Pd NPs.

Figure 4.27 shows a schematic diagram illustrating the reaction steps involved in the formation of the PL-G–Pd hybrids [87]. Due to the presence of large number of oxygen-containing functional groups, the GO nanosheets adsorbed Pd^{2+} ions from solution to form GO–Pd^{2+} complexes through physisorption, electrostatic interaction or charge-transfer processes [50]. The GO–Pd^{2+} complexes served as nucleation centers to further anchor metal atoms. Meanwhile, an appropriate pH value (about 10) was required to produce ionizable carboxyl groups and phenolic protons, which were necessary to ensure a uniform distribution of Pd^{2+} ions that would be reduced to Pd NPs. Therefore, the presence of oxygen-containing functional groups on the surfaces of the GO nanosheets was a critical factor to form the G–Pd hybrids.

Under the application of DBD plasma, charged particles (such as electrons, protons and ions) were generated and accelerated [90], and they interacted with the oxygen-containing functional groups on the GO nanosheets. As a result, these functional groups would be reduced, during which gaseous products, like H_2O and CO_2 gases, were generated. These gaseous items created pressures to exfoliate the GO during

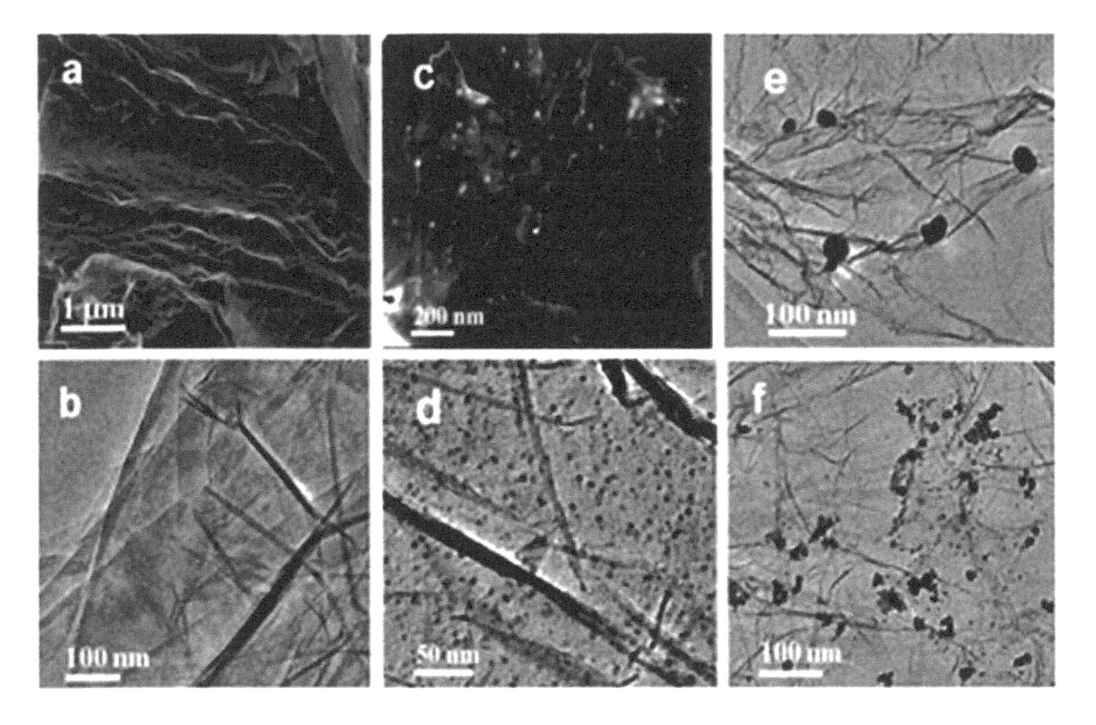

Fig. 4.26. FESEM (a, c) and TEM (b, d, e, f) images of the as-prepared graphene nanosheets and G–Pd hybrids: (a, b) blank graphene nanosheets exfoliated by using DBD plasma, (c, d) 2% PL-G–Pd, (e) 2% EG-G–Pd and (f) 2% H_2-G–Pd. Reproduced with permission from [87], Copyright © 2012, The Royal Society of Chemistry.

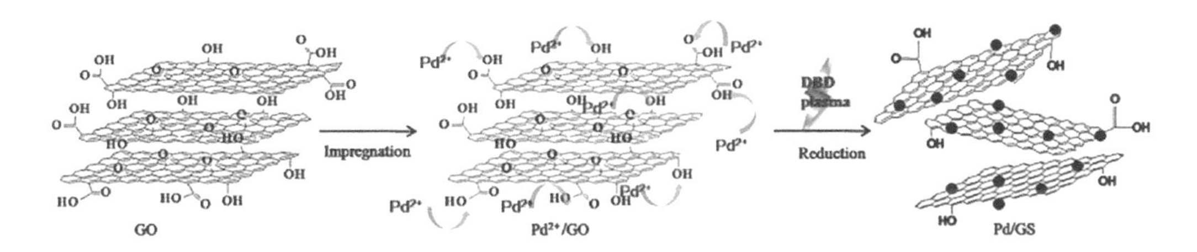

Fig. 4.27. Schematic diagram of preparation process of the G–Pd hybrids. Reproduced with permission from [87], Copyright © 2012, The Royal Society of Chemistry.

Color image of this figure appears in the color plate section at the end of the book.

the reduction. At the same time, $PdCl_2$ was reduced also during the plasma treatment. Such a simultaneous reduction ensured the formation of the G–Pd with a uniform distribution of Pd NPs as compared to the samples prepared by using other procedures.

More recently, a defect-engineered self-assembly process was reported to produce nanohole-structured and palladium-embedded 3D porous graphene (3D Pd-E-PG), with multifunctionalities, including ultrahigh hydrogen storage and CO oxidation [91]. The synthetic route was a one-pot process by using microwave irradiation. It could be used for fast mass production of 3D Pd-E-PG hetero-nanostructures, with highly activated nanoholes in the basal plane of the graphene nanosheets. Due to the low-power microwave radiation, Pd NPs were diffused into the graphene layers, which were then anchored to the functional groups on the surfaces of the GO nanosheets. As the power of the microwave radiation was increased, the Pd NPs agglomerated, and this produced physical nanoholes on the graphene nanosheets. The nanoholes in turn served to embed the Pd NPs. Because the carbon atoms along the edges of the nanoholes contained unsaturated bonds, they were highly reactive for hydrogen storage and CO oxidation.

A potential mechanism has been proposed to explain the formation of the 3D Pd-E-PG hybrid, with the presence of the nanoholes, as schematically shown in Fig. 4.28 [91]. Microwave irradiation heats a material by inducing the polar molecules in it to move and rotate. Firstly, microwave exfoliated graphene oxide (MEGO) with layered structure was produced as the graphite oxide was irradiated. Then the MEGO was mixed with palladium acetate [$Pd(O_2CCH_3)_2$] and ethanol (C_2H_5OH) in order to form a homogeneous suspension. It was then heated to completely evaporate the C_2H_5OH, leading to a powder. The powder was treated with low power microwave irradiation (700 W) for 30 s, during which Pd NPs were obtained due to the composition of $Pd(O_2CCH_3)_2$. The Pd NPs were deposited on the graphene nanosheets through the interaction with the oxygen-containing functional groups, thus forming Pd-D-G. After that, the

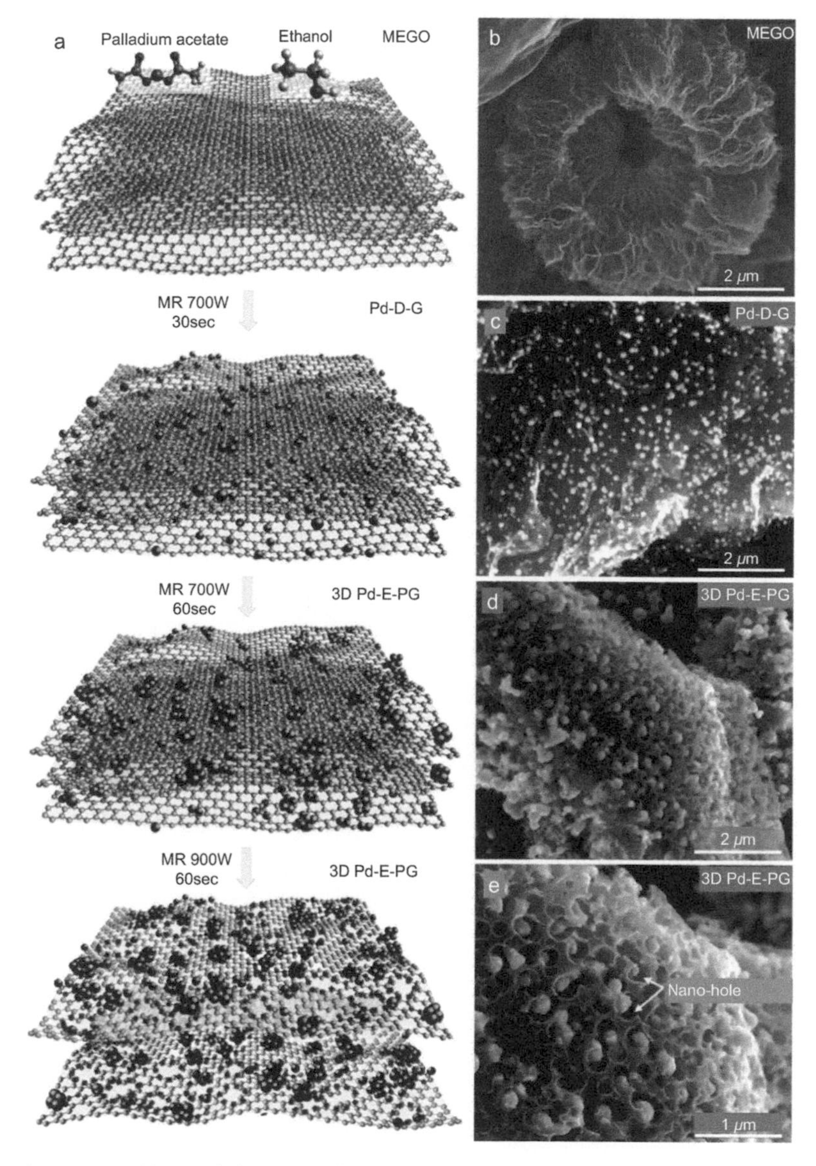

Fig. 4.28. Formation process of the nanohole-structured and palladium-embedded 3D porous graphene (3D Pd-E-PG) and SEM images of the samples obtained at different steps: (a) schematic diagram illustrating the microwave fabrication process towards the 3D Pd-E-PG, (b) SEM image of microwave-exfoliated graphene oxide, (c) SEM image of the uniform decoration of Pd NPs on graphene nanosheets after low power microwave irradiation, (d) SEM image of the aggregated Pd nanoparticles after successive high power microwave irradiation and (e) SEM image of nanohole generation and the perforated graphene structures after multistep microwave irradiation. Reproduced with permission from [91], Copyright © 2015, American Chemical Society.

Pd-D-G was further irradiated for a longer time (60 s), during which agglomeration of the Pd NPs occurred. After microwave irradiation at a high power of 900 W for 60 s, the agglomerated Pd NPs in the Pd-D-G diffused inside the graphene layers, generating nanoholes from the outer layers of the few-layered graphene nanosheets. The agglomeration of the Pd NPs was attributed to the high temperatures caused by the high power microwave irradiation. Finally, a 3D structure was obtained, in which large Pd particles were embedded in the nanoholes without detachment.

SEM images of the samples obtained at different steps are also shown in Fig. 4.28, illustrating the transition from MEGO to 2D Pd-D-G and finally to 3D Pd-E-PG [91]. The MEGO possessed porous, wrinkled and fluffy stacked graphene microstructures, due to microwave expansion of the graphite oxide in the c-axis direction, as shown in Fig. 4.28 (b). Figure 4.28 (c) revealed that the Pd NPs were homogeneously distributed on the MEGO nanosheets and entirely covered the graphene surfaces. As the microwave irradiation time was increased, nanoholes were generated inside the graphene layers, leading to the formation of 3D Pd-E-PG nanostructures, as shown in Fig. 4.28 (d). The Pd NPs became spheres and grew larger with increasing irradiation time. The 3D Pd-E-PG clearly shows a perforated graphene surface with a uniform and unique morphology. The nanopores on the surface of the MEGO were in the range of 10–100 nm.

Figure 4.29 shows TEM and HRTEM images of the 3D Pd-E-PG hybrids [91]. The graphene nanosheets in the 3D Pd-E-PG hybrids were transparent flakes with an average diameter at the micrometer scale. It was found that defects were produced on the local surfaces of the MEGO nanosheets by the NPs. The Pd NPs exhibited facet-like structures with different morphologies, as shown in Fig. 4.29 (b). Polycrystalline characteristics of the Pd NPs were demonstrated by the higher magnification HRTEM image, i.e., the inset of Fig. 4.29 (d). The lattice fringe of 0.225 nm was attributed to the Pd (111) crystalline planes of the Pd NPs.

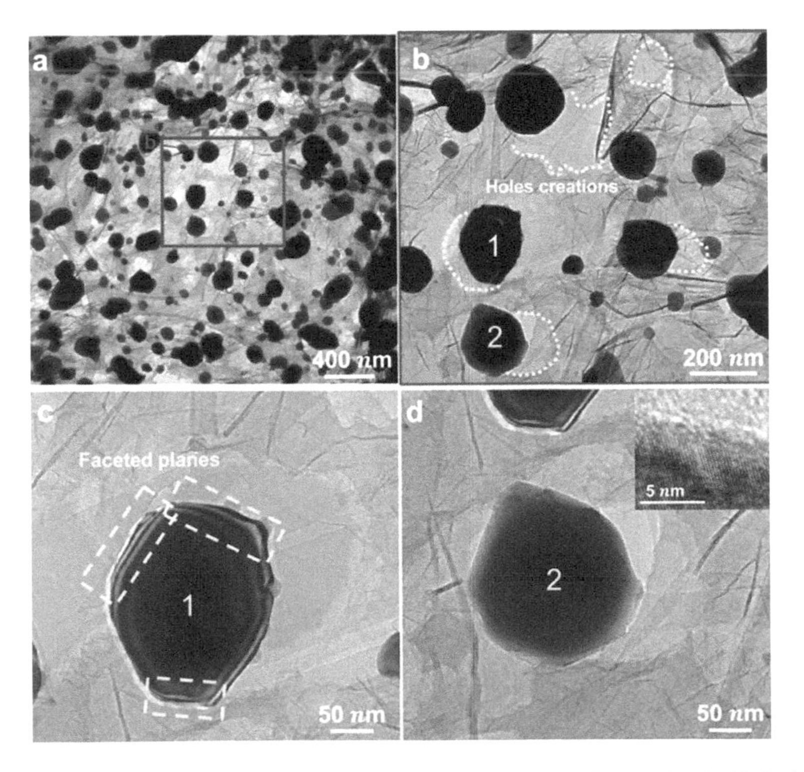

Fig. 4.29. TEM images of the 3D Pd-E-PG hybrids with nanoholes on graphene nanosheets and embedded Pd NPs: (a) TEM image of the 3D Pd-E-PG, with nanoholes produced by the Pd NPs attached on the amorphous graphene layer, (b) zoomed-in image of the box in (a) showing holes and (c, d) magnified TEM images around the Pd NPs in different areas, with the inset showing the lattice planes of the Pd NP in (d). Reproduced with permission from [91], Copyright © 2015, American Chemical Society.

4.2.4. Graphene-Pt

Pt nanocrystals have high catalytic activities, which have found a wide range of applications in various areas, such as fuel cells, electroanalysis and organic reaction catalysis [6]. Because Pt is a precious and rare element, various attempts have been made to reduce the absolute quantity in applications, by significantly increasing its activity. Incorporation of Pt with other materials, including metal oxides, activated carbons and polymers, has been an effective strategy to achieve this objective [92, 93]. Specifically, hybrids based on Pt nanostructures with graphene and its derivatives have been widely studied for such a purpose [94–109].

The processing techniques to fabricate G–Pt hybrids are similar to those discussed above [94–103]. The most commonly used precursor of Pt is K_2PtCl_4, which can be reduced by using various approaches, such as chemical reduction, photochemical synthesis, microwave assisted synthesis, electroless metallization and thermal evaporation [101–107]. For example, a GO–Pt nanohybrid was synthesized by refluxing K_2PtCl_4 and GO in a water–ethylene glycol solution [53]. As an alternative approach, graphene–Pt nanocomposite films have been successfully prepared on conductive indium tin oxide (ITO) electrodes by using an electrochemical synthetic route [110]. Pt nanoflowers decorated rGO nanosheets were obtained by using an electrochemical approach [102]. A new rGO–Pt hybrid for sensing applications has been developed by using photochemical reduction [94]. Selected examples are discussed in more detail in the following part.

A one-step microwave-assisted approach was reported to prepare hybrids of Pt NPs on graphene nanosheets (PNEGHNs) [99]. In these hybrids, the Pt NPs had a small size and a narrow size distribution, as well as a controllable loading level. The PNEGHNs/GCE exhibited more favorable electron transfer kinetics and much enhanced electrochemical reactivity than GNs/GCE, thus providing a more robust and advanced hybrid electrode material for electrochemical sensors and biosensors.

To synthesize the PNEGHNs, 5 ml GO ethylene glycol (EG) solution (1 mg/ml) was mixed with 5 ml water, into which 1 ml H_2PtCl_6 EG solution (77 mM) and 0.2 ml poly(methacrylic acid sodium salt) (PMAA) were added. After being quickly sonicated for about 2 min, the solution was treated in a microwave oven for 90 s at a power of 800 W. The product was collected by using centrifugation at 8000 rpm for 15 min, followed by consecutive washing and centrifugation cycles three times with water. The collected product was redispersed in a water-ethanol mixture (8 ml) with sonication to produce a colloidal suspension. During the formation of the PNEGHNs through microwave heating, EG was used as a reducing agent to reduce both the GO and Pt precursor. Meanwhile, PMAA had two functions, i.e., (i) stabilizing the GNs due to the strong hydrophobic interaction between PMAA and GNs and (ii) preventing Pt NPs from agglomeration. AFM analysis results indicated that thicknesses of the PNEGHNs and GNs were about 6.5 nm and 1.3 nm, respectively.

Figure 4.30 shows representative TEM and HRTEM images of the PNEGHNs [99]. It was observed that a large number of Pt NPs with a narrow size distribution were homogeneously distributed on the graphene nanosheets (Fig. 4.30 (A, B)). Due to the strong interaction between the graphene nanosheets and the Pt NPs, no isolated Pt NPs were observed. Figure 4.30 (C) revealed that the average size of the Pt NPs was about 2.6 nm. Individual Pt NPs could be clearly identified in the HRTEM image (Fig. 3.40 (D)). The *d*-spacing of 2.25 Å was attributed to the (111) planes of face-centered cubic (fcc) Pt. Density of the Pt NPs on graphene could be simply controlled by changing the synthetic parameters. When 2 ml of Pt precursor was used instead of 1 ml, the density of the Pt NPs in the PNEGHNs was increased, as demonstrated in Fig. 4.30 (E, F) (GPt3). In comparison, as the amount of Pt precursor was decreased to 0.5 ml, the density could be readily reduced (Fig. 4.30 (G–I) (GPt1). The high-quality GNs–Pt hybrids with Pt NPs of controllable density, small size and narrow distribution were attributed to the application of microwave treatment, as well as the use of PMAA. Therefore, this strategy could be extended to the synthesis of other NPs as well.

A simple *in situ* oxidation method was employed to prepare graphene nanoplate-Pt (GNP–Pt) hybrids [109]. Figure 4.31 shows a schematic diagram illustrating the process of synthesis of the GNPs–Pt hybrids. H^+ from the hydrolysis of platinum nitrate and NO_3^- were combined to introduce oxygen-containing groups onto the graphene nanosheets, on which Pt NPs were deposited. The GNPs obtained in this way contained only a very low level of defects, since the concentration of H^+ produced from hydrolysis was much lower, as compared with that generated in the conventional oxygenation processes, by using strong oxidation

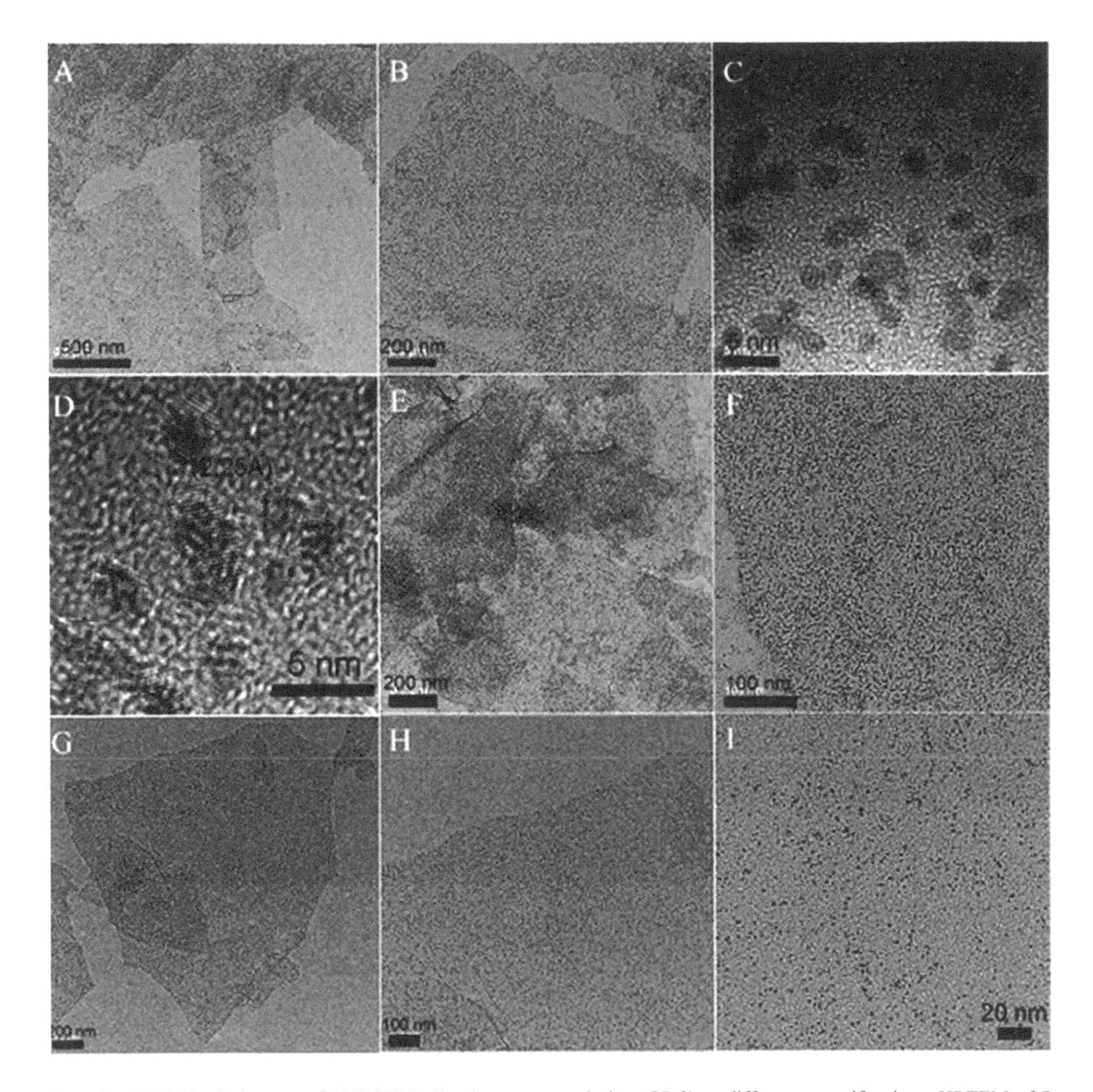

Fig. 4.30. TEM (A–C) images of PNEGHNs (1 ml precursor solution, GPt2) at different magnifications. HRTEM of Pt NPs (D). TEM (E, F) images of PNEGHNs with relative high density of Pt NPs (2 ml precursor solution, GPt3) at different magnifications. TEM (G–I) images of PNEGHNs with relative low density of Pt NPs (0.5 ml precursor solution, GPt1) at different magnifications. Reproduced with permission from [99], Copyright © 2010, American Chemical Society.

reagents, e.g., sulfuric acid and nitric acid. Such hybrids showed promising electrochemical performances, due to the intact crystal structure of graphene which offered very low charge transfer resistance.

GNPs were prepared by dispersing and exfoliating graphite in NMP at a concentration of 0.1 mg ml^{-1}. To synthesize the hybrids, ethylene glycol (EG) (30 ml) and 0.017 ml $Pt(NO_3)_2$ solution (0.45 M) were mixed with 50 ml of GNP dispersion under magnetic stirring for 15 min. The mixture was then solvothermally heated at 120°C for 12 h. After the solvothermal reaction, GNP–Pt hybrid powder was obtained after centrifugation, washing and drying. Conventionally produced GO was also used to prepare a hybrid for comparison, i.e., rGO–Pt.

Figure 4.32 shows XRD patterns of the GNP–Pt and rGO–Pt hybrid powders [109]. For the GNP–Pt sample, the broad peak at $2\theta \approx 25.0°$ was attributed to the (002) reflection of the GNPs, indicating the presence of few-layer graphene nanosheets in the hybrid. In rGO–Pt, the GO peak was suppressed, implying the recovery of the conjugated graphene network caused by the reduction process. At the same time,

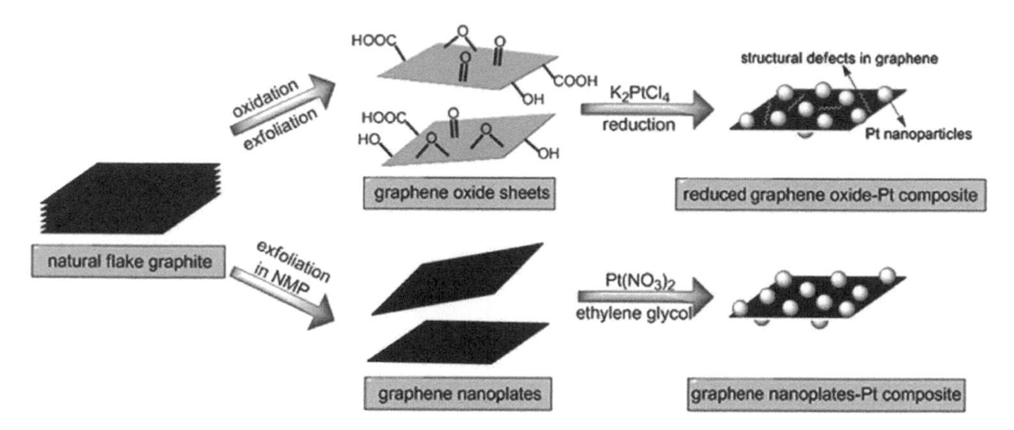

Fig. 4.31. Schematic diagram showing the synthetic procedures of the reduced graphene oxide-Pt composite and graphene nanoplate-Pt composite by the traditional oxidation–reduction method and the soft chemical method, respectively. Reproduced with permission from [109], Copyright © 2012, Elsevier.

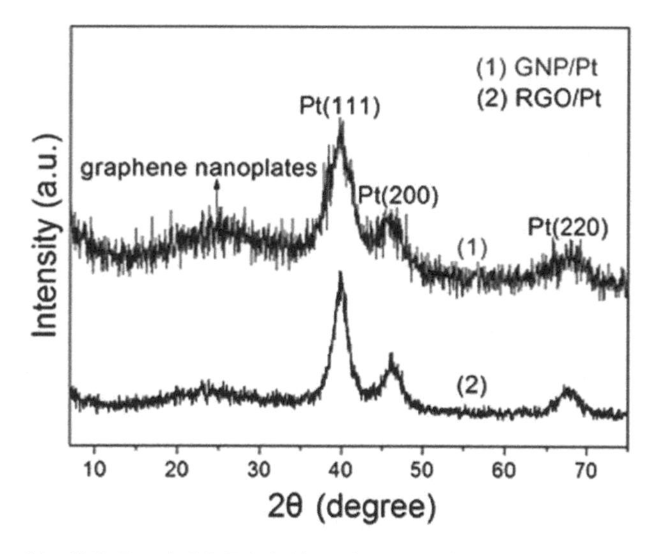

Fig. 4.32. XRD patterns of the GNP–Pt and rGO–Pt hybrid powders. Reproduced with permission from [109], Copyright © 2012, Elsevier.

diffraction peaks of the face-centered cubic (fcc) structured Pt could be observed clearly. The intensity of the Pt diffraction peaks of the rGO–Pt was stronger than that of the peaks in the GNP–Pt, which suggested that the Pt in rGO–Pt possessed higher crystallinity. This was attributed to the fact that the amount of oxygen-containing groups on the GO nanosheets was larger than that of the groups on the GNPs. Particle sizes of Pt NPs in GNP–Pt and rGO–Pt were 2.2 nm and 3.8 nm, respectively. The presence of finer Pt NPs in the GNP–Pt was closely correlated to the much softer synthetic conditions of the hybrid, in which the amount of functional groups was just sufficient.

Figure 4.33 shows TEM images and size distributions of the GNP–Pt and rGO–Pt hybrids [109]. The carbon nanosheets were nearly transparent, as shown in Fig. 4.33 (a), indicating the successful exfoliation of graphite flake into graphene nanoplates in NMP. The Pt NPs with diameters of 1–5 nm were uniformly dispersed on the GNPs. The d-spacing corresponding to the face-centered cubic (fcc) Pt (111) was clearly observed. In contrast, the Pt NPs in the rGO–Pt sample exhibited serious aggregation, as demonstrated in

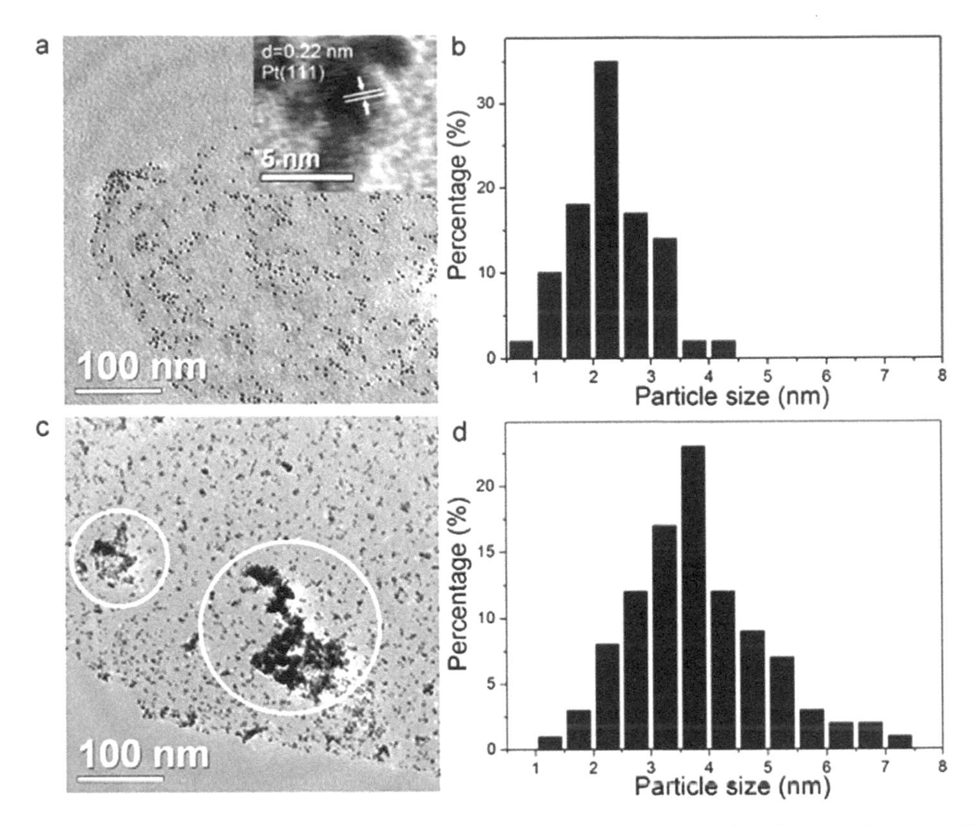

Fig. 4.33. TEM images and histograms of Pt particle size distribution of (a, b) GNP–Pt and (c, d) rGO–Pt. The inset of (a) is the HRTEM image of a GNP–Pt. Reproduced with permission from [109], Copyright © 2012, Elsevier.

Fig. 4.33 (c), which was due to the uneven distribution of the functional groups on the rGO nanosheets. The average sizes of Pt NPs were 2.3 nm and 3.7 nm in GNP–Pt and rGO–Pt, respectively. These values were almost the same as those estimated by using the Scherrer equation.

4.2.5. Graphene—Other Noble Metals

Besides the above discussed noble metals, other noble metallic NPs, including Ru, Rh, Ir and Os, have also been explored to develop hybrids with graphene. Ru, Rh and Ir are very strong catalysts in a wide range of applications, such as petroleum refining and the Monsanto acetic acid processing [2]. They are also high performance catalysts for the hydrogenation of unsaturated compounds [6]. Various processing routes have been developed to synthesize hybrids of these noble metals with graphene [111–118]. For instance, graphene-supported Ru NPs were synthesized through microwave irradiation in ionic liquids [113] or by using a wet-chemical method combined with thermal exfoliation of graphite [111]. More recently, G–Rh has been reported, by assembling Rh nanoclusters on graphene nanosheets [113]. One example is presented below to reveal the detailed reaction process of such hybrids.

A simple reduction method was used to prepare highly water dispersible G–Rh hybrids from Rh^{3+} on poly(ethylene oxide)/poly(propylene oxide)/poly(ethylene oxide) (PEO/PPO/PEO) triblock copolymer or pluronic-stabilized GO nanosheets with $NaBH_4$ as the reducing agent [117]. 15 ml GO suspension (0.5 mg ml^{-1}) was mixed with 10 ml 1×10^{-3} M solution of $RhCl_3 \cdot xH_2O$, which was sonicated for 20 min. A dilute aqueous solution of $NaBH_4$ was then added to mixture, followed by sonication at 60°C for 3 h. Finally, the precipitate was collected and redispersed in water.

Figure 4.34 shows a HRTEM image of the hybrid, demonstrating the homogeneous distribution of the Rh NPs on the graphene nanosheets, with an average size of 1–3 nm [117]. Polycrystalline nature of the Rh NPs was confirmed by SAED pattern (inset of Fig. 4.34). After reduction in the presence of pluronic, the graphene nanosheets were folded up, while the Rh NPs were uniformly distributed on the graphene nanosheets. The AFM images confirmed the homogeneous distribution of the Rh NPs over the 2D graphene nanosheets, with an average size of 3 nm. Thickness of the pluronic-stabilized graphene nanosheet was about 1.5 nm. Some graphene nanosheets contained nanopores after the reduction. The presence of the nanopores was attributed to the generation of unstable particles, which were removed during the sonochemical reduction of the GO–Rh hybrid. The G–Rh hybrid had a surface area of 285 $m^2 \cdot g^{-1}$ and a pore volume of 0.164 $cm^3 \cdot g^{-1}$. These properties made the hybrid a promising catalyst for various applications.

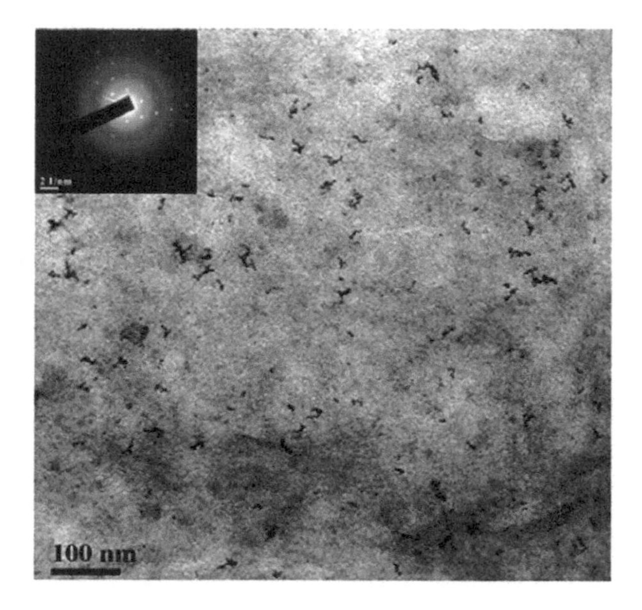

Fig. 4.34. HRTEM image of the G–Rh hybrid, with the inset to be the corresponding SAED pattern. Reproduced with permission from [117], Copyright © 2011, Springer.

Figure 4.35 shows a schematic diagram illustrating the formation process of the G–Rh hybrid [117]. Firstly, GO was stabilized by the PEO–PPO–PEO triblock copolymer, due to the hydrophobic interaction between the PPO chain fragment of the copolymer and the hydrophobic surface of the graphene nanosheets. In this case, the hydrophilic PEO chains remained outside the graphitic surfaces. As a result, the hydrophilic part of the triblock copolymer offered an extra stability of the graphene after the reduction, due to the hydrophilic interaction with water. Rh^{3+} was then combined with the pluronic-stabilized GO nanosheets and reduced by $NaBH_4$. During the sonication process, the particles having less π interactions with the graphene nanosheets were removed, thus leading to the formation of the nanopores in the final products.

A wet-chemical method has been used to prepare a range of metal nanoparticles, including Rh and Ru nanoparticles [111]. The hybrids were synthesized by using an impregnation method combined with heat treatment in H_2. $RuCl_3 \cdot xH_2O$ and $RhCl_3 \cdot xH_2O$ were used as the metal precursors of Ru and Rh. The metal precursors were mixed with graphene dispersed in acetone and ultrasonicated for 30 min. The resulting slurries were dried at 100°C for 10 h in order to evaporate the acetone. The dried powders were then treated in 4% H_2-N_2 at 250°C for 3 h. Figure 4.36 shows a schematic diagram to demonstrate the synthetic process of the hybrids. The presence of the metallic phases was confirmed by using XRD. Figure 4.37 shows TEM images, together with the average size and size distribution, of the two hybrids.

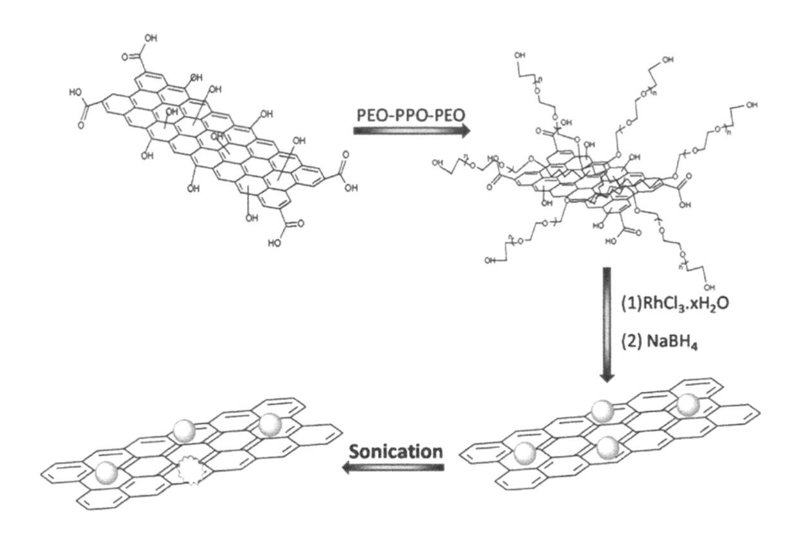

Fig. 4.35. Schematic diagram showing the reaction process to form the hybrids. Reproduced with permission from [117], Copyright © 2011, Springer.

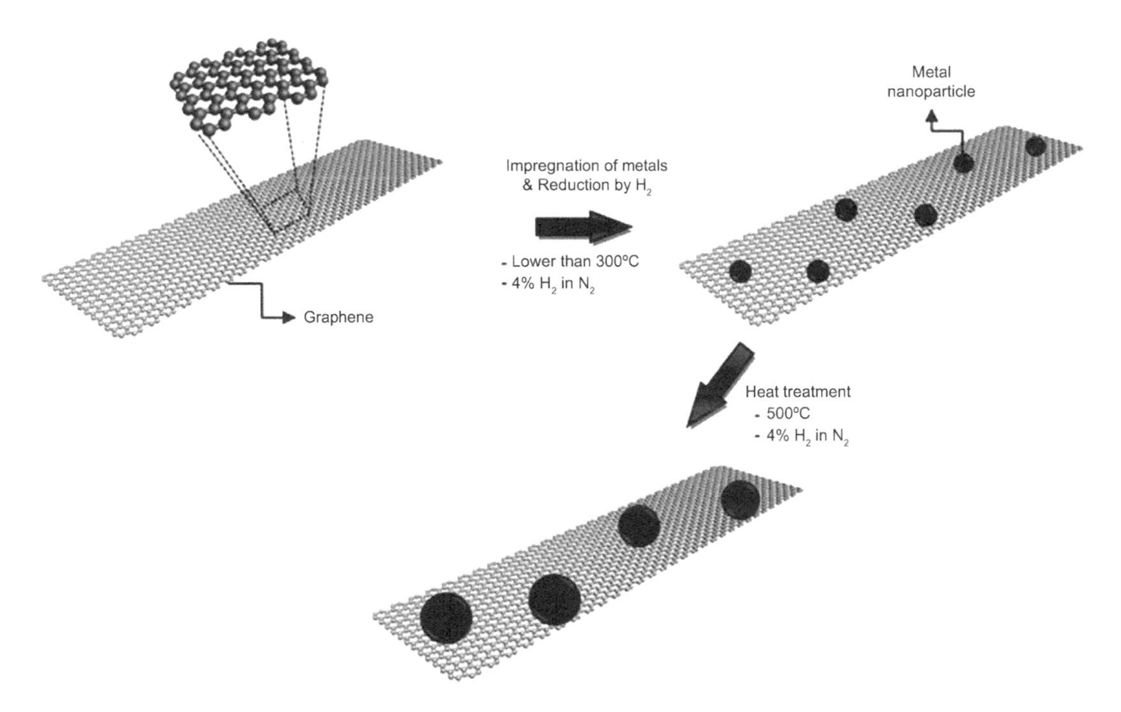

Fig. 4.36. Schematic diagram illustrating the synthesis process of the graphene-supported metal NPs and the influence of heat treatment. Reproduced with permission from [111], Copyright © 2011, Elsevier.

4.2.6. Graphene—Noble Bimetals

It has been shown that the catalytic, optical and electronic properties of noble metals can be significantly enhanced by forming bimetallic nanostructures [6]. As a result, the development of hybrids of bimetallic noble metals and graphene nanosheets has become a hot research subject all around the world [119–131]. Various combinations, including Au–Pt, Pt–Pd, Pt–Ru, Au–Pd, Pt–Ag, Pd–Ag and Pd–Ru, have been reported to form hybrids with graphene or its derivatives [119–132].

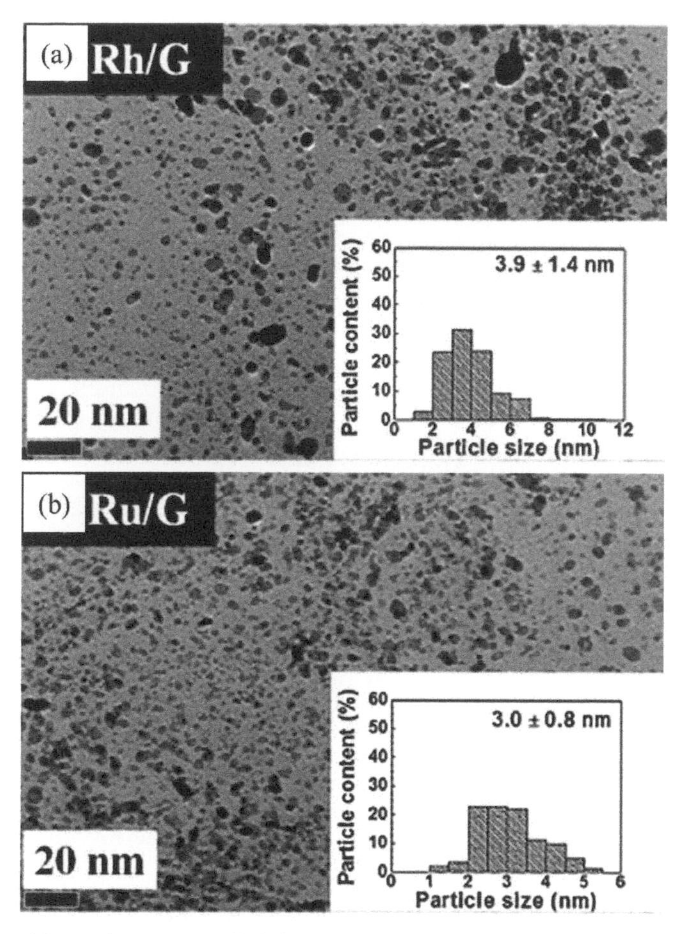

Fig. 4.37. TEM images of the graphene-supported hybrids with 60 wt% metal NPs: (a) Rh and (b) Ru, with the insets showing the histograms of the NPs. Reproduced with permission from [111], Copyright © 2011, Elsevier.

Pt–Pd hybrids of bimetallic nanodendrites loaded on graphene nanosheets possessed highly enhanced catalytic performances [120]. Pt–Ru NPs decorated graphene nanosheets with high catalytic activity towards methanol oxidation have been synthesized through the reduction of $PtCl_2$ and $RuCl_3$ in the presence of tetraoctylammonium hydrotriethylborate and subsequent thermal treatment in H_2/N_2 [122]. Through electrostatic interaction, pre-synthesized hollow flower-like Au–Pd NPs could be assembled on ionic liquid-grafted graphene nanosheets to obtain hybrids [123]. Au–Pd NPs-based hybrids have been synthesized by using a simple one-step *in situ* reduction method and the nanoparticles were well-distributed on the graphene nanosheets [130]. Graphene nanosheets decorated with Pd–Ag nanorings were reported through the galvanic displacement reaction between pre-synthesized Ag NPs and Pd ions [131].

A wet-chemical approach has been developed to synthesize 3D Pt-on-Pd bimetallic nanodendrites on graphene nanosheets (TP-BNGN) [120]. These hybrids demonstrated several advantages. Firstly, very fine single-crystal Pt nanobranches supported on Pd NPs with a porous structure and high dispersion could be directly deposited onto the graphene nanosheets, which a high electrochemically active area (81.6 m^2/g). Secondly, the number of the branches in the Pt-on-Pd bimetallic nanodendrite could be readily controlled by simply changing the processing parameters, so that tunable catalytic properties could be realized. Lastly, as compared with platinum black (PB) and commercial E-TEK Pt/C catalysts, the graphene/bimetallic nanodendrite hybrids displayed much a higher electrocatalytic activity in methanol oxidation reaction.

Figure 4.38 shows a schematic diagram demonstrating the synthetic process of the TP-BNGN hybrids [120]. The first step was a wet-chemical route to obtain poly(*N*-vinyl-2-pyrrolidone) (PVP)-functionalized graphene nanosheets through the reduction of GO with hydrazine. The average thickness of the graphene nanosheets was about 2 nm. Colloidal suspensions of the graphene nanosheets decorated with PVP had high stability in water, ethanol and dimethylformamide, due to the strong π-π interactions between graphene and PVP molecules. The second step was to deposit Pd NPs on the PVP-functionalized graphene nanosheets at room temperature. The Pd NPs were adsorbed on the surface of the graphene nanosheets, with an average size of about 3 nm. The G–Pd hybrids were used as seeds, on which dendritic Pt nanostructures were grown through the reduction of K_2PtCl_4 with ascorbic acid in an aqueous solution. The newly synthesized hybrids exhibited a rougher surface than the G–Pd hybrids.

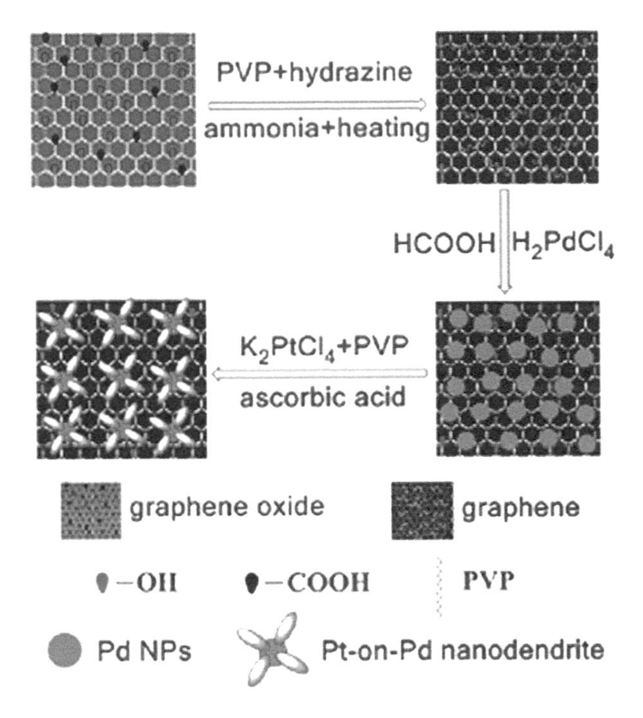

Fig. 4.38. Schematic diagram of the formation of the graphene nanosheet/Pt-on-Pd bimetallic nanodendrite hybrids. Reproduced with permission from [120], Copyright © 2010, American Chemical Society.

Color image of this figure appears in the color plate section at the end of the book.

Representative TEM images of the hybrids are shown in Fig. 4.39 (A–C) [120]. Clearly, Pt branches were formed on the Pd core, forming dendritic tendrils. Average size of the Pt-on-Pd bimetallic nanodendrites was 15 nm. HRTEM images of selected nanodendrites are shown in Fig. 4.39 (D, E), which revealed the overgrowth of Pt branches at multiple sites on the Pd seeds, as evidenced by the darker Pd center surrounded by lighter Pt branches in the nanodendrites. The diameter of the Pt branches was 35 nm. The Pt branches were single crystals, as confirmed by the highly ordered fringe patterns shown in Fig. 4.39 (E). (111) lattice plane of face-centered cubic (fcc) Pt, with lattice fringes of 0.23 nm, was clearly observed. Figure 4.39 (F) shows a fast Fourier transform (FFT) pattern of the HRTEM image (circled area in Fig. 4.39 (E)), further confirming the single crystal nature of the Pt nanobranches.

It was found that Pt nanodendrites on graphene nanosheets could not be synthesized without the presence of the Pd NPs, which indicated the indispensable role of the Pd seeds [120]. The growth of the Pt dendritic branches has been explained as high-rate reduction and (111) facet epitaxial growth facilitated by the Pd seeds with a large number of (111) facets [133]. In addition, the number of the Pt branches on the

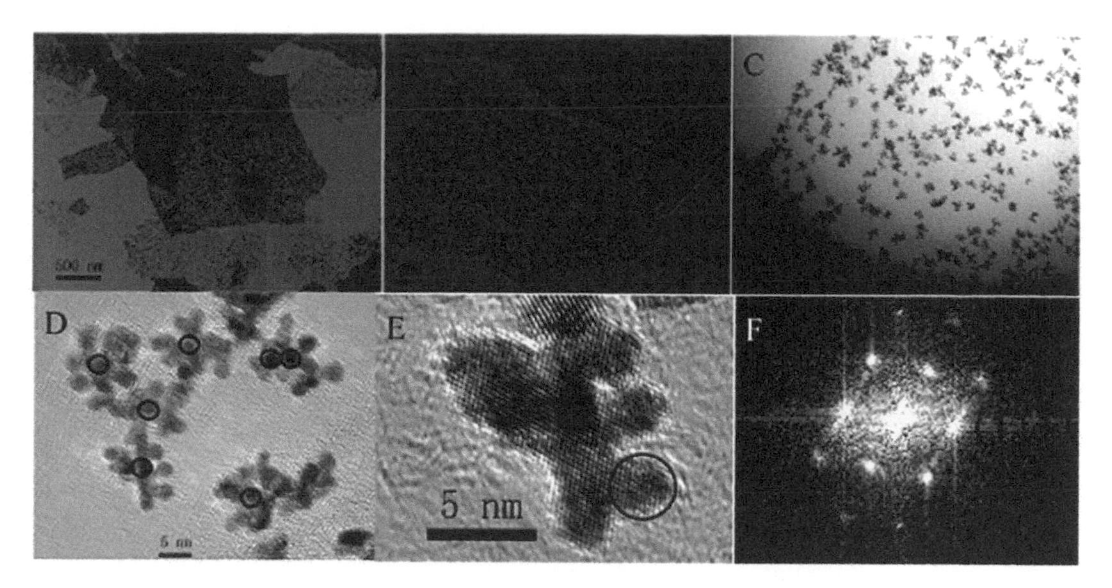

Fig. 4.39. TEM (A–C) and HRTEM (D, E) images of the TP-BNGN at different magnifications, with the circled parts in panel (D) to be Pd NPs. (F) FFT pattern of the HRTEM image shown in panel (E) (circled area). Reproduced with permission from [120], Copyright © 2010, American Chemical Society.

nanodendrites could be simply controlled by changing the reaction conditions. For instance, the number of the Pt nanobranches in the bimetallic nanodendrites could be increased by decreasing the level of graphene content, because less number of Pd NPs would be available to seed the growth of the Pt nanobranches. On the other hand, the microstructures could also be controlled by increasing the content of the Pt precursor. Nevertheless, by controlling the microstructures of the hybrids, the electrocatalytic activities of the TP-BNGN could be readily tuned.

A two-step approach has been reported to synthesize GNs–PdAg hybrids, with graphene nanosheets (GNs) to support the PdAg alloy nanorings [131]. The hybrids possessed three advantages: (i) uniform dispersion on the GN support, (ii) large surface area due to the availability of both the exterior and interior surfaces of a thin wall (~5.5 nm) and (iii) high efficiency of the Pd due to the low loading caused by the ring-like structure. Figure 4.40 shows a schematic diagram demonstrating the two-step synthetic process of the GNs–PdAg hybrids. At the first step, Ag NPs were deposited on graphene nanosheets to form GNs–Ag, by refluxing an aqueous mixture of $AgNO_3$ and GO with sodium citrate as the reducing agent. At the second step, nanocrystals with Ag core and Pd shell were produced through a galvanic replacement reaction between the Ag NPs and Pd^{2+}. Finally, ring-shaped PdAg alloy nanostructures were obtained due to the diffusion of the Ag atoms in the core to the shell after the reaction.

A TEM image of the GNs–PdAg is shown in Fig. 4.41 (a) [131]. Average diameters of the Ag NPs and PdAg nanorings were 27.3 ± 6.3 and 27.5 ± 6.2 nm, respectively, which implied that although hollow structures were formed after the galvanic replacement reaction, there was almost no change in the particle size. HRTEM images of the GNs–PdAg hybrids at different magnifications are shown in Fig. 4.41 (a, b). Average thickness of the ring walls was about 5.5 nm. Lattice fringes with an interplanar spacing of 0.22 nm were observed, corresponding to the (111) plane of the PdAg alloy. In the ring-shaped nanostructure, Ag and Pd were evenly distributed throughout the shells.

4.2.7. Graphene—Other Metals

There have also been reports on hybrids of various non-noble metallic NPs, such as Fe [134, 135], Ni [136–141], G-Cu [142–144] and Sn [145–147], for various applications, including electrode materials of energy storages, microwave absorbers and sensors. One example for each of the metals will be discussed in the following section, regarding details in processing and characterization.

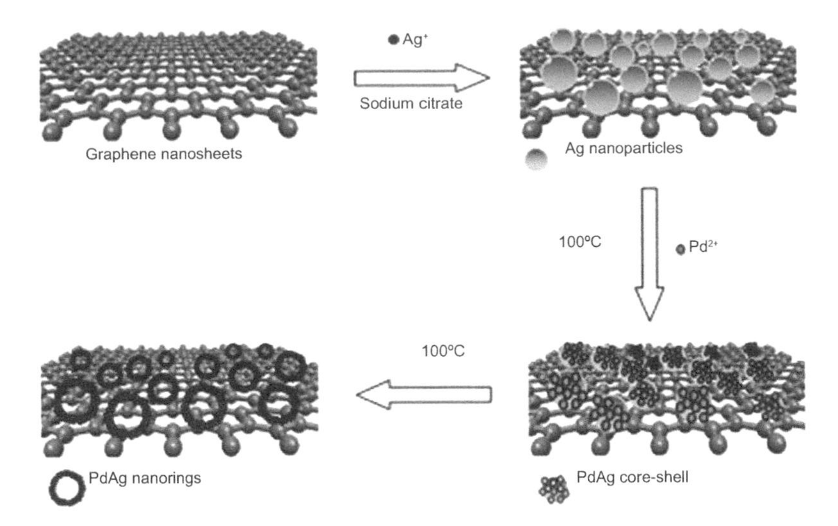

Fig. 4.40. Schematic illustrations of the procedure for preparing graphene nanosheets supporting PdAg nanorings. The preparation involves a galvanic replacement reaction between the Ag nanoparticles and Pd^{2+} ions. Reproduced with permission from [131], Copyright © 2013, Wiley-VCH Verlag GmbH & Co. KGaA, Weinheim.

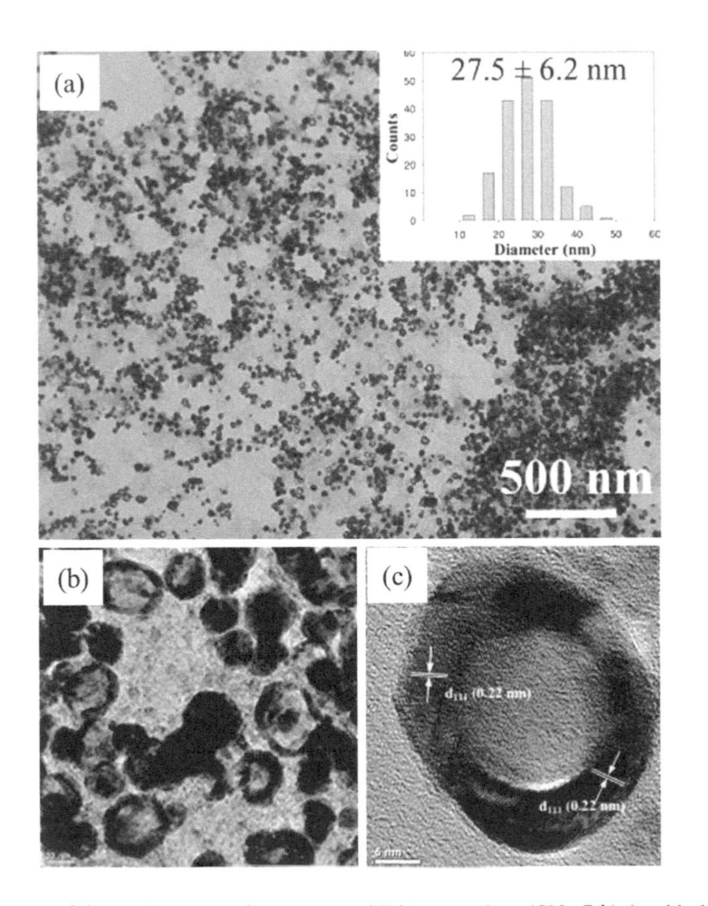

Fig. 4.41. (a) TEM image of the graphene nanosheet-supported PdAg nanorings (GNs–PdAg), with the inset showing the corresponding diameter histograms of the PdAg nanorings. (b, c) HRTEM images at different magnifications of the PdAg nanorings in the hybrids. Reproduced with permission from [131], Copyright © 2013, Wiley-VCH Verlag GmbH & Co. KGaA, Weinheim.

A magnetic hybrid has been synthesized through the intercalation of ferromagnetic Fe NPs into the interplanes of graphene nanoribbons (GNRs) [134]. The Fe-intercalated GNR stacks exhibited an anisotropic response to an external magnetic field, thus leading to an aligned architecture. Dispersibility of the GNRs could be enhanced by using edge functionalization, without affecting their magnetic properties and electrical conductivity. Iron(III) chloride ($FeCl_3$) is a well-known intercalator to form graphite intercalation compounds [148]. In this case, $FeCl_3$ was intercalated into the bundles of MWCNTs, as shown in Fig. 4.42. A two-compartment glass tube was used, where the MWCNTs were loaded into one compartment and the

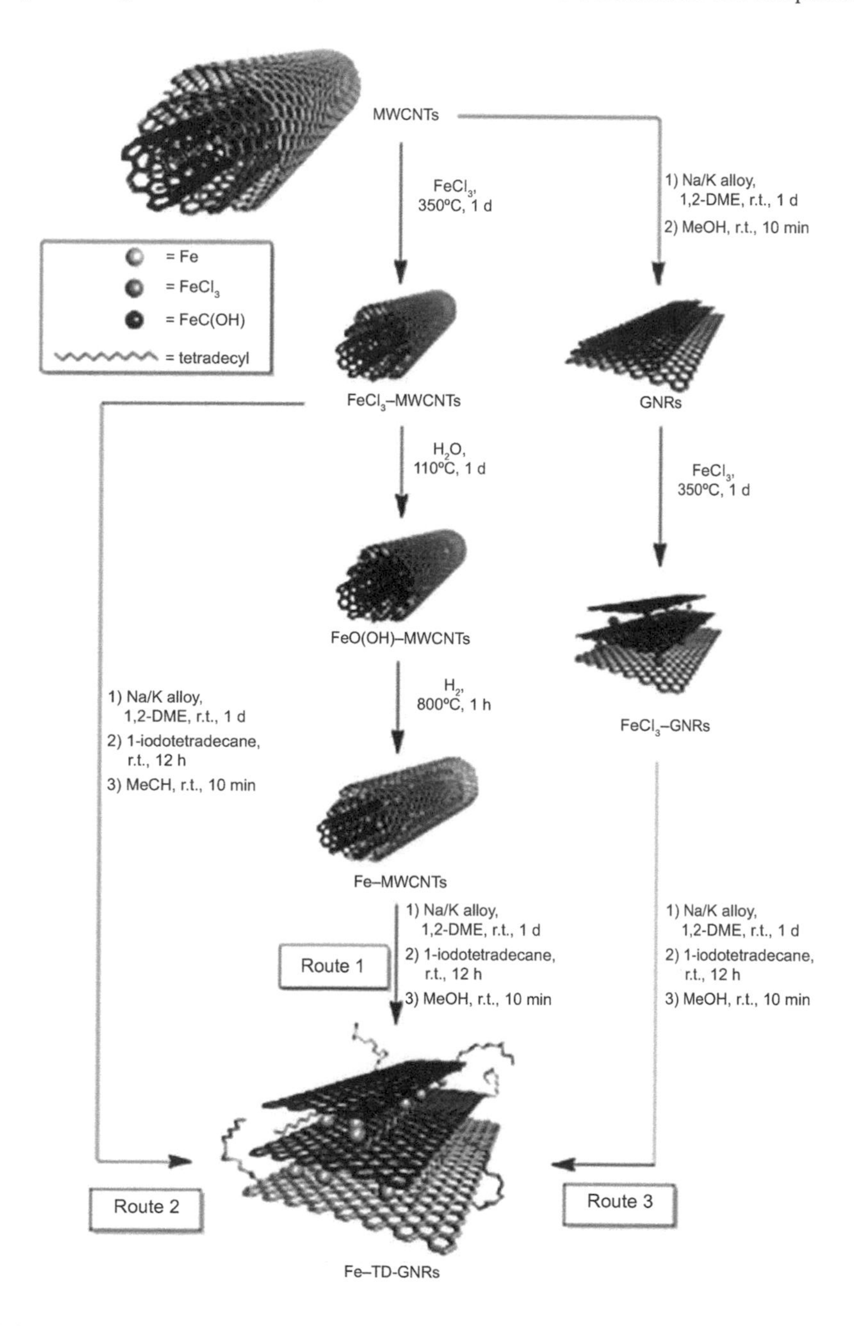

Fig. 4.42. Schematic diagram illustrating the synthetic route 1, route 2 and route 3 for the synthesis of iron-intercalated tetradecylated graphene nanoribbons (Fe–TD-GNRs). Reproduced with permission from [134], Copyright © 2012, American Chemical Society.

$FeCl_3$ into another. The tube was then vacuum-sealed and heated at 350°C for 1 day. The $FeCl_3$-intercalated MWCNTs ($FeCl_3$–MWCNTs) were hydrolyzed in water vapor at 110°C for 1 day. Fe-intercalated MWCNTs (Fe–MWCNTs) were obtained after reduction of the hydrolyzed products at 800°C for 1 h in H_2. One-pot splitting followed by edge functionalization led to iron-intercalated tetradecyledge-functionalized graphene nanoribbon stacks (Fe–TD-GNRs), which is shown in Fig. 4.42 (route 1).

Route 2 in Fig. 4.42 was used to optimize the procedure. Firstly, $FeCl_3$–MWCNTs were prepared similarly, which were then treated with an electride, i.e., Na/K alloy in 1,2-dimethoxyethane (1,2-DME), to simultaneously reduce Fe^{3+} to Fe, unzip the MWCNTs and functionalize the edges of the GNRs. The functionalization was achieved by quenching the reaction mixture with 1-iodotetradecane. The content of Fe in the product of route 2 was relatively low. As a result, route 3 was developed to increase the content of Fe. In this route, the MWCNTs were unzipped to yield GNR stacks first. The GNRs were then intercalated with $FeCl_3$ as a second step to form the $FeCl_3$–GNRs. Obviously, the intercalation of $FeCl_3$ was much easier in GNRs than in MWCNTs. Finally, Fe–TD-GNRs were obtained in a similar way. MeOH was used to quench all the Na/K reactions. Figure 4.43 shows TEM images of the Fe–TD-GNRs synthesized by using route 3, from which the edges of the functionalized hybrid were clearly visible.

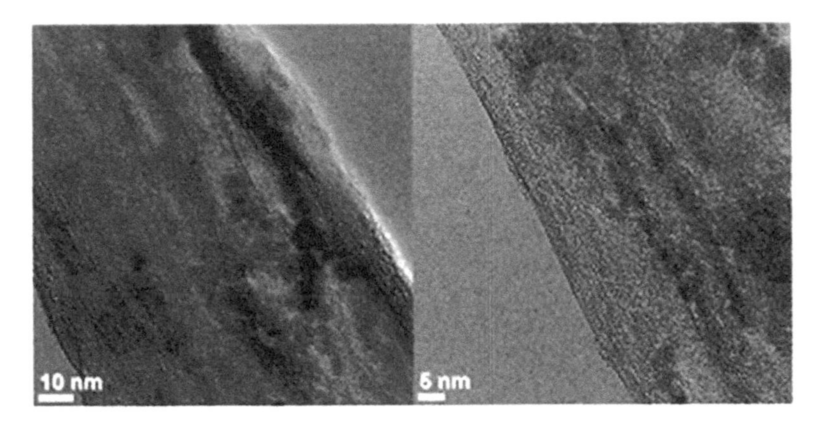

Fig. 4.43. Representative TEM images of the Fe–TD-GNRs synthesized through route 3. Reproduced with permission from [134], Copyright © 2012, American Chemical Society.

The Fe–TD-GNRs hybrids exhibited a high dispersion behavior in several organic solvents, such as chloroform and chlorobenzene, which made them highly processable for different applications [134]. The high solubility was attributed to the tetradecyl groups at the edges of Fe–TD-GNRs hybrids. The Fe–TD-GNRs in chloroform could be separated by using an external magnetic field (0.3 T) in 2 h, due to their strong magnetic properties. In contrast, no such a phenomenon was observed in the TD-GNRs samples without Fe. Meanwhile, the Fe–TD-GNRs hybrids had pretty high electrical conductivity, which is beneficial for various applications.

Due to their magnetic nature, electrical percolation could be improved when using the anisotropic Fe–TD-GNRs as conductive components, because they could be readily aligned by using a magnetic field. By using a special testing cell, resistances of a 5 wt% solution of Fe–TD-GNRs in a mixture of commercial hydrocarbons could be measured parallel and perpendicular to a magnetic field. The resistance related to electrical percolation was ~80 kΩ, in the direction parallel to the magnetic field. This was because the anisotropic Fe–TD-GNRs were aligned along the longer axis of the GNRs. When the measure was in the direction perpendicular to the magnetic field, the resistance of the suspension was increased to 275 kΩ, larger than that in the parallel direction by a factor of 3.5. Without the application of the magnetic field, the resistance was about 720 kΩ. Furthermore, the distribution of the measurement results parallel to the magnetic field was relatively narrow, as compared to those without a magnetic field, indicating once again the alignment effect of the magnetic field.

Figure 4.44 shows SEM images of the magnetically aligned and randomly arranged Fe–TD-GNRs samples derived from two MWCNTs, i.e., Mitsui and NanoTechLabs (NTL) [134]. They were both made by using route 3 and exhibited similar properties, such as XPS and Raman. Magnetic alignment of the particles was clearly observed in the two groups of samples.

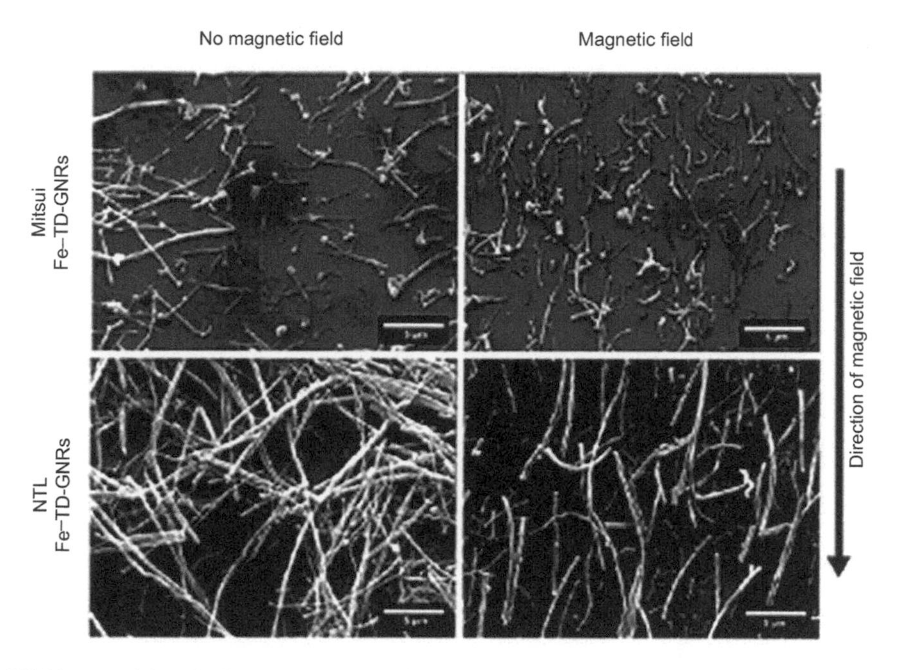

Fig. 4.44. SEM images of the Mitsui- and NTL-originated Fe–TD-GNRs. Suspensions of the Fe–TD-GNRs in chlorobenzene were drop-cast onto a SiO_2/Si substrate and dried with the presence of magnetic field (right two images), while the left two images were samples that were dried without the presence of magnetic field, as a comparison. All scale bars = 5 μm. Reproduced with permission from [134], Copyright © 2012, American Chemical Society.

A simple chemical method has been reported to prepare two types of hybrids based on Ni NPs, i.e., hexagonal close-packed (hcp) Ni nanocrystals and face-centered cubic (fcc) Ni nanospheres, by optimizing the reaction temperature and the type of solvents [140]. The hybrids were denoted as *h*-Ni–G and *c*-Ni–G, respectively. Finally, the Ni–G hybrids had very promising microwave absorbing properties, over 2.0–10.0 GHz.

To synthesize the *h*-Ni–G hybrids, GO was dispersed in 2-pyrrolidone to form a homogeneous brown solution through sonication for several hours. After that, $Ni(acac)_2$ and octadecylamine were added into the solution, which was then heated at 120°C for 30 min, followed by raising the temperature to the boiling temperature of 2-pyrrolidone, i.e., 245°C, for 2 h. The reaction was conducted in a flow of Ar gas. The color of the solution gradually changed from green to brown. After the reaction was finished, ethanol was injected into the reaction solution to cool it down as quickly as possible. The final product was collected by centrifugation, and was thoroughly washed with *n*-hexane and acetone and then dried at 40°C in vacuum.

The synthesis process of the sphere-like *c*-Ni–G hybrid was similar to that of the *h*-Ni–G hybrid. The only difference was that the second-step heating temperature was reduced from 245°C to 202°C. To prepare the cauliflower-like *c*-Ni–G hybrids, the 2-pyrrolidone was replaced with NMP as the solvent, whose boiling temperature was 202°C. Therefore, the second-step heating temperature was 202°C. Without the presence of GO in the reaction, *h*-Ni nanocrystals, *c*-Ni nanospheres and *c*-Ni nanoflowers, were obtained, denoted as Ni(245), Ni(202) and Ni(NMP). The corresponding hybrids were named *h*-Ni(245)–G, *c*-Ni(202)–G and *c*-Ni(NMP)–G, respectively. As mentioned previously, GO nanosheets were negatively charged, so that they absorbed Ni^{2+} ions through electrostatic interactions. Both the GO and the Ni^{2+} were reduced by octadecylamine in such a way that the Ni NPs were deposited on the graphene nanosheets, thus forming Ni–G hybrids. In this case, the Ni NPs were attached on the graphene nanosheets, without using any molecular linkers.

Figure 4.45 shows XRD patterns of the natural graphite, GO, graphene nanosheet (GN), *h*-Ni(245)–G, *c*-Ni(202)–G and *c*-Ni(NMP)–G hybrids [140]. The graphite was characterized by the (002) diffraction peak at $2\theta \approx 26.5^{\circ}$, with an interplanar distance of about 0.34 nm (Fig. 4.45 (a)). After oxidation, this peak shifted to $2\theta \approx 10.9^{\circ}$, corresponding to a larger basal spacing of 0.76 nm (Fig. 4.45 (b)), due to the introduction of a large amount of functional groups. After reduction, a weak and broad reflection peak at $2\theta \approx 23.5^{\circ}$ was observed, indicating the formation of amorphous carbon (Fig. 4.45 (c)). Figure 4.45 (d) revealed that the Ni(245)–G hybrid contained hexagonal close-packed Ni nanocrystals with lattice parameters of a = b = 2.651 nm and c = 4.343 nm. As shown in Fig. 4.45 (e, f), the Ni(202)–G and Ni(NMP)–G samples had face-centered cubic Ni NPs, with lattice parameters of a = b = c = 3.524 nm. It was found that the formation of the Ni NPs was not affected, no matter whether there was GO or not, as shown in Fig. 4.45 (c–f).

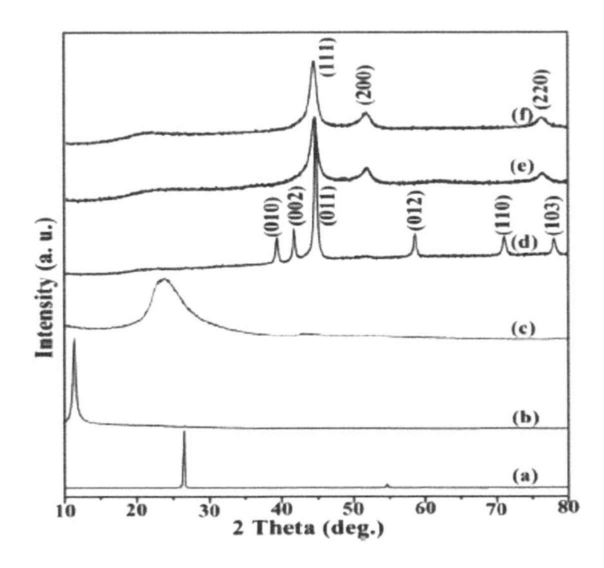

Fig. 4.45. XRD patterns of (a) graphite, (b) GO, (c) GN, (d) *h*-Ni(245)–G, (e) *c*-Ni(202)–G and (f) *c*-Ni(NMP)–G hybrids. Reproduced with permission from [140], Copyright © 2012, The Royal Society of Chemistry.

Figure 4.46 shows representative TEM images of the *h*-Ni(245) nanocrystals and *h*-Ni(245)–G hybrids [140]. Figure 4.46 (a) indicated that Ni nanocrystals with a relatively uniform size distribution were derived from the reduction reaction. The Ni NPs were attached on the two-dimensional graphene nanosheets, as clearly shown in Fig. 4.46 (b). It was further found that all the Ni nanocrystals were deposited on both sides of the graphene nanosheets. Figure 4.46 (c) revealed that the nanocrystals had a uniform rhombic shape with an average size of about 3 nm, which was confirmed by the HRTEM image (Fig. 4.46 (d)). The crystal structure of the Ni NPs was also evidenced by the SAED pattern. The *d*-spacing of 0.203 nm corresponded to the (011) plane of the hexagonal close-packed Ni, in agreement with the XRD results.

TEM images of the c-Ni(202)–G hybrids are illustrated in Fig. 4.47 [140]. On the graphene nanosheets, *c*-Ni nanospheres were randomly distributed, with diameters in the range of 50–200 nm (Fig. 4.47 (a, b)). A higher magnification TEM image of a *c*-Ni nanosphere–G hybrid indicated that *c*-Ni nanospheres were wrapped by the graphene nanosheets (Fig. 4.47 (c)). The submicrometer spheres consisted of *c*-Ni NPs with a diameter of about 200 nm, while the spheres were comprised of *c*-Ni nanocrystals with diameters in the range of 10–20 nm. HRTEM image revealed that crystal lattice fringes were present throughout the entire fcc Ni nanospheres, as shown in Fig. 4.47 (d)). The crystal lattice fringes exhibited a *d*-spacing of 0.203 nm, corresponding to the (111) plane of *c*-Ni. The crystalline characteristics of the graphene nanosheets were confirmed by the two-dimensional fast Fourier transform (FFT) analysis (the inset of Fig. 4.47 (d)).

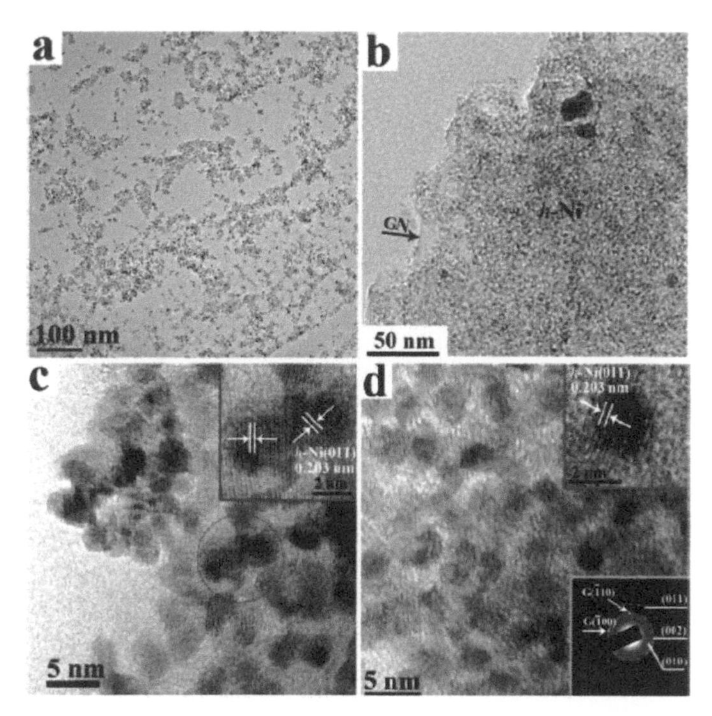

Fig. 4.46. TEM (a and b) and HRTEM (c and d) images of the *h*-Ni(245) nanocrystals and *h*-Ni(245)–G hybrids. Inset of (d) is the SAED pattern. Reproduced with permission from [140], Copyright © 2012, The Royal Society of Chemistry.

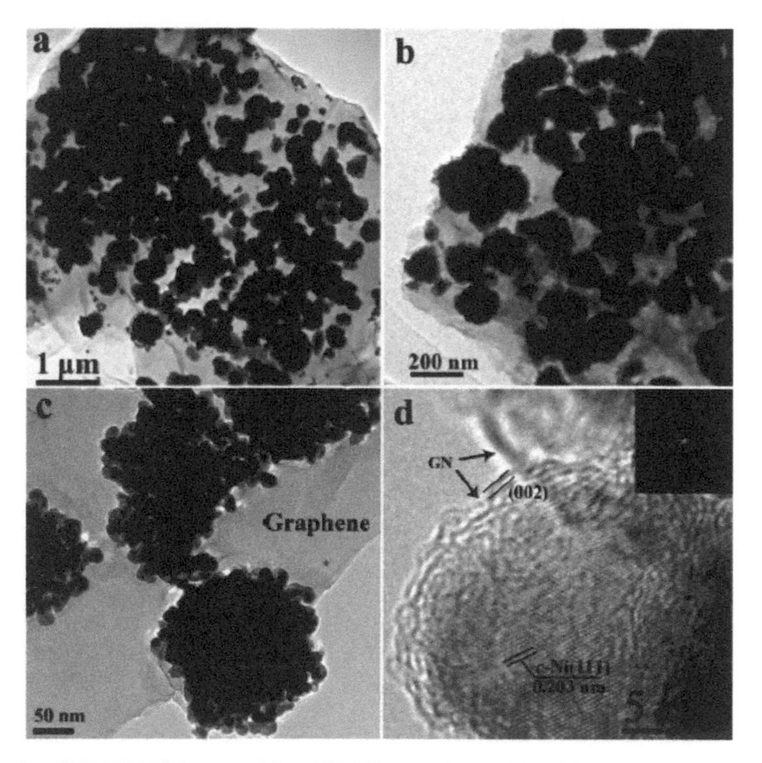

Fig. 4.47. TEM (a–c) and HRTEM (d) images of the *c*-Ni(202) nanospheres–G hybrid. Inset of (d) is the FFT analysis result. Reproduced with permission from [140], Copyright © 2012, The Royal Society of Chemistry.

The third type of Ni was cauliflower-like *c*-Ni nanostructure, which was obtained by using NMP as the solvent 202°C. The cauliflower-like *c*-Ni nanostructures were constructed with Ni NPs with sizes of to 10–20 nm. They were uniformly distributed on the graphene nanosheets, similar to the other two types of Ni NPs. The promising microwave absorbing behaviors of the Ni–G hybrids can be attributed to their unique magnetic and dielectric properties.

Composite nanowires, consisting of rGO and Cu nanoparticles, were fabricated to be used as gas sensors for NO_2 [142]. Figure 4.48 shows a schematic diagram of the fabrication process of the rGO–Cu hybrid. GO suspension was formed by mixing GO powders (150 mg) and DI water (500 ml) under ultrasonication for 1 h. Copper acetate monohydrate ($Cu(Ac)_2 \cdot H_2O$, 50 mg), 2 M-NaOH, and DI water (100 ml) were then added into the GO suspension. After the solution was heated at 80°C for 5 h, low-speed centrifugation at 4000 rpm was conducted to remove thick multilayer sheets until all the visible particles were removed. The supernatant was then further centrifuged at 8000–12,000 rpm for 45 min to remove the liquid solution, leading to solid GO–CuOx hybrid. The high-speed centrifugation at 12,000 rpm was repeated until a pH of 7 was reached. The products were then dried in vacuum at 90°C for 5 h. To obtain rGO–Cu hybrid powders, the dried powders were thermally treated at 400°C for 5 h in $Ar+H_2$.

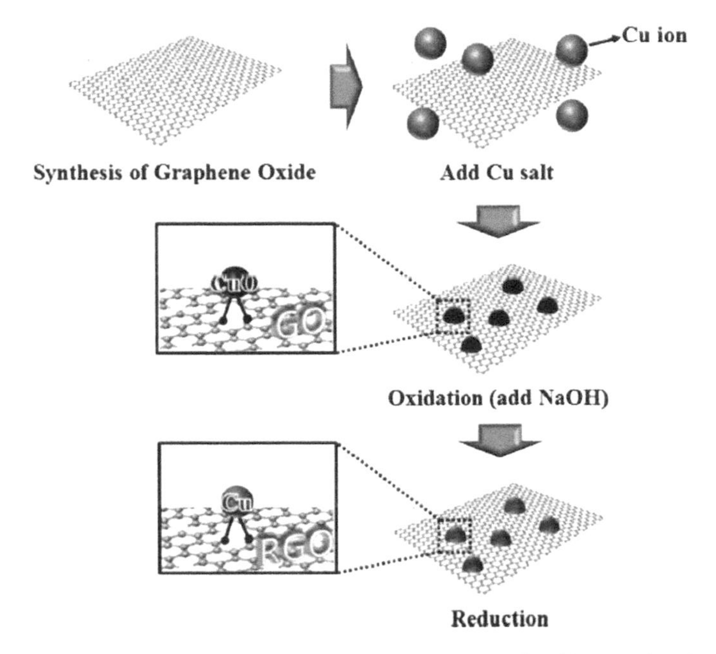

Fig. 4.48. Schematic diagram showing the fabrication process of the rGO–Cu hybrid. Reproduced with permission from [142], Copyright © 2015, Elsevier.

Figure 4.49 shows XRD patterns of Cu, rGO and rGO–Cu hybrid [142]. Figure 4.49 (a) indicated that the Cu was of a cubic phase, with lattice constant of a = 3.6150 Å. The rGO pattern was characterized by the (002) diffraction line at $2\theta = 24.81°$ (Fig. 4.49 (b)). The interlayer spacing was changed from 3.35 Å of graphite to 3.56 Å in the rGO, which could be readily attributed to the slight oxidation and interlamellar water content. However, Fig. 4.49 (c) indicated that the (002) diffraction line of the rGO corresponded to an interlayer spacing of 3.48 Å in the rGO–G hybrid. This value was slightly smaller than that of pure rGO, implying that the presence of Cu NPs prevented the further expansion of the rGO nanosheets.

Low-magnification TEM images of the rGO–Cu hybrid are shown in Fig. 4.50 (a, b) [142]. Cu NPs were present as dark dots on the surface of the rGO nanosheets. Figure 4.50 (c) shows a lattice-resolved TEM image of an individual dark Cu NP. The lattice fringes were attributed to the (111) lattice planes of cubic Cu, with spacing of 0.21 nm. The Cu NPs were polycrystalline, each of which contained several

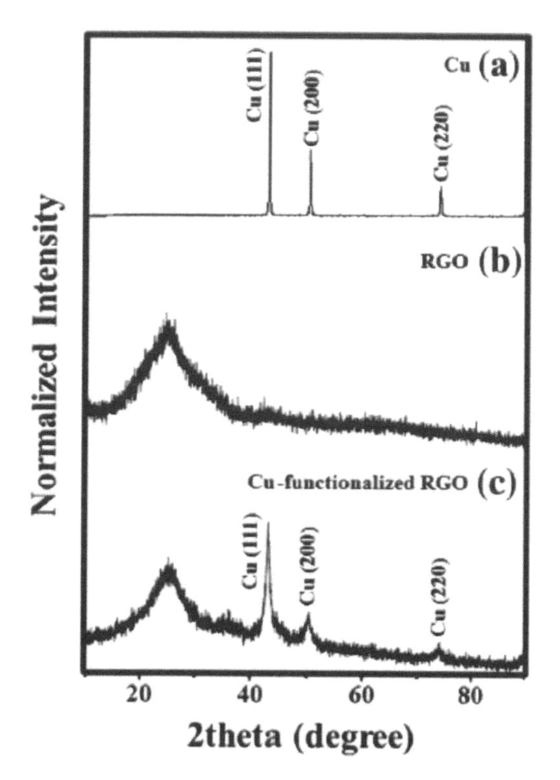

Fig. 4.49. XRD patterns of the samples: (a) Cu, (b) as-synthesized rGO and (c) rGO–Cu hybrid. Reproduced with permission from [142], Copyright © 2015, Elsevier.

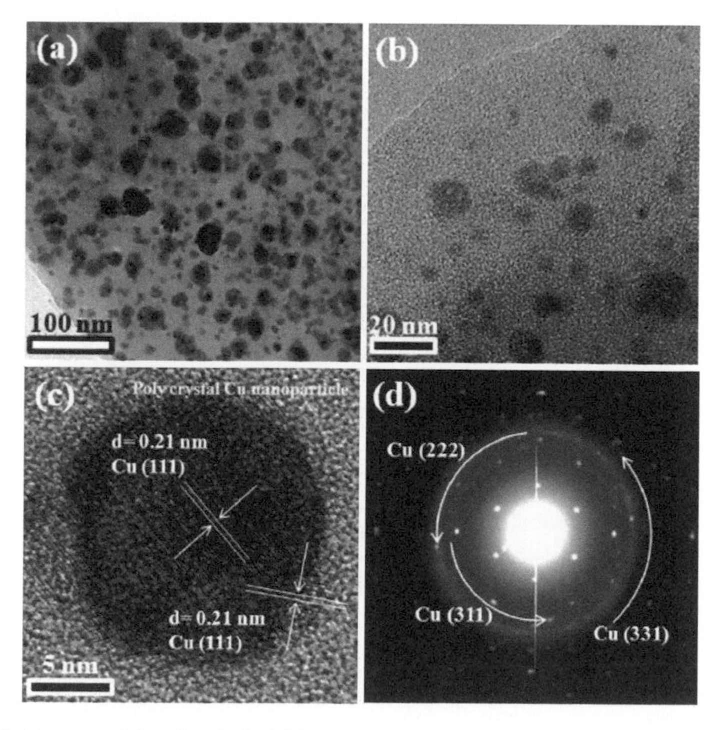

Fig. 4.50. (a, b) TEM images of the rGO–Cu hybrid. (c) Lattice-resolved TEM image of the Cu NPs on the rGO. (d) Corresponding SAED pattern. Reproduced with permission from [142], Copyright © 2015, Elsevier.

grains, as evidenced by the fact that the same lattice fringes could be observed at different locations in an oriented manner. In the SAED pattern (Fig. 4.50 (d)), both diffraction spots from the rGO and the diffraction rings corresponding to (311), (222) and (331) planes of the cubic Cu phase were observed.

It has been accepted that rGO is a *p*-type semiconductor [149, 150]. Because NO_2 is an electron-withdrawing molecule, it can extract electrons from the rGO nanosheets. As a result, the concentration of hole will be increased in the rGO, so as to increase the *p*-type conductivity or decrease the *p*-type resistivity. After the air is purged, the NO_2 species adsorbed on surface of the rGO nanosheets are removed by the newly adsorbed air molecules, so that the electrons taken by the NO_2 molecules will be released. In other words, the number of holes will be decreased and thus the *p*-type resistivity will be recovered.

Two possible mechanisms could be responsible for the enhancement in the sensitivity of the rGO–Cu hybrid [142]. On the one hand, the Cu sites could serve to adsorb the NO_2 molecules, so that bonds were formed between the NO_2 molecules and the Cu NPs, when the NO_2 molecules were adsorbed on the films. After that, the NO_2 molecules may migrate from the Cu cites to the surface of the rGO nanosheets. On the other hand, in the flow of NO_2 gas, Cu/CuO_x nanoarchitectures could enhance the sensitivity as well. While CuO and Cu_2O are *p*-type semiconductors with narrow band gaps, Cu in the Cu/CuO_x nanoarchitectures was an electron acceptor. In the presence of NO_2 molecules, electrons would be transferred from CuO to Cu_2O and then (from Cu_2O) to Cu, leading to the formation of holes on the surface of CuO or Cu_2O [151]. Because both CuO*x* and rGO are *p*-type semiconductors, electrons would be transferred conveniently from the rGO to the CuO_x, resulting in an increase in the density of holes in the rGO nanosheets. As a consequence, the gas sensitivity was further enhanced.

A chemical vapor deposition (CVD) method has been used to synthesize 3D porous graphene networks anchored with Sn nanoparticles encapsulated within graphene shells (Sn–G-PGNWs), which exhibited superior performances as the anodes of lithium ion batteries (LiBs), due to their high conductivity and large surface area [146]. Figure 4.51 shows a schematic diagram illustrating the synthetic process of the hybrids. It was a one-step *in situ* scalable technique with metal precursors as the catalysts and a 3D self-assembly of NaCl particles as the template, which could be readily removed due to its high solubility in water. The process combined (i) the self-assembly of water-soluble NaCl particles that were uniformly coated with an ultrathin film of $SnCl_2$–$C_6H_8O_7$ into a 3D structure by using freeze-drying and (ii) the calcination of the 3D $SnCl_2$–$C_6H_8O_7$/NaCl in H_2 to obtain Sn NPs. The Sn NPs then served as catalysts to grow graphene walls between the NaCl surfaces and graphene shells, thus leading to the formation of the 3D Sn–G-PGNW hybrids, with Sn NPs of 5–30 nm. Due to the presence of the CVD derived graphene shells, the encapsulated Sn NPs would not be exposed to the electrolyte and would be effectively isolated. The high elasticity of

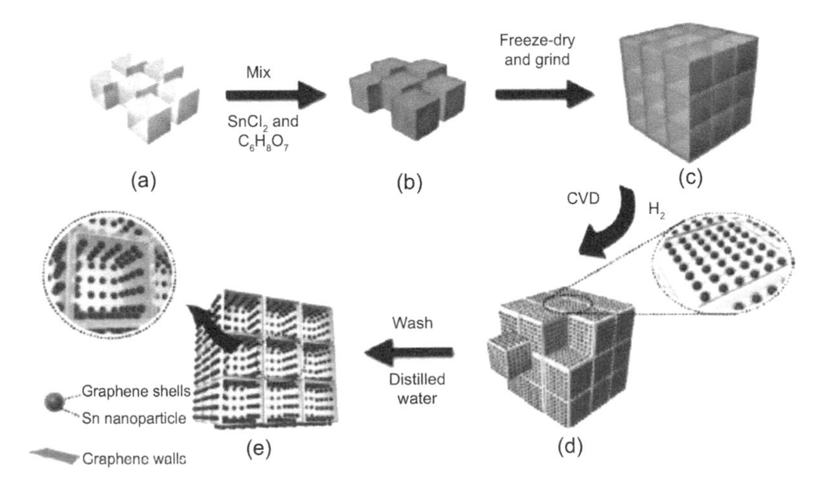

Fig. 4.51. Schematic illustration of the *in situ* CVD process for the one-step synthesis of 3D Sn–G-PGNWs using 3D NaCl self-assembly as a template. (a) NaCl particles. (b) $SnCl_2$–$C_6H_8O_7$-coated NaCl particles. (c) $SnCl_2$–$C_6H_8O_7$-coated NaCl self-assembly. (d) Sn–G-GNW-coated NaCl self-assembly. (e) 3D Sn–G-PGNWs. Reproduced with permission from [146], Copyright © 2014, American Chemical Society.

the graphene shells could help to buffer the volume expansion during the charge/discharge process. At the same time, the interconnected 3D porous graphene networks ensured high electrical conductivity, a large surface area and high mechanical flexibility, for application as an electrode [152].

XRD pattern of the hybrid is shown in Fig. 4.52 (a), indicating that the Sn NPs were of β-Sn crystal structure [146]. The sharp and narrow diffraction peaks implied that the Sn phase was highly crystallized. TGA and DTA results of the 3D Sn–G-PGNWs are shown in Fig. 4.52 (b), indicating that the content of Sn in the hybrid was about 46.8 wt%. Representative SEM images of the hybrid are shown in Fig. 4.52 (c, d). The interconnected macroporous networks were of a submicrometer size, while the walls of the interconnected 3D porous networks displayed a curved profile and a very low contrast, due to the ultrathin thickness (≤ 3 nm) and high mechanical flexibility of the graphene walls. Figure 4.52 (e) shows a high-magnification SEM image of the sample. The Sn–G NPs with sizes of 5–30 nm were uniformly deposited on the walls of the porous graphene networks.

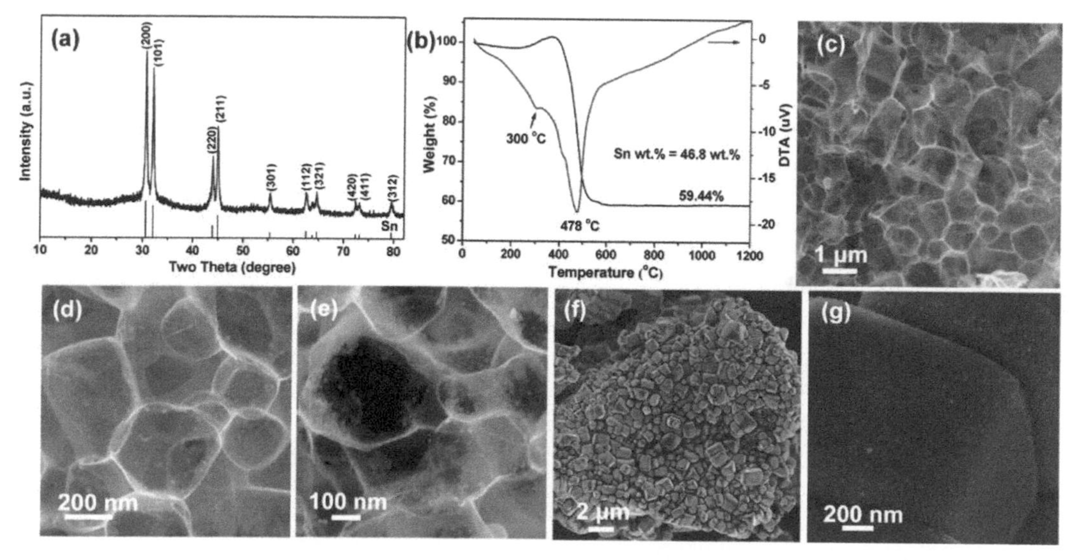

Fig. 4.52. (a) XRD pattern, (b) TGA and DTA profiles, and (c–e) SEM images of 3D Sn–G-PGNWs. (f, g) SEM images of the CVD synthesized products before eliminating NaCl. Reproduced with permission from [146], Copyright © 2014, American Chemical Society.

It was found that the presence of NaCl was a critical requirement for the synthesis of the 3D Sn–G-PGNWs. Without NaCl, the products derived from $SnCl_2$ and $C_6H_8O_7$ were 3D micrometer-sized amorphous carbon blocks, together with Sn particles with a large size of > 200 nm, without the presence of graphene. This is because Sn tended to form large particles without the NaCl template, due to its low melting point. The large Sn particles had a lower catalytic effect, so that graphene was not produced. As shown in Fig. 4.52 (f, g), before eliminating NaCl, the 3D structure of the Sn–G-PGNWs was formed on the surface of the 3D NaCl self-assembly. In other words, the 3D NaCl self-assembly provided a template to facilitate the growth of the 3D porous graphene networks.

Without the presence of the NaCl template, the mixture of citric acid was carbonized into amorphous carbon blocks, while Sn particles were formed from $SnCl_2$. In this case, porous graphene networks could not obtained, due to the three-dimensional cross-linked structure. With the presence of NaCl particles, they provided a large surface area to support the ultrathin $SnCl_2$–$C_6H_8O_7$ coating. When these NaCl particles coated with the $SnCl_2$–$C_6H_8O_7$ film were freeze-dried, they self-assembled into a 3D structure. This 3D structure ensured the formation of small and uniform Sn NPs, which in turn acted as a catalyst to produce the graphene walls. After the removal of NaCl through water washing, 3D porous graphene networks distributed with Sn–G NPs were obtained. Moreover, the pore size and wall thickness of the porous graphene and the size and content of Sn could be controlled by varying the content and size of the NaCl and the ratio of the metal precursor ($SnCl_2$) to the carbon precursor (citric acid).

Figure 4.53 shows representative TEM and HRTEM images of the 3D Sn–G-PGNWs [146]. The porous network deposited with uniform Sn–G NPs was constructed by continuous 3D ultrathin graphene walls of 5–30 nm, as shown in Fig. 4.53 (a, b). The interaction between the Sn–G NPs and the graphene walls was sufficiently strong to withstand ultrasonication. HRTEM images (Fig. 4.53 (c, d)) indicated that the graphene walls were composed of a few graphene layers with a thickness of ≤ 3 nm nevertheless. Some of the graphene walls had a thickness ≤ 1 nm. HRTEM images of the Sn–G NPs are shown in Fig. 4.53 (e, f). The NPs possessed a core-shell structure, in which the Sn NPs were entirely wrapped by the graphene shells. It was further observed that most of the Sn NPs also exhibited a core-shell structure, with the core section to be more crystallized and the shell section to be more amorphous-like. Such a unique microstructure was also confirmed by the SAED, STEM image (Fig. 4.53 (g)) and cross sectional composition line profiles (Fig. 4.53 (h)). The formation of such a crystalline-amorphous core-shell structure of the Sn NPs was closely related to the high cooling rate during the CVD process.

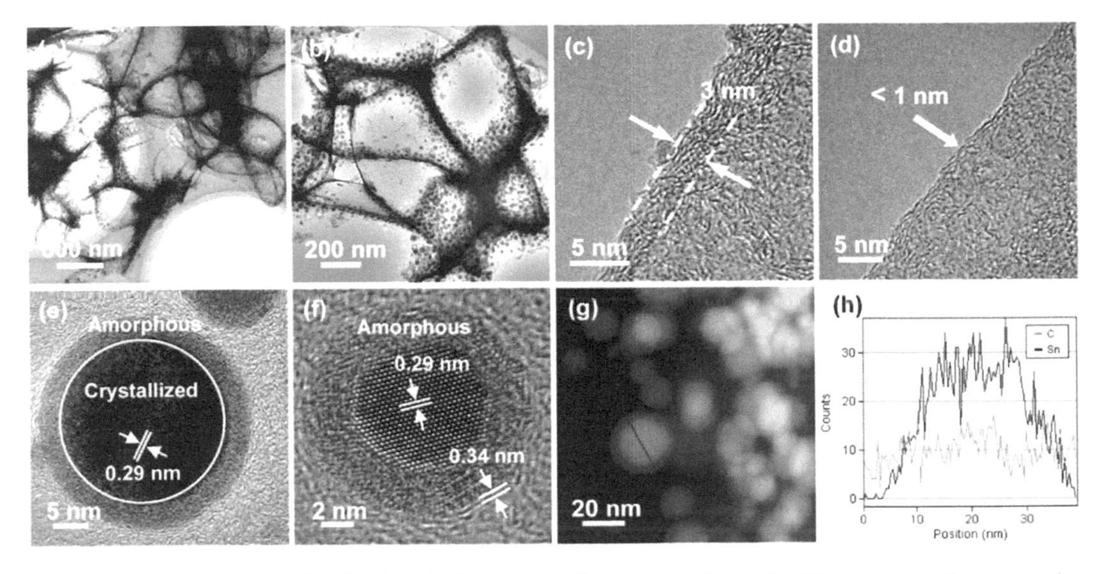

Fig. 4.53. (a, b) TEM images of the 3D Sn–G-PGNWs. (c, d) TEM images of the walls of the porous graphene networks. (e, f) HRTEM images and (g) STEM image of the Sn–G NPs. (h) Cross-sectional EDX elemental mapping of point 1 in (g). Reproduced with permission from [146], Copyright © 2014, American Chemical Society.

4.3. Graphene-Nanocarbons

Hybrids based on graphene (G) or graphene oxide (GO) and other nanocarbons, such as carbon nanotubes (CNTs) and carbon nanofibers (CNFs), have become a new group of carbon-based materials [14]. The incorporation of CNTs, including single-walled (SW), double-walled (DW) and multi-walled (MW), into G or GO, can be realized by using either chemical or physical methods. For example, CNTs–G/GO hybrid thin films are usually fabricated by using solution or suspension casting and layer-by-layer (LbL) deposition, free-standing sheets or papers are prepared by using vacuum filtration, while 3D hierarchical structures are produced by using chemical vapor deposition (CVD). CNTs–G/GO hybrids have also been used as fillers to fabricate polymer-based nanocomposites with synergistic effects of electrical and mechanical properties, which make them very useful for various potential applications, such as transparent electrodes replacing ITO, as electrode materials for supercapacitors, in lithium-ion batteries and dye-sensitized solar cells.

4.3.1. Graphene-CNTs

4.3.1.1. GO-CNTs

GO is dispersible in water, which is an advantage to be used to fabricate hybrids of GO–CNTs. In this section, representative GO–CNTs hybrids will be discussed to demonstrate their fabrication, characterization, properties and applications (brief description). In some cases, although the GO is reduced into rGO, the emphasis is on the GO-based hybrids. In addition, the following examples are presented without considering any specific order.

A simple mixing method was used to prepare CNTs–GO hybrid films [153]. It was found that the electrical conductivity of the films increased with the increasing content of CNTs because of the high conductivity of CNTs. Electrical conductivity also increased with the increasing thickness. However, the thin films were not optically transparent. Both chemically exfoliated GO nanosheets and commercial MWCNTs with OH groups were dispersed in DMF by using ultrasonic treatment. Figure 4.54 shows TEM images of the components. The two suspensions were mixed to form the precursor solutions. MWCNTs–GO thin films were prepared by casting the solutions on glass cover-slips. The ratio of MWCNTs to GO was controlled by the ratio of their precursor solutions, i.e., weight ratios of GONPs to MWCNTs-OH (*a*/*b*), labeled as G(*a*)/M(*b*). Thickness of the final films was simply adjusted by adjusting the volumes of the solutions used for depositions.

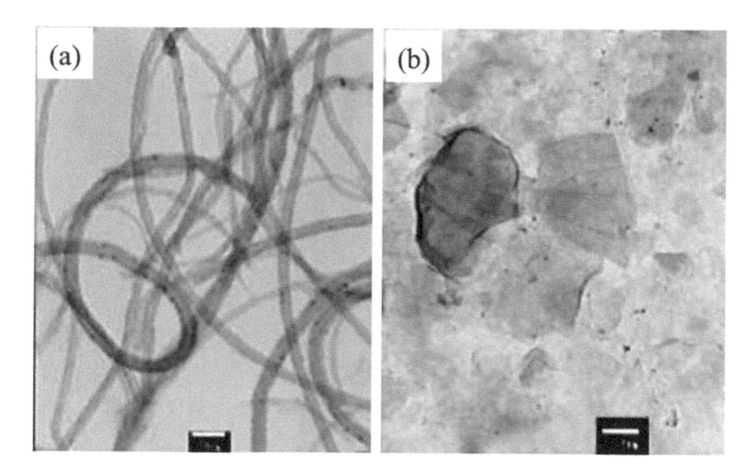

Fig. 4.54. TEM images of (a) exfoliated MWCNTs-OH (2.5 $mg \cdot g^{-1}$) and (b) GONPs (10 $mg \cdot g^{-1}$) in DMF. Reproduced with permission from [153], Copyright © 2008, WILEY-VCH Verlag GmbH & Co. KGaA, Weinheim.

Representative SEM images of the cross-section and natural surfaces of the MWCNTs–GO films are shown in Fig. 4.55. The films had multilayered microstructures and relatively homogeneous surface profiles. In the G(1)/M(0.5) film, most of the MWCNTs-OH (bright area) were covered by the GONPs (dark area), as shown in Fig. 4.55 (a, b). Percolated network of the CNTs was still not formed in this film sample. In contrast, in the G(1)/M(1) and G(1)/M(5) films, because the content of CNTs was increased, percolated CNTs network was observed, as shown in Fig. 4.55 (c, d) and Fig. 4.55 (e, f), respectively. This observation has a close effect on the square resistances of the hybrid films. The square resistance of the film for a given composition was also affected by the film thickness in a certain way. For example, the square resistance of the G(1)/M(0.5) film was significantly decreased from an unmeasurable value to 4.5 $\times 10^3$ $\Omega \cdot \square^{-1}$, as the thickness was increased from 2 μm to 3.5 μm. However, the square resistance was only slightly decreased to 1.3 $\times$ 10^3 $\Omega \cdot \square^{-1}$, when the thickness was further increased to 8 μm. Therefore, the percolation network of CNTs was the main factor that controlled the square resistances of the hybrid films.

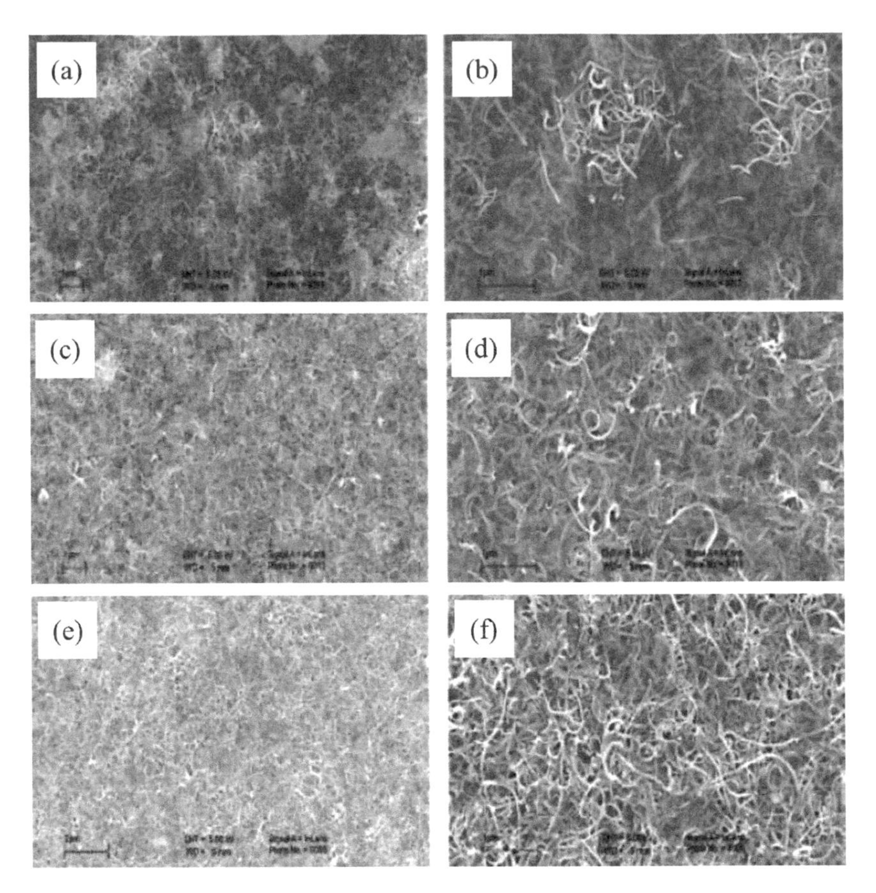

Fig. 4.55. Surface SEM images of the G(1)/M(0.5) film (a, b), G(1)/M(1) film (c, d) and G(1)/M(5) film (e, f). Reproduced with permission from [153], Copyright © 2008, WILEY-VCH Verlag GmbH & Co. KGaA, Weinheim.

Another example is the SWCNTs–GO hybrid thin film [154]. SWCNTs and GO with a weight ratio of 10:1 were dispersed with the aid of RNA. SWCNTs–GO films were obtained by using the filtration method. Careful study indicated that they could reduce the GO and remove the RNA by post-treatment with $NaBH_4$ and $NaOH/HNO_3$ consequently. This procedure resulted in significant reduction in the resistance of the films, making them highly promising for future applications. TEM images of the sample from the SWCNTs–GO solution, as well as that of the SWCNTs, are shown in Fig. 4.56. The SWCNTs were dispersed into thin bundles and the GO appeared as thin layers, which were mixed homogeneously. In comparison, the SWCNTs aggregated into thick bundles without the presence of GO nanosheets.

SWCNTs–GO hybrid films were prepared through the filtration method from the dispersions. The as-obtained films were post-treated successively with $NaBH_4$, NaOH and HNO_3. Sheet resistance of the samples decreased significantly after post-treatment. The SWCNTs–rGO hybrid film exhibited a high optical transmittance of 95.6% and a relatively low sheet resistance of 655 Ω per square after all the treatments. It was more conductive than the pure SWCNT film counterpart. The treatment of $NaBH_4$ was to reduce the GO to rGO and also to remove the RNA, so that the resistance of the film was effectively decreased. Similar effect was observed for the NaOH treatment. However, the successive application of these two treatments was more efficient than the single treatment of each compound. The function of HNO_3 included the removal of dispersants in the films and increasing the *p*-type conductivity of the SWCNTs and thus that of the films.

Figure 4.57 shows SEM images of the SWCNTs–GO films before and after the successive treatments. Obviously, the film was covered by RNA molecules before the post-treatment, which was responsible for

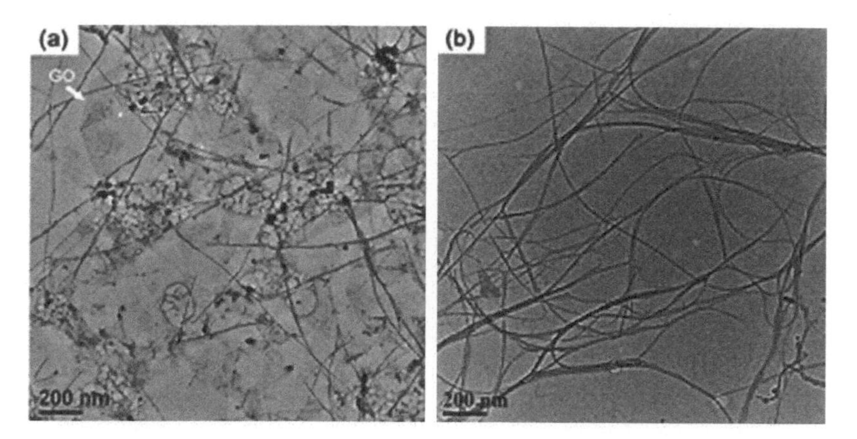

Fig. 4.56. TEM images of the SWCNTs–GO (a) and SWCNTs (b) from the respective solution dispersed with the aid of RNA. Reproduced with permission from [154], Copyright © 2011, The Royal Society of Chemistry.

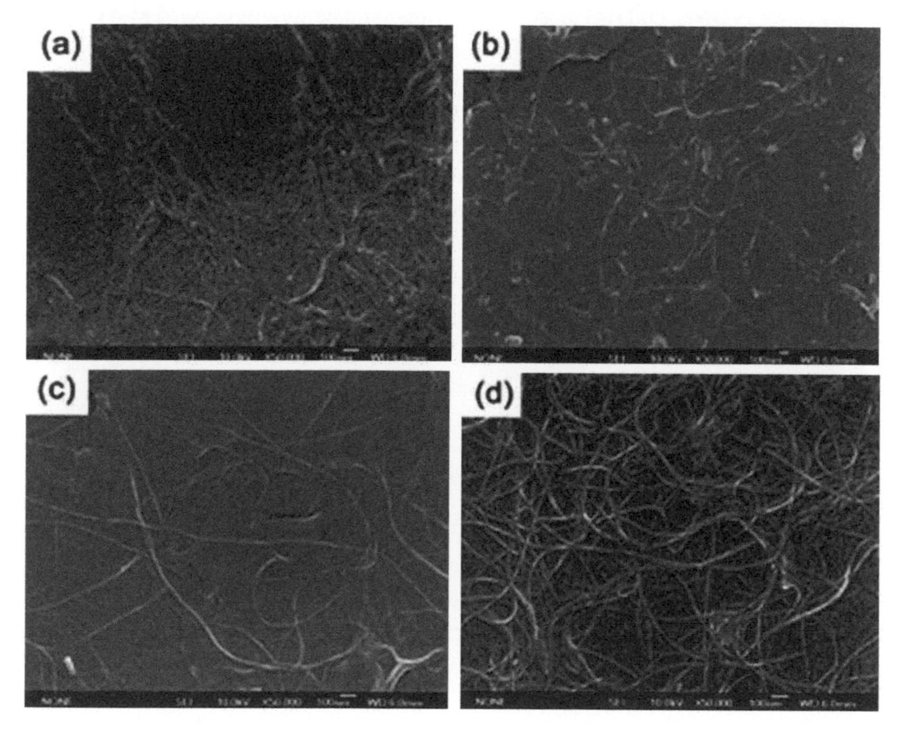

Fig. 4.57. SEM images of the SWCNTs–GO hybrid films before and after the treatment: (a) before treatment, (b) after treatment with $NaBH_4$ for 2 h, (c) after subsequent treatment with NaOH for 1 h and (d) after subsequent treatment with HNO_3 for 10 min. Reproduced with permission from [154], Copyright © 2011, The Royal Society of Chemistry.

the large sheet resistance of the film. After treatment with $NaBH_4$, some RNA molecules were removed and a SWCNT network could be observed. The SWCNT network became clearer after the treatment with NaOH solution, due to the further removal of RNA molecules. Finally, after the treatment with HNO_3, the SWCNT network became much denser, supporting the variation in the conducting properties of the films.

A double-layer structured GO–MWCNT hybrid film was created as a laser desorption/ionization (LDI) platform [155]. A layer of GO was first deposited on the substrates by immerging them in a GO aqueous solution. After drying, the sample was immerged in a MWCNT–NH_2 suspension to form a double layer of GO–MWCNT–NH_2. The films exhibited high salt tolerance, no sweet-spots for mass signals and good durability against mechanical stress. This platform could be used as an important tool for mass spectrometry-based biochemical analyses due to its outstanding performance, stability and cost effectiveness.

GO thin films were prepared by immersing APTES-functionalized glass substrate in an aqueous GO suspension of 1.5 mg·ml^{-1} for 1 h. After that, it was washed with water and ethanol, dried in N_2 and baked at 125°C for 10 min in N_2 flow. The substrate coated with GO thin film was then used to fabricate GO–MWCNT-NH_2 double layered films. The substrate was immersed in a MWCNT-NH_2 suspension of 120 μg·ml^{-1} for 1 h, followed by washing with water and ethanol and drying in N_2 flow. The GO–MWCNT-NH_2 double layered films were baked at 125°C for 10 min on a heated plate in N_2 flow. To make rGO and rGO–MWCNT-NH_2 double layered thin films, the GO and GO–MWCNT-NH_2 double-layer thin films were immersed in a 20% hydrazine monohydrate solution in DMF at 80°C for a day, followed by washing with DMF, water and ethanol, as well as drying in N_2 flow. For comparison, rGO–MWCNT and MWCNT-NH_2 films were prepared in a similar way.

LDI efficiency of surface immobilized carbon materials was systematically studied, in order to identify which nanomaterial played a key role, i.e., GO or MWCNT-NH_2. To do this, various types of substrates were prepared by utilizing the GO, rGO, MWCNT and MWCNT-NH_2, as well as their combinations. The substrates were divided into three groups, i.e., (i) substrates coated with only GO derivatives, (ii) those coated with only MWCNT-NH_2 and (iii) those coated with combinations of GO and MWCNT derivatives, as shown in Fig. 4.58 (a). GO derivative coated samples were prepared by immobilizing GO on aminated glass substrates, in which the surface immobilized GO was reduced with hydrazine monohydrate. The substrate coated with MWCNT-NH_2 was prepared by immobilizing MWCNT-NH_2 onto epoxide presenting glass. Three different substrates based on the combinations of GO and MWCNT derivatives were prepared by immobilizing MWCNT onto the rGO substrate (rGO–MWCNT) through π–π interaction, adsorbing MWCNT-NH_2 onto the GO substrate through electrostatic interaction to have GO–MWCNT-NH_2 and chemical reduction of the GO–MWCNT-NH_2 to form rGO–MWCNT-NH_2, as illustrated in Fig. 4.58 (b). SEM characterization results indicated all the samples possessed a high surface coverage of the immobilized carbon nanomaterials, as demonstrated in Fig. 4.58 (c).

Due to their oppositely charged surfaces, the GO and aminated MWCNT (i.e., MWCNT-NH_2) could be immobilized as a stable double-layered film in sequence on solid substrates. The roughness and surface area of the GO-coated layers were increased, after a layer of MWCNT-NH_2 was added on top. As a result, the analyte adsorption was enhanced and thus the LDI efficiency was greatly increased. The GO–MWCNT-NH_2 double layers exhibited high wettability in water due to their charged surface, were able to absorb UV light and could efficiently convert the absorbed energy into thermal energy, as expected from the properties of GO and MWCNT-NH_2 in the solutions [156, 157]. Therefore, these GO–MWCNT-NH_2 double layered films could be used as promising candidates for chip format LDI platforms.

It was found that GO–MWCNTs hybrids could be used as electrocatalysts with promising redox reversibility for VO^{2+}/VO_2^+ redox couples of vanadium redox flow batteries (VRFB) [158]. The GO–MWCNTs hybrids were produced by a unique electrostatic spray technique combined with an efficient ultrasonic treatment, as shown schematically in Fig. 4.59. It was discovered that the hybrid yielded a much higher electrocatalytic redox reversibility towards the positive VO_2^+/VO^{2+} couple, especially for the reduction.

Figure 4.60 shows representative TEM images of the GO nanosheets, MWCNTs and the GO–MWCNTs hybrid [158]. Figure 4.60 (a) revealed that the GO nanosheets were rough at the nanoscale level with the presence of wrinkles and thickness of about 3.18 nm. The MWCNTs were randomly dispersed and highly entangled with one another, as shown in Fig. 4.60 (b). As demonstrated in Fig. 4.60 (c, d), the MWCNTs were un-zipped into GO nanosheets and closely packed in the lateral direction. The gap between the GO nanosheets was bridged by the MWCNTs network. Due to the large lateral dimension, the GO nanosheets served as strong supporters, which were connected by the MWCNTs. As a result, a continuous network structure was formed together by the GO nanosheets and MWCNTs, which was attributed to the assembly of the two components during the ultrasonic treatment, thus forming the hybrid. The hybrid network made use of the high electrical conductivity of the MWCNTs and the abundant oxygen-containing groups of the GO simultaneously, which facilitated the promising electrocatalytic activity for the VO^{2+}/VO_2^+ redox couples in VRFB.

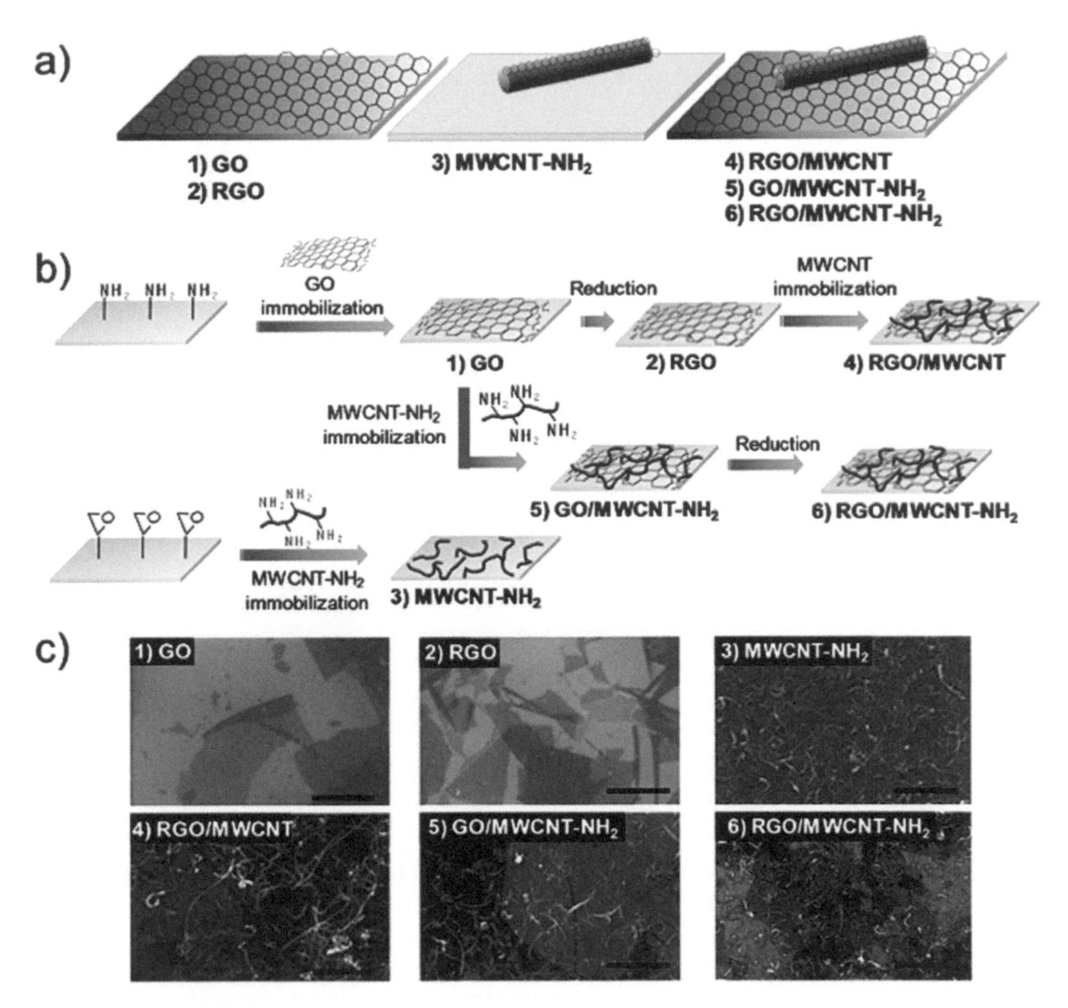

Fig. 4.58. Various carbon nanomaterial-based substrates fabricated for comparison of LDI efficiency. (a) Scheme of three substrates composed of GO derivative-coated, MWCNT-NH_2-coated, and GO and MWCNT derivative-coated as surface immobilized films. (b) Detailed scheme of preparation of six different graphene and/or MWCNT-based substrates. (c) SEM images of the prepared carbon nanomaterial substrates. Scale bars = 1 μm. Reproduced with permission from [155], Copyright © 2011, American Chemical Society.

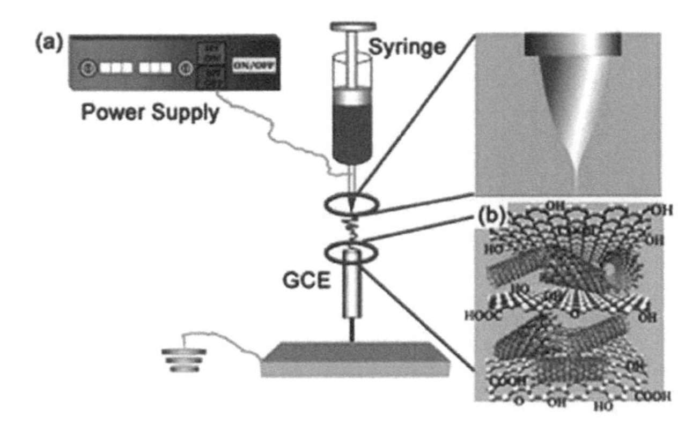

Fig. 4.59. (a) A schematic diagram of the e-spray technique used to fabricate the GO–MWCNTs hybrid on a GCE, (b) E-sprayed GO–MWCNTs on a GCE. Reproduced with permission from [158], Copyright © 2011, The Royal Society of Chemistry.

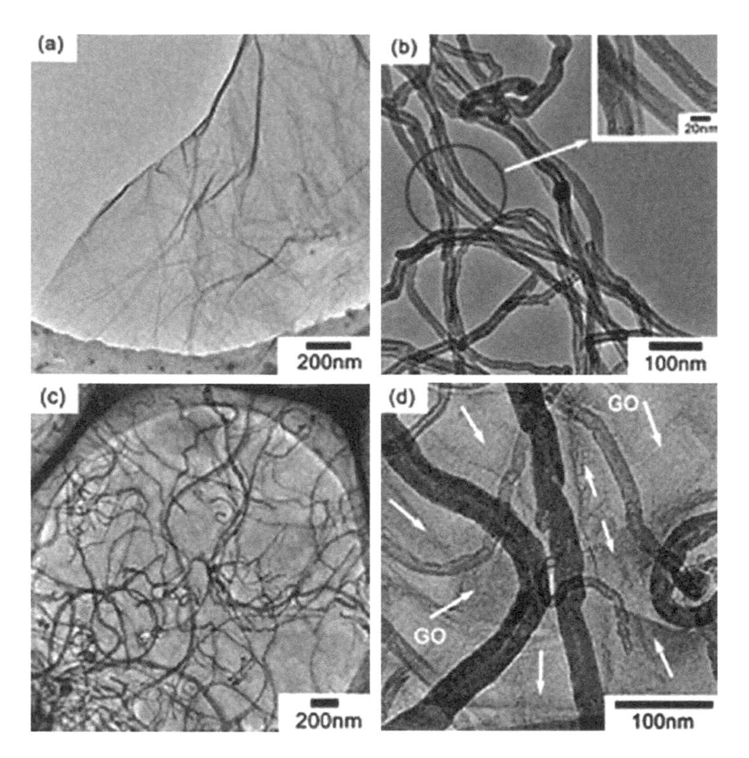

Fig. 4.60. TEM images of the GO nanosheets (a), MWCNTs (b), GO–MWCNTs (c) and a detailed view (d) of the red circle area in (c). Reproduced with permission from [158], Copyright © 2011, The Royal Society of Chemistry.

A stable aqueous dispersion, containing hydrophobic SWCNTs and GO nanosheets, was produced to prepare a GO–SWCNTs hybrid without using any surfactant [159]. The GO or GO–SWCNTs hybrid films were prepared by using the spray-coating method. To obtain the rGO, the samples were chemically reduced with hydrazine vapor at 80°C overnight. A superior performance in optical limiting and supercapacitance was exhibited in the hybrid, which suggested its potential application in optoelectronic devices and energy storage.

Figure 4.61 shows representative SEM and TEM images of the GO–SWCNTs hybrid [159]. The SWCNTs were in the form of heavily entangled bundles, with diameters of 20–30 nm, as demonstrated in Fig. 4.61 (a). Figure 4.61 (b) revealed that, in the GO–SWCNTs hybrid, the SWCNTs were disentangled, but with larger diameters of 50–80 nm. This observation could be attributed to the fact that the SWCNT bundles were connected by the GO nanosheets, thus forming a core–shell structure through π–π stacking. The presence of the core–shell structure in the GO–SWCNTs hybrid was confirmed by TEM observation. As illustrated in Fig. 4.61 (c), evident core–shell structures were observed, in addition to flat GO nanosheets. HRTEM image (Fig. 4.61 (d)) revealed the lattice fringes of the graphitic structure of the SWCNT bundles in the core of the hybrid. The fact that the SWCNTs were entirely covered by the GO nanosheets was also confirmed by the electrical conductivity of the hybrid.

Hybrids of CNTs and G/GO could also be used to fabricate composite fibers. One example is that of the GO–SWCNTs fibers prepared by using a coagulation spinning technique [160]. The hybrid fiber simultaneously offered high mechanical strength, high electrical conductivity and promising electrical actuation performance. Representative results are shown in Fig. 4.62. For instance, the fiber with a 2:1 volume ratio of SWCNT to GO achieved an increase in tensile strength and elastic modulus by 80% and 133%, respectively. The highly uniform dispersion of the nanocarbons and their stronger interfacial interactions were responsible for the synergetic property enhancement.

In this coagulation spinning process, acid was added to the PVA (poly(vinyl alcohol)) solution to assist the coagulation of GO. The optimized pH value of the PVA solution was below 2.0 for the GO fibers

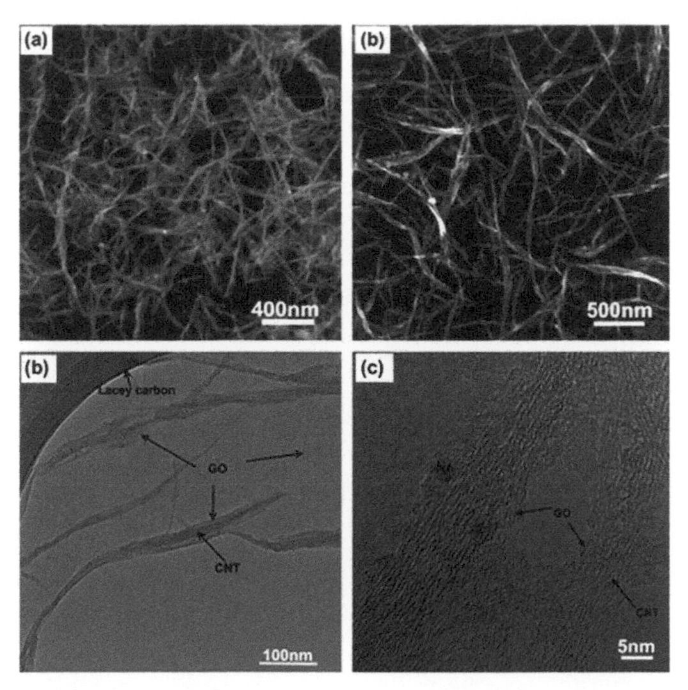

Fig. 4.61. (a) SEM image of SWCNTs. (b) SEM image of SWCNT–GO. (c) TEM image of SWCNT–GO–SWCNT. (d) High-resolution TEM image of SWCNT–GO. Reproduced with permission from [159], Copyright © 2011, Elsevier.

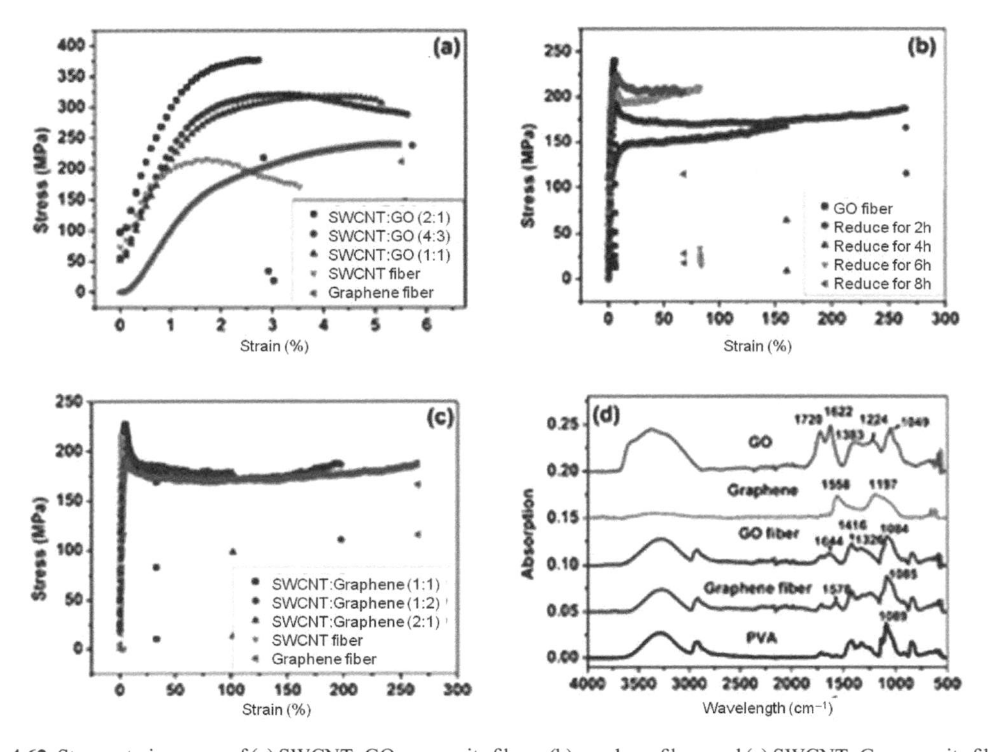

Fig. 4.62. Stress–strain curves of (a) SWCNT–GO composite fibers, (b) graphene fibers and (c) SWCNT–G composite fibers. (d) FTIR spectra of GO, graphene and their fibers. Reproduced with permission from [160], Copyright © 2011, The Royal Society of Chemistry.

to be spun continuously. The nanocarbons, GO or carbon nanotubes were prone to stacking or bundling together if only one of them was used in the fibers, as demonstrated in Fig. 4.63 (a, b). Generally, slippage occurs more likely between the GO nanosheets and carbon nanotube bundles, which has a negative effect on strength of the fibers. In contrast, a network was formed, when the GO and SWCNTs were dispersed together in the PVA matrix, as illustrated in Fig. 4.63 (c). Because GO is amphiphilic, it served as a dispersing agent to promote the dispersion of the SWCNTs [36], while the SWCNTs in turn prevented the restacking of the GO. The intercalation of the GO nanosheets into the SWCNT networks ensured an increased interfacial contact area, so that the stress could be readily transferred between the nanocarbons and PVA molecules. Also, GO contains a large number of carboxyl and hydroxyl groups, thus facilitating the formation of hydrogen bonds with the SWCNTs and PVA molecules and hence additionally strengthening the interfacial interactions.

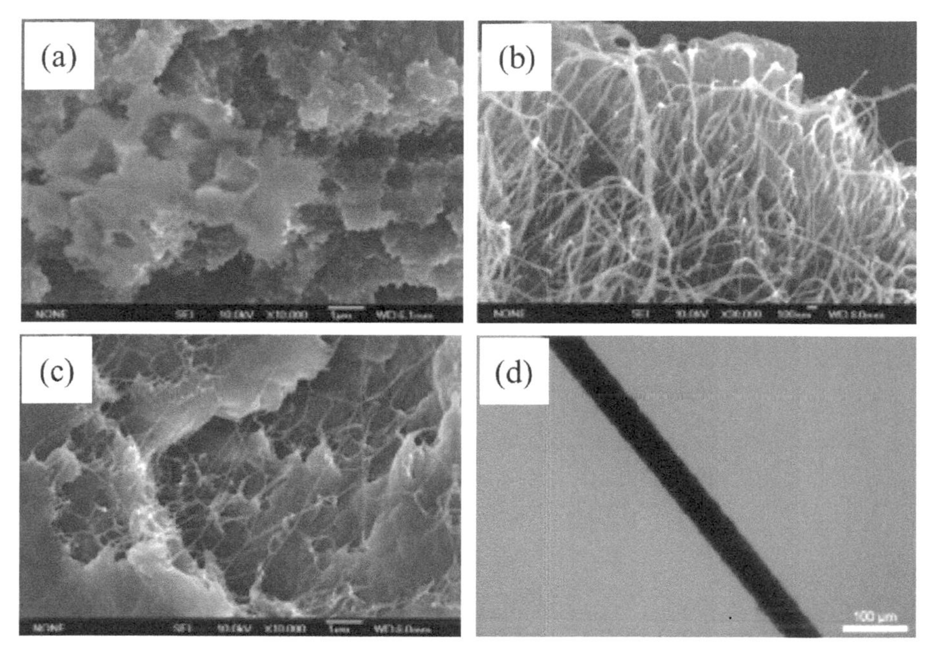

Fig. 4.63. SEM images of the cross-section of (a) GO fibers, (b) SWCNT fibers, (c) SWCNT–GO (2:1) composite fibers, (d) optical microscope image of the SWCNT–GO (2:1) composite fiber. Reproduced with permission from [160], Copyright © 2011, The Royal Society of Chemistry.

It was also found that the ratio of SWCNTs and GO could be used to adjust the mechanical properties of the fibers, which could be tuned to be from brittle to high toughness. As shown in Fig. 4.62 (c), the fibers possessed high toughness if GO was dominant, whereas brittle ones were obtained if the content of SWCNTs was dominant. This means that the toughness of the fibers could be varied, while their strength was only slightly affected. Interestingly, the G–SWCNT hybrid fibers possessed lower strength and conductivity than the GO–SWCNT fibers, which was attributed to the poor dispersion of the graphene nanosheets in the PVA matrix.

4.3.1.2. G-CNTs

Slightly different from GO, graphene nanosheets are not dispersible in water due to the absence of the functional groups. As stated previously, graphene nanosheets derived from GO through chemical reduction are usually called reduced GO or rGO. Various methods, including (i) solution processing/casting, (ii) layer-by-layer (LbL) deposition, (iii) vacuum filtration and (iv) chemical vapor deposition (CVD), have been employed to fabricate CNTs–G/rGO hybrids, which will described in a more detailed manner in this section.

Solution casting is the most convenient and cost-effective method to obtain CNTs–G/rGO hybrid thin films. An inexpensive method was used to develop rGO–CNTs hybrid thin films, from chemically converted graphene and pre-treated carbon nanotubes [161]. Both SWCNTs and GO were dispersed in anhydrous hydrazine, leading to a stable suspension. In this case, the GO was reduced at the same time. The ratio of SWCNTs to rGO could be readily controlled. The stable suspensions could be used to deposit thin films by solution casting and spin-coating. No surfactant was used, so that the intrinsic electrical and mechanical properties of the two components were retained. Optical transparency and electrical conductivity of the SWCNTs–rGO thin films were comparable with those of ITO on flexible substrates.

Figure 4.64 shows representative SEM and AFM images of single-component and the hybrid thin films. The hybrid film had a thickness of about 5 nm, with a relatively rough surface covered with CNT bundles. The presence of such bundles has been a problem; they could penetrate the active layers and thus cause shorting of devices. To address this problem, the G-SWCNT dispersions were sonicated sufficiently before they were used to deposit films. By using this treatment, the SWCNT bundles were broken up, so that the root mean square surface roughness could be reduced to about 1.49 nm, as illustrated in Fig. 4.64 (d).

A similar solution processing method was reported to fabricate rGO–SWCNTs hybrid films, with tunable work function (U_w), by using the doping of alkali carbonate salts with different properties [162]. The hybrid thin films were used to fabricate electrodes for inverted PVs to improve their efficiency. Such rGO–SWCNTs hybrid films could also be used as transparent electrodes for flexible optoelectronic devices by replacing ITO. To prepare hybrid suspensions of rGO and SWCNTs, dry GO and SWCNTs powders were dispersed directly in anhydrous hydrazine and aged by stirring for 1 week. As hydrazine contacted with the GO and SWCNT powders, bubbles were produced, after which a uniform dark gray suspension without visible precipitation was formed. The stable dispersions were centrifuged to separate any SWCNT bundles and aggregated rGO nanosheets. The homogeneity of the rGO-SWCNT dispersion was ensured by heating at 60°C under strong ultrasonication.

Figure 4.65 shows representative TEM images of the rGO, SWCNTs and their hybrids [162]. The rGO possessed a paper-like structure, with a few stacking layers of monatomic rGO nanosheets, as illustrated in Fig. 4.65 (a). SWCNTs can be well-dispersed in hydrazine through the formation of hydrazinium compounds, consisting of negatively charged SWCNTs capped with $N_2H_5^+$ counterions [161]. On precipitation was observed by the naked eyes, the aggregation and formation of SWCNT bundles occurred at nanoscales, as revealed by the TEM image in Fig. 4.65 (b). Fortunately, such SWCNT aggregates and bundles could be eliminated through redispersion by using prolonged sonication (e.g., several hours), as demonstrated in Fig. 4.65 (c, d). Due to its paper-like structure with a large surface area, the rGO served as a molecular surfactant to promote the dispersion of the SWCNTs [163, 164]. Due to the presence of the highly conjugated structures on their surface, the rGO nanosheets adhered firmly onto the cylindrical planes of the SWCNTs through π-π interactions. Figure 4.65 (e) indicated that many SWCNTs were laid over the surface of the rGO nanosheets, which ensured the formation of the conductive network in the rGO-SWCNTs hybrid.

It was also demonstrated that the electrical conductivity of the SWCNT films could be greatly increased by including a small amount of graphene [165]. A peak of DC conductivity was observed when 3 wt% graphene was used. rGO–CNT films deposited on a titanium substrate were prepared, which could be used as the binder-free electrodes of supercapacitors [166]. Supercapacitors with such hybrid electrodes possessed specific capacitance values much higher than those reported in the open literature. Besides G/rGO–CNTs two phase hybrids, multi-component hybrids have also been reported in the open literature. Either CNTs or G/rGO were modified [167, 168] or a third component phase was introduced to further improve the functionalities of the hybrid materials. For instance, a non-covalently dispersed *p*-type doped GO–SWCNT film was deposited by using a high-pressure air spray method [167]. Such a hybrid thin film displayed a low sheet resistance of 171 $\Omega \cdot \square^{-1}$ and a high transmittance of 84% at 550 nm. LCE cells fabricated with the *p*-GO–SWCNT thin films as the electrode demonstrated comparable performances with that of ITO electrodes. Pan et al. [168] used chitosan to graft rGO, so that it became soluble in water. Furthermore, it allowed for NWCNTs to be well-dispersed in acidic solutions with noncovalent interaction. The cast hybrid films exhibited significantly reinforced mechanical properties due to the presence of chitosan.

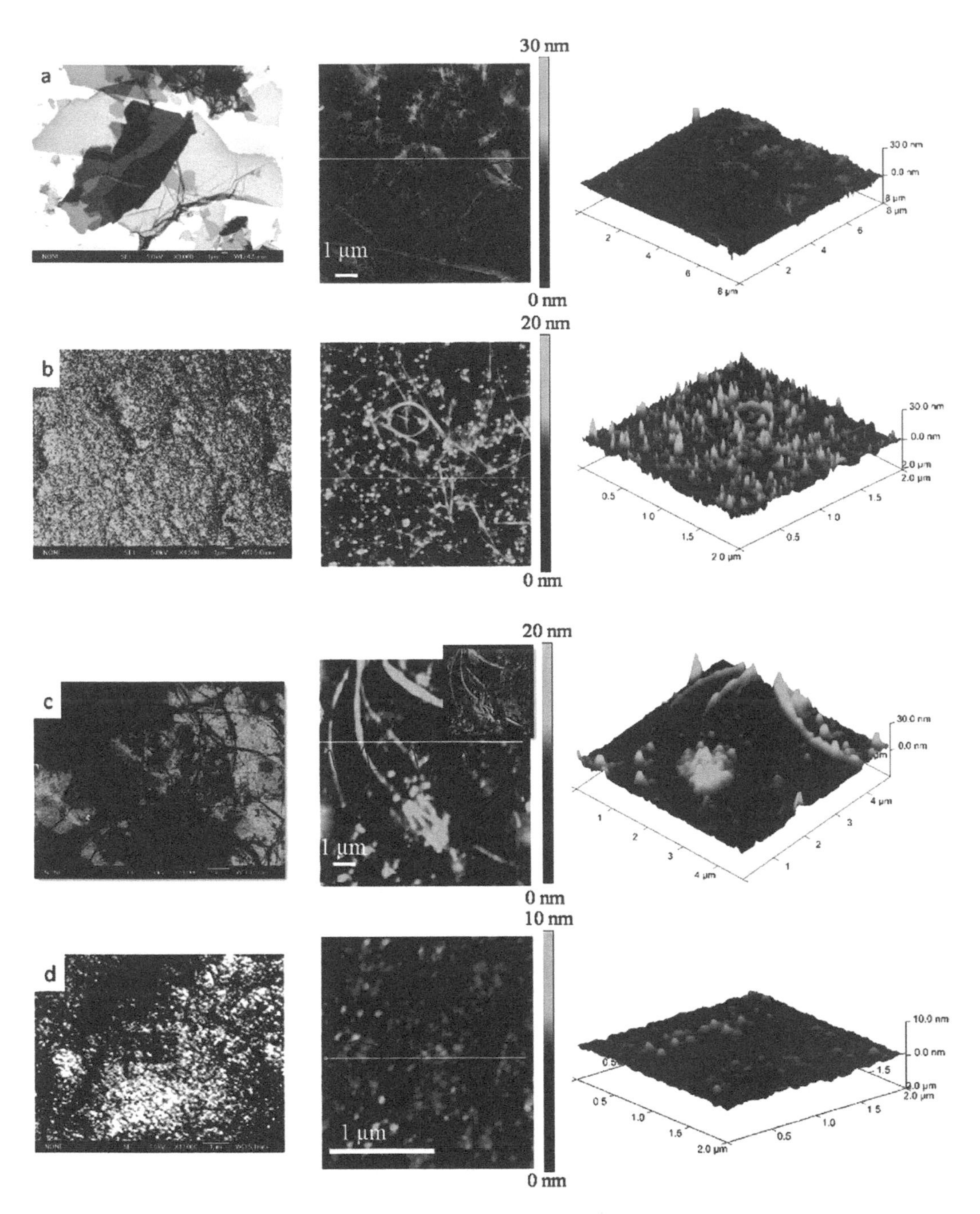

Fig. 4.64. SEM and AFM images of the G-CNT film, together with 3D topographies of (a) chemically converted graphene, (b) single-wall carbon nanotube network and (c) G-CNT hybrid film. The dense network of the G-CNT film exceeded the percolation threshold with an average surface roughness of 5–10 nm. Inset: phase image of the large sheet of chemically converted graphene and individual SWCNT. (d) G-CNT film after surface optimization. Height profile (blue curve) taken along the white solid line indicated an average surface roughness of 1.49 nm. Reproduced with permission from [161], Copyright © 2009, American Chemical Society.

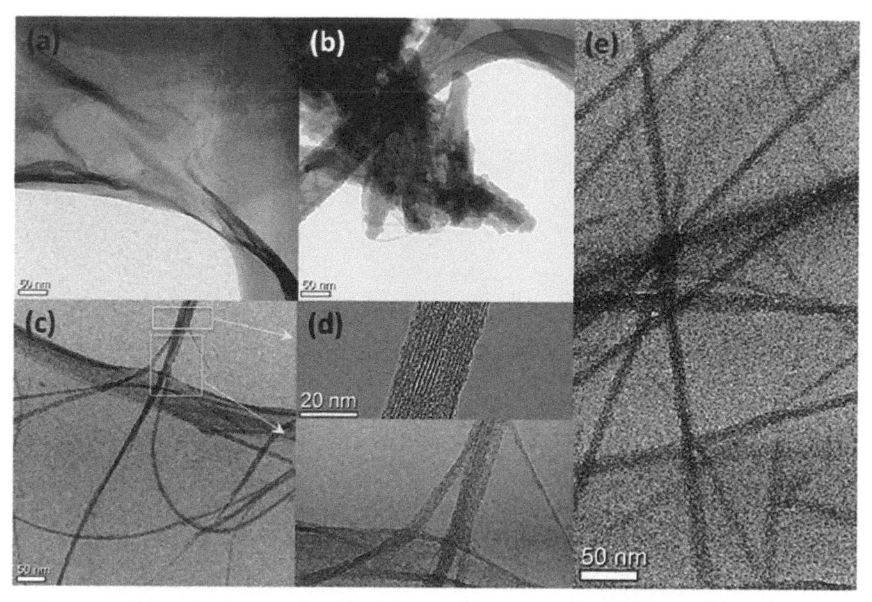

Fig. 4.65. TEM images of the samples from dispersed solutions. TEM images of (a) rGO, (b) SWCNTs and (c–e) rGO-SWCNTs hybrids. The low-magnification (scale bar = 50 nm) TEM images of the rGO-SWCNTs hybrids in (c) and (d) indicated that the rGO acted as a surfactant to disperse the SWCNTs through π-π interactions. The high-resolution (scale bar = 20 nm) TEM image of the rGO-SWCNTs hybrid in (d) indicated that the SWCNTs underwent debundling to be physically attached to the rGO nanosheets. (e) Central region of surface of the rGO-SWCNTs hybrid, revealing many SWCNTs extended across the rGO surface. Reproduced with permission from [162], Copyright © 2009, American Chemical Society.

G–MWCNTs hybrids processed through solution casting have been explored for application as the counter electrode of DSSC, in order to replace the Pt catalysts [169]. The catalytic effect of the G–MWCNTs hybrids was attributed to the presence of sharp atomic edges exposed to the ions in the electrolyte. An extremely thin G–MWCNTs hybrid sheet prepared by using a solar energy exfoliation technique has been demonstrated [170], where solar light was harvested and directed into the materials by using a convex lens with a diameter of 90 mm. The hybrid was then loaded with Pt nanoparticles, and was used to fabricate membrane electrodes for proton exchange membrane fuel cells (PEMFCs). By bridging the defects for electron transfer and increasing the basal spacing between the graphene nanosheets, the presence of the MWCNTs significantly improved the electrochemical performance of the electrodes.

A new type of 3D G–MWCNTs hybrid has been prepared by using a two-step solution method at room temperature [171]. There was an optimal proportion of MWCNTs to graphene nanosheets in terms of the overall performances of the hybrid films. The MWCNTs were evenly distributed on the surface of the graphene nanosheets, which in turn improved the dispersion behavior of the graphene nanosheets. Figure 4.66 shows a schematic diagram of the formation of the G–MWCNTs hybrids. The hybrids were incorporated with TiO_2 matrix to be used as the working electrode for dye-sensitized solar cell (DSSC). The hybrid displayed a higher degree of dye absorption and a lower level of charge recombination.

Both the pristine MWCNTs and the graphene were unstable in the solvent, so that aggregation and precipitation were soon observed, and were attributed to a lack of hydrophilic groups in their structure to have interactions with the polar solvent. Remarkably, the acid-MWCNTs and hybrids could be completely dispersed. However, aggregation and precipitation were observed in the acid-MWCNT suspension one week later, and were attributed to the attraction produced by the large hydrophobic specific surface of the CNTs, although they were treated with acid. In contrast, the hybrid dispersions were still stable without any precipitation at room temperature. In this case, there were π-stacking interactions between the π-conjugated aromatic surfaces of the graphene nanosheets and the walls of the acid-MWCNTs, while the hydrophilic surfaces of the acid-MWCNTs were responsible for the dispersion. Therefore, the incorporation of acid-MWCNTs tremendously enhanced the dispersion capability of graphene nanosheets in ethanol.

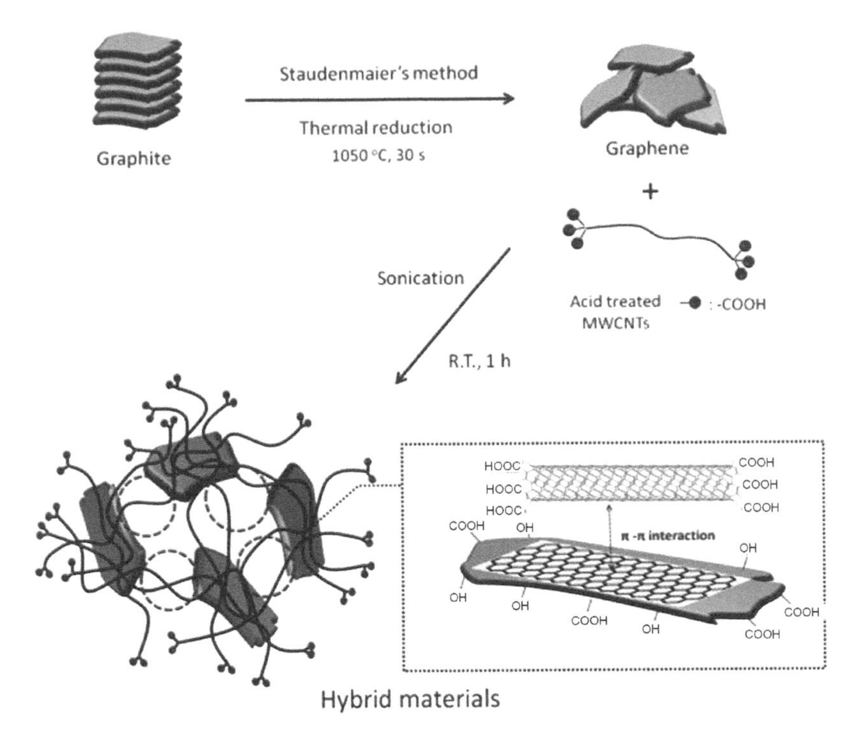

Fig. 4.66. Schematic diagram showing mechanism of the acid-MWCNTs in improving the dispersibility of graphene nanosheets. Reproduced with permission from [171], Copyright © 2011, Elsevier.

Hybrids with different ratios of acid-MWCNTs and graphene were prepared to optimize the dispersion behaviors of the hybrids. Figure 4.67 (a) revealed that the dispersion capability of the acid-MWCNTs was higher than that of the graphene by three orders of magnitude. The difference in dispersive behavior between the two nanocarbons was readily attributed to their difference in dimension and surface characteristics. The dispersion of the hybrids was strongly dependent on the ratio of acid-MWCNTs to graphene (inset of Fig. 4.67 (a)). At low concentrations of acid-MWCNTs, i.e., ratios of acid-MWCNTs to graphene from 0.5 to 6, the dispersion capability was increased with the increasing content of the acid-MWCNTs. Figure 4.67 (b) shows TEM image of the sample with ratio of acid-MWCNTs to graphene of 2:1. The acid-MWCNTs were evenly distributed on the surface of the graphene nanosheets, due to the interaction of the MWCNTs with the π-conjugated aromatic domain on the basal plane of the graphene nanosheets. A steric barrier was formed by the acid-MWCNTs to effectively separate the graphene nanosheets. The hydrophilic groups of the acid-MWCNTs ensured the dispersion of the graphene nanosheets through interaction with the polar solvent.

At higher concentrations (ratios of 6–12), the dispersion behavior was negatively affected by the content of the acid-MWCNTs. In this range, acid-MWCNTs bundles or clusters were formed, as illustrated in Fig. 4.67 (c). At the highest concentrations (ratios of 12–16), the high dispersion capability was recovered. However, Fig. 4.67 (d) indicated that the supernatants of these samples contained only acid-MWCNTs. All the graphene nanosheets were present in the precipitate after centrifugation, whose surfaces were coagulated with excessive MWCNTs clusters. This observation suggested that the dispersion capability cannot be solely used as a measure to evaluate the quality of the hybrids, while the ratio of the two components was more important. For example, that hybrids with ratios of 2, 10 and 14 exhibited similar dispersion behavior, but they had different microstructures. In this regard, the optimal ratio of acid-MWCNTs to graphene was 2:1.

LbL has been shown to be a powerful technique to fabricate multicomponent thin films with microstructures that can be finely tailored at nanometer scales. Therefore, it has been employed to prepare various G/GO–CNTs hybrids [172–174]. For example, a double layer structured ultrathin transparent

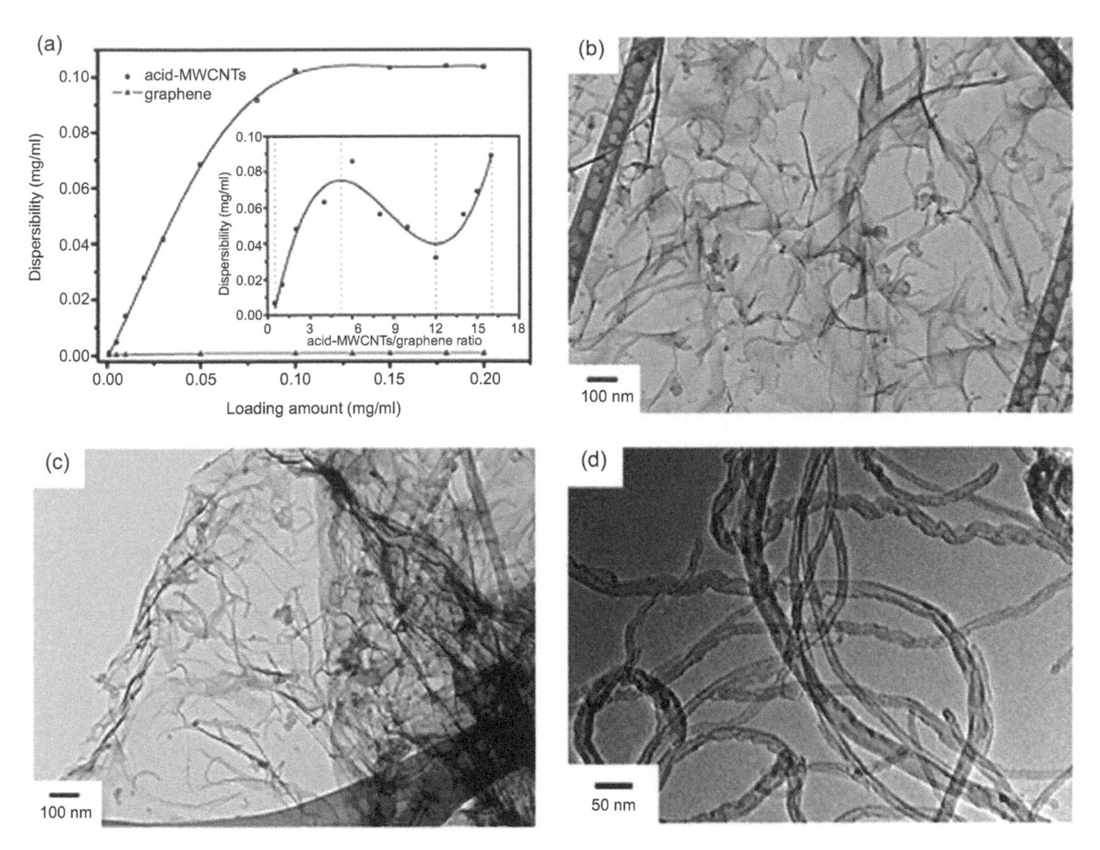

Fig. 4.67. (a) Limiting dispersion of the acid-MWCNTs and graphene. The inset is a plot of dispersibility as a function of the ratio of the hybrids at a concentration of 0.1 mg·ml^{-1} in ethanol after centrifugation at 3000 rpm for 15 min. TEM images of the hybrids with different proportions of acid-MWCNTs to graphene: (b) 2:1, (c) 10:1 and (d) 14:1. Reproduced with permission from [171], Copyright © 2011, Elsevier.

MWCNTs–rGO hybrid film was developed by using the LbL deposition method [172, 173]. A thin layer of rGO was first deposited on SiO_2/Si substrates through electrostatic adsorption, on which a layer of MWCNTs was deposited. The absorption of the MWCNTs onto the rGO film largely reduced the sheet resistance of the final films, while the optical transparency was not seriously influenced. Moreover, the double-layered films exhibited very high adhesion strength on the substrates, which is an important requirement for most applications with high durability.

A relatively simple and flexible LbL strategy was developed to fabricate rGO–MWCNTs hybrid multilayered thin films on silicon and quartz substrates [173]. The multilayered hybrid thin films were derived from a positively charged MWCNTs suspension and a negatively charged rGO suspension alternatively, as shown schematically in Fig. 4.68 (a). Electrical and optical properties of the films were controlled by controlling the number of the two layers. Electrical conductivity of the multilayered thin films was significantly increased by using the graphitization process. More significantly, the hybrid thin films could also be deposited on flexible substrates.

GO nanosheets were reduced into rGO nanosheets with hydrazine in the presence of ammonia. The rGO suspension prepared in this way exhibited a good stability (55 mV at pH 10), as demonstrated in Fig. 4.68 (b) [173]. AFM analysis results indicated that the rGO nanosheets had a relatively wide size distribution (0.5–2 μm). The positively charged MWCNTs were obtained through the oxidation of MWCNTs with strong acids to facilitate functionalization. After that, amine groups (NH_2) were introduced onto the surface of the functionalized MWCNTs-COOH, by using the reaction between the carboxylic acids and excess ethylenediamine mediated with *N*-ethyl-*N*'-(3-dimethylaminopropyl)carbodiimide methiodide

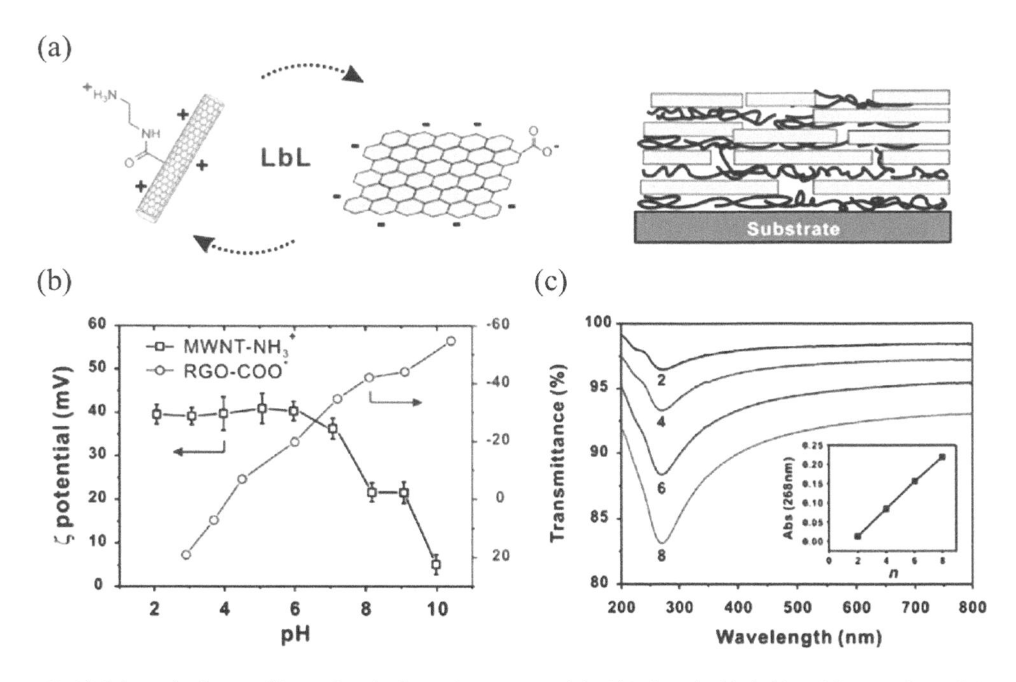

Fig. 4.68. (a) Schematic diagram illustrating the formation process of the LbL deposited hybrid multilayers of MWCNTs and rGO, (b) pH dependent ζ-potential of the chemically functionalized MWCNT-NH_3^+ and rGO-COO^- suspensions, (c) UV/vis curves of the $(MWCNT–rGO)_n$ multilayered films, with the inset to be absorbance at 268 nm as a function of the number of bilayers (*n*). The dimensions of MWCNTs and rGO are not to scale. Reproduced with permission from [173], Copyright © 2010, American Chemical Society.

(EDC). The positively charged MWCNTs suspension, consisting of MWCNTs-NH_3^+, was stable over a wide range of pH value (Fig. 4.68 (b)). The presence of the functional groups on the surface of the MWCNTs was confirmed by XPS results.

The multilayered thin films were deposited by using the spin-coating method on silicon or quartz substrates, with the suspensions of MWNT-NH_3^+ and rGO-COO^-, leading to films of $(MWCNTs–rGO)_n$, with n to be the number of bilayers. As shown in Fig. 4.68 (c), the characteristic UV/vis absorbance at λ_{max} = 268 nm of rGO within the films was gradually increased, implying the linear growth of the multilayered films and the reduction of the GO [173]. In addition, the transmittance of the hybrid films at 550 nm could be controlled by controlling the processing parameters and the concentrations of the precursors. For example, as the pH of the MWCNT suspension was decreased from 6.5 to 3.5, while that of the rGO was kept at pH = 10, the thickness of the 5-bilayer film was decreased from 14 nm to 11 nm. This was because the charge of the amine groups in the MWCNT-NH_3^+ was increased due to the decrease in the pH value, so that the quantity of the negatively charged rGO-COO^- was reduced. At the same time, the adsorption was also related to the effect of charge density on the MWCNTs.

Figure 4.69 shows SEM images of the hybrid multilayered films [173]. AFM results indicated that the one bilayer film consisted of a few overlapping rGO nanosheets, with MWCNTs distributed on the surface of the film. The final multilayered films were constructed with graphene nanosheets on which a percolating network was formed with randomly distributed MWCNTs, as demonstrated in Fig. 4.69 (c, d). A similar microstructure was observed in the unreduced samples, e.g., $(MWNTs–GO)_5$ film, as shown in Fig. 4.69 (c). The morphology of the film was almost not after the thermal reduction, whereas the edges of the graphene nanosheets were slightly blurred after thermal reduction, because the graphene nanosheets were restacked. However, chemical reduction with hydrazine led to significant change in the morphology of the graphene nanosheets, as illustrated in Fig. 4.69 (d).

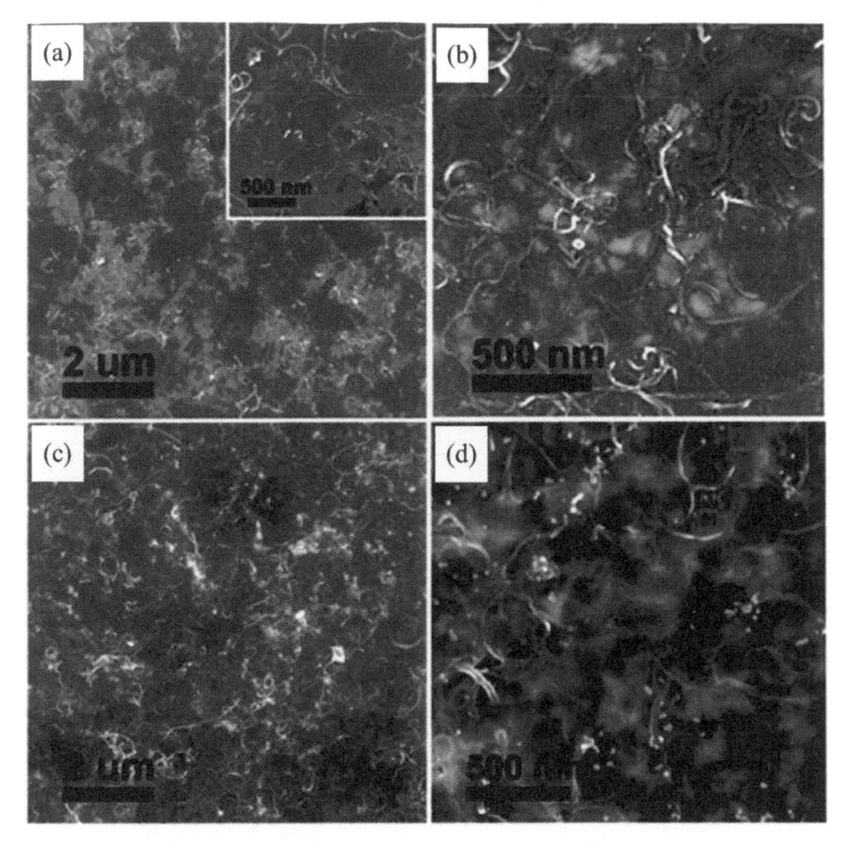

Fig. 4.69. SEM images of the (MWCNTs–rGO)$_n$ and (MWCNTs–GO)$_n$ hybrid multilayered films: (a) as-assembled (MWCNTs–rGO)$_5$ multilayered film, with the inset to be a high-resolution image, (b) (MWCNTs–rGO)$_5$ multilayered film after thermal reduction at 1000°C in H_2, (c) as-assembled (MWNTs–GO)$_5$ multilayered film and (d) (MWCNTs–GO)$_5$ multilayered film after chemical treatment with hydrazine. Reproduced with permission from [173], Copyright © 2010, American Chemical Society.

A similar LbL deposition method was reported to synthesize rGO–MWCNTs hybrid thin films, with potential applications as conductive electrode materials [174]. In this case, graphene nanosheets bound with polymer were dispersed in water to form a positively charged suspension. The graphene nanosheets were derived from *in situ* reduction of exfoliated GO in the presence of cationic poly(ethyleneimine) (PEI). Negatively charged MWCNTs suspension was prepared through acid oxidation. Hybrid thin films were coated on various substrates by using a multistep sequential assembly procedure, by immersing the substrates in the two suspensions alternatively. The obtained films had interconnected network carbon structures with well-defined nanopores, which facilitated fast ion diffusion and were promising for application as supercapacitor electrodes. Due to the presence of abundant -NH_2 groups, the surface of PEI-G was protonated to form -NH_3^+ over a certain range of pH value, pH = 2–9, when the PEI-GN was dissolved in DI water. The negative charge of the MWCNTs suspension was attributed to the formation of MWCNTs-COOH. The hybrid films were deposited on silicon or indium tin oxide (ITO) glass substrates, as shown schematically in Fig. 4.70 (a).

SEM images of the hybrid films are shown in Fig. 4.70 (b–e) [174]. In the initial film obtained by immersing in the PEI-G suspension, the graphene nanosheets had a quite high surface area coverage, with a wide size distribution of 0.5–5 μm, as demonstrated in Fig. 4.70 (b). After the immersion in the MWCNTs-COOH solution, randomly oriented CNTs with different lengths were deposited on the graphene layer (Fig. 4.70 (c)). Films with the desired thicknesses could be obtained by repeating the two alternative

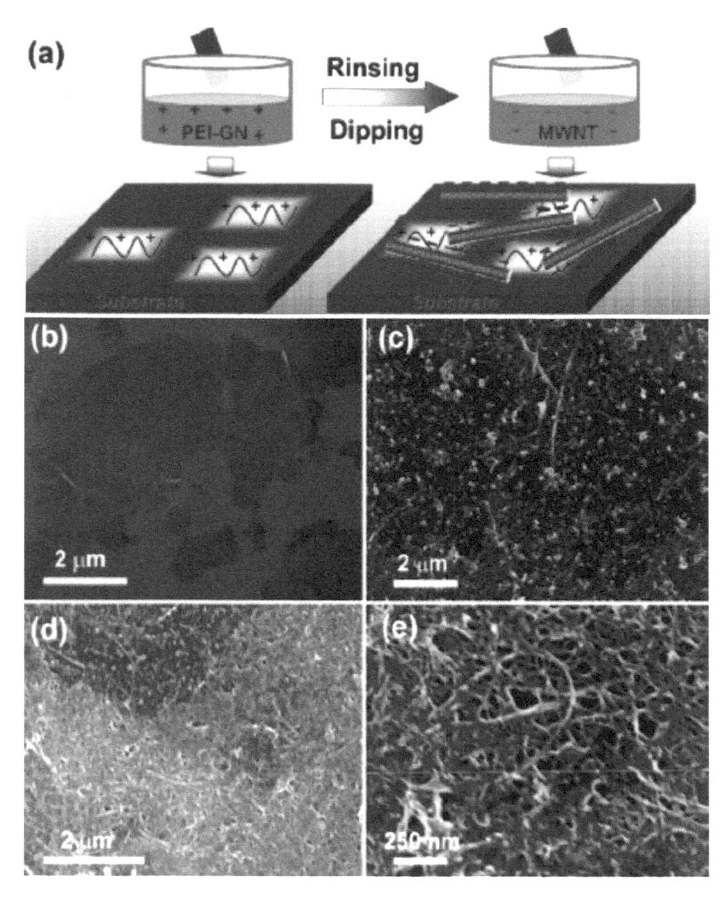

Fig. 4.70. (a) Schematic diagram illustrating the deposition process of the positively charged PEI-G and negatively charged MWCNT on substrates. (b, c) SEM images of the first layer of PEI-GN and the first bilayer $[PEI\text{-}GN\text{–}MWNT\text{-}COOH]_1$ film deposited on a silicon substrate, respectively. (d, e) SEM images of the $(PEI\text{-}G\text{–}MWNT\text{-}COOH)_9$ film with nine deposition cycles. Reproduced with permission from [174], Copyright © 2010, American Chemical Society.

steps. After thermal treatment at 150°C for 12 h in a vacuum, amide bonds between the PEI-modified graphene and the carboxylic acids on the acid-oxidized MWCNT surface were formed. SEM images of the $(PEI\text{-}G\text{–}MWNT\text{-}COOH)_9$ film with nine deposition cycles are shown in Fig. 4.70 (d, e). A dense and uniform network of carbon nanostructure with well-defined nanoscale pores was obtained, making the multilayered films potential candidates as electrodes for energy storage devices.

Vacuum filtration has been widely used to prepare free-standing G/rGO–CNTs hybrids with relatively large thicknesses [175, 176]. For example, a hybrid sheet was prepared from the dispersion of SWCNTs and G/nano-graphite in N-methyl pyrrolidone (NMP) [175]. The suspensions with different contents of G/nano-graphite were used to prepare hybrid sheets with different properties by using vacuum filtration, with film thickness in the range of 100–500 μm. It was found that the mechanical properties of the hybrids were better than those of either SWCNTs or G/nano-graphite sheets. The composite materials also had higher electrical conductivity.

A similar method was used to prepare a paper-like hybrid from few-layer graphene (FLG) and MWCNTs [176]. The ratio of FLG and MWCNTs in the hybrid paper was studied to optimize electrical conductivity. An overall synergetic effect was observed by combining the 2D FLG and 1D MWCNTs. Graphite powder was used to obtain FLG by thermal treatment at 800°C in N_2 for 2 min. FLG powder was dispersed in distilled water with the aid of surfactant poly(sodium 4-styrenesulfonate). MWCNT was suspended in distilled water with the surfactant Triton-X100. The mixture of FLG–MWCNTs suspension was heavily sonicated and then filtered through a 0.4 μm hydrophilic polycarbonate membrane to obtain paper-like sheets. Schematic diagram of the process and a photograph of the sample are shown in Fig. 4.71. After the filtration step, the papers were dried at 120°C for 2 h to completely remove water and the surfactants.

Figure 4.72 shows SEM images of the FLG–MWCNTs hybrid papers with different ratios of FLG/MWCNTs and a thickness of about 200 μm. The FLG sample possessed a very flat surface, together with small pores at the junctions of the FLG, as seen in Fig. 4.72 (a). In the FLG-dominated papers, both the

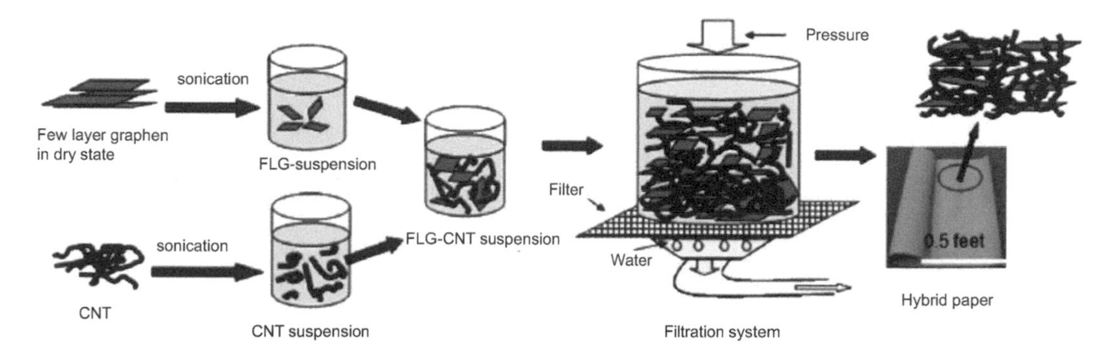

Fig. 4.71. Illustration of the process to make a FLG–MWCNT hybrid paper. Reproduced with permission from [176], Copyright © 2010, Elsevier.

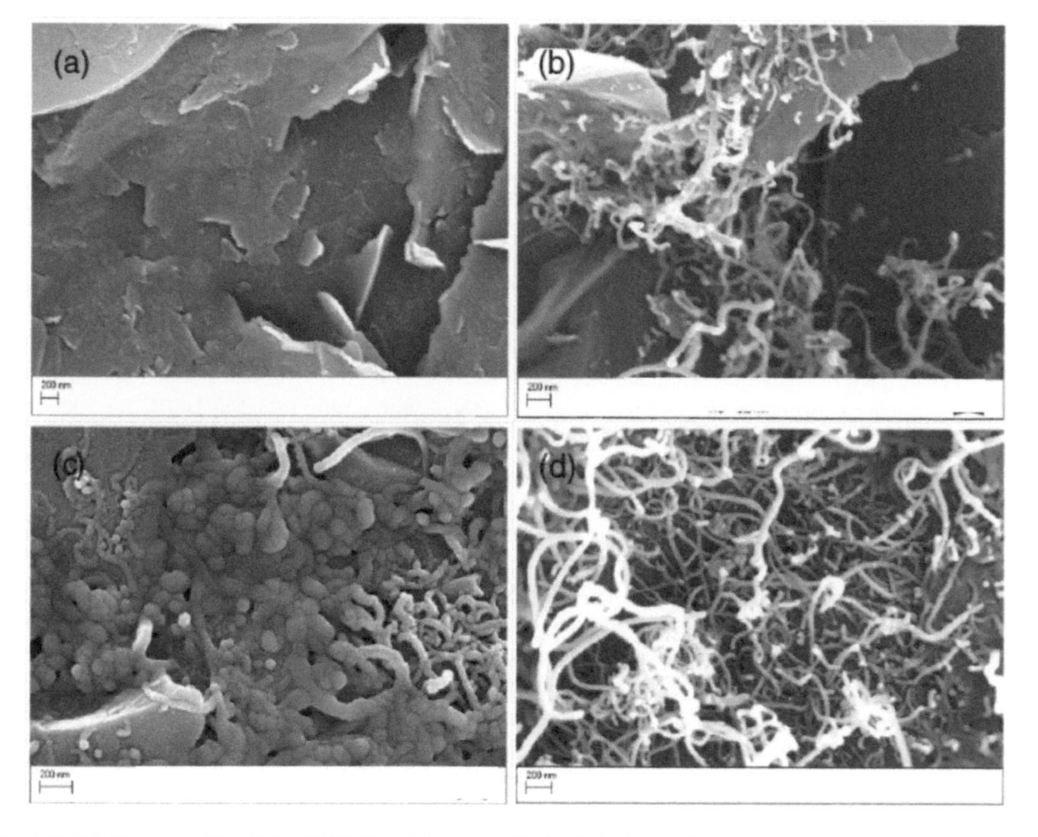

Fig. 4.72. SEM images of the FLG–MWCNTs hybrid papers (I): (a) FLG, (b) FLG/MWCNT = 5/1, (c) FLG/MWCNT = 1/1 and (d) FLG/MWCNT = 1/3. Reproduced with permission from [176], Copyright © 2010, Elsevier.

FLG and MWCNTs were uniformly dispersed with high entanglement. The flatness of the hybrid papers was related to the content of the MWCNTs, as shown in Fig. 4.72 (b, c). At a low content of MWCNTs, e.g., the sample with FLG/MWCNT = 5/1, the surface of the paper was rugged. The surface of the papers became more compact and flat as the content of MWCNTs was increased. A continuous network was developed, due to the close compact of a large amount of FLG and MWCNT in the lateral direction (Fig. 4.72 (c)). Such a compact was attributed to the self-assembly of the FLG and MWCNTs during the vacuum filtration process. The gap between adjacent FLGs was well-bridged by the MWCNTs network. The FLGs with a large lateral dimension served as the strong holders, while the MWCNTs functioned as wires to connect the FLG. Figure 4.72 (d) indicated that the MWCNT-dominated paper exhibited a different morphology as compared to the FLG-dominated paper. The inter-twined network was observed in the MWCNT-dominated paper.

Surface resistance increased with the content of MWCNTs up to FLG/MWCNTs = 5/1 and then decreased gradually to a minimum value of 687 mΩ/sq at FLG/MWCNTs = 1/3, which was followed by a gradual increase after that. The surface resistances of the FLG paper and the MWCNT paper were about 780 and 1571 $m\Omega \cdot \square^{-1}$, respectively. Obviously, surface resistance of the hybrid papers was determined by the conductive percolated network and the morphology of the papers. At a low content of MWCNTs, the straightness of the flexible MWCNTs was constrained by the rigid FLG, so that it was difficult to form a close-packed conducting network. Percolation was reached at a certain level of MWCNTs. Volume resistivity of the FLG-dominated papers decreased with the increasing content of MWCNTs. The volume resistivity was dependent on both in-plane and through-thickness conducting behaviors. In the MWCNTs dominated hybrid papers, there was a synergistic effect on the volume resistivity.

Various chemical vapor deposition (CVD) methods have been employed to fabricate G/GO/rGO–CNTs hybrids, with unique architectures that could not be obtained by other techniques [177–190]. Such unique 3D G/rGO–CNT multilevel structured hybrids could have various applications, such as in electron field emitters [180, 185], supercapacitors [181, 182], as anodes of lithium-ion batteries and dye-sensitized solar cells [183, 184, 187] and as electrodes for oxygen reduction reaction (ORR) [186]. Selected examples are described as follows.

3D G–CNT sandwich structures have been constructed by growing CNT pillars in between the graphene layers with the CVD technique. Paper-like substrate of graphene nanosheets was first prepared by using vacuum filtration of chemically derived graphene suspensions. Such hybrids exhibited promising electrochemical performances as electrode materials. Supercapacitors based on these electrodes possessed a specific capacitance of 385 F g^{-1} at 10 mV s^{-1} in 6 M KOH solution. After 2000 cycles, capacitance of the device was increased only by about 20% [177]. The CNTs grown in between the graphene nanosheets and distributed uniformly but sparsely on the whole sheet surface ensured high conductivity of the hybrids. The sandwich structures exhibited high electrical conductivity, low diffusion resistance to protons/cations and high electroactive areas as a candidate electrode material for high performance supercapacitors.

CNTs powder was purified in a mixture of H_2SO_4/HNO_3 with a ratio of 3:1 at 140°C for 1 h, which was then filtrated and washed with DI water. The treated CNTs powder was dispersed in GO suspension through ultrasonication for 1 h, with a mass ratio of CNT/GO of 1:10. After filtration and desiccation, the sample was thermally treated at 750°C for 1 h in Ar (99.999%) at a flow rate of 300 sccm. Samples based on pure graphene were also prepared similarly for comparison. Figure 4.73 shows a schematic diagram of the formation process of the 3D structured hybrids.

Figure 4.74 (a) shows an SEM image of the G–CNT hybrid sample. CNTs were grown well in between the graphene nanosheets, with a uniform distribution. The hybrid sample possessed a surface area of 612 $m^2 \cdot g^{-1}$, higher than that of graphene (202 $m^2 \cdot g^{-1}$), which was attributed to the intercalation and distribution of the CNTs inside the spaces formed by the graphene nanosheets. The distance between the CNTs was in the range of 100–200 nm, as seen in Fig. 4.74 (b, c). Figure 4.74 (d) shows a TEM image of the sample, revealing that the CNTs were mainly multi-walled, with inner diameters of 5–7 nm and outer diameters of 7–12 nm.

The concentration of the catalyst and the source of carbon exhibited a significant effect on morphology of the MWCNTs. If carbon monoxide was used as the carbon source, then the CNTs would be aligned, with sparse and short lengths. Otherwise, when methane was used, together with low catalyst concentration

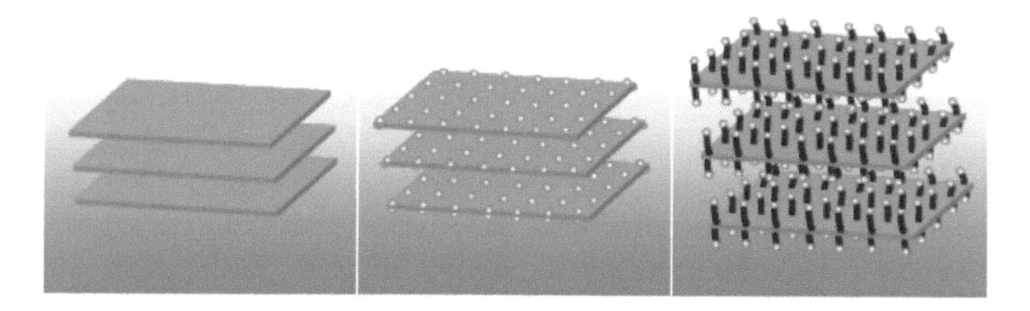

Fig. 4.73. Illustration of the formation of hybrid materials with CNTs grown in between graphene nanosheets, showing stacked layers of graphene oxide (left), catalyst particles adhered onto layer surface after deposition (middle), and CNTs in between graphene layers after growth (right). Reproduced with permission from [177], Copyright © 2010, WILEY-VCH Verlag GmbH & Co. KGaA, Weinheim.

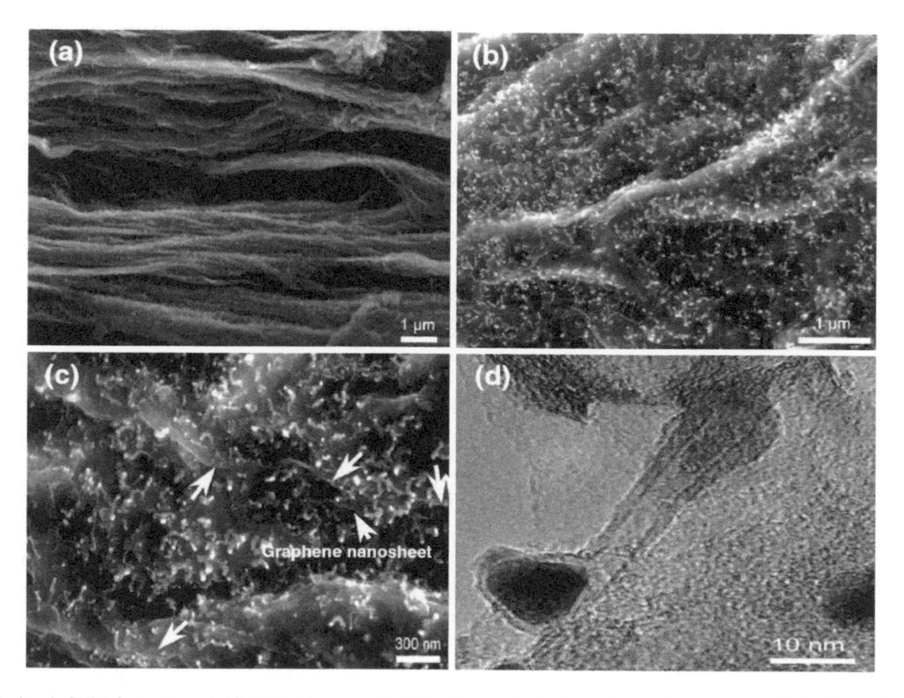

Fig. 4.74. (a–c) SEM images and (d) TEM image of CGS (Co catalyst: 16 wt%; carbon source: CO). Reproduced with permission from [177], Copyright © 2010, WILEY-VCH Verlag GmbH & Co. KGaA, Weinheim.

(e.g., 20 wt% Co), long CNTs of 1–2 μm would be produced, with a compact layer that randomly covered the surface of the graphene nanosheets, as illustrated in Fig. 4.75 (a, b). Additionally, if the concentration of the catalyst was too high, e.g., 60 wt% Co, the catalyst particles tended to aggregate, so that they would be wrapped inside the graphene nanosheets during the heat treatment, thus losing the catalytic activity to grow CNTs, as demonstrated in Fig. 4.75 (c, d). Therefore, both the source of carbon and concentration of catalyst should be optimized according to the requirements of the various applications.

A slightly different hybrid structure was derived by using CVD separately [178]. The heterostructure consisted of a graphene layer on top of the CNT arrays on substrates, which could be transferred to various polymer substrates and demonstrated observable sensitivity in terms of electrical conductivity in both tension and compression conditions. Fe thin film of 20 nm was deposited on four inch Si/SiO_2 substrates

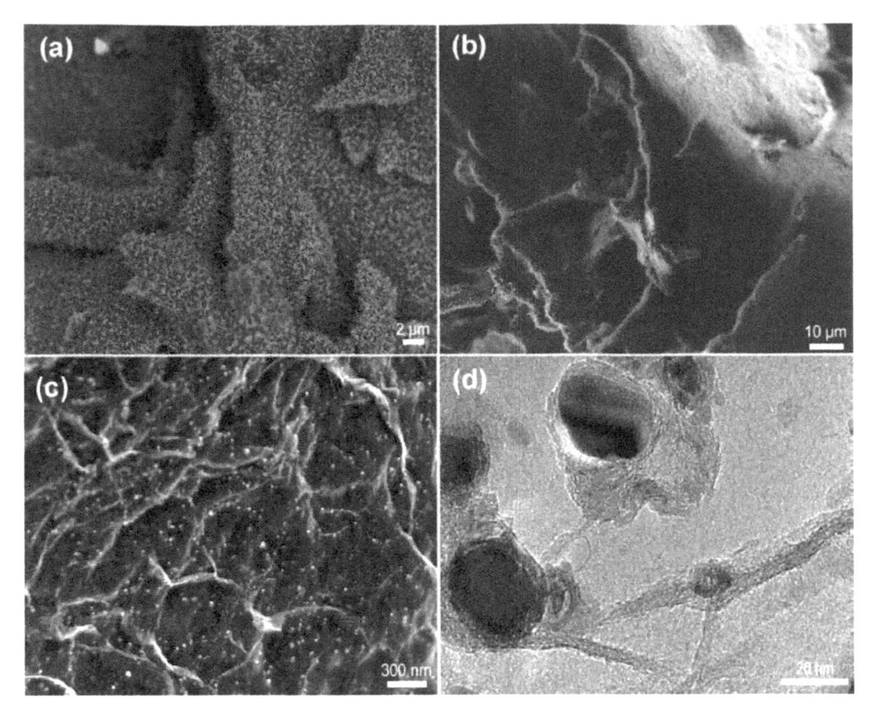

Fig. 4.75. (a, b) SEM images of the hybrid material with CNTs covering on the surfaces of graphene (Co catalyst: 16 wt%; carbon source: C_2H_4). (c) SEM and (d) TEM images of carbon/cobalt core/shell structures on the graphene sheets (Co catalyst: 60 wt%; carbon source: CO). Reproduced with permission from [177], Copyright © 2010, WILEY-VCH Verlag GmbH & Co. KGaA, Weinheim.

with SiO_2 thickness of 500 nm by using e-beam evaporation as the catalyst for the growth of CNTs. In addition, the Si/SiO_2 substrates were patterned by using photolithography, on which Fe catalyst layers with thicknesses of 10–30 nm were deposited. The CVD growth was conducted at 700°C using ethylene (C_2H_4) as the carbon source in Ar for 60 min.

Figure 4.76 (a) shows a representative SEM image of the hybrid structure, demonstrating the vertically aligned MWCNTs on Si/SiO_2 substrate and a multilayered graphene layer on top of the MWCNTs [178]. The MWCNTs had an average length of 30–40 µm. The formation mechanism of the MLG–CNTs hybrid structure is explained schematically by the diagram in Fig. 4.76 (b). In the beginning, a MLG film was formed on the Fe catalytic layer, after which MWCNTs started to grow below the graphene layer. Generally, MWCNT growth by using thermal CVD process is influenced by several factors, including the nature of the catalysts, transformation of metal catalyst particles into metal carbide particles, nucleation and growth of catalyst particles and diffusion of carbon atoms inside the catalyst matrix [191, 192]. For example, the morphology of the MWCNTs is closely related to the size of the catalytic NPs, which is determined by the thickness of the catalyst film. As a result, thickness of the catalyst thin film is an important factor. Experimental results indicated that the hybrid film was only obtained when the 20 nm catalyst film was used, while no hybrid film was available when the catalytic layers were 10 nm or 30 nm. It was found that the catalytic particles tended to agglomerate if the catalytic films were too thin or too thick.

Figure 4.77 (a) shows a SEM image of the MLG–CNTs hybrid, demonstrating clearly waved CNT arrays with a layer of graphene on top [178]. The graphene layer consisted of about 40 single-layers with a thickness of 15–20 nm, while each monolayer was about 3.4 Å, as revealed in Fig. 4.77 (b). Figure 4.77 (c) indicated that the average diameter of the MWCNTs was 10 nm. The catalyst was present in the form of iron carbide particles at the interface between the MLG and MWCNTs, supporting the tip growth mechanism of the MWCNTs. There was an interface between the MWCNTs and the MLG layer, as demonstrated in Fig. 4.77 (d).

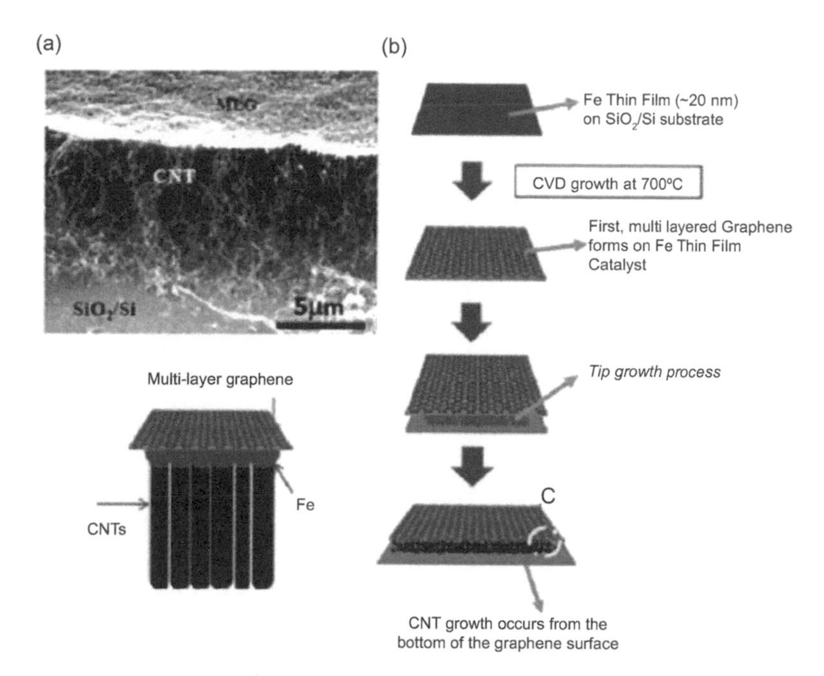

Fig. 4.76. (a) SEM image of the MLG–MWCNTs hybrid on Si/SiO_2 substrate. (b) Schematic diagram showing the growth process of MLG–MWCNTs hybrid. Reproduced with permission from [178], Copyright © 2011, The Royal Society of Chemistry.

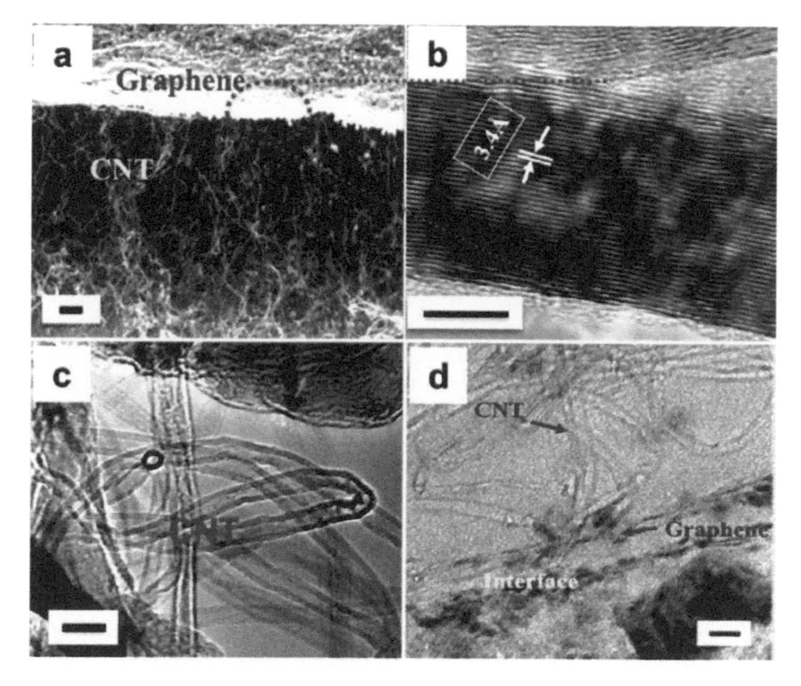

Fig. 4.77. (a) SEM image of the multilayered MLG–MWCNTs hybrid film (scale bar = 1 μm). (b) HRTEM image of the MLG film (40 graphene mono layer) as the top part of the nano-hybrid film as shown in panel (a) (scale bar = 5 nm). (c) Bright field HRTEM image of the multiwall carbon nanotubes (scale bar = 20 nm), with a diameter of about 10 nm. (d) TEM image demonstrating the interface between the graphene and MWCNT (scale bar = 20 nm). Reproduced with permission from [178], Copyright © 2011, The Royal Society of Chemistry.

Current-voltage (I–V) characteristics of the MLG–MWCNTs hybrids were linear. Uniform conductance over a large area of the hybrid films was observed, which was attributed to the Ohmic contacts between the graphene layer and the MWCNTs. In addition to good electrical properties, the MLG–CNT hybrid film has excellent electro-mechanical properties and good structural integrity when used to make flexible polymer films or stretchable electrodes. The process of transferring the MLG–MWCNT onto the PDMS substrate, leading to flexible MLG–MWCNT-PDMS, is shown in Fig. 4.78. The inset of Fig. 4.78 (b) shows a surface SEM image of the hybrid-polymer film. Tension and compression flexibilities of the MLG–CNT-PDMS composite films were evaluated. The resistance of the sample exhibited changes in the ranges of 0.85–2.3 kΩ and 0.9–5.5 kΩ, corresponding to a bending radius of flat to 3.3 mm in tension and flat to 1.4 mm in compression measurements, respectively. The conductivity was quickly recovered after the tension and compression loads were released. Therefore, such hybrid structures could be used for applications as mechanical (strain/pressure/touch) sensors. In addition, the MLG–MWCNTs hybrids could also be transferred onto flexible Cu foil [178]. The hybrid on Cu foil exhibited stable field emission performance for practical device applications.

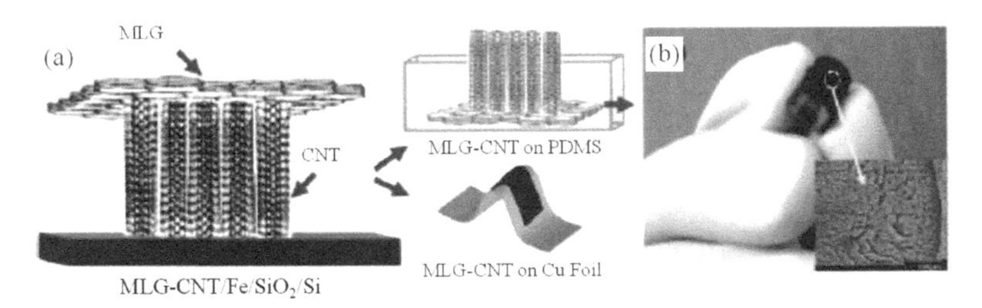

Fig. 4.78. (a) Schematic diagram showing the transferring of the MLG–MWCNTs hybrid film onto flexible polydimethylsiloxane (PDMS) and Cu foil. (b) Photograph of a transferred sample, demonstrating the flexibility of the MLG–CNT-PDMS composite film, with the inset showing a SEM image of the top surface of the composite film. Reproduced with permission from [178], Copyright © 2011, The Royal Society of Chemistry.

A new type of 3D hybrid architecture, consisting of vertically aligned carbon nanotubes (VACNT) and graphene, was developed by using CVD [181]. The VACNTs were grown by pyrolysing iron phthalocyanine (FePc), as illustrated in Fig. 4.79 (a)). In this case, thermally expanded highly ordered pyrolytic graphite (HOPG) was intercalated by the grown VACNTs through the pyrolysis of FePc. In these 3D pillared VACNT–G nanostructures, both the nanotube length (PL) and the inter-tube distance (MIPD) could be well-controlled, making them promising candidates as electrode materials for energy-related device applications. For example, pseudocapacitors with outstanding performances were demonstrated by electrodepositing nickel hydroxide on the 3D architecture.

Figure 4.79 (b) shows a schematic diagram of the procedure to fabricate the 3D pillared VACNT–G architectures [181]. In step 1, HOPG was first soaked in the mixture of sulfuric and nitric acids with a volumetric ratio of 3: 1 for 10 min at room temperature under sonication in a water bath. Step 2 was to heat the acid-treated HOPG at 900°C in Ar to vaporize the acid molecules, so that graphite was dramatically expanded along the c-axis. The thermally expanded HOPG was then thermally treated at 1000–1200°C in a flow of mixed Ar and H_2 through silicon tetrachloride ($SiCl_4$) to produce the SiO_2 coating in step 3, which was necessary to grow the VACNTs. The final step was to fabricate the 3D pillared VACNT–G architectures through the pyrolysis of FePc at 800–1000°C in Ar/H_2. The VACNT–G hybrids appeared as 3D networks, with a large increase in mass and hundred times volume expansion, as illustrated in Fig. 4.79 (c, d).

SEM images of the original HOPG and the acid-treated HOPG before and after the intercalated growth of VACNTs are shown in Fig. 4.80 (a–d) [181]. Figure 4.80 (a) indicated that the pristine HOPG already exhibited a well-defined layered structure. Due to the acid-treatment and thermal expansion, the layered structure was further expanded with a certain degree of distortion, as illustrated in Fig. 4.80 (b). The deposition of the SiO_2 also resulted in expansion and structure distortion (Fig. 4.80 (c)). As illustrated in Fig. 4.80 (d), VACNTs were incorporated into the gaps in between the thermally expanded graphene layers. It was observed that the VACNTs were grown on both sides of adjacent graphene nanosheets in the

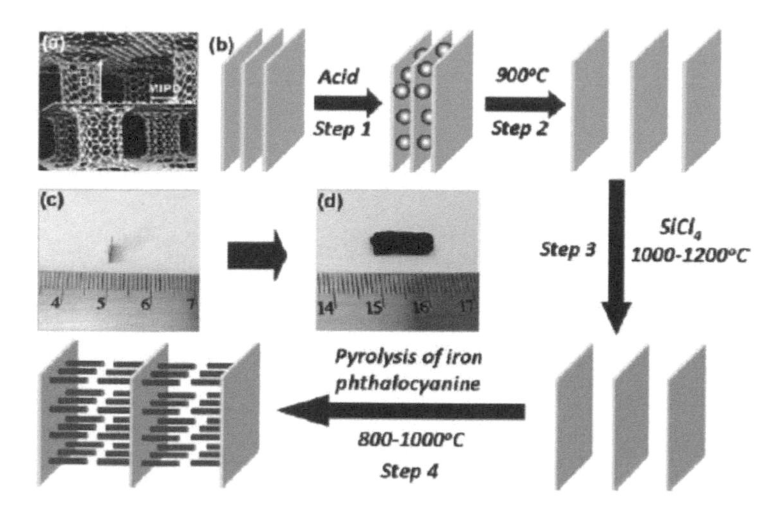

Fig. 4.79. (a) Schematic diagram of the 3D pillared VACNT–G hybrid nanostructure. (b) Schematic representation of the procedure to fabricate the 3D pillared VACNT–G hybrid architectures. Photographs of the original HOPG with a thickness of 80 μm (c) and the thermally expanded graphene layers intercalated with the VACNTs, revealing a huge expansion (d). Reproduced with permission from [181], Copyright © 2011, American Chemical Society.

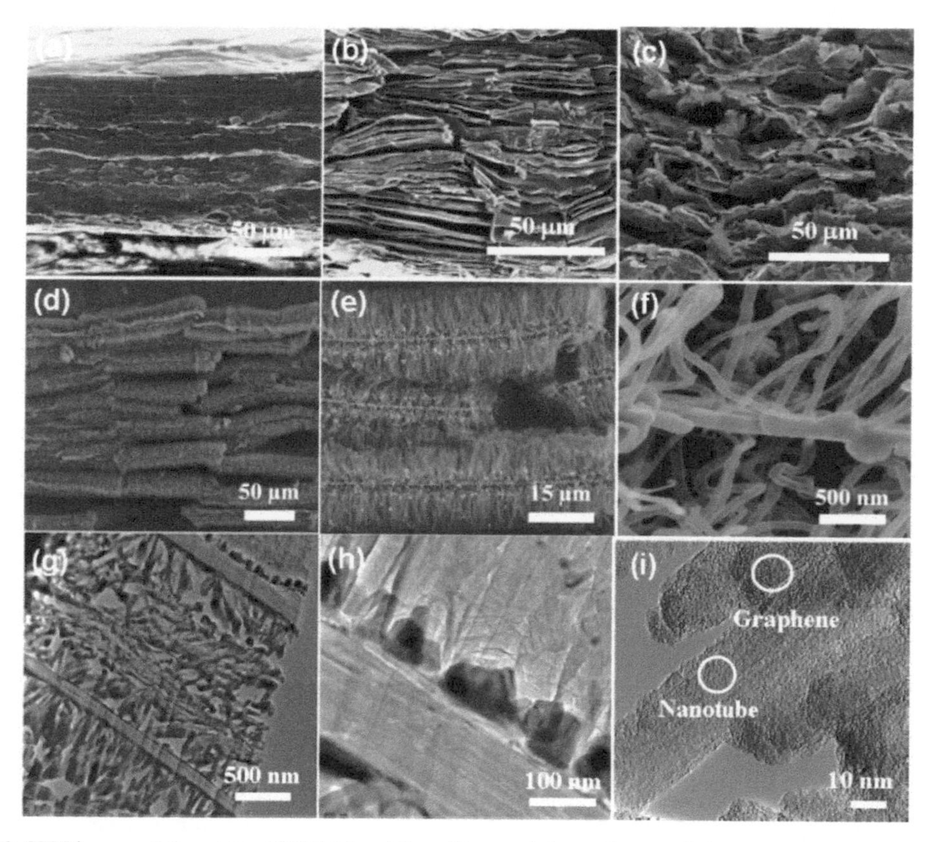

Fig. 4.80. SEM images of the pristine HOPG (a) and the acid-treated, thermally expanded HOPG without (b) and with (c) the SiO_2 coating layer. (d–f) Representative SEM images of the 3D pillared VACNT–G hybrid architectures at different magnifications. (g–i) Cross-sectional TEM images of the 3D pillared VACNT–G hybrid architectures at different magnifications. Reproduced with permission from [181], Copyright © 2011, American Chemical Society.

thermally expanded HOPG, as revealed clearly in Fig. 4.80 (e). High magnification image demonstrated that there was an intimate contact between the VACNTs and the graphene nanosheets (Fig. 4.80 (f)). Representative TEM images of the 3D VACNT–G hybrids are presented in Fig. 4.80 (g, h), confirming the structures observed in the SEM images. Iron catalyst particles at the VACNT–G interface became highly visible in the TEM image (Fig. 4.80 (h)). Figure 4.80 (i) shows a higher magnification TEM image taken at the VACNT–G interface, revealing a matched lattice orientation between the nanotube and the graphene layer.

The length of the VACNT pillars (PL) could be tailored by controlling the growth time [181]. SEM images of the samples grown for different times are shown in Fig. 4.81 (a–c). With increasing growth time, the length of the VACNT pillars was increased, as illustrated in Fig. 4.81 (d).

The 3D pillared VACNT–G hybrid architectures possessed BET surface areas of up to 213.7 ± 2.1 $m^2 \cdot g^{-1}$, with type-IV nitrogen adsorption isotherm. Therefore, these hybrids were classified into the category of mesoporous materials, with pore sizes in the range of 1–12 nm. These inherently porous 3D pillared VACNT–G architectures with a unique conductive carbon network exhibited high electrochemical performance as the electrode materials of pseudocapacitors.

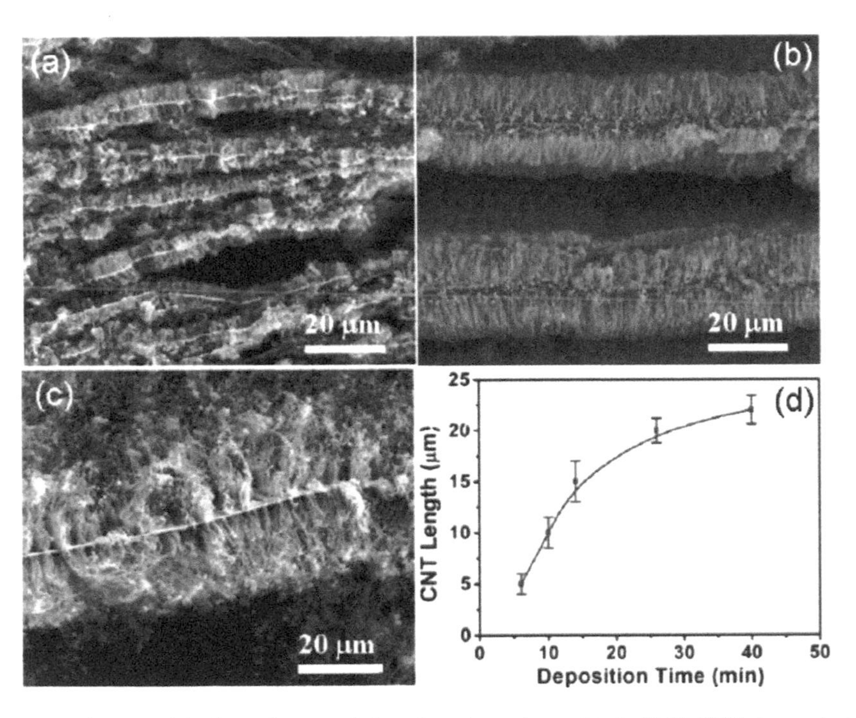

Fig. 4.81. (a–c) SEM images of the thermally expanded graphene layers intercalated with VACNTs for pyrolysis times of 5, 10 and 30 min, respectively. (d) Length of the VACNT pillars as a function of the deposition time of the VANCTs. Reproduced with permission from [181], Copyright © 2011, American Chemical Society.

4.3.2. Graphene—Other Nanocarbons

Other nanocarbons mainly include carbon black (AC) [193, 194], carbon nanoparticles [195–197] and microporous/mesoporous carbons [198–202], which have been combined with graphene or graphene oxide to form hybrids, with a wide range of applications.

A hybrid consisting of graphene nanosheets and carbon black (G–CB) was used for application as the electrodes of electrochemical double-layer capacitors (EDLCs) [193]. The presence of CB prevented the agglomeration of the graphene nanosheets, so as to facilitate electrolyte–electrode accessibility in order to achieve high capacitance. In this case, the CB particles acted as a spacer to separate the graphene nanosheets, so that the rapid diffusion path was provided with more edge plane participation in EDLCs.

The G–CB hybrids, with a G–CB weight ratio of 9:1, were prepared through *in situ* reduction with the aid of ultrasonication. Two approaches were adopted, as shown schematically in Fig. 4.82.

In approach 1, CB (acetylene black, 20–50 nm) was added during the reduction process of GO, while the GO was reduced to graphene by using hydrazine solution (20 wt% in water) at pH ≈10 adjusted with NaOH solution. The reaction system was refluxed for 24 h. G–CB-1 hybrid powder was obtained after filtration, washing and drying. Approach 2 was a simple physical mixing, i.e., GO was reduced without the presence of CB. Both graphene and CB powders were dispersed in distilled water ultrasonically. G–CB-2 was then derived in a similar way.

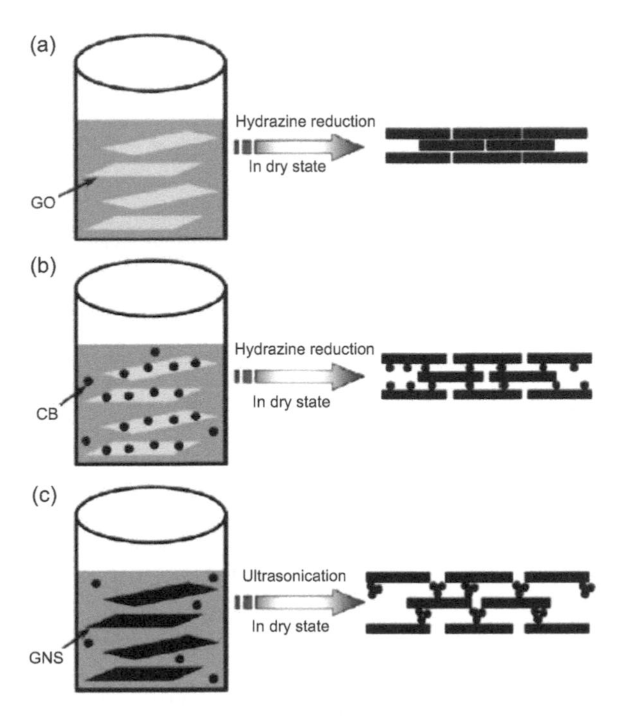

Fig. 4.82. Schematic diagram showing the synthesis procedure of graphene nanosheets (a) and G–CB hybrids (b, c) in dry state. Reproduced with permission from [193], Copyright © 2010, Elsevier.

The as-prepared GO exhibited wrinkles and folds at the edges with a thickness of 1 nm. The graphene nanosheets agglomerated to form a layered structure after drying, due to the van der Waals interactions between the graphene nanosheets. In G–CB-1, most of the CB particles were deposited on the basal surfaces of the graphene nanosheets, as shown in Fig. 4.83 (a, b). There were CB aggregates in the sample, because of the detachment of the CB particles from the graphene nanosheets, due to the removal of the oxygen-containing functional groups during the GO reduction. In comparison, the CB particles were only attached on the edge planes of the graphene nanosheets, simply because the remaining oxygen-containing groups are mainly at the edges of the graphene, as demonstrated in Fig. 4.83 (c, d). However, G–CB-2 exhibited a higher electrochemical performance than G–CB-1.

A solvothermal reduction method was reported to prepare rGO–AC (activated carbon) hybrid organogels, from GO–AC dispersion in propylene carbonate (PC) [194]. The rGO–AC hybrid organogels could be directly used as the electrodes of organic supercapacitors, without the use of conductive additives and binders. The rGO nanosheets served as a conductive network for the AC particles, so that the internal resistance was largely reduced and charge transfer distances in AC were significantly decreased. Supercapacitors based on the rGO–AC hybrid organic electrodes had higher specific capacitance, higher energy density and better rate performance than those made with conventional AC.

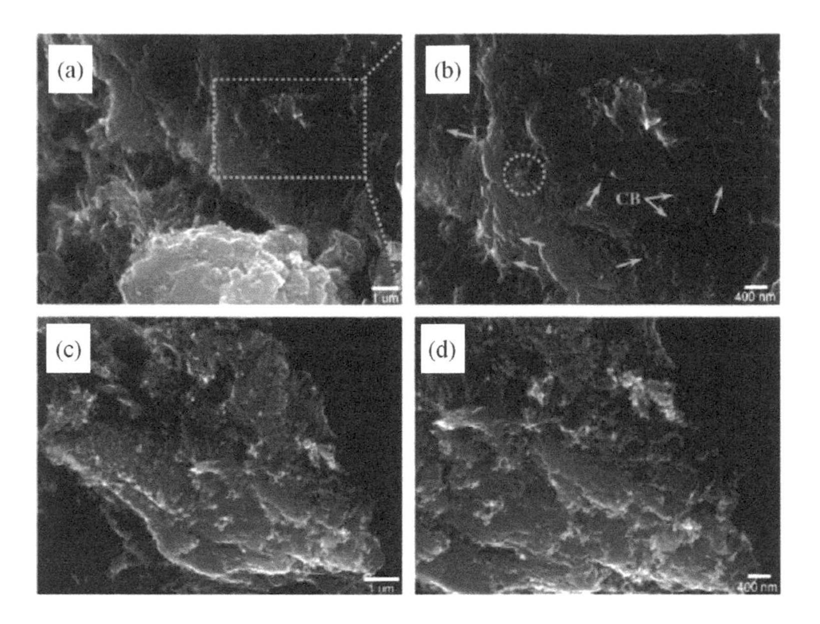

Fig. 4.83. SEM images of the G–CB-1 (a, b) and G–CB-2 (c, d). Reproduced with permission from [193], Copyright © 2010, Elsevier.

AC particles were pre-treated with nitric acid to increase their dispersion in the organic suspension of GO. Freeze-dried GO was dispersed in PC with the aid of sonication. The weight ratios of AC:GO were 0, 1 and 5. The mixtures were sonicated to form stable dispersions, which were then subjected to solvothermal reaction at 180°C for 14 h. After the solvothermal reduction, the weight loss of GO was about 50%, while the AC was not affected. The organogels derived from pure GO and GO–AC mixtures with weight ratios of 1 and 5 were named as SGO, GAC_2 and GAC_{10}, respectively.

Figure 4.84 (a) shows photographs of the three samples, indicating that the size of the cylindrical organogels increased with increasing content of the AC particles [194]. All the organogels exhibited high mechanical strength, so that they could be directly used to make electrodes of supercapacitors. SEM images of the freeze-dried SGO, GAC_2 and GAC_{10} are shown in Fig. 4.84 (b–d). The SGO had a 3D porous microstructure, with pore sizes at the scales from submicrometer to several micrometers, as illustrated in Fig. 4.84 (b). In the hybrid samples, the rGO nanosheets were also present as a 3D network, while the AC particles were homogeneously encapsulated within the framework, as demonstrated in Fig. 4.84 (c, d).

A new type of hybrid, made with chemically reduced graphene (CRG) and carbon nanospheres (CNS), was scalably synthesized for application as the anode material of LiBs [197]. In these special CRG–CNS hybrids, the CNSs were entirely encapsulated by the CRG nanosheets. The hybrids appeared as a 3D network, with unique pores and cavities, and offered promising electrochemical performances as the anodes of LiBs.

The CNSs were prepared by using a hydrothermal process, in which glucose was used as the carbon source and glutaric acid acted as the mineralization reagent. The hydrothermal reaction was conducted at 180°C for 5 h. To fabricate CRG–CNS hybrids, the synthesized CNSs and GO aqueous suspensions were thoroughly mixed with the aid of ultrasonication. Different weight ratios of GO to CNS, including 1/1, 5/2, 5/1, 10/1 and 20/1, were studied. Hybrids with different compositions were obtained after the complete evaporation of water. The GO–CNS hybrid powders were thermally treated at 900°C for 3 h in Ar gas flow containing 5% H_2, so that the GO was reduced to CRG and the CNSs were graphitized. The hybrids were denoted as CRG–CNS1-1, CRG–CNS5-2, CRG–CNS5-1, CRG–CNS10-1 and CRG–CNS20-1, corresponding to the samples of different compositions. In addition, CRG nanosheets and graphitized carbon nanospheres were also prepared similarly for comparison.

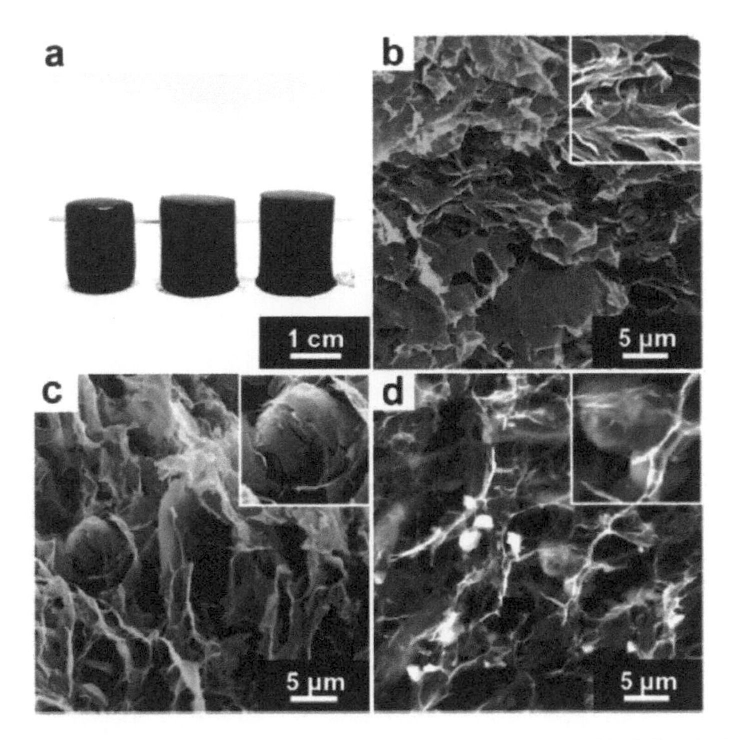

Fig. 4.84. (a) Photographs of SGO, GAC_2 and GAC_{10} (from left to right). SEM images of SGO (b), GAC_2 (c) and GAC_{10} (d) at low magnifications, with the insets to their SEM images at high magnifications. Reproduced with permission from [194], Copyright © 2010, The Royal Society of Chemistry.

Color image of this figure appears in the color plate section at the end of the book.

Figure 4.85 (a) shows an AFM image of the GO nanosheets, with an average thickness of 1 nm [197]. Figure 4.85 (b) indicated that the CNSs had diameters in the range of 200–250 nm, with a relatively narrow size distribution. Due to the use of glutaric acid as the mineralization reagent, the CNSs exhibited a high aqueous dispersion behavior, because surface oxygen-containing groups were produced from glutaric acid. Therefore, it was very convenient to prepare homogeneous GO–CNS hybrids by using simple mechanical mixing, because of the high aqueous dispersion capabilities of both the GO nanosheets and the CNSs.

Figure 4.86 shows SEM images of the GO–CNS hybrids and pure GO nanosheets [197]. As shown in Fig. 4.86 (a), all the CNSs were entirely encapsulated by the GO nanosheets, forming a 3D network. The formation of the 3D networks in the hybrids was attributed to several reasons. For instance, both the GO nanosheets and the CNSs contained oxygen-rich groups on their surfaces, which facilitated strong interactions between them through hydrogen bonding and electrostatic attraction. Also, homogeneous hybrids were formed due to their high water solubility. In addition, the flexible characteristics of the GO nanosheets with large lateral sizes naturally led to the encapsulating microstructure. Although the content of CNSs was decreased, the 3D network structure was retained. In these hybrids, the GO nanosheets were well-spaced by the CNSs, which was in a sharp contrast to the aggregating feature of pure GO, as seen in Fig. 4.86 (f).

Figure 4.87 shows SEM images of the CRG–CNS hybrids [197]. Obviously, the thermal reduction process did no damage to the 3D network structure of the hybrids. According to XRD patterns of the CNSs before and after the thermal annealing, it was found that the CNSs were graphitized, with crystallinities that were increased during the thermal treatment process. The CRG–CNS hybrids possessed a specific surface area of 325 $m^2 \cdot g^{-1}$, with an average pore size of 42 nm. Specifically, first cycle reversible capacity of the CRG–CNS hybrid anode was as high as 925 $mA \cdot h \cdot g^{-1}$ at a current density of 5 $A \cdot g^{-1}$, which was remained to be 700 $mA \cdot h \cdot g^{-1}$ after 200 charge–discharge cycles. This value is much higher than those of the anodes made of graphite and other carbonaceous materials.

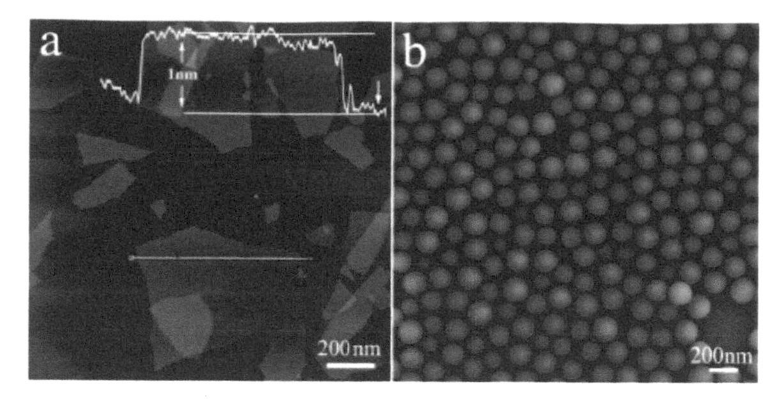

Fig. 4.85. (a) Tapping mode AFM image of the GO nanosheets deposited on the fleshly cleaved mica substrate, with the inset to be the corresponding height profile. (b) SEM image of the CNSs. Reproduced with permission from [197], Copyright © 2012, The Royal Society of Chemistry.

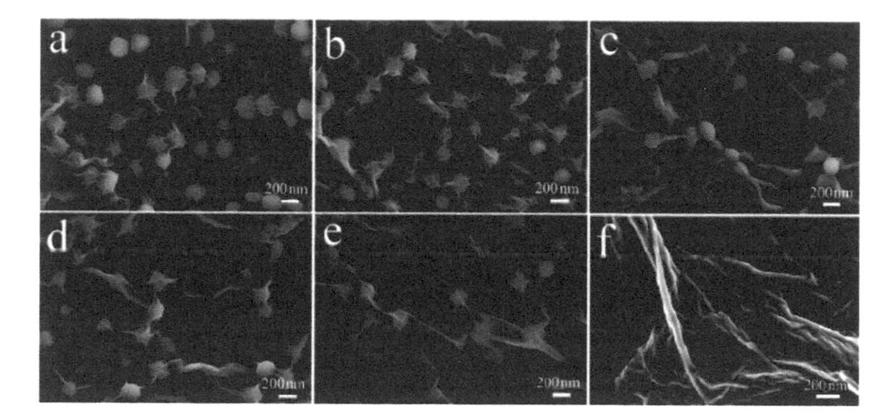

Fig. 4.86. (a–e) SEM images of the GO–CNS hybrids with GO to CNS weight ratios of 1/1, 5/2, 5/1, 10/1 and 20/1, respectively. (f) SEM image of pure GO nanosheets. Reproduced with permission from [197], Copyright © 2012, The Royal Society of Chemistry.

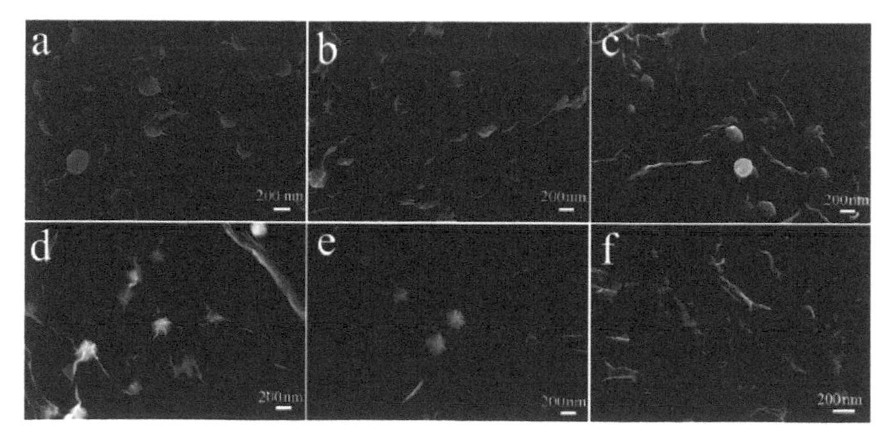

Fig. 4.87. (a–e) SEM images of the CRG–CNS hybrids: CRG/CNS1-1, CRG–CNS5-2, CRG–CNS5-1, CRG–CNS10-1 and CRG–CNS20-1. (f) SEM image of pure CRG nanosheets. Reproduced with permission from [197], Copyright © 2012, The Royal Society of Chemistry.

An electrostatic assembly method was described to prepare rGO–CMK-5 hybrids, which demonstrated excellent electrochemical performances as the electrodes of supercapacitors [198]. Firstly, negatively charged GO nanosheets and positively charged mesoporous carbon CMK-5 were assembled to form GO–CMK-5 hybrids, as schematically demonstrated in Fig. 4.88. Secondly, the GO–CMK-5 hybrids were thermally annealed in an inert atmosphere to reduce the GO into rGO. The derived hybrids possessed a hierarchical nanostructure with mesoporous carbon CMK-5 to be intercalated in between the rGO nanosheets. Because the CMK-5 platelets had an average size of about 500 nm, which was much smaller than that of the conventional CMK-5, e.g., several micrometers in length, the diffusion path of electrolyte ions was significantly shortened. CMK-5 possessed tubular pores, with a bimodal pore size distribution, which acted as straight channels for the ionic liquid to diffuse rapidly.

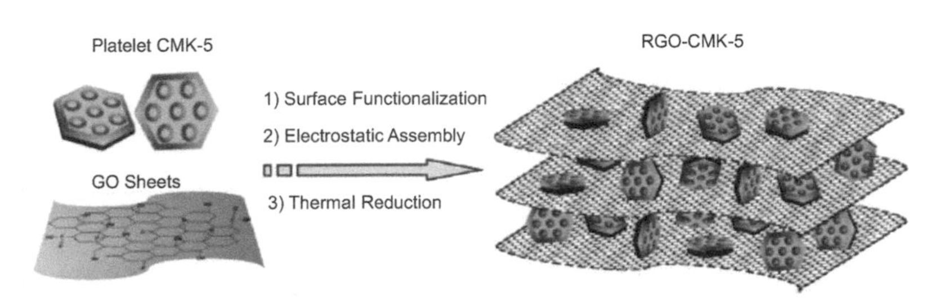

Fig. 4.88. Schematic diagram to describe the fabrication process of the rGO–CMK-5 hybrids. Reproduced with permission from [198], Copyright © 2013, The Royal Society of Chemistry.

The template used to synthesize the CMK-5 was mesoporous SBA-15, with a platelet morphology and short mesochannels [203]. With the platelet SBA-15, mesoporous carbon CMK-5 (silica–carbon) was deposited by using a CVD method [204]. The CVD products were annealed at 800°C for 90 min to realize further graphitization. CMK-5 carbon powder was obtained by etching the silica–carbon nanocomposite with 10% HF solution for 12 h to remove the silica. The CMK-5 powder was functionalized by soaking in poly(diallyldimethylammonium chloride) (PDDA, typical M_w of 100,000–200,000, 25% aqueous solution). The PDDA-CMK-5 dispersion was mixed with GO dispersion with nominated compositions to prepare the GO–CMK-5 hybrids. Finally, rGO–CMK-5 hybrids were synthesized through thermal annealing. Pure rGO sample was also prepared by annealing GO at 800°C, denoted as rGO-800.

Figure 4.89 shows a TEM image of the GO–CMK-5 hybrids [203]. The CMK-5 particles were present uniformly on the GO nanosheets. Figure 4.89 (b) shows a high-resolution TEM image of an individual CMK-5 particle, clearly demonstrating that the CMK-5 edge was partially encapsulated by the GO nanosheets. The straight mesochannels of the mesoporous carbon were well-retained. A representative TEM image of the rGO–CMK-5 hybrid is shown in Fig. 4.89 (c). Porous CMK-5 platelets dispersed on the surface of RGO sheets can be clearly seen. The rGO nanosheets had a multilayer structure, as shown in Fig. 4.89 (d). However, the pure sample, rGO-800, exhibited serious stacking phenomenon. Therefore, the CMK-5 particles served as spacers to effectively prevent the restacking of the rGO nanosheets in the rGO–CMK-5 hybrids.

Figure 4.90 shows SEM images of the rGO–CMK-5 hybrids and CMK-5 [203]. As seen clearly in Fig. 4.89 (a), the mesoporous CMK-5 particles had a platelet-like morphology with an average diameter of about 500 nm. Figure 4.90 (b) indicated that the rGO–CMK-5 hybrid exhibited a unique hierarchical structure. All the CMK-5 platelet particles were uniformly attached on the rGO nanosheets. The rGO nanosheets that wrapped the CMK-5 particles were slightly crumpled, as shown in Fig. 4.90 (c). Figure 4.90 (d) shows a cross-sectional SEM image of the rGO–CMK-5 hybrids, clearly illustrating their hierarchical microstructure. In such hybrids, the rGO layers with sufficient edge planes to be exposed would offer a large electrochemical surface area for charge storage, while the presence of the intercalated CMK-5 platelet particles effectively prevented the rGO nanosheets from restacking and facilitated a fast ion transportation with the straight and short mesochannels.

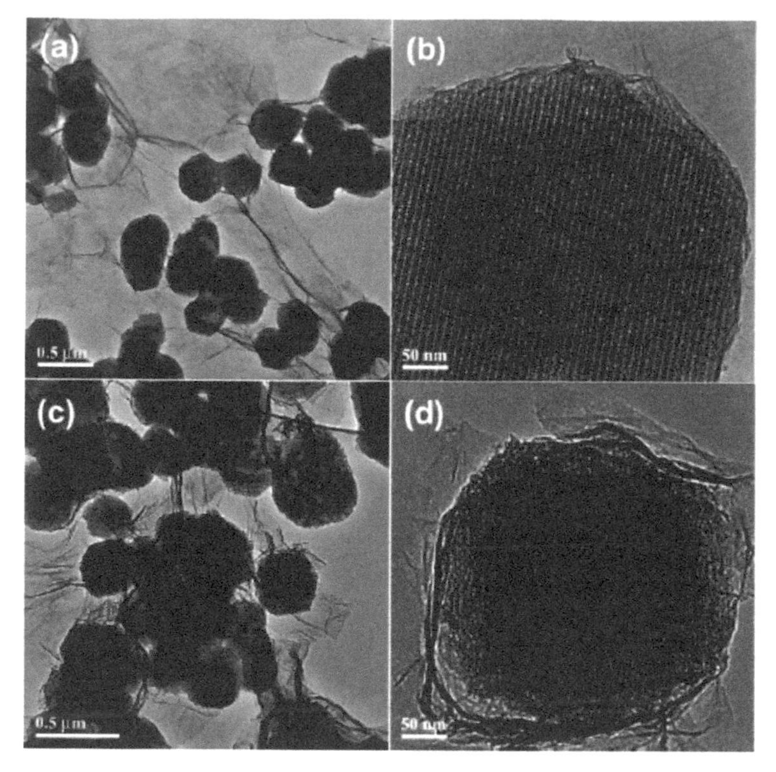

Fig. 4.89. TEM images of the GO–CMK-5 (a, b) and rGO–CMK-5 (c, d) hybrids. Reproduced with permission from [198], Copyright © 2013, The Royal Society of Chemistry.

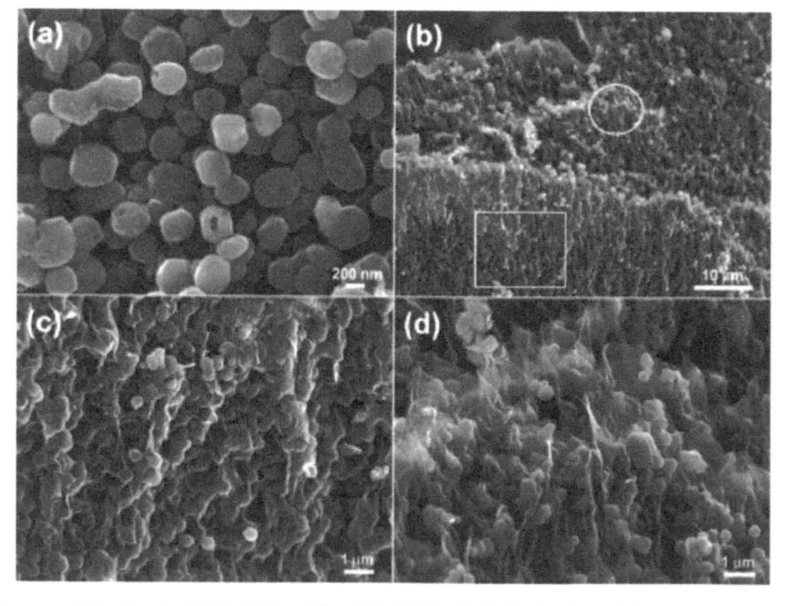

Fig. 4.90. SEM images of the platelet CMK-5 (a) and the rGO–CMK-5 hybrids (b–d) at different magnifications. Reproduced with permission from [198], Copyright © 2013, The Royal Society of Chemistry.

4.4. Summary

One important point of the graphene-based hybrids with nanostructured noble metals is to minimize their usage but maximize their efficiency, which however has not been systematically studied in the open literature. For all carbon-based hybrids, homogeneous distribution and combination of the graphene and the non-graphene components could be a critical issue, which should be paid more attention to in future studies. It is also suggested that the methods and approaches discussed in this contribution should be extended to components, so as to develop more materials with unexpected properties and performances.

References

[1] Daniel MC, Astruc D. Gold nanoparticles: Assembly, supramolecular chemistry, quantum-size-related properties, and applications toward biology, catalysis, and nanotechnology. Chemical Reviews. 2004; 104: 293–346.

[2] Xia YN, Xiong YJ, Lim BK, Skrabalak SE. Shape-controlled synthesis of metal nanocrystals: Simple chemistry meets complex physics? Angewandte Chemie-International Edition. 2009; 48: 60–103.

[3] Huang X, Qi XY, Huang YZ, Li SZ, Xue C, Gan CL et al. Photochemically controlled synthesis of anisotropic Au nanostructures: Platelet-like Au nanorods and six-star Au nanoparticles. ACS Nano. 2010; 4: 6196–202.

[4] Tao AR, Huang JX, Yang PD. Langmuir-Blodgettry of nanocrystals and nanowires. Accounts of Chemical Research. 2008; 41: 1662–73.

[5] Jones MR, Osberg KD, Macfarlane RJ, Langille MR, Mirkin CA. Templated techniques for the synthesis and assembly of plasmonic nanostructures. Chemical Reviews. 2011; 111: 3736–827.

[6] Guo SJ, Wang EK. Noble metal nanomaterials: Controllable synthesis and application in fuel cells and analytical sensors. Nano Today. 2011; 6: 240–64.

[7] Murray RW. Nanoelectrochemistry: Metal nanoparticles, nanoelectrodes, and nanopores. Chemical Reviews. 2008; 108: 2688–720.

[8] Yu KMK, Yeung CMY, Tsang SC. Carbon dioxide fixation into chemicals (methyl formate) at high yields by surface coupling over a Pd/Cu/ZnO nanocatalyst. Journal of the American Chemical Society. 2007; 129: 6360–1.

[9] Yin YD, Alivisatos AP. Colloidal nanocrystal synthesis and the organic-inorganic interface. Nature. 2005; 437: 664–70.

[10] Lee SS, Fan CY, Wu TP, Anderson SL. CO oxidation on Au_n/TiO_2 catalysts produced by size-selected cluster deposition. Journal of the American Chemical Society. 2004; 126: 5682–3.

[11] Huang X, Yin ZY, Wu SX, Qi XY, He QY, Zhang QC et al. Graphene-based materials: Synthesis, characterization, properties, and applications. Small. 2011; 7: 1876–902.

[12] Huang X, Qi XY, Boey F, Zhang H. Graphene-based composites. Chemical Society Reviews. 2012; 41: 666–86.

[13] Tan CL, Huang X, Zhang H. Synthesis and applications of graphene-based noble metal nanostructures. Materials Today. 2013; 16: 29–36.

[14] Kong HX. Hybrids of carbon nanotubes and graphene/graphene oxide. Current Opinion in Solid State & Materials Science. 2013; 17: 31–7.

[15] Xia YN, Xia XH, Wang Y, Xie SF. Shape-controlled synthesis of metal nanocrystals. MRS Bulletin. 2013; 38: 335–44.

[16] Zhou XZ, Huang X, Qi XY, Wu SX, Xue C, Boey FYC et al. *In situ* synthesis of metal nanoparticles on single-layer graphene oxide and reduced graphene oxide surfaces. Journal of Physical Chemistry C. 2009; 113: 10842–6.

[17] Ren WC, Fang YX, Wang EK. A binary functional substrate for enrichment and ultrasensitive SERS spectroscopic detection of folic acid using graphene oxide/Ag nanoparticle hybrids. ACS Nano. 2011; 5: 6425–33.

[18] Li J, Liu CY. Ag/graphene heterostructures: Synthesis, characterization and optical properties. European Journal of Inorganic Chemistry. 2010: 1244–8.

[19] Wen YQ, Xing FF, He SJ, Song SP, Wang LH, Long YT et al. A graphene-based fluorescent nanoprobe for silver(I) ions detection by using graphene oxide and a silver-specific oligonucleotide. Chemical Communications. 2010; 46: 2596–8.

[20] Pasricha R, Gupta S, Srivastava AK. A facile and novel synthesis of Ag-graphene-based nanocomposites. Small. 2009; 5: 2253–9.

[21] Lu GH, Mao S, Park SJ, Ruoff RS, Chen JH. Facile, noncovalent decoration of graphene oxide sheets with nanocrystals. Nano Research. 2009; 2: 192–200.

[22] Lu GH, Li HB, Liusman C, Yin ZY, Wu SX, Zhang H. Surface enhanced Raman scattering of Ag or Au nanoparticle-decorated reduced graphene oxide for detection of aromatic molecules. Chemical Science. 2011; 2: 1817–21.

[23] Zheng L, Zhang GN, Zhang M, Guo SH, Liu ZH. Preparation and capacitance performance of Ag-graphene based nanocomposite. Journal of Power Sources. 2012; 201: 376–81.

[24] Wang Y, Zhang S, Chen H, Li HX, Zhang P, Zhang ZY et al. One-pot facile decoration of graphene nanosheets with Ag nanoparticles for electrochemical oxidation of methanol in alkaline solution. Electrochemistry Communications. 2012; 17: 63–6.

[25] Sidorov AN, Slawinski GW, Jayatissa AH, Zamborini FP, Sumanasekera GU. A surface-enhanced Raman spectroscopy study of thin graphene sheets functionalized with gold and silver nanostructures by seed-mediated growth. Carbon. 2012; 50: 699–705.

[26] Guo YQ, Sun XY, Liu Y, Wang W, Qiu HX, Gao JP. One pot preparation of reduced graphene oxide (RGO) or Au (Ag) nanoparticle-RGO hybrids using chitosan as a reducing and stabilizing agent and their use in methanol electrooxidation. Carbon. 2012; 50: 2513–23.

[27] Lu XQ, Qi HT, Zhang XF, Xue ZH, Jin J, Zhou XB et al. Highly dispersive Ag nanoparticles on functionalized graphene for an excellent electrochemical sensor of nitroaromatic compounds. Chemical Communications. 2011; 47: 12494–6.

[28] Sun ST, Wu PY. Competitive surface-enhanced Raman scattering effects in noble metal nanoparticle-decorated graphene sheets. Physical Chemistry Chemical Physics. 2011; 13: 21116–20.

[29] Liu XW, Mao JJ, Liu PD, Wei XW. Fabrication of metal-graphene hybrid materials by electroless deposition. Carbon. 2011; 49: 477–83.

[30] Zhang ZY, Xu FG, Yang WS, Guo MY, Wang XD, Zhang BL et al. A facile one-pot method to high-quality Ag-graphene composite nanosheets for efficient surface-enhanced Raman scattering. Chemical Communications. 2011; 47: 6440–2.

[31] Chen JL, Zheng XL, Wang H, Zheng WT. Graphene oxide-Ag nanocomposite: *In situ* photochemical synthesis and application as a surface-enhanced Raman scattering substrate. Thin Solid Films. 2011; 520: 179–85.

[32] Zhu MS, Chen PL, Liu MH. Graphene oxide enwrapped Ag/AgX (X = Br, Cl) nanocomposite as a highly efficient visible-light plasmonic photocatalyst. ACS Nano. 2011; 5: 4529–36.

[33] Lu LH, Liu JH, Hu Y, Zhang YE, Chen W. Graphene-stabilized silver nanoparticle electrochemical electrode for actuator design. Advanced Materials. 2013; 25: 1270–4.

[34] Lerf A, He HY, Forster M, Klinowski J. Structure of graphite oxide revisited. Journal of Physical Chemistry B. 1998; 102: 4477–82.

[35] Gomez-Navarro C, Weitz RT, Bittner AM, Scolari M, Mews A, Burghard M et al. Electronic transport properties of individual chemically reduced graphene oxide sheets. Nano Letters. 2007; 7: 3499–503.

[36] Kim JY, Cote LJ, Kim F, Yuan W, Shull KR, Huang JX. Graphene oxide sheets at interfaces. Journal of the American Chemical Society. 2010; 132: 8180–6.

[37] Lotya M, Hernandez Y, King PJ, Smith RJ, Nicolosi V, Karlsson LS et al. Liquid phase production of graphene by exfoliation of graphite in surfactant/water solutions. Journal of the American Chemical Society. 2009; 131: 3611–20.

[38] Li XS, Cai WW, An JH, Kim SY, Nah JH, Yang DX et al. Large-area synthesis of high-quality and uniform graphene films on copper foils. Science. 2009; 324: 1312–4.

[39] Zhang YW, Liu S, Lu WB, Wang LH, Tian JQ, Sun XP. *In situ* green synthesis of Au nanostructures on graphene oxide and their application for catalytic reduction of 4-nitrophenol. Catalysis Science & Technology. 2011; 1: 1142–4.

[40] Huang X, Zhou XZ, Wu SX, Wei YY, Qi XY, Zhang J et al. Reduced graphene oxide-templated photochemical synthesis and *in situ* assembly of Au nanodots to orderly patterned Au nanodot chains. Small. 2010; 6: 513–6.

[41] Gong JM, Zhou T, Song DD, Zhang LZ. Monodispersed Au nanoparticles decorated graphene as an enhanced sensing platform for ultrasensitive stripping voltammetric detection of mercury(II). Sensors and Actuators B-Chemical. 2010; 150: 491–7.

[42] Huang X, Li SZ, Huang YZ, Wu SX, Zhou XZ, Li SZ et al. Synthesis of hexagonal close-packed gold nanostructures. Nature Communications. 2011; 2: 292.

[43] Ge SG, Yan M, Lu JJ, Zhang M, Yu F, Yu JH et al. Electrochemical biosensor based on graphene oxide-Au nanoclusters composites for L-cysteine analysis. Biosensors & Bioelectronics. 2012; 31: 49–54.

[44] Huang X, Li HB, Li SZ, Wu SX, Boey F, Ma J et al. Synthesis of gold square-like plates from ultrathin gold square sheets: The evolution of structure phase and shape. Angewandte Chemie-International Edition. 2011; 50: 12245–8.

[45] Vinodgopal K, Neppolian B, Lightcap IV, Grieser F, Ashokkumar M, Kamat PV. Sonolytic design of graphene-Au nanocomposites. Simultaneous and sequential reduction of graphene oxide and Au (III). Journal of Physical Chemistry Letters. 2010; 1: 1987–93.

[46] Huang X, Li SZ, Wu SX, Huang YZ, Boey F, Gan CL et al. Graphene oxide-templated synthesis of ultrathin or tadpole-shaped Au nanowires with alternating hcp and fcc domains. Advanced Materials. 2012; 24: 979–783.

[47] Hu YJ, Jin J, Wu PY, Zhang H, Cai CX. Graphene-gold nanostructure composites fabricated by electrodeposition and their electrocatalytic activity toward the oxygen reduction and glucose oxidation. Electrochimica Acta. 2010; 56: 491–500.

[48] Wang ZJ, Zhang J, Yin ZY, Wu SX, Mandler D, Zhang H. Fabrication of nanoelectrode ensembles by electrodepositon of Au nanoparticles on single-layer graphene oxide sheets. Nanoscale. 2012; 4: 2728–33.

[49] Huang J, Zhang LM, Chen B, Ji N, Chen FH, Zhang YB et al. Nanocomposites of size-controlled gold nanoparticles and graphene oxide: Formation and applications in SERS and catalysis. Nanoscale. 2010; 2: 2733–8.

[50] Muszynski R, Seger B, Kamat PV. Decorating graphene sheets with gold nanoparticles. Journal of Physical Chemistry C. 2008; 112: 5263–6.

[51] Zhang YZ, Jiang W. Decorating graphene sheets with gold nanoparticles for the detection of sequence-specific DNA. Electrochimica Acta. 2012; 71: 239–45.

[52] Kim YK, Na HK, Lee YW, Jang HJ, Han SW, Min DH. The direct growth of gold rods on graphene thin films. Chemical Communications. 2010; 46: 3185–7.
[53] Xu C, Wang X, Zhu JW. Graphene-metal particle nanocomposites. Journal of Physical Chemistry C. 2008; 112: 19841–5.
[54] Jasuja K, Berry V. Implantation and growth of dendritic gold nanostructures on graphene derivatives: Electrical property tailoring and Raman enhancement. ACS Nano. 2009; 3: 2358–66.
[55] Lu YH, Zhou M, Zhang C, Feng YP. Metal-embedded graphene: A possible catalyst with high activity. Journal of Physical Chemistry C. 2009; 113: 20156–60.
[56] Fu CP, Kuang YF, Huang ZY, Wang X, Du NN, Chen JH et al. Electrochemical co-reduction synthesis of graphene/Au nanocomposites in ionic liquid and their electrochemical activity. Chemical Physics Letters. 2010; 499: 250–3.
[57] Zhang ZJ, Chen HH, Xing CY, Guo MY, Xu FG, Wang XD et al. Sodium citrate: A universal reducing agent for reduction/decoration of graphene oxide with Au nanoparticles. Nano Research. 2011; 4: 599–611.
[58] Xiang GL, He J, Li TY, Zhuang J, Wang X. Rapid preparation of noble metal nanocrystals via facile coreduction with graphene oxide and their enhanced catalytic properties. Nanoscale. 2011; 3: 3737–42.
[59] Arif M, Heo K, Lee BY, Lee JY, Seo DH, Seo SA et al. Metallic nanowire-graphene hybrid nanostructures for highly flexible field emission devices. Nanotechnology. 2011; 22: 355709.
[60] Ko HH, Singamaneni S, Tsukruk VV. Nanostructured surfaces and assemblies as SERS media. Small. 2008; 4: 1576–99.
[61] Campion A, Ivanecky JE, Child CM, Foster M. On the mechanism of chemical enhancement in surface-enhanced Raman-scattering. Journal of the American Chemical Society. 1995; 117: 11807–8.
[62] Supriya L, Claus RO. Solution-based assembly of conductive gold film on flexible polymer substrates. Langmuir. 2004; 20: 8870–6.
[63] Mullins WW, Sekerka RF. Stability of planar interface during solidification of dilute binary alloy. Journal of Applied Physics. 1964; 35: 444–451.
[64] Langer JS. Instabilities and pattern-formation in crystal-growth. Reviews of Modern Physics. 1980; 52: 1–28.
[65] Xu QM, Wan LJ, Yin SX, Wang C, Bai CL. Effect of chemically modified tips on STM imaging of 1-octadecanethiol molecule. Journal of Physical Chemistry B. 2001; 105: 10465–7.
[66] Halder A, Ravishankar N. Ultrafine single-crystalline gold nanowire arrays by oriented attachment. Advanced Materials. 2007; 19: 1854–8.
[67] Li ZQ, Tao J, Lu XM, Zhu YM, Xia YN. Facile synthesis of ultrathin Au nanorods by aging the AuCl(oleylamine) complex with amorphous Fe nanoparticles in chloroform. Nano Letters. 2008; 8: 3052–5.
[68] Liu XH, Luo J, Zhu J. Size effect on the crystal structure of silver nanowires. Nano Letters. 2006; 6: 408–12.
[69] Kondo Y, Takayanagi K. Gold nanobridge stabilized by surface structure. Physical Review Letters. 1997; 79: 3455–8.
[70] Kitakami O, Sato H, Shimada Y, Sato F, Tanaka M. Size effect on the crystal phase of cobalt fine particles. Physical Review B. 1997; 56: 13849–54.
[71] Shen XS, Wang GZ, Hong X, Xie X, Zhu W, Li DP. Anisotropic growth of one-dimensional silver rod-needle and plate-belt heteronanostructures induced by twins and hcp phase. Journal of the American Chemical Society. 2009; 131: 10812–3.
[72] Caroff P, Dick KA, Johansson J, Messing ME, Deppert K, Samuelson L. Controlled polytypic and twin-plane superlattices in III–V nanowires. Nature Nanotechnology. 2009; 4: 50–5.
[73] Li Q, Gong XG, Wang CR, Wang J, Ip K, Hark S. Size-dependant periodically twinned ZnSe nanowires. Advanced Materials. 2004; 16: 1436–40.
[74] Li N, Wang ZY, Zhao KK, Shi ZJ, Xu SK, Gu ZN. Graphene-Pd composite as highly active catalyst for the Suzuki-Miyaura coupling reaction. Journal of Nanoscience and Nanotechnology. 2010; 10: 6748–51.
[75] Bong SY, Uhm SH, Kim YR, Lee JY, Kim HS. Graphene supported Pd electrocatalysts for formic acid oxidation. Electrocatalysis. 2010; 1: 139–43.
[76] Wu CH, Wang CH, Lee MT, Chang JK. Unique Pd/graphene nanocomposites constructed using supercritical fluid for superior electrochemical sensing performance. Journal of Materials Chemistry. 2012; 22: 21466–71.
[77] Huang HJ, Wang X. Pd nanoparticles supported on low-defect graphene sheets: for use as high-performance electrocatalysts for formic acid and methanol oxidation. Journal of Materials Chemistry. 2012; 22: 22533–41.
[78] Wang SY, Manthiram A. Graphene ribbon-supported Pd nanoparticles as highly durable, efficient electrocatalysts for formic acid oxidation. Electrochimica Acta. 2013; 88: 565–70.
[79] Wen ZL, Yang SD, Liang YY, He W, Tong H, Hao L et al. The improved electrocatalytic activity of palladium/graphene nanosheets towards ethanol oxidation by tin oxide. Electrochimica Acta. 2010; 56: 139–44.
[80] Li Y, Fan XB, Qi JJ, Ji JY, Wang SL, Zhang GL et al. Palladium nanoparticle-graphene hybrids as active catalysts for the Suzuki reaction. Nano Research. 2010; 3: 429–37.
[81] Srivastava V. Ionic liquid immobilized palladium nanoparticle - graphene hybrid as active catalyst for Heck reaction. Letters in Organic Chemistry. 2015; 12: 67–72.
[82] Zhao H, Yang J, Wang LH, Tian CG, Jiang BJ, Fu HG. Fabrication of a palladium nanoparticle/graphene nanosheet hybrid via sacrifice of a copper template and its application in catalytic oxidation of formic acid. Chemical Communications. 2011; 47: 2014–6.

[83] Hu ZL, Aizawa M, Wang ZM, Yoshizawa N, Hatori H. Synthesis and characteristics of graphene oxide-derived carbon nanosheet-Pd nanosized particle composites. Langmuir. 2010; 26: 6681–8.
[84] Lu LM, Li HB, Qu FL, Zhang XB, Shen GL, Yu RR. *In situ* synthesis of palladium nanoparticle-graphene nanohybrids and their application in nonenzymatic glucose biosensors. Biosensors & Bioelectronics. 2011; 26: 3500–4.
[85] Siamaki AR, Khder AERS, Abdelsayed V, El-Shall MS, Gupton BF. Microwave-assisted synthesis of palladium nanoparticles supported on graphene: A highly active and recyclable catalyst for carbon-carbon cross-coupling reactions. Journal of Catalysis. 2011; 279: 1–11.
[86] Moussa S, Abdelsayed V, El-Shall MS. Laser synthesis of Pt, Pd, CoO and Pd-CoO nanoparticle catalysts supported on graphene. Chemical Physics Letters. 2011; 510: 179–84.
[87] Xu WY, Wang XZ, Zhou Q, Meng B, Zhao JT, Qiu JS et al. Low-temperature plasma-assisted preparation of graphene supported palladium nanoparticles with high hydrodesulfurization activity. Journal of Materials Chemistry. 2012; 22: 14363–8.
[88] Giovanni M, Poh HL, Ambrosi A, Zhao GX, Sofer Z, Sanek F et al. Noble metal (Pd, Ru, Rh, Pt, Au, Ag) doped graphene hybrids for electrocatalysis. Nanoscale. 2012; 4: 5002–8.
[89] Yang J, Tian CG, Wang LH, Tan TX, Yin J, Wang B et al. *In situ* reduction, oxygen etching, and reduction using formic acid: An effective strategy for controllable growth of monodisperse palladium nanoparticles on graphene. Chempluschem. 2012; 77: 301–7.
[90] Chen MH, Chu W, Dai XY, Zhang XW. New palladium catalysts prepared by glow discharge plasma for the selective hydrogenation of acetylene. Catalysis Today. 2004; 89: 201–4.
[91] Kumar R, Oh JH, Kim HJ, Jung JH, Jung CH, Hong WG et al. Nanohole-structured and palladium-embedded 3D porous graphene for ultrahigh hydrogen storage and CO oxidation multifunctionalities. ACS Nano. 2015; 9: 7343–51.
[92] Li D, Mueller MB, Gilje S, Kaner RB, Wallace GG. Processable aqueous dispersions of graphene nanosheets. Nature Nanotechnology. 2008; 3: 101–5.
[93] He QY, Wu SX, Yin ZY, Zhang H. Graphene-based electronic sensors. Chemical Science. 2012; 3: 1764–72.
[94] Yin ZY, He QY, Huang X, Zhang J, Wu SX, Chen P et al. Real-time DNA detection using Pt nanoparticle-decorated reduced graphene oxide field-effect transistors. Nanoscale. 2012; 4: 293–7.
[95] Wu SX, He QY, Zhou CM, Qi XY, Huang X, Yin ZY et al. Synthesis of Fe_3O_4 and Pt nanoparticles on reduced graphene oxide and their use as a recyclable catalyst. Nanoscale. 2012; 4: 2478–83.
[96] Kou R, Shao YY, Wang DH, Engelhard MH, Kwak JH, Wang J et al. Enhanced activity and stability of Pt catalysts on functionalized graphene sheets for electrocatalytic oxygen reduction. Electrochemistry Communications. 2009; 11: 954–7.
[97] Li YM, Tang LH, Li JH. Preparation and electrochemical performance for methanol oxidation of pt/graphene nanocomposites. Electrochemistry Communications. 2009; 11: 846–9.
[98] Seger B, Kamat PV. Electrocatalytically active graphene-platinum nanocomposites. Role of 2-D carbon support in PEM fuel cells. Journal of Physical Chemistry C. 2009; 113: 7990–5.
[99] Guo SJ, Wen D, Zhai YM, Dong SJ, Wang EK. Platinum nanoparticle ensemble-on-graphene hybrid nanosheet: One-pot, rapid synthesis, and used as new electrode material for electrochemical sensing. ACS Nano. 2010; 4: 3959–68.
[100] Zhu CZ, Guo SJ, Zhai YM, Dong SJ. Layer-by-layer self-assembly for constructing a graphene/platinum nanoparticle three-dimensional hybrid nanostructure using ionic liquid as a linker. Langmuir. 2010; 26: 7614–8.
[101] Li YJ, Gao W, Ci LJ, Wang CM, Ajayan PM. Catalytic performance of Pt nanoparticles on reduced graphene oxide for methanol electro-oxidation. Carbon. 2010; 48: 1124–30.
[102] Yao ZQ, Zhu MS, Jiang FX, Du YK, Wang CY, Yang P. Highly efficient electrocatalytic performance based on Pt nanoflowers modified reduced graphene oxide/carbon cloth electrode. Journal of Materials Chemistry. 2012; 22: 13707–13.
[103] Gong F, Wang H, Wang ZS. Self-assembled monolayer of graphene/Pt as counter electrode for efficient dye-sensitized solar cell. Physical Chemistry Chemical Physics. 2011; 13: 17676–82.
[104] Kundu P, Nethravathi C, Deshpande PA, Rajamathi M, Madras G, Ravishankar N. Ultrafast microwave-assisted route to surfactant-free ultrafine Pt nanoparticles on graphene: Synergistic co-reduction mechanism and high catalytic activity. Chemistry of Materials. 2011; 23: 2772–80.
[105] Xin YC, Liu JG, Zhou Y, Liu WM, Gao J, Xie Y et al. Preparation and characterization of Pt supported on graphene with enhanced electrocatalytic activity in fuel cell. Journal of Power Sources. 2011; 196: 1012–8.
[106] Huang CC, Pu NW, Wang CA, Huang JC, Sung Y, Ger MD. Hydrogen storage in graphene decorated with Pd and Pt nano-particles using an electroless deposition technique. Separation and Purification Technology. 2011; 82: 210–5.
[107] Min SX, Lu GX. Dye-cosensitized graphene/Pt photocatalyst for high efficient visible light hydrogen evolution. International Journal of Hydrogen Energy. 2012; 37: 10564–74.
[108] He DP, Cheng K, Li HG, Peng T, Xu FG, Mu SC et al. Highly active platinum nanoparticles on graphene nanosheets with a significant improvement in stability and CO tolerance. Langmuir. 2012; 28: 3979–86.
[109] Huang HJ, Chen HQ, Sun DP, Wang X. Graphene nanoplate-Pt composite as a high performance electrocatalyst for direct methanol fuel cells. Journal of Power Sources. 2012; 204: 46–52.
[110] Liu S, Wang JJ, Zeng J, Ou JF, Li ZP, Liu XH et al. "Green" electrochemical synthesis of Pt/graphene sheet nanocomposite film and its electrocatalytic property. Journal of Power Sources. 2010; 195: 4628–33.

[111] Choi SM, Seo MH, Kim HG, Kim WB. Synthesis and characterization of graphene-supported metal nanoparticles by impregnation method with heat treatment in H_2 atmosphere. Synthetic Metals. 2011; 161: 2405–11.
[112] Sutter E, Albrecht P, Wang B, Bocquet ML, Wu LJ, Zhu YM et al. Arrays of Ru nanoclusters with narrow size distribution templated by monolayer graphene on Ru. Surface Science. 2011; 605: 1676–84.
[113] Sutter E, Wang B, Albrecht P, Lahiri J, Bocquet ML, Sutter P. Templating of arrays of Ru nanoclusters by monolayer graphene/Ru Moires with different periodicities. Journal of Physics-Condensed Matter. 2012; 24: 314201.
[114] Marquardt D, Vollmer C, Thomann R, Steurer P, Muelhaupt R, Redel E et al. The use of microwave irradiation for the easy synthesis of graphene-supported transition metal nanoparticles in ionic liquids. Carbon. 2011; 49: 1326–32.
[115] Liu X, Yao KX, Meng CG, Han Y. Graphene substrate-mediated catalytic performance enhancement of Ru nanoparticles: a first-principles study. Dalton Transactions. 2012; 41: 1289–96.
[116] Liu X, Meng CG, Han Y. Substrate-mediated enhanced activity of Ru nanoparticles in catalytic hydrogenation of benzene. Nanoscale. 2012; 4: 2288–95.
[117] Chandra S, Bag S, Bhar R, Pramanik P. Sonochemical synthesis and application of rhodium-graphene nanocomposite. Journal of Nanoparticle Research. 2011; 13: 2769–77.
[118] Tsai HM, Yang SJ, Ma CCM, Xie XF. Preparation and electrochemical activities of iridium-decorated graphene as the electrode for all-vanadium redox flow batteries. Electrochimica Acta. 2012; 77: 232–6.
[119] Zhang S, Shao YY, Liao HG, Liu J, Aksay IA, Yin GP et al. Graphene decorated with PtAu alloy nanoparticles: facile synthesis and promising application for formic acid oxidation. Chemistry of Materials. 2011; 23: 1079–81.
[120] Guo SJ, Dong SJ, Wang EK. Three-dimensional Pt-on-Pd bimetallic nanodendrites supported on graphene nanosheet: Facile synthesis and used as an advanced nanoelectrocatalyst for methanol oxidation. ACS Nano. 2010; 4: 547–55.
[121] Dong LF, Gari RRS, Li ZP, Craig MM, Hou SF. Graphene-supported platinum and platinum-ruthenium nanoparticles with high electrocatalytic activity for methanol and ethanol oxidation. Carbon. 2010; 48: 781–7.
[122] Bong SY, Kim YR, Kim I, Woo SH, Uhm SH, Lee JY et al. Graphene supported electrocatalysts for methanol oxidation. Electrochemistry Communications. 2010; 12: 129–31.
[123] Chai J, Li FH, Hu YW, Zhang QX, Han DX, Niu L. Hollow flower-like AuPd alloy nanoparticles: One step synthesis, self-assembly on ionic liquid-functionalized graphene, and electrooxidation of formic acid. Journal of Materials Chemistry. 2011; 21: 17922–9.
[124] Nethravathi C, Anumol EA, Rajamathi M, Ravishankar N. Highly dispersed ultrafine Pt and PtRu nanoparticles on graphene: formation mechanism and electrocatalytic activity. Nanoscale. 2011; 3: 569–71.
[125] Feng LL, Gao G, Huang P, Wang XS, Zhang CL, Zhang JL et al. Preparation of Pt Ag alloy nanoisland/graphene hybrid composites and its high stability and catalytic activity in methanol electro-oxidation. Nanoscale Research Letters. 2011; 6: 551.
[126] Liu JP, Zhou HH, Wang QQ, Zeng FY, Kuang YF. Reduced graphene oxide supported palladium-silver bimetallic nanoparticles for ethanol electro-oxidation in alkaline media. Journal of Materials Science. 2012; 47: 2188–94.
[127] Galal A, Atta NF, Hassan HK. Graphene supported-Pt-M (M = Ru or Pd) for electrocatalytic methanol oxidation. International Journal of Electrochemical Science. 2012; 7: 768–84.
[128] Zhang H, Xu XQ, Gu P, Li CY, Wu P, Cai CX. Microwave-assisted synthesis of graphene-supported Pd_1Pt_3 nanostructures and their electrocatalytic activity for methanol oxidation. Electrochimica Acta. 2011; 56: 7064–70.
[129] Hu YJ, Zhang H, Wu P, Zhang H, Zhou BX, Cai CX. Bimetallic Pt-Au nanocatalysts electrochemically deposited on graphene and their electrocatalytic characteristics towards oxygen reduction and methanol oxidation. Physical Chemistry Chemical Physics. 2011; 13: 4083–94.
[130] Chen HY, Li Y, Zhang FB, Zhang GL, Fan XB. Graphene supported Au-Pd bimetallic nanoparticles with core-shell structures and superior peroxidase-like activities. Journal of Materials Chemistry. 2011; 21: 17658–61.
[131] Liu MM, Lu YZ, Chen W. PdAg nanorings supported on graphene nanosheets: Highly methanol-tolerant cathode electrocatalyst for alkaline fuel cells. Advanced Functional Materials. 2013; 23: 1289–96.
[132] Neppolian B, Saez V, Gonzalez-Garcia J, Grieser F, Gomez R, Ashokkumar M. Sonochemical synthesis of graphene oxide supported Pt-Pd alloy nanocrystals as efficient electrocatalysts for methanol oxidation. Journal of Solid State Electrochemistry. 2014; 18: 3163–71.
[133] Lim BK, Jiang MJ, Camargo PHC, Cho EC, Tao J, Lu XM et al. Pd-Pt bimetallic nanodendrites with high activity for oxygen reduction. Science. 2009; 324: 1302–5.
[134] Genorio B, Peng ZW, Lu W, Hoelscher BKP, Novosel B, Tour JM. Synthesis of dispersible ferromagnetic graphene nanoribbon stacks with enhanced electrical percolation properties in a magnetic field. ACS Nano. 2012; 6: 10396–404.
[135] Long CL, Wei T, Yan J, Jiang LL, Fan ZJ. Supercapacitors based on graphene-supported Iron nanosheets as negative electrode materials. ACS Nano. 2013; 7: 11325–32.
[136] Li BJ, Cao HQ, Yin JF, Wu YA, Warner JH. Synthesis and separation of dyes via Ni@reduced graphene oxide nanostructures. Journal of Materials Chemistry. 2012; 22: 1876–83.
[137] Ji ZY, Shen XP, Zhu GX, Zhou HH, Yuan AH. Reduced graphene oxide/nickel nanocomposites: facile synthesis, magnetic and catalytic properties. Journal of Materials Chemistry. 2012; 22: 3471–7.
[138] Wang D, Yan W, Vijapur SH, Botte GG. Electrochemically reduced graphene oxide-nickel nanocomposites for urea electrolysis. Electrochimica Acta. 2013; 89: 732–6.
[139] Bajpai R, Roy S, Kulshrestha N, Rafiee J, Koratkar N, Misra DS. Graphene supported nickel nanoparticle as a viable replacement for platinum in dye sensitized solar cells. Nanoscale. 2012; 4: 926–30.

[140] Chen TT, Deng F, Zhu J, Chen CF, Sun GB, Ma SL et al. Hexagonal and cubic Ni nanocrystals grown on graphene: phase-controlled synthesis, characterization and their enhanced microwave absorption properties. Journal of Materials Chemistry. 2012; 22: 15190–7.
[141] Liu G, Wang YJ, Qiu FY, Li L, Jiao LF, Yuan HT. Synthesis of porous Ni@rGO nanocomposite and its synergetic effect on hydrogen sorption properties of MgH_2. Journal of Materials Chemistry. 2012; 22: 22542–9.
[142] Na HG, Cho HY, Kwon YJ, Kang ST, Lee CM, Jung TK et al. Reduced graphene oxide functionalized with Cu nanoparticles: Fabrication, structure, and sensing properties. Thin Solid Films. 2015; 588: 11–8.
[143] Wang SL, Huang XL, He YH, Huang H, Wu YQ, Hou LZ et al. Synthesis, growth mechanism and thermal stability of copper nanoparticles encapsulated by multi-layer graphene. Carbon. 2012; 50: 2119–25.
[144] Jia ZF, Chen TD, Wang J, Ni JJ, Li HY, Shao X. Synthesis, characterization and tribological properties of Cu/reduced graphene oxide composites. Tribology International. 2015; 88: 17–24.
[145] Luo B, Wang B, Li XL, Jia YY, Liang MH, Zhi LJ. Graphene-confined Sn nanosheets with enhanced lithium storage capability. Advanced Materials. 2012; 24: 3538–43.
[146] Qin J, He CN, Zhao NQ, Wang ZY, Shi CS, Liu EZ et al. Graphene networks anchored with Sn@graphene as lithium ion battery anode. ACS Nano. 2014; 8: 1728–38.
[147] Wang GX, Wang B, Wang XL, Park JS, Dou SX, Ahn HJ et al. Sn/graphene nanocomposite with 3D architecture for enhanced reversible lithium storage in lithium ion batteries. Journal of Materials Chemistry. 2009; 19: 8378–84.
[148] Dresselhaus MS, Dresselhaus G. Intercalation compounds of graphite. Advances in Physics. 2002; 51: 1–186.
[149] Deng SZ, Tjoa V, Fan HM, Tan HR, Sayle DC, Olivo M et al. Reduced graphene oxide conjugated Cu_2O nanowire mesocrystals for high-performance NO_2 gas sensor. Journal of the American Chemical Society. 2012; 134: 4905–17.
[150] Yoon HJ, Jun DH, Yang JH, Zhou ZX, Yang SS, Cheng MMC. Carbon dioxide gas sensor using a graphene sheet. Sensors and Actuators B-Chemical. 2011; 157: 310–3.
[151] Yang LX, Li L, Yang YG, Zhang G, Gong LH, Jing LQ et al. Facile synthesis of Cu/Cu_xO nanoarchitectures with adjustable phase composition for effective NO_x gas sensor at room temperature. Materials Research Bulletin. 2013; 48: 3657–65.
[152] Chen ZP, Ren WC, Gao LB, Liu BL, Pei SF, Cheng HM. Three-dimensional flexible and conductive interconnected graphene networks grown by chemical vapour deposition. Nature Materials. 2011; 10: 424–8.
[153] Cai DY, Song M, Xu CX. Highly conductive carbon-nanotube/graphite-oxide hybrid films. Advanced Materials. 2008; 20: 1706–9.
[154] Wang RR, Sun J, Gao LB, Xu CH, Zhang J, Liu YQ. Effective post treatment for preparing highly conductive carbon nanotube/reduced graphite oxide hybrid films. Nanoscale. 2011; 3: 904–6.
[155] Kim YK, Na HK, Kwack SJ, Ryoo SR, Lee YM, Hong SH et al. Synergistic effect of graphene oxide/MWCNT films in laser desorption/ionization mass spectrometry of small molecules and tissue imaging. ACS Nano. 2011; 5: 4550–61.
[156] Zhou Y, Bao QL, Varghese B, Tang LAL, Tan CK, Sow CH et al. Microstructuring of graphene oxide nanosheets using direct laser writing. Advanced Materials. 2010; 22: 67–71.
[157] Xu SY, Li YF, Zou HF, Qiu JS, Guo Z, Guo BC. Carbon nanotubes as assisted matrix for laser desorption/ionization time-of-flight mass spectrometry. Analytical Chemistry. 2003; 75: 6191–5.
[158] Han PX, Yue YH, Liu ZH, Xu WY, Zhang LX, Xu HX et al. Graphene oxide nanosheets/multi-walled carbon nanotubes hybrid as an excellent electrocatalytic material towards VO^{2+}/VO_2^+ redox couples for vanadium redox flow batteries. Energy & Environmental Science. 2011; 4: 4710–7.
[159] Dong XC, Xing GH, Chan-Park MB, Shi WH, Xiao N, Wang J et al. The formation of a carbon nanotube-graphene oxide core-shell structure and its possible applications. Carbon. 2011; 49: 5071–8.
[160] Wang RR, Sun J, Gao LB, Xu CH, Zhang J. Fibrous nanocomposites of carbon nanotubes and graphene-oxide with synergetic mechanical and actuative performance. Chemical Communications. 2011; 47: 8650–2.
[161] Tung VC, Chen LM, Allen MJ, Wassei JK, Nelson K, Kaner RB et al. Low-temperature solution processing of graphene-carbon nanotube hybrid materials for high-performance transparent conductors. Nano Letters. 2009; 9: 1949–55.
[162] Huang JH, Fang JH, Liu CC, Chu CW. Effective work function modulation of graphene/carbon nanotube composite films as transparent cathodes for organic optoelectronics. ACS Nano. 2011; 5: 6262–71.
[163] Kim F, Cote LJ, Huang JX. Graphene oxide: Surface activity and two-dimensional assembly. Advanced Materials. 2010; 22: 1954–8.
[164] Cote LJ, Kim F, Huang JX. Langmuir-Blodgett assembly of graphite oxide single layers. Journal of the American Chemical Society. 2009; 131: 1043–9.
[165] King PJ, Khan U, Lotya M, De S, Coleman JN. Improvement of transparent conducting nanotube films by addition of small quantities of graphene. ACS Nano. 2010; 4: 4238–46.
[166] Huang ZD, Zhang BL, Oh SW, Zheng QB, Lin XY, Yousefi N et al. Self-assembled reduced graphene oxide/carbon nanotube thin films as electrodes for supercapacitors. Journal of Materials Chemistry. 2012; 22: 3591–9.
[167] Jang WS, Chae SS, Lee SJ, Song KM, Baik HK. Improved electrical conductivity of a non-covalently dispersed graphene-carbon nanotube film by chemical p-type doping. Carbon. 2012; 50: 943–51.

[168] Pan YZ, Bao HQ, Li. Noncovalently functionalized multiwalled carbon nanotubes by chitosan-grafted reduced graphene oxide and their synergistic reinforcing effects in chitosan films. ACS Applied Materials & Interfaces. 2011; 3: 4819–30.
[169] Velten J, Mozer AJ, Li D, Officer D, Wallace G, Baughman R et al. Carbon nanotube/graphene nanocomposite as efficient counter electrodes in dye-sensitized solar cells. Nanotechnology. 2012; 23: 085201.
[170] Aravind SSJ, Jafri RI, Rajalakshmi N, Ramaprabhu S. Solar exfoliated graphene-carbon nanotube hybrid nano composites as efficient catalyst supports for proton exchange membrane fuel cells. Journal of Materials Chemistry. 2011; 21: 18199–204.
[171] Yen MY, Hsiao MC, Liao SH, Liu PI, Tsai HM, Ma CCM et al. Preparation of graphene/multi-walled carbon nanotube hybrid and its use as photoanodes of dye-sensitized solar cells. Carbon. 2011; 49: 3597–606.
[172] Kim YK, Min DH. Durable large-area thin flms of graphene/carbon nanotube double layers as a transparent electrode. Langmuir. 2009; 25: 11302–6.
[173] Hong TK, Lee DW, Choi HJ, Shin HS, Kim BS. Transparent, flexible conducting hybrid multi layer thin films of multiwalled carbon nanotubes with graphene nanosheets. ACS Nano. 2010; 4: 3861–8.
[174] Yu DS, Dai LM. Self-assembled graphene/carbon nanotube hybrid films for supercapacitors. Journal of Physical Chemistry Letters. 2010; 1: 467–70.
[175] Khan U, O'Connor I, Gun'ko YK, Coleman JN. The preparation of hybrid films of carbon nanotubes and nano-graphite/graphene with excellent mechanical and electrical properties. Carbon. 2010; 48: 2825–30.
[176] Tang YH, Gou JH. Synergistic effect on electrical conductivity of few-layer graphene/multi-walled carbon nanotube paper. Materials Letters. 2010; 64: 2513–6.
[177] Fan ZJ, Yan J, Zhi LJ, Zhang Q, Wei T, Feng J et al. A three-dimensional carbon nanotube/graphene sandwich and its application as electrode in supercapacitors. Advanced Materials. 2010; 22: 3723–8.
[178] Das S, Seelaboyina R, Verma V, Lahiri I, Hwang JY, Banerjee R et al. Synthesis and characterization of self-organized multilayered graphene-carbon nanotube hybrid films. Journal of Materials Chemistry. 2011; 21: 7289–95.
[179] Su Q, Liang YY, Feng XL, Muellen K. Towards free-standing graphene/carbon nanotube composite films via acetylene-assisted thermolysis of organocobalt functionalized graphene sheets. Chemical Communications. 2010; 46: 8279–81.
[180] Lee DH, Lee JA, Lee WJ, Kim SO. Flexible field emission of nitrogen-doped carbon nanotubes/reduced graphene hybrid films. Small. 2011; 7: 95–100.
[181] Du F, Yu DS, Dai LM, Ganguli S, Varshney V, Roy AK. Preparation of tunable 3D pillared carbon nanotube-graphene networks for high-performance capacitance. Chemistry of Materials. 2011; 23: 4810–6.
[182] Kim YS, Kumar K, Fisher FT, Yang EK. Out-of-plane growth of CNTs on graphene for supercapacitor applications. Nanotechnology. 2012; 23: 015301.
[183] Chen SQ, Chen P, Wang Y. Carbon nanotubes grown *in situ* on graphene nanosheets as superior anodes for Li-ion batteries. Nanoscale. 2011; 3: 4323–9.
[184] Li SS, Luo YH, Lv W, Yu WJ, Wu SD, Hou PX et al. Vertically aligned carbon nanotubes grown on graphene paper as electrodes in lithium-ion batteries and dye-sensitized solar cells. Advanced Energy Materials. 2011; 1: 486–90.
[185] Duc Dung N, Tai NH, Chen SY, Chueh YL. Controlled growth of carbon nanotube-graphene hybrid materials for flexible and transparent conductors and electron field emitters. Nanoscale. 2012; 4: 632–8.
[186] Ma YW, Sun LY, Huang W, Zhang LR, Zhao J, Fan QL et al. Three-dimensional nitrogen-doped carbon nanotubes/graphene structure used as a metal-free electrocatalyst for the oxygen reduction reaction. Journal of Physical Chemistry C. 2011; 115: 24592–7.
[187] Paul RK, Ghazinejad M, Penchev M, Lin J, Ozkan M, Ozkan CS. Synthesis of a pillared graphene nanostructure: A counterpart of three-dimensional carbon architectures. Small. 2010; 6: 2309–13.
[188] Kim UJ, Lee IH, Bae JJ, Lee SJ, Han GH, Chae SJ et al. Graphene/carbon nanotube hybrid-based transparent 2D optical array. Advanced Materials. 2011; 23: 3809–14.
[189] Lee DH, Kim JE, Han TH, Hwang JW, Jeon SK, Choi SY et al. Versatile carbon hybrid films composed of vertical carbon nanotubes grown on mechanically compliant graphene films. Advanced Materials. 2010; 22: 1247–52.
[190] Dong XC, Li B, Wei A, Cao XH, Chan-Park MB, Zhang H et al. One-step growth of graphene-carbon nanotube hybrid materials by chemical vapor deposition. Carbon. 2011; 49: 2944–9.
[191] Deck CP, McKee GSB, Vecchio KS. Synthesis optimization and characterization of multiwalled carbon nanotubes. Journal of Electronic Materials. 2006; 35: 211–23.
[192] Yoshida H, Takeda S, Uchiyama T, Kohno H, Homma Y. Atomic-scale *in-situ* observation of carbon nanotube growth from solid state iron carbide nanoparticles. Nano Letters. 2008; 8: 2082–6.
[193] Yan J, Wei T, Shao B, Ma FQ, Fan ZJ, Zhang ML et al. Electrochemical properties of graphene nanosheet/carbon black composites as electrodes for supercapacitors. Carbon. 2010; 48: 1731–7.
[194] Zhou QQ, Gao J, Li C, Chen J, Shi GQ. Composite organogels of graphene and activated carbon for electrochemical capacitors. Journal of Materials Chemistry A. 2013; 1: 9196–201.
[195] Zhou XS, Yin YX, Cao AM, Wan LJ, Guo YG. Efficient 3D conducting networks built by graphene sheets and carbon nanoparticles for high-performance silicon anode. ACS Applied Materials & Interfaces. 2012; 4: 2824–8.
[196] Wang MX, Liu Q, Sun HF, Stach EA, Zhang HY, Stanciu L et al. Preparation of high-surface-area carbon nanoparticle/graphene composites. Carbon. 2012; 50: 3845–53.

[197] Yang YQ, Pang RQ, Zhou XJ, Zhang YJ, Wu HX, Guo SW. Composites of chemically-reduced graphene oxide sheets and carbon nanospheres with three-dimensional network structure as anode materials for lithium ion batteries. Journal of Materials Chemistry. 2012; 22: 23194–200.
[198] Lei ZB, Liu ZH, Wang HJ, Sun XX, Lu L, Zhao XS. A high-energy-density supercapacitor with graphene-CMK–5 as the electrode and ionic liquid as the electrolyte. Journal of Materials Chemistry A. 2013; 1: 2313–21.
[199] Sun X, He JP, Tang J, Wang T, Guo YX, Xue HR et al. Structural and electrochemical characterization of ordered mesoporous carbon-reduced graphene oxide nanocomposites. Journal of Materials Chemistry. 2012; 22: 10900–10.
[200] Seredych M, Bandosz TJ. S-doped micro/mesoporous carbon-graphene composites as efficient supercapacitors in alkaline media. Journal of Materials Chemistry A. 2013; 1: 11717–27.
[201] Li M, Ding J, Xue JM. Mesoporous carbon decorated graphene as an efficient electrode material for supercapacitors. Journal of Materials Chemistry A. 2013; 1: 7469–76.
[202] Moon GH, Shin YS, Choi DW, Arey BW, Exarhos GJ, Wang CM et al. Catalytic templating approaches for three-dimensional hollow carbon/graphene oxide nano-architectures. Nanoscale. 2013; 5: 6291–6.
[203] Chen SY, Tang CY, Chuang WT, Lee JJ, Tsai YL, Chan JCC et al. A facile route to synthesizing functionalized mesoporous SBA-15 materials with platelet morphology and short mesochannels. Chemistry of Materials. 2008; 20: 3906–16.
[204] Lei ZB, Bai SY, Xiao YL, Dang LQ, An LZ, Zhang GN et al. CMK-5 mesoporous carbon synthesized via chemical vapor deposition of ferrocene as catalyst support for methanol oxidation. Journal of Physical Chemistry C. 2008; 112: 722–31.

CHAPTER 5

Graphene-Based Fibers

Ling Bing Kong,[1,a,*] *Freddy Boey,*[1,b] *Yizhong Huang,*[1,c] *Zhichuan Jason Xu,*[1,d] *Kun Zhou,*[2] *Sean Li,*[3] *Wenxiu Que,*[4] *Hui Huang*[5] and *Tianshu Zhang*[6]

5.1. Brief Introduction

Carbon fibers have various advantages, such as light weight, high mechanical strength and environmental stability, which make them be useful in various fields in our daily life. Since the discovery of carbon nanotubes (CNTs), CNTs-based fibers/yarns emerged as a new type of carbon fibers, which possessed high mechanical strength, high electrical and thermal conductivities, as compared to the conventional carbon fibers [1–8]. Graphene is a 2D layer of carbon atoms with a honeycomb lattice configuration, which is also the building block of carbon materials [9]. Similar to CNTs, graphene has high electron mobility [9–14], high thermal conductivity [13] and excellent elasticity and stiffness [14]. Therefore, the assembly of graphene and graphene related components has been a hot research topic. Currently, graphene has been assembled into 1D (fibers, yarns, wires), 2D (papers, sheets, membranes) and 3D (gels, foams, sponges), which will be discussed from this chapter onwards [15].

Graphene fibers (GFs) have formed a new group of materials, which have been acknowledged to possess special significances, due to their unique properties. By assembling individual graphene nanosheets into the macroscopic fibers, their properties are combined, so as to provide extraordinary performances, such as high mechanical flexibility, low density and potential functionalization capability [16]. This chapter aims to summarize the significant progress in the fabrication and characterization of GFs during the last several years. Their potential applications in unconventional devices, such as photovoltaic cells, supercapacitors, flexible fiber-type actuators and so on, will be briefly evaluated and discussed, with representative examples.

[1] School of Materials Science and Engineering, Nanyang Technological University, Singapore.
[a] E-mail: elbkong@ntu.edu.sg
[b] E-mail: mycboey@ntu.edu.sg
[c] E-mail: yzhuang@ntu.edu.sg
[d] E-mail: xuzc@ntu.edu.sg
[2] School of Mechanical & Aerospace Engineering, Nanyang Technological University, Singapore; E-mail: kzhou@ntu.edu.sg
[3] School of Materials Science and Engineering, The University of New South Wales, Australia; E-mail: sean.li@unsw.edu.au
[4] Electronic Materials Research Laboratory, School of Electronic and Information Engineering, Xi'an Jiaotong University, P. R. China; E-mail: wxque@xjtu.edu.cn
[5] Singapore Institute of Manufacturing Technologies (SIMTech), Singapore; E-mail: hhuang@SIMTech.a-star.edu.sg
[6] Anhui Target Advanced Ceramics Technology Co. Ltd., Hefei, Anhui, P. R. China; E-mail: 13335516617@163.com
* Corresponding author

5.2. Pure Graphene Fibers

5.2.1. Solution processing from graphene oxide (GO)

It has been found that soluble chemically oxidized graphene or GO nanosheets can form liquid crystals (LCs) in a twist-grain-boundary (TGB) phase-like model with simultaneous lamellar ordering and long-range helical frustrations, which can be continuously spun into macroscopic GO fibers [17–20]. Due to the presence of the lamellar structures, the GO LCs can reach concentrations that are sufficiently high for efficient alignment and effective coagulation. A simple syringe injection method was used to fabricate neat GO fibers, with 5 wt% NaOH-methanol solution as the coagulation bath [17]. GFs could be obtained by simply reducing the GO fibers in hydroiodic acid. The GFs exhibited promising mechanical strengths of 140 MPa at an ultimate elongation of 5.8% and high conductivity of 2.5×10^4 S m^{-1}. They had high mechanical flexibility for the design of patterns and complex textiles.

Figure 5.1 shows cryo-SEM images of the GO chiral LCs (CLCs) (a–f) [17]. When confined to a circular cavity, the GO CLCs displayed a freeze-fracture morphology as an annual ring-like structure, consisting of undulating bands with distinct boundaries, as seen in Fig. 5.1 (a). The undulating morphology was formed due to the presence of bands with neighboring orthogonal vectors along the radius, as the GO nanosheets were stacked up. One-half of the pitch ($p/2$) of the helix was used to describe the spacing between adjacent bands, as illustrated in Fig. 5.1 (e). Due to the micrometer scaled lateral width of the GO nanosheets, the pitches had a typical size of about 42 μm, which was larger than that of the giant TGB phases formed with

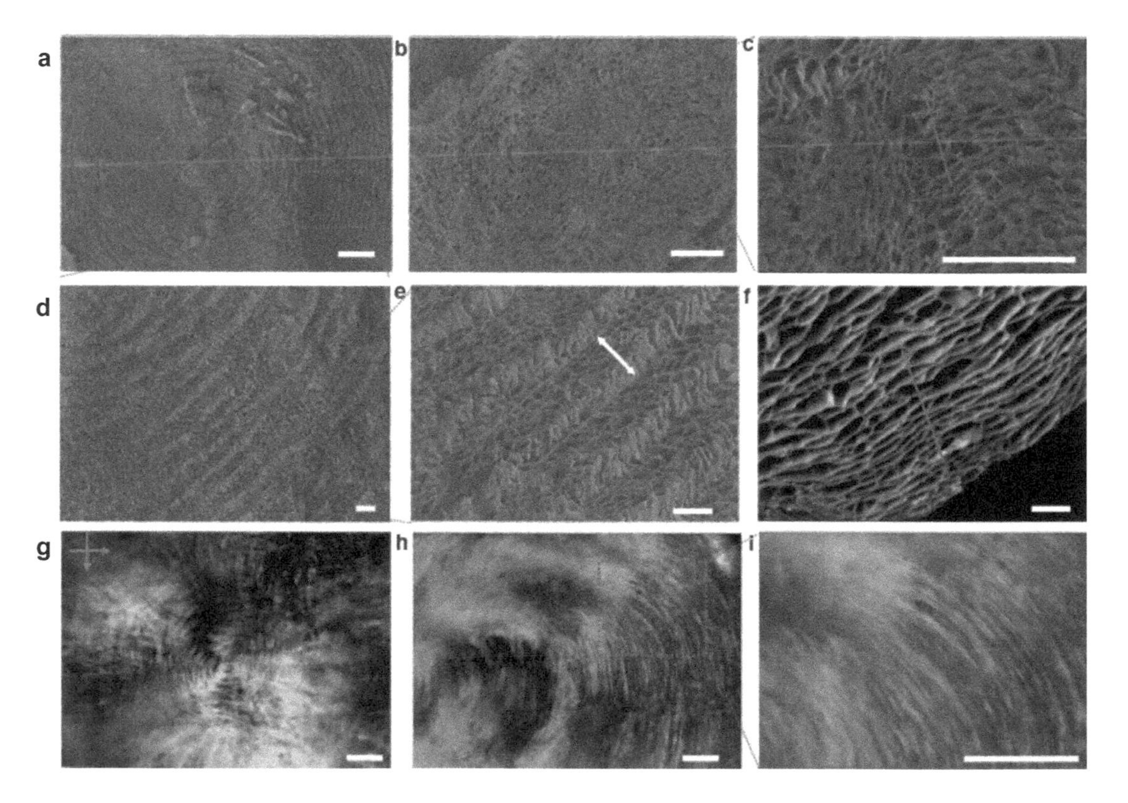

Fig. 5.1. Cryo-SEM images and POM textures of the GO CLCs. (a–e) Top-view SEM images showing the fracture morphology of the GO CLCs at a volume fraction φ = 0.98% confined in a circular cavity. The red crosses denote vectors (*ns*) into the paper and the orange arrows on the paper. (b) Central domain in (a). (c) Screw dislocation of neighboring GO lamellar blocks with twist vectors. (f) Side-view SEM image of the assembled GO blocks. (g–i) POM images between crossed polarizers of GO CLCs in the central (g) and lateral (h, i) domains. The regular ring textures (g–i) are in accordance with the annual-ring undulating fracture morphology (a–e) of the GO CLC. Scale bars = 200 μm (a), 25 μm (b–f) and 300 μm (g–i). Reproduced with permission from [17], Copyright © 2011, Macmillan Publishers Limited.

small molecules (0.1–1.2 μm) [21]. The lamellar feature with regular interlayer spacing of the assembled GO blocks is demonstrated in Fig. 5.1 (f). Polarized optical microscopy (POM) images of representative samples are shown in Fig. 5.1 (g–i), in which highly regular fingerprint-like and focal conic textures were present. The pitch of the GO nanosheet helix was decreased as the GO concentration was increased.

A TGB-like model has been proposed to explain the formation of the GO CLCs, as schematically illustrated in Fig. 5.2 [17]. GO nanosheets were assembled into lamellar blocks with a local vector (n) in normal direction, while the distance (d) between adjacent graphene nanosheets was decreased as the value of φ was increased, as illustrated in Fig. 5.2 (top left). Due to the negative charges at the boundaries and on the surfaces, the adjacent GO nanosheets would repel each other, thus leading to the formation of rotated configuration turned by an angle 7.0°, so as to minimize the free energy of the system (bottom right panel of Fig. 5.2). At the same time, the blocks were rotated successively in the same direction, either clockwise or anticlockwise, along the helical axis, so that a helical arrangement was formed, as demonstrated in Fig. 5.2 (center panel). Therefore, the formation of the orderly frustrated structures in the chiral phases was attributed to the negatively charged grain boundaries of the GO nanosheets which produced helical dislocations in the lamellar fluids. There were two twisted neighboring lamellar blocks, as shown in Fig. 5.1 (c), similar to that used to describe the TGB phase of conventional small molecules [22].

Concentration of the GO LCs played a crucial role in determining whether fibers could be spun out or not [17]. For example, the GO LCs with a concentration of 0.76 led to brittle fibers, while the 2.0% samples were collapsed belts. Continuous fibers were only available when the LC dispersions had a sufficiently high concentration, e.g., $\varphi = 5.7$ %. Diameters of the fibers could be controlled in the range of 50–100 μm, while the spinning rate was hundreds of centimeters per minute, by adjusting the size of nozzle and the drawing speed, with examples shown in Fig. 5.3 (a, b). Figure 5.3 (c) indicated that the fibers were not broken as the knots were tightened. As compared with the traditional carbon fibers, the GO fibers had much higher flexibility and torsion resistance.

The GO fibers exhibited a typical plastic deformation at room temperature, which was attributed to the possible stretching of the crumbled GO nanosheets, as well as the displacements of the GO nanosheets (Fig. 5.3 (d)). The fibers possessed a Young's modulus of 5.4 GPa at small deformation in the elastic region and demonstrated fracture elongations of 6.8–10.1%, which were greater than that of the filtrated GO papers

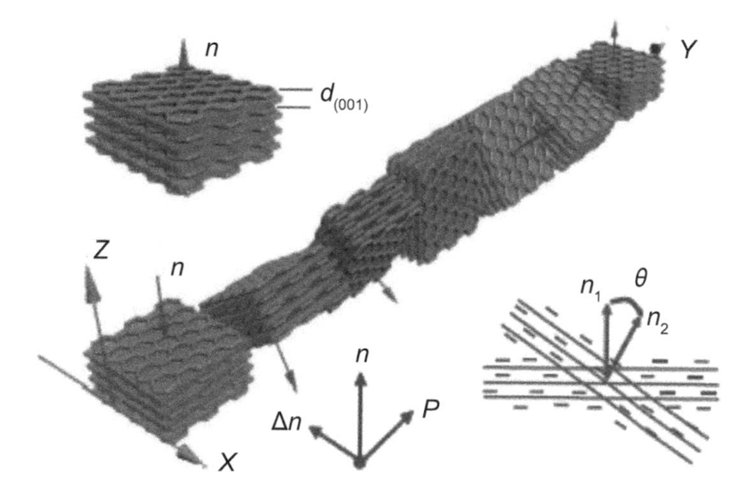

Fig. 5.2. Proposed model of one pitch of the GO CLCs. The vectors (*ns*) of the lamellar blocks are rotated anticlockwise along the helical axis. Top left: the lamellar structure of the organized GO block with an identical n and a regular spacing d, the semitransparent blue blocks represent the interlayer water and the GO nanosheets are indicated as pink nets (double bonds in the basal planes and pedant functional groups have been omitted for clarity). Bottom right: screw dislocation between two neighboring grain boundaries of the GO lamellar blocks (the red line denotes GO nanosheet boundaries). The negative charges (blue minus) spread on the surface and boundaries of the GO nanosheets, make two blocks repulse each other, with an angle θ between two vectors (n_1 and n_2). Reproduced with permission from [17], Copyright © 2011, Macmillan Publishers Limited.

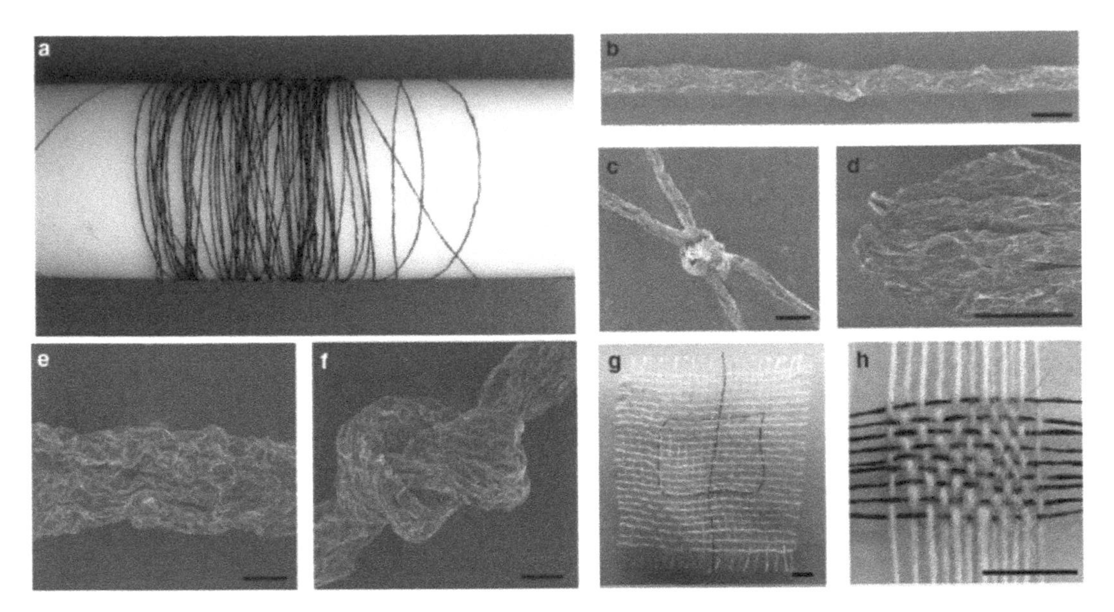

Fig. 5.3. Macroscopic GO fibers and chemically reduced graphene fibers. (a) Photograph of a four-meter-long GO fiber wound on a Teflon drum (diameter = 2 cm). (b) SEM image of the fiber and (c) typical tightened knots. (d) Fracture morphology of the GO fiber after tensile tests. The surface wrinkled morphology (e) and the tightened knot (f) of the graphene fiber. (j) A Chinese character ('中', Zhong) pattern knitted in the cotton network (white) using two graphene fibers (black). (k) A mat of graphene fibers (horizontal) woven together with cotton threads (vertical). Scale bars = 50 μm (b–f) and 2 mm (g, h). Reproduced with permission from [17], Copyright © 2011, Macmillan Publishers Limited.

(~0.4%) by more than one order of magnitude [23]. In addition, the GO fibers showed a fracture strength of 102 MPa, which was within the range of filtrated GO papers (70–130 MPa) [24].

After the GO fibers were reduced through chemical reduction in 40% hydroiodic acid, the fibers shrunk in diameter, accompanied by the presence of a gray metallic lustre. Meanwhile, the interlayer spacing was decreased from 8.9 Å ($2\theta = 9.86°$) to 3.7 Å ($2\theta = 24.25°$), according to the XRD analysis results, with the latter being close to the interlayer spacing of graphite (3.35 Å). The reduction led to an increase in both the Young's modulus (7.7 GPa) and fracture strength (140 MPa), while a fracture elongation of about 5.8% was also retained. The enhancement in mechanical strength was attributed to the stronger interactions between the graphene nanosheets, due to the much denser layer stacking. Similarly, crumbled graphene nanosheets were stretched when subjected to tensile stress, which resulted in the high fracture elongation of the graphene fiber, as illustrated in Fig. 5.3 (e). High flexibility of the graphene fibers was clearly demonstrated in Fig. 5.3 (f–h).

Although lyotropic LCs have been used to fabricate macroscopic fibers by using wet-spinning technology, successes were only achieved in linear rigid polymers (e.g., aromatic polyamides) and 1D nanostructures (e.g., CNTs) [6, 7, 25]. Such a phenomenon was ascribed to the noncovalent chemical bonds, including van der Waals forces and hydrogen bondings, strong mechanical chain entanglements and interconnection among different tubes. No fibers were obtained from the LCs of 2D colloids, mainly because the interactions between the solid-like flakes were too weak and the solubility of the colloids was not sufficiently high to support continuous spinning. In this respect, the GO and graphene fibers were the first examples reported in the open literature. This was because the LC dispersions of the GO nanosheets had high solubilities. The GFs showed strong mechanical strength as the nanosheet alignment of the intrinsic lamellar order of LCs was well-retained, so that the nanosheets had strong interactions. At the same time, the locally crumbled structures of the individual nanosheets ensured the flexibility of the GFs. Such GFS, with high electrical conductivity and promising mechanical strength, could find a wide range of applications in different fields, like chemical sensors and smart textiles.

The above technique was further extended for the fabrication of GO aerogel fibers with a "porous core-dense shell" structure, by combining spinning and ice-templating, from GO LCs [26]. Due to the interior uniform alignment of the GO nanosheets, the fibers possessed high tensile strength and high compression strength. The porous GO fibers could be reduced through chemical reduction and annealing process to obtain graphene fibers, without losing the porous structure, while gaining high electrical conductivity. This unique type of fibers had both high porosity and high mechanical strength and electrical conductivity, which is considered to be a conflict in most general materials.

The GO nanosheets used to prepare LCs had a thickness of 0.8 nm (t) and an average lateral width (w) of 2.0 μm, as shown in Fig. 5.4 (a). As a result, they possessed a huge aspect ratio (w/t) of > 2000, which ensured the formation of lyotropic LCs. With the increasing concentration of GO, there was a transition from nematic phase to lamellar phase [17, 20]. Figure 5.4 (b) shows that the GO gels loaded in a planar cell exhibited vivid banded textures between crossed polarizers under POM, demonstrating a typical lamellar optical texture with line disclinations. 2D synchronous small-angle X-ray scattering (SAXS) pattern of the GO gel, with a mass fraction f_m of 10.1%, corresponding to a volume fraction φ of 7.1%, is shown as the inset in Fig. 5.4 (b). The presence of the scattering loop revealed the lamellar structural character of the GO gels. The lamellar spacing (d_s) of two adjacent GO nanosheets was 11.2 nm, which was in a good agreement with the value calculated according to the 1D swelling model for sheet-shaped colloids, i.e., $d_s = t/\varphi = 11.3$ nm [27].

Figure 5.4 (c) shows a schematic diagram of the ice-templating process to produce porous GO fibers (GOPFs) [26]. The first step was to extrude the GO LC gels into liquid nitrogen, while the second step was the freeze-drying of the samples. To obtain continuous GOPFs, the concentration of the GO gels and the diameter of the nozzle were very important. GO gels with low concentrations and a small nozzle diameter could not lead to continuous gel fibers. It was found that the optimized concentration of the GO gel was at $f_m \approx 10.1\%$, while the diameters of the nozzles could be varied from 60 μm to 1 cm, thus resulting

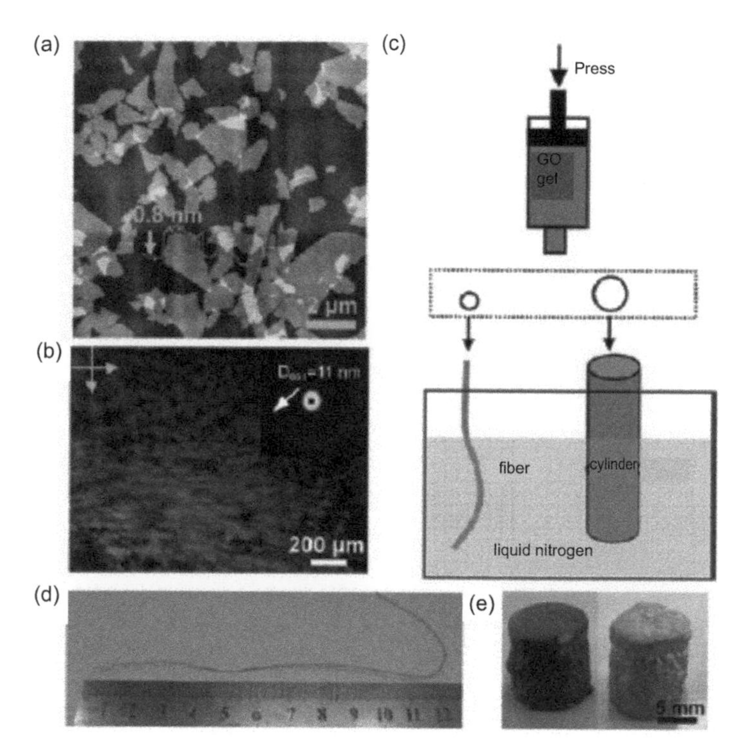

Fig. 5.4. (a) AFM image of the GO nanosheets deposited on mica. (b) POM microscopic image of the GO gels with f_m = 10.1% between crossed polarizers, with the inset to be the 2D SAXS pattern. (c) Schematic diagram showing the preparation process of the GO porous fibers (GOPFs) and the GO porous cylinders (GOPCs). Photographs of GOPF (d), GOPC (e, left) and rGPC (e, right). Reproduced with permission from [26], Copyright © 2011, American Chemical Society.

in GOPFs with variable diameters. Photographs of the GOPF with a diameter of ~100 μm and cut GO porous cylinders (GOPC) of about 1 cm in diameter are demonstrated in Fig. 5.4 (d, e). As illustrated in Fig. 5.4 (e), after the GOPFs and GOPCs with dull black features were reduced, the products of rGPFs and rGPCs exhibited metallic lustre. Meanwhile, the density of the samples was decreased from 110 mg·cm^{-3} of GOPFs to 71 mg·cm^{-3} of rGPFs and 56 of annealed rGPFs. Correspondingly, interlayer spacing was reduced from 0.828 nm to 0.363 nm and 0.342 nm, respectively. The rGO nanosheets showed more regular packing after annealing.

The rGPFs were characterized by a unique structure, which consisted of a wrinkled shell surface of densely packed graphene nanosheets and a porous part with interconnected graphene nanosheets enclosed by the dense shell [26]. Such rGPFs possessed very high flexibility. For example, they could be folded into a shape with an angle of 60° (Fig. 5.5 (c)), while the bent fibers were able to recover their original straight shape without fracture (Fig. 5.5 (d)). In the porous core of rGPF, the reduced graphene nanosheets were interconnected and aligned in the axial direction, thus leading to axially aligned empty cells, with a width of about 2–5 μm, as demonstrated in Fig. 5.5 (e–h). The high level orderings in the original lamellar GO LC gels were mainly responsible for the formation of the highly aligned porous microstructures. It was demonstrated in Fig. 5.5 (h) that the aligned pores were formed by the graphene nanosheets that were normal to the sheet plane and vertical to the fiber axis, i.e., the flowing direction. Figure 5.5 (i) shows a TEM image of the ribbon-like graphene nanosheets in porous cell walls of the rGPFs. HRTEM images (Fig. 5.5 (j, k)) revealed that the thickness of the strips was about 2 nm, corresponding to 3–5 graphene layers. Microstructure of the aligned porous graphene fibers was described by using the structural model schematically shown in Fig. 5.5 (l).

Because the fluid of the anisotropic 2D GO colloids was a shear flow, alignment was present along the flowing direction [5, 28]. Figure 5.6 shows a POM image of the GO gels in the form of powder LC, with lamellar ordered domains that were separated by line disclinations, as is schematically illustrated by

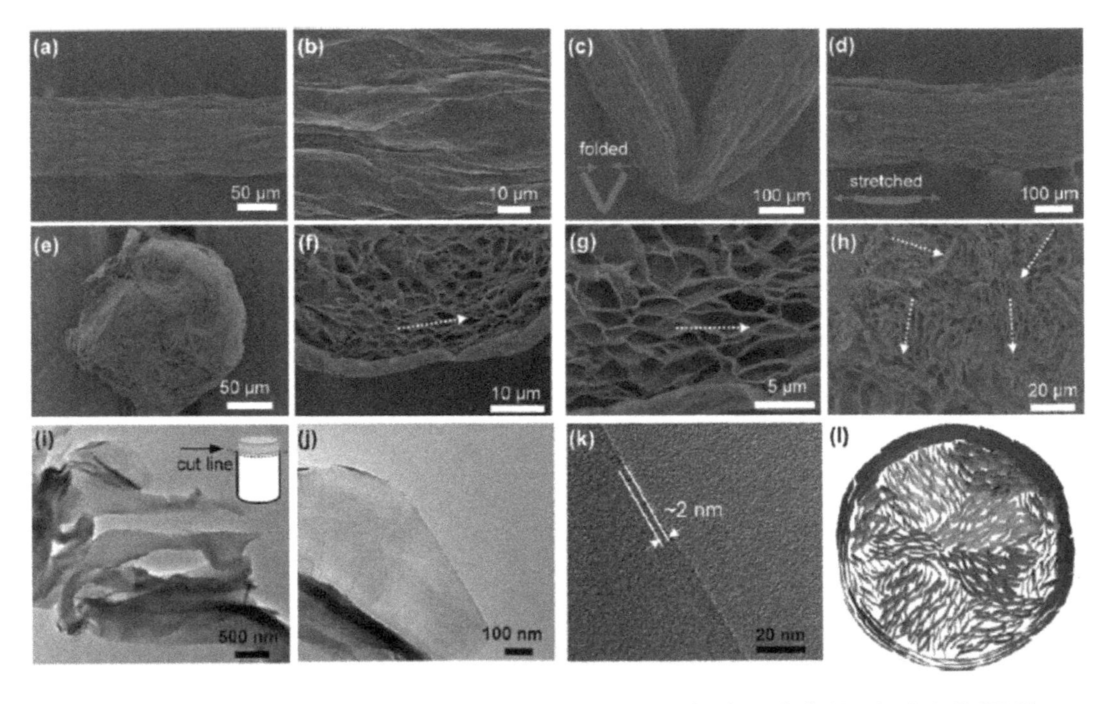

Fig. 5.5. Surface SEM images of the rGPF (a, b) and SEM images of the folded and stretched rGPFs (c, d). (e–h) SEM images of the fracture morphologies of the rGPFs. The arrows indicate the alignment direction of the rGO nanosheets. (i–k) TEM images of the fractured rGPFs embedded in epoxy resin. The inset in (i) shows the way to prepare TEM samples with cryo-cutting. (l) Schematic diagram of the core-shell structure model of the rGPFs and GOPFs. Regions of different colors represent GO lamellar regions with different director vectors in the core, while the red color denotes the dense shell. Reproduced with permission from [26], Copyright © 2011, American Chemical Society.

the structural model in Fig. 5.6 (b). During flowing of the GO gels, the GO nanosheets were aligned due to the capillary effect in the fluid direction. Such an alignment effect was well-evidenced by the uniform birefringence between crossed polarizers in the POM images in Fig. 5.6 (d–g). All the director vectors were perpendicular to the capillary surface. In the neighboring region of the inner capillary surface, the surface tension forced the GO nanosheets to form lamellar arrays that were parallel to the surface. Assembling of the GO nanosheets resulted in the tube-like shells, due to the dynamic templating of the inside wall of the spinning tube.

The graphene aerogel fibers exhibited a type IV nitrogen adsorption/desorption isotherm. The presence of a hysteresis loop at high relative pressure implied that there was a large fraction of mesopores in the aerogels. Sizes of the pores were in the range of 3.4–100 nm, with a pore volume of 5.85 $cm^3 \cdot g^{-1}$. Therefore, the aerogel fibers contained hierarchical pores. The fibers also had a quite high BET surface area that was up to 884 $m^2 \cdot g^{-1}$, nearly one-third of the theoretical surface area of monolayer graphene (2600 $m^2 \cdot g^{-1}$). Correspondingly, the average number of the graphene layers was close to 3, in a good agreement with the TEM results. As mentioned earlier, the mechanical properties of the fibers were not compromised by the highly porous structure, due to the unique core-shell configurations.

Plastic deformation of the GOPFs and rGPFs consisted of two stages, i.e., stretching and displacement. Representative fracture morphologies of the rGPFs are shown in Fig. 5.7 (a–f). As seen in Fig. 5.7 (a, b), the fracture surface exhibited a certain degree of elongation, implying that the rGPFs experienced an elastic breaking under tension. The stretching and displacement mechanisms were evidenced by the typical stretched fibrils at the fracture tips and distinct displacement lines of the slide graphene nanosheets, as demonstrated in Fig. 5.6 (c). Both cracks in the fracture tips (Fig. 5.7 (e)) and displacements (Fig. 5.7 (f)) were observed. It was suggested that sheet displacement was the dominating deformation in

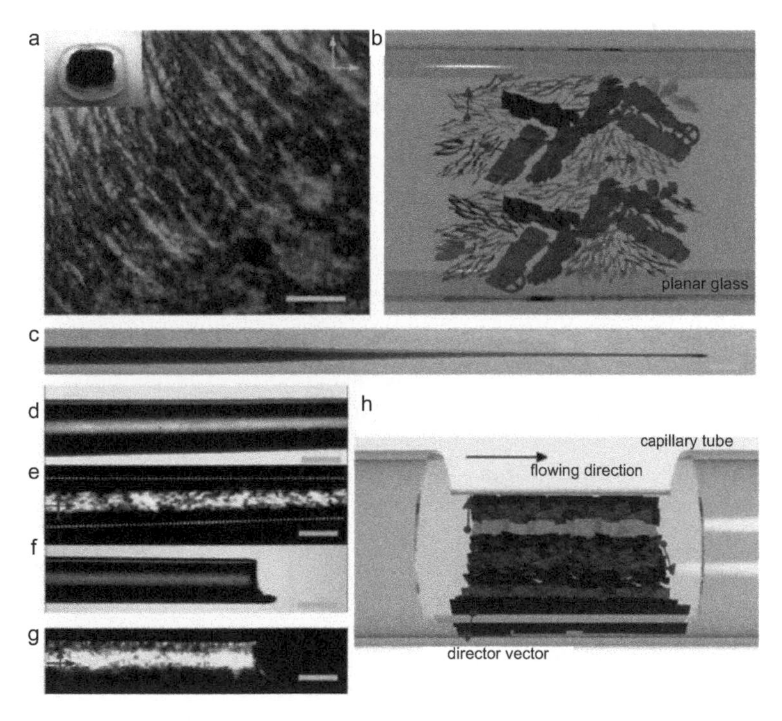

Fig. 5.6. (a) POM image of the GO gel powder between crossed polarizers, with the inset showing a photograph of the GO LC gel in a planar cell. (b) Structural model of the powder GO LC gels. The arrows and the crosses indicate director vectors on and into the observation plane, respectively. (c) Photograph of the flowing GO gel in a capillary tube. (d, f) Optical microscopic images of the flowing GO gels in a capillary tube at different sections. (e, g) POM images between crossed polarizers corresponding to panels (e) and (f), respectively. (h) Structural model of the flowing GO LC gels with flow-induced alignments of the GO nanosheets, demonstrating the formation of the core-shell structure under confined flowing. All the vectors are perpendicular to the capillary surface. Scale bars = 200 μm (a, d–g) and 1 cm (c). Reproduced with permission from [26], Copyright © 2011, American Chemical Society.

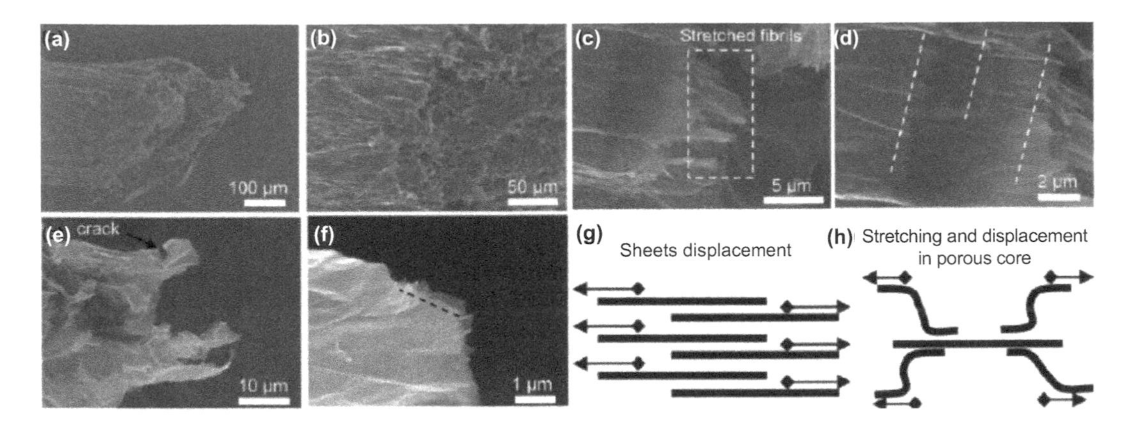

Fig. 5.7. SEM images of the rGPF surface (a, b) and failure morphology after tensile test (c, d). The lines in panels (d) and (f) indicate the slide lines of the rGO nanosheets. Deformation mechanism models of the shell (g) and the porous core (h) in the rGPFs, with the black lines representing the rGO nanosheets. Reproduced with permission from [26], Copyright © 2011, American Chemical Society.

the dense shells, while the porous cores experienced both stretching and displacement deformations, with models shown schematically in Fig. 5.7 (g) and (h), respectively. Due to the strong compression strength in the diametrical direction, all the porous cylinders, i.e., GOPCs, rGPCs and annealed rGPCs, could be cut into disks without crashing.

A special type of biomimetic composite fibers have been fabricated by using the wet-spinning method combined with the liquid crystal self-templating (LCST) approach [29]. In the composite fibers, guest compounds, such as polymers and inorganic nanoparticles, could be homogeneously dispersed in the inter-channels of the LC GO nanosheets. The nacre-mimetic fibers exhibited a highly ordered hierarchical structure. The concurrent effect of ultrahigh aspect ratio, well-preserved alignment of the GO nanosheets, uniform single-molecule interlayer of polymers and the presence of hydrogen bonding arrays ensured the formation of the macroscopic assembled artificial nacre, with a high σ of 555 MPa and a toughness of 18 $MJ\cdot m^{-3}$, which were 2–17 times higher than that of natural nacre (80–135 MPa and 0.1–1.8 $MJ\cdot m^{-3}$) [30, 31]. With an ultralow density of 1.0 $g\cdot m^{-3}$, the fibers possessed an optimal specific strength as high as 652 $N\cdot m\cdot g^{-1}$, which was nearly two times that of most metals and alloys. The reduced composite fibers possessed a high electrical conductivity of 5261 $S\cdot m^{-1}$.

Schematic diagram of the LCST strategy is shown in Fig. 5.8 (a–c). In this case, giant graphene oxide (GGO) nanosheets, with an average aspect ratio of as large as 1.6×10^4, were first synthesized. With such GGO nanosheets, lyotropic LCs could be formed at a very low concentration of 0.5 $mg\cdot ml^{-1}$. In aqueous solution of the GGO LCs, various guest compounds, such as polymers and inorganic nanoparticles, could be simply included, leading to host-guest composite LCs, in which the guest compounds were uniformed distributed in the interlayer channels of the GGO nanosheets. Typical LC behaviors of both the GGO and the composites were evidenced by POM images, as shown in Fig. 5.8 (d, e). The approach could be used to fabricate biomimetic fibers at large scale, with an example demonstrated in Fig. 5.8 (f). Since the guests were tightly trapped inside the interlayers of the GGO nanosheets, phase separation was effectively prevented. Due to their high flexibility, the fibers could be woven into textile items, as illustrated in Fig. 5.8 (g). Hyperbranched polyglycerol (HPG) was used to fabricate composite fibers.

Figure 5.9 (a) shows a photograph of the fluid near the nozzle under crossed polarizers. LC textures were observed both in the capillary and at the tip, as seen in Fig. 5.9 (b, c), where the degree of alignment of the GGO nanosheets was increased with the decreasing distance towards the tip of the nozzle, i.e., decreasing diameter. Once the fluid was injected into the coagulation solution, the diameter of the gel fibers was gradually reduced, while the alignment of the GGO nanosheets was still retained, so that the final solid fibers possessed a regularly compact brick-and-mortar (B&M) microstructure. Representative *in situ* POM and SEM images are shown in Fig. 5.9 (d–i). The solid fibers after wet-stretching showed

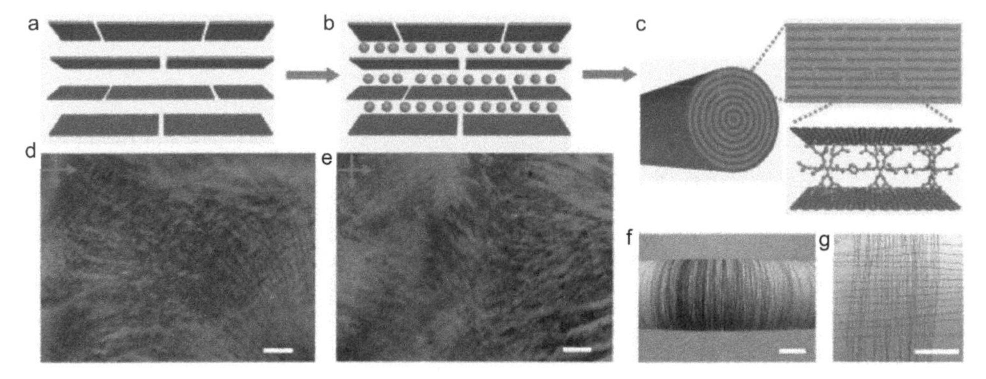

Fig. 5.8. Schematic diagram illustrating the LCST process and examples of the resulting composite fibers. (a–c) LCST protocol to prepare the host-guest layer-structured composites, including (a) formation of the host LCs of the 2D nanoplatelets with uniform nanochannels, (b) incorporation of guest compounds, e.g., polymers, biomacromolecules and nanoparticles, into the host nanochannels to form the host-guest complex LCs and (c) wet-spinning assembly of the complex LCs into nacre-mimetic fibers with hierarchical structures. (d) POM images of the host GGO LCs and (e) host-guest complex GGO-HPG LCs with a concentration of 4 mg·ml^{-1}. (f) Photographs of a thirty meters long GGO-HPG fiber and (g) a mat of GGO-HPG fibers. Scale bars = 300 mm (d, e) and 1 cm (f, g). Reproduced with permission from [29], Copyright © 2013, Macmillan Publishers Limited.

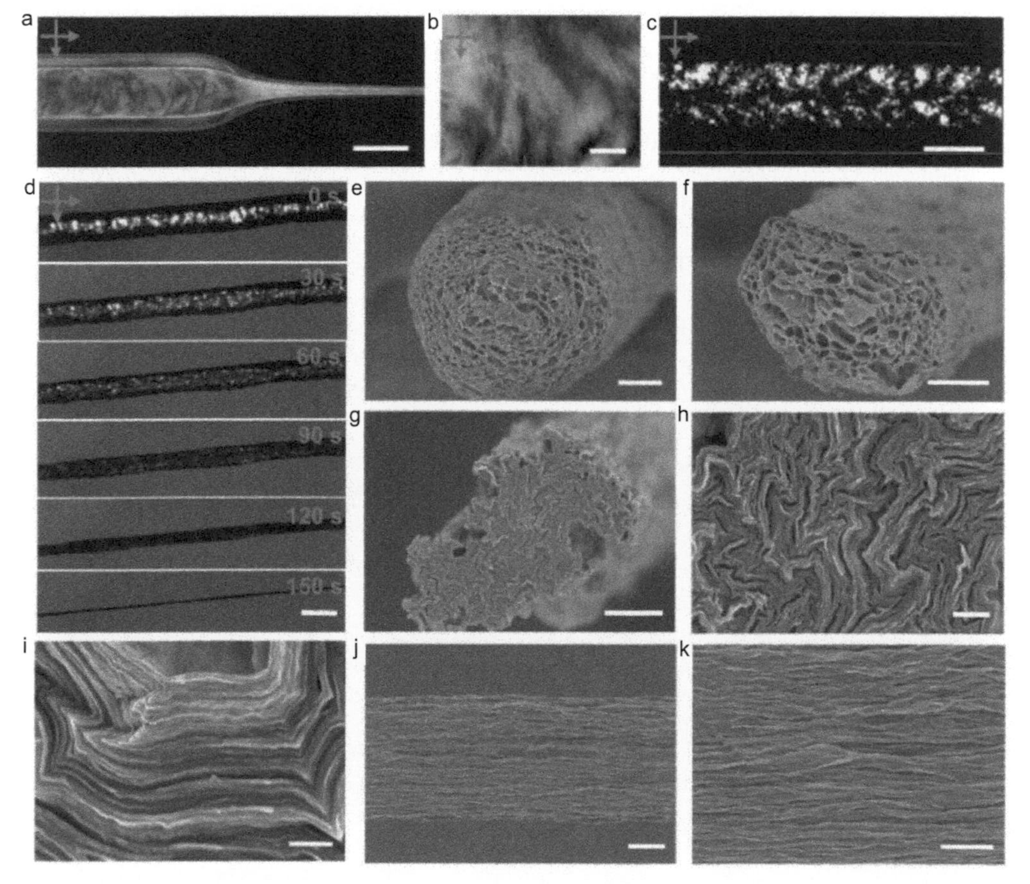

Fig. 5.9. Structural evolutions of the GGO-HPG suspensions during spinning. (a) Photograph of the GGO-HPG spinning solution inside the glass spinning head under crossed polarizers. (b–d) POM images of the GGO-HPG spinning solution in the capillary (b), at the spinning nozzle (c) and the drying procedure of the GGO-HPG fiber (d). (e–k) Cross-sectional SEM images of the GGO-HPG gel fibers at coagulation times of 0 min (e), 3 min (f) and the dried fiber (g–i). (j, k) Surface morphology of the resulting fibers. Scale bars = 5 μm (a), 500 μm (b), 250 μm (c), 100 μm (d), 30 μm (e), 25 μm (f), 3 μm (g, k), 500 nm (h), 250 nm (i) and 5 μm (j). Reproduced with permission from [29], Copyright © 2013, Macmillan Publishers Limited.

unhindered surfaces with spreading ridges along the axial direction (Fig. 5.9 (j, k)). The contraction in the radial direction led to the dense alignment of the GGO-HPG fibers. Figure 5.9 (h, i) indicated that the GGO nanosheets and the HPG molecules were piled alternatively.

The strong interactions between the functional groups contained in the GGO nanosheets and the HPG molecules resulted in high mechanical strength of the fibers. At low tensile loadings, the relatively weak layers of the HPG started to response. As the force was increased to a certain level, the action was transferred from the HPG layers to the GGO nanosheets. In this case, both the GGO nanosheets and the HPG interlayers would have shearing and sliding movement with respect to one another. After that, the hydrogen bonding networks between the adjacent GGO nanosheets and the HPG interlayers would be reformed and destroyed. Ultimately, failure of the fibers occurred by pull-out and fracture of the GGO nanosheets, as a result of the shear spreading, as shown in Fig. 5.10 (a–f). More importantly, the mechanical performances of the GGO-HPG fibers could be further enhanced by using glutaraldehyde (GA) to bridge the -OH groups of HPG and the GGO nanosheets. The GGO-HPG-GA fibers remained layered structures, with σ being increased by 17.5% to 652 MPa.

A one-step method to fabricate graphene fibers on a large scale, through *in situ* reduction in basic coagulation baths was reported [32]. Liquid crystallinity of the suspension was dependent on the concentration and size of the GO nanosheets, which had a direct effect on the processability of the LCs. Two types of fiber wet-spinning systems were employed to develop the graphene fibers. Spinnability of the GO dispersions was evaluated by using the "petri dish method", in which the GO dispersion was injected at flow rates of 5–10 $ml\cdot h^{-1}$ into a rotating petri dish with a coagulation bath that was rotated at 30–60 rpm. A custom-built wet-spinning apparatus was also used to study the continuous wet-spinning process. All the as-spun gel fibers were washed with 25 vol% ethanol-water and then air-dried at room temperature under tension, so as to obtain dried GO fibers. Two methods, i.e., (i) overnight annealing at 220°C in vacuum and (ii) hydrazine vapor treatment at 80°C for 3 h, were conducted to reduce the GO fibers to rGO fibers.

Ultra-large GO nanosheets, with lateral sizes of up to 100 μm, were necessary to support a continuous spinning (Fig. 5.11) [32]. Figure 5.11 (a) shows a representative AFM image of the GO nanosheets, confirming that the GO nanosheets were monolayer with an average thickness of about 0.81 nm. The highly-wrinkled feature of the GO nanosheets implied the formation of hydrogen bonds among the oxygenated functional groups, which in turn facilitated their full delamination in aqueous dispersions. In a polydisperse GO dispersion, large GO nanosheets generated excluded volume for small sheets, leading to entropic rearrangement to form a long-range ordering similar to a liquid crystalline state. There was a critical volume fraction (Φ) at which the transition from isotropic to nematic phase took place. For example,

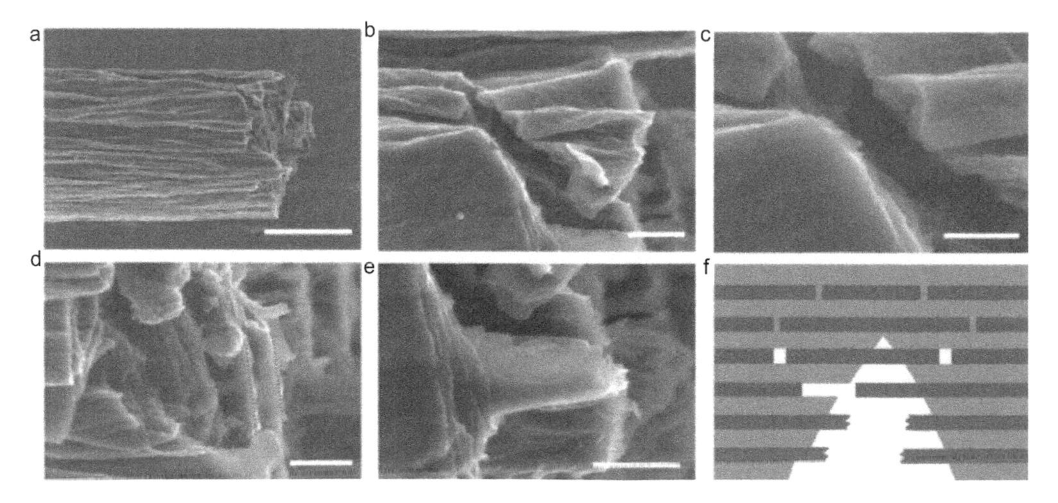

Fig. 5.10. Fractured sectional morphologies of the GGO-HPG fibers. (a–e) SEM images of the fractured surfaces of the GGO-HPG fibers at different magnifications. (f) Deformation mechanism model of the GGO-HPG fibers under tensile stress. Scale bars = 5 μm (a), 500 nm (b, d, e) and 250 nm (c). Reproduced with permission from [29], Copyright © 2013, Macmillan Publishers Limited.

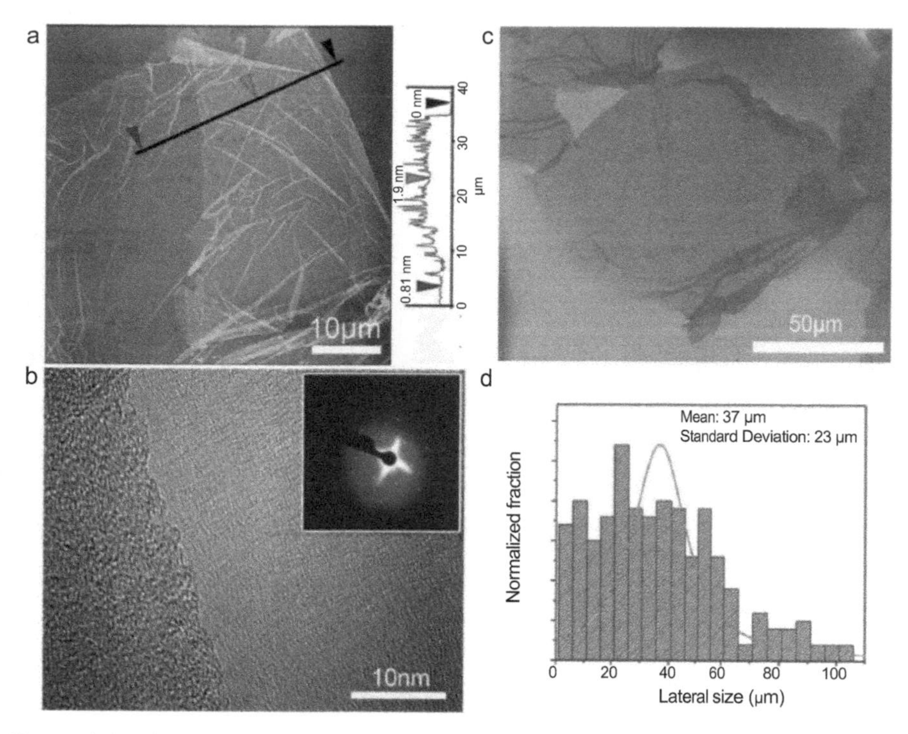

Fig. 5.11. Characteristics of the ultra-large GO nanosheets. (a) AFM image of a large GO nanosheet with a lateral size of > 75 μm, with the inset showing the height profile corresponding to the line in the AFM image. (b) HRTEM image of a monolayer GO nanosheet and the corresponding electron diffraction pattern (inset). (c) A representative SEM image of the GO nanosheets in dispersions with lateral sizes of up to 100 μm. (d) Lateral size distribution of the GO nanosheets (the diameter of an equal-area circle). Reproduced with permission from [32], Copyright © 2013, WILEY-VCH Verlag GmbH & Co. KGaA, Weinheim.

GO dispersions with concentrations below 0.25 $mg \cdot ml^{-1}$ were completely isotropic, and were unspinnable, with the GO nanosheets being simply spread into the coagulation bath. Transition from biphasic (i.e., isotropic and nematic phases co-existed) to fully nematic phase was observed in the suspensions at the concentrations of 0.25–0.75 $mg \cdot ml^{-1}$. In this concentration range, suspensions were partially spinnable, with the samples being characterized by weak cohesion among the GO nanosheets upon injection into the coagulation bath, so that only short segments were obtained. Not surprisingly, in this concentration range, the nematic phase volume fraction (φ_{nem}) was increased with the increasing GO concentration. As the GO concentration was increased to be ≥ 0.75 $mg \cdot ml^{-1}$, fully nematic phase was present, and continuous robust gel fibers could be spun out. The spinnability was affected by the GO concentration in the range of 0.75–5 $mg \cdot ml^{-1}$. Once the LC behavior is reached, high spinnability is always available, almost independent of coagulation methods. A wide range of coagulation methods can be used, such as (i) non-solvent precipitation, (ii) dispersion destabilization with acid, (iii) base or salt solutions, (iv) ionic cross-linking with divalent cations and (v) coagulation with amphiphilic or oppositely charged polymers.

Figure 5.12 (a) shows a POM image of the GO gel fibers displayed, with birefringence being clearly observed, confirming the highly ordered GO domains that were retained from the LC suspensions. It is well-known that the formation of the mono-domains of GO nanosheets requires a long timescale, if no external driving force is applied. One way to promote the alignment of LC GO is the application of external magnetic field. In this case, it took 5 h to form uniform LCs [32]. In contrast, macroscopic alignment during wet-spinning occurred almost immediately (within several seconds), which was ascribed to the action of the shear stress generated through the spinneret, because of the 2D feature of the GO nanosheets. Microstructure analysis indicated that, in the dry fibers, the GO nanosheets were stacked in a layer by layer way, with only a slight degree of folding. Meanwhile, they were orientated in the direction of the fiber axis, as revealed in Fig. 5.15 (b–d).

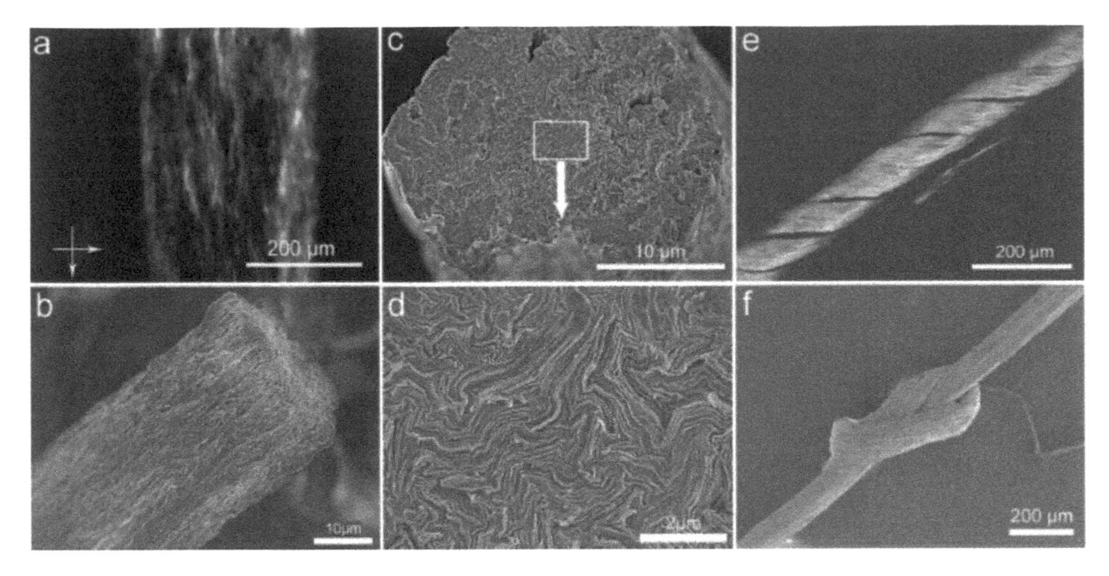

Fig. 5.12. (a) Polarized optical microscopy (POM) image of the as-spun GO gel fiber showing birefringence that confirmed the presence of ordered LC domains, with the arrows indicating the direction of the polarizers. SEM images of an as-spun GO fiber: (b) corrugated surface and (c) near-circular cross-section. (d) Close view SEM image of the cross-section of the GO fiber shown in (b) revealing GO nanosheet planes that were oriented along the fiber axis. (e, f) SEM images of the crumpled and knotted rGO yarns (reduced by annealing) showing their high flexibility. Reproduced with permission from [32], Copyright © 2013, WILEY-VCH Verlag GmbH & Co. KGaA, Weinheim.

A similar wet-spinning assembly approach was reported to fabricate graphene fibers with GO solutions combined with chemical reduction [33]. GO fibers were developed by spinning GO dispersions in a coagulation bath of hexadecyltrimethyl ammonium bromide (CTAB) solution. This is the first report in which the assembly of GO nanosheets into macroscopic fibers was realized at low GO concentrations. The assembly mechanism of the GO fibers was clarified. The graphene fibers exhibited promising mechanical strength, high electrical conductivity and high flexibility. Multi-functionality could be achieved through *in situ* or post-synthesis integration of various functional nanomaterials into the GO fibers.

The GO nanosheets had a thickness of 0.7 nm, and were used to prepare aqueous GO suspensions at a concentration of 10 mg·ml^{-1}. The suspensions were injected into a coagulation bath of 0.5 mg·ml^{-1} CTAB solution by using a simple double-jet pump, so that two pieces of GO fibers were spun out at the same time. The spinning process was continuous, as demonstrated in Fig. 5.13 (a, b). After drying, no change was observed in length, while the fibers were largely shrunk in diameter, as seen in Fig. 5.13 (c). The coil shape was retained after the GO fibers were taken off from the solid rods (Fig. 5.13 (d, e)). In addition, the dry GO fibers were highly flexible and could be easily knitted and woven. Representative SEM images of the GO fibers are shown in Fig. 5.13 (f–k). As revealed in Fig. 5.13 (f), the GO fibers exhibited high-order alignment, with the GO nanosheets being densely-stacked along the long axis. The GO nanosheets at the fiber outer surface were highly crumbled. Figure 5.13 (h) indicated that the assembly structures of the broken part were parallel to the axial direction. The uniformly layered structures of the GO nanosheets were further confirmed by the SEM images in Fig. 5.13 (i–k).

The diameter of the GO fibers could be controlled by using appropriate nozzles with desirable sizes or by adjusting the concentration of the GO suspensions. Graphene fibers could be obtained by reducing the GO fibers with hydroiodic acid as the reducing agent. The highly ordered microstructures of the GO fibers were entirely retained in the graphene fibers. The chemical reduction resulted in a decrease in the diameter of the fibers, due to the removal of oxygen-containing groups contained in the GO nanosheets.

CTAB was selected to form the coagulation bath due to its positive charge, which was important to ensure the assembly of the negatively charged GO suspensions. XRD analysis results indicated that the interlayer spacing of the GO fibers was 15.29 Å, which was much larger than those of the ordered GO

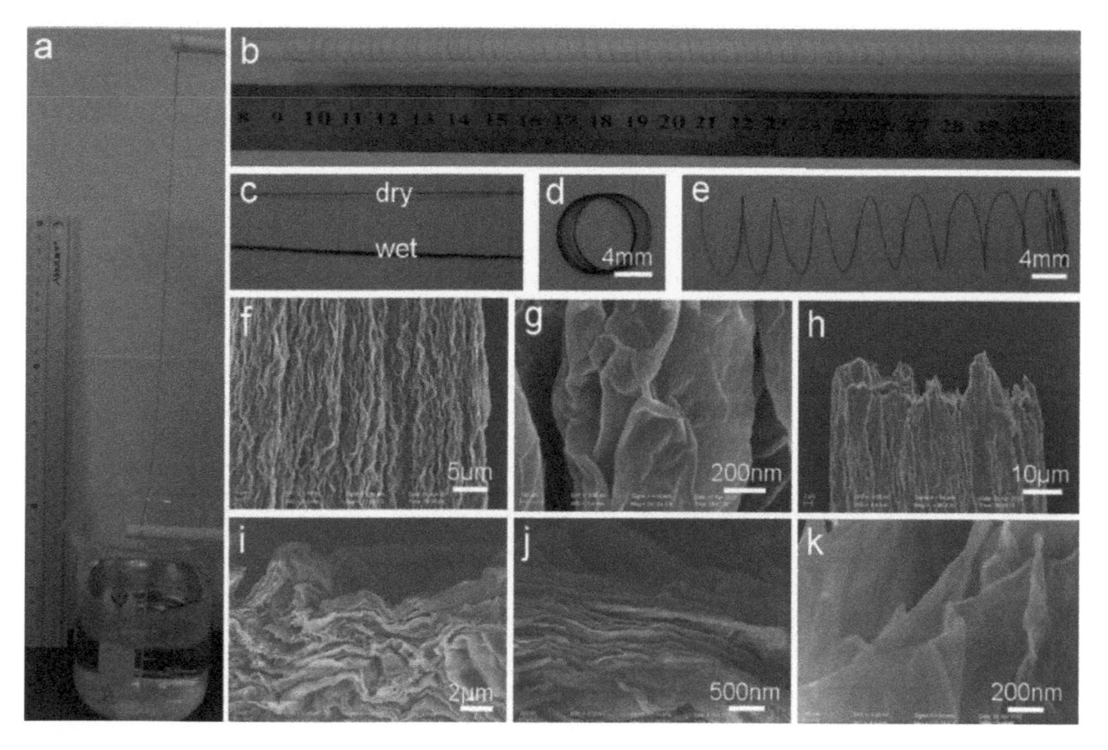

Fig. 5.13. The fabrication and microstructures of the GO fibers. Photographs of (a) 50 cm-long fiber spun out from the CTAB solution by wet spinning a 10 mg·ml^{-1} GO suspension in the coagulation bath of 0.5 mg·ml^{-1} CTAB solution and (b) 1.6 m-long fiber wound on a Teflon rod with a diameter of 8 mm, (c) dry and wet fibers, (d) fiber coil and (e) spring-like fiber by spreading out of coil in (d). (f, g) SEM images of the axial outer surface at different magnifications. (h–k) SEM images of the broken part at the end of the fiber at different magnifications. Reproduced with permission from [33], Copyright © 2012, Macmillan Publishers Limited.

assemblies reported in open literature (8–9 Å). The larger interlayer spacing observed in the GO fibers was attributed to the fact that lamellar structures were formed through the electrostatic interactions between the CTAB molecules and the GO nanosheets during fiber spinning. After chemical reduction, both the CTAB molecules and the functional groups were removed, so that interlayer spacing was reduced to 3.93 Å in the graphene fibers.

A low concentration CTAB solution (0.05 mg·ml^{-1}) was used to demonstrate the assembly process of the GO fibers. Representative SEM images and model of the fiber formation are shown in Fig. 5.14. Figure 5.14 (A (a)) shows a low-magnification SEM image of an unripe GO fiber, because the concentration of the CTAB solution was not sufficiently high. Therefore, the formation process of the GO fibers was revealed. There were four sections in the GO fiber marked with the rectangular frames, as illustrated in Fig. 5.14 (A (b–e)). As the GO suspension was injected into the coagulation bath, a multilayer film was formed from the GO nanosheets, with a width of about 245 μm, due to the strong repulsions between the negatively charged GO nanosheets (stage I). Edges of the GO film were curled and folded towards the center, because the negative charges of GO nanosheets were neutralized by the positive charges of CTAB, thus leading to a directionally aligned belt with a width of 130 μm (stage II). Gradually, the GO fiber started to appear, after repeated cycles of curling and folding, due to the continuous charge neutralizations. As a result, GO fibers were obtained with alignment of the GO nanosheets along the main axis (stages III and IV).

GO/epoxy resin composite fibers could be prepared by soaking the as-spun GO fibers in epoxy resin solution, followed by thermal solidification. The composite fibers obtained in this way exhibited good mechanical stability. The characteristic lamellar structures were well-retained in the GO/epoxy resin fiber. The incorporation of foreign items could also be carried out in the GO suspensions. For instance, if poly(N-isopropylacrylamide) (PNIPAM) was dissolved in a GO suspension, GO/PNIPAM fibers were

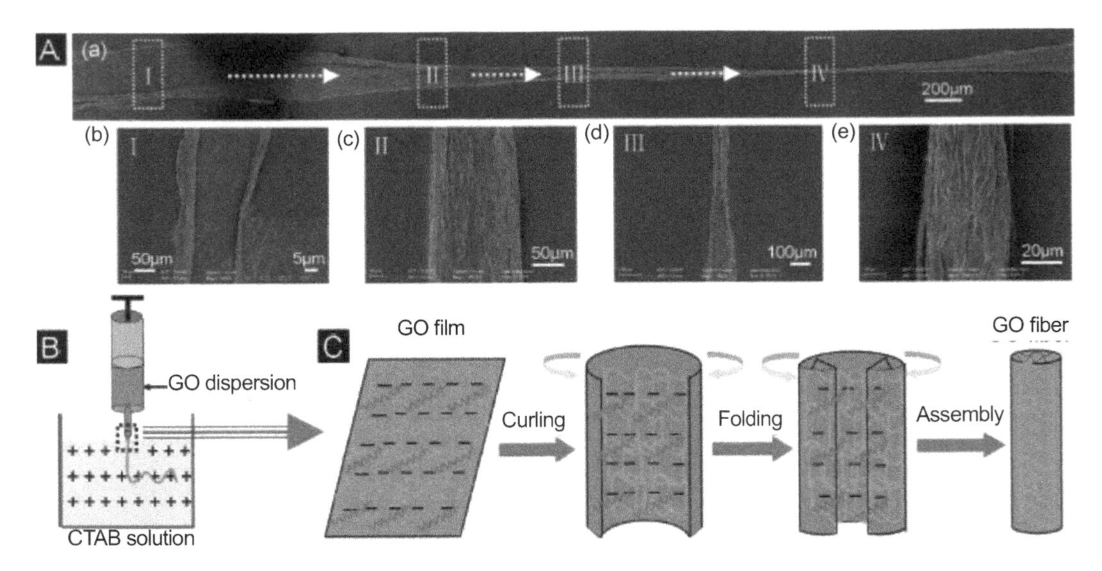

Fig. 5.14. Formation mechanism of the GO fiber. (A) SEM images illustrating the assembly process of the GO fiber by spinning 10 mg·ml^{-1} GO suspension into CTAB coagulation solution of 0.05 mg·ml^{-1}. (a) A single GO fiber. (b–e) Magnified SEM images corresponding to four parts of the single fiber of (a). (B) Schematic diagram of the wet-spinning apparatus. (C) Schematic diagrams showing the assembly mechanism of the GO fibers. Reproduced with permission from [33], Copyright © 2012, Macmillan Publishers Limited.

obtained, in which the ordered alignment along the long axis was not affected. More importantly, chemical reduction process could be conducted to form graphene composite fibers. GO–MWCNTs fibers have also been prepared similarly.

In a separate study, graphene fibers were fabricated by using the wet-spinning and coagulation process, with focus on the understanding of the processing-structure-property relationship [34]. Processing parameters include graphene dimension, spinning conditions, fiber density and orientation in achieving optimum properties. It was found that both mechanical strength and electrical conductivity of the graphene fibers were closely related to the degree of orientation of the graphene nanosheets in the fibers. The interaction between graphene nanosheets could be enhanced by using a solution of high GO concentration, which led to fibers with high electrical and mechanical properties. Also, the mechanical properties of the fibers were more strongly influenced by the fiber packing density related defects.

The wet-spinning was carried out by injecting condensed aqueous GO solutions into a coagulation bath with a 25 cm capillary tube [34]. The coagulation bath was 5 wt% NaOH ethanol solution. The gel-like GO solutions were extruded at rates of 0.8–5 m·min^{-1}, which were controlled by controlling the pressure of N_2 in the range of 0.8–3.0 MPa correspondingly. As the solutions were passed through the capillaries, the GO nanosheets in the solutions were subjected to shear force. Once the GO solutions were injected into the coagulation bath, continuous GO fibers were formed, which could be readily collected onto the spinning drum outside the bath. The as-spun GO fibers were washed with methanol to remove the residual NaOH and then dried for 24 h at room temperature. Graphene fibers were obtained through chemical reduction of the GO fibers in an aqueous solution of hydroiodic acid (40%) at 80°C for 8 h, followed by washing with methanol and vacuum drying for 12 h.

Diameters of the as-dried fibers were in the range of 40–150 μm, with linear densities of the reduced graphene fibers ranging from 0.20 to 1.56 tex (i.e., grams per kilometer fiber). The diameter and microstructure of the graphene fibers were closely related to the dimensions of the GO nanosheets and the processing parameters. Figure 5.15 shows representative SEM images of the GO fibers. Obviously, the diameter of the GO fibers was increased with the increasing diameter of the capillary tube. As shown in Fig. 5.15 (a–c), the diameters of the graphene fibers were in the range of 45–150 μm, corresponding to the capillary diameters of 0.35–0.5 mm. Similarly, the alignment of the GO nanosheets was along the long axis of the fibers, as confirmed by the cross-sectional SEM images in Fig. 5.15 (g–i).

As a comparison, graphene fibers from large GO (LGO) nanosheets were prepared, with a solution containing 3.8 wt% LGO, a capillary diameter of 0.40 mm and a spinning rate of 3 $m \cdot min^{-1}$. The LGO fibers were reduced to form rLGO fibers. Graphene fibers containing small graphene sheets were also prepared under the same processing conditions using GO solution. Figure 5.16 shows representative SEM images of the rLGO fibers. One observes that the LGO sheets are stacked densely with local alignments in the LGO fibers. Chemical reduction had almost no effect on the surface morphology of the LGO fiber, but the diameter was decreased from 54 μm to 43 μm, so that the rLGO fiber was more compact, as shown in Fig. 5.16 (f). XRD results indicated that the interlayer spacing was decreased from 0.88 nm to 0.37 nm, due to the removal of the oxygen-containing groups in GO.

It was found that the defects present in the fibers also influenced their mechanical properties. Nanoscale defects are mainly present in the graphene nanosheets, while voids are microscopic defects in the fibers. The microscopic voids could be prevented by using highly concentrated GO solutions. Packing density was used to characterize the defects and voids of the GO fibers. As the concentration of the GO solutions was increased from 2.8 to 5.2 wt%, the packing density of the GO fibers was increased from 0.20 to 0.33 $g \cdot cm^{-3}$. Tensile strength and modulus of the graphene fibers were in the ranges of 75–320 MPa and 3.1–11.6 GPa, respectively. With optimal processing conditions, graphene fibers with tensile strength and stiffness of up to 360 MPa and 12.8 GPa, respectively, could be developed.

As shown above, wet-spun GO fibers generally have a relatively low tensile modulus, due to the intrinsic alignment of the GO nanosheets along the long axis direction of the fibers. One of the strategies to address this problem is to use large GO nanosheets as the building blocks to assemble fibers [35]. By doing this, modulus could be increased by one order of magnitude. In this study, two types of GO

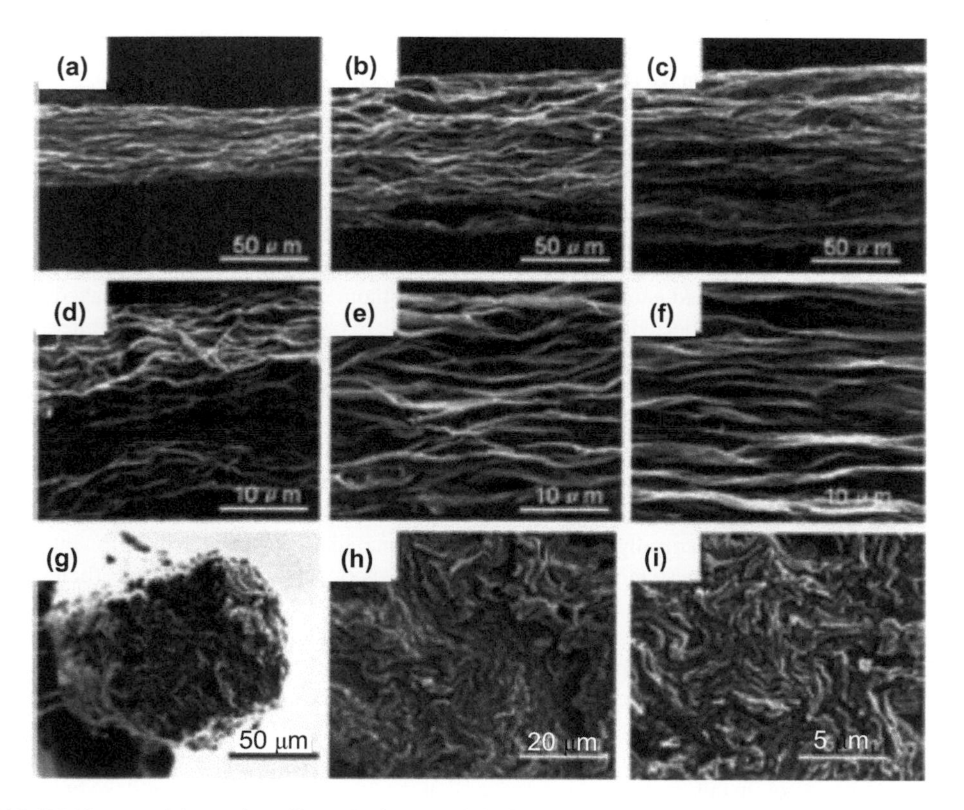

Fig. 5.15. SEM images of the graphene fibers. Surface morphologies of the graphene fibers by wet-spinning with nozzles of different sizes: (a, d) 0.30, (b, e) 0.45 and (c, f) 0.50 mm. (g–i) Cross-sectional SEM images of the graphene fibers at different magnifications. Reproduced with permission from [34], Copyright © 2013, The Royal Society of Chemistry.

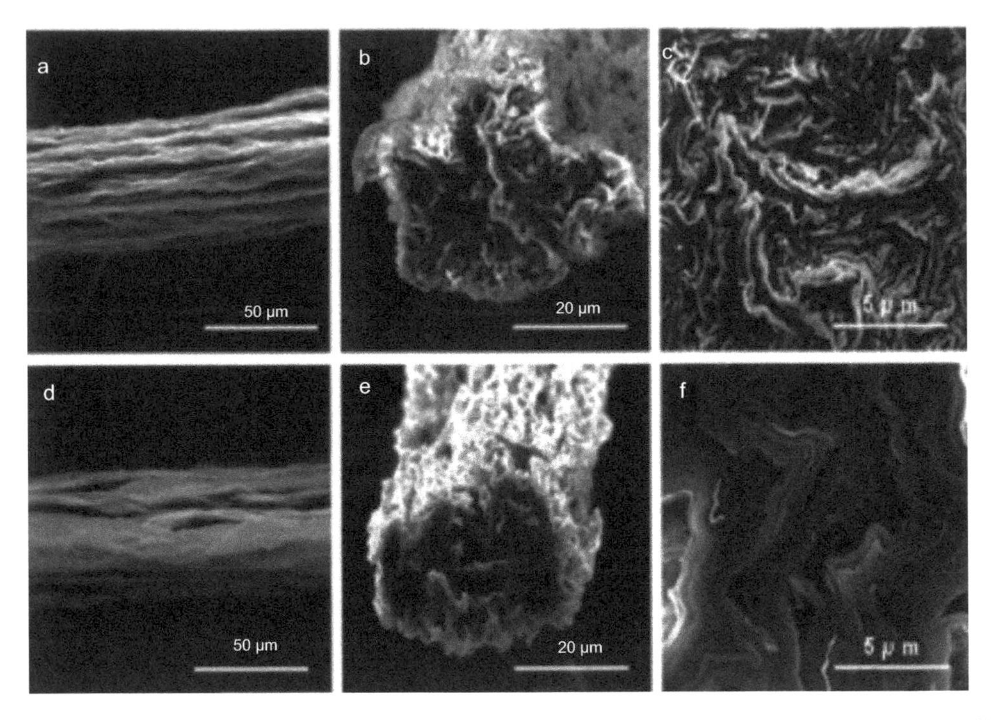

Fig. 5.16. SEM images of the LGO fibers: (a) surface and (b, c) cross-section. (d–f) SEM images of the rRLGO fibers. Reproduced with permission from [34], Copyright © 2013, The Royal Society of Chemistry.

nanosheets, including (i) large flake GO (LFGO) nanosheets, with an average diameter of 22 μm and (ii) small flake GO (SFGO) nanosheets, with an average diameter of 9 μm were used. Figure 5.17 (a) shows an optical microscope image of the LFGO dispersed in DI water. The LFGO possessed diameters in the range of 5 μm to 62 μm, while the SFGO had diameters from 1 μm to 21 μm. Both the SFGO and LFGO nanosheets were single layer.

Homogeneous gels with 5 wt% SFGO and LFGO were formed by dispersing them in DI, and were extruded through a 175 μm orifice at the rate of 0.1 $ml \cdot min^{-1}$, with ethyl acetate as the coagulation bath. The extruded fibers were injected into the coagulation bath without stretching, and were soaked for 5 min. The soaked fibers were then collected and laid on a Teflon plate to be dried at room temperature overnight. Figure 5.17 (b) shows specific stress-strain curves of the two groups of GO fibers. Obviously, the LFGO fibers demonstrated much higher mechanical performances than the SFGO fibers, in specific stress, specific modulus and elongation by 178%, 188% and 278%, respectively.

Due to the high flexibility, the LFGO fibers could be knotted into different types of knots. Figure 5.17 (c) shows an overhand knot. Knot efficiency is usually used to measure the flexibility of fibers, which is defined as the breaking stress of a knotted fiber divided by the breaking stress of a control fiber expressed in percentage [36]. Generally, KEs of nearly all polymer fibers are lower than 100%. Interestingly, we discovered that LFGO fibers spun without drawing show unconventional 100% knot efficiency (KE). Owing to the bending defects introduced by the high bending modulus, the stress of conventional fibers is degraded during the knotting process. As a result, breaking of knotted fibers always takes place at the knots. In sharp contrast, the breaking points observed in the LFGO fibers were far away from the knot, as revealed in Fig. 5.17 (d–f). This behavior is somehow similar to that observed in carbon nanotube yarns [36]. The high KE of the LFGO fibers was attributed to their relatively low bending modulus. Comparatively, the SFGO fibers possessed a lower KE, simply because the SFGO nanosheets were not sufficiently large to cover the entire knot. In this case, breaking occurred at the knot, due to bending at flake boundaries.

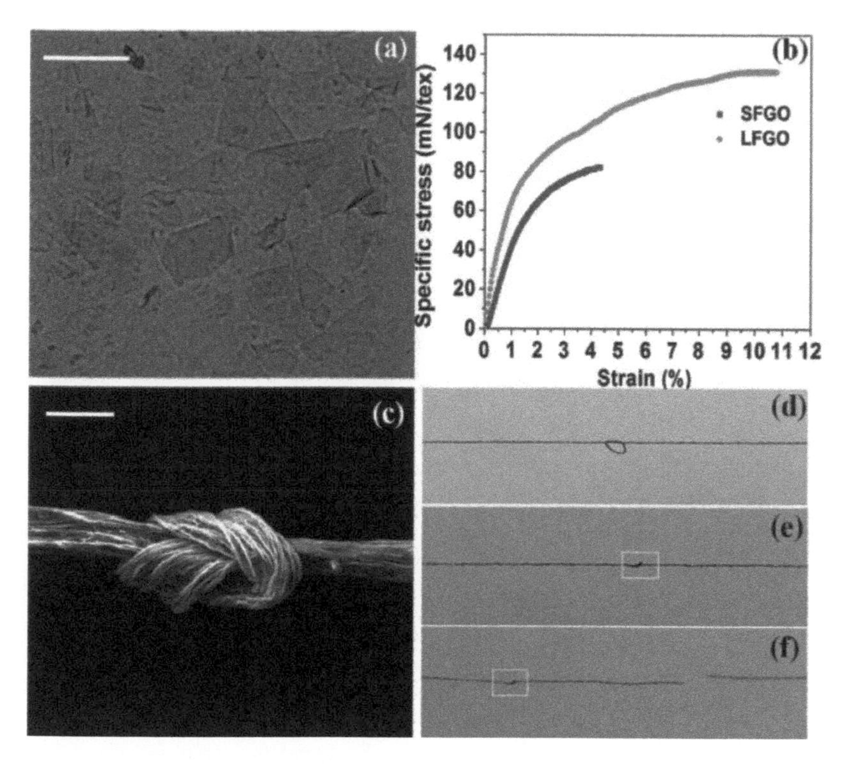

Fig. 5.17. (a) Optical microscope image of the 5 wt% LFGO dispersed in DI water, with scale bar = 100 μm. (b) Specific stress-strain curves of the SFGO and LFGO fibers spun without drawing. (c) SEM image of the LFGO fiber with an overhand knot, with scale bar = 100 μm. (d) A tied LFGO fiber. (e) Pulling the tied LFGO fiber to form a knot, with the yellow rectangle to highlight the position of the knot. (f) Continuing to pull the LFGO fiber causing breaking at a point far from the knot. Reproduced with permission from [35], Copyright © 2013, WILEY-VCH Verlag GmbH & Co. KGaA, Weinheim.

The SFGO nanosheets were used to achieve highly aligned GO fibers, because more pronounced aligned LC domains were formed in the SFGO suspension than in the LFGO suspension, when passing through the spinning orifice. In addition, due to the similar size, LFGO nanosheets tended to block the orifice. To spin SFGO fibers, methyl acetate or ethyl acetate should be used as the coagulation bath. Because there was a difference in the mass transfer rate between methyl acetate and the aqueous dope, the fibers from methyl acetate were of hollow structure. The solubility of methyl acetate in water is higher than that of ethyl acetate. As a result, much more water diffused out of the fibers than methyl acetate that diffused into the fibers, thus leading to hollow fibers. Therefore, ethyl acetate was used to obtain continuous solid fibers.

The GO fibers had an irregular morphology, with ripples on the surface and noncircle cross-section. The structure and thus mechanical properties of the SFGO fibers were closely related to the drawing parameters. Figure 5.18 shows representative SEM images of the fibers. The fibers obtained with draw ratios of 1.09, 1.27 and 1.45 showed a random orientation in the aligned direction, while the GO nanosheets were folded to a certain degree. As the draw ratio was increased to 1.82, the GO nanosheets in the fibers were uniformly aligned in one direction, so that their morphology was more like that of a GO paper. Mechanical performances, i.e., both the specific stress and modulus, were increased with the increasing draw ratio. The fibers exhibited very high flexibility.

A coaxial two-capillary spinning strategy was developed to continuously fabricate graphene-based hollow fibers (HFs) with well-controlled morphology [37]. Continuous GO-HFs and necklace-like HFs (nGO-HFs) could be spun out directly from concentrated GO suspensions. Graphene HFs (G-HFs) could be obtained through chemical reduction.

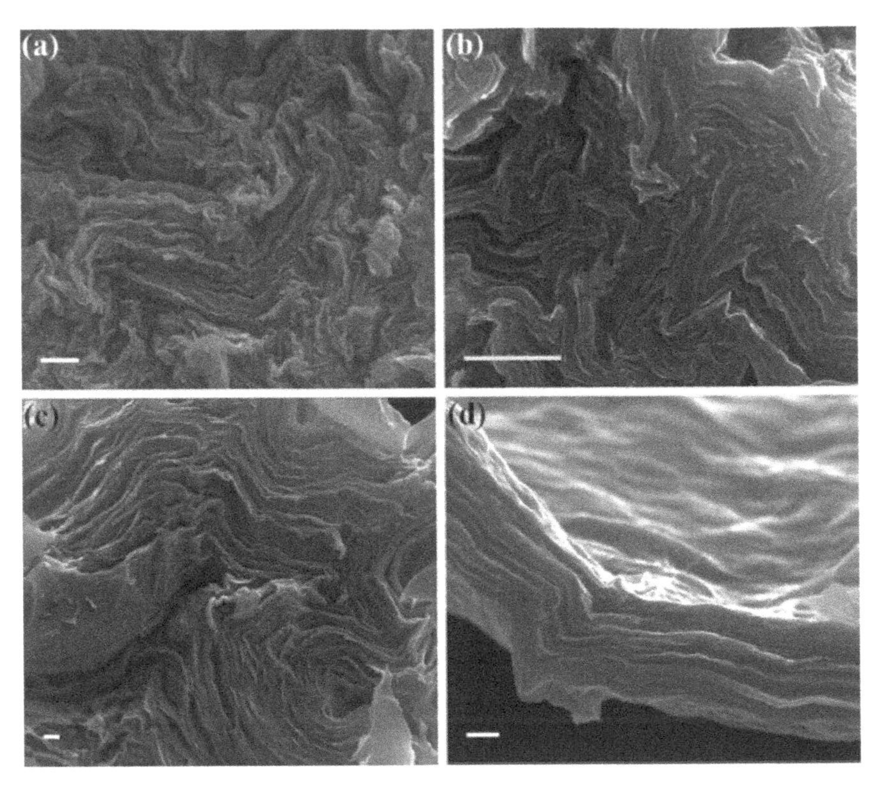

Fig. 5.18. Cross-sectional SEM images of SFGO fibers obtained with different draw ratios: (a) 1.09, (b) 1.27, (c) 1.45 and (d) 1.82. All scale bars = 1 μm. Reproduced with permission from [35], Copyright © 2013, WILEY-VCH Verlag GmbH & Co. KGaA, Weinheim.

Figure 5.19 (a) shows a schematic diagram of the setup used to spin GO-HFs. The coagulation bath was methanol solution containing 3 M KCl. A coaxial two-capillary spinneret was used to produce the GO-HFs. The spinneret was constructed by inserting a stainless steel needle, with an inner diameter of 300 μm and an outer diameter of 500 μm, into a branched glass tube with a capillary tip that had an inner diameter of 1000 μm. The coaxial two-capillary spinneret is shown in Fig. 5.19 (b). Figure 5.19 (c) shows a photograph of the GO solution with a concentration of 20 $mg \cdot ml^{-1}$ and a viscosity of 2.3×10^3 Pa·s, which was derived from a dilute GO solution through gradual evaporation of water. The GO nanosheets were of sizes in the range of 0.5–4 μm. Figure 5.19 (d) shows an experimental set-up, in which the stainless steel needle was connected to a syringe containing the coagulation bath, while the glass tube was filled with the GO suspension through the branch. Pressed air was applied to the branch to push the GO suspension out of the spinneret into the bath solution. Meanwhile the coagulation solution flowed out at a proper rate from the core capillary that was controlled by the injection pump. In this way, GO-HFs were continuously spun out at a production rate of 3.3 $cm \cdot s^{-1}$.

Figure 5.19 (e) shows a roll of the as-spun GO-HFs, which exhibited a semitransparent character, as seen in Fig. 5.19 (f). Hollow structure of GO fibers is demonstrated in the inset of Fig. 5.19 (f). Flexible and mechanically stable GO-HFs with a slight shrinkage in diameter were obtained after washing and drying (Fig. 5.19 (g)). Figure 5.19 (h) revealed that the GO-HFs had a tensile strength of 140 MPa. They exhibited a typical elongation at break of about 2.8%, which was attributed to the possible displacement of the GO nanosheets within the walls.

As shown in Fig. 5.20 (a), the as-spun GO-HFs possessed an incompact but highly cross-linking structure, in which the GO nanosheets were randomly interlaced within the wall. It was found that the GO nanosheets on the inner and outer surfaces were aligned along the tube axis, as illustrated in Fig. 5.20 (b, c), which was induced by the extrusion process. After drying, densely packed structures were observed

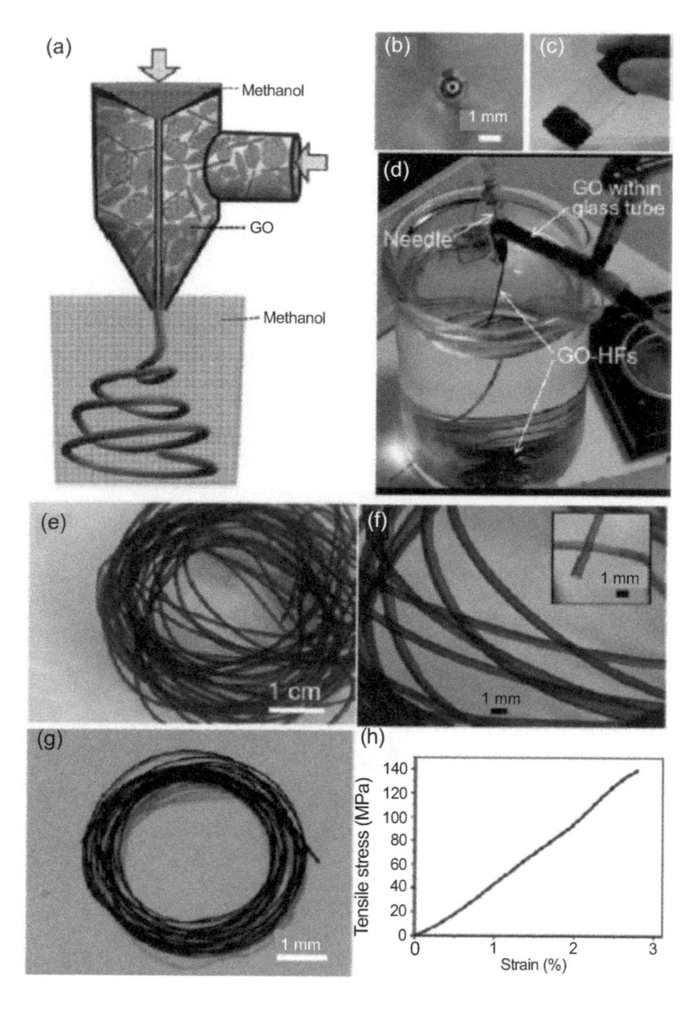

Fig. 5.19. (a) Schematic diagram of the setup consisting of a dual-capillary spinneret used to directly spin the GO-HFs. (b) A photograph of the coaxial two-capillary spinneret. (c) Photograph of the concentrated GO suspension with a concentration of 20 $mg \cdot ml^{-1}$. (d) Experimental spinning setup. (e, f) Photographs of the spun GO-HFs in methanol solution with 3 M KCl, with the inset in panel (f) showing the open tip of one GO-HF. (g) Photograph of the dried GO-HFs. (h) Typical stress-strain curve of a GO-HF. Reproduced with permission from [37], Copyright © 2013, American Chemical Society.

both on the surface and within the wall, as demonstrated in Fig. 5.20 (d–f). Due to the tight layer-by-layer stacking of GO nanosheets within the wall (Fig. 5.20 (g)), the GO-HFs were highly flexible. When the GO-HFs were bent, the walls were deformed but without the presence of any broken points, as seen in Fig. 5.20 (h, i). Both thickness of the wall and inner/outer diameter of the GO-HFs could be tailored by controlling the size/configuration of the spinneret and concentration of the GO suspensions.

Moreover, simultaneous functionalization of the GO-HFs could be achieved by introducing functional components directly into the core flow or mixing them with the initial GO suspension. For example, by mixing SiO_2 nanospheres of 200 nm into the flow medium of KCl/methanol in the inner capillary of the spinneret, the nanospheres were attached to the inner-walls of the GO-HFs, while the surface of the outer-walls was not affected. The functionalization had no effect on the mechanical properties of the GO-HFs.

By using thermal annealing, the GO-HFs could be converted into G-HFs, whereas their flexibility was not influenced while mechanical strength was improved. The conversion of GO-HFs could also be realized through chemical reduction in hydroiodic acid. The converted G-HFs possessed an electrical conductivity of 8–10 $S \cdot cm^{-1}$. TiO_2 functionalized G-HFs demonstrated fast photocurrent response with good repeatability.

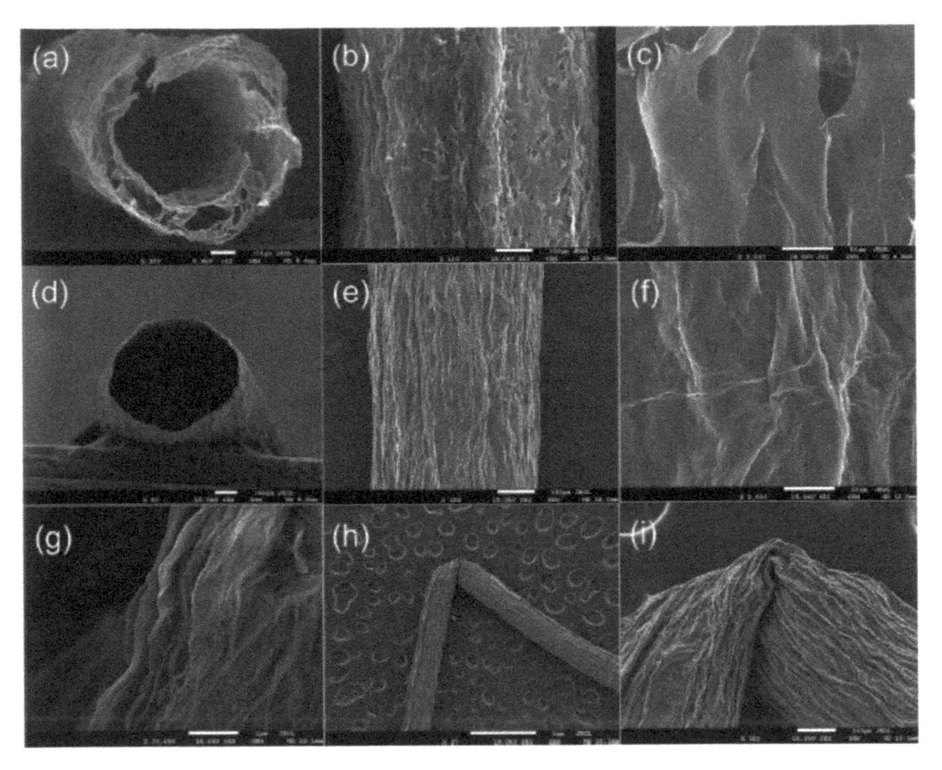

Fig. 5.20. (a–c) Cross-sectional and surface SEM images of the as-spun GO-HFs after rapid freeze-drying treatment. (d–f) SEM images of the naturally dried GO-HFs. (g) Cross-section view of the wall of the GO-HFs. (h, i) SEM images of a folded GO-HF. Scale bars = 100 μm (a, b, d, e, i), 10 μm (c, f), 1 μm (g) and 1 mm (h). Reproduced with permission from [37], Copyright © 2013, American Chemical Society.

As the inner fluid of KCl/methanol solution was replaced with compressed air, the GO-HFs exhibited a special necklace-like structure (nGO-HFs), in which the microspheres were connected by the fibers, as illustrated in Fig. 5.21 (a). It was observed that the diameter of the nGO-HFs was smaller than that of the GO-HFs. This was attributed to either the absence of the coagulation solution or the effect of the air pressure. The nGO-HFs were also highly flexible, and could be bent and folded, as seen in Fig. 5.21 (b) and the inset in it. Figure 5.21 (c) indicated that the microspheres had a diameter of about 700 μm. Surfaces of the nGO-HFs were constructed by compact GO films (Fig. 5.21 (d)). Figure 5.21 (e) shows an opened microbead, revealing a hollow interior. The shell consisted of densely stacked GO layers, with a thickness of about 200 nm (Fig. 5.21 (f)). As demonstrated in Fig. 5.21 (g, h), the hollow microsphere had a channel with a diameter of about 100 μm, which was connected with the outside fiber. The hollow channel was throughout the whole nGO-HFs. The nGO-HFs also exhibited promising mechanical strengths. As illustrated in Fig. 5.21 (i), a short nGO-HF was coated with a layer of polydimethylsiloxane (PDMS) and one end was sealed. With soap-suds, a soap bubble was generated from the tip of the hollow microsphere by pressing with fingers (Fig. 5.21 (j)). This process was highly reproducible, which implied that such nGO-HFs could be used as micropumps.

A modified wet-spinning method was reported to produce graphene fibers by applying shear stress [38]. The fibers fabricated in this way exhibited a macroscopic ribbon-like structure with high flexibility. Such ribbon-like graphene fibers could find a wide range of applications, such as elastic strain sensors, flexible counter electrodes for fiber solar cells and fabric electrodes for supercapacitors. Figure 5.22 (a) shows a schematic diagram of the spinning setup. The GO nanosheets with few-layers were several micrometers in size. Aqueous GO suspension had a concentration of 6 mg·ml^{-1}, which was injected by using a syringe pump into a chitosan solution placed on a stage that was rotated constantly at a speed of 10 rpm. The GO ribbons were collected by using a glass rod placed vertically in the chitosan solution.

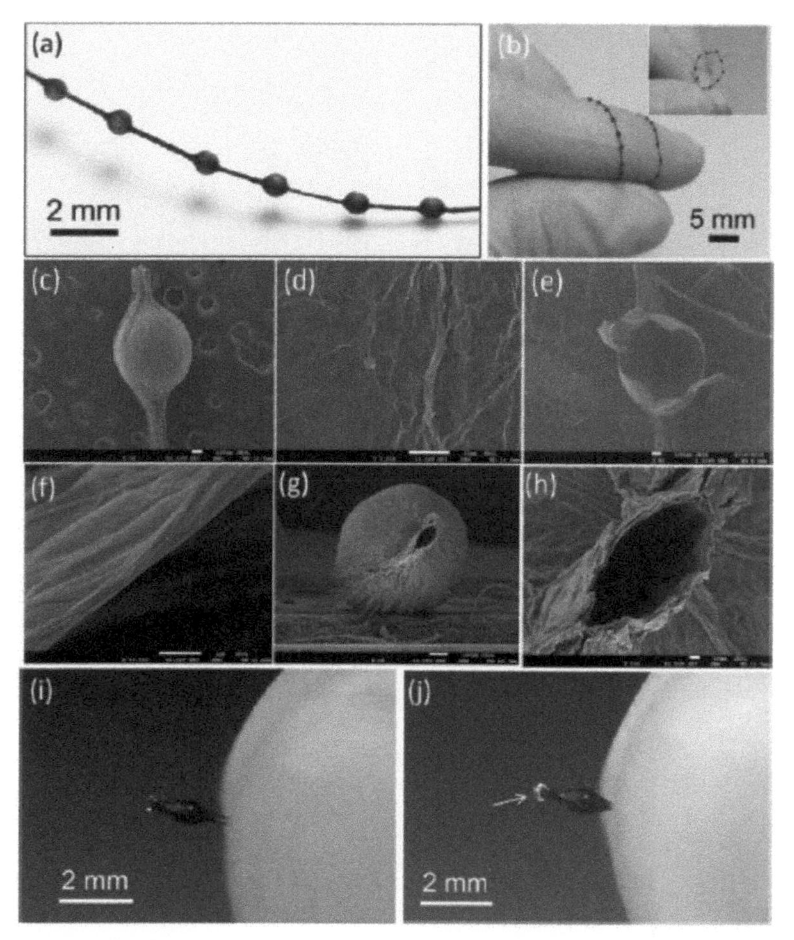

Fig. 5.21. (a) Photograph of the nGO-HF produced with air flow rate of 1.0 ml·min^{-1}. (b) Photograph of the nGO-HF coiled on finger and bent into a circle (inset). (c) An individual microsphere on the nGO-HF and (d) the corresponding enlarged surface image. (e, f) SEM images of a broken microsphere and its edge, respectively. (g, h) Cross-sectional view of the nGO-HF broken between microspheres. (i) Photograph of a short nGO-HF sucked with soap-suds and (j) a soap bubble produced on its tip by pressing the bubble of the nGO-HF. Scale bars = 100 μm (c, e, g), 1 μm (f) and 10 μm (d, h). Reproduced with permission from [37], Copyright © 2013, American Chemical Society.

During the spinning, the fluid flow direction was controlled to be perpendicular to the direction in which the GO suspension was injected (downward). Continuous ribbons were produced due to the presence of the shear stress. The ribbons exhibited a flat morphology, with a width-to-thickness ratio of up to 100, as seen in Fig. 5.22 (b). Photographs of the experimental setup, together with the glass rod, are shown in Fig. 5.22 (c–e). There were two reasons responsible for the formation of the ribbon structure, i.e., (i) electrostatic interaction between the GO nanosheets and chitosan in the coagulation bath and (ii) the larger shear fluid velocity compared with the injection rate of the GO suspension. The width of the ribbons in the range of 200–500 μm was determined by the diameter of the tilted opening of the needle. Other factors included the concentrations of the GO suspension and chitosan, injection rate of the GO suspension and rotation speed of chitosan, which could be optimized to tailor the morphology and property of the final ribbons.

Chemical reduction in hydroiodic acid was used to convert the GO ribbons into rGO ribbons, accompanied by a color change from brown to black, as illustrated in Fig. 5.22 (f, g). The rGO ribbons possessed conductivities of 100–150 S·cm^{-1}. Figure 5.22 (h) indicated that rGO ribbons showed very high flexibility. Wet ribbons could be wound onto a thin glass capillary to form a stable helical item after drying. When the spring-like structure was stretched, the twisted ribbon was not fractured and the shape could be recovered to a large extent.

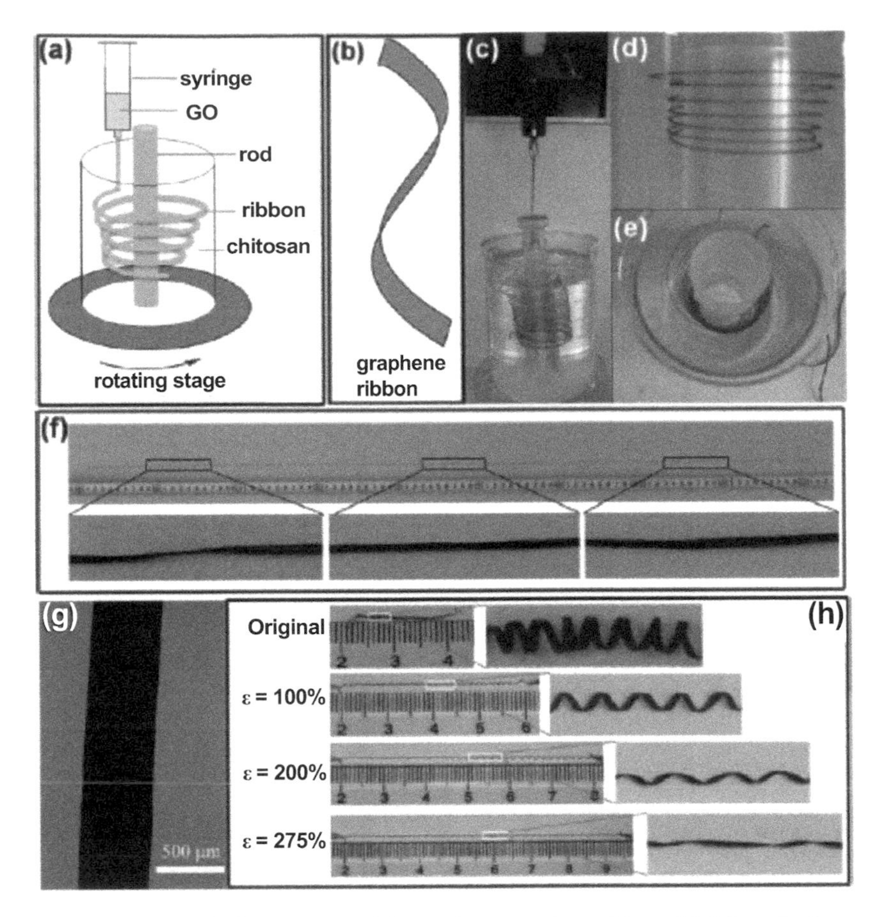

Fig. 5.22. Fabrication process and microstructures of the continuous graphene ribbons. (a) Schematic diagram of the wet-spinning setup. (b) Illustration of the flexible ribbon structure. (c) Photograph of the setup with a uniform GO ribbon spun in the chitosan solution. (d) Photograph of a single ribbon rotated around the collecting rod. (e) Top view of the setup. (f) Photograph of a 1 m long and 300 μm wide graphene ribbon after chemical reduction. (g) Optical image of a 500 μm wide ribbon. (h) A Ribbon made into spiral shape and then stretched to large strains. Reproduced with permission from [38], Copyright © 2013, American Chemical Society.

The orientation of wrinkles and morphology of the ribbons could be controlled by applying different loading conditions on the wet graphene ribbons during the drying process after the chemical reduction, as schematically illustrated in Fig. 5.23 (a). The buckling instability and wrinkles of the GO nanosheets were attributed to the effect of the surface tension forces that were generated during the drying process. Representative SEM images of the experiment results are shown in Fig. 5.23 (b–d). Ribbons with different widths were wound onto glass rods to make close contact, leading to strong interaction with the substrate to buffer the surface tension force, so that oriented wrinkles along the length direction of the ribbons were formed. Alternatively, a freestanding ribbon was held by fixing its ends vertically or horizontally, so as to have partially oriented wrinkles with two main orientations in the length (body part) and radial (surface layer) directions. Samples with random wrinkles without any specific orientation were obtained by drying the ribbons in a free state. As expected, the ribbons dried on the substrate exhibited the most pronounced orientation of wrinkles with a relatively smooth surface, while a freely dried sample was randomly wrinkled with a rough morphology. The graphene ribbons could be woven into spring-like loops with different pitches and simple butterfly knots, as seen in Fig. 5.23 (e–g). Even more complicated configurations could be produced, as demonstrated in Fig. 5.23 (h, i).

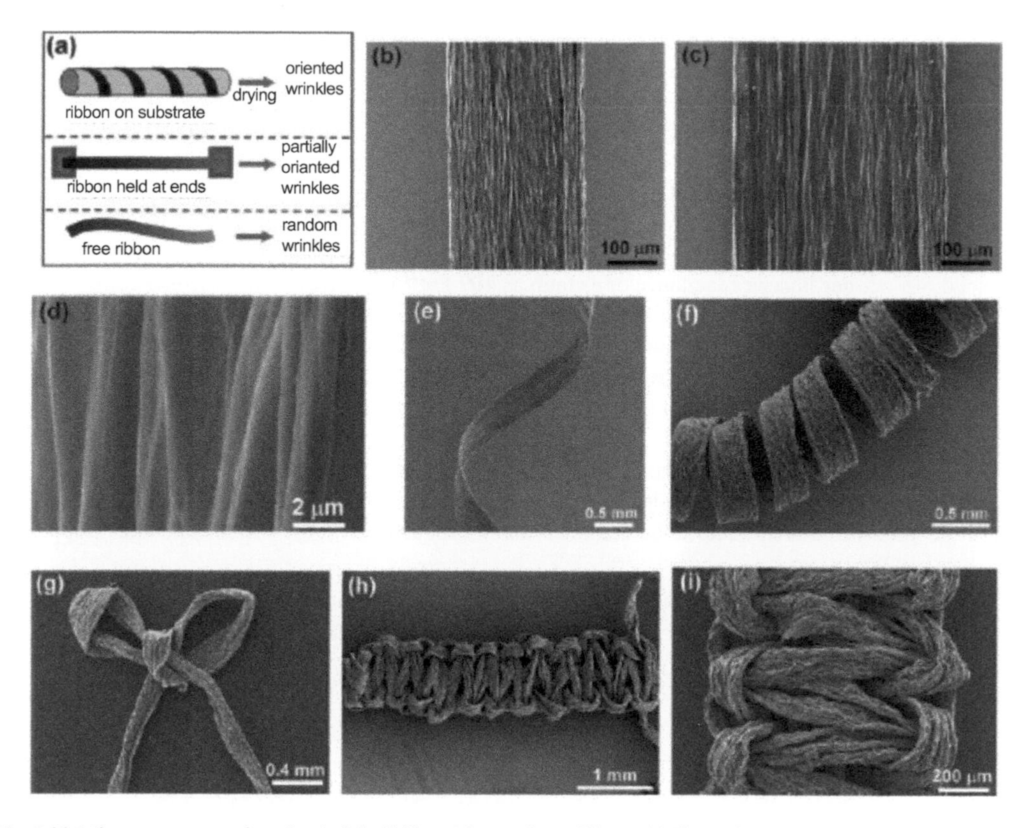

Fig. 5.23. Microstructures and mechanical flexibility of the graphene ribbons. (a) Illustrations of the different drying conditions to produce ribbons with oriented, partially oriented and random wrinkles, respectively. (b) SEM image of a 262 μm wide ribbon with oriented wrinkles. (c) SEM image of a 415 μm wide ribbon. (d) Close view showing a clean ribbon surface with wrinkles. (e) Flexible ribbon. (f) Spring-like loops made of a ribbon. (g) Graphene ribbon butterfly knot. (h) Graphene ribbon flat knot. (i) Close view showing highly twisted and complex structure in the flat knot. Reproduced with permission from [38], Copyright © 2013, American Chemical Society.

Mechanical properties of the graphene ribbons had a close relation to the orientation of the wrinkles. According to uniaxial tension testing results, oriented samples exhibited a relatively small failure strain of $\varepsilon < 4\%$, while those with random wrinkles demonstrated a larger failure strain of $\varepsilon > 9\%$. In the oriented samples, the presence of tensile stress during the drying process triggered the alignment of the graphene nanosheets, thus leading to a denser stacking between the graphene layers and hence higher strengths (> 100 MPa). In contrast, the highly wrinkled structure of the samples with random wrinkles possessed a lower strength due to the larger intersheet distance, corresponding to a moderate tensile strain (up to about 14%) before fracture. It was also found that the thickness of the ribbons was decreased with increasing fluid velocity, while the tensile strengths of the graphene ribbons were increased with the decreasing thickness over the range of 7–1.5 μm. This was simply because the thicker ribbons had more voids or defects.

The highly wrinkled ribbons were elastic over a modest range of strain ($\varepsilon < 5\%$). Elastic stretching was stabilized after several cycles. After 1000 cycles, the ribbon showed negligible deformation, showing high elasticity that was attributed to the presence of the random wrinkles in the ribbon. With such random wrinkles, the ribbons could be deformed reversibly within a certain degree. This was the first report on elastic graphene fibers in the literature. Variation in electrical resistance (R) was monitored during the cyclic tests, which was represented with a relative resistance change ($\Delta R/R$), subject to cycles under a stretching of $\varepsilon = 3\%$. The maximum change was as small as $\Delta R/R \approx 0.4\%$. As claimed by the authors, the conductivity variation could be further minimized by controlling the wrinkled structure of the ribbons. With these unique mechanical and electrical properties, the graphene ribbons could find potential applications as flexible electrodes and strain sensors. Potential application as the electrodes of fiber solar cells and supercapacitors has been demonstrated [38].

More recently, a specific study was reported to develop highly scalable graphene fibers and yarns with high supercapacitor performances [39]. It was demonstrated that the key to producing such graphene fibers and yarns was to preserve the large size of the nanosheets while simultaneously maintaining a high interlayer spacing between graphene nanosheets after the chemical reduction. To achieve high electrochemical performance, the number of covalently bonded carbon atoms per unit volume or mass should be maximized, while the number of other atoms present in the system and attached to the graphene nanosheets should be reduced. An extraordinary capacitance of as high as 409 $F{\cdot}g^{-1}$ by using a two-electrode configuration. Both the graphene fibers and yarns exhibited outstanding mechanical flexibility and strength. The 3D self-assembled binder-free aligned microstructures were achieved by using slightly acidic LC suspensions with pH $\approx$ 3 in acetone bath. The LC GO dispersions were spinnable by using various wet-spinning methods into both fibers and yarns, as shown in Fig. 5.24. The as-spun fibers were stable in structural integrity in water, as illustrated in Fig. 5.24 (b).

The acidic condition of the suspensions resulted in a high rate of fiber formation and solidification, thus leading to fibers with high porosity, as illustrated in Fig. 5.25. Also, stronger hydrogen bonding between the nanosheets facilitated in the acidic condition, together with the water molecules trapped in

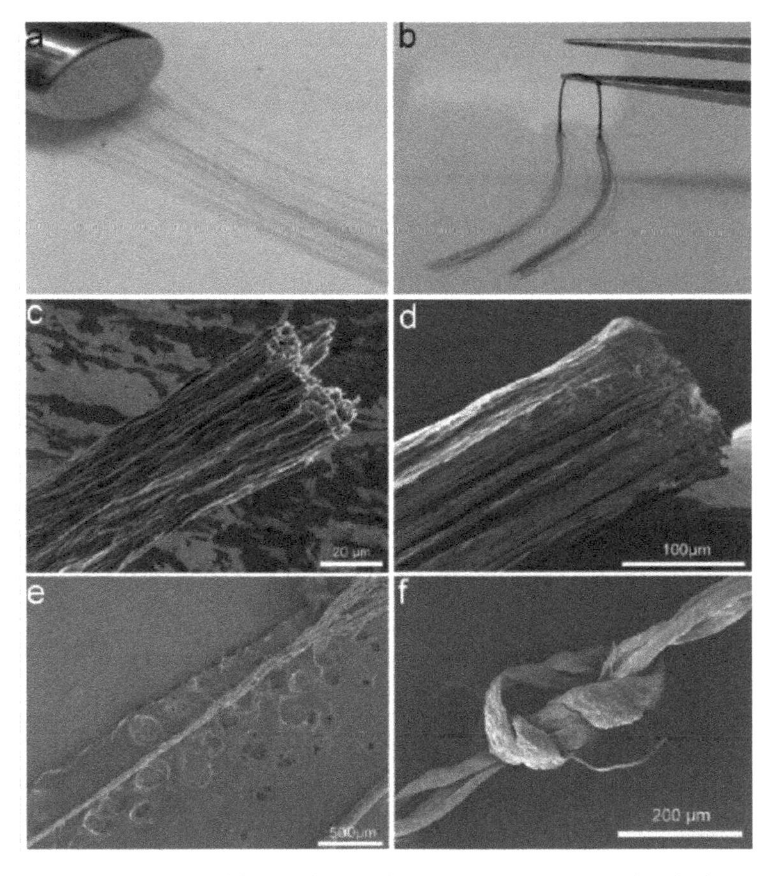

Fig. 5.24. Spinning and characterization of the GO fibers and yarns. (a) Photograph showing the formation process of the gel GO yarns produced with a multihole spinneret. As the LC GO was injected into the coagulation bath, GO filaments were coagulated instantly by the coagulation solution. (b) Dried GO yarns were easily separated to be present as individual filaments when they were immersed in water, while their structural integrity was still retained in the presence of water. SEM images of (c) an irregularly shaped GO fiber, (d) a GO yarn composed of many GO fibers, (e) an unwoven GO fiber yarn and (f) a loosely knotted GO yarn demonstrating the flexibility of the as-prepared GO fibers in an acetone bath. Reproduced with permission from [39], Copyright © 2014, American Chemical Society.

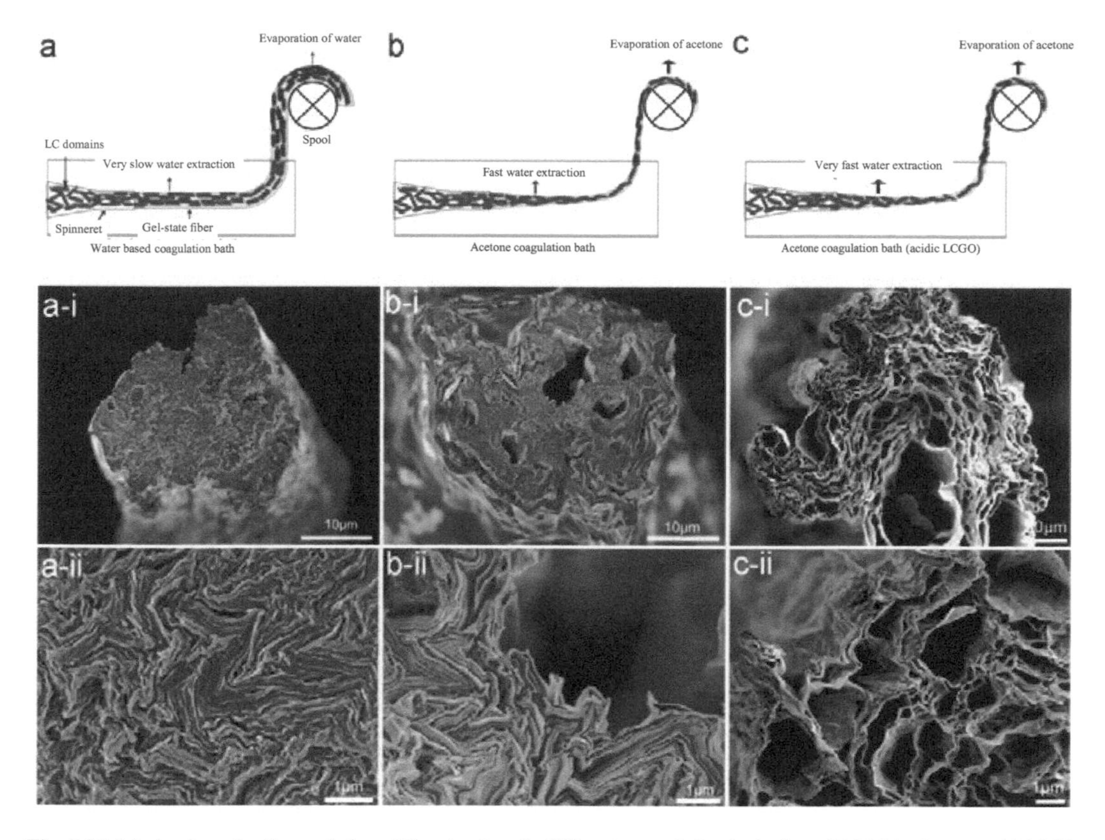

Fig. 5.25. Mechanisms for the evolution of the structure in different coagulation baths from (a) highly dense to (c) highly porous architectures. (a) Water-based coagulation bath resulted in slow expulsion of water from the as-injected gel fiber-like structure. Therefore, to collect such fibers, the length of the bath should be optimized to enable the formation of a solid-like sheath around the core of the fiber. The fiber could then be taken from the bath and transferred on a spool for the evaporation of water from the fiber, leading to a highly dense structure (a-i and a-ii). (b) Acetone coagulation bath resulted in a high rate of water extraction from the surface, due to the difference in the imbibition rate, thus leading to higher rate of solidification, so that porous fiber structure was obtained (b-i and b-ii). (c) Slight acidic condition of LC GO dopants (pH ≈ 3) further increased the difference in the imbibition rate, resulting in much higher water extraction rate, so as to produce more porous fibers (c-i and c-ii). Reproduced with permission from [39], Copyright © 2014, American Chemical Society.

between the inter-layers, effectively separated the nanosheets. By using acetone, due to its strong ability to expel water from the fibers, it was more likely to form highly porous structures, with was confirmed by the results in Fig. 5.25 (b, c).

It is well-known that the reduction of LC GO could trigger restacking of the nanosheets, which should be minimized to maintain the required interlayer spacing and porosity, in order to achieve high storage capacities. The optimal temperature range used to treat the as-prepared GO fibers was 200–220°C. XRD results indicated that d-spacing in the final architecture was from 0.3 to 0.65 nm. The relatively high d-spacing value was ascribed to the presence of some remaining functionalities on the surface of the rGO fibers after the thermal reduction. Furthermore, under this condition, the lateral size of the graphene nanosheets was not affected by the thermal reduction, which ensured the rGO fibers to have sufficiently high electrical conductivity and mechanical strength.

The rGO fiber yarns obtained in baths of acetone, alkaline (NaOH) and $CaCl_2$ (ionic cross-linking agents were all evaluated in terms of electrochemical performance based on their cyclic voltammogram (CV) responses at 10 $mV \cdot s^{-1}$. They showed similar EDLC performance, characterized by a near-rectangular CV curve representative. The yarns fabricated spun out in acetone bath were the best in terms of electrochemical response. It was suggested that, when using cross-linking agents, additional impurities were introduced into the fibers, which was responsible for the much lower EDLC capacitance.

A free-standing capacitor exhibited a specific capacitance of 394 $F\cdot g^{-1}$, corresponding to 0.99 $mF\cdot cm^{-1}$, at a scan rate of 10 $mV\cdot s^{-1}$, with a continuous capacitance of 160 at 100 $mV\cdot s^{-1}$. However, the resistivity was still relatively high, so that the performance of the device was limited at high scan rates, as evidenced by the large drop in capacitance. To address this problem, charge collectors had to be used. As a result, at 100 $mV\cdot s^{-1}$, a capacitance of close to 300 $F\cdot g^{-1}$ could be achieved. Energy density of the supercapacitors made with the rGO fibers could be further increased through the incorporation of metal oxides or conducting polymers.

5.2.2. Hydrothermal method

Hydrothermal treatment can facilitate spontaneous assembly and reduction of GO nanosheets, leading to graphene network via the strong interlayer π-π stacking between the graphene nanosheets [15]. A simple one-step dimensionally confined hydrothermal method was used to produce graphene fibers from aqueous GO suspensions [40]. A glass pipeline with an inter diameter of 0.4 mm was used as the reactor. GO suspension with a concentration of 8 $mg\cdot ml^{-1}$ was injected into the glass pipeline, with the two ends being sealed, and was then treated at 230°C for 2 h. Graphene fibers fit with the geometry of the pipeline were finally formed, with a diameter of 150 μm in wet state. The graphene fibers were collected from the pipeline by using N_2 flow and then dried in air. After drying, the length of the fibers was not changed, while they were shrunk in diameter to ~35 μm, due to the loss of water. The graphene fibers exhibited a very high flexibility, so that they could be woven into meshwork-like and cloth-like structures manually. The network of the graphene fibers could also be embedded into polydimethylsiloxane (PDMS) matrix by casting a mixed and degassed PDMS prepolymer on the graphene fiber meshes, followed by thermal curing.

Figure 5.26 (a) shows a graphene fiber with a length of 63 cm and a diameter of 33 μm. The process was very productive, where 1 ml GO suspension of 8 $mg\cdot ml^{-1}$ yielded graphene fiber of > 6 meter. Also, the diameter and length of the fibers could be readily controlled by either using pipelines with different lengths and inner diameters or adopting GO suspensions with different concentrations. The large shrinkage in the diameter of the fibers during the drying process would produce surface tension forces that could promote spontaneous orientation of the graphene nanosheets. The capillary force caused by the evaporation of water molecules led to a close packing of the porous graphene nanosheets. As a result, densely packed and certainly aligned fibers were formed as demonstrated in Fig. 5.26 (b–e). Figure 5.26 (e) shows SEM image of a broken fiber, with the individual graphene nanosheets and their packing entanglement being clearly observed. The dried graphene fibers had a Raman ratio of about 6:1, while the ratio of the wet fibers was close to 1:1, which further implied that the graphene nanosheets in the dried fibers were highly aligned parallel to the axis. Because no aggregation and preformed orientation were observed in the initial GO suspensions before and after they were injected into the pipeline, the alignment of the graphene nanosheets in the direction of the fiber axis was mainly ascribed to the effect of capillary-induced shear force and surface tension induced during the drying process. The graphene fibers could be curved into coils and enlaced in bundles in wet and dry states, as seen in Fig. 5.26 (f, g). The fiber was not broken as the knot was tightened (Fig. 5.26 (h)), while two-ply yarn could be made by twisting two fibers (Fig. 5.26 (i)).

The hydrothermally as-derived graphene fibers possessed tensile strengths of up to 180 MPa, which was increased to 420 MPa after they were thermally treated at 800°C for 2 h in vacuum. Due to their relatively high density of 0.23 $g\cdot cm^{-3}$, the fibers reached a density-normalized failure stress of 782 $MPa\cdot(g\cdot cm^{-3})^{-1}$. Typical elongations at beak of the graphene fibers were in the range of 3–6%, which was comparable with that of CNT fibers [3, 5] but much higher than that of graphite fibers (~1%). The elastic behavior of the graphene fibers at room temperature before breaking was closely related to the displacement of the graphene nanosheets. The graphene fibers exhibited an electrical conductivity of ~10 $S\cdot cm^{-1}$, which was close to those of wet-spun single-walled CNT fibers [5, 7]. The electrical conductivity was almost not affected when the fibers were bent over 1000 cycles.

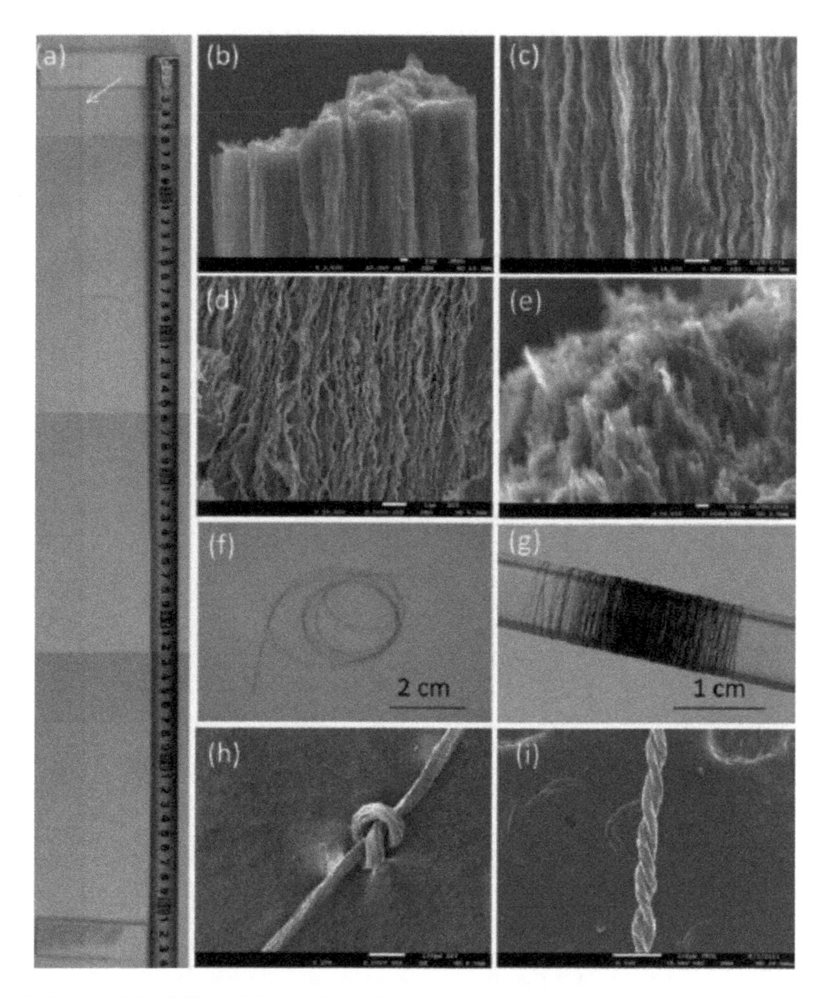

Fig. 5.26. Morphology and flexibility of the graphene fibers. (a) Photograph of a dry graphene fiber with a diameter of ~33 μm and a length of 63 cm. (b) SEM image of the broken part of a graphene fiber (scale bar = 1 μm). (c, d) Axial external surface and inner cross-section SEM images of the graphene fiber, respectively (scale bars = 1 μm). (e) High resolution SEM image of the broken section in (b) (scale bar = 100 nm). (f) Photograph of a wet graphene fiber coiled individually in water. (g) Photograph of dry graphene fibers coiled into bundle around a glass rod. (h, i) SEM images of the knotted and two-ply graphene fibers (scale bars = 100 μm). Reproduced with permission from [40], Copyright © 2012, WILEY-VCH Verlag GmbH & Co. KGaA, Weinheim.

Due to their high flexibility and mechanical strength, the graphene fibers could be shaped to various geometries. The configurations constructed with the wet fibers were well-retained after the drying process. As revealed in Fig. 5.27 (a), by wrapping the wet graphene fiber onto bars with different geometries, triangle and quadrangle configurations were obtained. The graphene fiber could be folded into 3D architectures, as in Fig. 5.27 (b–e). With these special mechanical and electrical properties, the graphene fibers could find applications in textile electronics (Fig. 5.27 (f, g)). Moreover, the graphene fibers could be incorporated into polymer composite films. Figure 5.27 (h) shows a graphene fiber network embedded in a polydimethylsiloxane (PDMS) matrix with tolerance that was mechanically flexible and optically transparent.

In addition, functional components could be incorporated into the graphene fibers to further amplify their functionality. For instance, the introduction of Fe_3O_4 nanoparticles led to the occurrence of magnetic graphene hybrid fibers. Fe_3O_4 nanoparticles were added in GO suspensions, with a weight ratio of Fe_3O_4:GO = 1:10. The magnetic graphene hybrid fibers were derived by using the same hydrothermal treatment. The G–Fe_3O_4 fibers exhibited similar mechanical flexibility, but with extra sensitive magnetic responses.

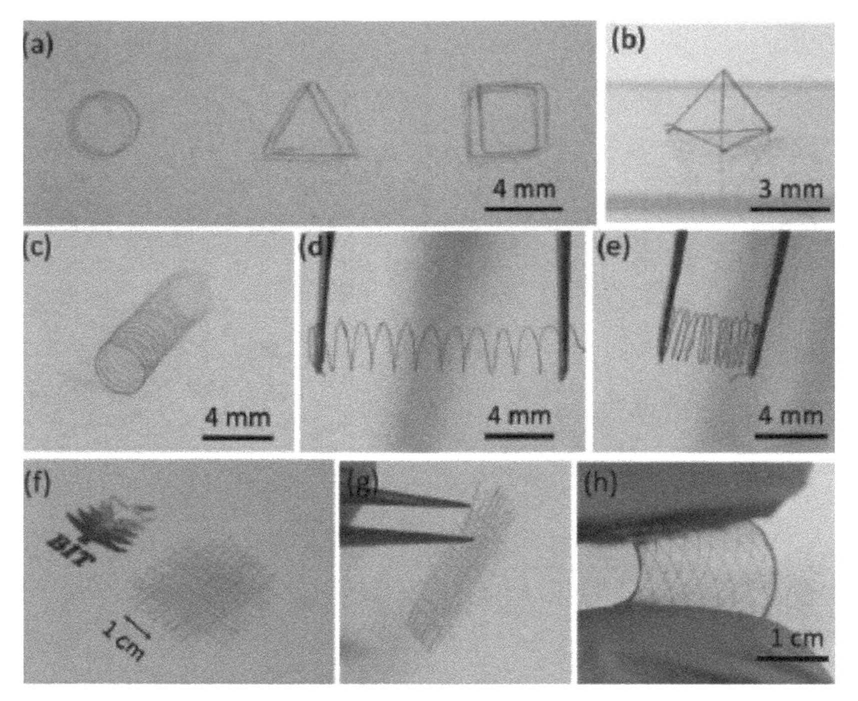

Fig. 5.27. Flexibility of the graphene fibers. (a, b) Handmade planar and 3D geometric structures with the graphene fibers. (c–e) Springs made of the graphene fiber at free, stretched and compressed status. (f, g) Photographs of the hand-knitted textile with the graphene fibers. (h) Photograph of the graphene fiber network embedded in PDMS matrix. Reproduced with permission from [40], Copyright © 2012, WILEY-VCH Verlag GmbH & Co. KGaA, Weinheim.

Because the as-obtained wet graphene fibers were highly porous, they can be used as hosts to incorporate other components during the drying process. For example, TiO_2 nanoparticles (NPs) were introduced into the framework of the graphene nanosheets by merging the wet graphene fibers in commercial TiO_2 aqueous suspension (10 mg·ml^{-1}) for 20 min with moderate stirring. The G–TiO_2 hybrid fibers with ~8 wt% TiO_2 NPs were produced after drying and annealing at 400°C for 30 min. The hybrid fibers showed promising photocurrent response with high repeatability, implying that there was a direct electron/hole injection between the TiO_2 NPs and the graphene nanosheets through the photoexcitation on TiO_2, which could be used in optoelectronic systems, such as photodetectors, photocatalysts and photovoltaic cells.

Further development of the hydrothermal method led to a dually geometric confinement approach, which enabled the synthesis of meter-long hollow GFs (hGFs) with tunable diameters [41]. The hGFs were microtubings (μGTs), which could find potential applications in fluidics, catalysis, purification, separation, sensing and environmental protection. The meter-long μGTs had diameters tunable in the range of 40–150 μm. The μGTs could be shaped to have hierarchical multichannels. In addition, selectively site-specific functionalization could be realized on the outer-wall, inner-wall, outer/inner-wall, and within-wall, in a well-controllable way, which made it possible to create μGTs with the desired properties for targeted applications, such as stimulus-responsive devices and self-powered micromotors.

To produce the μGTs, removable metal wires, e.g., Cu wire, were put inside the glass pipeline, as shown in Fig. 5.28 (a). GO suspensions were then filled into the glass pipeline, followed by thermal treatment at 230°C for 2.5 h. The Cu wire served as a support to induce the accumulation of the hydrothermally reduced GO, during the drying process, as illustrated in Fig. 5.28 (b–d). Therefore, μGTs with a diameter that was related to the diameter of the Cu wire were obtained after the Cu wire was etched in aqueous $FeCl_3$/HCl solution, as demonstrated in Fig. 5.28 (e).

Therefore, the method was called the dual geometric confinement approach. In this case, the glass pipeline defined the shape and outer diameter of the fibers, while the embedded Cu wires facilitated the formation of the tubular structure and determined the inner diameter of the fibers. The as-obtained

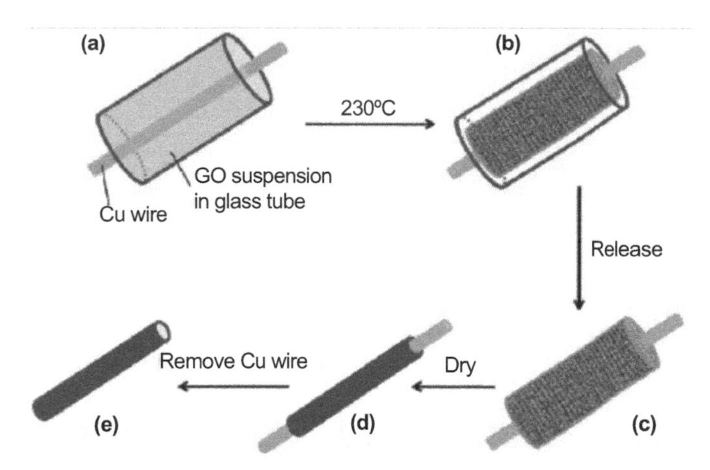

Fig. 5.28. Schematic diagram showing the fabrication process of the μGTs. (a) GO suspension was filled in the glass pipeline, in which a Cu wire was pre-intercalated. (b) The glass pipeline was heated at 230°C and hydrothermally reduced GO was deposited around the Cu wire. Two ends of the glass pipeline were sealed prior to the hydrothermal treatment. (c) The Cu wire enwrapped with hydrothermally converted graphene nanosheets was taken out from the glass pipeline. (d) Densely packed graphene layer on the Cu wire was formed after drying naturally. (e) The μGT was obtained after removing the Cu wire in aqueous 2.5 M $FeCl_3$ solution containing 0.5 M HCl. Reproduced with permission from [41], Copyright © 2012, American Chemical Society.

hydrothermally reduced GO fibers exhibited a highly 3D cross-linking porous structure, comprising of randomly entangled graphene nanosheets, as revealed in Fig. 5.29 (a, b). During the natural drying process, the capillary force induced due to the loss of water enabled the porous graphene nanosheets to be compacted along the Cu wire, as seen in Fig. 5.29 (c). The diameter of the sample was shrunk from 150 μm (Fig. 5.29 (a)) to about 100 μm (Fig. 5.29 (c)). Freestanding μGT was obtained, after the Cu wire was etched out, as demonstrated in Fig. 5.29 (d). The μGT possessed a wall of < 1 μm (Fig. 5.29 (e)), which was constructed by randomly packed graphene nanosheets (Fig. 5.29 (f)). Diameter of the μGTs was simply controlled by using Cu wires with different diameters. Two μGTs with diameters of about 150 and 40 μm are shown in Fig. 5.29 (g, h). Figure 5.29 (i) indicated that the μGT exhibited a tensile strength of 180 MPa, with an elongation at break of about 4% and a conductivity of ca. 10 $S \cdot cm^{-1}$.

The μGTs had stable flexibility and strong mechanical strength. The wet μGTs could have various predesigned configurations realized through the Cu wires. The predesigned structures were well-retained after drying. For instance, mechanically stable μGT springs that were highly stretchable and compressible could be obtained by enwinding the wet μGTs together with the Cu wire on a cylindrical bar, as illustrated in Fig. 5.30 (a, b). The spring shapes were maintained after drying and removal of the Cu wires. If a twist of two Cu wires (Fig. 5.30 (c), inset) was used as the templating core, then the μGT was helical in shape, as seen in Fig. 5.30 (c). Moreover, two-channel, three-channel and even four-channel μGTs could be developed by simply using two-, three- and four-ply Cu wires, as illustrated in Fig. 5.30 (d–f). The channels were well-separated by the graphene layers, making it possible to develop a multifunctional integrative microchannel system, with potential applications as vessels for macro/nanofluidic devices, multicomponent drug delivery and highly efficient catalyst supports.

The tubular structures of the μGTs offered opportunity to realize site-specific functionalization. Selective outer-wall, inner-wall, outer/inner-wall and within-wall modifications had been demonstrated in such fibers. With the presence of the Cu wires, the inner-walls of the fibers were protected, so that the outer surfaces could be selectively functionalized. For example, Pt nanoparticles with sizes of 200–300 nm were deposited onto the outer-wall of the μGTs, when the Cu wire supported μGTs were immersed in K_2PtCl_4 solution for 2 min, while the inner-wall was not affected. This functionalization was applicable to other metal nanoparticles as well.

To functionalize the inner-walls, Cu wires were pre-coated with a layer of Pt nanoparticles, with which the μGTs were prepared similarly. As the Cu wires were removed, μGTs with inner-walls distributed with

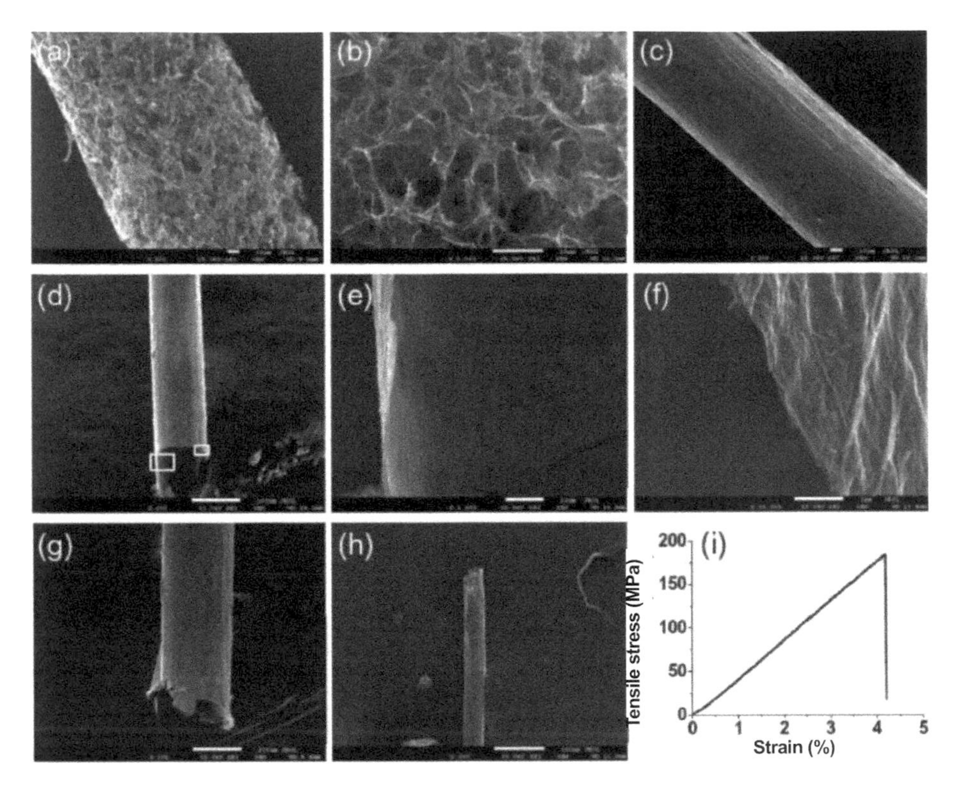

Fig. 5.29. Morphology and structural properties of the μGT. (a, b) SEM images of the hydrothermally converted graphene layer on a Cu wire with a diameter of about 100 μm, which was freeze-dried prior to the naturally drying process. (c) SEM image of the graphene on the Cu wire after naturally drying. (d) SEM image of a μGT after removing the Cu wire. (e, f) Close-view SEM images of the marked areas in (d). (g, h) μGTs generated by using Cu wires with different diameters (g: 150 μm, h: 40 μm). (i) Typical stress-strain curve of the as-prepared μGTs. Scale bars = 10 μm (a–c, e), 100 μm (d, g, h) and 1 μm (f). Reproduced with permission from [41], Copyright © 2012, American Chemical Society.

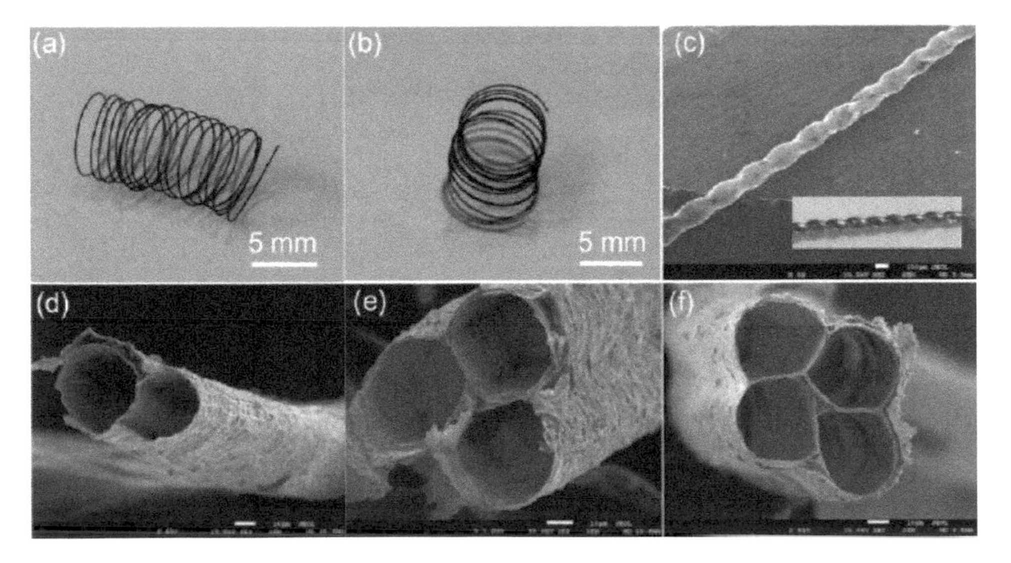

Fig. 5.30. Morphological controlling of the μGTs. (a, b) Photographs of a μGT spring lying and standing on a table. (c) SEM image of a helical μGT obtained by using a twist of two Cu wires of 100 μm in diameter (inset). (d–f) SEM images of the multichannel μGTs with a channel number of 2–4 (the Cu wire was 40 μm in diameter). Scale bars = 100 μm (c) and 10 μm (d–f). Reproduced with permission from [41], Copyright © 2012, American Chemical Society.

Pt nanoparticles were achieved. Density of the Pt nanoparticles on the inner-walls of the fibers was simply controlled by the density on the surface of the Cu wires. Both the inner-walls and outer-walls of the μGT could be functionalized, either symmetrically or asymmetrically. For instance, μGTs with inner-walls decorated with Pd nanoparticles were first obtained by using Pd nanoparticles-covered Cu wire, whose outer-walls were then deposited with Pt nanoparticles in the similar way as stated before.

If the dopants were introduced into the GO suspensions, the hydrothermally derived μGTs would have their walls thoroughly distributed with functional components. For example, when TiO_2 nanoparticles with an average size of 20 nm in diameter were dispersed into the GO suspensions, μGTs with TiO_2 homogeneously distributed in the tube walls were obtained. After drying, the TiO_2 nanoparticles were fully covered by the graphene nanosheets. In addition, μGTs with Fe_3O_4 magnetic nanoparticles had been prepared similarly. The TiO_2-modified μGTs possessed a fast photocurrent pulse upon exposure to light, while the Fe_3O_4-functionalized μGTs showed strong response to external magnetic fields.

In terms of real applications, the μGTs with inner-walls modified with Pt nanoparticles could be used as transporters, in which the Pt nanoparticles served as the catalysts to promote the decomposition of H_2O_2 to produce O_2. With one end being sealed, O_2 bubbles were ejected from the other side, thus driving the μGTs to move in the contrary direction, like a self-powered micromotor. Further improvements are expected to be made for more applications.

5.2.3. Chemical vapor deposition (CVD)

Chemical vapor deposition (CVD) graphene films have been used to fabricate graphene fibers through self-assembly [42]. The graphene fibers were porous and continuous fibers with tunable diameters and pore distribution. They were also mechanically flexible with high electrical conductivity. Graphene films were grown on Cu foils by a CVD method, with methane as the precursor. Four steps were involved in the 2D film to 1D fiber transformation process, as illustrated in Fig. 5.31 (a–d). Firstly, free-standing CVD graphene films were floated on water after the substrates were etched (Fig. 5.31 (a)). Secondly, the graphene films were collected and transferred onto the surface of ethanol. Once the graphene films were put on ethanol, their edges immediately scrolled up, thus leading to agglomerates, which sank into the liquid (Fig. 5.31 (b)). Thirdly, the graphene films were picked out (Fig. 5.31 (c)), and they then shrank into fibers after the evaporation of ethanol (Fig. 5.31 (d)). Eventually, graphene fibers were obtained after drying.

Figure 5.31 (e) indicated that the fibers obtained from 1 cm^2 films had a diameter of 20–50 μm. Geometric size of the fibers was mainly determined by the size of the original graphene films and could be controlled by the changing other parameters, e.g., the drawing rate. The graphene fibers were mechanically flexible, so that they could be conveniently manipulated. They exhibited an electrical conductivity of 1000 $S{\cdot}m^{-1}$, as demonstrated in the inset of Fig. 5.31 (e), due to the high quality of the CVD derived graphene nanosheets.

Figure 5.31 (f) shows SEM image of the fibers, which were wrinkled and highly porous. The porous structure forms upon ethanol evaporation during the drying process. The pores were formed by the scrolled graphene nanosheets, as illustrated in Fig. 5.31 (g). Figure 5.31 (h) shows typical Raman spectra of the graphene film and the fiber. There was only a small D peak at 1350 cm^{-1} in the pristine graphene film, while the fiber exhibited a higher D band intensity, due to the defected electronic structure related to the scrolling of the nanosheets. The G peaks at 1580 cm^{-1} of the film and the fiber were almost the same, which implied that the transformation had no effect on the C-C bonding structure. The 2D peaks at 2700 cm^{-1} were broadened, as the film was transformed into fiber, with an increase in the value of the full width at half maximum (FWHM) from 54.6 to 74 cm^{-1}, which was attributed to the overlapping of the Raman signals from different graphene nanosheets.

The self-assembly process experienced two stages, i.e., (i) scrolling of the film to form the fiber and (ii) shrinking of the fiber. The scrolling process was facilitated by the surface tension. Generally, because graphene films are relatively hydrophobic, they can be wetted by most organic solvents. On being contacted with ethanol, the graphene film became very flexible. At the same time, solvent evaporation played an important role in the formation of the graphene porous structure. After the fiber was formed, its surface

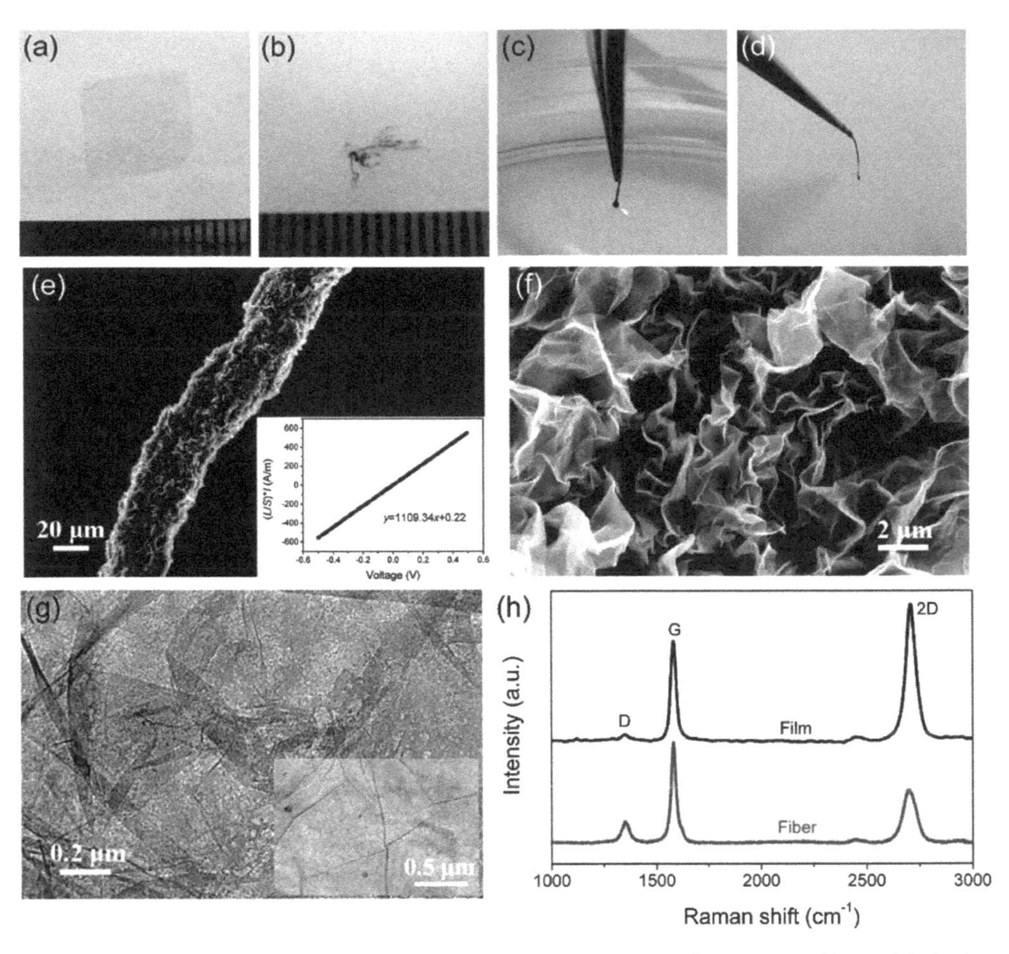

Fig. 5.31. Graphene fibers through the film to the fiber self-assembly. (a) Free-standing graphene film on deionized water surface. (b) Scrolled and wrapped graphene film when transferred onto ethanol. (c) Graphene fiber drawn out of ethanol. (d) Porous and wrinkled structure of the dried fibers. (e) Low-magnification SEM image of a graphene fiber, with the inset showing the current density-voltage curve, yielding a conductivity of 1000 $S \cdot m^{-1}$. (f) SEM and (g) TEM images of the graphene fiber, showing the wrinkled and porous structure, with the inset showing a TEM image of the pristine graphene film. (h) Raman spectra of the graphene film and the fiber. Reproduced with permission from [42], Copyright © 2011, American Chemical Society.

was still covered by a liquid layer of ethanol. Figure 5.32 (a) shows the formation process of the porous graphene fibers. Evaporation of ethanol triggered condensation of the droplets, so that graphene nanosheets were closely packed near the residue droplets. Porous structure was formed as the fibers were fully dried. Due to van der Waals force, the graphene nanosheets shrank and crumpled during the evaporation of ethanol, as shown in Fig. 5.32 (b).

Pore size of the porous graphene fibers could be controlled by controlling the rate of solvent evaporation. For example, the evaporation rate of ethanol is lower than that of acetone. Consequently, the fiber produced with ethanol was less porous than that obtained with acetone, as demonstrated in Fig. 5.32 (c) and (d), respectively. Figure 5.32 (e) indicated that quick shrinking was observed in the graphene fiber, when it was heated by using an infrared lamp. If a diluted ethanol solution with water was used, the fiber shrank very slowly, so that dense fiber was obtained. In porous graphene fibers, pore diameters were in the range of 500–2000 nm, with average values of 1170 nm for ethanol (Fig. 5.32 (f)), 800 nm for acetone (Fig. 5.32 (g)) and 1130 nm for ethanol plus heating (Fig. 5.32 (h)). Therefore, the higher the evaporation rate of solvent, the harsher the surface and the more porous the fiber would be. In addition, high evaporation rates led to a more uniform distribution of the pores.

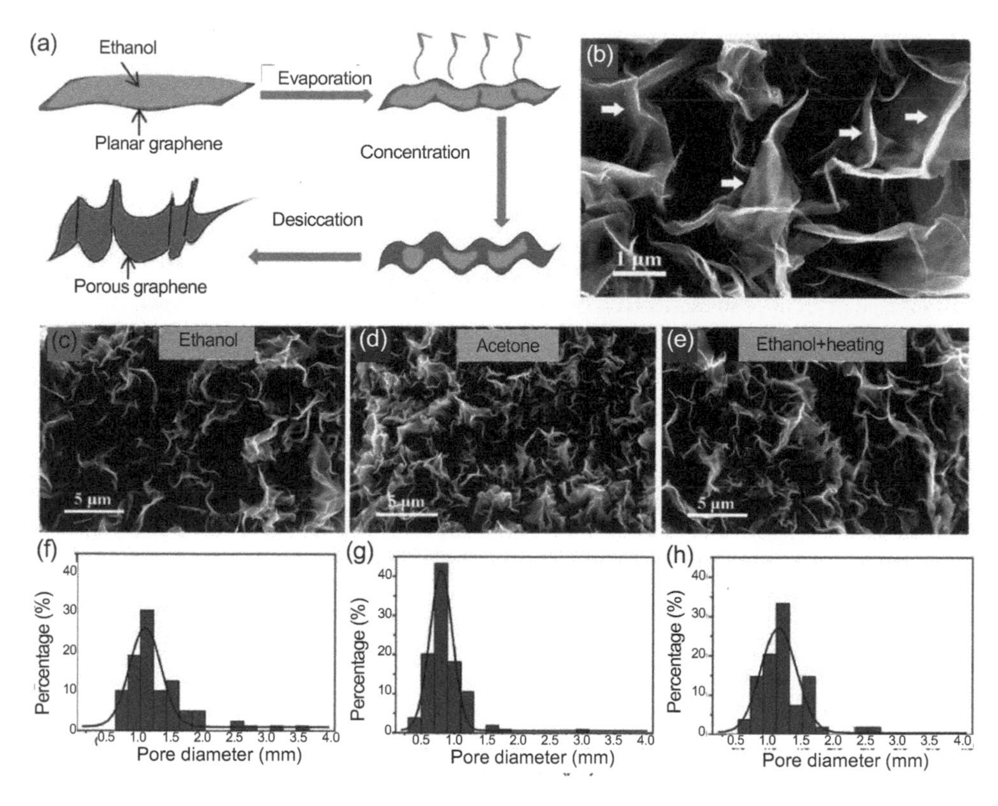

Fig. 5.32. Formation and characterization of the porous structure of the graphene fibers. (a) Schematic diagram showing the formation process of the porous structure. (b) SEM image of the porous graphene, with the arrows indicating the pinched and joint graphene nanosheets. SEM images of the graphene fibers obtained at different evaporations rates: (c) ethanol, (d) acetone and (e) ethanol with heating. Pore size distributions of the graphene fibers obtained at different evaporation rates: (f) ethanol, (g) acetone and (h) ethanol with heating, corresponding to average pore diameters of 1170, 800 and 1130 nm. Reproduced with permission from [42], Copyright © 2011, American Chemical Society.

Due to their porous structure, the graphene fibers could be used as the electrodes of electrochemical devices. They could also be further functionalized by incorporating other components. One example was to include MnO_2 nanoparticles in order to enhance the electrochemical properties. MnO_2 nanoparticles with contents of 1–3% were deposited onto the graphene fibers by using the chemical precipitation method, leading to G–MnO_2 hybrid fibers. Prototype supercapacitors made with the hybrid fibers exhibited specific capacitances of 1.4 $mF \cdot cm^{-2}$ at 10 $mV \cdot s^{-1}$ and 0.65 $mF \cdot cm^2$ at 500 $mV \cdot s^{-1}$.

The CVD method was further used to directly grow graphene layers on Cu meshes [43]. After deposition, the Cu wires were etched out with $FeCl_3$/HCl aqueous solution, so that hGFs with a network configuration were developed, which were named as graphene-based woven fabrics (GWFs). The GWFs could be transferred onto polydimethylsiloxane (PDMS) substrates, thus forming graphene-based woven fabric/PDMS hybrid films, which could find various applications. The GWFs possessed various special structural and characteristic features. Firstly, they exhibited high structural integrity, with much higher mechanical strength than polycrystalline graphene films. Secondly, there were micron holes present in the GWFs, so that they were highly permeable, as compared with graphene monolayers. Also, they had a unique exponential resistive response to external loadings, which could be used to develop strain sensors.

Two applications have been demonstrated by using the GWFs, i.e., (i) GWFs/polymer composites and (ii) GWFs/semiconductor solar cells. GWF/polymer composites were made by embedding aligned GMRs in polydimethylsiloxane (PDMS) to form 2D networks that possessed multi-joint conductive channels. GWF/semiconductor solar cells were constructed by using the GWFs as transparent electrodes in Schottky junctions. It was found that improved photovoltaic conversion efficiency was achieved by using the GWF/semiconductor solar cells. Furthermore, the special GWFs could be used directly as permeable membranes, while the periodic voids could be filled with other functional components to achieve enhanced functionality or multiple functionalities.

Figure 5.33 (a) shows a schematic diagram illustrating the steps used in the fabrication of the GWFs. Figure 5.33 (b) shows optical and plane view SEM images of a copper mesh before and after the graphene growth. Figure 5.33 (c) shows photographs of the as-obtained GWF films floating on water and deposited on glass and PET. Well-aligned arrays of GMR lines, with width and spacing of about 100 μm and 150 μm, were observed. The GWFs consisted of warp and weft GMRs through intersection, thus leading to precisely controlled dimension of the GWF pattern.

TEM image of a GMR and the corresponding electron diffraction pattern are shown in Fig. 5.33 (d). The GMR surfaces were clean and homogeneous. Typical six-fold symmetry of graphene and graphite was demonstrated by the electron diffraction pattern. Two sets of the hexagonal spots, with a rotation of 26°, were present in the diffraction pattern, coming from the front and back layers of the GMRs, due to the overlapping behavior. In other words, the graphene was not monolayered. Although monolayer graphene was grown on the Cu mesh, overlapping occurred as the Cu wires were etched out, due to the fragile nature of the graphene nanosheets. The presence of bilayered and few-layered graphene nanosheets was confirmed by Raman spectra.

GWFs could also be deposited on Ni meshes. However, the samples on Ni meshes were multilayered graphene, which in turn largely increased the etching process and residual nickel-carbon compounds were left inside the GMRs after etching. As a result, the as-obtained GWFs possessed higher structural integrity as compared with the GWFs on Cu meshes. The GWFs could be collected by using a quartz O-ring.

Optical image and magnified top view images of regions of the samples deposited on a silicon wafer are shown in Fig. 5.34 (a) and (b). The as-obtained GWFs exhibited a planar structure with two sets of GMRs that were intercalated, forming self-locked planar fibrous net-like configurations. Less transparent areas were observed in Fig. 5.34 (b), due to the overlapping or crumpling of the graphene layers. The lower and upper parts of the overlapped GMRs at the cross region were adhered and stuck together firmly,

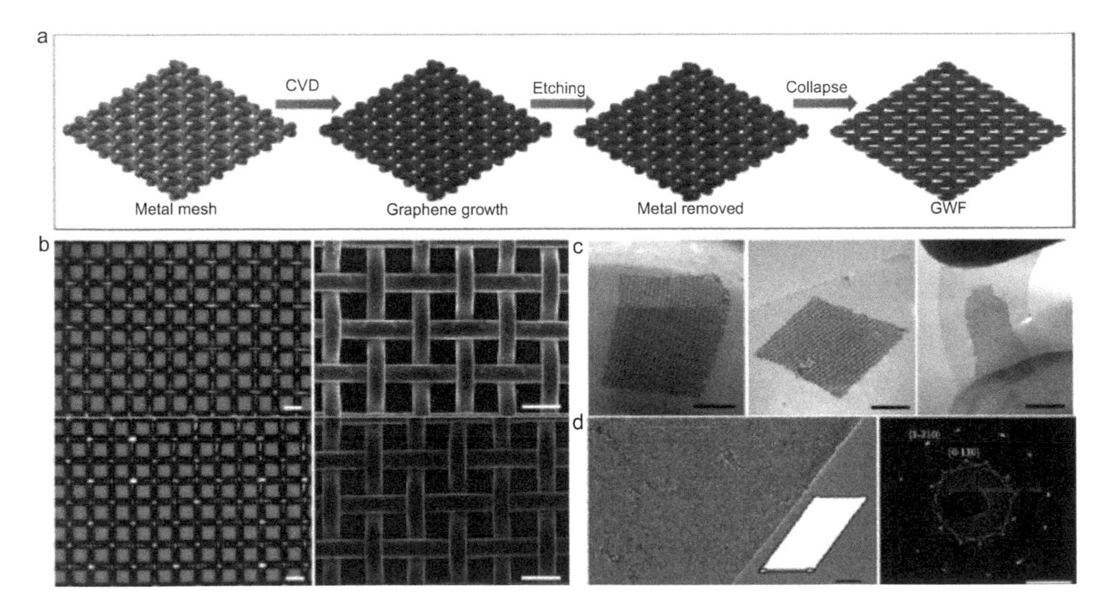

Fig. 5.33. Fabrication and characterization of the GWFs with Cu wire meshes as substrates. (a) Schematic diagram showing the steps to fabricate the GWFs. (b) Macroscopic optical images (left), top-view SEM images (right) of the Cu meshes before (top) and after (bottom) the growth of graphene. Scale bars = 200 μm. (c) Optical images of the GWF films floating on water and deposited on glass and PET substrates. Scale bars = 5 mm. (d) TEM image of a GMR and selected area electron diffraction pattern from the region marked with a yellow box. Scale bars = 50 nm (left) and 5 (1/nm) (right). Reproduced with permission from [43], Copyright © 2012, Macmillan Publishing Limited.

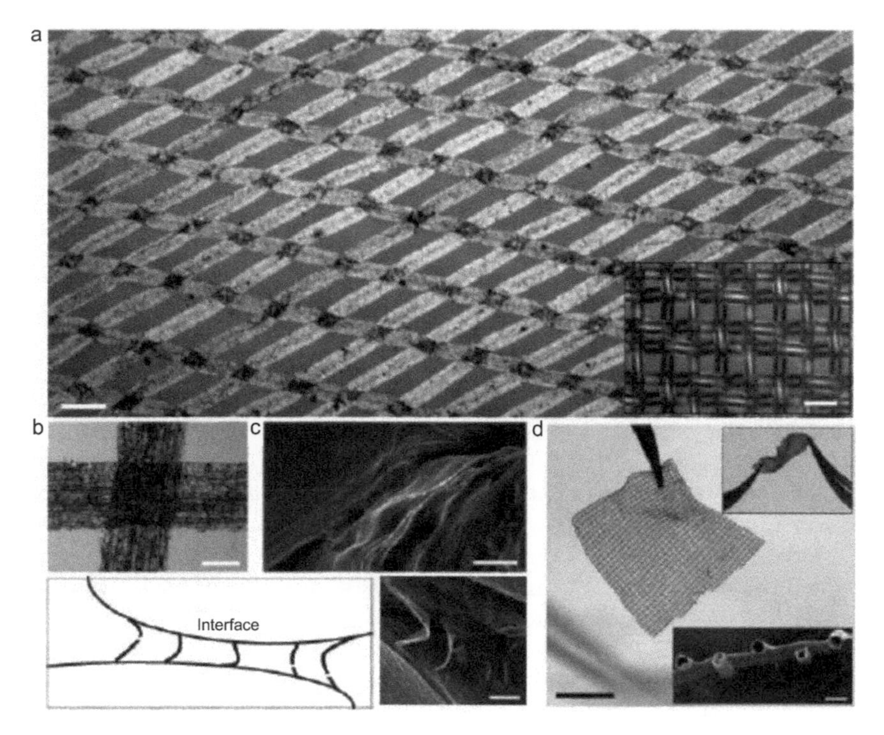

Fig. 5.34. Large-scale GWFs in supported and free-standing states. (a) Large-area optical image of a representative region of the GWF film. Scale bar = 100 μm. Inset shows optical image of a nylon woven fabric for comparison. Scale bar = 200 μm. (b) Top-view optical image (scale bar = 50 μm) and (c) high magnification SEM images of cross-sectional views of interlacing points of the GMRs. Scale bars = 5 μm (top) and 1 μm (bottom). (d) Optical images of the flexible GWF/PDMS composite films. Scale bar = 5 mm. The top inset shows the twisted GWF film held by tweezers. The bottom inset shows the cross-sectional SEM image of the composite film. Scale bar = 100 μm. Reproduced with permission from [43], Copyright © 2012, Macmillan Publishing Limited.

as illustrated in Fig. 5.34 (c). To fabricate GWF/PDMS composites, the samples before etching were covered by a thin layer of PDMS with controlled thicknesses, followed by crosslinking and curing. The PDMS layers acted as a support to prevent graphene networks from collapsing during the etching of the Cu meshes. Figure 5.34 (d) shows a free-standing GWF/PDMS composite film. The presence of the thin layer of PDMS coating did not compromise the optical and electrical properties of the composites, which were sufficiently flexible to be manipulated.

As expected, the structures of the GWFs were closely related to the diameter of the copper wires and density of the meshes. The packing density of the GMRs could be tailored through post treatments, such as controlling the flow of the liquid during the film transferring. Optoelectronic performance was characterized by sheet resistance (Rs) and transparency of the GWFs, which were determined by their geometrical parameters, such as GMR width and spacing between adjacent graphene nanosheets. It was found that film transparency was largely influenced by the density of cavity in the GWFs. As compared with the indium tin oxide (ITO), the performance of the GWFs needs to be further improved for practical applications. Nevertheless, this new type of flexible photovoltaic and composite materials offer an opportunity to develop flexible electronic devices.

3D graphene networks were prepared through substrate-assisted reduction and assembly of GO (SARA-GO) [44]. GO nanosheets were reduced and assembled into 3D networks on various substrates, including active metals of Zn, Fe and Cu, inert metals of Ag, Pt and Au, semiconducting Si wafers, nonmetallic carbon-based films and even indium-tin oxide (ITO) coated glass. Various graphene assemblies have been developed, including microtubes, four-way pipes, spiral tubes, multi-channel networks and micropatterns, which could be used to fabricate binder-free rechargeable lithium-ion batteries (LiBs).

Figure 5.35 (a) shows a photograph of the rGO layer on a Cu foil, which was obtained by merging the Cu foil in aqueous GO suspension with a concentration of 1 mg·ml^{-1} and pH = 4.0. The rGO appeared as a black layer covering the whole surface of the Cu foil, which was composed of a 3D network of rGO nanosheets as shown in Fig. 5.35 (b). The amount of the rGO deposited on the Cu foil was controlled simply by controlling the reaction time. Figure 5.35 (c) shows a TEM image of the as-obtained rGO nanosheets. There were nanoparticles present on the rGO nanosheets, which could be removed by washing with aqueous HCl solution, as demonstrated in Fig. 5.35 (d, inset). As the GO was reduced to graphene, the Cu substrate was oxidized into oxide. Other active metal foils, such as Al, Zn, Fe and Co, were also used to fabricate a 3D graphene network, as seen in Fig. 5.35 (e–h).

As shown in Fig. 5.36 (a, b), a graphene microtube (mGT) was formed through the reduction of GO on a Cu wire. The Cu wire served as a support to facilitate the aggregation of the rGO network. Diameter of the mGTs was dependent on the thickness of the Cu wire that was etched in aqueous $FeCl_3$/HCl solution (Fig. 5.36 (b)). Moreover, other configurations included spiral tubes, a four-way pipe (Fig. 5.36 (c)) and multi-channel network of graphene nanosheets (Fig. 5.36 (f)) with tunable length and diameter at large scale, which could be achieved similarly.

Figure 5.36 (c) shows the as-prepared four-way pipe, with the inserts illustrating the SEM images of the four nozzles. Hollow interior characteristic of the four-way pipe is shown schematically in Fig. 5.36 (d). Cu mesh (Fig. 5.36 (e)) and Fe mesh could induce the woven fabric of mGTs, as revealed in Fig. 5.36 (f–h). The region-defined patterns of rGO network were also available through the SARA-GO process. For example, Fig. 5.36 (i) indicated that rGO patterns on Pt foil were fabricated by pressing a Cu grid (200 μm in side length) on a Pt foil, which was put in a GO suspension. After that, the Cu grid was removed through etching. The region-confined 3D graphene network was completely retained after freeze-drying (Fig. 5.36 (j)).

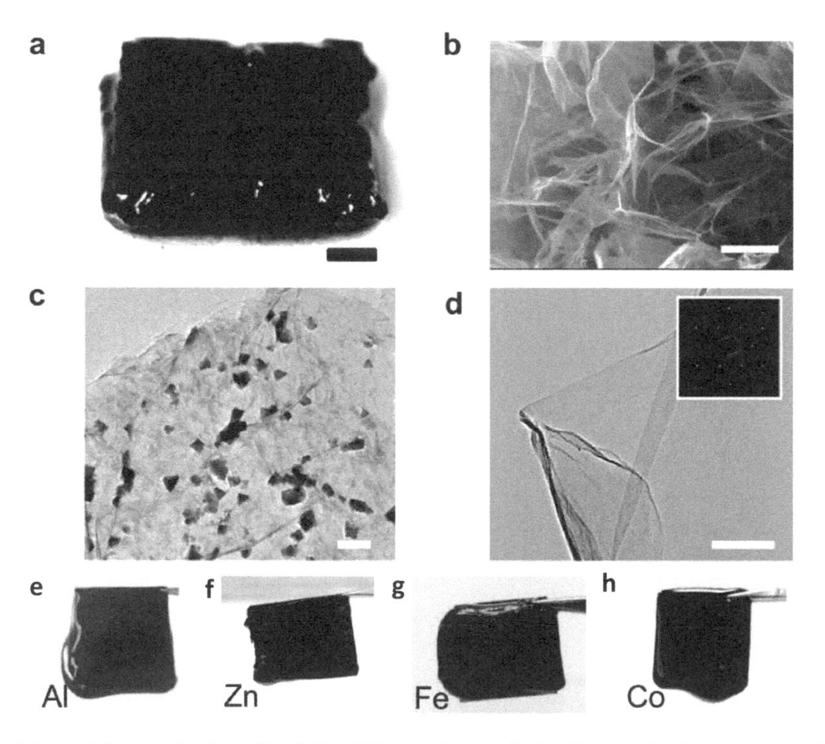

Fig. 5.35. Deposition and characterization of the SARA-GO on active metals. (a) Photograph of the as-deposited 3D rGO on a Cu foil. (b) SEM image of the freeze-dried sample of (a). (c, d) TEM images of the rGO on Cu foil before and after treatment with aqueous HCl solution, with the inset in (d) to be the corresponding electron diffraction pattern. (e–h) Photographs of the as-deposited rGO layers on Al, Zn, Fe and Co foils. Scale bars = 1 cm (a), 10 μm (b), 200 nm (c) and 100 nm (d). Reproduced with permission from [44], Copyright © 2013, Macmillan Publishing Limited.

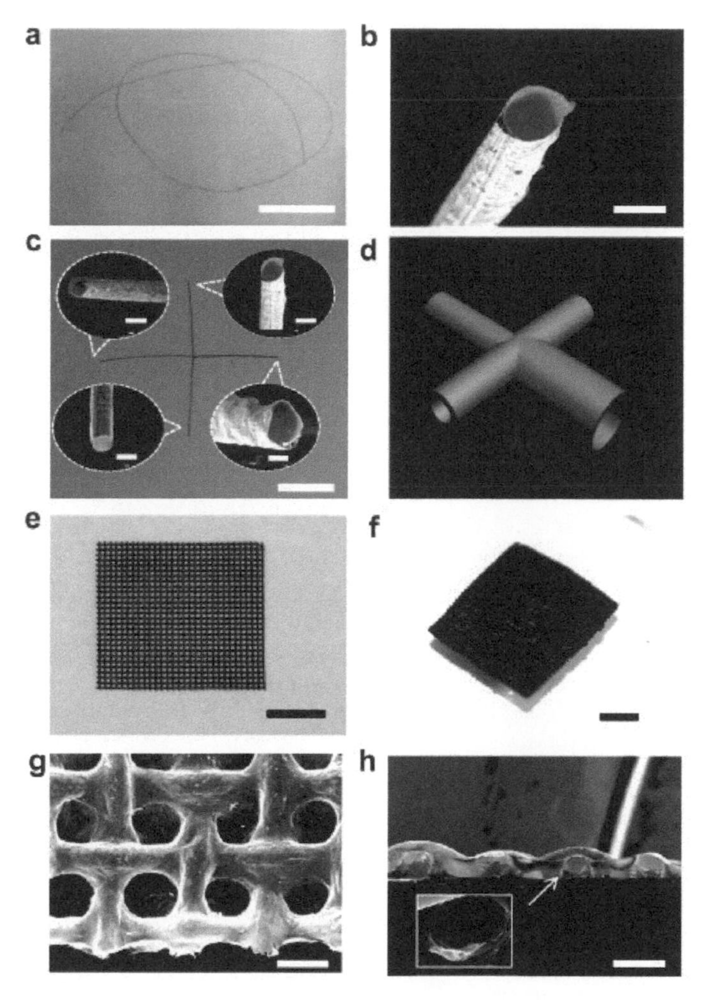

Fig. 5.36. Fabrication and characterization of the graphene microtubes, four-way pipe, multi-channel network and graphene micropatterns by using the SARA-GO process. (a, b) Photograph and SEM image of the graphene microtube produced by using SARA-GO on a Cu wire. (c) Photograph of a four-way tube with the corresponding SEM images of the nozzles. (d) Schematic diagram of (c). (e, f) Cu mesh and the derived graphene mesh through SARA-GO. (g, h) Top and side view SEM images of the graphene mesh in (f), respectively. Scale bars = 1 cm (a, c, e, f), 50 μm (b), 500 μm (g, h) and 100 μm (c (inset)). Reproduced with permission from [44], Copyright © 2013, Macmillan Publishing Limited.

5.2.4. Graphene ribbon fibers from unzipped-CNTs

Graphene nanoribbons with high aspect ratios can be obtained by unzipping CNTs [45–48]. These graphene ribbons have been used to prepare graphene fibers. A scalable method has been developed to fabricate long and narrow graphene nanoribbons, with which large graphene nanoribbon sheets were assembled and aligned first, while macroscopic neat graphene nanoribbon fibers were then obtained by twisting the sheets [49]. The graphene ribbon fibers showed high conductivity and good mechanical performance. The graphene ribbons were prepared by unzipping aligned nanotubes through oxidation, which were then reduced through thermal reduction. The graphene nanoribbons were highly aligned before and after the reduction process, which was confirmed by polarized infrared spectra. The graphene nanoribbon yarns exhibited a much higher electrochemical performance than the conventional twist-spun MWCNTs yarns, which was attributed to the incomplete reduction of the graphene nanoribbons. Most importantly, the conversion process could be scaled up for industrial production for applications as woven electrodes of fuel cells, supercapacitors and batteries.

Representative MWCNTs sheets that were drawn from spinnable nanotube forests are shown in Fig. 5.37 (a–c). The MWCNTs sheets were chemically oxidized and thoroughly washed to form free-standing graphene oxide nanoribbon sheets (GONSs). After the GONSs were withdrawn from the acidic aqueous solution containing hydrochloric acid, they collapsed into flexible gelated fibers. The conversion of MWCNTs sheets to GONSs was confirmed by XRD characterization results, as demonstrated in Fig. 5.37 (d). Diffraction pattern of the pristine MWCNTs sheets was characterized by a peak at 25.8°, corresponding to the typical inter-wall distance of 3.4 Å in MWCNTs. After oxidation, a broad peak located at 10° was observed, corresponding to the (002) reflection of ~9 Å between the graphene oxide planes.

TEM images of the MWCNTs and the GO ribbons are shown in Fig. 5.37 (e, f). The MWCNTs were completely (Fig. 5.37 (e)) converted into GO nanoribbons (Fig. 5.37 (f)). Figure 5.37 (g) shows FTIR absorption spectrum of the aligned GONS (black line), revealing the presence of stretching modes ~1736 cm^{-1}, which were attributed to edge terminal functional groups, such as carboxyls (-COOH) and carbonyls (–C=O). The MWCNTs had infrared inactive modes for 10 layers (Fig. 5.37 (g) red line). After being unzipped, the infrared active vibrational modes of sp^2C (C = C bond at 1590 cm^{-1}) appeared, indicating the formation of graphene oxide nanoribbons, with edge functional groups, such as carboxyl and carbonyl

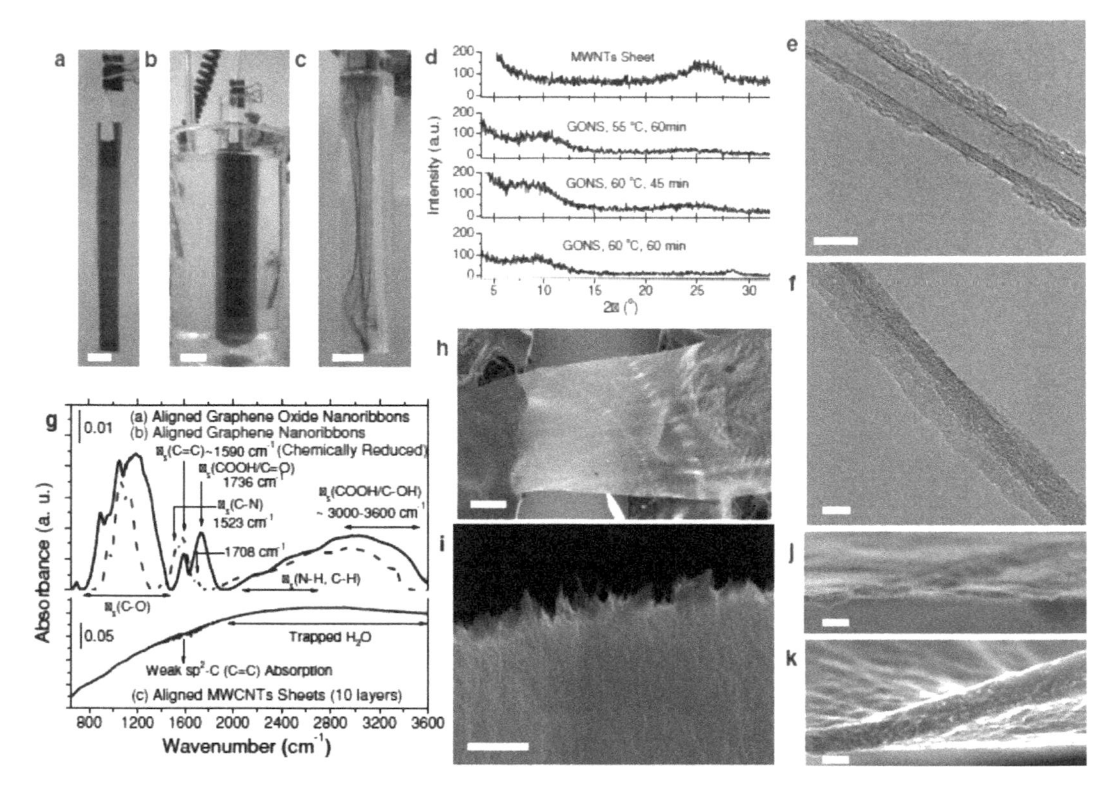

Fig. 5.37. Preparation and characterization of the aligned graphene oxide nanoribbon sheets (GONS). (a) 20 layers of MWCNTs sheets placed on a PTFE frame. Scale bar = 15 mm. (b) MWCNTs sheet immersed in the oxidizing bath for chemical unzipping. Scale bar = 15 mm. (c) Final GONS immersed in an acidic aqueous solution containing hydrochloric acid. Scale bar = 15 mm. (d) XRD patterns (Cu K α radiation) of the aligned MWCNTs sheets with 100 layers (top curve) and aligned GONS obtained from the 100 layers MWCNTs sheets through chemical unzipping. (e) HRTEM of an individual MWCNT. Scale bar = 5 nm. (f) HRTEM of an individual graphene oxide nanoribbon. Scale bar = 10 nm. (g) Infrared absorbance spectra of the aligned MWCNTs sheets (solid red curve). Transmission spectra of the unzipped MWCNTs sheet (solid black curve) and nanoribbons exposed to hydrazine vapor (blue dashed curve). (h) SEM image of the GONS obtained from unzipped MWCNTs sheets with 20 layers. Scale bar = 200 μm. (i) Top-view SEM image of the fractured section of the GONS. Scale bar = 400 nm. (j, k) SEM images of the GONS with thicknesses of about 100 nm and about 250 nm, respectively. Scale bars = 200 nm. Reproduced with permission from [49], Copyright © 2012, WILEY-VCH Verlag GmbH & Co. KGaA, Weinheim.

at 1708 cm^{-1}, 1151 cm^{-1} and 1076 cm^{-1} and hydroxyl at ~3000–3600 cm^{-1}. The broad infrared absorption region over 3000–3600 cm^{-1} was also partly contributed by the trapped physisorbed water molecules.

After the GO nanoribbons were reduced with hydrazine (N_2H_4) vapors, there was a sharp drop in the total infrared absorbance and IR intensities at 1708–1736 cm^{-1}, as illustrated in Fig. 5.37 (g) (blue dashed line), confirming the absence of carboxylic and carbonyl groups. A small red shift of the sp^2C (C=C) stretch band over 1585–1581 cm^{-1}, was observed after the chemical reduction process, due to the replacement of oxygen by nitrogen and H. At the same time, new stretches, such as C-N (weak band at about 1531 cm^{-1}), N-H and C-H (broad band at ~2100–3500 cm^{-1}), were present. Fabrication process of the GONS was demonstrated in Fig. 5.37 (h, i). They could also be fabricated by layering pristine MWCNTs sheets onto a flat rigid PTFE substrate. Thickness of the GONS was directly determined by the number of the pristine MWCNTs layers placed on the PTFE frame, as seen in Fig. 5.37 (j, k). Infrared polarization studies confirmed that most of the graphene nanoribbons were aligned and that the edge ether configuration was the most stable for the oxidized edge termination.

Graphene fibers or yarns can be obtained accordingly. The original MWCNTs sheets were first laid onto a PTFE frame, so that the unzipped free-standing graphene oxide nanoribbon sheets would collapse into graphene oxide nanoribbon yarns (GONYs), due to surface tension, when they were taken out from the solution, as illustrated in Fig. 5.38 (a–c). Figure 5.38 (a) shows a photograph of a MWCNTs sheet, with thirty layers of about 5 mm in width, which was wrapped vertically around a PTFE framework and immersed in an oxidizing bath and then a rinsing solution, thus leading to the GONS. The as-drawn fiber appeared as a gel because it contained a large amount of liquid, as demonstrated in Fig. 5.38 (b). H_2O_2 was used to remove the chlorine gas bubbles, so that no macroscopic structural defects were formed in the fibers. The formation of the graphene oxide gel phase was attributed to the large edge-to-surface-area ratio and edge functional groups in nanoribbons, which enabled their interaction with the solvent molecules in between the graphene nanoribbon bundles. After drying at room temperature, the GONY shrank from about 1 mm to about 25 μm in diameter, as seen in Fig. 5.38 (b–d).

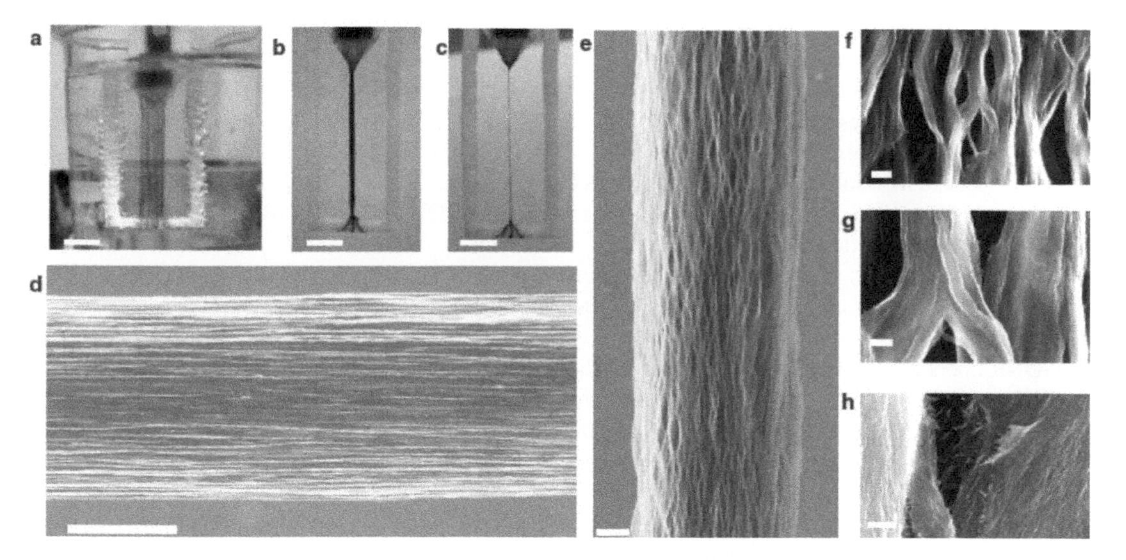

Fig. 5.38. Fabrication and characterization of the graphene nanoribbon yarns, illustrating the sequential steps of graphene nanoribbon yarn fabrication. (a) Photograph of a highly aligned and free-standing GONS in the acidic aqueous solution after the chemical unzipping process. Scale bar = 7.5 mm. (b) Collapse of the GONS into a GONY due to surface tension after being taken out from the liquid. Scale bar = 7.5 mm. (c) Photograph of the GONY after drying. Scale bar = 7.5 mm. (d) SEM image of the dry graphene nanoribbon yarn. Scale bar = 20 μm. (e) SEM images of the graphene nanoribbon yarn after chemical reduction in hydrazine vapors. Scale bar = 20 μm. (f–h) High magnification SEM images showing micrometer sized porosity and details of surface of the yarn and the graphene nanoribbons. Scale bars = 2 μm (f), 600 nm (g) and 200 nm (h). Reproduced with permission from [49], Copyright © 2012, WILEY-VCH Verlag GmbH & Co. KGaA, Weinheim.

Figure 5.38 (e) indicated that the GONY had a thickness of about 100 μm, after it was treated in N_2H_4 vapors at 95°C for 10–15 min. Micron-sized pores were distributed evenly in the yarn, as demonstrated in Fig. 5.38 (f–h). The porous structure would ensure the adsorption of gas or liquid electrolyte into the core of the yarns, thus enhancing catalytic and electrochemical reactions and ionic diffusion for certain applications. Figure 5.38 (h) indicated that the characteristics of the chemically reduced nanoribbons in the yarn were not affected during the thermal annealing process.

The yarns exhibited promising electrical, mechanical and electrochemical properties. Electrical conductivity of the yarns could be greatly increased after chemical reduction in 95°C N_2H_4 vapors, followed by thermal annealing at 300°C or 900°C for 2 h in high purity Ar or diluted hydrogen (5 vol%) in Ar. Specifically, the graphene nanoribbon yarns reduced by using only the thermal process at 300°C in pure Ar or 5 vol% H_2-Ar showed a large resistive behavior during the cyclic voltammetry measurement. In contrast, good supercapacitor performance was observed if the yarns were first reduced in N_2H_4 vapors and then thermally treated, which was attributed to the unique porous structure formed during the N_2H_4 reduction.

Similarly, GONRs and chemically reduced graphene nanoribbons can also be dispersed with high concentrations in chlorosulfonic acid to form anisotropic LC phases for wet spinning of GFs [50]. The graphene oxide nanoribbons (GONRs) were prepared by unzipping MWCNTs through oxidation. The GONRs fibers could be thermally reduced (tr) into trGNR fibers. Alternatively, GONRs could be chemically reduced (cr) with hydrazine into crGNRs, with which were crGNR fibers that were fabricated through spinning. As the crGNR fibers were further annealed (a), a-crGNR fibers were obtained.

Both the crGNRs and GONRs could be dispersed in chlorosulfonic acid to form anisotropic liquid crystal phases at room temperature. Diethyl ether (boiling point $b_p = 35$°C, viscosity = 0.224 cP at 25°C) was used as the coagulation bath solvent. The spinning apparatus included a spinning chamber with a piston. The piston was connected to a pressure controller at one end of the chamber, while a spinneret (capillary tube) was affixed to the other end of the chamber. The lyotropic materials were extruded, through the small spinneret, into the coagulation bath with an air gap to produce aligned ribbons within the as-spun fibers.

The ribbons had an average length of 4 μm and widths of > 100 nm. AFM analysis indicated that the height of the ribbon was about 1.2 nm, implying that the ribbons were single layer graphene. 2 wt% GONRs and crGNRs were studied by using a polarized optical microscope (POM). Figure 5.39 shows birefringent patterns of the samples, which were typical of the liquid crystal phase. Similar results were observed in the samples with 8, 12 and 15 wt% GONR and 8 wt% crGNR, indicating that all of them were liquid crystal phases. As a result, both the GONRs and crGNRs could be used for wet spinning. It was necessary to mention that large aggregations started to appear in the 15 wt % GONRs solution, which was not suitable to spin continuous and defect-free fibers. The GONRs and crGNRs were dispersed in the acid through a protonation mechanism.

Figure 5.40 shows photographs of the spinning process, including extrusion with a 12 cm air gap, coagulation and fiber collection. The extrusion rate was 0.0066 ml·min^{-1}, length (L) of the spinneret was 2.54 cm and the orifice diameter (D) was 125 μm, corresponding to L/D ratio of 203. More than 10 meters of GONR fibers were obtained in 1 h. The size of the air gap was very important to ensure the performance of the fibers. The optimized air gap for stable spinning was 15 cm. If the air gap was too long or too short, then the fibers would have inferior tensile properties. The alignment of the as-spun fiber could be further improved, if a rolling drum was used to stretch the fiber during the spinning. It was found that the setup could support a higher draw ratio.

It has been demonstrated that the cross-sectional shape of the wet-spun fibers is closely related to the coagulation conditions. A coagulation bath should meet certain requirements. For example, the GONRs should not be soluble in the coagulation bath. The cross-sectional shape of the wet-spun fibers is determined by the deformability of the coagulated layers and the mass transfer rate difference. It means that the coagulation rate should be sufficiently high, so that the coagulated layers will have low gradient at the interface near the surface layer. In this case, the coagulated outer layers will not collapse toward the core, thus leading to fibers with a circular shape. The mass transfer rate difference should be sufficiently low, in order to maintain a state where the solvent diffusion rate out of the fiber should be similar to the

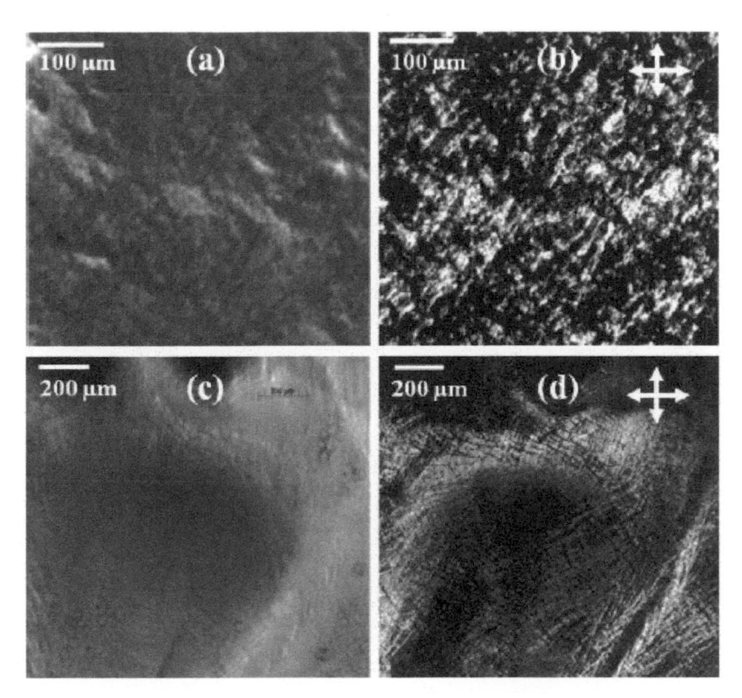

Fig. 5.39. Liquid crystals formed with the GONRs in chlorosulfonic acid as observed by using POM: (a) 2 wt% GONRs under transmission mode, (b) 2 wt% GONRs under cross-polarized mode, (c) 2 wt% crGNRs under transmission mode and (d) 2 wt% crGNRs under cross-polarized mode. Reproduced with permission from [50], Copyright © 2013, American Chemical Society.

Fig. 5.40. Dry jet wet spinning setup to fabricate the fibers. (a) Coagulation bath of ether covered with parafilm to slow down the evaporation. Continuous fibers were spun into the bath with a 12 cm air gap. The area framed by the red box was enlarged in the inset image, showing the fiber extruding from the spinneret. (b) Fibers spontaneously form a coil in the bottom of the coagulation bath. (c) Fibers collected on a Teflon drum. Reproduced with permission from [50], Copyright © 2013, American Chemical Society.

absorption rate of the nonsolvent. If the two rates are largely different, an irregular cross-sectional shape will likely be produced. Also, the viscosity of the bath should be as sufficiently low, in order to avoid surface etching of the fibers when going through the bath, thus leading to desirable surface morphology.

Water was first tested to prepare GONR solutions, due to the fact that it is easier to handle as compared with chlorosulfonic acid. Liquid crystal phase was observed in 5 wt% GONR aqueous solution. Ethyl acetate, methyl acetate and diethyl ether were used as the coagulation bath, which all could not result in fibers with promising morphologies. Therefore, water was not suitable to develop GONR fibers. Instead, chlorosulfonic acid had to be employed as the solvent.

Figure 5.41 shows SEM images of the GONR fibers spun from 8 wt% chlorosulfonic acid solution, with diethyl ether solvent as the coagulant bath, clearly indicating that the fiber possessed a nearly perfect circular cross-sectional shape. The GONR fibers spun with different air gaps had an average diameter of 54 μm, tensile strength of 33.2 MPa, modulus of 3.2 GPa and elongation of 1.64%. They were highly flexible. As shown in Fig. 5.41 (c), the fiber could be easily knotted into a loop with a minimum diameter of about 1 mm. The as-spun GONR fiber had a porosity, measured by using the BET method, corresponding to a surface area of 58 $m^2 \cdot g^{-1}$.

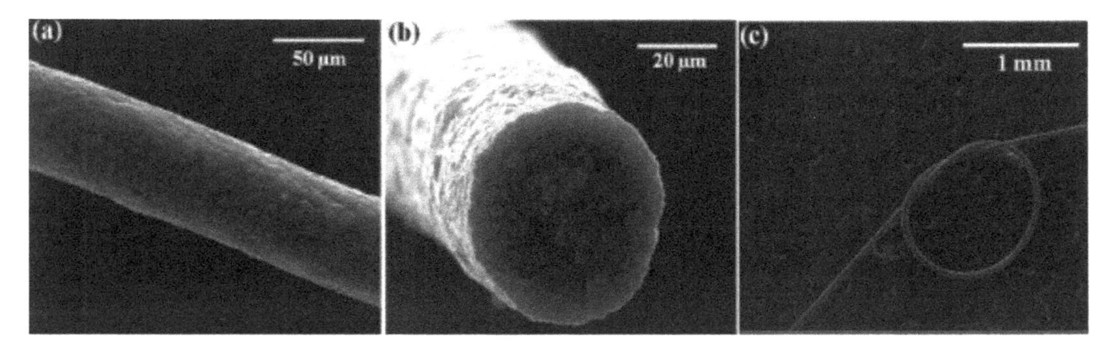

Fig. 5.41. Characterizations of the GONR fibers spun from the 8 wt% suspension. (a) Surface morphology of the as-spun fiber. (b) Transverse cross-section morphology. (c) Fiber knotted into a loop. Reproduced with permission from [50], Copyright © 2013, American Chemical Society.

Surface morphologies of the 1050°C trGNR with 1.3 g force (gf) pretension are shown in Fig. 5.42 (a, b). The diameter of the fiber was decreased because the oxygen functional groups and some voids were removed due to the thermal reduction. Optimized mechanical performances of the fiber included a tensile strength of 383 MPa, a modulus of 39.9 GPa and an elongation to break of 0.97%. A representative stress-strain curve is shown in Fig. 5.42 (c). The tensile strength and modulus were increased by nearly one order of magnitude, when compared with the as-spun GONR fibers. Higher molecular alignment leads to higher tensile modulus. The trGNR fibers showed higher tensile modulus than the graphene fibers fabricated by using other methods (~10 GPa), as discussed before [17, 33, 40]. Due to the presence of residual microvoids, the density of annealed trGNR fiber was 0.88 $g \cdot cm^{-3}$, which was less than half of that of conventional carbon fibers of 1.75–2.2 $g \cdot cm^{-3}$ [51]. As a result, specific strength of the trGNR fiber was 430 $kN^{*}m \cdot kg^{-1}$.

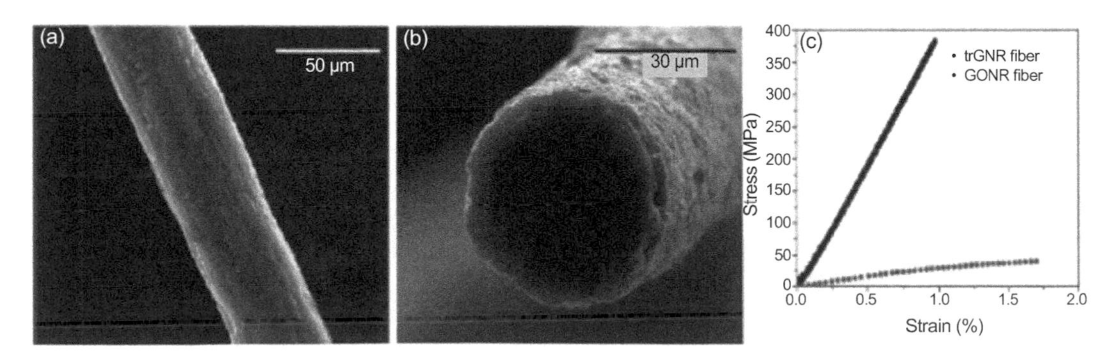

Fig. 5.42. (a) Surface morphology of the 1050°C annealed trGNR fiber. (b) Cross-section morphology of the 1050°C annealed trGNR fiber. (c) Typical stress-strain curve of the as-spun GONR fiber (12 cm air gap) and 1050°C annealed trGNR fiber (12 cm air gap) with 1.3 gf pre-tension. Reproduced with permission from [50], Copyright © 2013, American Chemical Society.

Microstructure of the fibers could be improved by using high temperature thermal annealing. In the as-spun fibers, the ribbons were poorly aligned, because the fibers were spun from a liquid crystal phase without significant alignment. After thermal treatment at 1300°C, large domains with partially ordered laminar structures were observed in the trGNR fibers. During the heat treatment, ribbons were transferred from limited ordering to regional ordering, as evidenced by XRD measurement with Herman's factor. The oxygen functional groups on the GONRs were removed during the heat treatment. The alignment could be further improved by applying pre-tensions during the thermal annealing process.

Electrical conductivity of the fibers could be significantly increased by removing the oxygen groups and aligning the graphene ribbons. For instance, the trGNR fiber treated at 1500°C possessed a conductivity of 285 $S \cdot cm^{-1}$, which was higher than that of the chemically reduced graphene fibers (35–250 $S \cdot cm^{-1}$) [17, 33] and the hydrothermally synthesized graphene fibers (10 $S \cdot cm^{-1}$) [40]. Although the GONRs with smaller sizes were supposed to have lower electrical conductivity than those of the GO flakes, this problem had been addressed by controlling the air gap during the spinning process and applying pre-tension during the thermal annealing step.

Spinning of the crGNR fibers was conducted similarly. Due to the high conductivity of crGNRs, the as-spun crGNR fibers already exhibited a pretty high conductivity of 5.6 $S \cdot cm^{-1}$. After being annealed at 1050°C without pre-tension, the derived a-crGNR fiber had a tensile strength of 90 MPa, which was lower than that of trGNR fiber. One possible reason was that the dispersion of crGNR in chlorosulfonic acid at high concentration was not very homogeneous, so that aggregations were formed. As a consequence, as-spun fibers were more discontinuous and contained more voids, which was evidenced by microstructural characterization results.

5.2.5. Other methods

An electrophoretic self-assembly method was used to fabricate polymer-free and surfactant-free fibers from rGO nanoribbons [52]. Because no additive was contained in the fibers, it was possible to monitor the state of oxidation as the graphene oxide nanoribbons (GONRs) were reduced. This state could be correlated to the property of the fibers for different applications. Electrical and field emission properties of the fibers as a function of the oxidation state of the GO nanosheets were evaluated. Especially for field emission, the rGONR fibers exhibited a low threshold electric field of 0.7 $V \cdot \mu m^{-1}$ and a giant current density of 400 $A \cdot cm^{-2}$. Additionally, the fibers were stable at a high current density of 300 $A \cdot cm^{-2}$.

The GONRs were prepared by unzipping CVD-produced carbon MWCNTs with solution of $KMnO_4$ in sulfuric acid. The optimal concentration of $KMnO_4$, in order to unzip the MWCNTs with a diameter of about 10 nm in the axis direction, was in the range of 800–900 wt%, with respect to the amount of the MWCNTs. For example, 30 mg of MWCNTs powder was dispersed in 30 ml of H_2SO_4, in which 250 mg of $KMnO_4$ was added. The mixture was then heated at 55°C for 30 min first. After that, the temperature was increased to 65°C so that the oxidation reaction was completed, and was then increased to 70°C, followed by cooling to room temperature. The GONRs were collected after thorough washing and drying. The dried GONRs were dispersed in a mixture solvent of DMF/H_2O with a volumetric ratio of 9:1. Chemical reduction of the GONRs was achieved with hydrazine monohydrate (1 μl for 3 mg GO) at 80°C for 12 h. Well-dispersed colloidal solutions were obtained after ultrasonication for 10 min.

A graphitic tip was used as a positive electrode which was immersed into the chemically reduced GONR colloidal suspension in a Teflon vessel in which a counter electrode was embedded. The immersed tip was separated about 5 mm away from the counter electrode. A constant voltage in the range of 1–2 V was applied between the electrodes during the withdrawal process of the graphitic tip, which was at 0.1 $mm \cdot min^{-1}$, at 20–25°C and 18–23% relative humidity. For thermal reduction, the chemically reduced GONR fibers were treated in Ar at atmospheric pressure, at temperatures of 200°C, 500°C and 800°C for 1 h. The rGONRs had widths of 15–35 nm and thicknesses of 2.5–6.0 nm, corresponding to few-layer stacks. Also, the rGONR fibers could be derived directly from the reduced GONR solution.

Figure 5.43 (A) shows the self-assembly method to fabricate chemically reduced GONR fibers by using electrophoretic assembling. The rGONR solution was injected into a cylindrical Teflon vessel with a diameter of 5 mm, so that the surface of the suspension appeared as a dome shape, in order to prevent

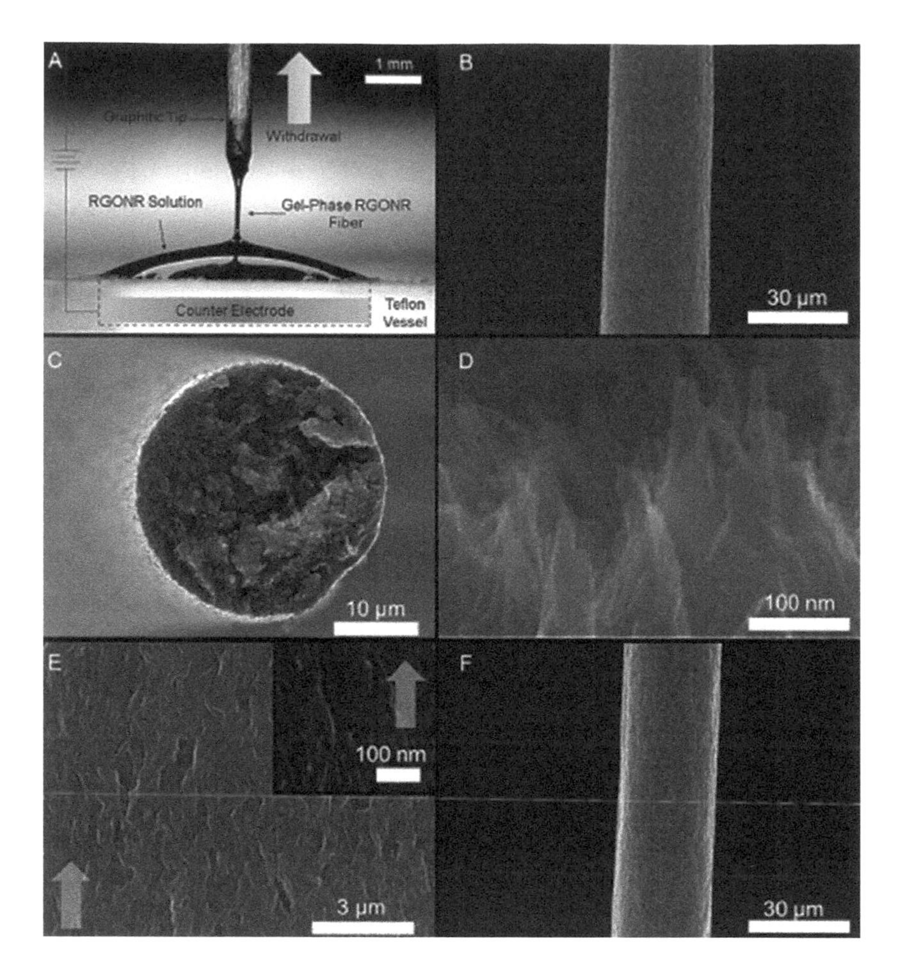

Fig. 5.43. (A) Photograph of a gel-phase rGONR fiber derived from chemically reduced GONR solution. (B) A solid rGONR fiber formed by using the electrophoretic self-assembly method. (C) Cross-section and (D) side-view SEM images of the fracture surface of the rGONR fiber. (E) Surface texture of the rGONR fiber and the partial alignment feature along the yarn longitudinal axis direction. The arrows indicate the axial direction of the rGONR fiber. The magnified image in the inset shows aligned nanoribbons. (F) SEM image of the fiber after thermal annealing at 800°C. Reproduced with permission from [52], Copyright © 2012, IOP Publishing Ltd.

the subsequently formed fibers from being attached to the inner side-wall of the vessel. The negative electrode was a copper plate that was mounted inside the Teflon vessel, while the positive electrode was a sharpened graphitic tip. As the tip was withdrawn from the suspension at a rate of 0.1 mm·min^{-1} with voltages of 1–2 V applied between the two electrodes, a gel-phase rGONR fiber was obtained suspended at end of the tip. Finally, solid rGONR fibers were formed, after the solvent in the gel fibers was evaporated.

The rGONR fibers had a cylindrical structure with diameters of 20–40 µm and a length of about 10 mm, as illustrated in Fig. 5.43 (B, C). The macroscopic fibers were constructed of graphene nanoribbons, with a highly dense packing, as revealed in Fig. 5.43 (C, D). Figure 5.43 (E) indicated that the rGONR fibers had a surface texture with the graphene ribbons aligned in the fiber axis direction. The longitudinal assembly of the rGONRs along the axis of the fiber (inset of Fig. 5.43 (E)) resulted from the effects of the electric field and surface tension-induced stresses during the withdrawing process. Electrical conductivity of the as-fabricated rGONR fibers was 0.37 S·cm^{-1}. After thermal annealing, the fibers retained the fibrous microstructure, while their diameter was only slightly decreased, as observed in Fig. 5.43 (F). The electrical conductivity was increased to 66 S·cm^{-1}, after the as-spun fibers were annealed at 800°C for 1 h.

A scalable self-assembly method at the liquid/air interface was reported to prepare GO fibers from aqueous GO suspensions, without using any polymer or surfactant [53]. The GO fibers were mechanically flexible, while their electrical conductivity could be significantly increased by using a low temperature hydrothermal treatment at 180°C for 5 h. When stable GO suspensions were kept in a beaker of 2000 ml at room temperature, GO fibers were assembled and grew at the liquid/air interface in two-weeks' time. The GO fibers were stable, free-standing and floating on the surface of the GO solution, and could be transferred onto arbitrary substrates. After washing and drying, dried GO fibers were obtained, which were intertwined into GO fiber films before they were transferred onto SiO_2/Si substrates. The GO fiber films were hydrothermally treated together with the SiO_2/Si substrates. The diameter and length of the fibers were controlled by either changing the size of the containers or controlling the duration of ultrasonic treatment and self-assembly process.

Figure 5.44 shows the self-assembly process of the GO fibers. The initial aqueous GO solution was light brown in color, with sedimentation at the bottom of the beaker, due to the precipitation of the multilayered GO nanosheets (Fig. 5.44 (a)). After being kept for two weeks at room temperature, white flecks were formed gradually at the liquid/air interface, as demonstrated in Fig. 5.44 (b). Figure 5.44 (d) indicated that intertwined fibers with diameters of 1–2 μm and lengths of hundreds of micrometers were present. These self-assembled GO fibers were much thinner than the graphene nanosheet fibers fabricated by other methods, probably due to the small dimension of the GO nanosheets, as seen in Fig. 5.44 (c).

The dispersive short fibers gradually assembled into a long one, with a bamboo-like structure as a foam on the surface. The individual GO fiber had a ring-like structure. According to the morphological evolution, from short fiber to long fiber, a scrolling-joining-stacking assembly mechanism was proposed. The water surface served as a 2D platform, on which the GO nanosheets with appropriate dimensions were floated. The stacking of the GO nanosheets was prevented by the repulsive electrostatic force, so that the GO solution was highly homogeneous. At the liquid/air interface, the GO nanosheets scrolled into short fibers with a ring-like structure, due to the presence of surface tensions. The short fibers were end-to-end joined together, leading to long fibers. The growth of the fibers was triggered by the π–π stacking and van der Waals forces in between the GO nanosheets.

The initially formed GO fibers had diameters of 1.42–1.62 μm, as shown in Fig. 5.45 (a), which were tangled together. Carbon particles were interspersed among the fibers, which were retained from the modified Hummers' process. Figure 5.45 (c) high magnified SEM image of the fibers, which were covered with GO nanosheets. The GO fibers could be cleaned by using ultrasonication, as demonstrated in Fig. 5.45 (d–f). At the same time, the diameter was slightly decreased to 1.63–1.99 μm. The GO fibers were not affected by the multiple twisting or folding, due to their high mechanical stability, as illustrated

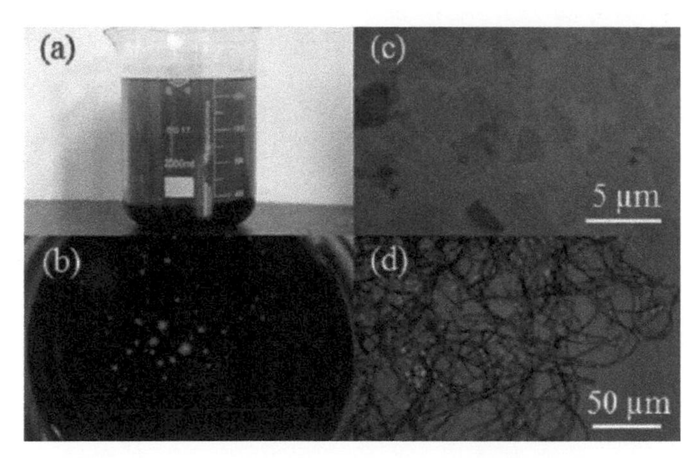

Fig. 5.44. Self-assembly process of the GO fibers: (a) optical photograph of the GO suspension, (b) optical photograph of the self-assembled fibers at the liquid/air interface, (c) optical microscope image of parent GO nanosheets and (d) optical microscope image of the GO fibers. Reproduced with permission from [53], Copyright © 2013, American Chemical Society.

in Fig. 5.46. The single GO fiber with a diameter of 1–2 μm had a Young's modulus of 1.1 ± 0.031 GPa, according to the measurement by using AFM nanoindentation. Therefore, the GO fibers demonstrated promising structural integrity, mechanical flexibility and resistance to torsion during the ultrasonic cleaning process. The alignment of GO nanosheets offered strong interactions, providing strength to the GO fibers.

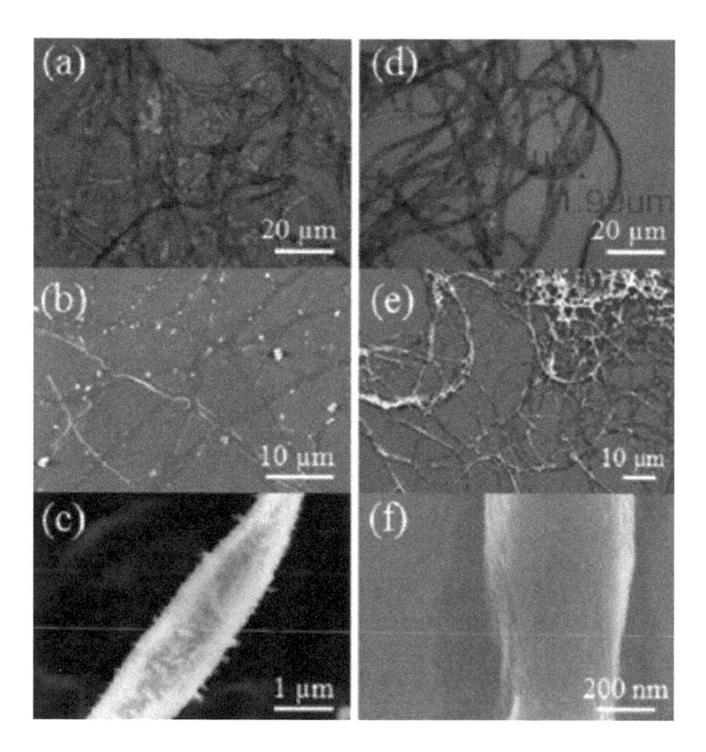

Fig. 5.45. Morphologies of the as-obtained GO fibers before and after ultrasonic cleaning. (a, d) Optical microscope images of the as-formed GO fibers before and after ultrasonic cleaning. (b, e) Low-magnification SEM images of the as-formed GO fibers before and after ultrasonic cleaning. (c, f) High-magnification SEM images of a single GO fiber before and after ultrasonic cleaning. Reproduced with permission from [53], Copyright © 2013, American Chemical Society.

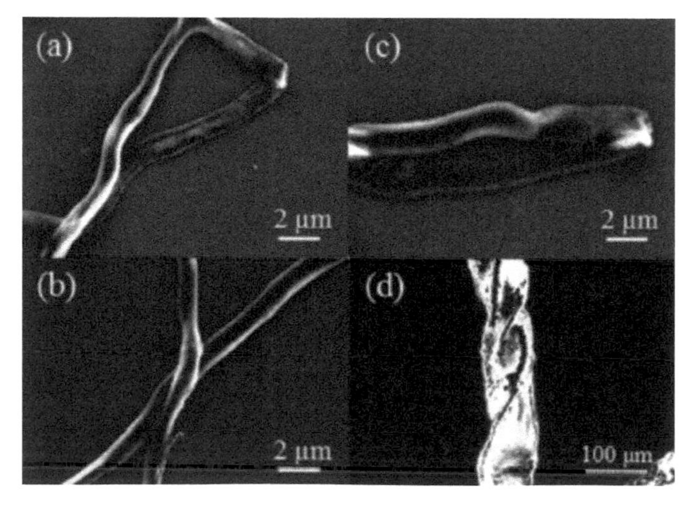

Fig. 5.46. SEM images of (a) the twisted, (b) inter-crossed, (c) folded GO fiber and (d) a bundle of the twisted GO fibers. Reproduced with permission from [53], Copyright © 2013, American Chemical Society.

5.3. Hybrid Graphene Fibers

5.3.1. Graphene–inorganic fibers

GFs could be used as a versatile platform to incorporate various functional components to achieve multiple functionalities [54, 55]. As discussed before, after Fe_3O_4 nanoparticles were *in situ* integrated into GFs, magnetic GFs were developed [40]. The Fe_3O_4 integrated GFs with high mechanical flexibility showed a strong magnetic response. Through post-functionalization, TiO_2 nanoparticles could be intercalated into the graphene nanosheets of the GFs, by soaking the wet as-prepared GFs in a commercial TiO_2 aqueous suspension [40]. The TiO_2 intercalated GFs exhibited promising photocatalytic performance. Moreover, the GFs could be selectively functionalized at specific sites [41].

To improve the electrical conductivity of GFs, Gao et al. [56] prepared graphene-metal hybrid fibers by wet spinning of GO LCs mixed with commercial Ag nanowires, followed by chemical reduction. The Ag-doped GFs exhibited a record high electrical conductivity of up to 9.3×10^4 S·m^{-1} and a high current capacity of 7.1×10^3 A·cm^{-2}. The combination of high conductivity and high mechanical strength and the flexibility of doped GFs renders them ideal stretchable conductors to be applied in soft circuits.

Continuous G-Ag hybrid fibers have been developed by using wet-spinning giant graphene oxide (GGO) liquid crystals mixed with commercial Ag NWs, followed by chemical reduction. The G-Ag hybrid fibers exhibited electrical conductivity as high as 9.3×10^4 S·m^{-1} and a high current density of 7.1×10^3 A·cm^{-2}, which were higher than those of the undoped fibers by more than three and fifteen times, respectively. Conductivity of the graphene fibers was dependent on the type of chemical reduction agents, with molecular doping mechanism for HI and mixing mechanism for vitamin C (VC). Due to their high electrical conductivity, strong mechanical strength and high flexibility, G-Ag hybrid fibers could find potential applications as stretchable conductors.

For the wet-spinning of the GGO-Ag NW fibers, GGO nanosheets were prepared first. As shown in Fig. 5.47 (A), the GGO nanosheets had an average lateral size of 20 μm, and a thickness of 0.8 nm. The giant lateral size and the atomic thickness of the GGO nanosheets ensured the formation of liquid crystals (LCs) in appropriate solvents, such as water, N,N-dimethylformamide (DMF) and N-methyl-2-pyrrolidone (NMP). The presence of Ag NWs had almost no influence on the liquid crystalline behavior of the GGO nanosheets in DMF, which was confirmed by the polarized optical microscopy (POM) image of the GGO dispersion with a concentration of 4 mg·ml^{-1}, as illustrated in Fig. 5.47 (D). In this case, the content of the

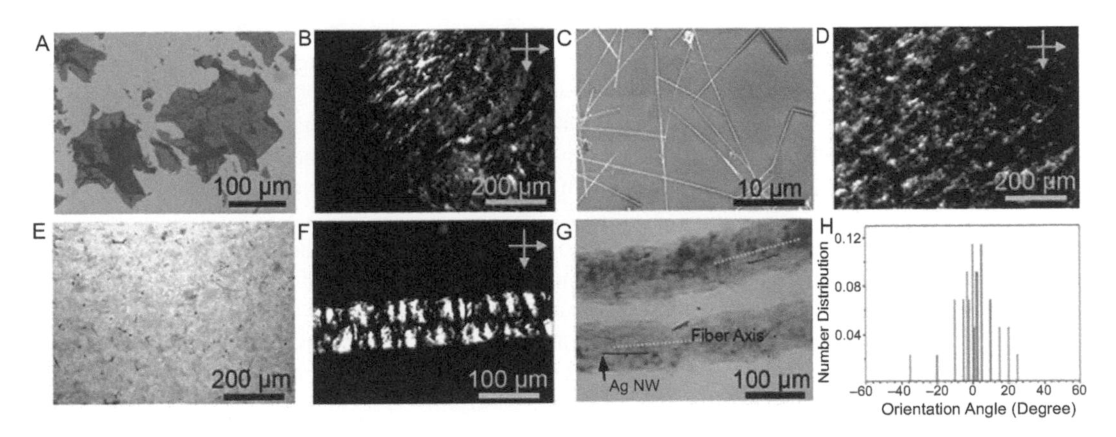

Fig. 5.47. (A) SEM image of the GGO nanosheets deposited on silica. (B) POM image of the GGO DMF dispersion (4 mg·ml^{-1}) between crossed polarizers. (C) SEM image of the commercial Ag nanowires. (D, E) POM images of the GGO-Ag NWs dispersions between crossed polarizers in the planar cell (D) and the corresponding optical image under natural light (E). (F, G) POM images between crossed polarizers (F) and optical image under natural light (G) of the GGO-Ag NWs gel fibers. The arrow in (E) indicates the Ag nanowire while the dashed lines in (G) indicate the gel fiber axis. (H) Orientation-angle distribution of the Ag NWs along the fiber axis in the gel fibers. Reproduced with permission from [56], Copyright © 2013, WILEY-VCH Verlag GmbH & Co. KGaA, Weinheim.

Ag NW was 5 wt%. It was found that the dispersibility of the Ag NWs was enhanced due to the presence of the GGO nanosheets. No precipitation was observed in the suspensions of GGO and Ag NWs for a quite long time, as demonstrated in Fig. 5.47 (E).

The GGO-Ag NWs hybrid fibers could be up to 100 m long, as seen in Fig. 5.48 (A, B). The spinning process triggered the alignment of the GGO nanosheets, as well as the alignment of the Ag NWs, along the fiber axis in the gel fibers. Figure 5.47 (F) shows POM image of the gel fiber, revealing that the nematic texture was more distinct and uniform than that of the spinning dopes in the planar cell. It implied that unidirectional flow during the spinning process ensured more regular alignment of the GGO nanosheets. The alignment of the Ag NWs in the gel fiber was confirmed by the POM image shown in Fig. 5.47 (G). Figure 5.47 (H) indicated that the distribution of the orientation angle between the Ag NWs and the fiber axis peaked in the range between −20° and 20°, while the average orientation angle was ±7.3°.

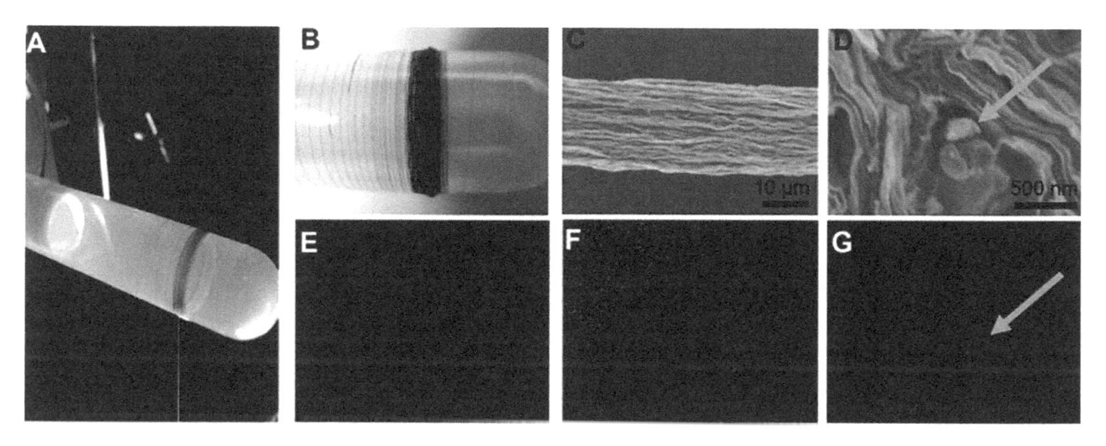

Fig. 5.48. (A, B) Digital photographs of the GGO-Ag NWs fibers collected on a plastic drum during the spinning process (A) and the dried fibers with a length of about 100 m (B). (C, D) SEM images of the GGO-Ag NWs fiber surface (C) and the cross-sectional morphology (D). The arrow in (D) denotes the section of the interior Ag NWs. EDS mapping images of carbon (E), oxygen (F) and Ag (G) in the area of (D). The arrow in (G) indicates the Ag NWs as seen in (D). Reproduced with permission from [56], Copyright © 2013, WILEY-VCH Verlag GmbH & Co. KGaA, Weinheim.

The average diameter of the dried fibers was about 10 μm, as compared with 100 μm of the gel fibers. The presence of the aligned wrinkles was attributed to the shrinkage in the single radical dimension of the fibers. Figure 5.48 (D) shows cross-sectional morphology of the GGO-Ag NWs fibers, with a compact origami-flower-like structure. The dentate bends with regular lamellar localized regions were caused by the orientation of the disclinations in the liquid crystalline gels. Energy-dispersive spectroscopy (EDS) mapping images of a section of the fiber are shown in Fig. 5.48 (E–G), indicating the presence of C, O and Ag elements.

It has been demonstrated that some divalent ions could serve as cross-linking agents to bridge the graphene nanosheets, so as to increase the mechanical strength of graphene fibers [57]. Similarly, giant graphene oxide (GGO) nanosheets, with a thickness (t) of ≈ 0.8 nm and an average lateral size (w) of 18.5 μm, corresponding to an aspect ratio (w/t) of as high as 2.3×10^4, were used to spin the graphene fibers. Due to the extremely high aspect ratio, the GGO could ensure the formation of LCs at very low concentrations.

Polarized optical microscope (POM) results indicated that stable GGO nematic LCs could be obtained at concentrations of 1–4 mg·ml^{-1}. With increasing GGO concentration, the orientational orderings were increased. The formation of local layered structure in the GGO LCs at 8 mg·ml^{-1} led to visible birefringence. The giant lateral size of the GGO nanosheets, together with sufficiently low concentration of the LCs, ensured stable liquid crystalline spinning dopes to possess sufficiently low viscosity, to enable the continuous spinning of the graphene fibers.

In the as-spun wet fibers, GGO nanosheets were aligned in a way similar to the pre-aligned orientational orderings in the LCs. However, the alignment was vulnerable to disruption, due to the different linear disclinations distributed. Dry GGO fibers with compact regularly aligned structures could be obtained by using two optimized processes, i.e., post-drying and wet-stretching. The GGO fibers without water washing had a flat ribbon-like shape, while those with water washing were compact fibers with circular sections. The fibers derived through stretching had nearly round sections and compact structures, as illustrated in Fig. 5.49 (a, b). As revealed in Fig. 5.49 (a), the diameter of the stretched GGO fibers was about 6 μm, a typical value of commercial carbon fibers. The GGO fibers without stretching exhibited crumbled morphologies, as illustrated in Fig. 5.49 (c, d).

The GGO fibers had a cross-section of compact origami-flower-like structures with dentate bends, as shown in Fig. 5.49 (e, f). The GGO nanosheets in the fibers were stacked densely with local alignments and the layered regions that contained orientational orderings around the bends. The dentate bends in the dry fibers were closely related to the orientational disclinations in the suspensions. There were spreading ridges on the surface of the GGO fibers, with a height of about 300 nm, as demonstrated in Fig. 5.49 (g, h). Due to the Eulerian buckling of the GGO nanosheets on surface of the gel fibers, dentate folds were formed through dehydration. Similar morphology was observed in the rGG fibers, suggesting that the chemical reduction with HI had no significant effect on the microstructure of the GGO fibers, as seen in Fig. 5.49 (i–k).

The GGO fibers spun under the optimized conditions with stretching in KOH coagulation bath possessed a tensile strength of 184.6 MPa at 7.5% ultimate elongation and a Young's modulus of 3.2 GPa. Such a promising mechanical performance was attributed to the size-induced enhancing effect of the GGO nanosheets. Meanwhile, the high mechanical strength of the fibers was also ascribed to the more efficient packing of the clean GGO nanosheets, because the oxidation debris on the GGO nanosheets were effectively eliminated when using the KOH coagulation bath [58, 59].

It has been acknowledged that divalent ions could serve to form interlayer and intralayer cross-linking to bridge the oxygen containing groups, so as to enhance the mechanical properties of GO papers [60, 61]. For such a purpose, $CaCl_2$ and $CuSO_4$ solutions were used as the coagulation baths to increase the strength of the graphene fibers. As shown in Fig. 5.50 (a, b), the Ca^{2+}-cross-linked GGO fibers possessed a tensile strength of 364.6 MPa at 6.8% ultimate elongation, while the fibers cross-linked with Cu^{2+} displayed a tensile strength of 274.3 MPa at 5.9% elongation. Mechanical performances of the Ca^{2+}-cross-linked GGO

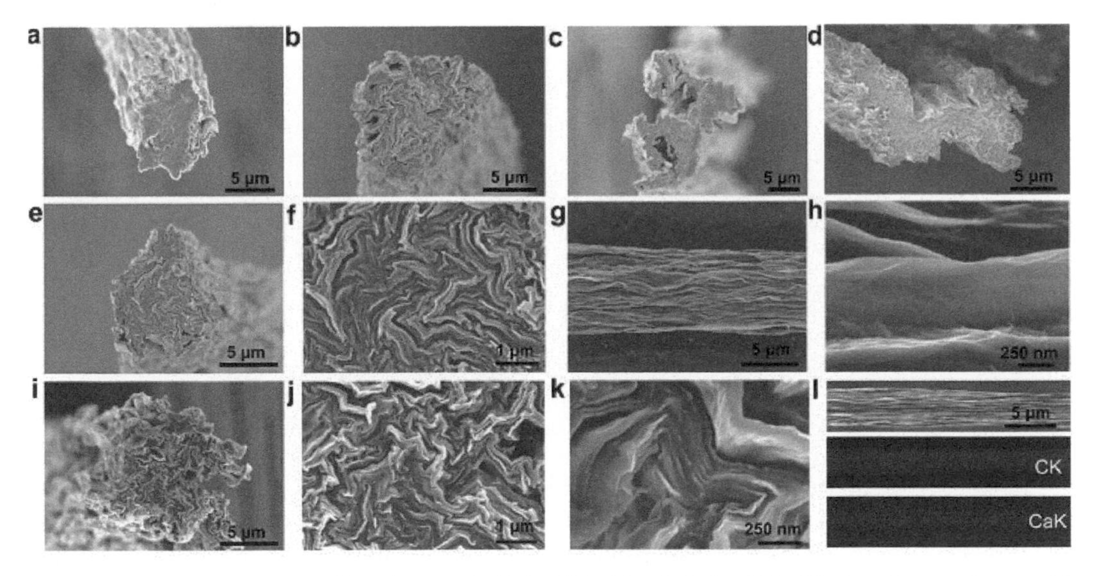

Fig. 5.49. SEM images of sections of the fiber spun in (a) 5 wt% $CaCl_2$ and (b) 5 wt% NaOH ethanol/water solutions with 1.3-fold stretching, and (c, d) their corresponding sectional SEM images without stretching. SEM images of (e, f) typical fracture and (g, h) surface morphology of the GGO fibers, and (i–k) sectional morphology of the rGG fibers. (l) SEM images of Ca^{2+}-cross-linked rGG fibers (top), together with C-element mapping (middle) and Ca-element mapping (bottom). Reproduced with permission from [57], Copyright © 2013, WILEY-VCH Verlag GmbH & Co. KGaA, Weinheim.

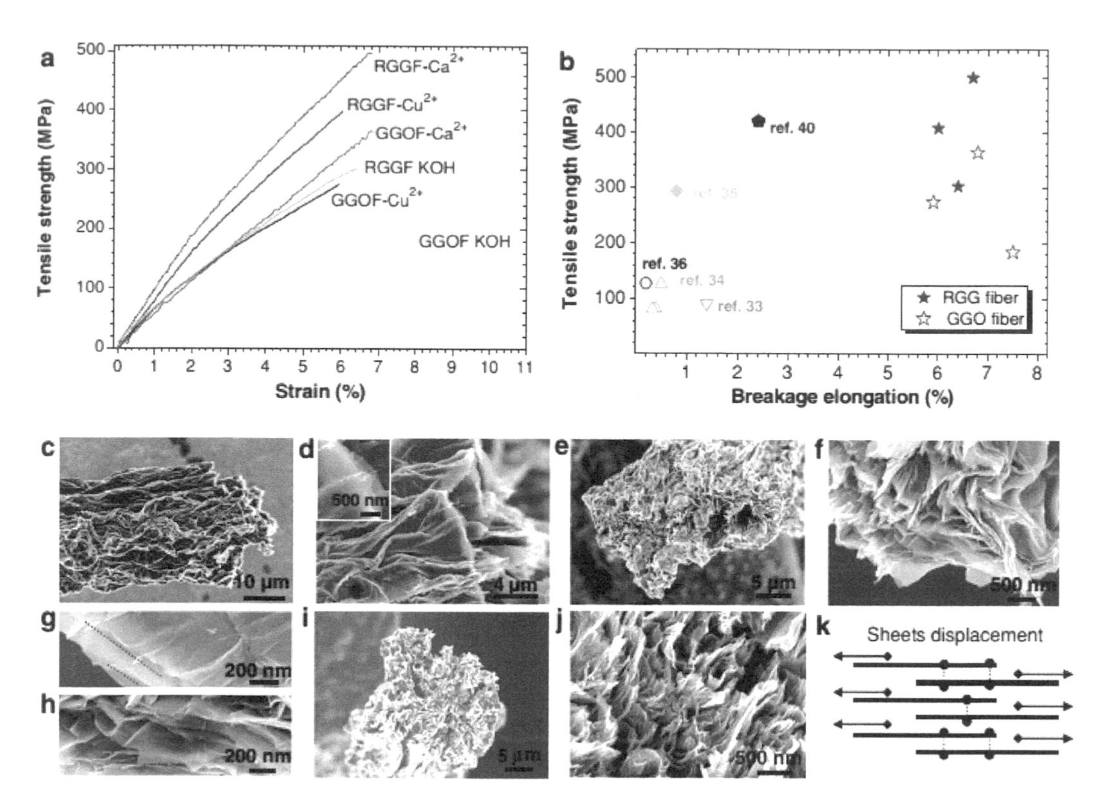

Fig. 5.50. (a) Typical mechanical measurements under tension for the GGO fibers and rGG fibers. (b) Diagram of mechanical performance data for the graphene-based neat papers and fibers in the literature and in this study. The hollow symbols represent the neat GO-based fibers/papers and the solid symbols stand for the reduced graphene-based fibers and papers. The red pentagrams denote GGO and rGG fibers in this study. SEM images of the fracture surfaces of (c, d) the GGO-KOH fiber, (e–h) the GGO-Ca^{2+} fiber and rGG-Ca^{2+} fiber. The dashed lines in (g) indicate either the boundary lines or the cracking lines of the pulled-out graphene nanosheets. (k) Deformation mechanism model for the GGO and rGG fibers under tensile stress. The dashed lines represent the hydrogen bonds and coordinative cross-linking bridges. Reproduced with permission from [57], Copyright © 2013, WILEY-VCH Verlag GmbH & Co. KGaA, Weinheim.

fibers were double those of the fibers spun in KOH solution. It was found that Ca^{2+} ions were homogenously distributed throughout the fibers, confirming that the enhancement in the mechanical properties of the fibers was mainly attributed to the cross-linking effect.

Deformation mechanism of the GGO and rGG fibers could be explained by using the tension-shear model [58, 62]. Generally, there are three dominant interactions between graphene nanosheets, i.e., (i) van der Waals interaction, (ii) hydrogen bonds and (iii) coordinative cross-linking. In the GGO fibers, because hydrogen bonds were dominant, they determined the mechanical strength of the fibers [63]. Chemical reduction resulted in an increase in the van der Waals interaction, due to the decrease in the interlayer space, thus leading to an enhancement in the mechanical properties, together with the hydrogen bonds between residual oxygen functional groups. With the presence of the divalent ions, oxygen-containing groups on the GGO nanosheets were bridged.

Representative SEM images of the cross-section and fractured sections of the fibers are shown in Fig. 5.50 (c–j). Tensile fractures of the GGO and rGG fibers exhibited elongation in the axes direction of the fibers, implying the presence of their elastic breakage behavior. Close inspection of the fracture tips indicated that the typical stretched fibrils were observed, with a fracture feature similar to the "pull-out" characteristic of CNT fibers under tensile stress. There were distinct boundary lines of pulled-out graphene nanosheets at the fracture tips, corresponding to the edges of the stacked graphene or the cracking lines of graphene nanosheets. Figure 5.50 (k) shows a schematic diagram of the tension shear model for the graphene fibers. The graphene nanosheets experienced a pulling force, so as to slide from the stacked graphene blocks due to the tensile force in the axial direction of the fiber. Therefore, the presence of the boundary lines of graphene nanosheets supported the tension-shear deformation mechanism. Due to their high mechanical flexibility, such graphene fibers could find applications in flexible and wearable sensors and energy storage devices.

Graphene-based biomimetic fibers, with graphene nanosheets as rigid platelets and hyperbranched polyglycerol (HPG) as elastic glue, have been fabricated by using wet-spinning technology [64]. With high solubility and abundant functional groups of HPG, HPG enveloped graphene nanosheets (HPG-e-Gs) were formed with sandwich-like building blocks, which were highly soluble in solvents, like N,N-dimethylformamide (DMF) and N-methyl pyrrolidone (NMP). At the same time, liquid crystals could be formed at high concentrations, thus leading to artificial nacre fibers (ANFs), with a brick-and-mortar (B&M) structure, consisting of graphene nanosheets and HPG binder. The fibers were constructed through hydrogen-bonding arrays, which exhibited mechanical and physical properties similar to those of nacre and bone.

The ANFs were developed using three steps: (i) synthesis of building blocks, (ii) formation of LC dopes and (iii) wet-spinning. Four HPGs with molecular weights of 4.4, 7.2, 87.6 and 125.8 kDa were used as the organic layers, corresponding to sandwich building blocks of HPG1-e-G, HPG2-e-G, HPG3-e-G and HPG4-e-G, respectively. Depending on their molecular weights, the fractions of the HPG attached onto graphene were 16, 34, 49 and 57 wt%. Due to the hydroxyl densities, all the HPG-eGs exhibited excellent solubility in polar solvents as mentioned above. It was found that the thickness of the HPG-e-Gs increases to 2.8 nm from 0.8 nm of the original GO nanosheets.

As shown in Fig. 5.51 (a), the HPG2-e-Gs could be used to spin continuous flexible fibers, while the fibers derived from HPG3-eGs were much less flexible with limited length, on the other hand HPG1-e-Gs and HPG4-e-Gs led to brittle fibers. The HPG2-e-Gs fibers were sufficiently flexible so that they could be knotted at micrometer scale, as demonstrated in Fig. 5.51 (b). It was found that the solution of NaOH in methanol was the most effective bath to spin continuous fibers with HPG2-e-Gs. As the suspensions were injected into the coagulation bath, gel fibers were formed, while the solvent (DMF) rapidly diffused out of the fibers into the methanol solution and the methanol was diffused into the fibers at the same time. The graphene nanosheets were quickly stacked with flow-induced alignment triggered by the NaOH solution as the suspensions were pushed out of the nozzle. The diameter of the fibers was in the range of 10–100 μm, which could be controlled by controlling the nozzle size, injection rate, flow conditions and coagulation process. The residual NaOH inside the fibers was moved by washing with methanol.

Figure 5.51 (c) shows surface SEM image of the dried fibers, with the graphene nanosheets being oriented along the fiber axis. Figure 5.51 (d) indicated that there was under contour in the cross-section, which was caused by shrinking of the dope during the coagulation process. HPG-e-G hybrid building blocks were present inside the fibers as densely ordered lamellar microstructures, similar to the B&M structure of nacre, as seen in Fig. 5.51 (e, f). As illustrated in Fig. 5.51 (e), curled and folded lamellae were observed, which were caused by the dislocation domains in the LC dopes. The incorporation of graphene in the ANFs were confirmed by using TEM and Raman spectroscopy analysis. Chemical structure of the graphene was not affected during the wet-spinning process.

Mechanical behavior of the fibers could be understood by using the shear lag model for nacre [30, 31]. At the tensile stress of < 50 MPa, the interface between two HPG adlayers started to yield in shear, while the HPG-e-G building blocks slided one another, thus leading to the destruction and reformation of the hydrogen bonding network. As the strain was increased, the load was transferred from the soft organic moieties to the hard graphene nanosheets, so that the shear was spread all over the fiber, resulting in the failure of the fiber by pulling-out of the HPG-e-G building blocks, as revealed in Fig. 5.51 (g, h). The network of the hydrogen-bonding between the HPG interlayers is schematically shown in Fig. 5.51 (i).

Further development of the strategy was the use of polymer to graft graphene oxide (PgG) nanosheets, in order to increase the mechanical performances of the graphene fibers [65]. Due to the covalent and uniform immobilization effect of the polymer chains on the surface of graphene oxide (GO) nanosheets, phase separation could be effectively prevented, while strong interfacial interaction was established at the same time. Most importantly, the PgG nanosheets were highly dispersible in organic solvents, thus leading to the formation of liquid crystals (LCs) to spin fibers. The PgG fibers with biomimetic features,

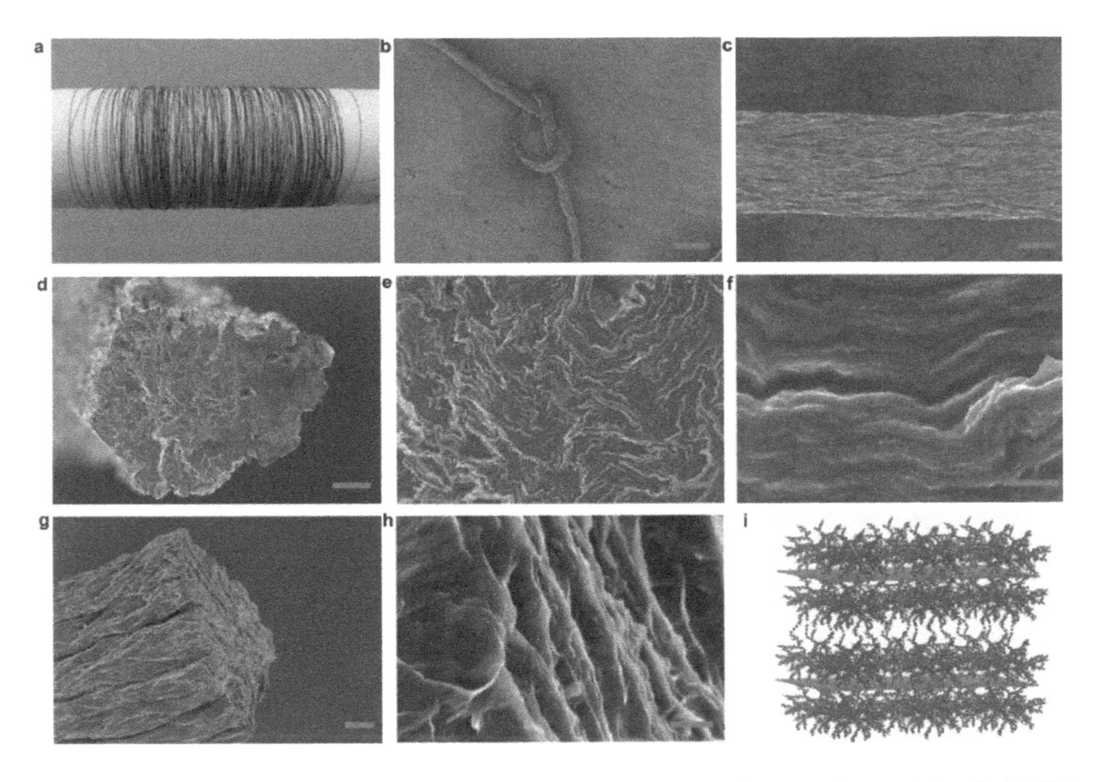

Fig. 5.51. Morphologies of the HPG2-e-G fibers. (a) Photograph of a HPG2-e-G fiber tens of meters in length. (b) SEM image of a knot of the HPG2-e-G fiber. (c) Wrinkled surface morphology of the HPG2-e-G fiber. (d–f) Cross-sectional SEM images of the HPG2-e-G fiber at different magnifications. (g, h) Side view of the fracture section of the HPG2-e-G fiber. The dislocated layered structure (h) indicates the displacement of adjacent HPG-e-G building blocks under tension. The dark gray area represents the left interspace after the "pulling-out" of some HPG-e-G nanosheets. (i) Schematic illustration of two adjacent HPG-e-G building blocks glued by the network of hydrogen bonding arrays between them. Scale bars = 200 μm (b), 20 μm (c), 10 μm (d), 3 μm (e), 500 nm (f), 10 μm (g) and 500 nm (h). Reproduced with permission from [64], Copyright © 2012, Macmillan Publishing Limited.

had promising mechanical performances, including a high strength of 500 MPa, a toughness of 7.8 $MJ \cdot m^{-3}$ and a Young's modulus of 18.8 GPa. The content of GO in the composite fibers could be controlled over a wide range of 25–88% by controlling the amount of the grafting polymer. Besides mechanical properties, the PgG fibers had very high chemical resistance.

Figure 5.52 shows the preparation protocol for the biomimetic composites [65]. The fiber spinning process involved three steps: (i) synthesis of building blocks of the PgG nanosheets, (ii) formation of colloidal LCs and (iii) spinning assembly. Polymer chains were covalently bonded onto the nanoplatelets through *in situ* free radical polymerization of vinyl monomers. The PgGs had an average thickness of 3.5 nm, much larger than that of the pristine GO nanosheets. The average width of the PgG nanosheets was 0.8 μm with polydispersity of 26%. Polymerization times were 1, 1.5, 5 and 18 h, so as to obtain PgGs with different contents of grafted polymer.

The PgG LCs were injected into a coagulation bath that was rotated by using a spinneret. The gel fibers showed birefringence under POM, as shown in Fig. 5.52 (a), due to the high alignment of the PgG building blocks. Figure 5.52 (b) indicated that after the gel fibers was dried into solid fibers, they experienced shrinkage in the cross-sectional direction. Figure 5.52 (c) shows a photograph of a dried PgG fiber with a length of 5 m. The fibers could be easily woven into textiles and tied into knots, due to their high flexibility, as illustrated in Fig. 5.52 (d).

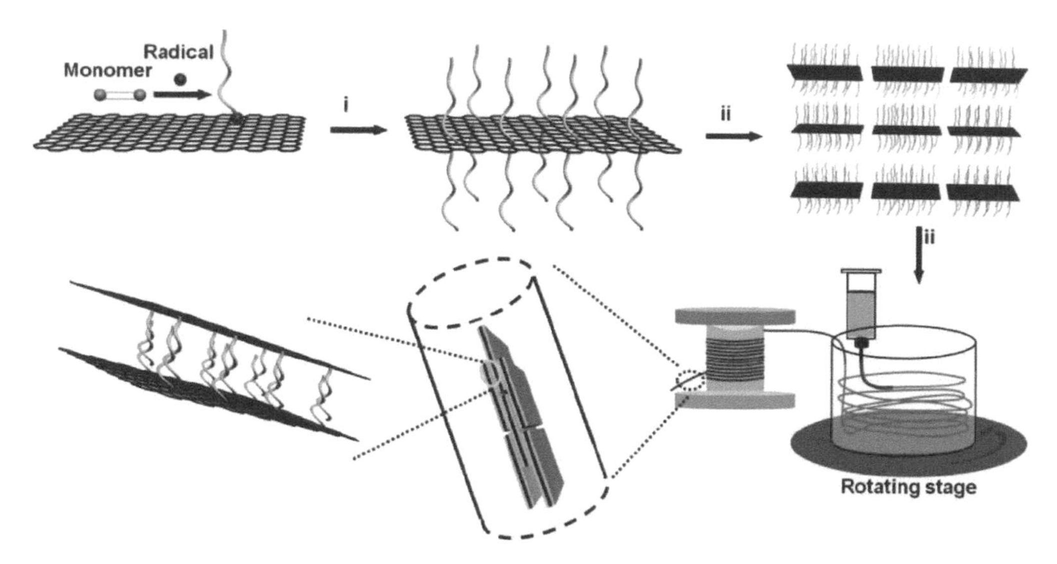

Fig. 5.52. Preparation of the PgG building blocks and wet assembly approach to spin the PgG fibers. (i) *In situ* free radical polymerization of GMA in the presence of GO to synthesize the PgG building blocks, followed by removal of the ungrafted polyGMA through cycles of centrifugation, decanting and redispersion. (ii) Formation of LC suspensions with spontaneous alignment of PgG nanosheets at above the critical concentration. (iii) Wet spinning of the PgG LCs into continuous fibers with B&M microstructures. Reproduced with permission from [65], Copyright © 2013, Macmillan Publishing Limited.

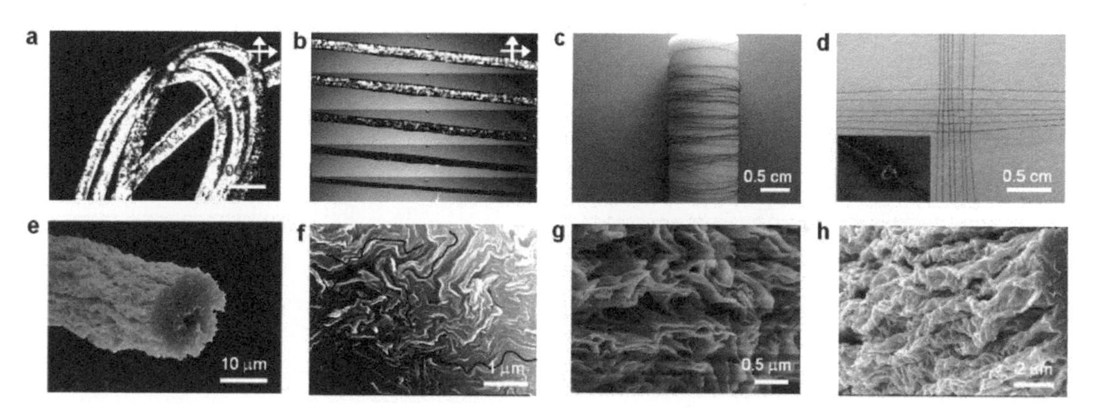

Fig. 5.53. Formation and structures of the PgG fibers. (a, b) POM images of the freshly spun gel PgG fibers and their evolution to solid fibers. (c) A five meter-long PgG fiber collected on a scroll. (d) Photograph of a hand-woven textile and SEM image (inset) of a knot made of the PgG fibers. (e–h) SEM images of the rounded morphology (e), fracture sections in a top-view direction (f) and in a tilt viewing direction (g) and the wrinkled surface (h) of the PgG fiber. Reproduced with permission from [65], Copyright © 2013, Macmillan Publishing Limited.

SEM images of the PgG fibers are shown in Fig. 5.52 (e–h). As seen in Fig. 5.52 (e), the fiber exhibited a rounded cross-sectional morphology, with a diameter of 15 μm. Highly uniform layered structures were observed in the fibers, confirming the B&M architecture of the PgG fibers. Staggered blocks with layered structure were observed in the cross-sectional SEM image, as highlighted by the blue lines in Fig. 5.52 (f). The staggered structures resulted from the multi-domains in the LC spinning doping. The formation of accordion-like structures was attributed to the liquid surface tension generated during the drying, which would enhance the out-of-plane interlocking and thus contributed to the enhancement in the mechanical strength of the fibers. Long-range ridges along the fiber axis were formed, reflecting buckling behavior of the PgG nanosheets, during the drying process, as demonstrated in Fig. 5.52 (h).

5.3.2. Graphene–nanocarbon fibers

Graphenes are typical 2D nanostructures with unique properties, while CNTs are important 1D nanomaterials, so that it has been naturally expected that their hybrids could have interesting properties. Attempts have been made to fabricate such hybrids recently. Several examples will be discussed in this section to demonstrate the synergistic effect when these two types of nanocarbon materials are combined.

Hybrid multiple-thread yarns of CNTs and graphene nanosheets (GNS), consisting of tens or hundreds of single-thread fibers of CNTs, were developed by using a chemical vapor gas flowing reaction, followed by post-treatment processing [66]. A CNT thick stick-like assembly was first fabricated by CVD processing in a horizontal reactor, on which the GNS were grown. CNT and GNS hybrid multiple-thread yarns could be drawn from the stretched thick stick-like CNT assembly. The as-prepared CNT multiple-thread yarns consisted of tens of thinner single-thread fibers and were several hundred micrometers in diameter. The single-thread fibers with a diameter of 20 μm contained double-walled CNTs and GNS. The twisted yarns exhibited a mechanical strength of 300 MPa and an electrical conductivity of 10^5 $S \cdot m^{-1}$. Potential applications of the CNT–GNS yarns as lamp thread and macroscopic body have been demonstrated.

The horizontal CVD reactor had a length of 1000 mm, consisting of a quartz tube with a diameter of 67 mm and a length of 1600 mm. Figure 5.54 (A) shows a schematic diagram of the setup used to fabricate the CNT assemblies. Both CNTs and GNS were derived from ethanol as the carbon source, with ferrocene as the catalytic agent and thiophene and water as the promoter agents. The reaction solution was prepared by dispersing 2.0 wt% ferrocene, 1.1 wt% of thiophene and 1.9 wt% of water into ethanol with the aid of ultrasonication. The reaction solution was then injected into the reaction gas at 1150°C and at a flowing rate of 10 $ml \cdot h^{-1}$. Continuous CNT assembly was formed after the injection, which was moved by the gas flow towards the end of the reactor. CNT and GNS hybrid assembly were obtained as the assembly was accumulated at the end of the reactor, as demonstrated in Fig. 5.54 (B). The reaction was maintained for 2 h.

TEM results indicated that the thread fibers consisted of high purity double-walled carbon nanotubes (DWNTs) and GNS nanosheets. Figure 5.55 (A) shows a TEM image of the CNTs taken from one end of a thread which is close to the low-temperature region, i.e., < 200°C. The CNT bundles exhibited diameters of 20–50 nm. Figure 5.55 (B) shows a TEM image of the CNTs at the center part of the thread, located at the middle temperature region (800°C). The CNT bundles were highly disordered and entangled, containing CNTs and graphene nanosheets. As illustrated in Fig. 5.55 (D), the graphene nanosheets contained two graphite layers. The GNS had dimensions of 50 × 50 nm, with a constituent of about 20% of the total carbon material.

Both content and dimension of the GNS were increased as the location was close to the high temperature region in the reactor. Figure 5.55 (C) show a TEM image of the CNTs and GNS formed at the location close to the high temperature region (1150°C). The hybrid possessed a cabbage-like structure. Dimension of the GNS was increased to 100 × 100 nm, while its content was increased to > 50 vol% of the total

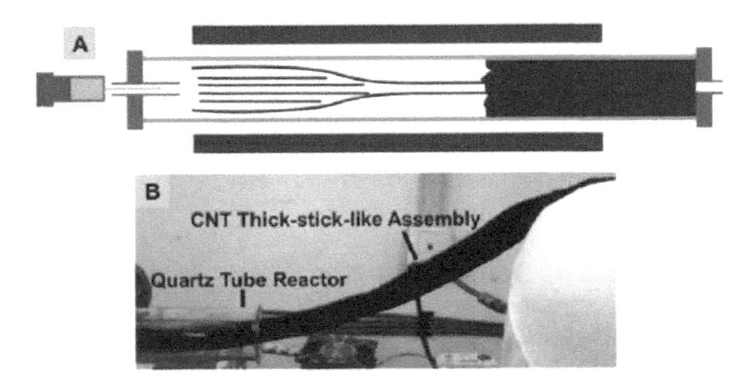

Fig. 5.54. Schematic diagram showing the synthesis of the thick stick-like CNT assembly from a chemical vapor gas flow reaction (A) and photograph of a thick stick-like CNT assembly drawn out from the end of the quartz tube reactor (B). Reproduced with permission from [66], Copyright © 2013, The Royal Society of Chemistry.

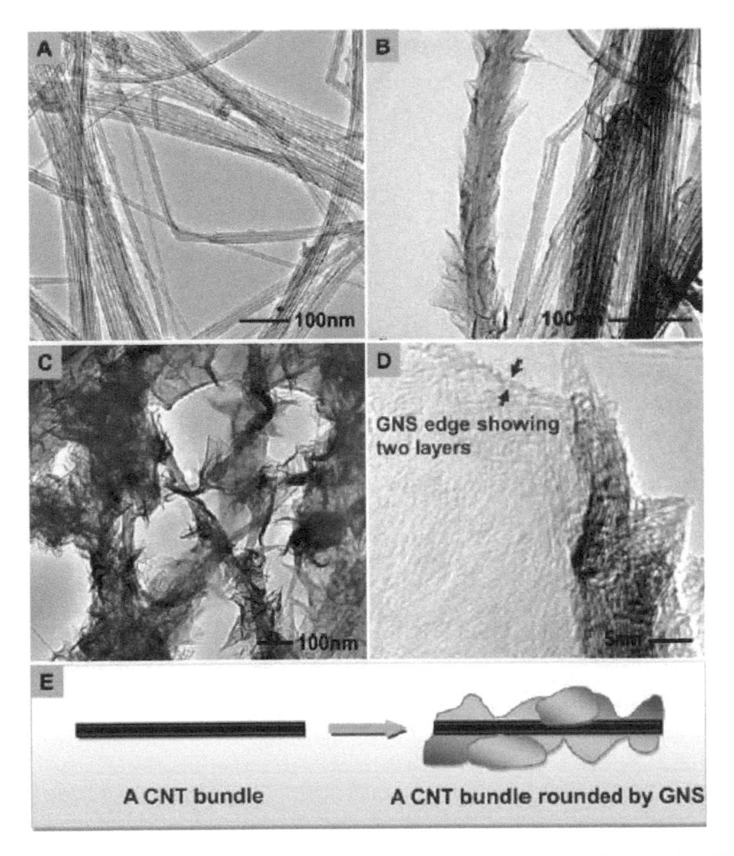

Fig. 5.55. TEM images of the CNT multi-threads and formation of the hybrid structure of CNT and GNS. (A) CNT bundles taken from the end of a single CNT thread close to the end of the quartz-tube reactor. (B) CNT bundles attached with graphene nanosheets. (C) CNT bundles rounded by GNS nanosheets. (D) A graphene nanosheet with 2 and 3 layers. (E) Schematic of formation of the single-thread fiber with hybrid structure of CNTs and GNS. Reproduced with permission from [66], Copyright © 2013, The Royal Society of Chemistry.

carbon materials. Therefore, high temperature supported the growth of GNS. This implied that it was possible to fabricate thread fibers with controlled composition of GNS and CNTs by simply controlling the reaction temperature.

A mechanism was proposed to explain the formation of the GNS nanosheets on the CNTS, as schematically demonstrated in Fig. 5.55 (E). The growth of the GNS required the support of the CNT bundles. This is because, according to the TEM images of the CNT and GNS hybrid multi-thread yarns, the GNS nanosheets were always observed on the CNT bundles, while no free GNS nanosheets were found somewhere else, as seen in Fig. 5.55 (B, C). Also, no GNS nanosheets were formed in the yarns deposited at the low-temperature region (Fig. 5.55 (A)). It therefore could be suggested that the presence of the GNS nanosheets was after the formation of the CNT bundles, which served as a support for the growth of the GNS. In other words, the CNT bundles formed first acted and then as a catalyst to facilitate the growth of the GNS nanosheets.

An alternative method was reported to fabricate CNT–G hybrid fibers, by growing CNTs on graphene fibers with embedded Fe_3O_4 nanoparticles as catalysts for chemical vapor deposition (CVD) of the nanotubes [67]. The CNT–G hybrid fibers have been used to build up a flexible electrochemical cell with textile electrodes. Fe_3O_4-containing graphene (G–Fe_3O_4) hybrid fibers were prepared by thermally treating the mixture of 8 mg·ml^{-1} GO and Fe_3O_4 NPs of 20 nm in a closed glass pipeline, as shown in Fig. 5.56. The weight ratio of GO suspension to Fe_3O_4 was 4:1. The G–Fe_3O_4 fiber was then put into a CVD chamber to grow the CNTs. The fabrication of the CNT–G fiber was carried out at 750°C in a flow of $C_2H_2/H_2/Ar$ (5/150/800 sccm) for 15 min.

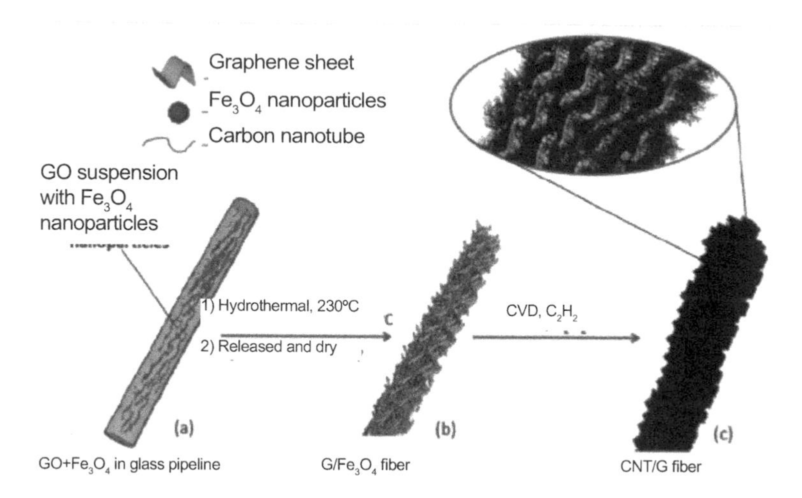

Fig. 5.56. Schematic diagram showing the fabrication process of the CNT–G fibers. (a) Aqueous GO suspension mixed with Fe_3O_4 NPs in a closed glass pipeline. (b) The G–Fe_3O_4 fiber released from the pipelines and dried after the hydrothermal process. (c) The as-prepared fluffy CNT–G fiber with enlarged schematic drawing of the CNTs grown in between the graphene nanosheets after the CVD process. Reproduced with permission from [67], Copyright © 2013, The Royal Society of Chemistry.

The G–Fe_3O_4 fiber, as shown in Fig. 5.57 (a, left), had a diameter of 34 μm, as seen in Fig. 5.57 (b). The fiber consisted of densely packed graphene nanosheets, as illustrated in Fig. 5.57 (b, c). Cross-sectional SEM image of the G–Fe_3O_4 fiber indicated that Fe_3O_4 NPs were homogeneously intercalated within graphene nanosheets (Fig. 5.57 (d)). EDS results revealed that O and Fe elemental mappings (Fig. 5.57 (f, g)) were consistent with the C mapping (Fig. 5.57 (e)), confirming the uniform distribution of the Fe_3O_4 NPs in the graphene fiber.

After CVD growth of the CNTs, there was almost no change in the length of the G–Fe_3O_4 fiber (Fig. 5.57 (a, right)), while diameter was greatly increased. Figure 5.57 (h) shows a CNT–G fiber with a diameter of 100 μm, which was nearly three times that of the G–Fe_3O_4 fiber. The surface of the CNT–G fiber was attached by highly entangled CNTs, as demonstrated in Fig. 5.57 (i). Cross-sectional SEM images of a broken CNT–G fiber are shown in Fig. 5.57 (j, k). The CNTs were grown not only on the fiber surface but also inside the fiber, thickening the fiber. The CNT–G fiber retained the flexibility of the graphene fiber, as illustrated in Fig. 5.57 (l, m). The expansion of the densely packed graphene nanosheets, due to the intercalation of the CNTs, resulted in a decrease in tensile strength, from 180 MPa to 24.5 MPa. Correspondingly, the G–Fe_3O_4 fiber possessed a large specific surface area of 79.5 $m^2 \cdot g^{-1}$, which was significantly higher than that of pure graphene fibers of 18 $m^2 \cdot g^{-1}$. In addition, the CNT–G fiber exhibited an electrical conductivity of about 12 $S \cdot cm^{-1}$ at room temperature, slightly higher than that of graphene fiber of 10 $S \cdot cm^{-1}$, due to the presence of the highly entangled CNTs.

A scalable method was developed to synthesize hierarchically structured carbon microfiber made of single-walled carbon nanotube (SWCNT) and nitrogen-doped rGO nanosheet, which were interconnected into network architecture, with extremely high electrical conductivity and high packing density, thus providing a large ion-accessible surface area for electrochemical supercapacitors [68]. Micro-supercapacitors (SCs) fabricated with the microfibers exhibited volumetric energy densities that were well comparable to those of thin-film lithium batteries, without compromising cyclability and rate capability.

A fused-silica capillary column was used as the hydrothermal microreactor to synthesize the SWCNT–rGO fibers, as schematically shown in Fig. 5.58 (a). Nitric-acid-treated SWCNTs and GO were used to prepare the hybrid microfibers. Ethylenediamine (EDA) was adopted as the nitrogen dopant and reducing agent to dope and reduce the GO through the hydrothermal reaction. In this case, GO acted as a surfactant to promote the dispersion of the SWCNTs, while the SWCNTs effectively prevented the restacking of the GO nanosheets [69], thus leading to homogeneous aqueous suspensions. Due to its two –NH_2 end groups, EDA served as a molecular 'end-anchoring' reagent to bridge the acid-oxidized SWCNTs and the GO

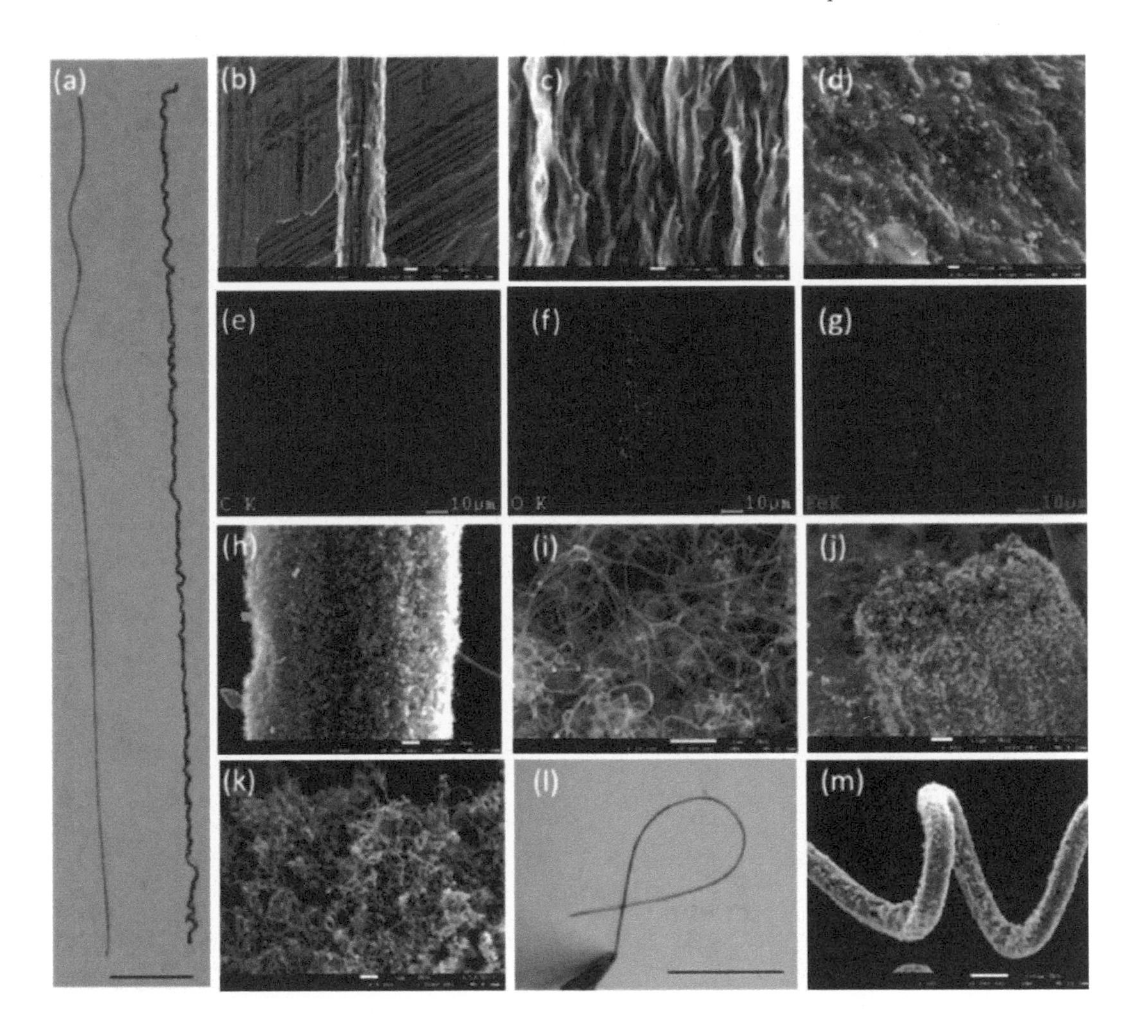

Fig. 5.57. (a) Photograph of a 5 cm long G–Fe_3O_4 fiber (left) and a CNT/G fiber (right). (b) SEM image of the G–Fe_3O_4 fiber. (c, d) Surface and cross-sectional SEM images of the G–Fe_3O_4 fiber, respectively. (e–g) EDS C, O and Fe elemental mappings of (b). (h, i) SEM images of the G–CNT fiber and enlarged surface, respectively. (j, k) Cross-sectional SEM images of the CNT–G fiber at different magnifications. (l) Photograph of a loop shaped CNT–G fiber. (m) Spring shape of the CNT–G fiber. Scale bars = 0.5 cm (a), 10 μm (b, h and j), 100 nm (c, d), 1 μm (i, k), 1 cm (l) and 100 μm (m). Reproduced with permission from [67], Copyright © 2013, The Royal Society of Chemistry.

nanosheets, leading to 3D pillared vertically aligned SWCNT–rGO architectures, though self-assembling inside the fibers. The pillared structures were highly oriented in the axial direction of the fiber, because of the shear force generated while the suspensions flew through the silica capillary columns, as shown in Fig. 5.59 (g–i). The aligned SWCNTs had a special contribution to the excellent electrical conductivity of the fibers, while the large-surface-area of the rGO nanosheets was responsible for high EDL capacitance.

Composition of the hybrid fibers in terms of micro-SC applications has been optimized. Fibers with SWCNT–GO mass ratios of 0:1, 1:8, 1:4 and 1:1 were denoted as rGO fiber, fiber-1, fiber-2 and fiber-3, respectively. Continuous fibers could not spun out as the mass fraction of GO was below 50 wt%. The as-synthesized wet fibers, with diameters of 260–300 μm, were dried in air for 4 h. The completely dried fibers had diameters of 40–60 μm, due to the water evaporation. They demonstrated tensile strengths of 84–165 MPa, yet with high flexibility, so that they could be bent into different shapes or woven into textile structures, as illustrated in Fig. 5.58 (d–f).

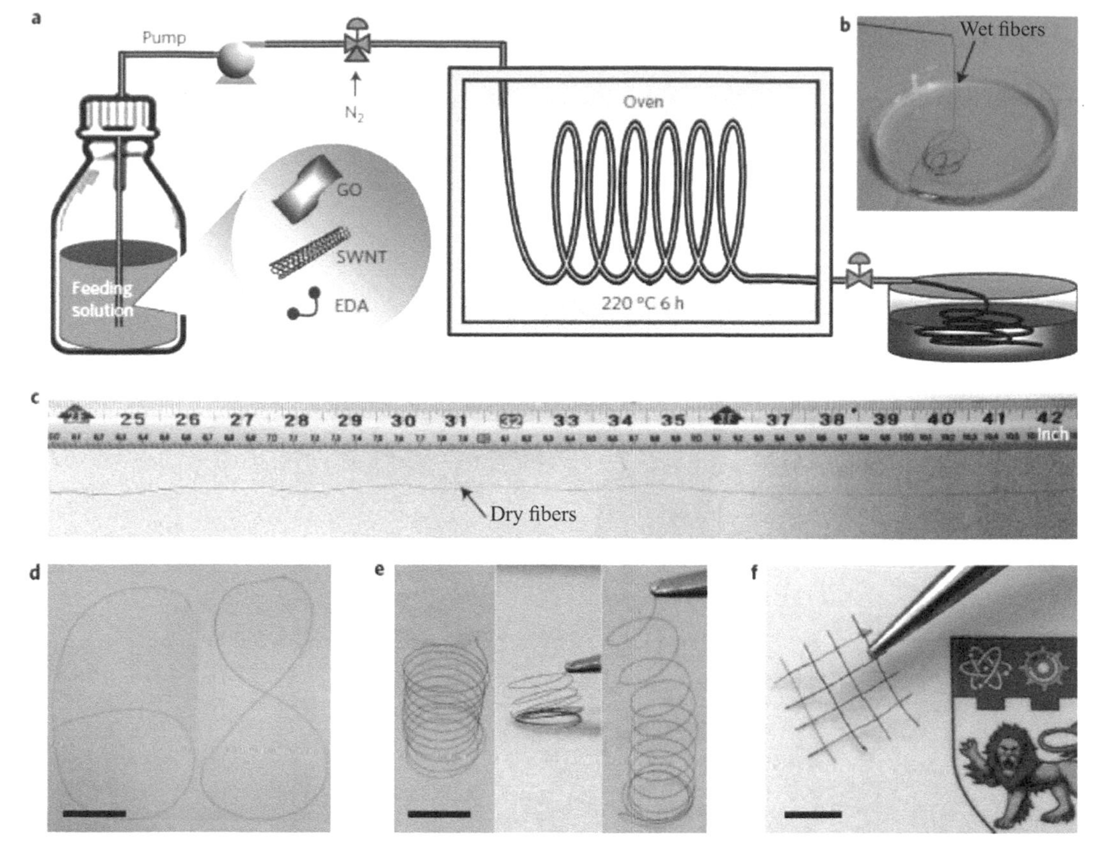

Fig. 5.58. Schematic diagram showing the synthesis process of the hybrid microfibers. (a) The fiber was synthesized by injecting a homogeneous solution containing acid oxidized SWCNTs, GO and EDA through a pump into a flexible silica capillary column, followed by *in situ* hydrothermal treatment in an oven at 220°C for 6 h before a continuous fiber was pushed into a water reservoir with a pressurized nitrogen flow. (b) Photograph of the as-prepared fibers collected in water. (c) A dry fiber with a diameter of 50 μm and length of 0.5 m. (d) Planar structures obtained by bending the fibers. (e) Compressed and stretched fiber springs. (f) A knitted textile fabricated from the fibers. All scale bars = 0.5 cm. Reproduced with permission from [68], Copyright © 2014, Macmillan Publishing Limited.

As shown in Fig. 5.59 (a–c), the rGO fiber was composed of densely stacked rGO nanosheets that were aligned in the axis direction. Due to the dense packing, the rGO fiber had a relatively low specific surface area. In comparison, specific surface areas of the hybrid fibers could be up to 396 $m^2 \cdot g^{-1}$ as the content of the SWCNTs was 50%. The hybrid fibers possessed an interconnected porous structure, as illustrated in Fig. 5.59 (d–f). The aligned SWCNTs were present in between the rGO nanolayers, which prevented the stacking of the rGO nanosheets and increased the porosity of the hybrid fibers, as demonstrated in Fig. 5.59 (g–i). Nitrogen adsorption–desorption isotherm indicated that the hybrid fibers exhibited a mesoporous structure. The pore sizes were in the range of 1.5–18 nm, peaked at 5 nm, which posed a positive effect on the electrochemical performance of the hybrid fibers.

G–CNT hybrid fibers have also been fabricated by using GO and pristine few-walled CNTs (FWCNTs) through wet-spinning [70]. In the above example, the mechanical properties of the hybrid fibers were degraded by the addition of oxidized-CNTs. Therefore, highly π-conjugated unfunctionalized FWCNTs were employed to form hybrid fibers with rGO, in order to achieve high ductility, strength and flexibility simultaneously. In addition, the presence of pristine FWCNTs tremendously increased the electrical conductivity of the hybrid fibers, with a maximum value of 210.7 $S \cdot cm^{-1}$. Flexible two-ply wire-shaped supercapacitors (WSSs), assembled with the as-fabricated hybrid fibers, demonstrated promising electrochemical performances.

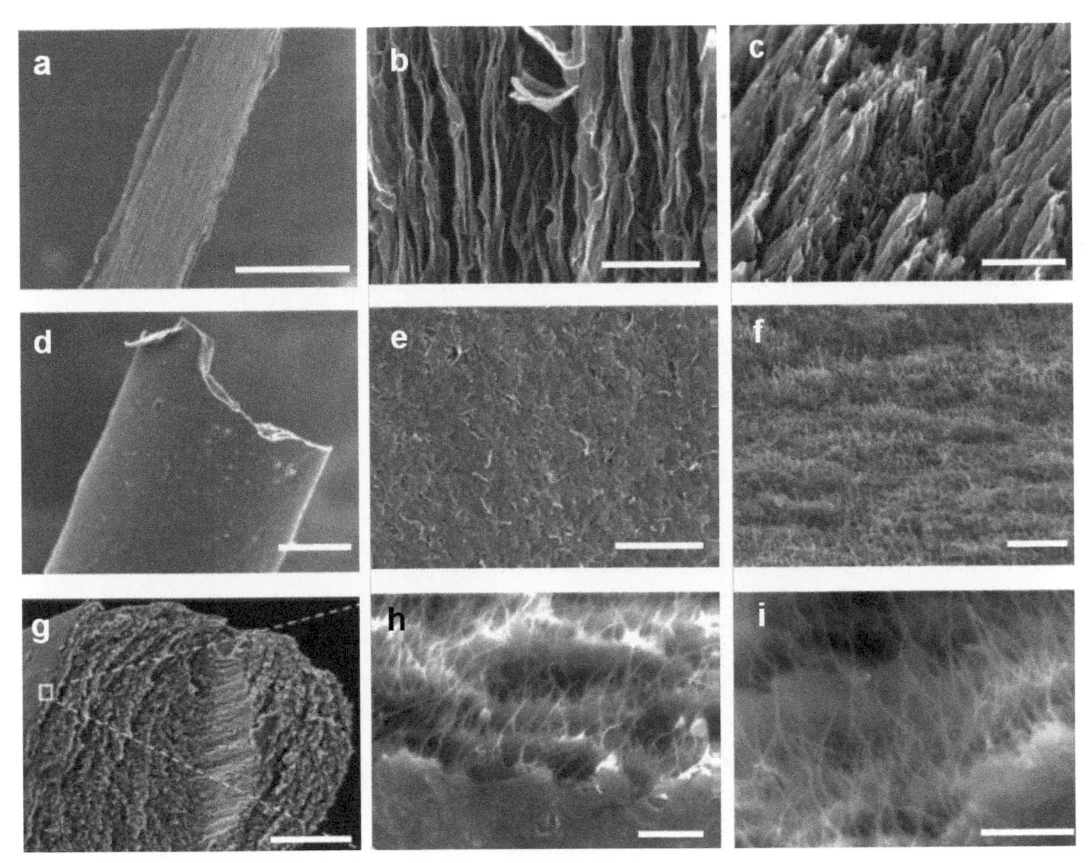

Fig. 5.59. Microstructures of the hybrid microfibers. SEM images of full view, outer surface and fracture end area of the rGO fiber (a–c) and fiber-3 (d–f). (g–i) Cross-sectional SEM images of the fiber-3. In (h), the square area in (g) is highlighted. In (i), SWCNT bundles are shown attached to the edges and surfaces of the rGO. Scale bars = 50 μm (a), 20 μm (d), 15 μm (g), 1 μm (c, f), 0.5 μm (b, e, h) and 300 nm (i). Reproduced with permission from [68], Copyright © 2014, Macmillan Publishing Limited.

GO nanosheets tens of micrometers in size were used. Unfunctionalized FWCNTs with CNT/GO mass ratios between 1:8 and 1:2 were dispersed to form aqueous solutions, corresponding to samples of CNT_1-rGO_8, CNT_1-rGO_6, CNT_1-rGO_4, CNT_1-rGO_3 and CNT_1-rGO_2. At CNT mass fractions of $\geq 50\%$, the CNT bundles were enlarged, which decreased the stability of the solutions, due to the electrostatic repulsion between the oxygen-containing groups on the GO nanosheets, thus leading to coagulation of the GO nanosheets. As long as well-dispersed CNT-GO solutions were formed, they exhibited nematic liquid crystal features under polarized optical microscopy (POM). Petridish wet-spinning approach was used to spin the hybrid fibers. It was found that continuous fibers could not be spun out from the solutions where the mass fraction of CNTs was $\geq 50\%$, because the content of the unfunctionalized FWCNTs was beyond the dispersion capability of the GO nanosheets.

The as-spun GO and CNT–GO hybrid fibers were reduced in hydroiodic acid aqueous solution to obtain rGO and CNT-rGO fibers. Figure 5.60 shows SEM images of the reduced fibers, with diameters in the range of 20–25 μm. The rGO nanosheets within the fibers were closely stacked and oriented in their axial direction and stacked together tightly. The FWCNTs showed partial alignment in the axial direction, especially in the fibers where the CNT/GO ratios were $< 1:3$, which was attributed to the presence of the GO nanosheets.

Mechanical properties of the hybrid fibers were increased after reduction, which exhibited a close relation to the content of the FWCNTs. Both the tensile strength and strain were increased first and then decreased, with the increasing FWCNT/RG ratio. Optimized mechanical performance was observed in the CNT_1-rGO_4 fiber, with an elongation of about 11.4% and a tensile strength of about 385.7 MPa. The

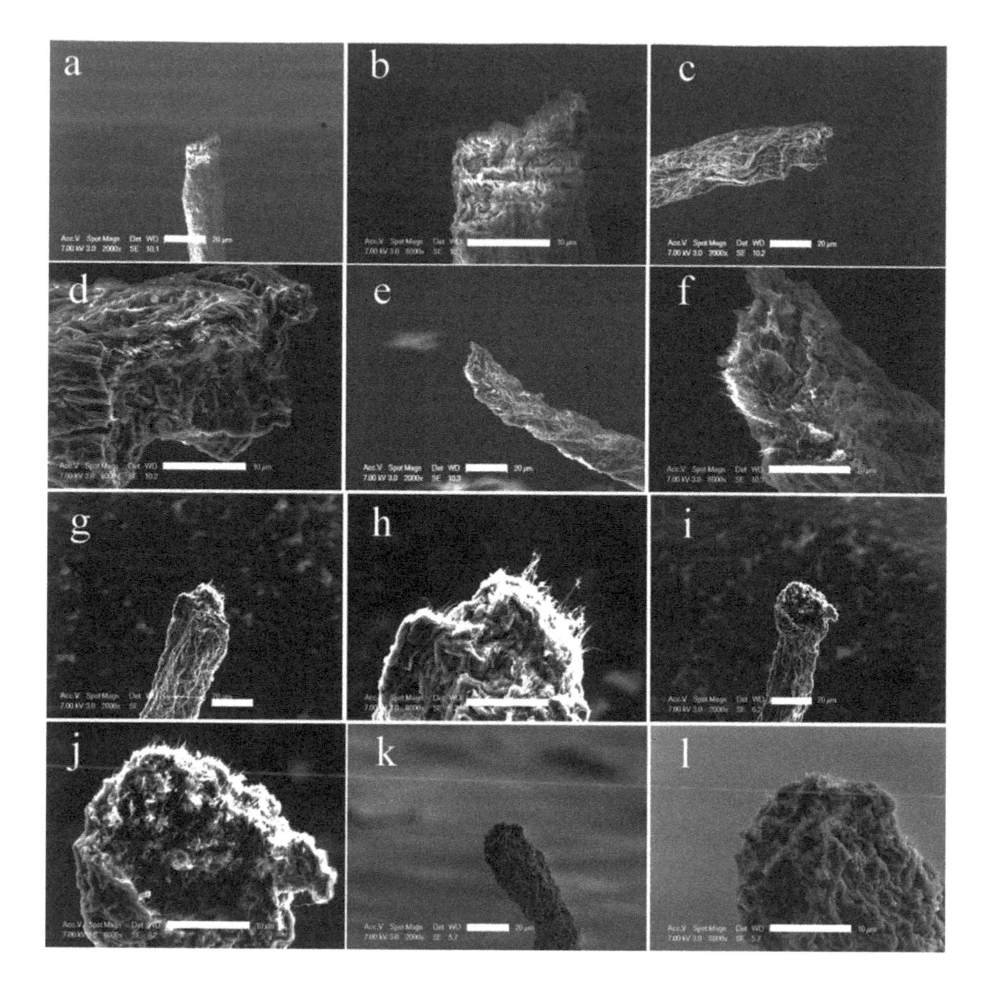

Fig. 5.60. Representative SEM images of the rGO and CNT-rGO fibers. (a, b) rGO, (c, d) CNT_1-rGO_8, (e, f) CNT_1-rGO_6, (g, h) CNT_1-rGO_4, (i, j) CNT_1-rGO_3, (k, l) CNT_1-rGO_2. Scale bars = 20 μm (a, c, e, g, i and k) and 10 μm (b, d, f, h, j and l). Reproduced with permission from [70], Copyright © 2015, American Chemical Society.

improvement in the strength and flexibility of the hybrid fibers was readily attributed to the synergistic effect due to the interconnection between the rGO nanosheets and the partially aligned FWCNTs. Because the unfunctionalized FWCNTs contained fewer defects than the functionalized FWCNTs, stronger van der Waals forces were formed between the FWCNTs and the rGO nanosheets in these hybrid fibers. Because CNTs tended to form bundles, the number of defects would be increased with the increasing content of the CNTs. As a result, there was an optimal composition in terms of the mechanical properties of the hybrid fibers. Additionally, the presence of the FWCNTs also had an influence on the electrical conductivity of the fibers. Similarly, there was an optimal content of the FWCNTs in terms of conductivity, which was observed in CNT_1-rGO_3, with a value of 275.1 $S{\cdot}m^{-1}$.

Instead of containing other nanocarbons, an all graphene core–sheath fiber has emerged recently, in which a core of GF was covered with a sheath of 3D porous network-like graphene framework [71]. The hierarchical hybrid structure was denoted as GF–3D-G. The core graphene fiber exhibited high electrical conductivity, while 3D graphene network created a high surface area, thus leading the great advantages as flexible electrodes for electrochemical supercapacitor applications, which has been demonstrated.

The GFs were prepared by using a one-step dimensionally-confined hydrothermal method with GO aqueous suspensions. The GF–3D-G fibers were obtained by using an electrochemical deposition, in GO aqueous suspension of 3 mg·ml^{-1} at pH = 3.62 containing 0.1 M $LiClO_4$. The GF with a diameter of 30–35 μm and a length of 2–10 cm was used as the working electrode, while Pt wire (1 mm in diameter) and Ag/AgCl (3 M KCl) were used as the counter and reference electrodes. The deposition was conducted at a constant potential of −1.2 V for 5 min. The as-deposited GF–3D-G fibers were thoroughly washed with distilled water and then freeze-dried. The dried fibers were highly flexible, as shown in Fig. 5.61 (a).

The GF–3D-G hybrid fiber exhibited a relatively higher tensile strength than the GF. The GO nanosheets were reduced into graphene and self-assembled into 3D networks on the GF fiber, as illustrated in Fig. 5.61 (b–f). The elimination of the oxygenated functional groups of the GO nanosheets was confirmed by the XPS results. The GF–3D-G had an electrical conductivity of 10–20 S·cm^{-1}, which was higher than that of the original GF. As seen in Fig. 5.61 (b, c), porous graphene layers were uniformly deposited on the fiber. The sizes of the pores were in the range of micrometers to ten micrometers, as demonstrated in Fig. 5.61 (d). Figure 5.61 (e, f) indicated that the graphene nanosheets encompassed the core of the GF, with the graphene planes nearly perpendicular to the surface of the GF, offering a high surface area for the access of ions to form electrochemical double-layers. The graphene nanosheets were strongly attached to the GF, which ensured mechanical stability of the GF–3D-G hybrid fibers.

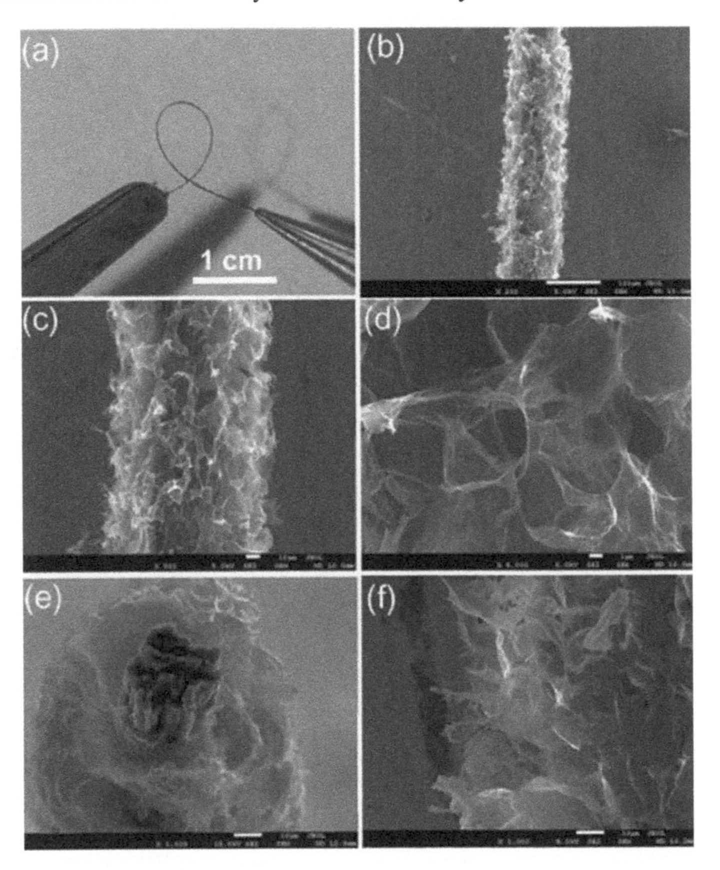

Fig. 5.61. (a) Photograph of a distorted GF–3D-G fiber. The observed particle-like bulging along the intersectant fiber resulted from the fiber-fiber attrition contact during the manual distortion. (b, c) SEM images of a GF–3D-G fiber. (d) An enlarged view of (c). (e) Cross-sectional SEM image of a GF–3D-G fiber showing the core GF surrounded by standing graphene nanosheets. (f) Edge view of a GF–3D-G fiber. Scale bars = 100 μm (b), 10 μm (c, e, f) and 1 μm (d). Reproduced with permission from [71], Copyright © 2013, WILEY-VCH Verlag GmbH & Co. KGaA, Weinheim.

5.3.3. Graphene-polymer composite fibers

In graphene-polymer composites, graphene is used as a filler to improve the mechanical properties of polymers and create new properties that the polymers do not have. For example, a very low content of graphene nanosheets can make the composites electrically conductive. Graphene nanosheets can be readily integrated with polymer matrix, due to the planar structure with large surface area. Furthermore, the carboxyl and hydroxyl functional groups on the basal planes and edges of the GO nanosheets act as linkers to bridge the graphene and the polymer. For example, graphene nanoribbon/carbon composite nanofiber yarns have been fabricated by using electrospinning from poly(acrylonitrile) containing GO nanoribbons (GONRs), combined with twisting and carbonization [72].

Poly(acrylonitrile) (PAN) and GONRs were dissolved in DMF to form spinning solutions. GONRs were added to 8 wt% PAN/DMF solution, with contents of up to 10 wt%. Figure 5.62 shows a schematic diagram of the device for electrospinning of the nanofibers. The polymer solutions were contained in a syringe with a stainless steel nozzle of 0.2 mm internal diameter. The nozzle was connected to a high-voltage regulated DC power supply. A constant volume flow rate was maintained by using a syringe-type infusion pump. A rotating disk, with a diameter of 250 mm and a width of 10 mm, was used to collect the aligned electrospun nanofibers. The distance between the nozzle tip and the collector was 100 mm, the applied voltage was 30 kV, the flow rate was 5 $\mu l \cdot min^{-1}$ and the rotating speed of the collector was 1000 rpm. The spinning experiment was conducted for 20 min.

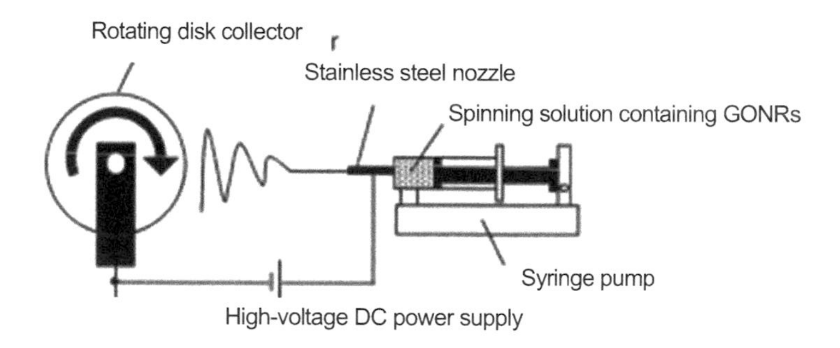

Fig. 5.62. Schematic diagram of the setup used to electrospin the composite fibers. Reproduced with permission from [72], Copyright © 2013, American Chemical Society.

Aligned GONR/PAN composite nanofibers, with compositions of GONR/PAN = 0/100, 0.1/99.9, 0.3/99.7, 0.5/99.5, 0.7/99.3, 1/99, 2/98, 3/97, 5/95 and 10/90 in wt/wt, were spun out from the corresponding spinning solutions. Figure 5.63 shows SEM images of the aligned GONR/PAN composite nanofibers. The fibers containing 0–2 wt% GONRs had an average diameter of about 200 nm, with bead-free and smooth morphologies, as seen in Fig. 5.63 (a–f). As the content of GONR was increased to 5 wt%, some rough surface regions were formed on the fibers, which was attributed to the serious agglomeration of the GONRs in the polymer matrix. Such phenomenon was evidenced by the TEM results in Fig. 5.64. Yarns with a thickness of about 50 μm could be fabricated by twisting the aligned nanofibers.

Figure 5.64 shows representative TEM images of the as-spun GONR/PAN fibers, in order to characterize the orientation of the GONRs in the composite. It was found that when the content of the GONRs in the composite fibers was 0.5 wt% and below, the GONRs were highly oriented along the fiber axis of the fibers, as illustrated in Fig. 5.64 (a, b). The orientation of the GONRs was induced by the electrified thin liquid jet during the electrospinning process. A directional shear force coupled with the external electric field to the flow of the spinning solution resulted in the orientation characteristics of the GONRs. In the samples with GONR contents of higher than 0.7 wt%, agglomeration of the GONRs was observed, i.e., the GONRs could not be well-dispersed in the fibers, so that orientation disappeared. The number and size of the agglomerates were increased with the increasing weight fraction of the GONRs in the composite nanofibers. Large agglomerates were frequently observed in the composite fibers with 5 wt% GONRs, as demonstrated in Fig. 5.64 (c, d).

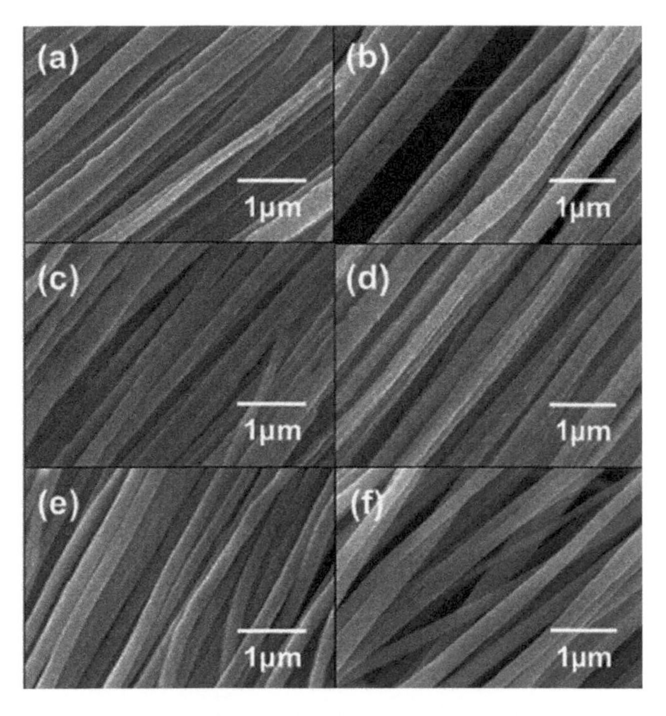

Fig. 5.63. Surface SEM images of the GONR/PAN composite nanofibers. (a) Pristine PAN nanofibers and the composite nanofibers containing different fractions of GONRs: (b) 0.1, (c) 0.3, (d) 0.5, (e) 1 and (f) 2. Reproduced with permission from [72], Copyright © 2013, American Chemical Society.

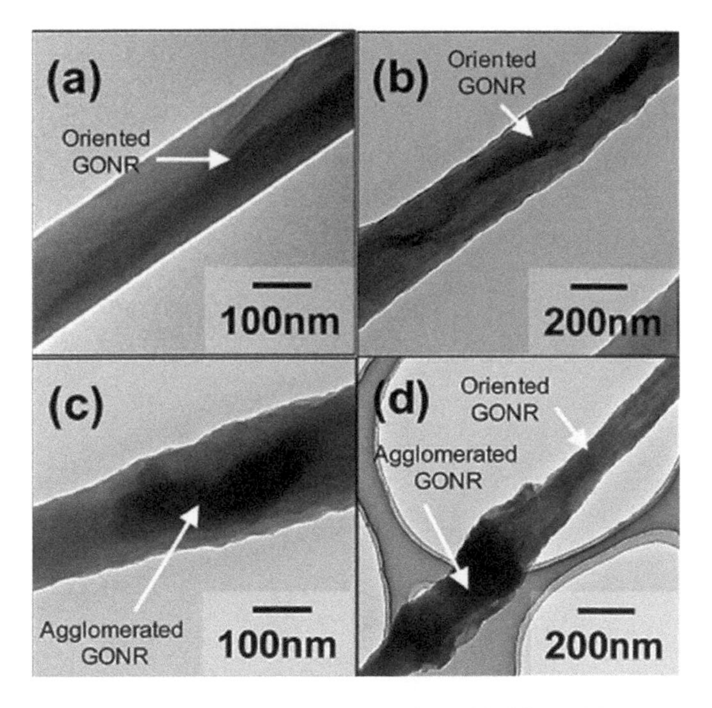

Fig. 5.64. TEM images of the as-spun GONR/PAN composite nanofibers with different GONR fractions: (a, b) 0.5 wt% and (c, d) 5 wt% GONR. Reproduced with permission from [72], Copyright © 2013, American Chemical Society.

The presence of the GONRs in the nanofiber retarded the movement of the polymer chains. Optimized tensile strength and Young's modulus were observed at the GONR fractions of 0.5 and 0.5–1 wt%, respectively, with maximum tensile strength of 179 MPa and maximum Young's modulus of 5.5 GPa, which were higher than those of the pristine PAN nanofiber yarns by 260% and 170%. The as-obtained GONR/PAN composite nanofiber yarns were thermally treated at 230°C for 3 h in air for stabilization and then at 1000°C for 1 h in N_2 for carbonization. During the carbonization, the GONRs were reduced at the same time. SEM observation indicated that the GNR/carbon composite nanofiber yarns exhibited no variation in morphology and shape after the carbonization treatment, as revealed in Fig. 5.65. Both the tensile strength and Young's modulus were significantly increased as compared to those of the pristine PAN nanofiber yarns and the PAN-based carbon nanofiber (CNF) yarns. Electrical conductivity of the GNR/carbon composite nanofiber was peaked at the GONR concentration of 0.5 wt.

During the electrospinning process, a directional shear force was generated to the flow of the spinning solution, which was further coupled with the high external electric field. As a result, as the fraction of the GONRs was sufficiently low, the GONRs could be highly aligned along the axial direction of the nanofibers in the electrified thin liquid jet. Due to the large surface area of the GONRs, together with the high orientation, their presence could greatly increase the mechanical properties of the composite nanofiber yarns. Other polymers used to fabricate graphene–polymer nanofibers by using electrospinning method include PVA [73, 74], poly(vinyl acetate) [75] and poly-acrylic acid [76].

It has been reported that the combination of CNTs with rGO could further result in solution-spun polymer composite fibers with further improved mechanical strength, as compared with those containing only CNTs [77]. The composite consisted of PVA, chemically reduced graphene oxide flakes (rGOFs)

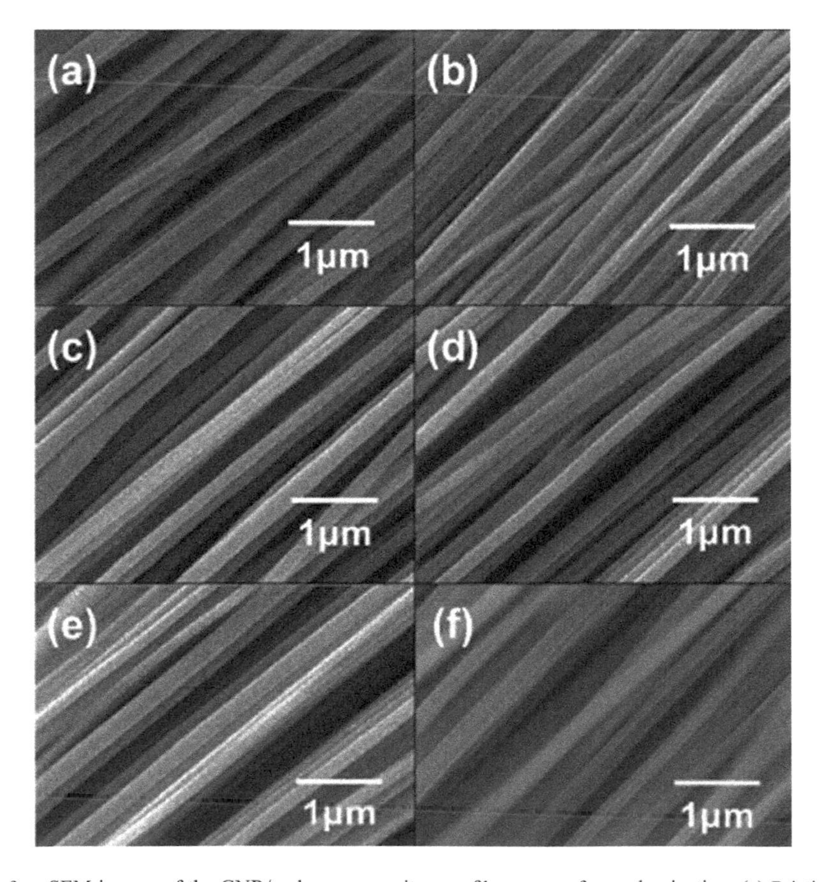

Fig. 5.65. Surface SEM images of the GNR/carbon composite nanofiber yarns after carbonization. (a) Pristine PAN-based carbon nanofiber yarn. Composite nanofiber yarns containing different GNR fractions: (b) 0.1, (c) 0.3, (d) 0.5, (e) 1 and (f) 2. Reproduced with permission from [72], Copyright © 2013, American Chemical Society.

and single-walled carbon nanotubes (SWCNTs). A network of the rGOFs–SWNTs nanoparticles was spontaneously formed in the fibers during wet spinning, if the solution composition and particle dispersion were well-optimized. The as-spun gel fibers already exhibited super-strong toughness, without the requirement of a subsequent drawing process. The high mechanical performance was attributed to the strong interactions between the two nanocarbon items, which were mutually self-aligned during the spinning process.

Fabrication process of the hybrid composite fibers is shown in Fig. 5.66 (a). Homogeneous dispersions of rGOFs and SWCNTs with various compositions were formed in aqueous solutions of sodium dodecyl benzene sulfonate (SDBS). The aqueous spinning solutions with pH = 7 contained 1 wt% SDBS and 0.3 wt% nanocarbon items. Dispersions were then injected into the flow stream of an aqueous solution of 5 wt% PVA, with molecular weight of 146,000–186,000 g·mol^{-1} and hydrolysis of ~99%. rGOF–SWCNT/PVA hybrid composite fibers were formed through the coagulation bath. The as-spun fibers were washed with DI water, followed by drying and methanol treatment to increase the crystallinity of the PVA. In such hybrid fibers, rGOFs and SWCNTs were dispersed in the PVA matrix, as demonstrated in Fig. 5.66 (b). PVA content in the fibers was about 30 wt%, while the ratio of rGOF to SWCNT was varied. In addition, PVA could be removed by thermally treating the rGOF/PVA fibers in vacuum at 600°C for 1 h or by soaking the rGOF/PVA fibers in 37% hydrochloric acid at room temperature for hours, so as to obtain rGOF fibers, as illustrated schematically in Fig. 5.66 (c). The polymer-free fibers were mechanically stable, due to the formation of the interconnected network by the rGOFs, as seen in Fig. 5.66 (d). If the rGOF–SWCNT suspensions were directly injected into 37% hydrochloric acid solution, then polymer-free fibers could also be developed.

It was shown that the strength and elongation at break of the rGOF–SWCNT/PVA hybrid composite fibers were higher than those of the SWCNT/PVA and rGOF/ PVA fibers by three and four times, respectively. Specifically, the volumetric toughness of the 1:1 rGOF–SWCNT/PVA hybrid composite fibers was 1,015–2,060 MJ·m^{-3}, with an average value of MJ·m^{-3}. The equivalent mass-normalized toughness was 480–970 J·g^{-1}. The ratio of rGOF to SWCNTs played an important role in determining the toughness performance of the hybrid composite fibers, which could be optimized experimentally.

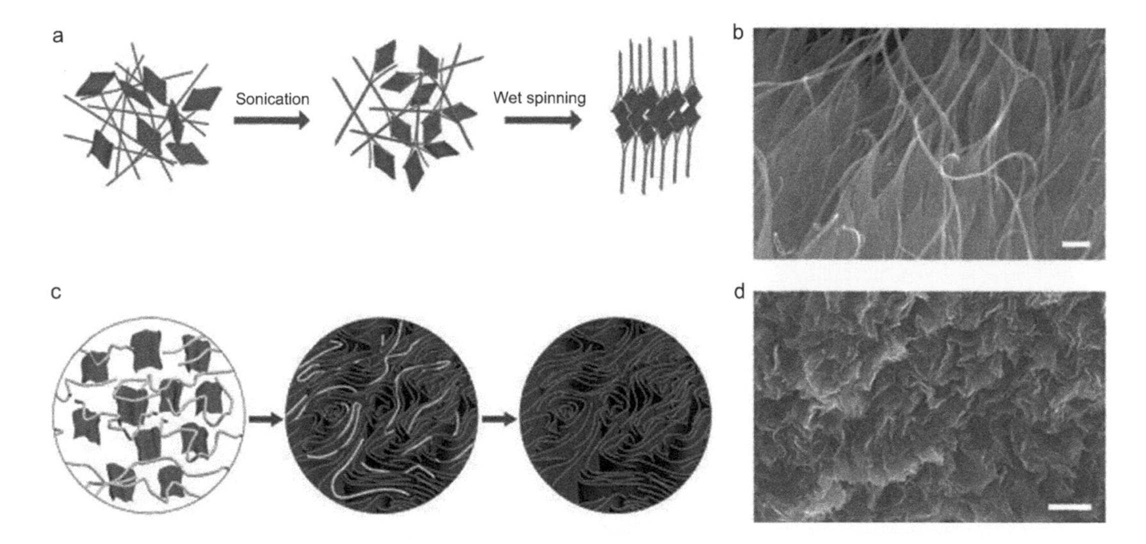

Fig. 5.66. Fabrication and microstructure of the graphene–CNT hybrid polymer composite fibers. (a) Schematic diagram showing the formation of the oriented interconnected network of rGOFs (curved rectangles) and SWCNT bundles (gray lines) as a result of sonication and subsequent wet spinning. (b) Cross-sectional SEM image of the rGOF–SWCNT/PVA hybrid composite fiber, with 1:1 weight ratio of rGOF to SWCNT, showing co-assembly of the rGOFs and SWCNTs. Scale bar = 100 nm. (c) Schematic illustration of the structural evolution between the coagulation-spun rGOF gel (left), polymer composite (middle) and polymer-free fibers (right). Yellow lines and blue curved rectangles represent PVA chains and rGOFs, respectively. (d) Cross-sectional image of the polymer-free rGOF fiber, indicating the presence of wrinkled rGOFs. Scale bar = 500 nm. Reproduced with permission from [77], Copyright © 2012, Macmillan Publishing Limited.

Figure 5.67 shows microscopic characteristics (SEM images) and a proposed model of the high toughness hybrid composite fibers. Before the fibers were stretched, they exhibited corrugations along the fiber longitudinal axis, due to the shrinkage encountered during the coagulation of the as-spun gel fibers, as demonstrated in Fig. 5.67 (a). After they were stretched to high strains, slight lateral corrugation was observed on surface of the fibers, as illustrated in Fig. 5.67 (b). Periodic wrinkles were observed on surfaces of the hybrid composite fibers, which could be the surface characteristics of metals that experienced a plastic deformation. Figure 5.67 (c) shows a possible deformation mechanism to explain the formation of the surface corrugations in the hybrid composite fibers.

The degree of orientation of the nanocarbon network was evaluated by observing the fiber fracture surfaces. Orientation of the nanocarbon items was very poor in the rGOF/PVA and SWCNT/PVA composite fibers. Figure 5.68 (a) shows fracture surface SEM image of the as-spun hybrid composite fiber with 1:1 ratio of rGOF to CNT, indicating that SWCNTs were highly aligned in the fiber longitudinal direction. As seen in Fig. 5.68 (b), the rGOFs were also preferably oriented in the same direction with the SWCNT bundles that were attached to edges and surfaces of the rGOF. In contrast, PVA-free hybrid fibers obtained by spinning into aqueous HCl directly exhibited a layered structure, consisting of wrinkled rGOFs and associated SWCNTs, as illustrated in Fig. 5.68 (c).

The hybrid composite fibers with 1:1 ratio of rGOF to SWCNT were highly flexible, so that they could be severely twisted without breaking by using an electric motor. Figure 5.68 (d) indicated that the deformation of the fibers was retained with a helix angle of about 50°, due to their high flexibility and toughness. They could be sewn into rubber bands and cloth, which repeatedly stretched to elongations of up to 20% (Fig. 5.68 (e)). After the hybrid composite fibers were coiled on a 1.2-mm diameter stainless steel tube and then annealed at 150°C for 1 h, a helical spring was produced, as illustrated in Fig. 5.68 (f). The fiber spring had a spring constant of about 41 $N \cdot m^{-1}$, which withstood both complete compression and elongation.

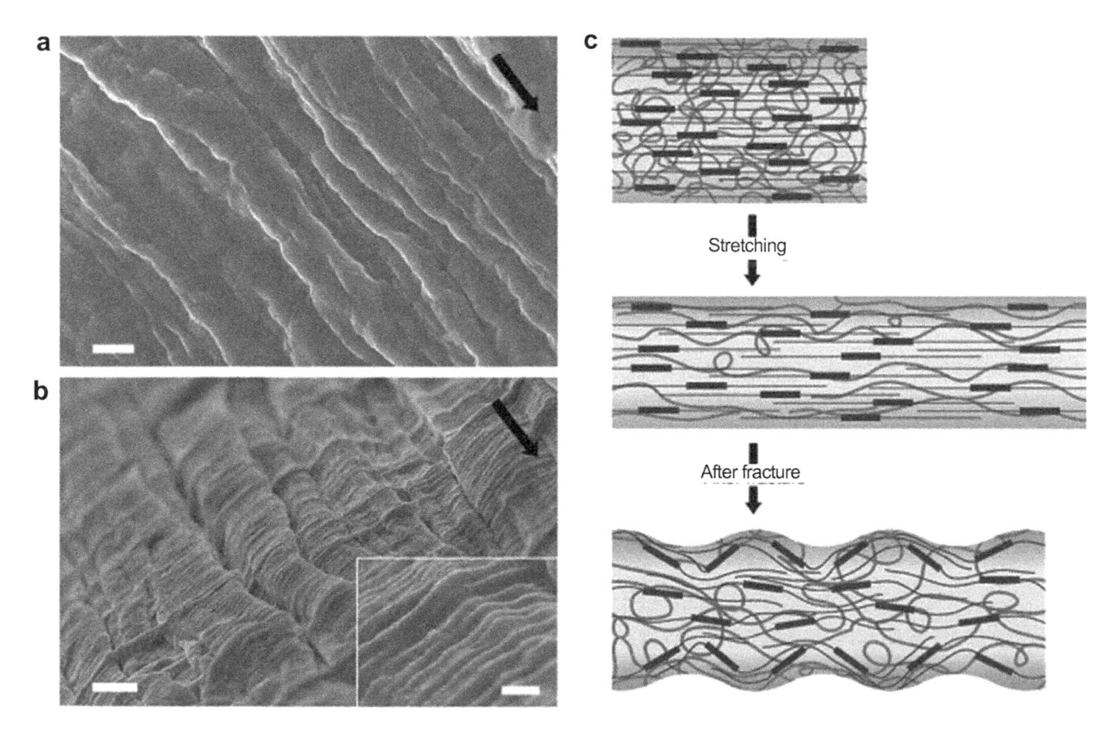

Fig. 5.67. Structural variation of the hybrid fibers during tensile tests. (a, b) SEM images showing the surface characteristics of the hybrid fibers before stretching (a) and after fracture (b). The arrows of (a) and (b) indicate the fiber axis. Scale bar = 1 μm and the inset scale bar in (b) = 200 nm. (c) Schematic illustrations showing the structural variation of the hybrid fibers during tensile tests. Black rectangles, black lines and blue curved lines represent rGOFs, SWCNTs and PVA chains, respectively. Reproduced with permission from [77], Copyright © 2012, Macmillan Publishing Limited.

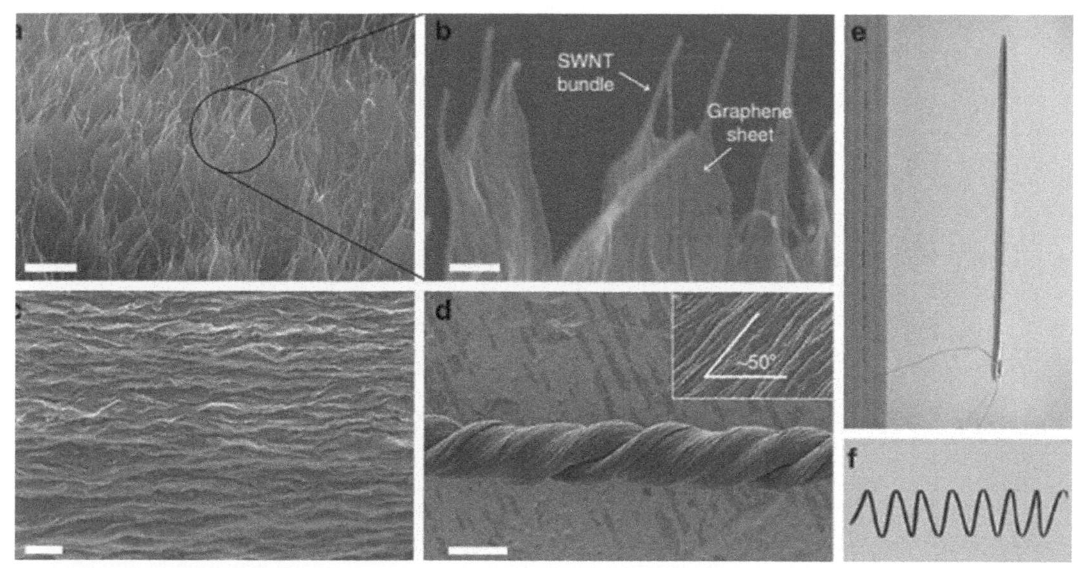

Fig. 5.68. Microstructure and mechanical flexibility of the hybrid composite fibers. (a) Fracture surface SEM image showing the well-oriented rGOFs and SWCNTs in the 1:1 rGOF/SWCNT hybrid fiber. Scale bar = 500 nm. (b) Close-view image of the fracture surface in (a). Scale bar = 100 nm. (c) SEM image of the layered structure on the surface of the 1:1 hybrid fiber with a polymer-free coagulation bath. Scale bar = 1 μm. (d) 1:1 hybrid composite fiber twisted by using an electric motor, with the inset showing the twist angle. Scale bar = 50 μm. The insert is the SEM image showing the orientation angle of the nanotubes with respect to the fiber axis. (e) Photograph showing a hybrid fiber sewn into a rubber band with a 3.5 cm needle. (f) A spring-shaped hybrid fiber that was formed by annealing at 150°C, with length of 6.5 mm, diameter of 1.2 mm and fiber diameter of 50 μm. Reproduced with permission from [77], Copyright © 2012, Macmillan Publishing Limited.

5.4. Potential Applications

5.4.1. Actuators and motors

Materials and structures with smart responses to environmental and external stimuli have potential applications in sensors, actuators and robots. In this respect, graphene-based materials and structures exhibited extraordinary properties that make them useful in actuation systems as actuators and motors [78–83]. For instance, electrochemical actuators based on unimorph and bilayer graphene films and 3D graphene skeletons have been reported in the open literature [84–87]. Moreover, GFs are mechanically flexible and electrically conductive, so that they can be readily used to replace the conventional carbon fibers, but with extra functionalities.

It is expected that GFs-based stimulus responsive structures have the additional advantage of flexibility. Electrochemical actuators based on GF/polypyrrole bilayer composite fibers have been reported [88]. The fiber actuators possessed high actuation performance and durability comparable to that of other graphene-based actuators. Furthermore, multi-armed tweezers and net actuators could be fabricated with the composite fibers, which could find potential applications in the fields of biological studies and microfabrication. Figure 5.69 shows a schematic diagram of the fabrication process of the GF/PPy composite fibers. Flexible GF of about 20 mm in length was fixed onto a Ni rod with conducting resins as the electrical contact to the potentiostat. The joint of the GF and the Ni rod was carefully placed about 5 mm above the electrolyte to ensure that only the GF was in contact with the electrolyte. To form the GF/PPy composite fiber, the free end of GF was floated on the surface of an aqueous solution of 0.1 M freshly distilled Py monomer and 0.2 M $NaClO_4$ in which PPy was partially deposited along the longitudinal direction of the GF through electropolymerization of Py at a constant potential of 0.8 V, in a standard three-electrode configuration with Pt wire and Ag/AgCl used as the counter and reference electrodes, respectively. Actuation response of the GF/PPy composite fiber was studied with the same three-electrode setup by cyclic voltammetry (CV) at a scan rate of 50 $mV \cdot s^{-1}$ between 0.8 V and −0.8 V in aqueous solution of 1 M $NaClO_4$.

A photograph of the as-prepared GF/PPy fiber which was 20 mm in length is shown in Fig. 5.70 (a), together with a schematic diagram of the GF/PPy fiber, with yellow tube and black corona to be the GF and PPy film, respectively. Cross-sectional SEM image of the GF/PPy structure is demonstrated in Fig. 5.70 (b), indicating that the GF had a thickness of about 30 μm, with a multilayer structure similar to that of the graphene nanosheets (inset in Fig. 5.70 (b)).

Electrochemical actuation performances of the GF/PPy fibers are illustrated in Fig. 5.71. The fiber was slightly off the vertical line perpendicular to the electrolyte surface at zero bias (Fig. 5.71 (b)), which was attributed to the asymmetrical structure of composite fiber. With CV scanning (Fig. 5.71 (d)), the GF/PPy fiber exhibited displacements rightwards and leftwards at positive and negative potentials, respectively (Fig. 5.71 (a, b)), with an angular velocity at the bias of near 0 V to 0.12 $rad \cdot s^{-1}$. Figure 5.71 (e) shows the bending angles of the GF/PPy fiber actuator as a function of applied potential, with a symmetrical bending displacement over a wide voltage range. In the durability test of over 100 cycles, the maximum displacement was retained by about 80% after 100 cycles, as seen in the inset of Fig. 5.71 (e). In addition, multiple section design could be applied to the GF/PPy composite fiber to realize complicated movement. As illustrated in Fig. 5.71 (f), two opposite PPys sections were deposited on the fibers. In this case, the combination of the two opposite bending forces resulted in an "S" shape actuation by the fiber, as demonstrated in Fig. 5.71 (g). The GF/PPy composite fibers could be used to construct more complicated actuators, such as bi-armed and tri-armed tweeters.

Graphene/GO (G/GO) asymmetric fibers with humidity response have been demonstrated through positioned laser reduction of as-spun GO fibers [89]. Remarkably, the G/GO fibers display complex, well-confined and predetermined motion and deformation once exposed to moisture. The excellent humidity

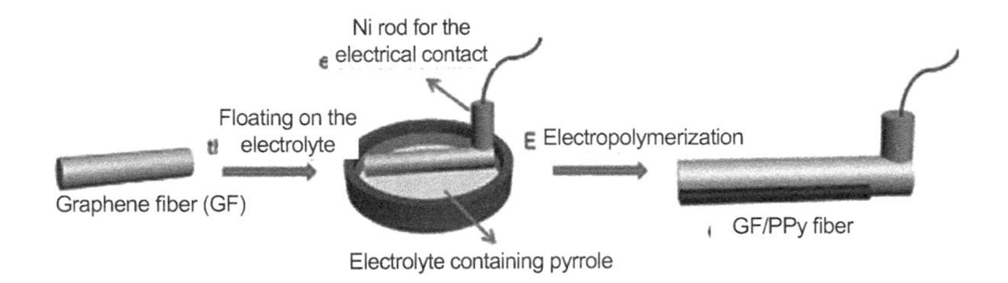

Fig. 5.69. Fabrication process of the GF/PPy composite fibers. Reproduced with permission from [88], Copyright © 2013, Elsevier.

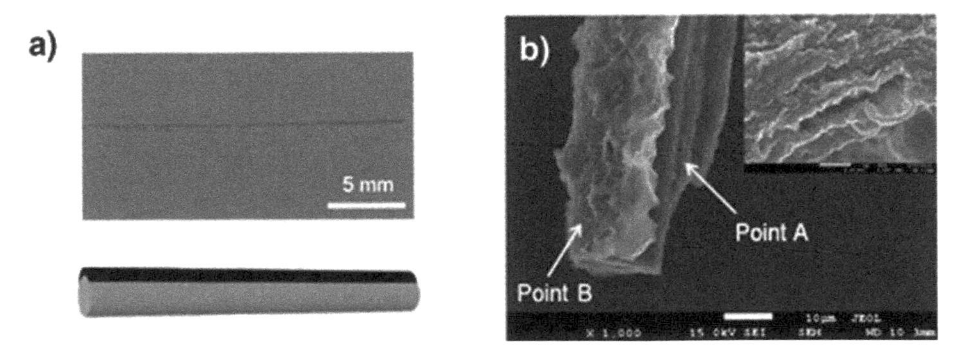

Fig. 5.70. (a) Photograph of the as-prepared GF/PPy fiber (upper panel) and a schematic diagram of the GF/PPy fiber with the yellow column and black corona to be the graphene fiber and the PPy film (bottom panel). (b) SEM image showing the microstructure of the GF/PPy composite fiber, with scale bar = 10 μm. The inset is an enlarged view of the GF showing the layered structure assembled with graphene nanosheets, scale bar = 1 μm. Reproduced with permission from [88], Copyright © 2013, Elsevier.

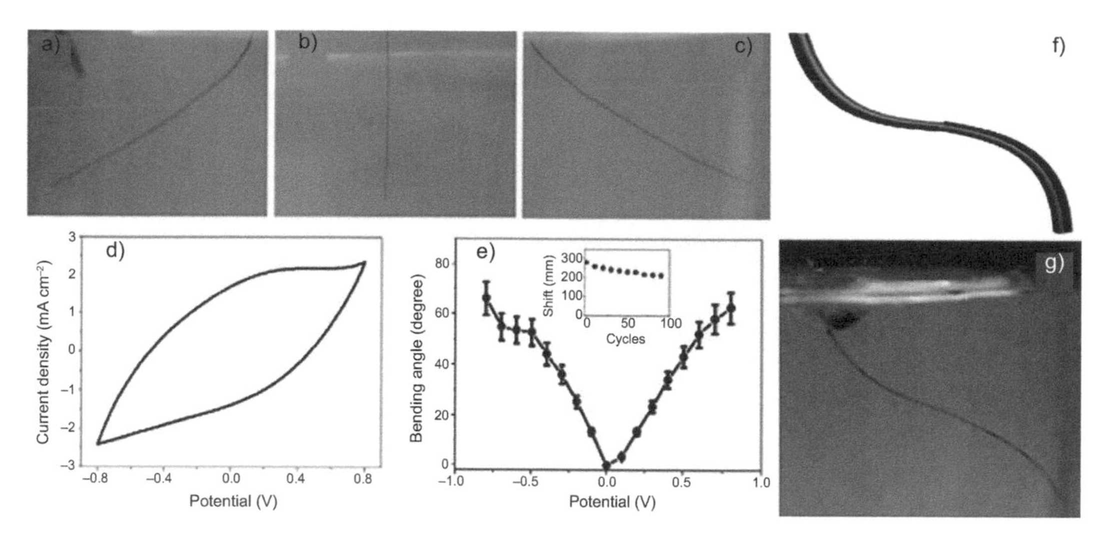

Fig. 5.71. (a–c) Snapshots of the bending status of the GF/PPy fiber actuator driven at electrochemical potentials of −0.8 V (a), 0 V (b), and +0.8 V (c), respectively. (d) Typical CV curve recorded during the actuation process. (e) Bending angle of the GF/PPy actuator as a function of the applied potential over the range of ±0.8 V. The inset is the durability test results over 100 cycles. (f) Schematic diagram illustrating the structure of the S-shaped GF/PPy composite fiber actuator with yellow and black zones to be graphene and PPy, respectively. (g) Photograph of the S-shaped actuation under CV scanning. Reproduced with permission from [88], Copyright © 2013, Elsevier.

responsive bending/unbending ability of the G/GO fibers enabled a new type of microfiber walking device. The G/GO fiber walker can travel in a narrow slit. Laser has been used to selectively reduce GO into graphene (G) in a specific area [90–92]. Figure 5.72 (a) shows an asymmetric G/GO fiber that was fabricated by scanning a laser beam along a preformed GO fiber. After the laser reduction, there was a color change from yellow-brown GO to black graphene in the scanned area on the GO fiber, as shown in Fig. 5.72 (b). The difference in feature between the scanned and unscanned areas was also confirmed by SEM, as seen in Fig. 5.72 (c).

Current–voltage (I–V) curves of the laser-reduced fiber and the unreduced fiber are shown in Fig. 5.72 (d). Obviously, the laser-reduced fiber exhibited a higher slope than that the GO fiber. The reduced area had a conductivity that was about two orders of magnitude higher than that of the GO surface. Side view photographs of the GO fiber and the asymmetric G/GO fiber are shown in Fig. 5.72 (e, f). A black layer of G region was observed along the G/GO fiber. The brighter feature of the GO areas in Fig. 5.72 (c, e, f) had a high content of O, with an O/C atomic ratio of about 1:3. In contrast, the laser-reduced areas had a much lower level of O, with an O/C of 1:8, confirming the reduction of GO by the laser irradiation.

The G/GO fibers were sufficiently flexible to withstand any deformation, showing a tensile strength of as high as 100 MPa. Due to the presence of the various oxygen-rich functional groups of GO, as well as the special stacking of the GO nanosheets, the GO layers would have a stronger affinity to water as compared with the graphene layers. As a result, the asymmetric G/GO fiber exhibited moisture-sensitive actuation behavior, as demonstrated in Fig. 5.73 (a).

For instance, the G/GO fiber bended towards the G side when it was exposed to moist air with a relative humidity (RH) of 80% (Fig. 5.73 (d)), while the fiber returned to its initial state as the humidity was recovered to the ambient conditions (Fig. 5.73 (c)). This process was fully reversible with an average motion rate of $8^{o}\cdot s^{-1}$. If the RH was decreased from 25% to 10%, the G/GO fiber experienced a displacement towards the GO side, as observed in Fig. 5.73 (b), which was attributed to the contraction of the GO layers caused by the loss of the adsorbed water molecules.

Figure 5.73 (e) indicated that the response of the G/GO fibers was almost linear as a function of relative humidity over a wide range of 10–80%. After 1000 cycles of repeated humidity variation between 10% and 80%, the response performance remained nearly unchanged, as illustrated in Fig. 5.73 (f), showing a high operation stability, thus making them promising candidates for practical applications. Three-wire moisture

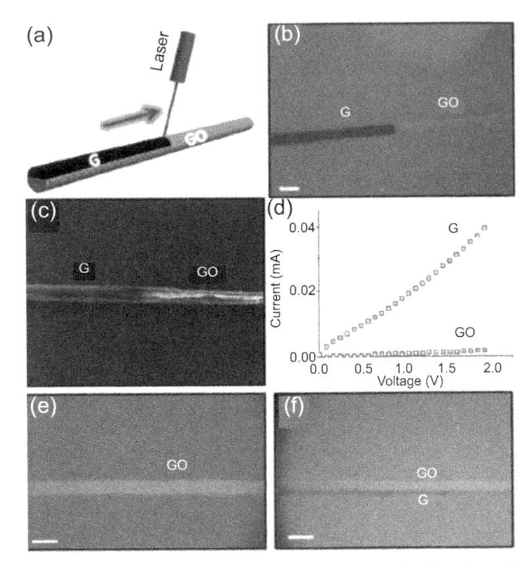

Fig. 5.72. (a) Schematic of the laser reduction on one side of a GO fiber. The black region was the laser-induced G region along the brown GO fiber. Photomicrograph (b) and SEM image (c) of the top surface of the as-reduced asymmetric G/GO fiber. (d) I–V curves of a laser-reduced G fiber and an untreated GO fiber with a length of 1 cm. (e, f) Photographs (side view) of the initial GO fiber and the asymmetric G/GO fiber, respectively. Scale bars = 50 μm. Reproduced with permission from [89], Copyright © 2013, Wiley-VCH Verlag GmbH & Co. KGaA, Weinheim.

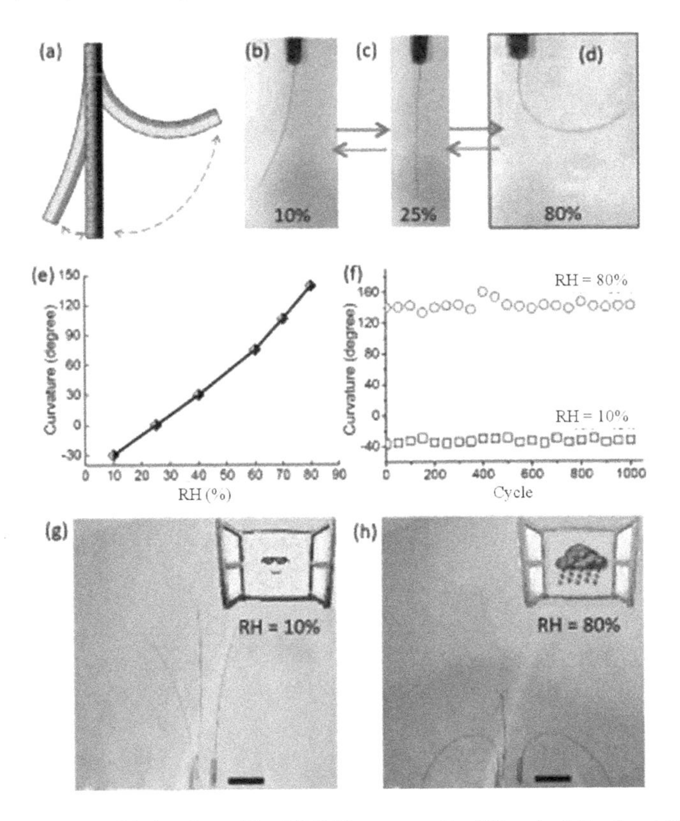

Fig. 5.73. (a) Schematic of the possible bending of the G/GO fiber exposed to different relative humidities. (b–d) Photographs of the G/GO fiber (2 cm in length) at different relative humidities. (e) Curvature of the G/GO fiber versus RH. (f) Durability test of the G/GO fiber. (g, h) Photographs of the three-wire moisture tentacles made of the G/GO fibers on a sunny and a rainy day, respectively. Scale bars = 5 mm. Reproduced with permission from [89], Copyright © 2013, Wiley-VCH Verlag GmbH & Co. KGaA, Weinheim.

tentacles have been constructed by using the G/GO fibers. As shown in Fig. 5.73 (g, h), the tentacles stood straight on sunny days, while they bended towards the opposite directions on rainy days, corresponding to a "close-open" action. The devices could be further modified or designed for other specific applications, as exemplified in Fig. 5.74.

By reconstructing the intrinsic configuration of the graphene nanosheets within the fiber, a new type of moisture-driven rotational motor has been prepared by rotary processing of freshly-spun GO hydrogel fibers [93]. The twisted GO fiber (TGF) with rearranged graphene nanosheets within the fiber exhibited very high performance as a reversible rotary motor, with rotary speeds of up to 5190 revolutions per minute and a tensile expansion of 4.7% under humidity variation. Moreover, the moisture-responsive actuation behavior of the TGF also led to the development of humidity switchers and moisture-triggered electric generators.

Figure 5.75 (a) shows schematic illustration of the TGF, which was fabricated by rotating the freshly-spun GO hydrogel fiber along the axis. During the rotating process, the fiber was shortened slightly, while its diameter was decreased correspondingly, due to the elimination of the solvent. The intrinsic structure of the GF was reconstructed during the twisting process. Originally, the GO nanosheets were aligned along the fiber axial direction (Fig. 5.75 (b)), whereas a helical configuration was formed after the twisting process (Fig. 5.75 (c, d)). At the same time, the compact graphene nanosheets on the surface of the TGF were conformed to the rotating direction, as seen in Fig. 5.75 (e). Cross-sectional view indicated that the graphene nanosheets inside the fiber were still densely packed along the axial direction of the fiber, so that the mechanical properties of the TGFs would not be significantly affected by the twisting process. The derived TGFs showed strong mechanical flexibility, while their properties could be adjusted by the varying the inserted twist to a certain degree.

Because the GO nanosheets contained a large amount of oxygen-rich functional groups, they experienced reversible expansion and contraction due to the adsorption and desorption of water molecules, respectively, as discussed above. The maximum deformation could reach as high as 5% [94]. Accordingly, the GO fibers with helical geometry could readily facilitate reversible torsional rotation as the humidity was changed, as illustrated in Fig. 5.76 (a). Therefore, a reversible variation in the helical angle experienced by the TGF was expected with varying humidity (Fig. 5.76 (a, b)). For example, the TGF had an initial helical angle (α) of 46.2° at a relative humidity (RH) of 20% (Fig. 5.76 (b), (1)), which was decreased to about 42°

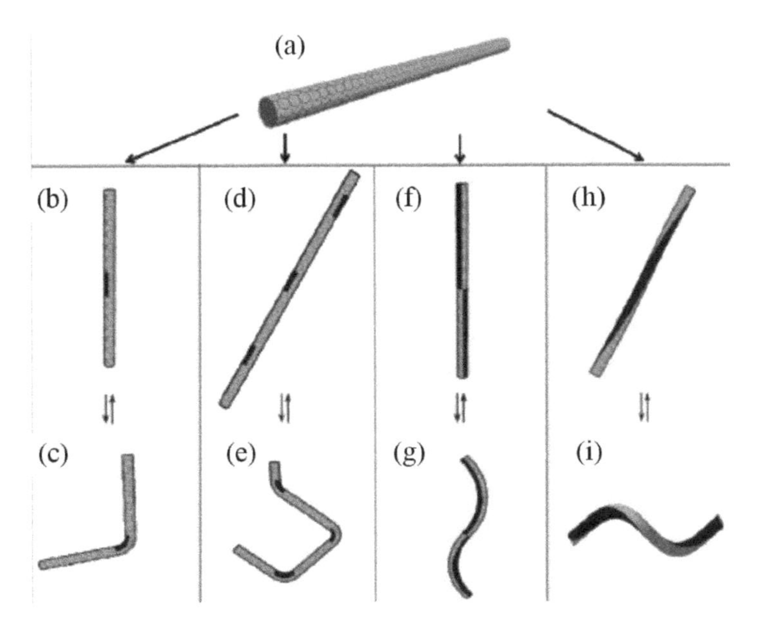

Fig. 5.74. Schematic diagrams of the G/GO fibers designed with different configurations and their deformations against humidity. (a) The original GO fiber. (b, d, f, h) GO fibers with region-selective laser reduction. (c, e, g, i) The corresponding predefined deformations upon exposure of the fibers to high humidity. Reproduced with permission from [89], Copyright © 2013, Wiley-VCH Verlag GmbH & Co. KGaA, Weinheim.

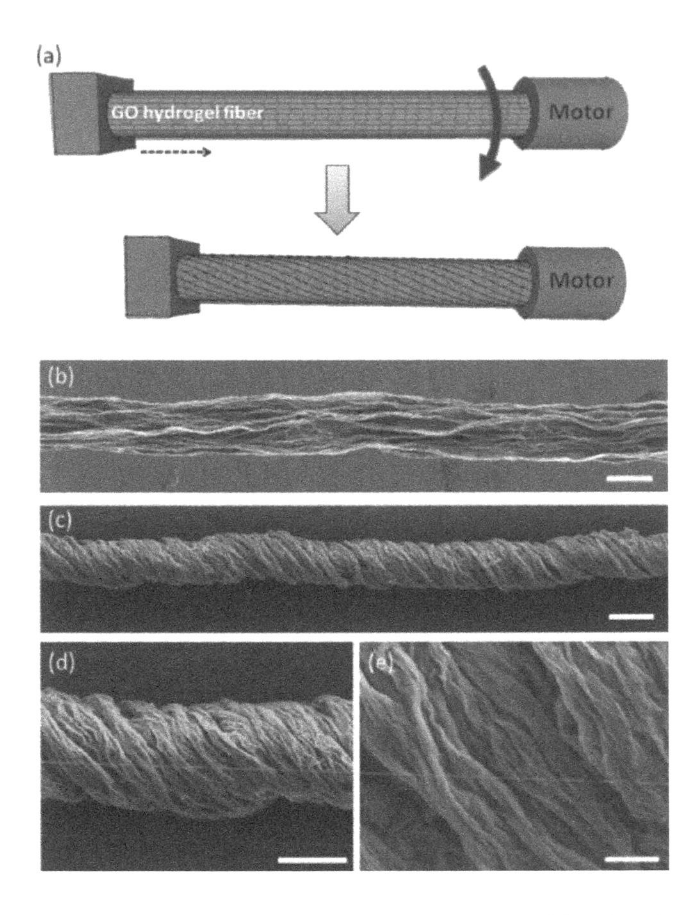

Fig. 5.75. Fabrication and characterization of the TGF: (a) scheme of fabrication of the TGF, with the arrow indicating the direction of rotation, (b, c) SEM images of the directly dried GO fiber and TGF formed with 5000 turns per meter, respectively, (d, e) enlarged view of the TGF and its surface. Scale bars = 50 μm (b–d) and 10 μm (f). Reproduced with permission from [93], Copyright © 2014, Wiley-VCH Verlag GmbH & Co. KGaA, Weinheim.

as the humidity was increased to RH = 85% (Fig. 5.76 (b), (2)). The TGF returned to its original state as the initial humidity was recovered (Fig. 5.76 (b), (3)). The cycle could be repeatedly conducted, as revealed in Fig. 5.76 (c). In addition, as demonstrated in Fig. 5.76 (d, e), both the rotation speeds and number of rotations were almost linearly proportional to the length of the TGFs and the degree of the inserted twist.

Micro-motors propelled by self-generated forces could find various potential applications to transport and deliver cargo in various environments [95–97]. One of the samples for such applications was demonstrated by using site-specific functionalized hGF, with which a self-powered graphene micromachine was created that could move in aqueous medium [41]. In that device, the inner-walls of the hGF were modified with Pt NPs. Similar to the tubular microjet, O_2 bubbles due to the decomposition of H_2O_2 were ejected from the open end, thus pushing the hGF to swim in the opposite direction. The Pt NPs acted as the catalyst to trigger the decomposition of H_2O_2 to produce O_2.

5.4.2. Energy related devices

Due to their high mechanical strength and electrical conductivity, GFs have been widely acknowledged to be a new type of electrode material of energy generation and storage devices, including solar cells, lithium ion batteries and supercapacitors. For instance, a wire-shaped photovoltaic cell has been developed by using G–Pt hybrid fibers as the counter electrodes and Ti wire impregnated with TiO_2 nanotubes as the working

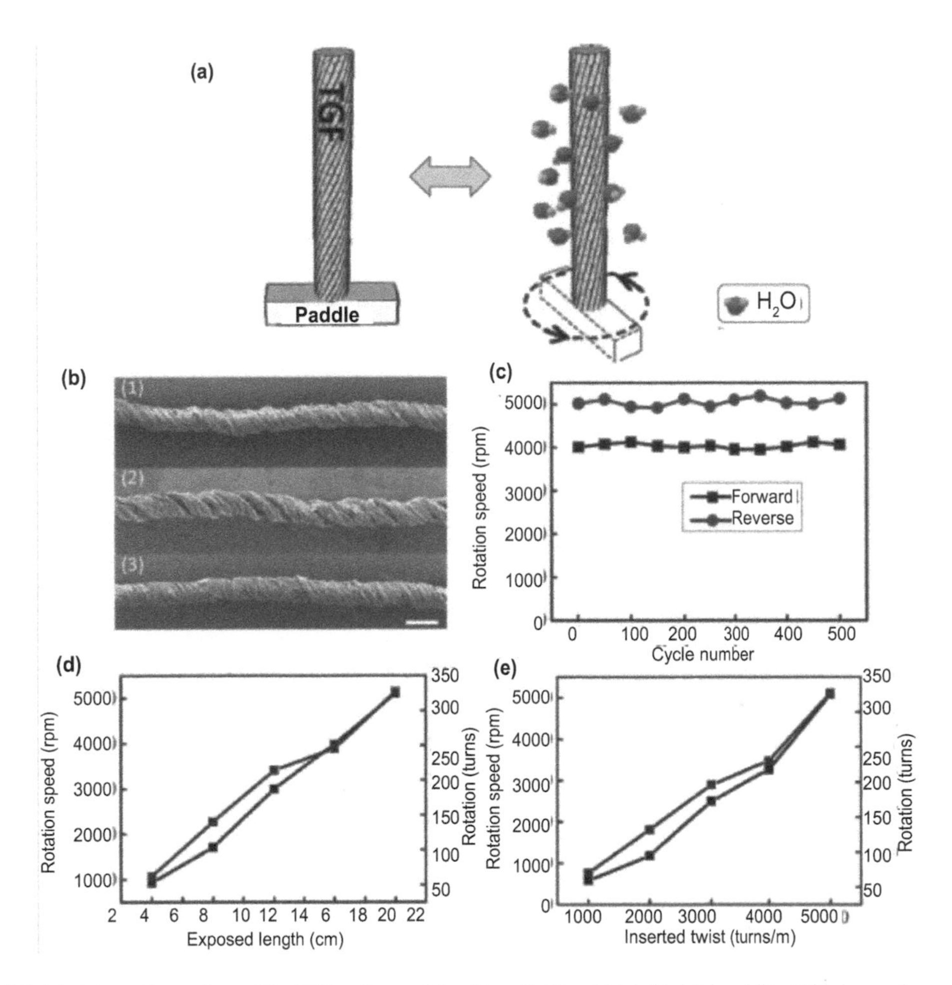

Fig. 5.76. (a) Schematic rotation of the TGF with a paddle at low (left) and high (right) humidity. With increasing moisture (right), the TGF drove the paddle rotating fast; then the paddle would reverse to the original state when the moisture was decreased (left). (b) SEM images of the initial TGF at RH = 20% (1), after exposure to high humidity of 85% (2) and the final state of TGF as the humidity was back to the initial RH = 20% (3). Scale bar = 100 μm. (c) Durability test of the TGF (5000 turns per meter) undergoing repeated RH changes, showing forward (the environment humidity varied from RH = 20% to 85%) and backward (RH = 85% to 20%) rotation speed versus cycle numbers. (d) Rotation speed and rotation numbers versus the TGF length exposed to RH = 85%. (e) Rotation speed and rotation numbers versus TGFs with different applied twist of turns per meter at a length of 20 cm. The environment humidity was changed from RH = 20% to 85%. Reproduced with permission from [93], Copyright © 2014, Wiley-VCH Verlag GmbH & Co. KGaA, Weinheim.

electrodes [98]. The high flexibility, mechanical strength and electrical conductivity of GFs resulted in a maximum energy conversion efficiency of 8.45%, which was much higher than that of other wire-shaped photovoltaic cells. It is expected that such GF fiber-based highly flexible photovoltaic devices should be easily incorporated with clothes, packages and other wearable devices to develop smart textile technologies.

The graphene fibers were spun from GO aqueous suspensions. Diameters of the GO fibers were controlled by varying the inner diameter of the nozzle and the drawing speed, as discussed previously. The GO fibers had a density of about 1.15 $g \cdot cm^{-3}$. Graphene fibers were obtained through the chemical reduction of the GO fibers with 40% hydroiodic acid. Representative SEM images of the graphene fibers are shown in Fig. 5.77 (a, b). The graphene fibers were highly flexible, with a tensile strength of about 330 MPa, due to the close packing of the graphene nanosheets, as confirmed by the cross-sectional SEM image of the broken fiber in Fig. 5.77 (c). The graphene fibers had a room temperature electrical conductivity of 200 $S \cdot cm^{-1}$.

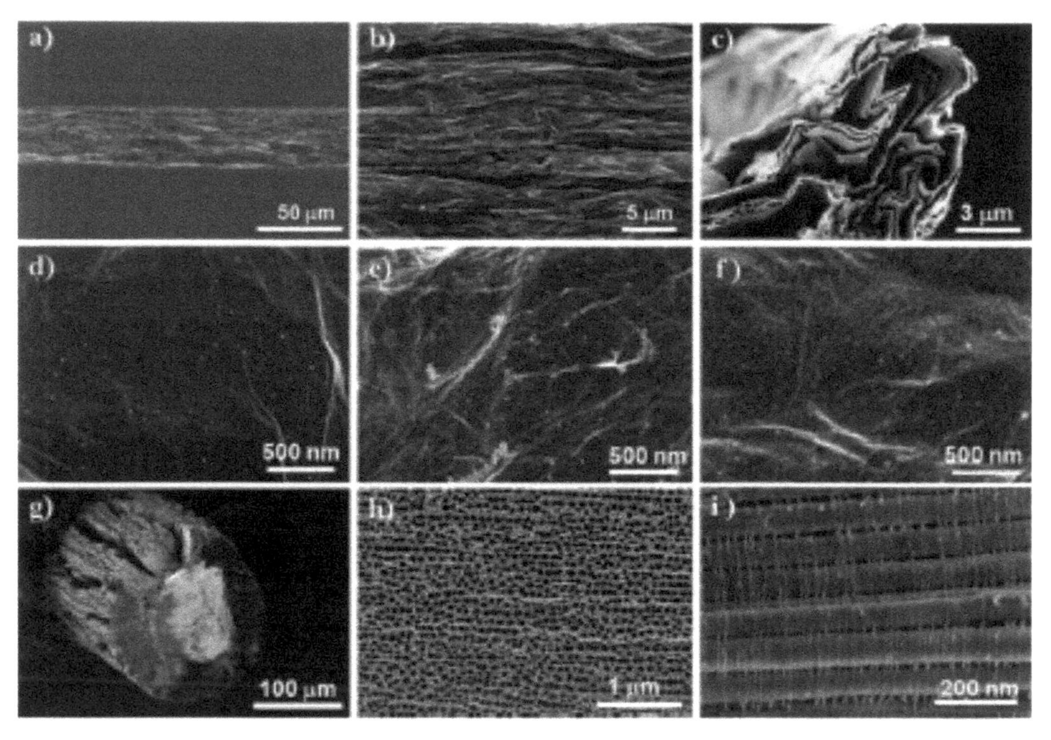

Fig. 5.77. SEM images of the graphene fibers, G–Pt hybrid fibers and TiO_2 nanotube-modified Ti wires: (a, b) graphene fiber at different magnifications, (c) cross-sectional view of the graphene fiber after failure in the mechanical test, (d–f) side views of the G–Pt hybrid fibers with Pt contents of 4.6%, 7.1% and 22.9%, (g) cross-sectional view of the Ti wire grown with TiO_2 nanotube arrays on the surface. (h, i) top and side views of the aligned TiO_2 nanotubes. Reproduced with permission from [98], Copyright © 2013, Wiley-VCH Verlag GmbH & Co. KGaA, Weinheim.

To be used as the electrodes of photovoltaic cells, the graphene fibers were modified with Pt NPs to have catalytic activity. Their high electrical conductivity ensured that the graphene fibers could be readily incorporated with Pt NPs by using electrodeposition. Surface morphologies of the G–Pt hybrid fibers with different contents of Pt are shown in Fig. 5.77 (d–f). With the increasing content of Pt, the electrical conductivity of the hybrid fibers was increased accordingly, while the mechanical performance was almost unaffected.

Dye-sensitized photovoltaic wires were prepared by using Ti wires, on which a layer of perpendicularly aligned TiO_2 nanotubes were grown, by using the anodizing method, as illustrated in Fig. 5.77 (g). Top and side view SEM images of the TiO_2 nanotubes are shown in Fig. 5.77 (h, i). The TiO_2 nanotubes had diameters of 70–100 nm, while their lengths could be controlled by controlling the anodizing time. After the adsorption of N719, the working electrode was twined with the G–Pt hybrid fiber, into which the electrolytes were introduced.

Figure 5.78 (a) shows a schematic diagram illustrating structure of the typical dye-sensitized photovoltaic wire, together with photographs of the two real cells, as seen in Fig. 5.78 (b–d). The photovoltaic device exhibited an open-circuit voltage (VOC) of 0.73 V, short-circuit current densities (JSC) of 12.67–17.11 $mA \cdot cm^{-2}$ and fill factors (FF) of 0.42–0.67, over the concentration range of Pt NPs of 0–7.1%. 7.1% was the optimized content of Pt NPs in terms of energy conversion efficiency (h = 8.41%).

Graphene-based materials have been extensively studied for potential applications in electrochemical capacitors [99–101]. Together with high electrical conductivity, the highly exposed surface area of the GFs with 3D graphene networks could be promising candidates for such applications. Also, they can be woven into textile electrodes to construct flexible supercapacitors with a high tolerance to repeated bending cycles [67]. Various flexible all-solid-state supercapacitors based on GFs have been reported in the open literature [102–104].

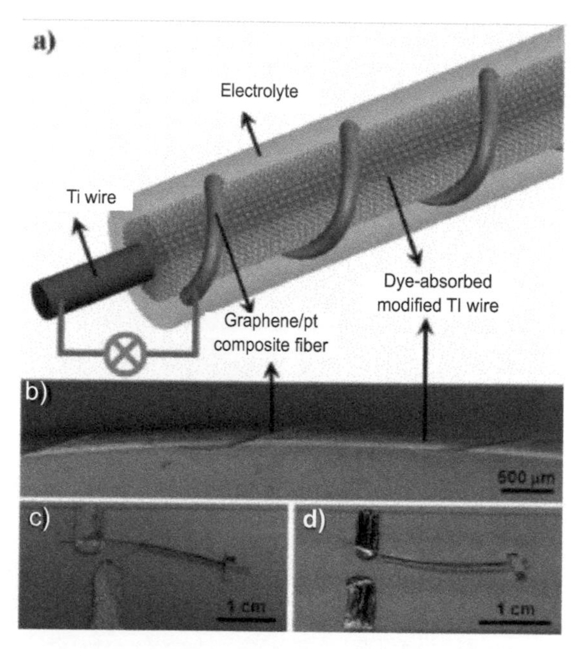

Fig. 5.78. Dye-sensitized photovoltaic wire fabricated by using the G–Pt hybrid fiber as the counter electrode and a Ti wire incorporated with TiO_2 nanotubes as the working electrode: (a) schematic diagram, (b) SEM image, (c, d) photographs of photovoltaic wires sealed in a capillary glass tube and flexible fluorinated ethylene propylene tube, respectively. Reproduced with permission from [98], Copyright © 2013, Wiley-VCH Verlag GmbH & Co. KGaA, Weinheim.

One example is that of the core–sheath GF/3D-G all carbon hybrid fibers, which were used as flexible electrodes to construct flexible electrochemical supercapacitors with high energy storage performances [71]. The fiber supercapacitors were assembled with two intertwined electrodes, both of which were solidified in H_2SO_4-PVA gel electrolyte, so as to develop all-solid-state fiber supercapacitors. The assembled fiber supercapacitors with GF/3D-G were highly flexible. The supercapacitors possessed an area-specific capacitance of 1.2–1.7 $mF \cdot cm^{-2}$ and a mass-specific capacitance of 25–40 $F \cdot g^{-1}$, both of which were comparable with those of other devices made with similar materials. The energy density and power density of the GF/3D-G supercapacitors were $0.4–1.7 \times 10^{-7}$ $Wh \cdot cm^{-2}$ and $6–100 \times 10^{-6}$ $W \cdot cm^{-2}$. These performances could be further improved by optimizing the materials and devices.

A hybrid fiber of MnO_2-modified graphene nanosheets on GF has been fabricated by directly depositing MnO_2 onto the graphene network to enclose the GF, leading to MnO_2–G/GF hybrid fiber supercapacitors [105]. By combining the electric double layer capacitance of the graphene network with the pseudocapacitance of the MnO_2 nanostructures, the all-solid-state fiber supercapacitor exhibited greatly enhanced electrochemical performances along with robust mechanical flexibility.

An aqueous solution consisting of 1 M Na_2SO_4 and 0.1 M $MnSO_4$ was used to deposit MnO_2. The electrodeposition of MnO_2 nanoflowers was conducted by using the three-electrode configuration, during which the G/GFs were immersed into the solution as the working electrode, while Ag/AgCl (3 M KCl) and Pt wire (1 mm in diameter) were employed as the reference electrode and counter electrode, respectively. A constant current of 400 $mA \cdot cm^{-2}$ was applied over periods of 5–40 min. The corresponding potential was 0.7 V during the electrodeposition process. After deposition experiments, the MnO_2–G/GFs were washed with DI water to remove the excessive electrolyte and then freeze-dried.

Flexible all-solid-state symmetric fiber supercapacitors were assembled by intertwining two electrodes of MnO_2–G/GFs that were deposited with either H_2SO_4-PVA gel polyelectrolyte or NaCl-PVA gel polyelectrolyte. In the asymmetric fiber supercapacitors, the MnO_2–G/GF was used as the positive electrode, while the G/GF was employed as the negative electrode. Figure 5.79 shows SEM images of G/GFs coated with MnO_2. After deposition for 5 min, the G/GF was only partly covered by the MnO_2 particles, as demonstrated in Fig. 5.79 (a). Figure 5.79 (b) indicated that the MnO_2 exhibited a flower-like

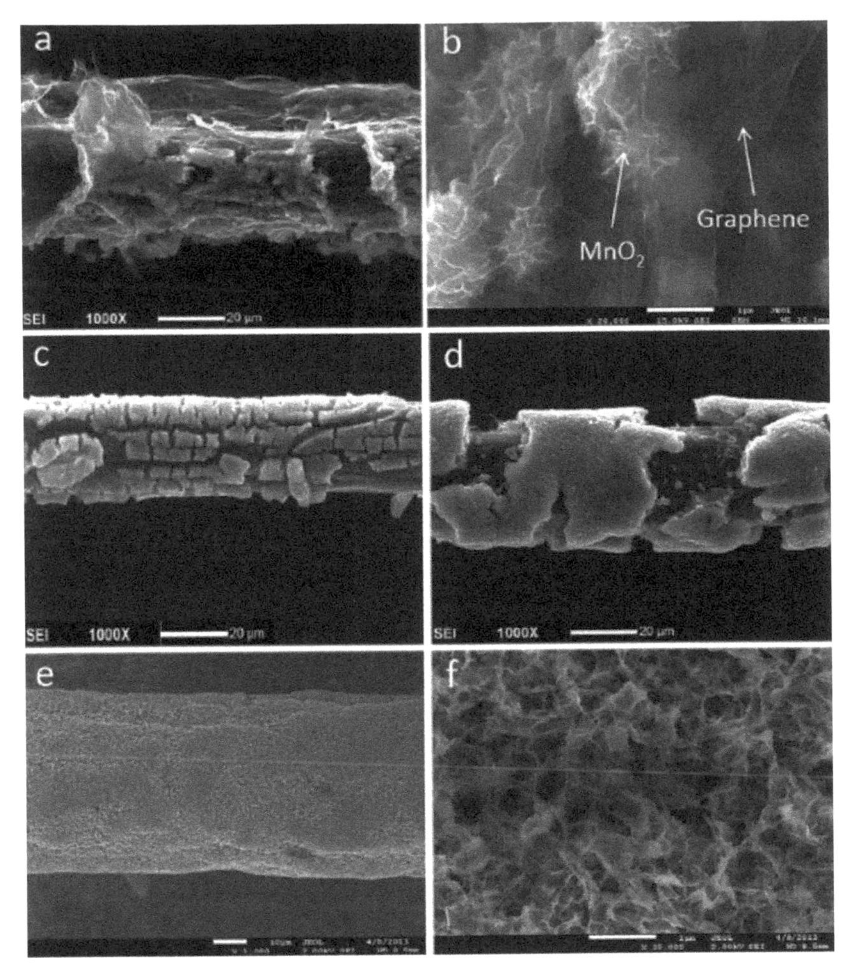

Fig. 5.79. SEM images of the MnO_2–G/GF with different MnO_2 deposition times: (a) 5 min, (c) 10 min, (d) 20 min, (e) 40 min, (b, f) enlarged views of (a) and (e). Reproduced with permission from [105], Copyright © 2014, Elsevier.

morphology, which were tightly attached on the graphene nanosheets. With increasing deposition time, the coverage of the MnO_2 particles on the G/GFs was gradually increased, as seen in Fig. 5.79 (c, d). Crack-free coating was formed after deposition for 40 min (Fig. 5.79 (e)). As illustrated in Fig. 5.79 (f), the presence of the MnO_2 particles had no effect on the porous structures of the fibers, thus ensuring the electrochemical performances.

Figure 5.80 (a) shows a schematic diagram of the flexible all-solid-state fiber supercapacitors assembled by intertwining two MnO_2–G/GFs. The H_2SO_4-PVA electrolyte had no negative effect on the performance of the MnO_2, due to the mediation effect of PVA. Bending test of the fiber capacitors was conducted by using the stage as schematically shown in Fig. 5.80 (b). As demonstrated in Fig. 5.80 (c, d), the supercapacitors could be twisted without losing their structural integrity and electrochemical performance. CV curves of the MnO_2–G/GF supercapacitors at different deformed states were almost the same, as revealed in Fig. 5.80 (e). All the CV curves showed ideal capacitive behavior with nearly rectangular shape without serious distortion. Similarly, the galvanostatic charge-discharge curves exhibited a typical triangular shape, as shown in Fig. 5.80 (f). All these observations confirmed that the MnO_2–G/GF supercapacitors possessed high flexibility and electrochemical stability.

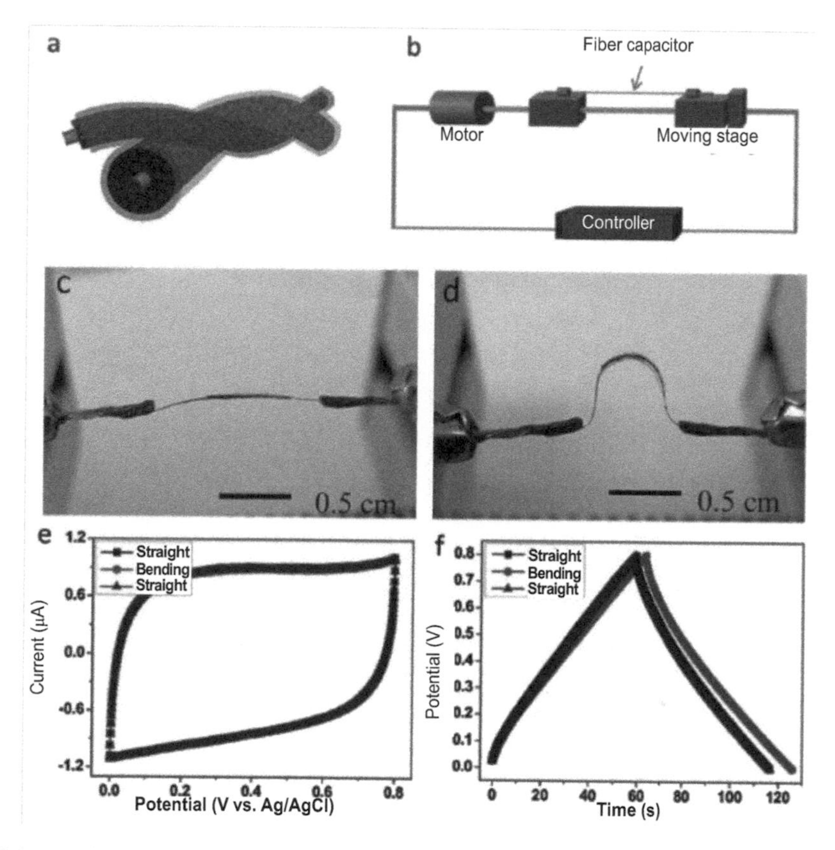

Fig. 5.80. (a) Schematic illustration of the fiber capacitor assembled with two entwined MnO_2–G/GFs containing polyelectrolyte. (b) Schematic illustration of the fiber capacitor for bending test. (c, d) Photographs of the fiber supercapacitor of MnO_2–G/GFs at straight and bending status. (e, f) CV and charge-discharge curves of the fiber capacitor with an effective length of 0.5 cm at straight and bending status. The scan rate in (e) was 10 $mV \cdot s^{-1}$, while the applied current in (f) was 1 mA. Reproduced with permission from [105], Copyright © 2014, Elsevier.

5.4.3. Other applications

Graphene-based fibers have also found applications in other interesting fields. For instance, graphene nanosheets could be incorporated into nanoscale composite fibers produced by using the electrospinning process, which have potential applications as the optical element in fiber lasers. It was found that electrospun graphene nanocomposites could be used as saturable absorbers with wideband absorption [75]. These efficient photonic materials are promising candidates in fiber lasers to generate ultra-short pulses.

Solid-phase microextraction (SPME) sorbents are important in analyzing trace organic contaminants in water [106]. They have simple operation and can be readily incorporated with other separation techniques, including gas chromatography (GC), liquid chromatography, capillary electrophoresis and ion chromatography [107–109]. Commercial fibers for SPME have been available in the market, e.g., PDMS, polyacrylate, divinylbenzene, carboxen, carbowax and their copolymers [106, 107]. These commercial fibers are cost-effective and have thermal and chemical stability, while they also have various problems to be addressed. It was found that graphene could be used as SPME coating for the extraction of pyrethroid pesticides [110], with higher extraction efficiencies and superior thermal and mechanical stabilities, as compared with the commercial counterparts. Similar advantages have also been observed in SPME based on other GFs [111, 112].

5.5. Concluding Remarks

Graphene-based macroscopic fibers have offered us a unique platform to create materials with special properties, which cannot be obtained from the conventional material systems. Various preparation methods have been made available to produce GFs at large scales, which is especially important for practical applications. Although significant progress has been made, we are still facing various challenges. For instance, the mechanical strength and electric conductivity of the GFs are still below the expected levels. The processing of the GFs should be standardized, while it is necessary to establish strategies that can be used to assemble graphene or GO nanosheets with variable dimensions, shapes and chemical compositions. There is expected to be a breakthrough in the near future.

References

[1] Zhang XB, Jiang KL, Teng C, Liu P, Zhang L, Kong J et al. Spinning and processing continuous yarns from 4-inch wafer scale super-aligned carbon nanotube arrays. Advanced Materials. 2006; 18: 1505–10.

[2] Jiang KL, Li QQ, Fan SS. Nanotechnology: Spinning continuous carbon nanotube yarns—Carbon nanotubes weave their way into a range of imaginative macroscopic applications. Nature. 2002; 419: 801.

[3] Zhang M, Atkinson KR, Baughman RH. Multifunctional carbon nanotube yarns by downsizing an ancient technology. Science. 2004; 306: 1358–61.

[4] Dalton AB, Collins S, Munoz E, Razal JM, Ebron VH, Ferraris JP et al. Super-tough carbon-nanotube fibres—These extraordinary composite fibres can be woven into electronic textiles. Nature. 2003; 423: 703–703.

[5] Vigolo B, Penicaud A, Coulon C, Sauder C, Pailler R, Journet C et al. Macroscopic fibers and ribbons of oriented carbon nanotubes. Science. 2000; 290: 1331–4.

[6] Ericson LM, Fan H, Peng HQ, Davis VA, Zhou W, Sulpizio J et al. Macroscopic, neat, single-walled carbon nanotube fibers. Science. 2004; 305: 1447–50.

[7] Davis VA, Parra-Vasquez ANG, Green MJ, Rai PK, Behabtu N, Prieto V et al. True solutions of single-walled carbon nanotubes for assembly into macroscopic materials. Nature Nanotechnology. 2009; 4: 830–4.

[8] Li YL, Kinloch IA, Windle AH. Direct spinning of carbon nanotube fibers from chemical vapor deposition synthesis. Science. 2004; 304: 276–8.

[9] Geim AK, Novoselov KS. The rise of graphene. Nature Materials. 2007; 6: 183–91.

[10] Novoselov KS, Geim AK, Morozov SV, Jiang D, Zhang Y, Dubonos SV et al. Electric field effect in atomically thin carbon films. Science. 2004; 306: 666–9.

[11] Novoselov KS, Geim AK, Morozov SV, Jiang D, Katsnelson MI, Grigorieva IV et al. Two-dimensional gas of massless Dirac fermions in graphene. Nature. 2005; 438: 197–200.

[12] Zhang YB, Tan YW, Stormer HL, Kim P. Experimental observation of the quantum Hall effect and Berry's phase in graphene. Nature. 2005; 438: 201–4.

[13] Balandin AA, Ghosh S, Bao WZ, Calizo I, Teweldebrhan D, Miao F et al. Superior thermal conductivity of single-layer graphene. Nano Letters. 2008; 8: 902–7.

[14] Lee CG, Wei XD, Kysar JW, Hone J. Measurement of the elastic properties and intrinsic strength of monolayer graphene. Science. 2008; 321: 385–8.

[15] Xu YX, Sheng KX, Li C, Shi GQ. Self-assembled graphene hydrogel via a one-step hydrothermal process. ACS Nano. 2010; 4: 4324–30.

[16] Cheng HH, Hu CG, Zhao Y, Qu LT. Graphene fiber: a new material platform for unique applications. NPG Asia Materials. 2014; 6: e113.

[17] Xu Z, Gao C. Graphene chiral liquid crystals and macroscopic assembled fibres. Nature Communications. 2011; 2: 571.

[18] Dan B, Behabtu N, Martinez A, Evans JS, Kosynkin DV, Tour JM et al. Liquid crystals of aqueous, giant graphene oxide flakes. Soft Matter. 2011; 7: 11154–9.

[19] Konkena B, Vasudevan S. Glass, gel, and liquid crystals: Arrested states of graphene oxide aqueous dispersions. Journal of Physical Chemistry C. 2014; 118: 21706–13.

[20] Xu Z, Gao C. Aqueous liquid crystals of graphene oxide. ACS Nano. 2011; 5: 2908–15.

[21] Fernsler J, Hough L, Shao RF, Maclennan JE, Navailles L, Brunet M et al. Giant-block twist grain boundary smectic phases. Proceedings of the National Academy of Sciences of the United States of America. 2005; 102: 14191–6.

[22] Goodby JW, Waugh MA, Stein SM, Chin E, Pindak R, Patel JS. Characterization of a new helical smectic liquid-crystal. Nature. 1989; 337: 449–52.

[23] Dikin DA, Stankovich S, Zimney EJ, Piner RD, Dommett GHB, Evmenenko G et al. Preparation and characterization of graphene oxide paper. Nature. 2007; 448: 457–60.

[24] Compton OC, Nguyen ST. Graphene oxide, highly reduced graphene oxide, and graphene: versatile building blocks for carbon-based materials. Small. 2010; 6: 711–23.

[25] Song WH, Kinloch IA, Windle AH. Nematic liquid crystallinity of multiwall carbon nanotubes. Science. 2003; 302: 1363–1363.
[26] Xu Z, Zhang Y, Li PG, Gao C. Strong, conductive, lightweight, neat graphene aerogel fibers with aligned pores. ACS Nano. 2012; 6: 7103–13.
[27] Gabriel JCP, Camerel F, Lemaire BJ, Desvaux H, Davidson P, Batail P. Swollen liquid-crystalline lamellar phase based on extended solid-like sheets. Nature. 2001; 413: 504–8.
[28] Gabriel JCP, Davidson P. New trends in colloidal liquid crystals based on mineral moieties. Advanced Materials. 2000; 12: 9–20.
[29] Hu XZ, Xu ZP, Liu Z, Gao C. Liquid crystal self-templating approach to ultrastrong and tough biomimic composites. Scientific Reports. 2013; 3: 2374.
[30] Yao HB, Fang HY, Wang XH, Yu SH. Hierarchical assembly of micro-/nano-building blocks: bio-inspired rigid structural functional materials. Chemical Society Reviews. 2011; 40: 3764–85.
[31] Wang JF, Cheng QF, Tang ZY. Layered nanocomposites inspired by the structure and mechanical properties of nacre. Chemical Society Reviews. 2012; 41: 1111–29.
[32] Jalili R, Aboutalebi SH, Esrafilzadeh D, Shepherd RL, Chen J, Aminorroaya-Yamini S et al. Scalable one-step wet-spinning of graphene fibers and yarns from liquid crystalline dispersions of graphene oxide: Towards multifunctional textiles. Advanced Functional Materials. 2013; 23: 5345–54.
[33] Cong HP, Ren XC, Wang P, Yu SH. Wet-spinning assembly of continuous, neat, and macroscopic graphene fibers. Scientific Reports. 2012; 2: 613.
[34] Chen L, He YL, Chai SG, Qiang H, Chen F, Fu Q. Toward high performance graphene fibers. Nanoscale. 2013; 5: 5809–15.
[35] Xiang CS, Young CC, Wang X, Yan Z, Hwang CC, Cerioti G et al. Large flake graphene oxide fibers with unconventional 100% knot efficiency and highly aligned small flake graphene oxide fibers. Advanced Materials. 2013; 25: 4592–7.
[36] Vilatela JJ, Windle AH. Yarn-like carbon nanotube fibers. Advanced Materials. 2010; 22: 4959–63.
[37] Zhao Y, Jiang CC, Hu CG, Dong ZL, Xue JL, Meng YN et al. Large-scale spinning assembly of neat, morphology-defined, graphene-based hollow fibers. ACS Nano. 2013; 7: 2406–12.
[38] Sun JK, Li YH, Peng QY, Hou SC, Zou DC, Shang YY et al. Macroscopic, flexible, high-performance Graphene ribbons. ACS Nano. 2013; 7: 10225–32.
[39] Aboutalebi SH, Jalili R, Esrafilzadeh D, Salari M, Gholamvand Z, Yamini SA et al. High-performance multifunctional graphene yarns: Toward wearable all-carbon energy storage textiles. ACS Nano. 2014; 8: 2456–66.
[40] Dong ZL, Jiang CC, Cheng HH, Zhao Y, Shi GQ, Jiang L et al. Facile fabrication of light, flexible and multifunctional graphene fibers. Advanced Materials. 2012; 24: 1856–61.
[41] Hu CG, Zhao Y, Cheng HH, Wang YH, Dong ZL, Jiang CC et al. Graphene microtubings: Controlled fabrication and site-specific functionalization. Nano Letters. 2012; 12: 5879–84.
[42] Li XM, Zhao TS, Wang KL, Yang Y, Wei JQ, Kang FY et al. Directly drawing self-assembled, porous, and monolithic graphene fiber from chemical vapor deposition grown graphene film and its electrochemical properties. Langmuir. 2011; 27: 12164–71.
[43] Li X, Sun PZ, Fan LL, Zhu M, Wang KL, Zhong ML et al. Multifunctional graphene woven fabrics. Scientific Reports. 2012; 2: 395.
[44] Hu CG, Zhai XQ, Liu LL, Zhao Y, Jiang L, Qu LT. Spontaneous reduction and assembly of graphene oxide into three-dimensional graphene network on arbitrary conductive substrates. Scientific Reports. 2013; 3: 2065.
[45] Kosynkin DV, Higginbotham AL, Sinitskii A, Lomeda JR, Dimiev A, Price BK et al. Longitudinal unzipping of carbon nanotubes to form graphene nanoribbons. Nature. 2009; 458: 872–6.
[46] Jiao LY, Wang XR, Diankov G, Wang HL, Dai HJ. Facile synthesis of high-quality graphene nanoribbons. Nature Nanotechnology. 2010; 5: 321–5.
[47] Jiao LY, Zhang L, Wang XR, Diankov G, Dai HJ. Narrow graphene nanoribbons from carbon nanotubes. Nature. 2009; 458: 877–80.
[48] Cano-Marquez AG, Rodriguez-Macias FJ, Campos-Delgado J, Espinosa-Gonzalez CG, Tristan-Lopez F, Ramirez-Gonzalez D et al. Ex-MWNTs: Graphene sheets and ribbons produced by lithium intercalation and exfoliation of carbon nanotubes. Nano Letters. 2009; 9: 1527–33.
[49] Carretero-Gonzalez J, Castillo-Martinez E, Dias-Lima M, Acik M, Rogers DM, Sovich J et al. Oriented graphene nanoribbon yarn and sheet from aligned multi-walled carbon nanotube sheets. Advanced Materials. 2012; 24: 5695–701.
[50] Xiang CS, Behabtu N, Liu YD, Chae HG, Young CC, Genorio B et al. Graphene nanoribbons as an advanced precursor for making carbon fiber. ACS Nano. 2013; 7: 1628–37.
[51] Minus ML, Kumar S. The processing, properties, and structure of carbon fibers. Jom. 2005; 57: 52–8.
[52] Jang EY, Carretero-Gonzalez J, Choi A, Kim WJ, Kozlov ME, Kim T et al. Fibers of reduced graphene oxide nanoribbons. Nanotechnology. 2012; 23: 235601.
[53] Tian ZS, Xu CX, Li JT, Zhu GY, Shi ZL, Lin Y. Self-assembled free-standing graphene oxide fibers. ACS Applied Materials & Interfaces. 2013; 5: 1489–93.

[54] Zhao Y, Song LP, Zhang ZP, Qu LT. Stimulus-responsive graphene systems towards actuator applications. Energy & Environmental Science. 2013; 6: 3520–36.

[55] Zhang J, Song L, Zhang ZP, Chen N, Qu LT. Environmentally responsive graphene systems. Small. 2014; 10: 2151–64.

[56] Xu Z, Liu Z, Sun HY, Gao C. Highly electrically conductive Ag-doped graphene fibers as stretchable conductors. Advanced Materials. 2013; 25: 3249–53.

[57] Xu Z, Sun HY, Zhao XL, Gao C. Ultrastrong fibers assembled from giant graphene oxide sheets. Advanced Materials. 2013; 25: 188–93.

[58] Rourke JP, Pandey PA, Moore JJ, Bates MA, Kinloch IA, Young RJ et al. The real graphene oxide revealed: Stripping the oxidative debris from the graphene-like sheets. Angewandte Chemie-International Edition. 2011; 50: 3173–7.

[59] Jeong HK, Lee YP, Lahaye RJWE, Park MH, An KH, Kim IJ et al. Evidence of graphitic AB stacking order of graphite oxides. Journal of the American Chemical Society. 2008; 130: 1362–6.

[60] Park SJ, Lee KS, Bozoklu G, Cai WW, Nguyen ST, Ruoff RS. Graphene oxide papers modified by divalent ions—Enhancing mechanical properties via chemical cross-linking. ACS Nano. 2008; 2: 572–8.

[61] Liu YL, Xie B, Xu ZP. Mechanics of coordinative crosslinks in graphene nanocomposites: A first-principles study. Journal of Materials Chemistry. 2011; 21: 6707–12.

[62] Miaudet P, Badaire S, Maugey M, Derre A, Pichot V, Launois P et al. Hot-drawing of single and multiwall carbon nanotube fibers for high toughness and alignment. Nano Letters. 2005; 5: 2212–5.

[63] Chen H, Muller MB, Gilmore KJ, Wallace GG, Li D. Mechanically strong, electrically conductive, and biocompatible graphene paper. Advanced Materials. 2008; 20: 3557–61.

[64] Hu XZ, Xu ZP, Gao C. Multifunctional, supramolecular, continuous artificial nacre fibres. Scientific Reports. 2012; 2: 767.

[65] Zhao XL, Xu Z, Zheng BN, Gao C. Macroscopic assembled, ultrastrong and H_2SO_4-resistant fibres of polymer-grafted graphene oxide. Scientific Reports. 2013; 3: 3164.

[66] Zhong XH, Wang R, Wen YY, Li YL. Carbon nanotube and graphene multiple-thread yarns. Nanoscale. 2013; 5: 1183–7.

[67] Cheng HH, Dong ZL, Hu CG, Zhao Y, Hu Y, Qu LT et al. Textile electrodes woven by carbon nanotube-graphene hybrid fibers for flexible electrochemical capacitors. Nanoscale. 2013; 5: 3428–34.

[68] Yu DS, Goh KL, Wang H, Wei L, Jiang WC, Zhang Q et al. Scalable synthesis of hierarchically structured carbon nanotube-graphene fibres for capacitive energy storage. Nature Nanotechnology. 2014; 9: 555–62.

[69] Cote LJ, Kim JY, Tung VC, Luo JY, Kim F, Huang JX. Graphene oxide as surfactant sheets. Pure and Applied Chemistry. 2011; 83: 95–110.

[70] Ma YW, Li PX, Sedloff JW, Zhang X, Zhang HB, Liu J. Conductive graphene fibers for wire-shaped supercapacitors strengthened by unfunctionalized few-walled carbon nanotubes. ACS Nano. 2015; 9: 1352–9.

[71] Meng YN, Zhao Y, Hu CG, Cheng HH, Hu Y, Zhang ZP et al. All-graphene core-sheath microfibers for all-solid-state, stretchable fibriform supercapacitors and wearable electronic textiles. Advanced Materials. 2013; 25: 2326–31.

[72] Matsumoto H, Imaizumi S, Konosu Y, Ashizawa M, Minagawa M, Tanioka A et al. Electrospun composite nanofiber yarns containing oriented graphene nanoribbons. ACS Applied Materials & Interfaces. 2013; 5: 6225–31.

[73] Qi YY, Tai ZX, Sun DF, Chen JT, Ma HB, Yan XB et al. Fabrication and characterization of poly(vinyl alcohol)/graphene oxide nanofibrous biocomposite scaffolds. Journal of Applied Polymer Science. 2013; 127: 1885–94.

[74] Wang CA, Li YD, Ding GQ, Xie XM, Jiang MH. Preparation and characterization of graphene oxide/poly(vinyl alcohol) composite nanofibers via electrospinning. Journal of Applied Polymer Science. 2013; 127: 3026–32.

[75] Bao QL, Zhang H, Yang JX, Wang SL, Tang DY, Jose R et al. Graphene-polymer nanofiber membrane for ultrafast photonics. Advanced Functional Materials. 2010; 20: 782–91.

[76] Jiang ZX, Li Q, Chen ML, Li JB, Li J, Huang YD et al. Mechanical reinforcement fibers produced by gel-spinning of poly-acrylic acid (PAA) and graphene oxide (GO) composites. Nanoscale. 2013; 5: 6265–9.

[77] Shin MK, Lee B, Kim SH, Lee JA, Spinks GM, Gambhir S et al. Synergistic toughening of composite fibres by self-alignment of reduced graphene oxide and carbon nanotubes. Nature Communications. 2012; 3: 650.

[78] Huang Y, Liang JJ, Chen YS. The application of graphene based materials for actuators. Journal of Materials Chemistry. 2012; 22: 3671–9.

[79] Zhu CH, Lu Y, Peng J, Chen JF, Yu SH. Photothermally sensitive poly(N-isopropylacrylamide)/graphene oxide nanocomposite hydrogels as remote light-controlled liquid microvalves. Advanced Functional Materials. 2012; 22: 4017–22.

[80] Wu CZ, Feng J, Peng LL, Ni Y, Liang HY, He LH et al. Large-area graphene realizing ultrasensitive photothermal actuator with high transparency: new prototype robotic motions under infrared-light stimuli. Journal of Materials Chemistry. 2011; 21: 18584–91.

[81] Zhang J, Zhao F, Zhang ZP, Chen N, Qu LT. Dimension-tailored functional graphene structures for energy conversion and storage. Nanoscale. 2013; 5: 3112–26.

[82] Lu LH, Liu JH, Hu Y, Zhang YE, Chen W. Graphene-stabilized silver nanoparticle electrochemical electrode for actuator design. Advanced Materials. 2013; 25: 1270–4.

[83] Liang JJ, Huang LP, Li N, Huang Y, Wu YP, Fang SL et al. Electromechanical actuator with controllable motion, fast response rate, and high-frequency resonance based on graphene and polydiacetylene. ACS Nano. 2012; 6: 4508–19.

[84] Xie XJ, Qu LT, Zhou CM, Li Y, Zhu JW, Bai HW et al. An asymmetrically surface-modified graphene film electrochemical actuator. ACS Nano. 2010; 4: 6050–4.
[85] Xie XJ, Bai HW, Shi GQ, Qu LT. Load-tolerant, highly strain-responsive graphene sheets. Journal of Materials Chemistry. 2011; 21: 2057–9.
[86] Liu J, Wang ZD, Xie XJ, Cheng HH, Zhao Y, Qu LT. A rationally-designed synergetic polypyrrole/graphene bilayer actuator. Journal of Materials Chemistry. 2012; 22: 4015–20.
[87] Liu J, Wang Z, Zhao Y, Cheng HH, Hu CG, Jiang L et al. Three-dimensional graphene-polypyrrole hybrid electrochemical actuator. Nanoscale. 2012; 4: 7563–8.
[88] Wang YH, Bian K, Hu CG, Zhang ZP, Chen N, Zhang HM et al. Flexible and wearable graphene/polypyrrole fibers towards multifunctional actuator applications. Electrochemistry Communications. 2013; 35: 49–52.
[89] Cheng HH, Liu J, Zhao Y, Hu CG, Zhang ZP, Chen N et al. Graphene fibers with predetermined deformation as moisture-triggered actuators and robots. Angewandte Chemie-International Edition. 2013; 52: 10482–6.
[90] Gao W, Singh N, Song L, Liu Z, Reddy ALM, Ci LJ et al. Direct laser writing of micro-supercapacitors on hydrated graphite oxide films. Nature Nanotechnology. 2011; 6: 496–500.
[91] Zhang YL, Guo L, Wei S, He YY, Xia H, Chen QD et al. Direct imprinting of microcircuits on graphene oxides film by femtosecond laser reduction. Nano Today. 2010; 5: 15–20.
[92] El-Kady MF, Strong V, Dubin S, Kaner RB. Laser scribing of high-performance and flexible graphene-based electrochemical capacitors. Science. 2012; 335: 1326–30.
[93] Cheng HH, Hu Y, Zhao F, Dong ZL, Wang YH, Chen N et al. Moisture-activated torsional graphene-fiber motor. Advanced Materials. 2014; 26: 2909–13.
[94] Nair RR, Wu HA, Jayaram PN, Grigorieva IV, Geim AK. Unimpeded permeation of water through helium-leak-tight graphene-based membranes. Science. 2012; 335: 442–4.
[95] Paxton WF, Sundararajan S, Mallouk TE, Sen A. Chemical locomotion. Angewandte Chemie-International Edition. 2006; 45: 5420–9.
[96] Sanchez S, Pumera M. Nanorobots: The ultimate wireless self-propelled sensing and actuating devices. Chemistry-An Asian Journal. 2009; 4: 1402–10.
[97] Mirkovic T, Zacharia NS, Scholes GD, Ozin GA. Fuel for thought: Chemically powered nanomotors out-swim nature's flagellated bacteria. Acs Nano. 2010; 4: 1782–9.
[98] Yang ZB, Sun H, Chen T, Qiu LB, Luo YF, Peng HS. Photovoltaic wire derived from a graphene composite fiber achieving an 8.45% energy conversion efficiency. Angewandte Chemie-International Edition. 2013; 52: 7545–8.
[99] Chen J, Li C, Shi GQ. Graphene materials for electrochemical capacitors. Journal of Physical Chemistry Letters. 2013; 4: 1244–53.
[100] Huang LP, Li C, Shi GQ. High-performance and flexible electrochemical capacitors based on graphene/polymer composite films. Journal of Materials Chemistry A. 2014; 2: 968–74.
[101] Li YR, Sheng KX, Yuan WJ, Shi GQ. A high-performance flexible fibre-shaped electrochemical capacitor based on electrochemically reduced graphene oxide. Chemical Communications. 2013; 49: 291–3.
[102] Li XM, Zhao TS, Chen Q, Li PX, Wang KL, Zhong ML et al. Flexible all solid-state supercapacitors based on chemical vapor deposition derived graphene fibers. Physical Chemistry Chemical Physics. 2013; 15: 17752–7.
[103] Huang TQ, Zheng BN, Kou L, Gopalsamy K, Xu ZP, Gao C et al. Flexible high performance wet-spun graphene fiber supercapacitors. RSC Advances. 2013; 3: 23957–62.
[104] Li YY, Sheng KX, Yuan WJ, Shi GQ. A high-performance flexible fibre-shaped electrochemical capacitor based on electrochemically reduced graphene oxide. Chemical Communications. 2013; 49: 291–3.
[105] Chen Q, Meng YN, Hu CG, Zhao Y, Shao HB, Chen N et al. MnO_2-modified hierarchical graphene fiber electrochemical supercapacitor. Journal of Power Sources. 2014; 247: 32–9.
[106] Dietz C, Sanz J, Camara C. Recent developments in solid-phase microextraction coatings and related techniques. Journal of Chromatography A. 2006; 1103: 183–92.
[107] Spietelun A, Pilarczyk M, Kloskowski A, Namiesnik J. Current trends in solid-phase microextraction (SPME) fibre coatings. Chemical Society Reviews. 2010; 39: 4524–37.
[108] Kaykhaii M, Dicinoski GW, Smedley R, Pawliszyn J, Haddad PR. Preparation and evaluation of solid-phase microextraction fibres based on functionalized latex nanoparticle coatings for trace analysis of inorganic anions. Journal of Chromatography A. 2010; 1217: 3452–6.
[109] Li QL, Ding YJ, Yuan DX. Electrosorption-enhanced solid-phase microextraction of trace anions using a platinum plate coated with single-walled carbon nanotubes. Talanta. 2011; 85: 1148–53.
[110] Chen JM, Zou J, Zeng JB, Song XH, Ji JJ, Wang YR et al. Preparation and evaluation of graphene-coated solid-phase microextraction fiber. Analytica Chimica Acta. 2010; 678: 44–9.
[111] Luo YB, Yuan BF, Yu QW, Feng YQ. Substrateless graphene fiber: A sorbent for solid-phase microextraction. Journal of Chromatography A. 2012; 1268: 9–15.
[112] Fan J, Dong ZL, Qi ML, Fu RN, Qu LT. Monolithic graphene fibers for solid-phase microextraction. Journal of Chromatography A. 2013; 1320: 27–32.

CHAPTER 6

Graphene-Based Papers and Films

Ling Bing Kong,[1,a,*] *Freddy Boey,*[1,b] *Yizhong Huang,*[1,c] *Zhichuan Jason Xu,*[1,d] *Kun Zhou,*[2] *Sean Li,*[3] *Wenxiu Que,*[4] *Hui Huang*[5] and *Tianshu Zhang*[6]

6.1. Brief Introduction

Graphene-based 2D materials have been reported in the open literature [1–3]. In this chapter, two parts are included, i.e., (i) thin films that are deposited on substrates and (ii) free-standing forms, which are known as papers, sheets, membranes and so on. There are slight overlappings between these two groups of materials in specific aspects. For example, vacuum membrane filtration can be used to fabricate both thin films and membranes. If the items are deposited on a given substrate, they are called thin films. Otherwise, if they are free-standing, they will be named as papers or membranes. For thin films, the thickness is usually at sub-micron level, whereas the membranes or papers could be as thin as micrometers or as thick as tens or even hundreds of micrometers. Although attempts have been made to standardize the names of these materials, it is difficult to do so in most cases, because of the high diversity of the sources that the references are coming from. Most likely, the nomenclature in this book follows the names of specific references, due to the consideration to match the context with the corresponding figures as much as possible.

6.2. Graphene-based Thin Films

Due to their high electrical conductivity and mechanical flexibility, graphene films are regarded as one of the most promising candidates that could be used for stretchable electronic and optoelectronic devices. With respect to those produced by using CVD approaches involving additional transferring process [4], the graphene films fabricated by using solution methods can avoid the use of expensive catalytic surfaces

[1] School of Materials Science and Engineering, Nanyang Technological University, Singapore.
[a] E-mail: elbkong@ntu.edu.sg
[b] E-mail: mycboey@ntu.edu.sg
[c] E-mail: yzhuang@ntu.edu.sg
[d] E-mail: xuzc@ntu.edu.sg
[2] School of Mechanical & Aerospace Engineering, Nanyang Technological University, Singapore; E-mail: kzhou@ntu.edu.sg
[3] School of Materials Science and Engineering, The University of New South Wales, Australia; E-mail: sean.li@unsw.edu.au
[4] Electronic Materials Research Laboratory, School of Electronic and Information Engineering, Xi'an Jiaotong University, P. R. China; E-mail: wxque@xjtu.edu.cn
[5] Singapore Institute of Manufacturing Technologies (SIMTech), Singapore; E-mail: hhuang@SIMTech.a-star.edu.sg
[6] Anhui Target Advanced Ceramics Technology Co. Ltd., Hefei, Anhui, P. R. China; E-mail: 13335516617@163.com
* Corresponding author

and can be operated under mild conditions. To obtain highly uniform graphene films, a good solution processability of GO is required. GO nanosheets in the form of thin films can be deposited on various substrates by using solution-based techniques, such as drop-casting [5, 6], dip-coating [7, 8], spraying [9–16], spin-coating [17–23], electrophoresis [24–30], Langmuir-Blodgett (LB) [31–38], Langmuir-Schaefer (LS) [39–42], layer-by-layer (LbL) assembly [43–51] and transfer via vacuum filtration [52–56]. In a typical case, conductive graphene films were produced by dip-coating an aqueous GO suspension on a quartz substrate followed by high temperature treatment [7]. The hydrophilic nature of GO is beneficial for a uniform deposition on the hydrophilic quartz surface.

6.2.1. Drop-casting and dip-coating

Obviously, the drop-casting [5, 6] and dip-coating [7, 8] methods to fabricate graphene films have several advantages, such as simple procedure, cheap chemicals and large-scale production. Gomez-Navarro et al. used a simple drop-casting method to deposit single- and multiple-layer GO nanosheets on various substrates to study the electronic transport characteristics of the individual nanosheets that were subjected to two types of chemical reduction methods [6]. The comparison with unmodified graphene indicated that the procedures enabled only partial recovery of the electrical conductivity, which was attributed to the fact that the oxidatively introduced point defects remained within the 2D carbon frameworks.

GO was dispersed in water with the aid of soft ultrasonication and the suspension was then deposited onto various substrates, such as mica, Si/SiO_2 and highly oriented pyrolytic graphite (HOPG). The suspension could be directly deposited onto mica and HOPG without the requirement of any functionalization, whereas the surface of the Si/SiO_2 substrates should be modified with 3-aminopropyltriethoxysilane (APTES) prior to the film deposition. Figure 6.1 (a) shows an AFM image of the GO nanosheet, with upper right edge

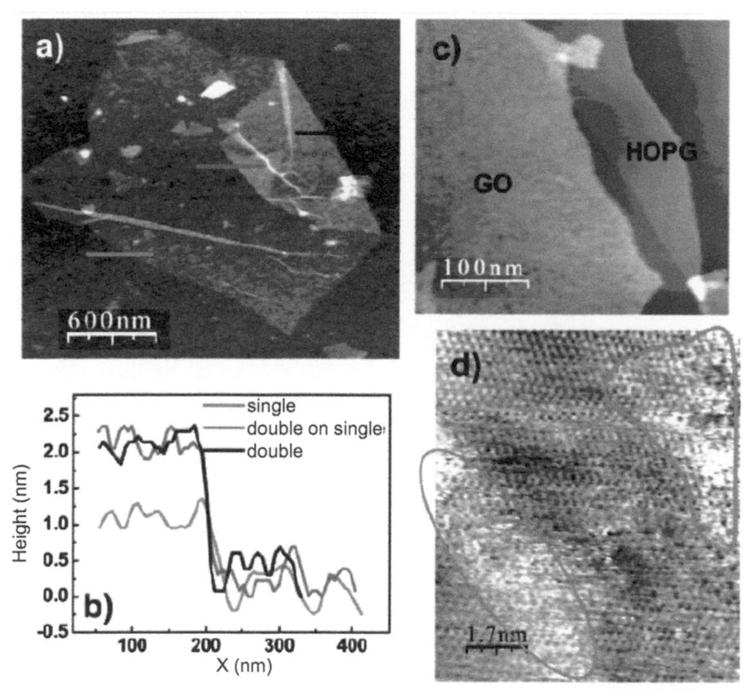

Fig. 6.1. (a) AFM image of the GO nanosheets with monolayer deposited on a SiO_2 substrate, showing a back-folded edge. (b) AFM section profiles along the three different lines in panel (a), revealing a mono-, bi-, and trilayer structure. (c) AFM image acquired from the GO monolayer on a HOPG substrate. (d) STM image of the GO monolayer on a HOPG substrate, taken under ambient conditions, with oxidized regions being marked with green contours. Reproduced with permission from [6], Copyright © 2007, American Chemical Society.

Color image of this figure appears in the color plate section at the end of the book.

being double-folded onto itself, which was confirmed by the cross-sectional profiles as demonstrated in Fig. 6.1 (b). Statistical analysis indicated that the GO nanosheets had lateral dimensions of 100–5000 nm and thicknesses of 1.1–15 nm. Most of the GO nanosheets (~80%) had a height of 1.1 ± 0.2 nm, while the rest exhibited multiples of that value. This thickness was slightly larger than the theoretical value of 0.8 nm, due to the presence of the oxygen-containing groups on two surfaces of the GO nanosheets. GO monolayers could be observed on the atomically flat surfaces of mica and HOPG, as illustrated in Fig. 6.1 (c). The GO nanosheets were examined by using scanning tunneling microscopy (STM) under ambient conditions. Figure 6.1 (d) shows a STM image of the GO monolayer, with the hexagonal lattice of the nanosheets being partially preserved. The STM images of the GO nanosheets were different from those of the pristine graphene by the presence of bright spots/regions, due to the formation of the disordered lattices, which were attributed to the presence of oxygenated functional groups.

A dip-coating method was reported to prepare transparent, conductive and ultrathin graphene films, which could be used as an alternative to the ubiquitously employed metal oxide window electrodes for solid-state dye-sensitized solar cells [7]. The graphene films were derived from exfoliated GO, combined with thermal reduction. The final films had an electrical conductivity of 550 $S \cdot cm^{-1}$ and an optical transparency of > 70% over the wavelength range of 1000–3000 nm. In addition, they demonstrated high chemical and thermal stabilities, with an ultrasmooth surface and tunable wettability.

GO nanosheets with lateral dimensions of several tens to hundreds of nanometers were used to deposit the GO films, as seen in Fig. 6.2 (A). The GO nanosheets were deposited on hydrophilic substrates, such as pretreated quartz, by using a dip coating method, from hot aqueous GO dispersion. The GO films were dried at controlled temperatures. The thickness of the films was controlled by adjusting the dipping repetition, combined with the temperature of the GO dispersion. For instance, two-fold dip coating of the GO dispersion at 70°C led to a film with a thickness of about 10 nm. Quasi-1D wrinkles with a length of 0.2–2.5 μm and a height of 5–20 nm were observed in SEM images in Fig. 6.2 (B), while AFM image indicated that they were formed due to the overlapping of the GO nanosheets, with some of the graphene edges being scrolled or folded during film fabrication, as revealed in Fig. 6.2 (C, D).

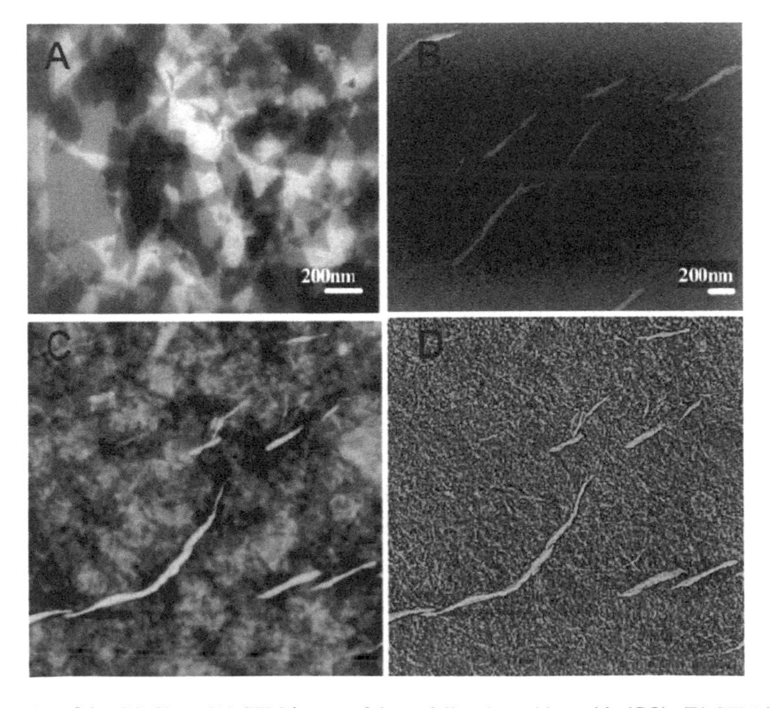

Fig. 6.2. Morphologies of the GO films. (A) SEM image of the exfoliated graphite oxide (GO). (B) SEM image of the GO film prepared through dip coating. (C) AFM height image (3.2 × 3.2 μm^2), with color scale (black to bright yellow) = 30 nm. (D) AFM phase image of the obtained GO film, with color scale (black to bright yellow) = 15°. Reproduced with permission from [7], Copyright © 2008, American Chemical Society.

Electrical conductivity of the graphene films was increased with the increasing annealing temperature over 550–1100°C. Sheet resistance (R_s) of the graphene film with a thickness of 10.1 ± 0.76 nm after annealing at 1100°C was 1.8 ± 0.08 kΩ·□$^{-1}$, with an average conductivity calculated to be 550 S·cm^{-1}. As the thickness of the graphene film was increased to 29.9 ± 1.1 nm, its conductivity was increased to 727 S·cm^{-1}, corresponding to R_s = 0.46 ± 0.03 kΩ·□$^{-1}$. In the films, graphene nanosheets could be stacked up to tens of layers, as shown in Fig. 6.3 (A).

Transmittance of the graphene films was closely related to their thickness. The films with a thickness of about 10 nm exhibited a transmittance of 70.7%, at the wavelength of 1000 nm, which was lower than those of the FTO (82.4%) and ITO (90.0%). With decreasing thickness, the transmittance of the films could reach 80.0%. However, in contrast to FTO and ITO, which showed strong absorptions in the region of near- (0.75–1.4 μm) and short-wavelength infrared (1.4–3 μm), the graphene films were always transparent in the whole regions, as demonstrated in Fig. 6.3 (B). Therefore, the graphene films could be used as window electrodes for optoelectronics with transparency over a wide range of wavelengths.

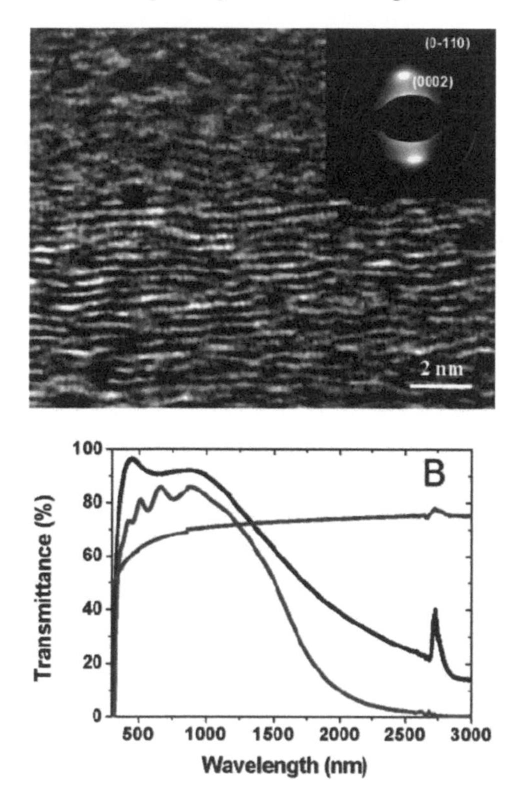

Fig. 6.3. Microstructure and optical transmittance of the graphene films. (A) HRTEM image of the graphene films with the corresponding SAED pattern (inset). (B) Transmittance curves of the graphene film with a thickness of about 10 nm (red), together with those of ITO (black) and FTO (blue). Reproduced with permission from [7], Copyright © 2008, American Chemical Society.

6.2.2. Spraying

The spraying method has been widely used as an economic technique to fabricate thin films or thin coatings for various applications. One of the advantages of this method is its capability to produce films over large areas with high homogeneities, so that it has been employed to deposit graphene-based thin films for various applications [9–16].

Availability of stable dispersions is a critical requirement to fabricate high quality graphene films by using the spraying deposition methods. High quality films should not have concentration gradients, sedimentation and agglomeration. In this case, the preparation of graphene suspensions with sufficiently high concentrations is a challenge. Usually, two strategies are used for such purpose, either the functionalization of the graphene nanosheets or the addition of surfactants into the dispersions. Any additive is added only when it has added value to the resultant films. Moreover, sufficiently strong adhesion between the films and the substrates should be maintained in order to have high mechanical and thermal stability and thus high performances.

A simple spray method has been used to deposit graphene thin films, with stable suspensions of graphene nanoflowers (GNFs) and multilayer graphene (MLG) flakes, which were dispersed in ethanol, N,N-dimethylformamide (DMF) and N-methyl-2-pyrrolidone (NMP) [12]. The suspensions of GNFs and MLG in DMF/ethanol were highly sprayable without addition of any surfactants. Micrometer-thick MLG/GNF films were deposited on glass substrates. Adhesion of the graphene films to the substrates was largely enhanced by coating a layer of methacrylic acid-methyl methacrylate (MA) copolymer on the substrates before the film deposition. With the presence of the MA coating layer, entire substrate coverage was achieved after spraying for 14 min, with the suspensions of 0.05 wt% GNFs and 0.1 wt% MLG. Graphene nanosheets were synthesized by inductively annealing silicon-carbon (Si-C) nanoparticles at 2600°C. The powders contained graphene nanoflowers (GNFs) and multilayer graphene (MLG) flakes, as shown in Fig. 6.4.

The schematic of the setup for film deposition is shown in Fig. 6.5 (a). An ultrasonic atomizer was connected to a continuous suspension feeding system consisting of a syringe pump and equipped with a high-frequency oscillator operating at 100 kHz, with an inlet for carrier gas. Piezoceramic components in the atomizer triggered mechanical oscillations to generate capillary waves, so that droplets were formed on the atomizer outlet, as illustrated in Fig. 6.5 (b). The droplets were ejected into the parabolic trajectories before they were detached from the peaks of the waves. Typical droplet size for aqueous solutions was 22 μm, with standard deviation of 1.7 μm, at a liquid flow rate of 60–1500 $ml \cdot h^{-1}$. The substrate was put on a hot plate and was perpendicular to the direction of the spray. When the droplets were moved downwards, the solvent was partially evaporated, while the droplets landed on the substrate. The hot plate temperature was set to be 150°C, while the actual temperature was slightly lower due to the heat dissipation. The average droplet velocity at a gas flow rate of 88 $l \cdot min^{-1}$ was 2.97 $m \cdot s^{-1}$. Air flow rate supplied into the atomizer was $l \cdot min^{-1}$, corresponding to an average droplet velocity of about 0.3 $m \cdot s^{-1}$, which carried the droplets to the substrate. A typical spray distance, i.e., the distance from the tip of the atomizer outlet to the substrate surface, was 10 cm, while the suspension feeding rate was 200 $l \cdot h^{-1}$.

The droplet residence time (t_r), i.e., the time taken to reach the substrate surface from the spray source, was 0.3 s. The spray setup was confined in a plexiglass tube with an inner diameter of 10 cm and wall thickness of 0.5 cm to prevent perturbations by the external air currents. The bottom edge of the plexiglass tube and the hotplate surface was separated by about 1 cm, in order to allow the carrier gas to

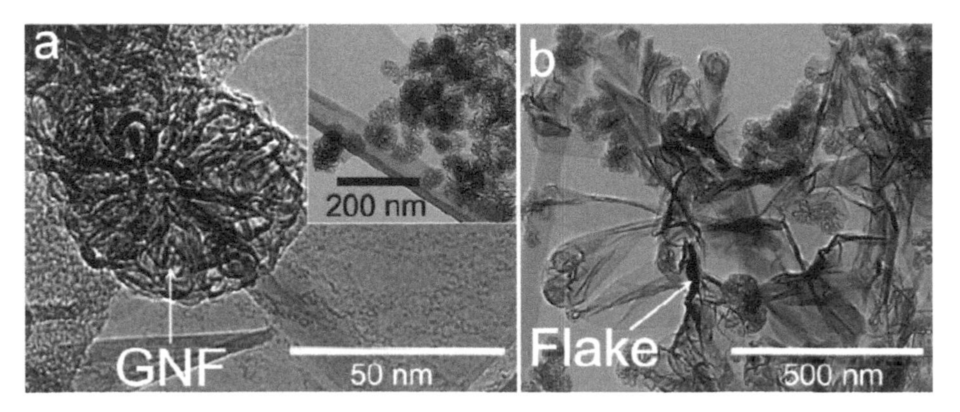

Fig. 6.4. TEM images of the graphene nanomaterials. (a) GNFs with the inset showing several GNFs on the TEM grid and (b) A MLG flake with some GNFs attached to the surface. Reproduced with permission from [12], Copyright © 2015, Taylor & Francis.

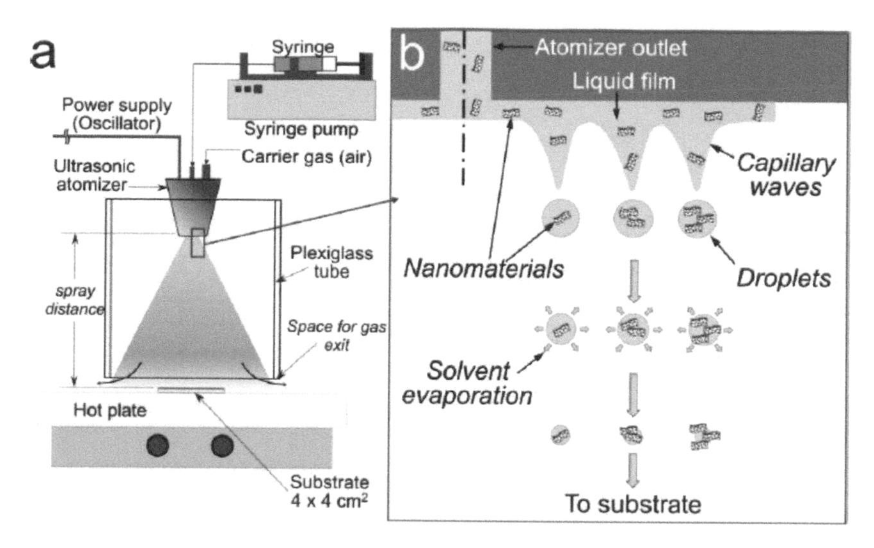

Fig. 6.5. (a) Ultrasonic spraying system used to deposit the graphene films (front view) and (b) description of droplet formation on the surface of the ultrasonic atomizer. Reproduced with permission from [12], Copyright © 2015, Taylor & Francis.

flow out of the substrate zone. The center of the substrate with a dimension of 4×4 cm^2 was aligned with the centerline of the atomizer and the plexiglass tube, so that the center of the spray plume was aligned to avoid possible edge effects on the substrate, as seen in Fig. 6.5 (a). The spraying time (t_s) was varied in the range of 7–14 min, while films were also collected by only spraying for 30 s to study the film formation process of the deposition. Slides of glass and MA-coated glass (MA-glass) were used as the substrates.

SEM images of the samples ultrasonically sprayed onto glass substrate, from the GNF suspensions with a concentration of 0.05 wt%, for 7 min and 14 min, are shown in Fig. 6.6 (a) and (b). The coverages of the substrate by the GNFs were 70% and 79% for 7 min and 14 min of spraying, respectively. MLG suspensions with concentrations of 0.05, 0.1 and 0.2 wt% were ultrasonically sprayed for 7 min onto the glass substrates, with their corresponding SEM images being shown in Fig. 6.6 (c–f). It was observed that the film from the 0.2 wt% MLG suspension exhibited agglomerates, as revealed in Fig. 6.6 (c). MLG flakes were scattered on the entire substrate, whereas the particles were not uniformly distributed, with the presence of uncovered areas, corresponding to a substrate coverage of about 79%. The samples from the suspensions of 0.1 wt% (Fig. 6.6 (d)) and 0.05 wt% (Fig. 6.6 (e)) exhibited spherical agglomerates, with diameters of ≤ 4 μm. Their substrate coverages were 73% and 64%, respectively. The flakes were entirely covered by the nano-flowers and their agglomerates. Figure 6.6 (f) indicated that, as the spraying time was increased from 7 min to 14 min, the 0.1 wt% MLG suspension led to the formation of larger deposits, but without significant variation in substrate coverage.

Although the substrates were not completely covered, the concentrations of the suspensions were sufficiently high to obtain percolated films in spraying times of minutes. Both the GNFs and MLG flakes were attached to the glass substrates through weak van der Waals forces. The films directly deposited on the glass substrates exhibited quite weak adhesion, because they were easily removed from the substrates. This was attributed to the fact that the graphene nanosheets have weak affinity with glasses due to the lack of functional groups on their surfaces.

The above problem could be addressed through functionalization. It was demonstrated that functional groups on MA polymeric chains could be employed to enhance the adhesion of the graphene nanosheets to the substrates. To do this, the GNFs and MLG suspensions were sprayed onto MA-glasses. SEM images of the samples obtained by spraying the GNF suspensions for 7 min and 14 min are shown in Fig. 6.7 (a, b). Obviously, the adhesion of the graphene nanosheets to the MA modified substrates was greatly enhanced, with the substrate coverages to be 80% and 96% after spraying for 7 min and 14 min, respectively. SEM images of the films from the MLG suspensions of 0.1 wt% sprayed on the MA modified glass substrates

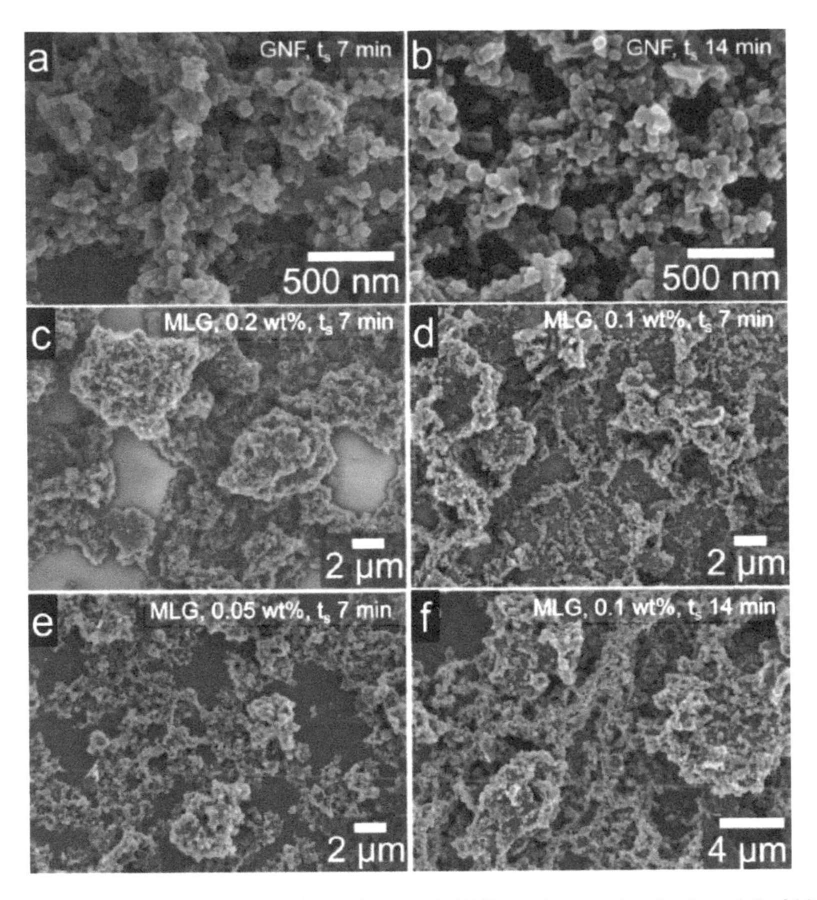

Fig. 6.6. SEM images of the graphene films on glass substrates. GNF films after spraying for times (t_s) of (a) 7 min and (b) 14 min. MLG films for t_s of 7 min from suspensions with concentrations of (c) 0.2 wt%, (d) 0.1 wt% and (e) 0.05 wt%. (f) Image of the film from the 0.1 wt% MLG suspension with a spraying time of 14 min. Reproduced with permission from [12], Copyright © 2015, Taylor & Francis.

for 7 min and 14 min are illustrated in Fig. 6.7 (c, d). The corresponding MLG flake coverages were 93% and almost 100%, for 7 min and 14 min spraying times. The MLG flakes were buried in the film and covered by GNFs, as demonstrated in the inset of Fig. 6.7 (d). The high coverage was also confirmed by the SEM images in Fig. 6.7 (e, f).

Figure 6.8 shows representative cross-sectional SEM images of the GNF and MLG flake films. Figure 6.8 (a) indicated that the MLG film on glass after spraying for 14 min exhibited a surface with irregular particles of varying sizes. With the MA modified glass substrate, the film thickness was also not uniform, ranging from 2 μm to 7 μm, as seen in Fig. 6.8 (b). The inset of Fig. 6.8 (b) shows SEM image of the MLG flake embedded in the film. The non-uniform thickness of the films was ascribed to the irregular shape of the flakes, leading to voids and pores. In comparison, the film deposited on the MA modified glass after spraying for 7 min was more uniform and dense, with an average thickness of 6.5 μm. Flakes were also clearly observed, as illustrated in the inset of Fig. 6.8 (c). Similarly, the GNF film on bear glass with a spraying time of 14 min (Fig. 6.8 (d)) was also non-uniform in thickness, with the presence of uncoated areas, while that on the MA modified glass (Fig. 6.8 (e)) was much more uniform and dense. Similar uniform film was observed in the sample sprayed for 7 min, as revealed in Fig. 6.8 (f).

After the glass substrates were modified with MA, the adhesion of the graphene coating was significantly improved. Functional groups, such as OH or COO, on the polymeric chains of MA, would interact with the graphene nanosheets through weak electrostatic bonds, e.g., hydrogen bonds. Also, the solvent vapor near the substrate surface could be combined with the softening of MA, due to the relatively

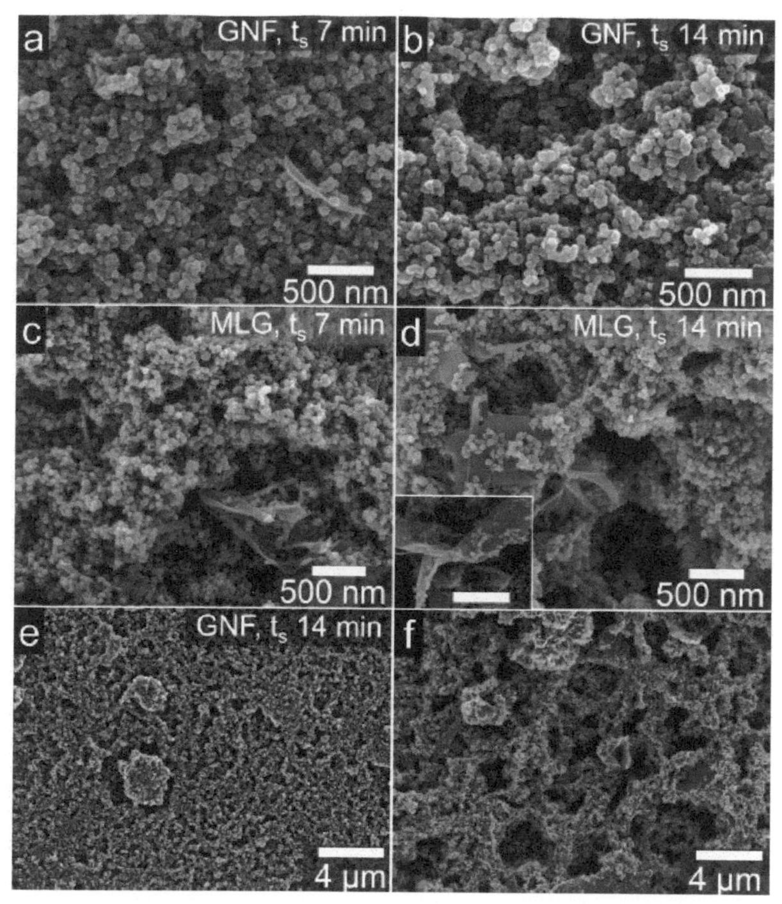

Fig. 6.7. SEM images of the graphene films on MA modified glass substrates. GNFs with spraying times (t_s) of (a) 7 min and (b) 14 min. MLG flakes with t_s of (c) 7 min and (d) 14 min, with inset scale bar = 1 μm. Low-magnification images of (e) GNFs and (f) MLGs sprayed for 14 min. Reproduced with permission from [12], Copyright © 2015, Taylor & Francis.

low glass transition temperature of 130°C, which contributed to the dissolution of the copolymer at the MA-air interface. The dissolved copolymer could confine the graphene nanosheets, so as to glue them to the substrate, so that the nanosheets would not be blown away by air flow from the atomizer or convection from the hot plate.

Highly efficient thin film counter electrodes (CEs) for dye-sensitized solar cells (DSSCs) were prepared by using chemically driven aqueous dispersible graphene nanosheets (GNS) [16]. The GNS thin films were fabricated by using an electro-spray (e-spray) method. The essential factors that influence the performance of CEs, including electrocatalytic activity, charge transfer resistance and electrical conductivity were studied, by focusing on the evolution of surface chemical structures of the graphene nanosheets. The efficiency of the CEs based on the GNS thin film (GNS-CE) was significantly enhanced by applying a post thermal annealing (TA), due to the reduction in oxygen concentration at the edges of the graphene nanosheets. Performance of the optimized e-sprayed GNS-CEs was comparable to that of the state-of-the-art CE materials, i.e., thermolytically prepared Pt.

Figure 6.9 shows the processing and characterization of the GNS films by using the e-spray method. The GNSs were prepared by the aqueous phase chemical reduction of GO nanosheets. Well-dispersed aqueous GNS solutions were prepared at concentrations of up to ~0.3 mg·ml^{-1} in DI water. GNS films with surface area and porosity, as well as low aggregation were essential requirements for high performances. Fluorine-doped tin oxide (FTO) coated glass plates were used to deposit the GNS thin films. Figure 6.9 (B) shows the setup of the e-spray experiment. The e-spray is an efficient film deposition

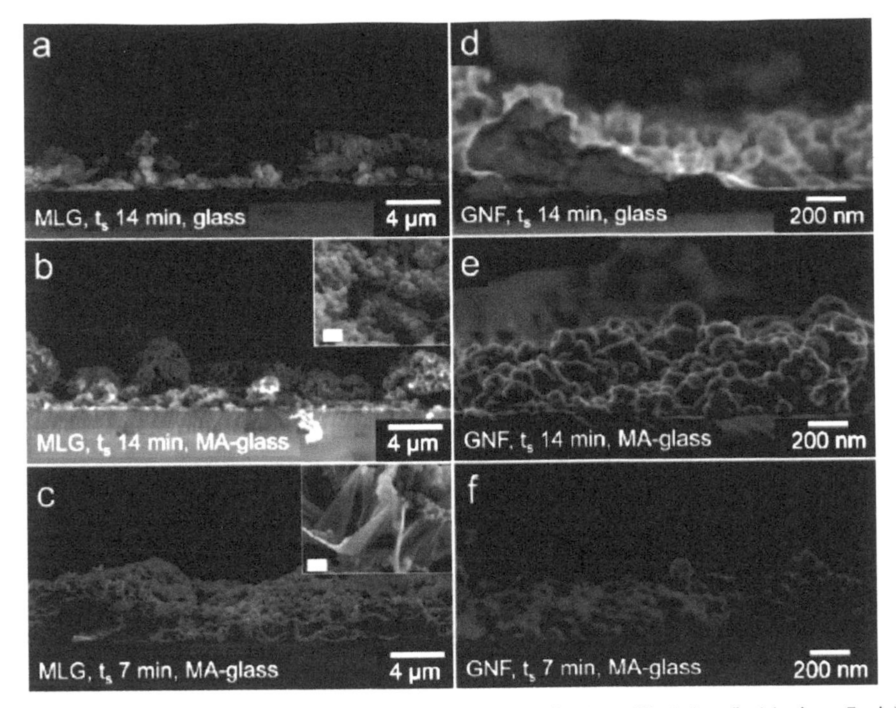

Fig. 6.8. Cross-sectional SEM images of the MLG on glass (a, 14 min) and MA modified glass (b, 14 min, c, 7 min) and GNF on glass (d, 14 min) and MA-glass (e, 14 min, f, 7 min). The insets show the MLG flakes embedded in the films, with scale bar = 200 nm. Reproduced with permission from [12], Copyright © 2015, Taylor & Francis.

method with uniform thickness and low material consumption. The GNS solution was sufficiently stable to deposit thin films by using the e-spray method. The deposition could be continuously carried out for >24 h without nozzle clogging.

The thickness of the resulting films was proportional to the deposition time. Figure 6.9 (A) shows a photograph of the well-dispersed GNS solution with a concentration of 0.16 $mg \cdot ml^{-1}$ in the mixture of DI water/ethanol (6/4). The GNS-CEs deposited on the FTO/glass substrate exhibited very high uniformity, as demonstrated in Fig. 6.9 (C). Representative SEM images of the GNS films are shown in Fig. 6.9 (D, E). The e-sprayed GNS layer had porous and layered microstructure, without obvious macro-aggregations, demonstrating the high efficiency of the e-spray method in fabricating GNS films. According to the Brunauer-Emmett-Teller (BET) measurement, the films possessed specific surface area and porosity of 276 $m^2 \cdot g^{-1}$ and 0.19 $cm^3 \cdot g^{-1}$, respectively.

6.2.3. Spin-coating

A spin-coating process was used to deposit large-area few-monolayer GO films, which were then converted into graphene films [17]. Such graphene films exhibited excellent mechanical properties and could be used to fabricate high Young's modulus ($\langle E \rangle = 185$ GPa) low-density nanomechanical resonators. Wafer-scale films with thicknesses of as thin as 4 nm were so robust that they could be delaminated and resuspended on a bed of pillars or field of holes. With these films, radio frequency resonators have been demonstrated, which possessed quality factors of up to 4000 and figures of merit ($f \times Q$) of $> 10^{11}$. The films could withstand high in-plane tension of up to 5 $N \cdot m^{-1}$, while their high Q-values indicated that the film integrity was enhanced due to the platelet-platelet bonding among the graphene nanosheets.

Exfoliated GO platelets were dispersed in methanol to form suspensions. Films were deposited on SiO_2/Si substrates with the suspensions by using a modified spin-casting method, as shown in Fig. 6.10 (a). It was found that solution evaporation could be accelerated by blowing dry N_2 during the spin-casting process, so that continuous films with GO platelets that were laid on the surface were obtained,

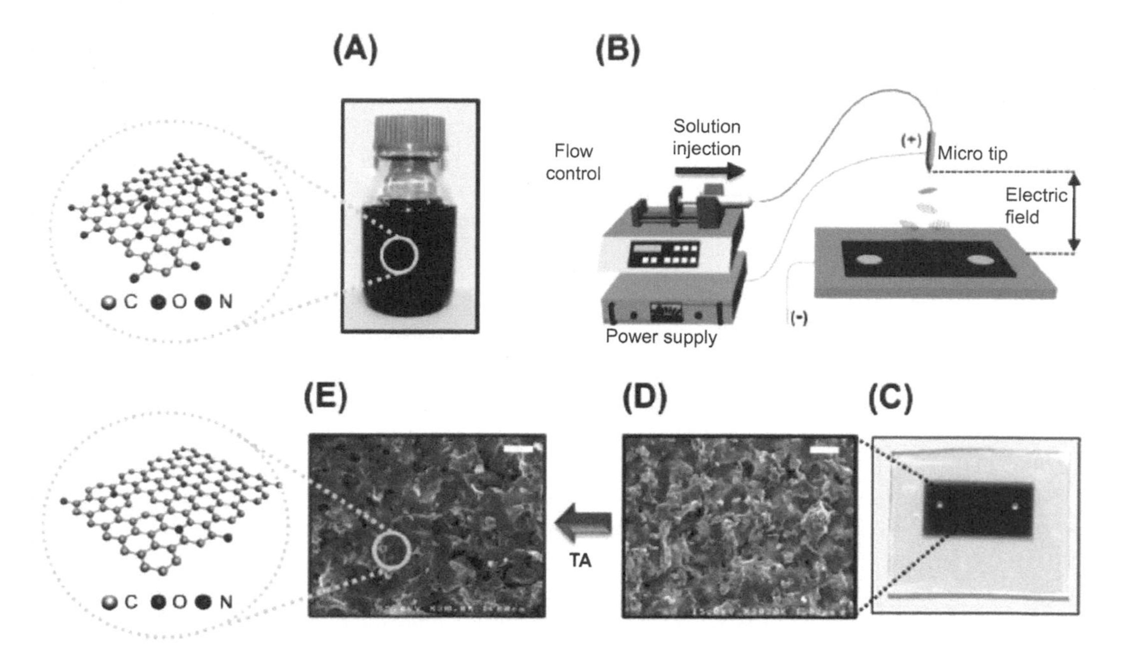

Fig. 6.9. The process and setup used to fabricate the GNS thin film CEs. (A) Photograph of the aqueous GNS solution, with a concentration of 0.3 mg·ml^{-1}. (B) Schematic diagram of the e-spray technique used to deposit the GNS thin film. (C) Photograph of the e-sprayed GNS thin film CE with a thickness of ~200 nm. (D, E) SEM images of the e-spayed GNS-based CEs: (D) as e-sprayed and (E) after TA. The scale bars at the up-right position of images are 500 nm. Reproduced with permission from [16], Copyright © 2012, American Chemical Society.

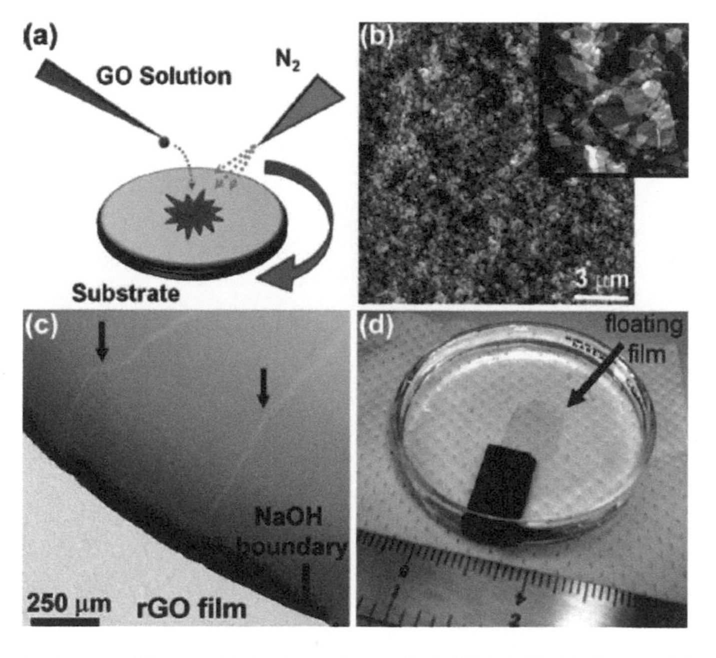

Fig. 6.10. (a) Schematic diagram of the modified spin-coating method. (b) AFM height image of the continuous GO film with an average thickness of 2 nm and roughness of 1 nm (z = 6 nm), with the inset image to be 2.75 × 2.75 μm^2. (c) Optical microscope image demonstrating the release of the 4 nm thick rGO film with sodium hydroxide. Arrows highlight two cracks in the delaminating film as the droplet propagated across the film surface. (d) Photograph of the rGO film transferred intact into water. The film had a thickness of about 8 nm with lateral dimensions same as the parent substrate (12 × 25 mm^2), submerged at bottom of the Petri dish. Reproduced with permission from [17], Copyright © 2008, American Chemical Society.

as revealed in Fig. 6.10 (b). After deposition, a low temperature thermal annealing and chemical reduction in hydrazine hydrate vapor were combined to form rGO films. To transfer the rGO films, delamination from the SiO_2/Si substrate was conducted by exposing the films to a basic solution of sodium hydroxide, as illustrated in Fig. 6.10 (c). As the samples were dipped into water, the entire delaminated films would be floating on the water surface, as seen in Fig. 6.10 (d). The regions of the films that were not exposed to the basic solution were retained on the substrate.

After the films were recaptured and dried in air on textured substrates, they were suspendable above the surface. Two rGO films with thicknesses of 20 and 4 nm are shown in Fig. 6.11 (a, b), which were suspended on a bed of Si pillars that were fabricated through a chemical nanomachining process. By using an escape route for water during the drying process, the films could suspended over areas of as large as 0.5 × 0.5 mm. The only limit was the extent of the support structure. To study elastic properties of the rGO films, drum resonators were constructed by transferring the films onto the prepatterned 250 nm SiO_2/Si substrates, with holes that were etched through the oxide having diameters ranging from 2.75 to 7.25 μm. Once the rGO films were dried over a field of holes, three final configurations were derived: (i) the film was pulled into the hole and in contact with the bottom, (ii) the film spanned the hole but was ruptured due to escaping of water molecules and (iii) the film spanned the hole intact, so that water beneath it was trapped, as illustrated in Fig. 6.10 (c, d). The trapped water could be released by using a focused ion beam (FIB) to mill a small hole (< ~300 nm) at the centers of the membranes (Fig. 6.11 (e)). In this case, the size and position of the holes had almost no effect on the fundamental drum acoustics.

A similar spin-coating approach was reported to deposit atomically thin highly uniform chemically derived graphene (CDG) films on 300 mm SiO_2/Si wafers [21]. The thin films could be lifted off to form uniform membranes. They were sufficiently strong to be either free-standing or transferred onto other substrates. Thickness of the films could be readily controlled in the range from 1–2 layers to 30 layers. Various methods have been used to reduce the CDG films. The transparent (70% transparency) rGO films were used to fabricate electrically active FET devices, with mobilities of 15 $cm^2 \cdot (V \cdot s)^{-1}$ and sheet resistance of 1 $k\Omega \cdot \square^{-1}$.

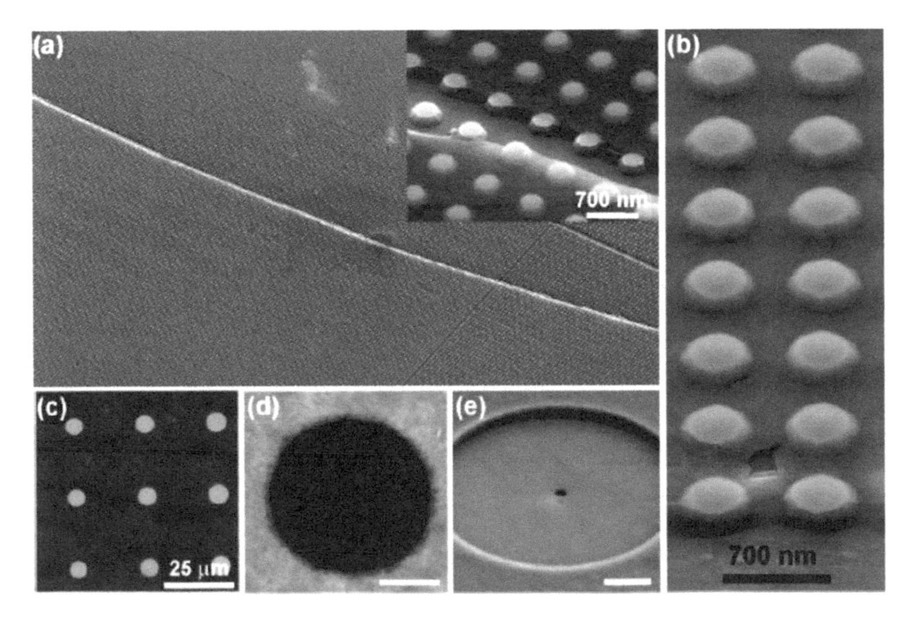

Fig. 6.11. Transfer and drying of the rGO films in air onto prepatterned substrates. SEM images of the rGO films with thicknesses of 20 nm (a) and 4 nm (b) suspended on a bed of Si pillars (pillar height = 100 nm). The inset in (a) shows a magnified image of the crack in the film that runs diagonally across the field. In (b), a small rupture in the electron transparent film can be observed near the pattern edge. (c) Optical microscope image showing nine rGO drum resonators. (d) AFM height image showing an intact drum resonator (thickness = 10 nm). The film was depressed about 10 nm below the top SiO_2 surface. (e) SEM image taken after a small hole was milled through a drum resonator (thickness = 15 nm). Scale bar = 1 μm in panels (d, e). Reproduced with permission from [17], Copyright © 2008, American Chemical Society.

Figure 6.12 (a) shows flowchart of the deposition method. The two key factors to deposit uniform graphene films on the 300 mm wafers were found to be the concentration of the CDG in the methanol/water suspension and the volume dispensed on the wafers for spin coating. For instance, if the concentration of the suspension was too high, the CDG flakes tended to aggregate, so that the film thickness would become non-uniform. If the suspensions were too diluted, the wafer surface could not be uniformly and continuously covered. It was also found that sufficient time was required for the dropped suspension to spread over the wafer surface before the stage was rotated. At the same time, during the deposition process, N_2 gas was blown at the center region of the SiO_2 (300 nm)/Si wafers, as demonstrated in Fig. 6.12 (a, iv). The formation of the CDG layer was reflected by the rapid change in color from purple to uniform light blue. The number of CDG layers was controlled by simply adjusting the spin coating speed. For example, films with 1–2 layers, 4–5 layers and 7–8 layers were obtained at 8000 rpm, 6000 rpm and 4000 rpm, respectively. Even thicker films could be fabricated at a rotation speed of 2000 rpm or by repeating the deposition step.

The lift-off and transfer procedures of the thin films to obtain free-standing membranes are also illustrated in Fig. 6.12 (a). By using this method, very uniform CDG thin films have been transferred onto 300 mm wafers with a yield of 100%. The lifted-off membranes were highly uniform, with atomically thin CDG films supported by polymethyl-methacrylate (PMMA). The suspended CDG membranes could be transferred onto any substrate by scooping, and then remained free-standing after the PMMA was removed by using acetone. The transferred films had no cracks. The fabrication of the free-standing CDG films implied that the individual flakes within the thin films were strongly adhered to form a continuous network. The structural integrity of the membranes could be well-retained even when the PMMA support was taken away. Photographs of the 150 mm and 300 mm wafers (6 and 12 in.) with atomically thin layers of uniform CDG and the transferred free-standing membranes are presented in Fig. 6.12 (b, c).

Figure 6.13 (a) shows a 300 mm membrane that was transferred onto PET film. The transferred membrane was highly flexible and optically transparent. Representative optical microscope image of the deposited graphene membrane without PMMA support is illustrated in Fig. 6.13 (b). Overlapping graphene flakes, with sizes in the range of 20–30 μm, were observed in the continuous membrane. The large size of the CDG flakes ensured the successful deposition of the CDG films with ultra-large areas. Figure 6.13 (c) shows AFM images of the thin films, indicating that the CDG flakes were lying flat on the substrate

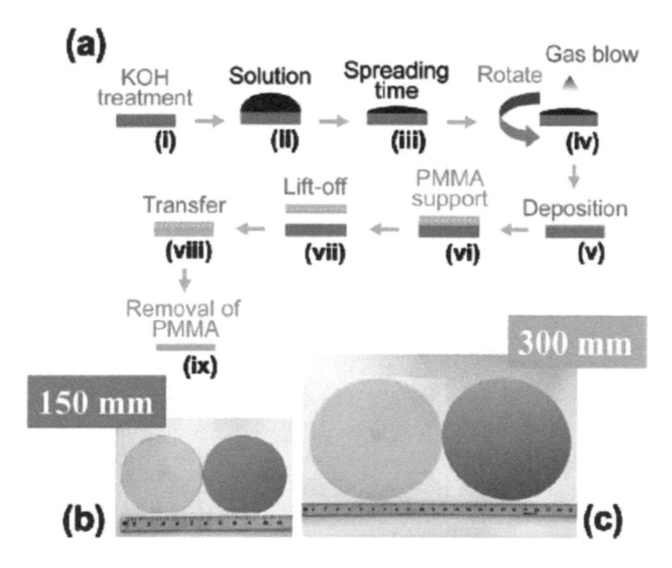

Fig. 6.12. (a) Schematic flow diagram of the deposition procedures for the CDG films on a 300 mm substrate. (iv) Deposition with spin coating while blowing N_2 gas at the center region of the SiO_2/Si wafer. Photographs of (b) 150 mm (6 in.) and (c) 300 mm (12 in.) wafers with atomically thin layers of uniform CDG films and the corresponding free-standing membranes after the transfer. Reproduced with permission from [21], Copyright © 2010, American Chemical Society.

Color image of this figure appears in the color plate section at the end of the book.

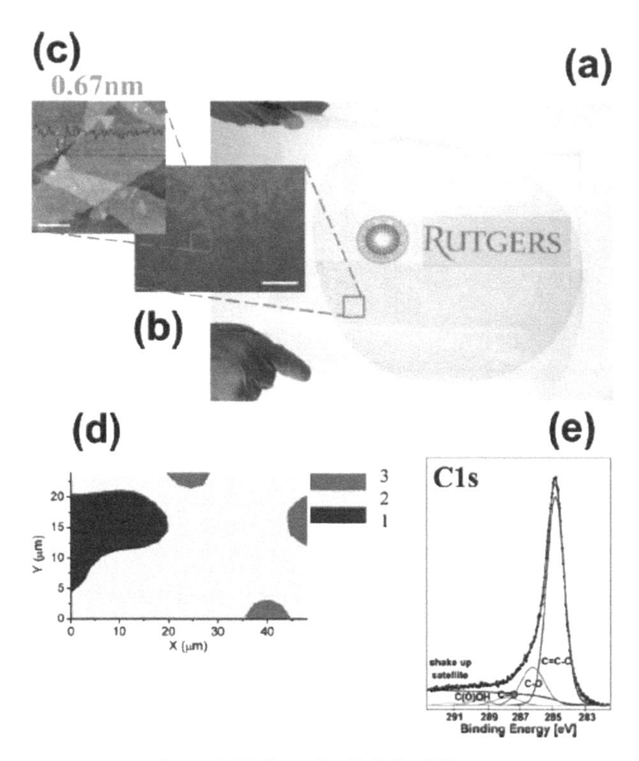

Fig. 6.13. (a) 300 mm membrane transferred onto PET film, with high flexibility and transparency. (b) Representative optical microscope image (scale bar = 50 μm) and (c) AFM image (scale bar = 1 μm) of the deposited graphene membrane without PMMA support. (d) Typical Raman map over 48 μm × 24 μm spatial region of the thinnest films, indicating that the deposited film consisted of mostly single to bilayer graphene nanosheets. (e) C1s peak of XPS spectra of the CDG thin films reduced through thermal annealing at 1100°C, with oxygen content to be reduced to 8 at.%. Reproduced with permission from [21], Copyright © 2010, American Chemical Society.

almost without any wrinkles, which was attributed to the combination of centrifugal force that was applied horizontally to the substrate during the spin coating process and the rapid evaporation of the solvent to confine the flakes before the formation of wrinkles. Height profile of the AFM image indicated that the step height was about 0.67 nm at the region corresponding to the edge of the flakes. More importantly, films with a thickness of as small as 0.7 nm could be obtained at optimized conditions, i.e., single-layer CDG membranes could be fabricated over 300 mm.

A spin self-assembly process was employed to form highly ordered layer-by-layer structures of GO nanosheets [22]. Due to their high aspect ratio, the GO nanosheets acted as building blocks to construct an ultra-high oxygen barrier system, together with polypropylene films. Because the GO layers were negatively charged, while the poly(allylamine hydrochloride) (PAH) layers were positively charged, a highly transparent and very efficient barrier layer with a thickness of < 300 nm could be obtained on a polyolefin film. The effect of the structural order of the nanosheets on the gas-barrier properties of the films was studied, by using different approaches, including spin-LbL assembly, dip-LbL assembly and PAH–GO mixture spin-coating. AFM results indicated that the films produced by using the spin-LbL assembly were flat and highly ordered, whereas those made by using the dip-LbL process possessed a rough surface and disordered structures, while the PAH–GO mixed films contained randomly oriented nanosheets in the polymeric matrix. Meanwhile, the spin-LbL films exhibited the highest barrier performance.

In the spin self-assembly method, the substrates were thoroughly cleaned first and then were treated with UV-ozone for 10 min. A droplet of aqueous solution of positively charged polyelectrolyte (PAH, 0.1 wt.%) was dropped on the PP substrate, which was rotated at 4000 rpm for 30 s and then the substrates were rinsed and dried. The negatively charged solution (GO, 0.1 wt.%) film was deposited on the PAH film. These two steps were repeated alternatively for *n* times to obtain the multilayer films of $(PAH/GO)_n$. Figure 6.14 shows a schematic diagram of the spin-coating process. In the dip self-assembly method, a given substrate was dipped into a positively-charged PAH solution for 5 min, then rinsed with deionized water for 30 s, followed by drying. Negatively-charged GO solution was processed similarly. Films with the desired thickness were prepared by repeating these two steps alternatively. To prepare the GO/PAH mixture films, the GO solution was mixed with completely deionized PAH solution at pH 12. Spin-coating was applied to the suspensions, in which a 50 nm thick layer on the PP film was produced at 4000 rpm for 30 s.

Figure 6.15 (a) shows AFM image of the GO nanosheets used to prepare the GO suspensions. AFM height profile in Fig. 6.15 (b) indicated that the GO nanosheets had an average thickness of about 0.8 nm, confirming the formation of single-layered GO nanosheets. The GO nanosheets had lateral areas widely ranging from 0.1 to 8 μm^2, with an average dimension of 1.6 μm^2. They exhibited an aspect ratio (length/thickness) of > 1000, which ensured the effectiveness of the spin self-assembly to obtain the multilayered films of $(PAH/GO)_n$.

Thickness and optical transparency of the multilayered PAH/GO films are shown in Fig. 6.16. The thickness was linearly proportional to the number of bilayers (*n*), with each repetition producing a thickness of 4.5 nm, which confirmed the regular and uniform deposition of the PAH and GO layers. Encouragingly, when the number of PAH/GO bilayers were increased from 5 to 30, the transmission of the films was only slightly and linearly decreased from 98.7% to 88.2% with respect to pristine PP films. The well-dispersed and regularly oriented GO nanosheets within the PAH matrix were responsible for the high optical transparency.

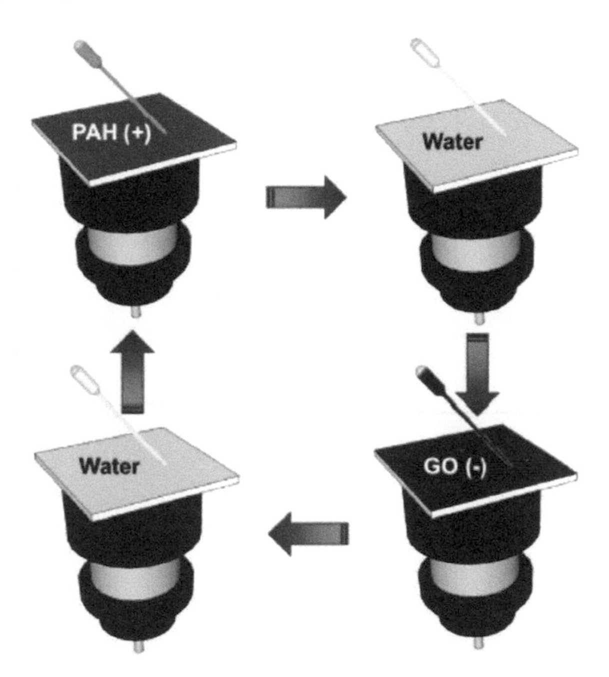

Fig. 6.14. Schematic diagram of the spin layer-by-layer assembly process with GO nanosheets as the negative charged building blocks and poly(allylamine hydrochloride) (PAH) as the positive charged polyelectrolyte. Reproduced with permission from [22], Copyright © 2015, Elsevier.

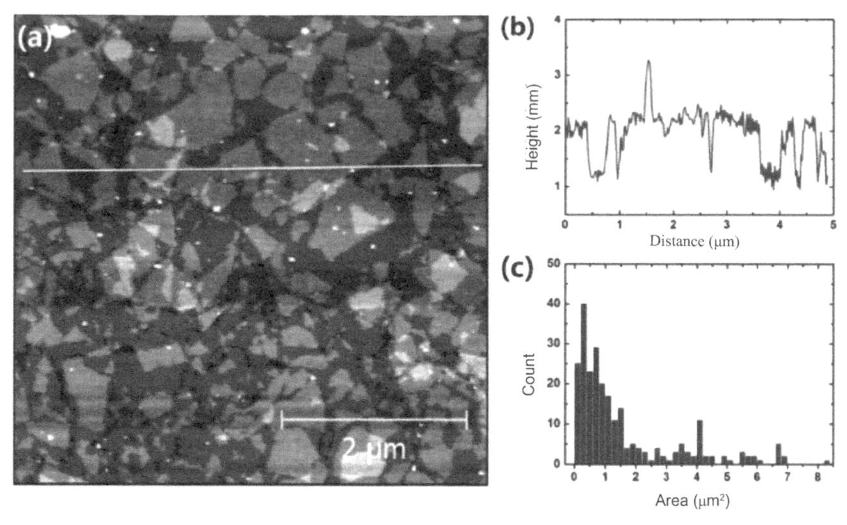

Fig. 6.15. (a) Tapping-mode AFM image of the GO nanosheets on an APTES-treated Si substrate. (b) Height profile of the AFM image and (c) corresponding histogram of the GO nanosheet size distribution. Reproduced with permission from [22], Copyright © 2015, Elsevier.

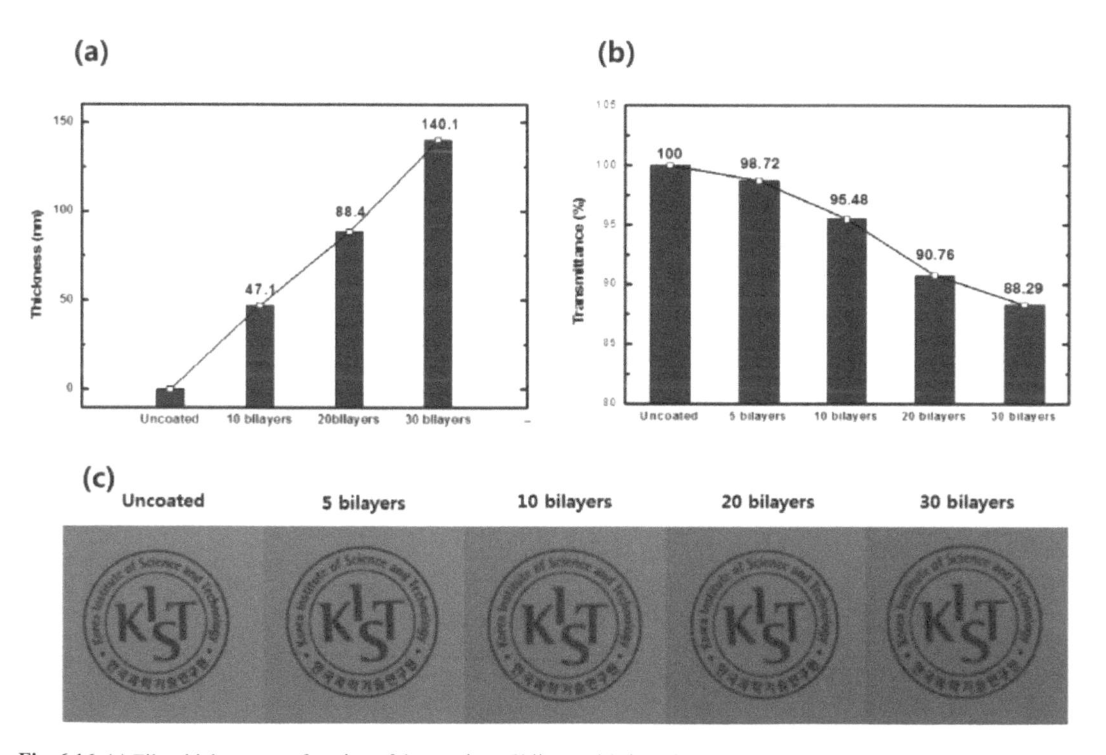

Fig. 6.16. (a) Film thickness as a function of the number of bilayers (*n*) deposited with the assembly of $(PAH/GO)_n$. (b) Light transmittance as a function of the number of bilayers. (C) Photograph of the $(PAH/GO)_5$, $(PAH/GO)_{10}$, $(PAH/GO)_{20}$ and $(PAH/GO)_{30}$ films on a 30 μm PP layer to demonstrate their optical transparency. Reproduced with permission from [22], Copyright © 2015, Elsevier.

6.2.4. Electrophoresis deposition

Electrophoretic deposition technology has various advantages as compared with other methods. Moreover, electrophoretic deposition of graphene nanosheets does not require the addition of any cross-linking molecules or binders, so that it has high film deposition rates. In addition, electrophoretic deposition results in a higher degree of packing and alignment of the graphene nanosheets when compared with other methods, such as spray-coating techniques.

A facile rapid scalable electrophoretic deposition approach has been reported to develop large-area graphene films on conductive substrates based on the electrophoretic deposition of GO and rGO nanosheets [25]. Conformal graphene films were fabricated by using two approaches. In the first method, GO nanosheets were electrophoretically deposited with an aqueous GO solution, which was formed through the oxidation of graphite to graphite oxide that was exfoliated into GO. The GO films were reduced by dipping them in an aqueous solution of hydrazine. In the second method, GO was reduced to graphene nanosheets in a strongly alkaline solution first and the rGO nanosheets were directly electrophoretically deposited onto the conductive substrates. The film thickness could be controlled by simply adjusting the deposition time.

Figure 6.17 shows photographs, optical microscopy images and SEM images of the graphene-based thin films on ITO substrates derived from the exfoliated GO and rGO suspensions by using the electrophoretic deposition method. The GO films exhibited colors from a light to a dark brown as the film thickness was increased. The brown color was related to the smaller size of intact conjugated domains within the GO nanosheets. Comparatively, the rGO suspensions were black in color, thus leading to black films. The films had thicknesses ranging from 9.6 to 146 nm. Very smooth films could be obtained when using the constant-current mode as compared with the constant-voltage mode. The voltage applied during the electrophoretic deposition was always < 20 V. Optical microscopy and SEM images indicated that the films were highly uniform over large areas without the presence of obvious cracking. Thicker films exhibited some crumpling of the graphene nanosheets at the boundaries between individual graphene platelets, as viewed in Fig. 6.17 (C).

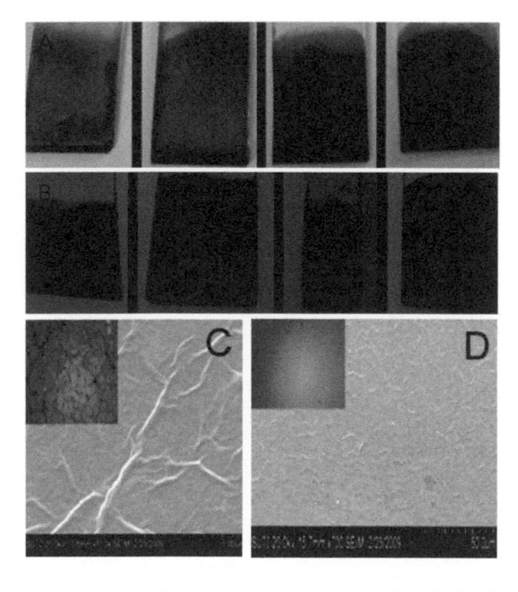

Fig. 6.17. (A) Digital photographs of the GO films. (B) Digital photographs of KOH/hydrazine-reduced graphene films, with film thicknesses obtained from spectroscopic ellipsometry for these films (from left to right) to be 9.6 ± 0.2 nm, 84.2 ± 1.1 nm, 108.6 ± 0.9 nm and 146.3 ± 1.6 nm. (C) SEM image of the KOH/hydrazine-reduced graphene film. (D) SEM image of the GO film. The insets show optical microscopy images. Reproduced with permission from [25], Copyright © 2000, American Chemical Society.

GO films could be directly reduced by using an electrophoretic deposition (EPD) process [27]. The oxygen functional groups on the GO nanosheets were removed during the EPD process. The as-deposited GO films possessed an electrical conductivity of 1.43×10^4 S·m^{-1}, which was much higher than that of the GO papers (0.53×10^{-3} S·m^{-1}). Without using any reducing agent, this method has demonstrated high potential for high yield, large area, low cost and environmentally friendly production of GO and rGO thin films.

Figure 6.18 (a) shows a schematic diagram of the single-compartment EPD setup. The GO nanosheets migrated toward the positive electrode, when a DC voltage was applied over the two electrodes. The deposition rate was dependent on several experimental parameters, such as the concentration of the GO suspensions, applied DC voltage and conductivity of the substrate. Various substrates, such as Cu, Ni, Al, stainless steel and *p*-type Si have been used to deposit the EPD-GO films. If a heavily doped *p*-type Si substrate was used, the deposition time was about five times as compared with that when using a stainless steel substrate. During the EPD process, gas bubbles were produced at the cathodes, while the deposition was formed on the anode. If a GO suspension with a concentration of 1.5 mg·ml^{-1} was used, a smooth film was deposited on a stainless steel substrate in less than 30 s at the applied potential of 10 V. The as-deposited samples were dried in air at room temperature for 24 h. Figure 6.18 (b) shows a cross-sectional SEM image of the air-dried EPD-GO film with a thickness of 4 μm that was obtained after deposition for 2 min. The film had a uniform thickness, with a packing morphology similar to that of the GO paper formed by using filtration. The thickness of the films could be varied in range between several hundreds of nanometers and tens of micrometers, by controlling the current and time.

More recently, a simple one-step EPD approach was developed to prepare Co_3O_4–graphene hybrid films [28]. The one-step EPD was combined with a subsequent thermal annealing. In this method, the use of harsh chemicals to reduce GO was effectively avoided, while polymer binders or conductive additives for fabricating lithium ion battery (LIB) electrodes were not required. The obtained Co_3O_4 NPs were homogeneously distributed on the surface and interlayer spacing of the graphene nanosheets, leading to a homogeneous hierarchical microstructure. Such Co_3O_4–graphene hybrid films demonstrated high reversible

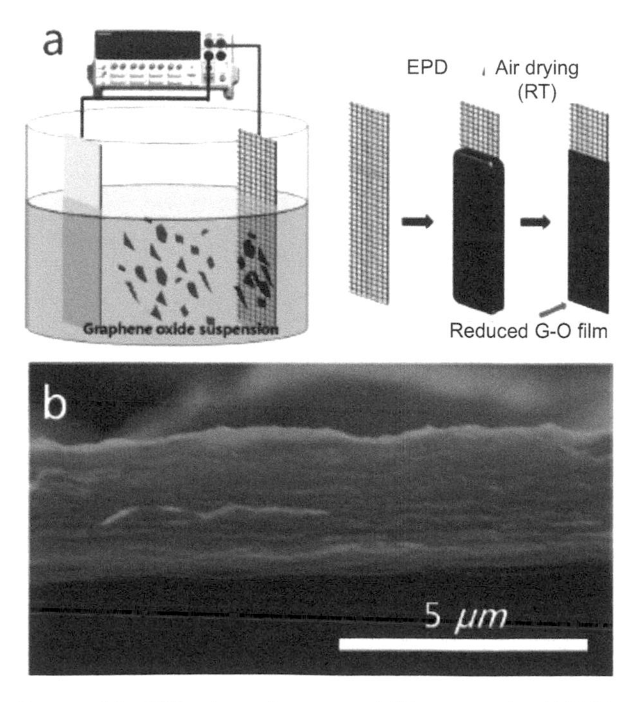

Fig. 6.18. (a) Schematic diagram of the EPD setup and process and (b) cross-sectional SEM image of the EPD-GO film. Reproduced with permission from [27], Copyright © 2010, American Chemical Society.

capacity, extraordinary cycling stability and excellent rate performance. The content of graphene in the hybrid films could be optimized in terms of their performances in lithium storage.

To prepare the EPD suspensions, 0.2 g GO and 0.5 g $Co(NO_3)_2 \cdot 6H_2O$ were dispersed in 200 ml of absolute ethanol with the aid of ultrasonication. EPD was performed in a standard two-electrode glass cell at 25°C, with copper foil as the working electrode and platinum foil as the counter electrode. The precursors of $Co(OH)_2$–GO hybrid films were deposited on the copper foil, with a deposition area of 25 mm × 35 mm. The DC voltage applied over the electrodes was 60 V, with a deposition time of 200 s. The as-deposited films were dried at room temperature and then thermally annealed in air at 200°C for 2 h and at 300°C for 2 h in Ar, in order to form the Co_3O_4–graphene hybrid film, which was denoted as Co_3O_4–graphene-1. The samples with the amount of $Co(NO_3)_2 \cdot 6H_2O$ increased to 1 g and 1.5 g were denoted as Co_3O_4–graphene-2 and Co_3O_4–graphene-3, respectively. Similar samples were prepared by using $Mg(NO_3)_2 \cdot 6H_2O$ with a much lower concentration, e.g., EPD 200 ml suspension contained 0.2 g GO and 0.025 g $Mg(NO_3)_2 \cdot 6H_2O$.

Figure 6.19 shows the fabrication process and reactions involved in the formation of the Co_3O_4–graphene hybrid films. Because the GO nanosheets dispersed in the ethanol suspension were positively charged, Co^{2+} ions were adsorbed by the GO nanosheets through electrostatic interaction, thus forming Co^{2+}–GO hybrids initially. During the EPD process, the Co^{2+}–GO hybrid items were driven by the electric field force, migrating towards the cathode, and were deposited on the copper foil. The deposited Co^{2+}–GO hybrids served as a micro-cathode, so that electrons could be transferred from the copper cathode to the outer surface of its sidewall, thus triggering the fast reduction of NO_3^- ions and H_2O molecules on the outer surface. Meanwhile, the OH^- ions were unable to diffuse into the suspension, due to the strong electro-migration of the Co^{2+} ions. Therefore, $Co(OH)_2$ NPs were formed as a uniform coating on the sidewalls of the GO nanosheets. Gradually, hierarchical $Co(OH)_2$–GO hybrids were obtained. Finally, the $Co(OH)_2$–GO

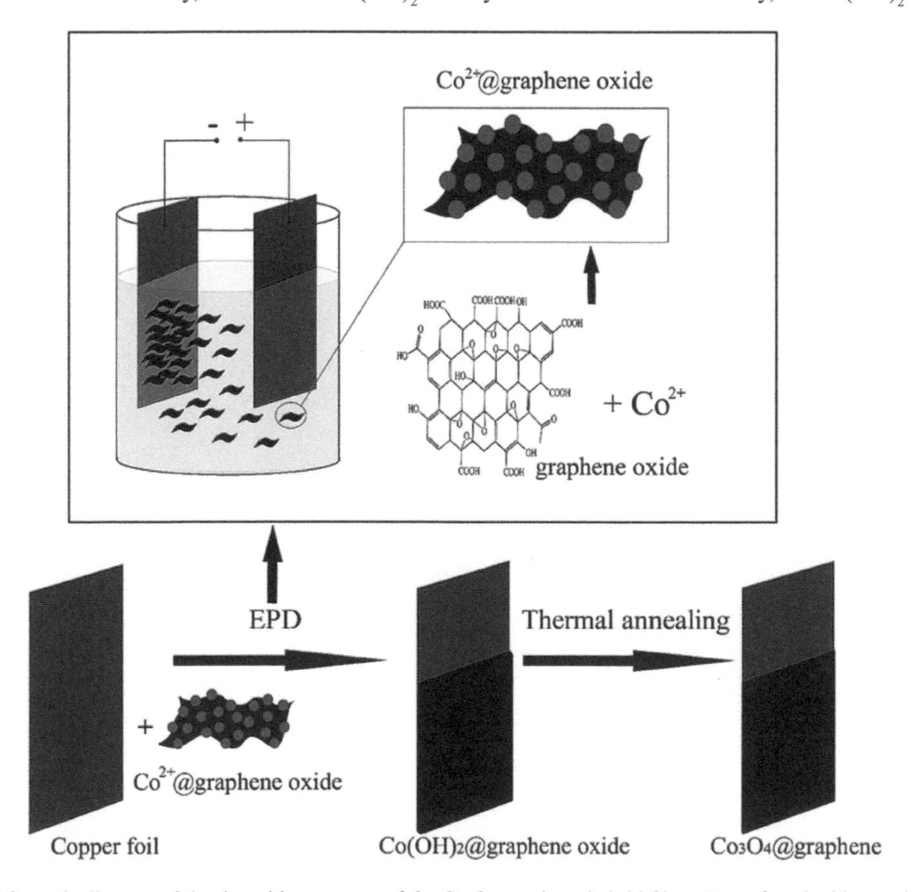

Fig. 6.19. Schematic diagram of the deposition process of the Co_3O_4–graphene hybrid films. Reproduced with permission from [28], Copyright © 2015, The Royal Society of Chemistry.

was converted into Co_3O_4–graphene through the subsequent two-step thermal annealing, i.e., at 200°C for 2 h in air and then at 300°C for 2 h in Ar. The phase evolution was confirmed by XRD measurement.

Figure 6.20 shows the morphology and microstructure characterization results of the Co_3O_4–graphene hybrid films. They possessed a unique hierarchical morphology, in which the Co_3O_4 NPs were attached on the surface and interlayers of layered graphene nanosheets, as demonstrated in Fig. 6.20 (a–c). In Co_3O_4–graphene-2, the Co_3O_4 NPs were densely dispersed on the graphene nanosheets, while those in Co_3O_4–graphene-1 were much more sparsely dispersed. However, in Co_3O_4–graphene-3, severe aggregation of the Co_3O_4 NPs was observed. This observation indicated that there was an optimized concentration for the Co_3O_4 NPs to be homogeneously distributed among the graphene nanosheets. Representative TEM images of the samples are shown in Fig. 6.20 (d–f). The Co_3O_4 NPs exhibited particle sizes in the range of 10–20 nm. At the same time, the particle size was gradually increased with the decreasing content of graphene in the hybrid films. Figure 6.20 (g) shows a cross-sectional SEM image of the Co_3O_4–graphene-2 hybrid films, confirming the presence of the hierarchical microstructure. The sample had a thickness of about 25 μm, with most graphene nanosheets lying horizontally on the substrate. The presence of the Co_3O_4 NPs effectively prevented the agglomeration of the graphene nanosheets, thus promoting the electrolyte diffusion

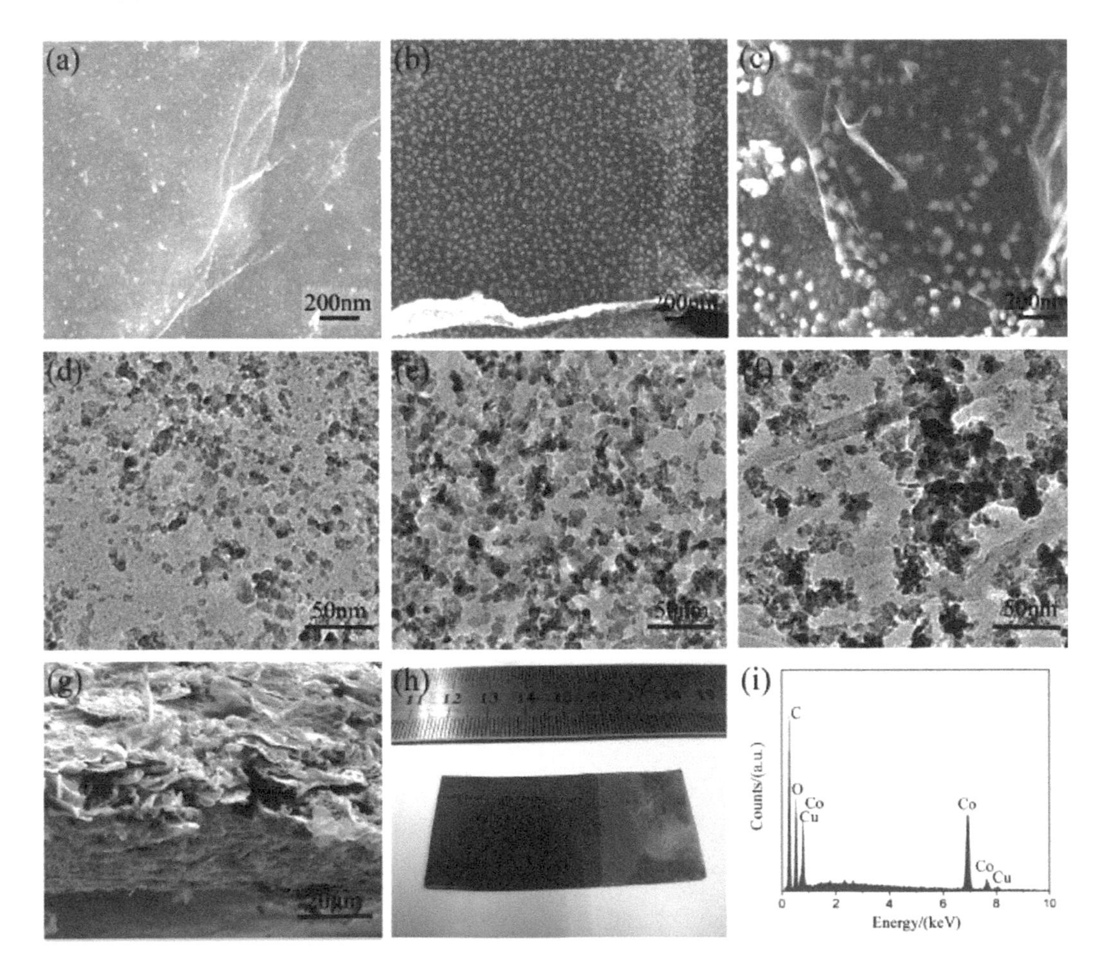

Fig. 6.20. SEM images of the hybrid films: (a) Co_3O_4–graphene-1, (b) Co_3O_4–graphene-2 and (c) Co_3O_4–graphene-3. TEM images of the hybrid films: (d) Co_3O_4–graphene-1, (e) Co_3O_4–graphene-2 and (f) Co_3O_4–graphene-3. (g) Cross sectional SEM image of the Co_3O_4–graphene-2 hybrid films. (h) Photograph of the Co_3O_4–graphene-2 hybrid films. (i) EDS spectrum of the Co_3O_4–graphene-2 hybrid films deposited on copper foil. Reproduced with permission from [28], Copyright © 2015, The Royal Society of Chemistry.

and improving the rate capability. The Co_3O_4–graphene-2 hybrid films were uniformly deposited on the copper foil, as seen in Fig. 6.20 (h). EDS in Fig. 6.20 (i) confirmed the phase composition of the films.

6.2.5. Langmuir-Blodgett (LB) and Langmuir-Schaefer (LS)

Hydrophilic interaction between GO nanosheets and solid substrate has been used as the main driving force to deposit GO or graphene thin films by using dip-coating, which however encountered a problem in precisely controlling the thickness and homogeneity of the graphene films. In this regard, LB assembly has been acknowledged to be an effective technique to produce graphene thin films, by floating a monolayer at the air-water interface due to the amphiphilic nature of GO. By using this method, large-area flat graphene films with controllable thicknesses can be readily fabricated [31, 32].

It has been found that the size of the GO nanosheets played a significant role in determining the self-assembly behavior during the LB assemblies [57–59]. First of all, because the ratio of hydrophobic basal planes and hydrophilic edges is dependent on size, the amphiphilicity of GO nanosheets can be controlled by controlling their size. Smaller nanosheets are more hydrophilic because of the higher charge density due to the ionizable carboxyl groups, which was evidenced by the fact that the zeta potential of GO suspensions was increased with the decreasing lateral size of the GO nanosheets. As a result, spontaneous interfacial size separation of GO nanosheets is observed due to the size-dependent amphiphilicity, i.e., the larger nanosheets float on the water surface, while the smaller nanosheets sink in the suspensions. Also, the properties of the LB monolayer films are influenced by the size of the GO nanosheets. Obviously, the larger the GO nanosheets, the more compact morphology, the fewer structural defects and the higher conductivity will be there in the final films.

There are two fundamental geometries of interacting single layers, i.e., (i) edge-to-edge and (ii) face-to-face. Such interactions could be studied at the air-water interface through the Langmuir-Blodgett (LB) assembly [31]. Due to the strong electrostatic repulsion between the 2D confined layers, stable monolayers of graphite oxide single layers could be formed without the need for any surfactant or stabilizing agent, by using the LB assembly technique. Because of the presence of repulsion, single layers were readily obtained without overlapping during compression. Different from molecular and hard colloidal particle monolayers, the single graphene layers have to fold and wrinkle at the edges to prevent the collapsing of the monolayers into multilayers. The monolayers could be transferred onto various substrates.

The graphite oxide single layer (GOSL) is a hexagonal network of covalently bonded carbon atoms, on which oxygen containing functional groups were attached to various sites, as illustrated in Fig. 6.21 (a, b). It is a special type of soft material, like the 2D membrane of single polymer molecule. The 2D membrane has the thickness of a single atomic layer and a lateral size that could be up to tens of micrometers, so that the GOSLs have a huge aspect ratio and surface area.

Colloidal stability is determined by the interaction between the colloidal particles. Figure 6.21 (f) shows the three classical types of DLVO stability of charged colloidal particles, given by total energy (U) versus particle separation (d) curves. If the electrostatic repulsion is dominant, the colloidal dispersion will be stable, for which the potential energy curve has a high energy barrier against potential flocculation or coagulation. If van der Waals attraction is dominant, the colloids become unstable, so that irreversible coagulation will be present, because the repelling barrier on the total energy curve disappears. If the combination of the repulsion and attraction forces produces a secondary minima, the colloids are stable kinetically. The flocculation, i.e., the state at the secondary minimum, can be reversed by agitation [31].

Due to electrostatic repulsion between the ionized carboxylic and phenol hydroxyl groups, GO can be well-dispersed in water. The charged groups are mainly at the edges of the nanosheets. Therefore, both electrostatic repulsion and van der Waals attraction are present, when two GOSLs approach each other. The total energy (U) is a sum of the two potentials. Due to their high anisotropicity, the total potential energy of two interacting GOSLs is dependent on their geometry. Two interacting geometries between the GOSLs are usually observed, edge-to-edge and face-to-face, as schematically demonstrated in Fig. 6.21 (c–e).

Water surface can be used as an ideal 2D platform to study the interactions of the GOSLs. The interface is geometrically similar to that of GOSLs, while the soft fluidic characteristic allows the GO nanosheets to freely move about, which supports both edge-to-edge and face-to-face interactions. This is the base of

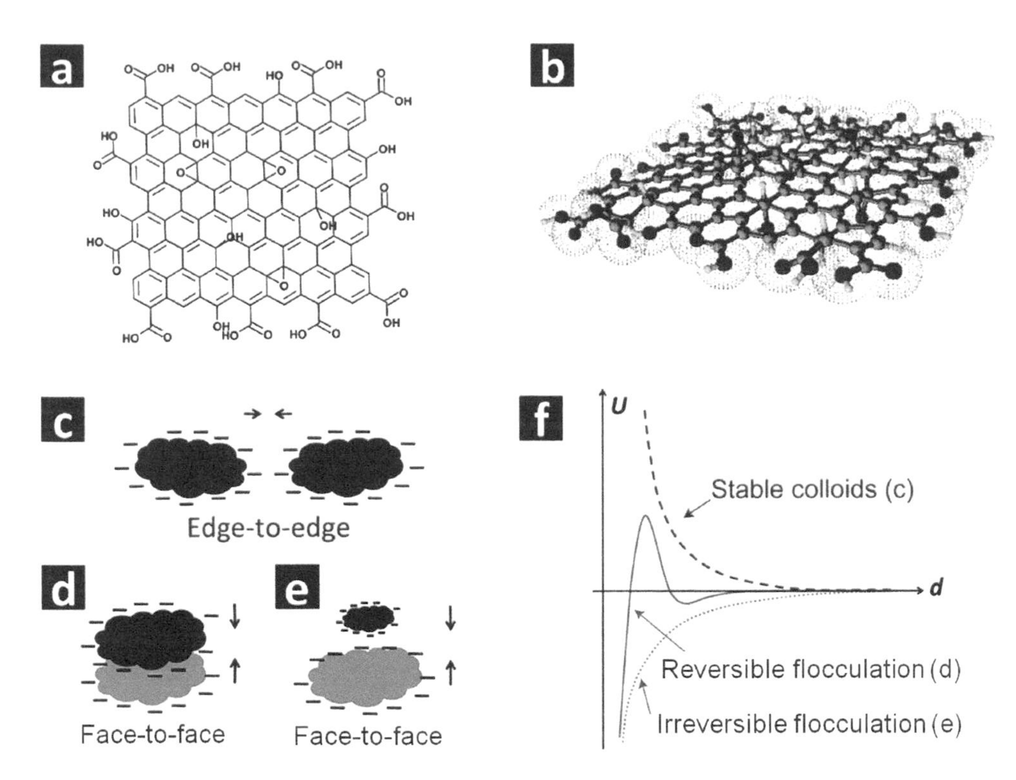

Fig. 6.21. (a) Structural model and (b) 3D view of the GOSL with carboxylic acid groups at the edge and phenol hydroxyl and epoxide groups at the basal plane. Two interacting geometries were observed for two single layers: edge-to-edge (c) and face-to-face (d, e). (f) Three classical types of DLVO colloidal stability, represented by the schematic total potential energy versus separation profiles: (i) dashed red line—strongly repelling colloids, (ii) solid blue line—kinetically stable colloids forming reversible flocculation and (iii) dotted green line—unstable colloid forming coagulation. Reproduced with permission from [31], Copyright © 2009, American Chemical Society.

the Langmuir-Blodgett (LB) assembly of GOSLs on water surface [31]. The assembled GOSL monolayers possessed a very high reversibility during the isothermal compression-expansion cycles. Single-layer GO nanosheets are similar to a cross-linked molecular monolayer, so that they can be collectively manipulated by the moving barrier, when they are floating on the water surface. The GO nanosheets will come together through edge-to-edge interaction by compression. Face-to-face interaction could be facilitated either *in situ* by applying an over compression, so that the GOSLs are forced to slide on top of one another, or *ex situ* through multiple step layer-by-layer dip-coating.

Experimental results indicated that methanol was a good solvent for the LB assembly, because the GO nanosheets could be well-dispersed in it and the suspensions were spread quickly on the water surface. In practice, DI water/methanol mixture with an optimized ratio of 1:5 was used for most LB assembly experiments. Average size of the GO nanosheets was controlled by varying the time of sonication. Two centrifugation steps were employed to further purify the GO nanosheets. The first centrifugation at 8000 rpm for 20 min was to further remove the smaller GO nanosheets and the by-products from the supernatant. The precipitates were collected and redispersed in 1:5 water/methanol mixture. The suspensions were then centrifuged at 2500 rpm for 10 min to remove aggregates and the large GO nanosheets. Finally, suspensions containing well-dispersed GO nanosheets with sizes in the range of 5–20 μm were obtained for LB experiments.

GO suspensions were gently spread onto the water surface, drop by drop with a glass syringe, at a speed of 100 $\mu l \cdot min^{-1}$, with total volumes of 8–12 ml. Surface pressure was monitored by using a tensiometer. GO films with a faint brown color were formed at the end of the compression, which were compressed by barriers at a speed of 20 $cm^2 \cdot min^{-1}$. The GO monolayers were transferred onto substrates at various points during the compression by vertically dipping the substrates into the trough and slowly pulling up at a rate of 2 $mm \cdot min^{-1}$. It was found that substrates with hydrophilic surfaces should be used to collect the GOSLs.

To confirm that the GO nanosheets were lying on the air-water interface rather than beneath the surface, the monolayers were monitored *in situ* by using surface-selective Brewster angle microscopy [60]. The presence of highly reflective shining pieces confirmed that the flat GO nanosheets were at the water surface. The density of the sheets could be well-controlled during the compression-expansion cycles. Figure 6.22 (e) shows isothermal surface pressure as a function of area, indicating that the surface pressure was gradually increased, as the barrier was closed. According to SEM observations, there were four types of GO assemblies. There was gas phase initially, where the surface pressure was constant during the compression, i.e., region "a" in Fig. 6.22 (e). The corresponding monolayer collected at this stage was composed of only isolated individual GO nanosheets, as seen in Fig. 6.22 (a).

With decreasing area, the surface pressure started to increase, while the GO nanosheets were pushed closer and closer. As the pressure entered region b in Fig. 6.22 (e), the GO nanosheets started to "touch" one another. As a result, a close-packed monolayer was formed, covering the entire 2D water surface, as revealed in Fig. 6.22 (b). The increase in surface pressure was attributed to the electrostatic repulsion between the adjacent GO nanosheets. The monolayers transferred onto the silicon wafer had thickness of about 1 nm. As the monolayer was further compressed, the surface pressure was further increased. Figure 6.22 (c) indicated that there was strong interaction between the GO nanosheets. Instead of overlapping, the GO nanosheets tended to fold at the touching points along their edges. Because the single layers were soft and flexible, the increased surface pressure was dissipated by the folding and wrinkling of the edges, so that the interior of nanosheets was flat, without the presence of buckling or wrinkling. According to AFM images, the folds were generally higher than 2 nm. At this stage, the GO coverage over the water surface was largely increased.

As the pressure was further increased, partial overlapping at the edges occurred, resulting in a complete monolayer with interlocked GOSLs, as illustrated in Fig. 6.22 (d). Due to the edge-to-edge interaction, the interior area of the GO nanosheets was always free of wrinkling, until a point at which there was no free space in the monolayer. Since the GO nanosheets were strongly interlocked one another, the monolayer

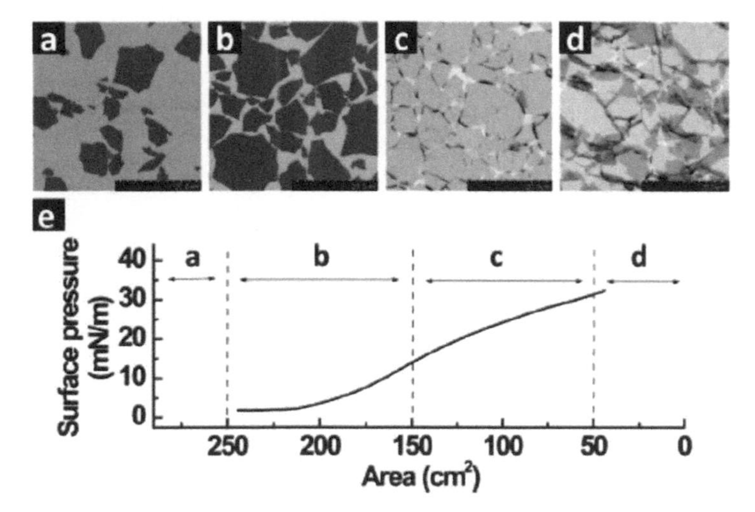

Fig. 6.22. Langmuir-Blodgett assembly of the GO single layers. (a–d) SEM images showing the collected GO monolayers on a silicon wafer in different regions of the isotherm. The packing density was continuously increased: (a) dilute monolayer of isolated flat nanosheets, (b) monolayer of close-packed GO nanosheets, (c) over packed monolayer with nanosheets to be folded at interconnecting edges and (d) over packed monolayer with folded and partially overlapped nanosheets interlocking with one another. (e) Isothermal surface pressure as a function of area, showing the corresponding regions of (a–d). Scale bars = 20 μm (a–d). Reproduced with permission from [31], Copyright © 2009, American Chemical Society.

behaved like a continuous thin film that could be free-standing. If macroscopic wrinkles at millimeter scale were formed, the monolayers would be collapsed, which implied that there was a critical point for the presence of stable monolayers.

According to the total potential energy analysis (Fig. 6.21 (c, f)), the LB films of the GOSLs would not experience flocculation or coagulation, as confirmed by the tiling behavior in Fig. 6.22. The strong edge-to-edge repulsion prevented stacking or overlapping of the GO nanosheets. The 2D GO monolayers were fully reversible after long compression-expansion cycles, as seen in Fig. 6.23. All the folds, wrinkles and partial overlapping completely disappeared, as the films were opened, as illustrated in Fig. 6.23. This observation implied that the face-to-face interaction was not stable, because the folding and overlapping would create partial face-to-face interaction.

Figure 6.23 (a) shows surface pressure curves of three compression-expansion cycles without sample collection. Almost the same shape and final pressure were observed in the three curves. As the cycles were continued, a small shift in the gas-liquid phase transition point was observed towards the smaller area, as shown in Fig. 6.23 (b). This was because a small amount of GO was lost from the monolayer in each cycle. As observed in Fig. 6.23 (c, d), double-layer structures were formed, which were composed of a small GO nanosheet (< 5 μm) on top of a larger one. The small nanosheets were most likely pushed onto the larger ones at high surface pressures. In this case, face-to-face type of interaction was introduced. The electrostatic repulsion between the edges of the two nanosheets would keep them in a complete overlapping state with a concentric-like arrangement, which was further stabilized through the van der Waals and residual π-π

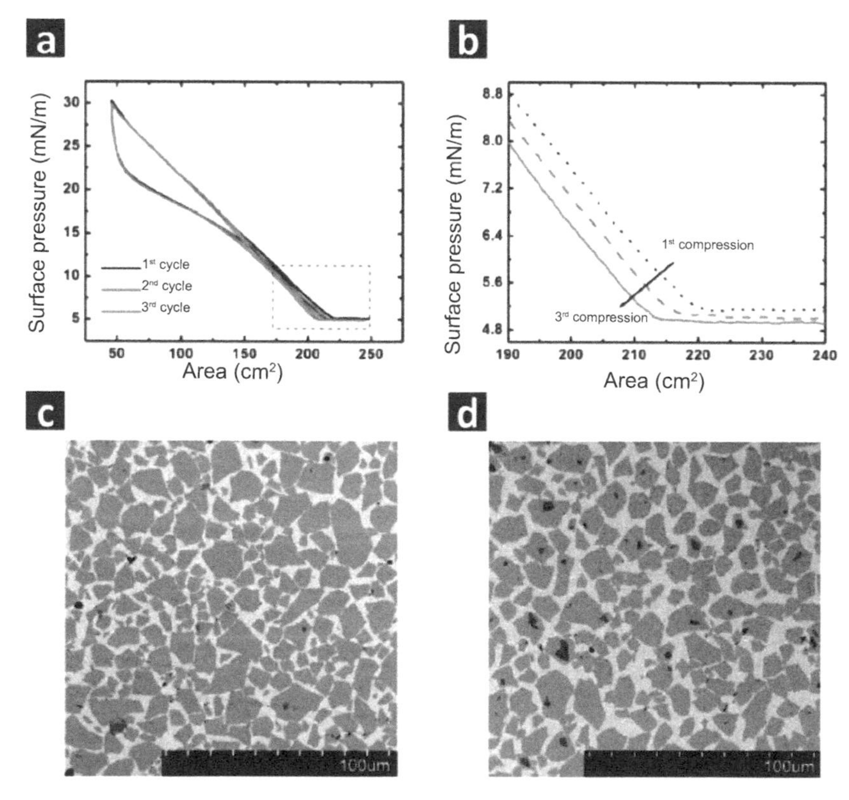

Fig. 6.23. GOSL monolayer with high reversibility and stability against compression. (a) Isotherm plots of three sequential compression-expansion cycles. The three plots essentially overlapped with one another, except in the early stage of compression, as indicated with the dotted-line box. (b) Close-up view of the initial stage of compression, revealing a shift of the plots to the lower area direction, indicating the loss of materials at the air-water interface after isotherm cycles. The SEM images of the monolayers (c) before and (d) after cycling showed that some smaller GO nanosheets were pushed onto larger ones, thus effectively reducing the amount of materials at the air-water interface. It also created double layers of GO nanosheets. Reproduced with permission from [31], Copyright © 2009, American Chemical Society.

Color image of this figure appears in the color plate section at the end of the book.

stacking between their contacting faces. However, double layers of nanosheets with similar size sheets and partially overlapped double layers were not present, after the surface pressure was released, which implied that the face-to-face interaction was favorable.

When the GO nanosheets were prepared during the oxidization of graphite particles, polydisperse in size was derived, which could be used to create double layers or multilayers of GO by using isothermal pressure-area cycling strategy. Double layers could also be obtained through sequential layer-by-layer dip-coating process. To enhance the adhesion to the substrates, the first layers were aged in air and annealed at a high temperature, onto which second layers were deposited to form double GO layers. Because the second layers of the GO nanosheets were repulsed by both the nanosheets in the same layer and those in the first layer, the second layer likely experienced wrinkling, especially when the density was high, as illustrated in Fig. 6.24 (c). It was also found that the second layer had a lower density than the first layer, if the deposition was conducted at the same surface pressure, as seen in Fig. 6.24 (b).

Homogeneous large-area hybrid transparent films, consisting of ultra-large graphene oxide (UL-GO) and functionalized single-walled carbon nanotubes (SWCNTs), were fabricated by using a layer-by-layer (LB) assembly strategy [38]. The GO–SWCNT hybrid thin films exhibited sheet resistances in the range of 180–560 $\Omega\cdot\square^{-1}$ and optical transmittances in the range of 77–86%, depending on the number of hybrid layers. Composition, structure and thickness of the hybrid films could be well-controlled.

Figure 6.25 shows a flow chart of the fabrication process. GO nanosheets with different lateral sizes were derived from natural graphite. The GO powder was diluted in DI water to a concentration of about 1 $mg\cdot ml^{-1}$ with gentle shaking. The as-prepared polydisperse GO nanosheets were grouped into four groups to obtain samples with different lateral sizes, by using three centrifugation runs. The as-prepared GO solution was initially centrifuged at 8000 rpm for 40 min. The supernatant led to small GO (S-GO). The precipitate was collected and dispersed for the second run of centrifugation at 8000 rpm for 25 min. The supernatant of this round resulted in large GO (L-GO). The second round precipitate was dispersed in water again to conduct the third run of centrifugation at 4000 rpm for 25 min, producing very large GO (VL-GO) (supernatant) and ultralarge GO (UL-GO) (precipitate). Representative SEM images and size distribution of the GO nanosheets with different sizes are shown in Fig. 6.26. High purity SWCNTs, with an outer diameter of about 2 nm and length of 5–30 mm, were refluxed using a mixture of sulfuric acid and nitric acid (3:1 volume ratio) for 24 h to activate the surface with oxygen functionalities.

The UL-GO and functionalized SWCNTs were transferred to a 1:5 and 1:3 water/methanol mixture, respectively. The solution was fed at a rate of 100 $ml\cdot min^{-1}$ to a total volume of 5–10 ml. The GO and SWCNT monolayers were stabilized for about 20 min before compression. The monolayers were then compressed by moving barriers at a speed of 10 $mm\cdot min^{-1}$. At the end of the compression, the GO and SWCNT films exhibited faint brown and black color. The monolayers were transferred onto a substrate at various stages of compression by using a dip-coating method, i.e., a quartz substrate was vertically dipped into the trough and pulled out at a speed of 0.1 $mm\cdot min^{-1}$. The films were transferred as the meniscus was

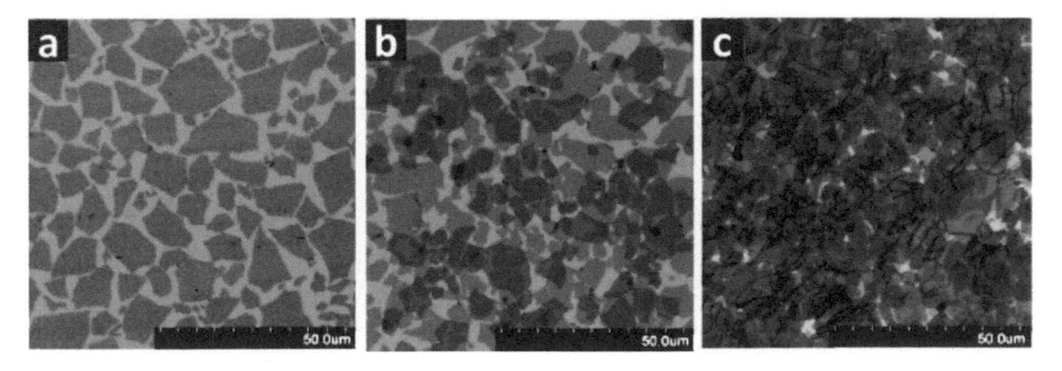

Fig. 6.24. SEM images showing the layer-by-layer assembly of the GO double layers with similar lateral sizes. (a) Close-packed single-layer GO monolayer as the first layer. (b) Double layers with a dilute top layer. (c) Double layers with a high-density top layer. The heavy degree of folding and wrinkling of the second layer in (c) implied the presence of strong repulsion between the two layers. Reproduced with permission from [31], Copyright © 2009, American Chemical Society.

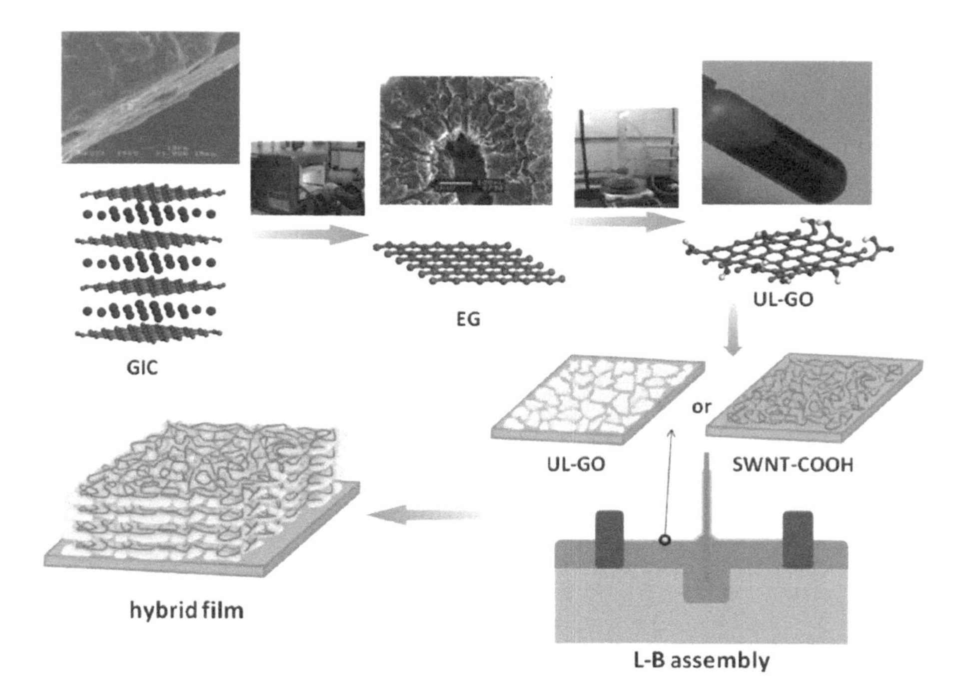

Fig. 6.25. Flow chart for the synthesis of UL-GO–SWCNT hybrid films. Reproduced with permission from [38], Copyright © 2012, The Royal Society of Chemistry.

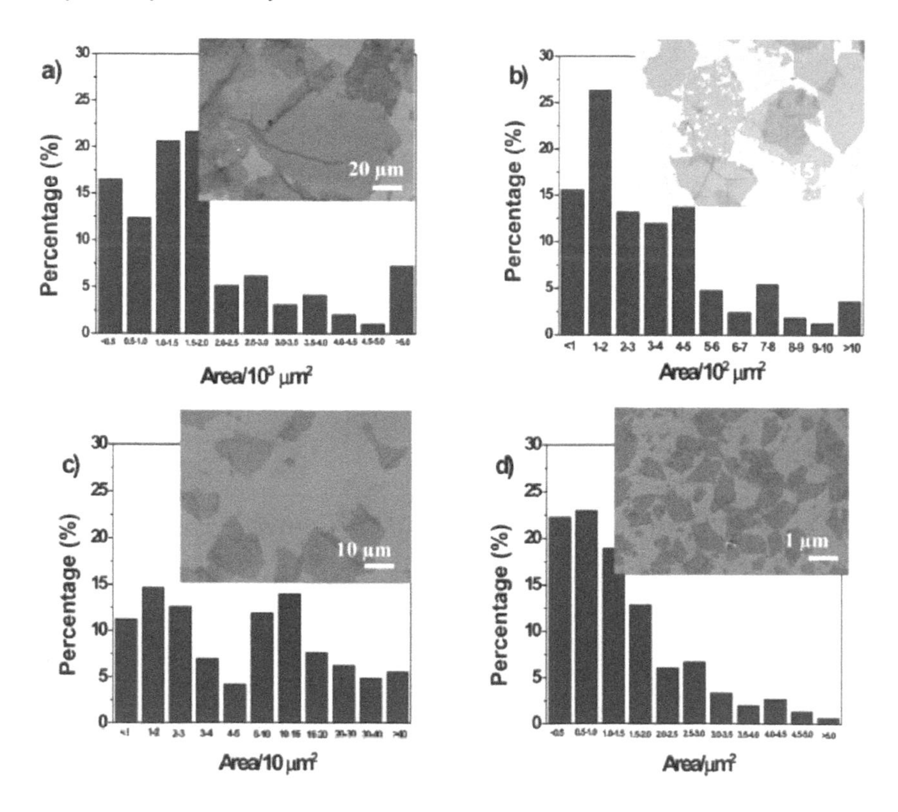

Fig. 6.26. Representative SEM images and size distributions of the UL-GO (a), VL-GO (b), L-GO (c) and S-GO (d). Reproduced with permission from [38], Copyright © 2012, The Royal Society of Chemistry.

spread on the substrate during the pulling out step. Similar to the statement above, the substrates should have a hydrophilic surface. To obtain hybrid transparent conductive films (TCFs) with uniform coverage and low optical scattering loss, the GO and SWCNT films were transferred onto the substrate by sequential layer-by-layer deposition method. After the deposition of each monolayer, the substrate was either dried in air overnight or baked in an oven at 80°C for 1 h to have stable adhesion between the conducting layers and the substrate.

The substrates with the deposited films were then thermally treated in controlled vacuum and gas flow. A vacuum pressure of 10^{-5} Torr was reached before the furnace was heated, while a continuous flow of ultra-pure Ar was maintained at a rate of 10°C·min^{-1}. The thermal annealing was conducted at 400°C for 1.5 h at 10^{-3} Torr. After that, the temperature was increased to 1100°C at a rate of 10°C·min^{-1} and kept for 0.5 h, so that rGO–SWCNT hybrid films were prepared.

UL-GO monolayers were deposited on the quartz substrate through a layer-by-layer step. The density and degree of wrinkling of the GO monolayers were accurately controlled by adjusting the LB processing parameters. Close-packed flat GO monolayers were produced at a surface pressure of about 10 mN·m^{-1}. The dispersion containing SWCNTs functionalized with carboxylic (COOH) groups was used to deposit the SWCNT monolayers.

Figure 6.27 shows schematic diagrams and microstructures of the GO–SWCNT hybrid films collected at different stages of the deposition process. As shown in Fig. 6.27 (b), partially wrinkled edges were formed, as the soft GO nanosheets were squeezed one another, during the pulling out of the substrate from the suspensions. Figure 6.27 (c) shows TEM image and the corresponding SAED pattern, indicating that the GO monolayer was of a predominantly amorphous nature. After the SWCNT layer was deposited on

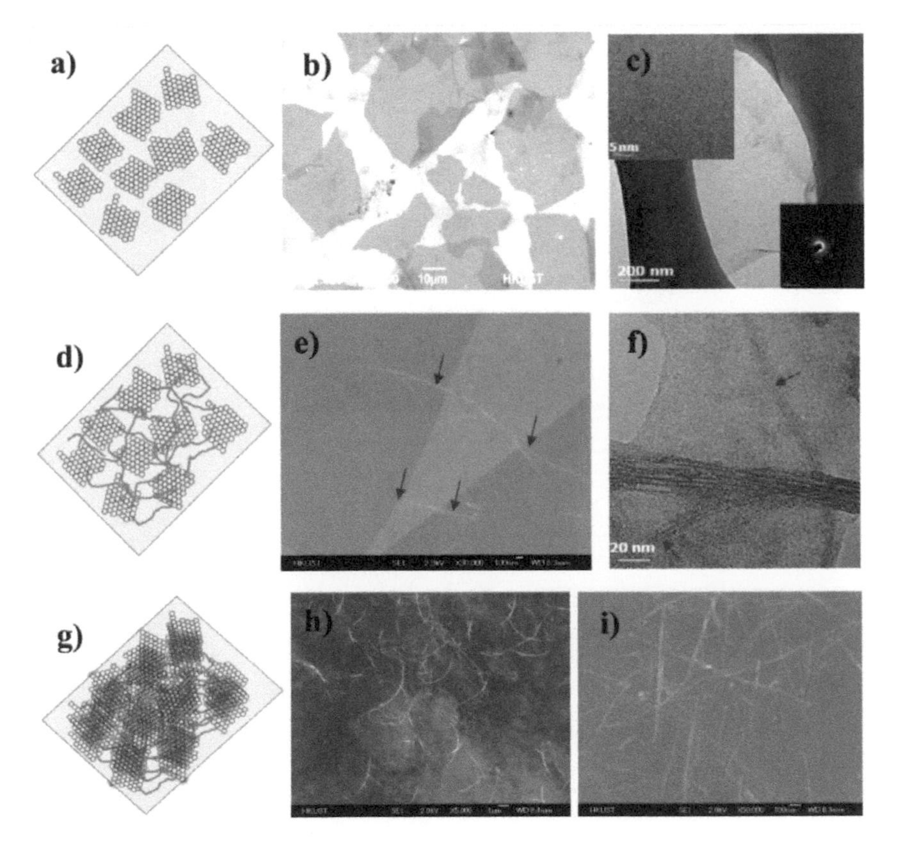

Fig. 6.27. Schematic diagrams and microstructures of the GO and GO–SWCNT hybrid films: (a) schematic, (b) SEM and (c) TEM images of a monolayer of GO, (d) schematic, (e) SEM and (f) TEM images of one bilayer of the GO–SWCNT hybrid film, with SWCNTs bridging the GO nanosheets, (g) schematic and (h, i) SEM images of the four bilayers of the GO–SWCNT hybrid film. Reproduced with permission from [38], Copyright © 2012, The Royal Society of Chemistry.

top of the GO layer, strong bonds were formed though *p–p* interactions and van der Waals forces, due to the amphiphilic nature of the surfaces of both the GO and SWCNT [33]. SEM images of the central regions of the GO–SWCNT hybrid films are shown in Fig. 6.27 (e, f), revealing that the SWCNTs served as conductive bridges to link the GO nanosheets. AFM observation confirmed that there were no aggregates in the SWCNT monolayer. The average AFM height of one bilayer GO–SWCNT hybrid film was only about 2 nm. The typical surface morphologies of 4 bilayers of the GO–SWCNT thin films are shown in Fig. 6.27 (h, i).

Surface roughness of the LB hybrid films was slightly increased with the increasing number of the GO–SWCNT layers. Small fluctuation in surface roughness was observed, which was caused by the deposition of the SWCNT layers. The thermal reduction usually led to decrease in the surface roughness with less fluctuation, which was closely related to the removal of the oxygenated functional groups and graphitization caused by the thermal annealing. For example, after thermal treatment, the 30 nm thick hybrid films had a surface roughness of about 10 nm, which was much smaller than those of the films fabricated by using other methods [61].

The Langmuir-Schaefer or LS method has been used to deposit single-layer graphene on arbitrary substrates in a controllable manner [39]. The deposition method combined with post-treatment is shown in Fig. 6.28. Water suspension of GO with a concentration of 5 ppm was used as a subphase for the LS deposition. The injection of the long-chain molecule octadecylamine (ODA) (Fig. 6.28 (g)) at the air–water interface facilitated the hybridization of the GO nanosheets by covalent bonding through functionalization with the amide functionality, as illustrated in Fig. 6.28 (b), thus leading to the formation of a mixed layer of ODA-GO. If an external pressure is applied to this hybrid Langmuir film through the movable barrier of a Langmuir–Blodgett (LB) apparatus, then a LB membrane will be obtained, as discussed above. It

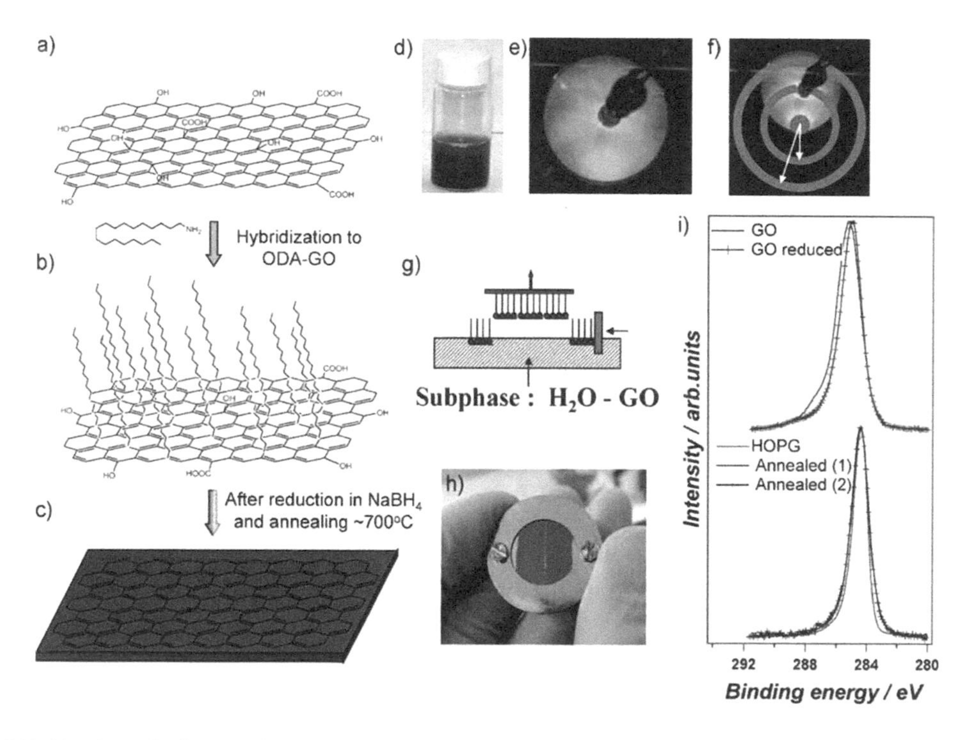

Fig. 6.28. (a) Schematic diagram of the GO nanosheet. (b) ODA-functionalized GO nanosheet. (c) GS on the substrate after reduction and annealing. (d) Photograph of a solution of the GO dispersed in water. (e) LEED pattern of GSs on the substrate after reduction and annealing. (f) The same with indication of the ring structure. (g) Schematic diagram of the deposition procedure. (h) Optical photography of the n^{++}-doped Si wafer with SiO_2 before (left) and after (right) the deposition of the GO nanosheets. (i) XPS curves of C1s core level region of the GO and rGO (top panel), annealed GS (bottom panel, curve 1), annealed GS after C_2H_4 treatment (bottom panel, curve 2) and HOPG (bottom panel, red curve). Reproduced with permission from [39], Copyright © 2010, The Royal Society of Chemistry.

the hybrid Langmuir film is transferred to a substrate by horizontally lowering the substrate to contact the ODA-GO–water interface, the transferring process is called the Langmuir–Schaefer (LS) method, as demonstrated schematically in Fig. 6.28 (g).

After the ODA-GO was transferred onto the desired substrates, chemical and thermal treatments were applied to convert the GO to rGO, by eliminating the majority of the functional groups, as confirmed by the XPS spectra in Fig. 6.28 (i, top panel). Electrical conductivity of the chemically reduced rGO was further increased by using post-thermal annealing at 700°C for 1 h in vacuum with the presence of C_2H_4. The deposited films exhibited high homogeneity, as seen in Fig. 6.28 (h). The substrate was n^{++}-doped Si wafer with a 300 nm SiO_2. Due to white-light interference between the dielectric layer and the deposited film, a clear contrast between the purple-violet coloration of the SiO_2 layer and the deposited GO layer was observed, corresponding to a green hue.

SEM images of the ODA-GO films transferred on Au are shown in Fig. 6.29 (a–c), demonstrating the dependence of coverage on surface pressures. Obviously, the graphene films had a high homogeneity over a relatively large area. Without pressure, the deposition resulted in a low coverage and well-isolated GSs (Fig. 6.28 (a)). The packing density was gradually increased with the increasing surface pressure, as illustrated in Fig. 6.28 (b, c). Although the coverage was increased, the substrate was covered by single layers, with very few overlaps between the nanosheets. Figure 6.28 (d) shows the lateral dimensional size distribution of the GS, which was in the range of 2.5–150 μm², with an average value of 30 μm² and a most probable value of 10 μm².

Another simple method was developed to generate graphene thin films by using the LS approach without the requirement of post-processing treatment [40]. To have a successful LS process, a mixed solvent was used, which ensured an efficient liquid phase exfoliation of graphene. At the same time, the ratio of the mixed solvent could be modified to produce different vapor pressures, which could be optimized to fabricate thin films with the desired properties. Under optimized conditions, the dispersions would facilitate self-assembly of the graphene nanosheets at the air–water interface, which were subsequently transferred onto solid substrates by using the LS technique. In this case, graphene films with large areas and controllable density could be readily obtained. In addition, the thickness of the films could be simply

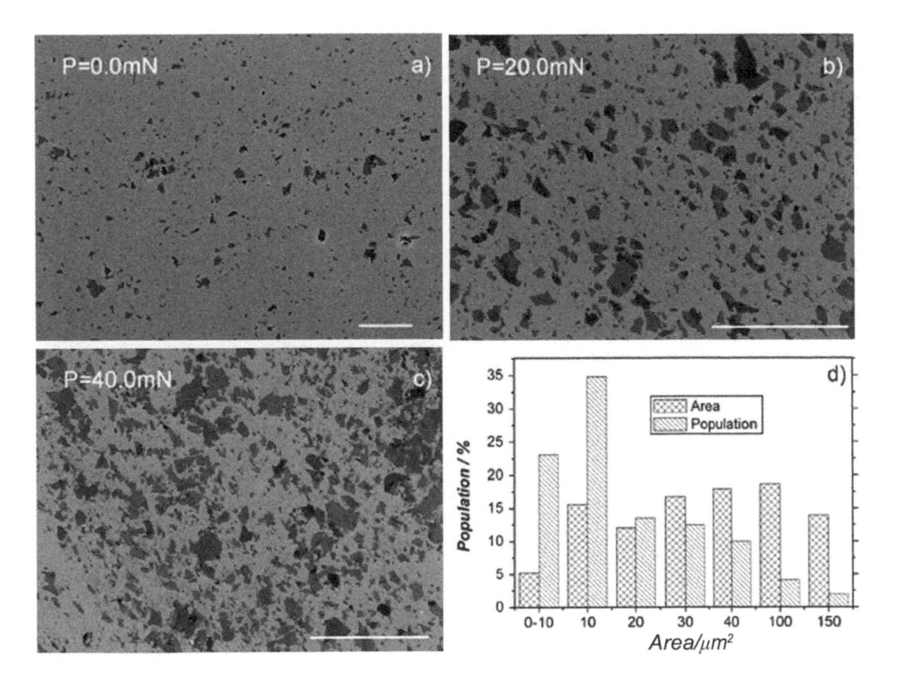

Fig. 6.29. SEM images of GSs obtained through deposition on Au/Si at various applied surface pressures: (a) 0 mN·m⁻¹, (b) 20 mN·m⁻¹ and (c) 40 mN·m⁻¹ (scale bar = 40 μm). (d) Distribution of the sizes of GSs deduced from the SEM images. Reproduced with permission from [39], Copyright © 2010, The Royal Society of Chemistry.

controlled by controlling the number of layers through repeated deposition. By doing this, it was possible to precisely tailor the electrical conductivity of the graphene thin films. These LS derived graphene thin films had potential applications in thin film flexible electronics, optoelectronics and sensors.

Because water has a high surface tension at room temperature, when the suspension of graphene dispersed in water-insoluble solvent is dropped on it, the droplets will quickly spread over the available area. As the solvent is evaporated, the graphene nanosheets will self-assemble into 2D thin films. If the number of graphene nanosheets per unit area is too low, it means that the adjacent graphene nanosheets were too far away, so that the interactions were weak. In this case, the monolayer is not continuous, which is called 2D gas. Because the graphene nanosheets have one interface with the aqueous phase and the other with air, they have almost no effect on the surface tension of water. If the available surface area of the graphene monolayer is reduced by sliding a barrier across the water surface, the graphene nanosheets will start to exert a repulsive force on each other, which results in a reduction in the surface tension. Due to the compression, the monolayer transients from the gas phase to the liquid phase and finally to the solid phase before collapse takes place. Such a process can be represented by the surface pressure-surface area (density) (Π-A) isotherm. Figure 6.30 shows a Langmuir compression isotherm for the graphene monolayer, with the phase evolution at a collapse pressure of above 39 $mN \cdot m^{-1}$.

The compressed graphene monolayer behaved like a 2D solid film, which was sufficiently strong and rigid so as not to collapse. Figure 6.31 shows photographs taken perpendicular to the optical axis, which indicated that the packing density of the graphene nanosheets could be manipulated by varying the surface pressure during the compression/expansion cycle. At $\Pi = 0$ $mN \cdot m^{-1}$, the graphene nanosheets were separated from one another, leading to the presence of monolayer islands with various sizes. The lateral dimension of the islands had a surface of a few μm^2. However, the suspension volume was too low (e.g., < 5 ml), and the graphene monolayer was not visible on the water subphase, as a result, it was difficult to observe optically, until large surface pressures were applied. By compressing the barriers, the monolayer was densified into multiple discrete islands which were connected, so that free space was decreased. As the pressure was increased to $\Pi = 40$ $mN \cdot m^{-1}$, the monolayer islands were connected entirely into a continuous closely packed monolayer.

There is a critical point for the interfacial area, at which the graphene nanosheets start to resist continuous packing into higher densities, because of the hard limit determined by the molecular cross-sectional area. Further compression after this critical point will damage the stability of the interface, which resulted in film collapse. As a consequence, the graphene nanosheets will be pushed from the interface into the film bulk. In worst case, an out-of-plane geometry will be formed.

As the surface pressure was ≤ 25 $mN \cdot m^{-1}$, the density of the monolayer could be tuned reversibly. Figure 6.31 (b) shows Π-A isotherms of two sequential compression-expansion cycles over 0–15 $mN \cdot m^{-1}$.

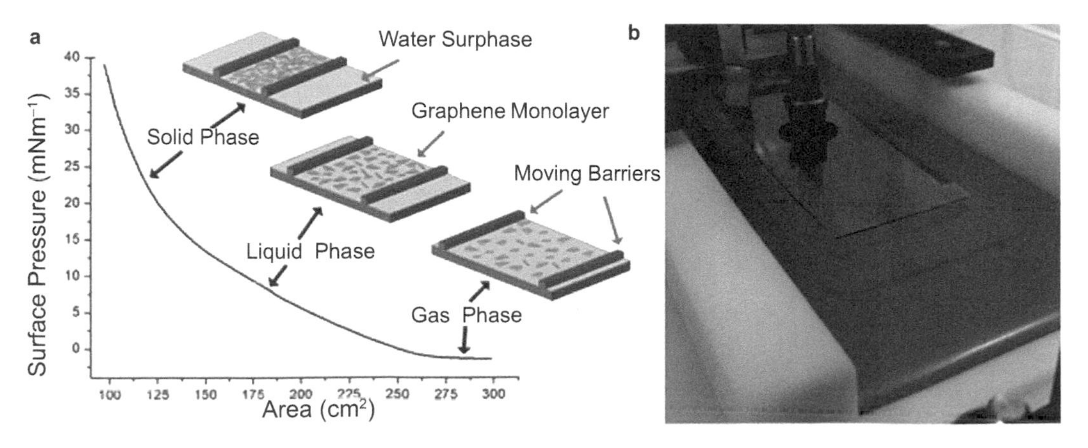

Fig. 6.30. (a) Representative surface pressure-area (P-A) isotherm of graphene monolayer deposited from chloroform/NMP mixture at the air–water interface. (b) A corresponding photograph of a Langmuir trough showing a dense graphene monolayer with a glass substrate above the surface ready for a Langmuir–Schaefer deposition. Reproduced with permission from [40], Copyright © 2013, Elsevier.

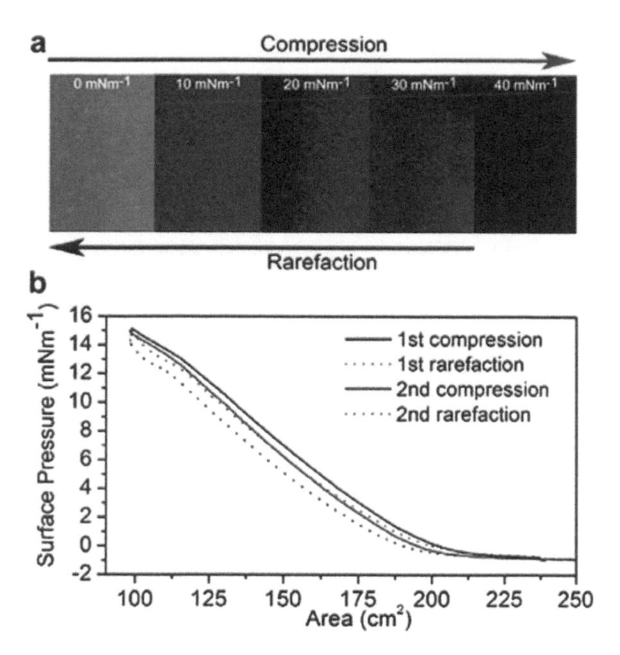

Fig. 6.31. (a) Optical photographs of the graphene monolayer films at the air/water interface deposited from chloroform/NMP mixture at different surface pressures taken perpendicular to the optical axis. (b) Π-A isotherms of two sequential compression-expansion cycles. Reproduced with permission from [40], Copyright © 2013, Elsevier.

Color image of this figure appears in the color plate section at the end of the book.

A weak hysteresis was observed in the first cycle of compression–rarefaction, which disappeared during the following recompression. The monolayers consisted of graphene nanosheets with various thicknesses in the range from single-layered to multilayered configurations. Because of that, overlapping was effectively prevented, so that the density of the monolayers became reversible. In addition, the presence of a small amount of NMP also could serve as a "molecular lubricant" to promote the spreading.

As the graphene monolayer was compressed to a desired density, the substrate was lowered onto the interface to contact the monolayer, after which the substrate was retracted. As a result, the graphene monolayer was transferred onto the substrate. As compared with LB, LS is more suitable to transfer highly rigid monolayers, because the floating monolayers experienced less disruptive forces. By following the Π-A isotherm, monolayers surface deposited on glass substrates at pressures in the range of 0–40 $mN \cdot m^{-1}$ are shown in Fig. 6.32. For instance, the film deposited from the gas phase (at $\Pi = 1$ $mN \cdot m^{-1}$) had a very low density with the presence of large voids. In contrast, the maximum surface coverage was obtained at 40 $mN \cdot m^{-1}$.

Possible mechanisms for the phenomena of GO wrinkling, in the system of octadecylamine (ODA)-GO Langmuir monolayers, processed with LS method, have been studied, as schematically illustrated in Fig. 6.33 [41]. There are two possible mechanisms that can be used to explain the GO wrinkling phenomena. Firstly, due to the formation of the hybrid, the ODA monolayer could have modified the flexibility of the assembled GO nanosheets. The ODA surfactant layers would occupy the areas without the GO nanosheets (colored in purple), so as to alter the flexibility of the GO film (Fig. 6.33 (a)). Once a compression was applied, the ODA molecules could entirely cover the whole area of the film, as illustrated in Fig. 6.33 (b). As a result, the hybrid GO–ODA layer behaved as a single-layer film, which was confirmed by AFM results, because the wrinkles were longer than the size of the GO nanosheets. Due to the increased flexibility, the hybrid layer tended to form wrinkles, in contrast to the individual flakes of a rigid small GO nanosheet.

Secondly, the ODA surfactant molecules on the GO nanosheets (colored in green as shown in Fig. 6.33 (b)) prevented the GO nanosheets from sliding one another during the compression. There were also surfactant molecules to be attached at the edges of the GO nanosheets, through the interaction between the positively charged surfactants and the negatively charged edges of the GO nanosheets. In this case, the

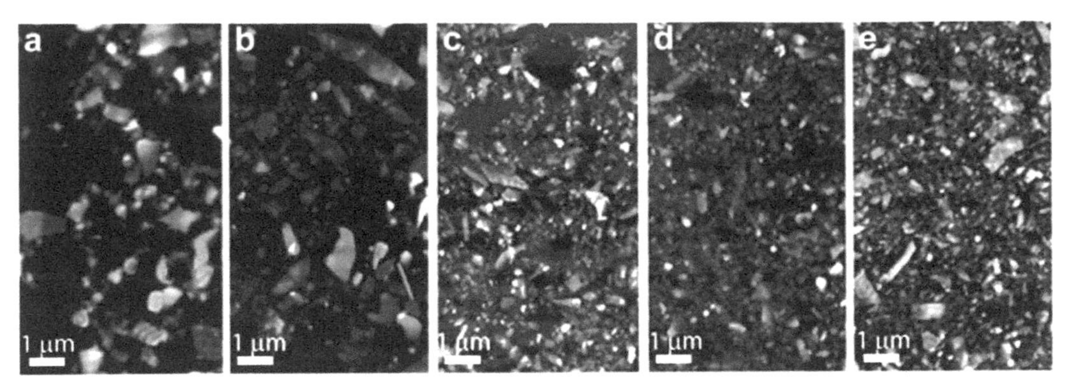

Fig. 6.32. AFM images of the graphene nanosheets deposited by using the Langmuir–Schaefer technique from NMP/chloroform mixture at different surface pressures: (a) 1, (b) 10, (c) 20, (d) 30 and (e) 40 $mN \cdot m^{-1}$. Reproduced with permission from [40], Copyright © 2013, Elsevier.

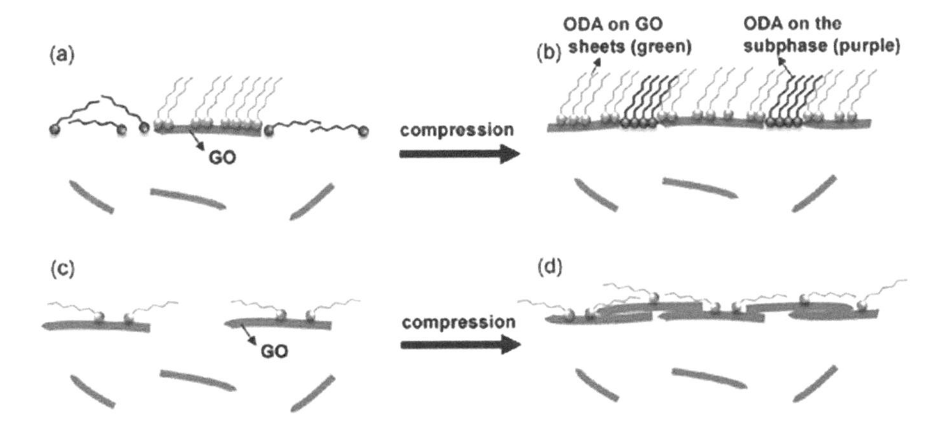

Fig. 6.33. Schematic diagrams illustrating the possible mechanisms during the formation of the GO–ODA hybrid films. (a) The number of ODA monolayers is sufficiently larger to spread at the air–GO suspension interface, so that the ODA molecules (colored in green) could be adsorbed on surface of the GO nanosheets, while the nonbound ODA molecules (colored in purple) would occupy the area without the GO nanosheets. (b) After compression, a GO monolayer was formed by the GO nanosheets with close-packing. The nonbound ODA molecules are also densely packed at the air–GO suspension interface. The GO–ODA hybrid monolayer exhibited no overlapping, because the adsorbed ODA molecules suppressed the sliding of the GO nanosheets during the compression. (c) If the amount of ODA is too low at the air–GO suspension interface, the ODA molecules (colored in green) would be only adsorbed on surface of the GO nanosheets. (d) Upon compression, the GO nanosheets could easily overlap one another at the interface. Reproduced with permission from [41], Copyright © 2014, American Chemical Society.

GO nanosheets would not slide due to the presence of the surfactant molecules on their surface, so that wrinkles were formed, while overlaps were prevented. This hypothesis has been supported by study, by changing the amount of ODA molecules. If the amount of ODA was sufficiently low, overlapping of the GO nanosheets was observed before the formation of wrinkling at a pressure of 25 $mN \cdot m^{-1}$, as shown in Fig. 6.33 (d). Comparatively, at the same pressure, when the amount of the ODA was sufficiently high, GO monolayer was formed without overlapping, which was evidenced by AFM examination. At low concentrations, the ODA molecules were only attached to the surface of the GO nanosheets, as seen in Fig. 6.33 (c). Therefore, to prevent the formation of sliding and overlapping of the GO nanosheets, the concentration of the ODA should be sufficiently high.

Graphene oxide films have been obtained by using both the BL and LS methods, which provided a comparison study on the two methods [42]. Figure 6.34 shows a schematic diagram of the two methods. As stated earlier, LB is based on the transfer process of Langmuir films from the air–water interface onto solids by vertical dipping of the substrate in the Langmuir monolayer, while LS involves horizontal

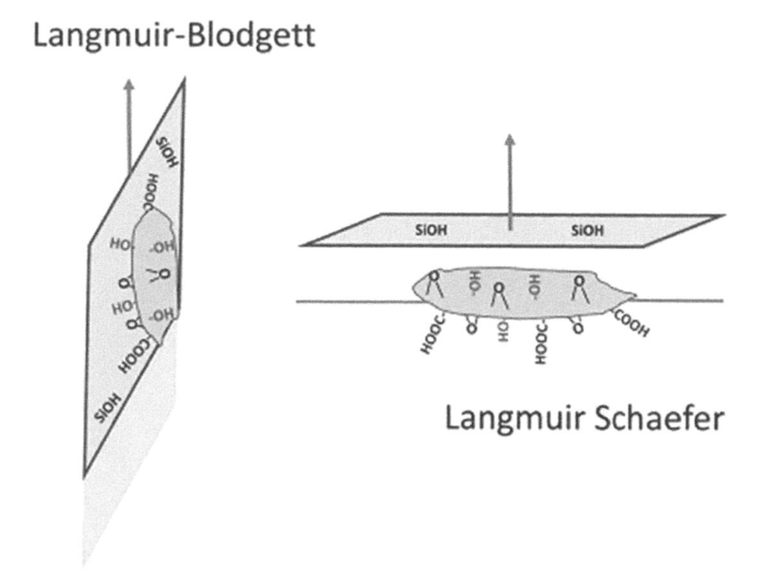

Fig. 6.34. Schematic diagram illustrating the functional groups involved during the contact between the solid substrate (Si/SiO_2) and the GO nanosheets in LB and LS deposition methods. Reproduced with permission from [42], Copyright © 2015, American Chemical Society.

dipping. Four types of samples were used in the study, including GO, purified GO (PGO), nanofiber GO (NGO) and purified nanofiber GO (PNGO).

Figure 6.35 shows SEM images of the LB films of different graphene oxide nanosheets. Selected high magnification images of some regions are presented as insets in the corresponding figures. The results indicated that the films from the GOs without purification had a relatively higher solid coverage. At the same time, the nanoplatelets of the nonpurified nanocarbon items were larger than those of the purified ones, while the nanoplatelets corresponding to the GO synthesized through the oxidation of fibrous nanocarbons were the smallest in size, which were consistent with the results of dynamic light scattering (DLS) measurements.

Figure 6.36 shows SEM images of GO films obtained by using the LS methodology. Their morphologies are quite similar to those of the LB films. It was found that the LB films exhibited a higher coverage than the LS ones. In addition, the solid coverage reached by the LB methodology was almost independent of the percentage of aromatic carbon, while it decreased as the Csp^2 percentage was increased for the films prepared by the LS technique.

Solid coverage of the LB films was slightly increased as the content of the C–O groups attached onto the nanosheets was increased, while it was increased with the level of the C–O groups more quickly for the films fabricated by using the LS method. Such a difference was readily attributed to the different orientation of the substrate with respect to the monolayer of the two deposition methods. For LS deposition, the solid substrate was horizontally dipped into the air–water interface, which was in entire contact with the monolayer. In this case, the solid substrate and the graphene nanoplatelets were linked through the epoxy and hydroxyl groups attached to the GO nanosheets and the silanol groups on the Si wafer. As a consequence, as the content of the C–O groups was increased, the interactions between the substrate and GO nanosheets were increased, thus leading to enhanced adhesion of the GO layers. However, for LB deposition, the solid was vertically dipped into the interface, so that the contact between the silanol groups on the solid substrate and the GO layer was through the O-groups on the GO nanosheets. Therefore, the substrate coverage was less dependent on the content of the C–O groups, as demonstrated in the schematic diagram of Fig. 6.34.

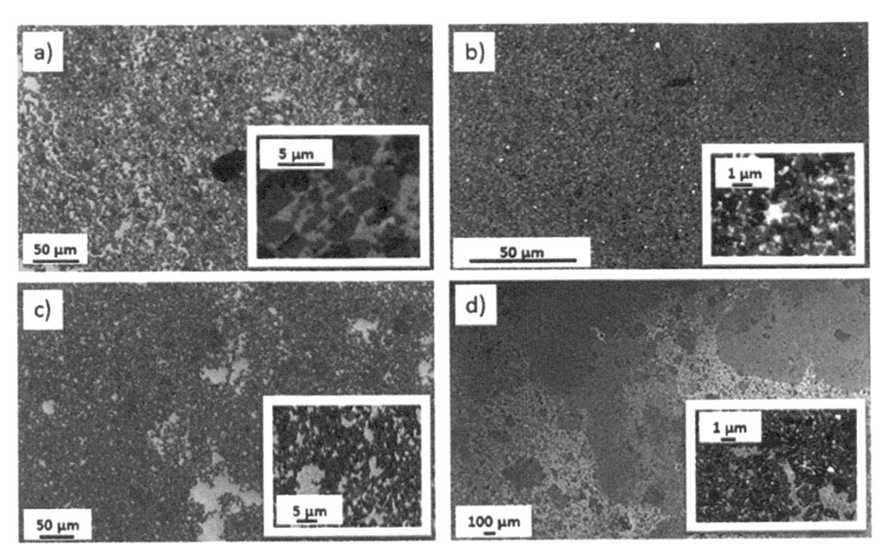

Fig. 6.35. SEM images of the LB films: (a) GO, (b) NGO, (c) PGO and (d) PNGO, which were fabricated at $\pi = 5$ mN·m^{-1}. Reproduced with permission from [42], Copyright © 2015, American Chemical Society.

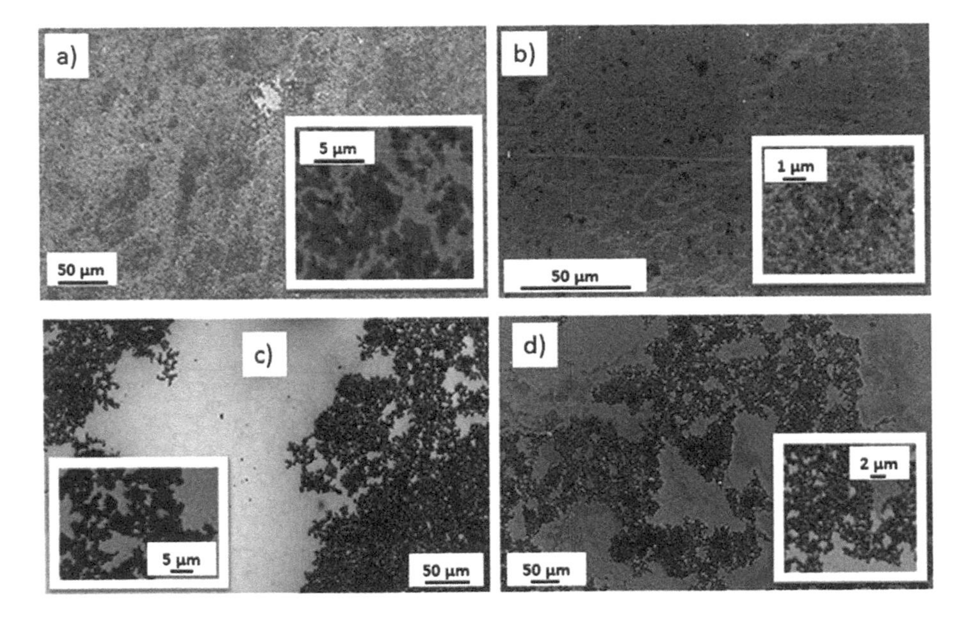

Fig. 6.36. SEM images of the LS films: (a) GO, (b) NGO, (c) PGO and (d) PNGO, which were obtained at $\pi = 5$ mN·m^{-1}. Reproduced with permission from [42], Copyright © 2015, American Chemical Society.

6.2.6. Layer-by-layer (LbL)

LbL self-assembly has been acknowledged to be a versatile strategy to prepare multilayer films, with enwrapped NPs of other nanomaterials. More recently, this technology has been employed to fabricate graphene-based hybrid thin films on flat substrates, through either ionic interactions or hydrogen bonding. In this technique, through the stepwise deposition of oppositely charged components containing GO/rGO nanosheets, together with other functional items, e.g., polymers, inorganic NPs, large area films, with multilayer graphene-based films containing alternative stacking of the components, could be developed [44–51]. The following part will be used to present several examples of LbL self-assembled graphene-based films.

It has been found that carboxylic acid-functionalized graphene nanosheets can be complexed with positively charged polyelectrolytes or nanoparticles [62–65]. Therefore, in order to make graphene nanosheets positively charged, one way is to attach a positively charged dispersing agent onto them, which together with the negatively charged polyelectrolytes can be used for the LbL assembly processing, as discussed later. The other way is attach positively charged amine functionality onto the surface of the GO nanosheets. In most cases, positively charged graphene nanosheets could only be dispersed in organic solvents, whereas their dispersion in aqueous condition has been a challenge [66–68].

There has been a report to synthesize water-dispersible positively charged graphene nanosheets through sequential functionalization processes, involving acyl-chlorination with thionyl chloride and amidation with ethylenediamine [44]. Since thionyl chloride could react strongly with oxygen, the epoxide groups on the basal plane of the GO nanosheets would be dissociated by the thionyl chloride during the acyl-chlorination, thus leading to the formation of surface defects on the GO nanosheets. Such structural defects could in turn prevent the aggregation of the nanosheets during the sample drying process. The positively and negatively charged GO nanosheets ensured the success of the LbL assembly through the electrostatic interactions, as schematically demonstrated in Fig. 6.37. The LbL assembled GO films were reduced to graphene films by using thermal annealing.

To prepare carboxylic acid-functionalized graphene dispersion (about 0.05 wt%), GO powder was mixed with hydrazine (35 wt%) at a ratio of 7:10, and was then heated at 80°C for 1 h to facilitate the chemical reduction. Meanwhile, amine-functionalized graphene dispersion was obtained by mixing expanded GO (not exfoliated) with acyl-chlorination, together with an excessive amount of thionyl chloride ($SOCl_2$), which was reacted with refluxing at 70°C for 12 h. After the reaction, the remaining thionyl chloride was removed by drying. For amidation, ethylenediamine was added to the acyl-chlorinated graphene suspension, which was reacted in anhydrous pyridine solution at 80°C for 1 day. The reaction products were thoroughly washed with methanol and DI water, followed by exfoliation with the aid of sonication for 30 min. The unexfoliated amine-functionalized items were removed by using centrifugation. The amine-functionalized graphene nanosheets were re-dispersed in DI water at a concentration of 0.05 wt%.

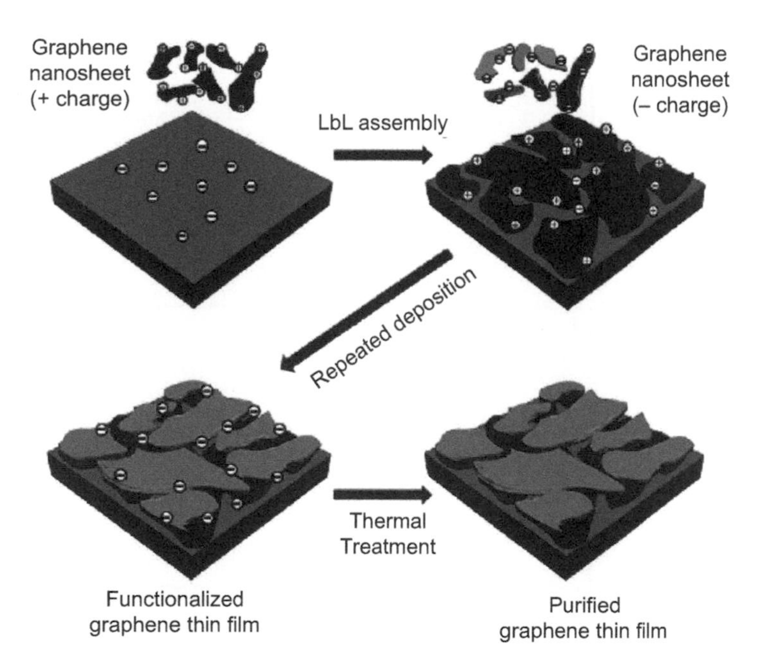

Fig. 6.37. Schematic diagram showing the fabrication process of the graphene thin films through the electrostatic LbL assembly with oppositely charged graphene nanosheets and subsequent thermal treatment. Reproduced with permission from [44], Copyright © 2011, American Chemical Society.

Before the LbL assembly deposition, quartz substrates were thoroughly cleaned with sonication in DI water and ethanol. The cleaned substrates were then plasma-treated for 1 minute to produce negatively charged surfaces. The LbL experiment was conducted by using a programmable slide stainer, with a deposition condition of eight minutes adsorption of charged graphene nanosheets, followed by three sequential washing steps in DI water bath. Considering the negatively charged surface of the substrate, positively charged graphene was first deposited, after which negatively charged graphene was coated. After that, the two layers were deposited alternatively according to the designed thicknesses, as illustrated in Fig. 6.37. The multilayered films were denoted as GS_X, with X referring to the layered pairs of graphene. Thermal reduction of the assembled graphene films were carried out in Ar at 300–900°C.

Figure 6.38 shows properties of the LbL assembled graphene thin films [44]. Figure 6.38 (A) indicated that the stepwise increment of the bilayer deposition was 3–4 nm. Note that the thickness of a single layer of graphene is about 1 nm, the thickness of a bilayer is about 2 nm. The presence of slightly thicker films

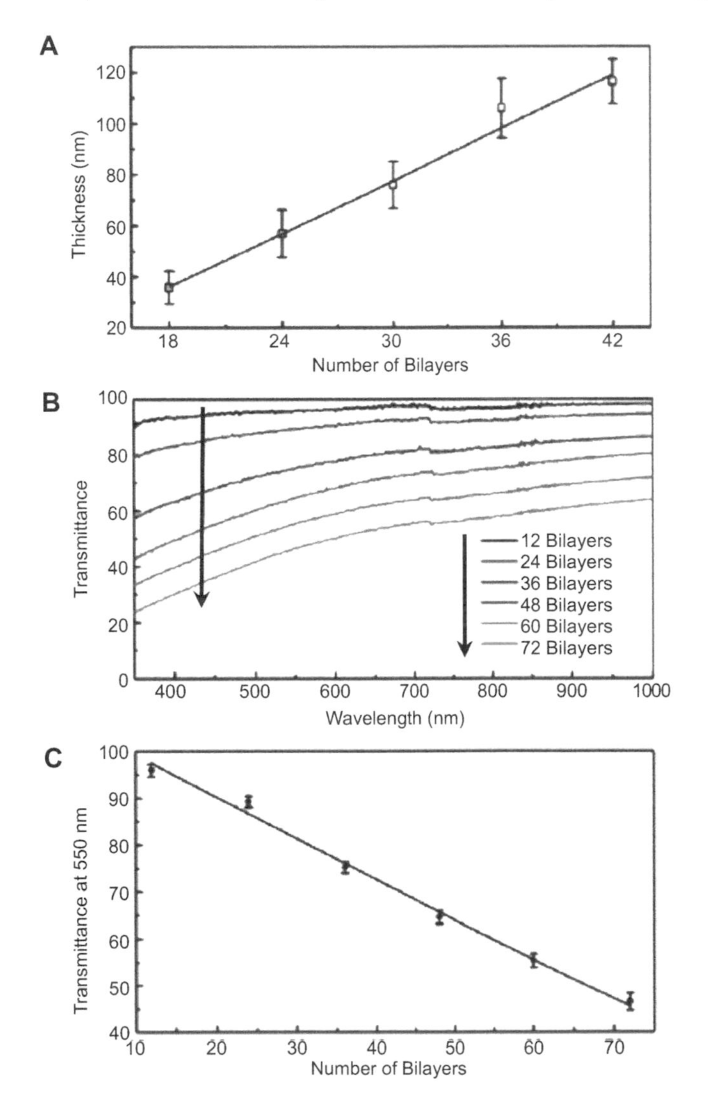

Fig. 6.38. Properties of the LbL-assembled graphene films. (A) Thicknesses of the graphene thin films as a function of the number of LbL depositions. The solid line is a linear fit and the error bars indicate the standard deviations. (B) Transparency spectra of the graphene assembled thin films. (C) Light transmittance of the graphene assembled thin films at a wavelength of 550 nm. The solid line is a linear fit and the error bars indicate the standard deviations. Reproduced with permission from [44], Copyright © 2011, American Chemical Society.

could be attributed to the functionalized graphene nanosheets. For example, in the amine-functionalized graphene nanosheets, structural deformations and disintegration, such as folding or texturing caused due to the generation of surface defects, could result in the increase in the stepwise thickness of the LbL deposition. As revealed in Fig. 6.38 (B), the films with LbL deposition numbers of less than 24 bilayers exhibited an optical transparency of $>80\%$. Corresponding thickness of the film was about 60 nm. Basically, the optical transmittance of the films linearly decreased with the increasing LbL deposition number, as illustrated in Fig. 6.38 (C).

The LbL method has been used to realize the bottom-up fabrication of highly ordered, free-standing, layered nanocomposites based on GO nanosheets [47]. Such nanocomposites exhibited excellent mechanical properties, with a toughness and elastic modulus of 1.9 $MJ \cdot m^{-3}$ and 20 GPa, respectively, for GO content as low as 8%. It was found the GO nanosheets were homogeneously distributed inside the LbL derived polyelectrolyte matrix, which resulted in a highly ordered stratification. The LbL assembly was carried out by using the LB method, in order to minimize the folding and wrinkling of the GO nanosheets, as shown in Fig. 6.39. The free-standing nanoscale (50 nm thick) multilayered nanomembranes with a monolayer of planar GO nanosheets with large lateral dimensions of up to a few centimeters across were sufficiently strong enough to be handled and transferred onto other substrates when necessary for given applications.

The concentration and size of the GO nanosheets were controlled by successive cycles of sonication combined with centrifugation. AFM analysis results indicated that the GO nanosheets had lateral dimensions of up to tens of micrometers, with a modest polydispersity in thickness of 0.96 ± 0.15 nn. Therefore, most of the exfoliated GO nanosheets were single-layered and bilayered. Poly(allylamine hydrochloride) (PAH) and poly(sodium 4-styrene sulfonate) (PSS) polyelectrolyte multilayers (PEMs) fabricated with spin-assisted LbL assembly were highly uniform, with microroughness being below 0.5 nm within 1×1 μm^2.

Since most applications require sufficiently high toughness along with a certain level of strength and elasticity, these unique GO-based nanocomposite membranes could meet the requirements, due to their outstanding toughness. In addition, the electronic and thermal properties of the GO nanosheets can be improved through chemical reduction or thermal annealing. The only problem that needs to be addressed could be the presence of the polymer matrix. The combination of electrical, thermal and optical properties, together with outstanding mechanical strength, has not been observed in most clay-based nanocomposites. Additionally, they could be used as heat sinks in electronic devices that are expected to produce more and more heat due to the rapid increase in the density of various components.

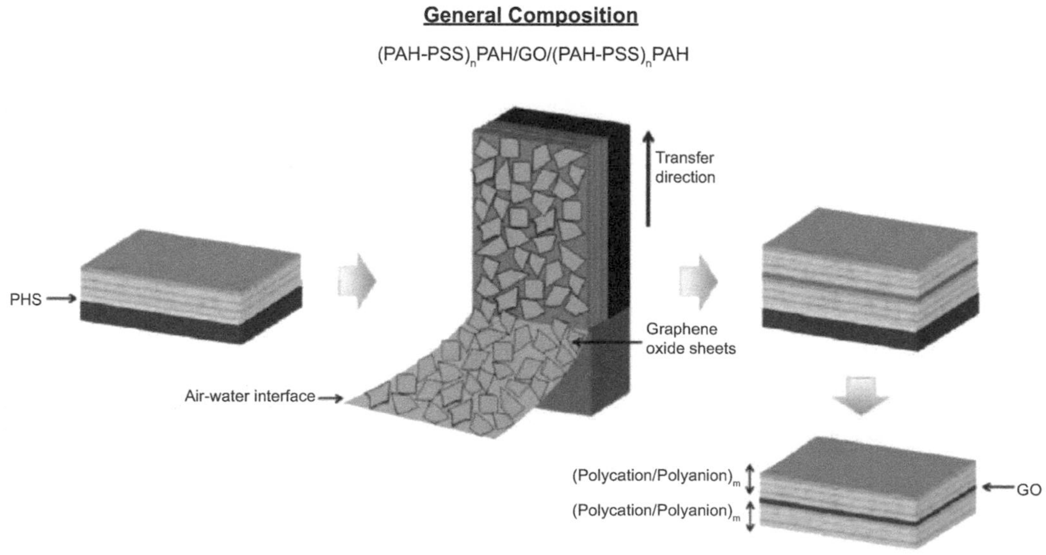

Fig. 6.39. Schematic diagram showing the fabrication and LbL-assembly of the free-standing multilayered membranes based on GO nanosheets. Reproduced with permission from [47], Copyright © 2010, American Chemical Society.

A simple LbL assembly method was developed to fabricate composite multilayer films of GO nanosheets and PANI (GO/PANI) [48]. The GO/PANI multilayer films could be readily reduced into graphene/PANI films by using chemical reduction with hydroiodic acid (HI). The graphene/PANI films exhibited largely increased conductivity and mechanical flexibility, which could be employed to make electrochromic devices without using ITO for the supporting electrodes.

The GO/PANI multilayer films were assembled on a pre-treated substrate by using the LbL assembly process, as demonstrated schematically in Fig. 6.40. The substrate was first dipped in the GO suspension for 10 min, was washed with DI water and then dried in N_2. The GO coated substrate was then dipped in the PANI solution, followed by washing with an HCl aqueous solution (pH = 3) and dried in a similar manner. Multilayer films were deposited by repeating the dipping in the negatively charged GO and positively charged PANI solutions alternatively. The derived films were denoted as $(GO/PANI)_n$, with n being the number of bilayers. The multilayer films were peeled off from the quartz substrates and then reduced in HI acid (55%) for 30 min. After that, the $(G/PANI)_n$ films were washed with acetone for 30 min two times, followed by drying in flowing N_2.

The $(G/PANI)_n$ films were systematically characterized, as shown in Fig. 6.41. There were two absorption bands at 320 nm and 630 nm, corresponding to the characteristic absorptions of PANI in the EB form, as seen in Fig. 6.41 (a). The band at 320 nm was ascribed to the π–π^* transition of benzoid rings, i.e., inter-band transition, while the broad absorption centered at 630 nm was related to the neutral PANI. With the increasing number of bilayers in the range of 2–15, the UV-visible absorbance of each band was increased gradually, illustrating a linear relation, as revealed in the inset of Fig. 6.41 (a). AFM images indicated that the thickness of the $(GO/PANI)_n$ film was also proportional to the number of the bilayers, with the average thickness of each bilayer to be about 3 nm. Figure 6.41 (c) shows photographs of the $(GO/PANI)_n$ films deposited on quartz substrates, which were smooth and transparent. As expected, the homogeneity and transparency of the films were decreased with the increasing deposition cycles. AFM image of the first bilayer of GO/PANI is shown in Fig. 6.41 (d), revealing that the negatively charged GO nanosheets were uniformly deposited on the positively charged surface of the silicon substrate. The PANI grains were adhered firmly onto the surface of the first GO layer, while the GO nanosheets could still be identified, as demonstrated in Fig. 6.41 (e).

The absorption peaks centered at 320 nm and 630 nm that were observed in the absorption spectra of the $(GO/PANI)_n$ films disappeared in the $(G/PANI)_n$ counterparts. Instead, two new absorption bands centered at 280 nm and 458 nm were observed. Similarly, their intensities were increased with the increasing number of bilayers. The presence of the weak broad band at 458 nm implied that the PANI was protonated, so that the EB form was changed to ES form [69]. The absorption peak at 280 nm was attributed to the presence

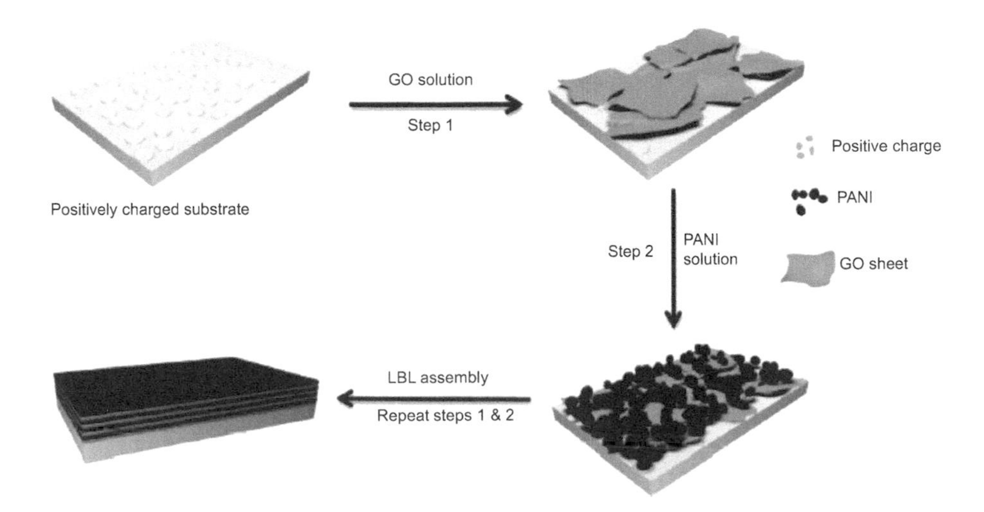

Fig. 6.40. Schematic diagram showing the LBL assembly process to synthesize the GO/PANI multilayer films. Reproduced with permission from [48], Copyright © 2014, Elsevier.

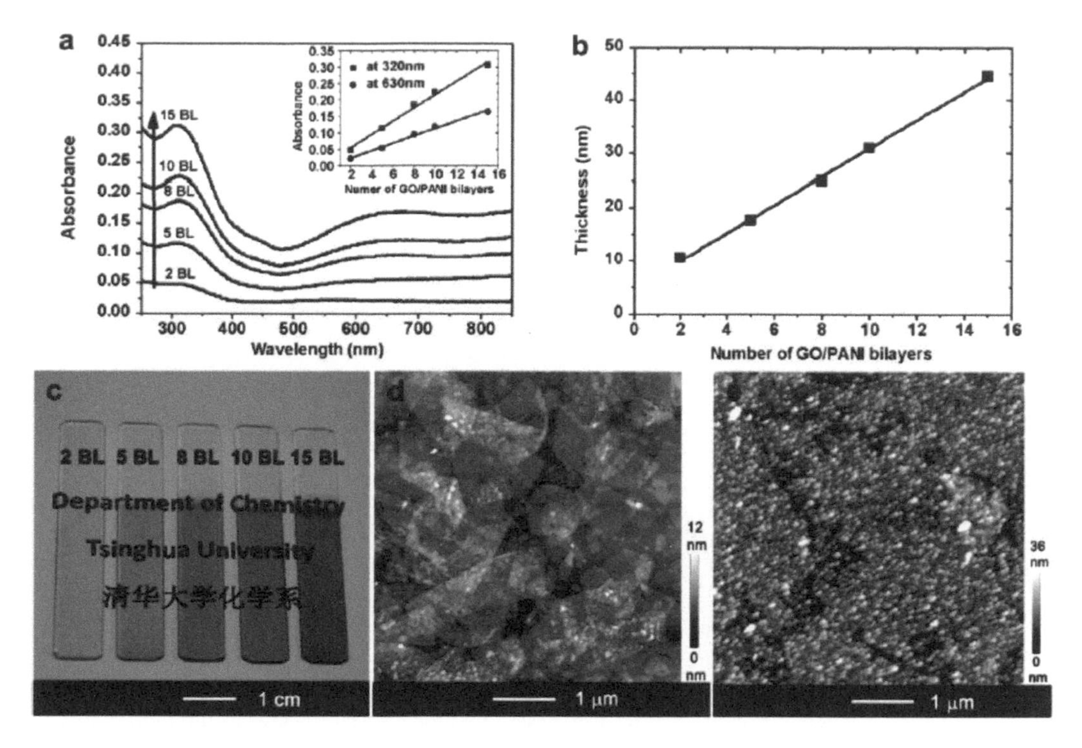

Fig. 6.41. (a) UV-visible spectra of the $(GO/PANI)_n$ films assembled on quartz substrates, with the inset to be the plots of absorption at 320 nm and 630 nm as a function of the number of bilayers. (b) Film thickness as a function of the number of bilayers. (c) Photographs of the $(GO/PANI)_n$ films deposited on quartz substrates, with 2, 5, 8 10 and 15 bilayers (BL). AFM images of the first GO layer (d) and $(GO/PANI)_1$ (e) deposited on silicon wafers. Reproduced with permission from [48], Copyright © 2014, Elsevier.

of rGO, implying partial recovery of the electronic conjugation in the GO nanosheets. The reduction of GO was also evidenced by the decrease in the resistances of the multilayer films. Resistance as a function of the number of cycles indicated that the formation of a continuous graphene film required at least five bilayers. SEM and AFM characterization results indicated that the $(G/PANI)_n$ films had a continuous and uniform surface, with PANI grains having diameters at nanometer scale.

A hybrid type of self-assembled LbL multilayer thin films consisting of alternating titania ($Ti_{0.91}O_2$) and GO nanosheets was reported [45]. After that, multilayered graphene-based films were derived from the corresponding GO films by exposing them to UV light which triggered the photocatalytic reduction. Therefore, the G–$Ti_{0.91}O_2$ hybrid materials could find potential applications in photocatalysis, capacitors and sensors.

LbL assembly fabrication procedure of the multilayer films, with alternating $Ti_{0.91}O_2$ and GO nanosheets, is schematically shown in Fig. 6.42. During the LbL assembly procedure, thoroughly cleaned quartz glass slides were used as the substrates. Specifically, a substrate was successively immersed in a protonic polyethyleneimine (PEI) aqueous solution, a colloidal suspension of negatively charged $Ti_{0.91}O_2$ nanosheets, a PEI aqueous solution and a negatively charged GO suspension. The dipping procedures were cycles for different times to obtain films with the desired thicknesses. The number of layers (circles), *n*, was indicated in the composition formula, i.e., $(PEI/Ti_{0.91}O_2/PEI/GO)_n$. The as-assembled multilayer films ranged from transparent yellow–brown in color. After being exposed to UV irradiation, the GO was reduced into graphene, with a corresponding change in the color of the films from transparent yellow-brown to black. More importantly, the PEI moiety could be removed by photocatalytic irradiation at the same time.

Similarly, a hybrid film consisting of rGO, poly(vinyl alcohol) (PVA) and Co–Al–NO_3 layered double hydroxide (LDH) could be fabricated by using the LbL assembly method [70]. The hydrogen bonding between rGO, PVA and Co–Al–NO_3 nanosheets was suggested to promote the assembly process on the quartz surface.

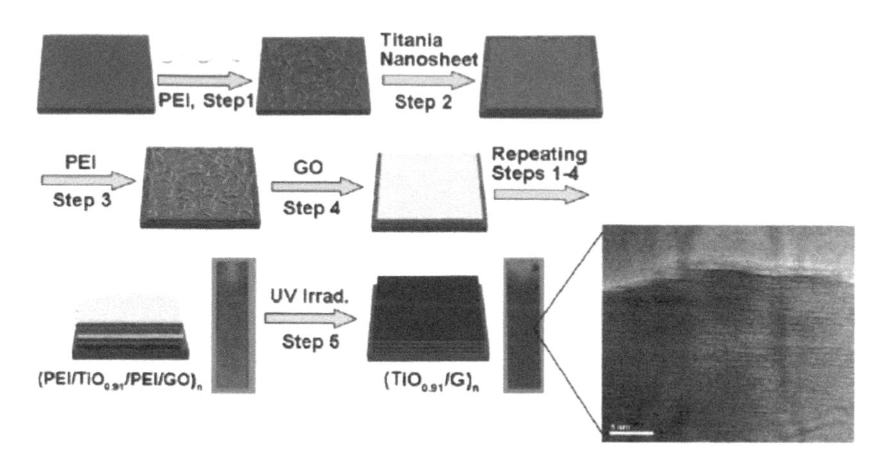

Fig. 6.42. Schematic diagram illustrating the formation of the LbL assembled multilayer films containing $Ti_{0.91}O_2$ and GO nanosheets, followed by subsequent UV-assisted photocatalytic reduction of the GO into graphene. Insets: multilayer hybrid films on glass slides turning dark after UV irradiation (bottom row) and HRTEM image of the G–$Ti_{0.91}O_2$ multilayer film (extreme right). Reproduced with permission from [45], Copyright © 2009, WILEY-VCH Verlag GmbH & Co. KGaA, Weinheim.

PVA was dissolved in DI water to form an aqueous solution with a concentration of 1 wt%. Quartz glass slides were used as the substrates, and were thoroughly cleaned with a mixture of H_2SO_4/H_2O_2 with a volumetric ratio of 3/1 for 1 h. The washed substrates were then washed with DI water and dried in N_2 flow. Figure 6.43 shows a schematic diagram of the LbL assembly process. Heterogeneous ultrathin films were deposited by repeating the following four steps: (i) dipping the substrate into exfoliated LDH suspension for 10 min, which was followed by thorough rinsing with DI water and drying in N_2 flow, (ii) dipping into the aqueous solution of PVA for 10 min, followed by rinsing with DI water and drying in N_2 flow, (iii) dipping into the suspension of GO suspension for 10 min, followed by water washing and N_2 drying and (iv) dipping into the solution of PVA again for 10 min, with similar post-treatment. After repeating the procedure for n times, multilayer films of $(LDH/PVA/GO/PVA)_n$ were produced. The as-prepared $(LDH/PVA/GO/PVA)_{50}$ hybrid films were immersed into 50% glutaraldehyde solution for 1 h for cross-linking reaction. Finally, $(LDH/PVA/G/PVA)_{50}$ films were obtained after reduction in hydrazine/DMF solution (0.5 ml 50% hydrazine + 30 ml DMF) at 85°C.

Figure 6.44 (a) shows the UV-vis spectra of the multilayer films with various deposition cycles. Figure 6.44 (b) illustrates the absorbance at 228 nm as a function of the number of deposition cycles. The intensity of the absorption peak was linearly increased with the increasing deposition cycles, implying the presence of a stepwise and regular growth in thickness. Such heterogeneous multilayer films with monolayer dispersed graphene and LDH nanosheets could be used for applications in both magnetic and electrical devices, due to their structural diversity and combined functionality.

Recently, a LbL process was demonstrated to fabricate photoactive hybrid films based on Keggin-type polyoxometalate cluster ($H_3PW_{12}O_{40}$, PW), GO, PEI and PAH [71]. Due to the photocatalytic activity of the PW clusters [72], GO could be transformed into rGO with UV irradiation, as illustrated in Fig. 6.45. As a result, uniform and large-area composite films based on rGO with precisely controlled thickness could be prepared on various substrates, such as quartz, silicon and plastic. The photo-reduced PW/rGO films with one and six layers, i.e., $(PAH/GO/PAH/PW)_1$ and $(PAH/GO/PAH/PW)_6$, had conductivities of 10 and 120 $S{\cdot}m^{-1}$, respectively, which were comparable to those of the rGO films produced by using the chemical reduction methods [73]. Therefore, photoreduction has been demonstrated as a green method to obtain graphene and graphene-based thin films.

The driving force to facilitate the LbL assembly process was electrostatic interaction between the negatively charged GO nanosheets and the positively charged PW clusters in the suspensions. The substrates were initially modified by coating a PEI/PW double layer, onto which PAH/GO/PAH/PW multilayers were deposited, as demonstrated in Fig. 6.45. In the multilayer films, the GO nanosheets and PW clusters were linked by the PAH layers. The PW layers were in close contact with the GO layer on both sides, through

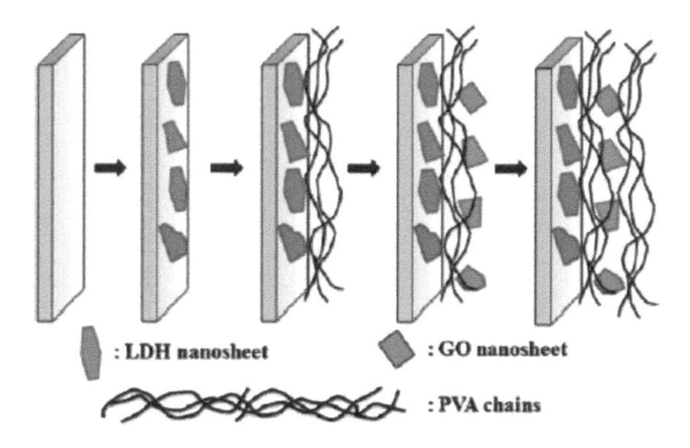

Fig. 6.43. Schematic diagram of the LbL assembly procedure. Reproduced with permission from [70], Copyright © 2010, American Chemical Society.

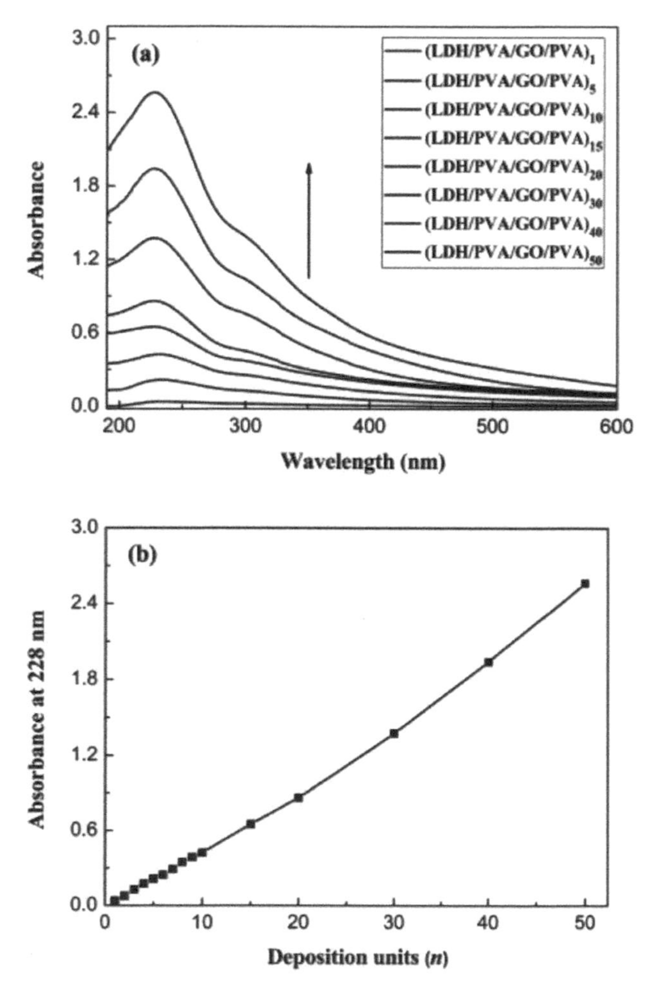

Fig. 6.44. (a) UV absorption spectra of the $(LDH/PVA/GO/PVA)_n$ multilayer films. (b) Absorbance at 228 nm as a function of the number of deposition cycles. Reproduced with permission from [70], Copyright © 2010, American Chemical Society.

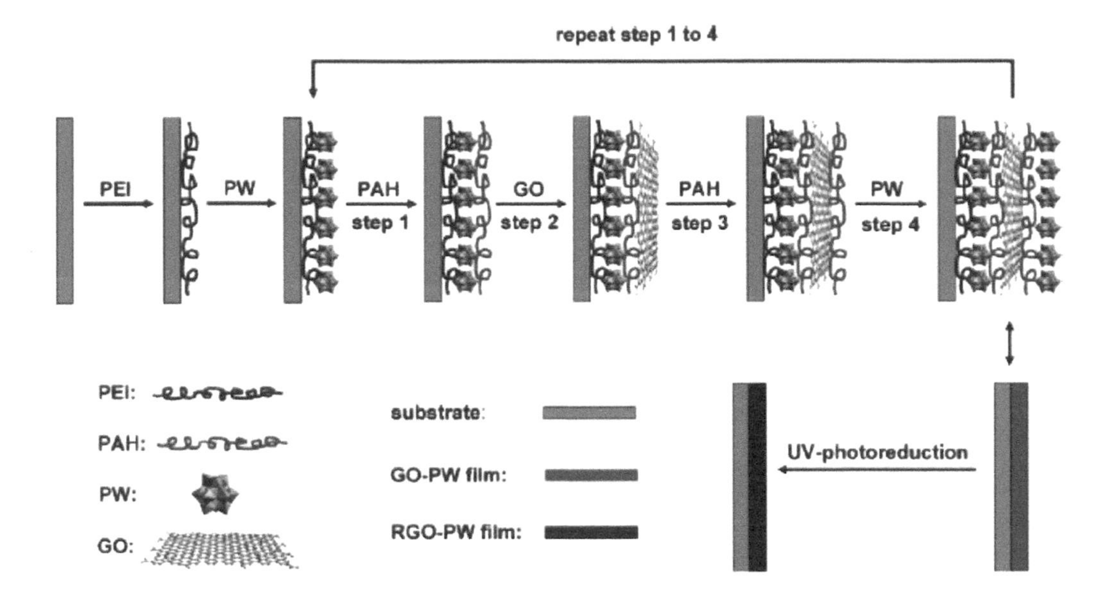

Fig. 6.45. Schematic diagram showing the fabrication procedure of the rGO-PW multilayer thin films, involving the LbL assembly of GO nanosheets and PW clusters using cationic polyelectrolytes PEI and PAH as the electrostatic linkers, and a subsequent *in situ* photoreduction to convert the GO into rGO. Reproduced with permission from [71], Copyright © 2011, American Chemical Society.

Color image of this figure appears in the color plate section at the end of the book.

the interaction between the oxygen-containing groups on the GO nanosheets and PW clusters. In this case, electrons were transferred from photoexcited PW to GO, so that GO was reduced in the PW-assisted photoreduction process.

Figure 6.46 (A) shows the UV-vis spectra of the GO and PW. There were two absorption bands at 230 nm and 300 nm in the GO curve, whereas the PW had an absorption band at 260 nm with an extinction coefficient of about 7.5×10^4 $M^{-1}\cdot cm^{-1}$, corresponding to the well-known ligand-to-metal charge-transfer (LMCT) transition. Figure 6.46 (B) illustrates the UV-vis absorption spectra of the $(PAH/GO/PAH/PW)_n$ multilayer films on quartz substrates. The absorption increased almost linearly with the number of the layers, as seen in the inset of Fig. 6.46 (B), which indicated that the LbL assembly was a highly reproducible process.

AFM image of the $(PAH/GO/PAH/PW)_1$ film on silicon substrate is shown in Fig. 6.47 (A). The film was covered by a monolayer of GO nanosheets with a high surface coverage, with a surface roughness of about 1.17 nm over 10–10 μm^2. Figure 6.47 (B) shows a magnified AFM image of the $(PAH/GO/PAH/PW)_1$ film, indicating that small PW domains were uniformly distributed on surface of the GO nanosheets. The uniform adsorption of the PW clusters ensured the smooth surface of the $(PAH/GO/PAH/PW)_1$ film. AFM images of the $(PAH/GO/PAH/PW)_6$ film are illustrated in Fig. 6.47 (C, D), demonstrating the close packing of the multilayer structure. At the same time, the size of the PW domains was increased to > 10 nm^2, while the surface roughness of the film also increased. Nevertheless, the $(PAH/GO/PAH/PW)_6$ films had a sufficiently smooth surface, which was one of the requirements for the fabrication of large-area thin-film electronic devices.

The GO-PW multilayer films could be deposited on various hydrophilic substrates, including quartz and oxygen plasma-treated flexible PET substrates, as shown in Fig. 6.48 (A, C). After *in situ* photoreduction, the GO was converted into rGO, which corresponded to a gradual change in color from light brown to dark, as illustrated in Fig. 6.48 (B, D). It is expected that multifunctional rGO devices can be fabricated by further incorporating other functional polyoxometalates, such as magnetic polyoxometalate clusters, for more device applications.

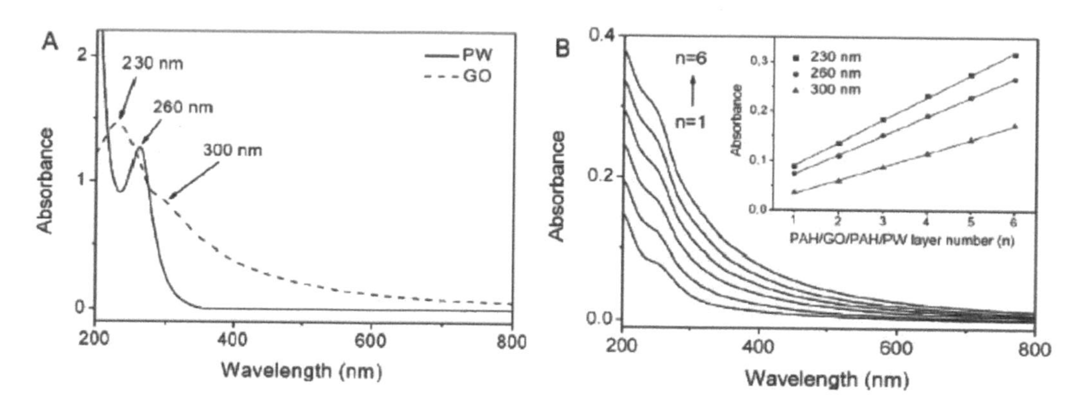

Fig. 6.46. (A) UV-vis absorption spectra of an aqueous 0.05 mg·ml^{-1} PW solution (black line, PW concentration is 1.7×10^{-5} M) and an aqueous 0.05 mg·ml^{-1} GO solution (red line). (B) UV-vis absorption spectra of the LbL-assembled $(PAH/GO/PAH/PW)_n$ multilayer films with layer number n = 1–6 on quartz substrates, which were modified with PEI/PW films. The blank substrate was used as reference. The curves, from bottom to top, correspond to n = 1–6. The inset shows the plots of the absorbance values at 230, 260, 300 nm as a function of the layer number. Reproduced with permission from [71], Copyright © 2011, American Chemical Society.

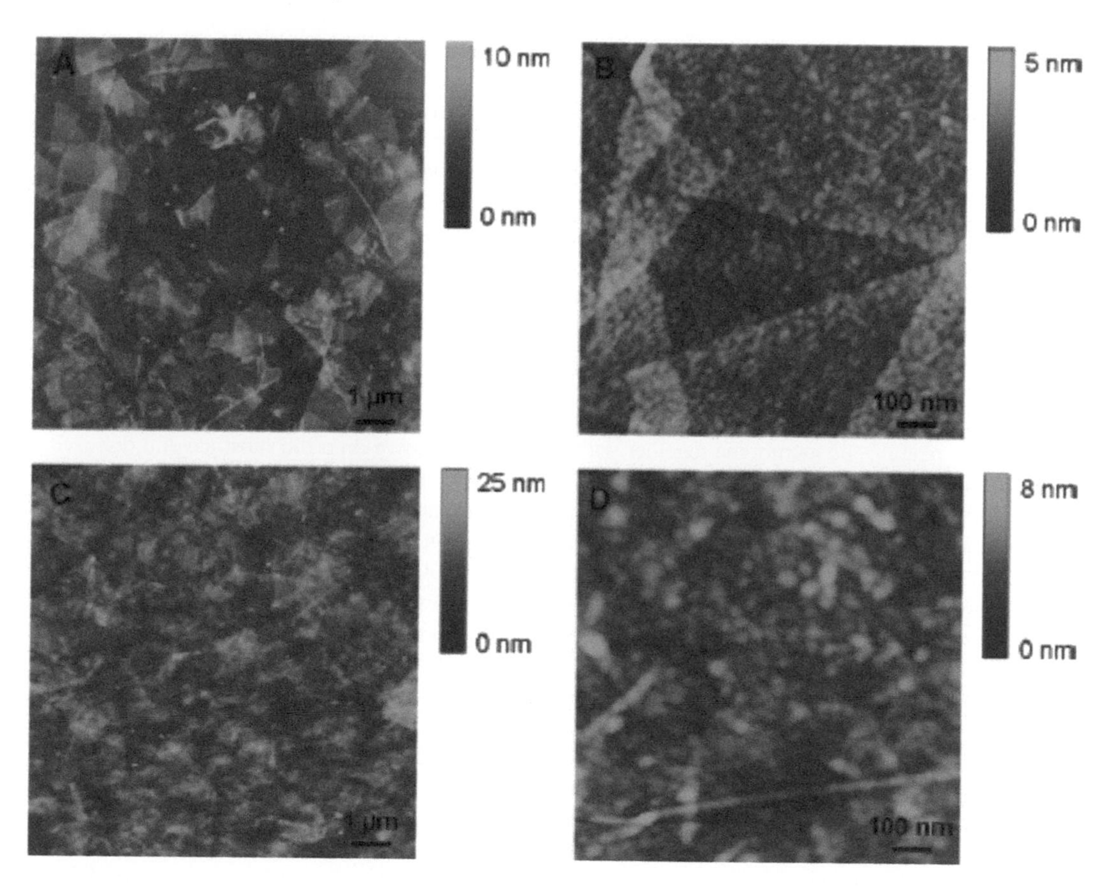

Fig. 6.47. AFM images of the $(PAH/GO/PAH/PW)_1$ film (A, B) and $(PAH/GO/PAH/PW)_6$ film (C, D), both on PEI/PW precursor film modified silicon substrates, with a layer of 300 nm thermal oxide. Reproduced with permission from [71], Copyright © 2011, American Chemical Society.

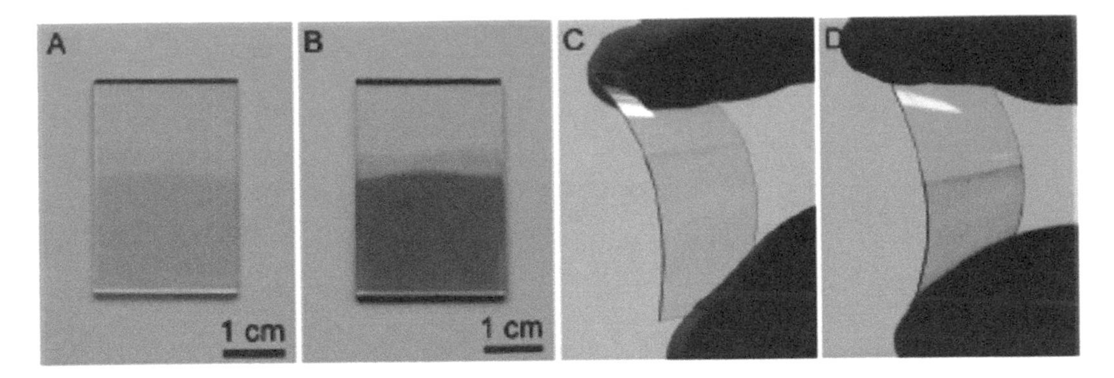

Fig. 6.48. Photographs of the $(PAH/GO/PAH/PW)_6$ multilayer film deposited on quartz substrates before (A) and after (B) UV photoreduction for 6 h and on a flexible PET substrate before (C) and after (D) UV photoreduction for 6 h. Both substrates were coated first by the PET/PW precursor film. Reproduced with permission from [71], Copyright © 2011, American Chemical Society.

6.2.7. Vacuum filtration

A solution-based vacuum filtration method was used to prepare rGO thin films with thicknesses ranging from a single monolayer to several layers over large areas [52]. Accordingly, the optoelectronic properties of the rGO thin films could be tailored over several orders of magnitude, thus offering potential applications as flexible and transparent semiconductors or semi-metals. The thinnest films exhibited graphene-like ambipolar transistor characteristics, while the thicker films behaved as graphite-like semi-metals.

Dilute GO suspensions, with concentrations of 0.33–2.64 $mg \cdot l^{-1}$, were vacuum-filtrated, with a commercial mixed cellulose ester membrane having pores of 25 nm. During the filtration process, the liquid was allowed to pass through the pores of the membrane, while the GO nanosheets were blocked. The rate of permeation of the solvent was related to the accumulation of the GO nanosheets on the pores. As the number of the GO layers was increased, the rate of filtration was decreased. The flow of the liquid facilitated the self-alignment of the GO nanosheets. The method offers control over the film thickness at a nanoscale level by simply varying either the GO concentration of the suspension or the volume of filtration.

After filtration, the GO thin films on the filter membrane could be conveniently transferred onto other substrates by placing the membrane together with the film side down onto the substrate and dissolving the membrane with acetone. The yield of the transfer process was almost 100% on any substrate, which was attributed to the sufficiently strong cohesive forces due to the van der Waals interactions among the GO nanosheets in the film and between the film and the substrate. Therefore, the as-deposited GO thin films could withstand lithographic processes, including rinsing, blowing with dry nitrogen and deposition of electrodes, without delamination or breaking. Figure 6.49 (a) shows an as-filtrated GO thin film with an area of 10 cm^2 on an ester membrane, which could be transferred onto glass and plastic substrates, as illustrated in Fig. 6.49 (b, c).

The GO thin films were efficiently reduced by using the combination of hydrazine vapor reaction and low-temperature thermal annealing treatment. The electrical and optical properties of the rGO thin films are shown in Fig. 6.50. It was found that the optoelectronic properties of the thin films could be tailored by controlling the amount of rGO on the film surface. As seen in Fig. 6.50 (a), sheet resistance of the hydrazine-treated GO thin films was almost independent on the filtration volume of below about 300 ml. After annealing at 200°C in N_2 or vacuum, the sheet resistance was significantly reduced to be $< 10^5\ \Omega \cdot \square^{-1}$. The saturation of the sheet resistance of the thin films in Fig. 6.50 (a) above the critical filtration volume was ascribed to the fact that only the uppermost layers were reduced. As expected, the chemically reduced and thermally annealed GO thin films exhibited a decrease in optical transparency.

Fig. 6.49. Photographs of the GO thin films on filtration membrane (a), glass (b) and plastic (c) substrates. Reproduced with permission from [52], Copyright © 2008, Macmillan Publishers Limited.

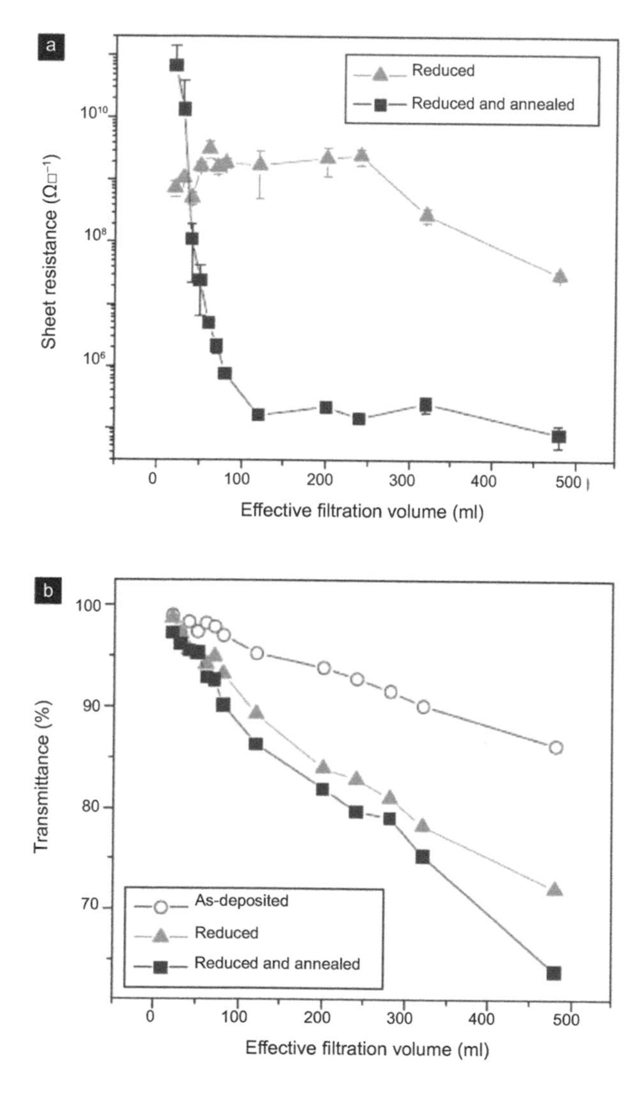

Fig. 6.50. Sheet resistance (a) and transmittance (b) at $\lambda = 550$ nm of the rGO thin films as a function of filtration volume. The plots are shown for thin films with different reduction steps, while the sheet resistance of the as-deposited GO thin films is out of scale. Reproduced with permission from [52], Copyright © 2008, Macmillan Publishers Limited.

6.2.8. Interface self-assembly

A graphene film was also produced at the oil/water interface from GO suspensions. With the presence of a small amount of ethanol, GO films with tunable transmittance and surface resistivity could be formed at the pentane/water interface, as the pentane was evaporated [74]. By using this method, GO–SWCNT hybrid films could also be prepared. After reduction with hydrazine vapor, the hybrid film showed improved conductivity as compared with the rGO films, which was attributed to the formation of conducting channels due to the presence of the highly conductive SWCNTs.

Figure 6.51 shows a schematic diagram demonstrating the fabrication process of the GO thin films [74]. 5 ml GO suspension was added into a 20 ml glass beaker, while 3 ml anhydrous pentane was layered on top of the GO suspension to create an interface between the organic and aqueous phases. 3 ml ethanol was then rapidly injected into the aqueous phase. After the injection of ethanol, the beaker was covered with a cap to minimize the effect of the pentane evaporation on the pentane-water interface. GO nanosheets started to self-assemble at the interface, whereas a GO layer was formed very quickly. The cap was removed 30 min later to speed up the evaporation of the pentane. After the pentane was depleted in 2 h, a GO film was floating on the surface of the aqueous solution. The film was collected on a quartz substrate, which was dried with N_2 blow. SWCNT aqueous dispersions were used as the source of SWCNTs to produce GO–SWCNT hybrid films.

Figure 6.52 (a) shows AFM image at edge of the GO film derived from 0.5 $mg{\cdot}ml^{-1}$ GO dispersion, deposited on a quartz substrate [74]. The nanometer scale uniform thin film had a thickness of ~7 nm, with a root-mean-square (rms) roughness of 3.23 nm, according to the height profile crossing its edge, as seen in Fig. 6.52 (b). Since a single-layer GO nanosheet is about 1 nm in thickness, the GO films obtained in this way consisted of multiple layers of GO nanosheets. Figure 6.52 (c) shows SEM image of the GO thin film, in which wrinkles were observed, suggesting the presence of overpacked layers. Both the GO and rGO films exhibited high transparency over a wide range of wavelength. Figure 6.52 (d) shows the transmittance curves of the GO and rGO films. They both possessed a flat optical transmittance profile, in the range over the visible and near-infrared regions. The peaks at 231 nm were attributed to the $\pi \rightarrow \pi^*$ transitions of the aromatic C–C bonds, whereas those at about 300 nm corresponded to the $n \rightarrow \pi^*$ transition of C=O bonds [75]. The rGO film was less transparent than the GO film, as illustrated in the inset of Fig. 6.52 (d). At the same time, the intensity of the absorption peaks from $\pi \rightarrow \pi^*$ transitions was increased, and the peak was also shifted to 270 nm, suggesting the recovery of the sp^2 carbon. Transmittances of the GO and rGO films at 550 nm were 91% and 83%, respectively.

Firstly, the rapid injection of ethanol into the GO aqueous dispersion could be a possible reason for the formation of the GO films, because it resulted in a vigorous mixing between ethanol and water. The liquid mixing created upward flows, so that the GO nanosheets were brought towards the interface. Secondly, ethanol acted as a nonsolvent for the GO aqueous suspension, which was responsible for the increased mass transfer of the GO nanosheets. Nevertheless, this method offered a simple way to fabricate GO and rGO thin films with potential for large-scale production.

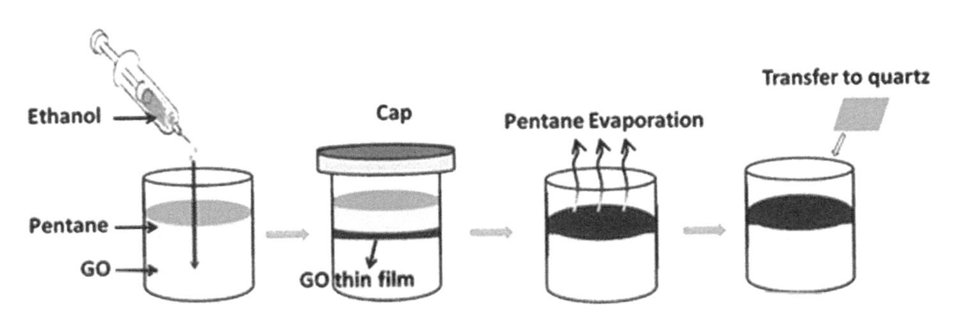

Fig. 6.51. Schematic diagram showing the preparation procedure of the GO thin film through the interface self-assembly. Reproduced with permission from [74], Copyright © 2011, American Chemical Society.

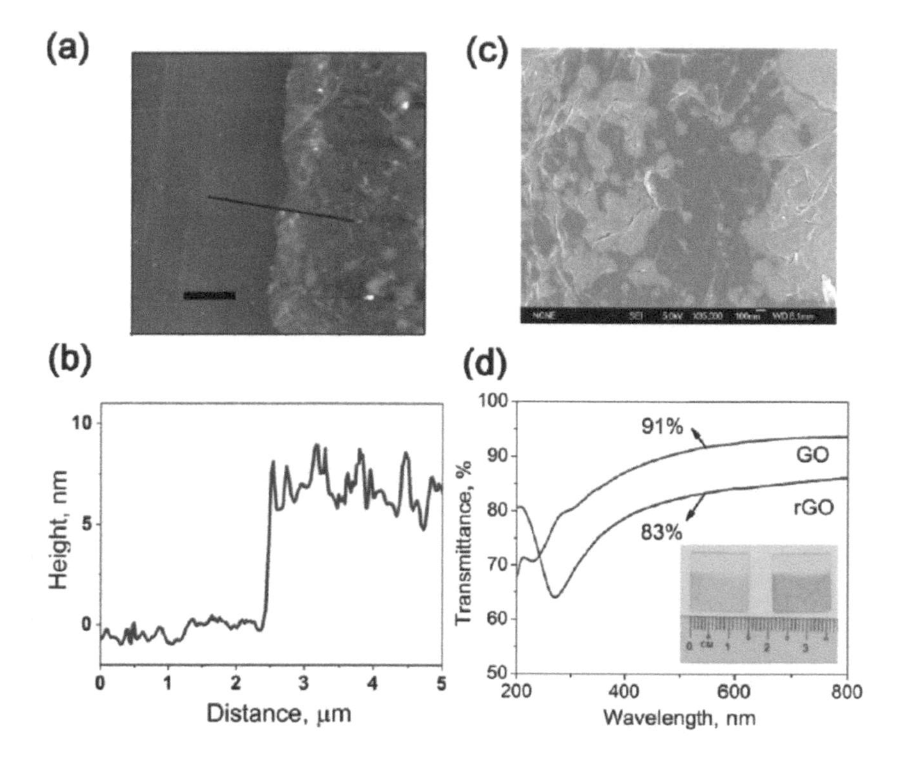

Fig. 6.52. (a) AFM image of the GO film on quartz substrate, which was derived from 5 ml 0.5 mg·ml^{-1} GO dispersion and 3 ml pentane with injection of 3 ml ethanol. (b) Height profile along the line in (a), crossing the edge of the GO film. (c) SEM image of the GO film. (d) Transmittances of the GO and rGO films. The inset shows the photographs of the GO (left) and rGO (right) films deposited on quartz substrates. Reproduced with permission from [74], Copyright © 2011, American Chemical Society.

6.3. Graphene-Based Free-Standing Papers

Free-standing papers or membranes based on graphene nanosheets have attracted much attention, due to their potential applications, especially in flexible electronic and energy storage devices [76–78]. Among various methods, vacuum filtration of G/GO suspensions through a porous membrane filter has been the most widely used technique to fabricate G/GO papers [79–129]. Other methods, including solution casting [130–136], tape casting [137], electro-spray deposition (ESD) [138], electrophoretic deposition (EPD) [139], interface self-assembly [140, 141], cryogel [142], pyrolyzed asphalt [143], hydrothermal synthesis [104], chemical vapor deposition (CVD) [144, 145], have also been employed to fabricate graphene papers.

6.3.1. Membrane vacuum filtration

Membrane vacuum filtration is the simplest and most straightforward method to prepare G/GO papers. Figure 6.53 shows a schematic diagram of the membrane vacuum filtration setup, which is connected to a pump during working process. Solvents are passed through the porous membranes, while G/GO nanosheets are blocked by the membranes as films. The films are then peeled off as free-standing papers after a certain degree of drying.

GO papers with thicknesses in the range of 1–30 µm have been fabricated by using this technique [77]. Structural characterization of GO paper indicated that the compliant GO nanosheets were interlocked/tiled together parallel in the horizontal direction, i.e., the formation of the ordered structures

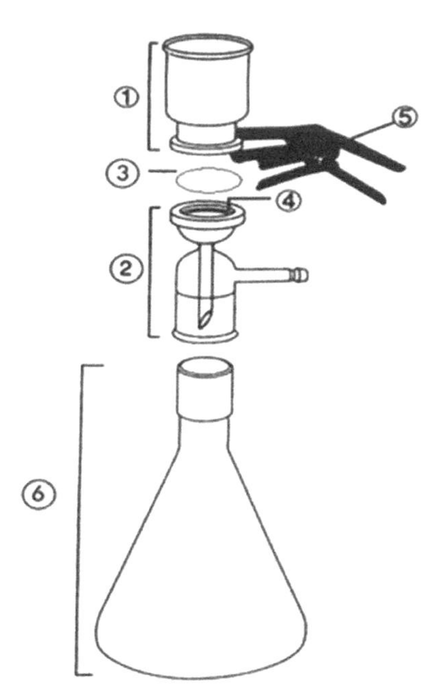

Fig. 6.53. Schematic diagram of the funnel filtering apparatus for vacuum membrane filtration: (1) filtering cup, (2) filtering head, (3) filtering membrane, (4) and core plate, (5) clamp and (6) conical flask.

was attributed to the flow-directed assembly process. The GO paper exhibited high flexibility and strong mechanical properties. As compared with the conventional carbon- and clay-based papers, the extraordinary mechanical properties of the GO paper originated from the strong van der Waals interactions and hydrogen-bonding within the GO nanosheets.

GO nanosheets were dispersed in water to form a GO suspension at a concentration of 3 $mg \cdot ml^{-1}$. The thickness of the GO papers was controlled by adjusting the volume of the colloidal suspension. The GO paper prepared in this way could be cut by using a razor blade. The papers were uniform, while they were dark brown under transmitted white light and almost black in reflection as the thickness was > 5 μm, as demonstrated in Fig. 6.54 (a–c). SEM images revealed that the highly oriented nanolayers were closely packed through almost the entire cross-section of the papers, with less densely packed 'wavy' skin layers with thicknesses in the range of 100–200 nm, as seen in Fig. 6.54 (e–g). The layered structure was also confirmed by XRD patterns. Layer-to-layer distance or *d*-spacing of the GO papers was about 0.83 nm, corresponding to the presence of one molecule thick layer of water interacting with the GO nanosheets through hydrogen bonding [146]. The average dimension of the ordered stack of GO nanosheets in the papers oriented perpendicularly to the diffracting plane was about 5.2 ± 0.2 nm, corresponding to 6–7 stacked GO nanosheets.

With three regimes of deformation in the stress–strain curves, i.e., straightening, almost linear (or 'elastic') and plastic, mechanical behavior of the GO papers was similar to that of most paper-like or foil-like materials, but with very high stiffness. Despite the presence of different levels of wrinkling and 'waviness' in the GO papers at different length scales, the initial straightening during the tensile loading was not very pronounced. During the rupture of the GO papers that were loaded beyond the 'elastic' regime, no pull-out of their lamellae was observed. Instead, nearly straight and flat fracture surfaces were formed, as seen in Fig. 6.54 (e–g). This was attributed to their high homogeneity and strong interlayer interaction. The GO papers exhibited an average modulus of 32 GPa, with the highest value of 42 ± 2 GPa.

Bending experiments were conducted for representative GO papers. Figure 6.55 (b) shows a strip of the GO paper that was bent to form a simple curve, which was then compressed between the two parallel plates until one kink or more was developed, as demonstrated in Fig. 6.55 (b, c). The radius of the curvature was measured before the samples lost their structural stability. From the solution for pure uniform bending

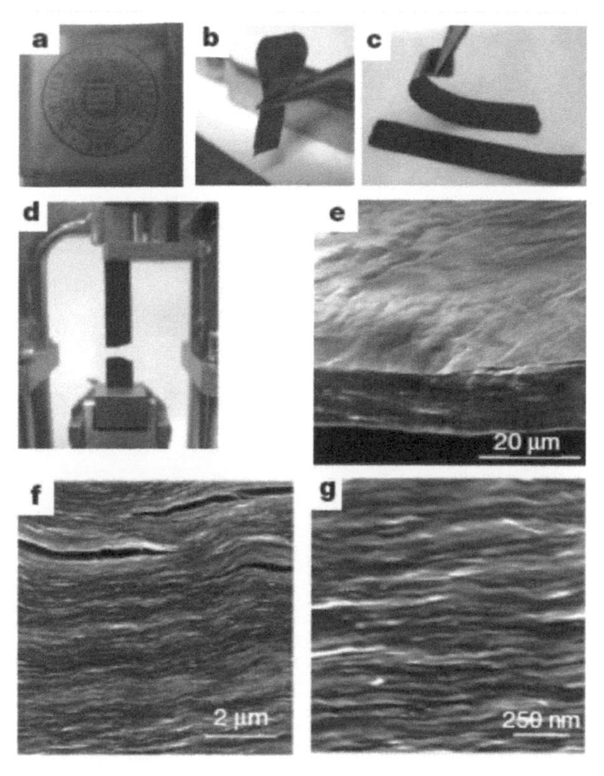

Fig. 6.54. Morphology and structure of the GO papers. (a–d) Photographs of the GO papers: (a) about 1 μm-thick film, (b) folded about 5 μm-thick semi-transparent film, (c) folded about 25 μm-thick strip and (d) strip after fracture after tensile loading. (e–g) Low-, middle- and high-resolution SEM side-view images of the ~10-μm-thick papers. Reproduced with permission from [77], Copyright © 2007, Macmillan Publishers Limited.

of a bar made of an isotropically homogeneous material, the positive or negative normal strain ε_x at the outer or inner surface of the bar is given by $|\varepsilon_x| = 0.5t/R$, with t and R to be thickness of the sample and radius of the curvature. Figure 6.55 (a) shows that the average normal strain value was $\varepsilon_x = 1.1 \pm 0.1\%$, according to the linear fitting of the experimental data (red line).

The mechanics of deformation at tension and bending were proposed, as schematically illustrated in Fig. 6.55 (e). A uniform distribution of stresses across the sample was formed under the uniaxial tension. The stresses were mainly transferred through the shear deformation of the interlamellar adhesive, i.e., the hydrogen-bonded water molecules, while the material bending resulted in only localized stresses at the surfaces of the bent papers. The stress at the outer surface was transferred between the adjacent layers by both the shear and pull-out of the water adhesive, thus leading to layer delamination, especially along the defects in the stacked structure, as seen in Fig. 6.55 (d, e). At the inner surface, the stress was compressive, which was responsible for the local shear and buckling of the layers, as illustrated in Fig. 6.55 (c, d). Under the critical uniaxial tension stress, the fracture propagated nearly straight across the sample without an obvious pull-out (Fig. 6.54 (e)). In comparison, under a bending, load delamination took place essentially along the microdefects, such as voids between some adjacent layers, as observed in Fig. 6.54 (f). Therefore, the GO papers consisted of in-plane stiff and out-of-plane compliant strongly interlocked GO nanolayers.

Because GO papers are nonconductive, they cannot be directly used for some applications, e.g., electrodes of LiBs and supercapacitors. Therefore, it is necessary to reduce GO papers into rGO papers. Various reduction methods have been employed to reduce GO papers, including hydrazine hydrate reduction [117], photothermal reduction [147], supercritical ethanol reduction [140] and thermal reduction [148, 149].

Thermal reduction has been applied to GO papers made of GO nanosheets with in-plane porosity that was generated on their basal planes by using a combination of ultrasonic vibration and mild acid oxidation,

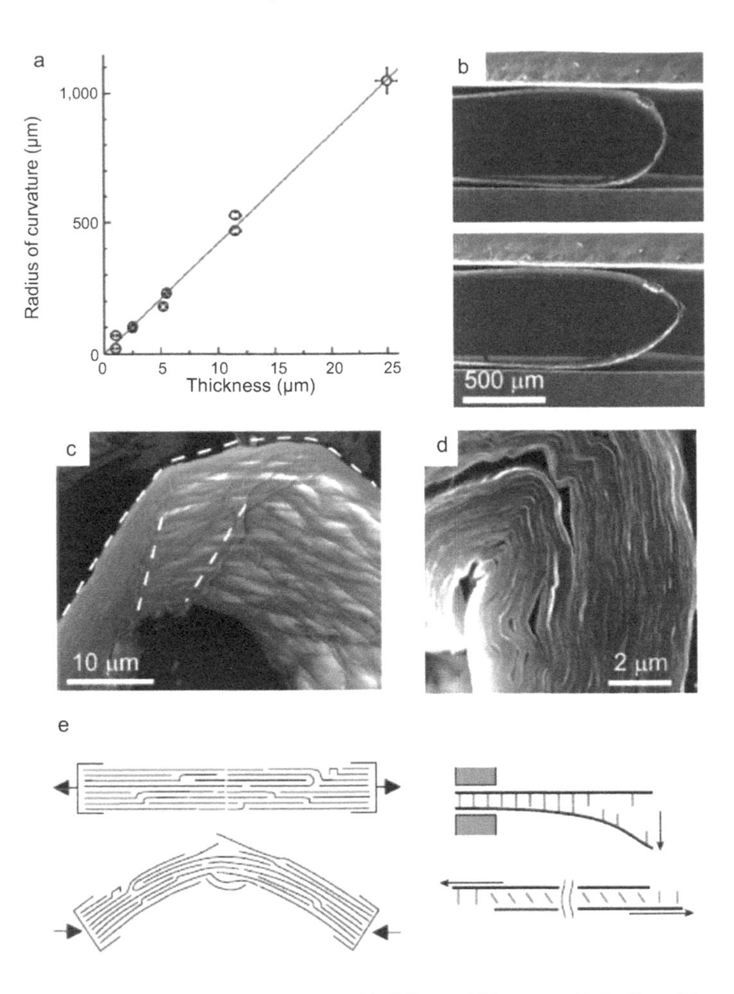

Fig. 6.55. Bending experimental results of the GO papers with different thicknesses. (a) Radius of the curvature at which the strip of paper lost mechanical stability (buckles) during bending, with the red line to be a linear fit of the experimental data. (b) Two low-magnification SEM images of the strip cut from membrane with a thickness of 5.2 μm when compressed between two parallel plates. The upper image was taken immediately before and the lower one after the sample buckled. (c) SEM image of the ~1 mm thick curved GO paper strip having an about 20 μm radius of curvature and showing two major creases (dotted white lines) due to the buckling. (d) High magnification SEM image of the 11 μm thick buckled strip. (e) Schematic diagrams of the uniaxial in-plane load-to-fracture and a bending-to-buckling test. The additional schematics represent the inter-lamellar water molecules (blue) to bridge the adjacent GO nanosheets. These interactions were broken upon bending of the stack or under tension, leading to fracture without pull-out. Reproduced with permission from [77], Copyright © 2007, Macmillan Publishers Limited.

as shown in Fig. 6.56 [149]. The milder condition was achieved by reducing the acid concentration and shortening the sonication time duration, in order to obtain control over the formation of defects more effectively. GO nanosheets were dispersed in water with a concentration of 0.1% w/w, into which a certain amount of 70% concentrated HNO_3 was added under continuous stirring. All the mixtures were sonicated at room temperature for 1 h. The high strain rates and frictional forces in cavitation bubbles were generated due to the ultrasound at sufficient acoustic pressures, which attached the carbonaceous surface and broke the framework. When a sufficient number of coordinatively unsaturated carbon atoms were formed at the damage sites and edge sites of the GO nanosheets, they would then react with HNO_3, thus leading to partial detachment and removal of the carbon atoms from the GO nanosheet, i.e., formation of holes. Four solutions with GO suspension/70% HNO_3 volume ratios of 1:5, 1:7.5, 1:10, and 1:12.5 were tested, corresponding to holey graphene oxide, HGO, denoted as HGO-I, II, III and IV, respectively.

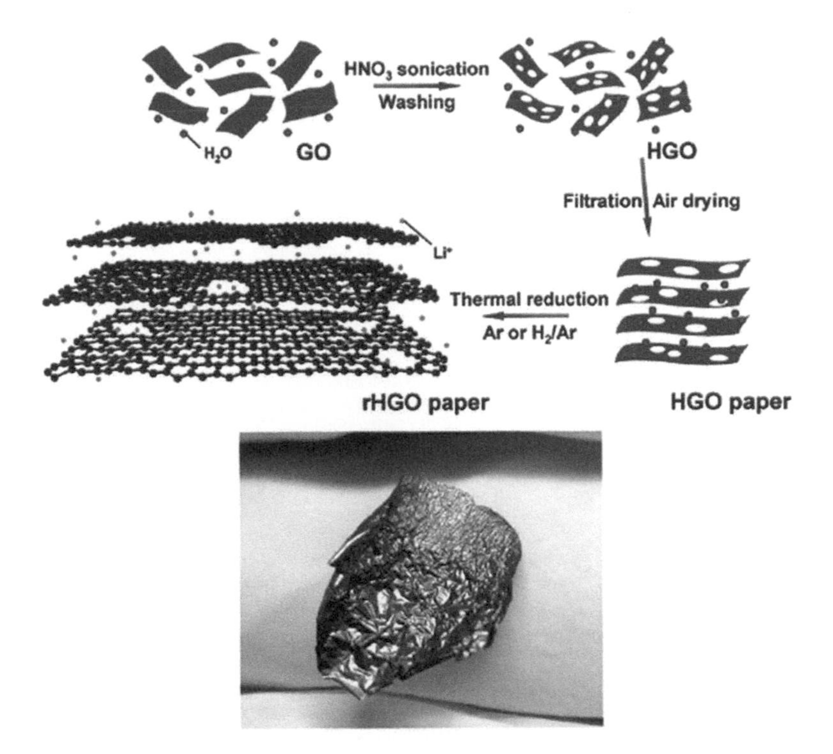

Fig. 6.56. Schematic diagram (not to scale) demonstrating the formation of the in-plane holes on the chemically exfoliated GO nanosheets and the holey GO (HGO) papers through filtration. Thermal reduction of the GO papers resulted in rHGO papers with in-plane porosity. Water molecules (blue spheres) were removed from the interlayers, leading to a 3D network with interconnecting graphitic domains having high electrical conductivity, structural integrity and disordered porous graphene regions for inter-plane diffusion. Li ions (green spheres) would diffuse throughout the structure rapidly through the in-plane pores and inter-plane channels. The interlayer spacing between the graphene sheets was greatly enlarged. The photograph shows an Ar-rHGO-III paper. Reproduced with permission from [149], Copyright © 2011, American Chemical Society.

GO and HGO papers were thermally reduced at 700°C in a flow of either pure Ar or mixture of H_2/Ar (10%/90%) to form rGO papers with high conductivity. The thermally reduced rGO papers had a three-dimensionally connected graphitic framework, consisting of partial overlap and coalescing of graphene nanosheets and pockets formed due to the rapid evaporation of water molecules that were trapped in between the GO nanosheets during the heating process. The in-plane porosity of the HGO was retained after thermal reduction, as seen in Fig. 6.57. The absence of most oxygen functional groups due to the thermal reduction was confirmed by the XPS analysis results of the reduced HGO (rHGO) papers. Comparatively, the reduction in the presence of H_2 was more effective than the reduction in pure Ar.

According to the SEM images, the size of the pores in the rHGO papers increased by increasing the content of the acid in the mild chemical treating process. The largest pores, observed in the Ar-rHGO papers I, II, III and IV, were 7, 20, 80 and 600 nm, respectively. The larger pores in the samples reduced in Ar were more easily identified than those reduced in the presence of H_2. For instance, the Ar-rHGO-IV had many 500–600 nm pores (Fig. 6.57 (H)), while the pore sizes in H-rHGO-IV were in the range of 200–300 nm (Fig. 6.57 (I)). The difference in pore size between the Ar- and H_2-reduced papers could be attributed to their different reduction mechanisms.

The reduction in Ar was conducted through the decomposition of the epoxy and carboxylate groups in the form of CO_2 and/or CO molecules. As a result, the samples would experience a large loss of carbon, together with the necessary atomic rearrangement in this case, so that the higher the content of oxygen, the severer the occurrence in the sample would be and it would be the most severe for a sample of the highest

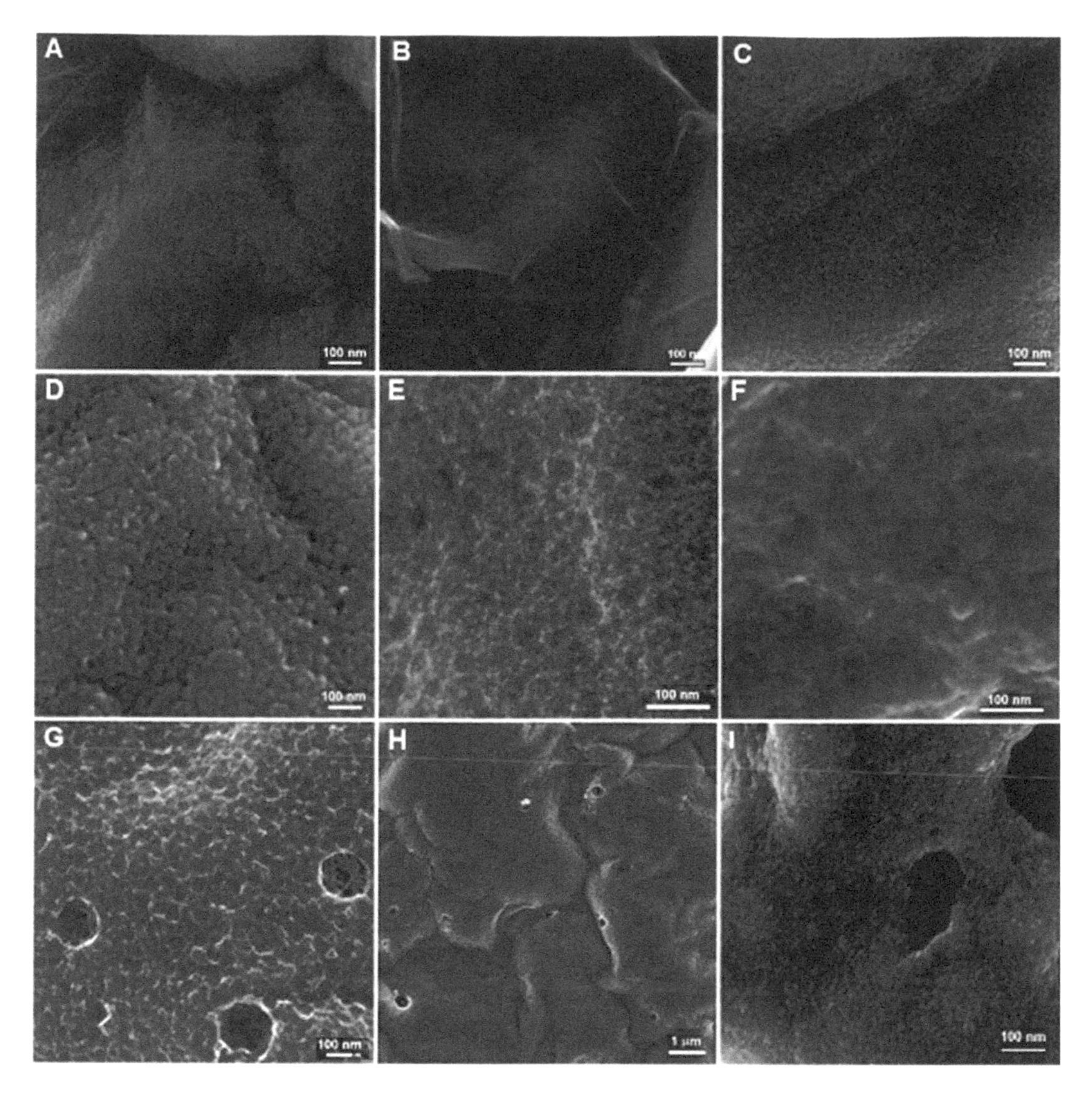

Fig. 6.57. Top surface SEM images of the rGO papers: (A) Ar-rGO, (B) H-rGO, (C) Ar-rHGO-I, (D) Ar-rHGOII, (E) Ar-rHGO-III, (F) H-rHGO-III, (G) Ar-rHGO-IV, (H) Ar-rHGO-IV at low magnification and (I) H-rHGO-IV. Reproduced with permission from [149], Copyright © 2011, American Chemical Society.

initial oxygen content [150]. However, in the presence of H_2, both the hydrogenation of the functional groups to form water and the returning of carbon atoms to the graphene nanosheets were present. The presence of these two competing reactions could reduce the size of the in-plane holes [150, 151].

Due to the highly conjugated structure, graphene and graphene-based materials have extraordinary electrical properties, which thus could be controlled by manipulating the degree of the in-sheet conjugation. This approach has been applied to graphene papers, whose conductivities could be varied over a wide range of 0.001–100 $S{\cdot}cm^{-1}$ [118]. The objective was achieved at the molecular level, through either covalent bonding or π–π stacking interactions by using either monofunctional or bifunctional molecules. Functional molecules, including monoaryl diazonium salts (MDS), bifunctional aryl diazonium salts (BDS) and bipyrene terminal molecular wire (BPMW), have been used for such a purpose. It was found that both MDS and BDS led to a decrease in the conductivity of the graphene papers, with BDS having a finer modification effect. In contrast, the graphene papers modified with BPMW had higher electrical conductivity than that of the other two groups of samples.

Chemically converted graphene obtained by reducing GO at 95°C in hydrazine solution at pH = 10 was used in this study. Covalent modifications with MDS and BDS molecules are shown in Fig. 6.58 (a, b), which disturbed the in-sheet conjugation of the graphene nanosheets [118]. As compared with MDS, BDS also created aryl linkages between the adjacent graphene nanosheets, thus leading to an enhancement in electronic coupling between nanosheets and thus in conductivity between the adjacent nanosheets. In contrast, as the graphene nanosheets were modified with BPMW, π–π stacking interactions were facilitated, without any damage to the in-sheet conjugation, as demonstrated in Fig. 6.58 (c). In this case, the inter-sheet electronic coupling was increased without any destruction of the in-sheet conjugation.

The GO papers were nearly insulating, with resistances of more than 1900 kΩ, corresponding to a conductivity of 5.3×10^{-6} $S \cdot cm^{-1}$. After thermal annealing reduction at 300°C, the derived graphene papers reached a conductivity of 50 $S \cdot cm^{-1}$, which was higher than that of the GO papers by almost seven orders of magnitude, but still much lower than that of high quality single graphene nanosheet. This observation was attributed to the relatively low electron mobility inside the graphene papers, due to the poor contacts between the adjacent graphene nanosheets. Also, graphene nanosheets tend to curve or corrugate, which hindered the close packing of the graphene nanosheets. In other words, in order to increase the electrical conductivity of graphene papers, it is necessary to enhance the inter-sheet electron coupling.

The purpose of the modification discussed above was to address this important issue. As clearly illustrated in Fig. 6.59 (a), the electrical conductivity was decreased constantly with the increasing molecular molar ratio of MDS to GDB [118]. However, if the content of MDS was too high the conductivity of the graphene papers would be negatively influenced, which could imply that the conjugation pathways for charge transfer were damaged. At the same time, it became more difficult to fabricate the graphene papers, which was attributed to the fact that the in-sheet aromatic conjugation of the graphene nanosheets was destroyed by the attached MDS molecules, so that the corresponding π–π interactions between the graphene nanosheets were lost. In addition, the aryl linkers were inserted between the graphene nanosheets made graphene papers to be expanded, which in turn weakened the electron transfer. Fortunately, the electrical conductivity of the as-modified graphene papers could be recovered through thermal annealing at 500°C, as demonstrated in Fig. 6.59 (a) [118]. Figure 6.59 (b) indicated that there was a sharp drop in the electrical conductivity of the graphene papers at a relatively low feeding ratio of BDS, which became the stable value as the content of BDS was > 10%. The BDS-modified graphene papers had higher conductivities than the MDS-modified samples, while it was easier to fabricate graphene papers when using MDS as the modifier.

To establish conductive links between the graphene nanosheets while not disrupting the in-sheet conjugation, so as to achieve high conductivity, π–π stacking interactions were explored, by using pyrene-functional precursors that could be attached onto aromatic macromolecules, including various nanocarbons

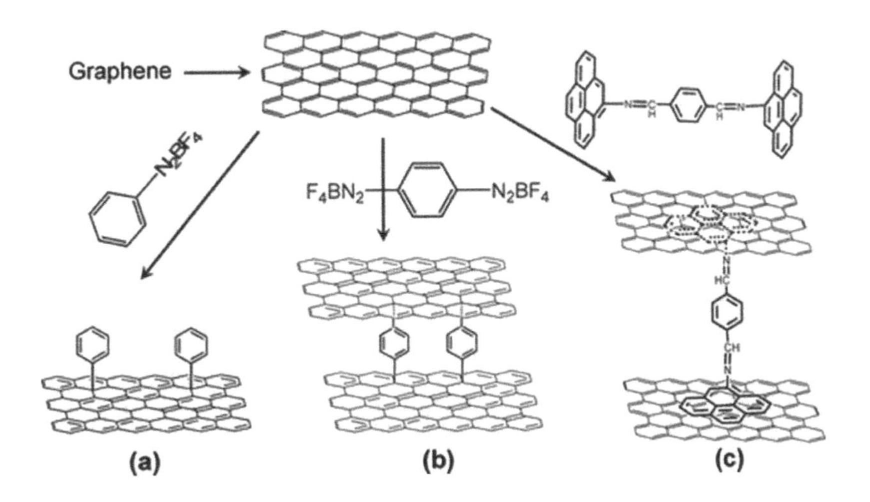

Fig. 6.58. Schematic diagram showing the modification of the graphene nanosheets with (a) MDS, (b) BDS and (c) BPMW. Reproduced with permission from [118], Copyright © 2012, American Chemical Society.

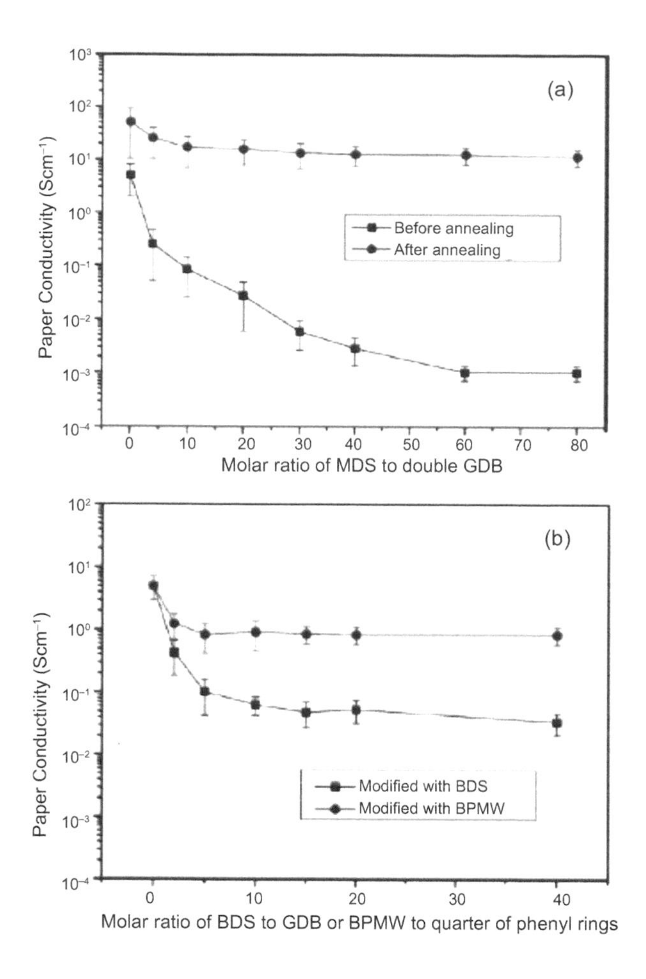

Fig. 6.59. Electrical conductivities of the graphene papers derived from the graphene nanosheets modified with (a) MDS at different ratios of MDS to GDB before (filled square) and after (filled circle) annealing and (b) with BDS at different molar ratios of BDS to GDB (filled square) and BPMW at different molar ratios of BPMW to a quarter of the phenyl rings (filled circle). Reproduced with permission from [118], Copyright © 2012, American Chemical Society.

[152–157]. BPMW was used to achieve this, which had no effect on the in-sheet electrical properties, but largely promoted the inter-sheet electrical conduction. This hypothesis was supported by the experimental results, as illustrated in Fig. 6.59 (b). Only a very slight drop in electrical conductivity was observed, while the stable conductivity was maintained at much a higher level as compared with other two groups.

A model has been proposed to describe the in-sheet and inter-sheet electrical conduction characteristics of the modified graphene papers, as illustrated schematically in Fig. 6.60 (a). Two paths have been responsible for charge transfer in graphene papers, i.e., in-sheet and inter-sheet. The in-sheet charge transfer is governed by the graphene conjugation structure (1), while the inter-sheet charge transfer is realized vie either electron hopping from one sheet to another when they are sufficiently close enough (2) [6, 158, 159] or through the molecular modifiers (3). It has been accepted that electronic coupling between the adjacent graphitic layers is very weak [160], so that constructing a charge transfer path by using a modifier becomes an effective way to increase the electrical conductivity of graphene papers.

When MDS was used to modify the graphene papers, the in-sheet conjugation was partially damaged, which led to the decrease in the conductivity. Also, the presence of the aryl linkers increased the spacing of the graphene nanosheets, which had a negative effect on the charge transfer. Similarly, BDS also destroyed the in-sheet conjugation and thus decreased the in-sheet conductivity. However, because the BDS molecule

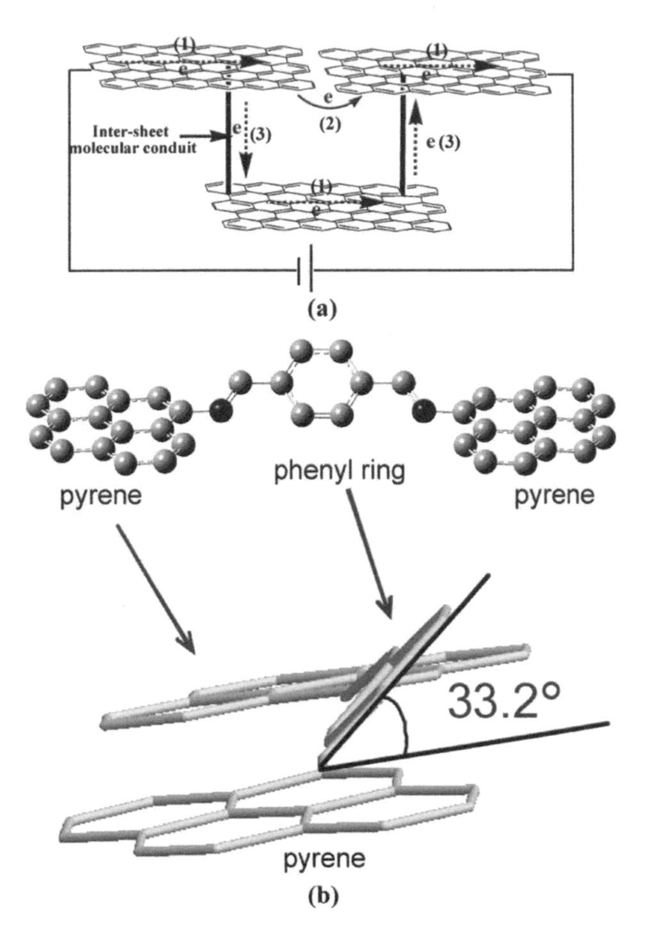

Fig. 6.60. (a) Model of the in-sheet and inter-sheet charge transfer of the graphene papers. (1) The in-sheet charge transfer, (2) the electron hopping from one sheet to another and (3) the charge transfer through the inter-sheet molecular conduits. (b) Calculated 3D structure of BPMW (without hydrogen atoms) using Gauss program (upper) and orientated 3D BPMW structure showing the plane-to-plane angle between the coplanar pyrene and the phenyl groups. Reproduced with permission from [118], Copyright © 2012, American Chemical Society.

had two diazonium functional groups, it would bridge every two adjacent nanosheets to form molecular conduit for inter-sheet charge transfer. As a consequence, BDS was more efficient in enhancing the electrical conductivity of the graphene papers.

Figure 6.60 (b) shows the most stable 3D structure of BPMW, according to density functional theory (DFT) study [118]. The central aryl moiety and the coplanar pyrene groups were not in a same plane, by deviating with a plane-to-plane angle of 33.2°. Due to this twisted 3D structure, the BPMW molecule could not link the graphene nanosheets through normal π–π stacking, so that ordered graphene structure was not obtained, which had been confirmed by the results of SEM observation.

Figure 6.61 shows cross-sectional SEM images of the graphene papers derived from graphene nanosheets modified with different functional precursors. The graphene papers with the modification of MDS were relatively denser, while the BDS and BPMW samples were much thicker with a net-like microstructure. As discussed above, the expansion of the BDS and BPMW papers was ascribed to the cross-linking effect of BDS and BPMW, while the 3D distorted structure of the BPMW was closely related to the molecular structure. Accordingly, the MDS graphene paper had a thickness of 6.5 μm, which was only slightly thicker than that of the pure graphene paper of 4 μm. In contrast, the BDS and BPMW graphene papers had thicknesses of 25 and 20 μm, respectively. The SEM results were supported by the XRD analysis results and were consistent with their mechanical properties.

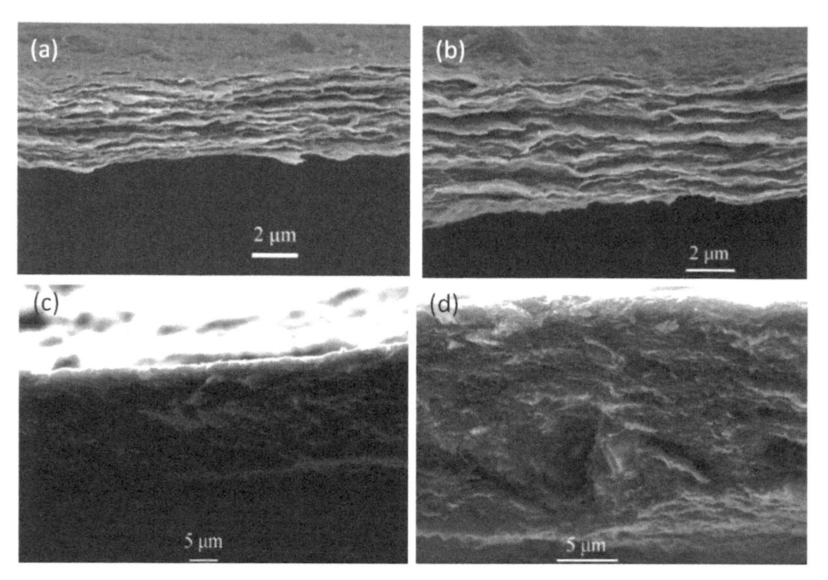

Fig. 6.61. Cross-sectional SEM images of the graphene papers made from the graphene nanosheets modified with different molecules: (a) pure graphene, (b) MDS, (c) BDS and (d) BPMW. Reproduced with permission from [118], Copyright © 2012, American Chemical Society.

Graphene papers consist of tightly stacked graphene nanosheets, thus leading to poor permeability, which could limit their applications where high permeability and tunable layer spacing are required. A simple method was developed to modulate the layer spacing of GO papers, by using the temperature-dependent decomposition reaction of ammonium nitrate (AN, NH_4NO_3) [119]. Different from the commonly used intercalation method, this approach could tailor the layer spacing of GO papers over a very large range of 123%–20,000%, but without the presence of any guest molecules. With such a wide range of expansion amplitude, the GO papers could find various applications, such as the highly efficient exclusion of small organic molecules, separation of ultrathin nanoparticles and loading of polar and nonpolar guest molecules. Moreover, once the GO papers were reduced to rGO papers, with high electrical conductivity, they could be used for flexible paper-based electrochemical devices.

AN has three decomposition reactions at different temperatures over 140–400°C. At 140°C, it decomposes slowly into ammonia (NH_3) and nitric acid (HNO_3). At 200°C, the decomposition products are N_2O and H_2O. At 400°C, it decomposes suddenly into N_2, O_2 and H_2O, with 1 mol AN to produce 3.5 mol gases. Therefore, it can be used to tailor the inter-layer spacing of GO papers by heating at different temperatures, as demonstrated in Fig. 6.62.

20 ml aqueous GO solution with a concentration of 0.5 mg·ml^{-1} was used to prepare the GO papers by using vacuum filtration, with 20 ml of GO solution through a mixed cellulose–ester membrane, having a diameter of 50 mm and pore size of 0.2 μm. After filtration, the wet GO papers together with the cellulose membrane were immediately immersed into AN solution with a concentration of 40 wt%. After being immersed overnight, the samples were dried at room temperature. Then, the GO papers could be easily peeled off from the membranes, and were heated at the temperatures of 140, 200 and 400°C, for 120 min, 120 min and 5 s, respectively.

Figure 6.63 (a) shows a GO paper with a thickness of 5.6 ± 0.3 μm. The GO paper exhibited a smooth surface (Fig. 6.63 (e)), while the GO nanosheets were nearly parallelly stacked against one another (Fig. 6.63 (i)). With the intercalation of AN, the weight of the papers was slightly increased by about 0.01 g, which was also confirmed by the SAXRD measurements, showing an increase in the periodicity from 2.3 to 2.7 nm, as revealed in Fig. 6.64.

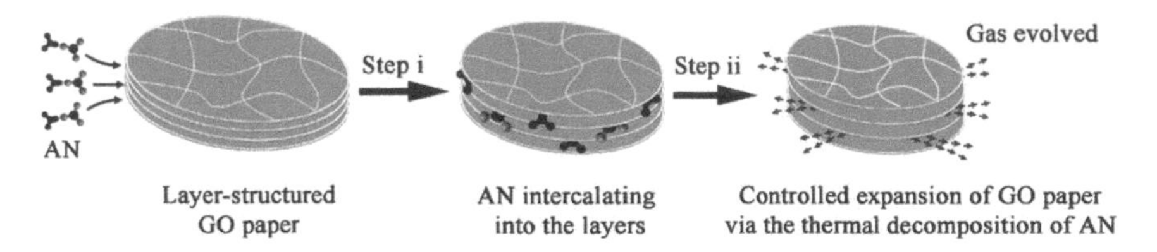

Fig. 6.62. Scheme diagram showing the selective expansion of GO papers. Reproduced with permission from [119], Copyright © 2012, American Chemical Society.

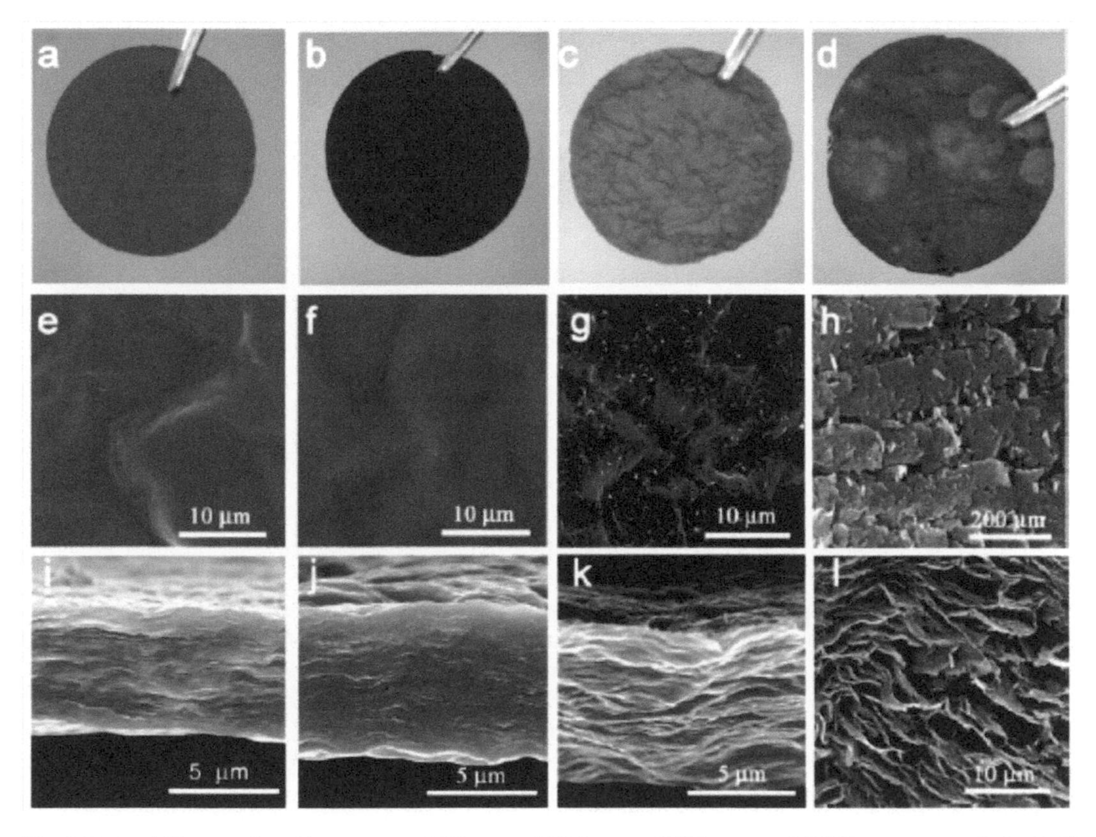

Fig. 6.63. (a–d) Photographs of the as-prepared GO paper, 140°C-paper, 200°C-paper and 400°C-paper, respectively. (e–h) Surface SEM images of the as-fabricated GO paper, 140°C-paper, 200°C-paper and 400°C-paper, respectively. (i–l) Cross-sectional SEM images of the as-obtained GO paper, 140°C-paper, 200°C-paper and 400°C-paper, respectively. Reproduced with permission from [119], Copyright © 2012, American Chemical Society.

After heating at 140°C, the color of the GO paper varied from translucent brown to opaque black, as seen in Fig. 6.63 (b). The decrease in transmittance and the variation in color were attributed to two reasons, (i) the increase in the thickness of the paper and (ii) the increment in the extinction coefficient caused by the thermal conversion. SEM image shown in Fig. 6.63 (j) indicated that the 140°C heated paper possessed a thickness of 6.9 ± 0.2 μm, which was expanded by 123% as compared to that of the unheated paper (5.6 ± 0.3 μm), while the near-parallel stacking of the GO nanosheets was not affected, probably due to the gentle decomposition of AN at 140°C. The ordered structure was also confirmed by the small angle X-ray diffraction (SAXRD) characterization, corresponding to a periodicity of about 2.5 nm (Fig. 6.64). Therefore, the interlayer spacing of the 140°C heated GO paper was increased by about 10% relative to that of the unheated GO paper, which was very close to the expansion ratio observed from SEM results.

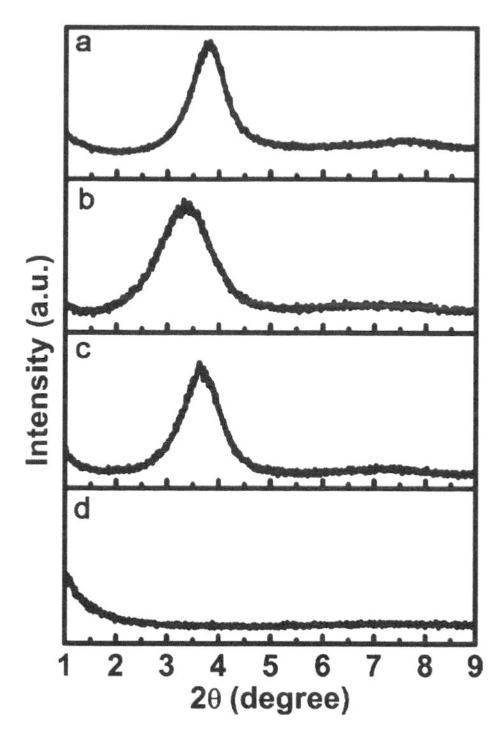

Fig. 6.64. SAXRD patterns of the GO papers: (a) the pristine GO paper, (b) GO paper intercalated by AN, (c) 140°C-paper and (d) 200°C-paper. Reproduced with permission from [119], Copyright © 2012, American Chemical Society.

The 200°C heated paper exhibited metallic luster, as viewed in Fig. 6.63 (c). It had physical thickness of 7.5 ± 0.2 μm, corresponding to an expansion ratio of 134%, as illustrated in Fig. 6.63 (k). Similarly, the 200°C-paper also possessed an ordered layer structure, while the long range ordering of the layer was weakened, so no diffraction peak was present in the SAXRD pattern in Fig. 6.64 (d). It was obviously attributed to the accelerated decomposition of AN at 200°C, so that the layer spacing was significantly increased, while the periodical structure of the GO nanosheets was damaged at the same time. However, according to the expansion ratio and the layer spacing of the unexpanded GO paper, the average layer spacing of the 200°C-paper was about 3.0 nm. Interestingly, the SEM image in Fig. 6.63 (i) revealed that in the 200°C-paper the edges of some GO nanosheets were lifted up by the gases evolved in the decomposition reaction.

Additionally, the GO nanosheets of the 200°C-paper were reduced to rGO, as evidenced by IR and Raman results. The reduction of GO in the 140°C-paper was due to the thermal effect, whereas the 200°C-paper experienced chemical reduction besides the thermal reduction. The chemical reduction was caused by the reducing gas of N_2O that was produced due to the decomposition of AN at 200°C. As a result, the 200°C-paper was entirely reduced, so as to exhibit metallic luster.

Because of the violent decomposition of AN, the GO paper heated at 400°C was loose and elastic only within 3–5 s, as illustrated in Fig. 6.63 (d). In this sample, the GO nanosheets were highly disordered by the gases, so that a large number of pores were created, as demonstrated in Fig. 6.63 (l). Thickness of the 400°C-paper was around 1.1 mm, which was thicker than the as-prepared GO paper by about 200 times. In addition, many cracks were observed in the paper, as seen in Fig. 6.63 (h). Such a microstructure could be beneficial to some applications requiring large surface area.

Besides wet-chemical exfoliated GO and graphene nanosheets, chemical vapor deposition (CVD) derived graphenes have also been used to fabricate graphene papers by using vacuum membrane filtration [108]. The graphene papers from the CVD-derived graphene exhibited superior electrochemical performances, with a capacity of 1200 $mAh \cdot g^{-1}$ at 50 $mA \cdot g^{-1}$ as electrodes of supercapacitors and exhibited excellent cycling characteristics as anodes of lithium ion rechargeable batteries. The CVD graphene nanosheets had a lateral size in the order of hundreds of micrometers, with a production at gram scale, by using a layered Fe containing natural mineral, expanded vermiculite, as the template. Due to the large sheet size of the CVD graphene nanosheets, the obtained graphene papers were highly flexible with a loose stacking state. Moreover, graphene–carbon nanotube (G–CNT) hybrids could also be synthesized in just one-step.

Vermiculite is a mica-like lamellar mineral that rapidly expands upon heating, so as to form a lightweight material, which is called expanded vermiculite. The expanded vermiculite was a golden powder with lamellar layers of about 500 μm and smooth surfaces, as shown in Fig. 6.65 (a), which served as substrates for the growth of graphene nanosheets. As illustrated in Fig. 6.65 (b, d, e), the as-synthesized graphene nanosheets had an extraordinarily large lateral size of 200–500 μm. The as-synthesized graphene powder was purified and dispersed in ethanol (0.4 g graphene in 40 ml ethanol) with the aid of ultrasonication for 10 min. After removing the sediments at bottom of the suspension, the suspension was used to obtain graphene papers through vacuum filtration by using a nylon filter membrane (0.2 μm). The wet papers were dried at 80°C for 2 h. Figure 6.65 (c) shows a photograph of the graphene papers, exhibiting a smooth surface. Tensile test results indicated that the as-obtained graphene papers had a mechanical strength of 36 MPa and a tensile ratio of up to 1.5%, demonstrating the high quality of the CVD-graphene papers.

Figure 6.65 (f) shows a high resolution TEM image of the products that were synthesized by using ethylene as the carbon source, which were multilayered graphene nanosheets with 3–8 layers, and

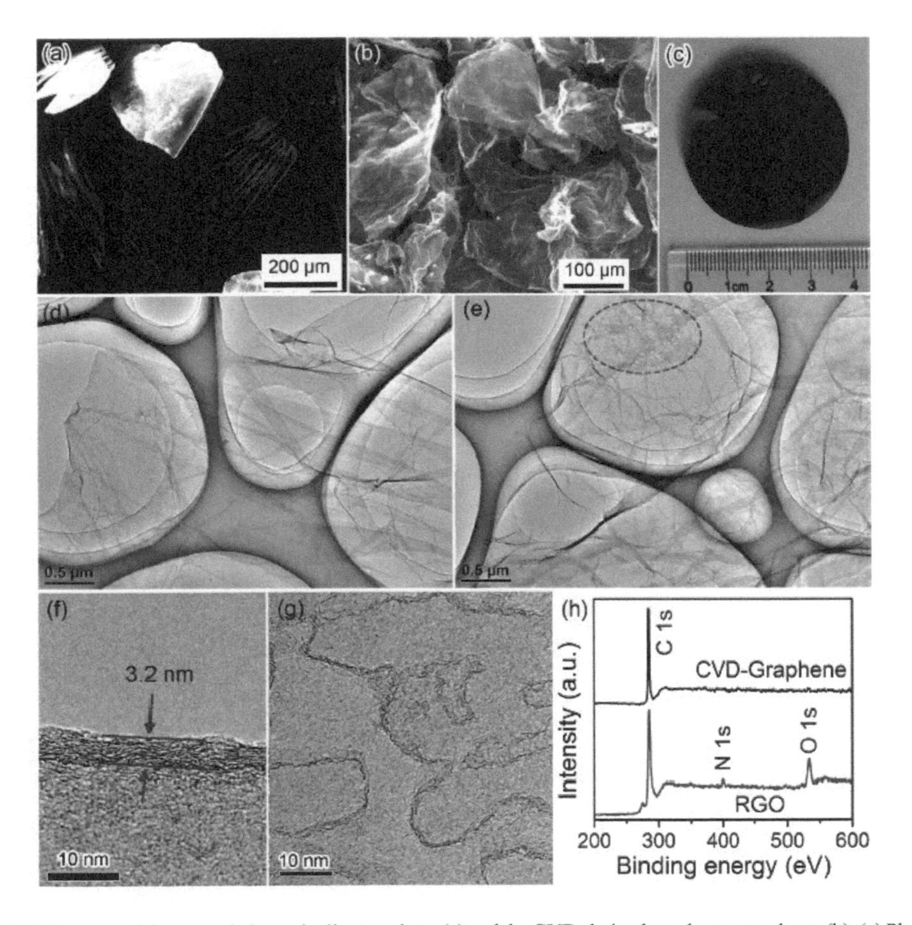

Fig. 6.65. SEM images of the expanded vermiculite templates (a) and the CVD-derived graphene nanosheets (b). (c) Photograph of the as-obtained graphene paper. (d, e) Low magnification and (f, g) high magnification TEM images of the CVD graphene nanosheets. The dotted ellipse in (e) indicates the pore structures of the graphene nanosheet shown in (g). (h) XPS spectra of the CVD-derived graphene and wet-chemically exfoliated rGO. Reproduced with permission from [108], Copyright © 2013, The Royal Society of Chemistry.

corresponding thicknesses of 1.5–4 nm. If methane was used as the carbon source at a reaction temperature of 900°C, graphene nanosheets with 1–2 layers would be obtained. Comparatively, the carbon yield of methane was quite low (< 1 wt%), so that it was not suitable for large-scale production. As seen in Fig. 6.65 (e), the graphene nanosheets contained some porous areas, which could be related to the incomplete crystallization at the surfaces of the vermiculite (Fig. 6.65 (g)). According to XPS analysis results, no oxygen-containing functional groups were observed in the CVD graphene nanosheets. In contrast, a sharp O1s peak was present in the wet-chemical derived rGO, as observed in Fig. 6.65 (h).

When H_2 was introduced into the reactant gas, Fe nanoparticles were formed on the vermiculite layers, which acted as the catalyst to facilitate the growth of CNT, together with graphene, thus leading to G–CNT hybrids. The CNTs were homogeneously distributed on the graphene nanosheets. The graphene to CNT ratio could be controlled by varying the H_2 to ethylene ratio. As the hydrogen to ethylene ratio was 1:2, only CNTs were obtained without the formation of graphene. To obtain G–CNT hybrids with an appropriate graphene to CNT ratio, the optimal hydrogen to ethylene ratio was 1:10.

Figure 6.66 (a) shows photographs of the flexible discs with a diameter of 13 mm cut from the CVD graphene papers, for use as the anodes of lithium ion batteries. Figure 6.66 (b) shows a surface SEM image of the graphene papers. The surface was very smooth, which was attributed to the fact that the papers were made of large-area graphene nanosheets overlapping one another. Thickness and surface density of the graphene papers could be easily controlled by controlling the volume of the graphene suspension. The arrangement of the graphene nanosheets was slightly disordered, as illustrated in Fig. 6.66 (c), also due to the large dimension of the graphene nanosheets.

Graphene quantum dots (GQDs) have been proposed as stabilizers to achieve aqueous dispersions of graphene nanosheets [104]. GQDs have both special atomic structure and surface chemistry, with one- to few-layered graphene nanosheets that have a lateral dimension smaller than 100 nm [161–163]. Due to the 2D sp^2 carbon structure, GQDs are expected to be able to strongly attach to the basal plane of graphene nanosheets through van der Waals attractions, i.e., π–π stacking. Also, GQDs contain the same hydrophilic surface groups, such as carboxyl, hydroxyl and epoxy groups, as GO, due to the similarity in their synthetic processing [164, 165]. Due to the presence of these hydrophilic groups, GQDs have high solubility in water, so that they could promote the dispersion of the intrinsically hydrophobic graphene nanosheets. With the incorporation of GQDs, commercially available graphene powder could be well-dispersed in water, which led to highly flexible and highly conductive graphene papers by using the vacuum filtration technique.

Homogeneous brown solution GQDs with a concentration of 0.5 $mg \cdot ml^{-1}$ could be readily prepared by ultrasonicating the mixture for only 10 min, with the presence of obvious Tyndall effect, as seen in Fig. 6.67 (a). In contrast, graphene powder could be dispersed in water, as demonstrated in Fig. 6.67 (b–d), because of the surface energy mismatch between graphene and water. According to XPS and FT-IR results in Fig. 6.67 (e–g), the GQDs had a high content of oxygen (27.8%), in the form of oxygen-containing (including carboxylic acid and hydroxyl) groups.

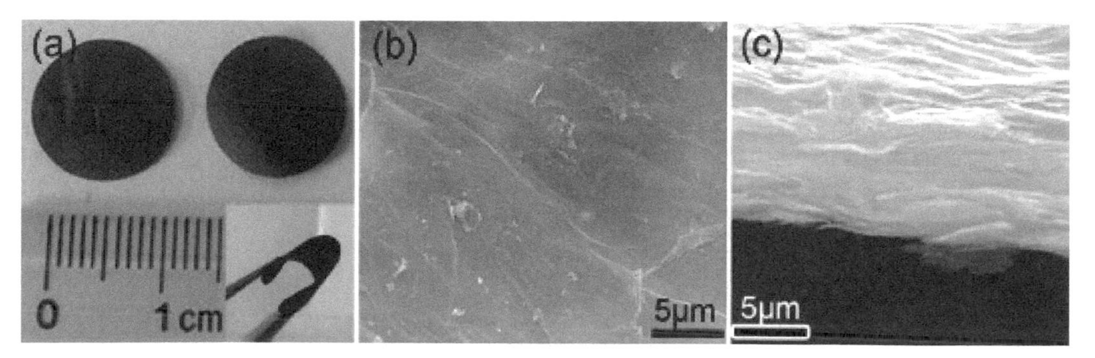

Fig. 6.66. Photographs (a) and SEM images (b: surface, c: side view) of the graphene discs used as the anodes of lithium ion battery measurement. Reproduced with permission from [108], Copyright © 2013, The Royal Society of Chemistry.

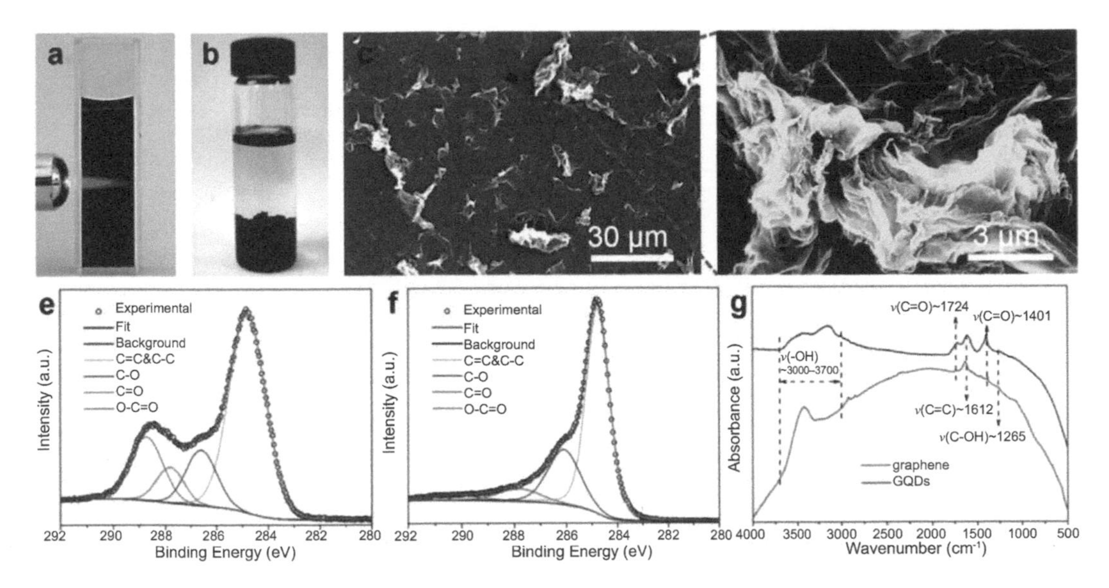

Fig. 6.67. Dispersion behaviors of the GQDs and graphene powder in water. (a) Photograph of the GQDs dispersed in water at 0.5 mg·ml^{-1} with obvious Tyndall effect with a red laser beam passing through, suggesting the formation of uniform colloidal dispersion. (b) Photograph of the graphene powder precipitated in water dispersion after 2 h of ultrasonication and 1 day standing. (c, d) SEM images of the precipitated graphene powder revealing the agglomerated graphene nanosheets. C 1s XPS spectra of (e) the GQDs and (f) graphene revealing different surface chemistry, together with their (g) FT-IR spectra. Reproduced with permission from [108], Copyright © 2015, American Chemical Society.

Color image of this figure appears in the color plate section at the end of the book.

Figure 6.68 shows a schematic diagram describing the detailed fabrication process. The graphene dispersions were prepared using three steps, including (i) dispersing graphene powder in the GQDs solution, (ii) removing the excess GQDs and (iii) re-dispersing the graphene slurry. In this case, the presence of the GQDs was a crucial requirement for stabilizing the graphene nanosheets in water. There could be two mechanisms governing the enhanced dispersion of graphene. Firstly, the surface of the graphene was modified due to the absorption of the GQDs, so that the graphene nanosheets became negatively charged and thus electrostatically stabilized in water. Secondly, the GQDs were uniformly distributed to surround the graphene nanosheets, so that they were blocked from one another. As a consequence, agglomeration of the graphene nanosheets was effectively prevented.

By removing the unnecessary GQDs, graphene slurry was obtained. Detailed analysis indicated that 7.8 mg of GQDs had been attached to 100 mg of graphene powder. The attached GQDs could not be removed with repeated washing and filtration. The resulting aqueous slurry with defined composition could be easily re-dispersed in water by sonicating for 30 min and diluted to dispersion with specific graphene concentrations. Figure 6.69 (a) shows a photograph of the re-dispersed graphene suspension, with a concentration of 0.2 mg·ml^{-1}, with colloidal behavior evidenced by the clear Tyndall effect phenomenon.

The concentration could be controlled in the range of 0.1–0.8 mg·ml^{-1} by varying the final volume of the dispersions. Figure 6.69 (b) shows the UV–vis absorption spectra of the graphene dispersion, solution of GQDs and graphene/GQDs complex dispersion. The graphene dispersion had an absorption peak at 276 nm, implying the presence of sp^2 carbon structure. Note the absorption peak of GQDs at 227 nm and the broad peak of graphene/GQDs between 227 and 276 nm, a sharp peak at 276 nm suggested that the unattached GQDs had been removed. Figure 6.69 (c) indicated that no significant loss in absorbance was observed in the dispersion with a concentration of 0.4 mg·ml^{-1} during the initial 60 h, whereas the 0.5–0.8 mg·ml^{-1} samples experienced a quick decrease in absorbance. Therefore, the stable concentration of graphene was 0.4 mg·ml^{-1}. As the graphene suspensions were deposited on Si substrates, a thin layer of

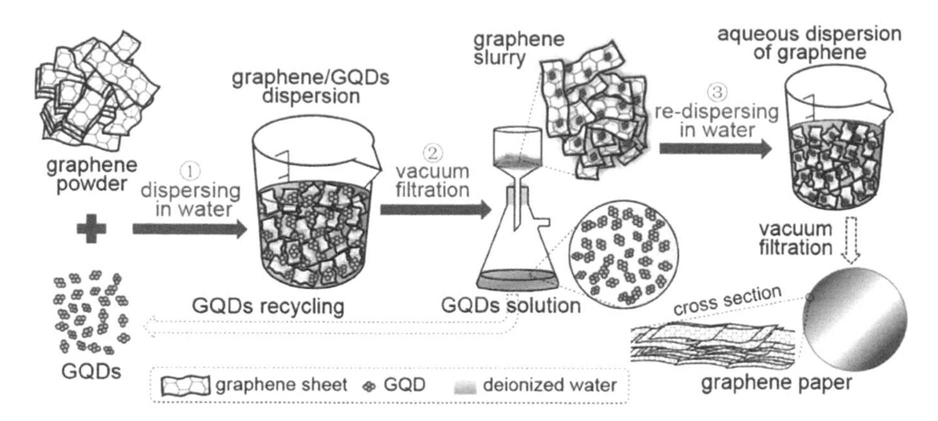

Fig. 6.68. Schematic diagram showing the fabrication process of the graphene papers from aqueous graphene dispersion: (1) dispersion of graphene powders and GQDs in water to obtain graphene/GQDs composite dispersion, (2) removal of excessive GQDs through vacuum filtration to get graphene aqueous slurry and (3) re-dispersion of the graphene slurry to form aqueous graphene dispersion for paper fabrication. Reproduced with permission from [108], Copyright © 2015, American Chemical Society.

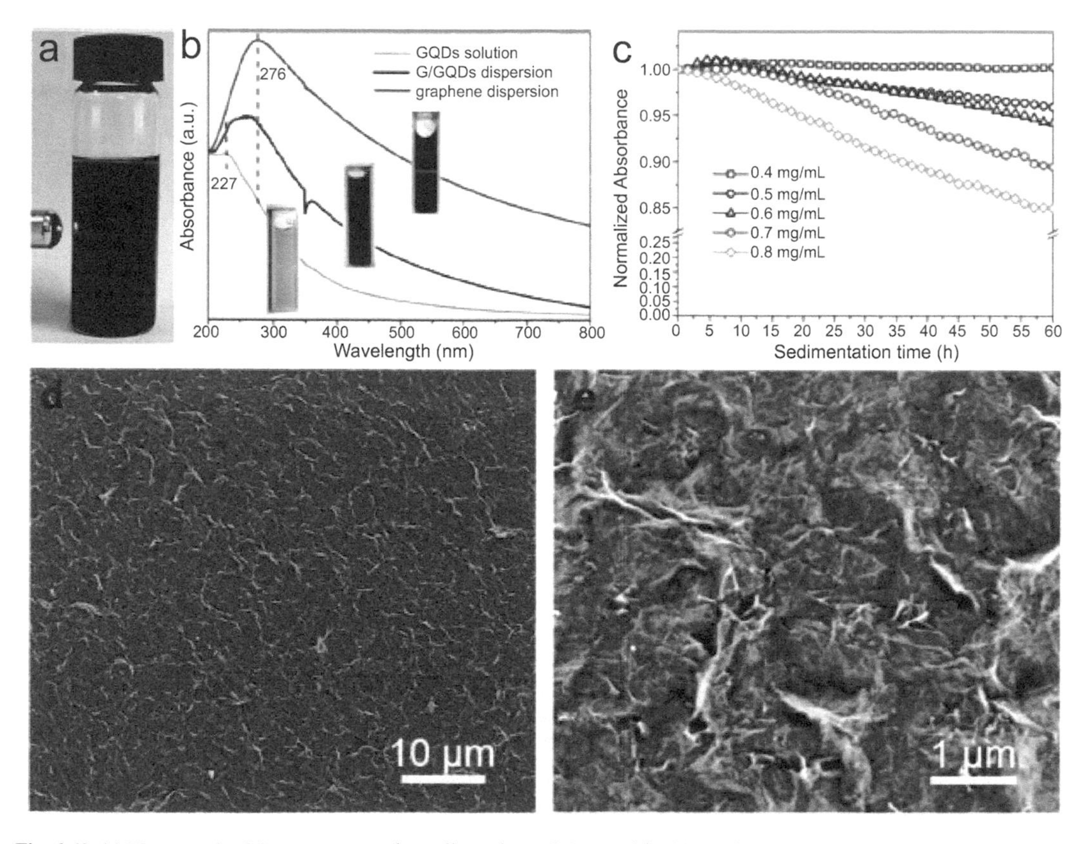

Fig. 6.69. (a) Photograph of the aqueous graphene dispersion at 0.2 mg·ml^{-1} with a red laser beam passing through, displaying evident Tyndall effect. (b) UV–vis absorbance spectra of the GQDs, GQDs/graphene and graphene water dispersions, with the insets demonstrating the dispersions in 1 mm quartz cuvettes. (c) Sedimentation behaviors of graphene dispersions with different concentrations (0.4–0.8 mg·ml^{-1}), indicating the stable concentration to be 0.4 mg·ml^{-1}. (d, e) Low- and high-magnification SEM images of the deposited graphene nanosheets on silicon wafer from the graphene dispersions, showing the outstretched nanosheets which confirmed the absence of aggregates. Reproduced with permission from [108], Copyright © 2015, American Chemical Society.

uniform film was formed, as illustrated in Fig. 6.69 (d, e). As confirmed in Fig. 6.69 (f, g), the dispersed graphene consisted of wrinkled nanosheets with a thickness of less than 3 nm, implying that the graphene nanosheets were not agglomerated in water due to the presence of the GQDs.

Generally, 100 ml graphene dispersion with a concentration of 0.2 $mg \cdot ml^{-1}$ was used to prepare graphene papers, followed by drying overnight at 60°C. Figure 6.70 (a) shows a photograph of the free-standing graphene papers with a diameter of about 3.8 cm, which were highly flexible and bendable. A free-standing and bendable paper with a diameter of about 3.8 cm (Fig. 6.70 (a)) was obtained after peeling off the membrane. Cross-sectional SEM images of the graphene papers are shown in Fig. 6.70 (b, c), at low- and high-magnifications. The graphene nanosheets were highly oriented in the plane of the papers, which suggested that the graphene nanosheets were parallelly deposited on the filter membrane. The completely dried graphene papers had apparent densities in the range of 1.22−1.51 $g \cdot cm^{-3}$.

The presence of partial π–π stacking between the adjacent graphene nanosheets was confirmed by the XRD results, as illustrated in Fig. 6.70 (d). The graphene papers had a relatively strong peak at 23.7°, corresponding to plane spacing of d_{002} = 0.37 nm. Electric conductivity of the graphene paper was up to 7240 $S \cdot m^{-1}$ at room temperature. It was found that the average conductivity was decreased from 7240 to 2506 $S \cdot m^{-1}$, as the content of the GQDs was increased from 7.0 to 28.8%. Therefore, the content of the GQDs in the graphene papers should be controlled at a sufficiently low level. For a graphene paper with a thickness of 18.0 μm, tensile strength, ultimate tensile strain and Young's modulus were 91.2 MPa, 1.7% and 5.3 GPa, respectively. In summary, this approach could be a new route to develop flexible graphene papers with controllable properties.

G–SWCNT hybrid papers have been prepared by using the membrane filtration method, which were studied as an adsorbent to adsorb various aromatic compounds from aqueous solutions [125]. The adsorbed molecules include 1-pyrenebutyric acid (PBA, $C_{20}H_{16}O_2$), 2,4-dichlorophenoxyacetic acid (2,4-

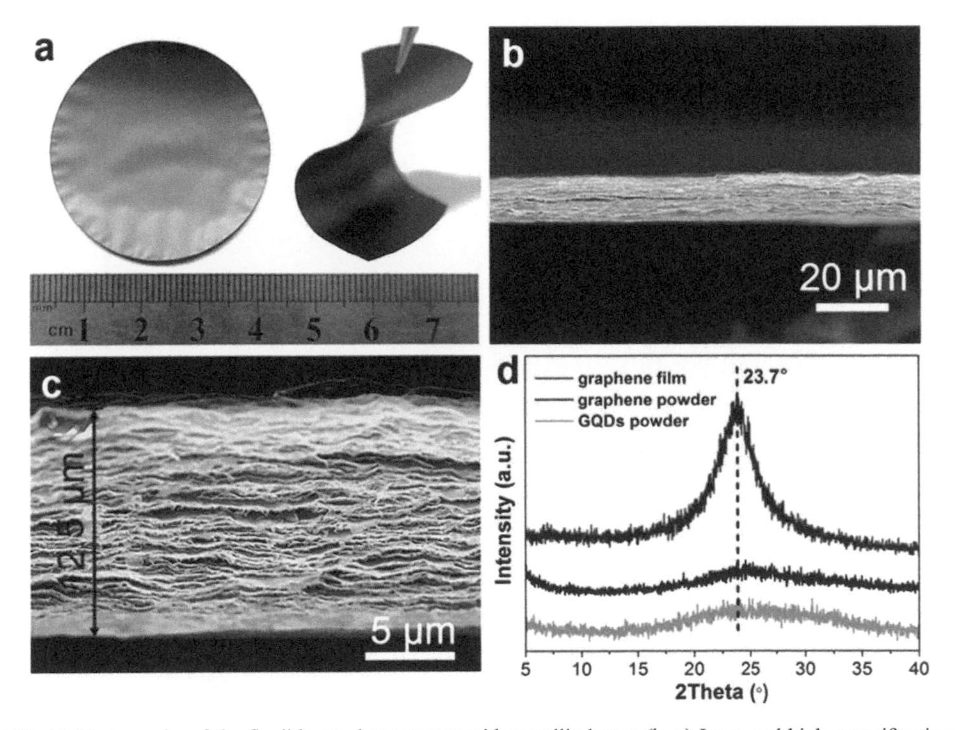

Fig. 6.70. (a) Photographs of the flexible graphene papers with metallic luster. (b, c) Low- and high-magnification cross-sectional SEM images of the graphene papers, illustrating the uniform thickness (12.5 μm) and layered structure. (d) XRD patterns of the graphene paper, graphene powder and GQDs, with a peak observed for the graphene paper confirming the π–π stacking of the graphene nanosheets formed during the paper filtration process. Reproduced with permission from [108], Copyright © 2015, American Chemical Society.

D, $C_8H_6Cl_2O_3$) and diquat dibromide (DqDb, $C_{12}H_{12}Br_2N_2$). PBA was selected as a model polyaromatic compound, while 2,4-D and DqDb were among the most extensively used agricultural herbicides to control broad-leaved weeds.

G–SWCNT hybrids papers were obtained from suspensions with various mass ratios of SWCNT:G, including 1:0, 2:1, 1:1, 1:2 and 0:1. The mixtures were soaked in hydrochloric acid at room temperature for 17 h to remove the metal particles. After that, the mixtures were then filtered through an aqueous membrane and thoroughly washed with DI water. The mixtures were dried and then dispersed in N,N-dimethylacetimide, with the aid of sonication for 45 min. Hybrid papers were obtained by filtering the suspensions through a 47 mm PTFE membrane (0.5 μm). The papers were first wetted with methanol, so as to be easily peeled off from the membrane. The as-obtained papers were annealed at 560°C for 10 min to remove the amorphous carbon.

It was found that the SWCNTs and graphene papers exhibited a slower adsorption rate, which could be attributed to hindered adsorbate diffusion due to the entangled SWCNTs and the closely stacked graphene nanosheets. Generally, the solution-phase adsorption of aromatic compounds is mainly governed by the available adsorption surface area instead of pore volume [166]. Due to their high aspect ratios, the SWCNTs were prone to form bundles and intertwine through the strong van der Waals attractions, as demonstrated in Fig. 6.71 (a). On the other hand, the graphene nanosheets were closely stacked into aggregates, which was attributed to the inter-planar π–π interactions, as seen in Fig. 6.71 (b). As a result, the access of the solute to the adsorption sites was hindered, thus leading to low adsorption rates.

In contrast, as the SWCNTs and graphene nanosheets were combined, aggregation was absent in the hybrid, as revealed in Fig. 6.71 (c). In this case, the SWCNTs were present as a random filamentous network, without the presence of large bundles, so that a large surface area structure was obtained. It was observed that the SWCNTs were attached to the graphene nanosheets at the edges and on the surface. As

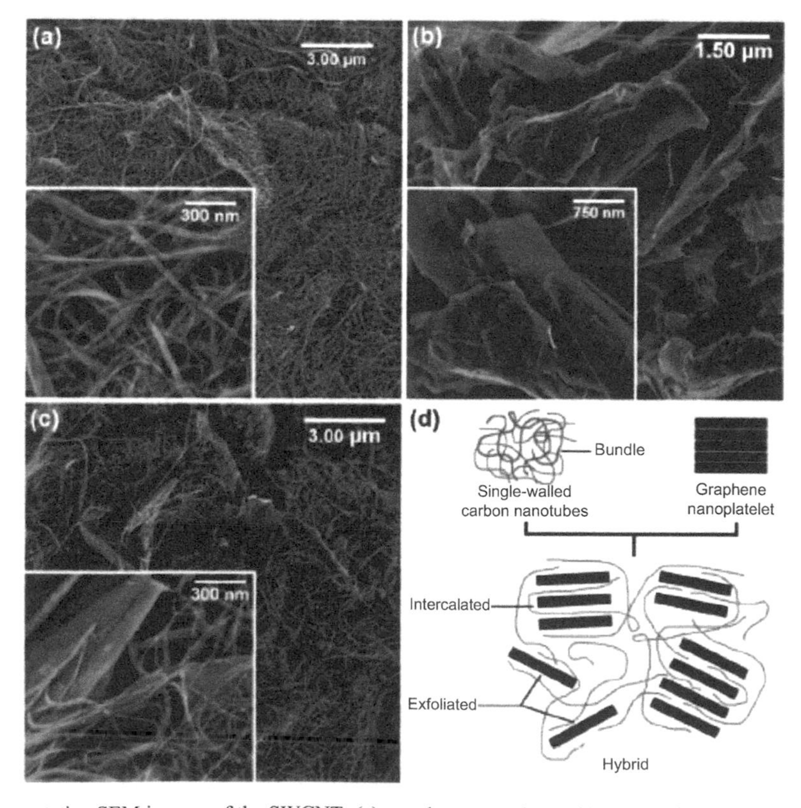

Fig. 6.71. Representative SEM images of the SWCNTs (a), graphene nanosheets (b) and 2:1 SWCNT–G hybrid papers (c). High magnification images are shown in the insets. (d) Schematic diagram of the SWCNT–G hybrid structure. Reproduced with permission from [125], Copyright © 2014, The Royal Society of Chemistry.

a result, the SWCNTs served as spacers to increase the spacing between adjacent graphene nanosheets and to even exfoliate the graphene nanosheets, as shown schematically in Fig. 6.71 (d). Because the adsorbate-accessible surface area was effectively enlarged, the contact between the solute and the carbon nanostructures was significantly increased, thus leading to enhanced adsorption rates and capacities.

A pillared-graphene network structure model has been proposed, which consisted of graphene nanosheets and carbon nanotubes (CNTs) to create a 3D network [167]. Molecular dynamics (MD) simulations predicted that the presence of the CNTs could largely enhance the interfacial thermal conductance between the adjacent graphene nanosheets. This full-carbon architecture potentially has high thermal transport properties. However, CNTs are too long to be used as the pillars to bridge the graphene nanosheets. In this regard, the super-short CNTs, i.e., carbon nanorings (CNRs), could be promising candidates for such a purpose [168]. Therefore, a 3D bridged CNR–graphene hybrid structure could be created, as schematically illustrated in Fig. 6.72 (a), in which the parallel graphene nanosheets acted as the building blocks, while the CNRs were vertically aligned within the inter-gallery spaces acting as pillars.

With this idea, 3D bridged CNR–graphene hybrid papers were fabricated by using the intercalation of polymer carbon source and metal particle catalyst, followed by *in situ* growth of CNRs in the confined space of the 2D interlayer galleries between the graphene nanosheets through thermal annealing [129]. A schematic diagram of the 3D CNT–graphene hybrid structure and a piece of hybrid paper are shown in Fig. 6.72 (c, b). The combination of graphene and CNR highly improved thermal conductivity in the normal direction of the graphene paper and the heat dissipation performance. The CNTs were synthesized on the

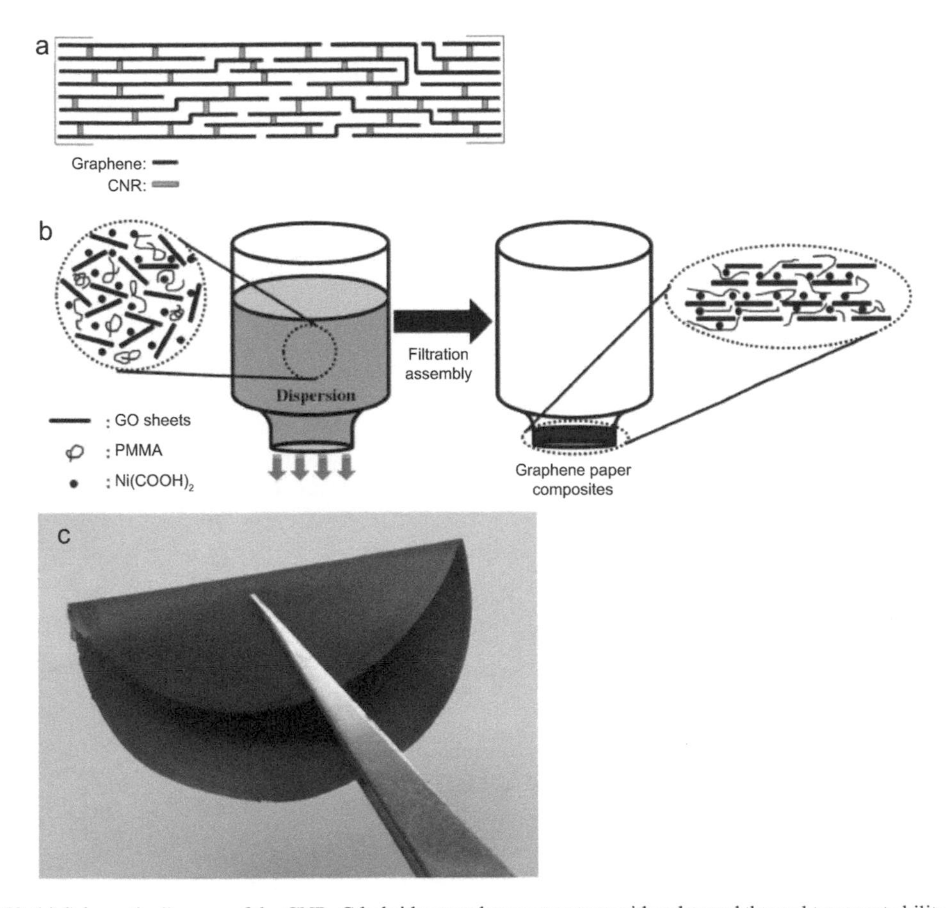

Fig. 6.72. (a) Schematic diagram of the CNR–G hybrid network nanostructure with enhanced thermal transport ability in the normal direction of graphene paper planes. (b) Schematic diagram of the vacuum-filtration fabrication process to make GO-PMMA-Ni$(COOH)_2$ papers. (c) Photograph of a hybrid GO-PMMA-Ni$(COOH)_2$ paper. Reproduced with permission from [129], Copyright © 2015, Wiley-VCH Verlag GmbH & Co. KGaA, Weinheim.

$Ni(COOH)_2$ particles, by using the CVD method with ethylene as the carbon source. Also, if poly methyl methacrylate (PMMA) was introduced into the $Ni(COOH)_2$ particles as a solid carbon source instead of a gas carbon source, CNTs could also be obtained.

As the hybrid papers were annealed, while $Ni(COOH)_2$ and PMMA decomposed, a carbon layer was formed on the metal particles that acted as the catalyst, as indicated by the solid arrows in Fig. 6.73 (a). A ring-shaped nanostructure was clearly observed, with the inter-planar distance measured by using fast Fourier transformation (FFT) to be about 0.334 nm, as illustrated in the inset in Fig. 6.73 (a). The ring-shaped nanostructures were multi-walled CNRs, with an outer diameter and wall thickness of about 12 and about 3 nm, respectively. The multi-walled CNTs were very short, so that the edge of the graphene underneath was clearly visible, as indicated by the solid curves in Fig. 6.73 (a). The dashed arrow indicated the atomic structure of the graphene-CNR junction regions, while the edges of the graphene nanosheets were connected to the outer walls of the CNRs. Figure 6.73 (b) shows a side view of two CNRs and the CNR root regions. The Ni particles were covered by the CNRs. Although the CNRs had few-walled ring-shaped structures, additional shells were observed at the CNR roots, as indicated by the dashed arrows in Fig. 6.73 (b). The CNTs were open-ended at the bottom and covalently bonded to the graphene layer at the junction area.

Cross-sectional SEM images of the rGP, rGP-10 and rGP-100 are shown in Fig. 6.74 (a–c). GP-10 was derived from 20 mg GO, 10 mg PMMA and 10 mg $Ni(COOH)_2$, while GP-100 was made with 100 mg PMMA and 100 mg $Ni(COOH)_2$. The rGP exhibited well-packed layers over the entire cross-section (Fig. 6.72 (a)), while voids were observed in the image of the rGP-10 (Fig. 6.74 (b)). However, the rGP-100 were more disordered, with white particles intercalated in between the graphene nanosheets (Fig. 6.74 (c)). Figure 6.74 (d) shows XRD patterns of the samples. The d-spacings of the adjacent GO nanosheets were 8.48, 10.15 and 10.78 Å, for pristine GP, GP-10 and GP-100, respectively. The increase in the d-spacing of GP-10 and GP-100 suggested that both PMMA and $Ni(COOH)_2$ had been intercalated into the graphene inter-gallery spaces.

After thermal annealing, the sharp XRD peaks of the GP disappeared, whereas new broad diffraction peaks appeared in the patterns of the rGP and rGP-10, which were close to the typical (002) diffraction peak of graphite, with d-spacing of 3.35 Å at $2\theta \approx 26.6°$. The diffraction peak of rGP-10 was much wider than that of rGP, due to its less ordered structure caused by the intercalations. Similarly, the rGP-100 had no distinguishable diffraction peak at $2\theta = 26.6°$, but with two sharp peaks, corresponding to d-spacing values of 3.12 and 3.04 Å, respectively. The two XRD peaks were attributed to the additional shells formed in the CNR root region.

A simple and efficient method has been reported to prepare flexible multifunctional free-standing montmorillonite-graphene (MMT-G) hybrid papers by using vacuum membrane filtration [106]. Aqueous colloidal dispersion of rGO, stabilized with MMT nanoplatelets, was obtained by directly reducing GO in the presence of exfoliated MMT nanoplatelets. The formation of stable rGO dispersion was attributed to

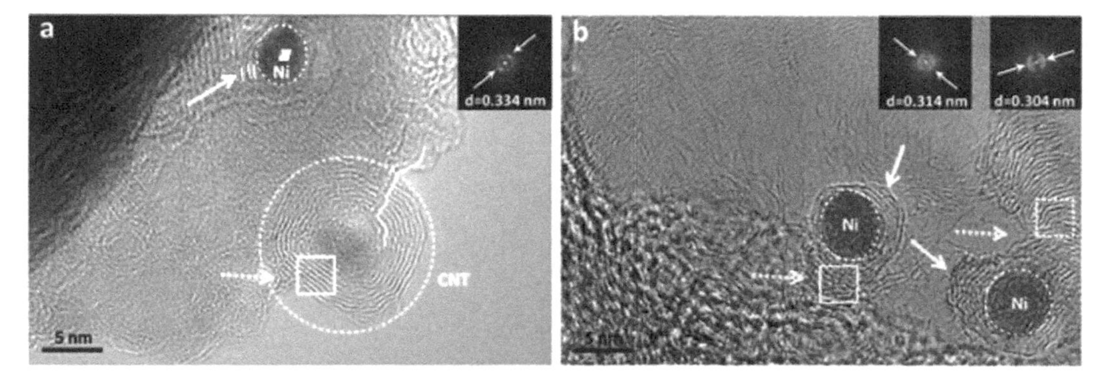

Fig. 6.73. (a, b) HRTEM images of the CNRs grown between graphene nanosheets. The corresponding FFTs in the inset of (a) and (b) were obtained from selected areas in (a) and (b), indicated by the squares. Reproduced with permission from [129], Copyright © 2015, Wiley-VCH Verlag GmbH & Co. KGaA, Weinheim.

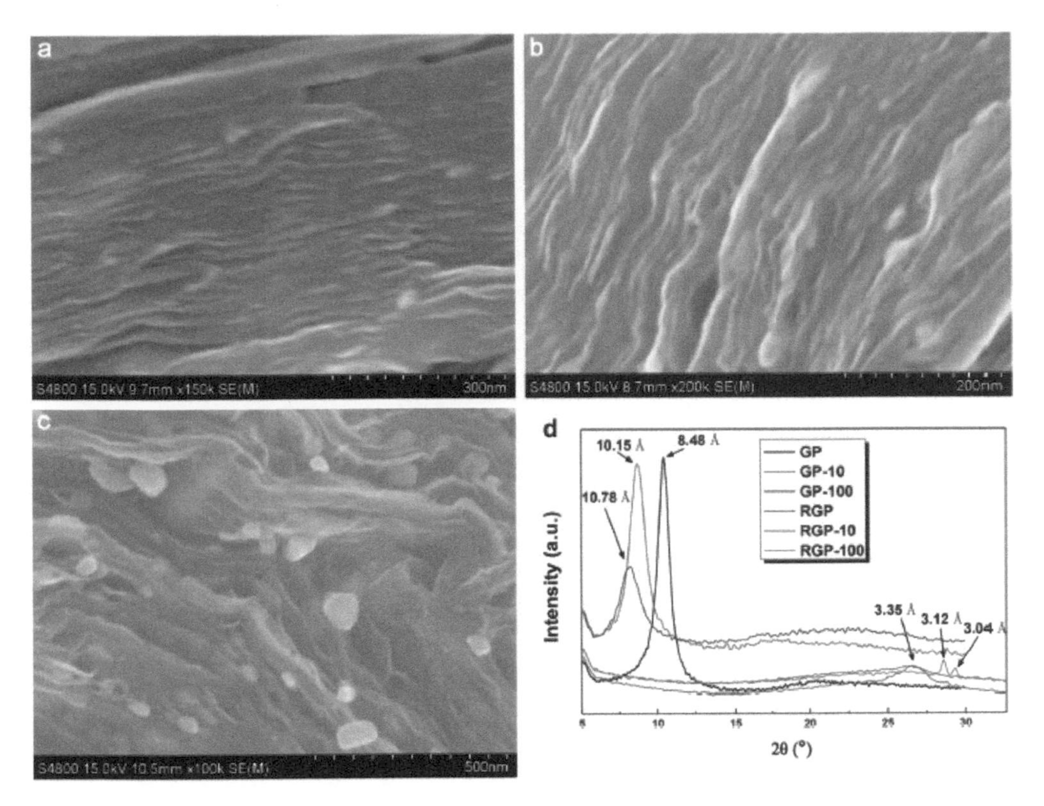

Fig. 6.74. Cross-sectional SEM images of the reduced GPs: (a) rGP, (b) rGP-10 and (c) rGP-100. (d) XRD patterns of the GPs and rGPs. Reproduced with permission from [129], Copyright © 2015, Wiley-VCH Verlag GmbH & Co. KGaA, Weinheim.

Color image of this figure appears in the color plate section at the end of the book.

two factors, i.e., (i) hydrogen-bonding interaction and (ii) the crosslinking effects due to the sodium ions that served as "cross-linkers" to bridge the rGO nanosheets and the MMT nanoplatelets. During filtration, the positively charged MMT and rGO were self-assembled to form highly oriented hybrid papers, in which the MMT and rGO were stacked with each other with an interlocking configuration. The MMT–rGO hybrid papers exhibited excellent flexibility, high electrical conductivity and fire-retardant properties.

A mechanism was proposed to explain the assistant effect of the MMT nanoplatelet on the dispersion of the rGO nanosheets in water, as demonstrated schematically in Fig. 6.75. The GO nanosheets were mixed with the exfoliated MMT nanoplatelets. The mixtures were subjected to reduction reaction from GO to rGO, during which the positively charged sodium ions (Na^+) acted as the linkers to bridge the two negatively charged nanosheets, i.e., MMT and rGO. At the same time, hydrogen-bondings were formed between the residual carboxyl groups on the rGO nanosheets and the hydroxyl groups on the MMT nanoplatelets. Due to these two types of interactions between rGO nanosheets and MMT nanoplatelets, MMT-G hybrids were formed in water, so that the rGO nanosheets could be dispersed in water to form stable suspensions.

Figure 6.76 shows the fabrication process of the highly oriented MMT–G hybrid papers by using the vacuum membrane filtration method. MMT papers were also prepared similarly for comparison. Due to the vacuum pressure during the filtration process, the MMT-G hybrid building blocks were deposited on the filter paper, with a preferential plane orientation (Fig. 6.76 (a)). The as-filtered papers adhered tightly to the filter paper, as shown in Fig. 6.76 (b). Thickness of the hybrid papers could be readily controlled by controlling the amount of the suspensions used for filtration. It was found that completely dried MMT-G hybrid papers were difficult to be peeled off from the filters, so that they should be transferred onto PTFE substrates, as demonstrated schematically in Fig. 6.76 (c). After drying at 60°C overnight, freestanding MMT-G hybrid papers could be obtained (Fig. 6.76 (d)). The hybrid papers were highly flexible and could be cut into different shapes as required (Fig. 6.76 (e, f)).

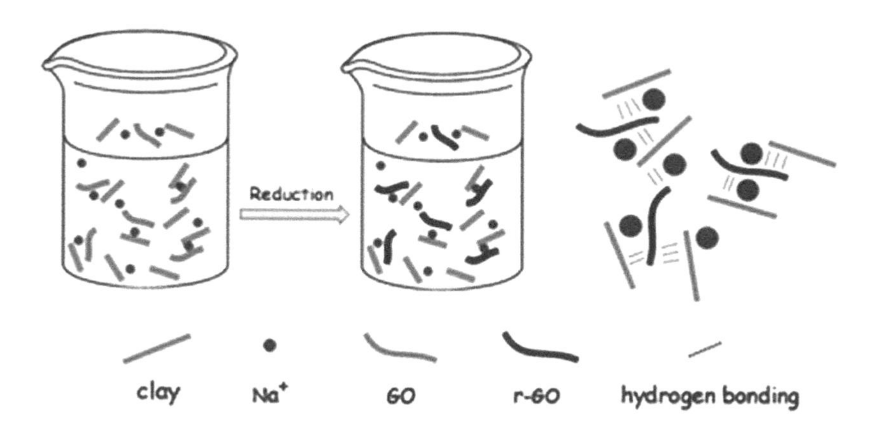

Fig. 6.75. Proposed mechanism for the assisted-dispersion of rGO nanosheets with exfoliated MMT nanoplatelets. Reproduced with permission from [106], Copyright © 2011, The Royal Society of Chemistry.

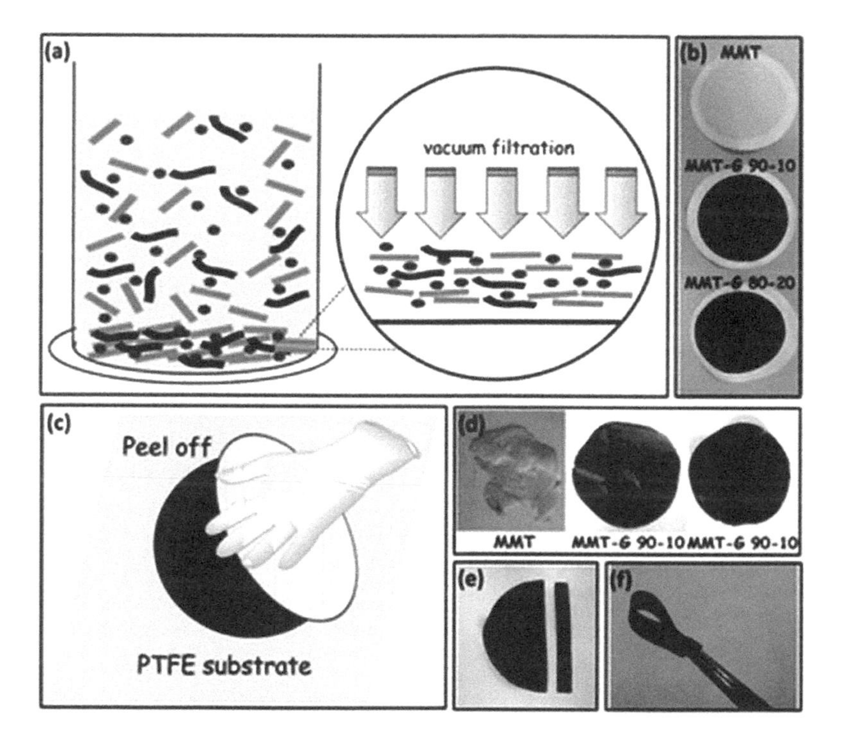

Fig. 6.76. Schematic illustration showing the fabrication of the free-standing MMT–G hybrid papers. (a) MMT–G hybrid dispersion and filtration process. (b) Photographs of the as-filtered MMT–G hybrid papers. (c) Papers peeled off from the membrane and transferred onto PTFE. (d–f) Free-standing hybrid papers. Reproduced with permission from [106], Copyright © 2011, The Royal Society of Chemistry.

The MMT-G hybrid papers exhibited flame retardant properties. When they were directly exposed to open flame, the hybrid papers were initially burned only in a flash, due to the presence of traces of residual oxygen-containing groups on the rGO nanosheets. The papers remained inert and retained their shape with only very slight shrinkage, even they were continuously exposed to a high-temperature torch flame for more than 10 min, as illustrated in Fig. 6.77 (a). The flame retardant behavior was attributed to the formation of the inorganic framework, consisting of a high content of interlocked MMT nanoplatelets and rGO nanosheets. Figure 6.77 (b) shows a photograph of the MMT–G hybrid paper after flame treatment,

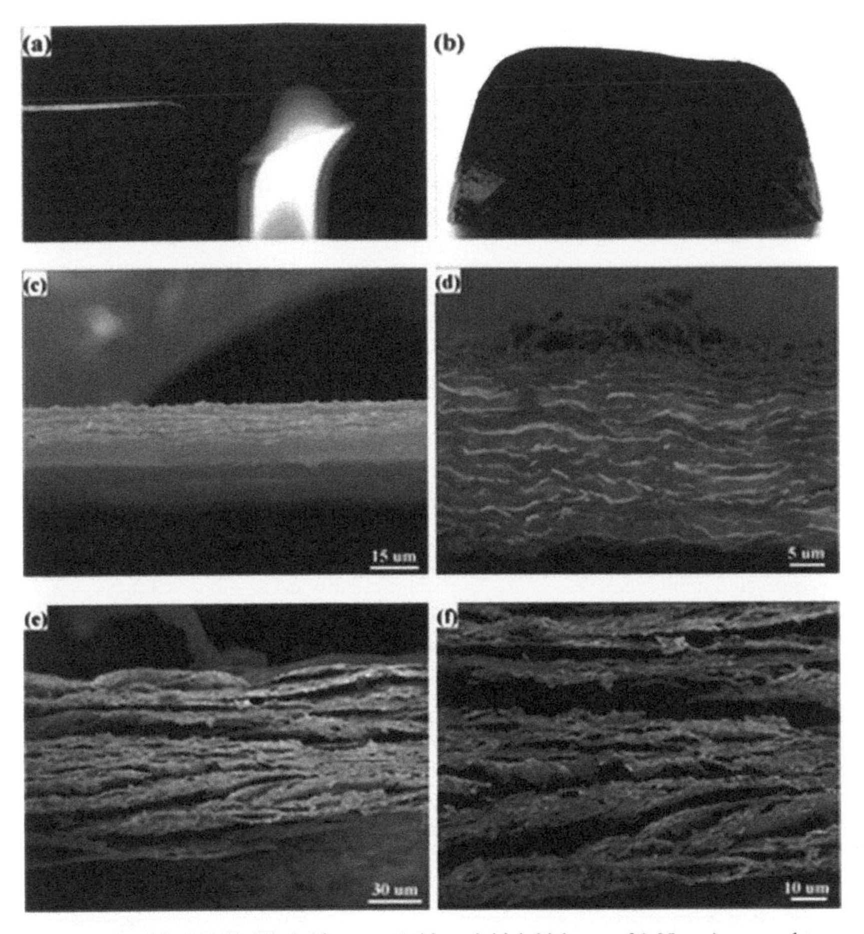

Fig. 6.77. (a) Photograph of the MMT–G hybrid papers (with an initial thickness of 0.05 mm) exposed to open flame. (b) Photograph showing the MMT–G hybrid paper after being flamed. Cross-sectional SEM images of the dried MMT–G (80–20) hybrid papers before (c, d) and after (e, f) the flame experiment at low and high magnifications, showing the presence of highly aligned anisotropic nanosheets. Reproduced with permission from [106], Copyright © 2011, The Royal Society of Chemistry.

with a well-retained shape. Cross-sectional SEM images of the MMT–G (80/20) hybrid paper are shown in Fig. 6.77 (c, d). The MMT and rGO nanosheets were highly oriented over the entire sample. SEM-EDX analysis results indicated that the papers had a homogeneous component distribution. Cross-sectional SEM images of the MMT–G (80/20) hybrid paper after flame treatment are shown in Fig. 6.77 (e, f). It had a tightly armored skin with the interior being a mesoporous layered structure. The fire retardant behavior was ascribed to the dense, multilayered inorganic micro/nanoporous structures of the hybrid papers. After the fire experiment, the paper was slightly expanded, but all the nanosheets were still highly oriented, which facilitated high heat dissipation efficiency. As a result, such MMT–G hybrid papers could be used for fire protection.

There are also reports on graphene-based papers that are incorporated with functional nanoparticles, with two cases to be discussed as examples. The first example is that of the rGO–MnO_2 paper used as a flexible electrode of supercapacitors [88]. The electrodes had a large areal mass, which combined flexible rGO with high conductivity and MnO_2 with large specific capacitance. The high flexibility of the paper made it possible for it to be integrated into any desired structure. Areal capacitance of as high as 897 $mF \cdot cm^{-2}$ was achieved by using the rGO–MnO_2 paper, corresponding to specific capacitance of 3.7 $mg \cdot cm^{-2}$.

$Mn(NO_3)_2$ was added into the GO solution, which was sonicated for 30 min. After that, $KMnO_4$ was added and then stirred at 85°C for 1 h. The solution was filtered with cellulose filter through vacuum filtration to obtain GO–MnO_2 papers. After drying, the paper could be peeled off from the cellulose filter paper. To make rGO–MnO_2 papers, the GO–MnO_2 papers were placed in a hydrazine solution, and were then sealed in a stainless steel autoclave and heated at 85°C for 24 h. The reduced paper electrodes were finally treated in HNO_3 solution and dried at 60°C for 24 h in air.

Cross-sectional SEM images of the rGO–MnO_2 papers are shown in Fig. 6.78 (a, b), clearly demonstrating well-separated and continuously aligned rGO nanosheets. The hybrid papers exhibited wrinkled and curved features upon reduction of the GO nanosheets (Fig. 6.78 (a)). Additionally, the rGO nanosheets had an open sheet arrangement, as revealed in the inset of Fig. 6.78 (a), which ensured the availability of a large accessible surface area and easy pathways for fast electrolyte ion diffusion, thus resulting in electrodes with low charge transfer resistance and high capacitance. Figure 6.78 (c) indicated that nanospherical MnO_2 particles were sandwiched within the rGO nanosheets. The content of the MnO_2 nanoparticles in the rGO–MnO_2 papers was about 20 wt%. High mechanical flexibility of the hybrid papers is demonstrated in Fig. 6.78 (d). According to dynamic mechanical analysis (DMA), the measured tensile strength and Young's modulus extracted from stress strain curve were about 8.79 MPa and 9.84 GPa, respectively.

The second example is that of the G–Fe_3O_4 hybrid paper, with both high electrical conductivity and strong magnetic properties for electromagnetic interference (EMI) [113]. Electrically conductive graphene nanosheets were used directly to grow Fe_3O_4 magnetic nanoparticles, without the use of additional chemical reduction or thermal annealing, so that no problem of GO was encountered. The free-standing G–Fe_3O_4 hybrid papers could find applications as conductive magnetically controlled switches.

Figure 6.79 shows a schematic diagram of the fabrication process, in which magnetic and conductive G–Fe_3O_4 hybrid papers were directly derived from hydrophilic and conductive graphene nanosheets through a hydrothermal treatment. Expanded graphite was used to obtain graphene nanosheets, which could be used to fabricate freestanding conductive graphene papers, with an electrical conductivity of $>20{,}000\ S \cdot m^{-1}$, without the requirement of chemical reduction or thermal annealing (Fig. 6.79 (c)). Due to the presence of a small amount of hydroxyl and carboxyl groups on the surface, the graphene nanosheets could be dispersed in water to form an aqueous suspension, so that Fe_3O_4 NPs could be introduced. The Fe_3O_4 NPs were incorporated by mixing $FeSO_4 \cdot 7H_2O$ solution that was precipitated with NaOH, as demonstrated in Fig. 6.79 (d). After a one-step hydrothermal treatment, the G–Fe_3O_4 hybrid was filtrated to form magnetic

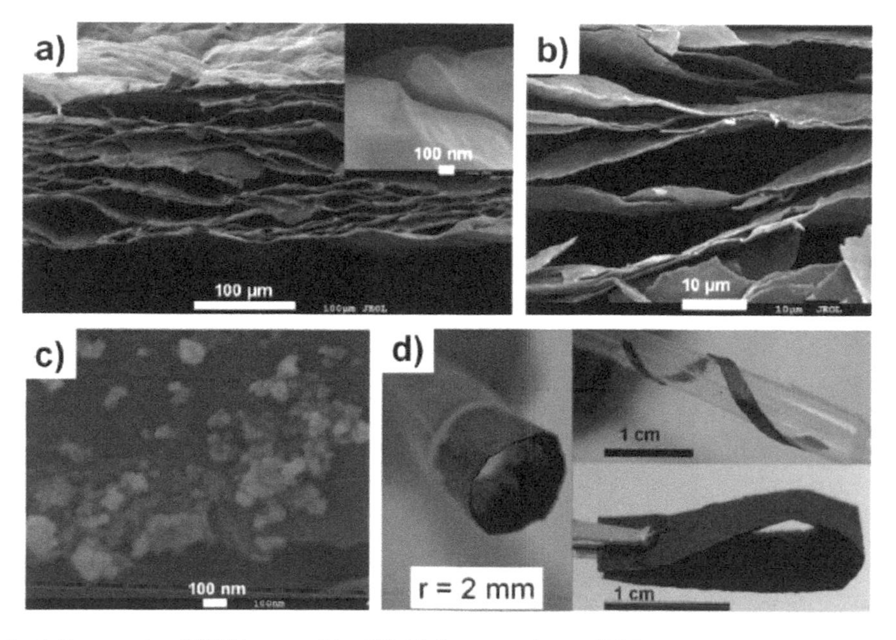

Fig. 6.78. (a, b) Cross-sectional SEM images of the rGO–MnO_2 paper at low and high magnifications, with the inset showing the edge of rGO–MnO_2 paper from the top view. (c) Image of rGO–MnO_2 paper showing the presence of MnO_2 nanoparticles. (d) Photographs of the rGO–MnO_2 paper. Reproduced with permission from [88], Copyright © 2013, Wiley-VCH Verlag GmbH & Co. KGaA, Weinheim.

and conductive G–Fe_3O_4 hybrid papers (Fig. 6.79 (e, f)). Content of the Fe_3O_4 NPs was simply controlled by controlling the amount of $FeSO_4 \cdot 7H_2O$ in the mixed solution before the hydrothermal treatments. In this study, Fe_3O_4 NPs were synthesized hydrothermally without the graphene nanosheets. For the G–Fe_3O_4 papers with different contents of Fe_3O_4 NPs, the amounts of $FeSO_4 \cdot 7H_2O$ and the corresponding samples were: G–Fe_3O_4-1, 35 mg $FeSO_4 \cdot 7H_2O$, G–Fe_3O_4-2, 70 mg $FeSO_4 \cdot 7H_2O$, G–Fe_3O_4-3, 105 mg $FeSO_4 \cdot 7H_2O$ and G–Fe_3O_4-4, 175 mg $FeSO_4 \cdot 7H_2O$.

Figure 6.80 shows representative SEM images of the samples. Particle size of the Fe_3O_4 NPs in the G–Fe_3O_4 hybrids (Fig. 6.80 (b)) was in the range of 100–200 nm, which was comparable with that of the Fe_3O_4 NPs synthesized without the presence of graphene nanosheets (Fig. 6.80 (a)). Cross-sectional images of the free-standing graphene and the G–Fe_3O_4 papers with the highest particle content (50 wt%) are shown in Fig. 6.80 (c) and (e), respectively. It was observed that the graphene nanosheets were well-aligned in the graphene papers (Fig. 6.80 (c)), but the orientation and alignment were disrupted to certain degrees as the Fe_3O_4 NPs were introduced, as seen in Fig. 6.80 (e, f). In addition, there was no significant different in the surface morphology between the graphene papers (Fig. 6.80 (d)) and the G–Fe_3O_4 hybrid papers (Fig. 6.80 (g)), although particle clusters could be observed in the G–Fe_3O_4 papers (Fig. 6.80 (g, h)).

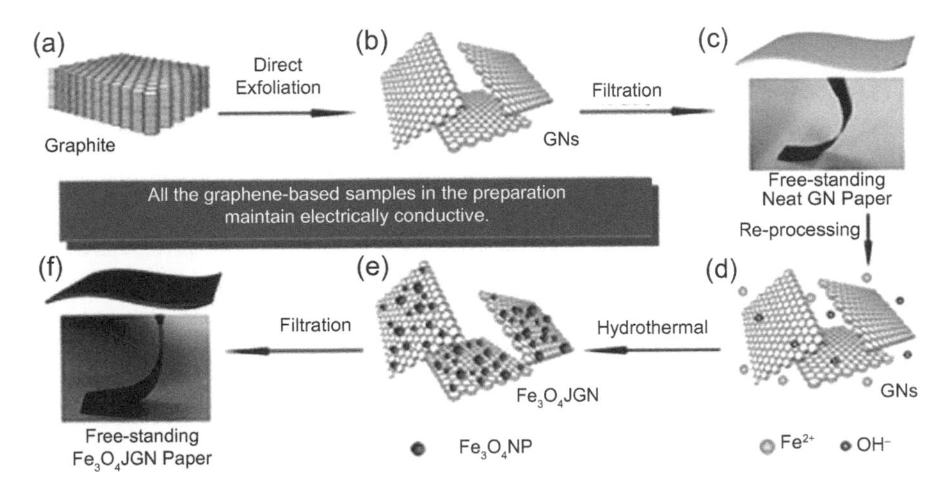

Fig. 6.79. Schematic diagram illustrating the fabrication process of the free-standing magnetic and conductive G–Fe_3O_4 hybrid papers. Reproduced with permission from [113], Copyright © 2015, The Royal Society of Chemistry.

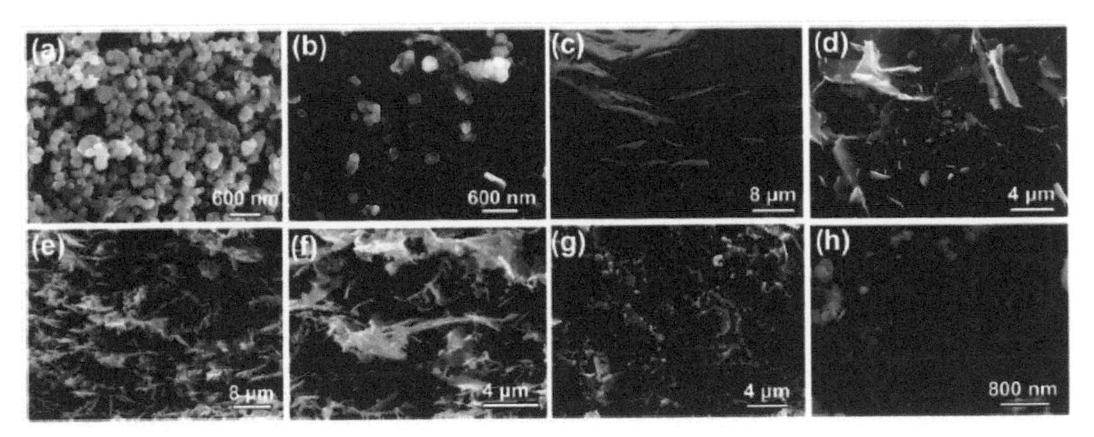

Fig. 6.80. Representative SEM images of the samples: (a) Fe_3O_4 NPs, (b) G–Fe_3O_4 hybrid, (c) cross-sectional view (c) and (d) top view of the pure graphene papers, (e, f) cross-sectional view and (g, g) top view of the G–Fe_3O_4-4 paper. Reproduced with permission from [113], Copyright © 2015, The Royal Society of Chemistry.

Because Fe_3O_4 NPs and their clusters were nearly 10- and 100-fold larger than the thickness of the single graphene nanosheet, the alignment of the graphene nanosheets was influenced, as illustrated schematically in Fig. 6.81 (a and b). Despite the considerable variation in the stacking of the graphene nanosheets (Fig. 6.80), free-standing G–Fe_3O_4 hybrid papers could still be formed, even when the content of Fe_3O_4 NPs was as high as 50 wt%. The formation mechanism of pure graphene papers was closely related to the hydrogen bonding due to the hydroxyl and carboxyl groups on the surface of the graphene nanosheets, as well as the compression of the air–water interface, as presented in Fig. 6.81 (c). The formation of the stable G–Fe_3O_4 hybrids was attributed to the fact that the un-nucleated polar sites, i.e., the hydroxyl and carboxyl groups, on surface of the graphene nanosheets facilitated the attachment of the Fe_3O_4 NPs [169]. For example, hydrogen bonding was formed between the graphene nanosheets and the Fe_3O_4 NPs in the aqueous solution, due to the high concentration of OH groups, as shown in Fig. 6.80 (d).

Graphene or GO papers have a nanoscale "brick-and-mortar" structure, in which the interaction between the adjacent nanosheets could be mediated by intercalated solvents [170–172]. There have been reports on the mechanical properties of the papers, together with strategies to improve their mechanical strength [170, 173]. However, in order to realize "materials design" for such special paper materials, it is necessary to clearly understand the formation mechanisms. As seen in the above discussion, vacuum membrane filtration has been the major technique to develop various G/GO papers, which involves the flow-directed assembly of graphene or GO nanosheets, as their suspensions are filtered over a supporting membrane. Obviously, the papers with well-aligned G/GO nanosheets are derived from suspensions in which all the nanosheets are entirely randomly distributed. Also, it has been observed that the layered structures could be swelled by several solvent molecules without destroying their integral structure [146, 174], while the swelling is reversible, i.e., after drying in air the original structures are completely recovered [103, 128].

It has been suggested that a gelation process of the graphene nanosheets took place at the solution–filter membrane interface during filtration, which was related to the formation of mechanically strong, highly conductive and anisotropic graphene papers [84]. More importantly, various organic or inorganic components could be incorporated into the graphene papers to form composites or hybrids [175, 176].

In a separate study, three formation mechanisms have been considered in order to explain the wide range of possible ordering sequences during the fabrication of GO papers through vacuum membrane filtration [87]. In brief, they were two extreme cases, i.e., highly ordered layering and complete disordering, together with the cases in between these extreme two. At the highly ordered layering end, order structure was developed as the nanosheets were deposited. At the other end, the GO nanosheets were randomly distributed during the filtration process, until they were aligned due to the geometric confinement caused by the removal of the solvents. The third mechanism was known as semi-ordered accumulation, which

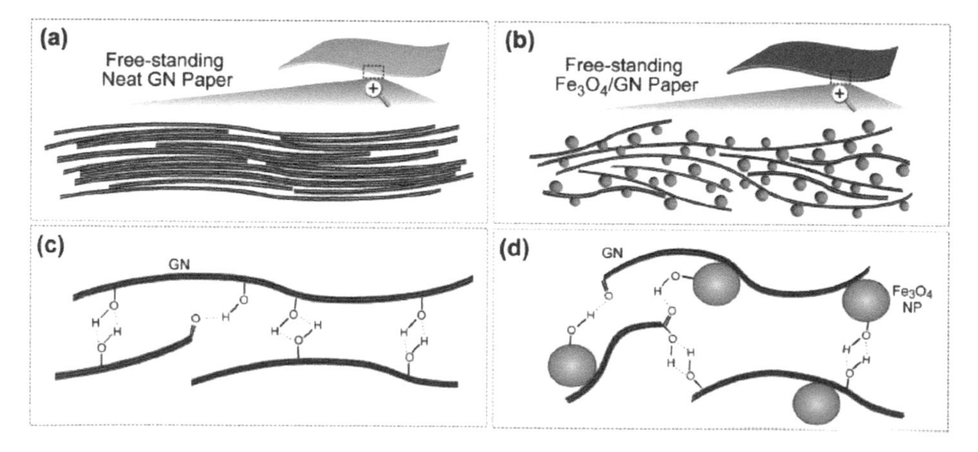

Fig. 6.81. (a, b) Schematic diagram showing the cross-sectional views of the free-standing GN paper and Fe_3O_4–GN papers. (c) Possible mechanism of the interactions between the GN layers in the neat GN paper and the (d) interactions between Fe_3O_4 and the GN nanosheets. Reproduced with permission from [113], Copyright © 2015, The Royal Society of Chemistry.

resulted in the formation of loosely stacked nanosheets at surface of the filter membranes. During the removal of the solvents, the semi-ordered layers were transferred into layered structures, due to the filtration induced compression. Experimental results suggested that the semi-ordered accumulation mechanism has governed the formation of GO papers through vacuum membrane filtration.

Mechanism 1 was characterized by highly ordered layering, in which the GO nanosheets were stacked one by one during the filtration process, as shown in Fig. 6.82 [87]. Similar mechanism was observed in the formation of ordered polymer-GO nanocomposites, when the GO nanosheets were deposited on the filter due to the hydrostatic forces as the solvent was removed [171]. A layer of assembled GO nanosheets was formed first, onto which more GO nanosheets were assembled, so that the thickness of the film continuously increased. Finally, paper structure was developed, after all of the excess solvent was removed. In this case,

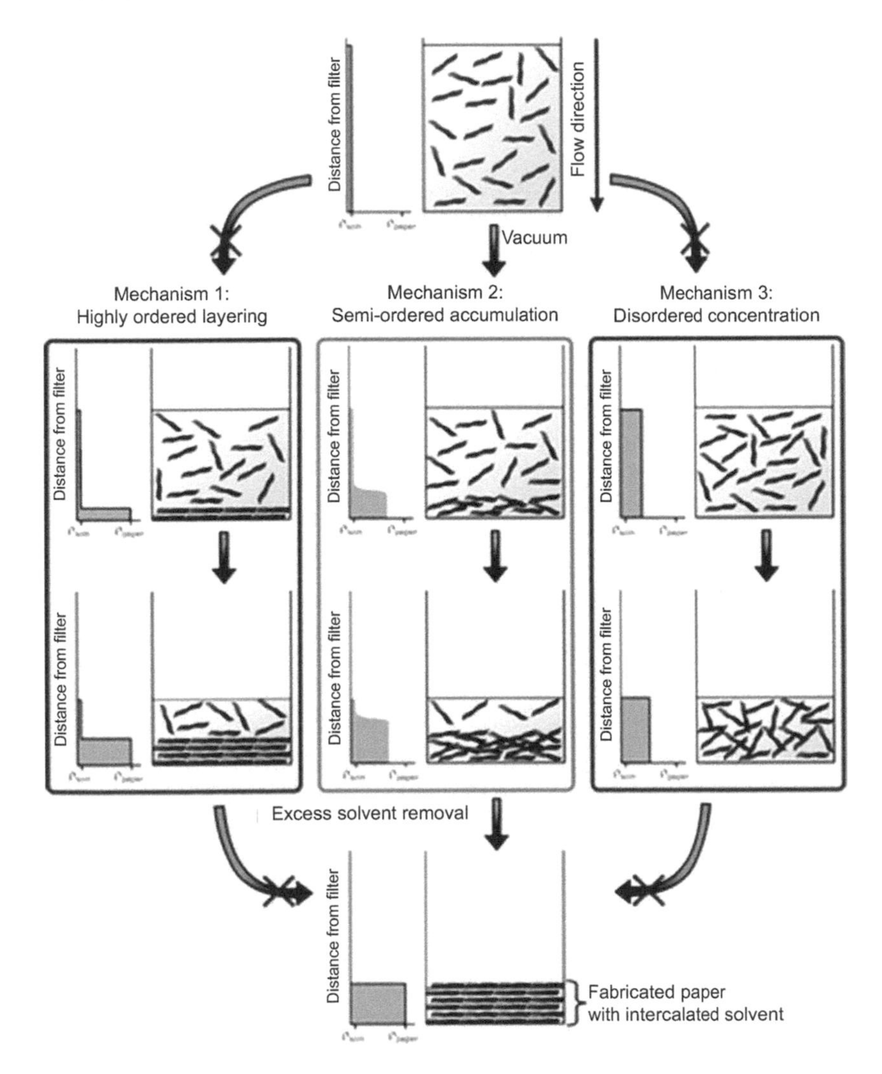

Fig. 6.82. Schematic diagram demonstrating the three possible mechanisms governing the formation of the GO papers (ρ = density). Highly ordered layering (mechanism 1) proceeded by the sequential stacking of the GO nanosheets in an ordered way with the density of the paper structure being constant throughout the filtration process. Semi-ordered accumulation (mechanism 2) characterized by the formation of a loose aggregate structure at a density in between the density of the dispersion and that of the final paper. After the solvent level was just at the top of the loose aggregate, compression simultaneously aligned the GO nanosheets perpendicularly to the flow direction and closed the gaps between the nanosheets. Disordered concentration (mechanism 3) characterized by a homogeneous density of GO nanosheets throughout the dispersion that was increased with time. Eventually, geometric constraints aligned the GO nanosheets into the final highly aligned papers. Reproduced with permission from [87], Copyright © 2011, American Chemical Society.

due to the multilayered structure, there was a spacing in between the nanosheets during the formation of the paper, which facilitated the solvent to flow through.

Mechanism 2 was called semi-ordered accumulation (Fig. 6.82). In this mechanism, as the solvent was removed through the filter, the GO nanosheets were brought into close contact with one another, because the local concentration of the GO nanosheets at the filter solvent interface was over their solubility in water. Due to weak interaction between the adjacent GO nanosheets, the structure was a loose aggregation, with the basal plane of the film consisting of partially oriented nanosheets. Therefore, the structure had a larger interlayer spacing as compared with the final paper. As the solvent level touched the top layer of the semi-ordered nanosheets, the second phase of paper formation was started. The lowering of the solvent level exerted compression force on the loose and weak aggregated structure, so that the spacing of the GO nanosheets was reduced, along with their orientation parallel to the solvent level. Finally, multilayered paper-like structure was formed with a higher degree of ordering throughout the sample.

Mechanism 3 was a completely disordered concentration, in which it was assumed that the GO nanosheets were randomly packed throughout the filtration process. Usually, the initial concentration of the dispersions is below the solubility of GO (> 15 $mg \cdot ml^{-1}$), so that the GO nanosheets that had been deposited on the filter could be re-dispersed into suspension [19]. As the solvent was removed, the GO concentration was gradually increased. When it approached 10 vol%, inter-sheet contacts would be established, due to their high aspect ratio [177]. Further removal of the solvent would lead to an increase in the alignment of the GO nanosheets, which were oriented along the surface plane of the membrane. As the air-water interface lowers, compression similar to that described for Mechanism 2 will produce increasingly aligned structures. This compression then continues until the final density of the self-supporting paper is reached.

However, both the mechanisms 1 and 3 have been excluded by experimental observations. On the one hand, the disordered concentration mechanism was denied, because the concentration of the GO nanosheets was kept constant during the filtration process, while filtration followed nonlinear dynamics. On the other hand, it was found that long chain polymers, e.g., poly(ethylene oxide) (PEO), could be intercalated through the GO aggregated structure during paper formation, while no XRD diffraction pattern was detected for the as-obtained wet GO papers, so that the highly ordered layering mechanism was not supported by experimental results. In contrast, a water-soaked GO paper sample experienced a complex evolution in XRD patterns during the drying process, which implied that a semi-ordered hybrid structure of GO nanosheets was formed and separated by residual water. Therefore, the experimental results supported the semi-ordered accumulation mechanism.

Figure 6.83 shows typical XRD patterns and representative SEM images of the experimental results designed to verify the paper formation mechanism [87]. For the GO-PEO nanocomposite papers, the adjacent nanosheets were separated by the polymer chains and water molecules concurrently. During the compression process, the water molecules were preferentially expelled first, while the polymer was retained. It was demonstrated that the mass of the retained polymer in the nanocomposite papers was increased linearly with the increasing molecular weight of the polymer, which implied that the nanosheet aggregated structure formed during the filtration exhibited a certain degree of ordering to accommodate the polymer. In other words, the semi-ordered accumulation mechanism was experimentally verified.

Moreover, the interactions between the solvents and the GO nanosheets during the vacuum membrane filtration process could be utilized as a tool to tailor the gallery spacing between the adjacent nanosheets in the final papers. According to the sum of the Hansen solubility factors, i.e., $\delta_{sum} = \delta_p + \delta_h$ [178], with δ_p and δ_h to be the polarity cohesion factor and hydrogen-bonding cohesion factor, respectively, the gallery spacing of the GO paper was contracted upon exposure to solvents with low δ_{sum} values (e.g., toluene), minor swelling took place in the presence of solvents with moderate δ_{sum} values (e.g., acetone) and complete loss of all orderings took place in solvents with high δ_{sum} values (e.g., water). This phenomenon could be very useful in the fabrication of GO-based nanocomposites, because the inter-sheet spacing within the structure could be tuned to accommodate molecules or nanoparticles with desired dimensions and morphologies.

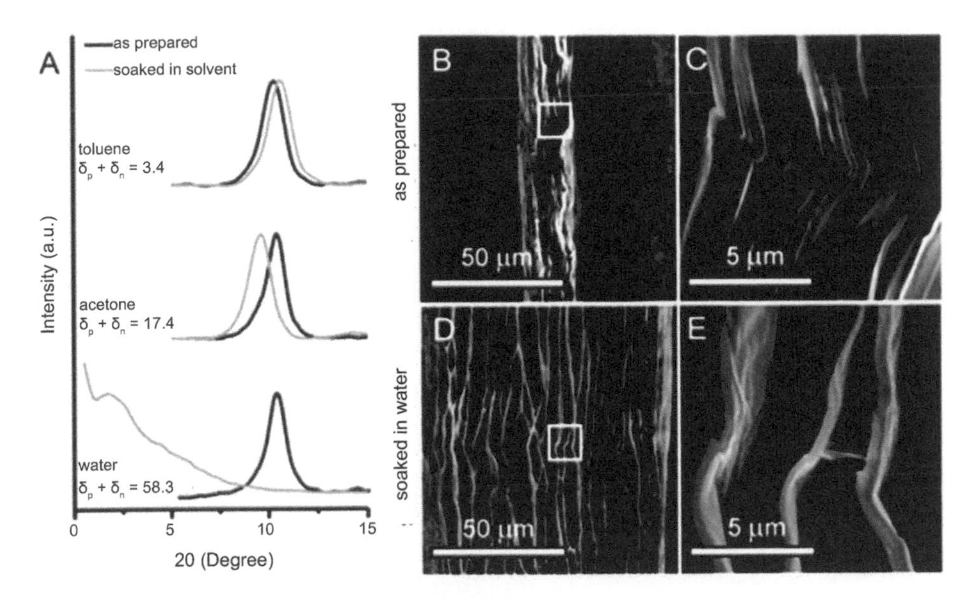

Fig. 6.83. (A) XRD patterns of the as-fabricated and solvent-immersed GO papers, demonstrating the effect of three different solvents, i.e., water, acetone and toluene, on inter-sheet spacings. (B) SEM image showing the entire width of the as-obtained GO papers with a thickness of about 11 μm. (C) Magnified SEM image of the boxed area in panel (B), revealing the inhomogeneous structure of the as-obtained GO paper. (D) SEM of the same paper swollen by water, with a thickness of about 100 μm. (E) Magnified SEM image of the boxed region in panel (D), revealing thin tendrils of GO stacks consisting of GO lamellae network structure. Both the untreated and the water-swollen samples were prepared from the same strip of GO paper. Reproduced with permission from [87], Copyright © 2011, American Chemical Society.

6.3.2. Other methods

6.3.2.1. Direction evaporation

A direct evaporation method was used to fabricate GO papers from GO suspension with mild heating [134]. The GO thin papers had an average thickness of about 8.4 μm, which exhibited excellent EMI shielding effectiveness of 20 dB and high in-plane thermal conductivity of 1100 $W \cdot m^{-1} \cdot K^{-1}$, as well as high mechanical flexibility and structure integrity withstanding repeated bending. Figure 6.84 (A) shows a schematic diagram describing the fabrication process of the GO thin papers through direct evaporation. GO suspensions were poured into Teflon dishes, which were heated gently at 50–60°C to evaporate the water, so that flexible dark brown GO papers were formed. Obviously, the thickness of the papers could be well-controlled from microns to ten microns by controlling the volume and concentration of the GO suspensions. Also, this method has no limit in terms of the size of the GO papers. In this respect, it was advantageous over the vacuum membrane filtration method discussed above.

Figure 6.84 (B) shows a photograph of the GO paper with a dimension of about 400 cm^2 made simply by using a mold of that size. The GO papers were sufficiently flexible, so that they could be rolled and folded, as seen in Fig. 6.84 (C). SEM observations indicated that the GO papers had smooth surface with a few thin ripples, as demonstrated in Fig. 6.84 (D). The multilayered structure was maintained through almost the entire cross-sections, as illustrated in Fig. 6.84 (E, F), without any difference from that of their vacuum filtration counterparts. During the evaporation of water, the concentration of the GO suspensions was gradually increased, so that the sheet-to-sheet interactions were facilitated, thus leading to self-alignment of the GO nanosheets due to their relatively large lateral dimension.

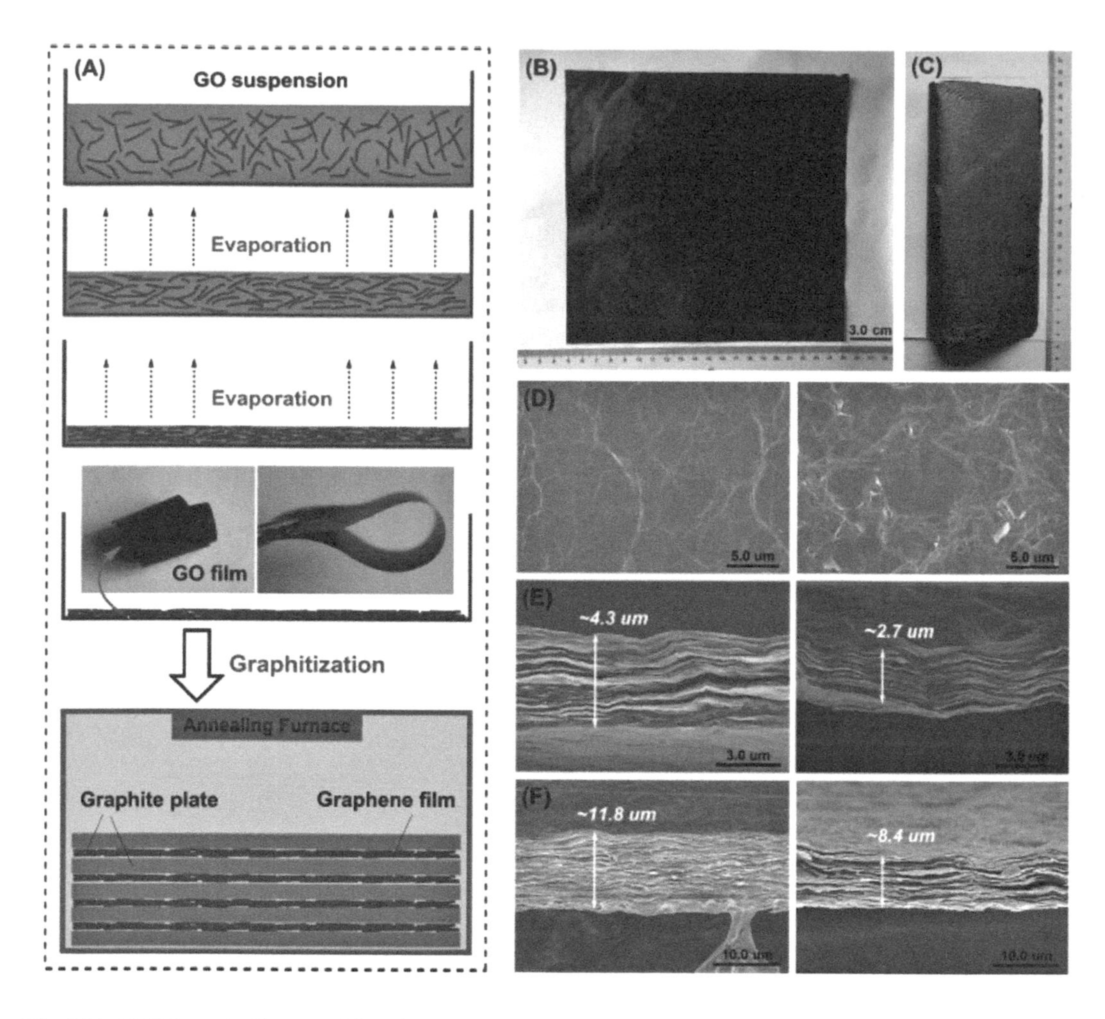

Fig. 6.84. (A) Schematic diagram of the self-assembly process to obtain the GO papers through evaporation and thermal graphitization treatment. The inset shows the photographs of the flexible GO papers. (B) Photograph of the free-standing and dark brown GO paper with a dimension of 20 cm×20 cm. (C) Photograph of the free-standing and shiny metallic GF-2000 (20 cm × 20 cm) folded on a paper. (D) Surface SEM images of the GO paper (left) and GF-2000 sample (right). (E, F) Cross-sectional SEM images of the GO paper and GF-2000 sample (with different thicknesses). Reproduced with permission from [134], Copyright © 2014, WILEY-VCH Verlag GmbH & Co. KGaA, Weinheim.

The GO papers were thermally annealed at 2000°C for further reduction and graphitization, which resulted in graphene papers (denoted as GF-2000) [179, 180]. The GF-2000 sample exhibited almost no variation in physical dimension as compared to the as-obtained GO papers, whereas it became darker and black with a shiny metallic luster, as observed in Fig. 6.84 (C). Figure 6.84 (D) shows SEM image of the GF-2000 sample, indicating its smooth surface and the presence of ripples. Cross-sectional SEM images of the GF-2000 sample are shown in Fig. 6.84 (E, F), illustrating the highly-oriented multilayer stacking of the graphene nanosheets. The thicknesses were decreased from 4.3 μm and 11.8 μm to 2.7 μm and 8.4 μm, respectively, due to the removal of the oxygen groups and water molecules.

Figure 6.85 (A) shows photographs of the graphene paper with a thickness of 8.4 μm, which could be folded and then pressed, without breaking and cracking. SEM images revealed that the graphene nanosheets were still continuously stacked and not cracked, as shown in Fig. 6.85 (B). Although the stress generated due to the bending was not sufficiently strong to break the graphene layers, it could make them slip out from the inter-lamination to accommodate the extension of the outer surface of the papers. The possible coalescence of the graphene nanosheets during the graphitization in the GF-

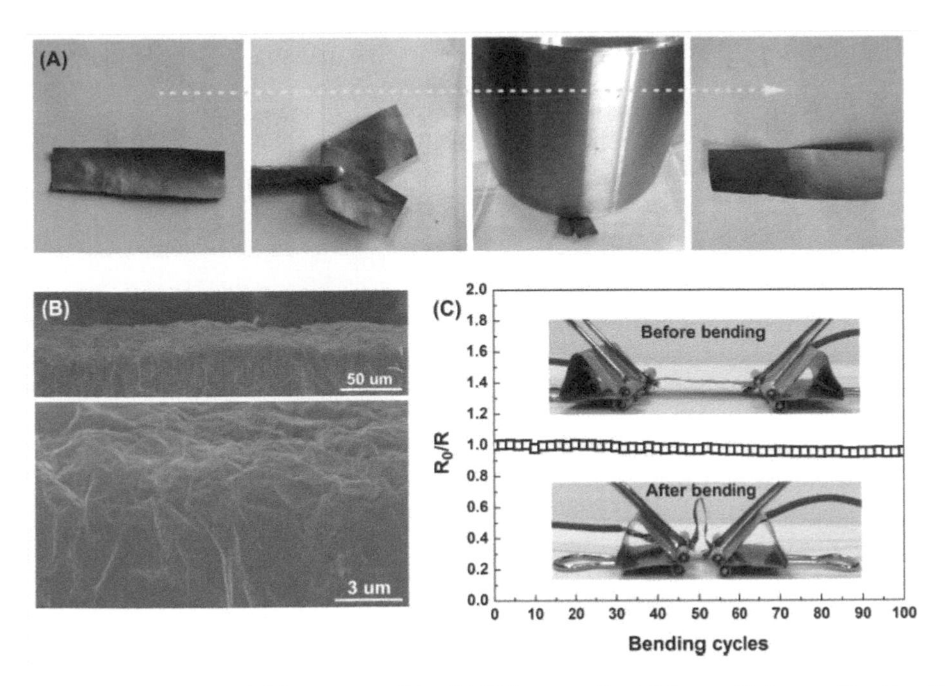

Fig. 6.85. (A) Photographs of the folding process of the GF-2000 paper with a thickness of 8.4 μm, without breaking and cracking after folding. (B) SEM images of the crease on the outer side of the folded GF-2000 (8.4 μm). (C) Resistance of the GF-2000 paper (8.4 μm) as a function of the bending cycle. Inset shows the unfolded (above) and folded sample (below). Reproduced with permission from [134], Copyright © 2014, WILEY-VCH Verlag GmbH & Co. KGaA, Weinheim.

2000 sample might have produced larger ones to withstand the slippage of the paper upon bending. Figure 6.85 (C) shows variation in the resistance of the GF-2000 sample as a function of the bending cycle, in order to check the bending effect on its electrical conductivity, with the insets showing the experimental setup. No visible variation in resistance was observed after 100 bending cycles, further confirming the flexible characteristics of the graphene papers.

6.3.2.2. Tape casting

Most of the methods that have been available in the open literature to fabricate G/GO papers are limited to the laboratory scale. Tape casting is a promising candidate that can be used to address this problem, as demonstrated in a recent study [137]. Polymer graphene composite tapes were prepared first, while the polymer was subsequently removed through thermolysis, which led to large-scale graphene papers with a network structure. The graphene tapes exhibited a high surface area, excellent electrical conductivity and promising mechanical strength.

Figure 6.86 shows the process of tape casting, in which the colloidal suspension was cast through a slit by moving the substrate relative to a doctor blade. The graphene used in the tape casting should have sufficiently large quantities, and was produced through the simultaneous thermal exfoliation and reduction of GO powder. Besides the oxygen-containing functionalities as stated earlier, the graphene network contained lattice defects and vacancies, as seen in the inset of Fig. 6.86. The functionalized graphene nanosheets (FGSs) could be distinguished according to their oxygen contents, which were expressed as the carbon-to-oxygen ratio (C/O in mol/mol). Wrinkles were formed on the graphene nanosheets, due to the presence of the defects and oxygen-containing groups, which decreased the contact area in between the adjacent graphene nanosheets and also prevented their re-stacking. This phenomenon could be an advantage for applications that require high surface area. Moreover, the chemical activity of the FGSs was increased because of the defect sites and the functional groups, which could find applications in electrochemical sensors, catalysts and fuel cells.

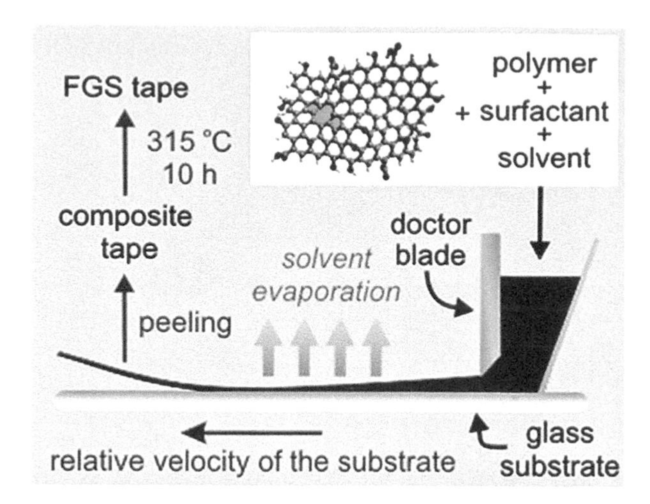

Fig. 6.86. Schematic diagram describing the fabrication of the FGS tapes with tape casting. The FGS schematic shows the defective and wrinkled structure of the graphene nanosheets. Oxygen functional groups are shown in red, with a 5-8-5 defect in pink and a 5-7-7-5 defect in yellow. Reproduced with permission from [137], Copyright © 2011, American Chemical Society.

To fabricate graphene tapes with high surface areas by using the tape casting method, the pre-requirement was to prepare highly dispersed suspensions of FGSs. A triblock copolymer surfactant, (ethylene oxide)$_{100}$-(propylene oxide)$_{65}$-(ethylene oxide)$_{100}$, or $EO_{100}PO_{65}EO_{100}$ in short, was used to disperse the FGSs in water. The amount of the surfactant needed to disperse the nanosheets was determined according to the concentrations of the FGSs in the supernatant of the suspensions containing different amounts of surfactants, which were compared with the initial overall suspension concentrations, as illustrated in Fig. 6.87 (a). The amount of FGSs in the supernatant was increased with the increasing concentration of the surfactant, corresponding to a continuous enhancement in the stability of the suspensions. As the surfactant-to-FGSs ratio reached 1 (g/g), the concentration of the FGSs was saturated, implying that they were sterically stabilized suspensions, i.e., the suspension stability was increased as the surface coverage of the particles was increased [181].

Hydrophobic PO chains of the $EO_{100}PO_{65}EO_{100}$ molecules were adsorbed on the surfaces of the graphene nanosheets, while their hydrophilic EO chains were extended into water [182]. Due to the high solubility of EO in water, the EO chains with strong repulsive interaction among themselves acted as a barrier which prevented the FGSs from aggregation. As the surfactant molecules could not fully cover the FGSs, the strong van der Waals forces would bring them together to form aggregates. Once the FGSs were entirely covered by the surfactant molecules, their dispersion concentration would be saturated. The decrease in supernatant concentration over time implied the sedimentation of the FGSs in the suspensions. This result suggests aggregation of the FGSs. This observation was mainly attributed to the van der Waals interactions if the weakly adsorbed surfactant molecules were desorbed. Because the time used for both the tape casting and drying processes was sufficiently short, the negative effect of desorption of the triblock copolymer from the FGSs was not observed.

Besides the requirement of high quality dispersions during the casting process, the suspensions should also have a good flowing behavior to pass through the gap between the blade and substrate under shear, as shown in Fig. 6.86. At the same time, the shape and thickness of the cast tape should be maintained as the shearing force disappeared after passing through the blade. Although these behaviors are closely related to the properties of the suspensions, the effects of the substrate wettability and the contact line pinning are also significant to a certain degree. It was found that, under the shear rates used in the casting, i.e., about 100 s^{-1}, the suspensions indeed exhibited shear thinning behavior, as seen in Fig. 6.87 (b). This was simply because both the PEO solutions and the FGS suspensions had shear thinning behavior, due to the disentanglement of the PEO molecules, the break-up of the FGS network and the shear-induced orientation of both the FGSs and the PEO molecules [183, 184].

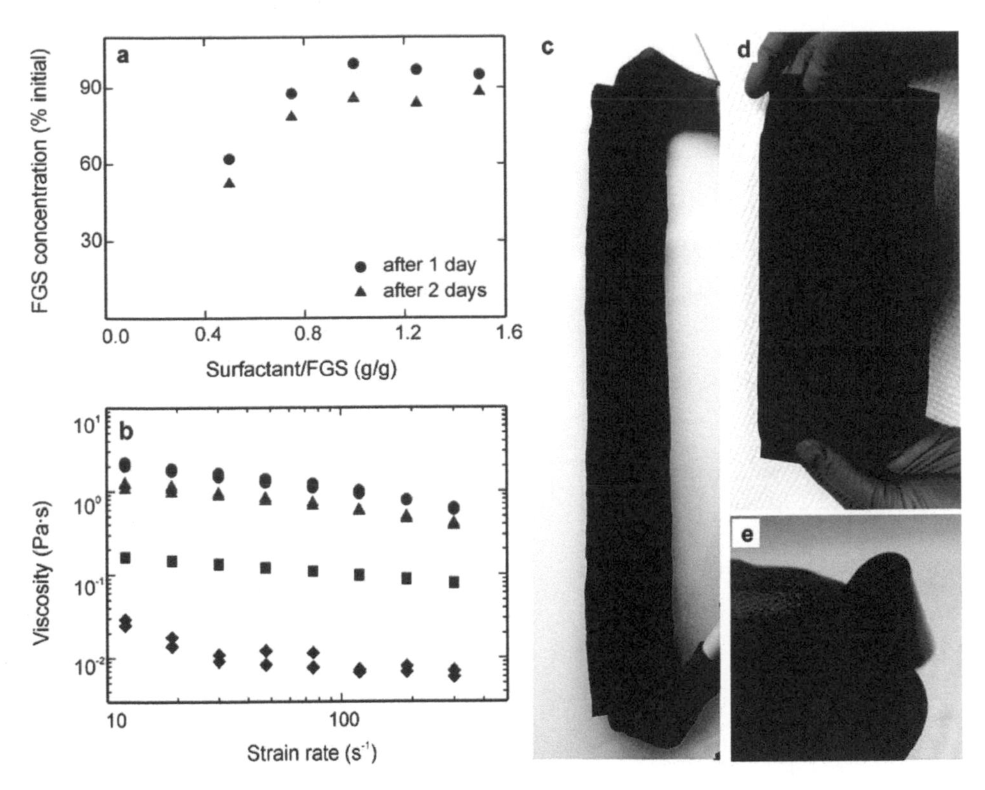

Fig. 6.87. (a) Supernatant concentration given as percentages of the initial concentration (0.5 mg·ml^{-1}) as a function of the surfactant to FGS ratio, 1 day and 2 days after the preparation of the suspensions with the aid of tip sonication. The linear relationship between the height of the absorbance peak at 280 nm and the FGS concentration was used to determine the concentrations of the aqueous FGS-surfactant suspensions. (b) Viscosity as a function of strain rate for the casting suspensions of the (●) 15 wt% FGS containing composite tape (2.8 wt% PEO, 0.6 wt% FGS, 0.6 wt% surfactant in the suspension), (▲) PEO solution (2.8 wt% PEO), (■) tape casting suspension used for casting of the 40 wt% FGS containing composite tape (0.68 wt% PEO, 1.36 wt% FGS, 1.36 wt% surfactant in the suspension) and (♦) FGS suspension (0.6 wt% FGS, 0.6 wt% surfactant). (c) Photograph of the self-supporting FGS/polymer/surfactant composite tape after being removed from the tape casting substrate. (d, e) Photographs of the FGS tapes after thermolysis to remove the polymer and surfactant. Reproduced with permission from [137], Copyright © 2011, American Chemical Society.

Viscosities of the FGS tape casting suspensions were variable over one order of magnitude at a given strain rate. The higher the concentration of FGSs, the lower the viscosity that the suspensions would have. The viscosity was also increased with the increasing content of PEO, so that the PEO solution with higher viscosity could be used as a thickening agent to adjust the viscosity of the suspensions. As the shear rate was decreased from 100 to 1 s^{-1}, the viscosity of the suspension required 1–2 s to return. As a result, the viscosity of the suspensions was increased suddenly in several seconds when they were cast. If nonwetting substrates were used, the contact line could not be effectively pinned [185]. Consequently, the suspension would be pulled away from the substrate, thus being condensed into a formless puddle. Therefore, it is necessary to use wetting substrates to maintain the shape of tapes during evaporation of the solvents. Figure 6.87 (c) shows a photograph of the free-standing continuous graphene composite tapes, which were easily peeled off from the substrates after drying. Polymer and surfactant could be removed by heating the composite tapes at 315°C. The resultant FGS tapes were flexible and mechanically robust, as illustrated in Fig. 6.87 (d, e).

Figure 6.88 shows representative SEM images of the FGS tapes, which indicated that they were highly uniform in thickness and free of defects in the microstructure. It was suggested that the thermolysis had been carried out at sufficiently low rates, so that the gases could escape from the tapes effectively without generating structural defects, thus leading to FGS tapes without the presence of microscale cracks. Also, the

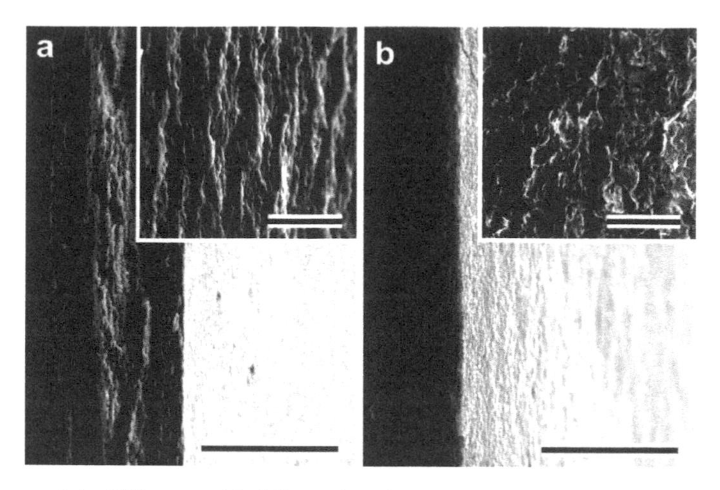

Fig. 6.88. SEM images of the FGS tapes with different densities: (a) 0.40 and (b) 0.15 g·cm^{-3}. Insets show their high magnification images. Scale bars = 50 μm (for the main images) and 5 μm (for the insets). Reproduced with permission from [137], Copyright © 2011, American Chemical Society.

density of the papers was almost linearly dependent on the initial concentrations of the FGS suspensions, implying that the polymer and surfactant were simply removed while the graphene frameworks remained intact.

The microstructure of the tapes exhibited a close relation to the content of the FGSs in the as-cast composite tapes. For the samples with high contents of FGSs, anisotropic structure was obtained, with striations parallel to the tape surfaces, i.e., the direction of the casting, as demonstrated in Fig. 6.88 (a). However, for those with lower densities, the anisotropic structure disappeared, so that the cross-section was more isotropic, as seen in Fig. 6.88 (b). The formation of the anisotropic structure was attributed to the preferred orientation of the graphene nanosheets, which was facilitated by the shear force during the casting process. However, the authors claimed that the anisotropic structure was not due to the shear effect, because there was sufficient time for the FGSs to randomize their orientation through rotational Brownian motion. Nevertheless, such FGS tapes (papers) exhibited excellent electrical and mechanical properties, which enabled them to be useful in various fields.

6.3.2.3. Electro-spray deposition

Another approach to produce large area free-standing graphene papers is called direct electro-spray deposition (ESD), which functions by combining a continuous roll-to-roll process and simple water exfoliation from highly hydrophilic aluminum substrates [138]. ESD has been widely used to deposit various thin films, because of its advantages, such as mass scale capacity, less material loss and high precision of property controlling [186–188]. Charged mono-dispersed fine droplets are generated due to the repulsion forces among them. The size of the droplets can be well-controlled by adjusting the flow rate and electric field applied to the injection nozzles and substrates, with smallest diameter to be hundreds of nanometers. The graphene papers fabricated in this way can have a wide range of thickness from hundreds of nanometers to hundreds of micrometers. They can be deposited on various substrates [189, 190].

The assembly and alignment of the graphene nanosheets in the final papers could be well-controlled by controlling the ESD processing parameters and further improved through mechanical compaction upon water exfoliation. In addition, thermal annealing was employed to heal the structural defects and remove the functional groups in the graphene nanosheets, so that their thermal and electrical properties were further improved [191, 192]. It was found that the optimal thermal annealing temperature was 2200°C, which led to defect-free, highly aligned and light-weight (density ≈ 2.1 g·cm^{-3}) graphene papers. The annealed graphene papers exhibited thermal and electrical conductivities of 1238.3 W·m^{-1}·K^{-1} and 1.57 × 10^{5} S·m^{-1}, respectively.

Well-dispersed graphene suspensions were used to deposit the graphene papers by using the ESD process. During the deposition, an electric voltage of 11.5 kV was applied between the injection nozzle and substrate which were separated by 2–3.5 cm to maintain a stable cone-jet mode. The graphene suspension was pushed through the nozzle at flow rates of 50–100 μl·min^{-1}, by using a syringe pump, so as to generate well-dispersed fine droplets that were then deposited on the heated substrate to form a uniform film. By controlling the solution concentration and the time of deposition, films with the desired thicknesses would be readily obtained. After deposition, the Al foils coated with the graphene films were immersed into water, so that free-standing graphene papers were peeled off from the substrate, as demonstrated in Fig. 6.89 (a–g). The graphene papers had well-retained geometry without visible fracture.

Due to the difference in surface properties and wettability, the graphene papers could be easily peeled off from the Al foils. The Al foil was more hydrophilic than the graphene papers, as illustrated in Fig. 6.89 (j, k). When the graphene coated Al foils were immersed in water, water molecules could penetrate the graphene films to reach the Al-graphene interface. Therefore, wet films were at the interface between the graphene layers and the hydrophilic Al substrate, so that the graphene papers were obtained when they were separated from the Al substrate.

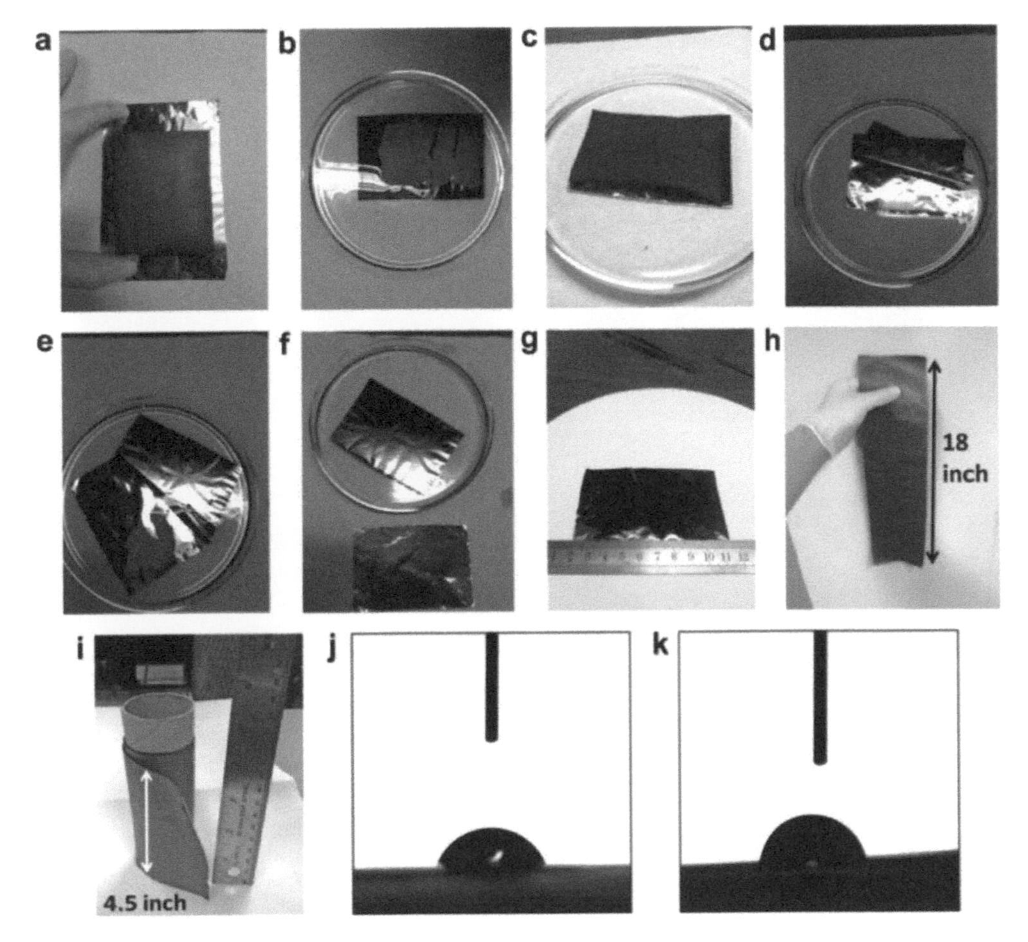

Fig. 6.89. (a) Electrospray deposited (ESD) graphene films on Al foils. (b–f) Photographs of the samples immersed in water for different time durations: (b) 30 s, (c) 2 min, (d) 4 min, (e) 7 min and (f) 9 min. (g) Photograph of a graphene paper (9 × 6 cm^2) peeled off from the Al foil and dried at room temperature. (h, i) Photographs of a 18 inch graphene paper strip (h) and a paper roll of 4.5 inch width (i) made by using the ESD integrated with roll-to-roll process. (j, k) Contact angle measurements of the Al substrate (64°, j) and the graphene paper (78°, k). Reproduced with permission from [138], Copyright © 2014, WILEY-VCH Verlag GmbH & Co. KGaA, Weinheim.

The ESD system could be integrated with a continuous roll-to-roll process [138]. Graphene strips and graphene rolls could be produced by using the integrated system. Although the large area free-standing graphene papers were derived without using polymer binders, they had pretty high flexibility, so that they could be rolled without the presence of visible damage, as demonstrated in Fig. 6.89 (h, i). Due to their high flexibility, the graphene papers could be processed into different shapes, with desired dimensions and geometries.

Representative SEM images of the graphene papers are shown in Fig. 6.90 (a, b), indicating the presence of a highly aligned structure [138]. The graphene nanosheets were stacked by overlapping one another in the horizontal direction parallel to the substrates, forming a layer-by-layer structure. To obtain such highly ordered and aligned structures, the flow rate, solution concentration and electric potential between the metallic nozzle and the substrate should be optimized in order to generate fine droplets during the ESD process. As the droplets were sprayed onto the substrates, the graphene nanosheets were stacked up layer-by-layer through the van der Waals forces when the solvents were evaporated.

The as-obtained graphene papers contained air gaps in between the graphene nanosheets, thus leading to a porous nanostructure, as seen in Fig. 6.90 (a, b). As a result, their thermal and electrical properties were relatively inferior. Mechanical compaction was employed to eliminate the air gaps, so as to achieve condense structure, as revealed in Fig. 6.90 (c). Thermal annealing was combined with mechanical press. After thermal annealing, the graphene papers exhibited highly ordered graphite layer texture, appearing as gray papers with silver luster, as illustrated in Fig. 6.90 (f). Moreover, large crystalline size and three-dimensional ordered microstructure were observed, after the samples were annealed at 1800°C (Fig. 6.90 (d)) and 2200°C (Fig. 6.90 (e)). The carbonized graphene papers had a crystallinity similar to that of pyrolytic graphite.

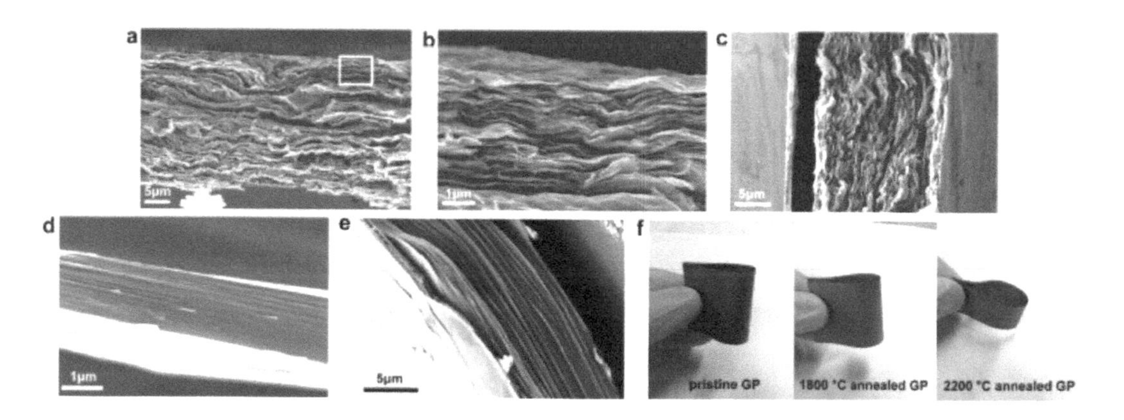

Fig. 6.90. (a) Cross-sectional SEM image of the as-obtained graphene papers. (b) High magnification SEM image of the square in (a) showing a layer-by-layer structure. (c) Cross-sectional SEM image of the graphene papers after mechanical pressing. (d, e) SEM images of the graphene papers thermally annealed at 1800°C (d) and 2200°C (e) for 30 min, followed by mechanical press, showing highly ordered and crystalline morphology. (f) Photographs of pristine papers (left), papers annealed at 1800°C (middle) and 2200°C (right). Reproduced with permission from [138], Copyright © 2014, WILEY-VCH Verlag GmbH & Co. KGaA, Weinheim.

6.3.2.1. Interface self-assembly

Interface induced self-assembly is also a promising method to construct graphene papers [57, 193–195]. In this respect, the amphiphilic characteristic of GO nanosheets is beneficial to form an oriented packing at the liquid–air interface. As the thickness is sufficiently large, free-standing GO papers can be obtained. This method has several advantages. Firstly, it is time-saving and low-energy consuming. For example, the GO membranes could be formed within 1 h. In contrast, the vacuum filtration method takes hours or even days to achieve similar thickness. Secondly, the GO suspensions can be repeatedly used, as long as the concentrations meet the requirement of membrane formation. Therefore, this process is sustainable and thus cost-effective. Thirdly, the thickness can be readily controlled by controlling the assembly time duration, while the size of the membrane is simply determined by the area of the liquid-air interface, i.e., the dimension of the reactors.

A facile self-assembly approach has been reported to obtain macroscopic GO membranes at a liquid-air interface by evaporating GO hydrosols [193]. In the experiment, 200 ml GO suspension in water with a concentration of 2 mg·ml^{-1} was treated ultrasonically for 30 min, followed by high-speed centrifugation at 5000 rpm for 20 min to remove impurities, as shown in Fig. 6.91 (a). By heating a stable hydrosol of GO to 353 K, a smooth and condensed GO thin film was formed at the liquid–air interface, as demonstrated in Fig. 6.91 (b). Further drying and reduction treatment yielded flexible, semi-transparent and free-standing graphene papers with thicknesses in the range of 0.5–20 μm, which could be tailored by controlling the evaporation time of the hydrosol, as seen in Fig. 6.91 (c). For example, GO papers with thicknesses of 5 μm and 10 μm were obtained after heating for 20 min and 40 min, respectively. The size of the flexible GO paper was limited only by the area of the liquid/air interface, so that large papers could be easily obtained by using large reaction vehicles (Fig. 6.91 (d)).

Figure 6.92 (a) shows a SEM image of the GO paper after heating for 40 min, which had a uniform thickness of 10 μm and a relatively smooth surface. Figure 6.92 (b) shows a surface SEM image, in which individual GO nanosheets can be clearly observed. Cross-sectional SEM image indicated the GO nanosheets were stacked in a layer-by-layer way, as illustrated in Fig. 6.92 (c). According to XRD analysis, the interlayer spacing was about 0.747 nm, which was almost the same as that of the parent GO. The self-assembled layer-by-layer macroscopic structure is presented in Fig. 6.92 (d). A mechanism was proposed to describe the formation process of the papers, as schematically presented in Fig. 6.92 (e).

As the hydrosol was heated, the Brownian motion of the GO nanosheets was sped up in the suspension. Due to the evaporation of water, the liquid level of the hydrosol gradually decreased, so that collision

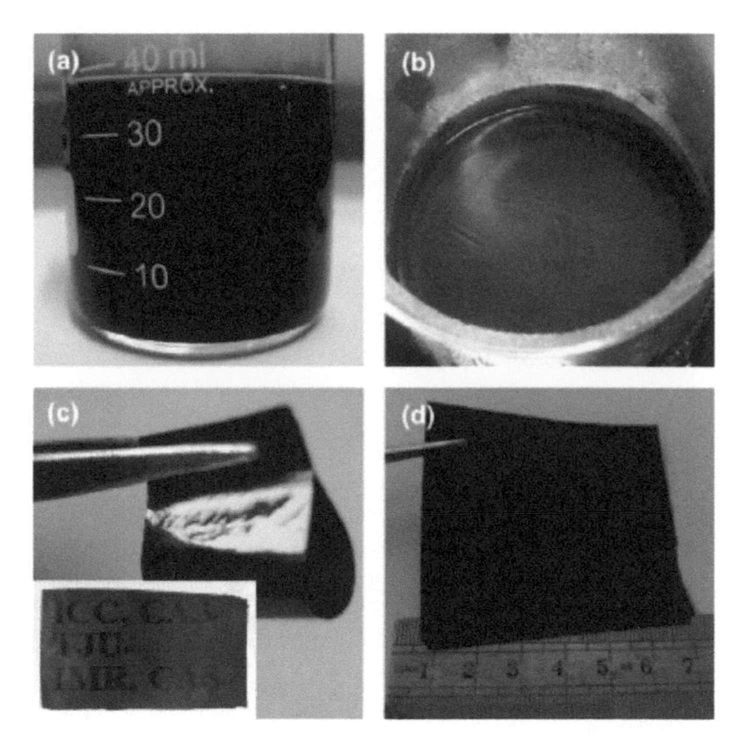

Fig. 6.91. (a) Photograph of the GO colloid suspension with a concentration of 2 mg·ml^{-1} kept stable for two weeks. (b) A film self-assembled at the liquid-air interface after heating at 80°C for 15 min. (c) Photograph of a flexible and semi-transparent GO paper (15 mm × 30 mm). (d) Photograph of a large-area (about 60 mm × 60 mm) GO paper. Reproduced with permission from [193], Copyright © 2009, WILEY-VCH Verlag GmbH & Co. KGaA, Weinheim.

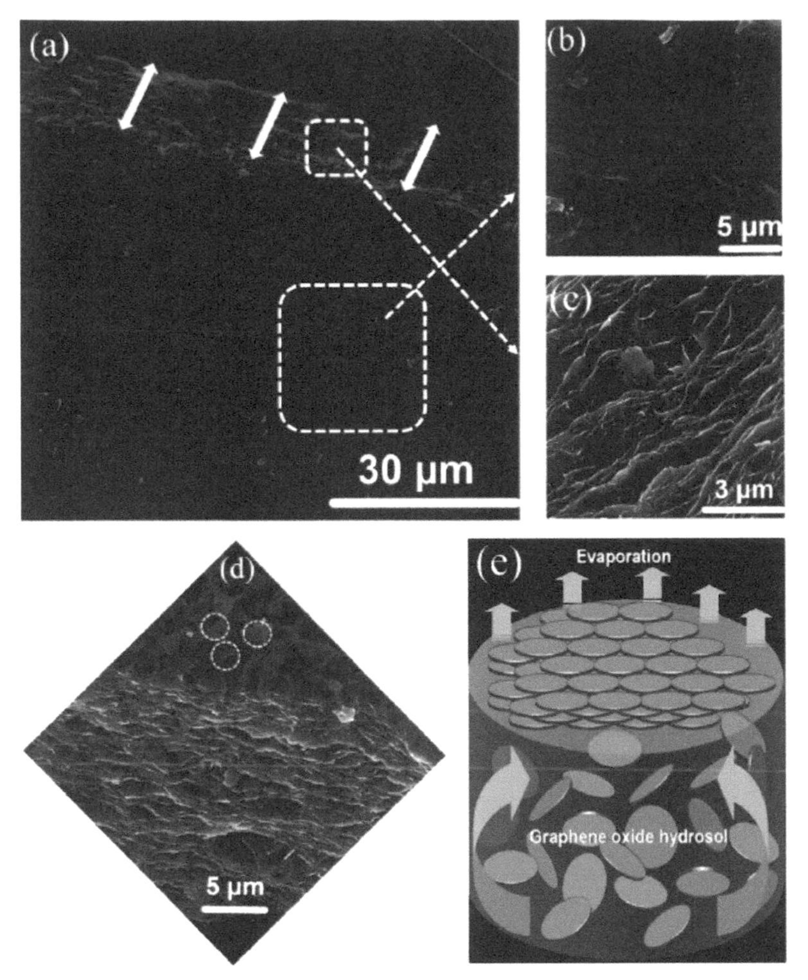

Fig. 6.92. (a–c) SEM images of the GO papers. (d) SEM image showing the multilayered structures. (e) Schematic diagram of the proposed formation mechanism. Reproduced with permission from [193], Copyright © 2009, WILEY-VCH Verlag GmbH & Co. KGaA, Weinheim.

and interaction among the GO nanosheets became more frequent and stronger. As a result, they migrated up to the liquid-air interface as the water molecules escaped from the hydrosol. The liquid-air interface offered a smooth space to host the 2D GO nanosheets. As the GO nanosheets reached the interface, they tended to aggregate to form a multilayered structure. This layer would become thicker and thicker, as new GO nanosheets were transferred near to the interface and accumulated, due to the strong interlayer van der Waals interactions. XRD results indicated that the average thickness of the ordered stack was 8.2 ± 0.1 nm, implying that the stacking structure consisted of 10–11 layers of graphene oxide nanosheets. Finally, macroscopic membranes with a thickness of several micrometers were develop over the whole liquid-air interface.

The interface self-assembled GO papers can be further reduced into graphene papers. For instance, supercritical ethanol could be used as an effective reductant to reduce the prefabricated GO papers [140]. The reduced GO papers were used as binder-free anodes, showing a capacity of 89 $mAh{\cdot}g^{-1}$, together with pretty high cycling stability.

Stable GO suspension was heated at 80°C for about 20 min in a water bath. At the liquid-air interface, a smooth GO thin film was formed. The suspension below the membrane was decanted, so that it was left at the bottom of the container. The dried GO paper was reduced by supercritical ethanol treatment at 260°C at 10 MPa for 5 h. The as-obtained bendable GO paper was almost black in reflection under white light, as seen in Fig. 6.93 (a). After treatment in supercritical ethanol, the paper exhibited metallic luster, as illustrated in Fig. 6.93 (b). It was still bendable and highly flexible. Figure 6.93 (c) shows a cross-sectional SEM image of the rGO paper after supercritical ethanol treatment, demonstrating well-packed layers throughout the entire cross-section of the paper.

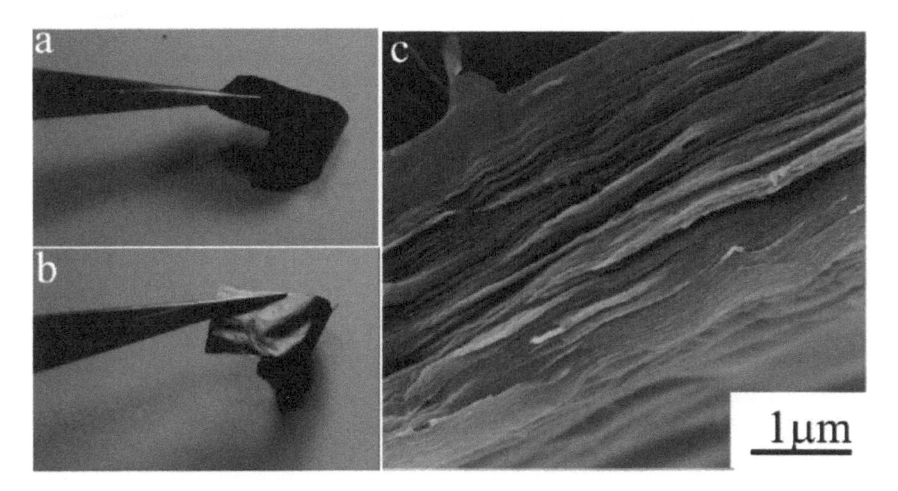

Fig. 6.93. Photographs of the GO paper (a) before and (b) after supercritical ethanol treatment. (c) Cross-sectional SEM image of the GO paper after supercritical ethanol treatment. Reproduced with permission from [140], Copyright © 2012, Elsevier.

6.3.2.4. Chemical vapor deposition

A chemical vapor deposition (CVD) process was used to fabricate graphene papers [144]. Firstly, graphene pellets were synthesized by the CVD method, with inexpensive nickel powder as the catalyst. The graphene pellets were then processed into graphene papers by using mechanical pressing. The graphene papers exhibited high electrical conductivity of up to 1136 ± 32 $S \cdot cm^{-1}$ and a breaking stress at 22 ± 1.4 MPa. Moreover, the graphene paper with a thickness of 50 μm demonstrated 60 dB electromagnetic interference (EMI) shielding effectiveness.

Nickel powder with an average particle size of 2–3 μm and specific surface area of 0.68 $m^2 \cdot g^{-1}$ was pressed into pellets of 6.4 cm in diameter. The nickel pellet was placed on a quartz platform inside a quartz tube to grow graphene nanosheets by using CVD. The nickel pellet was heated up to 1000°C in a tube furnace in Ar (1000 sccm). H_2 (325 sccm) was then purged into the tube for 15 min, to eliminate any metal catalyst oxide. After that, CH_4 was introduced for 5 min. Various hydrocarbon flow rates were tested, including 12, 15, 18, 25 and 28 sccm, corresponding to concentrations of 0.9, 1.1, 1.3, 1.9 and 2.1 vol%, respectively. The furnace was then cooled to room temperature at a rate of 100°C min^{-1} in Ar (1,000 sccm) and H_2 (325 sccm). The nickel pellet shrank by about 30% after the CVD reaction. The final 3D graphene structure in the form of the pellet was obtained by etching out the nickel from the graphene–nickel pellet with HCl (3 M) at 80°C for 10 h. The graphene pellet was washed with water to remove residual acid and then dried at room temperature. Graphene paper was obtained by compressing the graphene pellet with a presser between 2 flat steel plates. The thickness of the graphene papers was controlled by controlling the compression load.

Figure 6.94 (a) shows the procedure to obtain the graphene papers. The graphene paper was highly flexible, could be folded at 180° and then would recover its initial shape upon the release of the folding force, as shown in Fig. 6.94 (b). The graphene paper also exhibited high mechanical strength. Figure 6.94 (c) shows the stress–strain curve of the graphene paper synthesized with 1.9 vol% CH_4, possessing a breaking stress at 22 ± 1.4 MPa.

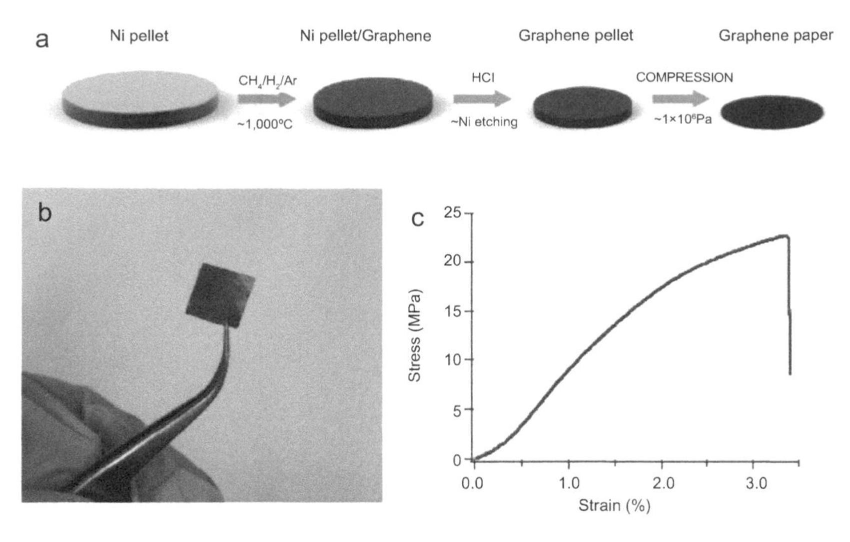

Fig. 6.94. (a) Schematic diagram showing the CVD procedure to synthesize the graphene pellet (3 cm in diameter) and its transfer into graphene paper. (b) Photograph of the graphene paper (1 cm × 1 cm). (c) Typical tensile stress strain curve of the graphene paper derived from the 3D graphene pellet synthesized with 1.9 vol% CH_4. (A color version of this figure can be viewed online.) Reproduced with permission from [144], Copyright © 2015, Elsevier.

Representative SEM images of the graphene paper are shown in Fig. 6.95 (a, b), revealing the wrinkles and ripples of the graphene nanosheets. This morphology could be attributed to the difference in thermal expansion coefficient between the nickel and graphene [196]. Gaps were left among the graphene nanosheets after the nickel powders were extracted from the random 3D samples. Due to the high mechanical strength, the cross section and morphology of the graphene paper could be identified in the SEM image of Fig. 6.95 (b). All the graphene nanosheets were tightly compacted, with a paper thickness of 35 μm. Figure 6.95 (c) shows a high-magnification TEM image of the graphene nanosheets, displaying a four-layer structure with an inter-layer spacing of 0.32 nm. The multilayered structure was also supported by the diffraction pattern, as illustrated in the inset in Fig. 6.95 (c). The graphene paper had a sharp 2D peak at 2707 cm^{-1}, indicating the presence of fewer layers of graphene nanosheets. Comparatively, the graphite, had a broad 2D peak at 2730 cm^{-1} in the Raman spectra, as seen in Fig. 6.95 (d). The suppressed D peak in the Raman spectrum implied the high graphene quality of the graphene papers.

Another example of CVD processing is the development of freestanding, flexible and transparent graphene paper (FFT-GP) from prism-like graphene (PLG) building blocks [145]. This freestanding paper possessed a large-area PLG structure, which could be transferred onto any other substrates. The PLG building blocks were connected in a face-to-face way and homogeneously aligned on silicon substrates. A single piece of FFT-GP was both highly transparent and flexible. It also maintained the electrical and physical properties of 2D graphene nanosheets, i.e., high electrical conductivity and large surface area. Such FFT-GPs could find potential applications in transparent and stretchable supercapacitors.

The FFT-GP was synthesized by using a microwave plasma-enhanced chemical vapor deposition (MPECVD) with NaCl powder as the template. Figure 6.96 shows a schematic diagram of the MPECVD process to synthesize the FFT-GP. The NaCl template was initially placed at the center of the substrate (Fig. 6.96 (a)). CH_4 as the carbon source was introduced into the chamber at microwave power of 750 W and 560°C, in order to obtain a thin graphene base layer on the silicon substrate (Fig. 6.96 (b)). The NaCl powder was fully molten, forming a thin layer on top of the graphene base layer (Fig. 6.96 (c)). At the same time, the NaCl nanocrystals were accumulated on the C/H plasma ball, through the carbon source/hydrogen radicals, due to the inelastic collisions between the radicals and NaCl nanocrystals. Meanwhile, nucleation and growth of the FFT-GP started on the thin NaCl layer. As the DC bias was turned off, prism-like NaCl crystals recrystallized in the thin NaCl layer coated with defective graphene fragments (Fig. 6.96 (d)). The prism-like NaCl crystals then served as templates for further growth of the PLG building blocks.

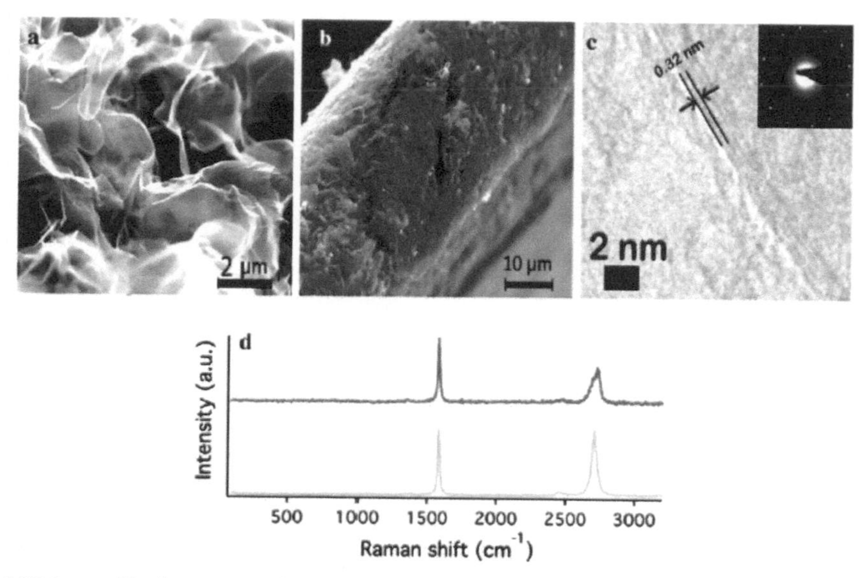

Fig. 6.95. (a) High magnification cross-sectional SEM image of the graphene paper. (b) Low magnification cross-sectional SEM image of the graphene paper, with a thickness of 35 μm. (c) TEM image of the graphene paper, showing a four-layer structure of graphene nanosheet with an inter-layer spacing of 0.32 nm. The inset is the corresponding electron diffraction pattern indicating the multilayer structure of the observed graphene flakes. (d) Raman spectra of pure graphite powder (upper/red) and the graphene paper (bottom/blue) that was synthesized with 1.9 vol% CH_4. Reproduced with permission from [144], Copyright © 2015, Elsevier.

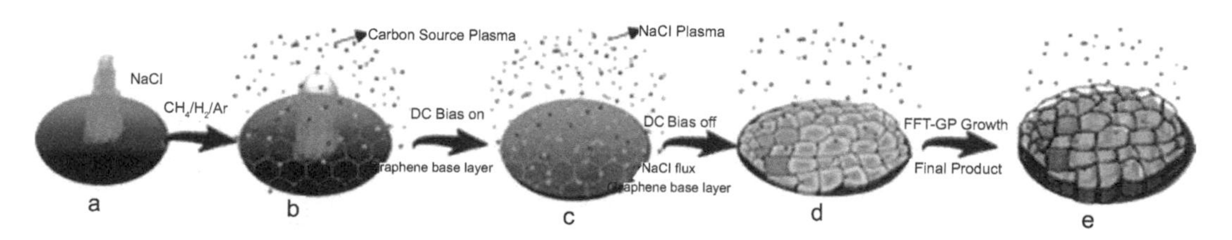

Fig. 6.96. Schematic diagram demonstrating the process to synthesize the freestanding, flexible and transparent graphene paper (FFT-GP). (a) Si substrate, with a bulk NaCl polycrystal at the center, placed into the reaction chamber. (b) Bulk NaCl gradually melted and the NaCl ions diffused into the mixed plasma. A graphene base layer was quickly formed on the Si substrate. (c) DC bias was applied, causing the NaCl to completely collapse and form a thin wet layer on the graphene base layer. (d) DC bias was turned off, so that prism-like graphene (PLG) started to form along with the recrystallization of the wet NaCl layer into crystal grains. (e) Final FFT-GP formed as a coating layer on the Si substrate. Reproduced with permission from [145], Copyright © 2015, American Chemical Society.

During the growth of the FFT-GP, NaCl was etched from the templates, while further growth of the PLGs started as the carbon source was precipitated onto the graphene seeds. The NaCl template was etched by using high-energy ion bombardment. During this process, some of the Na and Cl ions were incorporated into the plasma ball to form Na^+ and Cl^- plasma. Furthermore, some of the Na^+ and Cl^- plasma in the high-temperature zone of the plasma ball would combine with the carbon source to facilitate the growth of FFT-GP, while the rest in the low-temperature zone were recrystallized on the inside wall of the reactor or on the substrate. After reaction for 20 min, a thin graphene film base layer integrated with an interconnected PLG unit was formed on the silicon substrate, as seen in Fig. 6.96 (e). Due to its ductility, the interconnected graphene structure could be readily peeled off from the substrate. As a result, a freestanding, flexible and transparent graphene paper was developed, as demonstrated in Fig. 6.97 (a).

The freestanding graphene paper had wrinkles and light brown color, and was light, flexible and textural at the macroscale. Transparency of the FFT-GP is demonstrated in Fig. 6.97 (b), while its regularly arranged PLG microstructure is illustrated in Fig. 6.97 (c). Its flexible characteristic was also evidenced

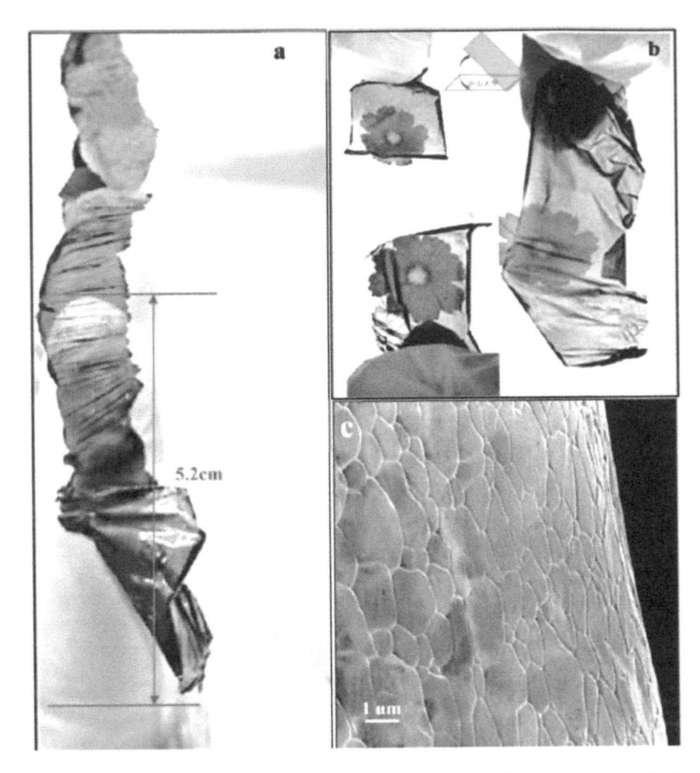

Fig. 6.97. (a) Photograph of a free-standing graphene sheet, 1.2 cm wide and 8.2 cm long, peeled off from a Si substrate. (b) Photographs taken with a 60° angle between the FFT-GP and the underlying logo, illustrating the transparency of the sheet from various angles. (c) Low-magnification SEM image of an FFT-GP sheet. Reproduced with permission from [145], Copyright © 2015, American Chemical Society.

by SEM observation. The ordered arrangement of numerous PLG building blocks was similar to that of an elastic net, as revealed by the high magnification SEM images, which indicated the morphological details of the upper side and bottom side, in Fig. 6.98 (a) and (b), respectively. Figure 6.98 (c) shows a high magnification cross-sectional SEM image of the FFT-GP, revealing its topological structure. The PLG units possessed various morphological blocks, including hexagonal, pentagonal and quadrangular prisms, which were connected in a face-to-face manner. The diameter of the PLG was less than 1 μm, while the thickness of the arris in the individual PLG was about 8.5 nm, which was larger than that of the facet of about 1 nm. The difference in size between the arrises and the facets was beneficial to the stability of the geometric structure.

A large number of pores were observed on the surface of each facet. Nitrogen adsorption/desorption isotherm of the FFT-GP belonged to type IV with an H_3-type hysteresis loop, implying the presence of slit-shaped pores, as seen in Fig. 6.98 (d). The sharp increase in N_2 uptake at low pressure and the hysteresis loop starting at $0.40P/P_0$ suggested the presence of meso- and microporous structures. The Brunauer–Emmett–Teller (BET) specific surface area of the FFT-GP was 909 $m^2 \cdot g^{-1}$. Figure 6.98 (e) shows the pore size distribution in the range of 0.9–110 nm, which was analyzed by using the density functional theory (DFT) method applied to the adsorption branch of the isotherm, with the assumption that the pores had a slit-shaped geometry. The pore dimensions were centered at 1.2 and 3.7 nm, in a good agreement with the micro- and mesoporous structures, respectively. There were also pores with dimensions centered at 9.2 and 75.5 nm, corresponding to small numbers of meso- and macro-porous structures, respectively. The formation of the pores on facets of the PLG structures was closely related to the growth process of the FFT-GP, as discussed before.

The detailed microstructure of the FFT-GP was revealed by using TEM observations. The ultrathin 2D graphene facets were integrated with interconnected graphitic carbon arrises to form a 3D graphene

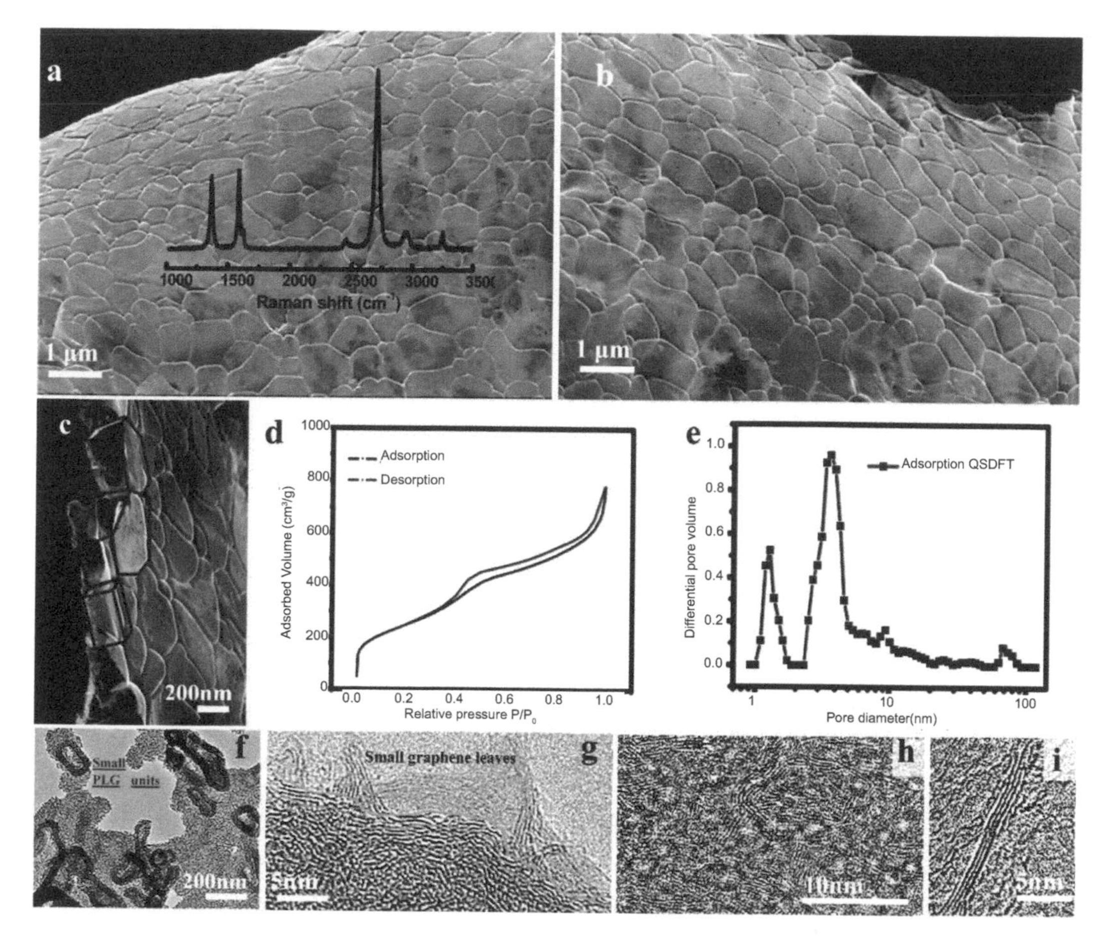

Fig. 6.98. (a, b) SEM images of the front side (a) and back side (b) of the FFT-GP sheet with a vertically aligned and regular arrangement of the connected PLG building blocks. The inset shows Raman spectrum of the FFT-GP. (c) Cross-sectional SEM image of the FFT-GP sheet, with the region marked in red to be the reconstructed topology of the corresponding connections of the PLG cells in the FFT-GP. (d) N_2 adsorption/desorption isotherms of the FFT-GP. (e) Pore-size distributions determined according to the original density functional theory with the slit porous model. (f) TEM image of one facet of the PLG. The sample was treated by ultrasonication before the TEM test to observe one single facet. (g) High resolution TEM (HRTEM) image of the small graphene leaves at the edge area of the PLG. (h) HRTEM images of three- to four-layered small graphene leaves at the center of one surface of the PLG. (i) HRTEM image of a four-layered graphene facet of the PLG. Reproduced with permission from [145], Copyright © 2015, American Chemical Society.

network. A hexahedron structure was observed in individual PLG. There were also small microstructures anchored on the facets, including small graphene leaves and hollow graphite cubes, as demonstrated in Fig. 6.98 (f). The small leaves consisted of 2–3 graphene nanolayers that were heavily distributed on the facets, similar to a loose and soft carpet, so as to effectively enlarge the surface area, as presented in Fig. 6.98 (g, h). The hollow graphite cubes were PLG units with a smaller size, while their crystalline arris frameworks and the graphene leaves were present in the high-resolution TEM images. The graphene facets consisted of 3–4 layers (Fig. 6.98 (i)).

6.3.2.5. Miscellaneous

A new technique was reported to fabricate multilayered graphene papers with large sheet sizes of > 10 cm^2 [143]. Carbon films with a metallic sheen were produced through the thermal decomposition of asphalt, forming fumes in a ceramic crucible that condensed onto a heated surface at a high temperature of 650°C.

A Fisher burner (natural gas/air) flame was used to heat the samples, while the crucible was covered but exposed to laboratory atmosphere. A 60 ml (70 mm) Coors casserole crucible with an inner 5 ml crucible was used as the reaction vessel to hold the starting material. The inner crucible was filled with 5 g of asphalt precursor and placed in the larger casserole crucible, as demonstrated in Fig. 6.99. The top of the setup was covered with a watch glass. In the thermolyzed asphalt reaction (TAR) process, the system was heated for 12–15 min, which was followed by cooling for 5–10 min. The yield of the TAR graphene based on the initial mass of starting material was 0.11% ($n = 3$, $\sigma = 0.08\%$) on the flat crucible target. This yield was dependent on the surface area, because it was increased to 0.93% ($n = 3$, $\sigma = 0.24\%$) when 5.0 g diatomite was used for the experiment.

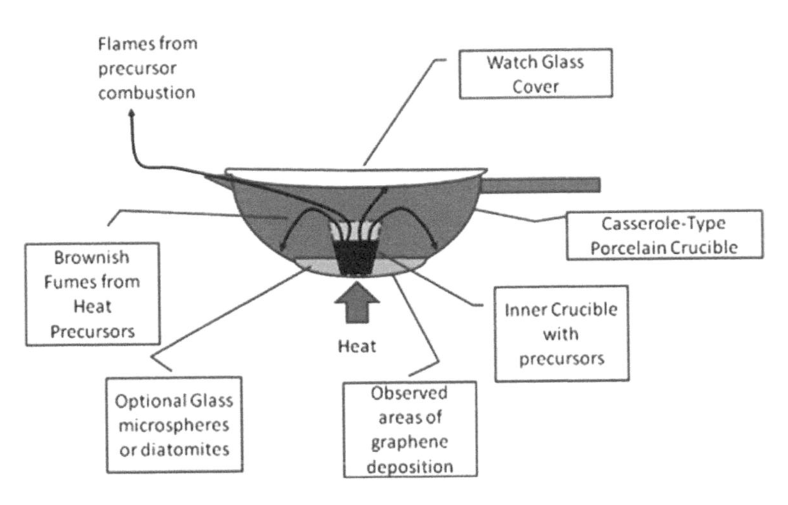

Fig. 6.99. Schematic of the graphene deposition setup. Reproduced with permission from [143], Copyright © 2011, Elsevier.

It was found that the carbon films could be strongly adhered to the ceramic substrates. In order to peel off the graphene papers from the ceramic crucible, the cooling rate should be sufficiently high. Figure 6.100 (A) shows a photograph of the graphene paper with a diameter of 25 mm. Obviously, larger papers could be obtained simply by using large crucibles, i.e., scaling up is not a problem of this method. The papers obtained in this way were relatively flat when compared with those made by using the vacuum filtration methods. Figure 6.100 (B) shows an optical micrograph of the graphene papers, illustrating their flat nature. Representative SEM images of the graphene papers are shown in Fig. 6.100 (C, D). The flatness of the papers was down to $\leq$ 10 nm. High magnification of TEM image indicated that (Fig. 6.100 (E)) the layered characteristics of the graphene papers were at the nanometer scale, as seen in Fig. 6.100 (E). The temperature required to support the deposition was estimated according to the melting point of aluminium (660°C). Because the aluminium foil was not molten and only just softening, the temperature for graphene deposition was estimated to be about 650°C.

More recently, graphene papers with a folded structure were derived from a freeze-dried GO cryogel, followed by subsequent thermal reduction [142]. This new type of porous graphene papers demonstrated significant enhancement in performances as electrodes of lithium ion batteries, as compared with the graphene paper fabricated by using the filtration method. A reversible capacity of 568 $mAh \cdot g^{-1}$ at 100 $mA \cdot g^{-1}$ was retained after 100 cycles.

Chemically reduced graphene (CRG) aqueous solution was obtained by reducing GO solution with hydrazine in ammonium solution. The as-prepared CRG solution was diluted to 1 $mg \cdot ml^{-1}$ and 0.4 $mg \cdot ml^{-1}$. CRG–GO hybrid solution was also prepared by mixing the GO and CRG solutions. The CRG/GO mass ratios were 1:5, 1:2 and 2:1. The resultant hybrid solution was transferred into a 50 ml beaker, to which a freezing-dry process was applied to form a cryogel, consisting of 8 h freezing at –30°C and 36 h freeze-

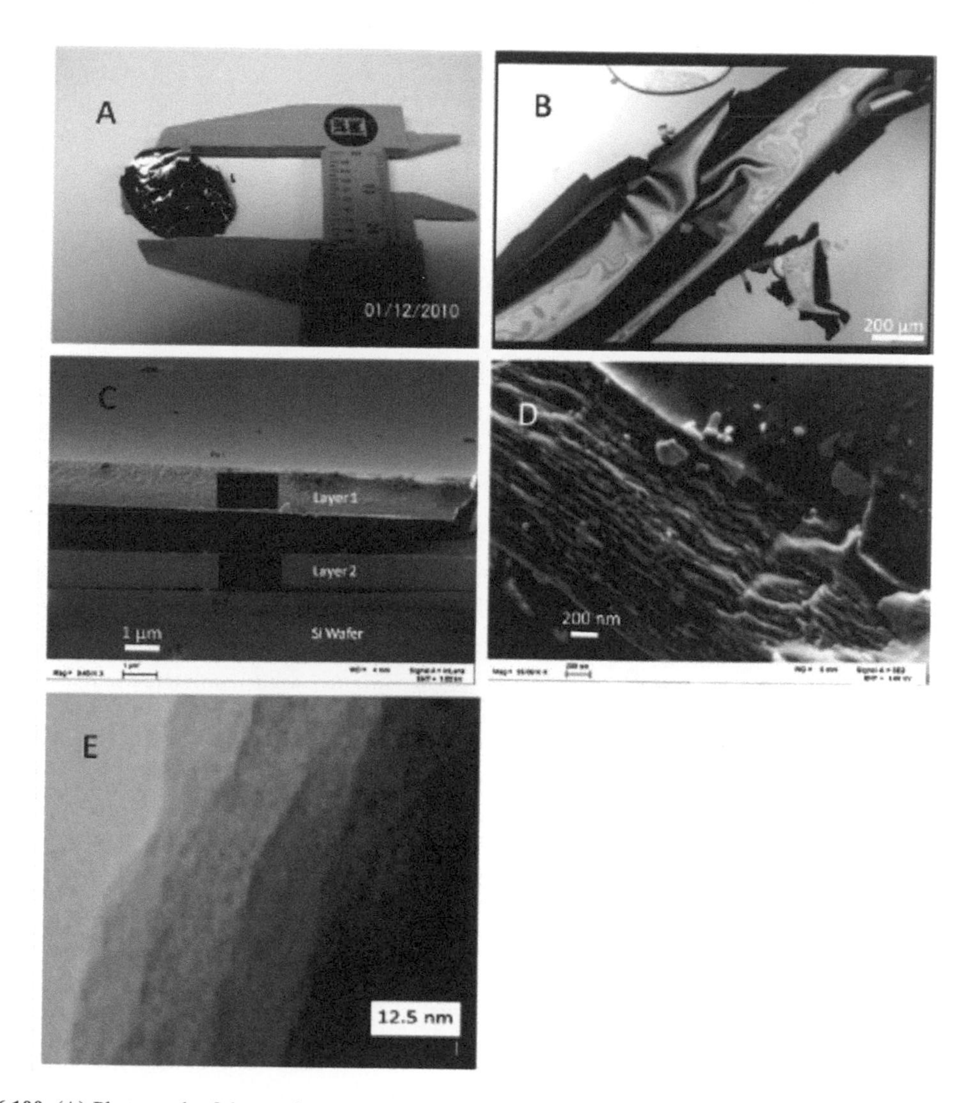

Fig. 6.100. (A) Photograph of the graphene paper with a diameter of 25 mm. (B) Optical micrograph (400×) of the free-standing graphene paper in water. Scale bar = 200 μm. (C) Cross-sectional SEM image of the folded graphene paper layers on Si. The distance bar = 1 μm. (D) Cross-sectional SEM image of the sample showing the layered characteristics. Scale bar = 200 nm. (E) TEM image showing the layered characteristics of the graphene paper. Reproduced with permission from [143], Copyright © 2011, Elsevier.

drying. CRG–GO hybrid papers were prepared by pressing the cryogels at 20 MPa, with GO–CRG-0.5, GO–CRG-2 and GO–CRG-5 to denote the corresponding CRG/GO mass ratios. The CRG–GO hybrid papers were reduced to rGO papers through thermal reduction at 220°C for 2 h in air. Accordingly, they were denoted as rGO–CRG-0.5, rGO–CRG-2 and rGO–CRG-5, respectively. A pure rGO paper was also prepared similarly for comparison.

Figure 6.101 (a) shows SEM image of the as-obtained cryogel, demonstrating an interconnected 3D porous network, with the pore wall consisting of graphene nanosheets. The formation of the pores was attributed to the sublimation of ice crystals during the freeze-drying process. The pores had a wide range distribution in size, from tens of micrometers to sub-millimeters. The graphene papers, with a thickness of about 20 μm, possessed a conductivity of 2–5 $S\cdot cm^{-1}$. All the cryogel papers had similar structures. Figure 6.101 (c) shows surface SEM image of the rGO– CRG-0.5 cryogel paper as a representative. The graphene

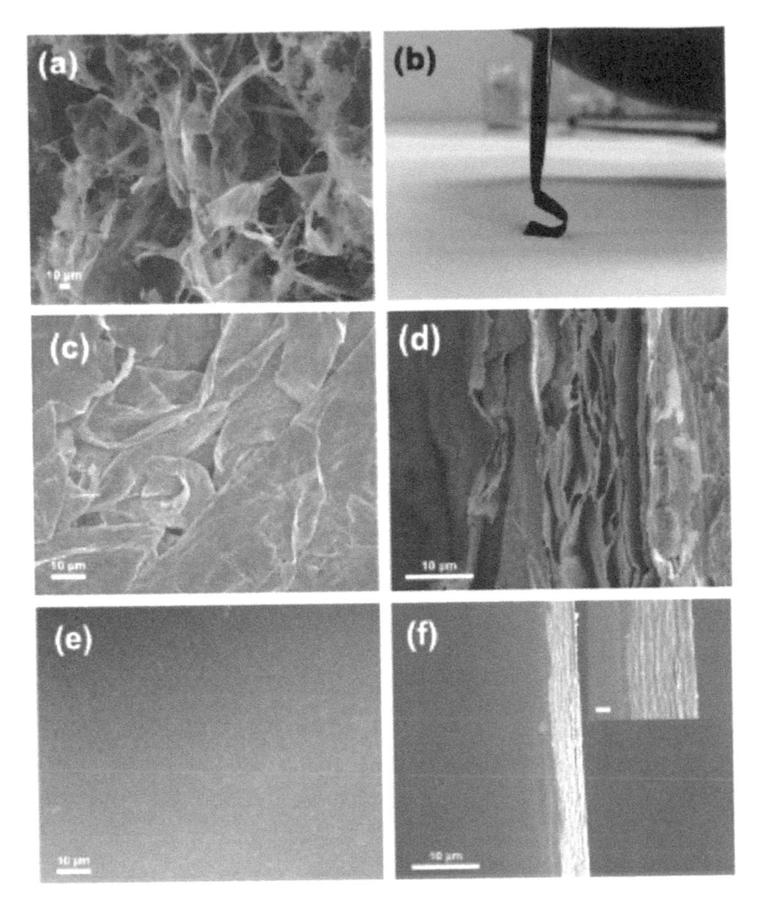

Fig. 6.101. (a) SEM image of the rGO–CRG-0.5 cryogel. (b) Photograph of the rGO–CRG-0.5 paper demonstrating its flexibility. (c, d) Surface and cross-sectional SEM images of the rGO–CRG-0.5 paper. (e, f) Surface and cross-sectional SEM images of the CRG paper made with vacuum filtration, with the inset to be the cross-sectional view of (f) at higher magnification. Scale bar = 1 μm. Reproduced with permission from [142], Copyright © 2014, The Royal Society of Chemistry.

papers had a relatively rough surface with folds of graphene nanosheets together with random size, as observed in Fig. 6.101 (c), which was different from the smooth surface of the graphene paper made with vacuum filtration (Fig. 6.101 (e)). They also exhibited a loosely packed layered structure (Fig. 6.101 (d)), in contrast to the tightly compact graphene layer structures of the filtration papers (Fig. 6.101 (f)). It was found that as the CRG/GO ratio in the precursor solution was increased, the mechanical properties of the graphene papers were tremendously enhanced.

6.3.3. Applications of G/GO papers

6.3.3.1. Energy related applications

As mentioned before, due to their excellent electrical and thermal conductivity, superior mechanical flexibility, a broad electrochemical window, together with the high specific surface area, graphene or graphene-based materials have become a promising candidate for energy related applications, especially as electrode materials of electrochemical energy devices, including batteries [149, 197–199], supercapacitors [200–203], fuel cells [204–208] and solar cells [7, 209, 210].

Specifically, the electrochemical interaction of lithium with graphene in lithium ion batteries (LIBs) has been the major research topic regarding graphene for such applications. It has been demonstrated that graphene-based electrodes can accommodate lithium more effectively than the common graphite anodes, due to the additional reaction mechanisms besides the intercalation, including fast lithium adsorption [211–213], defect trapping [214] and double layer or faradaic capacitance [215]. Currently, the main objective has been focused on graphene-based materials as electrodes in LIBs to replace the commonly used graphite.

More recently, free-standing rGO was also evaluated for the cathodes of LIBs [93]. It was found that the free-standing rGO cathodes exhibited an increase in capacity with the increasing content of oxygen, due to the role of oxygen functionalities in the faradaic redox reaction between the Li^+ ions and functionalized carbon. Moreover, the rGO cathodes showed excellent rate capability, due to the fast surface redox reaction. Therefore, rGO could be used for cathodes of high power LIBs.

As a special type of electrical energy storage devices, supercapacitors based on electrochemical double layer capacitance (EDLC) store and release electrical energy through charge separation at the nanoscopic scale at the electrochemical interface between an electrode and an electrolyte [216–220]. Because the energy stored is inversely proportional to the thickness of the double layer, supercapacitors have much higher energy density than the conventional dielectric capacitors. As compared with rechargeable batteries, the large amount of energy stored in supercapacitors can be delivered at very high power ratings. Supercapacitors can be used either as the primary power sources or in combination with other power sources, such as batteries and fuel cells.

Supercapacitors are superior to the conventional energy storage devices, because of several advantages, such as high power capability, long life-span, wide range of operating temperature, low weight, flexible packaging and less maintenance [221]. Besides EDLCs, there is another category of supercapacitors based on pseudocapacitance. While the charge storage mechanism of EDLCs is nonfaradic, that of pseudocapacitance is faradic, with redox reactions using electrode materials such as electrically conducting polymers and metal oxides. Although the energy densities of pseudocapacitors are higher than those of EDLCs, however, the phase changes within the electrode due to the faradic reaction limits their lifetime and power density.

The unit cell of supercapacitors is usually comprised of two porous carbon electrodes that are separated in order to avoid electrical contact with a porous separator [222]. Current collectors based on either metal foils or carbon impregnated polymers are used to conduct electrical current from the electrodes. Both the separator and the electrodes are impregnated with an electrolyte, which is an ionic conductor to facilitate ionic current to flow between the two electrodes, whereas electronic current is blocked to prevent discharging of the supercapacitor cell. Depending on the designed size and voltage for different applications, a packaged supercapacitor module could be composited of multiple unit cells.

CMG-based carbon electrode materials have been acknowledged to be a class of materials for the EDLC supercapacitor [223–226]. Figure 6.102 shows a schematic of the two-electrode supercapacitor test cell and fixture assembly [227]. The CMG-based supercapacitor cells were tested with three different electrolytes that are commonly used in commercial EDLCs, including KOH aqueous electrolyte (5.5 M) and two organic electrolyte systems, i.e., TEA BF4 in acetonitrile (AN) solvent and TEA BF4 in propylene carbonate (PC) solvent.

The performance of ultracapacitor cells is generally characterized by using cyclic voltammetry (CV), electrical impedance spectroscopy (EIS) and galvanostatic charge/discharge. The CV curves and galvanostatic charge/discharge are used to obtain specific capacitances of the CMG electrodes. The specific capacitance from the CV curves is integrated over the full CV curve to determine the average value. The specific capacitance determined from galvanostatic charge/discharge is calculated from the slope (dV/dt) of the discharge curves. The EIS data are analyzed by using the Nyquist plots, showing the frequency response of the CMG electrode/electrolyte system, which are plotted as the imaginary component (Z'') of the impedance versus the real component (Z'). Each point of the data corresponds to a frequency, with the frequency increasing from right to left along the Z'-axis. The intersection of the curve at the Z'-axis stands for the internal or equivalent series resistance (ESR) of the cell, which determines the rate at which the cell can be charged/discharged (power capability). The slope of the 45° portion of the curve is known as the Warburg resistance, which is attributed to the frequency dependence of ion diffusion/transport in the electrolyte.

The graphene material worked well in the commercial electrolytes. Due to its high electrical conductivity, very promising charge storage capability was achieved. CV curves were almost rectangular

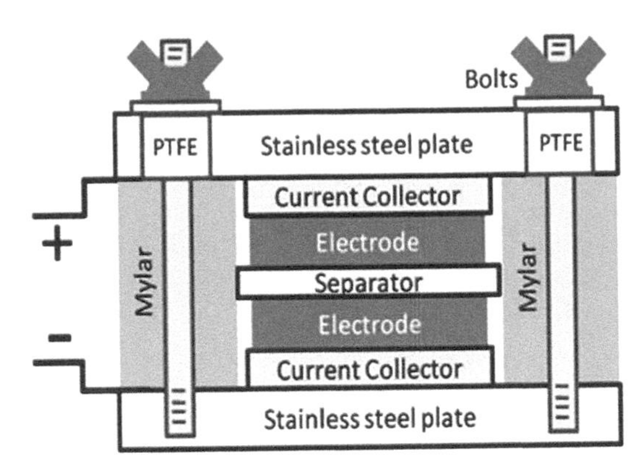

Fig. 6.102. Schematic of the supercapacitor cell assembly. Reproduced with permission from [227], Copyright © 2008, American Chemical Society.

in shape, implying that charge propagation within the electrodes was smooth. The CV curves of the CMG-based electrodes remained rectangular with only slight variance in specific capacitance, even at a scan rate of as high as 40 $mV \cdot s^{-1}$. Besides the low electrical resistance, the CMG material also provided a short and equal diffusion path length for the ions in the electrolyte, because the electrolyte was not penetrating into the particulate, so that only the graphene nanosheets at the surface were accessed. The high conductivity of the CMG material also resulted in low ESR of the cells. The internal cell resistance (real Z′ axis from Nyquist plots) was 0.15 Ω (24 kHz), 0.64 Ω (810 kHz) and 0.65 Ω (500 kHz) for KOH, TEABF4/PC and TEABF4/AN electrolytes, respectively.

The change in specific capacitance as a function of voltage was remained almost linearly at higher voltages. Test cells with KOH, PC and AN electrolytes were cycled to 1 V, 2.7 V and 2.5 V, respectively. A small fraction of pseudocapacitance was observed, which could be attributed to the presence of a low level of functional groups. Essentially, the current was increased nearly linearly with the increasing voltage, which suggested that the charge was basically nonfaradic. Specific capacitance of KOH electrolyte at 40 $mV \cdot s^{-1}$ scan rate during charging was very close to a constant of 116 $F \cdot g^{-1}$ over 0.1–0.9 V. Specific capacitance of the two organic electrolytes, AN and PC, were 100 $F \cdot g^{-1}$ and 95 $F \cdot g^{-1}$, during discharge at 20 $mV \cdot s^{-1}$, in the range from 1.5 V to fully discharged at 0 V and from 2.0 V to fully discharged at 0 V, respectively. It is expected that cost-effective supercapacitors could be available as a large-scale product once the price of graphene is significantly reduced.

6.3.3.2. Water treatment and molecule separation

Graphene-based membranes, either GO or rGO, have found potential applications in liquid and gas separation [228–230]. Because membranes containing nanochannels are formed between the 2D GO/rGO nanosheets, gases and ions with smaller sizes than those of the nanochannels are allowed to permeate, while all other larger species are blocked. Dry GO membranes have a void spacing of about 0.3 nm between the adjacent GO nanosheets, which is not permeable to most gases. However, water molecules can permeate freely if they are aligned in a monolayer, so that GO membranes can be used for the separation and selective removal of water. When GO membranes are immersed in aqueous solution, the spacing between GO nanosheets can be increased to 0.9 nm, due to the hydration effect, thus leading to decreased selectivity. Comparatively, although rGO membranes possess a similar layered structure to that of GO membranes, they are more stable in water because of the absence of the hydrated functional groups. One problem of rGO membranes for aqueous separation is the high transport resistance, which is attributed to the narrowed interlayer spacing. A potential way to increase the permeability of rGO is to decrease the membrane thickness. As a result, the development of ultrathin rGO membranes with the desired permeability together with superior selectivity is still a challenge in this research area.

Freestanding ultrathin rGO membranes made by using the vacuum filtration method have been evaluated for water treatment [91]. The permeability and selectivity of the freestanding ultrathin rGO membrane were assessed by using forward osmosis (FO) filtration, which is one of the osmotically driven processes. Because FO is driven by the osmotic pressure ($\Delta\pi$) of the draw solutions other than external hydraulic pressures, it consumes less energy than the conventional membrane processes [231]. It has been accepted that supported membranes are not suitable for FO processes, because there is severe internal concentration polarization (ICP) due to the porous support layers, so that low water flux is a serious problem [232, 233]. In this respect, freestanding rGO membranes become promising candidates for such applications, due to the absence of the support layers.

In an ideal FO process, the membrane should only experience the shear stress which is caused by the friction between the membrane and the fluids. The maximum shear stress in the FO system was about 5.04 $N{\cdot}m^{-2}$, corresponding to tensile strength of about 125 MPa for a membrane with a thickness of 40 nm and a dimension of 1×1 m^2 to resist the fluid shearing. It was found that the rGO membranes with thicknesses in the range of 20–40 nm were sufficiently strong enough to withstand the hydraulic pressures of 0.10–0.25 bar.

FO performance of the freestanding rGO membrane with a thickness of 100 nm was tested. Figure 6.103 (A) shows variations in the water fluxes of the freestanding rGO with a thickness of 100 nm and the commercial HTI CTA membranes, as a function of salt (draw solution) concentration, in which DI water and NaCl (2.0 $mol{\cdot}l^{-1}$) were used as the feed solution and the draw solution, respectively [91]. The 100 nm thick rGO membrane reached a water flux of 57.0 $l{\cdot}m^{-2}{\cdot}h^{-1}$, which was higher than that of the commercial

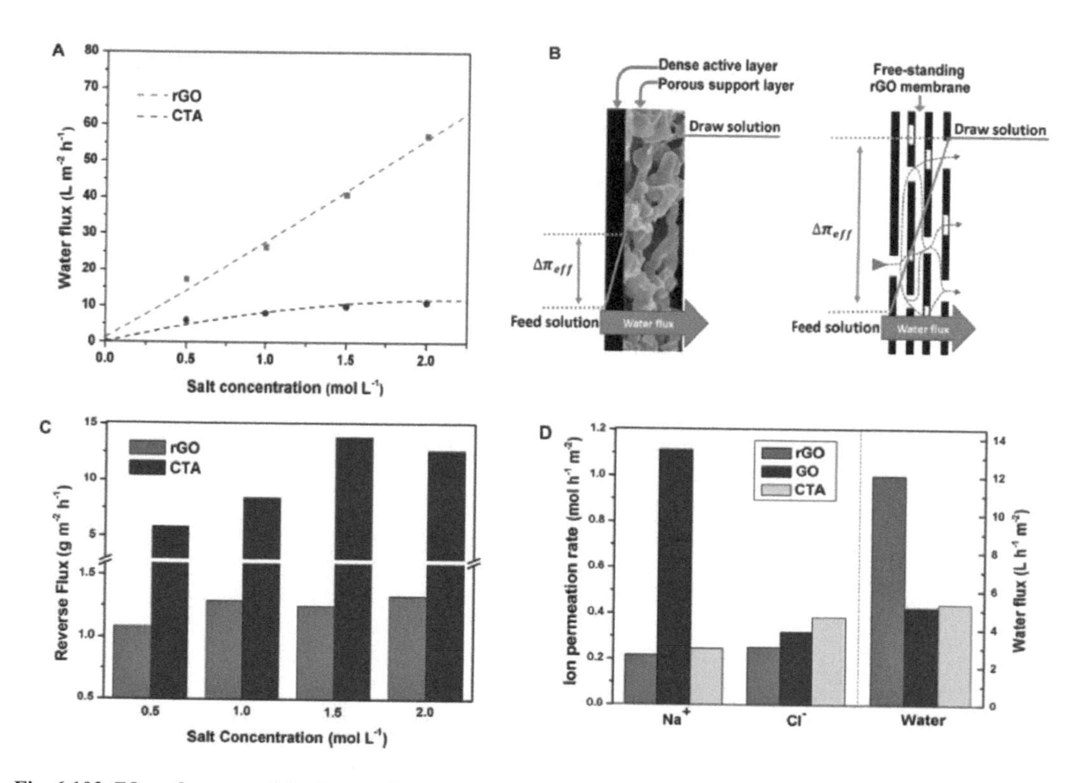

Fig. 6.103. FO performance of the freestanding rGO membranes. (A) Variations in the water fluxes of the freestanding rGO with a thickness of 100 nm and the commercial HTI CTA membranes, as a function of salt (draw solution) concentration. (B) Schematic drawings showing the difference between the osmotic pressure profiles for the conventional supported membrane (left) and freestanding rGO (right) membrane in FO process. For the rGO membranes, the intrinsic defects (e.g., pores and wrinkles) acted as the channels for mass transport. (C) Variations of reverse salt fluxes of the freestanding rGO (100 nm thick) and commercial HTI CTA membranes, as a function of salt (draw solution) concentration. (D) Permeation rates of Na^+ and Cl^- and water fluxes through the rGO, GO and CTA membranes, with NaCl as feed solution and ammonia as draw solution, respectively. Reproduced with permission from [91], Copyright © 2014, WILEY-VCH Verlag GmbH & Co. KGaA, Weinheim.

cellulose triacetate (CTA) FO membrane by nearly five times. Moreover, the commercial CTA membrane became saturated at high salt concentrations due to the effect of ICP, while the freestanding ultrathin rGO membrane exhibited a linear relationship between the water flux and the salt concentration in the range of 0.5–2.0 $mol \cdot l^{-1}$. This observation could be readily attributed to the absence of ICP in the freestanding ultrathin rGO membrane. As demonstrated in Fig. 6.103 (B), because of the ICP in the support layer, the conventional CTA FO membrane had a low effective osmotic driving force ($\Delta\pi_{eff}$). In contrast, the $\Delta\pi_{eff}$ loss was avoided in the freestanding rGO membrane.

After FO operation for 12 h, the conductivity of the feed solution was increased only slightly to 4 μS, although the draw solution had a NaCl concentration of as high as 2.0 $mol \cdot l^{-1}$, implying a low reverse NaCl flux of 1.3 $g \cdot m^{-2} \cdot h^{-1}$. In comparison, the CTA experienced a reverse NaCl flux that was nine times that of the rGO membrane, as seen in Fig. 6.103 (C). The NaCl rejection of the rGO membrane was assessed with 0.5 $mol \cdot l^{-1}$ ammonium bicarbonate (NH_4HCO_3) and 0.1 $mol \cdot l^{-1}$ NaCl as the draw solution and feed solution, respectively. As illustrated in Fig. 6.103 (D), both the water flux and NaCl rejection rate of the rGO membrane were higher than those of the CTA membrane. The higher NaCl rejection rate achieved by the rGO membrane was attributed to the smaller nanochannels formed by the rGO nanosheets. The membrane also exhibited pretty high selectivity, as shown in Fig. 6.103 (D).

Another example is the development of ultrafiltration nanostrand-channelled GO (NSC-GO) membranes, with nanochannels having diameters in the range of 3–5 nm, which exhibited excellent separation performance of small molecules and ultrafast water permeation [122]. The permeation rate was nearly ten times higher than that of normal GO membranes without compromising the rejections and more than 100 times higher than that of the commercial ultrafiltration membranes with similar rejections. Moreover, the separation behavior had an abnormal dependence on the applied pressure, which was distinctly different from the conventional porous polymer filtration membranes. It was believed that the observation was attributed to the mechanotunable effects [234–236]. Both experimental and molecular dynamics (MD) simulation results indicated that the abnormal behavior was caused by the reversible deformation of the nanochannels in the NSC-GO membranes. The shape and cross-sectional area of the nanochannels could be tuned by the application of pressure, which in turn affected the rejection rates of small molecules and water permeability. More importantly, the nature of water transportation through the NSC-GO membrane was in the form of classic viscous flow with substantially reduced boundary slip, in comparison with that in the plug-like flow confined in the hydrophobic graphene channels, such as single-walled carbon nanotubes (SWCNT) [237] and normal graphene membranes [229].

The NSC-GO membranes were prepared by using a multi-step method. Firstly, membranes were obtained from a mixed dispersion with positively charged copper hydroxide nanostrands (CHNs) and negatively charged GO nanosheets. Secondly, the obtained membranes were treated with hydrazine for 15 min. Finally, the CHNs were removed by using EDTA. During the filtration process, the CHNs were incorporated into the GO inter-layers. The GO/CHNs composite membranes possessed a nanofibrous surface structure. The CHNs were not only attached onto the surfaces of the GO nanosheets, but were also intercalated into the stacked GO layers. The incorporation of CHNs significantly improved the water permeation, when compared with the pure GO membranes.

Water permeability of the GO/CHN membranes was increased with the increasing volume of the CHN suspension, due to the increase in the number of nanochannels. The permeability was kept nearly constant as the ratio of V_{GO}/V_{CHNs} was over 3:12, because the negative charges on the surface of the GO nanosheets were all balanced by the positively charged nanostrands. The GO/CHN membranes exhibited high water permeability, while maintaining a low EB molecule rejection rate of only 57% (V_{GO}/V_{CHNs} = 3:12), as compared with that of the GO membrane (85%). The low rejection rate was ascribed to the enlarged inter-sheet spacing, due to the presence of the overlapped nanostrands. It was found that the membranes had relatively low stability and durability during the aqueous separation process, which was addressed by partially reducing the GO/CHNs membranes with hydrazine at room temperature before the removal of the CHNs. The treated membranes were very stable in aqueous separation applications.

Separation performance of the NSC-GO membranes as a function of applied pressure is shown in Fig. 6.104 (a). The rejection rate for EB molecules was the same (B83%) in zone I, but gradually decreased to 50% in zone II. As the pressure was further increased, the rejection rate recovered, reaching a value of as high as 95.4% in zone III. This observation was different from that observed in polymer-based membranes. For polymer membranes, the rejection rate was gradually increased with the pressure, due to the compression and/or densification of their porous structures [234–236]. According to the experimental pressure-dependent water flux through the NSC-GO membranes (black squares in Fig. 6.104 (a)), the flux was increased linearly, as the applied pressure was increased from 0.1 to 0.4 MPa at a rate of 101.1 $l \cdot m^{-2} \cdot h^{-1} \cdot bar^{-1}$, but increased nonlinearly as the pressure was increased over 0.4 MPa (zone II and III). Therefore, it could be concluded that no significant change occurred in the nanochannels in zone I. The rejection rate was decreased as the pressure was further increased from 0.4 to 0.9 MPa (zone II), implying that nanochannel pore was enlarged.

The decrease in permeability implied that the number of nanochannels was decreased. The underlying mechanism of the unconventional behavior was simulated with MD simulations by modeling a half cylindrical nanochannel in the NSC-GO membrane, as demonstrated in Fig. 6.104 (b), with an effective pore size of about 5 nm and a cross-sectional area of 31 nm^2. The oxygen-containing groups, including hydroxyl, epoxy and carboxyl groups, were generated by simulating a randomly assigned structure with oxygen-containing groups on both sides of the graphene nanosheet. Responses of the nanochannel to pressure exertion are shown in Fig. 6.104 (c–e). As a water-filled GO nanochannel was subject to pressure, it experienced a shape change from round into a flattened rectangle, as the pressure was increased to 0.75 MPa. The flattened rectangle was continuously compressed into round ripples, as the pressure was further increased to 1.5 MPa. The simulation results qualitatively explained the abnormal rejection rate of EB (the blue solid triangles in Fig. 6.104 (a)). As the channel was flattened, the cross-sectional area was increased from 31 to 41 nm^2, as illustrated in Fig. 6.104 (b). As a result, the permeability of the EB molecules was tremendously enhanced. With further compression, the nanochannels would be shrunk, so that the cross-sectional area was reduced and thus the rejection rate was increased again.

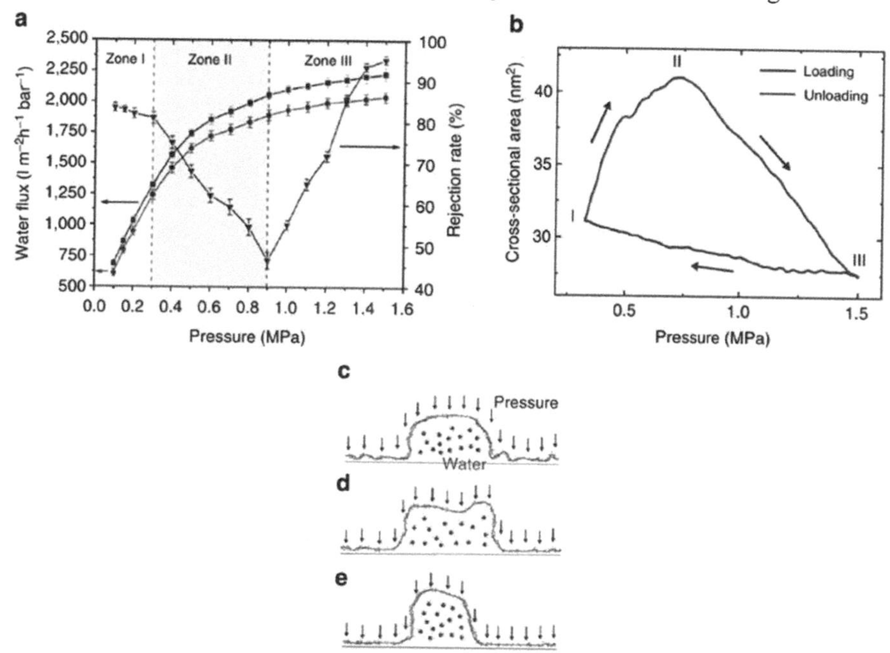

Fig. 6.104. Response of the NSC-GO membranes to applied pressure. (a) Pressure-dependent flux and rejection of EB molecules through the NSC-GO membranes. The black solid squares and red solid circles curve represent the flux variation during the first and third pressure-loading processes, respectively. The blue solid triangle curve denotes the rejection rate of EB during the first pressure-loading process. (b) Simulated cross-sectional area of the nanochannel as a function of the applied pressure. (c–e) Response of a half cylindrical GO nanochannel modeled with MD simulation. Reproduced with permission from [122], Copyright © 2013, Macmillan Publishers Limited.

6.3.3.3. Mechanical applications

It is believed that materials based on GO paper as a building block could mechanically respond to changes in humidity and/or temperature due to the changes in the amount of inter-lamellar water content between the GO nanosheets [128]. A general approach to fabricate mechanical actuators is to use two different building blocks to create asymmetric materials [238]. Bilayer paper composites were prepared with a layer of MWCNTs and a layer of GO. Both graphene and CNTs have the same basic structure, i.e., extended aromatic sp^2 carbon networks, so that this structural compatibility would allow for a stable interface between the two layers. The bilayer paper samples were obtained by using sequential filtration of COOH-functionalized MWCNTs and GO nanosheets both in aqueous suspensions. The GO layer was dark brown in color and electrically insulating, while the surface of the MWCNT layer was black and shiny and electrically conductive. SEM observations suggested that the surfaces of the GO layer and the MWCNT layer had no cross-contamination. The bilayer papers were highly flexible, with each layer to have a thickness of about 10 μm.

Because the GO layer swelled and shrank with increasing and decreasing relative humidity, respectively, the bilayer paper samples could provide actuation stimulated by the relative humidity at room temperature, as demonstrated in Fig. 6.105. At a low relative humidity (12%), the bilayer paper was rolled up with the MWCNT side facing outward, as seen in Fig. 6.105 (a). As the relative humidity was increased, the bilayer paper gradually unrolled, becoming almost flat at relative humidity values in the range of 55–60%, i.e., Fig. 6.105 (b, c). If the relative humidity was increased to above 60%, the bilayer paper was curled in the opposite direction, with the MWCNT side facing in and the GO side facing out, as revealed in Fig. 6.105

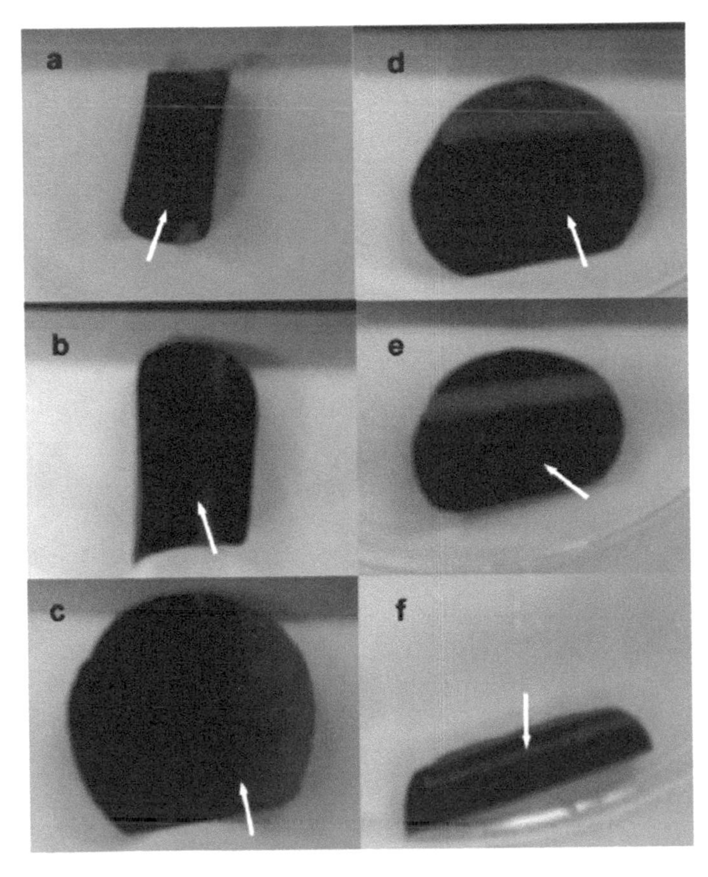

Fig. 6.105. Actuation of the bilayer graphene-based paper sample as a function of relative humidity (%): (a) 12, (b) 25, (c) 49, (d) 61, (e) 70 and (f) 90. White-arrowed side was the surface of the graphene oxide layer. Reproduced with permission from [128], Copyright © 2010, WILEY-VCH Verlag GmbH & Co. KGaA, Weinheim.

(d, e). The bilayer paper was completely curled at 85% (Fig. 6.105 (f)). Once the relative humidity was decreased, the bilayer paper was unrolled accordingly. Within about thirty seconds, the paper could reach a steady state. Similar actuation behavior was also triggered by temperature variation.

Moisture-responsive graphene papers have been prepared through focused sunlight-irradiation-induced photoreduction of GO [92]. The self-controlled photoreduction of GO was employed to fabricate GO/rGO bilayer structures by using sunlight as an irradiation source. Because both the sunlight penetration and the subsequent thermal relaxation were significantly suppressed after an expanded rGO layer was formed, only the surface layer of the thick GO paper was photoreduced. With this advantage, GO/rGO bilayer structures could be readily prepared in a self-controlled manner. The resultant GO/rGO bilayer papers exhibited strong moisture-responsive behaviors, because of the anisotropic water-molecule adsorption. With such unique GO/rGO smart papers, graphene-based moisture-responsive actuators have been developed.

Figure 6.106 (a) shows a schematic diagram illustrating the fabrication system that was used to prepare the GO/rGO bilayer papers. Pristine GO papers were prepared by using vacuum filtration of GO aqueous solution. Controllable photoreduction of the GO papers was realized by using focused sunlight irradiation. A piece of quartz coverslip was tightly pressed on the surface of the GO paper, so that it was isolated from ambient environment (O_2). The incident solar radiation was focused on the surface of the GO paper, with a 60 mm-diameter convex lens (focal length = 15 cm, focused spot diameter = 1 mm). A 2D movable platform with a movement rate of 0.2 $mm \cdot s^{-1}$ was used to guide the scanning path of the focal spot. The photoreduction of GO was observed as soon as the sunlight was focused on the GO paper, evidenced by the rapid color change from yellow–brown to black.

The photoreduced GO papers possessed a black color with a metallic luster, as illustrated in Fig. 6.106 (b). The back side of the rGO paper remained dark brown in color, due to the difference in reduction degree, as seen in Fig. 6.106 (c). As demonstrated in the insets of Fig. 6.106 (b, c), the irradiated side of the rGO paper showed a contact angle (CA) of about 90°, while the back side had a CA value of about 46°, which was almost the same as that of the pristine GO (about 45°). The variation in surface wettability was mainly attributed to the removal of hydrophilic oxygen-containing groups. Figure 6.106 (d) shows a cross-sectional SEM image of the rGO paper, confirming the difference between the two layers. The rGO layer possessed an expanded structure, with large gaps between the adjacent nanosheets. In contrast, the GO layer remained as a closely stacked layered structure. In other words, controllable photoreduction of the GO paper could be easily controlled by controlling the focused sunlight irradiation.

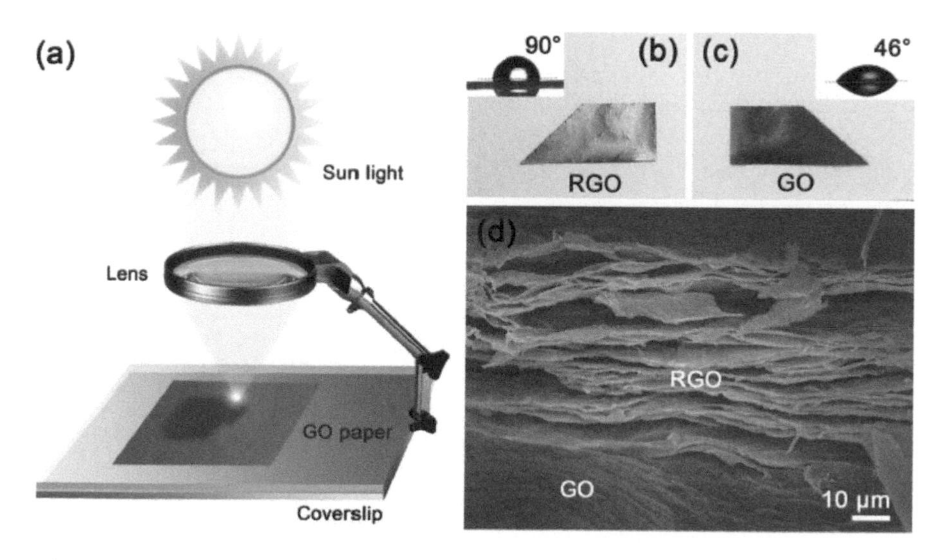

Fig. 6.106. (a) Schematic diagram showing the fabrication of the GO/rGO bilayer papers by using the focused sunlight reduction. (b, c) Photographs of the GO/rGO bilayer paper viewed from the front side (b) and the back side (c), respectively. The insets show water droplet contact angles of the two sides. (d) Cross-sectional SEM image of the GO/rGO bilayer paper. Reproduced with permission from [92], Copyright © 2015, WILEY-VCH Verlag GmbH & Co. KGaA, Weinheim.

The GO/rGO bilayer paper had a strong moisture-responsive bending behavior, because of the unique lateral anisotropy of the bilayer structure. Curvature degrees of a GO/rGO ribbon, with a width of 1.0 mm and a length of 12 mm, in different relative humidity (RH), are shown in Fig. 6.107. As the RH was increased from 24% to 86%, the curvature of the GO/rGO ribbon was increased from 0° to 168°. The corresponding photographs of the bended GO/rGO ribbon at different RHs are the insets in Fig. 6.107 (a). In comparison, pristine GO paper had no bending response to the variation in RH. Figure 6.107 (b) shows the response and recovery properties of the GO/rGO bilayer ribbon. As the RH was increased from 33% to 86% RH, the bending curvature was increased, taking only about 15 s to reach equilibrium. As the moisture was decreased back to the initial value of 33%, the bent GO/rGO ribbon became flat completely. The actuation behavior was repeatable, while the bending curvature was very stable.

As discussed previously, GO nanosheets have a strong interaction with water molecules through hydrogen bonding, while graphene nanosheets adsorb water molecules through the weaker van der Waals forces. Therefore, water molecules were mainly adsorbed by the GO layers, as demonstrated schematically in Fig. 6.107 (c). The selective adsorption/desorption of water molecules in the GO layers was responsible for the expansion/contraction behavior of the GO/rGO bilayers, corresponding to the reversible bending and straightening in moisture and dry conditions. The mechanism was confirmed by XRD results, as illustrated in Fig. 6.107 (d). At ambient humidity of ~30% RH, GO paper exhibited a diffraction peak

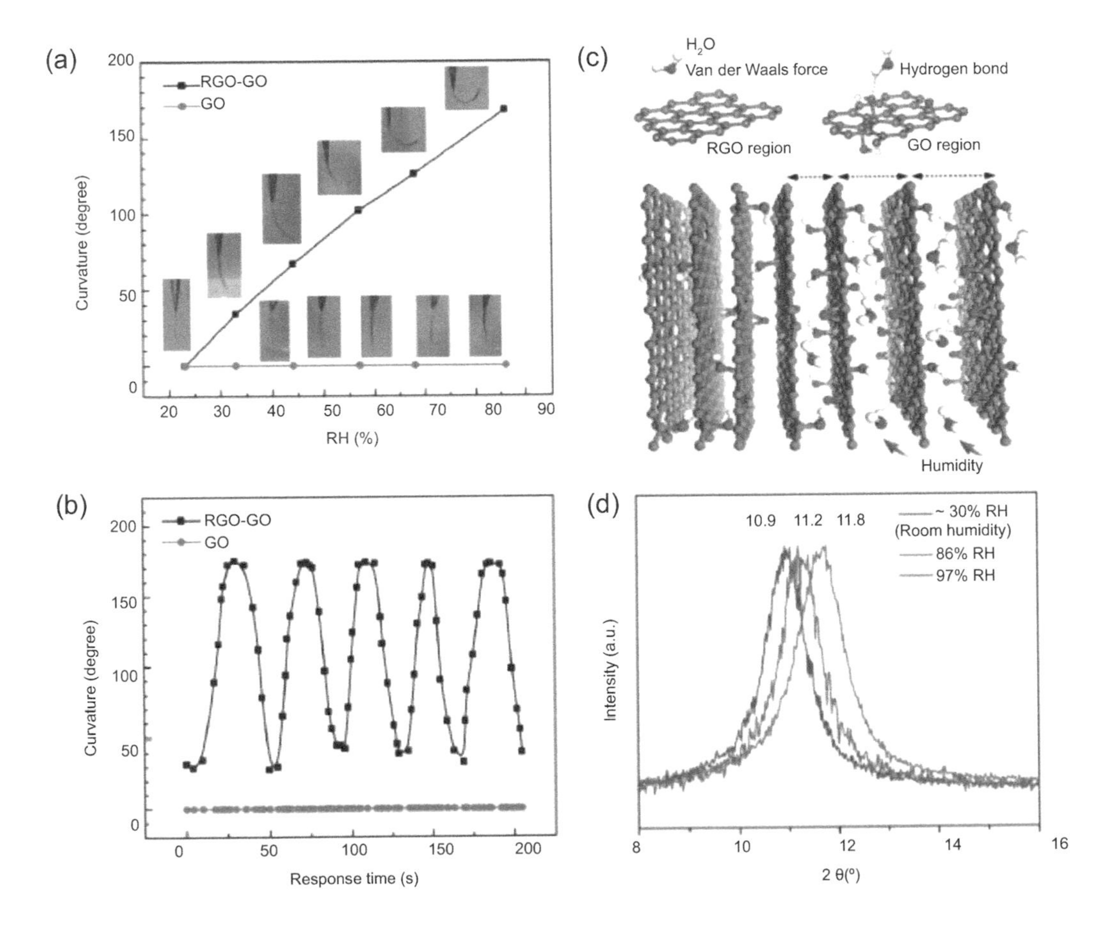

Fig. 6.107. Moisture-responsive properties and mechanism of the GO/rGO bilayer papers. (a) Curvature of the GO/rGO bilayer paper and pristine GO paper as a function of RH. (b) Responsive and recovery properties of the GO/rGO bilayer paper. The RH was switched between 33% and 86% five times. (c) Schematic diagram of the interaction between water molecules and graphene/GO nanosheets. (d) XRD patterns of the GO papers at different RHs. Reproduced with permission from [92], Copyright © 2015, WILEY-VCH Verlag GmbH & Co. KGaA, Weinheim.

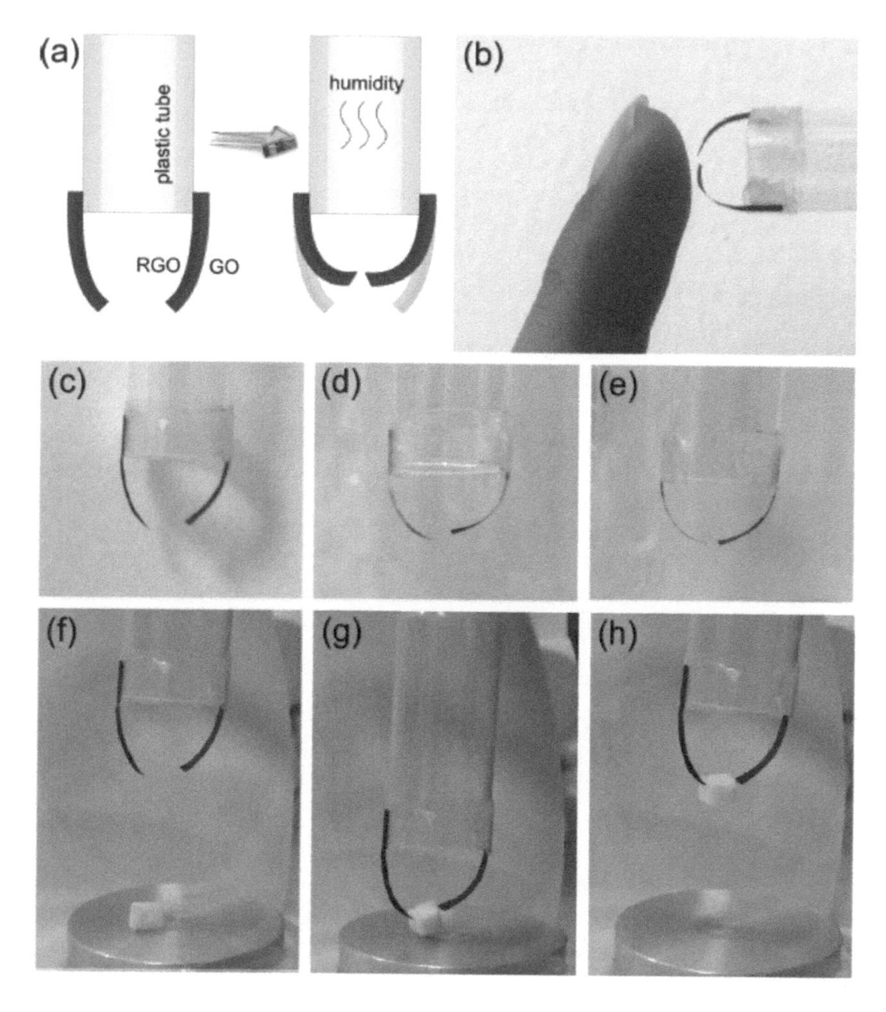

Fig. 6.108. Moisture-responsive performance of a smart claw based on the GO/rGO bilayer papers. (a) Schematic diagram showing the structure of the smart claw. (b) Photograph of the sensitive smart claw responding to a sweaty human finger. (c–e) Photographs of the moisture responsive claw manipulated by controlling the moisture. (f–g) Photographs of the smart claw grabbing a polymer foam block. Reproduced with permission from [92], Copyright © 2015, WILEY-VCH Verlag GmbH & Co. KGaA, Weinheim.

at about 11.8°, corresponding to an interlayer spacing of about 0.75 nm. As the humidity was increased to 86% and 97% RH, the diffraction peaks shifted to 11.2° and 10.9°, respectively, suggesting that the interlayer spacing was increased.

By using the moisture-responsive property of the GO/rGO bilayer papers, graphene-based smart actuators were designed and evaluated. Figure 6.108 (a) shows a schematic diagram for the fabrication of the moisture-responsive smart claw, two GO/rGO ribbons were stuck on the wall of a plastic tube, which was connected to a moisture supply system. The GO/rGO ribbons acted as smart fingers, as they responded to the variation in humidity. The GO/rGO ribbons demonstrated high sensitivity. For example, they even bent when approaching a sweaty human finger, as illustrated in Fig. 6.108 (b). By connecting the plastic tube to a humidity supply system, actuation behavior of the GO/rGO smart fingers could be evaluated systematically, as shown in Fig. 6.108 (c–e). More importantly, the smart fingers could pick up a 3 mm cubic polymer foam, as seen in Fig. 6.108 (f–h). In fact, there is no limit to use the GO/rGO bilayer papers to design actuators.

In another study, PANI was used as the interlayer spacer to fabricate a rGO/PANI nanocomposite electrode, due to its high specific capacitance, low cost and facial fabrication procedure [114]. In this case, the rGO nanosheets were decorated with the PANI nanoparticles on the surfaces through an *in situ* chemical polymerization from aniline monomer. Vacuum-assisted filtration method was used to fabricate the free-standing rGO/PANI hybrid electrode. The presence of the PANI nanoparticles prevented the restacking of the graphene nanosheets, which thus facilitated ion migration and accumulation in the electrode layers. Acid cross-linked PVA-based gel (H_2SO_4–PVA gel) was used as the electrolyte layer, due to its high ionic conductivity. Actuation of the composite actuator could be stimulated by using a very low voltage (0.1 V). With a square wave voltage of 0.5 V at 0.01 Hz, the actuator achieved a bending displacement of 4.3 mm, corresponding to a strain of 0.327%. The large actuation of the rGO/PANI actuator was attributed to the synergistic effect of the electric double-layer capacitor (EDLC) behavior of the rGO component and the pseudocapacitance of the PANI component. The study could be used as a guide for the design and preparation of ionic actuators based on nanostructured carbon materials, with high actuation strokes, large generated stress, long-term durability and low applied voltages.

Based on the electrochemical and actuation performance of the actuators, a mechanism was proposed, as shown in Fig. 6.109, which was attributed to the EDLC behavior of rGO and the pseudocapacitance of PANI. As a voltage was applied to the actuator, the H^+ and SO_4^{2+} ions were dissociated in the gel electrolyte layer, and were driven to migrate towards and accumulate at the cathode and anode, respectively, so as to maintain the electric field balance. For the rGO electrode, both the anode and the cathode would be expanded, due to the ion migration and accumulation, with the expansion of the anode being more significant than that of the cathode, because the SO_4^{2+} ions were larger than the H^+ ions. As a consequence, the actuator would be bending towards the cathode side. For the PANI actuator, the polymer also experienced dimensional changes, because of the ion insertion and deinsertion in the polymer backbone during the redox reaction process to balance the charge of the system.

The insertion of SO_4^{2+} ions into the PANI backbone caused swelling of the polymer chains, so that the anode was expanded, thus leading to a convex at a positive bias. After the rGO nanosheets were decorated with PANI, the actuator would have more significant bending deformation, because of the synergistic effect

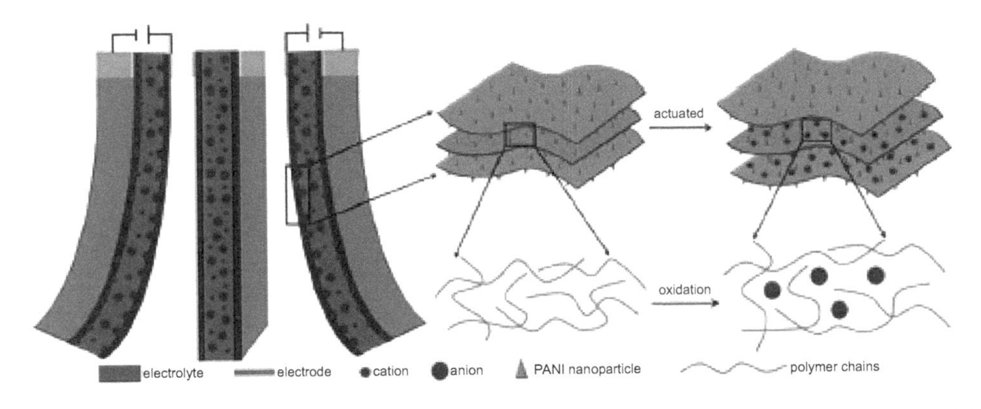

Fig. 6.109. Schematic diagram of the actuation deformation. Reproduced with permission from [114], Copyright © 2015, The Royal Society of Chemistry.

of the ion aggregation and the redox reaction. Because both the specific surface area and ion transportation channels in the loosely packed rGO/PANI nanocomposite film were increased, the ion migration resistance was decreased and ion transportation in the electrode was promoted. Therefore, the number of ions that could be inserted and accumulated in the electrodes was increased, so that the induced bending deformation was increased. Moreover, the redox reaction of PANI provided excessive charge transfer, so as to facilitate more ions to intercalate into the electrode and increase the ion storage capacity, which in turn enhanced the actuation performance of the composite electrodes.

6.3.3.4. Other applications

Because of the high electrical conductivity of graphene, graphene papers could be used for EMI applications. For instance, a graphene paper consisting of multilayer graphene nanosheets has been evaluated as a full-carbon flexible shielding material at radio-frequency [101]. It exhibited a shielding efficiency of as high as 50 dB at a thickness of as thin as 20 μm. The graphene paper was prepared by using vacuum filtration of graphene suspension. The paper had a DC electrical conductivity of > 1400 $S\cdot cm^{-1}$, without the requirement of high annealing temperature for post-synthesis graphitization. The shielding effectiveness performances of the graphene paper were assessed experimentally through measurements in the frequency range of 0.01–18 GHz.

Graphene films could have a highly hydrophobic and adhesive performance, with the wettability that was controllable by alternating UV irradiation and air storage [105]. Such graphene films could be used to design chemical engineering materials, microfluidic devices, devices to transport small amount of corrosive liquids. The graphene films showed a hierarchical micro-/nanotexture morphology, as shown in Fig. 6.110 (a). The surface microstructure consisted of 3 nm-thick graphene nanosheets with a size of 1 μm, while the nanostructure was composed of 1 nm-thick graphene nanoflakes. The multilayer graphene nanosheets were stacked randomly, while the few-layer graphene nanosheets were folded and agglomerated. This micro-/nano-textured surface was fundamentally important to the properties of the films.

The as-fabricated graphene films had a hydrophobic surface, with a CA of 140°, which was attributed to the micro-/nano-textured morphology of the films. To facilitate surface hydrophobicity, the microstructure and nanostructure should have sufficient roughness. The irregular folding and aggregation of the few-layer graphene nanosheets ensured the surface roughness, which was attributed to the presence of residual carboxyl and hydroxyl groups on surface of the graphene nanosheets.

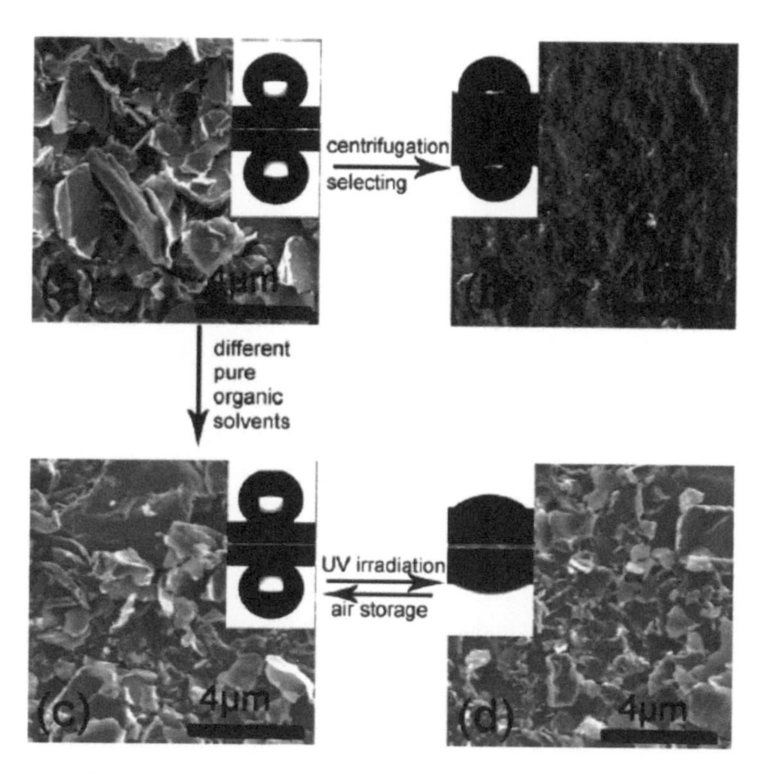

Fig. 6.110. SEM images of the films: (a) graphene–isopropanol film, (b) graphene–isopropanol film obtained through centrifugation at 16 krpm, (c) the graphene–ethanol film and (d) the graphene–ethanol film after UV irradiation, together with the shapes of water droplets on the corresponding surface under a tilt angle of 0° and 180°. Reproduced with permission from [105], Copyright © 2011, The Royal Society of Chemistry.

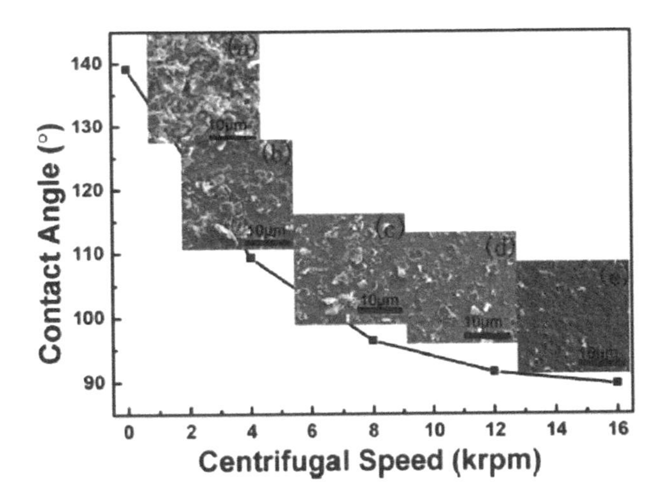

Fig. 6.111. CAs of the graphene–isopropanol films obtained from the suspensions at different centrifugal speeds. The insets shown SEM images of the films corresponding to different centrifugal speeds: (a) 0 krpm, (b) 4 krpm, (c) 8 krpm, (d) 12 krpm and (e) 16 krpm. Reproduced with permission from [105], Copyright © 2011, The Royal Society of Chemistry.

The relationship between the morphology and the hydrophobicity of graphene film was studied by using homogeneous graphene–isopropanol suspensions that were centrifuged at different centrifugal speeds. Figure 6.111 shows the correlation of the static CAs with the centrifugal speeds, together with their corresponding SEM images [105]. The CA was gradually decreased with the increasing centrifugal speed, which was closely related to the variation in the microtextured surface morphology. As the centrifugal speed was increased, the surface was gradually flattened, thus leading to the removal of the multilayer graphene nanosheets. The higher the centrifugal speed, the thinner the graphene nanosheets and thus the flatter the film surface would be. For example, at the centrifugal speed of 16 krpm, only nanostructures and few multilayer graphene nanoplates remained in the graphene film. Therefore, the CA of the film was 90°, which was close to that of the epitaxial graphene. In this case, the nanoscale folding and agglomeration have very weak effects on the surface hydrophobicity.

6.4. Summary

It is quite obvious that the properties of graphene thin films can be largely influenced by the properties and qualities of the substrates, besides the effects of the properties of the graphene building blocks, as well as the methods used to deposit the films. The use of different substrates for different applications is not avoidable. Therefore, it is necessary to have different standards for different films.

As for free-standing 2D graphene materials, the majority of them have been fabricated by using vacuum membrane filtration. However, while this method is most standardized, it has several serious disadvantages, such as a low deposition rate and limited sizes, which could restrict their practical applications, especially when large area samples are required. In this regard, solution casting or evaporation methods could have various advantages, while useful references are relatively scarce. It is therefore suggested that the research community should pay more attention to this issue.

References

[1] Wu DQ, Zhang FB, Liang HW, Feng XL. Nanocomposites and macroscopic materials: assembly of chemically modified graphene sheets. Chemical Society Reviews. 2012; 41: 6160–77.

[2] Eda G, Chhowalla M. Chemically derived graphene oxide: Towards large-area thin-film electronics and optoelectronics. Advanced Materials. 2010; 22: 2392–415.
[3] Du JH, Pei SF, Ma LP, Cheng HM. 25th anniversary article: Carbon nanotube- and graphene-based transparent conductive films for optoelectronic devices. Advanced Materials. 2014; 26: 1958–91.
[4] Kim KS, Zhao Y, Jang H, Lee SY, Kim JM, Kim KS et al. Large-scale pattern growth of graphene films for stretchable transparent electrodes. Nature. 2009; 457: 706–10.
[5] Schniepp HC, Li JL, McAllister MJ, Sai H, Herrera-Alonso M, Adamson DH et al. Functionalized single graphene sheets derived from splitting graphite oxide. Journal of Physical Chemistry B. 2006; 110: 8535–9.
[6] Gomez-Navarro C, Weitz RT, Bittner AM, Scolari M, Mews A, Burghard M et al. Electronic transport properties of individual chemically reduced graphene oxide sheets. Nano Letters. 2007; 7: 3499–503.
[7] Wang X, Zhi LJ, Muellen K. Transparent, conductive graphene electrodes for dye-sensitized solar cells. Nano Letters. 2008; 8: 323–7.
[8] Jumeri FA, Lim HN, Zainal Z, Huang NM, Pandikumar A. Titanium dioxide-reduced graphene oxide thin film for photoelectrochemical water splitting. Ceramics International. 2014; 40: 15159–65.
[9] Gilje S, Han S, Wang MS, Wang KL, Kaner RB. A chemical route to graphene for device applications. Nano Letters. 2007; 7: 3394–8.
[10] Li XQ, Zhang D, Yang C, Shang Y. Direct and efficient preparation of graphene transparent conductive films on flexible poly carbonate substrate by spray-coating. Journal of Nanoscience and Nanotechnology. 2015; 15: 9500–8.
[11] Duy-Thach P, Chung GS. Effects of rapid thermal annealing on humidity sensor based on graphene oxide thin films. Sensors and Actuators B-Chemical. 2015; 220: 1050–5.
[12] Modesto-Lopez LB, Miettinen M, Riikonen J, Torvela T, Pfuller C, Lehto VP et al. Films of graphene nanomaterials formed by ultrasonic spraying of their stable suspensions. Aerosol Science and Technology. 2015; 49: 45–56.
[13] Pirzado AA, Jouane Y, Le Normand F, Akilimali R, Papaefthimiou V, Ghimbeu CM et al. Electrical transport in "few-layer graphene" film prepared by the hot-spray technique: The effect of thermal treatment. Journal of Physical Chemistry C. 2014; 118: 873–80.
[14] Shi HF, Wang C, Sun ZP, Zhou YL, Jin KJ, Yang GZ. Transparent conductive reduced graphene oxide thin films produced by spray coating. Science China-Physics Mechanics & Astronomy. 2015; 58: 014202.
[15] Cruz R, Pacheco Tanaka DA, Mendes A. Reduced graphene oxide films as transparent counter-electrodes for dye-sensitized solar cells. Solar Energy. 2012; 86: 716–24.
[16] Jang SY, Kim YG, Kim DY, Kim HG, Jo SM. Electrodynamically sprayed thin films of aqueous dispersible graphene nanosheets: Highly efficient cathodes for dye-sensitized solar cells. ACS Applied Materials & Interfaces. 2012; 4: 3500–7.
[17] Robinson JT, Zalalutdinov M, Baldwin JW, Snow ES, Wei ZQ, Sheehan P et al. Wafer-scale reduced graphene oxide films for nanomechanical devices. Nano Letters. 2008; 8: 3441–5.
[18] Robinson JT, Perkins FK, Snow ES, Wei ZQ, Sheehan PE. Reduced graphene oxide molecular sensors. Nano Letters. 2008; 8: 3137–40.
[19] Becerril HA, Mao J, Liu ZF, Stoltenberg RM, Bao ZN, Chen YS. Evaluation of solution-processed reduced graphene oxide films as transparent conductors. ACS Nano. 2008; 2: 463–70.
[20] Pang SP, Tsao HN, Feng XL, Muellen K. Patterned graphene electrodes from solution-processed graphite oxide films for organic field-effect transistors. Advanced Materials. 2009; 21: 3488–91.
[21] Yamaguchi H, Eda G, Mattevi C, Kim HK, Chhowalla M. Highly uniform 300 mm wafer-scale deposition of single and multilayered chemically derived graphene thin films. ACS Nano. 2010; 4: 524–8.
[22] Lee KH, Hong JH, Kwak SJ, Park M, Son JG. Spin self-assembly of highly ordered multilayers of graphene-oxide sheets for improving oxygen barrier performance of polyolefin films. Carbon. 2015; 83: 40–7.
[23] Umar MIA, Yap CC, Awang R, Salleh MM, Yahaya M. The effect of spin-coated polyethylene glycol on the electrical and optical properties of graphene film. Applied Surface Science. 2014; 313: 883–7.
[24] Chen Y, Zhang X, Yu P, Ma YW. Stable dispersions of graphene and highly conducting graphene films: a new approach to creating colloids of graphene monolayers. Chemical Communications. 2009: 4527–9.
[25] Lee V, Whittaker, Jaye C, Baroudi KM, Fischer DA, Banerjee S. Large-area chemically modified graphene films: Electrophoretic deposition and characterization by soft X-ray absorption spectroscopy. Chemistry of Materials. 2009; 21: 3905–16.
[26] Dou YY, Luo M, Liang S, Zhang XL, Ding XY, Liang B. Flexible free-standing graphene-like film electrode for supercapacitors by electrophoretic deposition and electrochemical reduction. Transactions of Nonferrous Metals Society of China. 2014; 24: 1425–33.
[27] An SJ, Zhu YW, Lee SH, Stoller MD, Emilsson T, Park SJ et al. Thin film fabrication and simultaneous anodic reduction of deposited graphene oxide platelets by electrophoretic deposition. Journal of Physical Chemistry Letters. 2010; 1: 1259–63.
[28] Wu XY, Wang B, Li SM, Liu JH, Yu M. Electrophoretic deposition of hierarchical Co_3O_4@graphene hybrid films as binder-free anodes for high-performance lithium-ion batteries. RSC Advances. 2015; 5: 33438–44.
[29] Niu ZQ, Du JJ, Cao XB, Sun YH, Zhou WY, Hng HH et al. Electrophoretic build-up of alternately multilayered films and micropatterns based on graphene sheets and nanoparticles and their applications in flexible supercapacitors. Small. 2012; 8: 3201–8.

[30] Chavez-Valdez A, Shaffer MSP, Boccaccini AR. Applications of graphene electrophoretic deposition. A review. Journal of Physical Chemistry B. 2013; 117: 1502–15.
[31] Cote LJ, Kim F, Huang JX. Langmuir-Blodgett assembly of graphite oxide single layers. Journal of the American Chemical Society. 2009; 131: 1043–9.
[32] Li XL, Zhang GY, Bai XD, Sun XM, Wang XR, Wang E et al. Highly conducting graphene sheets and Langmuir-Blodgett films. Nature Nanotechnology. 2008; 3: 538–42.
[33] Cote LJ, Kim JY, Tung VC, Luo JY, Kim F, Huang JX. Graphene oxide as surfactant sheets. Pure and Applied Chemistry. 2011; 83: 95–110.
[34] Jia BP, Zou LD. Langmuir-Blodgett assembly of sulphonated graphene nanosheets into single- and multi-layered thin films. Chemical Physics Letters. 2013; 568: 101–5.
[35] Kim HK, Mattevi C, Kim HJ, Mittal A, Mkhoyan KA, Riman RE et al. Optoelectronic properties of graphene thin films deposited by a Langmuir-Blodgett assembly. Nanoscale. 2013; 5: 12365–74.
[36] Shao JJ, Lv W, Yang QH. Self-assembly of graphene oxide at interfaces. Advanced Materials. 2014; 26: 5586–612.
[37] Yang TM, Yang JH, Shi LF, Maeder E, Zheng QB. Highly flexible transparent conductive graphene/single-walled carbon nanotube nanocomposite films produced by Langmuir-Blodgett assembly. RSC Advances. 2015; 5: 23650–7.
[38] Zheng QB, Zhang B, Lin XY, Shen X, Yousefi N, Huang ZD et al. Highly transparent and conducting ultralarge graphene oxide/single-walled carbon nanotube hybrid films produced by Langmuir-Blodgett assembly. Journal of Materials Chemistry. 2012; 22: 25072–82.
[39] Gengler RYN, Veligura A, Enotiadis A, Diamanti EK, Gournis D, Jozsa C et al. Large-yield preparation of high-electronic-quatity graphene by a Langmuir-Schaefer approach. Small. 2010; 6: 35–9.
[40] Fahimi A, Jurewicz I, Smith RJ, Sharrock CS, Bradley DA, Henley SJ et al. Density controlled conductivity of pristine graphene films. Carbon. 2013; 64: 435–43.
[41] Kim HG, Jang YR, Yoo J, Seo YS, Kim KY, Lee JS et al. Morphology control of surfactant-assisted graphene oxide films at the liquid-gas interface. Langmuir. 2014; 30: 2170–7.
[42] Hidalgo RS, Lopez-Diaz D, Mercedes Velazquez M. Graphene oxide thin films: Influence of chemical structure and deposition methodology. Langmuir. 2015; 31: 2697–705.
[43] Lee TM, Min SH, Gu MS, Jung YK, Lee W, Lee JK et al. Layer-by-layer assembly for graphene-based multilayer nanocomposites: Synthesis and applications. Chemistry of Materials. 2015; 27: 3785–96.
[44] Park JS, Cho SM, Kim WJ, Park JH, Yoo PJ. Fabrication of graphene thin films based on layer-by-layer self-assembly of functionalized graphene nanosheets. ACS Applied Materials & Interfaces. 2011; 3: 360–8.
[45] Manga KK, Zhou Y, Yan YL, Loh KP. Multilayer hybrid films consisting of alternating graphene and titania nanosheets with ultrafast electron transfer and photoconversion properties. Advanced Functional Materials. 2009; 19: 3638–43.
[46] Ji QM, Honma I, Paek SM, Akada M, Hill JP, Vinu A et al. Layer-by-layer films of graphene and ionic liquids for highly selective gas sensing. Angewandte Chemie-International Edition. 2010; 49: 9737–9.
[47] Kulkarni DD, Choi I, Singamaneni S, Tsukruk VV. Graphene oxide-polyelectrolyte nanomembranes. Acs Nano. 2010; 4: 4667–76.
[48] Sheng KX, Bai H, Sun YQ, Li C, Shi GQ. Layer-by-layer assembly of graphene/polyaniline multilayer films and their application for electrochromic devices. Polymer. 2011; 52: 5567–72.
[49] Zhao X, Zhang QH, Hao YP, Li YZ, Fang Y, Chen DJ. Alternate multilayer films of poly(vinyl alcohol) and exfoliated graphene oxide fabricated via a facial layer-by-layer assembly. Macromolecules. 2010; 43: 9411–6.
[50] Li ZP, Wang JQ, Liu XH, Liu S, Ou JF, Yang SR. Electrostatic layer-by-layer self-assembly multilayer films based on graphene and manganese dioxide sheets as novel electrode materials for supercapacitors. Journal of Materials Chemistry. 2011; 21: 3397–403.
[51] Cong JJ, Chen YZ, Luo J, Liu XY. Fabrication of graphene/polyaniline composite multilayer films by electrostatic layer-by-layer assembly. Journal of Solid State Chemistry. 2014; 218: 171–7.
[52] Eda G, Fanchini G, Chhowalla M. Large-area ultrathin films of reduced graphene oxide as a transparent and flexible electronic material. Nature Nanotechnology. 2008; 3: 270–4.
[53] Mattevi C, Eda G, Agnoli S, Miller S, Mkhoyan KA, Celik O et al. Evolution of electrical, chemical, and structural properties of transparent and conducting chemically derived graphene thin films. Advanced Functional Materials. 2009; 19: 2577–83.
[54] Eda G, Lin YY, Miller S, Chen CW, Su WF, Chhowalla M. Transparent and conducting electrodes for organic electronics from reduced graphene oxide. Applied Physics Letters. 2008; 92: 233305.
[55] Xu YX, Bai H, Lu GW, Li C, Shi GQ. Flexible graphene films via the filtration of water-soluble noncovalent functionalized graphene sheets. Journal of the American Chemical Society. 2008; 130: 5856–7.
[56] Yi M, Liang SS, Liu L, Shen ZG, Zheng YT, Zhang XJ et al. Investigating the nature of graphene-based films prepared by vacuum filtration of graphene dispersions. Journal of Nanoscience and Nanotechnology. 2014; 14: 4969–75.
[57] Kim JY, Cote LJ, Kim F, Yuan W, Shull KR, Huang JX. Graphene oxide sheets at interfaces. Journal of the American Chemical Society. 2010; 132: 8180–6.
[58] Zheng QB, Ip WH, Lin XY, Yousefi N, Yeung KK, Li ZG et al. Transparent conductive films consisting of ultra-large graphene sheets produced by Langmuir-Blodgett assembly. ACS Nano. 2011; 5: 6039–51.
[59] Wang XL, Bai H, Shi GQ. Size fractionation of graphene oxide sheets by pH-assisted selective sedimentation. Journal of the American Chemical Society. 2011; 133: 6338–42.

[60] Danauskas SM, Ratajczak MK, Ishitsuka Y, Gebhardt J, Schultz D, Meron M et al. Monitoring x-ray beam damage on lipid films by an integrated Brewster angle microscope/x-ray diffractometer. Review of Scientific Instruments. 2007; 78: 103705.
[61] Huang JH, Fang JH, Liu CC, Chu CW. Effective work function modulation of graphene/carbon nanotube composite films as transparent cathodes for organic optoelectronics. ACS Nano. 2011; 5: 6262–71.
[62] Kovtyukhova NI, Ollivier PJ, Martin BR, Mallouk TE, Chizhik SA, Buzaneva EV et al. Layer-by-layer assembly of ultrathin composite films from micron-sized graphite oxide sheets and polycations. Chemistry of Materials. 1999; 11: 771–8.
[63] Cassagneau T, Fendler JH. Preparation and layer-by-layer self-assembly of silver nanoparticles capped by graphite oxide nanosheets. Journal of Physical Chemistry B. 1999; 103: 1789–93.
[64] Cassagneau T, Guerin F, Fendler JH. Preparation and characterization of ultrathin films layer-by-layer self-assembled from graphite oxide nanoplatelets and polymers. Langmuir. 2000; 16: 7318–24.
[65] Vickery JL, Patil AJ, Mann S. Fabrication of graphene-polymer nanocomposites with higher-order three-dimensional architectures. Advanced Materials. 2009; 21: 2180–4.
[66] Niyogi S, Bekyarova E, Itkis ME, McWilliams JL, Hamon MA, Haddon RC. Solution properties of graphite and graphene. Journal of the American Chemical Society. 2006; 128: 7720–1.
[67] Shen JF, Hu YZ, Li C, Qin C, Shi M, Ye MX. Layer-by-layer self-assembly of graphene nanoplatelets. Langmuir. 2009; 25: 6122–8.
[68] Fang YX, Guo SJ, Zhu CZ, Zhai YM, Wang EK. Self-assembly of cationic polyelectrolyte-functionalized graphene nanosheets and gold nanoparticles: A two-dimensional heterostructure for hydrogen peroxide sensing. Langmuir. 2010; 26: 11277–82.
[69] Ram MK, Salerno M, Adami M, Faraci P, Nicolini C. Physical properties of polyaniline films: Assembled by the layer-by-layer technique. Langmuir. 1999; 15: 1252–9.
[70] Chen DJ, Wang XY, Liu TX, Wang XD, Li J. Electrically conductive poly(vinyl alcohol) hybrid films containing graphene and layered double hydroxide fabricated via layer-by-layer self-assembly. ACS Applied Materials & Interfaces. 2010; 2: 2005–11.
[71] Li HL, Pang SP, Wu SD, Feng XL, Muellen K, Bubeck C. Layer-by-layer assembly and UV photoreduction of graphene-polyoxometalate composite films for electronics. Journal of the American Chemical Society. 2011; 133: 9423–9.
[72] Li HL, Pang SP, Feng XL, Muellen K, Bubeck C. Polyoxometalate assisted photoreduction of graphene oxide and its nanocomposite formation. Chemical Communications. 2010; 46: 6243–5.
[73] Stankovich S, Dikin DA, Piner RD, Kohlhaas KA, Kleinhammes A, Jia YY et al. Synthesis of graphene-based nanosheets via chemical reduction of exfoliated graphite oxide. Carbon. 2007; 45: 1558–65.
[74] Chen FM, Liu SB, Shen JM, Wei L, Liu AD, Chan-Park MB et al. Ethanol-assisted graphene oxide-based thin film formation at pentane-water interface. Langmuir. 2011; 27: 9174–81.
[75] Paredes JI, Villar-Rodil S, Martinez-Alonso A, Tascon JMD. Graphene oxide dispersions in organic solvents. Langmuir. 2008; 24: 10560–4.
[76] Baughman RH, Cui CX, Zakhidov AA, Iqbal Z, Barisci JN, Spinks GM et al. Carbon nanotube actuators. Science. 1999; 284: 1340–4.
[77] Dikin DA, Stankovich S, Zimney EJ, Piner RD, Dommett GHB, Evmenenko G et al. Preparation and characterization of graphene oxide paper. Nature. 2007; 448: 457–60.
[78] Hennrich F, Lebedkin S, Malik S, Tracy J, Barczewski M, Rosner H et al. Preparation, characterization and applications of free-standing single walled carbon nanotube thin films. Physical Chemistry Chemical Physics. 2002; 4: 2273–7.
[79] Chen H, Muller MB, Gilmore KJ, Wallace GG, Li D. Mechanically strong, electrically conductive, and biocompatible graphene paper. Advanced Materials. 2008; 20: 3557–61.
[80] Hu WB, Peng C, Luo WJ, Lv M, Li XM, Li D et al. Graphene-based antibacterial paper. ACS Nano. 2010; 4: 4317–23.
[81] He GY, Chen HQ, Zhu JW, Bei FL, Sun XQ, Wang X. Synthesis and characterization of graphene paper with controllable properties via chemical reduction. Journal of Materials Chemistry. 2011; 21: 14631–8.
[82] Lian M, Fan JC, Shi ZX, Zhang S, Li H, Yin J. Gelatin-assisted fabrication of graphene-based nacre with high strength, toughness, and electrical conductivity. Carbon. 2015; 89: 279–89.
[83] Gwon H, Kim HS, Lee KU, Seo DH, Park YC, Lee YS et al. Flexible energy storage devices based on graphene paper. Energy & Environmental Science. 2011; 4: 1277–83.
[84] Yang XW, Qiu L, Cheng C, Wu YZ, Ma ZF, Li D. Ordered gelation of chemically converted graphene for next-generation electroconductive hydrogel films. Angewandte Chemie-International Edition. 2011; 50: 7325–8.
[85] Cheng QF, Wu MX, Li MZ, Jiang L, Tang ZY. Ultratough artificial nacre based on conjugated cross-linked graphene oxide. Angewandte Chemie-International Edition. 2013; 52: 3750–5.
[86] Park SJ, Lee KS, Bozoklu G, Cai WW, Nguyen ST, Ruoff RS. Graphene oxide papers modified by divalent ions—Enhancing mechanical properties via chemical cross-linking. ACS Nano. 2008; 2: 572–8.
[87] Putz KW, Compton OC, Segar C, An Z, Nguyen ST, Brinson LC. Evolution of order during vacuum-assisted self-assembly of graphene oxide paper and associated polymer nanocomposites. ACS Nano. 2011; 5: 6601–9.
[88] Sumboja A, Foo CY, Wang X, Lee PS. Large areal mass, flexible and free-standing reduced graphene oxide/manganese dioxide paper for asymmetric supercapacitor device. Advanced Materials. 2013; 25: 2809–15.

[89] Tian Y, Cao YW, Wang Y, Yang WL, Feng JC. Realizing ultrahigh modulus and high strength of macroscopic graphene oxide papers through crosslinking of mussel-inspired polymers. Advanced Materials. 2013; 25: 2980–3.
[90] Guo W, Cheng C, Wu YZ, Jiang YN, Gao J, Li D et al. Bio-inspired two-dimensional nanofluidic generators based on a layered graphene hydrogel membrane. Advanced Materials. 2013; 25: 6064–8.
[91] Liu HY, Wang HT, Zhang XW. Facile fabrication of freestanding ultrathin reduced graphene oxide membranes for water purification. Advanced Materials. 2015; 27: 249–54.
[92] Han DD, Zhang YL, Jiang HB, Xia H, Feng J, Chen QD et al. Moisture-responsive graphene paper prepared by self-controlled photoreduction. Advanced Materials. 2015; 27: 332–8.
[93] Ha SH, Jeong YS, Lee YJ. Free standing reduced graphene oxide film cathodes for lithium ion batteries. ACS Applied Materials & Interfaces. 2013; 5: 12295–303.
[94] Hou ZL, Song WL, Wang P, Meziani MJ, Kong CY, Anderson A et al. Flexible graphene-graphene composites of superior thermal and electrical transport properties. ACS Applied Materials & Interfaces. 2014; 6: 15026–32.
[95] Li YF, Liu YZ, Shen WZ, Yang YG, Wang MZ, Wen YF. Free-standing optoelectronic graphene-CdS-graphene oxide composite paper produced by vacuum-assisted self-assembly. Applied Physics A-Materials Science & Processing. 2012; 106: 779–84.
[96] Xiang JL, Drzal LT. Thermal conductivity of exfoliated graphite nanoplatelet paper. Carbon. 2011; 49: 773–8.
[97] Valles C, Nunez JD, Benito AM, Maser WK. Flexible conductive graphene paper obtained by direct and gentle annealing of graphene oxide paper. Carbon. 2012; 50: 835–44.
[98] Park S, Suk JW, An JH, Oh J, Lee S, Lee W et al. The effect of concentration of graphene nanoplatelets on mechanical and electrical properties of reduced graphene oxide papers. Carbon. 2012; 50: 4573–8.
[99] Lee W, Lee JU, Jung BM, Byun JH, Yi JW, Lee SB et al. Simultaneous enhancement of mechanical, electrical and thermal properties of graphene oxide paper by embedding dopamine. Carbon. 2013; 65: 296–304.
[100] Kim DY, Kim MK, Kim DW, Suk JD, Park OO, Kang YK. Flexible binder-free graphene paper cathodes for high-performance Li-O_2 batteries. Carbon. 2015; 93: 625–35.
[101] Paliotta L, De Bellis G, Tamburrano A, Marra F, Rinaldi A, Balijepalli SK et al. Highly conductive multilayer-graphene paper as a flexible lightweight electromagnetic shield. Carbon. 2015; 89: 260–71.
[102] Kumar P, Shahzad F, Yu SG, Hong SM, Kim YH, Koo CM. Large-area reduced graphene oxide thin film with excellent thermal conductivity and electromagnetic interference shielding effectiveness. Carbon. 2015; 94: 494–500.
[103] Stankovich S, Dikin DA, Compton OC, Dommett GHB, Ruoff RS, Nguyen ST. Systematic post-assembly modification of graphene oxide paper with primary alkylamines. Chemistry of Materials. 2010; 22: 4153–7.
[104] He P, Sun J, Tian SY, Yang SW, Ding SJ, Ding GQ et al. Processable aqueous dispersions of graphene stabilized by graphene quantum dots. Chemistry of Materials. 2015; 27: 218–26.
[105] Zhang XQ, Wan SH, Pu JB, Wang LP, Liu XQ. Highly hydrophobic and adhesive performance of graphene films. Journal of Materials Chemistry. 2011; 21: 12251–8.
[106] Zhang C, Tjiu WW, Fan W, Yang Z, Huang S, Liu TX. Aqueous stabilization of graphene sheets using exfoliated montmorillonite nanoplatelets for multifunctional free-standing hybrid films via vacuum-assisted self-assembly. Journal of Materials Chemistry. 2011; 21: 18011–7.
[107] Zhang SD, Tao QH, Wang ZY, Zhang ZP. Controlled heat release of new thermal storage materials: the case of polyethylene glycol intercalated into graphene oxide paper. Journal of Materials Chemistry. 2012; 22: 20166–9.
[108] Ning GQ, Xu CG, Cao YM, Zhu X, Jiang ZM, Fan ZJ et al. Chemical vapor deposition derived flexible graphene paper and its application as high performance anodes for lithium rechargeable batteries. Journal of Materials Chemistry A. 2013; 1: 408–14.
[109] Shu KW, Wang CY, Li S, Zhao C, Yang Y, Liu HK et al. Flexible free-standing graphene paper with interconnected porous structure for energy storage. Journal of Materials Chemistry A. 2015; 3: 4428–34.
[110] Cheng C, Zhu JW, Yang XW, Qiu L, Wang YF, Li D. Dynamic electrosorption analysis: a viable liquid-phase characterization method for porous carbon? Journal of Materials Chemistry A. 2013; 1: 9332–40.
[111] Lee JU, Lee W, Yi JW, Yoon SS, Lee SB, Jung BM et al. Preparation of highly stacked graphene papers via site-selective functionalization of graphene oxide. Journal of Materials Chemistry A. 2013; 1: 12893–9.
[112] Jiang ZQ, Shi YL, Jiang ZJ, Tian XN, Luo LJ, Chen WH. High performance of a free-standing sulfonic acid functionalized holey graphene oxide paper as a proton conducting polymer electrolyte for airbreathing direct methanol fuel cells. Journal of Materials Chemistry A. 2014; 2: 6494–503.
[113] Song WL, Guan XT, Fan LZ, Cao WQ, Wang CY, Zhao QL et al. Magnetic and conductive graphene papers toward thin layers of effective electromagnetic shielding. Journal of Materials Chemistry A. 2015; 3: 2097–107.
[114] Liu Q, Liu LQ, Xie K, Meng YN, Wu HP, Wang GR et al. Synergistic effect of a rGO/PANI nanocomposite electrode based air working ionic actuator with a large actuation stroke and long-term durability. Journal of Materials Chemistry A. 2015; 3: 8380–8.
[115] Ping JF, Wang YX, Fan K, Tang WZ, Wu J, Ying YB. High-performance flexible potentiometric sensing devices using free-standing graphene paper. Journal of Materials Chemistry B. 2013; 1: 4781–91.
[116] Chen IWP, Saint Jhou SH, Chen W. Preparation of high-quality graphene sheets and their applications in highly conductive papers and a high-performance electromechanical actuator. Journal of Materials Chemistry C. 2013; 1: 5970–5.

[117] Abouimrane A, Compton OC, Amine K, Nguyen ST. Non-annealed graphene paper as a binder-free anode for lithium-ion batteries. Journal of Physical Chemistry C. 2010; 114: 12800–4.
[118] Liu JQ, Wang R, Cui L, Tang JG, Liu Z, Kong QS et al. Using molecular level modification to tune the conductivity of graphene papers. Journal of Physical Chemistry C. 2012; 116: 17939–46.
[119] Zhu JM, Zhu LW, Lu ZF, Gu L, Cao SL, Cao XB. Selectively expanding graphene oxide paper for creating multifunctional carbon materials. Journal of Physical Chemistry C. 2012; 116: 23075–82.
[120] Cetinkaya T, Ozcan S, Uysal M, Guler MO, Akbulut H. Free-standing flexible graphene oxide paper electrode for rechargeable Li-O_2 batteries. Journal of Power Sources. 2014; 267: 140–7.
[121] Yang MH, Ko SH, Im JS, Choi BG. Free-standing molybdenum disulfide/graphene composite paper as a binder- and carbon-free anode for lithium-ion batteries. Journal of Power Sources. 2015; 288: 76–81.
[122] Huang HB, Song ZG, Wei N, Shi L, Mao YY, Ying YL et al. Ultrafast viscous water flow through nanostrand-channelled graphene oxide membranes. Nature Communications. 2013; 4: 2979.
[123] Xu YX, Chen CY, Zhao ZP, Lin ZY, Lee C, Xu X et al. Solution processable holey graphene oxide and Its derived macrostructures for high-performance supercapacitors. Nano Letters. 2015; 15: 4605–10.
[124] Zhong D, Yang QL, Guo L, Dou SX, Liu KS, Jiang L. Fusion of nacre, mussel, and lotus leaf: bio-inspired graphene composite paper with multifunctional integration. Nanoscale. 2013; 5: 5758–64.
[125] Dichiara AB, Sherwood TJ, Benton-Smith J, Wilson JC, Weinstein SJ, Rogers RE. Free-standing carbon nanotube/graphene hybrid papers as next generation adsorbents. Nanoscale. 2014; 6: 6322–7.
[126] Huang XY, Zhi CY, Jiang PK, Golberg D, Bando Y, Tanaka T. Temperature-dependent electrical property transition of graphene oxide paper. Nanotechnology. 2012; 23: 455705.
[127] Bi H, Chen J, Zhao W, Sun SR, Tang YF, Lin TQ et al. Highly conductive, free-standing and flexible graphene papers for energy conversion and storage devices. RSC Advances. 2013; 3: 8454–60.
[128] Park SJ, An JH, Suk JW, Ruoff RS. Graphene-Based Actuators. Small. 2010; 6: 210–2.
[129] Zhang JW, Shi G, Jiang C, Ju S, Jiang DZ. 3D bridged carbon nanoring/graphene hybrid paper as a high-performance lateral heat spreader. Small. 2015; 11: 6197–204.
[130] Xiao F, Li YQ, Gao HC, Ge SB, Duan HW. Growth of coral-like PtAu-MnO_2 binary nanocomposites on free-standing graphene paper for flexible nonenzymatic glucose sensors. Biosensors & Bioelectronics. 2013; 41: 417–23.
[131] Xiao F, Song JB, Gao HC, Zan XL, Xu R, Duan HW. Coating graphene paper with 2D-assembly of electrocatalytic nanoparticles: A modular approach toward high-performance flexible electrodes. ACS Nano. 2012; 6: 100–10.
[132] Cui W, Li MZ, Liu JY, Wang B, Zhang C, Jiang L et al. A strong integrated strength and toughness artificial nacre based on dopamine cross-linked graphene oxide. ACS Nano. 2014; 8: 9511–7.
[133] Xiao F, Li YQ, Zan XL, Liao K, Xu R, Duan HW. Growth of metal-metal oxide nanostructures on freestanding graphene paper for flexible biosensors. Advanced Functional Materials. 2012; 22: 2487–94.
[134] Shen B, Zhai WT, Zheng WG. Ultrathin flexible graphene film: An excellent thermal conducting material with efficient EMI shielding. Advanced Functional Materials. 2014; 24: 4542–8.
[135] Zhang M, Huang L, Chen J, Li C, Shi GQ. Ultratough, ultrastrong, and highly conductive graphene films with arbitrary sizes. Advanced Materials. 2014; 26: 7588–92.
[136] Yang SJ, Kang JH, Jung HS, Kim TH, Park CR. Preparation of a freestanding, macroporous reduced graphene oxide film as an efficient and recyclable sorbent for oils and organic solvents. Journal of Materials Chemistry A. 2013; 1: 9427–32.
[137] Korkut S, Roy-Mayhew JD, Dabbs DM, Milius DL, Aksay IA. High surface area tapes produced with functionalized graphene. ACS Nano. 2011; 5: 5214–22.
[138] Xin GQ, Sun HT, Hu T, Fard HR, Sun X, Koratkar N et al. Large-area freestanding graphene paper for superior thermal management. Advanced Materials. 2014; 26: 4521–6.
[139] Wang M, Le Dai D, Oh JS, Nguyen Thi M, Kim SH, Hong SC et al. Large-area, conductive and flexible reduced graphene oxide (rGO) membrane fabricated by electrophoretic deposition (EPD). ACS Applied Materials & Interfaces. 2014; 6: 1747–53.
[140] Liu SY, Chen K, Fu Y, Yu SY, Bao ZH. Reduced graphene oxide paper by supercritical ethanol treatment and its electrochemical properties. Applied Surface Science. 2012; 258: 5299–303.
[141] Chen CM, Huang JQ, Zhang Q, Gong WZ, Yang QH, Wang MZ et al. Annealing a graphene oxide film to produce a free standing high conductive graphene film. Carbon. 2012; 50: 659–67.
[142] Shu KW, Wang CY, Wang M, Zhao C, Wallace GG. Graphene cryogel papers with enhanced mechanical strength for high performance lithium battery anodes. Journal of Materials Chemistry A. 2014; 2: 1325–31.
[143] Cheng IF, Xie YQ, Gonzales RA, Brejna PR, Sundararajan JP, Kengne BAF et al. Synthesis of graphene paper from pyrolyzed asphalt. Carbon. 2011; 49: 2852–61.
[144] Zhang L, Alvarez NT, Zhang MX, Haase M, Malik R, Mast D et al. Preparation and characterization of graphene paper for electromagnetic interference shielding. Carbon. 2015; 82: 353–9.
[145] Li N, Yang GZ, Sun Y, Song HW, Cui H, Yang GW et al. Free-standing and transparent graphene membrane of polyhedron box-shaped basic building units directly grown using a NaCl template for flexible transparent and stretchable solid-state supercapacitors. Nano Letters. 2015; 15: 3195–203.
[146] Lerf A, Buchsteiner A, Pieper J, Schottl S, Dekany I, Szabo T et al. Hydration behavior and dynamics of water molecules in graphite oxide. Journal of Physics and Chemistry of Solids. 2006; 67: 1106–10.

[147] Mukherjee R, Thomas AV, Krishnamurthy A, Koratkar N. Photothermally reduced graphene as high-power anodes for lithium-ion batteries. ACS Nano. 2012; 6: 7867–78.
[148] Hu YH, Li XF, Geng DS, Cai M, Li RY, Sun XL. Influence of paper thickness on the electrochemical performances of graphene papers as an anode for lithium ion batteries. Electrochimica Acta. 2013; 91: 227–33.
[149] Zhao X, Hayner CM, Kung MC, Kung HH. Flexible holey graphene paper electrodes with enhanced rate capability for energy storage applications. ACS Nano. 2011; 5: 8739–49.
[150] Bagri A, Mattevi C, Acik M, Chabal YJ, Chhowalla M, Shenoy VB. Structural evolution during the reduction of chemically derived graphene oxide. Nature Chemistry. 2010; 2: 581–7.
[151] Gao W, Alemany LB, Ci LJ, Ajayan PM. New insights into the structure and reduction of graphite oxide. Nature Chemistry. 2009; 1: 403–8.
[152] Liu Z, Sun XM, Nakayama-Ratchford N, Dai HJ. Supramolecular chemistry on water-soluble carbon nanotubes for drug loading and delivery. ACS Nano. 2007; 1: 50–6.
[153] Marquis R, Greco C, Sadokierska I, Lebedkin S, Kappes MM, Michel T et al. Supramolecular discrimination of carbon nanotubes according to their helicity. Nano Letters. 2008; 8: 1830–5.
[154] Mayo JD, Behal S, Adronov A. Phase separation of polymer-functionalized SWNTs within a PMMA/polystyrene blend. Journal of Polymer Science Part a-Polymer Chemistry. 2009; 47: 450–8.
[155] Liu Z, Robinson JT, Sun XM, Dai HJ. PEGylated nanographene oxide for delivery of water-insoluble cancer drugs. Journal of the American Chemical Society. 2008; 130: 10876–7.
[156] Yang XY, Zhang XY, Liu ZF, Ma YF, Huang Y, Chen YS. High-efficiency loading and controlled release of doxorubicin hydrochloride on graphene oxide. Journal of Physical Chemistry C. 2008; 112: 17554–8.
[157] Liu JQ, Tao L, Yang WR, Li D, Boyer C, Wuhrer R et al. Synthesis, characterization, and multilayer assembly of pH sensitive graphene-polymer nanocomposites. Langmuir. 2010; 26: 10068–75.
[158] Eda G, Chhowalla M. Graphene-based composite thin films for electronics. Nano Letters. 2009; 9: 814–8.
[159] Lee EJH, Zhi LJ, Burghard M, Muellen K, Kern K. Electrical properties and photoconductivity of stacked-graphene carbon nanotubes. Advanced Materials. 2010; 22: 1854–7.
[160] Faugeras C, Nerriere A, Potemski M, Mahmood A, Dujardin E, Berger C et al. Few-layer graphene on SiC, pyrolitic graphite, and graphene: A Raman scattering study. Applied Physics Letters. 2008; 92: 011914.
[161] Jin ZH, Owour P, Lei SD, Ge LH. Graphene, graphene quantum dots and their applications in optoelectronics. Current Opinion in Colloid & Interface Science. 2015; 20: 439–53.
[162] Zhou XJ, Guo SW, Zhang JY. Solution-processable graphene quantum dots. ChemPhysChem. 2013; 14: 2627–40.
[163] Ozfidan I, Korkusinski M, Hawrylak P. Electronic properties and electron-electron interactions in graphene quantum dots. Physica Status Solidi-Rapid Research Letters. 2016; 10: 13–23.
[164] Peng J, Gao W, Gupta BK, Liu Z, Romero-Aburto R, Ge LH et al. Graphene puantum dots derived from carbon fibers. Nano Letters. 2012; 12: 844–9.
[165] Yang SW, Sun J, Li XB, Zhou W, Wang ZY, He P et al. Large-scale fabrication of heavy doped carbon quantum dots with tunable-photoluminescence and sensitive fluorescence detection. Journal of Materials Chemistry A. 2014; 2: 8660–7.
[166] Zhang SJ, Shao T, Bekaroglu SSK, Karanfil T. The impacts of aggregation and surface chemistry of carbon nanotubes on the adsorption of synthetic organic compounds. Environmental Science & Technology. 2009; 43: 5719–25.
[167] Varshney V, Patnaik SS, Roy AK, Froudakis G, Farmer BL. Modeling of thermal transport in pillared-graphene architectures. Acs Nano. 2010; 4: 1153–61.
[168] Sun J, Liu HM, Chen X, Evans DG, Yang WS, Duan XF. Carbon nanorings and their enhanced lithium storage properties. Advanced Materials. 2013; 25: 1125–30.
[169] Brown GE, Henrich VE, Casey WH, Clark DL, Eggleston C, Felmy A et al. Metal oxide surfaces and their interactions with aqueous solutions and microbial organisms. Chemical Reviews. 1999; 99: 77–174.
[170] Medhekar NV, Ramasubramaniam A, Ruoff RS, Shenoy VB. Hydrogen bond networks in graphene oxide composite paper: Structure and mechanical properties. ACS Nano. 2010; 4: 2300–6.
[171] Putz KW, Compton OC, Palmeri MJ, Nguyen ST, Brinson LC. High-nanofiller-content graphene oxide-polymer nanocomposites via vacuum-assisted self-assembly. Advanced Functional Materials. 2010; 20: 3322–9.
[172] Yang XW, Cheng C, Wang YF, Qiu L, Li D. Liquid-mediated dense integration of graphene materials for compact capacitive energy storage. Science. 2013; 341: 534–7.
[173] Compton OC, Dikin DA, Putz KW, Brinson LC, Nguyen ST. Electrically conductive "alkylated" graphene paper via chemical reduction of amine-functionalized graphene oxide paper. Advanced Materials. 2010; 22: 892–6.
[174] Dekany I, Kruger-Grasser R, Weiss A. Selective liquid sorption properties of hydrophobized graphite oxide nanostructures. Colloid and Polymer Science. 1998; 276: 570–6.
[175] Park S, Mohanty N, Suk JW, Nagaraja A, An JH, Piner RD et al. Biocompatible, robust free-standing paper composed of a TWEEN/graphene composite. Advanced Materials. 2010; 22: 17361740.
[176] Liang JJ, Xu YF, Sui D, Zhang L, Huang Y, Ma YF et al. Flexible, magnetic, and electrically conductive graphene/Fe_3O_4 paper and its application for magnetic-controlled switches. Journal of Physical Chemistry C. 2010; 114: 17465–71.
[177] Dimarzio EA, Yang AJM, Glotzer SC. Mixing plate-like and rod-like molecules with solvent—A text of flory-Huggins lattice statistics. Journal of Research of the National Institute of Standards and Technology. 1995; 100: 173–86.

[178] Park SJ, An JH, Jung IH, Piner RD, An SJ, Li XS et al. Colloidal suspensions of highly reduced graphene oxide in a wide variety of organic solvents. Nano Letters. 2009; 9: 1593–7.
[179] Rozada R, Paredes JI, Villar-Rodil S, Martinez-Alonso A, Tascon JMD. Towards full repair of defects in reduced graphene oxide films by two-step graphitization. Nano Research. 2013; 6: 216–33.
[180] Ghosh T, Biswas C, Oh JS, Arabale G, Hwang TS, Nguyen Dang L et al. Solution-processed graphite membrane from reassembled graphene oxide. Chemistry of Materials. 2012; 24: 594–9.
[181] Almog Y, Klein J. Interactions between mica surfaces in a polystyrene-cyclopentane solution near the theta-temperature. Journal of Colloid and Interface Science. 1985; 106: 33–44.
[182] Alexandridis P, Hatton TA. Poly(ethylene oxide)-poly(propylene oxide)-poly(ethylene oxide) block-copolymer surfactants in aqueous-solutions and at interfaces—Thermodynamics, structure, dynamics, and modeling. Colloids and Surfaces a-Physicochemical and Engineering Aspects. 1995; 96: 1–46.
[183] Ebagninin KW, Benchabane A, Bekkour K. Rheological characterization of poly(ethylene oxide) solutions of different molecular weights. Journal of Colloid and Interface Science. 2009; 336: 360–7.
[184] Bihannic I, Baravian C, Duval JFL, Paineau E, Meneau F, Levitz P et al. Orientational order of colloidal disk-shaped particles under shear-flow conditions: a rheological-small-angle X-ray scattering study. Journal of Physical Chemistry B. 2010; 114: 16347–55.
[185] Decker EL, Garoff S. Contact line structure and dynamics on surfaces with contact angle hysteresis. Langmuir. 1997; 13: 6321–32.
[186] Jaworek A, Sobczyk AT. Electrospraying route to nanotechnology: An overview. Journal of Electrostatics. 2008; 66: 197–219.
[187] Pal M, Pal U, Gracia Y Jimenez JM, Perez-Rodriguez F. Effects of crystallization and dopant concentration on the emission behavior of TiO_2: Eu nanophosphors. Nanoscale Research Letters. 2012; 7: 1.
[188] Meng YN, Xin GQ, Nam JW, Cho SM, Chae HY. Electrospray deposition of carbon nanotube thin films for flexible transparent electrodes. Journal of Nanoscience and Nanotechnology. 2013; 13: 6125–9.
[189] Mustafa M, Awais MN, Pooniah G, Choi KH, Ko JB, Doh YH. Electrospray deposition of a graphene-oxide thin film, its characterization and investigation of its resistive switching performance. Journal of the Korean Physical Society. 2012; 61: 470–5.
[190] Lee CK, Park KW, Hwang SW, Lee SB, Shim JK. Direct electrospray deposition of graphene onto paper and effect of binder on its surface resistance. Journal of Nanoscience and Nanotechnology. 2013; 13: 7108–11.
[191] Jin MH, Kim TH, Lim SC, Duong DL, Shin HJ, Jo YW et al. Facile physical route to highly crystalline graphene. Advanced Functional Materials. 2011; 21: 3496–501.
[192] Song L, Khoerunnisa F, Gao W, Dou WH, Hayashi T, Kaneko K et al. Effect of high-temperature thermal treatment on the structure and adsorption properties of reduced graphene oxide. Carbon. 2013; 52: 608–12.
[193] Chen CM, Yang QH, Yang YG, Lv W, Wen YF, Hou PX et al. Self-assembled free-standing graphite oxide membrane. Advanced Materials. 2009; 21: 3007–11.
[194] Chen L, Huang LL, Zhu JH. Stitching graphene oxide sheets into a membrane at a liquid/liquid interface. Chemical Communications. 2014; 50: 15944–7.
[195] Kim F, Cote LJ, Huang JX. Graphene oxide: Surface activity and two-dimensional assembly. Advanced Materials. 2010; 22: 1954–8.
[196] Chae SJ, Guenes F, Kim KK, Kim ES, Han GH, Kim SM et al. Synthesis of large-area graphene layers on poly-nickel substrate by chemical vapor deposition: Wrinkle formation. Advanced Materials. 2009; 21: 2328–33.
[197] Yoo EJ, Kim JD, Hosono E, Zhou HS, Kudo T, Honma I. Large reversible Li storage of graphene nanosheet families for use in rechargeable lithium ion batteries. Nano Letters. 2008; 8: 2277–82.
[198] Bhardwaj T, Antic A, Pavan B, Barone V, Fahlman BD. Enhanced electrochemical lithium storage by graphene nanoribbons. Journal of the American Chemical Society. 2010; 132: 12556–8.
[199] Liu F, Song SY, Xue DF, Zhang HJ. Folded structured graphene paper for high performance electrode materials. Advanced Materials. 2012; 24: 1089–94.
[200] Wang DW, Li F, Zhao JP, Ren WC, Chen ZG, Tan J et al. Fabrication of graphene/polyaniline composite paper via *in situ* anodic electropolymerization for high-performance flexible electrode. ACS Nano. 2009; 3: 1745–52.
[201] Wu ZS, Ren WC, Wang DW, Li F, Liu BL, Cheng HM. High-energy MnO_2 nanowire/graphene and graphene asymmetric electrochemical capacitors. ACS Nano. 2010; 4: 5835–42.
[202] Zhao X, Sanchez BM, Dobson PJ, Grant PS. The role of nanomaterials in redox-based supercapacitors for next generation energy storage devices. Nanoscale. 2011; 3: 839–55.
[203] Wu XZ, Zhou J, Xing W, Wang GQ, Cui HY, Zhuo SP et al. High-rate capacitive performance of graphene aerogel with a superhigh C/O molar ratio. Journal of Materials Chemistry. 2012; 22: 23186–93.
[204] Seger B, Kamat PV. Electrocatalytically active graphene-platinum nanocomposites. Role of 2-D carbon support in PEM fuel cells. Journal of Physical Chemistry C. 2009; 113: 7990–5.
[205] Guo SJ, Dong SJ, Wang EK. Three-dimensional Pt-on-Pd bimetallic nanodendrites supported on graphene nanosheet: Facile synthesis and used as an advanced nanoelectrocatalyst for methanol oxidation. ACS Nano. 2010; 4: 547–55.
[206] Lu ZJ, Bao SJ, Gou YT, Cai CJ, Ji CC, Xu MW et al. Nitrogen-doped reduced-graphene oxide as an efficient metal-free electrocatalyst for oxygen reduction in fuel cells. RSC Advances. 2013; 3: 3990–5.

[207] Qu LT, Liu Y, Baek JB, Dai LM. Nitrogen-doped graphene as efficient metal-free electrocatalyst for oxygen reduction in fuel cells. ACS Nano. 2010; 4: 1321–6.
[208] He DP, Cheng K, Peng T, Mu SC. Graphene/carbon nanospheres sandwich supported PEM fuel cell metal nanocatalysts with remarkably high activity and stability. Journal of Materials Chemistry A. 2013; 1: 2126–32.
[209] Liu ZF, Liu Q, Huang Y, Ma YF, Yin SG, Zhang XY et al. Organic photovoltaic devices based on a novel acceptor material: Graphene. Advanced Materials. 2008; 20: 3924–30.
[210] Bonaccorso F, Sun Z, Hasan T, Ferrari AC. Graphene photonics and optoelectronics. Nature Photonics. 2010; 4: 611–22.
[211] Sato K, Noguchi M, Demachi A, Oki N, Endo M. A mechanism of lithium storage in disordered carbons. Science. 1994; 264: 556–8.
[212] Dahn JR, Zheng T, Liu YH, Xue JS. Mechanisms for lithium insertion in carbonaceous materials. Science. 1995; 270: 590–3.
[213] Xue JS, Dahn JR. Dramatic effect of oxidation on lithium insertion in carbons made from epoxy-resins. Journal of the Electrochemical Society. 1995; 142: 3668–77.
[214] Pan DY, Wang S, Zhao B, Wu MH, Zhang HJ, Wang Y et al. Li storage properties of disordered graphene nanosheets. Chemistry of Materials. 2009; 21: 3136–42.
[215] Yazami R, Deschamps M. High reversible capacity carbon-lithium negative electrode in polymer electrolyte. Journal of Power Sources. 1995; 54: 411–5.
[216] Fan XH, Phebus BD, Li LJ, Chen SW. Graphene-based composites for supercapacitor electrodes. Science of Advanced Materials. 2015; 7: 1916–44.
[217] Yang W, Ni M, Ren X, Tian YF, Li N, Su YF et al. Graphene in Supercapacitor Applications. Current Opinion in Colloid & Interface Science. 2015; 20: 416–28.
[218] Yu ZN, Tetard L, Zhai L, Thomas J. Supercapacitor electrode materials: nanostructures from 0 to 3 dimensions. Energy & Environmental Science. 2015; 8: 702–30.
[219] Zhang LL, Zhao XS. Carbon-based materials as supercapacitor electrodes. Chemical Society Reviews. 2009; 38: 2520–31.
[220] Zhang X, Zhang HT, Li C, Wang K, Sun XZ, Ma YW. Recent advances in porous graphene materials for supercapacitor applications. RSC Advances. 2014; 4: 45862–84.
[221] Kotz R, Carlen M. Principles and applications of electrochemical capacitors. Electrochimica Acta. 2000; 45: 2483–98.
[222] Pandolfo AG, Hollenkamp AF. Carbon properties and their role in supercapacitors. Journal of Power Sources. 2006; 157: 11–27.
[223] Palaniselvam T, Baek JB. Graphene based 2D-materials for supercapacitors. 2D Materials. 2015; 2.
[224] Shao YL, El-Kady MrF, Wang LJ, Zhang QH, Li YG, Wang HZ et al. Graphene-based materials for flexible supercapacitors. Chemical Society Reviews. 2015; 44: 3639–65.
[225] Vivekchand SRC, Rout CS, Subrahmanyam KS, Govindaraj A, Rao CNR. Graphene-based electrochemical supercapacitors. Journal of Chemical Sciences. 2008; 120: 9–13.
[226] Salunkhe RR, Lee YH, Chang KH, Li JM, Simon P, Tang J et al. Nanoarchitectured graphene-based supercapacitors for next-generation energy-storage applications. Chemistry-A European Journal. 2014; 20: 13838–52.
[227] Stoller MD, Park SJ, Zhu YW, An JH, Ruoff RS. Graphene-based ultracapacitors. Nano Letters. 2008; 8: 3498–502.
[228] Kim HW, Yoon HW, Yoon SM, Yoo BM, Ahn BK, Cho YH et al. Selective gas transport through few-layered graphene and graphene oxide membranes. Science. 2013; 342: 91–5.
[229] Nair RR, Wu HA, Jayaram PN, Grigorieva IV, Geim AK. Unimpeded permeation of water through helium-leak-tight graphene-based membranes. Science. 2012; 335: 442–4.
[230] Joshi RK, Carbone P, Wang FC, Kravets VG, Su Y, Grigorieva IV et al. Precise and ultrafast molecular sieving through graphene oxide membranes. Science. 2014; 343: 752–4.
[231] Klaysom C, Cath TY, Depuydt T, Vankelecom IFJ. Forward and pressure retarded osmosis: potential solutions for global challenges in energy and water supply. Chemical Society Reviews. 2013; 42: 6959–89.
[232] Song XX, Liu ZY, Sun DD. Nano gives the answer: Breaking the bottleneck of internal concentration polarization with a nanofiber composite forward osmosis membrane for a high water production rate. Advanced Materials. 2011; 23: 3256–60.
[233] You CF, Li Y. Desulfurization characteristics of rapidly hydrated sorbents with various adhesive carrier particles for a semidry CFB-FGD system. Environmental Science & Technology. 2013; 47: 2754–9.
[234] Persson KM, Gekas V, Tragardh G. Study of membrane compaction and its influence on ultrafiltration water permeability. Journal of Membrane Science. 1995; 100: 155–62.
[235] Ebert K, Fritsch D, Koll J, Tjahjawiguna C. Influence of inorganic fillers on the compaction behaviour of porous polymer based membranes. Journal of Membrane Science. 2004; 233: 71–8.
[236] Gin DL, Noble RD. Designing the next generation of chemical separation membranes. Science. 2011; 332: 674–6.
[237] Whitby M, Quirke N. Fluid flow in carbon nanotubes and nanopipes. Nature Nanotechnology. 2007; 2: 87–94.
[238] Ikuno T, Honda S, Yasuda T, Oura K, Katayama M, Lee JG et al. Thermally driven nanomechanical deflection of hybrid nanowires. Applied Physics Letters. 2005; 87: 213104.

CHAPTER 7

Graphene 3D Architectures

Ling Bing Kong,[1,a,*] *Freddy Boey,*[1,b] *Yizhong Huang,*[1,c] *Zhichuan Jason Xu,*[1,d] *Kun Zhou,*[2] *Sean Li,*[3] *Wenxiu Que,*[4] *Hui Huang*[5] and *Tianshu Zhang*[6]

7.1. Overview

In this chapter, recent progress in the fabrication and characterization of 3D graphene-based structures will be summarized. Although there have reports on this aspect [1, 2], a more extended discussion on each method has been not available, due to the space limitation in most cases. As stated earlier, graphene has attracted much attention, due to its unique structure, together with various extraordinary chemical and physical characteristics [3–5], thus leading to a wide range of potential applications in various fields [5, 6]. Due to its zero-gap semi-metal properties, pure graphene cannot be used to develop devices similar to those based on normal semiconductors [7–9]. This is the driving force to develop well-defined structures, including nanoribbons [10–12] and quantum dots [13, 14]. More specifically, graphene-based materials, which are used in energy, environment and biological applications, require 3D structures [15–18]. The 3D structured graphene materials could have high specific surface areas, strong mechanical properties and rapid mass and charge transport kinetics, due to the synergistic effect of the 3D porous structures and the intrinsic high conductivity of graphene. 3D graphene structures have also various other names, such as networks, foams, sponges, hydrogels and aerogels, which will not be specifically emphasized in this chapter and just follow the original references.

[1] School of Materials Science and Engineering, Nanyang Technological University, Singapore.
[a] E-mail: elbkong@ntu.edu.sg
[b] E-mail: mycboey@ntu.edu.sg
[c] E-mail: yzhuang@ntu.edu.sg
[d] E-mail: xuzc@ntu.edu.sg
[2] School of Mechanical & Aerospace Engineering, Nanyang Technological University, Singapore; E-mail: kzhou@ntu.edu.sg
[3] School of Materials Science and Engineering, The University of New South Wales, Australia; E-mail: sean.li@unsw.edu.au
[4] Electronic Materials Research Laboratory, School of Electronic and Information Engineering, Xi'an Jiaotong University, P. R. China; E-mail: wxque@xjtu.edu.cn
[5] Singapore Institute of Manufacturing Technologies (SIMTech), Singapore; E-mail: hhuang@SIMTech.a-star.edu.sg
[6] Anhui Target Advanced Ceramics Technology Co. Ltd., Hefei, Anhui, P. R. China; E-mail: 13335516617@163.com
* Corresponding author

7.2. Solution Processing of 3D Graphene Architectures and Related Chemistry

For practical applications, small graphene nanosheets usually need to be constructed or assembled into macroscopic materials [6, 19–23]. However, due to the extremely large aspect ratio, graphene nanosheets tend to aggregate or restack into the graphitic structure as they are brought together, which has been a critical issue encountered in this process. If this problem cannot be addressed, the advantageous structures and properties of graphene nanosheets cannot be well utilized as macroscopic materials. Therefore, it has been widely accepted that the integration of 2D graphene nanosheets into 3D macroscopic structures, such as porous films, scaffolds and networks, is the first step towards developing high-performance graphene-based materials in terms of their practical applications [15–18, 24]. In this section, the main strategies that have been used to fabricate 3D graphene architectures will be summarized.

7.2.1. Chemistry of solution processing

According to polymer science and physical chemistry, CMGs are 2D amphiphilic conjugated polyelectrolytes that consist of hydrophobic basal planes attached with hydrophilic oxygenated functional groups [25]. There is a hydrophilic–hydrophobic balance between the inter-planar van der Waals force and the electrostatic repulsion, which determines the properties and thus the self-assembly behaviors of the CMGs solutions or suspensions. As discussed in the previous chapters, in general, GO is soluble in water, while graphene (or rGO) nanosheets are insoluble. However, it has been found that rGO aqueous dispersion was stable at certain concentrations [26]. For example, there was a critical concentration of 0.5 $mg{\cdot}ml^{-1}$, below which stable rGO solutions were obtained, while gelation occurred once the concentration was raised above this critical value. Currently, solution based self-assembly, following the principles of colloid chemistry, has been one of the most versatile techniques reported in the open literature to develop 3D graphene architectures with hierarchical structures and special properties [24].

7.2.2. Gelation of GO

Gelation of GO dispersions has been acknowledged to be a simple yet feasible approach to fabricate porous 3D GO networks [27–34]. As stated earlier, GO nanosheets are soluble in water due to the edge-bound carboxyl moieties, together with the large amount of hydrophilic epoxy and hydroxyl groups on their basal planes [35]. GO stable aqueous dispersion can reach concentrations of up to 10 mg ml^{-1} [36].

To promote the gelation of GO nanosheets in water suspension, various additives have been explored. One example is poly(vinyl alcohol) (PVA), which is a water-soluble widely available polymer [27]. It acted as a cross-linking agent to boost the interactions of the 2D GO nanosheets, so as to enhance the GO gelation. GO/PVA hybrid hydrogels could be derived from the mixtures of GO and PVA aqueous solutions, which were heavily shaked for 10 s and then sonicated for 20 min. The strong hydrogen bonding interaction between the hydroxyl-rich PVA chains and oxygen-containing groups on the GO nanosheets, facilitated the formation of the cross-linking between the two species, as shown in Fig. 7.1. Because the number of cross-linking sites was sufficiently large, stable GO composite hydrogels were produced by forming GO networks.

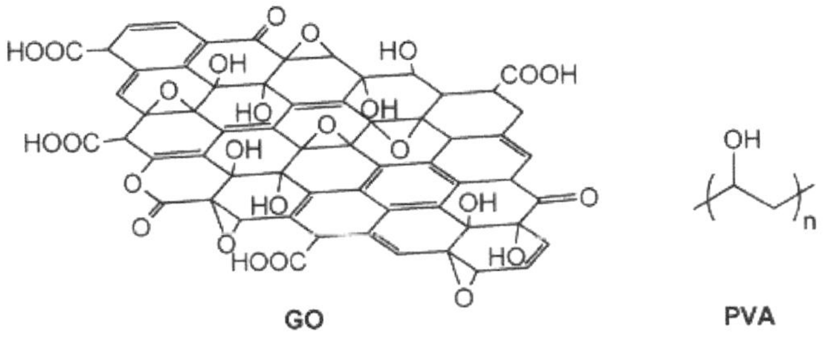

Fig. 7.1. Chemical structures of GO and PVA. Reproduced with permission from [27], Copyright © 2010, The Royal Society of Chemistry.

The gelation process was strongly influenced by the concentration of the cross-linking agent used in the experiments. GO/PVA mixtures containing 5 mg·ml^{-1} GO and PVA with various weight ratios, $r_{P/G}$, were studied. For PVA, there was a low and a high critical gelation concentration. A small amount of PVA, i.e., $r_{P/G}$ = 1:20, would be able to significantly increase the viscosity of the GO solution. For the GO/PVA mixtures with 5 mg·ml^{-1} GO, the gelation was only observed as the $r_{P/G}$ was in the range between 1:10 and 1:2. Further increasing the PVA content would facilitate a gel–sol transition. As $r_{P/G}$ = 1:1, the viscosity of the GO/PVA mixture was close to that of pure GO solution. This observation implied that the interactions between the GO nanosheets and the PVA molecules were different in the mixtures with low and high PVA concentrations.

Pure GO, GO/PVA mixed solutions and typical gels were lyophilized for morphological study. The freeze-dried sample derived from the 5 mg·ml^{-1} GO retained its original shape and volume, while the sample from the 1 mg·ml^{-1} GO demonstrated an obvious shrinkage caused by the lyophilization. Therefore, the 5 mg·ml^{-1} GO suspension was like a semi-diluted solution, so that the GO volume fraction was sufficiently large to form a GO network, after all water was removed. Figure 7.2 (A) shows SEM image of the GO network of sheets. Similarly, all the samples with PVA could retain their volumes after the lyophilization process. If only a small amount of PVA was added into the GO solution, e.g., $r_{P/G}$ = 1:20, the morphology of the lyophilized sample was not changed, i.e., the GO nanosheets were in extended states, as illustrated in Fig. 7.2 (B). The lyophilized gel derived from the mixture with $r_{P/G}$ = 1:5 had a similar morphology, as seen in Fig. 7.2 (C). As the $r_{P/G}$ was increased to 1:1, the sample would have a different morphology after lyophilization, as demonstrated in Fig. 7.2 (D). There were wire-like structures in the sample, implying that the GO nanosheets were rolled up either in the solution or during the lyophilization process.

In pure solution, the GO nanosheets could not form a gel network, due to the weak interaction force. Once the PVA polymer was added, the polymer molecules acted as a cross-linking agent. There was a strong interaction between the PVA molecules and the GO nanosheets through hydrogen bonding, due to the presence of a large amount of oxygen-containing groups on the GO and the hydroxyl-rich PVA polymer chains. One PVA chain could cater to two or more GO nanosheets, forming cross-linking bridges. As the number of cross-linking sites approached a critical value, GO composite hydrogel was produced, due to

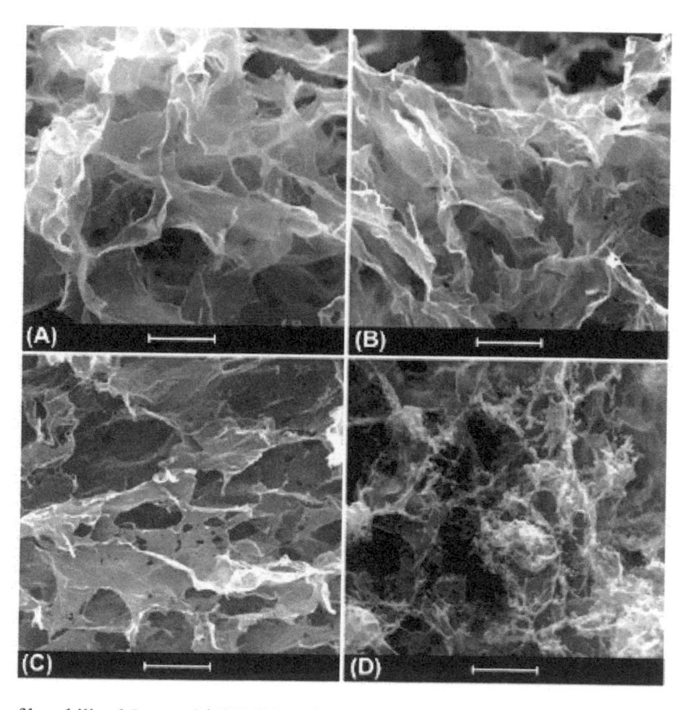

Fig. 7.2. SEM images of lyophilized 5 mg·ml^{-1} GO (A) and GO/PVA blends (B–D) with different ratios $r_{P/G}$: (B) 1:20, (C) 1:5 and (D) 1:1. Scale bar = 5 μm. Reproduced with permission from [27], Copyright © 2010, The Royal Society of Chemistry.

the formation of the GO network. However, if the PVA content was too high, e.g., $r_{P/G} = 1:2$, the number of PVA polymer chains adsorbed on the surfaces of the individual GO nanosheet was too large, so that the cross-linking effect was significantly weakened. Therefore, the PVA-coated GO nanosheets would exhibit behavior similar to that of pure PVA, i.e., gelation was prevented from happening. Also, at high concentrations of PVA, the GO nanosheets started to curl into wires in the composite solutions, as revealed in Fig. 7.2 (D). As a result, the opportunity for the GO nanosheets to contact was largely decreased, so that hydrogel could not be formed. Such GO hydrogels could find important applications in drug delivering and controlled releasing.

There have been other cross-linkers for GO gelation, such as DNA [28], protein [30], synthetic polymers with cationic charges and hydrogen bonding acceptors [29, 32], small quaternary ammonium salts and metal ions [29], through regulating the balance of the electrostatic repulsion, hydrophobic interactions and hydrogen bonding of GO-based colloidal suspensions.

DNA is an important category of biological building blocks, which could be used to non-covalently functionalize or promote the assembly of GO nanosheets. In GO/DNA composite assemblies, single-stranded DNA (ssDNA) chains would flatly lie on the surface of the GO nanosheets through π–π stacking interactions [37–39]. A simple and efficient approach was developed to trigger the 3D self-assembly of GO nanosheets into multifunctional hydrogels by using *in situ* formed ssDNA chains [28]. The GO/DNA composite self-assembled hydrogels (GO/DNA SH) had a high water content of 99%. They were mechanically strong, had self-healing ability and showed high environmental stability and dye-loading capacity.

The GO/DNA SH was obtained by simply mixing equal volumes of GO aqueous dispersion (6 $mg \cdot ml^{-1}$) and the aqueous solution of double-stranded DNA (dsDNA) (10 $mg \cdot ml^{-1}$), which was heated at 90°C for 5 min, as illustrated in Fig. 7.3. During the heating process, the dsDNA was unwound into ssDNA. The *in situ* formed ssDNA chains acted as a cross-linker to bridge the GO nanosheets, due to the strong non-covalent interactions, thus leading to the formation of stable SH based on the GO nanosheets and the ssDNA chains, as schematically demonstrated in Fig. 7.3. There was charge transfer from the nucleic acid bases of the DNA chains to the GO nanosheets [40, 41]. Also, the modulus of the aqueous mixture of the GO and dsDNA was increased tremendously, as the temperature was increased from 40°C to 90°C, which was still within the temperature range for dsDNA unwinding. Therefore, unwinding of the dsDNA was key to the formation of the SH, which supported the proposed mechanism described in Fig. 7.3.

SEM images of the as-prepared GO/DNA SH are shown in Fig. 7.4 (a, b), revealing the presence of well-defined and interconnected 3D porous network structure [28]. The samples had pore diameters ranging from submicrometer to several micrometers, with the pore walls consisting of ultra-thin layers of GO nanosheets. Figure 7.4 (c) shows XRD patterns of the samples, confirming the ssDNA bridging effect

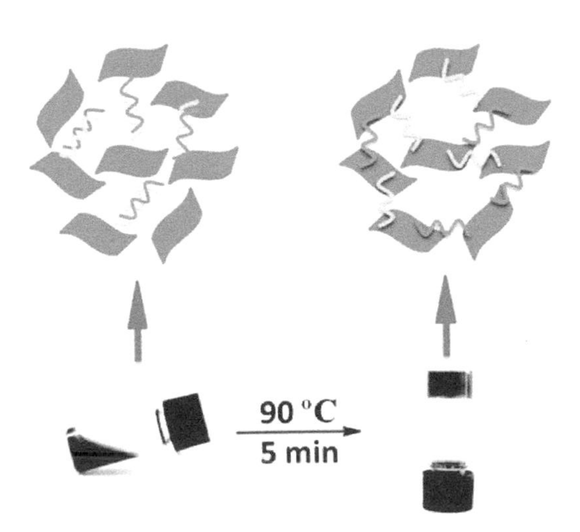

Fig. 7.3. The procedure for preparing the GO/DNA SH and schematic diagram of the proposed gelation mechanism. Reproduced with permission from [28], Copyright © 2010, American Chemical Society.

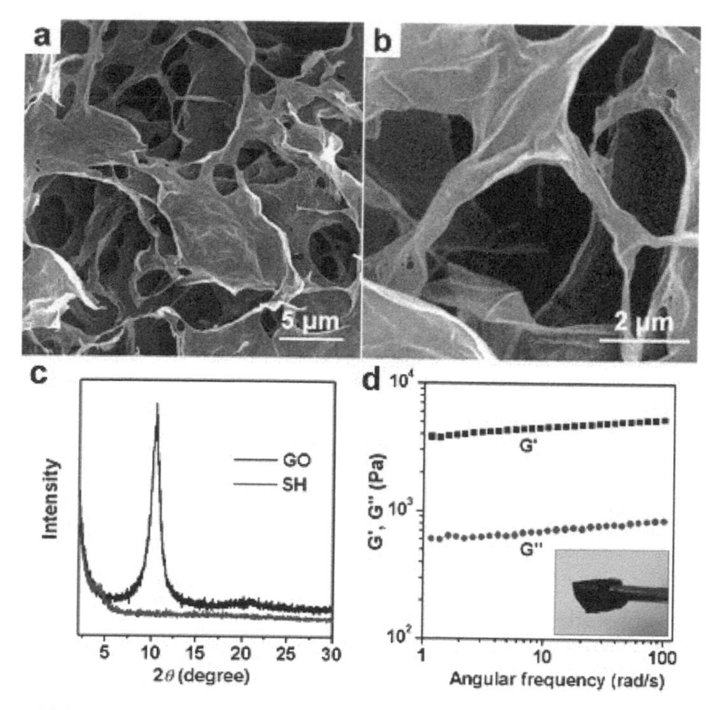

Fig. 7.4. SEM images with low (a) and high (b) magnifications revealing the interior microstructures of the GO/DNA SH. (c) XRD patterns of the freeze-dried GO and the GO/DNA SH. (d) dynamic rheological behavior of the SH prepared from the mixture of equal volume solutions of 6 mg·ml^{-1} GO and 10 mg·ml^{-1} dsDNA. The inset shows a photograph of the free-standing GO/DNA SH. Reproduced with permission from [28], Copyright © 2010, American Chemical Society.

Color image of this figure appears in the color plate section at the end of the book.

as the driving force for 3D self-assembly of the GO and DNA strains. The strong peak of the lyophiled GO at 10.6° was attributed to the layer-to-layer stacking spacing of 8.34 Å. However, the lyophiled GO/DNA SH sample had no XRD diffraction peak, which indicated that direct interaction or ordered stacking of GO nanosheets in the SH was absent. This was because the ssDNA had bound the GO nanosheets together to form 3D networks.

It was found that the storage modulus (G') of the SH had a substantial elastic response that was slightly dependent on the testing angular frequency. Moreover, the G' was always higher than the loss modulus (G") over 1–100 rad·s^{-1}, as illustrated in Fig. 7.4 (d). This observation suggested that the SH had a permanent network with a high degree of cross-linking. The SH possessed a typical G' value of about 4.6 kPa, so that free-standing SH was available, as seen in the inset of Fig. 7.4 (d). Due to the synergistic effect of the superiorly strong GO nanosheets and the robust binding between the ssDNA chains and the GO nanosheets, the free-standing SH samples exhibited a sufficiently high mechanical strength, even though their water content was as high as 99%. The gelation behavior of the mixtures and mechanical properties of the SH samples were strongly dependent on the concentrations of the GO and dsDNA. Generally, the higher the concentration of the two components, the stronger the hydrogels would be, due to the increased number of cross-linking sites between the GO nanosheets and the DNA chains.

The as-fabricated GO/DNA SH exhibited strong dye adsorption capability. Safranine O was used as a model dye in the form of aqueous solution (0.1 mg·ml^{-1}) with 0.6 ml to be mixed with 0.2 ml SH at room temperature. After 12 h, more than 80% of the dye was adsorbed by the SH, while it was almost completely absorbed after 24 h, as illustrated in Fig. 7.5 (a, b). The total dye-absorption capacity of the SH for safranine O was about 960 mg·g^{-1}. The high dye-loading capability of the SH could be partly attributed to the strong electrostatic interactions between the positively charged safranine O and the negatively charged GO nanosheets and the DNA molecules. At the same time, because of the porous structure, the SH had an extremely large specific surface area, which provided additional opportunity for the dye molecules and GO nanosheets to contact mutually.

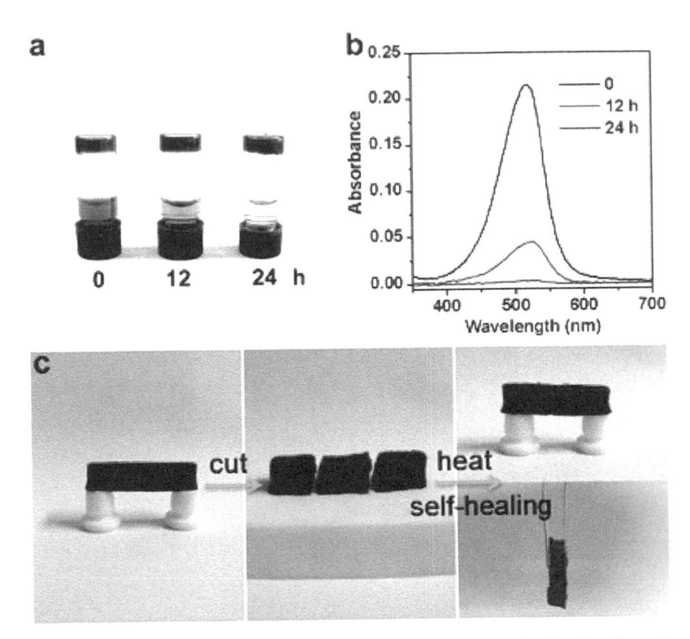

Fig. 7.5. Photographs (a) and absorption spectra (b) of the aqueous solution of safranine O after adsorption by the GO/DNA SH for different times at room temperature, where 0.6 ml of dye solution (0.1 mg·ml^{-1}) was added into 0.2 ml SH. (c) Self-healing process of the GO/DNA SH sample, which was cut into three small pieces first and then adhered by heating at 90°C in air for 3 min. Reproduced with permission from [28], Copyright © 2010, American Chemical Society.

More importantly, the GO/DNA SH had self-healing ability. For instance, a 2.5-cm long free-standing SH was cut into three pieces, which were then connected, as demonstrated in Fig. 7.5 (c). After the three pieces in contact surface-by-surface were then heated at 90°C in air for 3 min, the blocks were adhered into one whole piece. The self-healed SH was as strong as its original state. Because there were excessive ssDNA molecules in the SH, which were unbound and weakly bound to the GO nanosheets, they could reversibly be converted into dsDNA during the cooling process. The as-obtained SH also contained free dsDNA strains in water, which was confirmed by circular dichroism spectral examination. As the fresh surfaces of the cut SH were in close contact and heated, the free dsDNA would re-unwind to yield *in situ* ssDNA, which immediately linked the adjacent GO nanosheets at the interfaces, so that the three small pieces of SH were re-connected.

In a systematic study on GO gelation behavior, GO-based hydrogels were obtained by using different cross-linking approaches, including acidification, addition of small organic molecules, polymers and ions [29]. The effects of various driving forces, such as hydrogen bonding, electrostatic interaction and coordination, as well as the lateral dimensions of the GO nanosheets on GO gelation behavior were elucidated. The GO hydrogels were derived from GO dispersion that was mixed with the solutions of acid or other cross-linkers. The blend was then shaken heavily for several seconds to form hydrogels. The formation of the hydrogel was examined by using a tube inversion method, with a polypropylene tube that had an inner diameter of 8 mm.

Although GO nanosheets are similar to amphiphilic macromolecules, their lateral sizes could be at micrometer scale, so that they are much larger than the conventional polymers. As a result, they can contact one another even in the solutions with extremely low concentrations. For example, the freeze-dried sample of a 3 mg·ml^{-1} GO suspension could retain the original volume of the solution, whereas lyophilizing a 3 mg·ml^{-1} PVA (Fig. 7.6) led to samples with large shrinkage. Also, the lyophilized GO dispersion had a 3D network structure formed by the GO nanosheets, as illustrated in Fig. 7.7 (A). In this regard, it could be safely assumed that a loose dynamic GO network had been formed in the original GO suspension, due

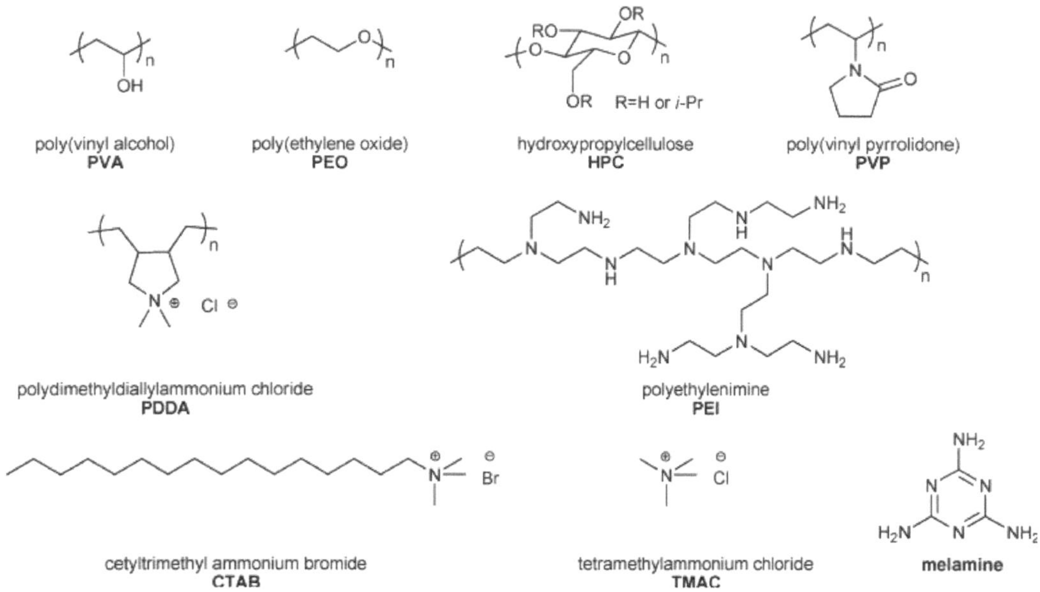

OH
poly(vinyl alcohol)
PVA
poly(ethylene oxide)
PEO
OR
RO
OR R=H or i-Pr
hydroxypropylcellulose
HPC
poly(vinyl pyrrolidone)
PVP
Cl ⊖
polydimethyldiallylammonium chloride
PDDA
NH2
H2N
polyethylenimine
PEI
Br
cetyltrimethyl ammonium bromide
CTAB
Cl
tetramethylammonium chloride
TMAC
melamine

Fig. 7.6. Chemical structures of several cross-linkers used to prepare the GO composite hydrogels. Reproduced with permission from [29], Copyright © 2011, American Chemical Society.

(A)
(B)
(C)
(D)

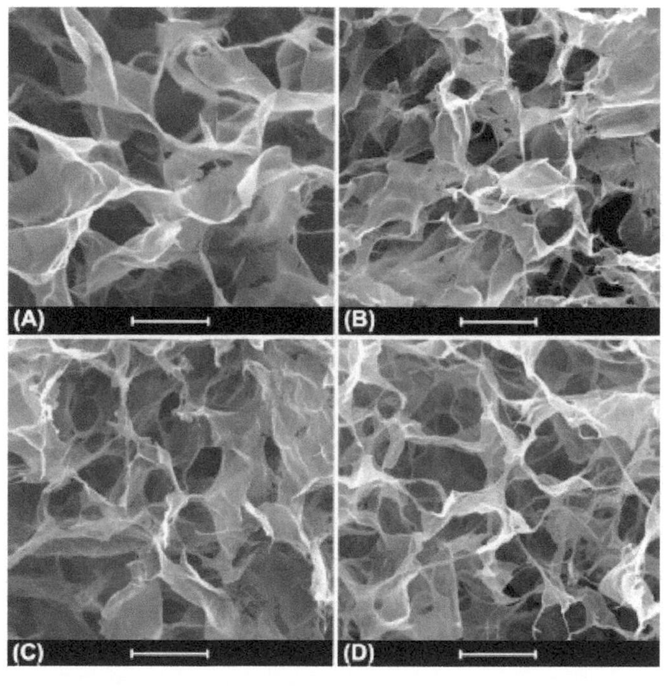

Fig. 7.7. SEM images of the lyophilized GO solution and three GO hydrogels: (A) GO solution, (B) GO/PVP hydrogel with 1 mg·ml^{-1} PVP, (C) GO/PDDA hydrogel with 0.1 mg·ml^{-1} PDDA and (D) GO/Ca^{2+} hydrogel with 9 mM Ca^{2+}. C_{GO} = 5 mg·ml^{-1}. Scale bar = 10 μm. Reproduced with permission from [29], Copyright © 2011, American Chemical Society.

to a force balance between electrostatic repulsion and binding interactions. As the GO network structure was reinforced, gelation took place accordingly. Acidification has been one of the effective ways to trigger this type of gelation.

Gelation and precipitation of GO nanosheets could be induced by the same reinforcing effect, but with different consequences. GO precipitate was the stacking of the GO nanosheets with certain degree of orientation, while GO hydrogel was comprised of the nanosheets in random distribution. Ordered stacking is energetically favorable, because there is a large contact area between every adjacent GO nanosheet. In real solutions or suspensions, the mobility of large GO nanosheets is highly limited, so that it is difficult to adjust their orientation to form ordered staking. Futhermore, the GO nanosheets with large conjugated basal planes are stiff and tend to form a stable network structure, which somehow prevented them from forming the energetically more stable state of precipitate. However, if GO nanosheets have sufficiently small lateral sizes, e.g., smaller than 1 μm, precipitation could be observed. This was the so-called size effect of gelation and precipitation. In addition, over-acidification of the suspension of large GO nanosheets would also result in the formation of irregular aggregates, because the repulsion forces were largely weakened.

Similar to PVA as discussed above, various other polymers have been shown to be able to trigger the gelation of GO suspensions. The polymers included poly(ethylene oxide) (PEO, M = 1,000,000), hydroxypropylcellulose (HPC) and poly(vinyl pyrrolidone) (PVP, M = 1,300,000), as schematically drawn in Fig. 7.6. PEO and PVP are hydrogen-bond acceptors, while HPC is a hydrogen-bond donor and acceptor. As a result, they can be attached onto the GO nanosheets by forming hydrogen bonds, so as to offer an additional bonding force to promote the GO gelation. All the GO/polymer composite hydrogels exhibited well-defined 3D network structures with morphologies similar to that of the pure GO hydrogel, as seen in Fig. 7.7 (B). This implied that the presence of the polymer chains had no negative effect on the assembly behavior of the GO nanosheets.

However, XRD analysis results indicated that the lyophilized GO/PVP and GO/PEO composite hydrogels demonstrated different features. The GO/PVP composites exhibited a peak at angles in the range of $2\theta = 10.5–12.5°$, which was related to the GO nanosheets and comparable with that of pure GO nanosheets at $2\theta = 11.5°$. It corresponded to an inter-planar distance of about 1.54 nm. This peak was weakened and broadened in the GO/PVP samples, which suggested that the stacking of the GO nanosheets was partly inhibited due to the presence of the PVP chains. No XRD peak was observed, which could be attributed to the fact that the PVP intercalated GO nanosheets had a large inter-layer spacing. It was therefore concluded that the PEO chains were sandwiched in between the GO nanosheets, while the PVP chains could form aggregates in the aqueous medium, due to their more hydrophobic characteristics. In summary, the formation of the GO/polymer composite hydrogels was influenced by several factors, including C_{GO}, pH value and the concentration and molar mass of the polymers.

Electrostatic interaction (ESI) is another driving force to induce the assembly of G/GO nanosheets induced to assembly [42, 43]. One example was polydimethyldiallylammonium chloride (PDDA), as demonstrated in Fig. 7.6. This alternative cross-linking agent functioned by forming hydrogen bonds. The ESI between the quaternary ammonium groups of PDDA and the carboxyl groups of GO sheets was the dominating driving force for gelation. PDDA had a very low critical gel concentration (CGC) of 0.1 $mg \cdot ml^{-1}$, which was much lower than those of PEO, PVP and HPC, implying that ESI was more effective than those hydrogen bonding cross-linkers. Figure 7.7 (C) shows a SEM image of the GO/PDDA hydrogel, sharing a similar microstructure to that of the lyophilized GO dispersion. Due to the significantly strong ESI between PDDA and the GO nanosheets, a high concentration of PDDA (e.g., 0.5 $mg \cdot ml^{-1}$) could trigger precipitation of the GO nanosheets, which was not observed in the GO/polymer gels obtained through hydrogen bonding.

The strong effect of ESI was also well-demonstrated by using polyethylenimine (PEI, Fig. 7.6). PEI chains contain primary, secondary and tertiary amines, most of which are protonated, thus facilitating both hydrogen bonding and ESI with GO nanosheets. PEI showed an even lower CGC of 0.2 $mg \cdot ml^{-1}$, while concentrations higher than 0.5 $mg \cdot ml^{-1}$ would cause GO precipitation. In addition, small quaternary ammonium salts, such as cetyltrimethyl ammonium bromide (CTAB, Fig. 7.6) and tetramethylammonium chloride (TMAC, Fig. 7.6), can also be used to promote GO gelation. This is different from the case of using small molecules as hydrogen-bonding cross-linkers. This could be attributed to the fact that ESI as a long-range force is generally stronger than that of a hydrogen bond.

Furthermore, CTAB had a lower CGC (0.3 $mg \cdot ml^{-1}$) than TMAC (1.9 $mg \cdot ml^{-1}$), suggesting that the longer the hydrophobic chain, the stronger the interaction with the GO nanosheets would be. It was also found that melamine (Fig. 7.6), that is a strong hydrogen-bond acceptor and multi-charged small molecule in water offered a similar cross-linking ability, with a CGC of as low as 0.3 mg·ml^{-1}. Cross-linking effect has also been observed by using cations, such as Ca^{2+} and Mg^{2+}. The microstructure of the ion-induced hydrogels was found to be similar to those of the GO/polymer hydrogels, as shown in Fig. 7.7 (D). Because the presence of metal ions almost has no effect on the pH value of the GO suspensions, the main driving force to facilitate the assembly of the GO nanosheets was ascribed to the coordination of the metal ions with the hydroxyl and carboxyl groups on the GO nanosheets. Accordingly, the higher the valence of the ions, the stronger the cross-linking effect would be, which has been confirmed by experimental observations.

A strategy has been develop to obtain ultralight chemically converted graphene aerogels with high compressibility [44]. In order to prevent the restacking of graphene nanosheets during the assembly process, a functionalization-lyophilization-microwave treatment approach was strategically adopted. GO nanosheets were functionalized in a controllable way, and were then assembled into monolithic functionalized graphene hydrogel (FGH) in ethylenediamine (EDA) aqueous solution. Finally microwave irradiation (MWI) was employed to remove the functional groups, in order to produce ultralight graphene aerogel (ULGA) that exhibited outstanding elasticity. Such aerogels could find potential applications in the areas of energy dissipation and vibration damping.

Figure 7.8 (a) shows the synthesis steps of the graphene aerogel with extremely low density. Firstly, a basic and weak reducing agent, i.e., EDA, was mixed into the GO colloidal solution to realize the functionalization of the GO nanosheets, so that they would self-assemble into a 3D network. The GO suspension initially became black, then turned sticky and finally converted into hydrogel, as illustrated in Fig. 7.8 (b). It was observed that the final hydrogel experienced almost zero volume shrinkage after gelation. After the lyophilization process, functionalized graphene aerogel (FGA) was derived from the FGH by removing all of the solvent. The FGA was further treated with MWI, thus leading to ULGA.

After being treated with MWI, the aerogel had changed in color from black to metallic gray, while the 3D network was well-retained, which implied that the ULGA was sufficiently strong to survive the MWI treatment. This was readily attributed to the strong interconnection of the graphene nanosheet building blocks. The ULGA had a foam-like structure with interconnected pores with sizes ranging from tens to hundreds of micrometers, as observed in Fig. 7.9 (a, b). The cell walls were comprised of well-assembled graphene nanosheets caused by the freezing process, during which the individual graphene nanosheets were brought together, due to the formation of ice crystals. Because of the large lateral size, the graphene nanosheets can be assembled into sheets with sizes of up to tens of micrometers, as illustrated in Fig. 7.9 (c), which was responsible for the elastic stiffness of the ULGA. Additionally, 1D wrinkles or folded regions along the surface of the assembled graphene nanosheets were observed, as revealed in

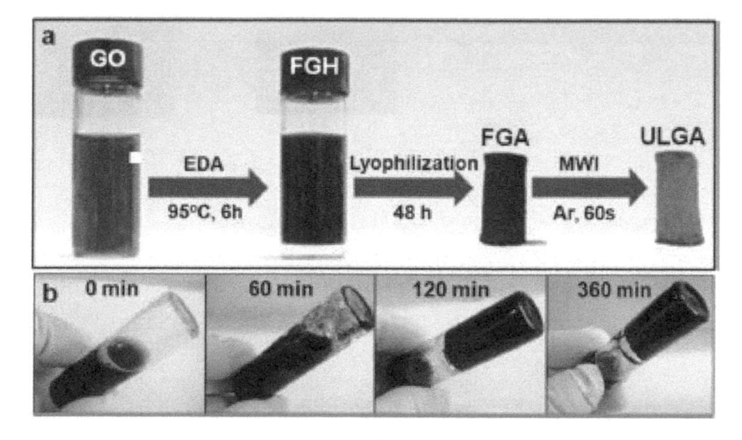

Fig. 7.8. (a) Illustration of the fabrication process of the ultralight graphene aerogel (ULGA). (b) Photographs of the time-dependent formation process of the functionalized graphene hydrogel (FGH). Reproduced with permission from [44], Copyright © 2013, WILEY-VCH Verlag GmbH & Co. KGaA, Weinheim.

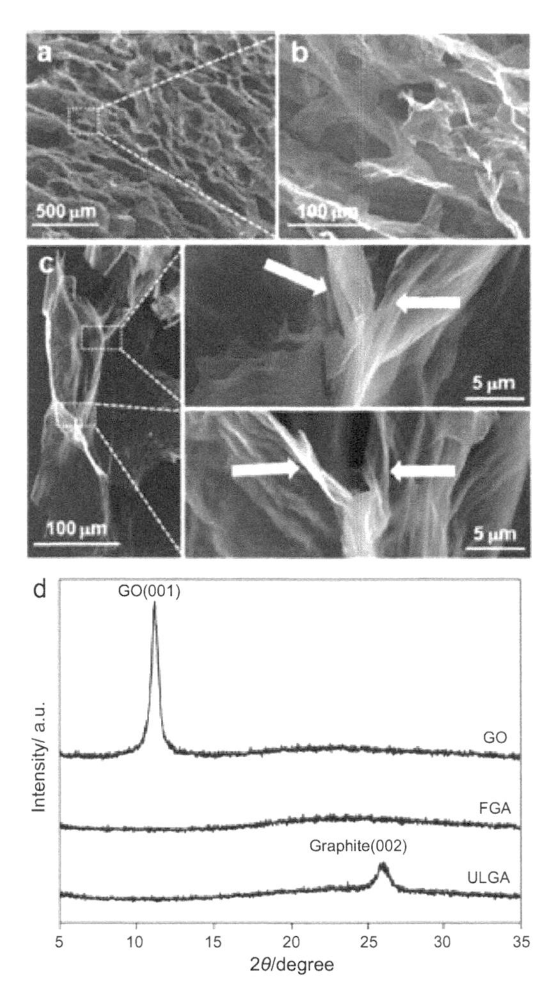

Fig. 7.9. (a, b) SEM images of the ULGA. (c) SEM images of the wall structure of the ULGA. (d) XRD patterns of the GO, FGA and ULGA. Reproduced with permission from [44], Copyright © 2013, WILEY-VCH Verlag GmbH & Co. KGaA, Weinheim.

Fig. 7.9 (c). TEM analysis results indicated the presence of graphitic structure in the folded and crumpled regions, which could be the additional factor resulting in the mechanical strength of the ULGA [45, 46].

Figure 7.9 (d) shows XRD patterns of the samples, which demonstrated the transformation from GO to the final aerogel, together with other analysis results including elemental composition, Raman spectra and XPS [44]. The GO shows a typical diffraction peak at about 11° [47]. The FGA sample had no diffraction peak, which suggested that the GO nanosheets were reduced due to the attachment of the EDA molecules. The ULGA showed a diffraction peak at 25.8°, corresponding to an interlayer spacing of 0.344 nm, which was an indication of the formation of graphitic structures [48]. It was found that the d-spacing in the final graphene aerogel was slightly larger than that of perfect graphite, which could be understood because there was bridge-linking in between the graphene nanosheets, due to the MWI induced release of amines. The initial GO demonstrated a C/O mole ratio of 1.9, while it was increased to 2.3 after the EDA-triggered assembly, implying the reduction of the GO nanosheets to a certain degree. After the MWI treatment, the C/O mole ratio was abruptly increased to 23.0 in the final ULGA sample, suggesting that the functional groups had been almost completely removed.

The as-obtained ULGA was ultralight. The weight loss during the transformation from FGA to ULGA was about 60%, which was much higher than those observed in other processes. In addition, the density of the ULGA could be tailored by controlling the EDA/GO ratio and the initial GO concentration, as plotted

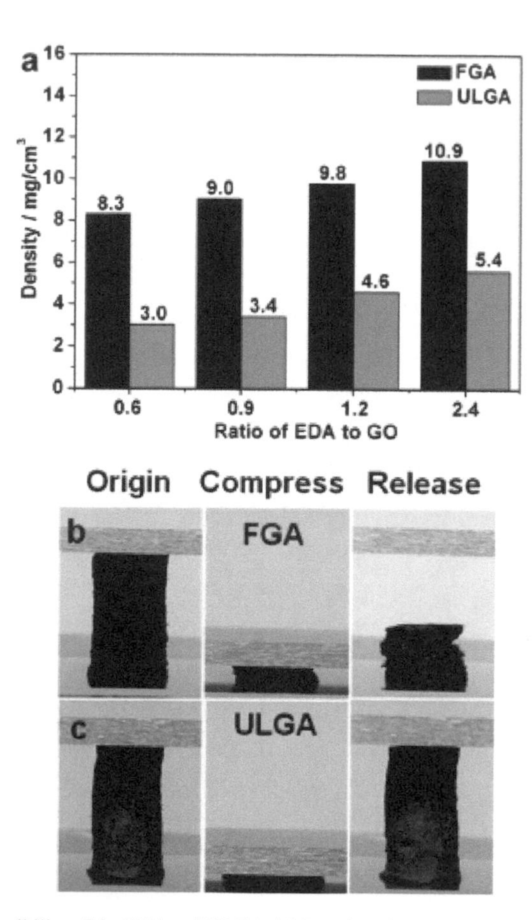

Fig. 7.10. Density and compressibility of the FGA and ULGA. (a) Density of the FGA and ULGA as a function of the EDA to GO ratio. (b, c) Photographs showing the compressibility of the FGA and ULGA, respectively. Reproduced with permission from [44], Copyright © 2013, WILEY-VCH Verlag GmbH & Co. KGaA, Weinheim.

in Fig. 7.10 (a). The density of the FGA can be varied from 8.3 to 10.9 mg·cm^{-3}, as the EDA/GO ratio was increased from 0.6 to 2.4. After the MWI treatment, the density of the aerogels largely decreased in range from 3 to 5 mg·cm^{-3}, which was nearly half of that of the FGA. The corresponding porosity was in the range of 99.7–99.8%. Compressive behaviors of the FGA and ULGA are demonstrated in Fig. 7.10 (b) and (c), respectively. Both FGA and ULGA could be compressed into a thin pellet. As the samples were released, the ULGA recovered nearly completely, whereas the FGA recovered by about 50%.

A special synthetic route was devised to fabricate 3D GO hydrogels and aerogels, through the self-assembly of 2D GO nanosheets that were cross-linked by using layered double hydroxides (LDHs) [49]. Recently, LDHs have formed a new class of multifunctional materials, with inorganic nanolamellar and positively charged frameworks, which have found various potential applications [50–55]. The LDH + GO hydrogels were fabricated through a one-step method, without using any hydrothermal processing or other additives. With an optimal ratio of organic and inorganic components, the Mg–Al LDHs served as a cross-linking agent, which effectively suppressed the restacking of the GO nanosheets. Instead, uniformly cross-linked GO nanosheets were assembled into 3D network structures, through various interactions, including strong hydrogen bonds, and relatively weak electrostatic and cation–π interactions. The GO-based hybrid aerogels possessed strong hydrophilic behaviors and demonstrated extraordinary structural stability in aqueous solutions, thus making them suitable for potential applications as absorbents. As a consequence, the LDH + GO hybrid aerogels exhibited a strong capability to remove dyes and heavy-metal pollutants from water. Due to their well-retained 3D monolithic structures, they could be easily separated, recovered and reused.

Two sets of experiments were conducted to synthesize the LDH + GO hydrogels. In the first experiment, the amount of LDH powder was controlled at 5 mg·ml^{-1}, while the concentration of the homogeneous GO suspension was decreased from 5 to 1 mg·ml^{-1}, so that five samples were obtained with LDH:GO ratios of 1:1, 2:1, 3:1, 4:1 and 5:1. In the second experiment, the GO concentration was fixed at 5 mg·ml^{-1}, whereas the concentration of the LDH was increased from 1 to 5 mg·ml^{-1}, which also led to five samples, with LDH:GO ratios of 1:5, 1:4, 1:3, 1:2 and 1:1. Once the LDHs were added to the GO suspensions, the mixtures were heavily shaken for homogenization and then sonicated for 1 h. The mixtures were kept at room temperature for 1 h to obtain LDH + GO hydrogels, as 3D dark-brown monoliths, which were further freeze-dried to form LDH + GO aerogels.

Photographs of the 3D LDH + GO monoliths before and after the ultrasonic treatment are shown in Fig. 7.11 (a, b). Obviously, the concentrations and the component ratios of GO and LDHs had a significant effect on the formation of the hydrogels. At the constant GO concentration of 5 mg·ml^{-1}, stable hydrogel was only available at LDH:GO ratios of ≥ 1:3, as illustrated in Fig. 7.11 (a). It was found that there was almost no volume shrinkage if a hydrogel could be formed. For the samples with LDH to GO ratios of less than 1:4, monolith gelation was not obtained. During the sonication process, the samples experienced a color change from the original brown to dark, suggesting that the assembly of the GO nanosheets

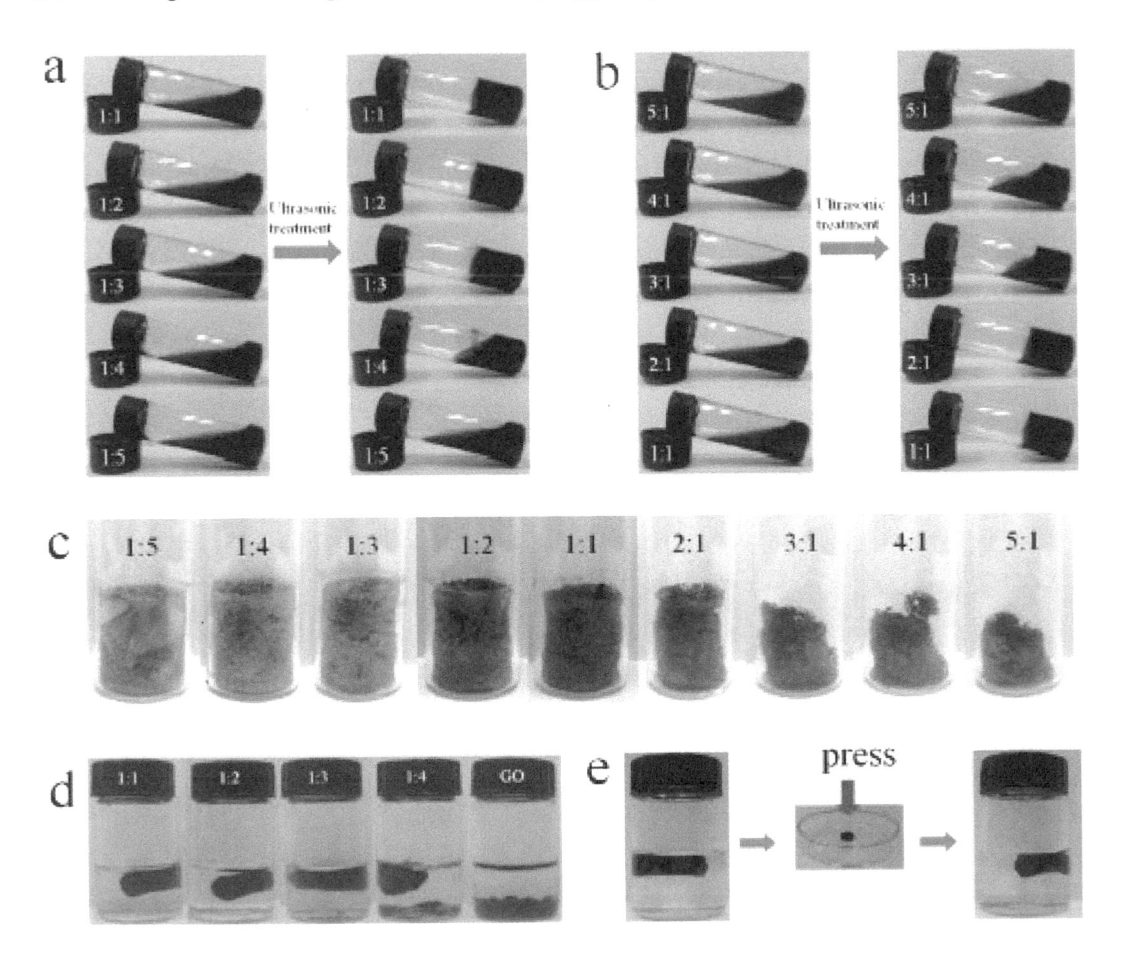

Fig. 7.11. Photographs of the samples demonstrating the formation process of the LDH + GO hydrogels and aerogels. (a) Hydrogels derived from the mixtures of the GO suspension (5 mg·ml^{-1}) and LDH solutions with different concentrations (5 mg·ml^{-1}/1 mg·ml^{-1} for 1:1/1:5 samples). (b) Mixtures of the LDH powder (5 mg·ml^{-1}) and GO dispersions with different concentrations (1 mg·ml^{-1}/5 mg·ml^{-1} for 5:1/1:1 samples). (c) Aerogels from the freeze-dried sonicated mixtures. (d) Aerogel stability in aqueous solutions with mechanical vibration. (e) Water swelling and shape recovery of the pressed 1:1 aerogel in an aqueous solution. Reproduced with permission from [49], Copyright © 2014, The Royal Society of Chemistry.

should be triggered by the interactions between the GO nanosheets and the LDHs. At the constant LDH concentration of 5 mg·ml^{-1}, the formation of the hydrogel followed a similar trend, as demonstrated in Fig. 7.11 (b). Monolithic hydrogel was only formed as the concentration of GO was ≥ 2.5 mg·ml^{-1}, i.e., only the samples with LDH:GO ratios of 2:1 and 1:1 could become hydrogels. In summary, the hydrogel formations in this case were closely related to the concentrations of both the GO and LDH suspensions.

Aerogels were obtained from the hydrogels by using freeze-drying to remove water. Figure 7.11 (c) shows photographs of the aerogels. As the LDH:GO ratio was increased from 1:5 to 5:1, the color of the aerogels varied from dark yellow to dark brown. The samples derived from the mixtures with the LDH:GO ratios in the range from 2:1 to 5:1 had a fluffy morphology and experienced obvious shrinkage. This implied that, if the GO concentration was less than 2.5 mg·ml^{-1}, the number of the GO nanosheets was not sufficient to assembly into a stable 3D network.

GO nanosheets have been assembled into hydrogels or aerogels by using different cross-linking interactions, due to their hydrophilic properties, together with the availability of their homogeneous suspensions. It is these properties that have made GO aerogels vulnerable in aqueous solutions, as seen in Fig. 7.11 (d). Therefore, it is desirable to convert GO gels into graphene aerogels, because the latter is hydrophobic, which can be used for oil and organic solvent sorption. Interestingly, the self-assembled GO aerogels cross-linked by using LDHs exhibited both high hydrophilic characteristics and high stability in aqueous solutions. For instance, the LDH + GO aerogels with ratios of 1:1 to 1:3 remained intact as they were immersed in the aqueous solution, whereas the 1:4 LDH + GO and pure GO aerogels were structurally weak, as displayed in Fig. 7.11 (d).

The LDH cross-linked GO aerogels exhibited high plasticity, as seen in Fig. 7.11 (e). Water-saturated LDH + GO aerogels could be deformed into circular flakes, by removing most of the water. More significantly, as the deformed sample was immersed in water solution, it swelled to the original shape without structural crashing. This observation implied that the GO nanosheets were linked so strongly that the linkage was not broken by the external compression, while the shrinkage of the aerogels was simply the deformation of the GO nanosheets. As a result, the original shape could be readily recovered in the aqueous solution.

Figure 7.12 shows representative SEM images of the hydrogels, indicating that they consisted of porous randomly oriented 3D GO frameworks. Micro-scale morphology of the LDH + GO aerogels had a close relation to the concentration of the LDHs. It was found that the higher the concentration of the LDHs, the

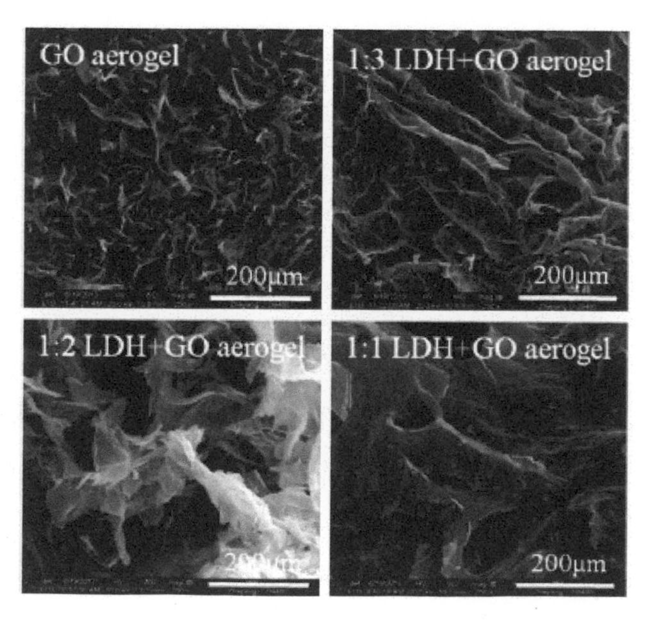

Fig. 7.12. SEM images of the pure GO aerogel and LDH + GO aerogels with various ratios of LDH and GO. Reproduced with permission from [49], Copyright © 2014, The Royal Society of Chemistry.

larger the assembled GO layers would be. For example, the pure GO aerogels contained numerous smaller GO pieces with small-sized fractals, whereas the presence of the LDHs obviously led to larger pieces and the size was increased with the increasing amount of LDHs. For instance, in the 1:1 LDH + GO aerogel, large pieces of flat sheets were observed. Therefore, it could be concluded that the LDHs had played an important role in the assembling of the GO nanosheets into hydrogels with strong 3D network structures.

Particles had been observed on the surface of the GO pieces in the LDH + GO aerogels, especially for those with high contents of LDHs, e.g., the 1:1 sample, as demonstrated in Fig. 7.13. High magnification image (Fig. 7.13 (a)) indicated that the particles were residual LDHs, whose morphology was same as that of the pure Mg–Al LDH powders. However, no impurities were observed in the 1:3 LDH + GO aerogel, due to the low content of LDHs, as illustrated in Fig. 7.13 (b). In other words, the excessive LDHs were simply agglomerated and deposited on the surface of the hydrogels. If all the LDHs were involved in the GO assembly, particles would be observed in the samples.

Additive-free GO hydrogels have been prepared by tuning the pH value [29] and using ultrasonication of the GO aqueous suspensions [31]. As mentioned earlier, the ionized carboxyl groups on the edges allowed the GO nanosheets to be dispersed in water through the electrostatic repulsion among them. In this case, acidification of the GO suspension would weaken the repulsion force, due to the protonation of the carboxyl groups. It has been confirmed that GO nanosheets became unstable in a strong acidic aqueous environment, due to the insufficient mutual repulsion forces [29, 56]. As the GO concentration was sufficiently high, hydrogel could be formed, instead of the presence of an amorphous precipitation [29]. More recently, it was found that ultrasonication could be used to convert aqueous GO suspensions into hydrogels, which had much smaller channels as compared with the aerogels derived from the parent GO suspension [31]. The sonication fractured the GO aggregates into smaller fragments, so that new sheet edges were present, but without the stabilizing carboxyl groups. This surface chemistry change triggered gelation of the GO suspensions. More importantly, such GO hydrogels exhibited very low critical gelation concentrations (CGC), in the range of 0.050–0.125 mg ml^{-1}, which was dependent on the power of the ultrasonic treatment.

Purified GO aqueous suspensions (50 ml) were diluted to a concentration of 2 mg ml^{-1} and sonicated with a probe sonicator set at 30% amplitude with alternating pulse and rest periods of 10 s each for different time durations (30–120 min). GO nanosheets are electrostatically stable in water due to the edge-bound carboxylic acid moieties. The aqueous dispersibility of GO nanosheets can be further enhanced by oxidizing the hydroxyl moieties into carboxyl groups on the basal planes. It was hypothesized that fracturing of the GO nanosheets to smaller fragments without the carboxylic acid groups along their freshly exposed edges

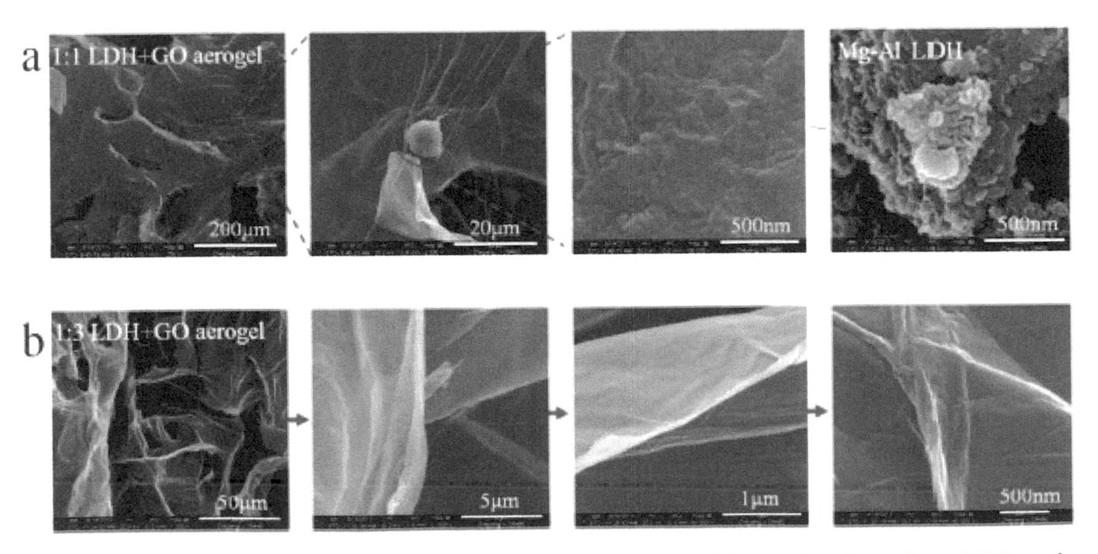

Fig. 7.13. SEM images of representative samples: (a) LDHs on the 1:1 LDH + GO aerogel surface and pure LDH powder and (b) 1:3 LDH + GO aerogel surface, without visible LDH particles. Reproduced with permission from [49], Copyright © 2014, The Royal Society of Chemistry.

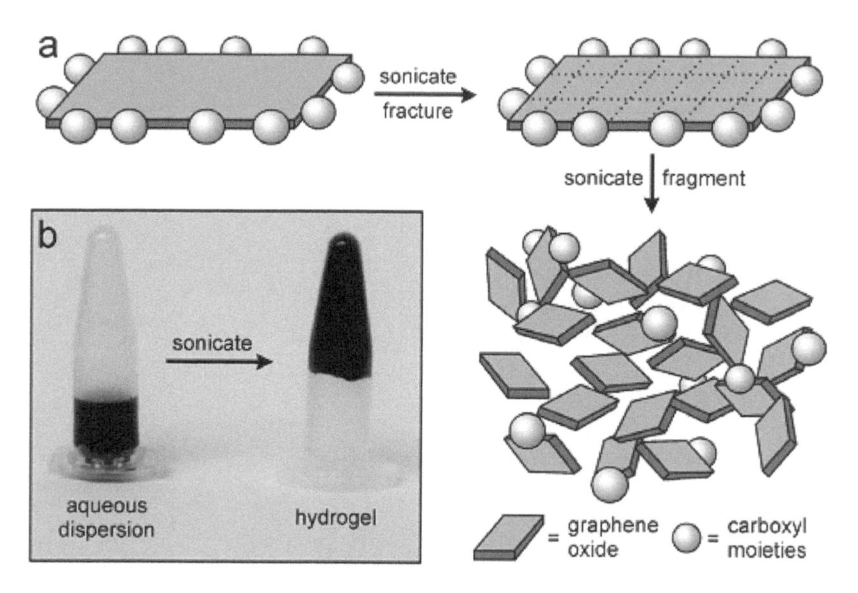

Fig. 7.14. (a) Schematic diagram illustrating the fracture and fragmentation of the GO nanosheet due to the ultrasonication, thus decreasing the coverage of carboxyl moieties (spheres) at the edges of the GO nanosheet fragments responsible for the gelation. (b) Photograph demonstrating the conversion of the GO aqueous suspension (left) into hydrogel (right) after the sonication treatment. Reproduced with permission from [31], Copyright © 2012, Elsevier.

would lead to a decrease in the dispersion capability, while irreversible aggregation was not induced, as schematically demonstrated in Fig. 7.14 (a).

After ultrasonication for 30 min, the dilute GO aqueous dispersion of 2 mg ml^{-1} was, gelated, as illustrated in Fig. 7.14 (b). Dynamic light scattering (DLS) results indicated that the average hydrodynamic diameter of the GO nanosheets was decreased from 590 to 250 nm after the ultrasonic treatment. As the sonication duration was increased to 120 min, stronger gel was formed. At the same time, hydrodynamic diameters of the GO nanosheets were further decreased to 80 nm. Although the lateral dimension of the GO nanosheets was decreased due to the ultrasonic treatment, zeta potentials of the GO suspension and the resultant hydrogels were kept almost constant at 56 ± 4 mV, which was still above the value of 30 mV required to prevent particle aggregation [26]. This suggested that the ultrasonic treatment caused negligible loss of the negatively charged functional groups. Therefore, it was more likely that the gelation was attributed to the fragmentation of the GO nanosheets that created new unfunctionalized edges which destabilized the suspensions.

In addition to the reduction in lateral size, surface chemistry of the GO nanosheets was also altered by the sonication treatment, which could be another factor to trigger the gelation of the suspensions. However, spectral characterization results, including XPS, FT-IR and NMR, as shown in Fig. 7.15 (a–c), indicated that no obvious change was observed in the chemical composition of GO after sonication for 120 min. This further confirmed that the destabilization of the GO nanosheets was caused by reduced carboxyl coverage at the edges of the newly formed fragments. Driving forces for the gelation included van der Waals attraction, hydrogen bonding between the functional groups and π–π stacking of the residual sp^2-hybridized regions on the GO nanosheet basal plane. The hydrogels were fragile in nature, suggesting the weak balance between the hydrophobic interactions and hydrogen bonding.

The GO hydrogels could be converted into aerogels through lyophilization. SEM images of the samples before and after sonication for 120 min are shown in Fig. 7.15 (d, e), illustrating their loose porous 3D networks of the bridged GO nanosheets. Due to the absence of the chemical cross-linkers, this ultrasonic method resulted in hydrogels and aerogels that could find potential applications in storage and delivery, especially as far as sensitive cargos are concerned.

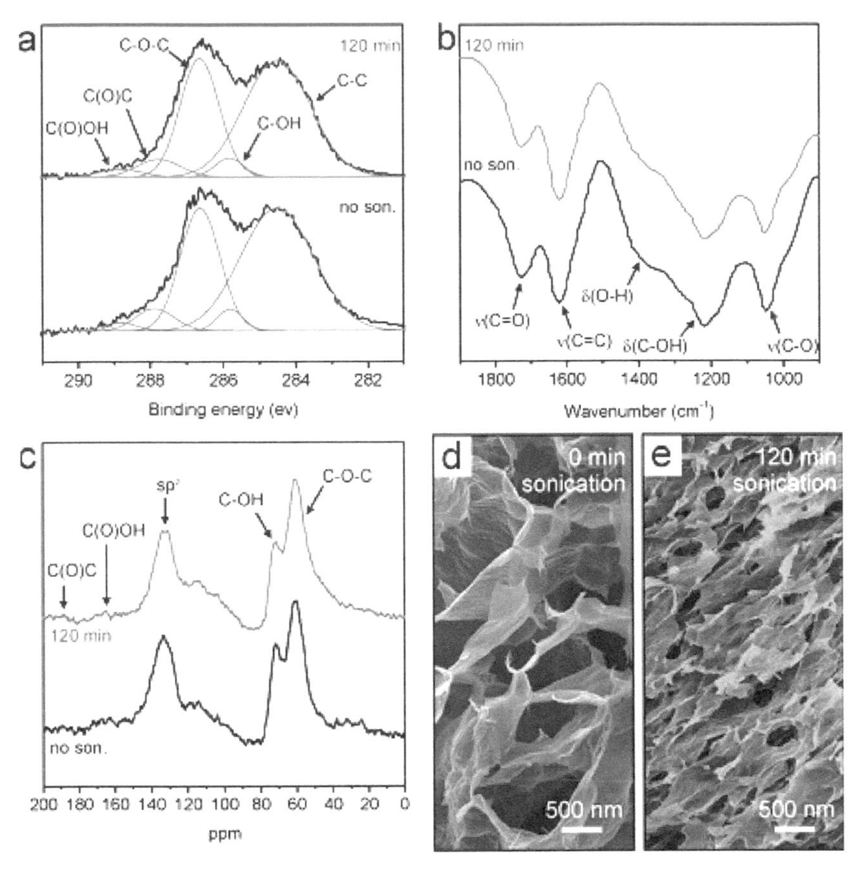

Fig. 7.15. XPS (a), FTIR (b) and solid-state ^{13}C NMR (c) spectra of the lyophilized GO before (black) and after (gray) ultrasonic treatment for 120 min. (d) SEM image of the lyophilized GO dispersion without sonication showing the porous structure of the cross-linked GO nanosheets. (e) SEM image of lyophilized hydrogel with 120 min of ultrasonication demonstrating smaller pores. Reproduced with permission from [31], Copyright © 2012, Elsevier.

7.2.3. Centrifugal evaporation-induced assembly of GO

Vacuum centrifugal evaporation has been demonstrated to be a strategy to fabricate additive-free GO hierarchical structures [57]. It can be used to produce large-scale GO sponges with a 3D interconnected network structure. The combination of vacuum evaporation and centrifugal force sped up the solvent evaporation while retaining the sample shape and morphology due to the effect of foaming or bumping. Once 1 ml GO colloidal suspension was formed in the 2 ml Eppendorf tube (Fig. 7.16 (a)), the evaporation process was initiated in a speed vacuum concentrator, so that possible sedimentation of the GO nanosheets was effectively prevented. The concentration of the GO colloidal suspension varied in range from 2 to 5 mg ml^{-1}, at temperatures of 40, 60 and 80ºC. At the low temperature of 40°C, the GO nanosheets were assembled into sponges with 3D porous networks. The driving forces for the assembly were the van der Waals forces, because during the water evaporation the outward centrifugal force was dominant over the upward evaporation force. After thermal annealing in H_2/Ar (50 cc·min^{-1}) at 800ºC for 12 h, the GO sponges were reduced to graphene sponges, with a large surface area and porous structure.

Figure 7.16 shows the formation process of the GO sponges and thin sheets. The experiments were started from homogeneous suspensions of single-layered GO nanosheets. The suspensions were placed on the rotor plate fixed in the vacuum chamber, with the vacuum centrifugal evaporator being operated at a rotating speed of 1800 rpm at a vacuum level of 10^{-4} torr. The as-formed GO sponges could directly collected

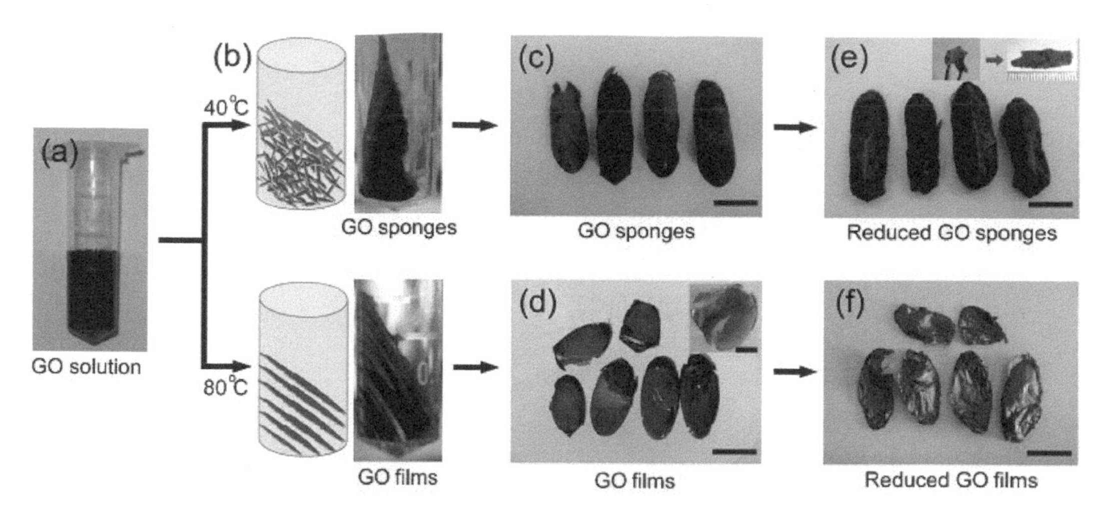

Fig. 7.16. Synthetic process of the fabrication of GO sponges and thin sheets by using the rapid vacuum centrifugation. (a) 1 ml GO solution (3 mg ml^{-1}) in a 2 ml microtube. (b) Schematic diagram showing the production of the GO sponges and films at different temperatures, each with a photograph of the sample. (c) Photographs of GO sponges. (d) Free-standing GO thin sheets, with the inset showing a photograph of the GO thin sheet with a diameter of 3 cm. (e) Photographs of the rGO sponges derived from the GO sponges after thermal annealing in H_2/Ar at 800°C for 12. (f) Free-standing graphene thin sheets. Scale bars = 1 cm (c–f). Reproduced with permission from [57], Copyright © 2010, WILEY-VCH Verlag GmbH & Co. KGaA, Weinheim.

from the microtube by using a tweezer, they were brown in color and of a cotton-like soft morphology, as demonstrated in Fig. 7.16 (c). Higher temperatures at 60 and 80°C led to a sponge-sheet mixtures and sheets, respectively. The sponge and film morphologies were well-retained after thermal reduction, with graphitic color and metallic luster, as seen in Fig. 7.16 (e, f). The rGO sponge exhibited high mechanical strength, as illustrated as the inset in Fig. 7.16 (e), which was obviously ascribed to the strong network structure formed by the GO nanosheets.

SEM images of the GO sponges are shown in Fig. 7.17 (a–d). Close-up view inside the GO sponges indicated that the individual GO nanosheets, with sizes in the range from 200 nm to 5 μm, were cross-linked to form GO assembled network with a size of about 30 μm. The network was constructed through mainly edge-to-edge and partially edge-to-surface interactions, as observed in Fig. 7.17 (b). In the near inner wall regions of the GO sponges, there were GO nanosheets that were parallelly stacked, with a relatively uniform thickness and narrow size distribution, caused by the centrifugal force (Fig. 7.17 (c)). After thermal reduction, porosity of the GO sponges was reduced, while the porous structure somehow collapsed, due to removal of the functional groups. Nevertheless, the randomly connected morphology of the GO sponges was still retained, while the large surface area was not destroyed, as illustrated in Fig. 7.17 (d).

Figure 7.17 (e) shows a HRTEM image of the graphene nanosheet isolated from the rGO sponges. Because the GO nanosheets were initially connected to one another mainly through edge-to-edge connection, the open side of GO nanosheets was prone to be scrolled during the thermal reduction process, thus leading to a highly folded structure at the edges. Figure 7.17 (f) shows the selected area electron diffraction (SAED) pattern of the graphene nanosheet. Ordered crystal structure of graphene was clearly observed, where the relative intensities of the inner peaks of the six-fold symmetry diffraction pattern were much stronger than those of the outer peaks, confirming that monolayer graphene was present in the rGO sponges. Such graphene sponges, with large surface area and highly porous structure, could find applications in various areas, including gas sensors, catalysis and composite materials.

7.2.4. *In Situ* gelation of rGO

3D rGO architectures would have more functions as compared with their GO counterparts, due to the recovered aromatic structure and thus enhanced conductivity. One of the simplest methods to develop 3D rGO architectures is the self-assembly of rGO *in situ*. The reduction of GO increases the van der Waals

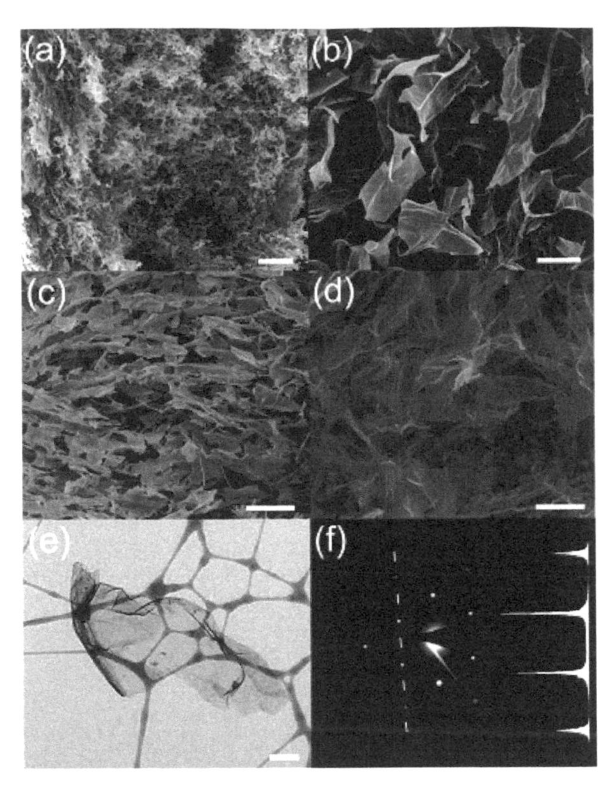

Fig. 7.17. (a) Low- and (b) high-magnification SEM images of the inner wall region of the GO sponges. (c) SEM of the outer region of the GO sponges. (d) SEM image of the rGO sponges. (e) TEM image of the single-layered graphene nanosheet isolated from the rGO sponges by using sonication. (f) Selected area electron diffraction (SAED) pattern taken at the periphery of the graphene nanosheet in (e). The scale bars = 100 mm (a, c), 10 mm (b), 50 mm (d) and 500 nm (e). Reproduced with permission from [57], Copyright © 2010, WILEY-VCH Verlag GmbH & Co. KGaA, Weinheim.

forces between adjacent graphene basal planes, resulting in rGO gelation or precipitation. As stated earlier, the force balance between electrostatic repulsion and inter-planar van der Waals interaction dominates the formation of rGO gels. Various reduction methods have been employed to reduce GO and trigger *in situ* gelation of rGO hydrogels.

7.2.4.1. Hydrothermal reduction of GO

Hydrothermal or solvothermal reduction has been the most widely used method for such a purpose [18, 58–61]. For example, 3D rGO interconnected networks were fabricated by using a hydrothermal process with the aid of noble-metal nanocrystals (Au, Ag, Pd, Ir, Rh, Pt, etc.) [59]. The hydrothermal treatment was conducted with GO suspensions (1.0 mg ml^{-1}), together with noble-metal salt and glucose in a Teflon-lined stainless-steel autoclave. The as-synthesized 3D microstructures were then freeze-dried in order to have good mechanical strength. Microstructures of the 3D structures, including porosity and pore size, were controlled by adjusting the GO concentration.

Figure 7.18 (a) shows AFM image of a representative GO nanosheet, with its corresponding height profile being shown in Fig. 7.18 (b). The GO nanosheet had a thickness of about 0.7 nm, while its edge was slightly thicker due to the presence of crinkles. Photographs of representative 3D self-assembled samples are shown in Fig. 7.18 (c–e). The cylindrical shapes of the samples resemble that of the vessel, which means that the size of the samples could be simply controlled by the volume of the container. More significantly, the size of the 3D self-assembled sample was only determined by the volume of the GO suspension, but not by the concentration. The 3D cylinder samples had a high mechanical strength, as illustrated in Fig. 7.18 (f–h). The exact mechanical properties of the sample were additionally tested

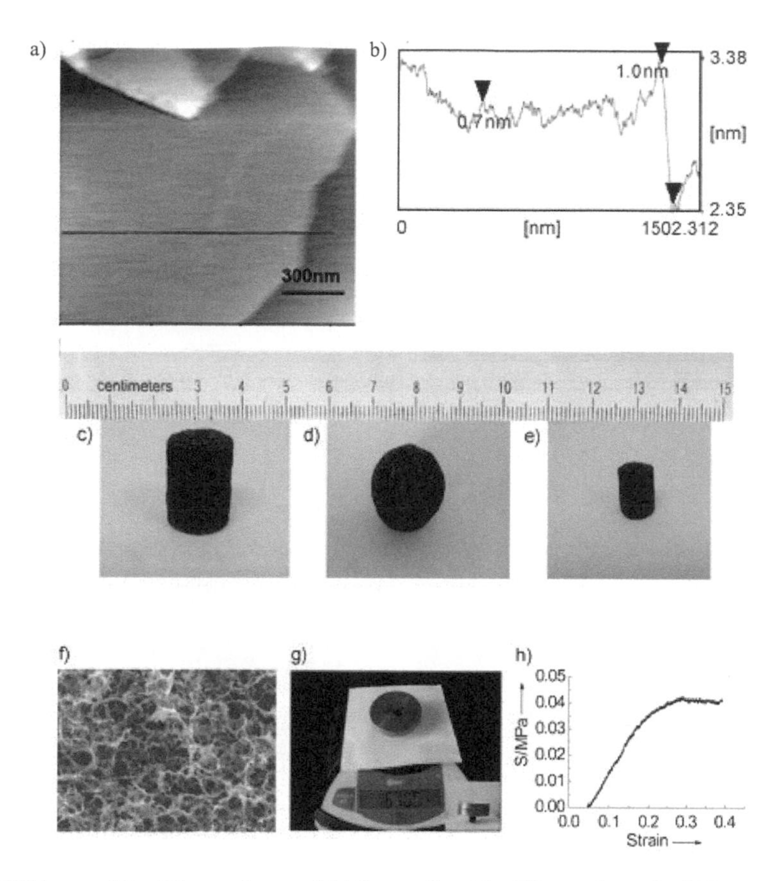

Fig. 7.18. (a) AFM image of the GO monolayer and (b) line profile of the GO monolayer. (c–e) Photographs of the freeze-dried 3D structures prepared by the self-assembly of GO with the aid of noble metals: (c) the front view and (d) bottom view of the sample derived from 35 ml GO suspension. (e) Front view of the sample from 20 ml GO suspension. (f) SEM image of the sample in (c). (g) Photograph showing an iron plate weighing 167 g placed on the 3D cylinder, with the inset showing the sample supporting two iron plates. (h) Strain–stress curve of the sample as determined from a compressive test (S = stress). Reproduced with permission from [59], Copyright © 2010, WILEY-VCH Verlag GmbH & Co. KGaA, Weinheim.

by a universal testing machine. Figure 7.18 (h) shows the strain–stress curve of the sample. It possessed compressive strength and compress modulus of about 0.042 MPa and 0.26 MPa, respectively. The sample was electrically conductive, with an average conductivity of about 2.5×10^{-3} S·cm^{-1}.

Representative SEM images of the samples with Pd are shown in Fig. 7.19 (a–d), demonstrating clearly their porous structure. Similar porous structure was observed in the samples with other noble metals, i.e., Ag, Au, Ir, Pt and Ru, as seen in Fig. 7.19 (e–h). The pores were uniformly arranged, with an average diameter of about 2 μm, as the GO suspension was 1 mg ml^{-1}. It was found that the pore diameter was increased tremendously to 6 μm, as the GO suspension was decreased to 0.6 mg ml^{-1}, as demonstrated in Fig. 7.19 (c, d), which meant that the pore structure and size of the hydrogel samples could be simply controlled by controlling the concentration of the GO suspensions.

If glucose was used to replace the noble metals, 3D assembly of the GO was not observed, instead, only powdery samples were obtained. This implies that the noble metals were indispensable to trigger the assembly process. The important role of the metal particles in forming the 3D self-assemble porous structure was further confirmed by dissolving Pd nanoparticles in aqua regia. Once the Pd nanoparticles were removed, the regular shape of the cylinder was damaged. Therefore, it was concluded that the Pd nanoparticles took part in the formation of the framework of the 3D cylinder.

A simple yet effective approach has been reported to produce nanoscale pores on graphene nanosheets by activating the graphene aerogels (GA) with H_3PO_4 [62]. Due to its mild characteristics, H_3PO_4 could be

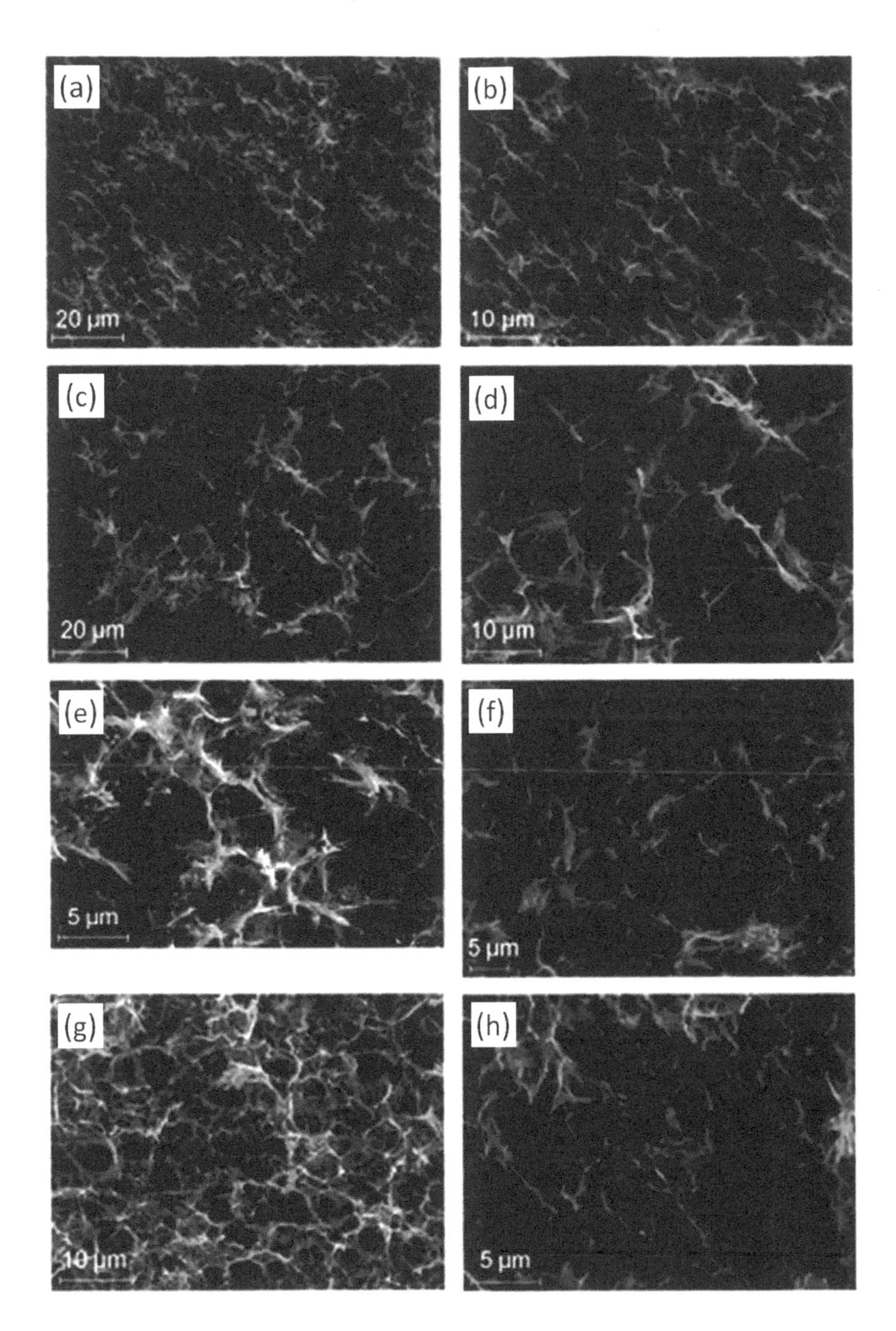

Fig. 7.19. (a, b) SEM images of the sample derived from the GO suspension with a concentration of 0.1 mg ml^{-1}. (c, d) SEM images of the sample obtained from the 0.07 mg ml^{-1} GO suspension. SEM images of the 3D structures made with different metals: (e) Au, (f) Pt, (g) Ir and (h) Ru. Reproduced with permission from [59], Copyright © 2010, WILEY-VCH Verlag GmbH & Co. KGaA, Weinheim.

used to generate nanopores, while preventing the severe corrosion to the experimental facilities and thus mitigating the environmental problems. As schematically depicted in Fig. 7.20 (a), the hydrothermally derived GAs in the presence of urea were impregnated with H_3PO_4 solution of different concentrations (Fig. 7.20 (b, c)). The GAs were then subject to the activation treatment at 800°C in flowing N_2 gas for

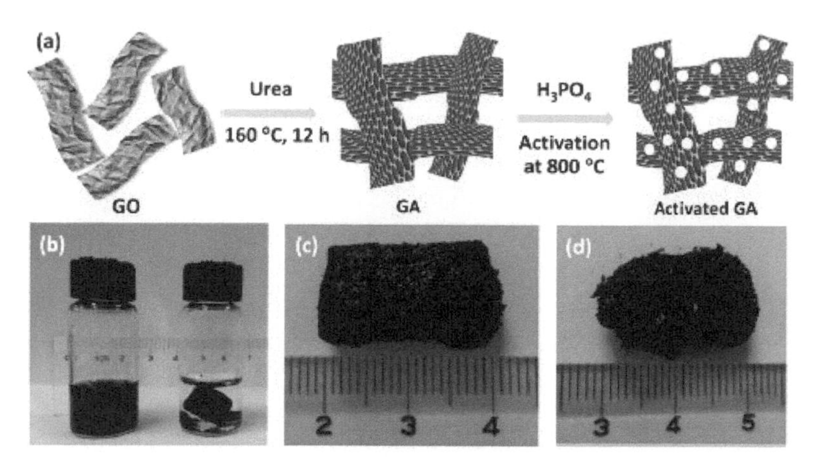

Fig. 7.20. (a) Schematic diagram showing the formation process of the aGA monoliths with H_3PO_4. (b) Photographs of the GO dispersion (1 mg ml^{-1}) before and after hydrothermal reduction with urea. Photograph of the as-obtained GA (c) and aGA-0.5 (d). Reproduced with permission from [62], Copyright © 2015, Elsevier.

90 min. The activated GA (aGA) (Fig. 7.20 (d)) retained the macroscopic morphology of the original aerogel (Fig. 7.20 (c)), while exhibiting an enormously enlarged specific surface area (SSA). The aGA possessed extraordinary electrochemical performance when used as the electrodes of supercapacitors. They were much superior in respect to the pristine GA, in terms of specific capacitance, rate capability and other parameters.

GO suspension with a concentration of 4 mg ml^{-1} was used for the experiments. Monolith GA was first obtained by hydrothermal gelation of GO with the aid of urea. For example, 0.85 g urea and 10 ml GO suspension (4 mg ml^{-1}) were added into 30 ml water to form a diluted GO dispersion of 1 mg ml^{-1}. After stirring for 1 h, the suspension was hydrothermally treated at 160°C for 12 h. The products were thoroughly washed with water until the pH value of the filtrate was close to 7. They were then freeze-dried for 24 h to obtain free-standing GA monoliths. The GA monoliths were impregnated with H_3PO_4 solution at concentrations of 0.2, 0.5, 1.0 and 1.2 M. After that, the samples were dried and then activated at 800°C for 90 min in flowing N_2. The activated GA samples were labeled as aGA-x, with x standing for the concentration of the H_3PO_4 solution used in the impregnation experiments. In comparison, the GA sample without the H_3PO_4 impregnation was thermally annealed at 800°C, and was denoted as aGA-800.

Because urea is an efficient reducing agent [63], the presence of urea in the hydrothermal treatment promoted the reduction of GO along with the gelation process [64]. At the same time, the urea also played a significant role in the formation of the monolithic GA at a low GO concentration, as confirmed in Fig. 7.20 (c). This was confirmed by the control experiment, in which the same hydrothermal reaction resulted in black and flocculated precipitates. Figure 7.21 (a) shows SEM image of the GA sample, which indicated that the GA exhibited a highly porous architecture, consisting of graphene nanosheets, with pore sizes in the range from sub-micrometer to several micrometers. SEM images of the aGA after activation with H_3PO_4 at different concentrations are shown in Fig. 7.21 (b–f). Obviously, the thermal activation had almost no effect on the porous microstructure of the GA monoliths. Furthermore, XRD broad diffraction peaks centered at near 25° suggested that the graphene layers in both the GA and aGA remained intact. It was observed that too high H_3PO_4 concentration (1.2 M) led to samples (i.e., aGA-1.2) with contaminations, as indicated by the arrows in Fig. 7.21 (f).

The GA monoliths consisted of multilayered graphene nanosheets that were partially overlapped or coalesced due to the deoxygenation during the hydrothermal reduction of GO with urea. TEM observation results indicated that the thermally activated graphene nanosheets had a large porosity, with pore sizes in the range of 1–5 nm. The oxygen-containing functional groups are thermally unstable, and can be easily taken away as H_2O vapor and CO_2 gas by thermal annealing at sufficiently high temperatures, so that pores are formed in between the compact graphene nanosheets [17]. After activation with H_3PO_4, the porosity was

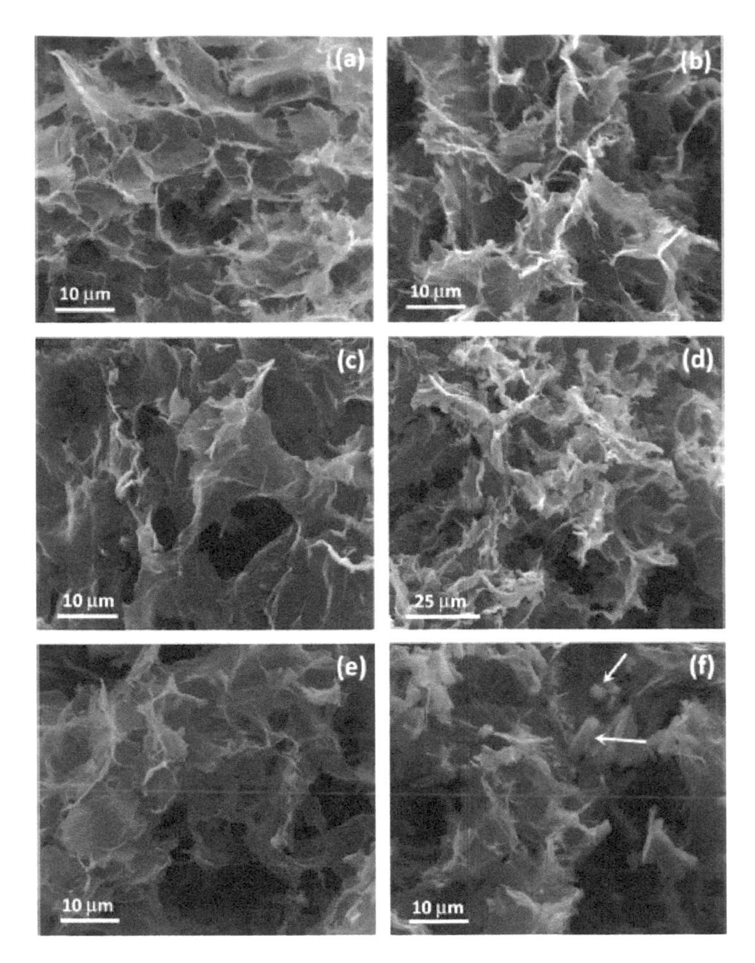

Fig. 7.21. SEM images of the as-synthesized GA (a) and aGA samples activated with H_3PO_4 at different concentrations: (b) 0 M, (c) 0.2 M, (d) 0.5 M, (e) 1.0 M and (f) 1.2 M. Reproduced with permission from [62], Copyright © 2015, Elsevier.

increased significantly, while the pore sizes were also slightly increased. The reaction of H_3PO_4 molecules with carbon atoms is given by: $4H_3PO_4 + 10C \rightarrow P_4 + 10CO + 6H_2O$ [48], which is thermodynamically favorable at temperatures of above 750°C. Therefore, etching of carbon atoms by the H_3PO_4 molecules led to the formation of the nanopores.

A simple hydrothermal method was developed to prepare rGO aerogels with variable and controllable surface wettability by using different amino acids as both cross-linkers and reducing agents [65]. For example, the cys-rGO aerogels obtained in the presence of L-cysteine (L-cys) were hydrophobic with extraordinary adsorption capacity to treat oil and organic solvent contaminated water, while the lys-rGO aerogels derived from L-lysine (L-lys) were hydrophilic, so as to be potential adsorbents to remove heavy metal ions from water.

The cys-rGO hydrogels were hydrothermally derived from GO suspension and L-cys with weight ratios of GO to L-cys ($r_{G/C}$) in the range of 1:0.25–1:2. Well-defined hydrogel was available as the $r_{G/C}$ was between 1:0.5 and 1:2. The hydrogels had a 3D cylindrical morphology which conformed to that of the vessel. The physical size of the obtained hydrogels was not affected by $r_{G/C}$. For instance, the sample with $r_{G/C}$ = 1:2 was about 15 mm in diameter and 25 mm in height, as shown in Fig. 7.22 (a). The dehydrated hydrogels after freeze-drying experienced almost no shrinkage, while the derived cys-rGO aerogels exhibited a well-retained 3D monolithic structure with a density of 13.5 mg·cm^{-3}. Figure 7.22

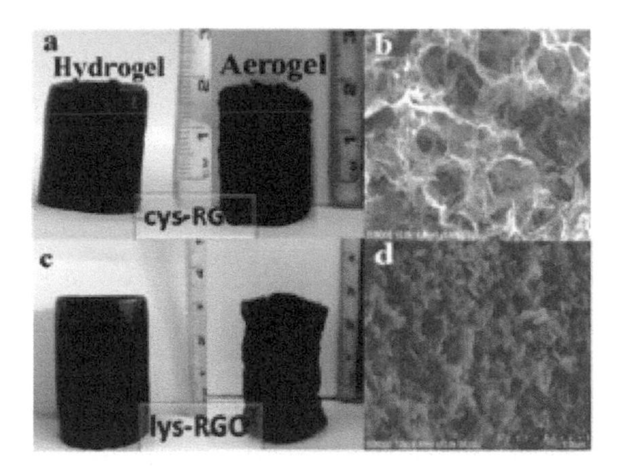

Fig. 7.22. (a) Photographs of the cys-rGO hydrogels and the corresponding aerogels. (b) SEM images of the cys-rGO aerogels. (c) Photographs of the lys-rGO hydrogels and the corresponding aerogels. (d) SEM image of the lys-rGO aerogel. Reproduced with permission from [65], Copyright © 2014, The Royal Society of Chemistry.

(b) shows SEM image of the cys-rGO aerogel. The GO nanosheets were interconnected to form the 3D porous network with pore sizes in the range of 5–20 μm. According to BET nitrogen adsorption isotherm, cys-rGO aerogel had a typical type II adsorption–desorption isotherm, with a BET specific surface area of 35.67 $m^2 \cdot g^{-1}$. XPS analysis results indicated that the rGO nanosheets in the cys-rGO aerogels were cross-linked through a disulfide bond.

Similar results were obtained when L-lysine was used as the cross-linker, with the sample of $r_{G/L}$ = 1:2 shown in Fig. 7.22 (c). The lys-rGO aerogels had an ultralow density of 3 $mg \cdot cm^{-3}$. Different from the cys-rGO aerogels, the lys-RGO aerogels consisted of interconnected and cross-linked crumpled rGO nanosheets, so that the 3D porous structure contained slit-shaped pores. As a result, the lys-rGO aerogels appeared to be fluffy as compared with the cys-rGO aerogel, as revealed in Fig. 7.22 (d). This difference implied that the microstructure of the rGO aerogels could be varied by using different amino acids as the cross-linkers. However, due to its fluffy morphology and low density, lys-rGO aerogel had a relatively low thermal stability that was confirmed by TGA results.

The formation mechanism of the cys-rGO aerogel and lys-rGO aerogel is schematically described in Fig. 7.23. The mixture of L-cys and GO had a pH ≈ 3.8, so that the L-cys was in the form of a cation with one positive charge with a protonated amino group, along with zwitterion in the solution. Most of the carboxyl groups on the GO nanosheets were deprotonated at this pH value, due to the low isoelectric point (IEP) of GO. In this case, the electrostatic force and the hydrogen bonds between L-cys and GO, together with the π–π stacking in between the GO nanosheets, triggered the gelation of the GO nanosheets, resulting in 3D hydrogels. During the hydrothermal treatment, GO was reduced to rGO by the L-cys. Comparatively, the L-lys was more basic, so that the mixture had a pH ≈ 10.8. Therefore, the L-lys was present as an anion with a deprotonated carboxyl group. Meanwhile, a small number of the L-lys molecules were in the form of zwitterion. Therefore, the electrostatic attraction between the protonated amino group on the side chain of lysine and the deprotonated carboxyl group on the GO surface, as well as the hydrogen bonds between the amino groups in lysine and the carboxyl groups of GO and the π–π stacking, could also facilitate the gelation of the GO nanosheets. Because L-lys was larger than L-cys, the lys-RGO aerogel was larger in physical dimension and had a lower density than the cys-RGO aerogel.

According to the test of contact angles, the cys-rGO aerogels were hydrophobic, whereas the lys-rGO samples were hydrophilic. Moreover, contact angles of the prepared cys-rGO aerogels were in the range of 98.8–153.0°, with different $r_{G/C}$ ratios. The sample with $r_{G/C}$ = 1:2 showed a superhydrophobic behavior, with a contact angle of 153.0°, as shown in Fig. 7.24 (a). For lys-RGO aerogels, the contact angle was decreased from 84.0° to 34.7°, as the $r_{G/L}$ was increased from 1:0.5 to 1:2, with an example demonstrated in Fig. 7.24 (b). In this regard, the surface wettability of the rGO aerogels could be readily determined by

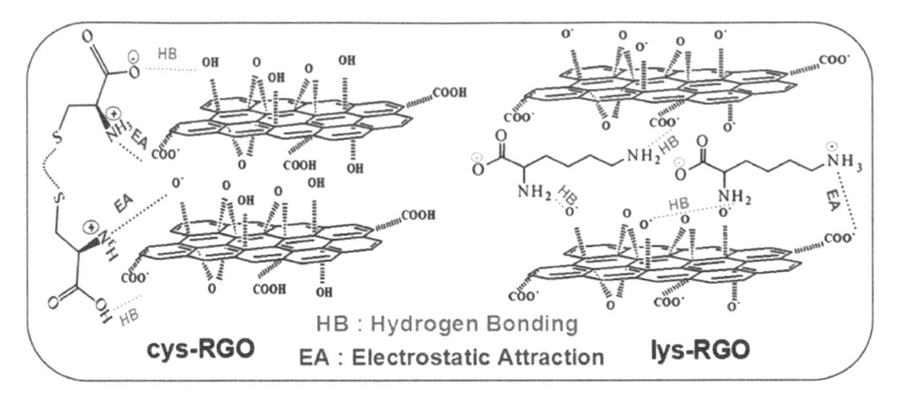

Fig. 7.23. Interactions between GO and L-cysteine or L-lysine. Reproduced with permission from [65], Copyright © 2014, The Royal Society of Chemistry.

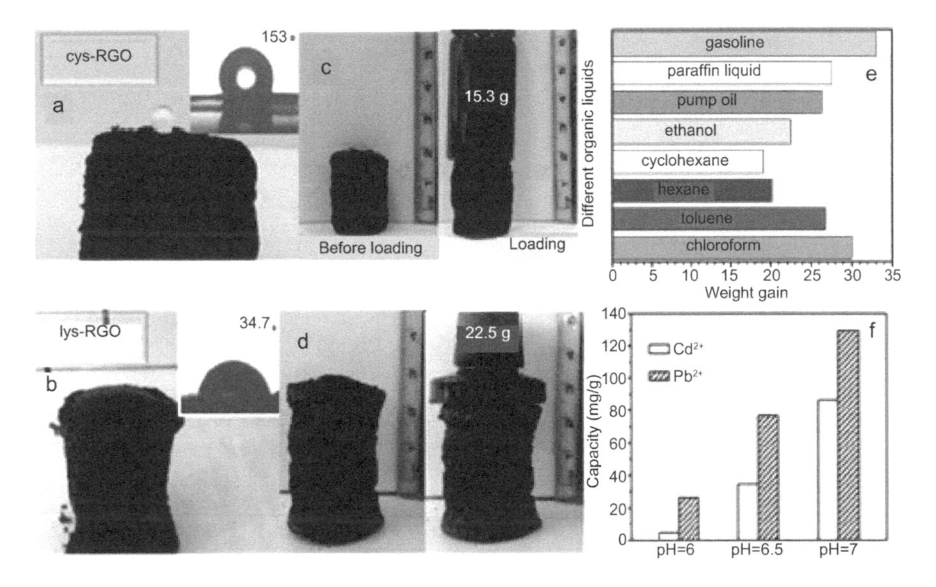

Fig. 7.24. Photograph of a water droplet supported on (a) cys-rGO and (b) lys-rGO aerogels, with the inset showing the water contact angle of the aerogel. Photographs of (c) cys-rGO and (d) lys-rGO aerogels. (e) Adsorption capacities of the cys-rGO aerogel towards various organic solvents in terms of their weight gain. (f) Adsorption toward Cd^{2+} and Pb^{2+} over the lys-rGO aerogels at different pH values. Reproduced with permission from [65], Copyright © 2014, The Royal Society of Chemistry.

the type and content of the amino acid cross-linkers. The difference in surface wettability between the two groups of aerogels was attributed to the difference in the functional groups of the two amino acids. The cys-rGO aerogels (L-cys) contained a hydrophobic thiol group, while thelys-RGO aerogel (L-lys) had a hydrophilic amino group. Additionally, they had high mechanical strength, due to the superior mechanical properties of the cross-linked rGO network, as demonstrated in Fig. 7.24 (c, d).

Due to their hydrophobic characteristics, the cys-RGO aerogels had a strong adsorption to various organic liquids in water, with an adsorption capacity of 19–33 times its weight for different oils and organic solvents, as illustrated in Fig. 7.24 (e). Furthermore, they exhibited very high recyclability. Because of their large surface area, the hydrophilic lys-rGO aerogels could be used to remove heavy metal ions from water. Figure 7.24 (f) shows the adsorption capability of the lys-RGO aerogels toward Cd^{2+} and Pb^{2+}. Metal ion absorption of the lys-RGO aerogels could be conducted in from low acidic to neutral pH conditions, with absorption capability dependent on the pH value. No metal hydroxide precipitation was observed during the entire adsorption process. The increase in adsorption capability with increasing pH value implied that the adsorption was attributed to the electrostatic interactions between the metal ions and the negatively charged GO, as well as the ion exchange between the proton and the metal ions.

It has been reported that rGO hydrogels could be directly obtained by hydrothermally treating GO aqueous suspension at appropriate concentrations without using any other reagents [18]. To prepare the self-assembled graphene hydrogels (SGH), a 2.0 mg ml^{-1} aqueous dispersion was sealed in a Teflon-lined autoclave and hydrothermally heated at 180°C for 12 h. Figure 7.25 (a) shows photographs of the sample before and after the hydrothermal experiment. The resulting rGO hydrogel contained 97.4 wt% water and 2.6 wt% rGO, with an interconnected 3D porous structure. The rGO hydrogels had an electrical conductivity of 5×10^{-3} S cm^{-1}, high thermal stability over 25–100°C and high mechanical strength with storage modulus of 470 ± 20 kPa. Furthermore, their structure and properties can be easily tuned by changing the concentration of GO dispersion and the hydrothermal reaction time.

Figure 7.25 (b) shows photographs of three SGH columns with a diameter of about 0.8 cm each, which were mechanically strong enough to support 100 g weight with almost no deformation. SEM images of the SGHs after freeze-drying are shown in Fig. 7.25 (c–e), demonstrating that they had a well-developed interconnected 3D porous network structure. The pores exhibited sizes in the range from submicrometer to several micrometers, while wall of the pores consisted of thin layers of tightly stacked graphene nanosheets. The formation of the physical cross-linking sites of the framework in the SGH was attributed to the partial overlapping or coalescing of the highly flexible graphene nanosheets, as seen in Fig. 7.25 (e). Therefore, it was the inherent flexibility of the graphene nanosheets that led to the porous 3D macrostructures.

I–V curve of the SGH is shown in Fig. 7.25 (f), with an Ohmic linear relationship and high electrical conductivity, due to the recovery of the π-conjugated system, because the GO nanosheets were hydrothermally reduced to rGO, as confirmed by XRD patterns. The freeze-dried SGH possessed an interlayer spacing of about 3.76 Å, which was much smaller than that of the GO precursor (6.94 Å) but slightly larger than that of natural graphite (3.36 Å). This observation implied that the graphene nanosheets in the SGH took π–π stacking, while there were still residual oxygenated functional groups. With these

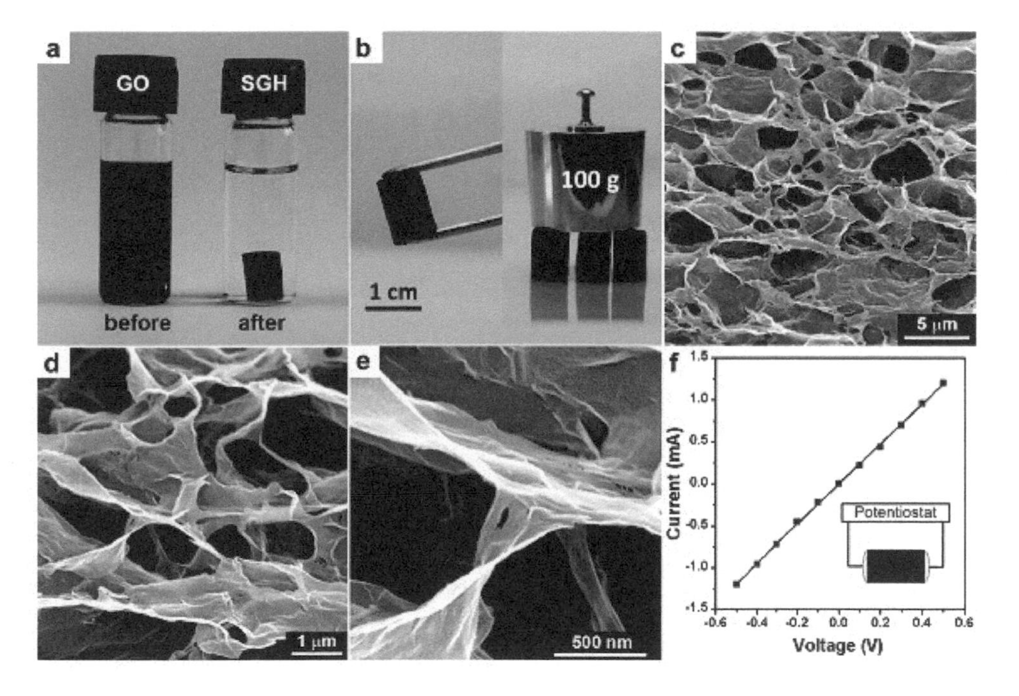

Fig. 7.25. (a) Photographs of the 2 mg ml^{-1} homogeneous GO aqueous suspension before and after hydrothermal reduction at 180°C for 12 h. (b) Photographs of the strong SGH that could support a heavy weight. (c–e) SEM images at different magnifications of the SGHs showing their interior porous microstructures. (f) Room temperature *I–V* curve of the SGH exhibiting Ohmic characteristics, with the inset showing the two-probe method for the conductivity measurement. Reproduced with permission from [18], Copyright © 2010, American Chemical Society.

residual hydrophilic oxygenated groups, the rGO nanosheets would encapsulate water molecules during the self-assembly process. The freeze-dried SGH exhibited a broad XRD peak, suggesting the decreased ordering of the graphene nanosheets and the presence of few-layer stacked graphene nanosheets.

It was found that the properties of the SGH were strongly dependent on the GO concentration (C_{GO}) and the hydrothermal reaction time. For instance, if the C_{GO} was too low, e.g., 0.5 mg ml^{-1}, the sample was black powder, as depicted in Fig. 7.26 (a). The samples with C_{GO} of 1 and 2 mg ml^{-1} were mechanically stable SGHs, whereas both the size and strength of SGHs were increased with the increasing C_{GO}. Figure 7.26 (b) shows photographs of the samples derived from the suspension with C_{GO} = 2 mg ml^{-1} for different time durations of hydrothermal reduction. Within the initial 6 h, the size of the as-obtained SGHs was decreased with the increasing reaction time and kept stable after that. In the starting suspensions, the GO nanosheets were randomly dispersed in water, due to their strong hydrophilicity and the electrostatic repulsion effect. During hydrothermal treatment, they became locally hydrophobic, because the conjugated domains were gradually recovered, due to the gradual elimination of the oxygenated functional groups, as confirmed by XRD and XPS analytical results. The synergistic action of the hydrophobic and π–π interactions triggered the flexible rGO nanosheets to be randomly stacked into a 3D network structure. Therefore, sufficiently high C_{GO} was required to maintain the cross-linking and partial overlapping of the flexible rGO nanosheets, as illustrated schematically in Fig. 7.27.

N-doped carbon nanomaterials and their composites or hybrids have become a hot research topic recently, due to their potential applications in energy-related applications, along with their extraordinary performances, as compared to their non-doped counterparts. Two examples of N-doped graphene 3D nanostructures synthesized by using the hydrothermal process are discussed in the following sub-section.

Marine mussels have shown very strong adhesive characteristics [66]. At the molecular scales, the mussels contain adhesive proteins that are densely decorated with Dopa (3,4-dihydroxy-L-phenylalanine), as well as a catecholic functionality [67]. Dopamine (DA), containing catechol and amine groups, is a unique molecule mimicking the adhesive proteins of the marine mussels, which can be spontaneously polymerized into polydopamine (PDA), so as to modify almost all types of surfaces of materials, without

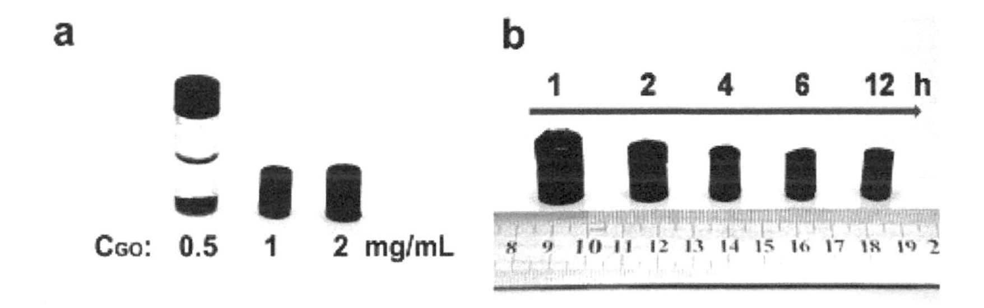

Fig. 7.26. (a) Photographs of the products derived from the GO suspensions with different C_{GO} after hydrothermal reduction at 180°C for 12 h. (b) Photographs of the products derived from the GO suspension with C_{GO} = 2 mg ml^{-1} after hydrothermal reduction at 180°C for different time durations. Reproduced with permission from [18], Copyright © 2010, American Chemical Society.

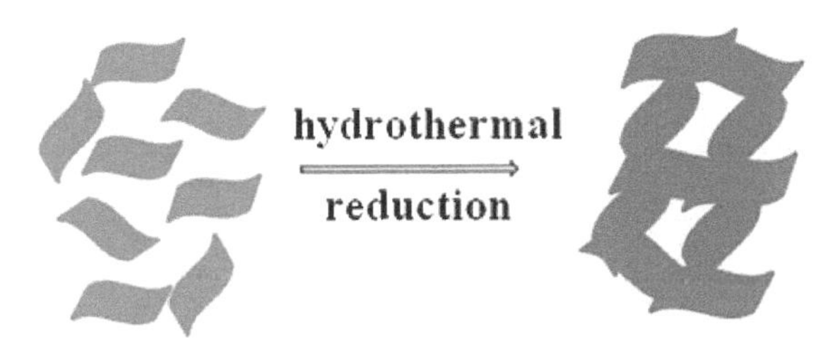

Fig. 7.27. Schematic diagram showing the formation mechanism of the SGH through the hydrothermal reduction. Reproduced with permission from [18], Copyright © 2010, American Chemical Society.

specific requirements on their chemical functionality or surface energy. As a result, its polymeric form can be used as an adhesive [68, 69]. A unique approach was used to develop 3D nitrogen (N) doped graphene aerogel (NGA), through the incorporation of mussel-inspired chemical motif of dopamine [70]. In this case, the DA served to facilitate oxidative polymerization, which thus reduced and functionalized the surface of graphene, while graphene nanosheets were assembled into 3D porous network structures at the same time. The DA was also employed as the source of nitrogen by offering N atoms to the graphene nanosheets during the pyrolysis process.

In the real experiment, 15 mg DA was added into 15 ml GO aqueous dispersion with a concentration of 1 mg ml^{-1} at pH $\approx$ 8.0, which was then sonicated for 40 min. The color of the solution gradually became pale brown. The mixture was then sealed in a 20 ml Teflon-lined autoclave to conduct hydrothermal treatment at 180°C for 12 h, in order to obtain NGA samples. The as-synthesized hydrogels were thoroughly washed and then freeze-dried. Ultralight NGAs were developed after the freeze-dried gel samples were calcined at 800°C for 3 h in Ar. The volume and density of the NGAs could be simply controlled by using GO suspensions with different concentrations in the range of 0.2–4.0 mg ml^{-1}, whereas the weight ratio of GO to DA was kept to be a consistent of 1:1. 3D GA without N-doping was also prepared similarly for comparison.

Figure 7.28 shows the synthetic procedure to produce the NGA samples, including two steps. The first step involved *in situ* hydrothermal cross-linking and polymerization of the mixture at 180°C for 12 h to obtain the 3D hybrid N-doped hydrogels. During this process, the DA was changed to the intermediate dihydroxyindole by liberating protons, so as to form PDA in between the individual GO nanosheets. Therefore, the intercalating PDA acted as a strong adhesive to cross-link the GO nanosheets. Meanwhile, the GO nanosheets were reduced, due to the oxidative polymerization of DA. Therefore, the 3D graphene network structure was formed from the rGO nanosheets induced by the DA. The second step was thermal annealing to realize the N-doping. The implied that the DA played a multi-functional role.

Figure 7.29 (a) shows a photograph of two NGAs standing on top of the red Calliandra haematocephala flower. The density of the NGA could be controlled in the range of 1.9–6.0 mg·cm^{-3}. The as-prepared NGAs showed a morphology similar to that of black cork. Figure 7.29 (b) shows SEM image of the sample in (a), in which twisted graphene nanosheets were randomly cross-linked to form a network structure with nano and micropores. Figure 7.29 (c) shows a TEM image of the sample, in which corrugated nanoflakes with a typical wrinkled and crumbled structure were observed. As shown in Fig. 7.29 (d), the HRTEM indicated that the graphene nanosheets in thin walls typically were of a few layers. The presence of high-quality graphene nanosheets was confirmed by XRD results.

Although the NGAs had an ultra-low density, they possessed strong mechanical strength. Figure 7.30 (a) shows a photograph of a 10 mg NGA sample, which was sufficiently strong enough to support a weight that was 5000 times its own weight. The NGAs were also fire-retardant, as demonstrated in Fig. 7.30 (b). Figure 7.30 (c) indicated that the 3D porous structure survived the firing and remained intact after being burnt repeatedly in ambient atmosphere. Due to the strong adhesion force formed between the DA and the graphene nanosheets, the NGAs had a reinforced mechanical strength as compared to the GA

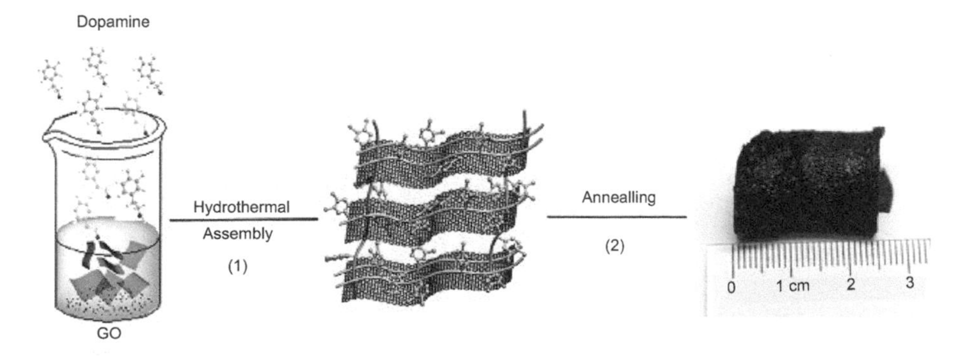

Fig. 7.28. Schematic diagram showing the production process of the ultralight 3D NGA, along with a photograph of the final NGAs. Reproduced with permission from [70], Copyright © 2014, Elsevier.

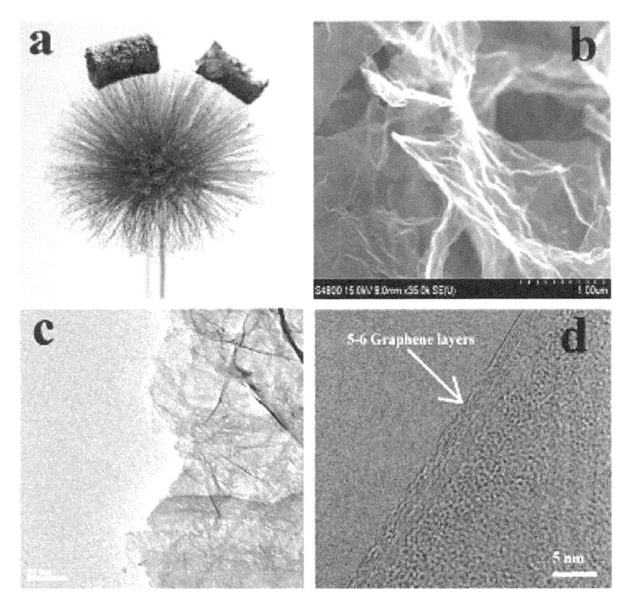

Fig. 7.29. (a) Photograph of two NGAs, with a cylinder shape of 1.9 cm in diameter and 1.2 cm in length, standing on a Calliandra haematocephala flower without deforming it. (b) SEM image of the sample in (a). (c) Representative TEM image of the NGA. (d) HRTEM image of the NGA. Reproduced with permission from [70], Copyright © 2014, Elsevier.

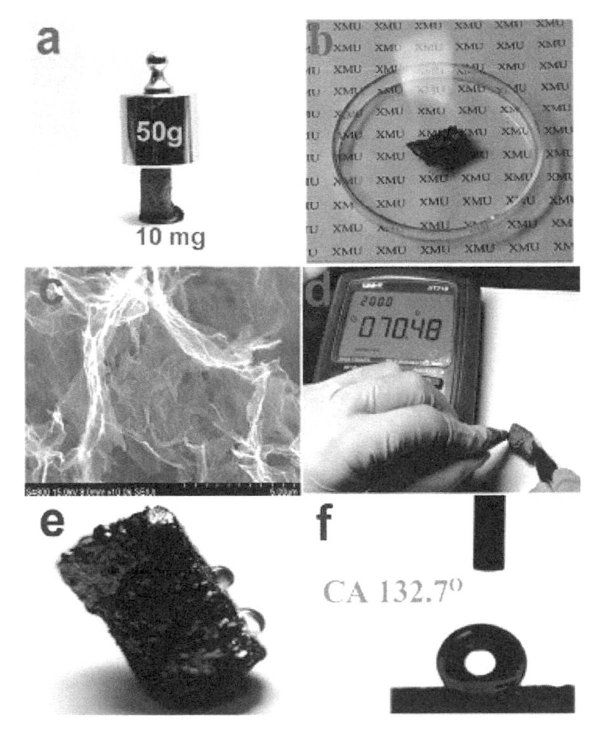

Fig. 7.30. (a) Photographs of a 10 mg NGA supporting a 50 g weight and (b) a burning NGA. (c) SEM image of the burned NGA without the loss of the porous network structure. (d) Demonstration of the electrical conductivity of the NGA. (e) Water droplets leaning on the surface of an NGA sample proving the hydrophobic nature of the NGAs. (f) Water contact measurement of the NGAs, with a contact angle of 132.7°. Reproduced with permission from [70], Copyright © 2014, Elsevier.

samples without DA. The NGAs were highly conductive to electricity, as demonstrated in Fig. 7.30 (d). In addition, the NGAs possessed a hydrophobic surface, as illustrated in Fig. 7.30 (e, f), which, together with the high porosity and the strong graphene skeleton, made them promising candidates for highly efficient separation/extraction of specific substances, such as organic pollutants and oils.

The second example of hydrothermally derived N-doped graphene hydrogels (NGHs) was synthesized by using urea [64]. The hydrothermal synthesis was conducted at 160°C for 3 h, with a low urea to GO mass ratio of 1:30. The as-synthesized NGHs were thoroughly washed to remove all the unreacted urea and were then fully dried. Finally, they were freeze-dried in vacuum. The NGHs exhibited high specific capacitance and excellent cycle stability when used as the electrode materials of supercapacitors.

Representative photograph of the NGHs prepared by using the hydrothermal process is shown as the inset in Fig. 7.31. The as-obtained NGHs contained 97.6 wt% water. Figure 7.31 shows the TGA curves of the parent graphite (PG), GO, wet and freeze-dried NGHs-4. PG was stable up to about 800°C with a weight loss of about 4%. The GO experienced three major weight losses, i.e., > 100°C, 120–180°C and 180–300°C, corresponding to evaporation of the adsorbed water, decomposition of the oxygen functionalities and desorption of the relatively stable oxygen functional groups, respectively. The dried NGHs-4 had no obvious weight loss over 180–300°C, suggesting that all the oxygen functional groups had been removed due to the hydrothermal treatment. The wet NGHs-4 exhibited a weight loss of about 95.4 wt% below 120°C, due to the evaporation of the adsorbed water.

Representative SEM images of the samples are shown in Fig. 7.32 (a–c). The GO nanosheets were loosely stacked with a morphology similar to that of foam or sponge. After the GO suspensions were hydrothermally reduced at 160°C for 3 h without the presence of urea, rGO hydrogels were obtained, as illustrated in Fig. 7.32 (b). With the presence of urea, the hydrothermally derived NGHs-4 possessed a well-developed interconnected 3D porous network structure, with pore diameters ranging from submicrometer to several micrometers and walls consisting of thin layers of stacked graphene nanosheets.

Figure 7.32 (d) indicated that the adsorption isotherm belonged to the type IV isotherm, with a sharp capillary condensation step at high relative pressures. A type 2 hysteresis loop was observed in the adsorption–desorption isotherms, implying the presence of mesopores. The majority of the pores had a specific volume of 0.27 $cm^3 \cdot g^{-1}$, with sizes in the range of 1.7–4.3 nm and a peak pore diameter of 3.81 nm. The sample had a BET specific surface area (SSA) of 300 $m^2 \cdot g^{-1}$, while its wet counterpart possessed a measured SSA of 1521 ± 60 $m^2 \cdot g^{-1}$. The NGHs-4 in wet state showed an excellent capacitive behavior of 308 $F \cdot g^{-1}$ at 3 $A \cdot g^{-1}$ and superior cycling stability of 92% retention after 1200 cycles in 6 $mol \cdot l^{-1}$ KOH.

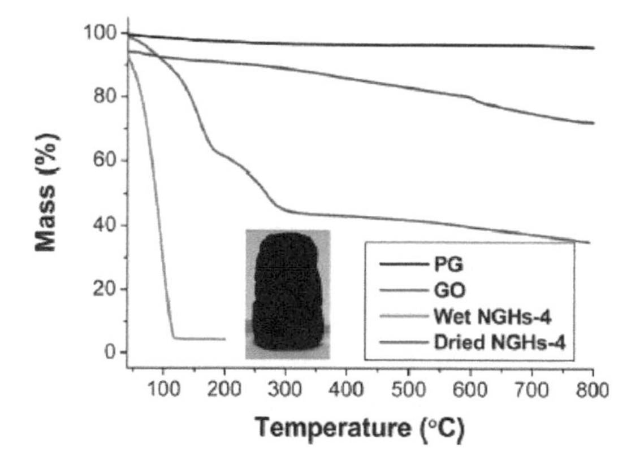

Fig. 7.31. TGA curves of PG, GO, wet and freeze-dried NGHs-4, with the inset showing a photograph of the as-obtained wet NGHs-4. Reproduced with permission from [64], Copyright © 2013, The Royal Society of Chemistry.

Color image of this figure appears in the color plate section at the end of the book.

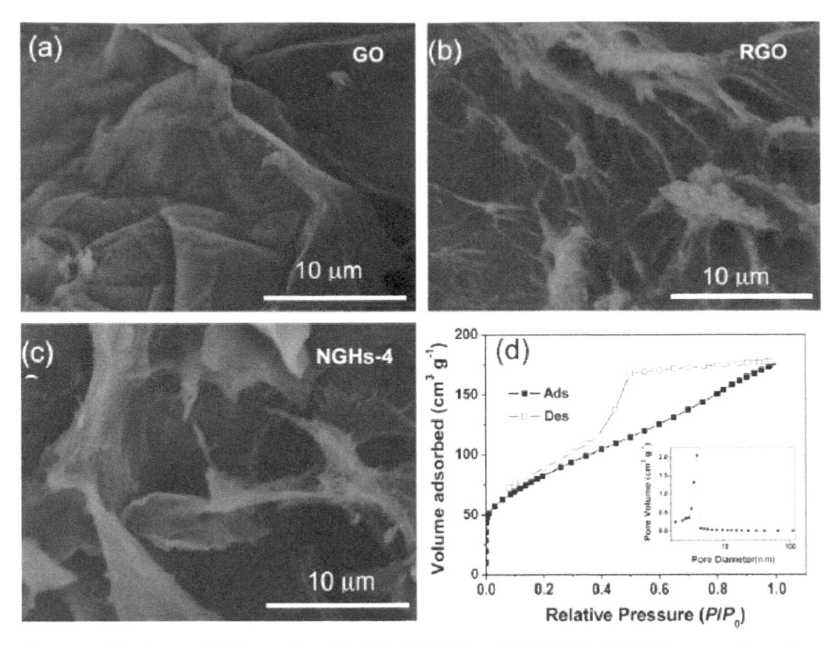

Fig. 7.32. SEM images of the freeze-dried samples: (a) GO, (b) RGO and (c) NGHs-4. (d) Nitrogen adsorption and desorption isotherms of the freeze-dried NGHs-4 with the inset showing the pore size distribution of NGHs-4. Reproduced with permission from [64], Copyright © 2013, The Royal Society of Chemistry.

7.2.4.2. Chemical reduction of GO

Self-assesmbly of *in situ* formed rGO through the chemical reduction of GO has been extensively studied. Various reducing agents, such as sodium ascorbate [71], L-ascorbic acid [72], hypophosphorous acid–iodine [73], $NaHSO_3$ [74, 75], Na_2S, HI, and hydroquinone [74], have been used to fabricate 3D graphene architectures through the *in situ* self-assembly of rGO nanosheets. Similar to the hydrothermal approaches, the chemical reduction of GO promoted hydrophobic and π–π interactions in between the graphene nanosheets, so as to trigger the self-assembly into 3D framework structures, with pore sizes ranging from sub-micrometer to several micrometers. Examples are discussed as follows.

High-performance 3D self-assembled graphene hydrogels (SGHs) have been fabricated through the chemical reduction of aqueous GO dispersions by using sodium ascorbate as the reducing agent [71]. Pure SGHs exhibited a high specific capacitance of 240 $F \cdot g^{-1}$, when they were used as the electrodes of supercapacitors. To prepare the SGHs, 6 ml GO aqueous suspension with a concentration of 2 mg ml^{-1} was loaded in a 15 ml glass vial, into which 24 mg sodium ascorbate was added. The sodium ascorbate was dissolved completely after sonication for 5 min, leading to a homogeneous yellow-brown solution, which was then heated at 90°C for 1.5 h to obtain the SGH. The excess sodium ascorbate was removed through dialysis.

Figure 7.33 (a) shows photographs of the GO aqueous suspension (2 mg ml^{-1}), with the addition of sodium ascorbate (left), along with a piece of SGH (right). The SGHs exhibited were mechanically strong, so that they appeared as free-standing cylinders. Interior microstructures of the as-prepared SGHs were characterized by using SEM. Figure 7.33 (b) shows a representative SEM of the graphene hydrogels, in which a well-developed cross-linked 3D porous structure was clearly observed, with pores having sizes in the range from submicrometer to several micrometers.

Both the microstructure and properties of the SGH were strongly dependent on several factors, including concentration (c_{GO}) of the starting GO suspensions, heating temperature and reaction time. Among them, c_{GO} played the most important role. As the concentration of the GO suspension was too low, e.g., c_{GO} = 0.5 mg ml^{-1}, only black-colored rGO powder was formed. At intermediate concentration, e.g., c_{GO} = 0.5 mg ml^{-1}, although SGHs could be obtained, they were mechanically weak. As the concentration of the

Fig. 7.33. (a) Photograph of an aqueous mixture of GO (2 mg ml^{-1}) and sodium ascorbate before (left) and after (right) chemical reduction at 90°C for 1.5 h. (b) SEM image of SGHs. Reproduced with permission from [71], Copyright © 2011, Elsevier.

GO suspensions was increased to 2 mg ml^{-1} and above, the SGHs would have sufficiently high mechanical strength. Rheological testing results indicated that the samples derived from the suspensions with c_{GO} = 2 and 3 mg ml^{-1} had almost the same mechanical strength. Therefore, the optimal c_{GO} was 2 mg ml^{-1}.

The heating temperature and reaction time could be combined to prepare mechanically strong SGHs. The heating temperature could vary from room temperature to 90°C. The higher the reaction temperature, the shorter the reaction time required. For instance, if the suspension was heated at 37°C, the reaction time was long as 30 h, while gelation occurred at 90°C in only 1.5 h. Figure 7.34 (a) shows photographs of the samples heated at 90°C for different time durations. Obviously, gelation was not started if the reaction time was shorter than 0.5 h. Once the gelation was initiated, the size of the SGHs was decreased gradually. Structural evolution of the SGHs was monitored by using the relative intensity of G (the E_{2g} mode of sp^2 carbon atoms) and D (the symmetry A_{1g} mode) bands of the Raman spectra, as shown in Fig. 7.34 (b). With the increasing heating time, the intensity ratio of D/G bands was gradually increased, which suggested that the GO nanosheets were reduced during the process. The evolution of the network structure was well reflected by the SEM observation, as illustrated in Fig. 7.34 (c–e). With increasing reaction time, the graphene nanosheets assembled more and more tightly, which implied that the reduction degree of the GO was the key factor to the formation of the SGHs.

A facile method was employed to produce chemically converted graphene xerogel (CCGX) through the chemical reduction of GO suspension with the presence of hypophosphorous acid and iodine (HPA-I) at 90°C for 12 h [73]. The reduction of GO in an aqueous solution of HPA-I resulted in a stable graphene hydrogel (GH), which was then freeze-dried to form the CCGX with very low density. The CCGX possessed a C/O atomic ratio of 14.7. Due to high degree of reduction, the CCGX exhibited extraordinary electrical properties, with an electrical conductivity (EC) of about 500 $S \cdot m^{-1}$. More importantly, by using HPA-I as the reducing agent, the hydrogelation could observed at a very low GO concentration of 0.5 mg ml^{-1}, so that ultra-low density xerogel could be obtained by using this method.

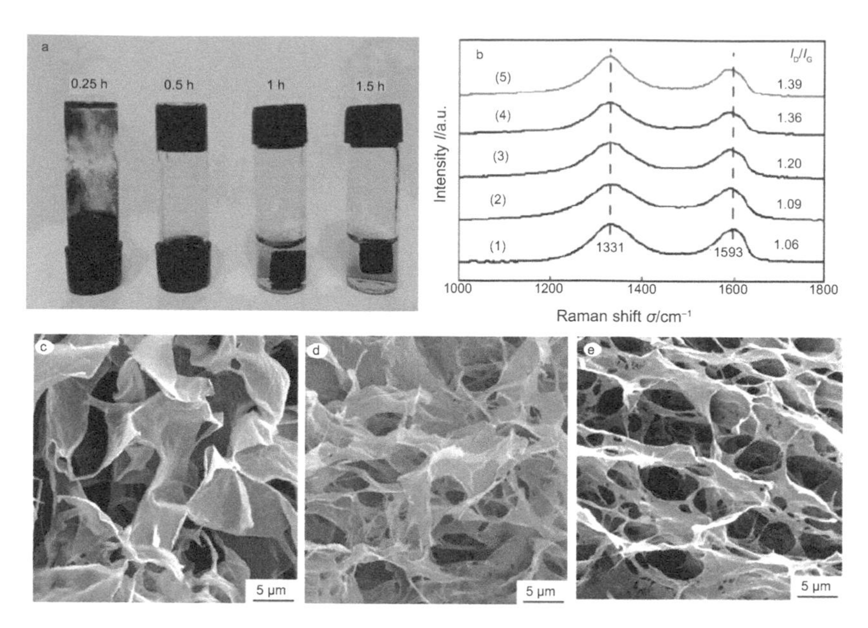

Fig. 7.34. Photograph (a) and Raman spectra (b) of the samples prepared after chemical reduction of the 2 mg ml^{-1} GO suspension at 90°C for different time durations: (1) 0 h, (2) 0.25 h, (3) 0.5 h, (4) 1.0 h and (5) 1.5 h. SEM images of the samples after reaction at 90°C for different time durations: (c) 0.25 h, (d) 0.5 h and (e) 1 h. Reproduced with permission from [71], Copyright © 2011, Elsevier.

Figure 7.35 (a) shows photographs of the GHs (right), which were derived from their corresponding aqueous mixtures of GO/HPA-I (left), with different GO concentrations. The as-synthesized GHs, monolith or slice (2 mm), were thoroughly washed with methanol and water (v/v = 1:1) in a soxhlet extractor for 12 h to remove all the residual HPA-I. It was observed that the volume of the GHs was closely related to the starting GO concentrations. The CCGX monolith, with a mass of 19 mg and a density of 13.0 mg·cm^{-3}, derived from the 3 mg ml^{-1} GO suspension, was strong enough to be able to support a 100 g weight with very little deformation. According to the stress–strain curves, the CCGXs with densities in the range of 6.8–13 mg·cm^{-3} demonstrated compressive yield stressed in the range of 24–50 kPa.

Self-assembly mechanism of the graphene nanosheets to form hydrogels due to the reduction of GO with the aid of HPA-I solution was proposed, as schematically illustrated in Fig. 7.35 (b, c). The mechanism involved two steps: (i) formation of GO agglomerate and (ii) transformation of the GO agglomerate into GH due to the reduction by HI. Figure 7.35 (c) indicated that the combination of hypophosphorous acid (or phosphorous acid) and iodine let to hydriodic acid [76, 77]. Furthermore, the presence of hypophosphorous acid lowered the pH value of the medium. At very low pH values, electrostatic repulsions were reduced, so that the GO nanosheets in the solutions could come together to form stable GO agglomerates. At the same time, the oxygen-containing functional groups, including epoxy, carbonyl and carboxyl groups, on the GO nanosheets were removed by the hydriodic acid, thus yielding CCG and iodine.

Representative SEM images of the CCGX are depicted in Fig. 7.36 (a, b), clearly revealing the macroporous 3D network structure that was constructed by the randomly oriented sheet-like structures with wrinkled texture. It is also observed that the microstructure of the CCGX was strongly dependent on the initial concentration of the GO suspension. Obviously, the pore size of the CCGX was decreased with the increasing GO concentration. Figure 7.36 (c) shows HRTEM image of the CCGX, in which wrinkled and folded paper-like textures of graphene nanosheets were clearly visible. Selected area electron diffraction (SAED) pattern of the CCGX is shown in Fig. 7.36 (d), presenting several bright hexagonal spots, with the intensity of the inner circle being relatively brighter than that of the outer one, suggesting the stacking characteristics of the graphene nanosheets [78, 79].

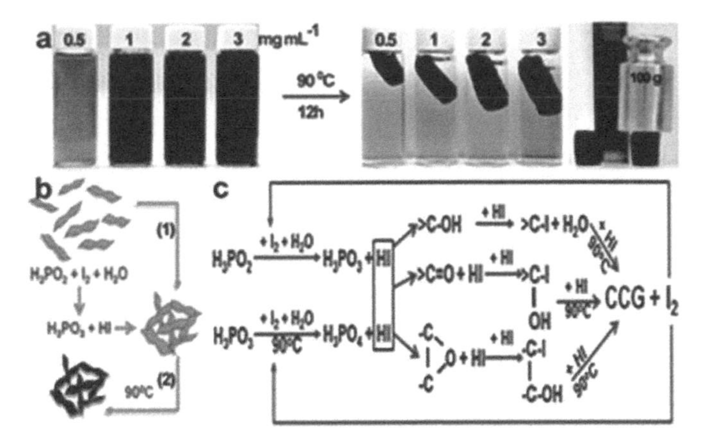

Fig. 7.35. (a) Photograph of the GHs prepared at different GO concentrations, (b) reduction of GO by using HPA-I solution and (c) mechanism of the GH formation under acidic conditions. Reproduced with permission from [73], Copyright © 2011, The Royal Society of Chemistry.

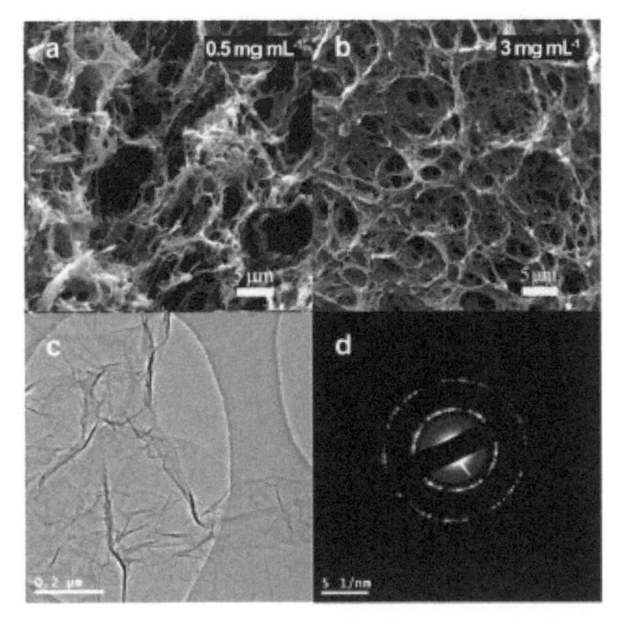

Fig. 7.36. SEM images of the CCGXs prepared at (a) 0.5 mg ml^{-1} and (b) 3 mg ml^{-1}, (c) HRTEM image of CCGX and (d) SAED of CCGX. Reproduced with permission from [73], Copyright © 2011, The Royal Society of Chemistry.

As illustrated in Fig. 7.36 (a), because the CCGX derived from the 0.5 mg ml^{-1} GO concentration contained quite a number of large pores with sizes of > 5 μm, it had a relatively low surface area of 35.64 $m^2 \cdot g^{-1}$, although its density was as low as 6.8 $mg \cdot cm^{-3}$. When high concentration GO suspensions were used, more cross-linking sites would be available, due to the partial overlapping of the randomly distributed GO nanolayers in the agglomerates, which however inhibited the layer-by-layer rearrangement of the graphene nanosheets during the reduction process. The CCGXs belonged to mesoporous structures, according to their Type IV nitrogen adsorption/desorption isotherm.

A method to prepare 3D architectures of graphene was developed, through *in situ* self-assembly of graphene nanosheets that were reduced from GO by using mild chemical reduction at 95°C and ambient pressure without stirring [74]. The reducing agents used to reduce the GO included $NaHSO_3$, Na_2S, Vitamin C, HI and hydroquinone. Both graphene hydrogels and aerogels with 3D architectures could be fabricated by the method, while their shapes could be controlled by simply using reactors with the desirable shapes. The graphene 3D architectures exhibited low density, high mechanical strength, thermal stability, electrical conductivity and specific capacitance, thus making them promising candidates for applications in supercapacitors, hydrogen storage and supports of catalysts.

Aqueous GO suspensions at concentrations of 0.1–2.0 mg ml^{-1} were added with various reducing agents, including $NaHSO_3$, Na_2S, Vitamin C, HI, and hydroquinone, at contents of 27–54 mmol ml^{-1}. The resultant suspensions were heated at 95°C for 30 min or 3 h without stirring to form graphene hydrogels. The as-obtained graphene hydrogels were dialyzed against deionized water for 3 days to remove the residual inorganic compounds. To obtain aerogels, the dried graphene hydrogels were freeze-dried to remove the absorbed water molecules.

$NaHSO_3$ can be used to reduce GO in aqueous suspension, and is considered to be a mild reducing agent [80]. If the reduction is conducted under stirring, the final products are aggregates or precipitates, due to the crude self-assembly of the quickly formed hydrophobic rGO nanosheets. Noting the steric hindrance effect of rGO nanosheets with micrometer sizes, if the reaction is not disturbed, it is expected to produce 3D rGO architectures. With the increasing degree of reduction, the partially reduced GO nanosheets would start to aggregate, due to the increase in hydrophobicity. Some rGO nanosheets assembled with a high degree of orientation, while others could be aggregated randomly. Finally, as more and more water molecules were driven out from the aggregates, compact 3D architectures were obtained consequently.

$NaHSO_3$ solution with a concentration of 41 mmol ml^{-1} was used to reduce GO in aqueous suspensions. Figure 7.37 (a) shows photographs of the samples after reaction for different time durations. The GO aqueous suspension had a concentration of 1.5 mg ml^{-1}. Figure 7.37 (b) shows a representative AFM image of the GO nanosheets deposited on the surface of newly cleaved mica, with a thickness of 0.84 nm. After reduction

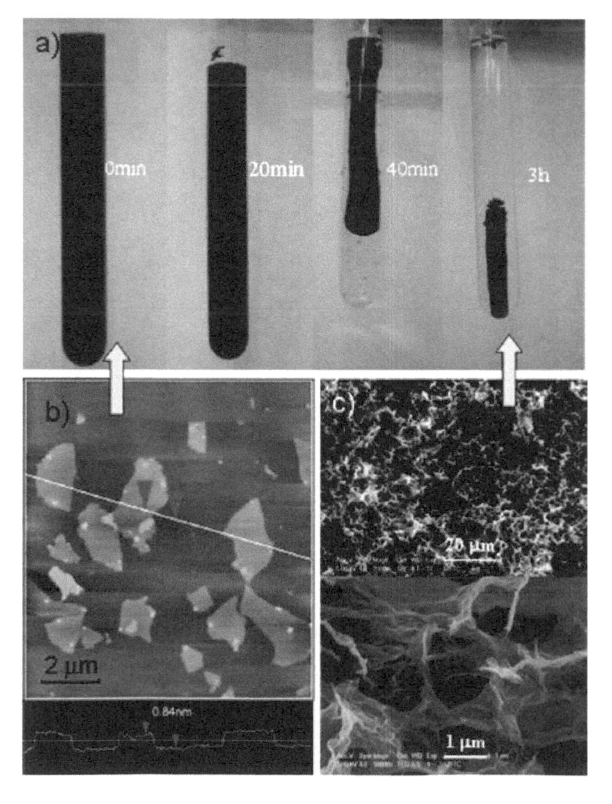

Fig. 7.37. (a) Photographs showing the formation process of the graphene hydrogels derived from an aqueous GO suspension (1.5 mg ml^{-1}), through the one-step self-assembly of the *in situ* reduced rGO for different reaction times using $NaHSO_3$ (41 mmol ml^{-1}) as the reducing agent. (b) AFM height image and height profile of the GO nanosheets deposited onto newly cleaved mica. (c) Cross-sectional SEM images of the as-prepared graphene hydrogel (3 h) at different magnifications. Reproduced with permission from [74], Copyright © 2011, The Royal Society of Chemistry.

reaction for 20 min, the aqueous suspension became dark because of the partial reduction of GO, while still remaining homogeneous. Reaction for 40 min resulted in the formation of a rod-shaped aggregate, floating on top of the suspension, while liquid at bottom of the tube became clear. The shrinkage of the black sample was attributed to the self-assembly of the rGO nanosheets. Finally, a small rod was developed at the end of the reduction (180 min). Due to its increased density, the aggregate sank to the bottom of the tube. The rod was hydrophilic, so that the water it contained inside could be squeezed out by pressing, i.e., graphene hydrogel was produced. Figure 7.37 (c) shows SEM images of rGO hydrogel, demonstrating a 3D architecture having homogeneously interconnected pores with sizes of 1–2 micrometers. The walls of the pores were very thin, suggesting the low level overlapping of the rGO layers.

The rGO hydrogels could be developed in different shapes by simply following the shapes of the used reaction vehicles. Figure 7.38 shows photographs of the graphene hydrogels with various shapes, including taper-like, cylindrical, pear-shaped and spherical shapes. This observation implied that the shrinkage of the hydrogels during the self-assembly was isotropic, thus offering a simple way to fabricate graphene hydrogels with the desired shapes for given applications. Figure 7.39 shows photographs of the graphene hydrogels prepared by using different reducing agents, including Na_2S, Vitamin C, HI and hydroquinone. Different reducing agents required different amounts of time to form the graphene hydrogels. For instance, $NaHSO_3$ took 30 min, while it needed only 10 min for Vitamin C and Na_2S. Furthermore, the graphene aerogels prepared using various reducing agents possessed different electrical conductivities. The HI hydrogels exhibited the highest electrical conductivity of 110 $S \cdot m^{-1}$.

More recently, a one-step strategy was developed to prepare macroscopic graphene–iron oxide hydrogels with 3D interconnected network structures, by using the synergistic effects of the self-assembly of GO nanosheets and *in situ* deposition of the metal oxide nanoparticles, including α-FeOOH nanorods and magnetic Fe_3O_4 nanoparticles [81]. The GO nanosheets were reduced by the ferrous ions that acted as reducing agents under mild conditions. The compositions of the graphene hydrogels could be readily controlled by adjusting the pH value of the starting GO suspensions. As the oil-saturated graphene–α-FeOOH superhydrophobic aerogel was burnt directly, 3D hematite α-Fe_2O_3 monoliths with porous microstructures were formed. More importantly, the separable and cost-effective graphene/metal oxide hydrogels and aerogels exhibited extraordinary capability to remove heavy metal ions and oils from water, thus making them potential candidates as efficient adsorbents in water purification and other waste-related applications.

Fig. 7.38. Photographs of the hydrogels with different shapes controlled by using different reactors. Reproduced with permission from [74], Copyright © 2011, The Royal Society of Chemistry.

Fig. 7.39. Photographs of the graphene hydrogels prepared using different reducing agents with a GO concentration of 1.5 mg ml^{-1} and a reducing agent concentration of 41 mmol ml^{-1}: (a) Na_2S, (b) Vitamin C, (c) HI and (d) hydroquinone. Reproduced with permission from [74], Copyright © 2011, The Royal Society of Chemistry.

In the experiments, a certain amount of $FeSO_4$ (0.0625–1 mmol) was quickly added into 10 ml GO aqueous suspension of 2 mg ml^{-1} in a 25 ml cylindrical sampler vial. The suspensions were then adjusted to pH = 3–11 with ammonia. All the reactions were carried out at 90°C in an oil bath for 6 h without stirring, leading to the appearance of 3D black monoliths. The samples were thoroughly washed with distilled water and freeze-dried into aerogels.

Figure 7.40 (a) shows photographs of the mixture of 10 ml GO suspension (2 mg ml^{-1}) and $FeSO_4$ (0.5 mmol) that was undisturbedly sealed in an oil bath at 90°C for different time durations. After reaction for 0.5 h, the GO suspension became black, due to the reduction of the GO nanosheets, which were still dispersible in water. Therefore, no aggregate was observed, but the sample was slightly floating from bottom of the vessel. As the reaction time was prolonged, a graphene monolith was formed and floated towards the top of the water. Finally, the black aggregate was further shrunk, leading to the formation of a columnar hydrogel. After reaction for 6 h, a well-developed black columnar hydrogel with a diameter of 10 mm and height of 7 mm was obtained, as demonstrated in Fig. 7.40 (b). No isolated graphene nanosheets were present elsewhere, so the transparent liquid was left in the vessel.

The as-synthesized graphene hydrogel contained about 95.8 wt% water, with a microstructure of well-defined and interconnected 3D network, as illustrated in Fig. 7.40 (c). The 3D network was uniformly dispersed with pores of several micrometers in size. A large number of nanoparticles were homogenously distributed on the graphene nanosheets, as seen in Fig. 7.40 (d). Figure 7.40 (e) shows TEM image of the nanoparticles, which had a rod-like morphology with a size of about 60 nm. XRD pattern in Fig. 7.40 (f) indicated that the nanoparticles were orthorhombic α-FeOOH.

The properties of the graphene–α-FeOOH hydrogels were closely dependent on the amount of the reducing agent Fe(II). The higher the content of $FeSO_4$ used in the GO suspension, the greater the number of α-FeOOH nanorods produced in the 3D network graphene hydrogels. Representative SEM and TEM images of the samples are shown in Fig. 7.41 (a–c). If the content was lower than 0.0625 mmol, the number of nanoparticles was very few, while the pores had a size of tens of micrometers. As a result, 3D gel cylinders would have a large physical size, suggesting the weak interaction in between the graphene nanosheets in the hydrogel network, as observed in Fig. 7.41 (d–f). When the graphene hydrogels with the content of $FeSO_4$ decreased from 1 to 0.0625 mmol were freeze-dried, the last two samples were collapsed instead of solid hydrogels, as demonstrated in Fig. 7.41 (g, h). In the powdery aerogel with the lowest content of $FeSO_4$ (0.0625 mmol), some GO nanosheets were not completely reduced. The microstructure and phase composition of the hydrogels were also affected by the pH value of the starting GO suspensions.

Figure 7.42 shows the formation mechanism of the graphene–iron oxide hydrogels. Firstly, the ferrous ions (Fe^{2+}) were adsorbed onto the GO nanosheets due to the electrostatic interactions, and were then oxidized into ferric ions (Fe^{3+}) by the oxygen-containing functional groups on the GO nanostheets. At low pH values, the resultant α-FeOOH nanorods were *in situ* deposited on the rGO surface due to hydrolysis of the Fe^{3+} ions. As the pH value of the starting GO suspension was sufficiently high, the Fe^{3+} ions and

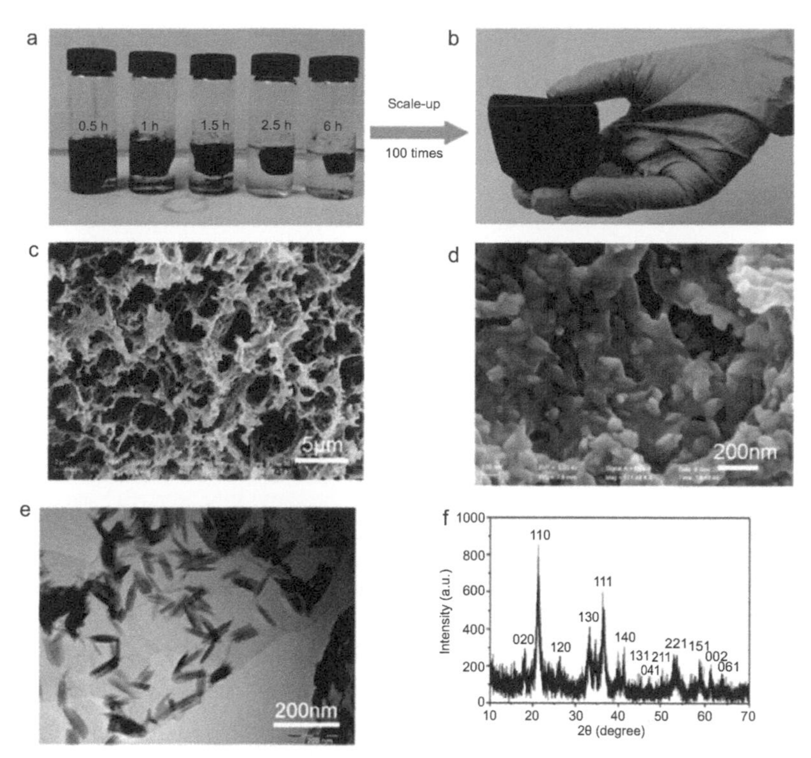

Fig. 7.40. (a) Photographs of samples after reaction for different times, showing the time-dependent formation process of the hydrogels derived from 10 ml of GO (2 mg ml^{-1}) suspensions at pH = 3 in the presence of 0.5 mmol $FeSO_4$. (b) Photograph of the hydrogels prepared by using 1000 ml GO suspension and 50 mmol $FeSO_4$. (c, d) Low- and high-magnification SEM images of the hydrogel. (e) TEM image of the freeze-dried graphene–α-FeOOH hydrogel. (f) XRD pattern of the graphene/α-FeOOH aerogel. Reproduced with permission from [81], Copyright © 2012, American Chemical Society.

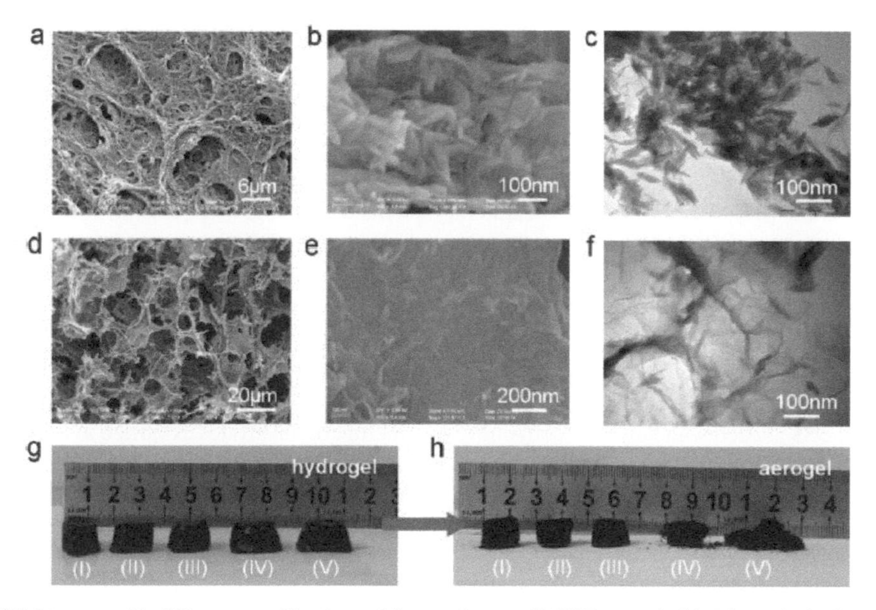

Fig. 7.41. SEM images with different magnifications of the graphene–α-FeOOH aerogels dried from the hydrogels derived from 10 ml GO suspension (2 mg ml^{-1}) at pH = 3, with different contents of $FeSO_4$: (a–c) 1 mmol and (d–f) 0.0625 mmol. Photographs of the graphene–α-FeOOH hydrogels (g) and the corresponding aerogels (h) dried from the hydrogels prepared using different amounts of $FeSO_4$: (I) 1, (II) 0.5, (III) 0.25, (IV) 0.125 and (V) 0.0625 mmol. Reproduced with permission from [81], Copyright © 2012, American Chemical Society.

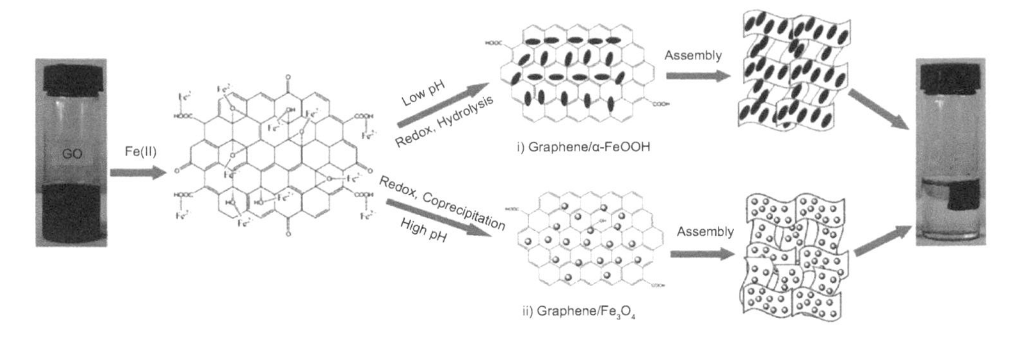

Fig. 7.42. Schematic diagram illustrating the formation mechanism of the graphene–iron oxide hydrogels. Reproduced with permission from [81], Copyright © 2012, American Chemical Society.

Fe^{2+} ions would be combined to the Fe_3O_4 precipitate. At the same time, the rGO nanosheets decorated with Fe_3O_4 nanoparticles were self-assembled into 3D hydrogel with interconnected networks, triggered by the synergistic effect of the hydrophobic and π–π stacking interactions, due to the decreased number of oxygenated groups on the graphene nanosheets. Moreover, the deposited nanoparticles on the rGO nanosheets acted as spacers to prevent the nanosheets from aggregation, so as to stabilize the graphene hydrogels. For example, if the graphene–α-FeOOH hydrogels were etched with HCl solution to remove the α-FeOOH nanorods, their 3D network structures were collapsed, because of the severe reaggregation of the graphene nanosheets. This strategy has been extended to other metal ions, such as Mn(II) and Ce(III), to form macroscopic graphene/Mn_2O_3 and graphene/CeO_2 hydrogels.

It was found that the graphene/α-FeOOH hydrogels were superhydrophobic and highly porous. When they adsorbed oils and nonpolar organic solvents, water was left, so that they could be used as a suction skimmer in marine oil-spill recovery. Figure 7.43 (a) shows photographs of the fast process to absorb a layer of gasoline labeled with Sudan III dye on the water surface, due to the superhydrophobic and capillary effects of hydrogels. More importantly, the aerogels could be recycled for many times, as demonstrated in Fig. 7.43 (b). The graphene–α-FeOOH hydrogels exhibited a high adsorption capacity after eight gasoline-absorbing and drying recycles, which could be readily attributed to their robust interconnected network and stable porous structure. Figure 7.43 (c) indicated that the hydrogels were able to absorb a wide range of nonpolar organic solvents and oils. The unusual superwetting and high selectivity of the graphene–α-FeOOH hydrogels resulted from the hydrophobic π–π stacking of the rGO nanosheets and the abundant inorganic nanoparticles that were distributed on the graphene nanosheets, which led to extremely high surface roughness [82, 83]. When an oil-saturated graphene–α-FeOOH aerogel was fired, a red cylinder was formed, as seen in Fig. 7.43 (d). XRD results indicated that the α-FeOOH was converted into hematite α-Fe_2O_3. The unique 3D porous microstructures were entirely retained after the burning, as illustrated in Fig. 7.43 (e, f). In other words, graphene–α-FeOOH was transferred into graphene–α-Fe_2O_3 aerogel.

7.2.4.3. Self-assembly of polymer grafted graphene

CMG nanosheets grafted with polymers containing special functional groups exhibited assembling ability in solutions, which led to the formation of 3D graphene architectures with unique microstructures and novel functionalities. It has been reported that, when 2D rGO nanosheets were noncovalently stabilized by using triblock copolymer PEO-*b*-PPO-*b*-PEO (F127), they could assemble into a composite hydrogel with the aid of α-cyclodextrin (α-CD). In this case, the dual roles of F127 in dispersing rGO and forming hydrogels were integrated [84]. The hydrophobic PPO segments were bound to the hydrophobic basal plane of the rGO nanosheets *via* the hydrophobic effect, while the hydrophilic PEO chains were extending into water, so that a supramolecular hydrogel was formed, due to the formation of a supramolecular pseudorotaxane structure between the PEO segments and α-CD. The hydrogel exhibited a gel–sol transition at an elevated temperature. Such rGO-based supramolecular hydrogels could find potential applications in drug delivery.

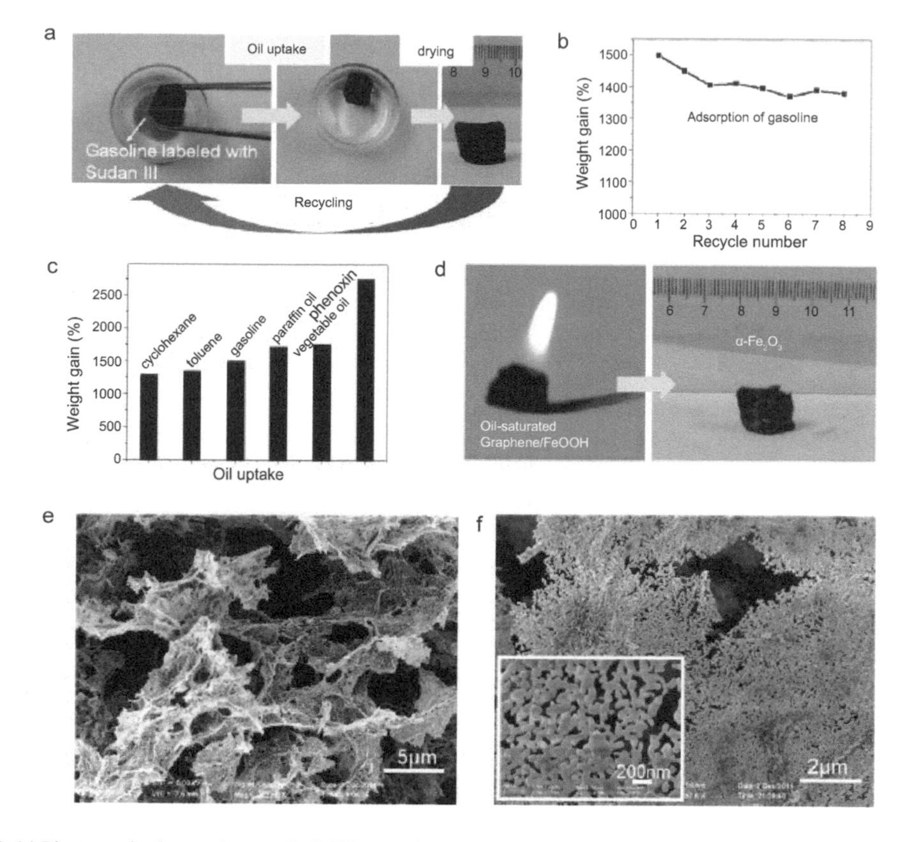

Fig. 7.43. (a) Photographs the graphene–α-FeOOH aerogels used to adsorb gasoline on water surface. The gasoline was labeled with Sudan III for clear presentation. (b) Regeneration capacity of the aerogels for adsorbing gasoline. Gasoline can be removed by heating the aerogel in an oven at 100°C for recycled use. (c) Adsorption capacities of the aerogels for a range of organic solvents and oils in terms of their weight gain. (d) Photographs of the oil-saturated graphene–α-FeOOH aerogel before and after burning. (e, f) SEM images of the 3D microstructures of the α-Fe_2O_3 by burning the oil-saturated graphene–α-FeOOH aerogel. Reproduced with permission from [81], Copyright © 2012, American Chemical Society.

To prepare the Pluronic copolymer functionalized graphene dispersion, 400 mg Pluronic copolymer was added to 15 ml aqueous GO suspension (1.0 mg ml^{-1}). After stirring for 10 min, the mixture was reduced with hydrazine monohydrate (400 μl) for 8 h with the aid of sonication at 40°C. After reduction, homogeneous black solution was obtained, which could withstand centrifugation at 22000 g for 15 min with almost no precipitation. The supramolecular hybrid hydrogel of triblock copolymer functionalized graphene was then prepared. To do that, α-CD and the Pluronic copolymer functionalized graphene solution were mixed, with the concentration of α-CD in the mixture to be 90 mg ml^{-1} and the molar ratio of the ethylene oxide units to α-CD to be 2:1.

Representative SEM images of the freeze-dried supramolecular hydrogels are shown in Fig. 7.44. It was observed that both the supramolecular native and the graphene containing hybrid hydrogels possessed a loose sponge-like structure in the form of cross-linked network, suggesting the elimination of the trapped water molecules during the freeze-drying process. No information indicated the effect of the graphene nanosheets on the inner cross-sectional morphology of the hydrogels. The Raman spectrum of the supramolecular hybrid hydrogel has almost no difference from that of the aqueous copolymer-coated graphene, confirming the incorporation of the graphene nanosheets into the supramolecular hybrid hydrogels.

Figure 7.45 (a) shows a schematic diagram of the structure model for the aqueous copolymer functionalized graphene solution. In this model, the hydrophobic PPO segments of the copolymer were attached to the hydrophobic surface of the graphene nanosheet due to the hydrophobic effect, while the hydrophilic PEO chains were freely drifting in water. The PEO chains would penetrate into the

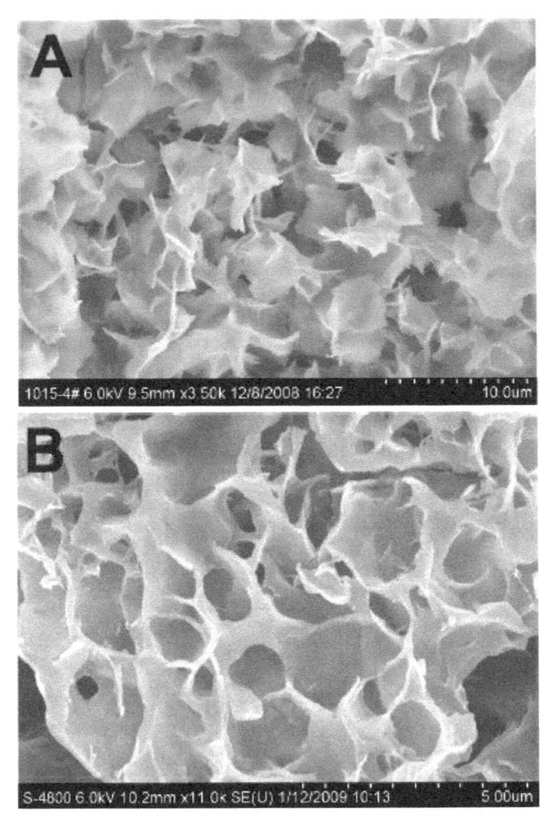

Fig. 7.44. SEM images of the supramolecular hydrogels and dispersion state of the graphene nanosheets: (A) hybrid hydrogel at low magnification and (B) hybrid hydrogel at high magnification. Reproduced with permission from [84], Copyright © 2009, American Chemical Society.

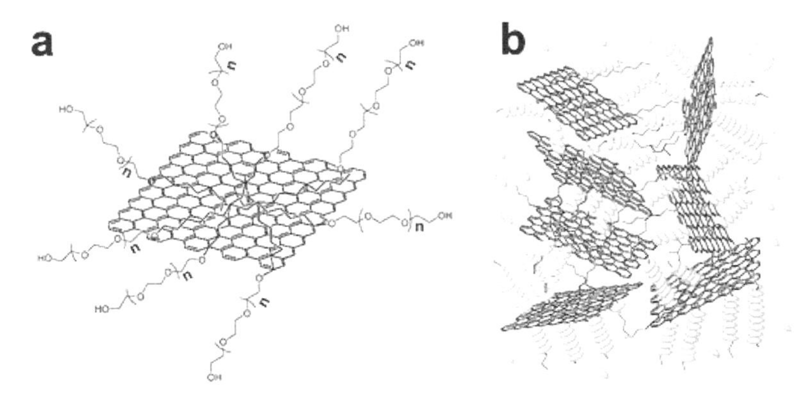

Fig. 7.45. Schematic diagram showing the structure of the copolymer coated graphene (a) and the supramolecular well-dispersed graphene nanosheet containing the hybrid hydrogel (b). Reproduced with permission from [84], Copyright © 2009, American Chemical Society.

cyclodextrin cavities, so that supramolecular hybrid hydrogel was formed. As compared with the native supramolecular hydrogel formed from the flexible chains of copolymer F68, the attached PPO segments and thus preorganized PEO chains in the copolymer-coated graphene solution ensured the formation of the hydrogel upon the presence of the α-CD, which acted as an accelerator to promote the gelation of the polymer grafted graphene solution. Meanwhile, because the PPO chains had been attached onto the graphene nanosheets, micellization or entangling of the copolymer, especially the PPO segments, to form a more complex structure, was effectively prevented. As a consequence, the final hydrogels had a well-defined lamellar structure, as depicted schematically in Fig. 7.45 (b).

It has been well-known that the threading of cyclodextrins onto polymer chains is entropically unfavorable, which is usually prompted by noncovalent weak force interactions, such as hydrogen bonding and hydrophobic interactions. An increase in temperature can induce dethreading of the cyclodextrin from the polymer chains. This gel-sol transition was observed, as the temperature of the supramolecular hydrogels was kept to be about 45°C for 20 min. Figure 7.46 shows photographs to demonstrate the gel-sol transition of the hybrid hydrogel triggered by the variation in temperature. The supramolecular hydrogel and the flowable sols are shown in Fig. 7.46 (A) and (B), respectively.

Another example is the preparation of stimuli-responsive GO composite hydrogel by covalently cross-linking GO nanosheets with poly(*N*-isopropylacrylamide)-*co*-poly(acrylic acid) (PNIPAM-*co*-PAA) microgels [85]. The highly efficient cross-linking reaction triggered the formation of a porous GO hydrogel network with high mechanical strength. Such hydrogels possessed thermal and pH responses with high responsive rate and excellent reversibility. The composite hydrogel had a hierarchical structure, consisting of polymer microgels with sizes of up to 10 nm and a cross-linked network of microgels and GO nanosheets. Such a unique microstructure would make the hydrogels useful to concurrently load and deliver small and large biomolecules.

The GO/PNIPAM IPN hydrogel was prepared by cross-linking GO nanosheets with PNIPAM-co-AA microgels in water, through the reaction between epichlorohydrin (ECH) and carboxyl groups, as illustrated schematically in Fig. 7.47 (a). The reaction was conducted in a sealed tube at 98°C (Fig. 7.47 (c)). The ECH could permeate into the aqueous phase to induce the cross-linking reaction, due to the formation

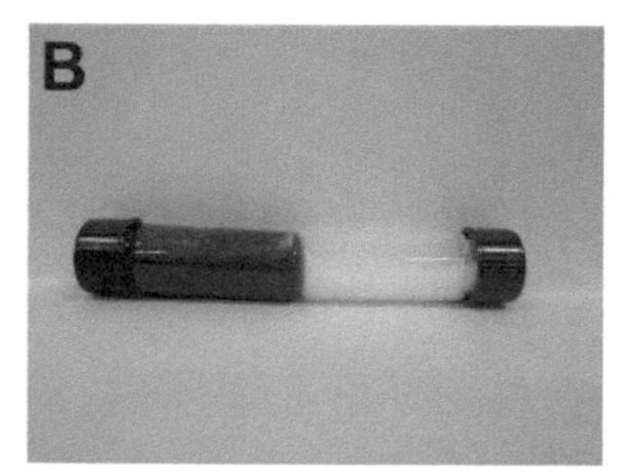

Fig. 7.46. Photographs of the supramolecular hydrogels (A) (left: graphene-containing hybrid hydrogel, right: native hydrogel) and the flowable sols (B) transited from the hydrogel triggered by raising the temperature. Reproduced with permission from [84], Copyright © 2009, American Chemical Society.

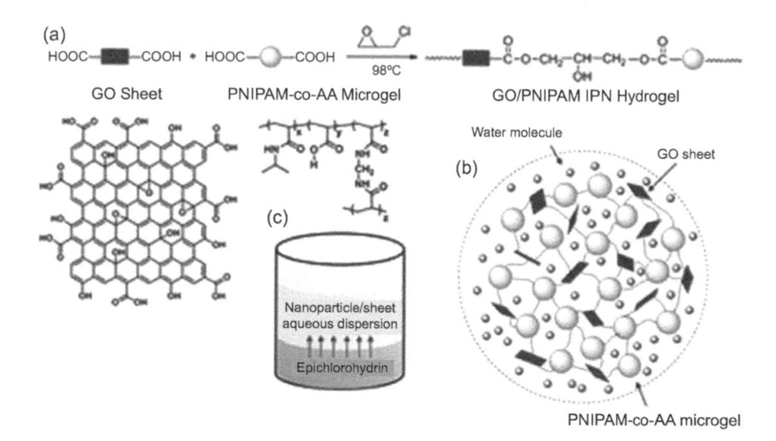

Fig. 7.47. (a) Formation of the GO/PNIPAM interpenetrating hydrogel networks due to the reaction between the ECH and the carboxyl groups on the GO nanosheets and the PNIPAM-co-AA microgels. (b) Structural sketch of the GO/PNIPAM IPN hydrogel. (c) Sealed reaction tube for the reaction at 98°C for incubation, in which the ECH was able to permeate into the aqueous phase to induce the cross-linking reaction. Reproduced with permission from [85], Copyright © 2011, American Chemical Society.

of ECH/water azeotrope at a sufficiently high temperature, noting the azeotropic temperature of 88°C, when containing 28 wt% water. Figure 7.47 (b) shows the schematic structure of the GO/PNIPAM IPN hydrogel. A two-level structural hierarchy was formed in both the GO/PNIPAM IPN hydrogel and the PNIPAM microgel network. The primary one was composed of cross-linked polymer chains inside each particle, with a mesh size of 1–10 nm, while the secondary one was a cross-linked system of nanoparticles and nanosheets, with a mesh size of 10–500 nm.

7.3. Interface Self-Assembly

Interface self-assembly has been shown to be an effective way to fabricate G/GO thin sheets, as discussed in the last chapter. In this section, porous graphene layered free-standing films, which have been classified as a new type of 3D architecture, are presented. These porous films have shown various advantages as compared to 3D porous scaffolds. For example, porous films can be directly used as active electrode materials for electrochemical applications without the requirement of tailoring their dimensions [86]. Various methods have been developed to fabricate graphene-based porous free-standing films, which are discussed as follows.

7.3.1. Breath-figure-templated assembly

Breath-figure, a widely used procedure for fabricating large-area porous polymer films, has been applied to the self-assembly of polystyrene-grafted GO (PS–GO) into macroporous films [87]. GO nanosheets were self-assembled into mechanically flexible, macroporous 3D free-standing large-scale films with tunable porous microstructures. Figure 7.48 (A) shows a schematic diagram of the breath-figure method. The polymer-grafted GO nanosheets were synthesized and dispersed in an organic solvent. The dispersion was then cast onto proper substrates, which were exposed to a stream of humid air flow. Due to the evaporation of the volatile organic solvent, spontaneous condensation and close packing of aqueous droplets occurred at the surface of the organic solution. Polymer-grafted GO-based macroporous films were developed on the substrates after completely drying them, from which flexible robust macroporous films consisting of rGO nanosheets were derived after pyrolysis. In addition, the electrical properties and chemical reactivity of the 3D rGO assemblies could be further increased by using nitrogen doping (N-doping).

The polymer-grafted GO nanosheets were prepared by using the surface-initiated atom transfer radical polymerization (ATRP) [88–90]. The polystyrene-grafted graphene oxide (PS-GO) nanosheets could be well-dispersed in benzene. Drop-casting the PS-GO/benzene dispersion with a concentration of 5 mg ml^{-1} onto a SiO_2 substrate resulted in a highly flexible macroporous film after the evaporation of the solvent, which could be transferred onto the flexible poly(ethylene terephalate) (PET) substrate, as demonstrated in Fig. 7.48 (B). The film had a thickness of 3.8 μm. After pyrolysis, the GO was thermally reduced to rGO, which thus became superhydrophobic, as illustrated in Fig. 7.48 (C).

SEM images of the rGO macroporous films are shown in Fig. 7.48 (D, E). The sample was characterized by closely packed macropores with a nanoscale rim thickness. The rGO assembled films were highly flexible and robust. Figure 7.48 (F) shows a plane-view SEM image of a concave film with a radius of curvature of about 2 mm, in which the pores were slightly elongated, but without damaging the macropore morphology. The porous structure was quickly recovered as the film was returned to a flat state, as seen in Fig. 7.48 (G).

Figure 7.49 (A) shows a high magnification SEM image of the interconnected rim structure of the films, which indicated that the rims were composed of stacked and overlapped rGO nanosheets, with highly wrinkled morphology along the boundary of the macropores. Representative TEM images are shown in Fig. 7.49 (B, C), confirming the presence of the wrinkled rGO nanosheets. Figure 7.49 (D) shows XRD pattern of the film, revealing an average interlayer spacing of about 3.5 Å, which was slightly larger than the interlayer spacing of the graphene layers in graphite (3.35 Å).

The size of the pores and the number of layers of the films could be controlled by controlling the concentration of starting solution and the chain length of the polymers, as shown in Fig. 7.50 (A). For instance, when 1 g initiator was used to modify GO with 5 ml and 2.5 ml styrene monomer, the samples PS-GO-5 and PS-GO-2.5 would contain 18 wt% and 58 wt% GO, respectively, after polymerization. As the

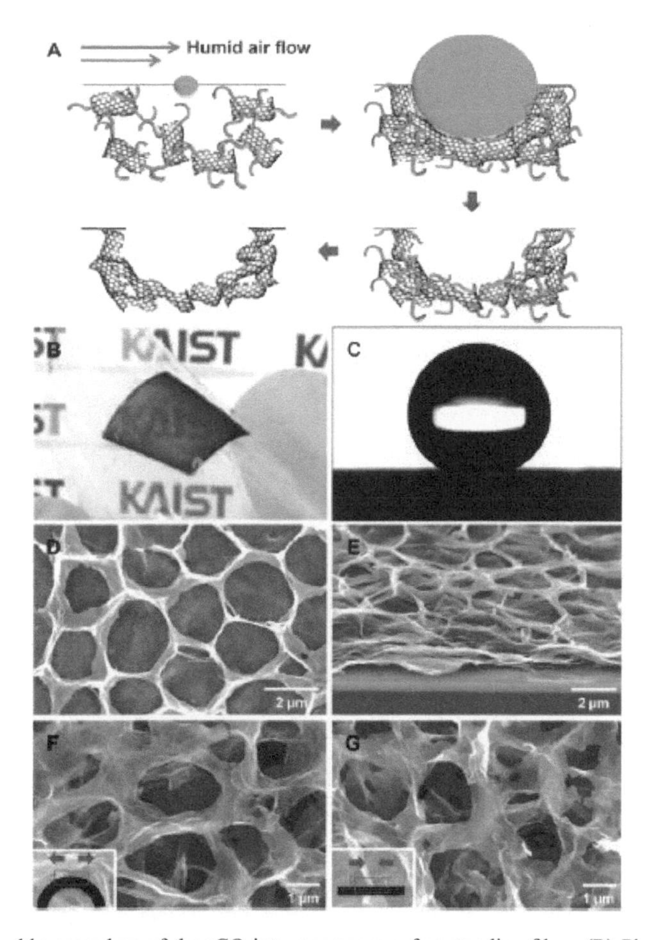

Fig. 7.48. (A) Self-assembly procedure of the rGO into macroporous free-standing films. (B) Photograph of the flexible semi-transparent macroporous rGO film on PET. (C) A water contact angle of 152° was observed for the superhydrophobic macroporous rGO film. (D) Plane-view and (E) 60°-tilted SEM images of the rGO film. (F, G) Plane-view SEM images of the porous rGO film upon (F) and after (G) deformation. Reproduced with permission from [87], Copyright © 2010, Wiley-VCH Verlag GmbH & Co. KGaA, Weinheim.

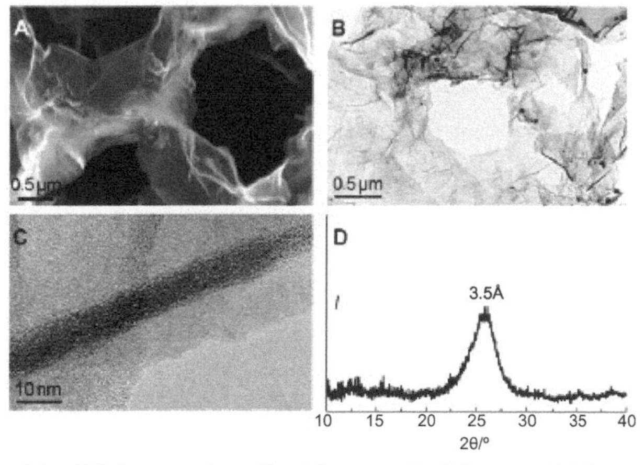

Fig. 7.49. (A) SEM image of the rGO sheets on the walls of the pores. (B, C) Low- and high-magnification TEM images of the porous rGO assembly. The rGO nanosheets surrounding the pores were pushed and wrinkled by the growth of the aqueous droplets and volume reduction due to the pyrolysis. (D) XRD pattern of the porous rGO film, showing an interlayer spacing of about 3.5 Å. Reproduced with permission from [87], Copyright © 2010, Wiley-VCH Verlag GmbH & Co. KGaA, Weinheim.

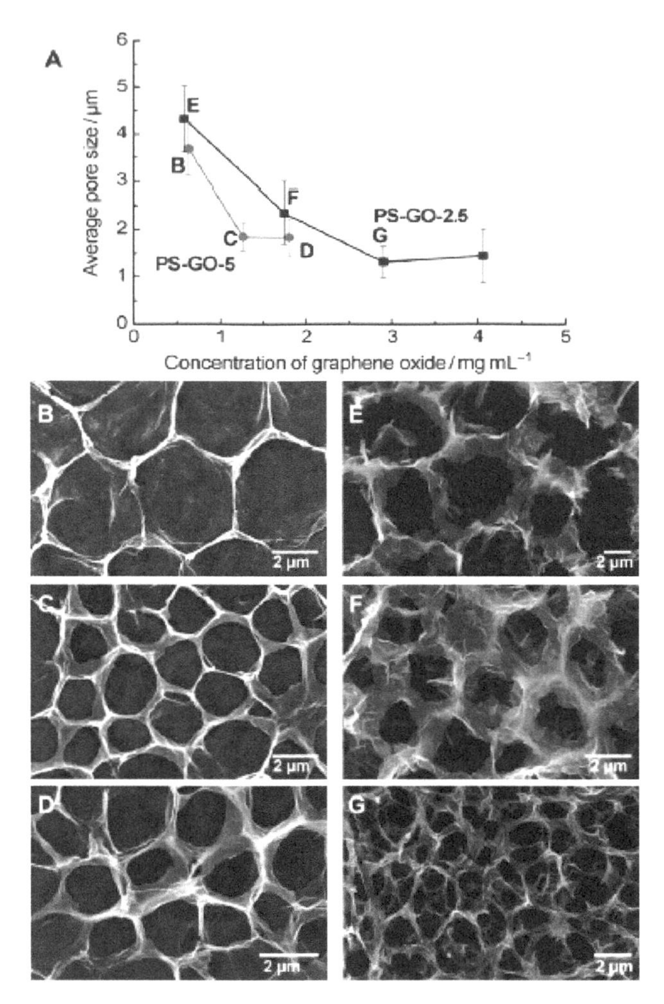

Fig. 7.50. (A) Average pore sizes versus the GO concentration in the PS-GO systems for different chain lengths of the grafted polymers. (B–D) SEM images of the porous derived films from PS-GO-5 at concentrations of 3.5 (0.63), 7 (1.26) and 10 (1.80) mg ml^{-1}. (E–G) SEM images of the porous films derived from the PS-GO-2.5 at concentrations of 1 (0.58), 3 (1.74) and 5 (2.90) mg ml^{-1}. The numbers in parentheses denote the concentration of the GO in the respective PS-GO solutions. Reproduced with permission from [87], Copyright © 2010, Wiley-VCH Verlag GmbH & Co. KGaA, Weinheim.

dispersion of PS-GO-5 in benzene had a concentration of 3 mg·ml^{-1}, the final film would have hexagonal-shaped pores with an average size of 3.7 µm, as illustrated in Fig. 7.50 (B). If 10 mg·ml^{-1} dispersion was used, the average pore size of the film was decreased to about 1.8 µm, as seen in Fig. 7.50 (C, D). Further increase in the GO concentration resulted in viscous solutions, which could be used to form aqueous droplets. SEM images of the PS-GO-2.5 samples are shown in Fig. 7.50 (E–G). As its concentration was increased from 1 mg·ml^{-1} to 3, 5 and 7 mg·ml^{-1}, the average pore size was monotonically decreased from 4.3 µm to 1.4 µm. The relationship between the average pore sizes and the content of GO is depicted in Fig. 7.50 (A).

7.3.2. Flow directed self-assembly

7.3.2.1. Leavening strategy

More recently, another approach based on the flow-directed self-assembly of GO followed by a leavening strategy was developed to prepare rGO films with 3D interconnected macroporous structures [17]. Leavening is usually a process applied to increase the volume of breads or cakes before or during baking or steaming by forming porous structures. Leavening has been demonstrated to be an effective approach to transfer compact graphene structures into porous ones, by finding a proper leavening procedure. Because unstable GO can be chemically reduced with mild heating to yield rGO, accompanied by the production of gaseous species such as H_2O and CO_2 [35, 91–93], if the gas released during the chemical reduction of the GO could be controlled properly, rGO foams could be obtained, similar to the formation of leavened bread. It was found that the rGO foams prepared in this way possessed structures with open pores and continuous cross-links of the rGO nanosheets. They could find potential applications as electrodes of supercapacitors and absorbents for selective absorption of oil and organic solvents for a clean environment.

Figure 7.51 (a) shows a schematic diagram of the experimental procedure. In the experiment, GO aqueous dispersions with a concentration of about 2.5 mg ml^{-1} were prepared, as shown in Fig. 7.51 (b). With the suspensions, nacre-like layer-structured films were fabricated by using the flow-directed assembly

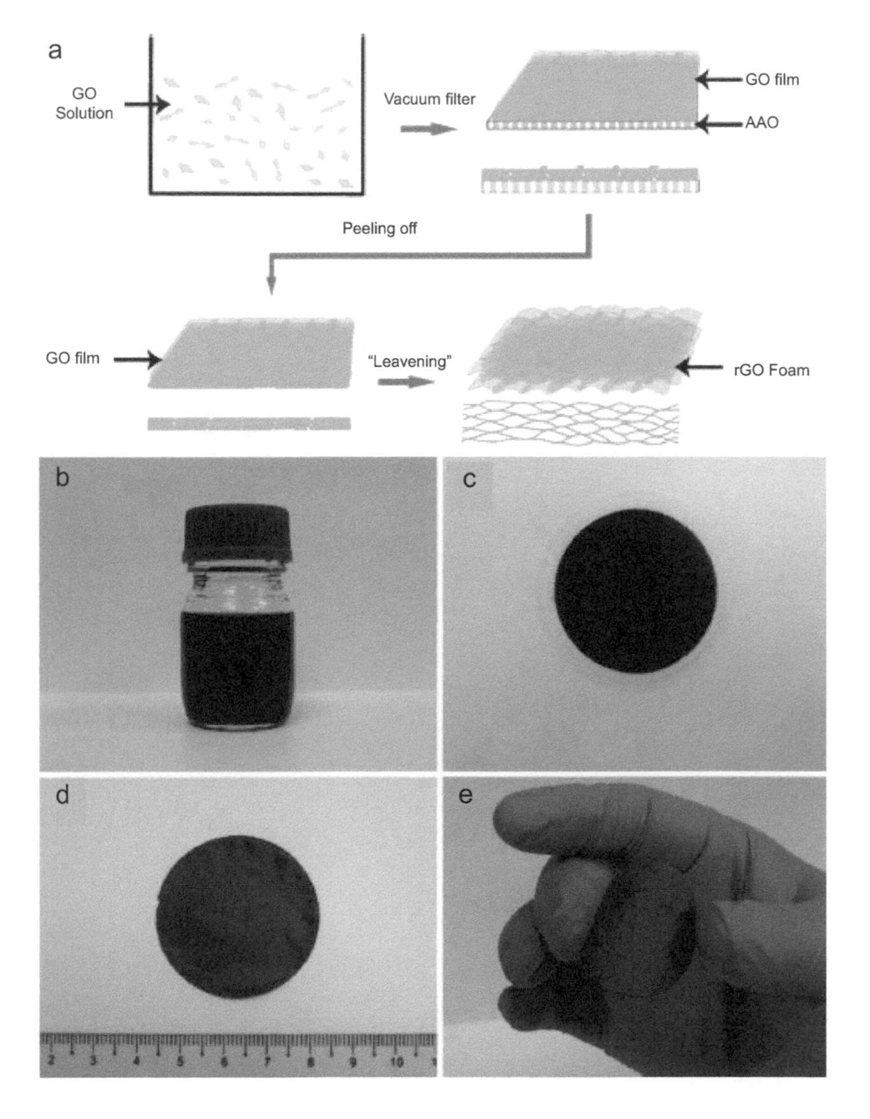

Fig. 7.51. (a) Schematic diagram showing the leavening process to prepare the rGO foams. (b) Photograph of the GO solution used for vacuum filtration. (c) GO film on AAO membrane. (d) Freestanding GO layered film. (e) Photograph of the freestanding paper-like rGO foam. Reproduced with permission from [17], Copyright © 2012, Wiley-VCH Verlag GmbH & Co. KGaA, Weinheim.

of the GO nanosheets. A porous anodized aluminum oxide (AAO) membrane with a pore diameter of 20 nm was used to filtrate the GO aqueous suspension to obtain samples with a shape as shown in Fig. 7.51 (c). Therefore, the thickness of GO films could be controlled by controlling concentration and volume of the GO suspensions. The films were peeled off from the AAO membranes, as illustrated in Fig. 7.51 (d). The freestanding GO films were put into a Teflon (polytetrafluoroethylene) vessel of 50 ml together with 80 μl hydrazine monohydrate. To avoid direct contact, the GO films were suspended well above the hydrazine monohydrate level. Hydrothermal treatment was carried out at 90°C for 10 h. Figure 7.51 (e) shows a photograph of the sample after the hydrothermal treatment. The foam structured film experienced a 50-fold volume expansion and 30% mass loss. During the reduction of GO to rGO by the hydrazine vapor, rapid evolution of gaseous species occurred, leading to the formation of the porous films.

Figure 7.52 (a) shows a cross-sectional SEM image of the film, demonstrating the presence of an open porous network with pore sizes in the range of sub-micrometer to several micrometers. The walls of the pores in the foams were continuously cross-linked, as illustrated in Fig. 7.52 (b). Because of the continuously cross-linked structures, the restacking of the graphene nanosheets in the films was prevented, while the films exhibited pretty high electrical conductivity, with a resistance of less than $100\ \Omega\cdot\square^{-1}$. The walls of the pores in the films were dense thin layers of stacked graphene nanosheets, as seen in Fig. 7.52 (c). The effective reduction of the GO to rGO was also confirmed by Raman and FTIR spectra, as depicted in Fig. 7.52 (d, e).

The role of hydrazine vapor in forming the rGO foams was further confirmed. For example, if 80 μl water was used to replace the hydrazine monohydrate, no change was observed in the morphology of the films. If the films were heated in a dry state or steamed in the presence of non-reducing solvents, e.g., toluene and benzene, the films were compact rather than porous, as observed in Fig. 7.53 (a). Therefore, the rapid evolution of the gases of H_2O and CO_2 could be the main reason for the formation of the porous structure. The pores were distributed evenly in the layer-structured films, thus forming the porous network. At the same time, the release of the gases also separated the compacted GO layers to produce the rGO foams, similar to the process of making bread from dough. As discussed in the last chapter, graphene papers made by using vacuum filtration contained a layer of water molecules in between the graphene layers, which provided diffusion paths for the hydrazine vapor to diffuse into the interior of the compact films.

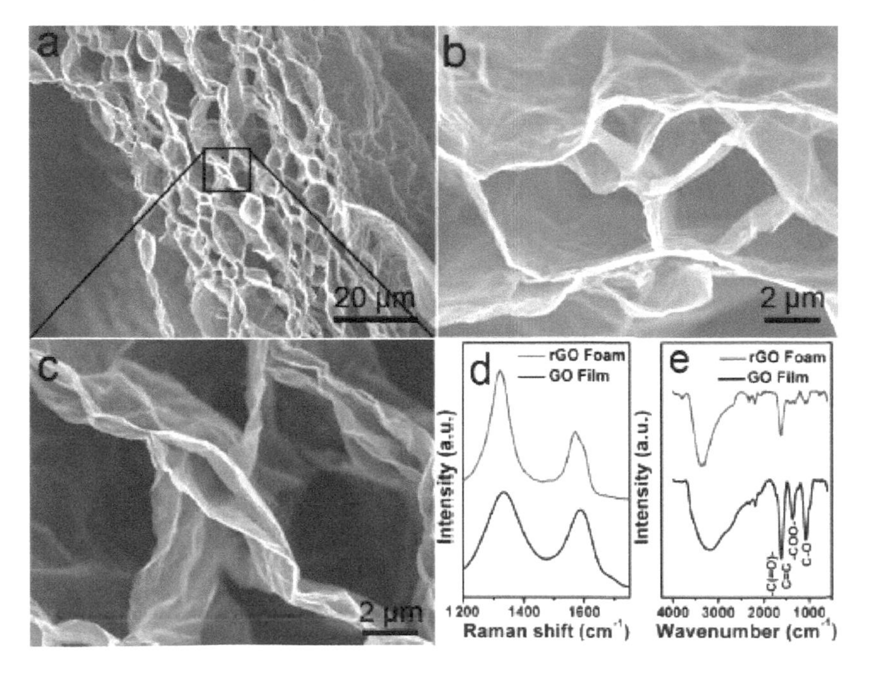

Fig. 7.52. (a–c) Cross-sectional SEM images of the rGO foams formed after 10 h of hydrothermal treatment at 90°C with 80 μl hydrazine monohydrate. Raman (d) and FTIR (e) spectra of the GO layered film (bottom) and rGO foam (top). Reproduced with permission from [17], Copyright © 2012, Wiley-VCH Verlag GmbH & Co. KGaA, Weinheim.

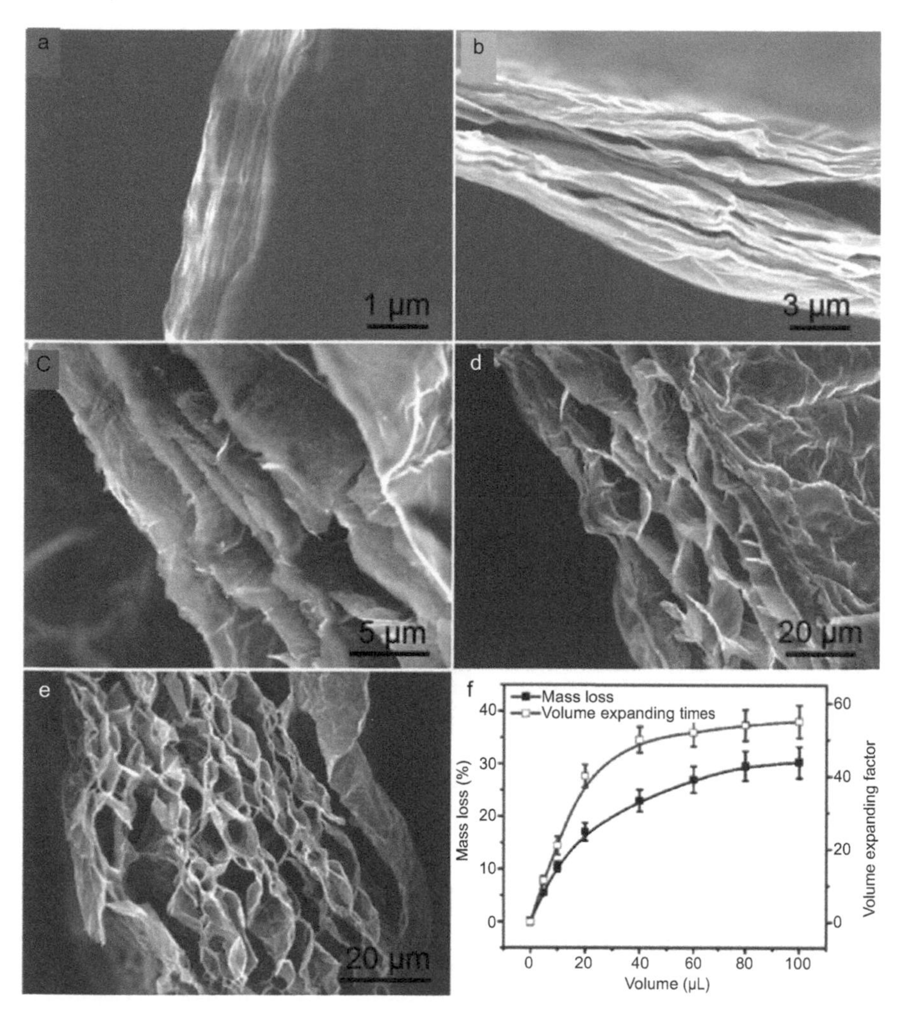

Fig. 7.53. (a–e) Cross-sectional SEM images of the rGO foams formed after 10 h of hydrothermal treatment at 90°C in the autoclave with different volumes of hydrazine monohydrate: (a) 0, (b) 5 μl, (c) 10 μl, (d) 20 μl and (e) 40 μl. (f) Volume expansion and mass loss of the rGO foams as a function of the volume of hydrazine monohydrate. Reproduced with permission from [17], Copyright © 2012, Wiley-VCH Verlag GmbH & Co. KGaA, Weinheim.

Porosity of the rGO foams could be controlled by controlling the amount of hydrazine used in the reduction reaction. Representative SEM images of the samples prepared at 90°C for 10 h with different amounts of hydrazine are shown in Fig. 7.53 (a–e). Without the presence of hydrazine, no rGO porous films could be obtained (Fig. 7.53 (a)). If the GO films were heated with 5 μl hydrazine monohydrate, the porosity was very low, as seen in Fig. 7.53 (b), with about a 11-fold volume expansion and about 5% mass loss. With the increasing amount of hydrazine in the autoclave, the porosity was increased gradually, as demonstrated in Fig. 7.53 (c–e) and 7.52 (a). However, too much hydrazine would not result in further increase in porosity. For instance, if more than 40 μl hydrazine monohydrate was used, no significant difference in the porosity of the samples was observed, as shown in Fig. 7.53 (f).

3.3.2.2. KOH activation of rGO porous structures

Chemical activation has been shown to be a useful procedure to produce porous carbon-based materials, which has also been employed to create highly conductive and porous rGO films [94, 95]. To realize the activation, GO suspensions were added with KOH at certain concentrations. The suspensions were then heated at 100°C to evaporate the water, thus leading to the formation of a thickened "ink paste". After that, the GO/KOH composite films were deposited by using vacuum filtration. After drying, the precursor films were activated in argon at 800°C for 1 h. The activated rGO films were free-standing with a smooth surface and uniform thicknesses. The flexible activated rGO porous films had a high in-plane electrical conductivity and high specific surface area. One example is discussed in a more detailed way as follows.

The activated reduced graphene oxide films were named as aGO [95]. The aGO films were mechanically flexible, physically free-standing and highly porous, with extremely high specific surface areas of 2400 $m^2 \cdot g^{-1}$ and a very high in-plane electrical conductivity of 5880 $S \cdot m^{-1}$. They had a very low H content of < 0.5 wt% and a high C/O atomic ratio of 14. The C/O atomic ratio was higher than that of the typical chemically converted graphene nanosheets. Figure 7.54 shows a schematic diagram to describe the processing steps of the film-like aGO porous thin sheets. Firstly, GO aqueous colloidal suspensions with a typical concentration of 1 $mg \cdot ml^{-1}$ were added with 1 M KOH dropwise, and were then heated in an oil bath at 100°C under constant stirring to evaporate the water, until ink pastes were obtained. The pastes contained GO nanosheets evenly dispersed with KOH together with a small amount of water, and were then used to deposit the films by using brief vacuum filtration with PTFE membrane.

To activate the GO/KOH films, the dry films were heated in a tube furnace in the flow of Ar at 800°C for 1 h at a pressure of 1 atm. Free-standing aGO films with thicknesses ranging from 7 to 30 μm were finally prepared, after thorough washing and drying. It was found that the concentration of KOH was a crucial factor that determined the homogeneity of the GO suspensions and hence the properties of the final aGO films. To ensure the layer-to-layer assembly, appropriate contents of KOH were required. Excessive KOH would lead to aggregation and precipitation of the GO nanosheets. With optimized conditions, the free-standing aGO films would be flexible, appearing with a metallic luster, as shown in Fig. 7.54. They had a smooth surface and uniform thickness. More importantly, they could be cut into different shapes and sizes, according to the requirements of the given applications. If KOH contents were not proper, dark cakes or delaminated papers would be formed, due to the inhomogeneous suspensions.

The degree of activation could be controlled by the content of KOH. Figure 7.55 shows SEM images of the aGO and the precursor GO/KOH films. The aGO films exhibited a flat and continuous sheet-like structure, as revealed in Fig. 7.55 (A, B), while the aMEGO (microwave exfoliated GO) had a chunk morphology, as demonstrated in Fig. 7.55 (D). The aGO14 film, derived from the suspension with a KOH/GO mass ratio of 14, had a continuous structure with a high porosity of micropores (Fig. 7.56), while the aGO16 film (KOH/GO ratio = 16) exhibited a rough surface that was heavily etched due to the high content of KOH. Microstructure of the final aGO films was directly related to the morphology of the precursor films. Figure 7.55 (C) demonstrates SEM image of the continuous precursor films, in which KOH was evenly intercalated in between the GO nanosheets.

It was found that the GO was partially reduced by KOH during deposition of the precursor film. TGA analysis results indicated that the GO/KOH precursor film undertook a slight weight loss (< 5 wt%) over 100–300°C in air, corresponding to the elimination of the oxygen-containing functional groups from the GO. Therefore, the removal of the functional groups was mainly realized during the evaporation process

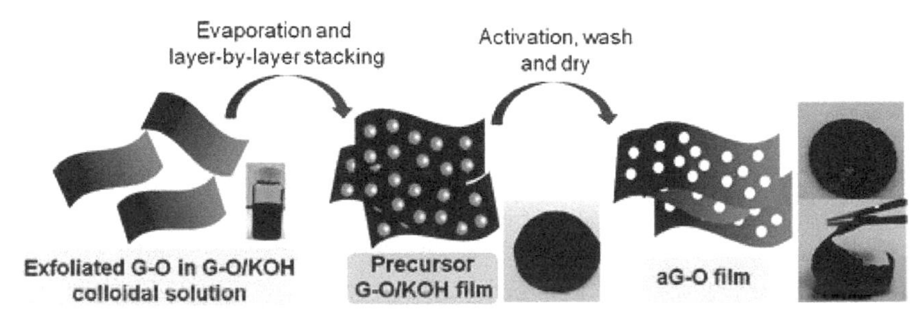

Fig. 7.54. Schematic diagram depicting the experimental steps to produce the aGO film. Reproduced with permission from [95], Copyright © 2012, American Chemical Society.

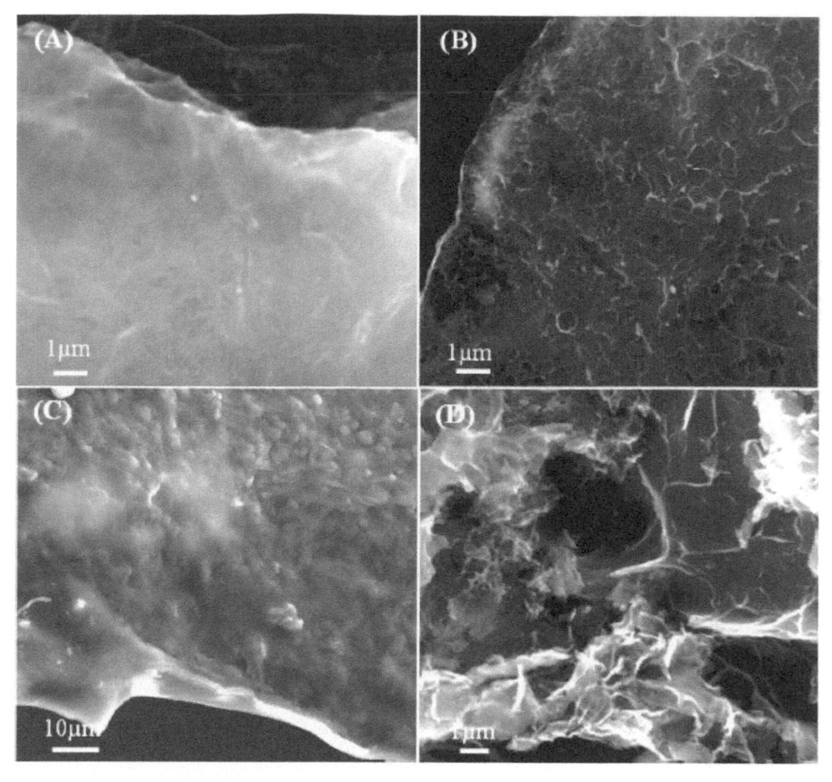

Fig. 7.55. Surface SEM images of the films: (A) aGO14 film, (B) aGO16 film, (C) the precursor GO/KOH film and (D) an aMEGO "chunk". Reproduced with permission from [95], Copyright © 2012, American Chemical Society.

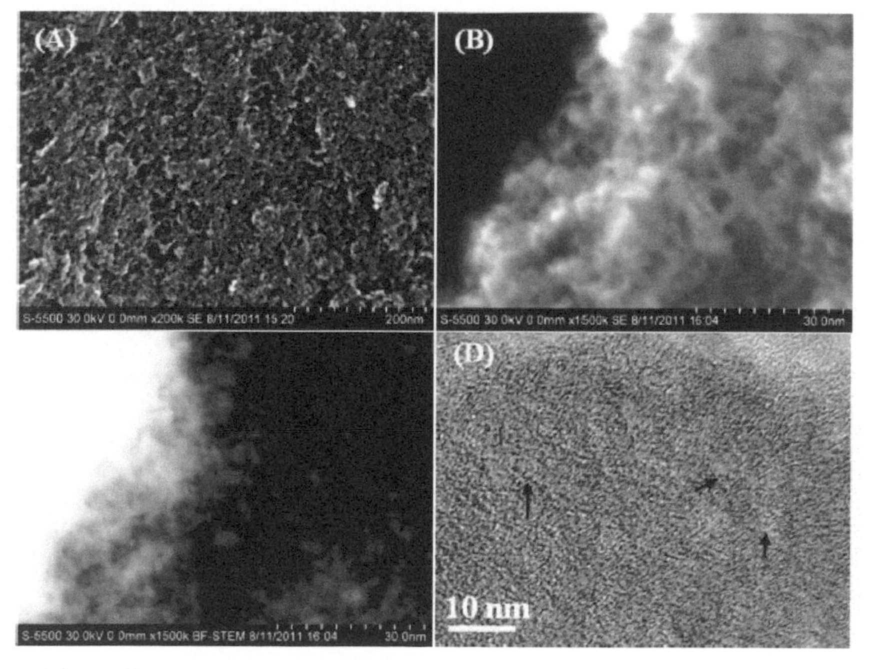

Fig. 7.56. (A) High-magnification SEM image of the aGO14 film with a highly porous structure. (B) SEM image of the aGO14 film and (C) the corresponding ADF-STEM image of the same region as in (B). (D) High-resolution TEM image of the aGO14 film, with the presence of uniformly distributed micropores of < 2 nm, as indicated by the black arrows. Reproduced with permission from [95], Copyright © 2012, American Chemical Society.

[96, 97]. C1s XPS spectra of the KOH reduced film and the as-deposited GO film indicated that the peak intensity of C=O was tremendously reduced after the KOH reduction, which further confirmed that effective deoxygenation of the GO nanosheets by KOH occurred mainly during the evaporation process.

High-magnification SEM images of the aGO14 film are shown in Fig. 7.56 (A, B), which revealed clearly that the sample possessed a highly porous and interconnected 3D microstructure. The corresponding annular dark field SEM (ADF-STEM) image is illustrated in Fig. 7.56 (C). A large amount of micro- and mesopores could be observed in the aGO14 film, which were homogeneously distributed throughout the highly porous structure. Figure 7.56 (D) shows high-resolution TEM (HRTEM) image of the sample, in which besides the densely present micropores, some small mesopores with sizes of < 5 nm could also be identified, as indicated by the black arrows.

7.4. Templating Approaches

Template-guided methods are also effective in developing 3D graphene architectures. This type of approach can be used to fabricate 3D porous graphene networks, with well-defined and shaped structures in the form of either 3D scaffolds or layered films [98–103]. The methods will be demonstrated and discussed by describing four groups of samples, including templated-directed CVD, ice-templating, PS colloidal temperate, and lithography method, in a detailed way.

7.4.1. Templated chemical vapor deposition (CVD)

Various rGO samples have been fabricated by different methods, some of which could be used to produce large-scale products. However, most of the rGO materials still have a relatively low electrical conductivity, which has limited their application in some areas, especially those that require high electrical conductivity. The main reason for that is the poor quality of the graphene nanosheets made by using those methods. Chemical vapor deposition (CVD) has been acknowledged to be an effective approach to produce high quality graphene nanosheets [104–110].

A 3D graphene macroscopic structure with a foam-like network, i.e., graphene foam (GF), has been fabricated by using a template-directed CVD method [99]. Unlike the structures formed with the chemically derived graphene nanosheets of relatively small lateral sizes, the GF fabricated by using the templated CVD method was a monolithic graphene 3D network, which facilitated the charge carriers to migrate with remarkably low resistance, due to the high-quality continuous CVD-derived graphene nanosheets. Therefore, this type of GF would have excellent electrical and mechanical properties.

Figure 7.57 shows the fabrication process of the GFs and their composites with polymers. Nickel foam, with a porous structure consisting of an interconnected 3D scaffold (Fig. 7.57 (a)), was selected as the template to grow the GFs. Methane (CH_4), used as the source of carbon, was used to deposit a thin layer of graphene into the nickel foam. High quality graphene layers were formed by decomposing CH_4 at 1000°C and ambient pressure [111–114]. The graphene layers contained a large amount of ripples and wrinkles, due to the difference in the thermal expansion coefficients between the nickel and graphene. The presence of these ripples and wrinkles offered mechanical interlocking with polymer chains, thus resulting in better adhesion with polymers when developing composite materials.

To prevent the graphene network from collapsing during the etching of the nickel skeleton, a thin layer of poly(methyl methacrylate) (PMMA) was deposited on top of the graphene films as a support. GFs were then obtained after the nickel foam was etched away with hot HCl or $FeCl_3$ solutions, as demonstrated in Fig. 7.57 (c, d). After that, the PMMA layer was gently removed with hot acetone, and monolithic GFs with a continuously interconnected 3D graphene network structure were produced, as seen in Fig. 7.57 (e). The graphene skeleton experienced a slight shrinkage. The interconnected 3D scaffold structure of the nickel foam template was well-retained by the GFs, so that all the graphene nanosheets in the GF were in good contact with one another. The application of the PMMA support layer was very important to create free-standing GFs, otherwise, the GFs would experience severe distortion and deformation. The final GFs were

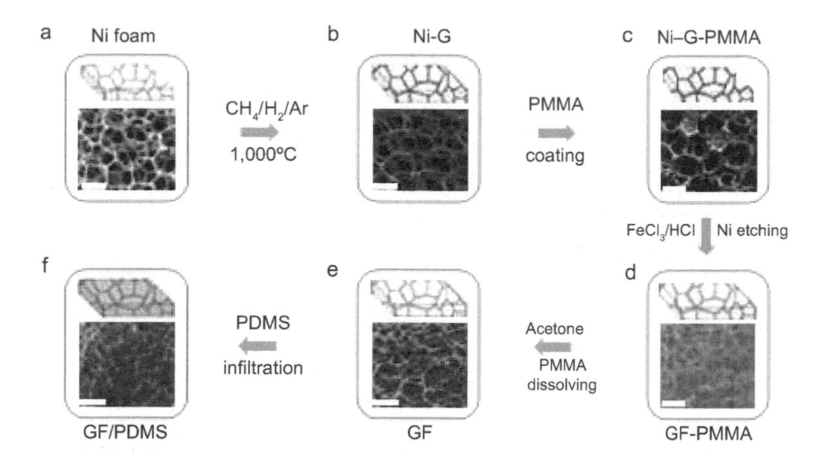

Fig. 7.57. (a, b) CVD growth of graphene films (Ni–G, b) by using nickel foam (Ni foam, a) as the template to fabricate the 3D graphene scaffold. (c) As-grown graphene film after coating a thin PMMA supporting layer (Ni–G-PMMA). (d) GF coated with PMMA (GF-PMMA) after etching the nickel foam with hot HCl or $FeCl_3$/HCl solution. (e) Free-standing GF after dissolving the PMMA layer with acetone. (f) GF/PDMS composite after infiltration with PDMS. Scale bars = 500 μm. Reproduced with permission from [99], Copyright © 2011, Macmillan Publishers Limited.

thinner than the original nickel foam, due to the liquid capillary force caused by the acetone evaporation. With the increasing number of graphene layers, the thickness of the GFs was increased from 100 μm to 600 μm, because thicker graphene sheets were stronger and thus had lower shrinkage.

The free-standing semi-transparent GFs were highly flexible and very light, with scaling-up capability by using large nickel foams, as shown in Fig. 7.58 (a). If the CH_4 concentration was 0.7 vol%, the final GFs had a density of as low as 5 $mg \cdot cm^{-3}$, corresponding to a porosity of as high as 99.7%. The GFs also possessed a quite high specific surface area of about 850 $m^2 \cdot g^{-1}$, corresponding to an average number of graphene layers of about 3. Figure 7.58 (c) shows a low-magnification TEM of the GFs, which clearly indicated that the ripples and wrinkles in the graphene building blocks of the GFs were well-retained after the nickel skeleton was removed. High-resolution TEM images revealed that the GFs consisted of single- to few-layer graphene nanosheets, as illustrated in Fig. 7.58 (d), which was in good agreement with the Raman spectra as depicted in Fig. 7.58 (e). The Raman spectra of free-standing GFs had a strongly suppressed defect-related D band, implying that the graphene in the GFs possessed pretty high quality.

Both macrostructure and the microstructure of the GFs could be controlled by controlling the processing conditions. For instance, by using nickel foams with different porous structures, the network and pore structures of the GFs could readily tailored. The average number of graphene layers, specific surface area and density of the GFs could be tuned by tuning the concentration of CH_4. The higher the concentration of CH_4 used, the larger the number of graphene layers would be. Accordingly, the properties of the GFs would be varied as a function of the number of graphene layers, as illustrated in Fig. 7.59 (c). If Cu foams were used as the template, the GFs had only monolayer graphene nanosheets due to the surface-catalyzed growth mechanism of graphene on Cu [115]. The single-layered GFs were too weak to be free-standing and to withstand the liquid capillary force caused by acetone evaporation.

Due to their special interconnected graphene network, the GFs could be used to fabricate composite materials with high electrical conductivity. For example, GF/PDMS composites have been prepared by infiltrating GFs with PDMS, as seen in Figs. 7.57 (f) and 7.59 (a). The composites were highly flexible and mechanically strong. Figure 7.59 (b) shows the fracture surface SEM image of the composites, indicating that the GFs were tightly adhered to the PDMS matrix. Because the interconnected 3D scaffold was not damaged, the composites had electrical conductivities close to those of the GFs, as observed in Fig. 7.59 (c). This observation is different from what had been observed in case of other graphene-based polymer composites, which have a critical concentration of graphene to beyond the percolation for high electrical conductivity.

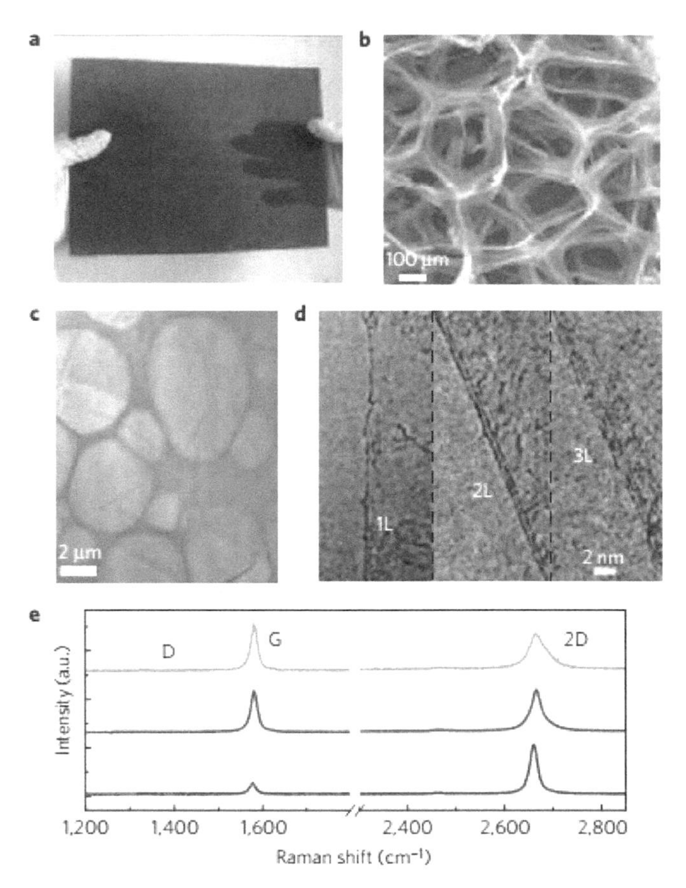

Fig. 7.58. (a) Photograph of the free-standing GF with a dimension of 170 × 220 mm^2. (b) SEM image of the GF. (c) Low-magnification TEM image of the GF. (d) High-resolution TEM images of the graphene nanosheets with different number of layers in the GF, with interlayer spacing of the bilayer (2L) and trilayer (3L) graphene to be about 0.34 nm. (e) Raman spectra of the GFs. The bottom spectrum and the other two correspond to monolayer and few-layer graphene, respectively, which were estimated from the intensity, shape and position of the G band and 2D band. The GF was prepared with a CH_4 concentration of 0.7 vol%. Reproduced with permission from [99], Copyright © 2011, Macmillan Publishers Limited.

7.4.2. Ice-templated unidirectional freezing

Unidirectional freezing, also described as ice segregation induced self-assembly (ISISA), is a well-known wet shaping technique to form porous materials, with unique porous microstructures [116–121]. This technique has also been applied to fabricate 3D graphene macroporous scaffolds, as well as their hybrids [98, 122, 123]. Several examples are presented in the following section to demonstrate the versatility of the technique.

A template-directed method has been used to prepare graphene–polymer nanocomposites with highly ordered 3D architectures, including sponge-like macroporous scaffolds and hollow micrometer-sized spheres [98]. The sponge-like macroporous scaffolds were obtained by using the ice-segregation-induced self-assembly (ISISA) of frozen aqueous dispersions, which were homogeneous mixtures of polystyrene sulfonate grafted graphene nanosheets (PSS-G) and poly(vinylalcohol) (PVA), as depicted in Fig. 7.60 (i). By directionally freeze-casting the PSS-G or GO aqueous dispersions, free-standing monoliths with a high surface area, constructed by internally aligned macro- and mesoscale pores were obtained. Also, PSS-G-coated polymer microparticles or hollow micrometer-sized PSS-G spheres were fabricated by depositing the negatively charged PSS-G dispersions on the surface of the positively charged polymer beads, as seen in Fig. 7.60 (ii). The PSS-G nanosheets had lateral sizes in the range from 500 nm to > 1 µm and a thickness of 1.5–2 nm.

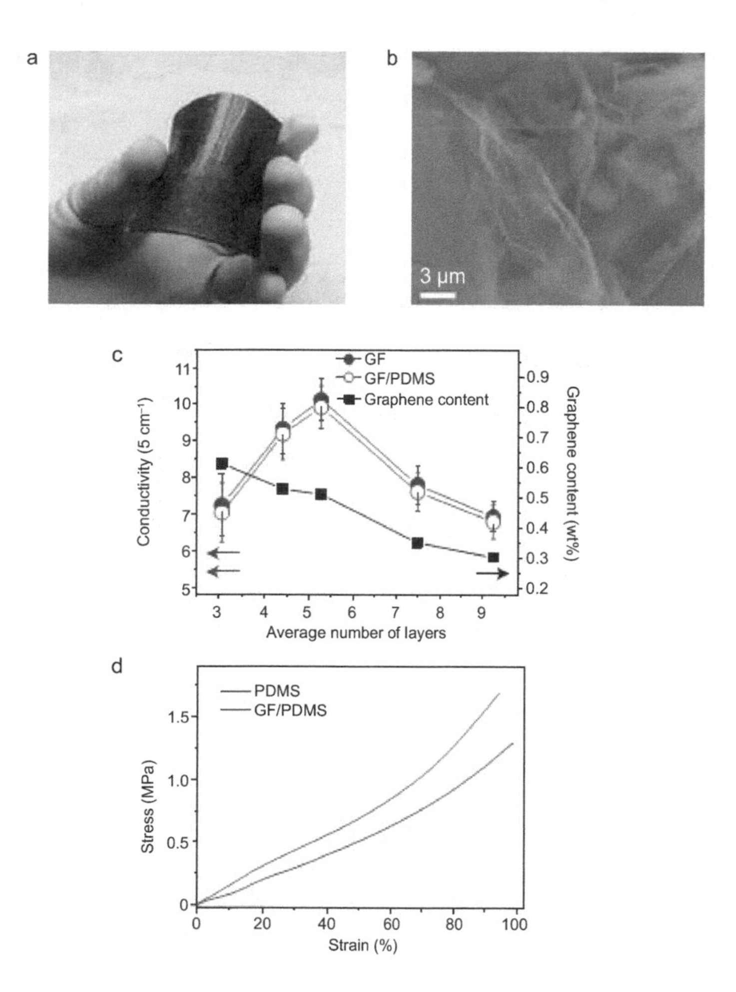

Fig. 7.59. (a) Photograph of the bent GF/PDMS composite, showing good flexibility. (b) Fracture surface SEM image of the composite. (c) Electrical conductivity of the GFs and GF/PDMS composites as a function of the number of graphene layers, corresponding to different weight fractions of the GFs in the composites. (d) Typical stress–strain curves of the PDMS and GF/PDMS composites with 0.5 wt% graphene content. Reproduced with permission from [99], Copyright © 2011, Macmillan Publishers Limited.

For the directional freeze-casting using the ISISA technique, a syringe containing homogenized mixtures of the PSS-G and PVA, or GO and PVA, with a typical PSS-G:PVA weight ratio of 1:10, was subjected to the unidirectional freezing in liquid nitrogen. The frozen samples were then freeze-dried to obtain porous monoliths, with a length of about 5 cm, as shown in Fig. 7.61. The as-fabricated scaffolds possessed a surface area of 37 $m^2 \cdot g^{-1}$. Morphologies of the longitudinal- and cross-sections of the samples in different regions, i.e., the immersion end of the syringe (X), the middle of the monolith (Y) and the far end (Z), were examined by using SEM, as illustrated in Fig. 7.62. According to the SEM images of the longitudinal sections of the PSS-G/PVA scaffolds (Fig. 7.62 (a, b)), the unidirectional freezing resulted in the growth of close-packed ice crystals, which compacted the PSS-G or GO nanosheets and the PVA components in between the boundaries of adjacent crystals. As a result, the subsequent high-vacuum sublimation of the ice produced highly ordered microchanneled structures became oriented along the freezing direction. Cross-sectional SEM images (Fig. 7.62 (c, d)) revealed that the walls of the scaffolds were highly textured and arranged in different motifs, which were dependent on their locations in the monolith. For instance, the region (X) was characterized by ordered lamellae with a spacing of about 5 mm, while the middle section (Y) contained mainly channel-like architectures with pores of 5–25 μm.

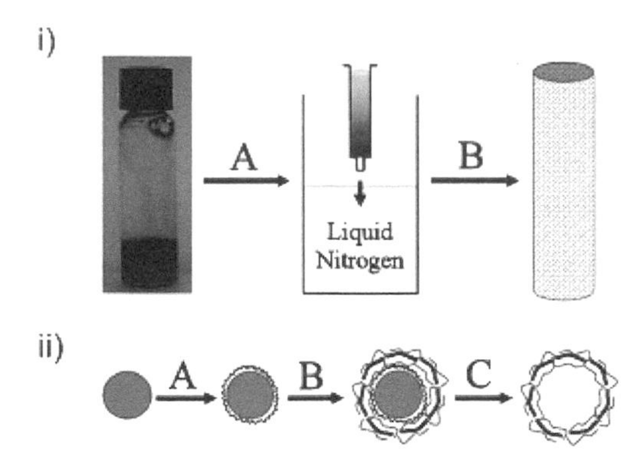

Fig. 7.60. Fabrication of the graphene–polymer nanocomposites with highly ordered 3D architectures. (i) Schematic diagram showing the formation process of the macroporous PSS-G/PVA monoliths by using the ice-segregation-induced self-assembly (ISISA): (A) homogeneous aqueous mixture of PVA and PSS-G (optical image) was transferred into an insulin syringe and unidirectionally frozen by immersion into liquid nitrogen at a dipping rate of 25.9 mm·min^{-1}, whereas (B) the frozen scaffold was then freeze-dried to sublime the ice crystals in order to produce a macroporous monolith. (ii) Schematic diagram showing the colloidal templating process to form the hollow PSS-G microspheres: (A) functionalization of the polystyrene beads with poly(allylamine hydrochloride) (PAH), (B) electrostatically induced templating of the negatively charged PSS-G nanosheets on the positively charged bead surface, and (C) removal of the core template by treatment with toluene to create hollow PSS-G microspheres. Reproduced with permission from [98], Copyright © 2009, WILEY-VCH Verlag GmbH & Co. KGaA, Weinheim.

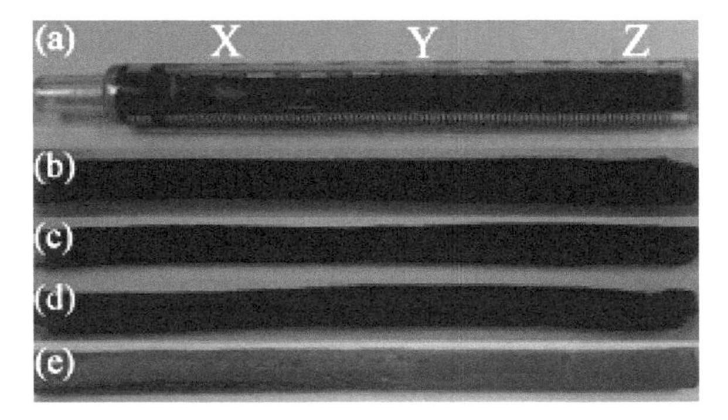

Fig. 7.61. Photographs of the freeze-dried monoliths fabricated by using the ISISA technique. (a) PSS-G/PVA monolith before removal from the syringe, with labeled sections, i.e., X (immersion end), Y (middle) and Z (far end). Photographs of the PSS-G/PVA monoliths with different weight ratios: (b) 1:5, (c) 1:10 and (d) 3:20. (e) Photograph of the GO/PVA monolith at a weight ratio of 3:20. Reproduced with permission from [98], Copyright © 2009, WILEY-VCH Verlag GmbH & Co. KGaA, Weinheim.

Figure 7.62 (e) shows high-magnification SEM image of the layered region. The walls had a thickness of about 1 μm, and were cross-linked by a secondary structure that traversed the inter-lamellar space. Figure 7.62 (e) clearly indicated that the walls were loosely packed with graphene nanosheets.

In a separate study, the ISISA approach was extended to fabricate macroporous rGO–Pt/Nafion hybrid films and 3D scaffolds, in which two steps were engaged [100]. Firstly, a porous network was derived from the aqueous suspensions of Nafion, GO and chloroplatinic acid, by using ice templating. Secondly, the porous network was reduced to result in graphene-supported Pt nanoparticles on Nafion scaffolds. Figure 7.63 shows a schematic diagram illustrating the formation process of the ice-templating technique. The porous hybrids exhibited both ionic and electronic conductivities along with catalytic activity, making them potential candidates for fuel cell and biosensor applications.

The scaffolds derived from the un-sheared Nafion dispersions consisted of well-defined ribbons periodically interconnected with transverse pillars, as shown in Fig. 7.64 (a). In contrast, the scaffolds made from the sheared viscous Nafion dispersions appeared as an interconnected porous 3D microstructure, as seen in Fig. 7.64 (b). Therefore, the more viscous and more sheared suspensions used, the larger the

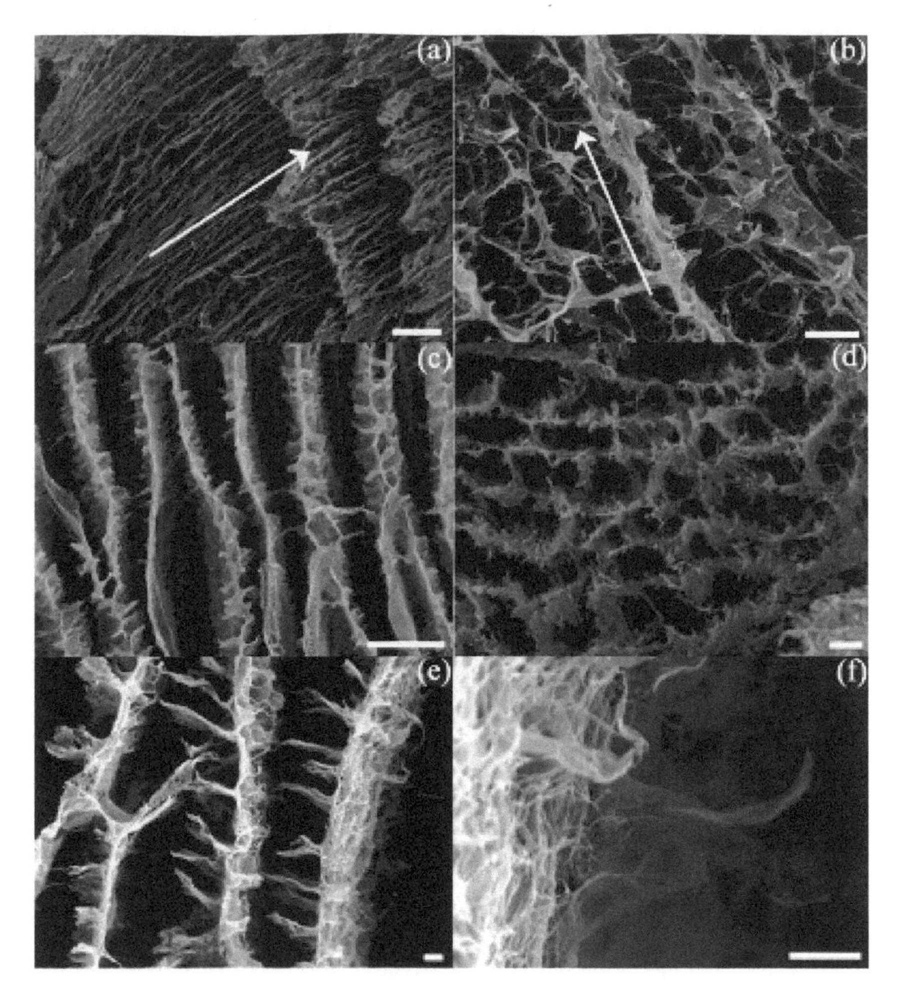

Fig. 7.62. SEM images of the PSS-G/PVA freeze-dried monolith (weight ration = 1:10) samples: (a, b) longitudinal-sections and (c, d) cross-sections. Scale bars = 10 μ (a, c, d) and 5 μm (b). Arrows in (a, b) indicate the direction of freezing. (a, c) Region X showing layered structure and (b, d) region Y with macroporous architecture. (e) Higher magnification SEM image of the cross-section in region X showing layered structure with inter-lamellar cross-links, with scale bar = 1 μm. (f) High magnification SEM image of the wall with loosely packed graphene nanosheets. Scale bar = 1 μm. Reproduced with permission from [98], Copyright © 2009, WILEY-VCH Verlag GmbH & Co. KGaA, Weinheim.

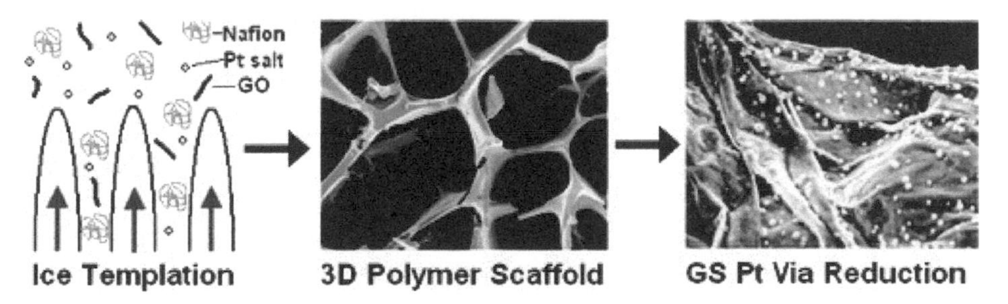

Fig. 7.63. Schematic diagram showing the ice templating method to synthesize the Nafion scaffold with Pt and graphene nanosheet precursors after chemical reduction. Reproduced with permission from [100], Copyright © 2011, American Chemical Society.

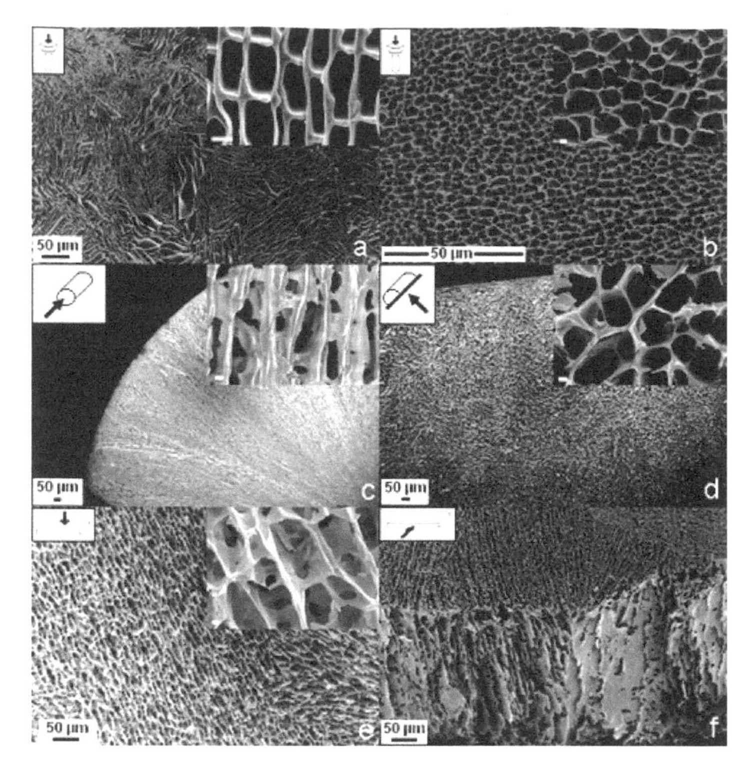

Fig. 7.64. SEM images of the cryostructured Nafion scaffolds. (a) Spherical samples derived from unsheared and (b) sheared aqueous suspensions. (c) Cylindrical samples viewed along the axis and (d) across the thickness. (e) Deposited films viewed across the thickness and (f) the edge plane. Insets at the top right of each image show the corresponding high-resolution SEM images. Scale bars = 1 μm. Reproduced with permission from [100], Copyright © 2011, American Chemical Society.

ice crystals would be formed, which promoted the branching behavior of the samples. Besides viscosity, there were also other factors that could influence the structure and morphology of the porous scaffolds, including the speed of the ice front, particle size of the species in the suspensions, osmotic pressure and surface tension of the suspensions [124, 125].

With sheared aqueous Nafion suspensions with concentrations of 10–15 wt%, robust crack-free scaffolds with different geometries could be obtained. In this study, the Nafion suspensions were immersed rapidly into the nitrogen slurry. The scaffolds exhibited highly anisotropic morphologies, as observed in Fig. 7.64, which was attributed to the presence of an intrinsic temperature gradient. Due to the cylindrical shape of the samples, the ice front travelled in the direction of the radius of the cylinder, i.e., from the periphery of the cylinder to the center, as demonstrated in Fig. 7.64 (c, d). The film samples possessed similar morphologies, with lamella arrays being perpendicular to the surface of the films, as observed in Fig. 7.64 (e, f).

The scaffolds had both nano- and macropores, which could be explained in terms of the presence of water. The nanopores of Nafion could absorb a large amount of water, which led to reversible swelling, while their size and connectivity were not affected by the processing parameters. On the other hand, the macropores were formed partly due to the ice templating effect, so that their size and morphology were closely related to the rate of crystallization of the ice and the particle size of the items in the suspensions.

To prepare the hybrid samples containing well-dispersed graphene nanosheets and Pt nanoparticles, substrates were immersed in homogeneous aqueous suspensions containing Nafion, GO and chloroplatinic acid, followed by freeze-casting. The resulting macroporous films were reduced with hydrazine or monosodium citrate in order to form GS and Pt hybrids. The outer surface, the middle region and the region near the substrate of the Nafion/GS–Pt films were examined by using SEM. It was found that both the nanosheets and the Pt nanoparticles were homogeneously distributed in the macroporous scaffolds. Figure 7.65 (a) indicated that individual graphene nanosheets on the surface of the Nafion were clearly visible. The Pt nanoparticles due to the reduction by hydrazine with sizes of 50–100 nm were attached onto the surface of the graphene nanosheets, as demonstrated in Fig. 7.65 (b). In comparison, if monosodium citrate was used as the reducing agent to replace hydrazine, the Pt nanoparticles became much smaller, with an average diameter of about 10 nm, as shown in Fig. 7.65 (d, e).

More recently, ice templating strategy has been employed to realize the uniform distribution of nitrogen doped graphene nanosheets, in order to achieve unique three-dimensional morphology of graphene-based materials, with enhanced electrochemical stability and ion transport network specifically for supercapacitor applications [122]. To synthesize nitrogen doped rGO (N-rGO), 20 ml GO solution with a concentration of 5 mg·ml^{-1} was diluted by adding 10 ml deionized water and was sonicated for 1 h to develop a homogeneous suspension. Melamine was used as the source of nitrogen, and was dissolved in deionized water by heating at 80°C for 30 min in order to form melamine solution. The GO and melamine solutions were mixed thoroughly with a mass ratio of 1:5 and then frozen in liquid nitrogen. After that, the samples were thawed following the phase diagram of water. To obtain N-rGO powder, the GO/melamine sponge type monoliths were annealed at 900°C for 30 min in Ar. Undoped samples of rGO were prepared similarly for comparison. During the annealing process, oxygen molecules were detached, this created active sites for nitrogen atoms to be incorporated into the honeycomb lattice of graphene. Figure 7.66 shows a schematic diagram to describe the synthetic procedure of the N-rGO.

SEM images of the rGO are shown in Fig. 7.67 (a, b), presenting an architecture caused by the interconnected ice-crystal growth assisted freeze-drying process. However, the N-rGO was characterized by a unique wrapped morphology, with a randomly oriented interconnected network exhibiting a continuous macroporous structure, as demonstrated in Fig. 7.67 (c, d). At the same time, a transparent wrinkled morphology was observed in the N-rGO samples. This could be attributed to the reaction between the oxygen containing functional groups on the GO nanosheets and the protonated amine groups of melamine, as the hot melamine solution was mixed with the GO solution [126, 127]. SEM images indicated that the samples without heat treatment possessed a similar structure, which suggested that the 3D morphology was retained from the pre-thermal treatment. There was only a slight shrinkage experienced by the samples after the thermal treatment. Representative TEM images are shown in Fig. 7.67 (e, f), indicating the presence of typical graphene nanosheets with marginally wrinkled profiles. Figure 7.67 (g) depicts the elemental mappings of C, N and O, which confirmed that N was uniformly distributed in the samples.

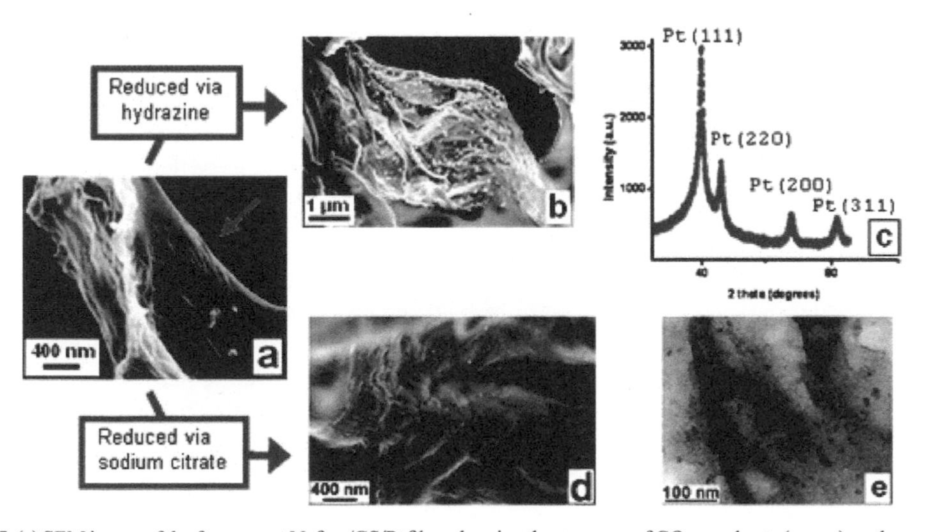

Fig. 7.65. (a) SEM image of the freeze-cast Nafion/GS/Pt films showing the presence of GO nanosheets (arrow) on the surface of the macropores. (b) Pt nanoparticles deposited on the surface of the graphene nanosheets, after reduction with hydrazine and (c) typical XRD pattern. (d, e) SEM and TEM images of Pt nanoparticles deposited on the surface of the graphene nanosheets, after reduction with monosodium citrate. Reproduced with permission from [100], Copyright © 2011, American Chemical Society.

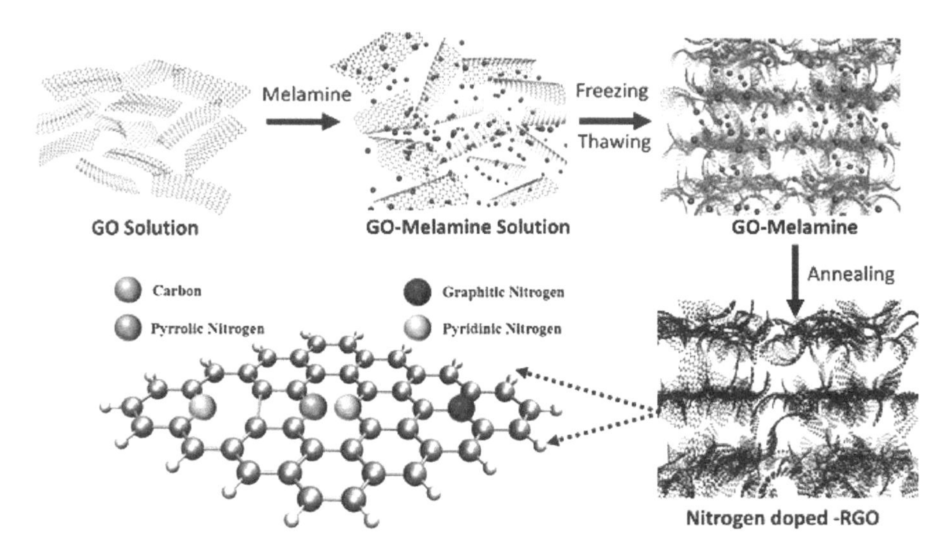

Fig. 7.66. Schematic diagram illustrating the procedure to synthesize the nitrogen doped graphene, with different nitrogen doping configurations in graphene lattice. Reproduced with permission from [122], Copyright © 2016, Elsevier.

Color image of this figure appears in the color plate section at the end of the book.

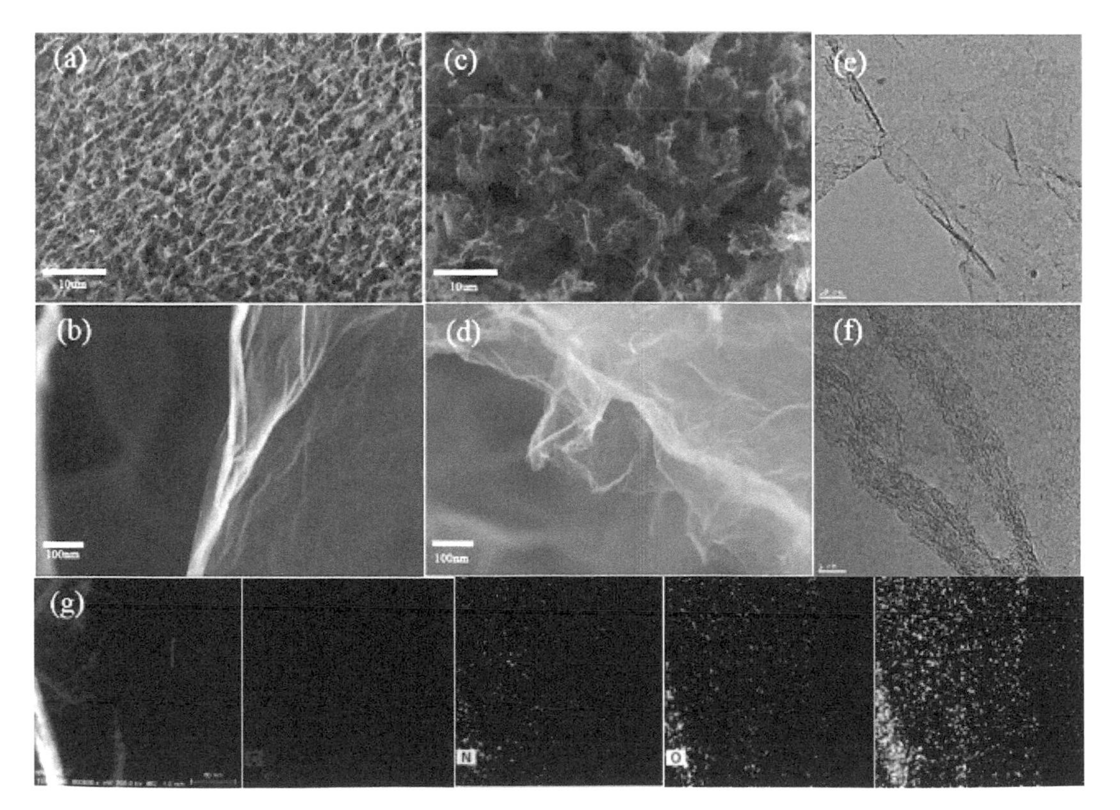

Fig. 7.67. (a, b) Low and high magnification SEM images of pure rGO. (c, d) Low and high magnification SEM images of the N-rGO. (e, f) TEM images of the N-rGO. (g) Dark field STEM image and C, N, O and overlaid elemental mapping images. Reproduced with permission from [122], Copyright © 2016, Elsevier.

The self-assembled monoliths had a specific surface area of 190 $m^2 \cdot g^{-1}$. The content of N was about 6.2% in the N-rGO, with a predominant pyridinic N-type configuration. When used as the electrodes of supercapacitors, a specific capacitance of 217 $F \cdot g^{-1}$ was achieved, at a scan rate of 5 $mV \cdot s^{-1}$, which was about three times higher than that of the pure rGO counterpart, due to the pseudocapacitive behavior of the N-rGO. The high rate of capability of the N-rGO electrode was closely related to its high electronic conductivity with low charge transfer resistance, which was confirmed by the results of electrochemical impedance spectroscopy. Moreover, the N-rGO electrode also demonstrated excellent cyclic stability with almost zero decay in capacitance after 5000 cycles at a scan rate of 100 $mV \cdot s^{-1}$.

The last example in this section is the preparation and characterization of interstacked $VOPO_4$–graphene hybrids, vertically porous 3D microstructures, by using the ice-templated self-assembly technique [123]. The new type of hybrid materials exhibited extraordinary performances as electrodes of supercapacitors, because of their superior characteristics, including micrometer-sized pores, vertically porous structures and high electrical conductivity. The vertically porous 3D structure of the graphene-based framework especially, had a high surface area, with vertical microchannels as pathways for the ions to be smoothly transported in the electrolyte. Meanwhile, in the inter-stacked structures of the 3D $VOPO_4$–graphene hybrids, the $VOPO_4$ and the graphene nanosheets were strongly bound through hydrogen bonding. Therefore, the plane contact of the graphene nanosheets was maximized, which in turn ensured a rapid charge transfer and structural stability of the electrode during the multiple redox reactions [128–130]. Coin-shaped asymmetric supercapacitor (ASC) assembled with the vertically porous 3D $VOPO_4$–graphene hybrids as electrodes achieved a high energy density of 108 $Wh \cdot kg^{-1}$.

Figure 7.68 shows a schematic diagram of the preparation procedure to prepare the vertically porous 3D nanohybrids of $VOPO_4$ and rGO nanosheets. The GO nanosheets had an average lateral area of about 6 μm^2 and an average thickness of ~1 nm, i.e., they were monolayer nanosheets. $VOPO_4$ nanosheets were derived from V_2O_5 powder and HPO_4, and were exfoliated with the aid of sonication in isopropanol [131]. They had an average lateral area of about 0.4 μm^2 and average thickness of about 5 nm, with 6–7 layers. Phase composition of the $VOPO_4$ was confirmed by XRD results.

During the ice-templated self-assembly process, because the liquid suspension was frozen, the perpendicularly growing ice crystals would expel the dispersed $VOPO_4$ and GO nanosheets, so that they were accumulated radially in between the ice crystals, as seen in Fig. 7.68. $VOPO_4$–GO nanosheets were inter-stacked in between the ice crystals, resulting in their directed self-assembly. Due to the directional segregation, 2D plane–plane interactions among the two types of nanosheets were largely promoted. The hydrazine vapor treatment reduced the GO to rGO, without chemical effect on the $VOPO_4$ in 3D nanohybrids, which was evidenced by XPS and Raman spectroscopy. The $VOPO_4$–rGO hybrid had an electrical conductivity of 1.3×10^2 $\Omega \cdot cm^{-1}$, which was five orders of magnitude higher than that of pure $VOPO_4$ (3.0×10^7 $\Omega \cdot cm^{-1}$). The nanohybrid had a high specific surface area of 325 $m^2 \cdot g^{-1}$.

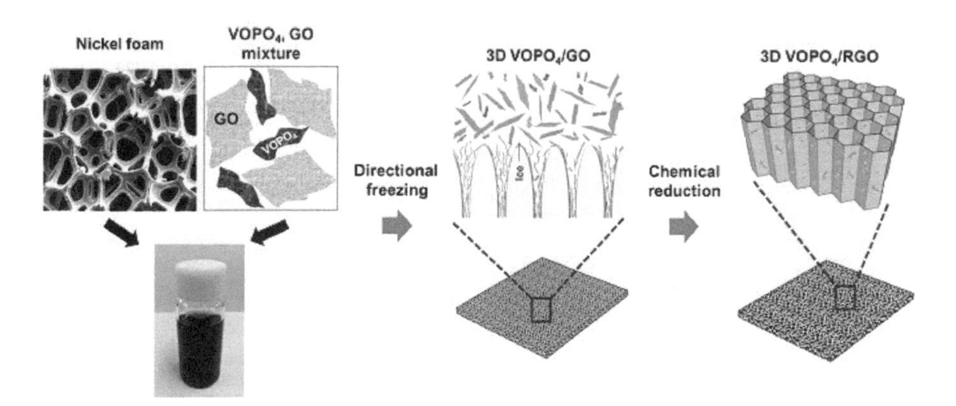

Fig. 7.68. Schematic diagram illustrating the fabrication procedure of the $VOPO_4$–rGO nanohybrids with vertically-aligned porous 3D structures realized through ice-templated self-assembly. Reproduced with permission from [123], Copyright © 2015, Macmillan Publishers Limited.

Figure 7.69 shows SEM images of the ice-templated 3D $VOPO_4$–rGO nanohybrids that were deposited inside nickel foam, as a current collector and a structural support for supercapacitor performance characterization. The 3D $VOPO_4$–rGO nanohybrid had a weight ratio of 1:1, which could be well-incorporated into the nickel foam to form an interconnected 3D porous network, as demonstrated in Fig. 7.69 (a). Figure 7.69 (b) shows a top-view SEM image of the 3D $VOPO_4$–rGO nanocohybrid, indicating that the pores with micrometer sizes were regularly packed, while the $VOPO_4$–rGO nanohybrid items were entirely in the form of a framework. Cross-sectional side and tilt SEM images are shown in Fig. 7.69 (c, d), from which the vertically-aligned microchannel structures oriented all the way along the freezing direction could be clearly identified. This was attributed to the unidirectional freezing and growth of the close-packed ice crystals. Such vertically-aligned microchannels were directly responsible for the high rate of ion transport in the electrolyte. High-magnification SEM and representative TEM images of the 3D $VOPO_4$–rGO nanocohybrid are depicted in Fig. 7.69 (e, f). The larger-sized rGO nanosheets were assembled to form the porous frameworks, onto which the smaller $VOPO_4$ nanosheets were attached to form a sandwich-like interstacked microstructure, through hydrogen bonds. This interstacked structure was especially important for electrochemical reactions, due to its structural stability and the minimized contact resistance between the $VOPO_4$ and graphene.

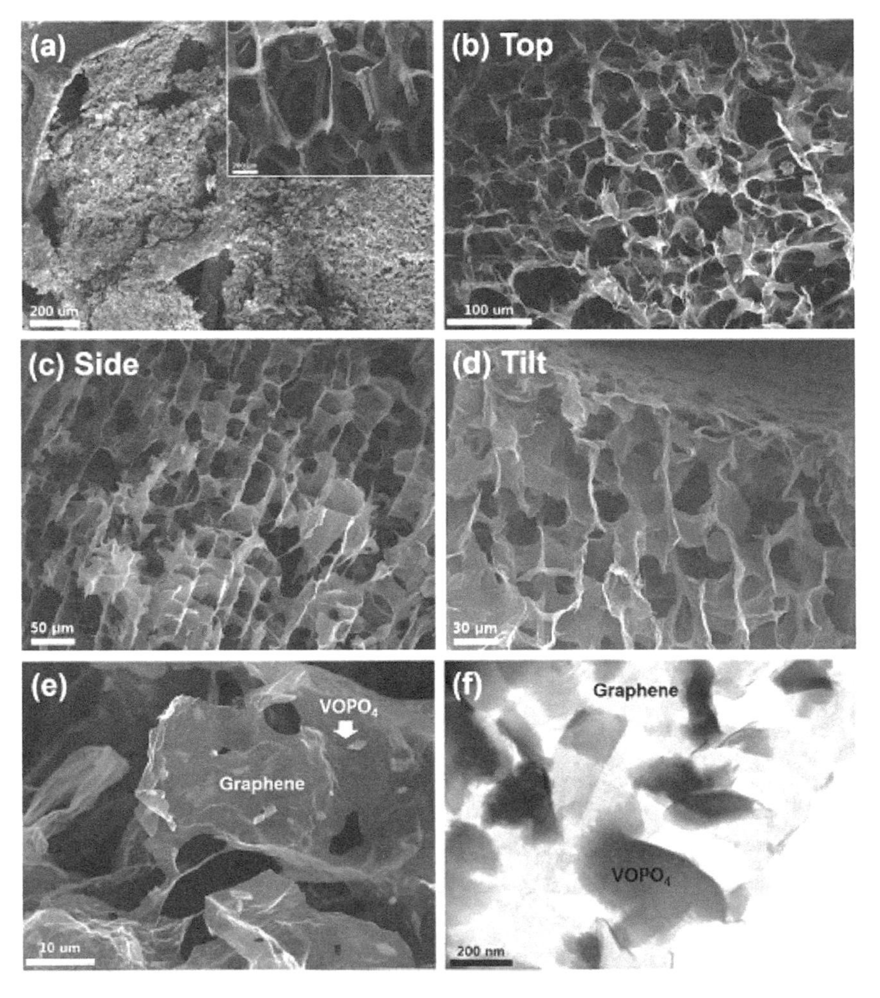

Fig. 7.69. (a) SEM image of the porous 3D microstructures of the $VOPO_4$–rGO nanohybrid incorporated inside nickel foam, with the inset showing the pristine nickel foam. (b) Cross-sectional top view, (c) cross-sectional side view and (d) tilt view of the porous 3D microstructures of the $VOPO_4$–rGO nanohybrid. (e) High-magnification SEM image and (f) TEM image of the porous $VOPO_4$–rGO nanohybrid. Reproduced with permission from [123], Copyright © 2015, Macmillan Publishers Limited.

7.4.3. PS colloidal templating

As stated earlier, polystyrene (PS) colloidal particles, which are a common sacrificial template, have been used as templates to promote the formation of hollow micrometer-sized rGO spheres and 3D graphene architectures by electrostatically induced assembly [98]. A slightly different process was reported to fabricate a 3D rGO macroporous structure templated by PS colloidal particles, which involved two steps, i.e., (i) the formation of free-standing PS/rGO films by using the vacuum filtration of a mixed aqueous colloidal suspension containing rGO nanosheets and PS particles with a diameter of 2.0 μm, (ii) the elimination of the PS template particles to obtain 3D macroporous structures, as shown in Fig. 7.70 [101]. The 3D rGO macroporous structures had a very high surface area and extraordinary mechanical strength, with the porous network constructed by interconnected rGO nanosheets (Fig. 7.70). The unique structure could be used for electrodes that were expected to have high ionic and electronic conductivities, as well as the potential to incorporate with metal oxides, such as MnO_2. Asymmetric supercapacitors were assembled by using the rGO and rGO–MnO_2 as the negative and positive electrodes, respectively.

Toluene was used to remove the PS template particles in the PS/CMG composites, thus leading to the open porous structures, as demonstrated in Fig. 7.71 (a, b). The porous structure remained intact after the PS particles were completely removed, since the interconnected walls formed by the multilayered rGO nanosheets were sufficiently strong to withstand the impact caused by the removal of the PS particles, as illustrated in Fig. 7.71 (c, d). Therefore, due to the high mechanical strength of the graphene nanosheet-based framework and the mild impact of the evaporation of toluene, the 3D porous structure could be well-retained. It was the presence of the interconnected rGO nanosheets in the porous networks that made them highly conductive, with an electrical conductivity of 1204 $S \cdot m^{-1}$.

MnO_2 was incorporated into the rGO porous structures by immersing them into 0.1 M $NaMnO_4$/0.1 M Na_2SO_4 solution at neutral pH value, for various dipping times (10–60 min), as schematically depicted in Fig. 7.70 [132, 133]. Figure 7.72 (a) shows a high-angle annular dark-field scanning TEM (HAADF-STEM) image, which indicated that the 3D macroporous structure was well-retained after the deposition of MnO_2. The presence of C and O, together with Mn that was homogeneously distributed on the walls of the 3D porous frameworks, was confirmed by the EDS elemental mapping results, as shown in Fig. 7.72 (a). The amorphous characteristics of MnO_2 were also observed in the high resolution HAADF image, as revealed in Fig. 7.72 (b). The MnO_2 layer fully covered the rGO nanosheets, with coating thicknesses of 20–30 nm. Figure 7.72 (c) shows XRS spectrum of the hybrid sample, confirming the entire coverage of the MnO_2. It was found that, with the PS templating process, the MnO_2 was only deposited over the cross-section of the dense samples.

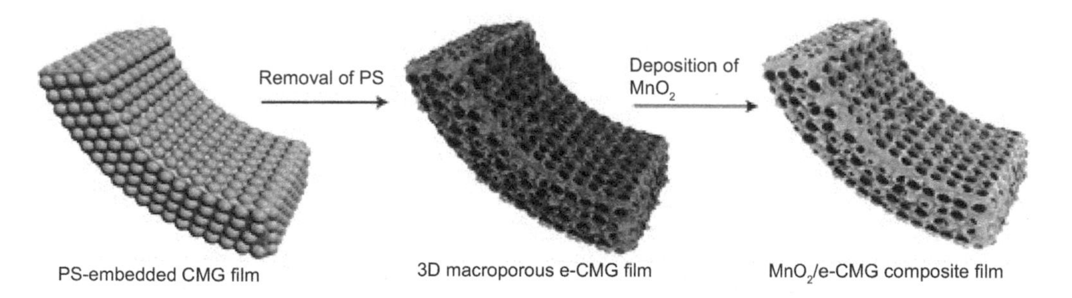

Fig. 7.70. Schematic diagram illustrating the procedure to prepare the 3D macroporous structures through the embossing process with PS particles as templates, followed by the deposition process of MnO_2 to produce rGO–MnO_2 hybrid porous frameworks. Reproduced with permission from [101], Copyright © 2012, American Chemical Society.

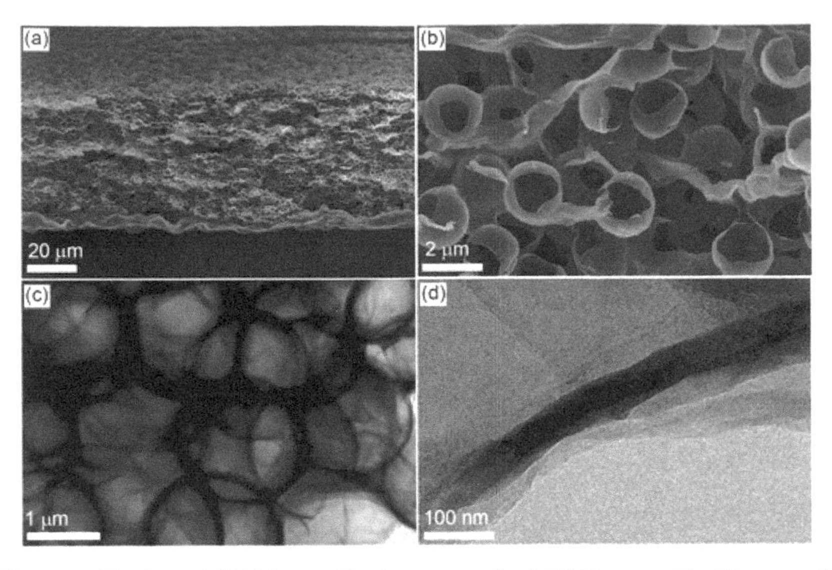

Fig. 7.71. (a) Low-magnification and (b) high-magnification cross-sectional SEM images of the 3D porous rGO framework structures. (c, d) TEM images at different magnifications of the porous rGO samples. Reproduced with permission from [101], Copyright © 2012, American Chemical Society.

7.4.4. Lithography method

Multilayered microporous graphene structures with dimensions that were two orders of magnitude smaller than those derived from nickel foam templates were prepared through the conversion of pre-defined 3D pyrolyzed photoresist films (PPFs) [103]. Three steps were involved in the fabrication process. The first step was the generation of 3D porous carbon structures with a face-centered cubic-like configuration, by using the interference lithography technology. In the second step, the 3D amorphous carbon structures were sputtered with a thin layer of nickel, and were then thermally annealed at 750°C for 50 min in H_2/N_2 (5%/95%), so that the carbon was converted into 3D graphene monoliths. The nickel was eventually etched out with acidic solutions, thus leading to 3D graphene structures, with pore sizes at the scale of 500 nm. The whole process is demonstrated schematically in Fig. 7.73.

Figure 7.74 (A) shows SEM image of the 3D porous carbon with a face-centered cubic-like structure, which was created by using the interference lithography. The structures had five interconnected layers, which were constructed with nanometer-sized carbon arms and nodes that were arranged in an in-plane triangular configuration. The as-produced 3D porous structures contained mainly sp^3 carbon, together with microdomains containing sp^2 carbon. Figure 7.74 (B) indicated that a thin layer of Ni was deposited entirely on the porous carbon structures, due to their open porous configuration. After thermal annealing, the carbon was converted into graphene, due to the catalytic effect of Ni. After that, the Ni layer was removed by etching in a 2 M H_2SO_4 solution for 8 h. As a result, 3D porous graphene structures were finally formed, with typical wrinkles and ripples, as demonstrated in Fig. 7.74 (C, D).

Figure 7.75 shows cross-sectional SEM images of the 3D porous graphene structures, revealing the presence of hollow structural characteristics. Because the amorphous carbon was prone to diffusion into Ni at sufficiently high temperatures, graphitization would occur at the surface of the nickel layer [134]. The formation of the hollow structure could result from two possible factors. On the one hand, the carbon could be completely reordered at the surface of the nickel. On the other hand, the carbon could also be incompletely diffused through the Ni layer, so as to be deposited in the interior of the Ni layer. In both cases, as the Ni layer was etched, 3D porous structures were obtained, as illustrated in Fig. 7.73. According to the SEM images, the thickness of the walls that connected the arms was less than 5 nm (Fig. 7.75 (A, B)), while that of the spherical nodes was in the range of 20–30 nm (Fig. 7.75 (C, D)). The presence of the two

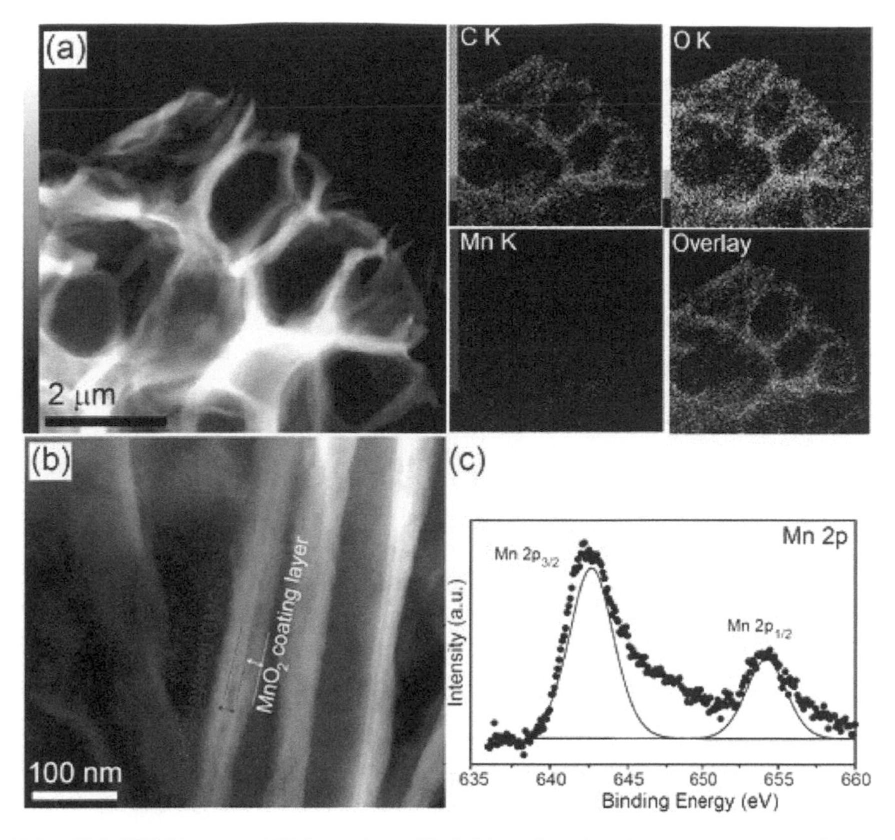

Fig. 7.72. (a) HADDF-STEM image and EDS mappings of C, O, Mn and overlay elements on a segment of the porous rGO–MnO_2 structures. (b) High-resolution HADDF-STEM image of the porous rGO–MnO_2 structures. (c) Mn 2p XPS spectrum of the porous rGO–MnO_2 structures. Reproduced with permission from [101], Copyright © 2012, American Chemical Society.

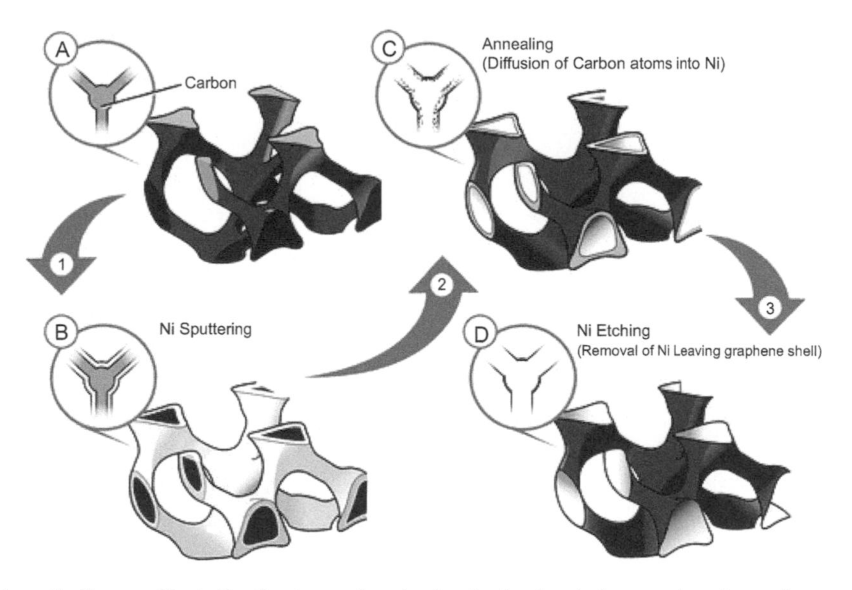

Fig. 7.73. Schematic diagrams illustrating the steps and mechanism for the chemical conversion of amorphous porous carbon into 3D porous graphene: (A) porous carbon, (B) conformal Ni coating, (C) diffusion of carbon into Ni top surface during thermal annealing and (D) hollow 3D graphene after the etching of the Ni layer. Reproduced with permission from [103], Copyright © 2012, American Chemical Society.

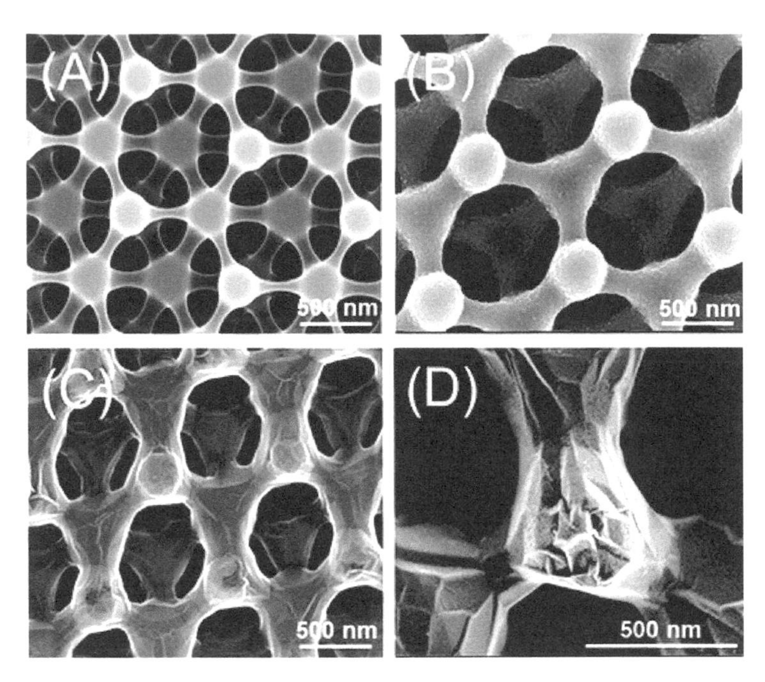

Fig. 7.74. SEM images of the porous carbon structures (A), porous carbon structures coated with nickel (B) and 3D porous graphene structures (C, D). Reproduced with permission from [103], Copyright © 2012, American Chemical Society.

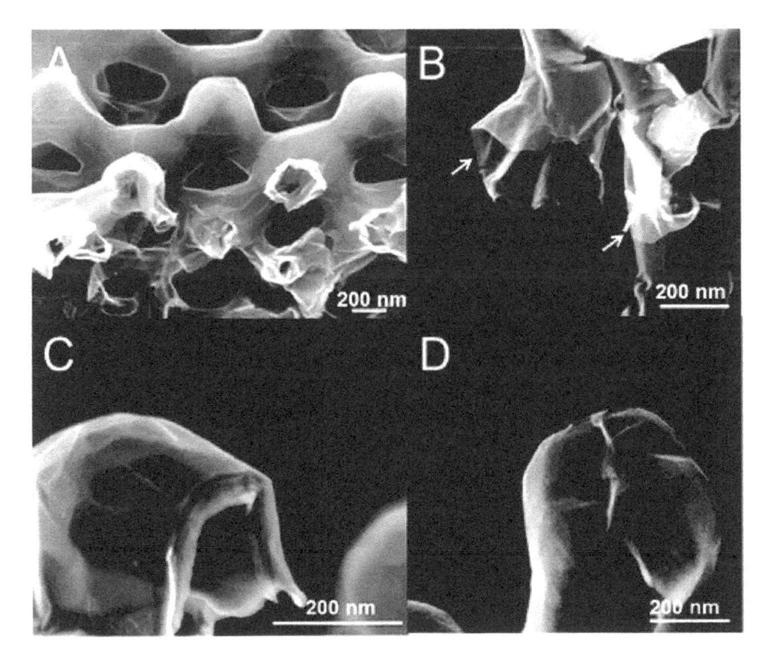

Fig. 7.75. Cross-sectional SEM images of 3D graphene structures, showing the details of the hollow thin arms (A, B) and thick nodes (C, D). The arrows point to the thin layered graphene nanosheets. Reproduced with permission from [103], Copyright © 2012, American Chemical Society.

distinct regions with different graphene thicknesses was further confirmed by the Raman analysis results. These porous 3D graphene structures could find potential applications as the electrodes of supercapacitors, due to their excellent electrochemical performances.

7.5. Solvothermal Assembly

Solvothermal treatment has been used to assemble porous GO/G-based frameworks (GOFs), for various applications [135–137]. A potential problem can be encountered when using the boronic acids because of their self-condensation to form anhydride moieties [138–140]. In this regard, it was suggested that diboronic acids should be used as the pillars to prepare the GOFs. For example, highly porous GO frameworks (GOFs) with a large specific surface area were developed by pillaring GO nanosheets with linear aromatic boronic esters through the well-known reaction between boronic acids and diols under hydrothermal treatment [135]. The solvothermal reaction occurred between the GO nanosheets and 1,4-phenyldiboronic acid at temperatures in the range of 80–150°C for 48 h in methanol. Figure 7.76 shows a schematic diagram of the hydrothermal reactions. GOF structures exhibited controllable pore size and volume, by controlling the length of the pillaring unit and the reaction temperature. If 100°C solvothermal temperature and 1,4-phenyldiboronic acid linker were used, then the GOFs possessed the optimized interlayer space and maximized BET surface area of 470 m^2 g^{-1}.

Grand canonical Monte Carlo (GCMC) simulations were carried out to evaluate the potential of the GOF materials for H_2 storage. A series of idealized systems with different diboronic acid linker concentrations, i.e., different pore size, pore volume, and surface areas, were evaluated. The optimized structure should have an interlayer spacing of about 1.1 nm. Absolute hydrogen adsorption isotherms at 77 K of the representative GOF structures derived from the simulations are demonstrated in Fig. 7.77 [135]. The simulation predictions were quite close to the experimental results. Reasonably, the higher the concentration of the linker, the denser the GOF structures, and thus the less accessible the pores would be. However, too low concentrations of the linker would weaken the structural stability and reduce the interlayer spacing in between the graphene nanosheets. For instance, in a pure graphene structure, the interlayer distance is about 0.34 nm, so that it has no capability to adsorb H_2. Although pure GO has a relatively larger interlayer spacing of 0.7 nm, it is unable to host H_2 as there are O and OH groups to fill the spaces. Therefore, there was a tradeoff that had to be considered when optimizing the structures of the GOFs, so as to maximize their H_2 storage capacity.

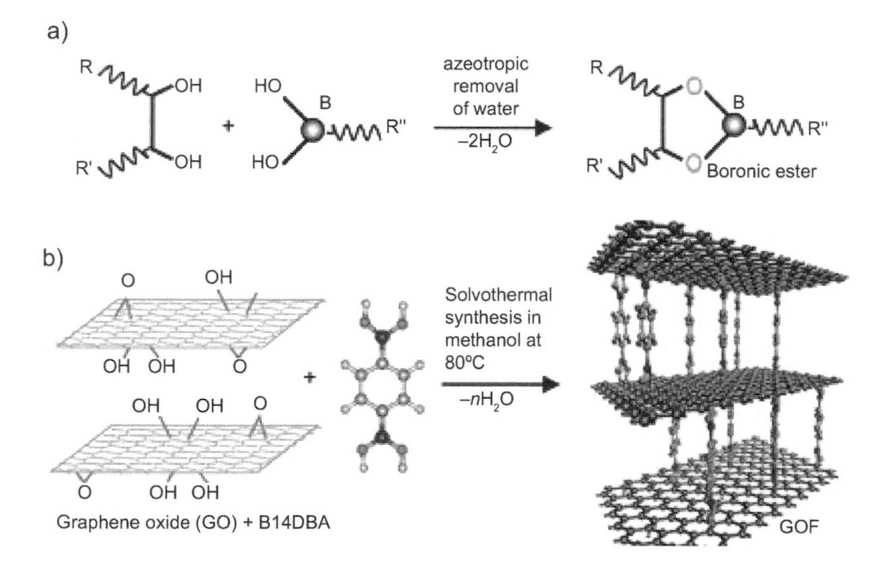

Fig. 7.76. Schematic diagram of the solvothermal reactions: (a) boronic ester and (b) GOF formation. Idealized GO frameworks (GOFs) were composed of layers of GO nanosheets connected by the benzenediboronic acid pillars. Reproduced with permission from [135], Copyright © 2010, Wiley-VCH Verlag GmbH & Co. KGaA, Weinheim.

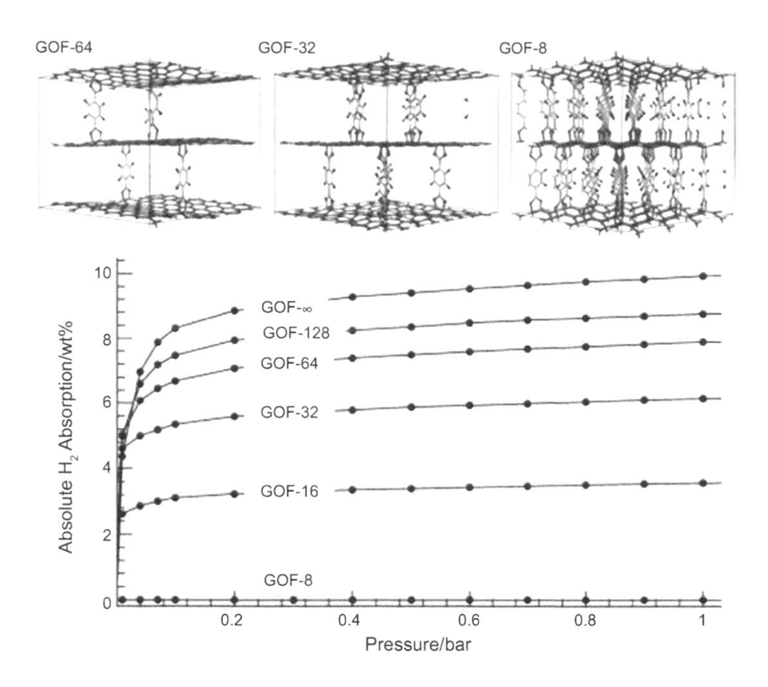

Fig. 7.77. Grand canonical Monte Carlo (GCMC) simulations to optimize the structures of the GOF-n with n graphene carbons per linker, together with the structures of three examples with n = 64, 32 and 8. Reproduced with permission from [135], Copyright © 2010, Wiley-VCH Verlag GmbH & Co. KGaA, Weinheim.

The above strategy has been extended to other porous GOFs by pillaring GO layers with various linear boronic acids [136]. Figure 7.78 shows a schematic diagram to illustrate the structures of the GOFs with the boronic acids. Due to the formation of underlying strong boronate-ester bonds in between the GO layers, an improved stability over GO nanosheets and significantly enlarged porosity had been achieved after the evacuation. As a result, the H_2 storage capacity of the GOFs at low pressures was increased by a factor of twice, because of their increased specific surface areas, as compared with those of the commonly available porous carbon materials. In addition, it was observed that the type and length of the cross-linkers, together with the reaction temperature, had a great impact on both the accessible surface area and framework stability of the GOFs.

A facial one-step solvothermal synthetic route was reported to prepare a graphene–metal organic framework (GOF) composite by using functionalized rGO nanosheets as the building blocks [137]. Solvothermal treatment of pyridine-functionalized rGO with 5,10,15,20-tetrakis(4-carboxyl)-21*H*, 23*H*-porphine (TCPP) and $FeCl_3$ in a DMF–methanol mixed solvent led to the formation of a graphene–metalloporphyrin GOF with specific surface area that was higher that of zeolite and porphyrin-based GOFs, or $(Fe{-}P)_n$ GOFs. Figure 7.79 shows a schematic diagram demonstrating the chemical structure and formation mechanism of the GOFs. The presence of rGO altered the crystallization process of the iron–porphyrin in the MOFs, thus resulting in an increase in the porosity and an enhancement in the electrochemical charge transfer rate of iron–porphyrin. G-dye was used to represent the rGO nanosheets with functionalization by the donor-π-acceptor dyes, which were terminated in pyridinium moieties, i.e., electron-withdrawing group. The solubility of the systems was improved by the pyridine ligands, due to the stabilization of the electron-rich phenylethyl group and the inhibited aggregation.

The combination of G-dye and $(Fe{-}P)_n$ MOF led to the formation of composites named as $(G\text{-}dye\text{-}FeP)_n$ MOF. Various weight percentages of G-dye (5, 10, 25 and 50 wt%) were combined with the chemical precursors of $(Fe{-}P)_n$ GOF to obtain (G-dye 5, 10, 25 and 50 wt%-FeP)$_n$ GOF composites, in order to clarify the structure–composition relationship. As the content of G-dye in the composites was increased from 5 to 50 wt%, the lattice distortion of the GOFs was increased gradually, so that a transition from crystalline into amorphous state was observed. In this case, the G-dye acted as an impurity intercalant

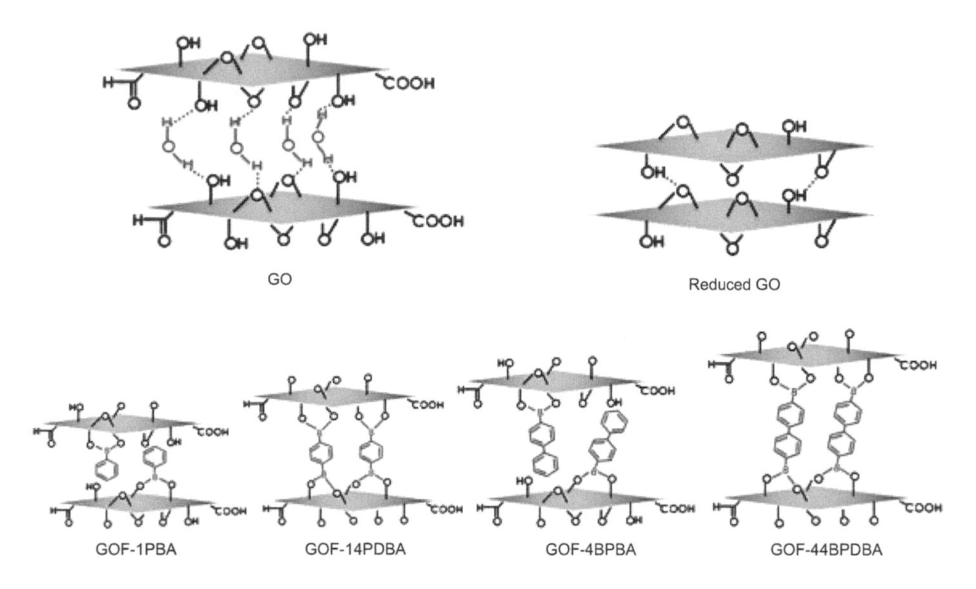

Fig. 7.78. Schematic diagram describing the structures of the GOFs with different types of phenylboronate pillaring units. Reproduced with permission from [136], Copyright © 2011, The Royal Society of Chemistry.

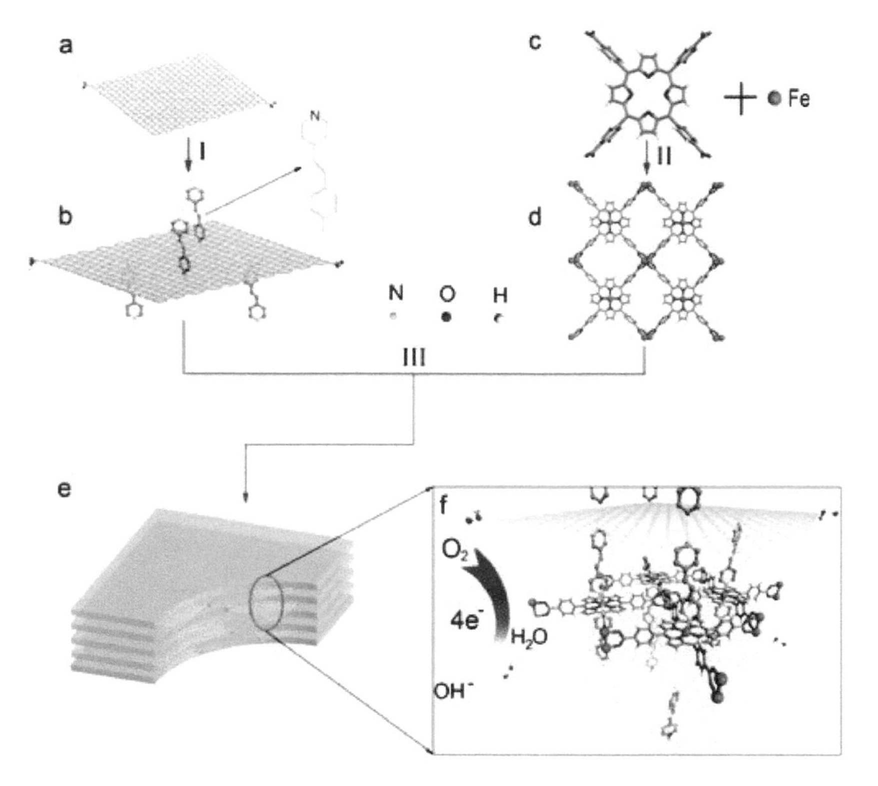

Fig. 7.79. Schematic diagram of the chemical structures of the components and composites: (a) rGO, (b) G-dye, (c) TCPP, (d) $(Fe-P)_n$ MOF, (e) $(G\text{-dye-FeP})_n$ GOF and (f) magnified view of the layers inside the framework of the $(G\text{-dye-FeP})_n$ GOF, showing the way the graphene nanosheets were intercalated by the prophyrin networks. Synthetic process to prepare the graphene–porphyrin GOFs: (I) G-dye derived from the rGO nanosheets through the diazotization with 4-(4-aminostyryl) pyridine, (II) $(Fe-P)_n$ GOF prepared through the reaction between TCPPs and Fe ions and (III) $(G\text{-dye-FeP})_n$ GOF synthesized through the reaction between $(Fe-P)_n$ GOF and G-dye. Reproduced with permission from [136], Copyright © 2012, American Chemical Society.

in the basic frameworks of $(Fe{-}P)_n$ GOFs. Figure 7.80 shows representative SEM images of the composite samples. Accompanying the variation in crystal structure, the morphology of the $(Fe{-}P)_n$ GOFs was changed in shape from plate to rod, as the content of the G-dye was increased from 5 and 25 wt%.

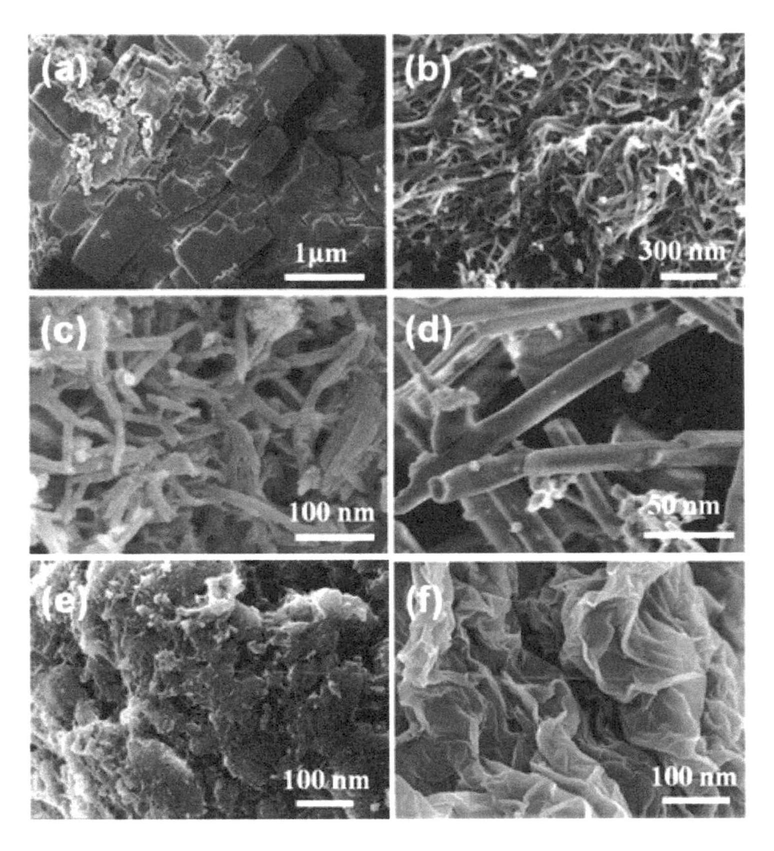

Fig. 7.80. SEM images of the samples: (a) $(Fe{-}P)_n$ MOF, (b) (G-dye 5 wt%-FeP)$_n$ MOF, (c, d) (G-dye 10 wt%-FeP)$_n$ MOF, (e) (G-dye 25 wt%-FeP)$_n$ MOF and (f) G-dye. Reproduced with permission from [136], Copyright © 2012, American Chemical Society.

7.6. Sol–Gel Polymerization

Ultra-low density graphene aerogels have been synthesized by using the sol–gel polymerization of resorcinol and formaldehyde (R/F), with sodium carbonate as the catalyst in an aqueous GO suspension, followed by pyrolysis of the GO–R/F gels [141, 142]. Firstly, GO–R/F gels were prepared by curing an aqueous suspension containing 4 wt% R/F and 1 wt% GO in sealed glass vials at 85°C, with a molar ratio of R to F of 200:1. The resulting GO–R/F wet gels were then thoroughly washed with acetone to remove the water inside the pores of the gel network, and were then dried with supercritical CO_2 to produce the GO–R/F aerogels. Finally, the dry GO–R/F aerogels were pyrolyzed at 1050°C in N_2 for 3 h, thus leading to the formation of monolithic graphene aerogels. These graphene aerogels possessed a high electrical conductivity of 1.0 $S{\cdot}cm^{-1}$, a large specific surface area of 584 $m^2{\cdot}g^{-1}$ and a pore volume of 2.96 $cm^3{\cdot}g^{-1}$. Such graphene aerogels could find potential applications in energy storage, catalysis and sensors.

SEM images of the graphene-based aerogels are shown in Fig. 7.81 [142]. All the graphene-based aerogels had a sheet-like microstructure. The content of RF exhibited a significant effect on the microstructure of the graphene aerogels. Although there were RF polymers in the precursors, no carbon nanoparticles were derived accordingly, even when the content was as high as 4 wt%. As the content of

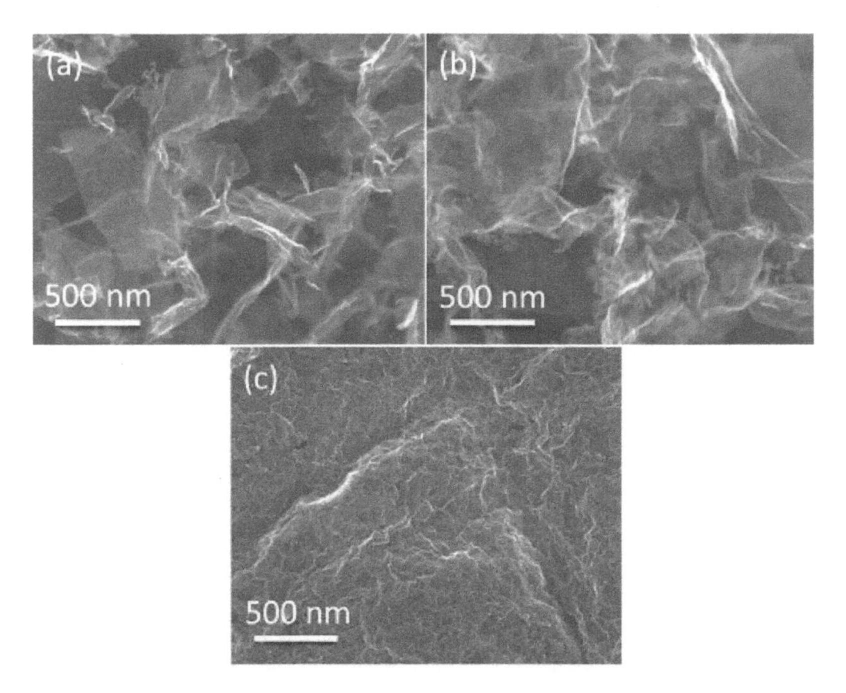

Fig. 7.81. SEM images of the graphene aerogels with different contents of the initial RF: (a) 4 wt%, (b) 2 wt% and (c) 0 wt%. Reproduced with permission from [142], Copyright © 2011, American Chemical Society.

RF was decreased, the microstructure of the aerogels became finer and finer. For example, some sheets in the aerogels were so thin that they were transparent. At the same time, the size of the pores in between the sheets became smaller and smaller.

An ultrafast acid-catalyzed sol-gel process in acetonitrile to prepare wet GO-based gels in 2 h at temperatures below 45°C has been reported [143]. Supercritical drying and pyrolysis were used to reduce the GO gels into aerogels without the presence of residual carbon-nitrogen bonds from the acetonitrile or its derivatives. This rapid synthesis method could be easily scaled-up to fabricate graphene aerogels with various applications, for example as sorbents for environmental toxins, support materials for electrocatalysis and as the high-performance electrodes of supercapacitors and solar cells. Figure 7.81 outlines the sol-gel process used to prepare the graphene-based gels and aerogels. GO-acetonitrile dispersion was first homogenized with the aid of sonication, into which RF-catalyst reaction mixture was added, followed by a second sonication, so as to achieve rapid gelation. Both RF aerogels (RFA) and GO-loaded RF aerogels (GO-RFA) were obtained through carbonization, which were denoted as RFA-P and GO-RFA-P, respectively.

Low and high magnification SEM images of the GO-RFA-P are shown in Fig. 7.82 (a), from which graphene nanosheets could be clearly identified. The randomly-oriented graphene nanosheets possessed a lateral size at the micron scale, and were embedded in a network of carbonized chains derived from the RF. The carbonized RF chains had a relatively large spacing of > 100 nm. Small RF particles were mainly observed on the surface of the GO nanosheets, which could be attributed to the fact that the solvent was on the surface of the GO nanosheets in the precursor suspensions, because of the negatively charged O^- and COO^- groups. Figure 7.82 (b) shows SEM images of the RFA-P sample. Without the presence of the GO nanosheets, a fine sponge-like structure was formed, while the filamentous morphology was not observed. The RF network was denser and exhibited finer pores and features. Higher magnification image indicated that the RFA-P was extraordinarily rich in uniformly sized (< 20 nm) bead-like particles, which were joined to form a disordered array.

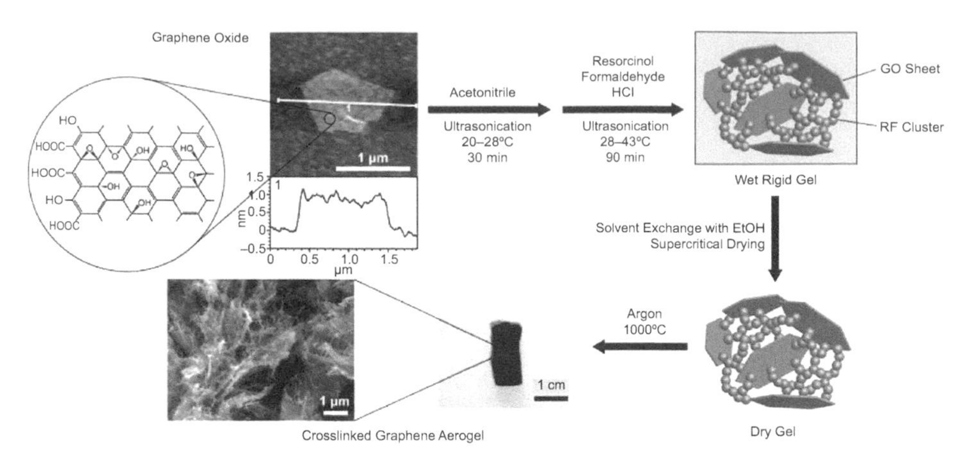

Fig. 7.82. Diagram showing the fabrication process, microstructure and representative samples of the GO–RF aerogels. Reproduced with permission from [143], Copyright © 2015, Elsevier.

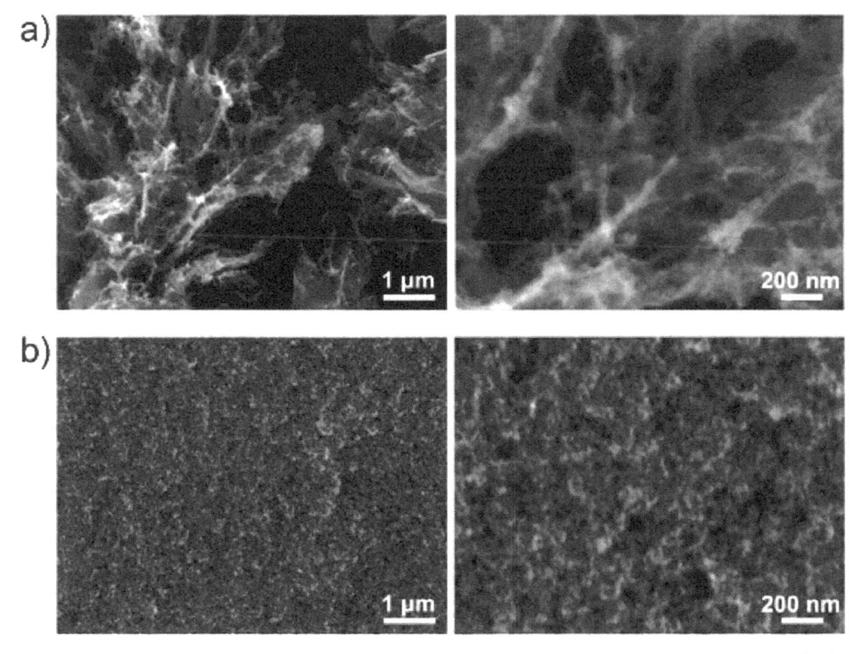

Fig. 7.83. SEM images of the samples at different magnifications: (a) pyrolyzed 5.5 wt% RF aerogel loaded with 1.7 wt% GO and (b) pyrolyzed 5.5 wt% RF aerogel. Reproduced with permission from [143], Copyright © 2015, Elsevier.

7.7. 3D Printing

Conventional manufacturing is known as a subtractive process, in which an unwanted portion is removed from a work piece in order to fabricate an item with the desired shapes. In contrast, additive manufacturing (AM), also called 3D printing, creates objects from 3D model data by joining materials, generally in a layer-by-layer way [144]. Therefore, 3D printing is superior to the traditional subtractive manufacturing [145–150]. For example, it can be used to produce complex items and parts that are made of expensive materials. Complex items are those that have special characteristics, such as complicated geometries, internal features, engineered porosity and material gradients. Also, it is suitable for a high level of customization with small-scale production. More recently, 3D printing has been employed to produce graphene-based complicated items with special functionalities [151–156]. Representative examples are discussed in the following section.

More recently, electrically conductive, mechanically resilient and biocompatible scaffolds with a high graphene content of 60 vol% in solid have been developed by using 3D printing, with printable graphene inks [154]. The 3D graphene (3DG) inks were synthesized with graphene nanosheets and the biocompatible, biodegradable and hyperelastic polyester polylactide-co-glycolide (PLG). The 3D printing was conducted at room temperature by using an extrusion method to obtain self-supporting user-defined structures with high fidelity and precision. The graphene-based structures were mechanically strong while maintaining high flexibility. Moreover, the structures had quite high electrical conductivity, and could be used as electrically conducting scaffolds for potential applications in tissue regenerative engineering. Figure 7.84 shows a schematic diagram of the whole process for the fabrication and characterization of the 3D graphene-based bioactive structures.

The 3DG inks were derived from commercially available materials by using the standard mixing method in an ambient environment. This simple process, together with their high stability, made the inks suitable for scalable production. To prepare the 3DG inks, PLG was first dissolved in dichloromethane (DCM), while graphene powder was dispersed in a mixture of DCM and a small amount of 2-butoxyethanol and dibutyl phthalate. The graphene power contained nanosheets with thicknesses of 3–8 atomic layers. Volume ratios of graphene to PLG suspensions include 3:2 (60 vol% or 75 wt% graphene), 2:3 (40 vol% or 56 wt% graphene) and 1:4 (20 vol% or 32wt % graphene), corresponding to weight ratios of 3:1, 3:2, and 3:7, respectively. Excessive DCM was initially used to promote the dispersion and mixing of the components, and was then evaporated during sonication. Desired 3DG inks were developed until the quasi-static shear rate viscosity approached about 30 Pa·s. The obtained 3DG inks should be stored in well-sealed glass containers at a sufficiently low temperature. Due to the shear-thinning nature of the inks and the graded volatility of the solvents, room temperature extrusion-based printing to produce self-supporting structures could be realized.

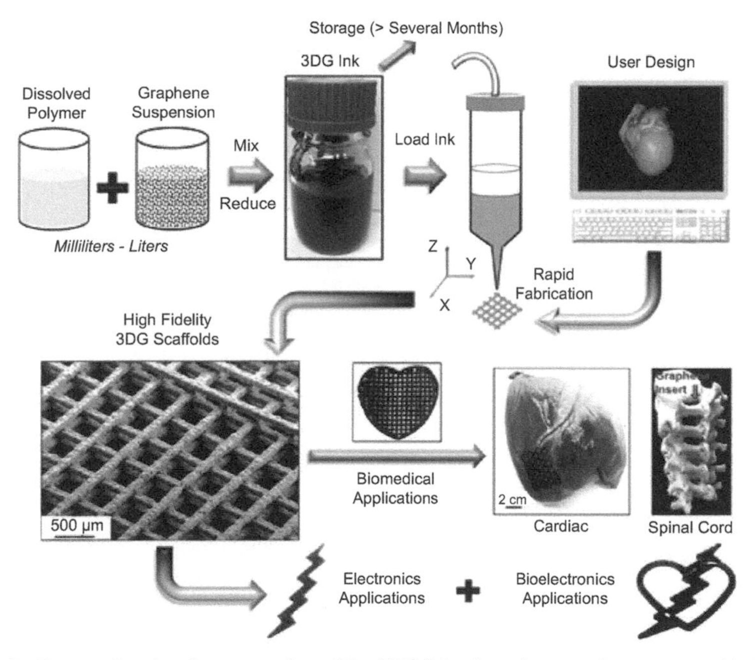

Fig. 7.84. Schematic diagram showing the preparation of the 3DG inks, from the extrusion process with the inks to the 3D printed graphene-based structures and their potential applications. The 3DG inks were obtained through simple combination and mixing of elastomer solutions with the dispersion of graphene powder in a graded solvent, followed by volume reduction and thickening. User-defined architectures 3D-printed with the 3DG could find various potential applications, for example as the electrodes of energy storage devices, and in the fields of bioelectronics, and tissue and organ engineering. Reproduced with permission from [154], Copyright © 2015, American Chemical Society.

The 3DG ink with 60 vol% graphene and 40 vol% PLG was used to carry out the 3D printing. Due to the 2D nature of the graphene nanosheets and the uniaxial nature of the printing process, the objects exhibited anisotropic microstructures and properties. The 3DG ink was a viscous dispersion, with randomly-oriented graphene nanosheets suspended in the dissolved elastomer solution. During the extrusion process, the shear forces triggered the reorientation and alignment of the graphene nanosheets along the direction of the flow, as illustrated in Fig. 7.85 (a). As a result, filament microstructures with the graphene nanosheets oriented along the fiber surface were formed, as seen in Fig. 7.85 (b–e). In this case, the content of the elastomer was critical to form a continuous interconnecting matrix in between the graphene nanosheets. At the same time, evaporation of the DCM upon extrusion should be sufficiently fast enough to ensure the formation of self-supporting fibers without significant deformation after the deposition, as observed in Fig. 7.85 (a). Because of the high vapor pressure of DCM, together with the high surface area to volume ratio of the extruded objects, rapid evaporation of the DCM was achievable.

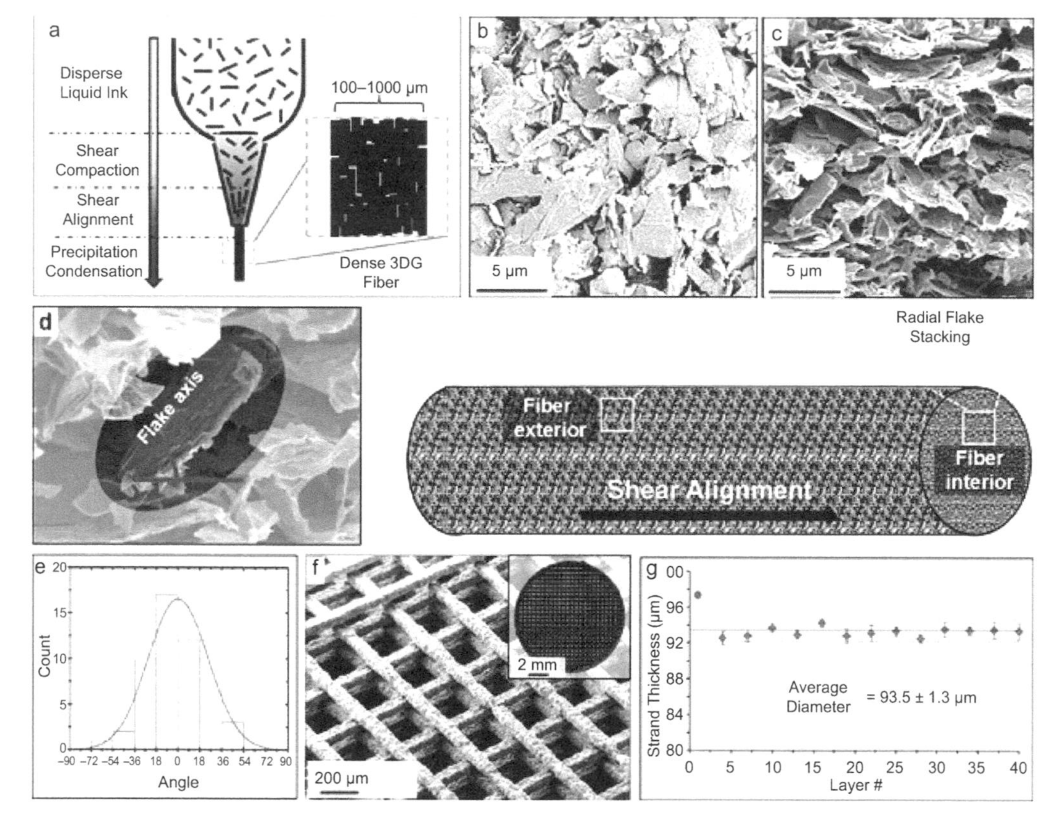

Fig. 7.85. (a) 3DG inks (60 vol% graphene + 40 vol% PLG) were liquid before extrusion. At pressure, the ink flew into the narrowing diameter nozzle, while shear forces drove the graphene nanosheets into alignment. Upon exiting the nozzle, the DCM was rapidly evaporated, while the fiber was solidified, with a slight diameter reduction. (b, c) SEM images of the exterior and cross-section of the fiber, respectively, showing the alignment of the graphene nanosheets along the exterior of the fibers and the nanosheet stacking within the fibers. (d) Measurement of the graphene nanosheet orientation in an end-on cross-sectional view of the 3DG fiber. (e) Histogram of the graphene nanosheet orientations with respect to the horizontal direction. The distribution around 0° indicated that there was a preferential alignment of the nanosheet within the fiber interior. (f) SEM image and photograph (inset) of the 3DG printed object with a 100 μm tip, displaying a high degree of regularity. (g) Uniformity of the 3DG structure quantified by fiber thickness in a 40 layer construct printed with a 100 μm tip. While the first layer deviated from the mean due to the spreading on the substrate, the subsequent 39 layers showed narrow standard deviations within particular layers (error bars on individual points), as well as across many layers (dashed bounding lines on either side of the solid, mean line), with average diameter and deviation. Reproduced with permission from [154], Copyright © 2015, American Chemical Society.

Meanwhile, the presence of the low vapor pressure solvents, i.e., 2-butoxyethanol and dibutyl phthalate, served to ensure a seamless merging of the subsequent printed layers, thus leading to physically smooth transitions between the adjacent layers. The 3DG inks could allow for extrusion from the smallest tip with a diameter of 100 μm, at speeds of > 40 mm·s^{-1}, so that highly uniform multilayered structures could be readily obtained, as shown in Fig. 7.85 (f). The individual strands had a diameter which was smaller than the tip diameter by only about 6%, due to the shrinkage of the items during drying. Additionally, the dimensions of the printed fibers were consistent throughout the objects, as depicted in Fig. 7.85 (g).

Although the content of graphene was quite high, the composite possessed high mechanical stability and flexibility, so that thinner structures could be freely manipulated. For example, as shown in Fig. 7.86 (a), a complex 2D hexagonal array could be rolled to form a macrotube-like 3D object. The highly flexible behavior of the printed objects was attributed to the presence of the PLG as the polymer binder. The 3D-printed graphene-PLG composites could be strained to > 210% and 81% prior to failure, for 20 vol% and 60 vol% graphene loadings, respectively, as shown in Fig. 7.86 (b). The elastic modulus was initially increased from pure PLG to the composites with 20 vol% graphene and then decreased significantly as the content of graphene was further increased, as seen in Fig. 7.86 (c). This is because the tensile loads were primarily carried by the PLG elastomer in the samples with low contents of graphene. As the content of graphene was increased to ≥ 40 vol%, the binding effect of the PLG on the graphene nanosheets was reduced, due to its insufficient amount. As a result, both the strength and elastic moduli of the composites with high contents of graphene were decreased.

It was believed that the shape of the graphene nanosheets had a significant effect on the stability and flexibility of the 3D-printed objects. Control experiments indicated that the inks derived from other carbon-based nanomaterials, such as carbon nanotubes, could not be used to print structures with similar mechanical performances as those printed from the 3DG inks. Instead, the carbon nanotube-based structures were very brittle. Due to the presence of dense micron-scale clusters, carbon nanotubes cannot translate upon loading, leading to brittle fractures. As shown in Fig. 7.86 (d), the 3DG cylinders behaved like typical plastic foam under compressive loads. Figure 7.86 (d) also indicated that the tensile and compressive properties were retained after the samples were annealed in air at temperatures up to 100°C, while they were decreased quickly if the annealing temperature was over 150°C because of the decomposition of PLG. The samples annealed at different temperatures exhibited similar microstructures (Fig. 7.86 (e)), while their mechanical properties were significantly different, as demonstrated in Fig. 7.86 (f).

The ability to surgically implement the 3DG at clinically relevant scales was demonstrated. For instance, uni- and multichannel nerve conduits could be printed by using 3DG inks, as illustrated in Fig. 7.87 (a, b) and (c–e), respectively. The inks allowed the production of not only nerve conduits with various diameters (Fig. 7.87 (b)), but also objects with high aspect ratios and hundreds of layers (Fig. 7.87 (f)). In addition, items with large sizes could be printed first, and then could be cut in one with any size of interest for the given applications. For example, a 2 cm section was cut from a 12 cm cylinder, and was sutured around and to the dorsal branch of the ulnar nerve in an unfixed human cadaver through multiple surgical steps, as observed in Fig. 7.87 (g). After the ulnar nerve was exposed, the 2 cm tubular 3DG nerve conduit was longitudinally cut with surgical shears, so that it was wrapped around the nerve bundle. After wrapping, the longitudinal 3DG cut was sutured shut along the length of the nerve conduit. These results indicated that the 3D printed graphene-based structures could be potentially used for pre- and intraoperative precision surgical procedures on the scales relevant to humans.

A 3D printing strategy has been demonstrated to fabricate 3D graphene aerogels with designed macroscopic architectures [155]. The approach was based on the precise deposition of GO ink filaments on a pre-defined tool path. Two key challenges were encountered in this process, i.e., (i) the development of printable graphene-based inks and (ii) retaining the intrinsic properties of single graphene nanosheets in the final 3D printed structures, such as large specific surface area, strong mechanical properties and high electrical conductivity.

To address the first challenge, the composition and rheology of the printable inks should be tailored to meet the requirements, e.g., they should be able to reliably flow through the fine nozzles and the final products should be self-supporting. As a result, the inks should have corresponding physical properties, such as a sufficiently high viscosity and non-Newtonian fluid behavior. Generally, most GO-based graphene

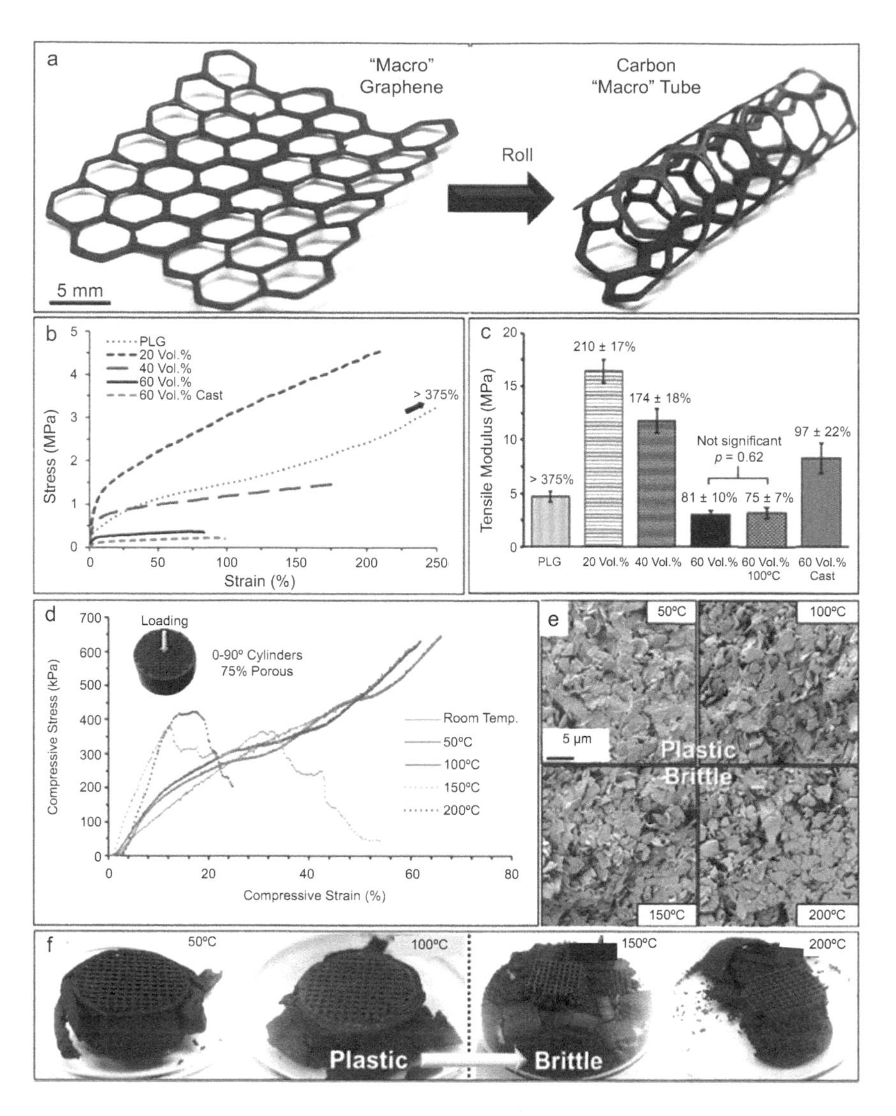

Fig. 7.86. (a) Photographs of the thin 3DG sheets with high flexibility and a sample rolled into a 3D form. (b) Quasi-static tensile measurements of the pure 3D-printed PLG, 3D-printed PLG-graphene composites and cast 60 vol% graphene composite. (c) Elastic moduli obtained from the tensile results and average percent strain-to-failure for each sample group (n = 4), which included the as-printed 60 vol% graphene (3DG), 100°C annealed for 30 min 3DG and cast 60 vol% ink. The moduli and strain-to-failure of all groups were significantly different from one another ($p \ll 0.05$), except for the 60 vol% and 60 vol% 100°C. (d) Compression measurement of the 75% porous 3DG (60 vol%) cylinders, without and with annealing at different temperatures, illustrating the characteristic ductile-to-brittle transition between 100 and 150°C. (e) SEM images of the thermally annealed 60 vol% samples, showing similar particle network microstructure. (f) Photographs of the 60 vol% cylinders after compression testing, illustrating the brittle failure of the samples annealed above 150°C. Reproduced with permission from [154], Copyright © 2015, American Chemical Society.

aerogels are derived from dilute precursor GO suspensions, e.g., with concentrations of < 5 mg·ml^{-1} GO. Such suspensions cannot be used for 3D printing, due to their too low viscosity. In order to make them printable or processable, their rheological behavior must be tailored [157]. Figure 7.88 (a) shows apparent viscosity of the high-concentration GO suspensions as a function of shear rate. The GO suspension at a concentration of 20 mg·ml^{-1} was a shear-thinning non-Newtonian fluid, which met the requirements of printable inks. As the concentration of GO was further increased to 40 mg·ml^{-1}, the apparent viscosity was increased by another order of magnitude, so that the printability was increased accordingly.

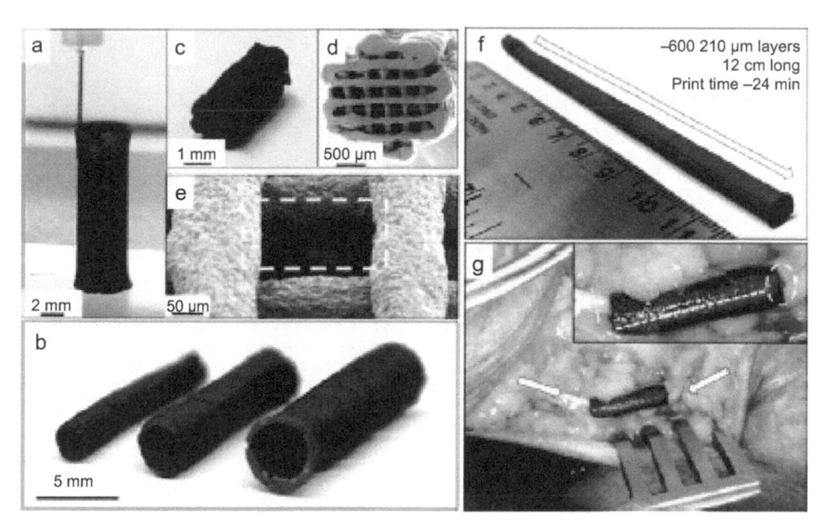

Fig. 7.87. (a) Photograph of the self-supporting tubular structures (140 layers) from the 3DG (60 vol% graphene) ink. (b) Photographs of the samples with different sizes. (c, d) Uniaxial, multichannel nerve guides, with uniaxial channels achieved by significantly reducing the z-spacing of progressive layers. (e) SEM image of the multichannel 3DG nerve conduit with every other layer close to contact (yellow box), minimizing or eliminating pores orthogonal to the major axis of the nerve guide. (f) Photograph of a long tube composed of hundreds of layers, which could be cut into smaller ones with desired sizes. (g) Photograph of tubular 3DG nerve conduit cut from (f) that was implanted into a human cadaver via longitudinal transection and wrapping around the ulnar nerve (white arrows). The 3DG nerve conduit was then sutured closed along the longitudinal transection (white dotted line), as well as the surrounding epinerium and nerve tissue (inset, yellow circle). Excessive 3DG nerve conduit length was then cut with surgical shears to expose additional nerve tissue. Reproduced with permission from [154], Copyright © 2015, American Chemical Society.

Moreover, the apparent viscosity could be further increased by adding hydrophilic fumed silica powders to the GO suspensions. The presence of silica filler brought both shear thinning behavior and a shear yield stress to the GO suspensions. Figure 7.88 (b) shows the storage and loss moduli of the pure GO suspensions and the representative GO inks, with different compositions. It was found that the pure GO suspensions with a concentration of 20 mg·ml^{-1} without fillers possessed a plateau in their elastic modulus (G') of about 1,000 Pa and a yield stress (τ_y) of about 40 Pa, respectively. As 20 wt% silica powder was added into the pure 20 mg·ml^{-1} GO suspension, both the elastic modulus and yield stress were increased by nearly one order of magnitude. Additionally, 10 wt% silica could be used to increase the elastic modulus and yield stress of the 40 mg·ml^{-1} GO suspensions by over an order of magnitude.

As discussed earlier, aerogels are special porous solids with ultralow density, derived from hydrogels after the removal of the liquid in the pores. To convert 3D printed GO structures into aerogels, the GO inks must remain wet during the printing and gelation, after which the liquid in the GO gels is removed by using supercritical or freeze-drying, so as to prevent the gels from the capillary force-induced collapsing. In this case, the items were printed into a bath of liquid, which had a density lower than that of water and was immiscible with the GO inks. Figure 7.88 (c) shows the fabrication steps of the 3D printing. The GO inks were loaded into a syringe barrel and extruded through a micronozzle to print the 3D structures. If the inks were dried in air, then the tip of the printing apparatus would be clogged or the pores in the printed structures would be collapsed. Therefore, the printing was conducted in isooctane, an organic solvent that was immiscible with the aqueous inks. After the structures were printed out, the silica filler was etched in order to get periodic 3D graphene aerogel microlattices. A woodpile with a primary cubic-like lattice consisting of multiple orthogonal layers of parallel cylindrical filaments was printed in a layer-by-layer way [155]. The 3D simple cubic lattice had an in-plane center-to-center filament spacing (L) of 1 mm and a filament diameter (d) of 0.25 mm, corresponding to a spacing-to-diameter ratio (L/d), as demonstrated in Fig. 7.88 (c, 4). Geometric densities of the 3D printed graphene structures could be controlled, by simply controlling the filament spacing and the diameter.

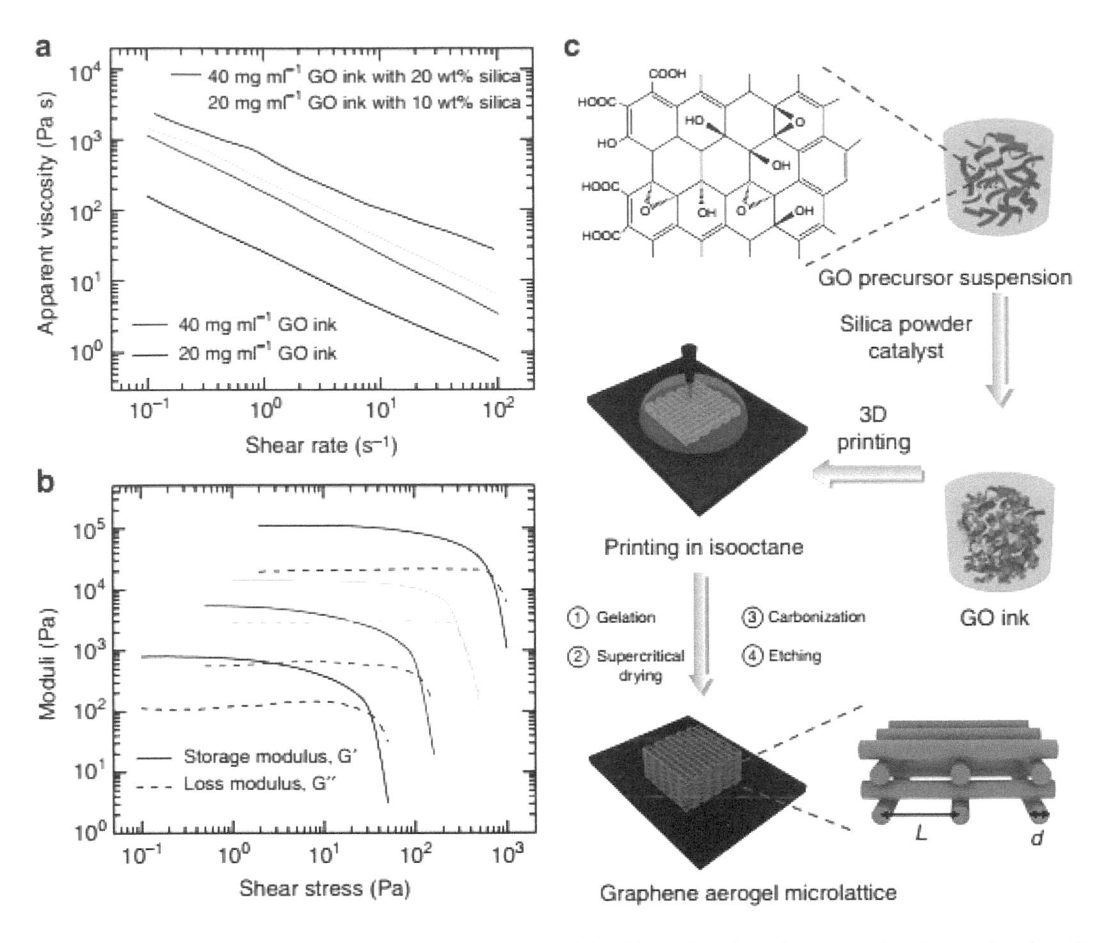

Fig. 7.88. Fabrication steps and rheological properties of the GO inks. (a) Log–log plots of apparent viscosity as a function of shear rate and (b) storage and loss modulus as a function of the shear stress of the GO inks with and without silica fillers. (c) Schematic of the fabrication process. Following the arrows, fumed silica powders and the catalyst were added into the as-mixed aqueous GO suspensions. After mixing, a homogeneous GO ink with the desired rheological properties was prepared. The GO ink was extruded through a micronozzle immersed in isooctane to prevent drying during the printing. The printed microlattice structure was supercritically dried to remove the liquid. Then, the structure was heated at 1050°C in N_2 for carbonization. Finally, the silica filler was etched with HF acid. The in-plane center-to-center rod spacing is defined as *L*, while the filament diameter is defined as *d*. Reproduced with permission from [155], Copyright © 2015, Macmillan Publishers Limited.

Color image of this figure appears in the color plate section at the end of the book.

The printed 3D graphene aerogel microlattices exhibited extraordinary structural stability and micro-architecture accuracy, as observed in Fig. 7.89 (a, b), which suggested that the inks possessed high quality for the 3D printing process. After the removal of the silica fillers, randomly distributed large-sized pores were left in the graphene aerogels, as seen in Fig. 7.89 (c, d). This also indicated that the microstructure of the 3D printed graphene aerogels could be tailored by modifying the compositions of the GO inks. It has been accepted that gelation chemistry has a significant effect on the microstructures of the final hydrogels [142]. To confirm this in the 3D printed structures, two gelation reactions were examined, i.e., (i) basic solution of $(NH4)_2CO_3$ to directly cross-link the graphene nanosheets through the functional groups and (ii) resorcinol (R) and formaldehyde (F) with sodium carbonate to catalyze and link the graphene nanosheets. With the organic sol–gel chemistry (R–F solution), the GO networks had more open pores (Fig. 7.89 (d)), as compared with those obtained through the gelation based on the native functionality of the GO nanosheets (Fig. 7.89 (c)). A series of graphene aerogel microlattices with varying thicknesses and a large area graphene aerogel honeycomb have been printed to demonstrate the versatility of the technique, as illustrated in Fig. 7.89 (e, f).

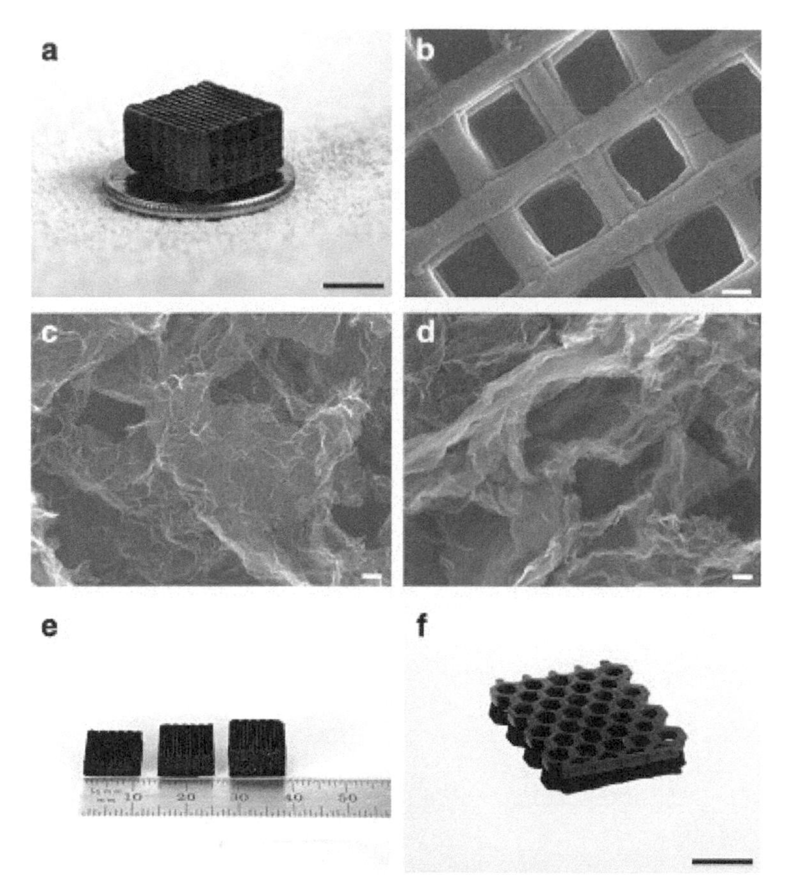

Fig. 7.89. Morphology and structure of the graphene aerogels. (a) Photograph of the 3D printed graphene aerogel microlattice. SEM images of the samples: (b) 3D printed graphene aerogel microlattice, (c) graphene aerogel without R–F after etching and (d) graphene aerogel with 4 wt% R–F after etching. (f) Photograph of the 3D printed graphene aerogel microlattices with varying thicknesses and (f) the 3D printed graphene aerogel honeycomb. Scale bars = 5 mm (a), 200 μm (b), 100 nm (c, d) and 1 cm (f). Reproduced with permission from [155], Copyright © 2015, Macmillan Publishers Limited.

3D printing technique has been used to fabricate highly stretchable and conductive carbon nanomaterials-based stretchable conductive materials (SCMs) [156]. Split-level and aligned 3D porous patterns based on polylactic acid (O-PLA) were first prepared by using 3D printing, and were then used as the templates to develop 3D porous poly(dimethylsiloxane) (O-PDMS). After the CNTs and graphene were introduced into the O-PDMS, the 3D skeleton became electrically conductive. As a result, split-level O-PDMS/CNTs–graphene (S-OPCG) and aligned O-PDMS/CNTs–graphene (A-OPCG) porous structures with high mechanical stretchability and electrical conductivity could be developed. More importantly, due to their extraordinary split-level structure, the S-OPCG structures were significantly superior to the A-OPCG ones, in terms of both the electrical and mechanical properties. Under 100% uniaxial stretching, conductivity retentions of the S-OPCG and A-OPCG were 40% and 25%, respectively. Also, the electrical conductivity of the S-OPCG remained almost constant after 5000 bending cycles, while only showing a slight decrease after 100 cycles of a stretching–releasing process by a 50% magnitude.

Figure 7.90 shows a schematic diagram of the process to prepare the S-OPCG [156]. PLA was selected as the basic material to print the 3D skeletons, because it is eco-friendly and easily removed. For the 3D printing, a 200-μm syringe needle was used to print the O-PLA, with the spacing between the adjacent skeletons to be 200 μm. O-PLA scaffolds with both split-level and aligned structures could be printed out, by controlling the printing parameters. Representative cross-sectional SEM images of the aligned and split-level O-PLA structures are shown in Fig. 7.91 (a, c). The skeletons with the O-PLA aligned in different

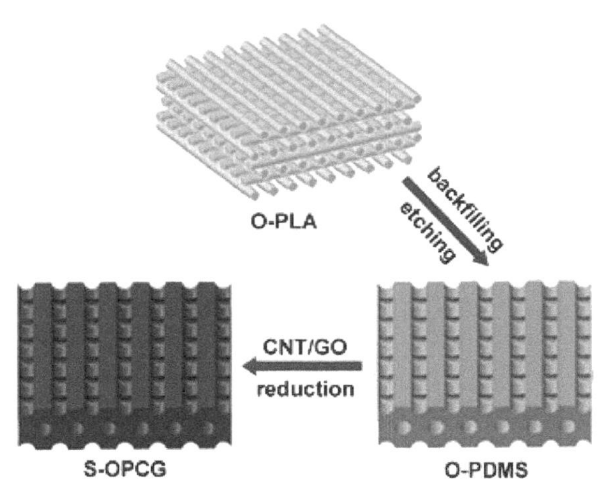

Fig. 7.90. Schematic diagram showing the procedure to print the 3D porous S-OPCG. Reproduced with permission from [156], Copyright © 2016, American Chemical Society.

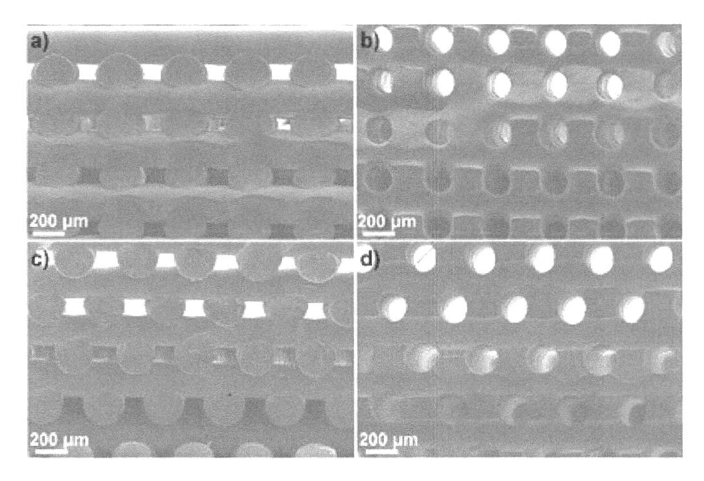

Fig. 7.91. Cross-sectional SEM images of the samples: (a) aligned O-PLA, (b) aligned O-PDMS, (c) split-level O-PLA and (d) split-level O-PDMS. Reproduced with permission from [156], Copyright © 2016, American Chemical Society.

layers were in a straight line, while those of the split-level O-PLA were interlaced. After the structures were filled with PDMS in the pores, the PLA was removed through solvent dissolving. Therefore, PDMS would replicate the porous structures of the O-PLA, thus leading to aligned and split-level O-PDMS. SEM images of the O-PDMS samples are shown in Fig. 7.91 (b, d). Continuous pores were distributed in the PDMS structures, while the surface of the O-PDMS was smooth.

To make conductive PDMS structures, CNTs and graphene were introduced [156]. In order to do that, a suspension containing CNTs and GO was filled into the O-PDMS frameworks there were stretched by 50%. The samples were then treated for vacuum degassing, so that the CNTs and GO nanosheets were coated onto the inner surface of the O-PDMS scaffolds. Hydroiodic acid was used to reduce GO into rGO in order to increase the electrical conductivity. With 1 wt% CNTs–graphene, conductivity of the S-OPCG could reach 5.12 $S \cdot m^{-1}$, while that of the A-OPCG was only 0.34 $S \cdot m^{-1}$. The difference in electrical conductivity between the two samples could be attributed to their difference in structure. Due to their interlaced porous structure, the conductive solution could entirely permeate the skeleton of the split-level O-PDMS more slowly, so that there was time to coat the structure. SEM images of the S-OPCG are shown in Fig. 7.92 (a, c), which indicated that a dense conductive layer was formed on the O-PDMS skeletons. Moreover,

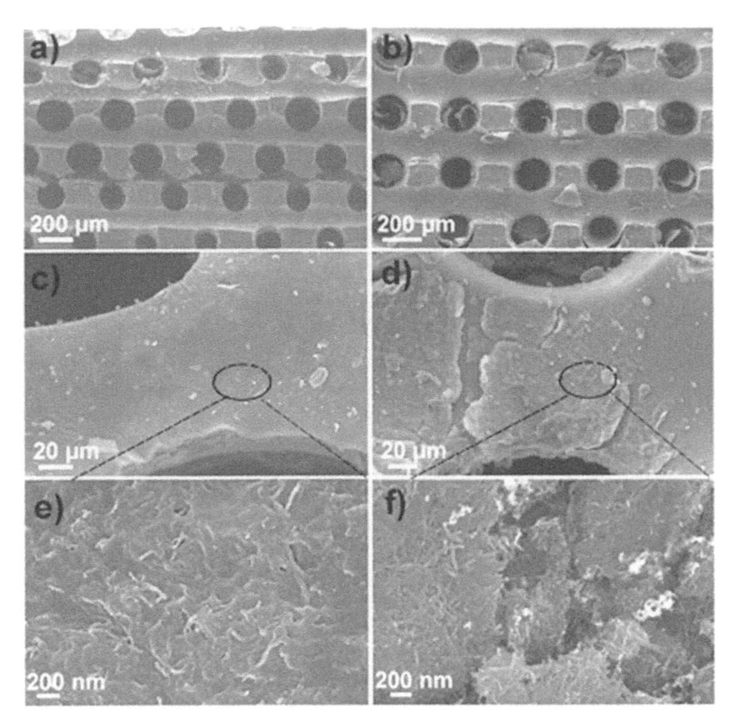

Fig. 7.92. Cross-sectional SEM images of the samples: (a, c) S-OPCG and (b, d) A-OPCG. (e, f) High magnification SEM images of the parts in (c) and (d), respectively. Reproduced with permission from [156], Copyright © 2016, American Chemical Society.

both the graphene nanosheets and CNTs were uniformly distributed, without the presence of agglomeration or stacking, due to the synergistic effect of the CNTs and graphene nanosheets with an optimized mass ratio, as illustrated in Fig. 7.92 (e).

In contrast, the pores of the aligned O-PDMS in different layers were connected, so that the conductive solution flowed through it in a relatively quick way, thus leading to an incomplete and unconsolidated coating, as seen in Fig. 7.92 (b, d). As a consequence, the distribution of the CNTs and graphene nanosheets would be non-uniform, as revealed in Fig. 7.92 (f). Therefore, the A-OPCG structures had lower conductivity than the S-OPCG ones. Additionally, the geometric parameter of the O-PDMS also played an important role in determining the electrical conductivity of the S-OPCG. With the same content of CNTs–graphene, the thinner the PDMS skeleton and the smaller the pitch, the higher the electrical conductivity would be. Also, the prestretching and the slow addition of the conductive solution were necessary to achieve high electrical conductivity.

Figure 7.93 shows SEM images of the S-OPCG after 100% stretching. Figure 7.93 (a) indicated that the conductive layer on the A-OPCG was still attached to the framework tightly without falling off to be observed. However, cracks were present inevitably, as demonstrated in Fig. 7.93 (b). If the cracks were too large, the connections of the conductive layers were broken, while if the cracks were small, the connections of the conductive components were still retained, as seen in Fig. 7.93 (c, d). This could be the reason why the S-OPCG composites could retain electrical conductivity under a large degree of stretching.

7.8. Miscellaneous

To further demonstrate the richness of 3D graphene-based structures, two more examples are discussed in this section. The first example is a layer-by-layer (LbL) self-assembly of GO nanosheets into porous 3D macrosized beads [158]. The entirely different assembly behavior was observed when GO was complexed with branched polyethyleneimine (b-PEI). In contrast to chitosan, b-PEI is able to penetrate the GO/b-PEI complex interface, which allows the layer to grow continuously into 3D structures with thicknesses

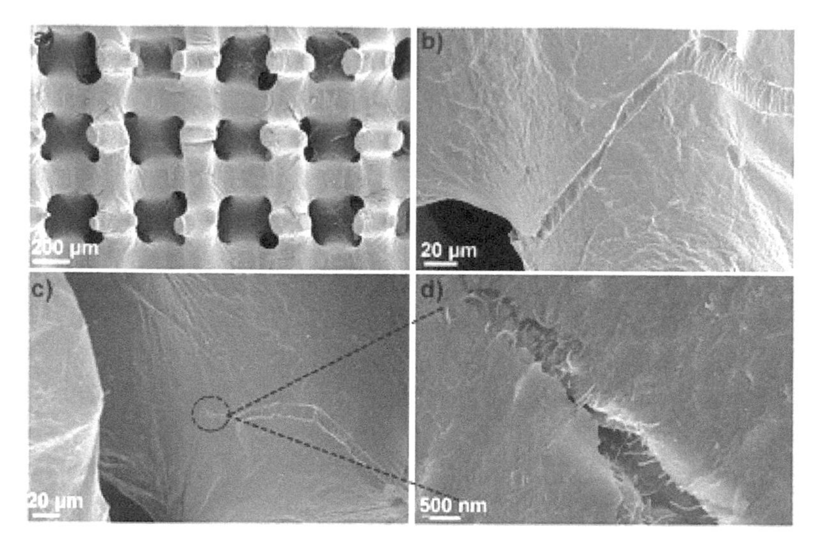

Fig. 7.93. (a–d) SEM images of the S-OPCG after 100% stretching. Reproduced with permission from [156], Copyright © 2016, American Chemical Society.

at the millimeter scale. More importantly, the assembled structures possessed a foam-like highly porous microstructure. The formation of GO/b-PEI structures followed the LbL reassembly characteristics. It contrast to the conventional LbL assemblies, the assembly was driven by the diffusion of b-PEI, so that it was achieved through multiple deposition steps. As a consequence, once the assembly was started, it would be continued without the requirement of external force or stimuli. Therefore, this was called diffusion driven layer-by-layer (dd-LbL) assembly. This assembly process has been used to (i) prepare various porous frameworks with tunable porosity, (ii) promote the growth of patterned multilayered films on rigid and flexible substrates, and (iii) fabricate free-standing 3D structures with shapes that could be controlled by using templates.

In order to examine the interaction of b-PEI and GO at a liquid–liquid interface, droplets of about 12 ml b-PEI solution, with $M_w = 750,000$, $M_n = 60,000$, 25 wt.% in water, were added into a GO suspension, with a concentration of 8 $mg \cdot ml^{-1}$ in a mixture of dimethylformamide (DMF) and water with a ratio of 9:1. The mixture was then kept for several hours, as schematically shown in Fig. 7.94 (a). Complexation of the GO and b-PEI occurred at the surface of the droplet, thus forming a dark brown layer on the droplet, as demonstrated in Fig. 7.94 (b). The diameter of the spheres was increased from 2.8 mm of the initial b-PEI droplet to 6.5 mm. The layer was very stable, so that the spherical shape was well-retained after the spheres were taken out from the liquid. The spheres were called GO/b-PEI beads, with shell thicknesses of up to several millimeters, as seen in Fig. 7.94 (c). SEM images of the GO/b-PEI complex layer are shown in Fig. 7.94 (d–f). The beads were freeze-dried in order to avoid structural change. Highly porous networks were formed by the GO nanosheets, which implied that the wet spheres contained a large amount of water. The GO nanosheets were assembled into an onion-like structure. Figure 7.94 (g) shows XRD patterns of the GO/b-PEI beads and pure GO, with peaks of $2\theta = 5.3^{\circ}$ and 10.57°, corresponding to interlayer spacings of d = 1.64 nm and 0.84 nm, respectively. The large interlayer spacing indicated that the b-PEI had been incorporated in between the GO layers during the assembly process.

Figure 7.94 (h) shows a schematic diagram of the formation mechanism of the GO/b-PEI beads. As the b-PEI droplet was immersed into the GO suspension, electrostatic interactions facilitated complexation between the polyelectrolyte and nanosheets at the interface. Therefore, a thin complex layer was formed, so that the droplet was encapsulated by the negatively charged GO nanosheets. However, the b-PEI could penetrate the barrier to continuously diffuse out through the interlayer space in between the GO nanosheets, which was most likely driven by the osmotic pressure. Once the negatively charged outer surface was attached with b-PEI, more GO nanosheets would be deposited onto the shell. Meanwhile, the

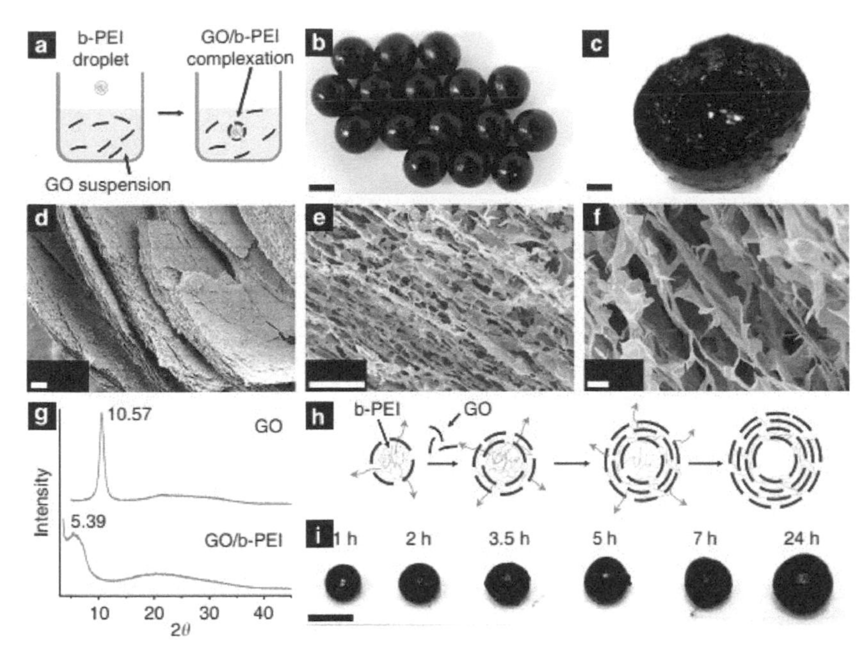

Fig. 7.94. Diffusion driven layer-by-layer (dd-LbL) assembly of GO nanosheets. (a) Schematic diagram showing the experimental setup. As the droplets of the branched polyethylenimine (b-PEI) solution were dropped into the GO suspension, complexation between b-PEI and GO took place at the droplet surface, leading to the formation of stable bead-like structures (b). (c) Cross-section vies of a bead, with the layer of the GO/b-PEI complex to be a few millimeters thick. (d–f) SEM images of the beads, indicating the layered and porous network formed by the GO nanosheets. (g) XRD patterns of the GO/b-PEI and GO, demonstrating the increase in interlayer spacing, due to insertion of the b-PEI in between the GO nanosheets. (h) Schematic diagram of assembly mechanism. (i) Photographs of the samples with different reaction time durations. Scale bars = 5 mm (b), 1 mm (c), 100 μm (d), 10 μm (e), 1 μm (f) and 5 mm (i). Reproduced with permission from [158], Copyright © 2014, Macmillan Publishers Limited.

concentration of GO in the vicinity was decreased accordingly, so that more b-PEI would diffuse out. As a result, a new interface was formed, which led to the formation of the large space between the layers. The cycle of complexation–diffusion–complexation was responsible for formation of the GO shells covering the b-PEI droplet. Finally, layered and porous 3D macrostructures were obtained, as seen in Fig. 7.94. Because the reactants were continuously consumed during the assembly process, a decreased frequency of complexation was expected, so that the growth of the assembled structure was gradually slowed. This was the reason why the size of the GO/b-PEI beads was not linearly proportional to the reaction time, as observed in Fig. 7.94 (i). It has been confirmed that the critical factor for the continuous assembly of the GO nanosheets was that the polyelectrolyte should be able to penetrate the initial complexation layer.

The dd-LbL assembly strategy was also used to fabricate GO/b-PEI films on solid/flexible substrates, by first casting the b-PEI solution on substrates, such as glass, as shown in Fig. 7.95 (a) [158]. A green dye was used for visualization. After the glass slide coated with b-PEI was immersed in the GO suspension (8 mg·ml^{-1}) for several hours, a thick film was formed, due to the diffusion of b-PEI from the deposited droplet, as demonstrated in Fig. 7.95 (b). The GO/b-PEI films exhibited properties and microstructures similar to those of the beads. After freeze-drying, the shape of the film was well-retained. Figure 7.95 (c) shows a cross-sectional SEM image of the film, revealing a similar porous network. If the film was dried by using normal heating, the sample experienced a serious shrinkage, thus leading to the presence of a tightly packed multilayered structure, as seen in Fig. 7.95 (d), due to the surface tension of water bringing the GO nanosheets together.

The GO/b-PEI films were very robust, and could be deposited on substrates with various types of surfaces and sizes. Figure 7.95 (e) shows a film that was deposited on a filter paper. In this case, the b-PEI solution was first dropped onto the filter paper with a dimension of 8 × 10 cm^2. Then, the filter paper

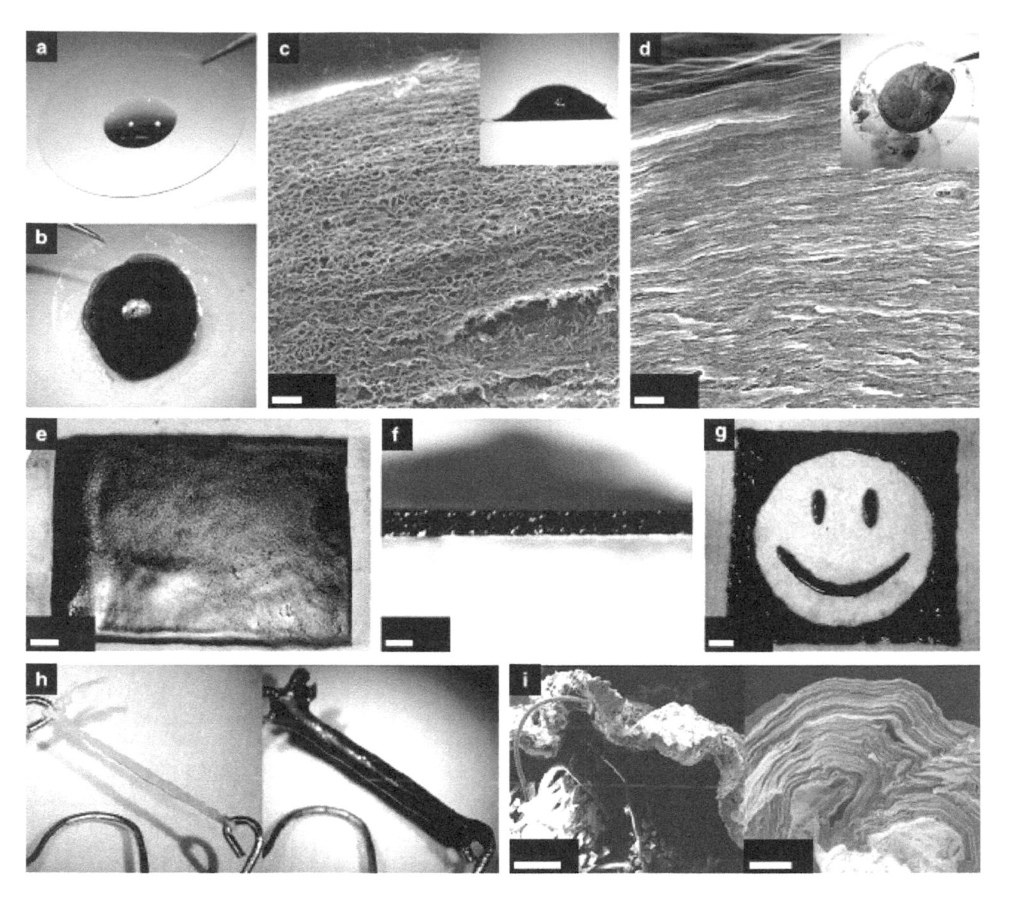

Fig. 7.95. Deposition of the GO/b-PEI films on solid and flexible substrates. (a–d) Films on a glass slide. (a) A drop of b-PEI solution was dropped onto the substrate. (b) After the glass slide was immersed in the GO suspension for several hours, a thick film was formed over that area. Freeze-drying the sample resulted in a porous foam-like structure (c), whereas oven-drying led to a tightly packed multilayered film (d). (e) Large area (8 cm × 10 cm) film formed on a filter paper. The film had a uniform thickness (f). (g) GO/b-PEI film with pre-determined patterns. (h, i) Films deposited on the surface of a thin thread. (i) Cross-sectional SEM confirming the formation of multilayer coating on the thread. Scale bars = 5 μm (c), 1 μm (d), 1 cm (e), 1 mm (f), 1 cm (g), 100 μm (i, left panel) and 10 μm (i, right panel). Reproduced with permission from [158], Copyright © 2014, Macmillan Publishers Limited.

was immersed into the GO suspension, so the GO/b-PEI film was formed over the same area. The film had a uniform thickness over the whole area, as illustrated in Fig. 7.95 (f). Because the formation of the GO/b-PEI film was confined only to the area covered with b-PEI it was possible to create patterned films on a given substrate on pre-determined areas, as demonstrated in Fig. 7.95 (g). Moreover, the dd-LbL assembly approach could be readily extended to non-flat substrates. For instance, a thick GO/b-PEI layer was deposited onto a thin thread, as seen in Fig. 7.95 (h, i).

Additionally, the shapes of the assembled products could be controlled by simply using templates with the desired shapes, as demonstrated in Fig. 7.96. A polydimethylsiloxane slab with a hole in the middle was placed on a glass slide, so that a well shape was formed. A small amount of b-PEI solution was dropped into the well to cover its bottom (Fig. 7.96 (a)). After that, the substrate was immersed in GO suspension for several hours, so that complexation of the GO and b-PEI was triggered within the well due to diffusion of the polyelectrolyte (Fig. 7.96 (b)). The GO/b-PEI structures could be easily detached from the polydimethylsiloxane, with the shapes that could be well-retained, as illustrated in Fig. 7.96 (c). This simple yet versatile approach could find potential applications in various areas.

Fig. 7.96. Tailoring the assembled GO/b-PEI structures by using templates. (a) A polydimethylsiloxane slab with a hole placed on a glass slide to form a well, into which a small amount of b-PEI solution was dropped. (b) After the slide was immersed in GO suspension for several hours, complexation between GO and b-PEI was carried out within the confined space to form a disk-like macrostructure. (c) Photographs of free-standing pieces, with shapes of their templates. Reproduced with permission from [158], Copyright © 2014, Macmillan Publishers Limited.

Spark plasma sintering (SPS) has recently been widely used to consolidate nanostructured ceramics, metals and CNTs reinforced composites [159–163]. SPS is a sintering method that applies high pressure, high temperature, along with a pulsing current through the material during sintering [164]. The current leads to the formation of plasma in the form of micro-discharge, due to impurities on the particle surfaces. The plasma provides an enhanced heat transfer effect, so as to produce better bonding of the sintered materials. This technique has various advantages over the conventional sintering technologies, including high densification rate, short sintering times, controlled grain growth and enhanced bonding between the composite components. More recently, SPS has been used to consolidate graphene reinforced composites based on aluminum oxide [165–167] and silicon nitride [168], at pressures of 35–60 MPa and temperatures of 1000–1650°C. It was found that the SPS processing had no any negative effect on the graphene nanosheets, which were almost defect-free in the composites.

Inspired by these observations, a bulk sample of graphene nanosheets has been prepared by using the SPS technique at an extremely high temperature of 1850°C and a pressure of 80 MPa [169]. The structure of the graphene nanosheets remained unchanged after the SPS processing at such an extremely high temperature. Graphene nanosheets were less vulnerable, as compared with CNTs, probably due to their special structure [170–173]. Graphene nanosheets are 2D-like, while CNTs are tubular-like structures. Comparatively, tubular-like structured CNTs are vulnerable to buckling and collapse at high pressures.

The consolidation of the graphene nanosheets was conducted, with graphite die of 20 mm in diameter, at a pressure of 80 MPa and temperature of 1850°C, for 10 min. The die was in an environment of Ar at a pressure of 3–6.5 Pa. The graphene powder was wrapped in a graphite foil to improve the current flow, with a peak current of 2500 A during the 10 min hold at 1850°C and a current of 1110 A at 1100°C for 3 min. The heating rate used was $200°C \cdot min^{-1}$. Figure 7.97 shows a photograph of the sintered graphene pellet. The starting graphene powder had a density of $1.82\ g \cdot cm^{-3}$, which was increased to $2.11\ g \cdot cm^{-3}$ after the SPS processing. Low magnification fracture surface SEM images of the SPS graphene pellet are shown in Fig. 7.98. The pellet had dense structure near the top and bottom surfaces, while the intermediate part consisted of highly oriented stacked graphene nanosheets.

Figure 7.99 shows fracture surface SEM images of the graphene pellet. Unique bending, folding and curling structures were observed, which were responsible for energy dissipation during the fracture of the pellet. For example, bending of a single stack can be identified in Fig. 7.99 (a, b, d). Figure 7.99 (c) shows

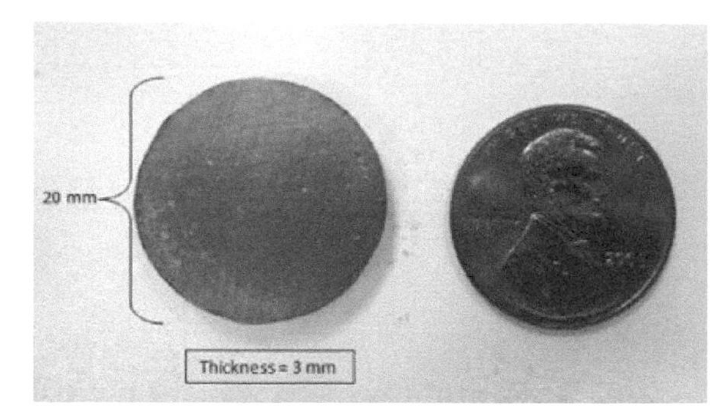

Fig. 7.97. Photograph of the bulk graphene pellet, together with a penny for size comparison. Reproduced with permission from [169], Copyright © 2012, Elsevier.

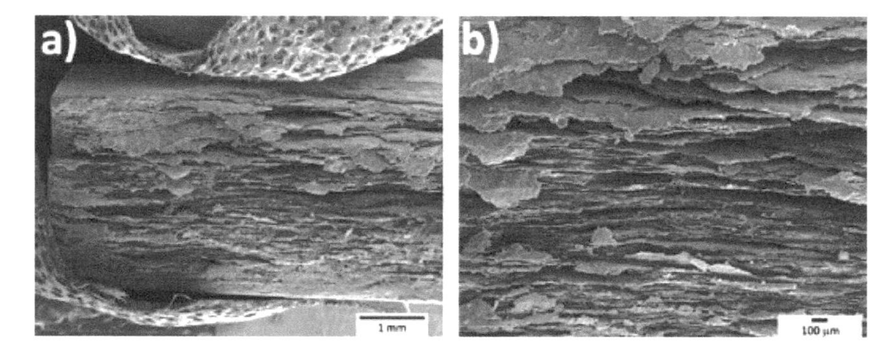

Fig. 7.98. Fracture surface SEM images of the SPS consolidated graphene pellet, revealing stacks of uniformly oriented graphene nanolayers at (a) low magnification and (b) high magnification. Carbon tape used to hold sample for SEM is seen in (a). Reproduced with permission from [169], Copyright © 2012, Elsevier.

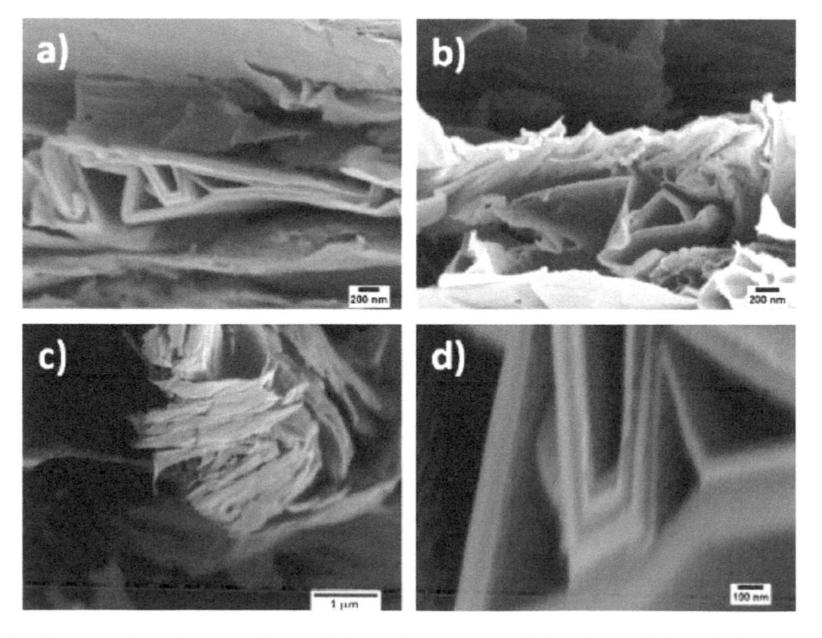

Fig. 7.99. (a) Bending at high angles across the multiple graphene stacks. (b) Bending at high angles in a single stack. (c) In-plane view of the bent sample. (d) Bending of multiple graphene stacks, with some stacks to be offset because of the interlayer sliding. No fracture was observed at high bending angles, indicating high strength and flexibility of the bulk graphene pellet. Reproduced with permission from [169], Copyright © 2012, Elsevier.

wrapping phenomena, with minimal damage on the surface. Figure 7.99 (d) indicated that, although the bending angles exceeded 90° and occasionally reached 180°, the graphene nanosheets were not fractured, leading to the presence of U-shaped structures. The graphene layers concurrently underwent sliding during the bending, especially between adjacent stacks, as observed in Fig. 7.99 (d). The results suggested that graphene nanosheets could provide unique energy dissipating mechanisms, when used as reinforcements in ceramic matrix composites. Tribological studies suggested that the graphene nanosheets sheared off and welded at a higher load, so as to offer a lubricating effect.

7.9. Applications of 3D Graphene Architectures

With the above discussion, it can be concluded that graphene-based materials with 3D architectures have various special properties, such as large accessible specific surface areas, interconnected conductive networks and a special microenvironment. As a consequence, they have found a wide range of applications, including energy storage, catalysis, environmental protection, stretchable electronics and so on, with enhanced performances as compared to their powdery counterparts. Examples of the applications of 3D graphene architectures are discussed in this section.

7.9.1. Energy storage

The applications of the 3D architectures for energy storage are mainly in supercapacitors. Supercapacitors, also known as ultracapacitors and electric double layer capacitors (EDLCs), have formed a special group of energy storage devices, with high power densities, long life and high rate capability [174]. Previous studies indicated that graphene-based supercapacitors could achieve capacitances of 135 and 99 F g^{-1} in aqueous and organic electrolytes, respectively [94, 95, 175]. However, these values are much lower than the theoretical prediction of 550 F g^{-1} for single-layer graphene [176]. It is well-known that the capacitance of an EDLC is generated due to the accumulation of electrostatic charge at the electrode/electrolyte interfaces, which is generally proportional to the effective specific surface area of the electrode materials. CMGs with 3D microstructures have large surface areas that are accessible to form electric double layers, while providing facile routes for electron and electrolyte molecules to transport because of the interconnected conductive networks [18]. Therefore, the performances of supercapacitors could be largely enhanced by using the 3D porous graphene-based materials as active electrodes.

A one-step hydrothermal self-assembled rGO hydrogel [18], was used for the electrodes of supercapacitors, with a 3D porous network and a high specific surface area of 964 $m^2 \cdot g^{-1}$ [177]. Due to the unique 3D porous structure, most of the 2D graphene nanosheets could be accessed by the electrolytes to form electric double layer charges. The EDLCs based on the rGO hydrogel exhibited a high specific capacitance of 152 $F \cdot g^{-1}$, which was higher than that of the supercapacitor based on rGO agglomerate particles by about 50% [175]. After the rGO hydrogel was reduced with hydrazine, the specific capacitance was increased to 222 F g^{-1} at 1 $A \cdot g^{-1}$ [177]. More importantly, the EDLCs based on chemically reduced rGO hydrogels possess excellent rate performance and a highly stable cycling life. For example, the EDLCs could maintain a high specific capacitance of 165 F g^{-1} at a fast discharge rate of 100 A g^{-1} in aqueous KOH electrolyte. In addition, specific capacitance of the EDLC could be kept to be 92% after 2000 charge–discharge cycles over 0–1 V at a discharge current density of 4 A g^{-1}. The performances could be further improved by using various strategies [178–184].

7.9.2. Stretchable electronics

Rubbery and stretchable conductive materials have attracted much attention in recent years, because of their potential applications in stretchable electronics and smart textiles [114, 185]. One of the examples is the composite based on graphene foams (GFs) and poly(dimethyl siloxane) (PDMS), which was prepared by infiltrating the GFs with PDMS. The 3D continuous and interconnected network of GFs conducted electricity, while the PDMS ensured mechanical flexibility, as discussed above [99].

7.9.3. Catalytic applications

Graphene or GO has been extensively studied for catalytic applications, as either catalysts or supports for other catalytic materials [186–189]. In this regard, 3D structured graphene-based architectures could be ideal candidates for such applications, because of their unique porous structures, excellent mechanical strength and electrical properties. Specifically, the highly porous interconnected network would ensure a high rate of ion diffusion and enhanced transfer kinetics. They also could provide a special reaction microenvironment and conductively multiplexed pathways for rapid charge transfer and conduction, especially when they are incorporated with other active components [190–194].

7.9.4. Hydrogen storage

The development of hydrogen storage materials with high capacity and reversibility has been one of the hot research topics related to renewable energies, due to their potential applications in clean and environmentally benign hydrogen-based fuel cells [195–201]. It has been demonstrated that GO frameworks with a 3D network nanostructure had hydrogen sorption capacity, which was predicted by theoretical simulation and supported by experimental results [135, 202]. Until now, the experimental H_2 uptake is still less than the theoretical prediction for ideal GO framework structure, which has been attributed the presence of oxygenated groups in the GO frameworks. Therefore, the hydrogen storage capacity of GO/G-based nanomaterials could be improved, by eliminating the unreacted oxygenated groups as completely as possible through chemical reduction.

7.9.5. Environmental remediation

It has been acknowledged that the recyclability of the absorbents that are used to remove organic pollutants or oil spills from water is an important indicator of their performances [203]. It was observed that a GO/DNA composite hydrogel could be used to efficiently extract and remove the model dye safranine O from water [28]. After incubating 0.6 ml aqueous safranine O solution at a concentration of 0.1 mg ml^{-1} with 0.2 ml GO/DNA hydrogel for 24 h, the model dye could be absorbed by the hydrogel almost completely. The loading capacity for safranine O was up to 960 mg g^{-1} GO, which was attributed to the concurrent effect of the electrostatic and π–π interactions between the dye molecules and the GO nanosheets and the DNA molecules. The large specific surface area of the 3D porous structure of the hydrogel could be an additional factor responsible for the high loading capacity.

7.9.6. Sensors

Graphene-based materials in general have been employed to fabricate various sensing devices, because graphene nanosheets have a huge specific surface area and unique optoelectronic properties [23, 204–209]. More recently, GFs with 3D networks have been found to be more promising candidates to develop sensors, e.g., chemiresistor-type sensor, which could be used to detect environmental pollution gases with sensitivity at the ppm level [16]. Usually, the resistance change ($\Delta R/R$) as a function of time was used to characterize the sensitivity of the chemiresistors. In this case, $\Delta R/R$ of the active layer of the GFs was decreased from about 30% in 1000 ppm to about 5% in 20 ppm NH_3. The GFs-based sensor was also highly sensitive to NO_2, with $\Delta R/R$ of about 4% in 20 ppm NO_2. In addition, the GF device could be operated at room temperature and atmospheric pressure, with a very low power consumption [210]. It is expected that more and more sensors will be developed by using the 3D graphene-based highly porous architectures.

7.10. Concluding Remarks

Significant progress has been made in developing 3D graphene-based architectures with controlled micro- and nano-structures, which are expected to find various applications in flexible electronics, supercapacitors, catalysis, hydrogen storage, sensors and environmental remediation. Various strategies and approaches have been developed to fabricate 3D graphene-based architectures, such as self-assembly, templated growth, sol–gel synthesis and lithography patterning technology. The 3D structures not only have the intrinsic properties of graphene or graphene oxide, but also demonstrate additional functionalities originating from their unique microstructures. They are specifically characterized by conductive frameworks with high specific areas, which make them promising candidates for loading catalysts, adsorbing organic or inorganic molecules, forming double layer charges and facile routes for charge or ion transportation.

Despite significant progress, there are still challenges to be addressed in the near future for the 3D graphene-based structures for practical applications. For example, because the structures and properties of the 3D graphene architectures are closely related to their building blocks, i.e., graphene or graphene oxide nanosheets, it is important to be able to synthesize nanosheets with the desired physical and chemical properties. Other issues, include restacking of the single-layer graphene building blocks, porosity and mechanical strength, purity and electrical conductivity, formation mechanisms of the 3D structures, especially for those derived from solutions or suspensions. More applications of such materials should be further explored. Nevertheless, there will be more and more studies and reports on this type of special graphene-based materials in the future.

References

[1] Li C, Shi GQ. Three-dimensional graphene architectures. Nanoscale. 2012; 4: 5549–63.

[2] Zeng M, Wang WL, Bai XD. Preparing three-dimensional graphene architectures: Review of recent developments. Chinese Physics B. 2013; 22: 098105.

[3] Novoselov KS, Geim AK, Morozov SV, Jiang D, Zhang Y, Dubonos SV et al. Electric field effect in atomically thin carbon films. Science. 2004; 306: 666–9.

[4] Rao CNR, Sood AK, Subrahmanyam KS, Govindaraj A. Graphene: The new two-dimensional nanomaterial. Angewandte Chemie-International Edition. 2009; 48: 7752–77.

[5] Allen MJ, Tung VC, Kaner RB. Honeycomb carbon: A review of graphene. Chemical Reviews. 2010; 110: 132–45.

[6] Bai H, Li C, Shi GQ. Functional composite materials based on chemically converted graphene. Advanced Materials. 2011; 23: 1089–115.

[7] Novoselov KS, Geim AK, Morozov SV, Jiang D, Katsnelson MI, Grigorieva IV et al. Two-dimensional gas of massless Dirac fermions in graphene. Nature. 2005; 438: 197–200.

[8] Meric I, Han MY, Young AF, Ozyilmaz B, Kim P, Shepard KL. Current saturation in zero-bandgap, topgated graphene field-effect transistors. Nature Nanotechnology. 2008; 3: 654–9.

[9] Szafranek BN, Fiori G, Schall D, Neumaier D, Kurz H. Current saturation and voltage gain in bilayer graphene field effect transistors. Nano Letters. 2012; 12: 1324–8.

[10] Biro LP, Nemes-Incze P, Lambin P. Graphene: nanoscale processing and recent applications. Nanoscale. 2012; 4: 1824–39.

[11] Jia XT, Campos-Delgado J, Terrones M, Meunier V, Dresselhaus MS. Graphene edges: a review of their fabrication and characterization. Nanoscale. 2011; 3: 86–95.

[12] Bai JW, Huang Y. Fabrication and electrical properties of graphene nanoribbons. Materials Science & Engineering R-Reports. 2010; 70: 341–53.

[13] Zhu SJ, Tang SJ, Zhang JH, Yang B. Control the size and surface chemistry of graphene for the rising fluorescent materials. Chemical Communications. 2012; 48: 4527–39.

[14] Shen JH, Zhu YH, Yang XL, Li CZ. Graphene quantum dots: emergent nanolights for bioimaging, sensors, catalysis and photovoltaic devices. Chemical Communications. 2012; 48: 3686–99.

[15] Chen QW, Zhang LY, Chen GH. Facile preparation of graphene-copper nanoparticle composite by *in situ* chemical reduction for electrochemical sensing of carbohydrates. Analytical Chemistry. 2012; 84: 171–8.

[16] Yavari F, Chen ZP, Thomas AV, Ren WC, Cheng HM, Koratkar N. High sensitivity gas detection using a macroscopic three-dimensional graphene foam network. Scientific Reports. 2011; 1: 166.

[17] Niu ZQ, Chen J, Hng HH, Ma J, Chen XD. A leavening strategy to prepare reduced graphene oxide foams. Advanced Materials. 2012; 24: 4144–50.

[18] Xu YX, Sheng KX, Li C, Shi GQ. Self-assembled graphene hydrogel via a one-step hydrothermal process. ACS Nano. 2010; 4: 4324–30.

[19] Wan XJ, Huang Y, Chen YS. Focusing on energy and optoelectronic applications: A journey for graphene and graphene oxide at large scale. Accounts of Chemical Research. 2012; 45: 598–607.

[20] Huang X, Qi XY, Boey F, Zhang H. Graphene-based composites. Chemical Society Reviews. 2012; 41: 666–86.

[21] Luo B, Liu SM, Zhi LJ. Chemical approaches toward graphene-based nanomaterials and their applications in energy-related areas. Small. 2012; 8: 630–46.

[22] Sun YQ, Wu Q, Shi GQ. Graphene based new energy materials. Energy & Environmental Science. 2011; 4: 1113–32.

[23] Yang WR, Ratinac KR, Ringer SP, Thordarson P, Gooding JJ, Braet F. Carbon nanomaterials in biosensors: Should you use nanotubes or graphene? Angewandte Chemie-International Edition. 2010; 49: 2114–38.
[24] Xu YX, Shi GQ. Assembly of chemically modified graphene: methods and applications. Journal of Materials Chemistry. 2011; 21: 3311–23.
[25] Li D, Kaner RB. Materials science—Graphene-based materials. Science. 2008; 320: 1170–1.
[26] Li D, Mueller MB, Gilje S, Kaner RB, Wallace GG. Processable aqueous dispersions of graphene nanosheets. Nature Nanotechnology. 2008; 3: 101–5.
[27] Bai H, Li C, Wang XL, Shi GQ. A pH-sensitive graphene oxide composite hydrogel. Chemical Communications. 2010; 46: 2376–8.
[28] Xu YX, Wu Q, Sun YQ, Bai H, Shi GQ. Three-dimensional self-assembly of graphene oxide and DNA into multifunctional hydrogels. ACS Nano. 2010; 4: 7358–62.
[29] Bai H, Li C, Wang XL, Shi GQ. On the gelation of graphene oxide. Journal of Physical Chemistry C. 2011; 115: 5545–51.
[30] Huang CC, Bai H, Li C, Shi GQ. A graphene oxide/hemoglobin composite hydrogel for enzymatic catalysis in organic solvents. Chemical Communications. 2011; 47: 4962–4.
[31] Compton OC, An Z, Putz KW, Hong BJ, Hauser BG, Brinson LC et al. Additive-free hydrogelation of graphene oxide by ultrasonication. Carbon. 2012; 50: 3399–406.
[32] Sahu A, Choi WI, Tae GY. A stimuli-sensitive injectable graphene oxide composite hydrogel. Chemical Communications. 2012; 48: 5820–2.
[33] Zhang J, Cao YW, Feng JC, Wu PY. Graphene-oxide-sheet-induced gelation of cellulose and promoted mechanical properties of composite aerogels. Journal of Physical Chemistry C. 2012; 116: 8063–8.
[34] Wang X, Lu LL, Yu ZL, Xu XW, Zheng YR, Yu SH. Scalable template synthesis of resorcinol-formaldehyde/graphene oxide composite aerogels with tunable densities and mechanical properties. Angewandte Chemie-International Edition. 2015; 54: 2397–401.
[35] Dreyer DR, Park S, Bielawski CW, Ruoff RS. The chemistry of graphene oxide. Chemical Society Reviews. 2010; 39: 228–40.
[36] Lin YR, Ehlert GJ, Bukowsky C, Sodano HA. Superhydrophobic functionalized graphene aerogels. ACS Applied Materials & Interfaces. 2011; 3: 2200–3.
[37] Mohanty N, Berry V. Graphene-based single-bacterium resolution biodevice and DNA transistor: Interfacing graphene derivatives with nanoscale and microscale biocomponents. Nano Letters. 2008; 8: 4469–76.
[38] Patil AJ, Vickery JL, Scott TB, Mann S. Aqueous stabilization and self-assembly of graphene sheets into layered bio-nanocomposites using DNA. Advanced Materials. 2009; 21: 3159–64.
[39] He SJ, Song B, Li D, Zhu CF, Qi WP, Wen YQ et al. A graphene nanoprobe for rapid, sensitive, and multicolor fluorescent DNA analysis. Advanced Functional Materials. 2010; 20: 453–9.
[40] Erdem A, Papakonstantinou P, Murphy H. Direct DNA hybridization at disposable graphite electrodes modified with carbon nanotubes. Analytical Chemistry. 2006; 78: 6656–9.
[41] Das B, Voggu R, Rout CS, Rao CNR. Changes in the electronic structure and properties of graphene induced by molecular charge-transfer. Chemical Communications. 2008: 5155–7.
[42] Kotov NA, Dekany I, Fendler JH. Ultrathin graphite oxide-polyelectrolyte composites prepared by self-assembly: Transition between conductive and non-conductive states. Advanced Materials. 1996; 8: 637–41.
[43] Xu YX, Zhao L, Bai H, Hong WJ, Li C, Shi GQ. Chemically converted graphene induced molecular flattening of 5,10,15,20-tetrakis(1-methyl-4-pyridinio)porphyrin and its application for optical detection of cadmium(II) ions. Journal of the American Chemical Society. 2009; 131: 13490–7.
[44] Hu H, Zhao ZB, Wan WB, Gogotsi Y, Qiu JS. Ultralight and highly compressible graphene aerogels. Advanced Materials. 2013; 25: 2219–23.
[45] Qiu L, Liu JZ, Chang SLY, Wu YZ, Li D. Biomimetic superelastic graphene-based cellular monoliths. Nature Communications. 2012; 3: 1241.
[46] Luo JY, Jang HD, Sun T, Xiao L, He Z, Katsoulidis AP et al. Compression and aggregation-resistant particles of crumpled soft sheets. ACS Nano. 2011; 5: 8943–9.
[47] Zhu YW, Murali S, Stoller MD, Velamakanni A, Piner RD, Ruoff RS. Microwave assisted exfoliation and reduction of graphite oxide for ultracapacitors. Carbon. 2010; 48: 2118–22.
[48] Compton OC, Dikin DA, Putz KW, Brinson LC, Nguyen ST. Electrically conductive "alkylated" graphene paper via chemical reduction of amine-functionalized graphene oxide paper. Advanced Materials. 2010; 22: 892–6.
[49] Fang QL, Chen BL. Self-assembly of graphene oxide aerogels by layered double hydroxides cross-linking and their application in water purification. Journal of Materials Chemistry A. 2014; 2: 8941–51.
[50] Fan GL, Li F, Evans DG, Duan XF. Catalytic applications of layered double hydroxides: recent advances and perspectives. Chemical Society Reviews. 2014; 43: 7040–66.
[51] Li CM, Wei M, Evans DG, Duan XF. Layered double hydroxide-based nanomaterials as highly efficient catalysts and adsorbents. Small. 2014; 10: 4469–86.
[52] Li W, Yan D, Gao R, Lu J, Wei M, Duan X. Recent advances in stimuli-responsive photofunctional materials based on accommodation of chromophore into layered double hydroxide nanogallery. Journal of Nanomaterials. 2013: 586462.

[53] Qu J, Zhang QW, Li XW, He XM, Song SX. Mechanochemical approaches to synthesize layered double hydroxides: a review. Applied Clay Science. 2016; 119: 185–92.
[54] Wang Q, O'Hare D. Recent advances in the synthesis and application of layered double hydroxide (LDH) nanosheets. Chemical Reviews. 2012; 112: 4124–55.
[55] Zhang K, Xu ZP, Lu J, Tang ZY, Zhao HJ, Good DA et al. Potential for layered double hydroxides-based, innovative drug delivery systems. International Journal of Molecular Sciences. 2014; 15: 7409–28.
[56] Wang XL, Bai H, Shi GQ. Size fractionation of graphene oxide sheets by pH-assisted selective sedimentation. Journal of the American Chemical Society. 2011; 133: 6338–42.
[57] Liu F, Seo TS. A controllable self-assembly method for large-scale synthesis of graphene sponges and free-standing graphene films. Advanced Functional Materials. 2010; 20: 1930–6.
[58] Cao JY, Wang YM, Xiao P, Chen YC, Zhou Y, Ouyang JH et al. Hollow graphene spheres self-assembled from graphene oxide sheets by a one-step hydrothermal process. Carbon. 2013; 56: 389–91.
[59] Tang ZH, Shen SL, Zhuang J, Wang X. Noble-metal-promoted three-dimensional macroassembly of single-layered graphene oxide. Angewandte Chemie-International Edition. 2010; 49: 4603–7.
[60] Sun YQ, Wu Q, Shi GQ. Supercapacitors based on self-assembled graphene organogel. Physical Chemistry Chemical Physics. 2011; 13: 17249–54.
[61] Zhao JP, Ren WC, Cheng HM. Graphene sponge for efficient and repeatable adsorption and desorption of water contaminations. Journal of Materials Chemistry. 2012; 22: 20197–202.
[62] Sun XX, Cheng P, Wang HJ, Xu H, Dang LQ, Liu ZH et al. Activation of graphene aerogel with phosphoric acid for enhanced electrocapacitive performance. Carbon. 2015; 92: 1–10.
[63] Lei Z, Lu L, Zhao XS. The electrocapacitive properties of graphene oxide reduced by urea. Energy & Environmental Science. 2012; 5: 6391–9.
[64] Guo HL, Su P, Kang XF, Ning SK. Synthesis and characterization of nitrogen-doped graphene hydrogels by hydrothermal route with urea as reducing-doping agents. Journal of Materials Chemistry A. 2013; 1: 2248–55.
[65] Liu WJ, Wang YK, Li ZH. Tuning of surface wettability of RGO-based aerogels for various adsorbates in water using different amino acids. Chemical Communications. 2014; 50: 10311–4.
[66] Lee HS, Dellatore SM, Miller WM, Messersmith PB. Mussel-inspired surface chemistry for multifunctional coatings. Science. 2007; 318: 426–30.
[67] Lee BP, Messersmith PB, Israelachvili JN, Waite JH. Mussel-inspired adhesives and coatings. In: Clarke DR, Fratzl P, editors. Annual Review of Materials Research 2011. p. 99–132.
[68] Lee HS, Rho JS, Messersmith PB. Facile conjugation of biomolecules onto surfaces via mussel adhesive protein inspired coatings. Advanced Materials. 2009; 21: 431–4.
[69] Ryu JK, Ku SH, Lee HS, Park CB. Mussel-inspired polydopamine coating as a universal route to hydroxyapatite crystallization. Advanced Functional Materials. 2010; 20: 2132–9.
[70] Song XH, Lin LP, Rong MC, Wang YR, Xie ZX, Chen X. Mussel-inspired, ultralight, multifunctional 3D nitrogen-doped graphene aerogel. Carbon. 2014; 80: 174–82.
[71] Sheng KX, Xu YX, Li C, Shi GQ. High-performance self-assembled graphene hydrogels prepared by chemical reduction of graphene oxide. New Carbon Materials. 2011; 26: 9–15.
[72] Zhang XT., Sui ZY, Xu B, Yue SF, Luo YJ, Zhan WC et al. Mechanically strong and highly conductive graphene aerogel and its use as electrodes for electrochemical power sources. Journal of Materials Chemistry. 2011; 21: 6494–7.
[73] Hai Dinh P, Viet Hung P, Tran Viet C, Thuy-Duong NP, Chung JS, Shin EW et al. Synthesis of the chemically converted graphene xerogel with superior electrical conductivity. Chemical Communications. 2011; 47: 9672–4.
[74] Chen WF, Yan LF. *In situ* self-assembly of mild chemical reduction graphene for three-dimensional architectures. Nanoscale. 2011; 3: 3132–7.
[75] Chen WF, Li SR, Chen CH, Yan LF. Self-assembly and embedding of nanoparticles by *in situ* reduced graphene for preparation of a 3D graphene/nanoparticle aerogel. Advanced Materials. 2011; 23: 5679–83.
[76] Hicks LD, Han JK, Fry AJ. Hypophosphorous acid-iodine: a novel reducing system. Part 1: Reduction of diaryl ketones to diaryl methylene derivatives. Tetrahedron Letters. 2000; 41: 7817–20.
[77] Gordon PE, Fry AJ. Hypophosphorous acid-iodine: a novel reducing system. Part 2: Reduction of benzhydrols to diarylmethylene derivatives. Tetrahedron Letters. 2001; 42: 831–3.
[78] Gao W, Alemany LB, Ci LJ, Ajayan PM. New insights into the structure and reduction of graphite oxide. Nature Chemistry. 2009; 1: 403–8.
[79] Moon IK, Lee JH, Ruoff RS, Lee HY. Reduced graphene oxide by chemical graphitization. Nature Communications. 2010; 1: 73.
[80] Chen WF, Yan LF, Bangal PR. Chemical reduction of graphene oxide to graphene by sulfur-containing compounds. Journal of Physical Chemistry C. 2010; 114: 19885–90.
[81] Cong HP, Ren XC, Wang P, Yu SH. Macroscopic multifunctional graphene-based hydrogels and aerogels by a metal ion induced self-assembly process. ACS Nano. 2012; 6: 2693–703.
[82] Shieh J, Hou FJ, Chen YC, Chen HM, Yang SP, Cheng CC et al. Robust airlike superhydrophobic surfaces. Advanced Materials. 2010; 22: 597–601.

[83] Zou RJ, Zhang ZY, Yu L, Tian QW, Wu JH, Sun YG et al. Oriented free-standing ammonium vanadium oxide nanobelt membranes: Highly selective absorbent materials. Chemistry-A European Journal. 2010; 16: 14307–12.
[84] Zu SZ, Han BH. Aqueous dispersion of graphene sheets stabilized by pluronic copolymers: Formation of supramolecular hydrogel. Journal of Physical Chemistry C. 2009; 113: 13651–7.
[85] Sun ST, Wu PY. A one-step strategy for thermal- and pH-responsive graphene oxide interpenetrating polymer hydrogel networks. Journal of Materials Chemistry. 2011; 21: 4095–7.
[86] Yang XW, Zhu JW, Qiu L, Li D. Bioinspired effective prevention of restacking in multilayered graphene films: Towards the next generation of high-performance supercapacitors. Advanced Materials. 2011; 23: 2833–8.
[87] Lee SH, Kim HW, Hwang JO, Lee WJ, Kwon J, Bielawski CW et al. Three-dimensional self-assembly of graphene oxide platelets into mechanically flexible macroporous carbon films. Angewandte Chemie-International Edition. 2010; 49: 10084–8.
[88] Ayres N. Atom transfer radical polymerization: A robust and versatile route for polymer synthesis. Polymer Reviews. 2011; 51: 138–62.
[89] Matyjaszewski K. Atom transfer radical polymerization: From mechanisms to applications. Israel Journal of Chemistry. 2012; 52: 206–20.
[90] Matyjaszewski K, Xia JH. Atom transfer radical polymerization. Chemical Reviews. 2001; 101: 2921–90.
[91] Stankovich S, Dikin DA, Piner RD, Kohlhaas KA, Kleinhammes A, Jia YY et al. Synthesis of graphene-based nanosheets via chemical reduction of exfoliated graphite oxide. Carbon. 2007; 45: 1558–65.
[92] An SJ, Zhu YW, Lee SH, Stoller MD, Emilsson T, Park SJ et al. Thin film fabrication and simultaneous anodic reduction of deposited graphene oxide platelets by electrophoretic deposition. Journal of Physical Chemistry Letters. 2010; 1: 1259–63.
[93] Gao XF, Jang J, Nagase S. Hydrazine and thermal reduction of graphene oxide: Reaction mechanisms, product structures, and reaction design. Journal of Physical Chemistry C. 2010; 114: 832–42.
[94] Zhu YW, Murali S, Stoller MD, Ganesh KJ, Cai WW, Ferreira PJ et al. Carbon-based supercapacitors produced by activation of graphene. Science. 2011; 332: 1537–41.
[95] Zhang LL, Zhao X, Stoller MD, Zhu YW, Ji HX, Murali S et al. Highly conductive and porous activated reduced graphene oxide films for high-power supercapacitors. Nano Letters. 2012; 12: 1806–12.
[96] Fan XB, Peng WC, Li Y, Li XY, Wang SL, Zhang GL et al. Deoxygenation of exfoliated graphite oxide under alkaline conditions: A green route to graphene preparation. Advanced Materials. 2008; 20: 4490–3.
[97] Park S, An JH, Piner RD, Jung I, Yang DX, Velamakanni A et al. Aqueous suspension and characterization of chemically modified graphene sheets. Chemistry of Materials. 2008; 20: 6592–4.
[98] Vickery JL, Patil AJ, Mann S. Fabrication of graphene-polymer nanocomposites with higher-order three-dimensional architectures. Advanced Materials. 2009; 21: 2180–4.
[99] Chen ZP, Ren WC, Gao LB, Liu BL, Pei SF, Cheng HM. Three-dimensional flexible and conductive interconnected graphene networks grown by chemical vapour deposition. Nature Materials. 2011; 10: 424–8.
[100] Estevez L, Kelarakis A, Gong QM, Da'as EH, Giannelis EP. Multifunctional graphene/platinum/nafion hybrids via ice templating. Journal of the American Chemical Society. 2011; 133: 6122–5.
[101] Choi BG, Yang M, Hong WH, Choi JW, Huh YS. 3D macroporous graphene frameworks for supercapacitors with high energy and power densities. ACS Nano. 2012; 6: 4020–8.
[102] Sun HM, Cao LY, Lu LH. Bacteria promoted hierarchical carbon materials for high-performance supercapacitor. Energy & Environmental Science. 2012; 5: 6206–13.
[103] Xiao XY, Beechem TE, Brumbach MT, Lambert TN, Davis DJ, Michael JR et al. Lithographically defined three-dimensional graphene structures. ACS Nano. 2012; 6: 3573–9.
[104] Batzill M. The surface science of graphene: Metal interfaces, CVD synthesis, nanoribbons, chemical modifications, and defects. Surface Science Reports. 2012; 67: 83–115.
[105] Huang PL, Lin SC, Yeh CY, Kuo HH, Huang SH, Lin GR et al. Stable mode-locked fiber laser based on CVD fabricated graphene saturable absorber. Optics Express. 2012; 20: 2460–5.
[106] Kibena E, Mooste M, Kozlova J, Marandi M, Sammelselg V, Tammeveski K. Surface and electrochemical characterisation of CVD grown graphene sheets. Electrochemistry Communications. 2013; 35: 26–9.
[107] Kyle JR, Guvenc A, Wang W, Ghazinejad M, Lin J, Guo SR et al. Centimeter-scale high-resolution metrology of entire CVD-grown Graphene sheets. Small. 2011; 7: 2599–606.
[108] Munoz R, Gomez-Aleixandre C. Review of CVD Synthesis of Graphene. Chemical Vapor Deposition. 2013; 19: 297–322.
[109] Tsen AW, Brown L, Havener RW, Park J. Polycrystallinity and stacking in CVD graphene. Accounts of Chemical Research. 2013; 46: 2286–96.
[110] Yamada T, Kim J, Ishihara M, Hasegawa M. Low-temperature graphene synthesis using microwave plasma CVD. Journal of Physics D-Applied Physics. 2013; 46: 063001.
[111] Yu QK, Lian J, Siriponglert S, Li H, Chen YP, Pei SS. Graphene segregated on Ni surfaces and transferred to insulators. Applied Physics Letters. 2008; 93.
[112] Reina A, Jia XT, Ho J, Nezich D, Son HB, Bulovic V et al. Large area, few-layer graphene films on arbitrary substrates by chemical vapor deposition. Nano Letters. 2009; 9: 30–5.

[113] Chae SJ, Guenes F, Kim KK, Kim ES, Han GH, Kim SM et al. Synthesis of large-area graphene layers on poly-nickel substrate by chemical vapor deposition: Wrinkle formation. Advanced Materials. 2009; 21: 2328–33.
[114] Kim KS, Zhao Y, Jang H, Lee SY, Kim JM, Kim KS et al. Large-scale pattern growth of graphene films for stretchable transparent electrodes. Nature. 2009; 457: 706–10.
[115] XS, Cai WW, An JH, Kim SY, Nah JH, Yang DX et al. Large-area synthesis of high-quality and uniform graphene films on copper foils. Science. 2009; 324: 1312–4.
[116] Li WL, Lu K, Walz JY. Freeze casting of porous materials: review of critical factors in microstructure evolution. International Materials Reviews. 2012; 57: 37–60.
[117] Deville S. Ice-templating, freeze casting: Beyond materials processing. Journal of Materials Research. 2013; 28: 2202–19.
[118] Medri V, Sciti D, Fabbriche DD, Piancastelli A, Landi E. Ice templating of ZrB_2-SiC systems. Ceramics International. 2015; 41: 10324–30.
[119] Papa E, Medri V, Benito P, Vaccari A, Bugani S, Jaroszewicz J et al. Synthesis of porous hierarchical geopolymer monoliths by ice-templating. Microporous and Mesoporous Materials. 2015; 215: 206–14.
[120] Deville S, Saiz E, Tomsia AP. Ice-templated porous alumina structures. Acta Materialia. 2007; 55: 1965–74.
[121] Gutierrez MC, Ferrer ML, del Monte F. Ice-templated materials: Sophisticated structures exhibiting enhanced functionalities obtained after unidirectional freezing and ice-segregation-induced self-assembly. Chemistry of Materials. 2008; 20: 634–48.
[122] Kota M, Yu X, Yeon SH, Cheong HW, Park HS. Ice-templated three dimensional nitrogen doped graphene for enhanced supercapacitor performance. Journal of Power Sources. 2016; 303: 372–8.
[123] Lee KH, Lee YW, Lee SW, Ha JS, Lee SS, Son JG. Ice-templated self-assembly of $VOPO_4$-graphene nanocomposites for vertically porous 3D supercapacitor electrodes. Scientific Reports. 2015; 5: 13696.
[124] Deville S. Freeze-casting of porous ceramics: A review of current achievements and issues. Advanced Engineering Materials. 2008; 10: 155–69.
[125] Deville S, Maire E, Bernard-Granger G, Lasalle A, Bogner A, Gauthier C et al. Metastable and unstable cellular solidification of colloidal suspensions. Nature Materials. 2009; 8: 966–72.
[126] Lin ZY, Song MK, Ding Y, Liu Y, Liu ML, Wong CP. Facile preparation of nitrogen-doped graphene as a metal-free catalyst for oxygen reduction reaction. Physical Chemistry Chemical Physics. 2012; 14: 3381–7.
[127] Lu ZJ, Xu MW, Bao SJ, Tan K, Chai H, Cai CJ et al. Facile preparation of nitrogen-doped reduced graphene oxide as a metal-free catalyst for oxygen reduction reaction. Journal of Materials Science. 2013; 48: 8101–7.
[128] Wu CZ, Lu XL, Peng LL, Xu K, Peng X, Huang JL et al. Two-dimensional vanadyl phosphate ultrathin nanosheets for high energy density and flexible pseudocapacitors. Nature Communications. 2013; 4: 2431.
[129] Low JX, Cao SW, Yu JG, Wageh S. Two-dimensional layered composite photocatalysts. Chemical Communications. 2014; 50: 10768–77.
[130] Dai K, Lu LH, Liu Q, Zhu GP, Wei XQ, Bai J et al. Sonication assisted preparation of graphene oxide/graphitic-C_3N_4 nanosheet hybrid with reinforced photocurrent for photocatalyst applications. Dalton Transactions. 2014; 43: 6295–9.
[131] Yamamoto N, Hiyoshi N, Okuhara T. Thin-layered sheets of $VOHPO_4$-$0.5H_2O$ prepared from $VOPO_4$-$2H_2O$ by intercalation-exfoliation-reduction in alcohol. Chemistry of Materials. 2002; 14: 3882–8.
[132] Fischer AE, Pettigrew KA, Rolison DR, Stroud RM, Long JW. Incorporation of homogeneous, nanoscale MnO_2 within ultraporous carbon structures via self-limiting electroless deposition: Implications for electrochemical capacitors. Nano Letters. 2007; 7: 281–6.
[133] Lee SW, Kim JH, Chen S, Hammond PT, Yang SH. Carbon nanotube/manganese oxide ultrathin film electrodes for electrochemical capacitors. ACS Nano. 2010; 4: 3889–96.
[134] Anton R. On the reaction kinetics of Ni with amorphous carbon. Carbon. 2008; 46: 656–62.
[135] Burress JW, Gadipelli S, Ford J, Simmons JM, Zhou W, Yildirim T. Graphene oxide framework materials: Theoretical predictions and experimental results. Angewandte Chemie-International Edition. 2010; 49: 8902–4.
[136] Srinivas G, Burress JW, Ford J, Yildirim T. Porous graphene oxide frameworks: Synthesis and gas sorption properties. Journal of Materials Chemistry. 2011; 21: 11323–9.
[137] Jahan M, Bao QL, Loh KP. Electrocatalytically active graphene-porphyrin MOF composite for oxygen reduction reaction. Journal of the American Chemical Society. 2012; 134: 6707–13.
[138] Rambo BM, Lavigne JJ. Defining self-assembling linear oligo(dioxaborole)s. Chemistry of Materials. 2007; 19: 3732–9.
[139] Cote AP, Benin AI, Ockwig NW, O'Keeffe M, Matzger AJ, Yaghi OM. Porous, crystalline, covalent organic frameworks. Science. 2005; 310: 1166–70.
[140] Severin K. Boronic acids as building blocks for molecular nanostructures and polymeric materials. Dalton Transactions. 2009: 5254–64.
[141] Worsley MA, Pauzauskie PJ, Olson TY, Biener J, Satcher JH, Jr., Baumann TF. Synthesis of graphene aerogel with high electrical conductivity. Journal of the American Chemical Society. 2010; 132: 14067–9.
[142] Worsley MA, Olson TY, Lee JRI, Willey TrM, Nielsen MH, Roberts SK et al. High surface area, sp^2-cross-linked three-dimensional graphene monoliths. Journal of Physical Chemistry Letters. 2011; 2: 921–5.

[143] Lim MB, Hu MJ, Manandhar S, Sakshaug A, Strong A, Riley L et al. Ultrafast sol-gel synthesis of graphene aerogel materials. Carbon. 2015; 95: 616–24.
[144] Slotwinski JA, Garboczi EJ. Metrology needs for metal additive manufacturing powders. Jom. 2015; 67: 538–43.
[145] Deckers J, Vleugels J, Kruthl JP. Additive manufacturing of ceramics: A review. Journal of Ceramic Science and Technology. 2014; 5: 245–60.
[146] Giannitelli SM, Mozetic P, Trombetta M, Rainer A. Combined additive manufacturing approaches in tissue engineering. Acta Biomaterialia. 2015; 24: 1–11.
[147] Gmeiner R, Deisinger U, Schoenherr J, Lechner B, Detsch R, Boccaccini AR et al. Additive manufacturing of bioactive glasses and silicate bioceramics. Journal of Ceramic Science and Technology. 2015; 6: 75–86.
[148] Studart AR. Additive manufacturing of biologically-inspired materials. Chemical Society Reviews. 2016; 45: 359–76.
[149] Travitzky N, Bonet A, Dermeik B, Fey T, Filbert-Demut I, Schlier L et al. Additive manufacturing of ceramic-based materials. Advanced Engineering Materials. 2014; 16: 729–54.
[150] Yang S, Zhao YYF. Additive manufacturing-enabled design theory and methodology: a critical review. International Journal of Advanced Manufacturing Technology. 2015; 80: 327–42.
[151] Kim JH, Chang WS, Kim DH, Yang JR, Han JT, Lee GW et al. 3D printing of reduced graphene oxide nanowires. Advanced Materials. 2015; 27: 157–61.
[152] Lin D, Jin SY, Zhang F, Wang C, Wang YQ, Zhou C et al. 3D stereolithography printing of graphene oxide reinforced complex architectures. Nanotechnology. 2015; 26.
[153] Yang ZX, Yan CZ, Liu JH, Chabi S, Xia YD, Zhu YQ. Designing 3D graphene networks via a 3D-printed Ni template. RSC Advances. 2015; 5: 29397–400.
[154] Jakus AE, Secor EB, Rutz AL, Jordan SW, Hersam MC, Shah RN. Three-dimensional printing of high-content graphene scaffolds for electronic and biomedical applications. ACS Nano. 2015; 9: 4636–48.
[155] Zhu CF, Han TYJ, Duoss EB, Golobic AM, Kuntz JD, Spadaccini CM et al. Highly compressible 3D periodic graphene aerogel microlattices. Nature Communications. 2015; 6: 6962.
[156] Duan SS, Yang K, Wang ZH, Chen MT, Zhang L, Zhang HB et al. Fabrication of highly stretchable conductors based on 3D printed porous poly(dimethylsiloxane) and conductive carbon nanotubes/graphene network. ACS Applied Materials & Interfaces. 2016; 8: 2187–92.
[157] Naficy S, Jalili R, Aboutalebi SH, Gorkin RA, III, Konstantinov K, Innis PC et al. Graphene oxide dispersions: tuning rheology to enable fabrication. Materials Horizons. 2014; 1: 326–31.
[158] Zou JL, Kim F. Diffusion driven layer-by-layer assembly of graphene oxide nanosheets into porous three-dimensional macrostructures. Nature Communications. 2014; 5: 5254.
[159] Munir ZA, Anselmi-Tamburini U, Ohyanagi M. The effect of electric field and pressure on the synthesis and consolidation of materials: A review of the spark plasma sintering method. Journal of Materials Science. 2006; 41: 763–77.
[160] Olevsky EA, Kandukuri S, Froyen L. Consolidation enhancement in spark-plasma sintering: Impact of high heating rates. Journal of Applied Physics. 2007; 102: 114913.
[161] Saheb N, Iqbal Z, Khalil A, Hakeem AS, Al Aqeeli N, Laoui T et al. Spark plasma sintering of metals and metal matrix nanocomposites: A review. Journal of Nanomaterials. 2012: 983470.
[162] Wang LJ, Zhang JF, Jiang W. Recent development in reactive synthesis of nanostructured bulk materials by spark plasma sintering. International Journal of Refractory Metals & Hard Materials. 2013; 39: 103–12.
[163] Yurlova MS, Demenyuk VD, Lebedeva LY, Dudina DV, Grigoryev EG, Olevsky EA. Electric pulse consolidation: an alternative to spark plasma sintering. Journal of Materials Science. 2014; 49: 952–85.
[164] Lu K. Sintering of nanoceramics. International Materials Reviews. 2008; 53: 21–38.
[165] He T, Li JL, Wang LJ, Zhu JJ, Jiang W. Preparation and consolidation of alumina/graphene composite powders. Materials Transactions. 2009; 50: 749–51.
[166] Wang K, Wang YF, Fan ZJ, Yan J, Wei T. Preparation of graphene nanosheet/alumina composites by spark plasma sintering. Materials Research Bulletin. 2011; 46: 315–8.
[167] Fan YC, Wang LJ, Li JL, Li JQ, Sun SK, Chen F et al. Preparation and electrical properties of graphene nanosheet/ Al_2O_3 composites. Carbon. 2010; 48: 1743–9.
[168] Walker LS, Marotto VR, Rafiee MA, Koratkar N, Corral EL. Toughening in graphene ceramic composites. ACS Nano. 2011; 5: 3182–90.
[169] Nieto A, Lahiri D, Agarwal A. Synthesis and properties of bulk graphene nanoplatelets consolidated by spark plasma sintering. Carbon. 2012; 50: 4068–77.
[170] Apak B, Sahin FC. C-CNT produced by spark plasma sintering. Acta Physica Polonica A. 2015; 127: 1029–31.
[171] Li JL, Bai GZ, Feng JW, Jiang W. Microstructure and mechanical properties of hot-pressed carbon nanotubes compacted by spark plasma sintering. Carbon. 2005; 43: 2649–53.
[172] Sato Y, Nishizaka H, Sawano S, Yoshinaka A, Hirano K, Hashiguchi S et al. Influence of the structure of the nanotube on the mechanical properties of binder-free multi walled carbon nanotube solids. Carbon. 2012; 50: 34–9.
[173] Yamamoto G, Sato Y, Takahashi T, Omori M, Hashida T, Okubo A et al. Single-walled carbon nanotube-derived novel structural material. Journal of Materials Research. 2006; 21: 1537–42.
[174] Winter M, Brodd RJ. What are batteries, fuel cells, and supercapacitors? Chemical Reviews. 2004; 104: 4245–69.
[175] Stoller MD, Park SJ, Zhu YW, An JH, Ruoff RS. Graphene-based ultracapacitors. Nano Letters. 2008; 8: 3498–502.

[176] Xia JL, Chen F, Li JH, Tao NJ. Measurement of the quantum capacitance of graphene. Nature Nanotechnology. 2009; 4: 505–9.
[177] Zhang L, Shi GQ. Preparation of highly conductive graphene hydrogels for fabricating supercapacitors with high rate capability. Journal of Physical Chemistry C. 2011; 115: 17206–12.
[178] Sheng KX, Sun YQ, Li C, Yuan WJ, Shi GQ. Ultrahigh-rate supercapacitors based on eletrochemically reduced graphene oxide for ac line-filtering. Scientific Reports. 2012; 2: 247.
[179] Miller JR, Outlaw RA, Holloway BC. Graphene double-layer capacitor with ac line-filtering performance. Science. 2010; 329: 1637–9.
[180] Xu YX, Lin ZY, Huang XQ, Wang Y, Huang Y, Duan XF. Functionalized graphene hydrogel-based high-performance supercapacitors. Advanced Materials. 2013; 25: 5779–84.
[181] Wu XZ, Zhou J, Xing W, Wang GQ, Cui HY, Zhuo SP et al. High-rate capacitive performance of graphene aerogel with a superhigh C/O molar ratio. Journal of Materials Chemistry. 2012; 22: 23186–93.
[182] Van Hoang L, Huynh Ngoc T, Le Thuy H, Nguyen Thi Minh H, Oh ES, Chung JS et al. Synthesis of a highly conductive and large surface area graphene oxide hydrogel and its use in a supercapacitor. Journal of Materials Chemistry A. 2013; 1: 208–11.
[183] Hong JY, Yu X, Bak BM, Pang CH, Park HS. Bio-inspired functionalization and redox charge transfer of graphene oxide sponges for pseudocapacitive electrodes. Carbon. 2015; 83: 71–8.
[184] Tian WQ, Gao QM, Tan YL, Zhang YL, Xu JD, Li ZY et al. Three-dimensional functionalized graphenes with systematical control over the interconnected pores and surface functional groups for high energy performance supercapacitors. Carbon. 2015; 85: 351–62.
[185] Rogers JA, Someya T, Huang YG. Materials and mechanics for stretchable electronics. Science. 2010; 327: 1603–7.
[186] Machado BF, Serp P. Graphene-based materials for catalysis. Catalysis Science & Technology. 2012; 2: 54–75.
[187] Radovic LR, Mora-Vilches C, Salgado-Casanova AJA. Catalysis: An old but new challenge for graphene-based materials. Chinese Journal of Catalysis. 2014; 35: 792–7.
[188] Zhang XP, Liu D, Yang L, Zhou LM, You TY. Self-assembled three-dimensional graphene-based materials for dye adsorption and catalysis. Journal of Materials Chemistry A. 2015; 3: 10031–7.
[189] Wang XB, Qin YL, Zhu LH, Tang HQ. Nitrogen-doped reduced graphene oxide as a bifunctional material for removing bisphenols: Synergistic effect between adsorption and catalysis. Environmental Science & Technology. 2015; 49: 6855–64.
[190] Hou Y, Wen ZH, Cui SM, Ci SQ, Mao S, Chen JH. An advanced nitrogen-doped graphene/cobalt-embedded porous carbon polyhedron hybrid for efficient catalysis of oxygen reduction and water splitting. Advanced Functional Materials. 2015; 25: 872–82.
[191] Yu MH, Huang YC, Li C, Zeng YX, Wang W, Li Y et al. Building three-dimensional graphene frameworks for energy storage and catalysis. Advanced Functional Materials. 2015; 25: 324–30.
[192] Zhu SH, Wang JG, Fan WB. Graphene-based catalysis for biomass conversion. Catalysis Science & Technology. 2015; 5: 3845–58.
[193] Dong XC, Xu H, Wang XW, Huang YX, Chan-Park MB, Zhang H et al. 3D graphene-cobalt oxide electrode for high-performance supercapacitor and enzymeless glucose detection. ACS Nano. 2012; 6: 3206–13.
[194] Ding Y, Wang Y, Su L, Bellagamba M, Zhang H, Lei Y. Electrospun Co_3O_4 nanofibers for sensitive and selective glucose detection. Biosensors & Bioelectronics. 2010; 26: 542–8.
[195] Yang ZX, Xia YD, Mokaya R. Enhanced hydrogen storage capacity of high surface area zeolite-like carbon materials. Journal of the American Chemical Society. 2007; 129: 1673–9.
[196] Alfonso Alonso J, Cabria I, Jose Lopez M. Simulation of hydrogen storage in porous carbons. Journal of Materials Research. 2013; 28: 589–604.
[197] Durbin DJ, Malardier-Jugroot C. Review of hydrogen storage techniques for on board vehicle applications. International Journal of Hydrogen Energy. 2013; 38: 14595–617.
[198] Niaz S, Manzoor T, Pandith AH. Hydrogen storage: Materials, methods and perspectives. Renewable & Sustainable Energy Reviews. 2015; 50: 457–69.
[199] Sevilla M, Mokaya R. Energy storage applications of activated carbons: supercapacitors and hydrogen storage. Energy & Environmental Science. 2014; 7: 1250–80.
[200] Ghosh AK, Subrahmanyam KS, Krishna KS, Datta S, Govindaraj A, Pati SK et al. Uptake of H_2 and CO_2 by graphene. Journal of Physical Chemistry C. 2008; 112: 15704–7.
[201] Srinivas G, Zhu YW, Piner R, Skipper N, Ellerby M, Ruoff RS. Synthesis of graphene-like nanosheets and their hydrogen adsorption capacity. Carbon. 2010; 48: 630–5.
[202] Zhang ZW, Li JC, Jiang Q. Density functional theory calculations of the metal-doped carbon nanostructures as hydrogen storage systems under electric fields: A review. Frontiers of Physics. 2011; 6: 162–76.
[203] Yuan JK, Liu XG, Akbulut O, Hu JQ, Suib SL, Kong J et al. Superwetting nanowire membranes for selective absorption. Nature Nanotechnology. 2008; 3: 332–6.
[204] Schedin F, Geim AK, Morozov SV, Hill EW, Blake P, Katsnelson MI et al. Detection of individual gas molecules adsorbed on graphene. Nature Materials. 2007; 6: 652–5.
[205] Bogue R. Graphene sensors: a review of recent developments. Sensor Review. 2014; 34: 233–8.

[206] Deng XH, Tang H, Jiang JH. Recent progress in graphene-material-based optical sensors. Analytical and Bioanalytical Chemistry. 2014; 406: 6903–16.
[207] Gao L, Lian CQ, Zhou Y, Yan LR, Li Q, Zhang CX et al. Graphene oxide-DNA based sensors. Biosensors & Bioelectronics. 2014; 60: 22–9.
[208] Meng F-L, Guo Z, Huang X-J. Graphene-based hybrids for chemiresistive gas sensors. TRAC-Trends in Analytical Chemistry. 2015; 68: 37–47.
[209] Wu SX, He QY, Tan CL, Wang YD, Zhang H. Graphene-based electrochemical sensors. Small. 2013; 9: 1160–72.
[210] Shimizu Y, Egashira M. Basic aspects and challenges of semiconductor gas sensors. Mrs Bulletin. 1999; 24: 18–24.

Index

N

P

R

S

Z

Color Plate Section

Chapter 2

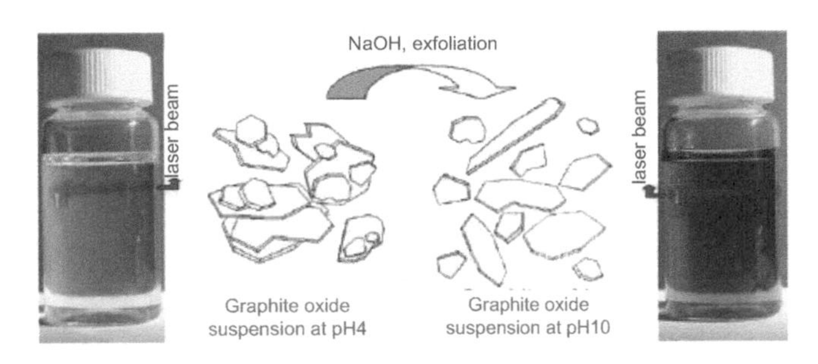

Fig. 2.2. Please refer to Page 7 for figure caption

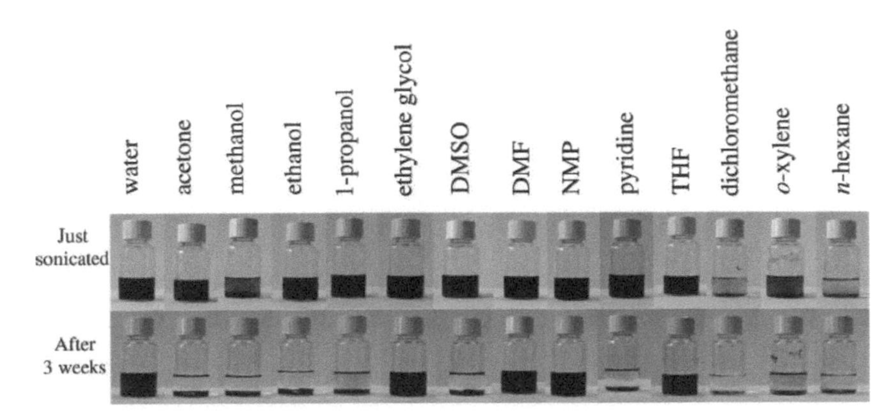

Fig. 2.7. Please refer to Page 11 for figure caption.

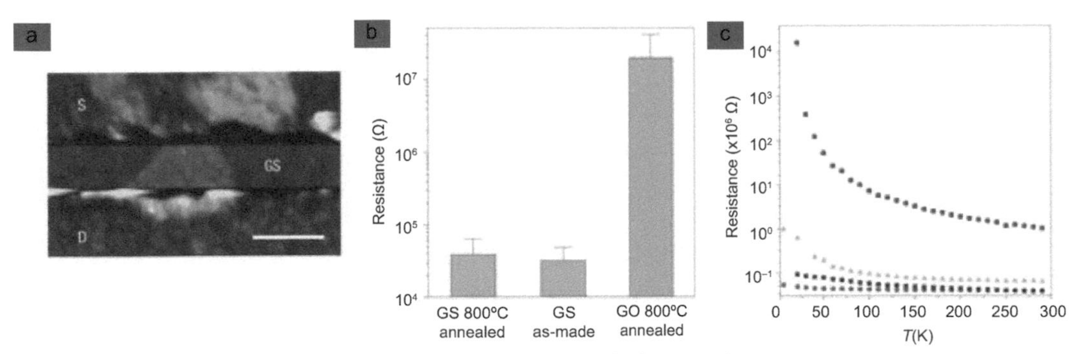

Fig. 2.29. Please refer to Page 30 for figure caption.

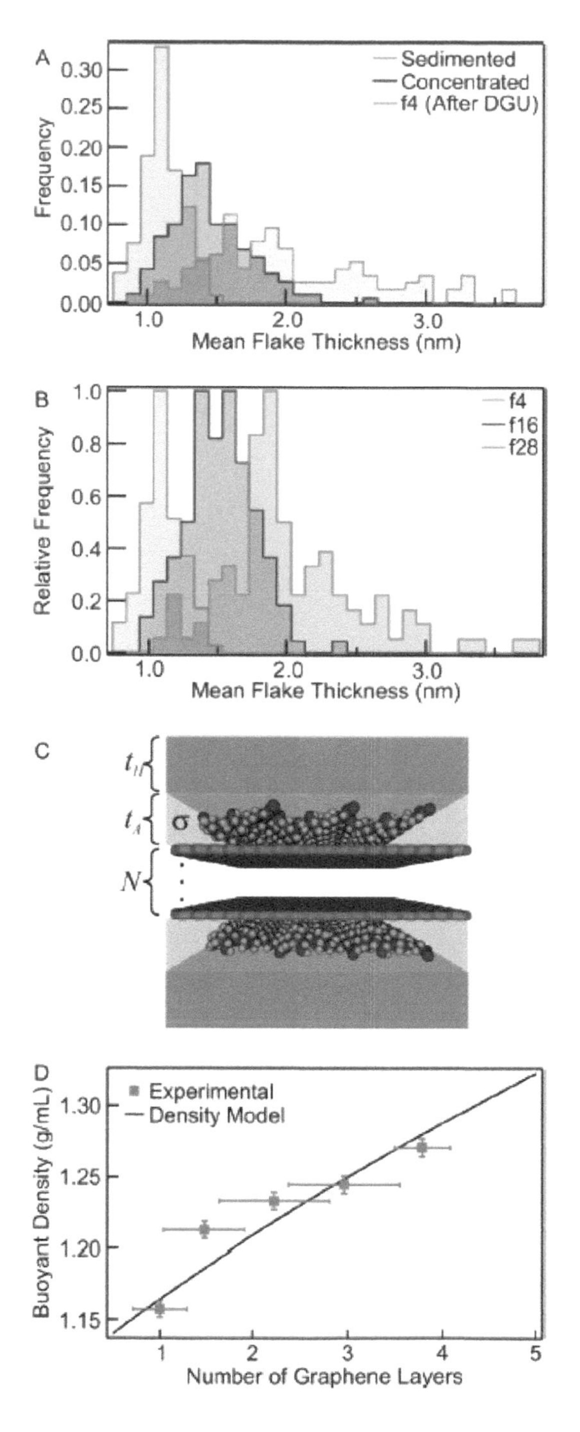

Fig. 2.73. Please refer to Page 66 for figure caption.

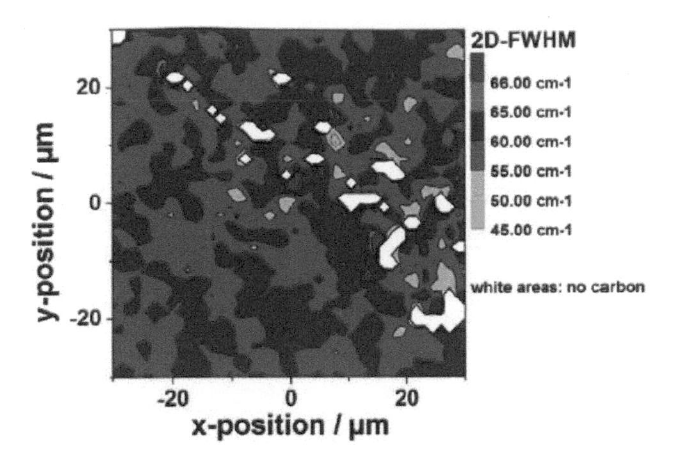

Fig. 2.81. Please refer to Page 72 for figure caption.

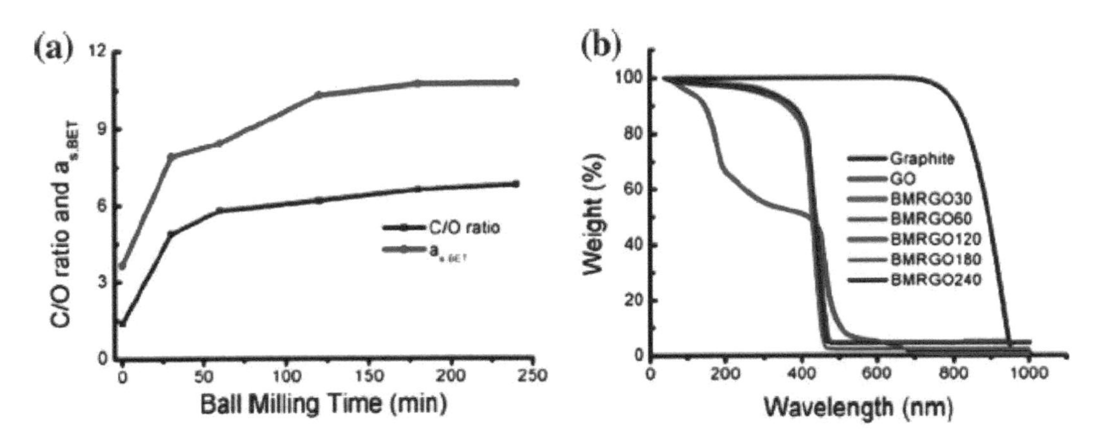

Fig. 2.83. Please refer to Page 73 for figure caption.

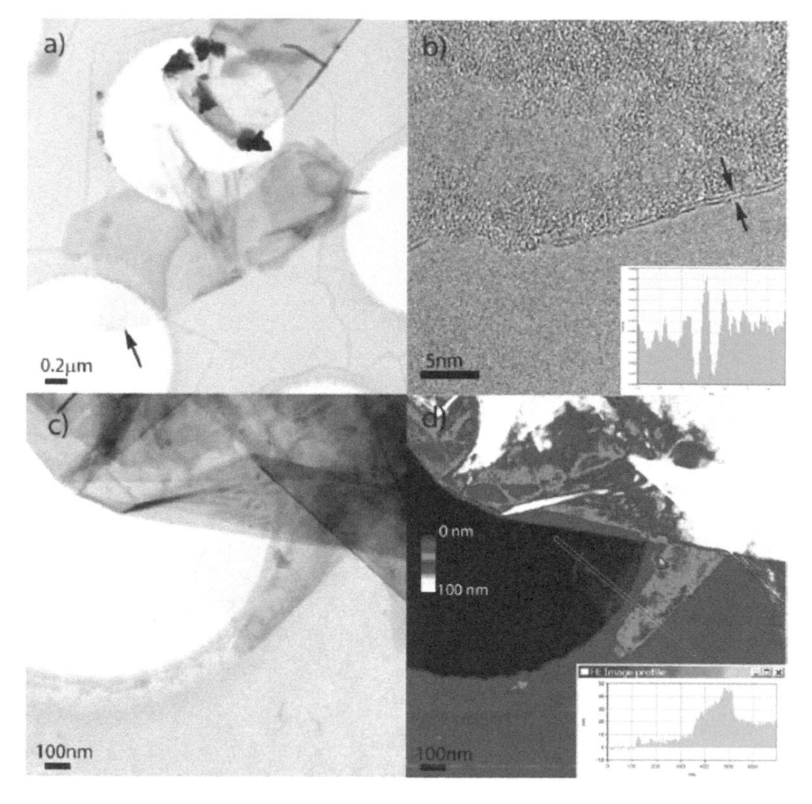

Fig. 2.87. Please refer to Page 77 for figure caption.

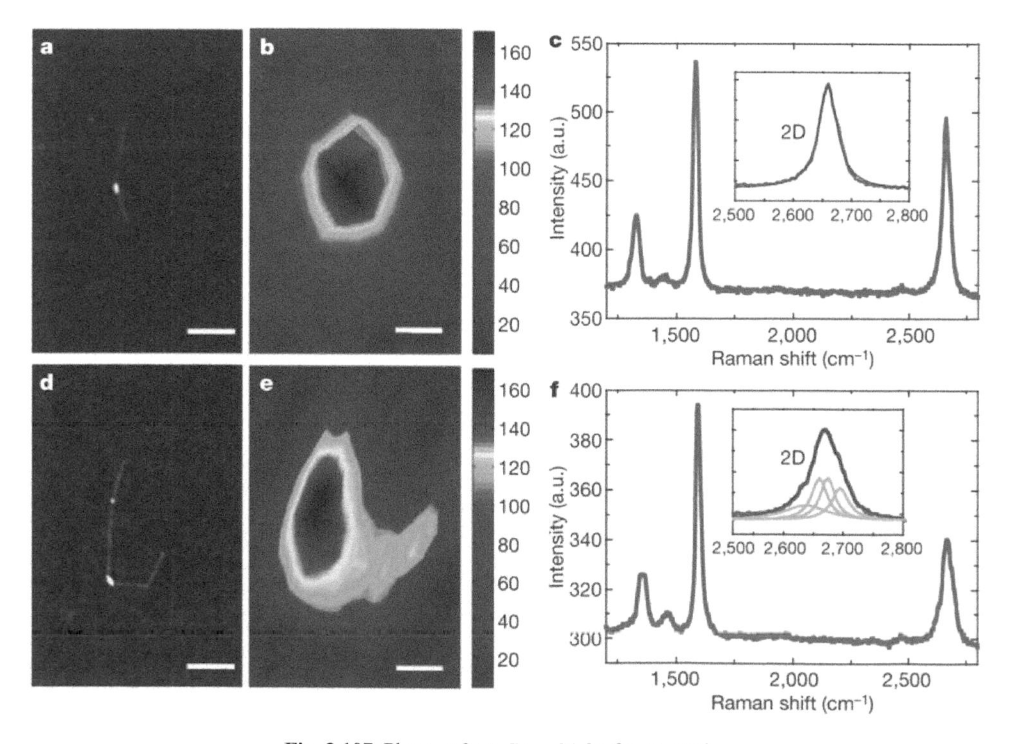

Fig. 2.107. Please refer to Page 94 for figure caption.

Chapter 3

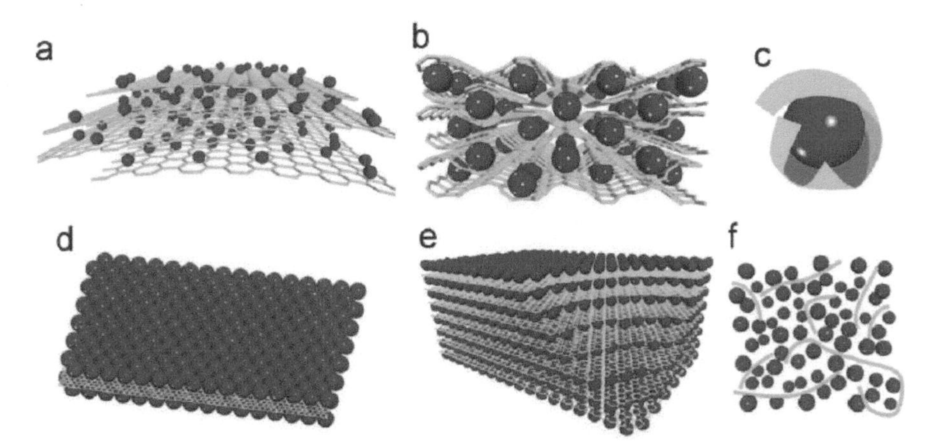

Fig. 3.1. Please refer to Page 112 for figure caption.

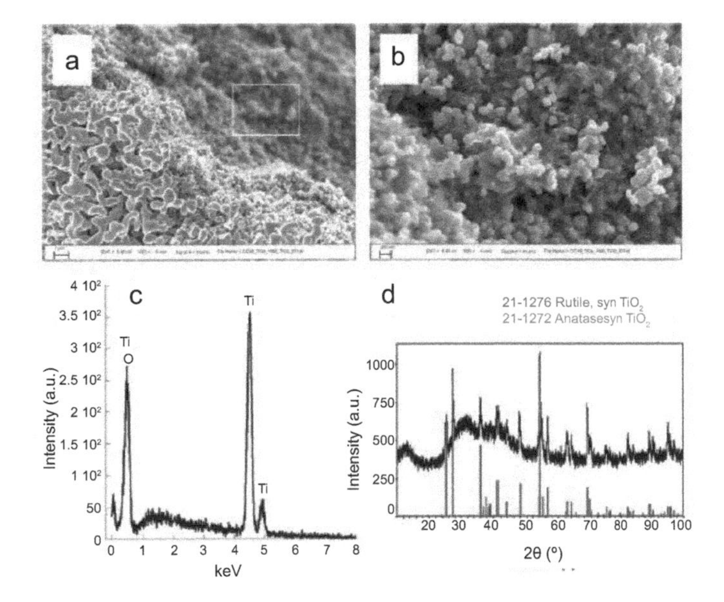

Fig. 3.13. Please refer to Page 123 for figure caption.

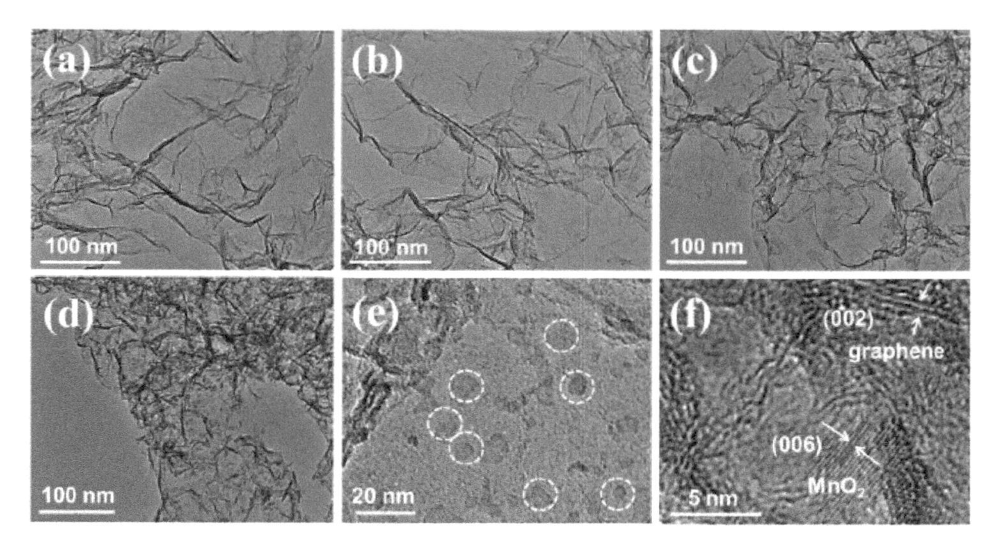

Fig. 3.43. Please refer to Page 147 for figure caption.

Chapter 4

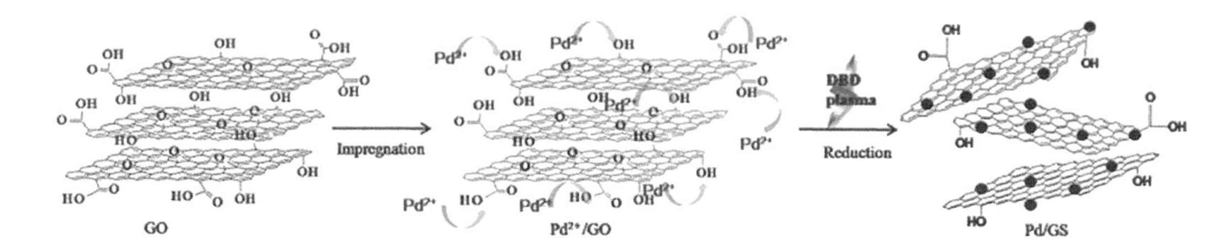

Fig. 4.27. Please refer to Page 242 for figure caption.

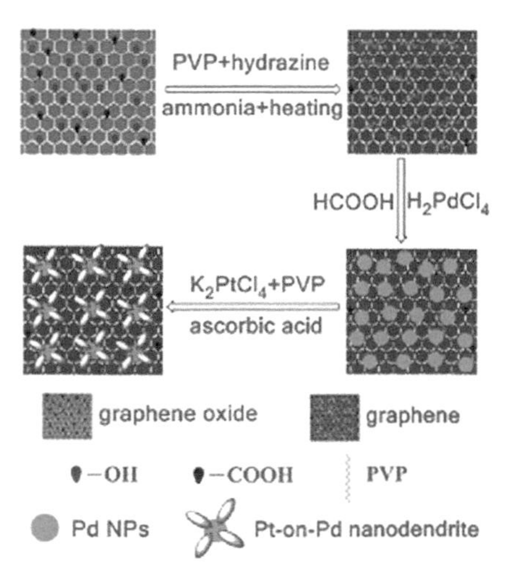

Fig. 4.38. Please refer to Page 252 for figure caption.

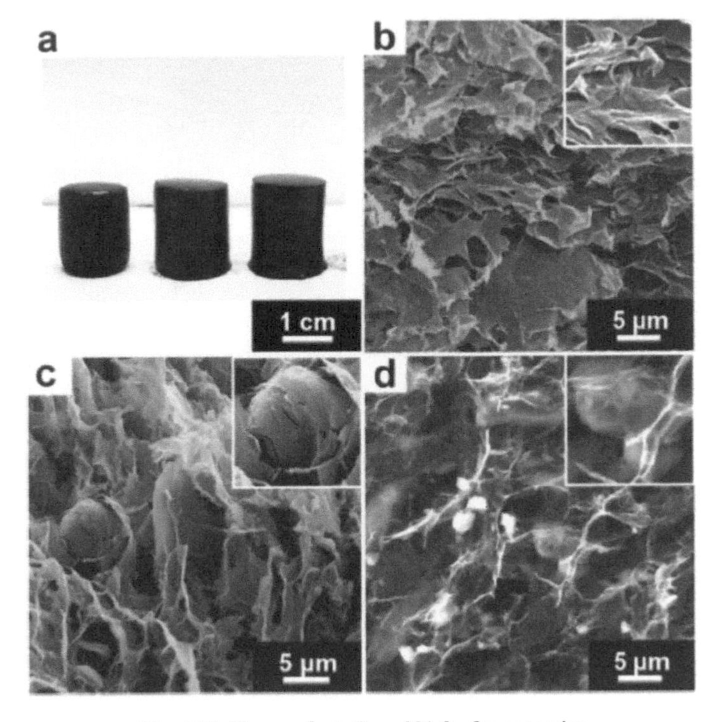

Fig. 4.84. Please refer to Page 291 for figure caption.

Chapter 6

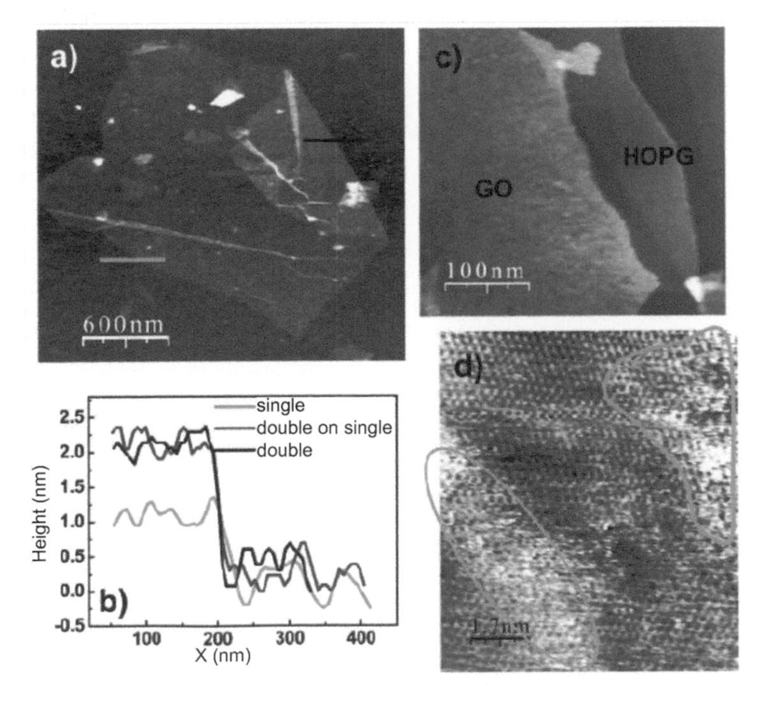

Fig. 6.1. Please refer to Page 385 for figure caption.

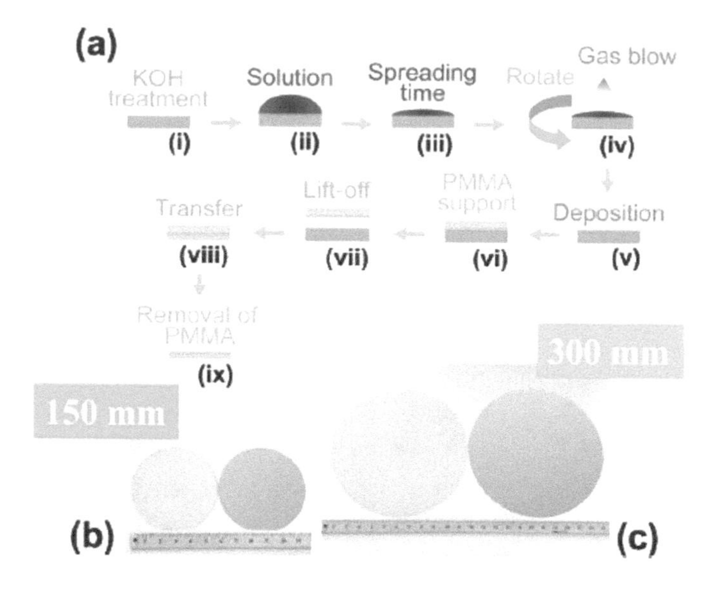

Fig. 6.12. Please refer to Page 395 for figure caption.

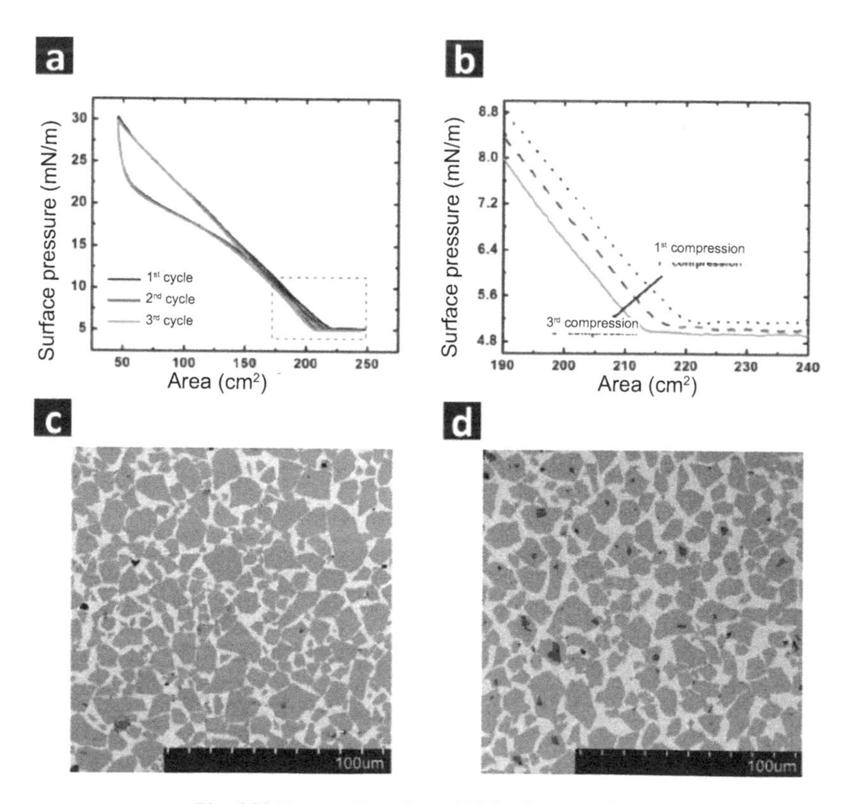

Fig. 6.23 Please refer to Page 406 for figure caption.

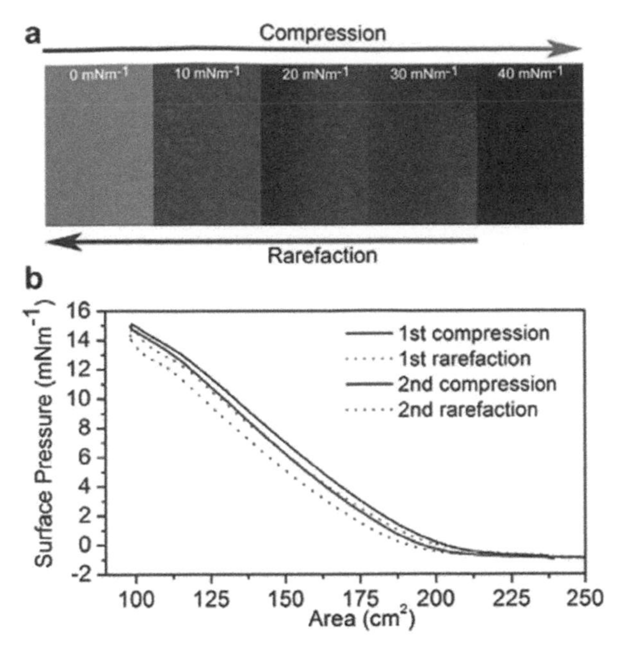

Fig. 6.31. Please refer to Page 413 for figure caption.

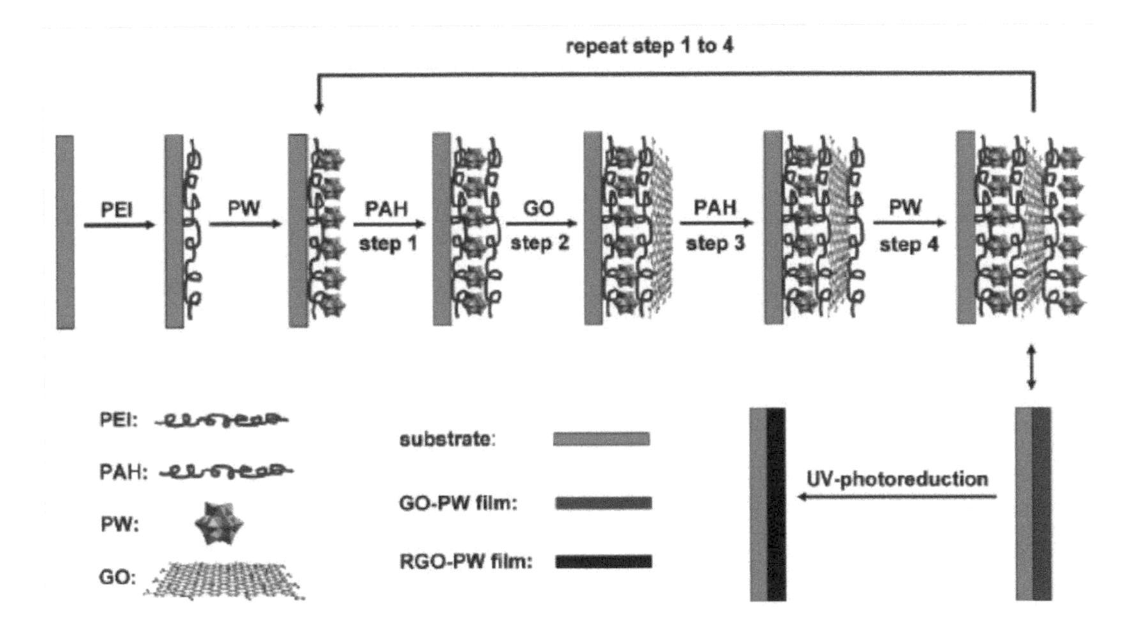

Fig. 6.45. Please refer to Page 424 for figure caption.

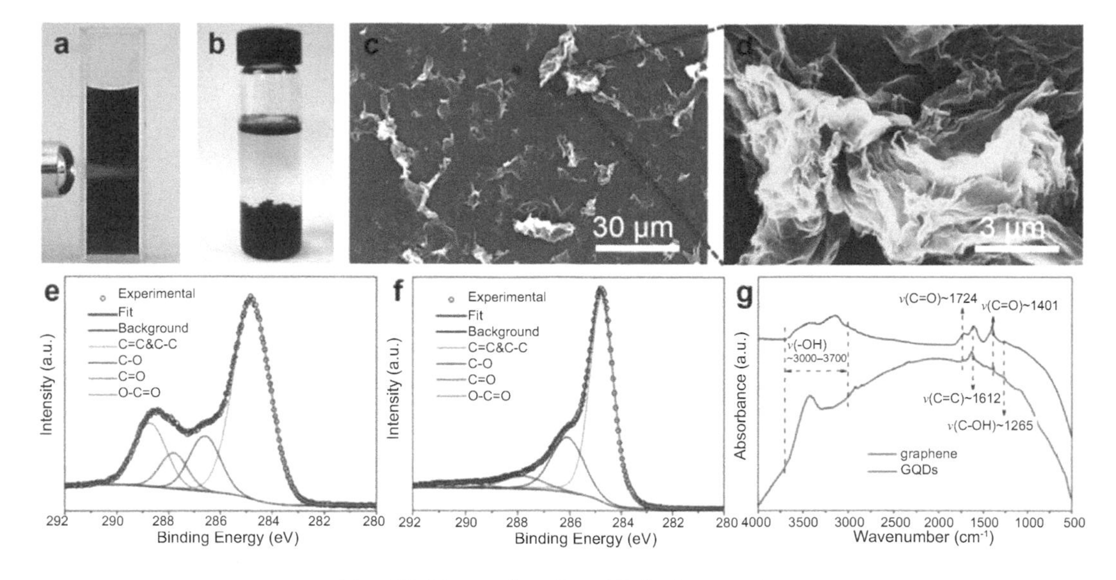

Fig. 6.67. Please refer to Page 443 for figure caption.

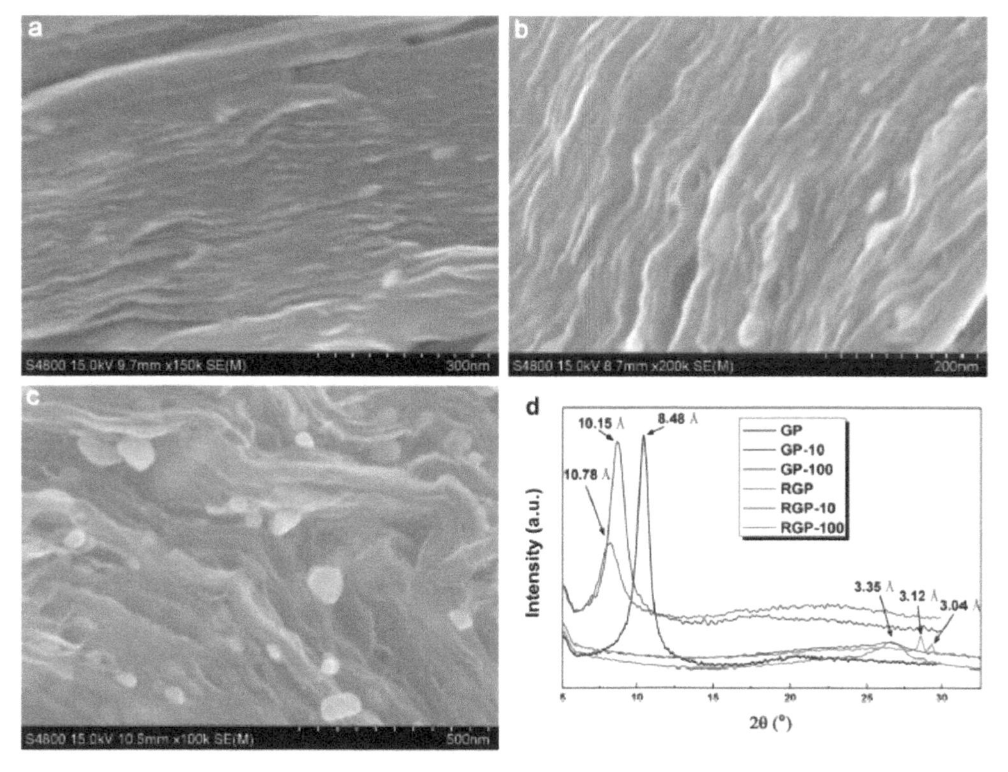

Fig. 6.74. Please refer to Page 449 for figure caption.

Chapter 7

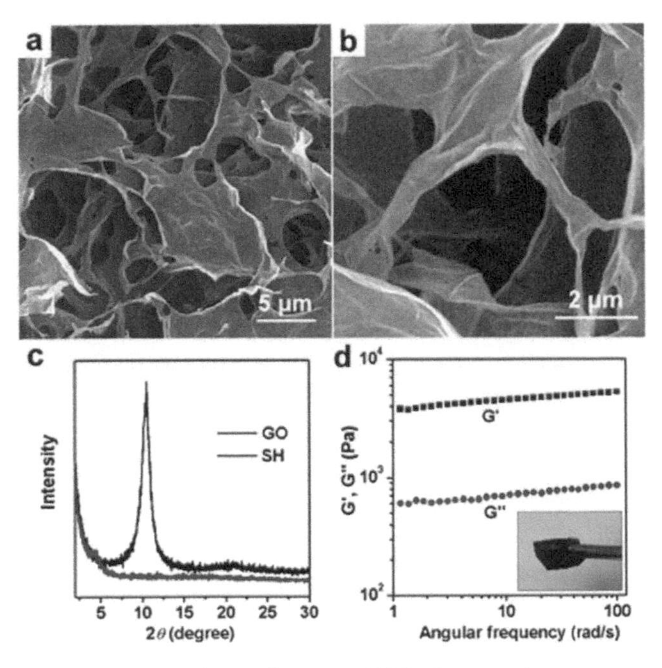

Fig. 7.4. Please refer to Page 499 for figure caption.

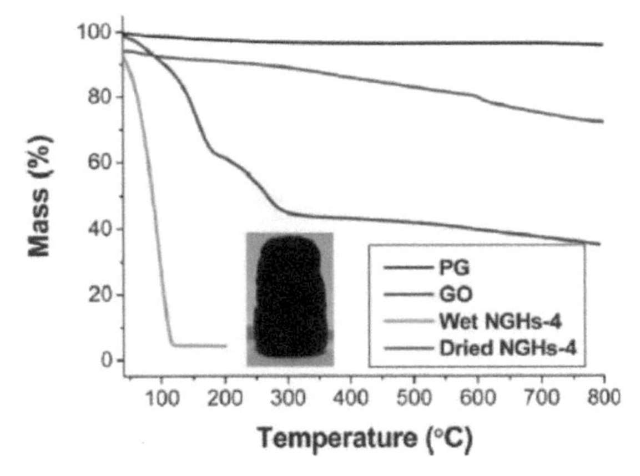

Fig. 7.31. Please refer to Page 523 for figure caption.

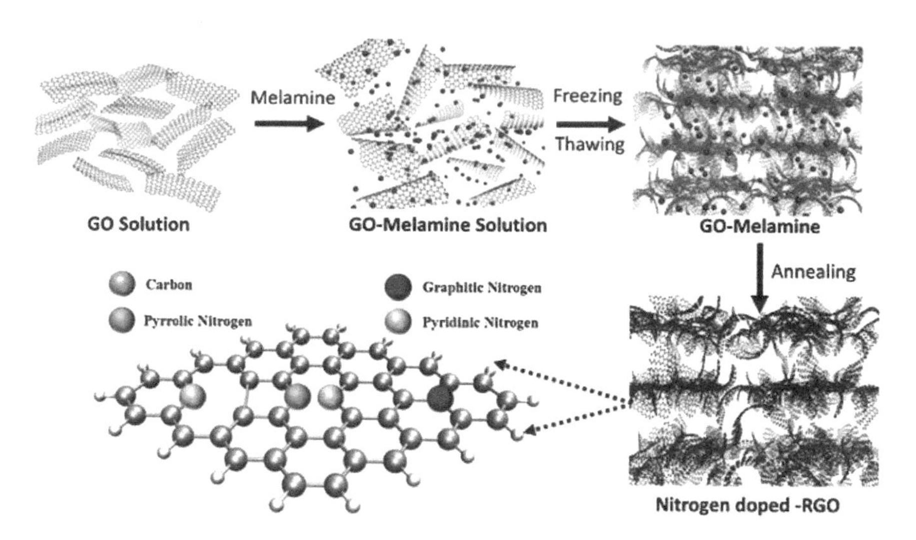

Fig. 7.66. Please refer to Page 552 for figure caption.

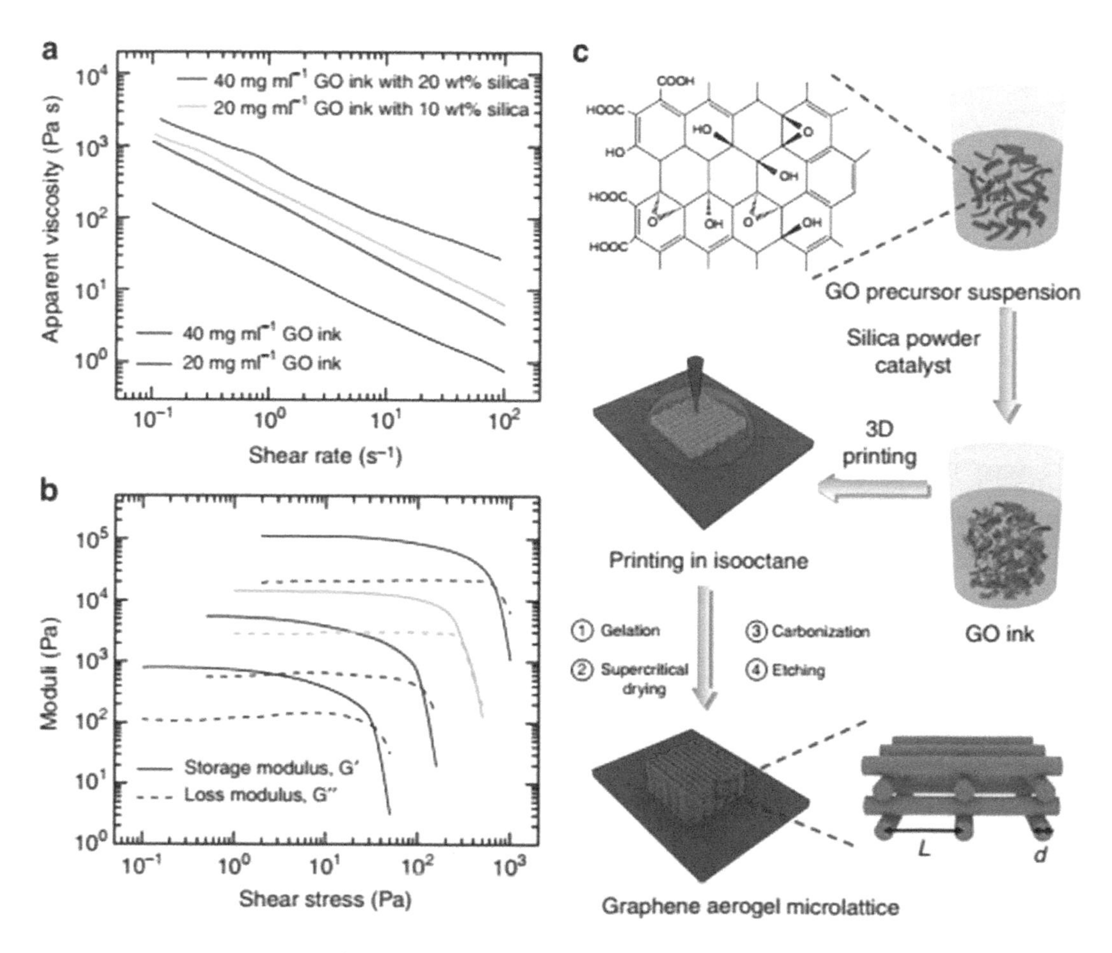

Fig. 7.88. Please refer to Page 570 for figure caption.